Legal Aspects of Architecture, Engineering, and the Construction Process

Third Edition

Legal Aspects of Architecture, Engineering, and the Construction Process

Third Edition

Justin Sweet
Professor of Law
University of California (Berkeley)

West Publishing Company
St. Paul   New York   Los Angeles   San Francisco

Printed in the United States of America

Library of Congress Cataloging in Publication Data

Sweet, Justin.
    Legal aspects of architecture, engineering, and the construction process.

    Includes index.
    1. Building—Contracts and specifications—United States.   2. Engineers—Legal status, laws, etc.—United States.   3. Architects—Legal status, laws, etc.—United States.   I. Title.
KF902.S93 1985   343.73'078624   84–25643
ISBN 0–314–77827–6   347.30378624

1st Reprint—1986

To Sheba

Preface to the Third Edition

Why a new edition? This is a question any author must be prepared to answer. Since the second edition was published in 1977, *some* statutory and common law has changed. In a book about law, such changes can justify a new edition.

Indemnification continues to be a topic for legislative activity. Completion statutes that bar claims against participants in the construction process after a specified period of time are still being challenged. Legislative changes have affected public contract dispute resolution, particularly federal contracts. The Copyright Act passed by Congress in 1976 went into effect in 1978. Bankruptcy law changed in 1979 and 1984.

Court decisions demonstrate an increased recognition of adhesion contracts. Unconscionability as a control over contractual overreaching and implication of good faith and fair dealing between contracting parties have achieved greater respectability. Yet few clauses in contracts in the business world have been upset and courts are struggling with good faith and fair dealing. Arbitration—no longer thought of as a panacea—continues to generate court decisions. Also, changes in the American Arbitration Association's Construction Industry Arbitration Rules, reproduced in Appendix F, have been made since the second edition. Courts have had to deal increasingly with tort claims, both claims by one party to a contract against the other and claims against persons with whom the claimant has no contractual relationship. Tort law expansion has led to more powerful remedies and increased recognition of claims for economic losses.

These changes, however, in and of themselves, could not justify a new edition. In a treatise designed to teach and explain, they are only at the margins. What did lead to this edition was my belief that the treatise could be improved. The second edition did not extend sufficient attention to delay claims and defects. A core common law concept—restitution—needed specific coverage. A more refined analysis of owners was needed to understand many legal doctrines. Financial sources for construction deserved special treatment.

Some core material dealing with basic elements of laws relating to real property deleted from the second edition has been restored. New "players" such as construction managers have become more prominent. Though not new, techniques such as fast tracking and design/build have received greater and more publicized use. Licensing laws under heavy attack, particularly those requiring contractors to be licensed, have been given more analysis. More complex problems for the classroom have been added at the request of academic users.

Speaking bluntly, I have learned much in seven years through my contact with those in the field, both guests of my Berkeley law school seminar and those with whom I have participated in other seminars and lectures.

Better cases have replaced those that did not work well as teaching tools or were not adequately representative. I have concentrated on cases in areas of greatest difficulty, such as professional liability, cost limitations, defects, and delays.

My basic objective is still to explain how law interacts with construction. My years of contact with those in the construction world who are *not* legally trained convince me that they can understand legal doctrines that are described clearly, particularly if examples are given and reasons for the doctrines are explored. I continue to emphasize proper planning to get projects built with a minimum of disputes and litigation. I have taken the liberty to make my own views on certain issues explicit, taking care to separate them from my description of the law and debates over the law.

Textual footnotes have been reduced to a minimum. Footnote references have been limited to documenting specific cases described and backing up (where possible or where necessary) textual legal conclusions. Occasionally references have been made to material that may be useful to a reader who wishes to explore a problem in greater depth than can be done in this treatise.

Description of American law must look to a federal jurisdiction, fifty states, the District of Columbia, and various territories. No one can describe the law in all these jurisdictions. As a result, I emphasize issues, debates, generalizations, and trends. Legal advice in actual disputes requires a competent lawyer.

I make many references to documents published by the American Institute of Architects (AIA). Such documents often reflect customary practices and signal problem areas. They are the invaluable connectors for many participants in the Construction Process and may be more important than statutes and case decisions.

All references, unless otherwise indicated, are to A101 (1977), A201 (1976), A401 (1978), and B141 (1977). Each (along with A311 and A312) is reproduced in appendices. AIA plans to revise its documents in 1986. References to AIA Documents are designed to facilitate instruction. They should not be taken as necessarily representing *current* AIA Documents.

The term *owner* tends to dominate construction contract usage, particularly if we judge by standard contracts such as those published by AIA. However, when I want to emphasize the *professional* relationship created when a person or entity retains a professional adviser to design, advise, and administer the project, I use the term *client*. When the client begins the actual construction process, I shift to the term *owner*. Since the shift is to a degree often imperceptible, the two terms are interchangeable.

As noted in extracts from the Preface to the Second Edition, I made stylistic choices which recognized the increased entry of women into design and construction. However the awkwardness and inelegance of "he or she" or "his or her" or other generally neutral phrases has led me to eliminate them. Instead I have used neuter pronouns where appropriate. Where sexual pronouns must be used I have alternated, male in odd chapters, and female in even ones.

Some acknowledgements are in order. Judy Brown, currently a third-year law student at the University of California, checked references and did invaluable editing. Jeffrey Chu, another third-year student, assisted. Robert McLeod, Esq., and Carl Seneker, III, Esq., of San Francisco each read a chapter and made suggestions. I owe much to the students who have taken my course in Construction Law at Berkeley. They have used this treatise and the special supplement prepared for law students. They pin-pointed weaknesses and made suggestions for improvement as well as prepared valuable seminar papers.

Justin Sweet

May 1985

From the Preface to the Second Edition

The extract from the First Edition contains a brief description of citations to court decisions. Many significant reported appellate opinions come from the Federal Circuit Courts of Appeal. The citation system for these opinions was not described in the First Edition.

Suppose the case is *Moorhead Construction Co. v. City of Grand Forks*, 508 F.2d 1008 (8th Cir.1975). After the name of the parties, an abbreviation of the Report in which the opinion is contained, (F.2d standing for Federal, Second Series), preceded by the volume number of the Report (508) and followed by the page where the case begins, (1008) is given. The particular circuit court deciding the case follows, (8th) and is followed by the year of the decision, (1975).

* * *

In the first edition, citations to cases discussed were given in the text. However, for a number of reasons, case citations and other materials are placed in footnotes in this edition. Case citations in the text slow down reading. More important, users have requested more references to original and secondary source material. Lawyers who have used the first edition have pointed to the absence of an authoritative legal text dealing with construction problems. To respond to this vacuum, more citations to legal authorities dealing with construction disputes have been given.

Also, an unexpected number of my architectural and engineering students have requested additional legal material. Many have done formal research papers dealing with legal problems. Others desired to explore legal problems more deeply. This is an additional reason for more footnote references to original sources in this edition. (One of the unanticipated effects of this treatise has been increased student interest in obtaining law degrees.)

* * *

Finally, the treatise has been "desexed." Increased number of women entering architecture, engineering, and construction has made it even more important to avoid sexual role stereotypes.

* * *

Finally, no words can begin to express my indebtedness to my wife Sheba.

Justin Sweet

June 1977

From the Preface to the First Edition

From 1963 to 1965 I gave a short series of informal lectures to architecture students at the Berkeley campus of the University of California. These lectures dealt with selected legal aspects of professional practice. Because of student and faculty interest, the Department of Architecture asked me to give a regular academic course which would treat the important legal aspects of architecture, engineering, and the construction process.

The course was given in 1967, 1968, and 1969. Because I found existing published materials did not meet my needs, I prepared teaching materials. Retaining what worked well in class and eliminating what did not, I revised the materials each year. In 1968 I added reported appellate cases. This treatise is an expanded and revised version of the 1969 materials.

In planning the teaching materials and the approach to the course, I remembered a remark made by a Berkeley architect which related to the transition from school to practice. He referred to the "cold bath" a young architect receives when he moves from the world of teachers and books to the world of clients, contractors, developers, building inspectors, and loan officers. An understanding of law and legal institutions may not eliminate the "bath," it should reduce the shock and make the transition more tolerable.

Another objective of this treatise is to dispel some of the mistrust and even hostility some design professionals feel toward law and legal institutions. These architects and engineers view law as an unreasonable interference with their professional work, a device to take money out of their pockets unjustly, and a means by which unscrupulous clients avoid paying for design services.

Too often hostile attitudes are the result of limited personal contact with the law. Persons with such attitudes rarely are aware of the role allocated to law by society. The law must select from competing and often conflicting social, political, and economic goals. Also, law must adjust and choose from the demands of competing interest groups.

This role is a difficult one. Even if the reader believes the legal solution to a problem to be unjust (laws and legal institutions, being created and operated by humans, are far from perfect), the reader should recognize the difficulty of the task. This recognition should reduce the suspicion and hostility with which some design professionals view law.

* * *

Some mention should be made of the legal citation system. When specific cases are reproduced in whole or in part or cited, the full legal citations are given. This is to enable the reader, whether he be an architect, engineer, or lawyer, to find the complete case if he so desires.

While legal citations often appear complicated, they are, in reality, quite simple. There are four elements to a citation. A typical citation would be *Sniadach v. Family Finance Corp.*, 395 U.S. 337 (1969). First, the name of the case is given, usually with the plaintiff (the person starting the lawsuit) first, followed by the defendant. "U.S." is the abbreviation of the reporter system from which the case is taken, in this case the United States Supreme Court Reports. The number preceding the abbreviation of the reporter system (395) is the volume in which the case is located. The number following the abbreviation of the reporter system (337) indicates the page on which the case commences. Finally, the citation concludes with the year that the court announced the decision.

Most states have official reports of the appellate court decisions of the state. Also, most state court decisions are collected in regional reports, published by the West Publishing Company. A typical state court citation would be *Weiner v. Cuyahoga Comm. Coll. Dist.*, 19 Ohio St.2d 25, 249 N.E.2d 907 (1969). "Ohio St.2d" is the abbreviation for the Ohio State Reports, Second Series. This abbreviation is preceded by the volume in the reports (19) and followed by the page number of the volume where the case begins (25). This is followed by the citation to the regional reports, Northeast, Second (N.E.2d), with the volume preceding (249) and page number following (907). Finally, the citation is concluded with the year of the decision. A list of abbreviations of the reports will be given following the Table of Cases.

Rather than employ the terms "architect-engineer", "architect or engineer", or "architects and engineers", generally I combine my reference to the two professions by employing the term "design professional". Also, I use the term "prime" contractor rather than "general" contractor. "Supplier" is favored over "materialman", but the latter term is employed occasion.

Justin Sweet

June 1970

Summary Table of Contents

Table of Contents

Table of Cases

Principal cases are in Italic type. Cases cited or discussed are in Roman type.

Table of Abbreviations

The Federal Supplement (F.Supp.) contains opinions of Federal (U.S.) District Courts. These courts are located throughout the United States. In some states they are located in one place. As an example, a reference in the F.Supp. to (D.Neb.) indicates an opinion of *the* Federal District Court for Nebraska. Others are located in different parts of the state. As an example, a reference to (E.D.Pa.) indicates the Eastern District of Pennsylvania; (M.D.) Middle District, (C.D.) Central District, (N.D.) Northern District, (S.D.) Southern District and (W.D.) Western District.

A.	Atlantic Reporter [3]
A.2d	Atlantic Reporter, Second Series [3]
AAA	American Arbitration Association
AGC	Associated General Contractors
A.B.A.J.	American Bar Association Journal [6]
A.C.	Appeal Cases (English)
A.D.2d	Appellate Division, Second Series (New York)
AIA	American Institute of Architects
AIA Doc.	American Institute of Architects Document
A.L.R.2d	American Law Reports, Second Series [7]
A.L.R.3d	American Law Reports, Third Series [7]
A.L.R.4th	American Law Reports, Fourth Series [7]
ASBCA	Armed Services Board of Contract Appeals [10]
Ala.	Alabama Reports [1]
Am.Jur.2d	American Jurisprudence, Second Series [9]
Ariz.	Arizona Reports [1]
Ariz.App.	Arizona Appeals Reports

---

1. Highest Court of State
2. Highest Court of U.S.
3. Regional Reporter
4. All New York Cases
5. All California Cases
6. Periodical
7. Annotation to important cases
8. Statute
9. Encyclopedia
10. Federal Agency Appeals Board
11. All Illinois Cases

Ariz.L.Rev.	Arizona Law Review [6]
Ariz.Rev.Stat.Ann.	Arizona Revised Statutes Annotated [8]
Ark.	Arkansas Reports [1]
Baylor L.Rev.	Baylor Law Review [6]
B.C.L.Rev.	Boston College Law Review [6]
BCA	Board of Contract Appeals (Federal)
BNA	Bureau of National Affairs (Publisher)
B.R.	Bankruptcy Reporter
C.F.R.	Code of Federal Regulations
CI Rules	American Arbitration Association, Construction Industry Arbitration Rules
CM	Construction Manager
Cal.	California Reports
Cal.2d	California Reports, Second Series [1]
Cal.3d	California Reports, Third Series [1]
Cal.App.	California Appellate Reports
Cal.App.2d	California Appellate Reports, Second Series
Cal.App.3d	California Appellate Reports, Third Series
Cal.Civ.Code	California Civil Code [8]
Cal.Exec.Order	California Executive Order
Cal.Stats.	California Laws [8]
Calif.L.Rev.	California Law Review [6]
Cal.Rptr.	California Reporter [5]
Cal.W.L.Rev.	California Western Law Review [6]
Clev.St.L.Rev.	Cleveland State Law Review [6]
Colo.	Colorado Reports [1]
Colo.App.	Colorado Appeals Reports
Conn.	Connecticut Reports [1]

---

1. Highest Court of State
2. Highest Court of U.S.
3. Regional Reporter
4. All New York Cases
5. All California Cases

6. Periodical
7. Annotation to important cases
8. Statute
9. Encyclopedia
10. Federal Agency Appeals Board
11. All Illinois Cases

Conn.App.	Connecticut Appeals Reports
Conn.Gen.Stat.Ann.	Connecticut General Statutes Annotated [8]
Conn.Sup.	Connecticut Supplement Reports
Corn.L.Rev.	Cornell Law Review [6]
Ct.Cl.	Court of Claims (Federal)
Dauph.	Dauphin County Reporter (Pennsylvania)
D/B	Design/Build
D.C.App.	Appellate Court, District of Columbia
Del.Super.	Delaware Superior Court
DOTCAB	Department of Transportation, Civil Aeronautics Appeal Board [10]
Det.C.L.Rev.	Detroit College Law Review [6]
Del.Supr.	Delaware Supreme Court [1]
Duke L.Rev.	Duke Law Review [6]
Exec.Order	Executive Order (Federal)
Eng.Rep.	English Reports
Eng.B.C.A.	Corps of Engineers, Board of Contract Appeals [10]
EPA	Environmental Protection Agency (Federal)
F.	Federal Reporter
F.2d	Federal Reporter, Second Series
FAACAP	Federal Aviation Agency, Contract Appeals Board [10]
Fed.Cir.	United States Circuit Court, Federal Circuit
Fed.Reg.	Federal Register
Fed.Rules	Federal Rules Reporter
Fla.	Florida Supreme Court [1]
Fla.App.	Florida Appellate Court
Fla.Stat.Ann.	Florida Statutes Annotated [8]
Forum	Forum [6]

---

1. Highest Court of State
2. Highest Court of U.S.
3. Regional Reporter
4. All New York Cases
5. All California Cases

6. Periodical
7. Annotation to important cases
8. Statute
9. Encyclopedia
10. Federal Agency Appeals Board
11. All Illinois Cases

GMC	Guaranteed Maximum Cost
GMP	Guaranteed Maximum Price
GSBCA	General Services Board of Contract of Contract Appeals [10]
Ga.	Georgia Reports [1]
Geo.L.Rev.	Georgia Law Review [6]
Gonz.L.Rev.	Gonzaga Law Review [6]
Hastings L.J.	Hastings Law Journal [6]
HUD	Department of Housing and Urban Development (Federal)
Hawaii	Hawaii Reports [1]
IBCA	Interior Board of Contract Appeals [10]
Idaho	Idaho Reports [1]
Ill.	Illinois Reports [1]
Ill.2d	Illinois Reports, Second Series [1]
Ill.Ann.Stat.	Illinois Annotated Statutes [8]
Ill.App.2d	Illinois Appellate Court Reports, Second Series
Ill.App.3d	Illinois Appellate Court Reports, Third Series
Ill.Dec.	Illinois Decisions [11]
Ill.L.For.	Illinois Law Forum [6]
Ill.L.Rev.	Illinois Law Review [6]
Ind.	Indiana Reports [1]
Ind.App.	Indiana Appellate Court Reports
Ins.Coun.J.	Insurance Counsel Journal [6]
Iowa	Iowa Reports [1]
Iowa Code Ann.	Iowa Code Annotated [8]
Iowa L.Rev.	Iowa Law Review [6]
J.Pat.Off.Soc.	Journal of Patent Officials Society [6]
J.Mar.L.Rev.	John Marshall Law Review [6]

1. Highest Court of State
2. Highest Court of U.S.
3. Regional Reporter
4. All New York Cases
5. All California Cases

6. Periodical
7. Annotation to important cases
8. Statute
9. Encyclopedia
10. Federal Agency Appeals Board
11. All Illinois Cases

Kan.	Kansas Reports [1]
Kan.App.2d	Kansas Appellate Reports, Second Series
Ky.	Kentucky Reports [1]
Ky.App.	Kentucky Appellate Court
Ky.Rev.Stat.	Kentucky Revised Statutes [8]
L.Ed.	Lawyer's Edition [2]
L.Ed.2d	Lawyer's Edition, Second Series [2]
L.W.	United States Law Week
La.	Louisiana Reports [1]
La.App.	Louisiana Appellate Court
Law & Contemp.Probs.	Law and Contemporary Problems [6]
Loy.L.Rev.	Loyola Law Review [6]
Loy.L.A.L.Rev.	Loyola (Los Angeles) Law Review [6]
M.L.D.C.	Model Land Development Code
Marq.L.Rev.	Marquette Law Review [6]
Mass.	Massachusetts Reports [1]
Mass.App.	Massachusetts Appeals Reports
Mass.Gen.Laws Ann.	Massachusetts General Laws Annotated [8]
McKinney's N.Y.Gen.Obligation Law	McKinney's General Obligations Law (New York) [8]
Md.	Maryland Court of Appeals [1]
Md.App.	Maryland Court of Special Appeals
Me.	Maine Reports [1]
Mich.	Michigan Reports [1]
Mich.App.	Michigan Appeals Reports
Mich.Comp.Laws Ann.	Michigan Compiled Laws Annotated [8]

1. Highest Court of State
2. Highest Court of U.S.
3. Regional Reporter
4. All New York Cases
5. All California Cases

6. Periodical
7. Annotation to important cases
8. Statute
9. Encyclopedia
10. Federal Agency Appeals Board
11. All Illinois Cases

Misc.2d	Miscellaneous New York Reports, Second Series
Minn.	Minnesota Reports [1]
Miss.	Mississippi Reports [1]
Miss.L.J.	Mississippi Law Journal [6]
Mo.	Missouri Supreme Court [1]
Mo.App.	Missouri Appellate Court
Mo.L.Rev.	Missouri Law Review [6]
Mont.	Montana Reports [1]
N.C.	North Carolina Reports [1]
N.C.App.	North Carolina Appeals Reports
N.D.Lawyer	Notre Dame Lawyer [6]
N.E.	North Eastern Reporter [3]
N.E.2d	North Eastern Reporter, Second Series [3]
N.H.	New Hampshire Reports [1]
N.H.Rev.Stat.Ann.	New Hampshire Revised Statutes Annotated [8]
N.J.	New Jersey Reports [1]
N.J.L.	New Jersey Law Reports
N.J.Stat.Ann.	New Jersey Statutes Annotated [8]
N.J.Super.	New Jersey Superior Court Reports
N.L.R.B.	National Labor Relations Board (Federal)
N.M.	New Mexico Reports [1]
N.S.P.E.	National Society of Professional Engineers
N.W.	North Western Reporter [3]
N.W.2d	North Western Reports, Second Series [3]
N.Y.	New York Reports [1]
N.Y.2d	New York Reports, Second Series [1]
N.Y.Admin.Code	New York Administrative Code [8]

1. Highest Court of State
2. Highest Court of U.S.
3. Regional Reporter
4. All New York Cases
5. All California Cases

6. Periodical
7. Annotation to important cases
8. Statute
9. Encyclopedia
10. Federal Agency Appeals Board
11. All Illinois Cases

N.Y.Gen.Obl.	New York General Obligations Act [8]
N.Y.S.	New York Supplement [4]
N.Y.S.2d	New York Supplement, Second Series [4]
Neb.	Nebraska Reports [1]
Nev.	Nevada Reports [1]
O.F.P.P.	Office of Federal Procurement Policy
O.S.H.C.	Occupational Safety and Health Commission
Ohio App.	Ohio Appellate Reports
Ohio App.2d	Ohio Appellate Reports, Second Series
Ohio Misc.	Ohio Miscellaneous Reports
Ohio No.U.L.Rev.	Ohio Northern University Law Review [6]
Ohio Ops.	Ohio Opinions
Ohio St.	Ohio State Reports [1]
Ohio St.2d	Ohio State Reports, Second Series [1]
Ohio St.3d	Ohio State Reports, Third Series [1]
Okl.St.Ann.	Oklahoma Statutes Annotated [8]
Or.	Oregon Reports [1]
P.	Pacific Reporter [3]
P.2d	Pacific Reporter, Second Series [3]
P.L.	Public Law (U.S.) [8]
Pa.	Pennsylvania State Reports [1]
Pa.Commw.	Pennsylvania Commonwealth Court Reports
Pa.State	Pennsylvania State Reports
Pa.Super.	Pennsylvania Superior Court Reports
P.S.B.C.A.	Postal Services Board of Contract Appeals
Pub.Contract L.J.	Public Contract Law Journal [6]
Pub.Law	Public Law (U.S.) [8]

1. Highest Court of State
2. Highest Court of U.S.
3. Regional Reporter
4. All New York Cases
5. All California Cases

6. Periodical
7. Annotation to important cases
8. Statute
9. Encyclopedia
10. Federal Agency Appeals Board
11. All Illinois Cases

R.I.	Rhode Island Reports [1]
Real Est.L.J.	Real Estate Law Journal [6]
Rev.Code Mont.	Montana Revised Code [8]
Rev.Code Wash.Ann.	Washington Revised Code Annotated [8]
S.C.	South Carolina Reports [1]
S.Cal.L.Rev.	Southern California Law Review [6]
S.Ct.	Supreme Court Reporter (U.S.) [2]
S.D.	South Dakota Reports [1]
S.E.2d	South Eastern Reporter, Second Series [3]
Smith-Hurd Ill.Stat.Ann.	Smith-Hurd Illinois Statutes Annotated [8]
So.	Southern Reporter [3]
So.2d	Southern Reporter, Second Series [3]
Stan.L.Rev.	Stanford Law Review [6]
Stat.	Statutes-at-Large (U.S.) [8]
S.W.	South Western Reporter
S.W.2d	South Western Reporter, Second Series [3]
Tenn.	Tennessee Reports [1]
Tenn.App.	Tennessee Appeals Reports
Tex.Civ.App.	Texas Civil Court of Appeals
Tex.Tech.L.Rev.	Texas Tech. Law Review [6]
U.C.C.	Uniform Commercial Code
U.C.C.Rep.	Uniform Commercial Code Reporter
U.C.L.A. L.Rev.	University of California, Los Angeles Law Review [6]
U. of Chi.L.Rev.	University of Chicago Law Review [6]
Univ. of Cinn.L.Rev.	University of Cincinnati Law Review [6]
Univ.Pa.L.Rev.	University of Pennsylvania Law Review [6]
U.S.	United States Supreme Court Reports [2]

1. Highest Court of State
2. Highest Court of U.S.
3. Regional Reporter
4. All New York Cases
5. All California Cases

6. Periodical
7. Annotation to important cases
8. Statute
9. Encyclopedia
10. Federal Agency Appeals Board
11. All Illinois Cases

U.S. Arbitration Act	United States Arbitration Act [8]
U.S.C.A.	United States Code Annotated [8]
U.San Fran.L.Rev.	University of San Francisco Law Review [6]
Utah	Utah Reports [1]
Va.	Virginia Reports [1]
Vt.	Vermont Reports [1]
Vt.L.Rev.	Vermont Law Review [6]
W.Va.	West Virginia Reports [1]
Wn.	Washington Reports [1]
Wn.2d	Washington Reports, Second Series [1]
Wash.	Washington Reports [1]
Wash.2d	Washington Reports, Second Series [1]
Wash.App.	Washington Appeals Reports
Wash.L.Rev.	Washington Law Review [6]
Wash. & Lee L.Rev.	Washington and Lee Law Review [6]
West's Ann.Calif.Bus. & Prof.Code	West's Annotated California Business and Professions Code [8]
West's Ann.Calif.Civ. Code	West's Annotated California Civil Code [8]
West's Ann.Cal.Govt. Code	West's Annotated California Government Code [8]
West's Ann.Calif. Health and Safety Code	West's Annotated California Health and Safety Code [8]
West's Ann.Calif.Ins. Code	West's California Insurance Code [8]
West's Ann.Calif. Labor Code	West's California Labor Code [8]
West's Ann.Calif.Pen. Code	West's Annotated Penal Code [8]

1. Highest Court of State
2. Highest Court of U.S.
3. Regional Reporter
4. All New York Cases
5. All California Cases

6. Periodical
7. Annotation to important cases
8. Statute
9. Encyclopedia
10. Federal Agency Appeals Board
11. All Illinois Cases

West's Ann.Calif.Pub. **Contract Code**	West's Annotated California Public Contract Code [8]
Wis.	Wisconsin Reports [1]
Wis.2d	Wisconsin Reports, Second Series [1]
Wis.L.Rev.	Wisconsin Law Review [6]
Yale L.J.	Yale Law Journal [6]

1. Highest Court of State
2. Highest Court of U.S.
3. Regional Reporter
4. All New York Cases
5. All California Cases

6. Periodical
7. Annotation to important cases
8. Statute
9. Encyclopedia
10. Federal Agency Appeals Board
11. All Illinois Cases

Legal Aspects of Architecture, Engineering, and the Construction Process

Third Edition

Part One

The Substructure: Legal Institutions and Core Legal Concepts

Introduction

This treatise discusses legal doctrines organized around the principal participants and activities in the Construction Process, an inclusive term to describe ownership, design, and construction. Part I sets the stage for the balance of the treatise by describing legal institutions and law central to the Construction Process. Other legal doctrines are described in Parts II through IV.

1

Sources of Law:
Varied and Dynamic

SECTION 1.01 RELEVANCE

Law consists of coercive rules created and enforced by the state to regulate the citizens of the state and provide for the general welfare of the state and its citizens. Law is an integral part of modern society and plays a major role in the Construction Process. Since this treatise examines the intersection between law and the Construction Process, it is important to be aware of the various sources of law and the characteristics and functions of the law.

Many illustrations can be provided, but suppose a man who owns property wishes to build a house to provide shelter for his family. Without assurance that stronger persons will not use force to seize materials with which he is building or throw him out of the house after it is built, it would take an adventurous or powerful person to invest time and materials to build the house. Similarly, workers would be reluctant to pound nails or pour concrete if they were fearful of being attacked by armed gangs. Here, criminal law protects both the property owner from those who might take away his property and workers from those who might harm them.

Similarly, contractors would hesitate to invest their time or money to build houses if they did not believe they could use the

civil courts to enforce their contracts and help them collect for their work if owners do not pay them. Workers would be less inclined to work on the house if they did not have confidence that they could use the civil courts to collect for work that they had done or to provide compensation for them if they were injured on the job.

Finally, some would be unwilling to engage in construction activity if they were not confident that an impartial forum would be available if disputes arose over performance. Were the state not to provide such a forum participants might settle their disputes by force.

Today various sources of law seek to provide these needs. These sources of law are spotlighted in this chapter. Chapter 2 focuses on the American judicial system.

SECTION 1.02
THE FEDERAL SYSTEM

Very large countries, such as the United States, Canada, Australia, Nigeria, and India, employ a federal system of government. Even smaller countries, particularly those with distinct religious, linguistic, ethnic, or national communities (for example, Switzerland), may choose to live under a federal system. This gives local entities limited autonomy to deal with cultural and other activities. The alternative would be domination by the majority or small, often economically inefficient or weak nation-states.

In a federal system power is shared; the exact division of power between the central government and constituent members vary. For example, Canada has a looser federal system than the United States with Canadian provinces having more autonomy than American states.

A federal system may be created in a large country where persons living in one part of the country hold political, social, or economic views distinct from the rest of the country. For example, New York might choose to provide greater social benefits for its citizens than Texas. Similarly, some states may wish to execute murderers while others may not. In a federal system, diverse views may be accommodated within one country.

Also, a federal system, depending upon its characteristics, can bring citizens closer to those who govern them. A rancher in Wyoming may resent being controlled by a legislature in Washington but may be more willing to submit to laws that come from Cheyenne. When the entities of a federal system are themselves large, such as Canadian provinces or American states, citizens may see even the state capital as being too remote and prefer to be governed by those they elect in their cities or towns.

But the federal system recognizes the need for a central government to deal with certain issues on behalf of *all* citizens. For example, the American federal government controls currency, foreign relations, and defense, to mention a few of the areas controlled exclusively by the federal government. The constitutional framers did not want each state to have its own currency, diplomatic service, and army.

Other functions can be shared. For example, the federal government enacts tax laws, as do states. Similarly, federal and state laws deal with crime, labor relations, and, to use an illustration more germane to construction, worksite safety.

To understand how the law regulates the Construction Process it is important to be aware of the American federal system. For example, under protectionist pressures, Congress can require that under certain circumstances designated equipment procured by the federal government be made in America. But a state that enacts a "Buy America" statute would be invading the exclusive power of the federal government to conduct foreign policy.[1] Although states can choose their own procurement policies, they may not interfere with the exclusive central government's power to conduct foreign policy.

Mention has been made of concurrent or shared power to deal with workplace safety. To avoid duplication of enforcement efforts and to relieve construction contractors from inconsistent regulations, the federal government, though dominant, can delegate workplace safety standards and their enforcement to the states so long as the states meet federal standards. Similar delegations are found in environmental protection laws.

Despite general federal supremacy over the states, the U.S. Constitution reserved some authority to the states. For example, except for contracts made by the federal government or those affected with a strong federal interest, state law determines which contracts will be enforced, the remedies granted for breach of contract, the conduct that gives rise to civil liability, and the laws that relate to the ownership of property. These are core legal concepts in the Construction Process. As a result, most law

---

1. *Bethlehem Steel Corp v. Board of Commissioners*, 276 Cal.App.2d 221, 80 Cal.Rptr. 800 (1969) (invalidating California Buy American Act).

that regulates construction is determined by the state in which the project is located or in which the activities in question are performed.

This can and sometimes does lead to variations in legal rules that relate to construction. However, the dominance of nationally created standard contract forms, the willingness of courts in one state to look at and often follow the decisions from another state, and the unification of areas of private law, such as the sale of goods, have minimized the actual variation in state laws that relate to construction. However, some laws vary greatly, such as mechanics' lien laws and licensing, both of which are regulated by state statutes.

Power-sharing arrangements endemic to a federal system are not static. For example, before 1978, states could protect common law copyrights—the right to determine if and when written matter will be disclosed to the public. This right existed despite the federal government's having a constitutionally created monopoly over patents and copyrights granted by the U.S. Constitution. But in 1976, Congress enacted legislation which, effective in 1978, for the most part abolished the power of states to protect common law copyright in order to unify the law that relates to intellectual property.

SECTION 1.03 CONSTITUTIONS

When organizations are created, whether political, economic, or social, they usually attract members because of the goals they seek to achieve, the methods they will choose to achieve the goals, and the procedures by which they operate. Often these factors are set forth in a basic set of principles or rules providing a constitutional framework that induces persons to join and regulates the operations of the organization. Although this basic set of rules may be referred to by different labels, it provides a constitutional framework intended to be durable and, although not immutable,

amendable only with the consent of most of the members of the organization.

Most organizations need rules to govern day-to-day operations. These rules, though created with less formality than constitutional rules and more responsive to changing conditions, must stay within the basic framework of the constitutional rules. For example, a business corporation will usually have Articles of Incorporation that operate as a constitution of the corporation and are agreed upon by those who create the corporation and available to those who wish to purchase shares. The corporation is also likely to have bylaws that regulate corporate activities and that are often more detailed and more easily changed. In addition, a large corporate organization may need standard operating procedures and hierarchical authority mechanisms to control the corporate operation.

The U.S. Constitution has regulated the power among the branches of the federal government, the federal government and the states, and all governments and their citizens for two centuries. Despite universal admiration for this durable document, the wealth of federal laws that have been enacted attests to the importance of legislation as a method to deal with changing circumstances and the varying allocation of political power.

States also have constitutions. On the whole they are longer than the federal Constitution and change more frequently. While not identical, they share common characteristics with the U.S. Constitution, mainly the separation of powers among the legislative, executive, and judicial branches of government, and the importance of protecting citizens from abuse of power by the states.

Constitutional law does not play a significant role in the Construction Process. Yet a few areas exist where the federal Constitution or state constitutions have been invoked. For example, constitutional law has been used to challenge laws requiring that minority contractors be given a certain portion of federal construction contracts.[2] It has, along with state constitutions, been invoked to challenge the validity of state legislation that relieves certain parties who engage in design and construction from liability after a designated number of years have elapsed following completion of a project.[3]

Finally, the federal Constitution, with little success, has been used as the basis for challenging land-use legislation enacted by local entities such as cities.[4] But reference in this treatise to constitutions, whether federal or state, is rare.

SECTION 1.04   LEGISLATION

Legislation, mainly at the state level, plays an increasingly larger role in the Construction Process. Those who wish to understand how law affects design and construction must pay increasingly careful attention to legislation.

Legislation is the most democratic law-making process. It expresses the will of the majority of the citizens of the state. (Perhaps the only source of law *more* democratic is the law created by contracting parties to regulate their private rights and duties.)

Legislatures are political instrumentalities. Persons and organizations seek to influence lawmakers. Though often denigrated as an area where power and money control, legislation does, as a rule, reflect popular attitudes. Lawmakers who vote in ways not popular with their constituents are likely to be removed from office. Also, legislation can be enacted quickly to respond to what the legislature believes to be important social and economic needs of its citizens.

The legislative process functions at different governmental levels. Most are aware of the federal, state, county, and city legislative bodies. But special districts, such as school and sewerage districts, cannot be ignored. They often do not attract the attention of the voters and the media, and those that are elected are often unknown community figures. This can mean a lesser degree of voter responsiveness to their activities and, unfortunately, greater opportunities for corruption. But these special districts are often important to the Construction Process as they commission construction work.

Legislators do not have absolute freedom to enact legislation. The checks and balances so central to the American political process bar legislators from absolute power even though the legislatures do represent the political will of the citizens. Laws must be constitutionally enacted and the constitutionality of a law is determined by courts. Courts in their role of interpreting legislation can indirectly control the legislature and the political desires of the majority. Similarly, local units such as counties, cities, or special districts are limited by the legislation that has created them.

As stated, state legislation affects design and construction. Licensing and registration laws determine who can design and, increasingly, who can build. State incorporation laws determine the forms of organization that can be used to accomplish particular objectives.

Other state statutes affect design and construction. Dispute resolution has been increasingly affected by legislative activities encouraging arbitration. Similarly, construction litigation has been affected by state laws that preclude certain types of

---

2. *Fullilove v. Klutznick,* 448 U.S. 448, 100 S.Ct. 2758, 65 L.Ed.2d 902 (1980) (upheld as a valid exercise of legislative power).

3. See Section 27.03(G).

4. See Section 11.15.

indemnification provisions in construction contracts and by legislation that cuts off liability after the expiration of a designated time. Finally, local legislative bodies exercising their power to protect the public and regulate land use have had an increasingly important affect on construction. Local authorities must approve many types of projects. Through housing and building codes, they also control the quality of construction.

Mention must be made of the Uniform Commercial Code (U.C.C.), particularly Article 2, which deals with transactions in goods. The Code was developed by the American Law Institute and the Commissioners on Uniform State Laws, both private organizations devoted to unification of private law. At present all states but Louisiana have adopted the Code. (Louisiana influenced by its civil law tradition has adopted only some parts of the Code.) Article 2 regulates the sale of materials and supplies used in a project. While it does not regulate construction itself, inasmuch as it does not govern services, it can be influential in a construction dispute. Reference is made to it in this treatise.

SECTION 1.05
THE EXECUTIVE BRANCH

The separation of powers central to the American political system has been mentioned. It is designed to prohibit any of the three branches—executive, legislative, and judicial—from having dominant power. While laws come out of the legislature, the executive and judicial branches participate in lawmaking. The executive branches— the President at the federal level, the governor at the state level, and the mayor in some municipal systems—are elected by all the citizens. Federal and state chief executives and sometimes local chief executives can veto legislation. To override a veto requires more than a simple majority of the legislature, usually two-thirds.

In this treatise the executive branch plays a limited role. The chief executive can issue executive orders to those under his control. For example, the first prohibition of discrimination in employment resulted from a presidential executive order.[5] Also, an executive order by the governor of California in 1978[6] changed the dispute resolution system for California state public contracts and led to legislative change.[7]

SECTION 1.06
ADMINISTRATIVE AGENCIES

One great change in the American governmental structure has been the emergence of administrative regulatory agencies at every level of government, but particularly at federal and state levels. Such agencies developed for a variety of reasons. First, special activities and industries were thought best regulated by experts in those activities and industries. Second, legislatures were often unable or unwilling to involve themselves in the details of regulation. Third, regulation through agencies was thought to be a better alternative than no regulation (this has changed in the 1980s with some industries being deregulated) or government operation of these activities and industries.

The constitution for a regulatory agency is the legislation creating it. In that sense the agency is a creation of the legislature. However, the chief executive usually appoints, often with the advice and consent of the legislature, key agency officials. Legis-

---

5. The first was Exec. Order No. 8802, 6 Fed. Rules 3109 (1941). The history is collected in *Farmer v. Philadelphia Electric Co.*, 329 F.2d 3 (3d Cir.1964).

6. Cal. Exec. Order B50–78 (1978).

7. West's Ann. Cal. Civ. Code §1670. See Section 34.20(B).

latures, particularly the Congress, monitor agency performance through exercising an oversight function. This is accomplished through committee hearings during which complaints are heard and agency officials are asked to give explanations.

Agencies operate through issuance of regulations and through disciplinary actions. Such activities are subject to judicial review. During the period when many of these agencies were created, courts extended considerable and almost total deference to such agencies because of presumed agency expertise. As the agencies became more active and as complaints about their aggressiveness and preoccupation with problems that some thought trivial became more vocal, agencies' powers and activities were looked at more carefully by the courts and legislatures.

Regulation through administrative agencies has generated intense controversy. Some agencies are attacked as being under the control of those they are supposed to regulate. This is asserted by pointing to agency employees being selected from the regulated industries or agency employees looking forward to being hired by members of the industry after they leave agency employment. On the other hand, agencies have sometimes been attacked for overzealous regulation, and such attacks have had some success in the 1980s.

Administrative agencies play an important role in some aspects of the Construction Process. They are often given responsibility for implementation of laws enacted by the legislatures. For example, licensing and registration of professionals are largely in the hands of state regulatory agencies, which determine levels of enforcement, the nature of examinations, the requirements for licensing and registration, and other matters relating to such professions. Also, the preventive aspects of workplace safety and environmental protection are largely within the control of federal and state agencies. Social insurance for workplace injuries is operated by state workers' compensation agencies. At local levels, administrators play important roles in land-use control and construction quality. In areas affecting public safety, the most important source of law is regulations issued by administrative agencies.

SECTION 1.07
COURTS: THE COMMON LAW

Courts are an important part of the law-making process. Exercising their power of judicial review, they determine whether legislation is constitutional and interpret the legislative enactments. They have the principal responsibility for granting remedies under legislative systems. Courts also pass upon and interpret administrative regulations and grant remedies for violations.

The sense in which courts are a source of law central to the construction process is, however, principally based on the dual function of American appellate courts. On one hand, the appellate courts review decisions made by trial courts by providing a forum for those litigants who are dissatisfied with judgments of the trial courts. More important, the process by which appellate courts judge the correctness of decisions made by the trial courts has had an immense impact upon design and construction. American courts follow precedent to resolve disputes. An understanding of how courts make law and resolve disputes requires an understanding of the precedent system, discussed in Section 2.14.

SECTION 1.08
CONTRACTING PARTIES

Sources of law usually are public bodies. Yet contract law, with the broad autonomy granted the contracting parties to determine the terms of their exchange, grants lawmaking power to those who make contracts.

Even if autonomy is looked upon as an inherent liberty in a free society, the state still plays a role—enough of a role to create a "partnership" between contracting parties and the state. Despite broad autonomy given the parties, the state still determines who can contract, creates the formal requirements, and provides remedies when contracts are not performed.

If the contracting parties can make their contract a source of law, this is the most *democratic* form of lawmaking. The contracting parties freely determine the terms of their exchange and limit their freedom of action voluntarily by agreeing to perform in the future, with the coercive arm of the state operating if they do not.

Chapter 5 is devoted to some aspects of contract law and Chapter 6 to contract remedies. At this point, it is sufficient to mention the emergence of the bipolar analysis of contract law, one that draws a sharp distinction between *negotiated* contracts and contracts of *adhesion*. The former emphasize autonomy or freedom of contract based upon consent being freely given by the contracting parties. The latter involve contracts made with the terms, or at least most of them, largely dictated by *one* of the parties and merely adhered to by the other. This distinction is discussed in Section 5.04(C).

SECTION 1.09 PUBLISHERS OF STANDARDIZED DOCUMENTS

Adhesion contracts, represented by pre-prepared printed contracts, are prepared by *one* party to the contract largely if not exclusively with its interests in mind. Yet many pre-prepared printed forms are used in contracts for design and construction that should not be considered adhesion contracts in the sense that the term is used in

Sections 1.08 and 5.05. Often contracts for design and construction services are based largely upon documents published by professional associations such as the American Institute of Architects (AIA), the National Society of Professional Engineers (NSPE), and the Associated General Contractors (AGC), to mention some of the most important.

These associations have no official status and some would contend that they cannot be considered sources of law. Certainly contracting parties need not use the standardized documents or, if they do use them, are free to make drastic changes. But the frequency with which these documents are used largely unchanged justifies classifying these associations as sources of law. Even if in a *technical* sense they should not be placed along with courts, the realities of construction contracting make them even *more* important.

SECTION 1.10 SUMMARY

Understanding how law bears upon design and construction requires a recognition of the wide variety of law sources. Those involved in design and construction should be aware of the laws and legal institutions that have placed their mark on design and construction. Trends as to which sources of law predominate and the forms that legal intervention take must be appreciated. A half century ago it could be confidently stated that the principal sources of law in design and construction were the contracting parties and the courts. Increasingly, other instrumentalities such as legislatures and regulatory agencies are leaving their imprint upon design and construction.

2

The American Judicial System: A Forum For Dispute Resolution

SECTION 2.01 STATE COURT SYSTEMS: TRIAL AND APPELLATE COURTS

Each state has its own judicial system. Courts are divided into the basic categories of trial and appellate courts. Within each category, there may be a subclassification based upon the amount or type of relief sought or upon the nature of the matter being litigated.

The basic trial court, frequently called the court of general jurisdiction, hears all types of cases. Depending upon the state, it may be called a superior, district, or circuit court. The bulk of the work before such courts consists of criminal cases, personal injury cases, commercial disputes, domestic matters (divorce, custody, and adoption), and probate (transfer of property at death).

A court of general jurisdiction may review determinations of administrative agencies (zoning, employment injuries, licensing, etc.). The presiding official will be a judge, and in certain matters there may also be a jury. The division of functions between judge and jury is discussed in Section 2.11. Courts of general jurisdiction are usually located at county seats.

Many states have established subordinate courts of limited jurisdiction. Municipal or city courts often have jurisdiction to try matters that involve less money than a minimum figure set for courts of general jurisdiction. There can also be a jury in such cases. As a rule, the dockets are less congested in these courts of limited jurisdiction than are the dockets in the courts of general jurisdiction. The procedures in municipal or city courts are essentially the same as those in the courts of general jurisdiction.

Some state legislatures have established small claims courts that provide expeditious and inexpensive procedures for disputes involving small sums. Usually the party seeking a remedy from such a court will be able to begin the action by paying a small filing fee and filling out a form provided by the clerk of the small claims court. The procedures are usually informal. There are no juries. While the judges are lawyers, attorneys are less important in such disputes, and in some states attorneys are not permitted to represent the parties.

Some states still use justices of the peace, a vestige of a rural, dispersed society. These judges often have had little or no legal training. Although they are being phased out in favor of small claims, municipal, or city courts, they still exist in sparsely populated areas of some states. The future may witness their elimination.

At least one appeal is usually possible from a decision of a trial court. Generally the appeal is to the next highest court. For example, a party losing a decision in a municipal court may have a right to appeal to the court of general jurisdiction or to an appellate division consisting of judges of the court of general jurisdiction. Appeals from courts of general jurisdiction are made to an appellate court. In the more populous states, usually an intermediate court must review a case if the trial court decision is appealed. In such states, the state supreme court generally is given the discretion to decide whether it will hear an appeal from an intermediate appellate court. In states without intermediate appellate courts, an appeal is made from the court of general jurisdiction to the supreme court of the state.

State court judges are either appointed by the governor or elected. Even states that elect judges have many judges who were initially appointed. Generally judges are reelected and leave office only when they retire or die. Such vacancies are filled by interim appointments, and appointees are usually elected when they go before the voters.

Some states use an elective process in which all judges—or at least appellate judges—periodically submit their records to the electorate but do not run against other candidates.

SECTION 2.02
FEDERAL COURT SYSTEM

The American judicial system includes two systems—federal and state—which to a degree exist side by side. Each state has its own judicial system, while the federal courts also operate in each state. The federal courts have jurisdiction to decide federal questions, that is, disputes involving the federal Constitution or federal statutes. They also have jurisdiction to hear civil disputes between citizens of different states, often called "diversity of citizenship" jurisdiction. While in theory the amount in controversy in each type of case must exceed $10,000, the amount in controversy is

important principally in diversity of citizenship cases.

Many claims brought before federal courts can also be brought in state courts, creating concurrent jurisdiction. Some matters, however, can be brought only in the federal courts. The principal areas of exclusive federal jurisdiction are as follows:

1. admiralty
2. bankruptcy
3. patent and copyright
4. actions involving the United States
5. violations of federal criminal statutes

In exclusive jurisdiction matters, no "amount in controversy" requirement exists.

The federal courts operate under more modern and less formal procedural rules than do most state courts. With regard to substantive law (laws that establish legal rights and duties), the federal courts use the substantive law of the state in which they sit or some other applicable state law, unless the case involves federal law, such as a federal constitutional provision or a federal statute.

The basic trial court in the federal court system is the district court. Each state has at least one, and the populous states have a number of district courts located in the principal metropolitan centers.

The district court is presided over by a federal district court judge, and provisions for juries exist in certain cases. A party may appeal from a decision of the district court to a circuit court of appeals. Generally, each circuit court of appeals reviews decisions of district courts that are located in states in a designated geographical area.

A party dissatisfied with the result of a decision by the circuit court of appeals may ask that the U.S. Supreme Court review the case. In general, the Supreme Court determines which cases it will review. If it decides not to hear the case, the matter is ended. If it decides to hear the case, briefs and oral arguments are presented to the

Supreme Court. The Supreme Court usually rejects petitions for hearings unless the matter involves either a conflict between federal circuit courts or a very important legal issue.

Federal court judges are appointed by the President of the United States and are confirmed by the Senate. In essence the judges serve for life or until voluntary retirement.

Within the federal system, a few courts deal with specialized matters. For example, the Claims Court hears claims against the U.S. government, appeals from which are taken to the U.S. Court of Appeals for the Federal Circuit. Usually these cases relate to tax disputes or to disputes between government contractors and government agencies that award government contracts. Other examples of specialized courts are the Tax Court, which hears disputes between taxpayers and the government, and the Patent Court, which reviews disputed patents.

SECTION 2.03
STATUTE OF LIMITATION: TIME TO BRING THE LAWSUIT

Statutory provisions require that legal action be commenced within a specified period of time. Such statutes in the context of construction are discussed in Sections 17.08(B) and 27.03(G).

SECTION 2.04
HIRING AN ATTORNEY: ROLE AND COMPENSATION

Usually a person with problems that may involve the law or the legal system will consult an attorney. After an inquiry into the facts and a study of the law, the attorney advises a client of its legal rights and responsibilities. While the attorney may also give an opinion on the desirability of instituting legal action or defending against any action that has been asserted against a client, the litigation choice is usually made by the client.

Unless the law sets the fee, which is rare,

attorney and client can determine the fee. Sometimes the fee is a specified amount for the entire service to be performed, such as uncontested divorces or simple incorporations. The fee may be a designated percentage of what is at stake, such as in the probate of an estate. Often the fee is based on time spent by the attorney, typically computed on an hourly or daily basis. If no specific agreement on the fee is reached, the client must pay a reasonable amount.

Hourly rates charged depend upon the attorney's skill, the demands on her time, the amount involved, the complexity of the case, what the client can afford, the locality in which the attorney practices, and the outcome in the event of litigation. Although hourly rates vary greatly, at present an experienced attorney located in a large city is likely to charge from $100 to $150 an hour. (The second edition published in 1977 states the fee at $50–75.)

A potential client should ask a prospective attorney the likely charges for legal services. Such advance inquiry can avert possible later misunderstandings or disputes over the hourly rate. The client may wish to set a maximum figure for a particular legal service. Attorneys are reluctant to accept maximum figures because it is often difficult to predict the amount of time needed to provide proper legal service.

Commonly in personal injury or death cases and occasionally in commercial disputes, attorneys use contingent fee contracts. The lawyers are not paid for legal services if they do not obtain a recovery for the client. Usually a client agrees to reimburse the attorney for out-of-pocket costs for such items as deposition expenses, filing fees, and witness fees. For taking the risk of collecting nothing for time spent, the attorney will receive a specified percentage of any recovery. In personal injury cases, the percentage can range from 25% to 50%, depending upon the locality, the difficulty of the case, and the reputation of the attorney. In some metropolitan areas, attorneys will charge 25% to 33% if they obtain a settlement without trial and 33% to 40% if they go to trial.

The contingent fee system is much criticized, especially where the attorney obtains an astronomical sum in a tragic injury or death case. Such a system gives the attorney an entrepreneurial stake in the claim. Some feel this is unprofessional and may influence the attorney's advice as well as raise questions when attorney and client disagree over settlement. Yet some defend the contingent fee as the only method by which poor clients can have their claims properly presented. States are beginning to regulate contingent fees.

A retainer is an amount paid by a client to an attorney either at the commencement of or periodically during the attorney-client relationship. What function does the retainer serve? Is it an advance on fees, a nonrefundable payment made to have a call upon the attorney's services or an agreed value on fees for certain services? Put another way, if the value of the attorney's services is less than the amounts paid as a retainer, must the attorney refund the balance?

While the retainer payment can be an agreed-upon value for the legal services performed by the attorney or an amount to pay for having a call upon the attorney's services unrelated to fees for services, the retainer payment is more commonly an advance payment on any fees that the client is obliged to pay the attorney. Sometimes it covers routine services but does not include extraordinary services such as litigation. The attorney and the client should agree upon the function and operation of any retainer.

In the past, bar associations published schedules of recommended or minimum fees. Such schedules violate the antitrust laws and are no longer used. For a discussion of architectural and engineering fee schedules see Section 14.03.

The high cost of legal services and an increasing use of prepaid group health plans has led to the suggestion of prepaid group legal service plans.

The law recognizes the need for a client to be as candid in communicating with an attorney as with a member of the clergy, a doctor, or a spouse. For this reason, the attorney must keep confidential any communication made by the client, unless the communication was to plan or commit a crime, or if the client challenges the competence of the attorney's performance.

SECTION 2.05
JURISDICTION OF COURTS

Jurisdiction—the power to grant the remedy sought—means power over the person being sued and over the subject matter in question. Since federal courts have been discussed in Section 2.02, discussion in this section centers upon state court jurisdiction.

The U.S. Constitution requires that state and federal governments insure that a defendant receives due process, the opportunity of knowing what the suit is about, who is suing, and where and when the trial will take place. Usually this is accomplished by having a process server hand a summons ordering the defendant to appear and a complaint stating the reasons for the lawsuit to the defendant in the state where the trial is to take place. The power of the state to compel the defendant to appear does not extend beyond the border of the state. If the defendant does not come within the state, traditionally the state court in which the lawsuit has been brought would not have jurisdiction over the person or entity against whom the claim has been made. The plaintiff would have to sue in the state where the defendant actually could be handed the legal papers.

The requirement of physical presence in the state where the court sits can place great hardship on a plaintiff who may be forced to begin a lawsuit in a state that is far from the evidence and witnesses. For this reason, most states have enacted long-arm statutes that permit a plaintiff in certain cases to sue in her home state despite the fact that she cannot hand the legal papers to the defendant within the plaintiff's state.

Long-arm statutes are often used to sue defendant motorists who reside in a state other than the one where the accident occurred or where the injured party resides. Suppose an Iowa driver were involved in an accident in Illinois that injured an Illinois resident. The Iowa driver returns to Iowa after the accident. The injured Illinois resident can always sue in Iowa. Usually, the injured party would also be given the opportunity to sue in Illinois by mailing a notice of the lawsuit to the other driver in Iowa or by filing legal papers with a designated state official in Iowa or Illinois.

Sometimes long-arm statutes are used to sue businesses that have their legal residence in a state other than that of the plaintiff. Suppose a New York company sells, directly or indirectly, electric drills to Wyoming purchasers. A Wyoming buyer is likely to be able to sue the New York seller in a Wyoming court for injuries resulting from purchase and use of a defective drill.

Jones Enterprises, Inc. v. Atlas Service Corp.[1] involved a claim brought in the federal court in Alaska by a prime contractor against an out-of-state supplier, an out-of-state subsupplier to the supplier, and an out-of-state structural engineer retained by the supplier to design precast concrete work. The claim resulted when the structure constructed in Alaska by the prime contractor and made with materials designed and shipped to Alaska by the out-of-state defendants collapsed. The defendants were not served legal papers in Alaska, since they were not physically present there. But the plaintiff relied upon an Alaska long-arm statute.

None of the out-of-state defendants were qualified to do business in Alaska, none had

agents in Alaska, and all the work had been performed by the defendants outside Alaska. However, the court held that all the defendants were subject to the jurisdiction of the Alaska federal court in which the action had been brought. Although the supplier and the subsupplier to the supplier had had some minimal contact with Alaska, the structural engineer had *never* even been in Alaska in connection with the project. But since he knew that the building was to be constructed in Alaska and he sent his design into the stream of commerce aimed at Alaska, the court concluded that he, along with all of the other out-of-state defendants, was subject to the Alaska court.

The tendency has been to expand jurisdiction of a court over persons who are not personally served with the legal papers within the state where the court is located.

Jurisdiction may also involve the subject matter of the lawsuit. For example, if land is located in State A and there is a dispute as to the ownership of the land, only State A will have jurisdiction, even though persons who may claim an interest in the land may not be residents of State A and may not be served with the legal papers in State A.

States usually limit their jurisdiction to matters of concern to that state. A state may have jurisdiction because property is located in that state, the injury occurred in that state, the plaintiff or defendant is a resident of that state, or a contract was made or performed within that state.

Another aspect of jurisdiction relates to the type of remedy or decree sought. Early in English legal history two sets of courts existed—law courts and equity courts. Juries were used in the former but not in the latter. In addition, equity court judges could award remedies not available in the law courts. Parties seeking equitable relief had to show that their remedy in the law courts was inadequate. As a result of this dual set of courts, two sets of procedural

---

1. 442 F.2d 1136 (9th Cir.1971).

and substantive rules emerged. Generally, the procedures and remedies tended to be more flexible in the equity courts.

The division between law and equity was incorporated into the U.S. judicial system. However, a gradual merger of the two systems has occurred in most states. For all practical purposes, in most states only one set of courts exists.

Some differences remain, however. The most important is that certain remedies may be given only by an equity court. The most important of these remedies are specific performance (under which one party is ordered by a court to perform as promised in a contract), injunctions (orders by the court that persons do or do not do certain things), and reformation (rewriting of a written document to make it accord with the parties' actual intention).

As Section 2.10 explains, there are no juries when an equitable remedy is requested.

SECTION 2.06
PARTIES TO THE LITIGATION

Generally the party commencing the action is called the plaintiff, while the party against whom the action is commenced is called the defendant. In construction disputes it is common for the defendant to assert a counterclaim against the plaintiff or to make claims against third parties arising from the same transaction. For example, suppose an employee of a subcontractor sues the prime contractor based upon a claim that the prime contractor has not lived up to the legal standard of conduct and this caused an injury—a loss to the plaintiff. The prime contractor may, in addition to defending the claim made by the employee of the subcontractor, sue the architect and the owner, claiming that the latter was responsible for failure of the architect to live up to the legal standard, or sue the subcontractor employer of the claimant based upon indemnification.

The prime contractor in asserting these

claims is called a cross-complainant in most state courts and a third-party plaintiff in the federal courts. Those against whom these claims are made would be called cross-defendants in most state courts and third-party defendants in the federal courts.

Cross claims by defendants were difficult to maintain in the law courts in England and under early American procedural rules. However, equity courts freely permitted a number of different parties to be involved in one lawsuit, so long as the issues did not become too confusing. They sought to resolve disputes relating to the same transaction in one lawsuit. Ultimately, the rules developed in the equity courts prevailed, and it is generally possible to have multiparty lawsuits if jurisdiction can be obtained over all the parties and if litigation can proceed without undue confusion or difficulty.

SECTION 2.07
PREJUDGMENT REMEDIES

Not uncommonly the defendant against whom a judgment has been obtained has insufficient assets to pay the judgment. A judgment may be uncollectible if assets are hidden or heavily encumbered by prior rights of other creditors. Most states have statutes creating prejudgment, or as they are sometimes called, provisional remedies. Plaintiffs (or defendants who are asserting a counterclaim) may be able, in advance of litigation, to seize or tie up specific assets of the opposing party in such a way to insure that if they prevail, the judgment can be collected. These assets may be attached if in the possession of the defendant, or garnished if in the hands of third parties. It is not uncommon for a plaintiff to attempt to tie up assets such as bank accounts or other liquid assets in advance of the litigation.

The party whose assets are tied up can dissolve (have removed) the attachment or garnishment by posting a sufficient bond. Parties whose assets have been tied up may ultimately prevail in the litigation. If so, they may have unfairly suffered damage from the seizure or tying up of their assets. To protect against the risk of not finding assets out of which the prevailing party can be indemnified for such losses, the law frequently requires that the attachment or garnishment be accompanied by a bond.

Even if a party can avoid having assets tied up in advance of litigation by posting a bond, such a protective device is often not feasible where the party whose assets are being tied up or seized is poor. As a result, great hardship can occur to poor people and their families if the wage earner's wages are seized or if the sheriff repossesses a car, a television set, or furniture being purchased under a conditional sales contract before the court hearing. Under such arrangements the buyer takes possession, but the seller retains ownership until the payments have been made. As a result of this, the U.S. Supreme Court has looked closely at prejudgment remedies to insure that they give the defendant wage earner or debtor due process.[2]

Another prejudgment remedy is a preliminary injunction, an equitable decree ordering a defendant to cease doing something that the plaintiff claims is causing or will wrongfully cause injury to the plaintiff. Such an injunction can be issued before a full trial if the plaintiff can show that it is needed to preserve the status quo until the trial, that the plaintiff would suffer irreparable loss if it were not issued, and that there is a strong likelihood that the plaintiff will prevail at the trial.

Suppose an executive of a company that makes scientific instruments has threatened or begun to violate a promise made in her employment contract not to compete with the employer after leaving the company. If such a promise not to compete is reasonable, a court of equity will enforce it by ordering the executive to live up to the promise. The employer would very likely seek a preliminary injunction because irreparable harm can be caused the employer between the time the executive starts to work for a competitor and the issuance of a court order prohibiting the executive from doing so after a full trial. If the employer can show a strong likelihood that the clause will be enforced, the court should award a preliminary injunction.

The employer must post a bond to protect the executive from any damage caused by the issuance of the preliminary injunction if the executive should prevail after a full trial.

SECTION 2.08 PLEADINGS

The lawsuit is usually begun by handing a summons and complaint to the defendant, called service of process. The defendant has a specified time to respond. The defendant's answer may assert that the plaintiff has not stated the facts correctly or that defenses exist even if the allegations of the plaintiff's complaint are true. In some states, the legal sufficiency of the complaint may be attacked by filing a demurrer, a statement that even if the plaintiff's facts are true, the plaintiff has no valid claim.

In addition to answering the plaintiff's complaint, the defendant in some cases may assert a claim against a third party by filing a cross-complaint or may file a counterclaim against the plaintiff asserting that the defendant has a claim against the plaintiff.

In some states the plaintiff will be required to submit a reply to the answer. Most states require an answer to a counterclaim. In early American legal practice, there were a substantial number of other pleadings that might be filed. The tendency in American procedural law today is to reduce the number of pleadings.

The pleadings should inform each party of the other party's contentions and eliminate from the trial matters upon which

2. *Fuentes v. Shevin,* 407 U.S. 67, 92 S.Ct. 1983, 32 L.Ed.2d 556 (1972).

there is no disagreement. By an examination of the pleadings, an attorney or a judge should be able to determine the salient issues to be explored in the litigation. Pleadings should streamline the litigation and avoid the proof of unnecessary matters. Trial preparation should be more efficient and settlement expedited if each side knows the issues upon which the other intends to present evidence at the trial.

The complaint should be a concise statement of facts upon which a plaintiff is basing a claim and the specific remedy sought. Unfortunately, complaints are often prolix and contain factual statements and serious charges that the plaintiff's attorney may not be able to prove or which have not been checked out carefully. The inexperienced litigant should understand some reasons for this.

Frequently the lawsuit is commenced before the plaintiff's attorney has had sufficient time to investigate thoroughly and use the discovery procedures described in Section 2.09. Unsure of the facts, an attorney may plead a number of different legal theories as a protective measure.

The attorney wants to be certain that the complaint will be legally sufficient, that is, that the judge will not uphold a claim made by the opposing attorney that no legal claim has been stated. The plaintiff's attorney may decide to use the identical language of pleadings that have been held to be sufficient in the past. Such language may not be tailor-made to the particular case in which it is used. If the language is taken verbatim from an old pleading form book it may contain archaic and exaggerated language.

The plaintiff's attorney may believe that the only way to get the defendant or the defendant's attorney to seriously consider a client's claim is to start a lawsuit immediately, make strong accusations, and ask for a large amount of money.

Unfortunately, exaggeration and overstatement are frequently used legal weapons. But when litigants are not made aware of the reasons for such language, its use may simply increase the already existing hostility between the litigants and make settlement more difficult.

If the defendant has been served with a summons and complaint, failure to answer or to receive an extension of time to answer within the time specified in the statute allows the plaintiff to take what is called a default judgment and the case is lost by default.

The plaintiff who obtains a default judgment can collect on this judgment in the manner described in Section 2.12. Most states permit the judge to set aside the default judgment. To do so, the defendant must present cogent reasons why it would be unfair to enforce the judgment despite failure to respond to the pleadings in the designated time. Modern courts are *more* willing than courts a generation ago to set aside default judgments if the defendant can show a valid defense, that failure to answer was the result of an excusable mistake and the amount of the judgment is substantial. Yet default judgments are rarely set aside. For this reason, a defendant who wishes to contest a claim made in a complaint should have an attorney answer the complaint within the specified time.

SECTION 2.09
PRETRIAL ACTIVITIES

The parties and their attorneys wish to discover all facts that are material to the lawsuit. "Discovery" is the legal process to uncover information in the hands of the other party. One method of discovery is written interrogatories. One party's attorney sends a series of written questions to the other party, who has a specified period of time in which to respond.

The other method is to take a deposition, that is, to compel the other party or its agent to appear at a certain time and place and answer questions asked by the attorney seeking discovery. The attorney is also likely to demand that the person being questioned bring all relevant documents relating to the matter in dispute.

The person being questioned usually brings an attorney. Questions are asked under oath. Questions and answers are transcribed by a reporter, and a transcript is made available to both parties.

Generally the transcript does not substitute for testimony at the trial. However, the transcript can be used to impeach (that is, to contradict) the witness if the testimony at the trial is inconsistent with the statements made under oath in the discovery process.

A proper use of discovery should avoid surprise at the trial, reduce trial time, and encourage settlement. However, the sweeping range of inquiry and the wide latitude given to demand that documents be produced has made the process time consuming and very costly.

Witnesses who will be out of jurisdiction at the time of trial or who are very sick and may not be alive when the case is tried can also be examined under oath and their testimony recorded and transcribed. This deposition process can substitute for testimony that would otherwise be unavailable at the time of trial.

In many states, the pretrial conference is another step in the litigation process. Usually such a conference is conducted in the judge's chambers in the presence of the judge, the attorneys, and sometimes the parties. The main purposes of the pretrial conference are, like those of the pleadings, to narrow the issues, avoid surprise and encourage settlement.

SECTION 2.10   THE JURY

A dispute not settled or abandoned before trial will be submitted to a judge and sometimes to a jury. In many cases the right to a jury trial is constitutionally guaranteed. As stated in Section 2.05, one exception is actions where decrees given by an equity judge are sought. The parties can agree to try the case without a jury. The principal present use of juries is in criminal and personal injury trials. Their use in commercial disputes, although available, is less common.

Historically juries consisted of twelve lay persons selected from the community to pass upon guilt and sometimes sentence in criminal matters and to decide disputed factual questions in civil matters. Juries have been attacked as inefficient. Jury trials take longer, cost more, and have a higher degree of reversals by appellate courts. Some states reduce jury size from twelve to six members. In criminal matters, the decision must be unanimous. Some states require only a five-sixths majority decision in civil disputes.

In many state courts, attorneys are allowed to question prospective jurors to determine their impartiality. An attorney dissatisfied with a particular juror can ask that the juror be excused for cause. The judge rules on whether there is proper cause to excuse the juror. Usually the attorneys can strike (that is, excuse) a designated number of jurors peremptorily (that is, for no reason at all). In the federal courts, jury examination is typically conducted by the judge.

Jurors need give no reason for their decision. While the judge instructs the jury as to the law, jurors are, for all practical purposes, free to do as they choose. Some criticize this, while others feel that it is a useful safety valve, especially in criminal matters, to allow members of the community to excuse law violators.

Until recently jurors did not make the details of their deliberations public. But of late, jurors, especially in controversial cases, openly describe the jury process to representatives of the media and anyone else who may be interested or willing to pay.

Some feel that juries are easily manipulated by clever and persuasive attorneys. Others feel that jurors are sensible and generally come to the right result. The jury system will always be controversial.

SECTION 2.11 TRIALS: THE ADVERSARY SYSTEM

The trial is usually conducted in public and is begun with an opening statement by the plaintiff's attorney. Sometimes the defendant's attorney also makes an opening statement. In the opening statement, the attorney usually states what she intends to prove and also seeks to convince the jury that the client's case is meritorious. Opening statements are less common in trials without a jury.

After opening statements, the plaintiff's attorney may call witnesses who give testimony. Also, physical exhibits and documents can be offered into evidence. In civil actions, the other party can be called as a witness.

An attorney cannot ask leading questions of the witnesses, except to establish less important preliminary matters or where the witness is of low mentality or is very young. A leading question is one that tends to suggest the answer to the witness. Such questions can be and usually are asked witnesses called by the other party.

Witnesses are supposed to testify only to those matters that they have perceived through their own senses. Usually, witnesses cannot express opinions upon technical questions unless they are qualified as experts. Expert testimony is discussed in greater detail in Section 17.06.

Under the hearsay rule, witnesses cannot testify what they have been told if the purpose of the testimony is to prove the truth of the statement. The danger in hearsay testimony is that the party who made the statement is not in court and cannot be cross-examined as to the basis upon which the statement was made. However, there are many exceptions to the hearsay rule.

After the attorney has finished questioning the witness, the other party's attorney will cross-examine and try to bring out additional facts favorable to her client or to discredit the testimony given on direct examination. Cross-examination can be an effective tool to catch a perjurer or show that a witness is mistaken. However, when used improperly, it may create sympathy for the witness or reinforce testimony of the witness.

Documents play a large role in litigation. Under modern rules of evidence, it is relatively easy to introduce into evidence documents to be considered by the court. Documents are hearsay testimony. They are usually writings made by persons who are not in court. However, there are many exceptions to the hearsay rule that permit the admission of documents such as official records and business entries. Often the attorneys will stipulate to the admissibility of certain documents.

After the plaintiff's attorney has presented her client's case, the defendant's attorney will present the defendant's case. After this is done, the plaintiff is given the opportunity to present rebuttal evidence. After all the evidence has been presented, the judge submits most disputed matters to the jury (when one is used). The judge instructs the jury on the law using instructions often difficult for a juror to understand. The jury meets in private, discusses the case, takes ballots, decides who prevails and how much should be awarded to the winning litigant. (In controversial cases this is often followed by a press conference. Refer to Section 2.10.)

The adversary system, though much criticized because of its expense, its often needless consumption of time, and the hostility it can engender, is central to the American judicial process. Each party, through its attorney, determines how its case is to be presented and can present its case in a vigorous and persuasive manner. Also, each party, mainly through cross-examination and oral argument, has considerable freedom to attack the other's case. Generally the judge acts as an umpire to see that the procedural rules are followed.

One justification for the adversary system is that when the smoke has cleared the truth will emerge. This assumes well-trained advocates of relatively equal skill and a judge who insures that procedural rules are followed.

Another justification for the adversary system is that it gives the litigants the feeling that they are being honestly represented by someone of their choosing and are not simply being judged by an official of the state. It can give the litigants a sense of personal involvement in the process rather than simply being passive recipients of decisions handed them by the state. But excessive partisanship can generate bitter wrangles and delays. Also, it can mean that a trial resembles military combat rather than a reasoned pursuit of the truth.

Even under the adversary system there are limits on advocacy. The law prescribes rules of trial decorum administered by the judge. The judge can punish those who violate these rules by citing them for contempt. The person cited can be fined and, in some cases, imprisoned. The legal profession can discipline its members who violate professional rules of conduct. As obvious illustrations, an attorney must not bribe witnesses, encourage or permit perjured testimony, or mislead the judge on legal issues. An attorney is the champion for the client yet part of the system of the administration of justice and must not do anything that would subvert or dishonor that system.

Attorneys frequently make objections in hotly contested litigation. This is always irksome to participants, but it is often, though not always, necessary. If objections are not made at the proper time, the right to complain of errors may be lost. The judge should be given the chance to correct judicial mistakes on the spot.

The possibility of drama and tension in the litigation process can increase if a jury is used. Some attorneys employ dramatic tactics to influence the jury. Although some enjoy the combat of litigation, litigation is at best unpleasant and at worst traumatic for the litigant or witness who is not accustomed to the legal setting, the legal jargon, or the adversary process. This can be intensified if the trial attracts members of the public or is reported in the media.

A trial is an expensive way to settle disputes. In addition to attorneys' fees, witness fees, court costs, and stenographic expenses, there are less obvious expenses to the litigant. Much time must be spent preparing for and attending the trial. The litigant may have to disrupt business operations by searching through records. For these as well as other reasons, most lawsuits are settled.

SECTION 2.12 JUDGMENTS

A judgment is an order by the court stating that one of the parties is entitled to a specified amount of money or to another type of remedy. Some judges rule immediately after the trial. Usually, the judge takes the matter under consideration. Often judges write opinions giving the reasons for their decisions.

SECTION 2.13 ENFORCEMENT OF JUDGMENTS

If the plaintiff obtains a money award, the defendant should pay the amount of money specified in the judgment. However, if the defendant does not pay voluntarily, the plaintiff's attorney will deliver the judgment to a sheriff and ask that property of the defendant in the hands of the defendant or any third party be seized and sold to pay the judgment. It is often difficult to find property of a defendant. In some states, defendants can be compelled to answer questions about their assets.

Even if assets can be found, exemption laws may mean that certain assets may not be taken by the plaintiff to satisfy the judgment. Legislatures declare certain property exempt from seizure. Statutes usually contain a long list of items of property that cannot be seized by the sheriff to satisfy a judgment because they are considered necessary to basic existence. Such items are automobiles, personal clothing, television sets, tools of trade, the family Bible, and other items that vary depending upon the state and the time in which the exemption laws were passed.

Homestead laws exempt the house in which the defendant lives from execution to satisfy judgments. Financial misfortunes notwithstanding, a debtor and those dependent on the debtor should have shelter. Sometimes homestead rights are limited in amount. Under certain circumstances a homestead can be ordered sold.

In many cases, judgments are not satisfied because the defendant has no property or the property in the defendant's possession is exempt. Collection of a judgment when the defendant does not pay voluntarily may be difficult, uncertain, and costly—another reason for settlement.

SECTION 2.14
APPEALS: THE USE
OF PRECEDENT

The party who appeals the trial court's decision is called the appellant. The party seeking to uphold the trial court decision is called either the appellee or the respondent.

The attorney for each party will submit a printed brief to the appellate court. The court can permit a typewritten brief. The brief of the appellant seeks to persuade the appellate court that the trial judge has made errors, while the brief for the respondent seeks to persuade the appellate court to the contrary or that any errors that have been committed were not serious enough to warrant reversal.

Usually the attorneys make brief oral arguments before the appellate court. Sometimes these arguments consist of a summary of the briefs. Sometimes the attorneys are questioned by the judges on specific points that appear in the briefs or trouble the judge. Introduction of evidence is not permitted in appeals.

In the American system most appellate courts write a decision stating their reasons for the way they have decided the case.

Decisions vary in length. Usually, the legal basis for a decision in a trial court or appellate court is a statute, a regulation, or a prior case precedent. The latter requires amplification.

All decision makers seek guidance from the past. Nonjudicial decision makers, such as committees, boards of directors, and organizations of all types, often seek to determine what they have done in the past as a guide for what they should do in the present. However, in such decision making it is not likely that the precedents of the past *must* be followed. English and American courts must follow precedent, subject to an exception that is described later in this section. Within a particular judicial system judges must decide matters in accordance with earlier decisions of higher courts within the system.

For example, all judges in the State of California must follow the precedents set by the highest member of the system—the state supreme court. Trial judges and members of the intermediate appellate courts must follow decisions of the intermediate appellate courts.

As stated, appellate courts give reasons for their decisions when they write written opinions. All members of the system are expected to follow the reasoning as well as the results in earlier cases decided by a higher court. However, California judges are not compelled to follow decisions from other state courts, although they may look to decisions of other state courts for guidance. Nor are they compelled to follow the decisions of federal courts except the decisions of the U.S. Supreme Court.

Despite reverence for precedent, some European countries forbid following precedent and some decision makers in this country, such as labor and commercial arbitrators, avoid it. This suggests that arguments can be made for and against following precedent.

A legal system should be reasonably predictable. Persons and organizations wish to know the law in advance to plan their activities. If they can feel confident that future judges will follow earlier decisions in similar cases, they should be able to predict and plan more efficiently. Knowing in advance what a judge or a court will do should encourage settlement of disputed matters. A fair judicial system should treat like cases alike.

Another argument for following precedent relates to conservation of judicial energy. If a matter has been thoroughly reviewed, analyzed, and reasoned in an earlier decision, there is no reason for a later court to replow the same ground. This reason for following precedent assumes great confidence in the decision makers of the past and a relatively static political, economic, and social order in which the system operates.

Those who attack the precedent system maintain that the sought-after certainty is illusory because courts can avoid precedent. They also argue that a precedent system tends to become rigid and unresponsive to changing needs. Finally, they contend that the precedent system can be an excuse for judges avoiding hard questions that should be examined.

Precedents must be followed only in similar cases. If the facts are different, precedents can be distinguished by judges and held not to control the current case. Even under a system of following precedents, there are times when earlier precedents become so outmoded that they are overruled. When American judges cannot accept the result that applying precedent compels, they create exceptions to the precedent. The greater degree of dissatisfaction, the greater the number of exceptions likely to be created. When the exceptions seem to swallow up the precedent, an activistic court may recognize the realities and overrule the precedent.

Good courts seek to avoid unthinking rigidity at one extreme and whimsical decision making at the other. The important feature of a precedent system is reasonable predictability that can accommodate change. The skillful attorney reads the precedents but also looks at the facts and any relevant political, social, and economic factors that may bear upon the likelihood that the precedent will be distinguished or overruled.

Usually all the members of the appellate court agree. Sometimes a dissenting opinion gives reasons for not agreeing with the majority. Sometimes the majority of the court that agrees on the disposition of the case cannot agree on the reasons for the decision, and one or more majority judges may write a concurring opinion.

SECTION 2.15
INTERNATIONAL CONTRACTS

Increased worldwide competition exists for building and engineering contracts. As a result, those who design and construct may be engaged by a foreign national, often the sovereign itself, in a foreign country where the work is to be performed. These international contracts raise special problems that are discussed in Sections 9.06 and 34.21. One topic deserves mention here, however.

Legal systems vary not only as to substantive law but also as to the independence of their judiciary and dispute resolution processes. In some countries, the judiciary is simply an arm of the state that follows to a large degree the will of the head of state or the agency with whom the contract has been made. A foreigner may not have confidence in the impartiality of such a dispute resolution process.

This can even be a problem in the United States or within a particular state. The U.S. Constitution recognized this by allowing the removal of a case from a state court to the federal court if there is diversity of citizenship. This problem can be particularly difficult when the designer or contractor is a company from an industrialized country doing business in a lesser developed country.

Lack of confidence in the judiciary often

necessitates a contract clause under which disputes will be resolved by some international arbitration process. Even where there is more confidence in the independence of the judiciary, such as in a transaction where disputes would normally be held before a court in the United States or a western European country, unfamiliarity with the processes and the possibility of the application of unfamiliar law may lead to contractual provisions dealing with disputes. For example, European legal systems do not use the adversary system.

They use a more inquisitorial system under which the judge is likely to be a professional civil servant who plays a more active role in resolving disputes. Similarly, laws in western European countries are often based upon brief yet comprehensive civil codes that are quite different from those an American lawyer may encounter in domestic practice. Under such conditions, it is common for the parties to provide by contract that disputes will be resolved by international arbitration with the applicable law that of a neutral and respected legal system.

3

FORMS OF ASSOCIATION: ORGANIZING TO ACCOMPLISH OBJECTIVES

SECTION 3.01 RELEVANCE

People engaged in design or construction should have a basic understanding of the ways in which individuals associate to accomplish particular objectives. The professional in private practice should know the basic elements of those forms of associations professionally available. Anyone who becomes an employee or executive of a large organization should understand the basic legal structure of that organization. The design professional who deals with a corporate contractor on a project should understand corporate organization. Those who are shareholders in or transact business with a corporation should understand the concept that insulates shareholders from almost all liabilities of the corporation.

The following are the most important organizational forms:

1. sole proprietorship
2. partnership
3. corporation
4. joint venture
5. unincorporated association
6. "loose" association

SECTION 3.02
SOLE PROPRIETORSHIP

The sole proprietorship, although not a form of association as such, is the logical business organization with which to begin the discussion. It is the simplest form and is the form used by many private practicing design professionals.

The creation and operation of the sole proprietorship is informal. By its nature, the sole proprietor need not make arrangements with anyone else for the operation of the business. Generally, no state regulations apply, except for those requiring registration of fictitious names or for having a license in certain businesses or professions. Sole proprietors need not maintain records relating to the operation of the business except those necessary for tax purposes. Sole proprietors have complete control over the operation of the business, taking the profits and absorbing the losses. They rent or buy space for the operation of the business, hire employees, and may buy or rent personal property used in the business operation.

The proprietorship continues until aban-

23

donment or death of the sole proprietor. Continuity of operation in the event of death sometimes can be achieved through a direction in the proprietor's will that his executor continue the business until it can be taken over by the person to whom it is sold or given by will. This insures that the business is a continuing operation during the handling of the estate and that there is no costly hiatus in operation.

The proprietor may hold title to property in his own name or in any fictitious name. Interest in the business can be transferred, the only exception being that in some states the spouse of the proprietor may have an interest in certain types of property used in the enterprise.

Capital must be raised by the proprietor or by obtaining someone to guarantee the indebtedness. This differs from a corporation, which can issue shares as a method of raising capital.

SECTION 3.03 PARTNERSHIP

(A) GENERALLY:
UNIFORM PARTNERSHIP ACT

A partnership is an association of two or more persons to carry on a business for profit as co-owners. Unlike a corporation it is not a legal entity. All states except Georgia and Louisiana have adopted the Uniform Partnership Act, which sets forth the rights and duties between the partners themselves if the partners have not specified to the contrary in the partnership agreement. It also deals with claims of third parties.

(B) CREATION

While it is not necessary that the partnership agreement be evidenced by a sufficient written memorandum, it is advisable to do so. In addition to reasons for having any agreement in writing, other reasons exist for expressing partnership agreements in writing. The transfer of an interest in land must be evidenced by a written memoran-

dum as must agreements that by their terms cannot be performed within a year. If the creation of the partnership is accompanied by a transfer of land, that portion of the partnership agreement dealing with the land should be evidenced by a written memorandum.

If the duration of the partnership expressly exceeds a year, the partnership agreement should be evidenced by a written memorandum. Agreements to answer for the debts of another must, under certain circumstances, be evidenced by a written memorandum. Partnership agreements with provisions of this type should be evidenced by a written memorandum.

In stating that certain agreements must be evidenced by a written memorandum, a distinction is frequently misunderstood. Suppose a partnership is created orally for a period of more than a year. The rights and duties of the partners will still be governed by the oral agreement and by the Uniform Partnership Act. However, partners have no legally enforceable obligation to perform in the future, although they will have to perform those obligations that accrued before the decision to terminate the oral partnership.

Third parties will still have whatever rights the law would give them against the partners despite the partnership agreement being required to be evidenced by a written memorandum. The fact that the law required the partnership agreement to be so evidenced because it was to last over a year will have a limited impact upon the partners and, in most cases, no impact upon third parties.

(C) OPERATION

Generally, the partners decide who is to exercise control. Most matters can be decided by majority vote. However, certain important matters, such as amendment of the partnership agreement, must be by unanimous vote. Sometimes partners wish to have something other than the majority rule apply. In a large partnership, a small group of partners may be given the authority to decide certain matters without requiring that a majority of the partners approve these decisions.

(D) FIDUCIARY DUTIES

As fiduciaries, the partners have an obligation to act in good faith toward each other. A partner must account to the partnership for any secret profits made. A partner must not harm the partnership because of any undisclosed conflict of interest. For example, where the partnership represents the owner in dealing with a construction company, it would be improper for one of the partners to have a financial interest in the construction company unless this were disclosed in advance to the partners and to the owner-principal.

Since the fiduciary obligation is of great importance, it may be helpful in the partnership agreement to spell out the permissible scope of outside activities of individual partners. In the absence of an understanding among the partners as to permissible outside activities, any activity that raises a conflict of interest or competes in any way with the partnership would not be proper. For further discussion see Section 14.04(B).

(E) PERFORMANCE OBLIGATIONS, PROFITS, LOSSES, WITHDRAWAL OF CAPITAL AND INTEREST

In the absence of any agreement to the contrary, the partners are to devote full time to the operation of the business and share profits and losses equally even if one works more than the others. If a specified proportion of the profits is allocated by partnership contract to specified partners, that same percentage will apply to losses. If some arrangement other than this is desired, it must be expressed in the partnership agreement.

Normally, partners cannot withdraw capital during the life of the partnership unless permitted by the partnership agreement. A partner may collect interest for money or property lent to the partnership, as well as for capital contributions advanced upon request of the partnership.

(F) AUTHORITY OF PARTNER

Section 9(1) of the Uniform Partnership Act states:

> Every partner is an agent of the partnership for the purpose of its business, and the act of every partner, including the execution in the partnership name of any instrument, for apparently carrying on in the usual way the business of the partnership of which he is a member binds the partnership, unless the partner so acting has in fact no authority to act for the partnership in the particular matter, and the person with whom he is dealing has knowledge of the fact that he has no authority.

Clearly the authorized acts of the partner will bind the partnership. The more difficult problems arise when the partner's acts are not authorized.

Here the doctrine of apparent authority can charge the partnership with unauthorized acts by a partner. The partnership can create apparent authority by making it appear to third parties that a partner has authority he does not actually possess (See Section 4.06.). Certain extraordinary acts, such as criminal or other illegal acts, are not charged to the partnership. However, because of the vast range of authority given to the partners, both actual and apparent, and the fiduciary obligation owed by each partner to the other, the character and integrity of prospective partners are of great importance.

(G) LIABILITY OF GENERAL PARTNERSHIP AND INDIVIDUAL PARTNERS

This section treats general partnerships. The liability of limited partners is discussed in Section 3.03(J).

The liability of the partnership and the partners for debts of the partnership and individual debts of the partners is complicated and confusing. Clearly, creditors of the partnership can look to specific partnership property and, if this is insufficient, can look to the property of the individual partners. Creditors of individual partners who obtain a court judgment may satisfy the judgment out of the partner's interest in the partnership property, but the partnership can avoid losing the property by paying the judgment.

Even more complicated is the liability of an incoming partner for obligations created before becoming a partner and of an outgoing partner for obligations incurred after leaving the partnership. Generally, the partner is not liable for obligations incurred before becoming a partner. The partner can eliminate liability for obligations that occur after leaving the partnership by informing those who have dealt with the partnership of his departure. But because of the uncertainty of the law in this area, it is best for those who enter existing partnerships to determine what obligations exist at the time they enter the partnership and those who leave partnerships to notify those who have dealt with the partnership of their departure.

Because of the ease in which partnerships can be dissolved and reconstituted, considerable confusion exists as to the right of creditors of predecessor partnerships to hold successor partnerships for the obligations of the predecessor. Sections 17 and 41 of the Uniform Partnership Act should be consulted.

A partner who incurs liability or pays more than a proper share of a debt should receive contribution from the other partners. Similarly, a partner who incurs liability or pays a partnership debt should be indemnified by the partnership. Usually these are expressly dealt with in the partnership agreement.

(H) TRANSFERABILITY OF PARTNERSHIP AGREEMENT

A partner has some interests that are transferable. Profits can be assigned, but not specific partnership property or management and control.

(I) DISSOLUTION AND WINDING UP

Partnerships do not have the legal stability of corporations. Section 31 of the Uniform Partnership Act lists a formidable number of events that will dissolve the partnership. Section 32 lists an equally formidable number of events that will give a court the power to dissolve a partnership on application of a partner or a purchaser of a partner's interest. Sections 37–42 are rules for winding up a partnership.

(J) LIMITED PARTNERSHIP

Most states permit the creation of a limited partnership made up of one or more general partners and one or more limited partners. Such an organization permits investors to receive profits but limits their liability to their investment without having to use the corporate form. Generally, the general partner manages or controls the corporation while the limited partners are investors.

Investments cannot be services if the limited partnership status is to be maintained. Designers who work for developers and are paid in shares may find they are general partners with greater liability.

Nevertheless, investors who wish to protect their investment need not take absolutely passive roles. Section 10 of the Uniform Limited Partnership Act, effective in all states except Louisiana, permits a limited partner to inspect and copy records, to demand information on matters affecting the partnership, or to apply to a court to dissolve and wind up the limited partnership. Section 7 states that a limited partner will not be liable as a general partner unless he takes part in control of the business.

SECTION 3.04 PROFIT CORPORATION

(A) USE

The corporate form is used by most large and medium-sized businesses in the United

States and has become the vehicle by which many small businesses are conducted. In addition, there is an increasing tendency on the part of practicing design professionals to choose the corporate form, where possible, for their business organization. The corporate form takes on even greater significance in light of the increasing number of design professionals who are employed by corporations. Because many complexities exist in corporation law that cannot be discussed in this treatise, the discussion here must be brief and simple.

(B) GENERAL ATTRIBUTES

While the partnership is merely an aggregate of individuals who join together for a specific purpose, the corporation is itself a legal entity. It exists as a legal person. It can take hold and convey property and sue or be sued in its corporate name. As shall be seen, the other important corporation attributes are centralization of management in the board of directors, free transferability of interests, and perpetual duration. In addition, the corporation offers the advantage of limiting shareholder liability for the debts of the corporation to the extent of the obligation to pay for the corporate shares purchased. Although not all of these attributes are available to all types of corporations, as a general introductory statement, these attributes set the corporation apart from sole proprietorship and partnership.

(C) PREINCORPORATION PROBLEMS

Often persons called promoters set into motion the creation of a corporation. Although they may continue in control, sometimes they merely organize the corporation and turn control over to others. Usually they are compensated by the corporation after the corporation has been organized. This compensation may be in the form of cash, shares of stock, stock options, or positions within the corporation.

During the promotion phase, promoters often make contracts with third parties.

Third parties should consider the possibility that the corporation may not be formed or may not adopt the contract made by the promoter. Consideration of these possibilities may induce the third party to insist that the promoter be individually liable if the corporation is not formed or does not adopt the promoter's contracts. A third party who has doubts about the finances of the prospective corporation may wish to hold the promoter even if the contract is adopted by the corporation.

(D) SHARE OWNERSHIP

The corporation is a separate legal entity owned by the shareholders, who do not own any part of specific corporation property. Shareholders have a right against the corporation that is governed by statutes, the articles of incorporation, and the wording of the shares. One of the great strengths of modern corporation law is the variety of types of shares that can be employed by the corporation—preferred and common, par and no-par, and voting and nonvoting.

Ordinarily shares of stock in a corporation are freely transferable (one of the major advantages of incorporation). Usually restrictions exist on transferability of shares in closely held corporations whose shares are not sold to the general public. This enables the shareholders to keep control of the company by eliminating the possibility of outsiders becoming shareholders. Restrictions on transferability sometimes accompany shares purchased by executives or employees as part of a stock option plan.

(E) PIERCING THE CORPORATE VEIL

One of the principal functions of the corporation is to shield shareholders from corporate obligations. Ordinarily the liability of shareholders is limited to paying for shares they have purchased. This limitation of liability can operate unfairly where a third party relies upon what appears to be a solvent corporation. The corporation thought to be solvent may be merely a shell. The assets of the corporation may not be sufficient to pay the corporate obligations. Someone injured by the acts or failure to act of a corporation may find that the corporation is unable to pay for damages because the amount of capital paid into the corporation was very small.

The corporate form can be disregarded and, to use a picturesque phrase, the corporate veil pierced and shareholders held liable. A court may do so if unjust or undesirable consequences would result by interposing the corporation as an entity between the injured party (or a creditor) and the shareholders. This is more likely to be done where a person has suffered physical harm through corporate activities; where a creditor could not reasonably have been expected to check on the credit of the corporation; and where the one-person, family, or closely held corporation is used. The latter types of corporations have been popular because they combine control with limitation of liability. Ordinarily such corporations are valid and protect the shareholders from liability. However, this is so only if the corporation is used for legitimate purposes and the business is conducted on a corporate basis. Also, the enterprise must be established on an adequate financial basis so that the corporation is able to respond to a substantial degree for its obligations. If not, circumstances such as those mentioned can result in "piercing the corporate veil." Similar problems can arise when a corporation organizes a subsidiary corporation and holds all the shares of the subsidiary.

(F) ACTIVITIES, MANAGEMENT, AND CONTROL

The state law under which the corporation was created determines the outer limits of corporate activities and organization. Since state laws generally extend considerable latitude to the corporation on these matters, as a rule activities and organization are governed by the articles of incorporation.

The articles of incorporation are the constitution of the corporation. They generally set forth the permissible activities of the corporation, organization and management of the corporation, and the rights of the shareholders. Sometimes these matters are phrased in general terms in the articles of incorporation and articulated more specifically in the corporate bylaws. In large corporations, a set of corporate documents is likely to exist that delineates the chain of command and states the authority of corporate officers and employees to handle particular corporate matters. The ultimate power within the corporation lies with the shareholders, who delegate this power to an elected board of directors. In theory, the board of directors controls long-term corporate policies, while the day-to-day operations are handled by the corporate officers.

This model of corporate control may vary depending upon the type of corporation. In a smaller corporation, the board of directors, or even large shareholders, may exert an influence upon day-to-day operations. If the corporation is a large, publicly held corporation with thousands of shareholders, the actual power is largely with the management. Even though theoretically the shareholders can displace the board of directors, the diffusion of share ownership often makes it difficult for this to be done and gives the board of directors and the officers of the corporation effective power, with the shareholders having little control over policy or corporate acts. However, this does not make the corporation immune from "takeover" bids by other corporations. In the 1980s there were many well-publicized takeovers by soliciting shares from shareholders at above-market prices.

Some corporations have cumulative voting of shareholders for directors. A shareholder asked to choose a slate of seven members of the board may cast all seven votes for one candidate. Cumulative voting is required in some states and has the effect of protecting minority interests in the corporation.

The directors choose the officers of the corporation. Usually they will select a president, vice president, secretary, and treasurer. Larger corporations might also, through the board of directors, designate persons to serve as general counsel or controller or as other corporate officers.

The officers are in charge of the day-to-day running of the corporation. The larger the corporation, the more likely it is that many of the details will be delegated to other employees of the corporation.

There has been a marked development in American law towards extending fiduciary obligations of fair dealing to directors and to officers. Directors and officers owe each other fiduciary duties of fair dealing and a fiduciary duty to the shareholders. It is not proper, for example, for a member of the board or for an officer to use inside information to purchase shares of corporate stock from shareholders who are not aware of the inside information known by the director or officer. Directors and officers are not permitted to take advantage of economic opportunities that should be made available to the corporation.

The articles of incorporation usually provide for annual shareholder meetings, as well as for periodic meetings of the board of directors. Minutes must be kept of all board proceedings. The administrative burden of operating a corporation can be formidable.

(G) PROFITS AND LOSSES

Usually the board of directors determines the disposition or distribution of profits made by the corporation. Sometimes control is limited by the articles of incorporation, especially the rights of preferred shareholders. In addition, it may have obligations to the creditors of the corporation based upon contracts or upon agreements made between the corporation and lenders such as shareholders or banks. Profits may

be reinvested in the corporation and used for corporate purposes.

Subject to statutory regulations the board determines when (and how large) a share of the profits is to be paid as dividends to the shareholders. The board may pay dividends only out of certain specified funds. Sometimes the board issues a stock dividend instead of a cash dividend. In such cases, the shareholders receive additional shares in the corporation instead of money. Statutory or other limitations may exist relating to the redemption of shares by the corporation and to the repurchase by the corporation of its own shares.

If the board unlawfully issues dividends, the board members are liable to the corporation and to those creditors of the corporation harmed by the unlawful declaration of dividends.

Normally, individual shareholders are not responsible for losses of the corporation because of the insulation from personal liability given corporate shareholders. A shareholder who paid for shares in accordance with the purchase agreement with the corporation is not liable for any obligations of the corporation.

(H) LIFE OF CORPORATION

One advantage of the corporation is its perpetual life. Sole proprietorships end with the death of the sole proprietor. Often partnerships end with the death of any of the partners. The corporation will continue despite the death of shareholders. A court can dissolve a corporation in the event of a deadlock among the board of directors or under other circumstances.

(I) DISSOLUTION

The dissolution of a corporation is complicated and is governed largely by statute. The bankruptcy laws play a large role in disposing of assets of the corporation upon dissolution. This treatise does not discuss dissolution.

SECTION 3.05
NONPROFIT CORPORATION

Nonprofit corporations are similar in organization and operation to profit corporations. The major difference is that no profits can be distributed to shareholders by a nonprofit corporation. Examples of nonprofit corporations are hospitals, most educational institutions, and charities.

Capital for a nonprofit corporation is raised by donations, grants, and, occasionally, the sale of shares. Usually there are members instead of shareholders. The members elect the board of directors or trustees; the board selects officers to run the corporation. There are articles of incorporation, bylaws, meetings, and other similarities to a profit corporation. Generally, the shareholders are insulated from personal liability for the debts of the corporation. Such corporations are exempt from taxes since they have no profit. Tax exemptions can be lost if certain types of political or profit-making activities are engaged in by nonprofit corporations.

SECTION 3.06
PROFESSIONAL CORPORATION

While design professionals traditionally have practiced as sole proprietors or partnerships, increasingly states have permitted them to practice through a professional corporation. The corporation form has been used mainly to take advantage of tax laws that allow employees of corporations to receive fringe benefits without having them included within taxable income.

Some states that have allowed professionals to incorporate have made it clear that they cannot use the corporate form to shield their individual assets, a principal reason why business corporations are used.

Herkert v. Stauber [1] demonstrates the interrelationship of legal issues in construction disputes as well as the ambit of professional corporation statutes. The dispute involved a claim by an owner against a professional corporation that had agreed to design and build an apartment house for the elderly. The owner claimed that the contract required the professional corporation to obtain approval of a federal agency that would assist in funding the project. One of the issues involved the extent to which individual shareholders in the professional corporation were liable for damages caused by breach of this obligation by the corporation.

The applicable professional incorporation law did not allow the corporate form to limit liability for negligent performance of professional services. Did failure to obtain funding from a federal agency constitute negligent performance of *professional* services?

The court did not have to determine whether the breach was negligent because it concluded, citing this treatise, that obtaining financing for construction is not a professional service. It limited professional architectural services to *design* and *supervision*. In addition to pointing to the absence of any professional training in this field, it pointed to the absence of standards to determine whether failure to obtain financing was negligent.

SECTION 3.07 JOINT VENTURE

Joint ventures can be created by two or more separate entities who associate together usually to engage in one specific project or transaction. Such arrangements are contractual and can be expressed by written contracts or implied by acts. In large construction projects, two contractors may find that they cannot handle a particular project individually but can do so if they associate themselves in a joint venture.

Usually the agreement under which such

---

1. 106 Wis.2d 545, 317 N.W.2d 834 (1982).

a joint venture is created is complex and sets forth in detail the rights and duties of the joint venturers. Joint ventures created informally by acts have gaps as to specific rights and duties. These gaps are filled in by the law. When this must be, it is likely that principles of partnership will be applied. Refer to Section 3.03.

Suppose there is a dispute as to the existence of a joint venture. Since joint ventures are considered very much like partnerships for one transaction, the relationship is often examined to determine whether partnershiplike attributes are present. For example, sometimes it is stated that there must be a community of interest, a common proprietory interest in the subject matter, the right of each to govern policy, and a sharing of profits and losses.[2]

If a joint venture has some of the attributes of a partnership as between the venturers themselves, such attributes can relate to the rights of third parties who deal with a joint venture. If a joint venture has been created, each venturer can bind the other. For example, in *Medak v. Cox,*[3] the plaintiff contracted to perform design services with a president of one of two companies that had associated together to build a development. The court held that the president of one venturer could bind the other venturer and both were held liable to the plaintiff.

In the *Medak* case, the joint venture agreement stated that *both* venturers had to consent before the joint venture could enter into any obligations. However, the plaintiff was not aware of this agreement and was not bound by it.

In *Leo F. Piazza Paving Co. v. Foundation Constructors Inc.,*[4] the court had to determine whether one joint venturer could obtain enforcement of an indemnification provision given to the other by a subcontractor. Inasmuch as the case also introduces the complex topic of indemnification (a topic covered extensively later [5]) the facts of the case will be set forth.

The two joint venturers, Piazza and Jones, entered into a prime contract as a joint venture with a public entity for road construction work. Shortly thereafter, each of the joint venturers entered into a joint venture agreement. The agreement divided the work so that Jones performed specific bid items while Piazza performed the balance. Payments were to be divided so that each joint venturer would receive payment for the items of work that it did. Profits and losses were to be strictly divided between the joint venturers as if each were a single unit. Neither was entitled to share in the overall profit of the other, nor was either responsible for the losses of the other. The insurance premiums were to be paid based upon the proportion of the work performed by each, and each was to carry its own workers' compensation insurance.

Jones entered into a subcontract for performance of some of the work for which it was responsible. The prime contract between the public entity and the joint venture was incorporated by reference. The subcontract designated Jones as the contractor and did not mention Piazza. The subcontractor agreed to indemnify "the owner and contractor" for certain losses.

A subcontractor's employee was injured and instituted legal action against each of the joint venturers. The subcontractor employer would defend Jones but not Piazza. A number of witnesses testified as to the intention of the parties when the contracts were made. Representatives of Piazza testified that the joint venture agreement made each venturer responsible only for its own work. They also testified that they did not

---

2. *Richton v. Farina,* 14 Ill.App.3d 697, 303 N.E.2d 218 (1973).

3. 12 Cal.App.3d 70, 90 Cal.Rptr. 452 (1970).

4. 128 Cal.App.3d 583, 177 Cal.Rptr. 268 (1981).

5. See Chapter 36.

feel responsible for any of the work performed by Jones's subcontractors, that they were clearly under Jones's supervision. (This contention is incorrect. While the joint venturers may have sought to segregate the liability between *themselves* for their acts or those of their subcontractors, any claims by the public entity or a third party would be maintainable against the joint venture, including each joint venturer.)

Witnesses for the subcontractor stated that there had been *no* representations that the subcontract was made on behalf of the joint venture and that it entered into *no* agreement with Piazza.

The court noted that its objective was to determine the intention of the contracting parties—Jones and the subcontractor. The court pointed to the joint venture agreement, which seemed to make each responsible for its own work and not for the work of the other. Each was allowed to subcontract provided that each venturer remained fully responsible. As a result, Jones's intention when it entered into the subcontract was to protect only *itself* and not Piazza. Had the subcontractor, according to the court, examined the prime contract and the joint venture agreement, it might have recognized that "contractor" encompassed the two joint venturers. Here the subcontractor had only *general* knowledge of the prime contract. According to the court, this did not constitute an express recognition that the prime contract was a joint venture agreement and that the term *contractor* included both joint venturers. The subcontractor did not have to defend and indemnify Piazza.

SECTION 3.08 UNINCORPORATED ASSOCIATION

Individuals sometimes band together to accomplish a collective objective without using any of the forms of association described thus far, such as partnerships or corporations. Instead they may organize an unincorporated association such as a fraternal lodge, a social club, a labor union, or a church. Design professionals may perform professional services for these associations or may become involved in the associations themselves.

Generally, unincorporated associations are not legal entities. For this reason, early American law did not allow them to hold property in the association name, make contracts, sue in the name of the association, or be sued as a group. They were merely a group of individuals who banded together to accomplish a particular purpose.

In most states, statutes have removed many of the former procedural difficulties. While they are still not entities, unincorporated associations are often permitted to contract, to hold property, and to sue or be sued in the name of the association.

Usually such groups have constitutions, bylaws, and other group-created rules that govern the rights and duties of the members as between themselves. They elect officers who have specified authority, such as to hire employees and run the activities of the association.

Who is liable for the contractual obligations of the association? All the members generally are responsible for those contracts entered into by the officers that were authorized by the members. Liability rests upon agency principles. See Sections 4.05, 4.06.

Frequently it is difficult to establish the authority of the officers, since formalities are often disregarded in these organizations. It may be necessary to show evidence of which members voted for resolutions authorizing the officers to make a particular contract. While the officers may have certain inherent authority by virtue of their positions, important projects, such as engaging a design professional or a contractor, usually are beyond the scope of their inherent authority.

Officers who make the contract may be

individually liable if the contract was clearly made by them as individuals. It may be possible to hold the officers for having misrepresented their authority if the contract was not authorized. Because of the potential risk to officers and to members, many contracts contain provisions that limit liability to certain designated property held in trust for the association in those states where they are not permitted to own property in the association name. (See Section 10.07.) Where the association is permitted to hold property in its own name, liability is often limited to that property.

SECTION 3.09 "LOOSE" ASSOCIATION: SHARE–OFFICE ARRANGEMENT

Sometimes design professionals use the term *association* to describe an arrangement under which they are independent sole proprietors but they join together to share offices, equipment, and clerical help. They may also do work for one another. Sometimes these are known as share-office arrangements.

During the sharing arrangement, legal problems can develop. First, suppose one associate performs services for another and the latter is not paid by his client. In the absence of any specific agreement or well-documented understanding dealing with this risk, the associate who performs services at the request of another should be paid.

Second, suppose the associate performing services at the other associate's request does not perform properly and causes a loss to the client. The client would have a claim against the associate who did the work and the associate with whom it dealt. If the latter settled the claim or paid a court judgment, it would very likely have a valid

claim for indemnification against the associate who did not perform properly. See Chapter 36.

Third, if those in the share-office arrangement create an impression that they are a partnership (or corporation) to the outside world, they will be treated as if they were a partnership. For example, if the stationery, building directory, and telephone directory listed them as Smith, Brown and Jones, such facts may have created an apparent partnership. If so, each member of the association can create partnershiplike contracts and tort obligations.

When a loose association terminates or one participant withdraws, all associates in the first case and the withdrawing associate in the case of withdrawal can freely compete with former associates or those remaining. However, the associates in the first case and the withdrawing associate in the second cannot do any of the following:

1. Take association records that belong to another associate.

2. Represent that they still are associated with former associates if this is not the case.

3. Wrongfully interfere with contractual or stable economic relationships existing at the time the association ended or one associate withdrew.

If the associates were members of a professional association such as the American Institute of Architects or the National Society of Professional Engineers, any disputes may be governed by rules of the professional association that do not violate antitrust laws.[6]

SECTION 3.10 PROFESSIONAL ASSOCIATIONS

In Section 3.09, mention was made of the

---

6. In *Corrigan v. Cox*, 254 Cal.App.2d 919, 62 Cal.Rptr. 733 (1967), a dispute of this type was controlled by the rules of the American Dental Association to which both disputants belonged.

professional associations to which those who design and construct belong, such as the AIA, the NSPE, and the AGC. These associations have had a substantial impact on design and construction, and although they are not organizations like the others mentioned in this chapter created for the purposes of engaging in design and construction, some mention should be made of their activity and some of the legal restraints upon them.

These professional associations engage in many activities. They speak for their members and the professions or industries associated with them. In addition, they seek to educate their members in matters that relate to their activities. One of the most important activities is publishing standard forms for design and construction that are used not only as the basis for contracts for design and construction but also to implement the construction administrative process. It must be kept in mind that membership in these professional associations is not required for one to design or build.

By and large, associations can choose their members. However, as these associations have become more important, the law particular in professional associations is be-ginning to set limits on their power to determine who will be admitted to membership [7] and what the legitimate grounds are for discipline. Restraints placed on professional associations have related to attempts by public authorities to encourage competition in the professional marketplace. This has resulted in attacks by public officials upon what were called the rules of ethical conduct that determine whether an applicant would be admitted to the association and can be disciplined for violation of ethical rules of the association.

Often ethical rules limit the power of one professional to supplant another and bar or sharply limit competition based upon price. At one time, professional associations had minimum or recommended fee schedules. These have been largely abandoned after attacks by public officials based upon the contention that they tended to suppress competition. In addition, attempts to discipline members based upon "ungentlemanly conduct such as competing with a fellow member on the basis of price" have been found to violate agreements between an association and federal antitrust officials that the association would not limit competition.[8]

PROBLEMS

1. A and B were partners in an architectural firm. B's brother C was a struggling young architect, hardly able to pay his bills. B was approached by a prospective client to build a $50,000 residence at a $5,000 commission. B suggested that the client go to C, since C needed the business much more than A and B. A is unhappy with this. Does he have any legal recourse against B?

2. Smith and Jones were young architects who decided to share expenses and office space. Smith was wealthy, while Jones was barely able to make ends meet. Their offices were in the Atlas Building, and the door was painted with the inscription "Smith and Jones, Architects." This was also the listing in the telephone directory and in the office directory of the Atlas Building.

Actually, each had his own clients and neither did any work for the other. They hired a secretary who also

---

7. *Pinsker v. Pacific Coast Society of Orthodontists*, 12 Cal.3d 541, 116 Cal.Rptr. 245, 526 P.2d 253 (1974).

8. *United States v. American Society of Civil Engineers*, 446 F.Supp. 803 (S.D.N.Y.1977).

acted as a bookkeeper. Each income item was allocated to the person who had performed the services, and expenses were allocated based upon who used the services or supplies. Each had a separate checking account, and there was a third account in the name of Smith and Jones. The latter account was used to pay for the rent, the cost of the secretary, and other office expenses.

On occasion each would consult the other with regard to professional design matters. It was not uncommon for one to call in the other while a client was present to get informal advice on a design question.

a. Jones ordered a copying machine that cost $2,000. The order was placed in the name of Smith and Jones. Jones did not pay for the machine and the seller claims that they have a legal right to collect from Smith as well as from Jones. Do you agree? Give your reasons. Would your answer depend in any way upon whether the copier machine company knew of Smith's wealth and supplied the machine based upon his good credit?

b. Jones designed an exclusive residence for a client. His negligence in design required that certain work be redone at an expense of $2,000 to the client. The client's lawyer ascertained that Jones had no assets or professional liability insurance and that any judgment against him would be uncollectible. The client's lawyer now asserts that he has the right to recover from Smith because Jones and Smith appear to the world as a partnership. Is he correct? Would your conclusion be changed or reinforced if it could be shown that during a conference between Jones and his client, Smith was called in to offer some suggestions relating to design? (Assume that the suggestions did not relate to matters that ultimately were the basis for the claim and that Smith was not negligent in any way).

4

The Agency Relationship: A Legal Concept Essential to Contract Making

SECTION 4.01 RELEVANCE

Agency rules and their application determine when the acts of one person bind another. In the typical agency problem, there is a principal, an agent, and a third party. The agent is the person whose acts are asserted by the third party to bind the principal. There can be legal problems relating to the rights and duties of the principal and agent as between themselves as well as disputes between the agent and the third party. But because the third party v. principal part of the agency triangle is most important—and most troublesome—these problems will be used to demonstrate the relevance of agency law.

The design professional, whether in private practice or working as an employee, may be in any of the three positions of the agency triangle.

For example, suppose the design professional is a principal (partner) of a large office. The office manager orders an expensive computer. In this illustration, the design professional, as a partner, is a principal who may be responsible for the acts of the agent office manager.

Suppose the design professional is retained by an owner to design a large structure. The design professional is also engaged to perform certain functions on behalf of the owner in the construction process itself. Suppose the design professional orders certain changes in the work that will increase the cost of the project. A dispute may arise between the owner and the contractor relating to the power of the design professional to bind the owner. Here the design professional falls into the agent category, and the issue is the extent of her authority.

Suppose X approaches a design professional in private practice regarding a commission to design a structure. X states that she is the vice-president of T Corporation. The design professional and X come to an agreement. Does the design professional have a contract with T Corporation? Here the design professional is in the position of the third party in the agency triangle.

The agency concept is basic to understanding the different forms by which persons conduct their business affairs, such as partnerships and corporations.

SECTION 4.02 POLICIES BEHIND AGENCY CONCEPT

(A) COMMERCIAL EFFICIENCY AND PROTECTION OF REASONABLE EXPECTATIONS

As the commercial economy expanded beyond simple person-to-person dealings, commercial necessity required that persons be able to act through others. Principals needed to employ agents with whom third parties would deal. Third parties will deal with an agent if they feel assured that they can look to the principal. The agent may be a person of doubtful financial responsibility. The concept of agency filled the need for giving third persons some assurance that they can hold the principal.

Agency exposes the principal to risks. The principal may be liable for an unauthorized commitment made by the agent. Suppose the principal authorizes its agent to make purchases of up to $1,000, but the agent orders $5,000 worth of goods. From the third party's standpoint, this $5,000 purchase may be reasonable in light of the position of the agent or what the latter had been ordering in the past. The law protects the principal from unauthorized commitments but not at the expense of reasonable expectations of the third person. That problems such as this can develop did not destroy the unquestioned usefulness of the agency concept. Such problems required the law to create rules and solutions to handle such questions that would be in accord with commercial necessity and common sense.

(B) RELATIONSHIPS BETWEEN PRINCIPAL AND AGENT

As a rule, the relationship of principal and agent is created by a contract expressed in words or manifested by acts. The agent may be a regular employee of the principal or an independent person hired for a specific purpose and not controlled as to the details of requested activities. Several aspects of the agency relationship make it different from the ordinary commercial, arm's-length relationship.

An arm's-length transaction is one whereby the parties are expected to protect themselves. In such a transaction, no general duty is imposed upon one party to protect the other party, nor is any duty imposed to disclose essential facts to the other party. Although there are some exceptions, generally commercial dealings are at arm's-length. On the other hand, principal and agent have a fiduciary relationship, one of trust and loyalty. (See Section 14.04(B).)

Often the arrangement under which an agent performs services is sketchy and does not specifically delineate all the rights and duties of principal and agent. In such a case terms often must be implied by law.

SECTION 4.03 OTHER RELATED LEGAL CONCEPTS

Confusion often occurs because of the overlapping nature of terms such as *principal-agent, master-servant,* and *employee-employer.* This chapter deals only with the principal-agent relationship and the power of agency under which one person may bind another to a contractual obligation.

SECTION 4.04 CREATION OF AGENCY RELATIONSHIP

Generally agency relationships are created by a manifestation by the principal to the agent that the agent can act on the principal's behalf and by some manifestation by the agent to the principal that she will so act. However, these elements of consent are often informal. They need not be in writing, except in cases where statutes require that the agent's authority be expressed in writing if the transaction to be consummated by the agent is one that would also require a writing.

SECTION 4.05
ACTUAL AUTHORITY

The agent is ordinarily authorized to do only what it is reasonable to believe the principal wants done. In determining this, the agent must look at the surrounding facts and circumstances. If, for example, the principal has authorized the agent to purchase raw materials to be used in a particular manufacturing process and the agent then learns that the principal has decided not to proceed with the project, it is unreasonable for the agent to believe that the authority to buy the materials still exists.

The agent must consider the situation of the principal, the general usages of the business and trade, the object the principal wishes to accomplish, and any other surrounding facts and circumstances that would reasonably lead to a belief that the agent can or cannot do something. This is largely a matter of common sense. Sometimes the agent is given specific authority to do those incidental acts that are reasonably necessary to accomplish the primary act. For example, when an agent is given authority to purchase a car for the principal, it is likely that the agent also has authority to buy liability insurance for the principal. The agent, when possible, should seek authorization from the principal to perform those acts that are not expressly authorized.

Sometimes emergencies arise that make it difficult or impracticable for the agent to communicate with the principal. In such cases, the agent is given authority to do those necessary acts to prevent loss to the principal with respect to the interests committed to the agent's charge. Such situations may arise in a construction contract. Frequently the design professional is given authority to act in emergencies. Even without this express authority, the design professional should have authority to bind the principal if the acts appear to be reasonably necessary and inferable from other authority. But as indicated in Section 15.08, the tendency has been to *reduce* the authority of the design professional in order to limit liability exposure.

SECTION 4.06　APPARENT AUTHORITY: *FRANK SULLIVAN CO. v. MIDWEST SHEET METAL WORKS*

Most difficult principal-agency cases arise under the doctrine of apparent authority. Apparent authority exists when the principal's conduct reasonably leads a third party to believe that the principal consents to acts done on its behalf by the person purporting to act for it.

It may be useful to draw some initial distinctions that are essential to an understanding of apparent authority problems. First, suppose the asserted agent neither is an employee of the principal nor has been hired to perform any specific task by the principal. This can be illustrated by *Amritt v. Paragon Homes, Inc.*[1] The plaintiff homeowner sued for damages caused by faulty construction of a home built for her by Romero, a contractor. Clearly Romero was liable, but evidently Romero could not pay a court judgment. The homeowner sued Paragon, the manufacturer of the prefabricated home erected by Romero and Sewer, the manufacturer's local distributor.

The homeowner had arranged through Sewer to purchase the materials, plans, and drawings for such a home. Evidently Sewer represented himself to the homeowner as the agent of Paragon and implied that Paragon would not be responsible unless the homeowner used Romero to construct the home. The agreement was on a form supplied by Paragon and filled in by Sewer. The homeowner borrowed money from and gave a mortgage to Paragon for the construction costs. The homeowner made pay-

---

1. 474 F.2d 1251 (3d Cir.1973).

ments by depositing funds in escrow with Paragon, and Paragon demanded a completion certificate before disbursing the funds. Sewer inspected the construction work for Paragon.

Paragon, Sewer, and Romero were all separate legal entities. Romero worked for himself, as did Sewer. The court, however, held that Sewer and Romero were agents for Paragon, since both had "the outward trappings of apparent authority to be Paragon's agents." [2] The court concluded that the homeowner was "reasonably led to believe they were Paragon agents." [3]

The court quoted § 27 of the Restatement of Agency,[4] which states:

> apparent authority to do an act is created as to a third person by written or spoken words or any other conduct of the principal which, reasonably interpreted, causes the third person to believe that the principal consents to have the act done on his behalf by the person purporting to act for him.[5]

The court pointed to Paragon's conduct— supplying all materials and forms through Sewer, requiring that the house be inspected by Sewer and built to its plans, and disbursing all construction payments—the combination of these activities could reasonably have led the homeowner to believe Paragon had authorized Sewer and Romero to be its agents. Although the court spoke in terms of apparent authority, it actually found an apparent agency.

Because of the many legal entities involved in a construction project, one party not infrequently seeks to bind an entity other than the one with whom it has dealt on the same theory as that used in *Amritt*

v. Paragon Homes, Inc. This is especially likely when consumers find that the persons with whom they have dealt are unable to respond when those persons have not performed in accordance with their legal obligation. The *Amritt* case illustrates that courts frequently look to a more solvent party, especially if that solvent party has organized the transaction, controlled it, and sought to eliminate responsibility for a risk legitimately a part of its enterprise by the use of a separate legal entity that lacks the financial capability to respond for product defects. The conclusion reached in the *Amritt* case is also more justified if the party dealing with the contractor is a consumer of limited commercial experience.

A second problem can stem from a client's engaging a professional person. For example, the owner may engage a design professional to perform services relating to design and contract administration for a particular project. The latter is rarely an employee of the owner but has certain designated authority. This matter is treated more fully in Section 21.05(B).

The third and perhaps most difficult problem relates to whether an admitted employee of an employer has the authority to engage in certain acts and thereby to bind the employer. Here there is an agency relationship in that the agent has some authority to bind the principal. But the problems that develop relate to the extent of that authority.

The hazards of this problem are illustrated by *School District of Philadelphia v. Fram-*

---

2. Id. at 1252.

3. Ibid.

4. The Restatements of the Law are collections of rules, comments, and illustrations about a particular legal subject. They are published by the American Law Institute, a private organization made up of lawyers, judges, and legal scholars. The Institute's function is to collect case law from all the states and distill it into rules. Unless adopted by a legislature or followed by a court, it is *not* law. The principal Restatements for the purposes of this treatise are those of Torts, Contracts, and Agency.

5. 474 F.2d at 1252.

lau Corp.[6] In the middle of the trial the attorney for the defendant school district notified the judge and the other party that the school district's president had agreed to a settlement. The plaintiff contractor agreed to the settlement; the judge, after stating that the case had been settled, discharged the jury. The school district board did not agree and refused to consummate the settlement. The plaintiff contractor sued the school district, based upon the settlement; but the court held that neither attorney nor the school board president had authority to settle the dispute. This case seems harsh and perhaps can be explained by the reluctance of courts to apply the apparent authority concept to public entities.

The following case also deals with apparent authority of an employee in the construction contract context.

FRANK SULLIVAN COMPANY v. MIDWEST SHEET METAL WORKS
United States Court of Appeals, Eighth Circuit, 1964.
335 F.2d 33.

BLACKMUN, Circuit Judge.[7] Midwest Sheet Metal Works, a Minnesota partnership, instituted this diversity suit against Frank Sullivan Company, a Boston contractor, to recover damages for breach of contract. The jury returned a verdict in favor of Midwest for $85,000.

. . . The controversy arises out of the project for the extension and remodeling of the United States Post Office and Customs House at Saint Paul. Minnesota law controls. Although Sullivan asserts twenty separate points on its appeal, these come down essentially to four primary issues:

1. The identity of Midwest's Exhibit 5 as the agreement between the parties, and its effectiveness as a contract.

. . . [second issue omitted]

3. The authority of Sullivan's agent to sign the agreement.

. . . [fourth issue omitted]

The prime contractor on the project was Electronic & Missile Facilities, Inc., of New York City (EMF). On December 4, 1961, EMF and Sullivan executed a lengthy and detailed subcontract whereby Sullivan undertook all plumbing, heating apparatus, air conditioning, ventilation work on the job for an agreed price of $1,650,000. This contract was executed on behalf of Sullivan by Francis J. Sullivan (Frank) as its president.

In the fall of 1961, before the formal execution of the agreement between EMF and Sullivan, Frank had contact with Michael J. Elnicky, the dominant partner of Midwest, about Midwest's taking on the sheet metal and air conditioning portion of Sullivan's subcontract. Sullivan had even invited a quotation from Midwest for this work. On November 1 Midwest quoted a figure in excess of a million dollars. Frank by telephone told Elnicky that this bid was about $200,000 too high. Elnicky indicated he might reduce

---

6. 15 Pa.Cmwlth. 621, 328 A.2d 866 (1974). Note the name of the case. Although most courts list the plaintiff first, some, such as Pennsylvania, place the appealing party first in the case name.

7. Judge Blackmun has since been appointed to the U.S. Supreme Court.

his price somewhat but could not approach Sullivan's suggested figure. Frank testified that he then told Elnicky, "Well, look, we have got a fellow going out there, and I will show you that these are not prices that we dreamed up, these are prices that we used in making our bid, and we got them confirmed by letters by reputable people."

Near the close of 1961 EMF told Sullivan that it was imperative that the work in Saint Paul be started. Sullivan promised EMF that it would get a superintendent, a foreman, and men and material on the job by the end of January. Sullivan sent John Sullivan (Jack) from Boston to Saint Paul in early January. On this trip he conferred with EMF's superintendent on the project. Jack was back in Minnesota later in the same month with Byers who was to be Sullivan's general superintendent on the job. Before he left Boston on this second trip Jack had been instructed by Frank to look over the labor situation in the area, to check in with prospective subcontractors, to see Elnicky and give him quotations which "will back up the reduction of his bid," and to "get the job started." Frank gave Jack the job estimates which had been prepared by Sullivan but he was not given and had not seen the prime contract.

Upon their arrival Byers called upon local union business agents, purchased material and tools, received other equipment from Boston, and placed four steamfitters on the job.

On Monday, January 22, Jack came to Midwest's office. This was the first time Elnicky saw him. Elnicky and some of his employees testified that Jack told him on this visit that he was "part of the [Sullivan] organization" and had "a piece of it." Jack denied that he made any such statement. Elnicky conceded that he made no attempt to check Jack's authority with the Sullivan home office and that he did not ask for written evidence of it.

Elnicky also testified that Jack early in this first meeting suggested that Midwest price off "the whole works"; in any event, Elnicky indicated that he was interested in taking over the entire Sullivan job. Jack did not object and said that he would try to get a copy of the prime contract for him. The two men met again on Tuesday when Jack permitted Elnicky and his people to review the bids Sullivan had received. By Wednesday Jack obtained a copy of the prime contract from EMF's office on the job. He gave it to Elnicky who kept it overnight. Meanwhile, Jack talked with other prospective subcontractors. Elnicky and Jack met further, and sometimes socially, during the same week. Jack told him that Sullivan would have to have a minimum of $100,000 if Elnicky took over. On Thursday Jack told Frank by telephone of the discussions he was having with Midwest. As to this conversation Frank testified that he told Jack that this could not be done, that EMF wanted Sullivan on the job, and that Jack should "pick up what you got and come home." Midwest finished its estimating on Saturday, January 27. That evening Jack was at Elnicky's home with Elnicky and two of the latter's men. They discussed costs and what Elnicky might offer to do the job but no conclusion was reached.

Early in the afternoon of the next day, Sunday, Elnicky came to Jack's hotel room. Jack was planning to return to Boston. Elnicky arrived with a fifth of Scotch. The men were together for three and one-half hours, discussing the job and drinking the entire fifth. Byers was present but left for a time to get the copy of the prime contract which had been left elsewhere. Elnicky made an offer of $1,550,000 to perform the work. This was discussed as was the question of what to do with the equipment and materials which Sullivan already had on the job. Jack then started to write something out. Elnicky dictated part of it. Several drafts were made. Later Jack dictated a draft to a hotel typist and he and Elnicky signed it. Elnicky then took Jack to the airport. The typed draft is Midwest's Exhibit 5 * and is the document in controversy. It was admitted in evidence over Sullivan's objection.

The testimony as to the execution of the exhibit is in sharp conflict. Jack testified that he told Elnicky that this agreement was subject to approval by Frank and EMF, that there was no sense in working out other details until this was done, that it was his intention that Exhibit 5 be merely a proposal by Elnicky to Sullivan, and that he did not intend thereby to turn the Sullivan contract over to Midwest. Byers testified that Jack told Elnicky that it had to be approved by Frank and EMF; that Elnicky acknowledged this; and that he, Byers, had expressed a hope that

---

*

"Hotel Saint Paul
St. Paul, Minnesota
January 28, 1962

"The Midwest Sheet Metal Co.
340 Taft St.
Minneapolis, Minnesota

"The Midwest Sheet Metal Company of 340 Taft St., Minneapolis, Minnesota, agrees to take over Frank Sullivan Company's contract with Electronic Missile Facilities, Inc., in the amount of One Million Five Hundred Fifty and 00/100 Dollars ($1,550,000).

"The cost of the bond will be paid for by the Frank Sullivan Company.

"The Midwest Sheet Metal Company will man the above mentioned project on January 29, 1962, to show good faith regarding this contract.

The above agreement is made between

Midwest Sheet Metal
Midwest Sheet Metal
M.J. Elnicky
Frank Sullivan Co.
John Sullivan"

The discrepancy between the words and the figures is readily apparent.

Frank would approve it so he "could go home." Elnicky flatly denied that Jack had said Exhibit 5 was subject to approval by Frank and EMF.

The next day, Monday the 29th, Jack called Byers from Boston to see if Elnicky had someone on the job as the writing provided. Byers called Elnicky who told him he would have someone there on Tuesday. On Tuesday Jack and Elnicky conferred by telephone. Byers then sent Elnicky's men away from the job. On the same day, January 30th, Jack, signing on behalf of the Sullivan Company, wrote Midwest that "the agreement made on January 28th, 1962, between the Midwest Sheet Metal Company and the Frank Sullivan Company is hereby cancelled" and that they would try to arrange for Midwest to quote on the job's ventilating, air conditioning, and refrigeration. Jack testified that he wrote this letter without discussion with Frank. Midwest's receipt of that letter led to the present suit.

Jack Sullivan's status is obviously of vital importance. In January 1962 he was 29 years of age. He was a high school graduate and had had one year of "night college." He held a plumber's union card and a journeyman plumber's license. He had worked for Sullivan for ten years. He had started there as a plumber and, by 1959, was a job superintendent. About mid-1960 he became an "outside superintendent." In this capacity he traveled to various jobs, examined labor and material situations and reported back to Sullivan. He did no hiring or firing. He did not order materials. He did make recommendations. He was not an officer, director, or shareholder of Sullivan. He was not related to Frank. He had nothing to do with obtaining or negotiating subcontracts. He had had no experience in sheet metal or air conditioning. He had done no estimating. He had been given no specific authority to sign any contract for Sullivan or to assign Sullivan's subcontract with EMF.

In January 1962 Elnicky was about 52. He had been in sheet metal and similar work for many years and had run his own business since 1947.

A. The contract. Sullivan's basic argument here is . . . that Exhibit 5 . . . is not sufficiently clear and definite to be valid and to constitute an enforceable contract.[8] . . .

[I]f

". . . substantial terms are left open and subject to further agreement, which is never reached, there is not only no complete agreement but no contract at all. . . . The next inquiry is whether, there being an agreement which is asserted to be a contract, it is complete. It is not unless it is in all essential terms definite and certain or capable of being made so by the aid of competent evidence and permissible

---

8. [Ed. note: The certainty requirement for a valid contract is discussed in Section 5.06(D).]

interpretation. If as a contract it be incomplete, a court can no more complete it for the parties than it could make it for them in the beginning." Wells Const. Co. v. Goder Incinerator Co. . . .

"There is no contract where there is no mutual and final assent to all the essential terms of a bargain." New England Mut. Life Ins. Co. v. Mannheimer Realty Co. . . . Vagueness and indefiniteness may affect the validity of an alleged agreement. "If an alleged contract is so uncertain as to any of its essential terms that it cannot be carried into effect without new and additional stipulations between the parties, it is not a valid agreement." Ames-Brooks Co. v. Aetna Ins. Co. . . . And "where substantial and necessary terms are specifically left open for future negotiation, the purported contract is fatally defective." King v. Dalton Motors, Inc., . . . But

"A proper administration of justice does not permit an over-zealous quest for subtle ambiguity to destroy the intent of the parties when the court, despite some incompleteness and imperfection of expression, can reasonably find that intent by applying the words used, with all their reasonable implications, to the subject matter as the parties themselves, under all the surrounding circumstances, must have applied, used, and understood them. This court is reluctant to invoke the principle that indefiniteness prevents the creation of a contract where a just result, consistent with a reasonably expressed intent of the parties, can be reached by upholding the agreement." Hartung v. Billmeier. . . .

. . . Reasonable certainty is thus the standard. . . .

In the light of the foregoing authorities, we affirm on this point. We do so for the following reasons:

*   *   *

2. While Exhibit 5 may have been born with somewhat of an alcoholic background, this fact alone does not make it any less a contract. Evidently both Elnicky and Jack desired or were content to negotiate in that kind of atmosphere. It is not now claimed that there was duress or that Elnicky took advantage of Jack through alcohol.

3. Exhibit 5 on its face is clear and complete. It succinctly states that Midwest "agrees to take over" Sullivan's contract with EMF. It states the figure at which it does so. It leaves a $100,000 gross profit margin, less the cost of the bond, for Sullivan. It calls for immediate manning of the job by Midwest. It contains no ambiguity and Sullivan so concedes.

4. Although, as Sullivan observes, Exhibit 5 consists of but one page, as contrasted with the many pages of the subcontract between EMF and Sullivan, this difference is understandable. The

document could have been more formal but it was formulated by two laymen and it in effect incorporated the detailed EMF subcontract by reference. By so doing it achieved certainty as to underlying details. Sullivan itself was originally content in this respect or it would not have made the subcontract with EMF.

5. The discrepancy between the words and the figures of Exhibit 5 is not fatal. It is true that the normal rule is that, where there is a discrepancy of this kind, the words and not the figures control. . . . This usually does not invalidate a contract. The cited case also discloses that there are even situations where resort may be made to the figures. In any event, despite the facts that the difference here was a half million dollar one and that it should have been noticed by either Jack or Elnicky, there was no confusion in the minds of the two men. Each knew that the proper amount was $1,550,000. No one was misled.

6. Exhibit 5 does not have that vagueness or indefiniteness or incompleteness which, in the several Minnesota cases cited above, has served to defeat alleged contracts.

* * *

8. Jack's own letter of January 30, 1962, is almost persuasive in itself. It refers to "the agreement" of January 28 "between Midwest . . . and . . . Sullivan" and reaches out to cancel it.

* * *

C. Jack's authority. Midwest does not contend that the record supports a finding that Jack possessed actual authority to act on behalf of Sullivan. The issue is one of apparent authority. Sullivan asserts that the evidence as a matter of law was insufficient to support a finding of apparent authority and that the court's submission of this issue to the jury and its rulings and instructions consistent therewith were prejudicially erroneous.

Jack certainly assumed the mantle and the posture of responsibility and authority. There is evidence to the effect that he professed an ownership interest in Sullivan, possessed and produced the bids and the cost estimates the company has assembled, permitted Elnicky to review them, obtained a copy of the EMF-Sullivan subcontract for Elnicky, mentioned to others than Elnicky his interest in contracting out the entire Sullivan portion of the job, demonstrated a permissive attitude toward Midwest's interest in taking on the full subcontract, was in contact with Elnicky's performance bond man and asked him to confirm his comments by letter, accepted Elnicky's entertainment favors, bargained continuously for a week, and even wrote the letter of cancellation.

But apparent authority must be founded on something more than the conduct and statements of the agent himself. Liability can be imposed upon a principal only "for that appearance of authority caused by himself." 2 Williston on Contracts (3d Ed.

1960), § 277A, pp. 222–24; . . . Of course, a degree of reasonableness and of diligence is required of one who deals with the agent. . . .

We find in this record adequate support for the submission of the issue of apparent authority to the jury. Accepting the evidence, as we must, in the light most favorable to the prevailing plaintiff, we have, apart from and in addition to Jack's own acts and statements, all the following: (1) Sullivan sublet part of every job it had for there were certain types of work (air conditioning and sheet metal, for example) which it never performed; (2) Sullivan sent Jack to Saint Paul to get the job started; (3) Sullivan instructed Jack to get in touch with area people in the construction industry and to obtain subcontract offers; (4) Sullivan placed Jack in possession of the breakdown of costs it had prepared and of the bids it had received; (5) Frank told Elnicky that he had a man going out to Minnesota; (6) Jack possessed the Sullivan name; (7) Jack was the highest person in authority in Sullivan's employ on the job; (8) so far as the Saint Paul job was concerned, Byers followed Jack's instructions; (9) Jack, while he was in Saint Paul the week of January 22, was in telephone communication with Boston, and, specifically, talked with Frank; and (10) Frank knew that Jack was discussing with Elnicky a complete takeover by Midwest and yet did nothing to disavow his status to Elnicky. And, for what it is worth, Jack was still with the Sullivan Company at the time of the trial.

Of course, there is an opposing factual argument, namely, that Sullivan had given Jack no instructions to turn over the entire job; that Jack was not supplied with a copy of the EMF-Sullivan subcontract and had to obtain it from the EMF man on the job; that the preliminary conversations between Sullivan and Elnicky had only to do with a limited area of work; that Elnicky knew who Frank was and was in communication with him; and that Elnicky did not inquire of Jack or of Frank as to Jack's authority. But this is just another argument for the trier of fact. The jury was not persuaded.

Sullivan places great emphasis on Elnicky's failure to make inquiry as to Jack's authority, and it urges the important nature of the contract as demonstrated by the amount involved and the time required for its performance. This argument, however, cuts both ways. If the job was so large and so important, a jury might properly infer that Sullivan's top man on the project was there with workable authority. We feel that the cases Sullivan cites in support of its argument are distinguishable on their facts. Hill v. James, supra, concerned a traveling salesman, a fact which the court stressed, and a questionably completed contract. The court readily recognized that a principal may clothe even such a salesman with apparent authority. Dispatch Printing Co. v. National Bank of Commerce, supra, concerned a Saint Paul newspaper's Minneapolis advertising solicitor-collector and his indorsing checks drawn in favor of the newspaper and depositing them in his own account. Mooney v. Jones . . . concerned a mortgagor's representation, claimed to be made on behalf of the mortgagee, to a contractor that the latter's lien waiver would not affect his lien rights for future work. Language in these opinions relative to an agent's actual authority and a third person's duty to inquire is accepted law but the Minnesota court has consistently recognized the established principles of apparent authority. Illustrative is the court's comment in Mooney itself . . . that "the record is insufficient to show that National Guardian did anything to hold Gustafson out as its agent". . . . And in Sauber v. Northland Ins. Co., supra, Chief Justice Knutson, in speaking for the court, said, "Apparent authority exists by virtue of conduct on the part of the principal which warrants a finding that a third party, acting in good faith, was justified in relying on the assumption that the agent had authority to act." [footnote omitted]

The situation here strikes us as one where, as the negotiations developed, Jack sensed the opportunity to bring his company out of the project at a convenient profit of $100,000, without additional cost or participation, and further sensed that this would be a feather in his cap if he could bring it about. Whether Sullivan sensed this, desired it, and encouraged it, we shall, of course, never know with positive assurance. But the record supports just such an inference by the jury. We are not at liberty to overturn that body's conclusion.

. . . Affirmed.

The *Sullivan* opinion reflects the differing roles of trial and appellate courts. The opinion notes arguments and evidence for and against the conclusion that Elnicky could have reasonably believed that Jack Sullivan had authority to make the contract. A reasonable inference is that the appellate court or at least some judges of that court might have come to a different conclusion. But where the issue is factual, that is, what Elnicky might have reasonably believed, and where there is conflicting testimony and evidence at the trial, the appellate court is likely to affirm the judgment of the trial court.

As seen in the *Sullivan* case, apparent authority can protect a party who has relied upon appearances. However, a third party unsure of an agent's authority should check with the principal where feasible. The application of the apparent authority doctrine is uneven, and the doctrine's existence cannot justify carelessness by third parties when dealing with agents.

SECTION 4.07
TERMINATION OF AGENCY

If authority is conferred for a specified period, it will terminate at the end of that period. If no time is specified, authority continues for a reasonable time. If authority is limited to performing a specified act or to accomplishing a certain result, it terminates when the act or result is completed.

Sometimes the terms of the authorization specify that the authorization is to continue until a certain event occurs. If so, the occurrence of the event will terminate the agency unless the event would not come to the attention of the agent. Sometimes the loss or destruction of certain subject matter terminates the agency. For the effort of bankruptcy upon an agency relationship see Section 38.04(C).

Sometimes the agency terminates when principal and agent consent to terminate the relationship. This can occur if either principal or agent manifests to the other that the relationship will no longer continue. In some cases, the agency can also be terminated by death of the principal, and almost always by death of the agent. However, agency relationships sometimes are created by contract for fixed terms. The principal may, despite a fixed-term contract, terminate the agency and terminate the power of the agent to bind it, but such a termination will be a breach of contract with the agent.

These events effect *actual* authority of the agent. If termination of authority has taken place but manifestations to third parties that the agent has authority still exist, the agent may be able to bind the principal by the doctrine of apparent authority. Apparent authority will continue only so long as the third party should not realize that the agent no longer has authority to bind the principal.

SECTION 4.08
DISPUTES BETWEEN
PRINCIPAL AND THIRD PARTY

In most cases, the issue is whether the principal is *bound* by the acts of its agent. In such cases, the third party seeks to hold the principal and the principal contends that there was no agency, that the act was not within the agent's authority, or that there was no apparent authority.

Sometimes the principal *wishes* to be bound by a transaction between the principal's agent and the third party. The latter may refuse to deal with the principal, claiming it did not know it was dealing with an agent. The agent may not have informed the third party that she was an agent, and the third party may not have had reason to know that the agent was acting on behalf of someone else. The undisclosed principal may disclose its status and hold the third party, unless the facts would make it unjust to permit the principal to assert its status. (Once the principal is disclosed, the third party can hold the principal.)

Suppose the agent did not have actual authority to enter into a transaction but the principal discovers the transaction. In such a case, the principal may ratify the transaction within a reasonable time and bind the third party (and itself) by notifying the third party and affirming the agent's unauthorized acts.

SECTION 4.09
DISPUTES BETWEEN
AGENT AND THIRD PARTY

Disputes between an agent and a third party are relatively rare. If the agent is acting on behalf of an undisclosed principal, the third party has the right to sue the agent individually. The third party may lose this right if it pursues its remedy against the principal, once the principal becomes disclosed. If the agent has misrepresented her authority and the third party is unable to hold the principal, the third party may have an action against the agent for misrepresentation of authority.

Klepp Wood Flooring Corp. v. Butterfield [9] demonstrates the liability exposure of an architect who acts as an agent for a client. The owner commissioned the defendant architect to design an expansion for a private school and play an undefined role in procuring contractors. The architect asked the plaintiff, a flooring contractor, to submit a

---

9. 176 Conn. 528, 409 A.2d 1017 (1979).

bid for the floor installation work. The architect told the contractor his bid was accepted. No formal contract was prepared. The contractor performed and billed the *architect.*

The trial judge held the architect liable as a general contractor for an undisclosed principal. The contractor relied upon the reputation of the architect and expected payment from him.

The appellate court found evidence to sustain the trial court's decision that the contractor did not realize he was dealing with an agent and had no knowledge of the identity of any principal. The court held that to avoid *personal* liability, the agent must disclose that he is acting as an agent and must disclose the identity of the principal, the third party having no duty to discover these facts.

The confusion here may have been traceable to the unclear relationship between architect and school. Also, the architect may have a claim for restitution against the school, the value of which depends upon the school's solvency. But note that the trial judge found that the architect had acted as general contractor, a conclusion that would place responsibility upon the architect. If at all in doubt design professionals should make their agency status clear to those with whom they deal. Also, it is better to have the contract made with the owner-client.

PROBLEMS

1. A has authority to buy normal office supplies for an engineering firm. She enters into a written contract with T under which T will furnish an intercom system for the office at a cost of $825. Can T enforce this contract against the firm? What additional facts would be helpful in deciding this question, and why would these facts be helpful?

2. Here are the salient facts in a court case: On June 18, 1953, R.J. McDonald was the owner of a 1952 Hudson automobile. On that date he procured an insurance policy on the car from defendant. Among other things, the policy covered damages caused by collision or upset. The policy ran for two years, and the premium was paid for that length of time.

   On November 20, 1953, McDonald sold the car to his brother-in-law, John E. Sauber. The transfer was completed in a bank at Farmington. After transferring the title card to Sauber, McDonald handed him an envelope containing the insurance policy.

Sauber then called Northland Insurance Company, defendant herein, on the telephone about the insurance. His testimony is that a woman answered the telephone. She inquired whether she could help him, and his testimony in that regard was as follows:

> I was informed, naturally, it was the Northland Insurance Company; I didn't know her name or whether she said this was Northland Insurance Company, but she knew I was talking to the right place; the purpose was, I told her I had purchased the car and it was transferred to me and I was the new owner of the car and I had the insurance policy and I wanted to know if it was all right I would drive the car with this insurance and she said it is perfectly all right, go ahead and that is about the summary of the whole deal; I was the new owner of the car and it was insured by them people.

A Northland clerk corroborated the phone call but insisted she did not state the policy would be transferred to him. She stated she told him to bring in the policy and fill out some forms. Assume the car was driven, an injury resulted, and Northland denied coverage, claiming first that no assurance had been given and second, any assurance was not authorized. Would the insurance company be liable if a jury believed Sauber's testimony?

What facts are relevant to the issue of apparent authority?

5

Contracts and Their Formation: Connectors for Construction Participants

SECTION 5.01   RELEVANCE

The private design professional will make contracts with, among others, clients, consultants, employees, landlords, and sellers of goods. In addition, contracts are made between owners and prime contractors, prime contractors and subcontractors, contractors and suppliers, employers and employees, buyers and sellers of land, and brokers and property owners.

Contracts for design and for construction are analyzed in other parts of the treatise. Sections 5.01 through 5.10 provide a framework for such analysis. Sections 5.11 and 5.12 present a brief treatment of employment contracts and contracts incident to the acquisition of land ownership, respectively, both relevant to the Construction Process.

SECTION 5.02   THE FUNCTION OF ENFORCING CONTRACTS: FREEDOM OF CONTRACT

The principal function of enforcing contracts in the commercial world is to encourage economic exchanges that lead to economic efficiency and greater productivity. This is accomplished by protecting the reasonable expectations of contracting parties that each will perform as promised. While many and perhaps most contracts are and would be performed without resort to court enforcement, the availability of legal sanctions plays an important role in obtaining performance.

Generally, American law gives autonomy to contracting parties to choose the substantive content of their contracts. Since most contracts are economic exchanges, giving parties autonomy allows each to value the other's performance. To a large degree autonomy assumes and supports a marketplace where market participants are free to pick the parties with whom they will deal and the terms upon which they will deal. In addition to the parties being in the best position to determine terms of exchange, the alternative of state-prescribed rules for economic exchanges not only would lead to rigidity but would place a heavy burden on the state. Also, parties are more likely to perform in accordance with their promises if they have participated freely in making the exchange and determined its terms. Finally, such autonomy, often called freedom of contract, fits well in a free society that encourages individual enterprise.

However, broad grants of autonomy assume contracting parties of relatively equal bargaining power and a relatively free marketplace. Those conditions should preclude overreaching. A party who believes the other party's terms to be unreasonable can deal with others. If the parties do arrive at an agreement under such conditions, the give-and-take of bargaining should insure a contract that falls within the boundaries of reasonableness.

The development of mass-produced contracts and the emergence of large blocs of economic power have made this earlier model of the negotiated contract the exception. If the state, through its courts, enforces adhesion contracts (contracts presented on a take-it-or-leave-it basis,) the state is according almost sovereign power to those who have the economic power to dictate contract terms. For this reason, many inroads have been made upon contractual freedom by federal and state legislation, regulations of administrative agencies, and courts through their power to interpret contracts and to determine their validity.

SECTION 5.03
PRELIMINARY DEFINITIONS

The promisor is a person who makes a promise, and the promisee is the person to whom the promise is made. In most two-party contracts, each party is a promisee and promisor, both making and receiving promises.

The offeror is a person who makes an offer, and the offeree is a person to whom the offer is made. All other definitions will be given as the particular term is discussed.

SECTION 5.04
CONTRACT CLASSIFICATIONS

(A) EXPRESS AND IMPLIED

Sometimes contracts are classified according to the method by which they are created. Using this classification, there are express contracts and implied-in-fact contracts. In express contracts, the parties manifest their assent or agreement by oral or written words. In implied-in-fact contracts assent is manifested by acts rather than by words.

A recent case illustrates this in a construction project context.[1] The employee of a subcontractor brought equipment he had rented to the job site to replace equipment that a third party had supplied but had removed. The employer admitted that the employee's equipment was on the job site and was used but contended that there was no express agreement under which he would pay the employee rent for the equipment.

In ruling for the employee the court stated:

> . . . even if there was no express oral agreement for equipment rental, an implied contract may be inferred from the conduct of the parties. Implied contracts arise . . . where there is circumstantial evidence showing that the parties intended to make a contract. Where one performs for another a useful service of a character that is usually charged for, and such service is rendered with the knowledge and approval of the recipient who either expresses no dissent or avails himself of the service rendered, the law raises an implied promise on the part of the recipient to pay the reasonable value of such service.[2]

Unless the agreement is required to be in writing, the implied-in-fact agreement is as valid as an express contract.

(B) SUBJECT MATTER

Sometimes contracts are classified by the transaction involved, such as sales of land, sales of goods, loans of money, leases, service contracts, professional service contracts, insurance and family contracts. While traditionally American contract law has been thought of as a unitary system, different transactions are treated differently. For example, the seller of *goods* is in many instances held to a *successful out-*

---

1. *United States v. Young Lumber Co.,* 376 F.Supp. 1290 (D.S.C.1974).

2. Id. at 1298. Where such an arrangement is considered nonconsensual, recovery can be based upon unjust enrichment. See Section 8.04.

come standard, usually referred to as implied warranty of fitness, while those who sell professional *services* are usually held to a less strict standard,[3] one that compares what was done to what *others* would have done. Subject matter variations are manifested by regulatory legislation over certain types of contracts. Judicial opinions often treat one transaction differently from another.

(C) BARGAIN AND ADHESION

Contracts are sometimes classified as negotiated or adhered to. The latter are referred to as contracts of adhesion. A negotiated contract arises when two parties with reasonably equivalent bargaining power enter into negotiations, give and take, and *jointly* work out a mutually satisfactory agreement.

The adhesion contract has no or minimal bargaining. The dominant party hands the contract to the weaker party on a take-it-or-leave-it basis. At its extreme the weaker party who wishes to enter into the transaction must accept all the terms of the stronger party.

Sometimes important terms can be negotiated, although the balance will be dictated by one party with little opportunity for bargaining. For example, the purchaser of a new automobile may be able to bargain on price but will have to accept the standardized terms of the dealer as to all or almost all other aspects of the purchase. The rare buyer who reads the standardized terms and objects to them will have great difficulty in persuading the dealer to change the terms.

Sometimes the adhesion contract is accompanied by monopoly power. In a competitive economy a weaker party who does not want to accept harsh terms may deal with others. However, in many transactions the weaker party will find the same terms used by the competitors of the person whose terms were unpalatable or will find no competitor. Frequently such adhesion contracts are printed, and the person with whom the weaker power is dealing (such as a salesperson or clerk) lacks authority to vary the printed terms of the contract. Modern courts recognize the difference between the negotiated contract and the contract of adhesion.

The increasing use of standardized forms can lead not only to one-sided contracts but also to misunderstandings. For example, in *Impastato v. Senner* [4] the parties orally agreed that the architect would not be paid until completion of his work, but they signed standardized forms of the American Institute of Architects which provided for interim fee payments. The court did not have to face the conflict because the parties agreed in court that the oral agreement was not to be affected by the subsequent signing of the standard form. But the fact pattern reflects the importance of making certain that no conflict exists between the oral agreement and the standardized form contract.

SECTION 5.05
CAPACITY TO CONTRACT

Capacity usually relates to age and mental awareness. Since they rarely affect contracts for design or construction, they are not discussed here.

SECTION 5.06
MUTUAL ASSENT

(A) OBJECTIVE THEORY OF CONTRACTS: MANIFESTATIONS OF MUTUAL ASSENT

Early English contract law stated there had to be a meeting of the minds—actual agree-

---

3. See Section 17.05.

4. 190 So.2d 111 (La.App.1966).

ment as to the existence of a contract before there could be a valid contract.

As a rule each party believes it has made a contract, though perhaps differing as to the exact nature of performance. Suppose one person harbors a secret intention not to be bound and yet manifests to the other party an intention to be bound. The party who relied upon the objective manifestation should be protected. The law protects the reasonable expectations of the innocent party through the objective theory. A party is bound by what it manifests to the other party. Secret intentions are not relevant. If one party, innocently or otherwise, misleads the other into thinking that it has serious contractual intention, a contract exists despite the lack of actual agreement of the parties. The same principle holds true if one party is only joking when entering into negotiations and making an agreement with the other party. Unless the other party should reasonably have realized that the negotiations were not serious, the party who is not serious will be held to the agreement.

(B) THE ASSENT PROCESS: OFFER AND ACCEPTANCE

Typically the process by which contracting parties make an agreement involves communications, oral and written, that culminate with one party making an offer and the other party accepting it.

Where parties exchange communications at the same time and place and an agreement is reached, rarely is it necessary to determine whether there has been an offer and acceptance. But where parties communicate with each other at a distance, it is often necessary to determine whether the parties have arrived at an agreement, and this necessitates an examination of whether there has been an offer and acceptance.

Offers are differentiated from preliminary negotiations or proposals for offers. While a number of abstract definitions of an offer exist, it is more useful to consider the *effect* of an offer. An offer creates a "power of acceptance in the offeree." The offeree can create a legally enforceable obligation without any further act of the offeror. How is it determined whether the offeree has a reasonable belief that he has a power of acceptance?

Some cases scrutinize the language of the offer, especially a written offer, looking for definite words of commitment on the part of the offeror, such as "I offer" or "I promise." However, parties only rarely express themselves in legal terms.

The entire written proposal is examined to see if a reasonable person receiving it would think he can "close the deal." Factors include the certainty of the terms, any indication that the proposer will not have to take further action, the past dealings between the parties, and the person to whom the offer is made. For example, if nothing is stated on essential terms such as price, quantity, or quality, the bargaining is probably in a preliminary stage. The first proposal is intended merely to start the negotiating mechanism, and more negotiating will take place before agreement is reached.

If the proposer is negotiating with a number of other persons at the same time and the person to whom a particular proposal is directed knows of this, it is likely that the latter realizes, or should realize, that the proposer wishes to have the last word rather than risk being obligated to a number of persons. In such a case, the communication is probably not an offer.

Even if stated to be irrevocable, the common law holds that offers are generally revocable. At any time before acceptance an offeror can withdraw the offer, provided the offeror communicates the revocation directly or indirectly to the offeree. This rule of revocability has generated many exceptions, the most important being reliance upon the offer and the firm offer under § 2–205 of the Uniform Commercial Code dealing with the sale of goods. The latter makes a firm written offer in a signed writing by a merchant irrevocable for the time stated or, if not stated, for a reasonable time not to exceed three months.

Suppose a written offer states that it will be open for a specified period, say ten days. At the expiration of this period, the power of acceptance terminates. If the offer is made by letter, it may not be clear when the time period begins. Does it begin on the date of the letter offer, at the time the letter is mailed, at the time when it would normally be received, or at the time when it is actually received? Usually it begins when it is actually received. It is better to use a specific terminal date such as "You have 10 days from May 1, 1985." (See (C) later in this section.)

If the duration of the offer is not stated specifically, the offer remains open for a reasonable time. Facts considered are the state of the market for the goods or services in question, the need of the offeror to be able to deal with others in the event the offeree does not decide to accept, and custom and usage in the particular transaction. For example, the reasonable time to accept an offer for the sales of shares of stock in a fluctuating market will usually be shorter than the reasonable time to accept an offer for the sale of land. The time to accept the offer to sell perishable goods will be shorter than the time to accept an offer to sell durable goods. All of this is judged from the viewpoint of the offeree, whose reasonable belief as to the duration of the offer governs.

An offer can be revoked by the offeror. In most states, revocation must be communicated to the offeree. A few states, notably California, make a revocation effective when *placed* in the means of communication, such as post or telex, similar to an acceptance. See Section 5.06(C). An offeree who is still interested must make another proposal to the original offeror. This power of revocation frequently exists despite a specified time limit in the offer.

Generally an offer terminates if it is rejected by the offeree. Sometimes the rejection is explicit. The offeree may communicate a lack of interest. In such a case, the offeror is free to deal with others. Usually rejection is implied if the offeree makes a counteroffer. The normal expectation of the offeror in the event of a counteroffer is that the offeree is no longer interested in doing business on the basis of the original offer. This implication may be negated if the counter-offer makes clear that the offeree is still considering the original offer. A counteroffer both creates a power of acceptance in the original offeror and usually terminates the power of acceptance of the original offeree.

The acceptance must be sent or communicated while the power of acceptance still exists. Any acceptance after that time does not operate to create the contract without a further act by the offeror.

The acceptance must be unequivocal and must not propose new or different terms. The following case illustrates this rule and provides an illustration of a negotiation over the supply of construction material. It also illustrates a subcontractor "quote" used by the prime and their failure to agree on terms, a topic explored in Section 32.02.

WESTERN CONTRACTING CORP. v. SOONER CONSTRUCTION CO.

United States District Court, Western District of Oklahoma, 1966. 256 F.Supp. 163.

[Ed. note: Footnote renumbered.]

DAUGHERTY, District Judge. This is an action by the plaintiff, Western Contracting Corporation, against the defendant, Sooner Construction Company, for breach of an alleged subcontract between the parties on a runway project at Tinker Air Force Base, Oklahoma. There was no written subcontract signed by the parties.

. . . From the evidence the Court finds that Sooner orally quoted certain unit prices on asphalt paving to Western prior to Western submitting its bid on the project for the prime contract. Western was successful on its bid. Thereafter, Sooner confirmed its orally quoted prices to Western by a letter dated March 25, 1963.[5] The oral quotes and the letter quotes were the same. Regarding the item of hot mix surface the price quote of Sooner was 17,220 tons at $8.32 per ton less a 50¢ per ton discount if payment is made by the 10th of the month. Sooner asked for the subcontract during these activities but the request was denied. During the period from March 25, 1963, the date of the above

---

5.

Received after Bidding
Haskell Lemon Construction Co.
Road Building and Paving
Phone WIndsor 6–3357   P.O. Box 7118
OKLAHOMA CITY, OKLAHOMA

Western Contracting Corp.
March 25, 1963   Sioux City, Iowa
RECEIVED
Mar 28 1963

Western Contracting Corporation
400 Benson Building
Sioux City, Iowa

Gentlemen:

Congratulations on receiving the Tinker Field contract. I hope that it will prove to be a most successful job for you.

We would like very much to make a contract with you to do your asphalt paving work on this job. This letter is to confirm the prices which we quoted you at Fort Worth.

Item 9. Prime 48,150 gallons @ .19 Furnish, deliver and apply .. no brooming or blotting.

Item 10. Tack coat, 260 gal. @ .37

Item 11. Tack coat, 318 gal. @ .37

Item 13. Hot mix surface 17,220 T. @ $8.32, less 50¢/ton discount for payment by 10th of month

Item 14. Asphalt (85–100) 215,350 gal. @ .13

No quotation on item 12, 15, 16, 17 as they may be deleted.

We will be happy to help you in any way possible on this contract. We would appreciate your contacting us when you establish your job office here. We hope the job will prove to be both pleasant and profitable and that we will have the opportunity of working with you.

Yours very truly,
/s/ Haskell Lemon
Haskell Lemon
Haskell Lemon Construction Co.

mentioned letter, until July 15, 1963, Western opened an office at Tinker and the parties had various contacts, telephone calls, and discussions regarding the possibility of a subcontract. Western was obtaining asphalt quotes elsewhere. On July 15, 1963, at Tinker a meeting was had attended by a Mr. Hastie for Western, a Mr. Lemon for Sooner, a Mr. Pybas, a superintendent of Sooner, and a representative or representatives of the United States Corps of Engineers. At the meeting discussions were had regarding the specifications, equipment and rolling stock. Hastie testified that after this meeting he for Western and Lemon for Sooner reached an oral agreement on the subcontract following which a form of subcontract, unsigned by Western, was forwarded by Hastie to Sooner for execution and return to Western for execution by Western at its home office in Iowa. On the hot mix surface this written subcontract submitted by Hastie contained a price of $7.82 per ton thereon but did not provide for payment by the 10th of the month. Rather, it provided for partial payments to Sooner, less a retained percentage of 10%, as Western was paid on estimates by the owner and final payment to Sooner upon complete performance of the subcontract within 45 days after final payment is received from the owner by Western. Lemon denied that an oral subcontract was agreed upon on July 15, 1963, with Hastie and denied that any discussions were even had whereby Sooner would agree to the $7.82 price with the retainage provision and final payment provision as above set out instead of payment for the hot mix surface by the 10th of the month. Pybas, who testified that he was with Lemon at all times going to, at and from the Tinker meeting on July 15, 1963, also denied any oral agreement on the subcontract or any discussions about the discounted price of $7.82 being agreeable without payment by the 10th of the month. Lemon testified that shortly after receiving the written subcontract from Hastie he called Hastie on the phone several times and objected to the lower price of $7.82 per ton without payment being provided for by the 10th of the month in accordance with his quoted terms. Hastie acknowledged several telephone conversations after July 15, 1963, with Lemon regarding the retainage and that Hastie suggested in one of these conversations a reduction of the retainage to only 50% of the work. Hastie further testified that Lemon never gave him an answer to this suggestion.

* * *

In late September, 1963, certain developments took place. Sooner sent a signed subcontract to Western to which it attached certain amendments, six in number, one of which called for full payment for each of the three phases of the work to be done by Sooner within 45 days of completion by Sooner of each phase in lieu of Sooner's requirement in its written confirmation of payment by the 10th of the month and Western's requirement in the written subcontract it prepared and submitted of the 10%

retainage and the final payment in 45 days. Western wrote Sooner a letter advising Sooner that it was delinquent in the performance of its subcontract and that if Sooner did not correct this default in performance within five days Western would exercise its rights under the subcontract. Then on September 29 or 30, 1963, a meeting was held in Oklahoma City which brought a Mr. Shaller down from Iowa for Western. Shaller as manager of heavy construction for Western was over Hastie. At this meeting the six amendments were discussed one by one and Shaller disapproved the amendment about full payment in 45 days following each phase of completion as well as two other amendments and in his own hand wrote "out" opposite each of the three amendments so disapproved. Shaller approved the other three amendments. In the language of Shaller, finally at this meeting "things were terminated." Under date of October 2, 1963, Western made a subcontract with Metropolitan Paving Company for larger unit prices as to all items (the price for hot mix surface—the largest item—was $8.53 per ton) and sues herein for the difference amounting to $16,957.08 plus interest, overhead and profit and other expenses.

*   *   *

15 Oklahoma Statutes, Section 71 provides:

"An acceptance must be absolute and unqualified, or must include in itself an acceptance of that character, which the proposer can separate from the rest, and which will include the person accepting. A qualified acceptance is a new proposal."

Anderson v. Garrison . . . provides:

"In order that a counter-offer and acceptance thereof may result in a binding contract, the acceptance must be absolute, unconditional, and identical with the terms of the counter-offer."

*   *   *

It is true that the parties may have a binding agreement, provided they have reached one, even though they have the understanding that the agreement should be formally drawn up. This depends upon the intention of the parties.

Fry v. Foster . . . holds:

"Where parties to an agreement make its reduction to writing and signing a condition precedent to its completion, it will not be a contract until this is done, and this is true although all the terms of the contract have been agreed upon. But, where parties have assented to all the terms of the contract, and they are fully understood in the same way by each of them, the mere reference in conjunction therewith to a future contract in writing will not negative the existence of a present contract." . . .

Thus, the importance of what actually took place on July 15, 1963, between Hastie and Lemon—whether they reached an oral agreement or not—becomes apparent. Hastie says they reached an oral agreement. Lemon and Pybas say they did not. Hastie does not claim that the alleged oral agreement was reached in the presence of the representatives of the Corps of Engineers who attended the meeting, therefore, these representatives are not able to give any assistance to the problem. The Court is of the opinion and finds and concludes that Hastie and Lemon did not on July 15, 1963, reach an oral agreement on the subcontract or discuss and settle the price differential on the hot mix surface with reference to the two alternative prices quoted by Sooner and the effect of the method of payment on the same. It is believed that when Western prepared the subcontract shortly after July 15, 1963, it sought to take advantage of the lower quoted price without meeting the condition attached to the same regarding payment. To this Sooner promptly objected and Sooner did not sign and return the subcontract as requested by Western. Hastie admits that several telephone conversations immediately followed his mailing the subcontract he prepared, these telephone conversations coming from Lemon and that the subject matter of the calls had to do with the retainage provision of the submitted subcontract as it affected the price of the hot mix surface.

* * *

The Court, therefore, finds and concludes from the evidence that Sooner quoted a price of $7.82 a ton on hot mix surface provided payment was received for the same by the 10th of the month—otherwise the price would be $8.32 a ton; that the weight of the evidence indicates that this alternative quote and payment condition of Sooner was not changed or discussed on July 15, 1963; that Western in submitting a subcontract to Sooner set out the $7.82 per ton price but did not meet the payment condition attached to the same; that Sooner immediately and admittedly by several telephone conversations objected to this feature of the subcontract as submitted; . . . that while a period of several weeks lapsed before Sooner submitted its subcontract with amendments such lapse of time transpired in the face of and after objections were made by Sooner to the subcontract submitted by Western and particularly with reference to the use of the lower quoted price on hot mix surface without complying with the requested method of payment in the use of such price; that at the meeting on September 30, 1963, Western would not agree to the originally quoted alternative price of Sooner or its modification as later proposed by Sooner and terminated the matter . . . The Court is of the opinion that the parties never reached a meeting of the minds on the price and method of payment for the hot mix surface, the principle item involved. . . . In simple summary, Sooner quoted an alternative price on hot mix surface to Western, Western submitted a subcontract containing a price and method of payment different from each alternative, Sooner submitted an amended contract with still a different price and method of payment, this was not acceptable to Western and the matter was terminated by Western.

Plaintiff is, therefore, not entitled to the judgment it seeks.

As demonstrated in the preceding case when the offeree, that is, the person to whom the offer is made, proposes different terms, it usually means no agreement as yet has been reached. The increased use of standardized forms and stationery with printed provisions along the margins can create problems.

Suppose a letter of acceptance contains printed provisions along the margins that differ from the offer. A leading case held that such a letter was not an acceptance because it proposed different terms.[6] (The printed terms on the stationery probably were not intended to be part of the agreement.) The mirror image rule, that is, the acceptance must be a "mirror" of the offer, often frustrated the common intention of the parties.

To remedy this problem and to deal with transactions where buyer and seller in the sale of goods will sign only their own forms that rarely agree, the Uniform Commercial Code adopted § 2–207. Inclusion of additional or different terms does not necessarily preclude the communication from being an acceptance if it appears that the offeree is accepting the terms of the offer and is not conditioning his acceptance on the offeror agreeing to additional or different terms. Those additional terms are considered proposals for additions to the contract that are sometimes binding without further communication of the offeror. Section 2–207 also

provides that if the conduct of the parties indicates that they have made a contract although the writings of the parties do not match, a contract has been concluded on the terms that do match with the supplemental terms being supplied by law.

Since offerors usually wish to know whether their offers have been accepted, acceptance, as a rule, must be communicated to the offeror. The offeror can deprive himself of the right to receive actual notice of acceptance by stating, "If you accept, sign the letter and you need not communicate any further with me."

(C) CONTRACTS BY CORRESPONDENCE

Generally, to be valid, offers, revocations, and rejections must be communicated to the other party. Since contract law protects reasonable expectations, it is essential that contracting parties know where they stand. However, a special rule has been developed for acceptances.

When parties began to make contracts by correspondence, English courts adopted the mailbox or dispatch rule. If the offeree used the same means of communication as the offeror had employed, the acceptance would be effective when placed in the means of communication. It did not have to be actually received. For example, if the offeree posts an acceptance to the offeror in

6. *Poel v. Brunswick-Balke-Collender Co. of New York*, 216 N.Y. 310, 110 N.E. 619 (1915).

response to an offer received by post, the address on the letter of acceptance is correct, and the proper postage is placed on the letter, the contract is formed at the time the letter is mailed. This places the risk of a delayed or lost letter upon the offeror and protects the offeree's expectation that when the letter has been posted a contract has been formed.

While early cases scrutinized the means of communication used by the offeree, generally the mailbox rule applies if a reasonable means of communication is employed.

One aspect of this rule is often ignored. The rule applies only if the offeror has not *specifically* required that the communication actually be *received.* For example, if the offeror makes an offer by mail but states that word *must* be received by a specified date as to whether the offeree accepts, the contract is not formed unless the offeror receives the offeree's acceptance by the stated date.

(D) REASONABLE CERTAINTY OF TERMS

As mentioned in Section 5.07(B) one test for determining whether an offer has been made is the clarity and completeness of the terms of the proposal. Even when an agreement on terms has been reached and no issue exists as to the existence of an offer, agreed terms that lack reasonable certainty of meaning preclude a valid contract.

Why require that even agreed terms be reasonably clear in meaning before a valid contract can exist? First, vagueness or incompleteness of terms may indicate that the parties are still in the bargaining stage. Second, a third party, such as a judge or an arbitrator, should be able to determine without undue difficulty whether the contract has been performed. But *reasonable*

certainty does not require terms so clear that there can be *no* doubt as to their meaning.

Janzen v. Phillips [7] involved an action by a landscape architect to recover his fee. The architect and his client were somewhat hazy in their initial dealings. There had been no definite discussion of costs and the defendant stated that he wanted a "first class job" and work went ahead on this basis.

Difficulties developed when the interim billings exceeded what the client thought he would have to pay. After discussions regarding costs, the architect sent a letter to the client in which he stated that he would "substantially landscape most of your property" for $6,500.00, or "nearly so." [8] The court was concerned with the terms *substantially* and *nearly so.* Without these terms, clearly assent to this letter would have created a valid contract. The court stated that where it appears the parties have intended to make a contract, courts will tread lightly in concluding that their efforts were not successful because of the absence of reasonable certainty. These terms, according to the court, have a definite meaning and are frequently used to indicate that the cost would be *about* $6,500 and the work would be essentially completed for that price. Some latitude was expected and the architect would have some leeway from the $6,500 figure.

Another court, though, was less willing to be flexible when an event occurred that neither party anticipated. In *Goebel v. National Exchangors, Inc.,* [9] the contract stated that the architect would be paid $500 per apartment for each apartment constructed. Because of costs climbing substantially beyond the budget of the owner, the project was abandoned but not until the architect had performed substantial services. The

7. 73 Wash.2d 174, 437 P.2d 189 (1968).

8. 437 P.2d at 191.

9. 88 Wis.2d 596, 277 N.W.2d 755 (1979).

court held the $500 figure to be of no help in determining the compensation because the testimony of the parties revealed that the number of apartments was "totally indefinite, ranging anywhere from 68 to 100 or more units." This, according to the court, precluded a valid contract from having been formed. However, the court also held that the architect should have been given the opportunity to establish his claim based upon having made an implied contract allowing him to recover the reasonable value of his services.

While the court's analysis is not clear, it could have been based upon the parties' having *impliedly* agreed that if the project did not go forward the architect would be paid a reasonable value for the services that he had performed. Alternatively, if the services were found to have benefited the client, recovery could have been based upon unjust enrichment, the client being required to pay the architect for the benefit that the architect's services had conferred on the client. (See Section 8.04(B).)

Relevant factors in resolving "certainty" disputes are the importance of the term in question, whether performance has commenced, how far it has proceeded, and whether it appears that the party claiming that the contract is invalid appears to be looking for an excuse to avoid contractual obligations.

(E) A MORE FORMAL DOCUMENT

Suppose an informal writing is concluded but the parties expect that a more formal document will be drawn up. Have they intended to bind themselves, with the formal written document only a memorial of their actual agreement, or did they not intend to bind themselves until assent to the more formal writing? This depends upon the intention of the parties.

An expression of their intention will control. But suppose specific statements of intentions do not exist. If the transaction is complicated and nonroutine and the basic agreement is sketchy in form, it is likely that the parties do not intend to bind themselves until the more formal writing is prepared and signed. Commencement of performance by both parties very likely means that a contract has been formed without a formal agreement. Performance by one party is more ambiguous.

Sometimes parties begin performance on the basis of a letter of intention or order to proceed. If no final agreement is made, the letter often states that compensation will be based upon a specified formula or reasonable compensation. If the letter of intent does not deal with payment in the event no contract is concluded, the party receiving the other party's performance must restore the benefit where possible, or pay the value of the benefit conferred based upon the principle of unjust enrichment. (See Section 8.04(B).)

Modern courts tend to find a contract at the earliest possible stage if the court is convinced that the parties intended to be bound and only minor gaps need to be filled.[10]

(F) AGREEMENTS TO AGREE

Frequently parties to a contract are unable to specifically define every aspect of their relationship in the contract, yet they may wish to create a legally enforceable agreement. To accomplish these two objectives they may specify that certain terms will be agreed upon by the parties at a later date. For example, a long-term supply of goods contract may specify that the shipping arrangements or delivery dates will be agreed upon at a later date. The parties have *agreed to agree.*

Older cases refused to enforce contracts that contained such provisions. Judges saw

---

10. *Saul Bass & Associates v. United States,* 205 Ct.Cl. 214, 505 F.2d 1386 (1974) (contract found on basis of letters of intent).

themselves as simply enforcing agreements that the parties made rather than making agreements for the parties.

Modern courts are more willing to enforce contracts that contain agreements to agree as long as the facts do not indicate that the parties are still in the negotiation stage and so long as the matters to be agreed upon are not essential. For example, § 2–204 of the Uniform Commercial Code states that a contract for sale is valid even though certain terms are left open, "if the parties have intended to make a contract and there is a reasonably certain basis for giving an appropriate remedy."

Where courts have faced contracts that contain agreements to agree, as a rule the parties are not disputing the to-be-agreed-to provisions. One party contends that it is not bound because there is a fatal defect that rendered the contract invalid. If performance has begun and the matter to be agreed to is not so essential that it indicated the parties were still in the bargaining stage, courts are likely to enforce the agreement. The agreed-to provision will be supplied by a standard of reasonableness that will often take into account custom and usage.

Sometimes the parties do not state that they will agree but state that they will negotiate in good faith on certain matters. While some courts will not enforce these provisions, there is an increasing tendency to enforce them by requiring the parties to negotiate in good faith. If the parties in good faith seek to agree but cannot, further performance should not continue and any performance that has been rendered should be dealt with by restitution. However, if one of the parties does not negotiate in good faith, the court may determine what good-faith negotiation would have been produced by way of agreement.

In *Purvis v. United States*,[11] prime contractor and subcontractor agreed to work

out one part of the subcontract later but were not able to agree. The court could have concluded that each party attempted to negotiate in good faith and decided the disputed matter on the basis of restitution. But the court completed the agreement for the parties by deciding what they would have agreed to had they negotiated in good faith.

SECTION 5.07
DEFECTS IN THE
MUTUAL ASSENT PROCESS

(A) FRAUD AND
MISREPRESENTATION

The negotiation process frequently consists of promises and factual representations made by each party. If important factual representations or promises are made to deceive the other party, the deceived party may cancel the transaction or receive damages if it has reasonably relied upon the representations or promises.

Sometimes such representations are made not with the intention of deceiving but are negligent or even innocent. As the conduct becomes less morally reprehensible, the remedies to the deceived party will be fewer. For example, in the case of fraudulent representation, the deceived party may receive punitive damages, designed to punish the conduct and not to compensate the victim. While the victim of negligent misrepresentation will be able to recover damages to compensate for the loss or cancel the transaction, the only relief likely to be accorded someone who relies upon an innocent misrepresentation is cancellation of the contract.

Suppose one party to a contract wishes to cancel the transaction because of a claim that the other party should have disclosed important facts, which, had they been known, would have persuaded the former not to enter into the transaction. While

---

11. 344 F.2d 867 (9th Cir.1965).

early cases rarely placed a duty to disclose on contracting parties, there is an increasing tendency to do so where the matter that is not disclosed is important and where the party knowing of the facts should have realized that the other party was not likely to ascertain them.

(B) ECONOMIC DURESS

A valid contract requires consent freely given. Consent cannot be obtained by duress or compulsion. Early common law duress was physical, such as obtaining a deed or a contract by threatening to kill or injure the other party.

Modern duress in the commercial world principally involves economic duress. Economic duress or (as it is sometimes called) business compulsion can exist if one party exerted *excessive* pressure beyond permissible bargaining and the other party consented because it had no real choice. Courts have found economic duress to exist in very limited cases. Their hesitance undoubtedly relates to the inherent pressures involved in the bargaining process and the fear that many contracts could be upset if the economic duress concept were used too frequently.

In *Austin Instrument, Inc. v. Loral Corp.,*[12] a prime contractor under a federal procurement contract contended that a modification it had made with a supplier was obtained by economic duress because the supplier had threatened to stop deliveries unless the prime contractor agreed to pay more money.

The prime contractor was in no position to obtain the goods from anyone else. It would have suffered substantial liquidated damages for delay and might have had its contract terminated by the government. A substantial portion of its business was with the government, and failure to deliver might have jeopardized its chances for future contracts. The Court of Appeals of New York concluded· that it had no real choice. As a result the prime contractor was relieved from its promise to pay more money for the goods and was able to recover excessive payments made.

The case is a close one because both parties were experienced businessmen, the prime contractor might have obtained relief from the government had it been sought, and there was no indication that the price was outrageously high. Some courts would have considered this simply a matter of hard bargaining.

Selmer Co. v. Blakeslee-Midwest Co.[13] faced duress in the context of a subcontractor who sought to avoid a settlement made with a prime contractor. The Court held that the financial difficulty of the subcontractor was insufficient in and of itself to avoid the settlement. Had the prime contractor acknowledged the correctness of the subcontractor's claim yet forced the subcontractor to accept a significantly lower amount, this, coupled with the subcontractor's financial difficulties, might have justified a finding of economic duress.

(C) MISTAKE

The doctrine used most often in attacking the formation process is mistake. Here again, there are no fixed or absolute rules. There may be mistakes as to the terms of the contract. One party may not read the agreement, or because of mistake or fraud the agreement may not reflect the earlier understanding of the parties. Also, parties are often mistaken as to the basic assumptions upon which the contract was made. Everyone who makes a contract has certain underlying assumptions that, if untrue, make the contract undesirable.

A few generalizations can be made. While there is always some carelessness in the making of a mistake, the greater the degree of carelessness, the less likely it is that the party making the mistake will be

---

12. 29 N.Y.2d 124, 324 N.Y.S.2d 22, 272 N.E.2d 533 (1971).

13. 704 F.2d 924 (7th Cir.1983).

given relief. A comparison of the values exchanged is relevant. If one party is getting something for almost nothing, mistake is more likely to be employed to relieve the other party from the contract. The question of when the mistake is uncovered is important as well. If the mistake is discovered before there has been reliance by the other party or before there has been performance by the other party, there is a greater likelihood that the contract will not be enforced.

In extreme cases courts will relieve a party from certain risks. However, the law also seeks to prevent parties from avoiding performance of their promises by dishonest claims of mistake when it proves economically advantageous to do so. Steering a course between these two policies, as well as supporting the policy that persons should be able to rely upon agreements, has meant that there are few fixed legal rules that can be used to predict probable results in such cases.

A determination of whether relief will be granted requires a careful evaluation of the facts, with principal scrutiny paid to the relative degree of negligence, any assumption of risk, the disparity in the exchange values in the transaction, and the likelihood that the status quo can be restored.

(D) UNCONSCIONABILITY

Equity courts refused to enforce contracts that were unconscionable—contracts that shocked the conscience of the judge. Borrowing this concept, the Uniform Commercial Code in § 2–302 gave the trial judge the power to strike out all or part of an unconscionable contract for the sale of goods. Gradually the concept is being accepted in *all* contracts. But contracts that have been struck down (and few have) have tended to:

1. Be consumer rather than commercial transactions.
2. Involve sharp dealing in the bargaining process rather than harsh terms.
3. Involve clauses that exculpate a party

from obligation or make vindication of legal rights very difficult rather than price terms.

As yet the doctrine has not made much of a direct impact upon contracts for design and construction except for anti-indemnity statutes to be discussed in Section 36.05(D). Its development has, however, encouraged courts to examine all the circumstances that surround the making of a contract and to intervene more frequently in the bargain.

SECTION 5.08 CONSIDERATION AS A CONTRACT REQUIREMENT

(A) FUNCTIONS OF CONSIDERATION

No legal system enforces all promises. Some promises are not important enough to enforce nor intended to create enforceable rights. For these reasons, the law will not enforce a promise to attend a social event.

Enforcement of some promises can disrupt important institutions. Suppose an eight-year-old boy can obtain judicial relief to enforce a promise by his father to take him to the zoo. This can disrupt the order needed for a family unit.

Some promises are made hastily or impulsively, without appreciation for the consequences. Altruistic promises are often made without regard for the burden they can place on the promisor. Some promises are best left to other sanctions. For example, the law generally does not use contract law to hold politicians to their campaign promises. Instead, the ballot box and public opinion are mechanisms to pressure public officials to live up to their promises.

Some contracting parties have used their bargaining power to impose harsh obligations on weaker parties, promising little or nothing of value in return. In extreme cases, this overreaching may be restrained by denying enforcement of the weaker party's promise.

The common law has made the consideration doctrine the chief vehicle for determining which promises will be enforced by the courts. It operates as a limit on complete contract freedom.

(B) DEFINITIONS: EMERGENCE OF BARGAIN CONCEPT

Early law required benefit to the promisor (the one who had made the promise). Later decisions, however, enforced a promise where the promisor did not receive pecuniary benefit but the promisee suffered a detriment.

In this century, the concept of bargain developed. Rather than look for benefit to the person making the promise or detriment to the person to whom the promise had been made, the courts examined the transaction to see whether there had been a bargain. If so, usually there is consideration. Fairness of the bargain is not a requirement, except in certain consumer transactions and where equitable remedies are sought.

Most commercial contracts involve bargains, and consideration is rarely a problem. However, sometimes a guarantee or surety promise, that is, one to answer for the obligations of another, can raise consideration questions because it is often made after the contract that is being guaranteed. For example, in *Northern State Construction Co. v. Robbins*,[14] the contractor made a construction contract with a corporate owner of questionable financial responsibility on February 27, 1962. On March 1, 1962, several shareholders in the corporation executed a written guarantee of the corporate owner's performance under the construction contract.

The contractor's fears about the financial instability of the corporate owner proved well grounded. Upon completion of the work, the corporate owner was unable to pay the outstanding balance of the contract price. The contractor brought an action against the guarantors.

The action was unsuccessful because the court concluded that the guarantee occurred *after* the construction contract had been made. This, according to the court, meant that the guarantors received nothing for their promise inasmuch as the owner was already committed to perform the contract. Had the construction contract and the guarantee been part of one transaction and the agreement by the prime contractor induced in part by the guarantee, the guarantee would have been enforceable. A different result would be reached in other states.[15]

(C) RELIANCE

Acts of a substantial nature in justifiable reliance upon a promise may enforce the promise. Beginning in gift promises and promises to charities, courts began to use the doctrine of reliance to supplement the bargain concept.

Reliance has extended into commercial areas and has been used to enforce promises of bonuses made by employers to employees and promises made by manufacturers to franchised dealers. In bonus cases, the reliance is the employee's remaining on the job to obtain the bonus. In the dealer franchise cases, reliance frequently consists of expansion of the dealer's facilities and investment of capital. While it is possible to find a bargained-for exchange in these cases, some courts have avoided the often tortured reasoning necessary to find a bargain and instead have used the reliance concept to make the promise enforceable.

The reliance concept has entered the construction process. Section 32.02 describes its use as a means of enforcing bids made by subcontractors to prime contractors. *Bethlehem Fabricators, Inc. v. British Overseas Airways Corp.* held the owner to its promise to a prospective subcontractor that the prime contractor would be required to

---

14. 76 Wash.2d 357, 457 P.2d 187 (1969).

15. Restatement (Second) of Contracts § 88 (1981).

post a payment bond because of the subcontractor's reliance by its having entered into the subcontract.[16]

While reliance is not as accepted as is the bargain concept, it is acquiring greater respectability and recognition by the law.

(D) PAST CONSIDERATION

By and large the promisor who has not received anything in exchange for its promise or whose promise has not been reasonably relied upon will not have its promise enforced. In this sense it is often stated that promises motivated by benefits having been received in the past are not enforceable. But there are inroads in this rule. In some states a promise based upon past benefits not conferred as a gift may be enforced "to the extent necessary to prevent injustice." [17]

In the commercial world there are exceptions to the traditional denial of nonenforceability. Promises to pay a debt that has been either discharged by bankruptcy [18] or barred by the Statute of Limitations are enforced.[19]

SECTION 5.09
PROMISES UNDER SEAL

When most people were unable to write, a method was needed to accomplish conveyances and other legal acts. The method adopted in early English legal history was the seal, an impression placed upon soft wax that had been placed on the document. The impression was made by a ring or other similar instrument. It emphasized the seriousness of the act being performed. Consideration was not necessary. A sealed instrument was a powerful document.

In the United States, the seal began to lose its power when, instead of being a

formal impression in wax, the seal was reduced to either putting the word *seal* on the document or using some abbreviation such as *LS.* This did not have the trappings of formality possessed by the old seal. As a result, in many states, the seal was denied any function in the preparation of the documents. In other states, a document under seal creates enforceability or creates a presumption of consideration.

SECTION 5.10
WRITING REQUIREMENT:
STATUTE OF FRAUDS

(A) HISTORY

In 1677 the English Parliament passed the Statute of Frauds. The statute required that a sufficient written memorandum be signed by the defendant before there can be judicial enforcement of certain specified transactions. One reason for the writing requirement is to protect litigants from dishonest claims and protect the courts from the burden of hearing claims of questionable merit. Another is that the requirement of a writing acts as a cautionary device. It warns persons that they are undertaking serious legal obligations when they assent to a written agreement. The Statute of Frauds has been adopted in all American jurisdictions in one form or another. There has been a tendency to expand the classifications that require a writing.

(B) TRANSACTIONS
REQUIRED TO BE EVIDENCED
BY A SUFFICIENT MEMORANDUM

Transactions singled out in the original statute were promises by an executor or administrator to pay damages out of his own estate, promises to answer for the debt,

---

16. 434 F.2d 840 (2d Cir.1970).

17. Restatement (Second) of Contracts § 86 (1981).

18. Id. at § 83. Formal requirements to pay debts barred by bankruptcy are contained in 11 U.S.C.A. § 524.

19. Restatement (Second) of Contracts § 82. Many states require a written promise.

default or miscarriages of another, agreements made upon consideration of marriage, contracts for the sale of land or an interest in land, agreements not to be performed within a year from their making, and contracts for the sale of goods over a specified value. Some states require contracts for the performance of real estate brokerage services and contracts to leave property by will be evidenced by a written memorandum.

Classification of the transaction has been one device by which courts have cut down the effectiveness of this statute. For example, a contract to construct a twenty-story building will not require a writing if the court concludes that the contract *could* be performed within a year. Many other limitations and exceptions upon the various transactions have been made by the courts.

(C) SUFFICIENCY OF MEMORANDUM

The memorandum must be signed by the party to be charged. It must contain the basic terms but need not express the entire agreement. It need not have been signed at the time the agreement was made but can be signed later. For example, a letter written and signed by a party after a dispute has arisen will provide the needed memorandum if the letter contains sufficient terms. In goods transactions, the statute is satisfied by an admission in the course of litigation.[20] As technology changes, the law will have to decide whether audio or visual tapes will satisfy the requirement.[21]

(D) AVOIDING THE WRITING REQUIREMENT

Sometimes oral agreements that generally require a memorandum are enforced de-spite the absence of a memorandum. Section 2–201 of the Uniform Commercial Code states that acceptance of or part payment for the goods is sufficient to make enforceable an oral contract for *those* goods.

In cases involving the sale of land, the part performance doctrine developed as a substitute for a writing. For example, if the buyer took possession and made improvements, the court could order the seller to convey despite the absence of a writing. However, part payment by the purchaser was not sufficient part performance.

Some jurisdictions enforce oral agreements if one party has reasonably relied on and changed its position based upon the promised written agreement, or if the other party has been unjustly enriched by the performance.[22]

The Statute of Frauds has become eroded in many states. If a claim seems genuine and if the claimant has relied reasonably on the oral agreement, a strong likelihood exists that the claimant will be permitted to prove the agreement.

SECTION 5.11 EMPLOYMENT CONTRACTS

(A) INTRODUCTION

This chapter has described principles of contract formation and remedies. Generally this treatise discusses legal doctrines within the context of the participants and activities related to design and construction. Application of general legal doctrines to design and construction contracts are discussed ahead. However, two other types of contracts important in the construction process—those that relate to employment and those that relate to acquiring ownership of

---

20. U.C.C. § 2–201(3b). But the contract is enforced only to the quantity admitted. This was not adopted in California.

21. *Ellis Canning Co. v. Bernstein,* 348 F.Supp. 1212 (D.Colo.1972) held that an audio tape satisfied the Statute of Frauds.

22. *Monarco v. Lo Greco,* 35 Cal.2d 621, 220 P.2d 737 (1950).

real property by purchase—have particular rules that were developed to meet specific needs and contractual practices. With this in mind, Sections 5.11 and 5.12 touch upon some of the salient legal characteristics of those contracts. This section deals with employment contracts, and Section 5.12 deals with contracts that relate to the acquisition of interests in real property by purchase.

(B) CONTRACTS FOR INDEFINITE PERIODS

The common law classified employment agreements for an indefinite period as agreements at will. Such contracts can be terminated at any time by either party. In essence, the contract governed the rights and duties of the parties so long as they were both willing to continue the relationship.

The at-will rule was based in part on the difficulty the employer would have enforcing any contract against the employee because it was likely that the employee would not be financially able to respond to any court judgment. Courts would not *order* an employee to go back to work for an employer and only rarely would order an employee not to compete with the employer even if the employee promised not to compete. The at-will rule may be in accord with the intention of the parties. Increasingly, though, wrongful dismissals, either bad faith discharges[23] or those made without cause[24] may justify compensatory and even punitive damages.[25] However, in a recent case an employer was allowed to discharge a high-level employee who had hired an attorney to handle a wage dispute with his employer. The discharge letter stated that the employer could not function if executives had to deal through attorneys.[26]

(C) MODIFICATION

Suppose an employee receives an offer of a better job during the term of employment and informs the employer, and the employer promises to pay more money for the same work for the balance of the term. Generally the employer's promise is not enforceable. The employer received nothing for the promise other than the performance which the contract entitled the employer, and the employee did no more than what was required under the employment contract. Denial of enforcement was based upon what was called the "preexisting duty rule."

Perhaps the real basis for such decisions was the presence of actual or subtle duress on the employer. Yet in many situations, there was no duress or unfairness. As a result, exceptions developed to the normal rule in employment cases. For example, if the employee's duties are changed in the slightest, the promise to pay more money would be enforced. Also, if the old contract has been cancelled by the parties, the preexisting duty rule does not apply.

In some states, statutes were passed stating that the modifications did not require additional consideration to be enforceable if the modifications were in writing. Perhaps these exceptions reflect a judicial and statutory recognition that employees' rights to try to improve their positions are more important than being held to their contracts.

(D) WRITING REQUIREMENT

The Statute of Frauds requires that con-

---

23. *Monge v. Beebe Rubber Co.*, 114 N.H. 130, 316 A.2d 549 (1974) (discharged for not dating foreman). See Annot., 9 A.L.R.4th 314 (1981).

24. *Pugh v. See's Candies, Inc.*, 116 Cal.App.3d 311, 171 Cal.Rptr. 917 (1981) (thirty-two-year employee).

25. *Tameny v. Atlantic Richfield Co.*, 27 Cal.3d 167, 164 Cal.Rptr. 839, 610 P.2d 1330 (1980).

26. *Kavanagh v. KLM Royal Dutch Airlines*, 566 F.Supp. 242 (N.D.Ill.1983).

tracts that by their terms cannot be performed within a year be evidenced by a written memorandum signed by the party from whom performance was legally demanded. But a writing is not required if the work can be performed within a year. The employment of an engineer for a large project that could take two years need not be in writing if the contract *could* be performed within a year.

(E) PUBLIC EMPLOYMENT

Even without civil service rules requiring a fair hearing, some public employees cannot be discharged without a hearing because of expanded application of federal constitutional obligations to states and public agencies.

Public employees are increasingly joining unions or employee associations which has led to legislation permitting collective bargaining by some employees. When negotiations break down, strikes often result. Although strikes by public employees are many times forbidden, public officials are often unwilling to invoke legal sanctions against such strikes.[27]

(F) GOVERNMENT REGULATION OF PRIVATE EMPLOYMENT

Private law is of diminished importance in employment relations. Collective bargaining is, to a certain degree, controlled by federal and state legislation. Increasingly, state and federal governments regulate private employment, mainly in the area of wages, hours, working conditions, and employee selection. Laws limit the work hours of minors and require extra pay for work over a stated minimum number of hours per day or week. Minimum wage laws set a wage floor.

During the 1950s, many states enacted fair employment practices acts. The policy of nondiscrimination was given a substantial impetus by the enactment of the Federal Civil Rights Act of 1964. This act applies, with some exceptions, to employers engaged in interstate commerce. Given the interstate characteristics of most businesses, it realistically affects almost all employers who have eight or more employees. The Act makes it an unfair employment practice:

> to fail or refuse to hire or to discharge any individual, or otherwise to discriminate against any individual with respect to his compensation, terms, conditions, or privileges of employment, because of such individual's race, color, religion, sex, or national origin.[28]

In 1967 the federal government enacted comparable legislation prohibiting discrimination on the basis of age unless age is a bona fide occupational qualification or unless there is a bona fide seniority or retirement plan.[29]

(G) REMEDIES

(See Section 6.09.)

SECTION 5.12 CONTRACTUAL ACQUISITION OF INTERESTS IN LAND: SPECIAL PROBLEMS

(A) BROKER CONTRACTS

Function

While it is not necessary to hire a broker to assist a seller or buyer, typically a person wishing to buy land or to sell land will contact a real estate broker to assist in find-

---

27. A California intermediate appellate court held that this particular strike by employees of a municipal sanitary district was illegal. It cited and discussed authorities and commentaries. *County Sanitary Dist. No. 2 v. Los Angeles County Employees Ass'n,* 147 Cal.App.3d 990, 195 Cal.Rptr. 567 (1983) mod. in 148 Cal.App.3d 838d, 195 Cal.Rptr. 567 (1983).

28. 42 U.S.C.A. § 2000e–2.

29. 29 U.S.C.A. § 623.

ing a buyer or seller. Ordinarily, the broker is hired to bring prospective buyers or sellers to the principal. The broker is not given the authority to actually sell the land.

Sometimes brokers try to find prospective buyers or sellers without any advance agreement as to commissions. More often, a listing or exclusive listing is made by owners of land with the real estate broker, or similar arrangements are made by prospective buyers. Because it is more common for the seller than the buyer to obtain a real estate broker, the discussion centers on the use of the broker by the prospective seller of property.

Listings

Generally the owner of land promises to pay the broker a specified commission if the broker procures a buyer ready, willing, and able to purchase the property. The owner of the property is not obliged to sell to the person making an offer to purchase. Federal, state, or local open occupancy laws may limit the seller's right if the reason for refusing to sell relates to race, color, religion, sex, or national origin. To avoid payment if the sale is not consummated, owners can condition their obligation to pay the fee upon the consummation of a contract for purchase, the actual transfer of title, or the receipt of a specified payment.

Listings fall into three categories: general, exclusive agency, and exclusive right to sell. A general listing, usually for an unspecified time, permits owners the greatest freedom. Owners can authorize any number of agents, sell the land themselves, or withdraw before a purchaser is found. An exclusive agency gives the chance to one agent, but owners can sell the land themselves. The tightest listing is an exclusive right to sell, with the agent receiving a commission no matter who makes the sale.

Multiple listings allow a number of agents to find a buyer, with the fee being divided between the agent who obtains the listing and the one who finds the buyer.

Fees

Traditionally brokers have received a specified percentage of the sale price as their compensation. This method of computing the fee, like a similar method traditionally used by professional designers, has been criticized. Often it does not relate to the amount of time spent by the broker. Some owners resent having to pay a full fee when it took little effort to sell the property at the asking price, not an uncommon phenomenon in a rising market. Also, such a method can induce the broker to ask a low asking price which, while producing a lower fee, can result in a quick sale with minimal effort.

As the trend continues toward alternative methods of providing professional services and continued effort by governmental officials to encourage competition, it is likely that alternative fee methods will develop. Some brokers provide a service that is not conditioned upon a sale but is compensated for by a fee based upon the time spent or by a flat fee. Some states have enacted legislation requiring that brokers tell prospective clients that it is possible to negotiate a fee different from the one customarily used in the community. (Most indications are that such laws have very little effect upon actual fees.)

Legal Problems

The problems that arise most frequently relate to the one-sidedness of the agreement. In the typical brokerage arrangement, the broker does not promise to do anything. The promise is made by the owner, and for that reason, the contract is typically classified as unilateral. If it is unilateral, earlier cases allowed the listing to be revoked by the owner at any time prior to full performance. However, doctrines have been developed to protect the broker from expending substantial amounts of effort and spending money in performance of the contract only to find that the agreement has been revoked before the broker's completion of performance.

Some brokers protect themselves today from revocation by language in the listing agreement that binds brokers to use their best efforts to secure a buyer ready, willing, and able to buy the property. This technique is intended to convert the contract into a two-sided contract (a bilateral contract) where *both* parties have obligations. Another technique employed by brokers is the inclusion of a provision allowing the owner to withdraw the property but to condition this right on payment of the broker's fee.[30]

Formalities

In some states agreements by which the owner of land promises to pay a commission if a broker procures a buyer or a renter must be evidenced by a written memorandum. In most categories of transactions required to be evidenced by a sufficient written memorandum, the courts have been astute in finding exceptions that would in effect enforce an oral transaction.[31] In contracts involving brokerage services, courts have generally refused to employ exceptions.

Fiduciary Relationship

Hiring a broker creates an agency relationship between the client and the real estate broker with mutual fiduciary obligations. The parties must disclose pertinent information and any conflict of interest to one another, not make profits at the other's expense, and behave fairly toward one another. There has been a tendency to hold brokers liable to the buyer for secret profits made by telling the buyer the seller will take only a price higher than the seller wishes, buying at the lower price from the seller and reselling at a higher price to the buyer.[32]

(B) CONTRACTS TO PURCHASE AN INTEREST IN LAND

When Are Parties Bound?

One problem that recurs in contracts for the purchase of land is whether the parties intend to wait until a more formal writing is signed before they bind themselves. Ordinarily the history of a contractual transaction begins with preliminary negotiations. One party puts forth a proposal to which the other party may either agree or make a counter proposal. At some stage, there is general agreement as to the basic terms of the transaction. For example, the parties may agree as to the land that is to be sold, the price, the financing terms, and the date of transfer of possession.

Many other elements exist to this contract that are not likely to be considered by the parties at this stage. These elements will be agreed upon when the agreement is transformed into a more formal written contract. The type of deed, the allocation between the parties of the risk of loss between the time the contract is formed and the execution of the deed, the furnishing of title insurance, and provisions for an escrow arrangement are not likely to be discussed at the time the preliminary agreement is reached.

Sometimes the parties may make a notation informally on a piece of paper as to the basic terms to which there *has* been agreement. They may, in addition, sign this informal writing. Have they intended to bind themselves by this informal writing so that neither party can withdraw at its own discretion? Is the informal writing only a tentative agreement, with the binding contract to be formed when both parties have

---

30. *Blank v. Borden,* 11 Cal.3d 963, 115 Cal.Rptr. 31, 524 P.2d 127 (1974).

31. See Section 5.10(D).

32. *Harper v. Adametz,* 142 Conn. 218, 113 A.2d 136 (1955); *Ward v. Taggart,* 51 Cal.2d 736, 336 P.2d 534 (1959).

assented to a more formal writing that will contain details omitted in the informal writing?

Whether the parties are to be bound at the preliminary stage or only upon their assent to a more formal document depends upon the intention of the parties. However, criteria exist to determine such intention. Objective criteria are useful if the matter is disputed and both parties are likely to state that they had contrary intentions. The law examines the complexity of the transaction, whether either or both parties had acted in reliance upon the earlier informal agreement, and whether the law provides sufficient guidelines to fill in the gaps not yet resolved.

On the whole, the law finds that the parties in land transactions do *not* intend to bind themselves until there is assent to a more formal writing. A transaction involving the sale of an interest in real property usually involves much money and is not routinely made by one or both parties. It is better to delay binding the parties until they or their advisers have had an opportunity to examine provisions dealing with the problems not anticipated at the time the informal agreement is negotiated. However, parties *can* bind themselves at the time of the informal document, particularly if standard provisions accepted by the community can satisfactorily fill in the gaps.

Agreement to Agree

(See Section 5.06(F).)

Options and Contingencies

In these contracts purchasers may not be willing to bind themselves before inquiring into such matters as financing, costs of construction, and the uses to which the property can be made. They may, instead of binding themselves, wish to obtain an option under which the seller will agree to sell,

while the buyer retains the freedom of choosing to buy.

As noted in Section 5.06(B), most offers are revocable even if the offer is stated that the offer will not be revoked for a specified period of time. To make the offer irrevocable and to create a binding option, the person giving the option—in this case the seller—must receive something for the option. Usually this is accomplished by the buyer's paying a nominal amount, such as one dollar or ten dollars. If a written option is made for a reasonable period for a purchase at an adequate price, no inquiry will be made into the question of whether this nominal amount was bargained for or if this was a price actually paid for the option. Some options are bought for an amount of money that reflects the value of the option. Although the payment of the nominal amount does not always reflect the value of the option, payment of this amount usually will make the option irrevocable.[33]

The seller may not be willing to grant a one-sided option. If the buyer cannot obtain an option, the buyer may agree to buy the land, subject to the ability to obtain financing of a specified amount and at a specified rate. Buyers may want to condition their obligations to purchase upon obtaining a variance from the zoning laws, upon selling their own homes, or upon other important events occurring before they are obligated to complete the purchase. Sometimes the buyer's promise is phrased in this way: "I will buy if I obtain financing of 80% of the purchase price for a loan not to exceed 20 years at a rate not to exceed 12%." (The first edition in 1970 stated $7\frac{1}{2}$%.) In such a situation, the buyer may not have made a promise at all but has given a conditional acceptance. Neither party is bound to perform. However, more likely the buyer has made an absolute promise to perform but inserted a number of contingencies in the contract.

---

33. Restatement (Second) of Contracts § 87 (1981).

Sometimes the occurrence of contingencies that condition the obligation to perform is largely within the control of the purchaser. Suppose the contingency is the buyer selling his own home. Buyers may not try hard to sell their house if they have second thoughts about a house they may have contracted to buy. They may not make attempts to obtain financing if financing is a contingency spelled out in the contract to purchase the land.

Since occurrence of such conditions is to a substantial degree within the buyer's power, does this destroy the two-sidedness of the contract? It may appear that the seller is bound *absolutely* to perform, while the buyer can retain the power to determine whether or not to perform. For this reason, some courts have held the contract too one-sided for enforcement. Most, though, enforce the contract by implying promises of best efforts or of good faith on the part of the buyer to cause the conditions to occur.[34] Contingencies are common, and generally the parties intend to arrive at a valid agreement.

Sometimes sellers wish to protect themselves by the use of contingencies. They may not wish to sell their house unless able to purchase another house or be admitted to a community for senior citizens. The same principles discussed in connection with protective contingencies for the buyer apply if the seller obtains this protection.

Formal Requirements

(See Section 5.10.)

Destruction Before Conveyance or Transfer of Possession

Often a substantial period of time must elapse before the title is actually conveyed or possession is transferred. Suppose the house is destroyed before transfer of either title or possession? Does the buyer still have to pay? Is a seller who cannot deliver the premises in the state promised by the contract guilty of breach of contract?

Unless covered in the contract, the result varies from state to state. Most states force the buyer to complete the purchase even if the house has been destroyed. Some relieve the buyers even if they have gone into possession. Some have adopted the Uniform Vendors and Purchasers Risk Act, which places the risk of loss upon the buyer if *either* the title has been conveyed or the purchaser has gone into possession.

The parties should consider this matter at the time the contract is made. If the contract allocates the loss to one party, that party should protect himself against this loss by insurance.

Delay

Usually the contract states a date for the payment of money and the issuance of the deed. Yet performance is often delayed, usually because of delay in obtaining the loan or clearing title.

Suppose no contract provision deals with delay. The injured party can accept delayed performance and receive damages for the delay. But it is possible that delay can terminate the other party's obligation to perform. However, the frequency of delay in such transactions is likely to preclude such a drastic result in actual practice. If a claim is made for the land or money *itself*, that is, for specific performance, in the absence of a contract provision to the contrary, time would not be "of the essence." Although delay is a breach, it does not terminate the other party's obligation to perform.

Contracts usually state that time is of the essence. Suppose the party guilty of delay has relied upon promised performance by the other party, either by making other

---

34. *Mattei v. Hopper*, 51 Cal.2d 119, 330 P.2d 625 (1958).

commitments or by paying money under the contract. In such a case the law may not permit termination, particularly in the case of delay in making payments by the buyer if the latter tenders the amount due with interest.[35] Decisions, though, will vary, making generalizations precarious. About the best that can be said is that if the delay is truly minimal and not caused by fault and if the innocent party is not damaged by the delay, termination will very likely be refused.

Even though late performance *may* not end the contract, the party guilty of unexcused delay must pay any damages caused by the delay. In the case of delayed performance by the buyer, interest is usually the measure of recovery; by the seller, it is usually loss of use or rental value.

Implied Warranty

Suppose the buyer after taking possession finds defects in the house or terrain. In the sale of goods, courts often imply a warranty of fitness or merchantability. Sellers upon whom buyers relied are responsible if the goods did not accomplish the buyer's known purpose. It was not necessary to show an *express* warranty or fault.

Historically, implied warranty was not used in sales of real property. One reason was that the deed transferred ownership and strong policies existed in favor of making the deed final and complete. Not only were warranties not implied in deeds, the common law held that warranties made *specifically* were merged into the deed. Even express warranties were lost if not contained in the deed.

Denial of implied warranties was based upon *caveat emptor* —let the buyer beware. Buyers can inspect the premises and should

not be given judicial protection when they could have protected themselves.

Today the seller of a new home (and possibly of a used one) impliedly warrants to the buyer and, in some states, to *subsequent* purchasers that the house is fit for the intended purpose of the buyer, that it was built in a workmanlike manner, and that it is suitable for human habitation.[36]

Remedies

(See Section 6.06.)

(C) ESCROW

In many transactions involving the purchase of real property an escrow arrangement is used. The seller deposits the deed and any other documents that are ultimately to go to the buyer with a third party, commonly a title company or a lending institution. The instructions to the third party escrow holder may be to deliver these documents to the buyer when the buyer has delivered to the escrow holder the purchase price (and perhaps has met other conditions agreed to by the parties). The escrow holder acts as a clearance house to insure that the buyer does not have to pay without getting the deed and that the seller does not have to deliver the deed without being paid. Usually, the escrow holder is a stranger, although it may be possible to use the attorney of one of the parties as the escrow holder.

Escrows are used for a number of reasons. If both parties begin their performance by opening an escrow and making a deposit into the hands of the escrow holder, it is less likely that the deal will fall through. The advance execution of the deed makes it less likely that there will be delay and difficulty if the seller dies, as compared with those situations where the

---

35. *MacFadden v. Walker*, 5 Cal.3d 809, 97 Cal.Rptr. 537, 488 P.2d 1353 (1971).

36. *Humber v. Morton*, 426 S.W.2d 554 (Tex.1968). See Section 28.09.

seller does not make out the deed until the purchaser has put up the money. The escrow assures each party that when it performs it takes no risk that the other party may not perform. Escrows are also useful when there are a number of interested parties to the transaction. In many transactions involving the sale of land, there will be a prior lender who will want the debt paid off and a new lender who is advancing money to the purchaser. A substantial portion of the new loan will be used to pay off the prior lender. A number of mechanical details can be taken care of by escrow holders, especially when the escrow holders are professional escrow companies.

Sometimes problems develop because the escrow instructions frequently on standard printed forms vary from the contract of sale. The documents *as a whole* must be examined to find the common intention of the parties. The escrow instructions and the contract for the sale should be consistent.

(D) TITLE ASSURANCE

By Seller

Purchasers of land want to be reasonably certain that they will acquire undisputed ownership of the land. They want the title, usually marketable title, promised by the seller. This is a title sufficiently free from dispute that it would be accepted by most buyers in the community. Buyers may be concerned with the possibility that someone will assert a claim against the land that they will either have to buy off or litigate. If the seller does not have the title promised, the buyer has an action against the seller for breaching the warranty to deliver marketable title free of encumbrances other than those specified. However, the seller may be out of the jurisdiction or unable to pay any judgment awarded to the buyer.

By Attorney

The buyer often wants better protection than a legal claim against the seller. For this reason, the buyer may require that the seller furnish a title abstract (a summary of all the recorded transactions that pertain to the land). The buyer asks his attorney to review the abstract to see if the seller has the title claimed and to find any encumbrances on the land. This is a system used in small communities in the United States.

The principal defect of such a system is that any recovery against the attorney must be based upon the attorney's negligence, which may be difficult to show. Even if the attorney were negligent, it may be that the damages caused by the attorney's negligent performance are substantially greater than his ability to pay. The examining attorney does not assume responsibility for the correctness of the abstract that has been compiled by an abstract company. Any action for an improper copying would have to be brought against the abstract company. The abstract is not a reproduction of the original documents on record but is a condensed statement of the key facts in each transfer.

By Title Company

For these reasons, much of the title assurance work in larger cities and in many states is handled by title insurance. The title insurance company issues a policy at the time of the transfer, insuring the title against certain defects that appear in the land records. The title insurance company checks through the recorded documents pertaining to the property. If it makes a mistake, the title insurance company is usually capable of paying damages.

Certain defects or "clouds on title" cannot be discovered from a search of the land records. Such title defects usually are not covered by title insurance. However, most buyers are satisfied by the title policy because such defects are rare.

Title insurance is not absolute insurance. The policy only warrants that an expert company has checked the records and can pay if it makes a mistake. A buyer who wishes to know the scope of the insurance should read the title insurance policy.

(E) SECURITY INTERESTS

Purpose

Land transactions frequently involve substantial sums. Usually the purchaser is unable to pay the entire purchase price and frequently borrows money from a lender to buy the property.

Lenders choose borrowers who are likely to repay the loan. However, lenders want a security interest in the property out of which they can be repaid in the event the borrower is unwilling or unable to pay back the debt. For example, a purchaser who wishes to buy property for $100,000 may have only $10,000 to put down on the property and may wish to borrow the balance from a bank. If the bank will lend $90,000, the buyer will usually sign a note for the $90,000 with an agreement to make specified payments at specified intervals. The buyer will also give a security interest to the lender. The security interest may be created by a mortgage or, in some states, by what is called a deed of trust.

Operation

Generally the buyer (borrower) of the land takes title and the lender has a security interest in the land. If the borrower does not pay in accordance with the loan obligation, the holder of the security interest has the right to foreclose on the security. In some states, this can be done without judicial sale.

In most cases, the lender will go to court and ask for a judicial sale of the property. The lender will be paid out of the proceeds of the property to the extent of the unpaid debt. If the sale does not net enough money to pay off the debt, a deficiency judgment is awarded by the court to the lender against the borrower. In some states, defi-

ciency judgments are prohibited by law in certain transactions. In these states the lender can recover only the amount obtained by the judicial sale.

Secured transactions can become very complicated. Many parties may demand the proceeds of the sale which can result in difficult priority problems. Sometimes there are both a first and a second mortgage—sometimes even a third. There may be construction loans on the property, and mechanics' liens and tax liens may be asserted. In such cases, litigation is often necessary to unravel the competing claims.

Equity of Redemption

Usually the mortgagor-debtor has an "equity of redemption." The foreclosure sale will not be effective to transfer ownership permanently until a specified period set by statute has elapsed. During this statutory period, the mortgagor-debtor may pay the amount owing and redeem the property. During the statutory period the purchaser at a foreclosure sale does not have good title. This can affect the amount the buyer is willing to bid at the foreclosure sale.

Land Contract

A land contract is a security device under which the transfer of ownership or the delivery of the deed is withheld until all payments have been made. It may be used when the buyer does not have enough money to make a substantial down payment. Suppose the buyer pays for a long period and then is unable to pay. The buyer may have paid much more than the value received by occupying the property. The buyer may have also made improvements. It seems unfair to have to forfeit the interest in the land for failure to make some payments. For this reason, a few states treat the land contract as a security device with an equity of redemption. Most allow the seller to retake and keep all payments.

PROBLEMS

1. A is a president of Acme Corporation and has two years left on a written four-year contract at an annual salary of $100,000. He is approached by a competitor and offered $125,000 a year. A tells this to the board of directors of Acme who raise his salary to $125,000. A agrees. Is this modification of the original agreement legally enforceable? If not, how can it be made enforceable.

2. The following is a statement of the facts in a recent court case. After reading it carefully, be prepared to answer questions a through f.
   a. What would A argue to support its claim?
   b. What would B argue to support its claim?
   c. Was there a contract? If so, what were its terms?
   d. Suppose there were no contract. Can A recover from B for the work A performed? Can B recover the payments it made?
   e. How could this misunderstanding have been avoided?
   f. How does this case differ from *Western Contracting v. Sooner Construction Co.* set forth in Section 5.06(B)?

A is an established firm specializing in a form of foundation work known as grouting, which consists of pressurized injection of a cement-based mixture into the soil underlying a building for the purpose of arresting subsidence and in some cases actually raising foundation walls. In the summer of 1980, B, a small construction firm, entered into a contract with Brazilian Embassy to stabilize and partially reconstruct a building in the District of Columbia known as the Brazilian Annex. The Annex was a relatively old building constructed in large part on filled ground. The structure had

sunk on all four sides with the result that the floors bowed in the middle. B's task, among other things, was to stabilize the structure to prevent further sinking and to raise certain parts of the foundation, particularly the northeast corner, in order to partially alleviate the unevenness of the floors. This lawsuit arises from B's decision to ask A to perform this aspect of the job using A's grouting technique in lieu of alternative methods available.

During early August 1980, Davis, a vice-president of A, and Downey, B's president, discussed the project in a number of conferences and telephone calls. On August 14, 1980, Davis submitted a written proposal to which was annexed a standard set of conditions. The proposal, to the extent here material, made no guarantee that efforts either to stabilize or to lift the building would be successful, made no commitment as to time of completion of the job, and contemplated that the work would be billed at a per diem rate without any stated limitation on the total price of the job.

After further conversations with Davis—the precise contents of which are hotly disputed—Downey sent Davis a telegram on August 25 indicating that A's proposal was accepted subject to "verbally agreed changes" and that a signed revision would follow. The next day Downey prepared and signed an edited version of A's written proposal to be mailed to Davis. The purport of the revision was to indicate that A was committed to stabilize the building and to lift the northeast corner by at least one and one-half inches and that the job would be undertaken in approximately ten days with a maximum payment of $20,000. A never received this document or inquired as to why it had not been received as promised.

Both parties proceeded on the assumption that they had come to some type of agreement and work on the site began August 28. Although A was eventually able to stabilize the perimeter of the building, it was unable, despite protracted effort, to achieve the desired rise in the northeast corner of the building. As the work proceeded, A reported in writing daily to B, and B therefore had full knowledge that A was proceeding without marked success. At the end of eleven days of work, A billed B at the per diem rate in A's proposal, and the work was paid for in the total amount of $9,936.75.

B at no time indicated to A that it should stop working. Downey constantly reiterated, however, that B had only $20,000 to pay for the work. A never consented to $20,000 cap and urged B to seek an adjustment in the contract price from the Brazilian Embassy which B consistently refused to do. After approximately twenty-five days of continuous work, A concluded that it would not be possible to lift the building one and one-half inches and, requiring the equipment for another job, informed B that it was terminating work. It did no more work.

6

Remedies for Contract Breach: Emphasis on Flexibility

SECTION 6.01 AN OVERVIEW

Determining the remedies available for breach of contract is much more difficult than determining whether a valid contract has been made. It is exasperatingly difficult to predict *before* trial what the law will require that a breaching party must pay. Even after all the evidence has been produced at the trial, it can be very difficult to determine the precise remedy that the party entitled to relief should receive. This sometimes leads to a trial procedure under which remedies are considered *before* going into the issues of whether a valid contract has been formed and whether it has been breached. If very little can be accomplished remedially, it makes little sense to have a long, protracted lawsuit to establish the claim.

While the function of awarding a remedy for contract breach is to compensate the injured party (the exceptions to this being discussed in Section 6.04), measuring the losses incurred and gains prevented can be very difficult. There may be agreement on general objectives, yet the implementation of these objectives can be difficult because of the great variety of fact situations, the different times at which a breach can occur in the history of a contract, the variety of causes that may generate the breach, and

the different judicial attitudes toward breach of contract itself.

The common law has developed conventional formulas that are applied in particular cases designed to implement the basic compensation objective. These formulas may not appear to achieve a just result when they are applied. Borderline cases make it difficult to determine which formula should be applied. The range of remedies, both as to type and amount, can give discretion to juries, trial judges, and appellate courts. For example, if one party seems to have taken unfair advantage of the other, doubts may be resolved in favor of the latter. The reason for the breach, though perhaps not exculpating the breaching party, can be influential in measuring the award. The relative abilities of plaintiff and defendant to bear the loss can be an influential factor, although often an unstated one.

Generally when juries are used they are given considerable latitude to determine the award. The jury award is likely to be upheld if it is based on substantial evidence or reasonable inferences from the evidence, unless the award seems to result from passion or prejudice. While compromise can play a strong part in jury determinations, the failure to require a jury to be specific as to the reasons for its award can give the jury a mechanism for achieving what it believes to be a just result even if this would differ from what the law appears to require.

SECTION 6.02 RELATIONSHIP TO OTHER CHAPTERS

As indicated in the introductions to Parts II, III, and IV, the legal doctrines discussed in Part I are designed to be building blocks for the balance of the treatise. They are elemental legal doctrines that should be mastered before examining more detailed problems that develop in the course of the Construction Process. This chapter explores *generally* the judicial remedies awarded for breach of contract that are applied *specifically* to Construction Process problems in succeeding chapters dealing with remedies for breach of contract.

Usually claims by a design professional are for services rendered for which payment has not been made. Claims by the owner against the design professional usually relate to a delay caused by the design professional, defective design, or a failure to monitor the contractor's performance. These topics are considered in greater detail in Section 17.10.

Claims by owners against contractors usually involve losses asserted because of unexcused contractor delay (covered in Chapter 30) or failure by the contractor to build the project as required (covered in Chapter 28). Claims by the contractor against the owner are usually based upon payment for work performed that has not been received or upon increased cost of performance attributable to the owner (discussed in Chapters 27 and 29).

This chapter, in addition to providing a broad sketch of contract remedies generally, looks briefly in Sections 6.09 and 6.10 at special remedial problems relating to employment contracts and contracts to purchase an interest in real property. These contracts were discussed in Sections 5.11 and 5.12, respectively.

SECTION 6.03 MONEY AWARDS AND SPECIFIC DECREES: DAMAGES AND SPECIFIC PERFORMANCE

Judicial remedies can be divided into judgments that simply state that the defendant owes the plaintiff a designated amount of money (sometimes called the money award) and judgments that specifically order the defendant to do something (specific performance) or to stop doing something (injunction). It is important to recognize the essential differences between money awards and specific decrees.

The ordinary court judgment (the money award) is not a specific order to the defendant to pay this amount to the plaintiff. In the absence of voluntary compliance by the defendant with the court decree, the plaintiff must take the initiative to ask law enforcement officials to seize property of the defendant, now a judgment debtor, that the law does not exempt. If this is done—often a costly and frustrating process—the property is sold to satisfy the judgment. Any amount remaining after payment of the judgment and costs of sale is paid to the defendant. Enforcement of court judgments in this fashion is often costly where successful and often is unsuccessful.

The specific decree is a much more effective remedy. Failure by the defendant to comply with the court order can be the basis for citing the defendant for contempt of court. The defendant is brought before the judge to explain why it has not complied with the decree. If the explanation is not satisfactory, the judge can punish the defendant by fine or imprisonment or coerce the defendant into performing by stating that the defendant must pay a designated amount or stay in jail until it performs.

By and large, the claims that are central to this treatise will, if successful, result in money awards. It is the principal remedy sought when claims are made that someone has not performed a contract for design or construction services. Courts have been very reluctant to issue orders requiring designers or contractors to perform in accordance with their contractual obligations. By the time the dispute reaches the courts, continued contact between the disputing parties is likely to generate additional disputes that can place a substantial burden on the judge. (A specific decree is a personal order by the judge.)

The most economic and efficient way to resolve such a dispute may be to issue a court judgment that will provide funds that can enable the owner to retain another designer or contractor. This is often better than compelling performance when designer or contractor has made it clear she no longer wishes to perform. Specific decrees can be considered judicially supervised slavery in violation of the Thirteenth Amendment. Although there has been some expansion of this remedy[1] it is still not seen much in design and construction disputes.

Despite heavy emphasis upon the money award, some specific decrees are awarded in Construction Process disputes. An unpaid contractor or designer may have a right to assert a mechanics' lien against property improved by the contractor or designer. The lien is an equitable remedy that requires that the property be sold and the holder of the lien be paid out of the proceeds. As shall be seen in Sections 6.09 and 6.10, specific equitable decrees are often awarded to restrain an employee from competing with the employer or from divulging trade secrets or to compel a seller of real property to convey in accordance with the contract. In New York an arbitrator with broad remedial powers ordered a contractor to perform, an award confirmed by the court.[2]

SECTION 6.04
COMPENSATION AND PUNISHMENT: EMERGENCE OF PUNITIVE DAMAGES

As stated in Section 6.01, the basic purpose in awarding damages for breach of contract is to compensate a party who has suffered losses or who has been prevented from making gains. The common law did not look upon contract breach as an immoral act that might justify punishment. This may have been based upon the importance in a market-oriented society of persons engaging in economic exchanges.

Excessive sanctions may discourage persons from making contracts. In the past few years courts have been willing to award punitive damages for certain types of contract breach, justifying this by concluding that a particular breach was tortious. Wrongful dismissals claims discussed in Section 5.11 have begun to generate punitive damages. Punitive damages are awarded to the claimant to prevent the conduct from being repeated by the defendant or to make an example of the defendant to deter others.[3]

Punitive damage awards have been quite rare in construction disputes. However, a few recent cases have upheld punishing the defendant. In *F.D. Borkholder Co. v. Sandock*,[4] the contractor deviated from the plans. The court justified the punitive damages award by concluding that the defendant was guilty of intentional and wrongful acts that constituted fraud, misrepresentation, deceit, and gross negligence in its dealings. The court stated:

The Court of Appeals cited our decision in *Hibschman Pontiac, Inc. v. Batchelor*, 266 Ind.

1. Restatement (Second) of Contracts §§ 359–360 (1981).

2. *Grayson-Robinson Stores, Inc. v. Iris Construction Corp.*, 8 N.Y.2d 133, 202 N.Y.S.2d 303, 168 N.E.2d 377 (1960). (See Section 34.12.)

3. *Tameny v. Atlantic Richfield Co.*, 27 Cal.3d 167, 164 Cal.Rptr. 839, 610 P.2d 1330 (1980).

4. 274 Ind. 612, 413 N.E.2d 567, 570–571 (1980).

310, 362 N.E.2d 845 (1977) for the proposition that punitive damages are recoverable in breach of contract actions only when a separate tort accompanies the breach or tort-like conduct mingles in the breach. Here, prior to the execution of the contract, Sandock representatives expressed their concern about moisture on the walls. Under the terms of the contract, they were to pay $200 for plans to be drawn up by Borkholder's architect. The contract provided that all labor and material would be furnished in accordance with specifications. Sandock was given a copy of the plans. However, contrary to these plans, the top and bottom courses of block forming the one wall were not filled with concrete, thus constituting latent variances. Furthermore, the roofline was shortened which represented an additional deviation from the plans.

There was testimony that the cut-off roofline enabled water to leak down into the top of the block wall. Other evidence indicated that the wetness problem resulted from this water percolating down through the inside of the wall, collecting at the bottom, and then rising again by capillary action. Sandock made numerous complaints but was constantly reassured by several Borkholder representatives that the problem was caused by simple condensation, a theory ultimately disproved by an on-site test conducted by the Borkholder firm. Sam Sandock testified that Freeman Borkholder, president of the company, promised that the situation would be remedied whereupon Sandock tendered all but $1,000 of the contract price. The problem was never corrected. The Borkholder people knew, of course, that the blocks in the wall were not filled with concrete. Also, Borkholder himself conceded that the roofline adjustment increased the likelihood of water running down into the core of the wall.

We believe that there is cogent and convincing proof that the Borkholder firm engaged in intentional wrongful acts constituting fraud, misrepresentation, deceit, and gross negligence in its dealings with Sandock. *Hibschman Pontiac, Inc., supra.* Accordingly, we agree with the Court of Appeals that the trial court could have concluded that separate torts accompanied the breach. Next, relying on *Hibschman,* the Court of Appeals attempted to identify the public interest to be served by imposing punitive damages. However, the majority could not perceive any such interest and refused to let the award stand. We disagree. As Judge Garrard stated in his dissent:

"I have no problem identifying the public interest to be served in requiring that the builders of public buildings be deterred from fraudulently disregarding building code requirements or those contained in the plans and specifications they have agreed to comply with." *Sandock v. Borkholder, supra,* at 959.

The purpose of punitive damages generally is to punish the wrongdoer and to deter him and others from engaging in similar conduct in the future. . . . An award of such damages is particularly appropriate in proper cases involving consumer fraud. . . .

The building contractor occupies a position of trust with members of the public for whom he agrees to do the desired construction. Few people are knowledgeable about this industry, and most are not aware of the techniques that must be employed to produce a sound structure. Necessarily, they rely on the expertise of the builder. Here, the builder has been found to have engaged in fraudulent or deceptive practices by constructing a building with latent deviations from the plans which resulted in damage to the owner. Further, the builder has attempted to disclaim responsibility for such damage when it may be inferred that it knew or should have known that its work was the cause. Under these circumstances, certainly the imposition of punitive damages furthers the public interest.

In *Jeffers v. Nysse,*[5] a purchaser made a claim against a developer-builder. The jury had found that the builder had misrepresented the insulation and heating costs when he sold the house, knowing those misrepresentations to be untrue and with the intention of deceiving the buyers. Yet the jury did not find that the seller's conduct had been malicious or vindictive. Nevertheless, the court affirmed a judgment of punitive damages since the conduct had been wanton, willful, or reckless despite the absence of malice. The court concluded that a requirement of malice

would shield the defendants from any liability beyond the costs of compensating the [buyers] for their costs in putting the house in the condition it was represented to be in originally. But "putting the cookies back in the jar" when caught is not enough. If that result were reached, sellers could make any misrepresenta-

---

5. 98 Wis.2d 543, 297 N.W.2d 495 (1980).

tion necessary to make a sale. If it was not discovered, or was discovered or not pursued, the seller would make a windfall gain. If the fraud were discovered and successfully proven, the seller would only be liable to make good on his representations. He would suffer no punishment nor would he be deterred from similar conduct in the future.[6]

Both of these cases involve fast-selling builders willing to do anything to obtain a sale. Awarding punitive damages will be very rare in most construction disputes, but damages may be awarded where it seems needed to protect unsuspecting consumers from those who would prey upon them.

SECTION 6.05
PROTECTED INTERESTS

Increasingly, law looks upon the varying ways in which the party who makes a contract can be protected from breach of the other party. This protection is frequently described as encompassing expectation, reliance, and restitution.

The most protected of the three interests, *restitution* seeks to restore the status quo that existed before the contract was made by awarding the plaintiff any benefit the plaintiff has conferred on the defendant. It is most protected because it is relatively easy to establish and it involves not only a loss to the plaintiff but a gain to the defendant. (It is part of the legal concept of restitution discussed in greater detail in Chapter 8.) Its importance in measuring claims by designers and contractors for services performed justify its inclusion in Section 31.02.

A plaintiff protects its *reliance interest* by obtaining reimbursement of expenses from the defendant that have been incurred either in reliance upon the contract or expenses incurred before the contract was made that have become valueless because of the breach. Such a remedy looks backward to a point in time even before the contract was made (something similar to restitution) but also looks forward. If the defendant can establish that the expenditure sought would never have been reimbursed in the venture engaged in by the plaintiff, the plaintiff cannot recover from the defendant.

Suppose a contractor failed to build a building in which the owner intended to manufacture Nehru jackets, a sartorial meteor of the 1960s. Suppose further that the owner invested a large amount of money to promote these jackets. If the contractor can establish that market saturation or changing fashions meant not *one* jacket would have been sold, the owner could not recover the marketing costs, as they would not have been reimbursed from sale proceeds.[7]

The third protected interest, *expectation*, looks forward and seeks to place the plaintiff in the position it would have found itself had the defendant performed. One way of accomplishing this objective is to order the defendant to perform as promised, a remedy generally unavailable in contracts to perform design or construction services. The same objectives can be achieved by a money award, however.

Suppose a designer or contractor refuses to perform its contract. Awarding an amount that would enable the owner to hire someone to perform identical design or construction services would protect the owner's expectation interests. Since this is a formula used very frequently in claims by owners, it is discussed in greater detail in Chapter 31.

---

6. 297 N.W.2d at 499. But punitive damages were not awarded where the issue of breach was a close one, *Quedding v. Arisumi Brothers, Inc.,* 661 P.2d 706 (Hawaii 1983), nor where there was no evidence the contractor intended not to perform when the contract was made, *Century Prop. Inc. v. Machtinger,* 448 So.2d 570 (Fla.App. 1984).

7. *L. Albert & Son v. Armstrong Rubber Co.,* 178 F.2d 182 (2d Cir.1949).

One issue that arises frequently in construction disputes, although it is commented upon in greater detail in Sections 26.06 and 31.03, should be mentioned briefly here. Suppose a contractor breaches a contract by installing roofing shingles that are slightly discolored and do not match. One way of protecting the expectation interest of the owner is to require that the entire roof be replaced with the proper roofing shingles. This may cost an amount substantially greater than the diminished value of the house resulting from the failure of the contractor to properly perform. Suppose the cost of completely reroofing the house would be $15,000 and the diminution in the value of the house, that is, the difference between what the house would have been worth had it been built properly and what it is worth now, is $2,000. Under such circumstances, some courts would award the lower amount if persuaded that it would be economically wasteful to replace the entire roof, that the nonperformance was not deliberate, and that if the amount were awarded, the owner would not use it for that purpose.[8]

SECTION 6.06
LIMITS ON RECOVERY

Using rules described in this section, the law limits recovery by a claimant who has suffered losses that can be connected to the other contracting party's failure to perform in accordance with the contract. These rules often provide insurmountable obstacles to the claimant, a phenomenon that has prompted some to contend that the law does not give adequate protection to those who suffer losses because of contract breach. Undoubtedly these obstacles can make recovery difficult. Yet they can also be looked upon as rules that encourage parties to make contracts without inordinate fear that their nonperformance will expose them to unpredictable or devastating damage claims.

These doctrines *can* limit recovery but do not *invariably* do so. To the extent that any generalization can be made, modern law seems *more* willing to grant greater compensatory damages than earlier periods in American legal history.

(A) CAUSATION

The claimant must show that the defendant's breach has caused the loss. Losses may be caused by more than one actor, and on occasion, causal factors can include other events and conditions. For example, the contractor's performance may be delayed by the owner, by strikes, by material shortages, *and* by the contractor's poor planning. It may be difficult if not impossible to establish the amount of the loss caused by contributing causes or conditions.

The defendant will be responsible for the loss if its breach was a substantial factor in bringing about the loss. It need not be the sole cause of the loss.[9] Generally this question is determined by the finder of fact, sometimes the jury, and sometimes the trial judge. Any judicial finding regarding causation not based upon guesswork will be upheld.

Difficult causation problems can develop in construction contracts. Suppose the specifications for a street overpass requires one hundred bars of steel to be installed. The contractor installs eighty bars, and a failure results. Suppose the contractor then points to structural flaws in the overpass *generally* that would have caused the entire overpass to collapse within a short period of time. This contention would be seeking to negate the element of causation

---

8. *Salem Towne Apartments, Inc. v. McDaniel & Sons Roofing Co.*, 330 F.Supp. 906 (E.D.N.C.1970). But cost of correction was awarded in *O.W. Grun Roofing & Construction Co. v. Cope*, 529 S.W.2d 258 (Tex.Civ.App.1975).

9. *Krauss v. Greenbarg*, 137 F.2d 569 (3d Cir.1943), cert. denied 320 U.S. 791, 64 S.Ct. 207, 88 L.Ed. 477 (1943).

that is an essential part of the owner's claim. The owner must show that "but for" the contractor's breach, the loss would not have occurred. Put another way, the contractor contends its breach did not cause the loss, as it would have happened anyway. Such a defense will rarely succeed, the likelihood being that *each* will be chargeable with the entire loss.[10]

One common fact pattern involves tracing responsibility for a defect both to defective design and improper workmanship. Similarly, delayed performance by the contractor may be traceable to poor management by the contractor and excessive design changes made by the design professional. These difficult problems are treated in Chapters 28 and 30.

(B) CERTAINTY

A claimant must prove the extent of losses with *reasonable* certainty. One court described the certainty rule in these terms:

> Courts have modified the "certainty" rule into a more flexible one of "reasonable certainty." In such instances, recovery may often be based on opinion evidence, in the legal sense of that term, from which liberal inferences may be drawn. Generally, proof of actual or even estimated costs is all that is required with certainty.
>
> Some of the modifications which have been aimed at avoiding the harsh requirements of the "certainty" rule include: (a) if the fact of damage is proven with certainty, the extent or the amount thereof may be left to reasonable inference; (b) where a defendant's wrong has caused the difficulty of proving damage, he cannot complain of the resulting uncertainty; (c) mere difficulty in ascertaining the amount of damage is not fatal; (d) mathematical precision in fixing the exact amount is not required; (e) it is sufficient if the best evidence of the damage which is available is produced; and (f) the plaintiff is

entitled to recover the value of his contract as measured by the value of his profits.[11]

In construction litigation, certainty questions are legion. Court decisions take up pages deciding whether adequate proof has been made of the following:

1. Contractor profits on performed or unperformed work.

2. Owner lost profits for delayed or faulty construction.

3. Contractor cost of performing the work.

4. Owner cost of completing the work.

5. Added costs of performance made necessary by the other party's breach.

Proving lost profits is discussed in Section 6.06(E). However, it may be useful to look at the certainty requirement in establishing the value of labor and materials that have gone into the project.

Clearly, record keeping is essential. The contractor seeking to prove the actual or reasonable value of the labor and materials furnished who can produce accurate, detailed records that establish this will be in a sound settlement or litigation position. Likewise, a contractor who can produce sound expert testimony on diminished productivity, accounting records, and expert testimony that establishes on-site and office overhead or expert testimony on profit margin will be in an advantageous position. But contractors and subcontractors often seek to prove these items without complete and meticulous records or competent expert testimony.

As to cases, a contractor who kept time records but who did not segregate them by jobs was able to prove how much should have been attributed to the defendant's pro-

---

10. This was the result in *City of Reno v. Ken O'Brien & Associates, Inc.*, an unpublished opinion of the U.S. Court of Appeals for the Ninth Circuit, No. 74–2094, March 23, 1978 (design and construction combined to create indivisible loss).

11. *M & R Contractors & Builders, Inc. v. Michael*, 215 Md. 340, 138 A.2d 350, 355 (1958), approved in *Certain-Teed Products Corp. v. Goslee Roofing & Sheet Metal, Inc.*, 26 Md.App. 452, 339 A.2d 302 (1975); *Elte, Inc. v. S.S. Mullen, Inc.*, 469 F.2d 1127 (9th Cir.1972).

ject by testimony of his foreman.[12] Similarly, subcontractors are sometimes relieved from strict certainty proof.[13]

On the other hand, a prime contractor was limited to the amount that he could prove by time cards for labor work performed after the defendant subcontractor had defaulted.[14] Similarly, an estimate by the owner as to the cost of correction was held inadequate to prove damages.[15]

Court decisions are not always consistent as to the amount of certainty required to establish a loss. Some courts will not hold the claimant to a high standard of certainty if they are convinced that a loss has occurred, that it has been caused by the defendant's breach, and that the claimant has marshalled the best available evidence. Others, particularly in cases in which the issue of breach is a close one and it appears that the defendant has performed in good faith, will use the certainty requirement to either limit or bar recovery.

(C) FORESEEABILITY: FREAK EVENTS AND DISPROPORTIONATE LOSSES

A series of improbable events can combine to lead to large losses. In the contract context, suppose a contractor is hired to build a high-rise office building. Shortly after the work is completed and accepted, the electrical system fails because of improper workmanship. This causes the elevators to be out of service for two hours. During those two hours, the building owner has scheduled an interview with a prospective tenant who is thinking of leasing three floors of office space at a rental very attrac-

tive to the owner. Because the elevators are out of service, the prospective tenant decides to rent elsewhere. Should the contractor be liable to the building owner for the extraordinary lease profits that the owner lost?

Clearly the contractor's breach, perhaps a minor one, set off a chain of events that culminated in the loss of extraordinary rental profits. The improper installation caused an electrical failure, causing the elevators to malfunction. The prospective tenant was in the building at that very time. The prospective tenant's decision not to rent may have been caused by the tenant's irascibility or ignorance. This series of possible but improbable events has caused the building owner to suffer a loss. Is it fair to transfer this loss to the contractor?

The loss was an extraordinary loss, and one that could not be reasonably foreseen at the time the contract was made. Contrast this with an electrical malfunction that causes a small fire or an inconvenience to the tenants. Such a loss falls more easily into the type of loss that can be reasonably foreseen at the time a contract is made.

The second type of loss, one vastly disproportionate to the price paid for performance, can be demonstrated by *Hadley v. Baxendale*,[16] the leading English case that gave rise to the requirement that losses be reasonably foreseeable as a *probable* result of the breach. A shipper claimed against a carrier when the latter's delay in returning a shaft that had been sent to the factory for repair caused the shipper's plant to be shut down. The amount paid for shipping the shaft was disproportionately small compared to the cost of an entire plant shut-

---

12. *Don Lloyd Builders, Inc. v. Paltrow,* 133 Vt. 79, 330 A.2d 82 (1974).

13. *St. Paul-Mercury Indemnity Co. v. United States,* 238 F.2d 917 (10th Cir.1956); *McDowell-Purcell, Inc. v. Manhattan Construction Co.,* 383 F.Supp. 802 (N.D.Ala.1974), affirmed 515 F.2d 1181 (5th Cir.1975); *Certain-Teed Products Corp. v. Goslee Roofing & Sheet Metal, Inc.* supra note 11.

14. *Welch & Corr Construction Corp. v. Wheeler,* 470 F.2d 140 (1st Cir.1972).

15. *Gross v. Breaux,* 144 So.2d 763 (La.App.1962).

16. 156 Eng.Rep. 145 (1854). See Restatement (Second) of Contracts § 351 (1981).

down. The court concluded that the carrier is liable for losses *naturally* resulting from the breach and other losses, the possibility of which is brought to the carrier's attention at the time a contract is made.

Advance awareness of the risk gives the carrier the chance to adjust its rates for performance or decide to forgo the transaction. Similarly, is it fair to place the responsibility for a building shutdown on an electrician who is called to make a minor repair in a high-rise building but who fails to perform in accordance with the contract obligations? This may be too onerous in light of the small profit earned.

In a construction context it may be foreseeable that inadequate insulation may cause some tenants to move out and generate lost rentals but not cause bankruptcy to the owner when there is a wholesale evacuation of tenants.[17]

Many foreseeability cases involve lost—extraordinarily *high*—profits. (Because these cases also involve certainty problems, please review (B).) Exposure to potentially great risks has generated attempts by parties whose performance under a contract may cause large losses to insure or develop risk management techniques, such as exculpation, remedial limitation, and indemnification such as those discussed in Chapter 18.

(D) AVOIDABLE CONSEQUENCES (THE CONCEPT OF MITIGATION)

The rule of avoidable consequences is another limitation. A claimant cannot recover those damages that the claimant could have reasonably avoided. Sometimes this rule is expressed as one that requires the victim of a contract breach to do what is reasonable to mitigate or reduce the damages. This limiting rule relates to the requirement that the breaching party is responsible only for those losses that its breach has caused. If the loss could have been reasonably avoided or reduced by the claimant, the claimant cannot transfer that loss to the breaching party.

This limitation has not been a favored one. The law not only has placed the burden of establishing that the loss could have been avoided by the claimant upon the breaching party but also has been hesitant to give the rule much scope. The cases illustrate this.

In *C.A. Davis, Inc. v. City of Miami*,[18] a defendant contractor sought to reduce the recovery by the owner by asserting that the owner spent more than was necessary to correct the contractor's work. The court held that the contractor could challenge the completion cost only if he could show waste, extravagance, or lack of good faith. It will be unusual for a court to conclude that expenses incurred by the claimant to correct defective work by the contractor were out of line and not recoverable. Yet one court did reduce an award because it concluded that overtime was not necessary and the claimant's refusal to allow the designer-builder to provide free engineering services was not justified.[19]

In *Great Lakes Gas Transmission Co. v. Grayco Constructors, Inc.*,[20] a breaching construction contractor contended that the owner, a public utility, could pass its loss to its customers through rate increases. In addition to the uncertainty as to whether this *could* be done, it seems as if denial of this assertion was based in part on the "chutzpah" of the contract-breaker in seeking relief by putting such a burden on the utility and its customers.

---

17. *R.E.T. Corp. v. Frank Paxton Co.*, 329 N.W.2d 416 (Iowa 1983) (court avoided foreseeability hurdle by concluding contractor's breach was tortious, based upon negligence).

18. 400 So.2d 536 (Fla.App.1981).

19. *First National Bank of Akron v. Cann*, 503 F.Supp. 419 (1980), affirmed 669 F.2d 415 (6th Cir.1982).

20. 506 F.2d 498 (6th Cir.1974).

This doctrine was invoked, strangely, in *S.J. Groves & Sons Co. v. Warner Co.*, by a subcontractor in default who claimed the prime contractor should have fired him.[21] The prime contractor sought damages against the subcontractor who had failed to supply ready-mixed concrete.

Trouble developed almost at the outset. The owner was forced repeatedly to reject the concrete supplied by the subcontractor. In addition, the subcontractor frequently failed to make deliveries in accordance with the prime contractor's instructions. The prime contractor considered using other sources but felt it had no real alternatives. It would cost too much to build its own plant. The only other concrete source had not been certified to do state work, and its price was higher than that of the subcontractor. In addition, the only alternative source had limited facilities and trucks. Despite the difficulties, the subcontractor continued to assure the prime contractor that things would improve.

Despite these promises, the subcontractor's performance continued to be erratic and the public entity ordered all construction halted until the subcontractor's service could be discussed at a conference. Again after renewed assurances that things would improve, the public entity allowed work to resume. For succeeding months, the subcontractor's performance continued to be uneven and unpredictable.

During performance, the prime contractor approached the alternate source, which by then had been certified by the state. The alternate source agreed to reduce its price to the same price as the subcontractor, but the prime contractor continued to use its subcontractor as its sole supplier.

Had the prime contractor acted reasonably in continuing to use the subcontractor despite its poor performance? The trial court concluded that the prime contractor

had not, but the appellate court did not agree.

After noting that the burden of proving that the losses could have been avoided by reasonable effort was upon the breaching party, the court looked at the alternatives available to the prime contractor. One alternative was to simply terminate the subcontractor, an alternative the court did not find realistic. Another alternative was for the prime contractor to set up its own cement batching plant, an alternative the court found impractical because of time and expense. Another alternative was to accept the subcontractor's assurances that performance would be satisfactory in the future, the alternative selected. Another alternative was to use the alternate supplier as a supplemental source or as a substitute.

The appellate court concluded that *all* the alternatives had their drawbacks. Even if the alternate supplier had been engaged as a supplemental source, there was still no guarantee that the subcontractor would perform properly. The use of two suppliers might raise other problems, and there was a question as to whether the alternate supplier would have been able to perform.

The court concluded that, confronted with these choices, the prime contractor's decision to stay with the subcontractor may not have been the best choice. The test is, however, whether the course chosen was *reasonable*, not whether it was necessarily the best. The court was not willing to engage in hypercritical examination of the choice made. It concluded that staying with the subcontractor may have been not only reasonable but the best choice under the circumstances. The court noted that the breaching subcontractor, who sought to second-guess the choice made by the prime contractor, could also have engaged a supplemental supplier and that, where each party had the equal alternative to reduce the damages, the defendant was in no posi-

---

21. 576 F.2d 524 (3d Cir.1978).

tion to contend that the plaintiff failed to mitigate. (The avoidable consequences doctrine raises special problems in employment contracts and is discussed in Section 6.09.)

(E) LOST PROFITS

Claims for lost profits, extraordinary or ordinary, have caused particular difficulty. They involve the preceding limitations on contract recovery, causation, certainty, foreseeability, and avoidable consequences.

Profits on the *very* contract in question, such as claims by a contractor for lost profits on the construction contract, are part of the routine measurement used in contractor claims. (This aspect of lost profits is discussed in Chapter 31.)

This subsection comments upon claims by contractors for profits on *other* contracts it asserts it would have obtained. (Depriving the owner of lost profits, though similar to those made by the contractor for profits on other contracts, is covered in Section 31.02.)

Suppose a contractor is unjustifiably terminated, either before starting performance or during performance. Suppose further that the contractor claims that completion of this contract would have earned additional profits through the award of other contracts. Recovery of such profits requires that a formidable number of hurdles be overcome.

First, even if the contractor can establish the loss of other contracts, were they lost because of the owner's refusal to allow the contractor to complete this contract or for other reasons? Second, how certain is it that the contractor would have received not only the other contracts but also the amount of profit that might have been earned? Third, was it reasonably foresee-able at the time the contract was made that unjustified termination of the contract would have caused the contractor to lose other contracts? Fourth, could the contractor by reasonable effort have avoided losing any contracts that were lost?

Some jurisdictions deny recovery of lost profits for new businesses, proof being too uncertain that it would have earned profits. The modern tendency has been to treat this issue as any other factual issue. Even a new business can *try* to prove it would have earned additional profits from other contracts.

In *Texas Power & Light Co. v. Barnhill*,[22] the owner terminated Barnhill's contract. A jury awarded substantial damages to Barnhill based upon Barnhill's testimony that he was unable to secure bonding necessary to obtain another contract or bid on two other jobs because he was terminated. The appellate court held that lost profits from collateral contracts are recoverable if they were within the contemplation of the parties at the time they entered into the contract. Although the profits need not be proved to exact calculations, there must be competent evidence with a reasonable degree of certainty and exactness as to the amount. The evidence, though not direct, supported a strong inference that the owner was aware of these collateral contracts. In the owner's twenty-two-year relationship with the contractor, the contractor continually worked on outside contracts. The court found evidence from which a jury could have calculated with a reasonable degree of certainty the amount of lost profits, the appellate court sustaining the jury's finding.

(F) COLLATERAL SOURCE RULE

Suppose the defendant has breached but the

---

22. 639 S.W.2d 331 (Tex.Civ.App.1982). Similarly, see *Eastern Tunneling Corp. v. Southgate Sanitation District*, 487 F.Supp. 109 (D.Colo.1979). For cases failing in proof, see *John W. Johnson, Inc. v. J.A. Jones Construction Co.*, 369 F.Supp. 484 (E.D.Va.1973); *Liberto v. Villard*, 386 So.2d 930 (La.App.1980); and *United States v. Mountain States Construction Co.*, 588 F.2d 259 (9th Cir.1978).

plaintiff has been compensated by a third party. Can the defendant show that the plaintiff has not suffered a loss?

If the loss is caused by a tortious defendant, that is, by one who has not lived up to the legal standard of care, the defendant cannot reduce or eliminate its liability by pointing to the loss having been compensated by a third party. The third party in such cases is called a "collateral source." Any acts of that source are collateral and cannot be taken into account. Typically, tort cases involve harm to persons, and the collateral source is an insurance company, an employer who provides medical benefits, or the state that provides replacement for lost earnings or medical losses. Because the tortious defendant is considered a wrongdoer, the collateral source rule deprives the defendant of a windfall created by contract protection purchased by the injured party or benefits awarded by the state.

Whether the collateral source rule deprives parties who have breached contracts of the right to point to the loss having been compensated by a third party is not clear. For example, in *Industrial Development Board v. Fuqua Industries, Inc.,*[23] a local development board contracted for the construction of a plant that was to be used by a manufacturer. The development board planned to execute a long-term lease of the plant to the manufacturer. The development board made the contract to enable the manufacturer to obtain better financing terms. In effect the plant was designed by the manufacturer for the manufacturer's use. The defects in the plant were corrected by the manufacturer, and this, according to the court, precluded the development board from having any contract claim against the contractor. (However, the court would allow the manufacturer to recover from the contractor to preclude unjust enrichment.)

New Foundation Baptist Church v. Davis[24] came to a different conclusion. A church sued the contractor for defects in the sanctuary floor that caused collapse during a funeral three years after the church was completed. A church member carpenter donated his labor and repaired the damage for a total cost to the church of $3,000. Yet the jury's award of $6,500 to the church was upheld. The court noted that the church member had no wish to benefit the contractor, and it would be unfair for the contractor to receive the advantage of the church member's generosity.

The issue arose in *Huber, Hunt & Nichols, Inc. v. Moore,*[25] where a contractor claimed against the owner and the architect. The contractor settled with the owner before trial. One of the reasons given for denying recovery against the architect was that the contractor had been paid substantial amounts by the owner. This was a tort claim, there being no contract between contractor and architect. Although this would make the compensation collateral, the court held that the owner and the architect were sufficiently associated with each other to preclude the owner from being considered a collateral source.

There seems to be a trend toward treating tort and contract claims similarly and basing the application of the rule on other criteria.[26] Because of the variety of sources, and types of claims, the results are likely to vary from state to state.

(G) CONTRACTUAL CONTROL: A LOOK AT THE U.C.C.

The preceding discussion has assumed an absence of any controlling contract clauses

---

23. 523 F.2d 1226 (5th Cir.1975).

24. 257 S.C. 443, 186 S.E.2d 247 (1972).

25. 67 Cal.App.3d 278, 136 Cal.Rptr. 603 (1977).

26. Fleming, *The Collateral Source Rule and Contract Damages,* 71 Calif.L.Rev. 56 (1983).

regulating the remedy. Yet it is becoming more common for contracts to regulate the remedy. Some contracts specify or liquidate the damages when establishing actual damages would be difficult if not impossible to prove. Some contracts contain language insisted upon by one of the contracting parties to either exculpate itself from responsibility or limit its exposure.

When remedies are specified, it is common to state in the contract that they are not exclusive. For example, the American Institute of Architects (AIA) provides remedies for termination. Yet A201, ¶ 7.6.1, states that remedies specified are not exclusive. Similarly, evidence must be clear that a specified remedy is to be exclusive before it is given that effect.

Contractual control of remedies are treated in other parts of this treatise. Section 30.09 discusses liquidated damages. Chapter 18 dealing with risk management goes into other contractual attempts to regulate the remedy. At this point attention is directed to the Uniform Commercial Code, which governs transactions in goods. It has been having increasing impact on construction contract disputes. Section 2–719 allows the parties to provide for remedies by contract but states:

> Where circumstances cause an exclusive or limited remedy to fail of its essential purpose, remedy may be had as provided in this Act.

Frequently sellers of goods seek to limit their obligation for breach to repair and replacement of defective goods. In *Coastal Modular Corp. v. Laminators, Inc.*,[27] a contractor was allowed to recover damages from a supplier of defective panels for a construction project despite a contract providing that the remedy was limited to repair and replacement. The court held that the remedy had failed of its "essential purpose" under § 2–719. The contractual remedy would apply only if the defect had been discovered while the work was in progress. Here the defect had been discovered after completion. Repair and replacement would not have been a viable remedy.

The limited remedy was applied in *Price Brothers Co. v. Charles J. Rogers Construction Co.*[28] in a slightly different context. The pipe subcontract stated that the supplier-installer would pay only for damages relating to *above-ground* repair and replacement. The pipe failed and the contractor incurred large expenses removing and replacing the defective pipe *below* ground. It sought to recover this expense from the supplier-installer.

The court held the contractual remedy did not fail under § 2–719. The event, though rare, was foreseeable. Also, the contractor received the benefit of a lower price. The stated remedy in the contract *would* have failed if the replacement pipe had *also* been defective.

(H) NONECONOMIC LOSSES

To this point, with the exception of the discussion relating to punitive damages, this chapter has focused upon compensatory economic losses, losses that can be established precisely or roughly in the marketplace. Increasingly claims for breach of contract, sometimes tied with tort claims, are made for noneconomic losses, such as emotional distress, caused by a breach of contract.

Generally contract law does not grant recovery for such losses. Denial has usually been based on the lack of foreseeability. However, a more acceptable rationale is that contracting parties should not be liable for potentially open-ended and freak losses that are extremely difficult to measure in economic terms. A homeowner who cannot take possession of a new residence when promised can certainly suffer emo-

---

27. 635 F.2d 1102 (4th Cir.1980).

28. 104 Mich.App. 369, 304 N.W.2d 584 (1981).

tional distress. But the likelihood and gravity of such distress is generally dependent upon the emotional and psychological makeup of the homeowner.

Some decisions that have classified the contract breach as tortious have allowed recovery for emotional distress. *McCune v. Grimaldi Buick-Opel, Inc.*[29] allowed for recovery where an employer had failed to perform his contractual obligation to provide medical insurance for an employee, exposing the employee to emotional distress caused by bill-collection tactics of the hospital. Such a contract was found to be not simply a commercial one but also one geared to provide mental solicitude to one of the parties. Increasingly, though modestly, noneconomic losses are beginning to be awarded in contracts or conduct related to the construction process.

Randa v. United States Homes, Inc.[30] upheld a jury verdict for *intentional* infliction of emotional distress against a contractor. There were many defects, causing the wife of the buyer of a $160,000 home to spend time in the hospital for a nervous breakdown. She was "petrified" when told she could not put mirrors in the bedroom. Filing of liens shocked the buyers, with the wife getting "sick" and starting "to cry."

Other decisions have allowed recovery for noneconomic losses without the need to establish intentional infliction of emotional distress. *B & M Homes, Inc. v. Hogan*[31] involved a claim by the Hogans against Morrow (the owner of B & W Homes) based upon the contract under which the Hogans agreed to buy a lot and a house to be built by Morrow. During construction, Mrs. Hogan discovered a hairline crack in the concrete slab. Morrow told her such cracks were common and not to worry. After the Hogans moved in, the crack widened and caused extended damage. Morrow was notified and dealt with the damage, but did not attempt to repair the slab himself. The claim was based in part upon mental anguish that the Hogans suffered. The Hogans introduced evidence that they were concerned over their safety because they believed the house to be structurally defective, that the condition of the house might cause gas and water lines to burst, and that they were forced to live in a defective house because they could not afford to move.

The court noted that as a general rule, damages for mental anguish are not recoverable for breach of contract. However, an exception is made for contracts that involve "mental concern or solicitude." The court held that this contract fell into that category and that it was reasonably foreseeable that faulty construction of a house would cause the homeowners to suffer severe mental anguish. The court noted the home as the largest single investment for most families and one that places the family in debt for many years. The court pointed to an earlier decision that had allowed recovery for mental anguish when a builder performed improperly under a contract to build a home, emphasizing the homeowner's view of her home as her castle and a place to protect her against the elements and shelter her belongings.

Another recent case, *Orto v. Jackson,*[32] allowed recovery for "aggravation and inconvenience." In upholding the award, the court stated:

> . . . for over two years the Jacksons have suffered partial loss of use and enjoyment of their basement, which still leaks after heavy rains, and the total loss of use and enjoyment of their backyard due to the cesspool located there. . . . The aggravation and inconvenience suf-

---

29. 45 Mich.App. 472, 206 N.W.2d 742 (1973).

30. 325 N.W.2d 905 (Iowa App.1982). See also *Alsteen v. Gehl*, 21 Wis.2d 349, 124 N.W.2d 312 (1963) (dictum).

31. 376 So.2d 667 (Ala.1979). See Annot., 7 A.L.R.4th 1178 (1981).

32. ___ Ind.App. ___, 413 N.E.2d 273, 278 (1980).

fered by the Jacksons certainly arose naturally from the builders' breach of the contract and it could hardly be said that these damages were not in the parties' minds at the time they entered the contract.

These recent cases do not suggest that noneconomic losses will be freely awarded in construction contract disputes. They do suggest that recovery for such losses will no longer be automatically denied. The law, as part of consumer protection, may compensate those who suffer emotional distress when homes built for them are defective and the builders refuse to stand behind their work. While such claims are susceptible to abuse, the *possibility* of recovery may cause builders to take their contractual obligations seriously.

SECTION 6.07 COST OF DISPUTE RESOLUTION: ATTORNEYS' FEES

This treatise has suggested that contract remedies should compensate the injured party. Yet full compensation is rarely achieved. The rules that determine whether and how much damages are avoided sometimes make it difficult to achieve full compensation. Even more important, the winner cannot recover its cost of litigation,—an item that has risen to staggering proportions. American law does not transfer the prevailing party's attorneys' fees to the other party unless the claim is based upon a contract that contains a provision providing for a recovery or recovery is based upon a statute granting attorneys' fees. Under extraordinary circumstances, such as commission of an *intentional* tort [33] or the losing party's claim having been vex-

atious or frivolous, such costs can be recovered.

Denial of attorneys' fees is based upon the reluctance of American law to discourage citizens from using the legal system. While an unsuccessful claimant may have to bear the cost of its own fees, the law absolves the claimant of the responsibility of bearing the other party's costs.

This rule has led to increased legislation authorizing attorneys' fees as part of costs in consumer and civil rights cases. It has led to reciprocal attorneys' fees legislation in some states. These statutes provide that if one party can recover attorneys' fees under a contract, the other party can do so even though the latter is not specifically granted this in the contract. Some states grant attorneys' fees to the prevailing party where a mechanics' lien has been sought. Some give attorneys' fees when claims are made on surety bonds. A few states allow attorneys' fees *generally* for contract actions.

Drafters often seek advantages, as shown in *P & C Thompson Brothers Construction Co. v. Rowe*.[34] The subcontract provided that the prime contractor recovered its attorneys' fees if it obtained *a* judgment but the subcontractor recovered its attorneys' fees only if it recovered *all* it sought.

Documents published by the AIA do not provide for attorneys' fees, each party bearing its litigation costs.

AIA documents require arbitration under the Construction Industry Arbitration Rules administered by the American Arbitration Association. These rules do not specifically provide for attorneys' fees. They deal mainly with other costs of arbitration, including the arbitrator's fee. While the arbitrator would very likely have freedom to

---

33. *Waldinger Corp. v. Ashbrook-Simon-Hartley, Inc.*, 564 F.Supp. 970 (C.D.Ill.1983) (specialty contractor recovered from engineer who rigged the specifications and refused to consider a substitution based upon intentional interference with contractor's contract with a manufacturer).

34. 433 So.2d 1388 (Fla.App.1983) (court allowed fees to the subcontractor *without* compliance with the contract condition).

award attorneys' fees, arbitrators customarily do not. However, *Harris v. Dyer,*[35] involving a claim under a mechanics' lien statute, awarded the successful litigant the costs it incurred in arbitration based upon the Oregon statute providing for attorneys' fees to those who establish mechanics' liens being preserved in the contract.

Local law, currently in a state of ferment, must be consulted by those who draft contracts or those who must counsel as to the availability of attorneys' fees to the prevailing party. While discussion has centered upon attorneys' fees, the other formidable costs of construction litigation—particularly complex litigation that involves defects and impact claims—cannot be ignored. These claims generate immense costs to reproduce documents, classify them, analyze them, and store them, as well as the staggering costs involved in conducting pretrial discovery. To these costs must be added the costs of preparing exhibits and retaining expert witnesses and the nonproductive costs incurred by personnel in preparing for the lawsuit. Those who draft contracts may wish to deal with these costs. Such costs should remind the parties they must use every effort to avoid litigation.

SECTION 6.08   INTEREST

Many cases have involved recoverability of prejudgment interest. Claimants often seek interest from a time earlier than entry of court judgment. Perhaps this is due to the ease with which a claim for interest can be tacked onto a claim for work performed or defective work. In any event, states vary considerably in their treatment of claims for prejudgment interest.

The varying legal rules dealing with pre-judgment interest are caused by the differing judicial attitudes toward the desirability of *complete* compensation on one hand and on the other hand punishing a defendant when the latter exercised a good-faith judgment not to pay that turned out to be incorrect.

While case holdings and statutes have many slight variations, they follow three principal rules. One limits prejudgment interest to liquidated (specific) amounts or unliquidated amounts that are easily determinable by computation with reference to a fixed standard contained in the contract without reliance on opinion or discretion.[36] A party against whom a claim is made should be able to avoid prejudgment interest by tendering the amount due. Unless the amount is known or easily determined, tender cannot be made. This ignores the likely possibility that payment is not made because of a dispute over the validity of the claim, not the amount due were the claim held to be valid.

Some courts and statutes give the judge or jury discretion to determine whether interest should be awarded.[37] Often discretion takes into account whether the refusal by the defendant to pay was vexatious, what the inflation rate is, and how important money use is to the party entitled to the payment.

Some jurisdictions, either by court decision or increasingly by statutes, give prejudgment interest unless special circumstances would make it unjust to award it.[38] Undoubtedly this reflects increased legislative recognition that delay in payment causes a serious loss that should be compensated through interest. Again, local law must be consulted.

Frequently special rules deal with claims

---

35. 292 Or. 233, 637 P.2d 918 (1981).   See Section 34.13.

36. *E.C. Ernst, Inc. v. Koppers Co.,* 626 F.2d 324 (3d Cir.1980).   See Annot., 60 A.L.R.3d 487 (1974).

37. West's Ann.Civ.Code § 3287(b) (grants judge discretion to award interest from date no earlier than date legal action commenced).

38. See *Hedla v. McCool,* 476 F.2d 1223 (9th Cir.1973).

against public authorities. Until recently, interest could not be recovered against the federal government. This changed in 1978.[39]

With costs of financing now more recognizable as an important element in construction costs, contracts should specify that payments will bear interest from the time they are due. As to claims that are difficult to evaluate, the issue is not so clear. Suppose each party has a good faith belief in the merit of its position and for that reason refuses to settle. Ultimate determination that refusal to pay was unjustified permitting recovery of prejudgment interest can place a heavy burden on the party who has asserted an honest reason for not paying. On the other hand, undoubtedly the loss should have been paid, and not awarding prejudgment interest denies full compensation to the party whose position was ultimately vindicated. Balancing compensation against punishment makes resolution of this issue difficult.

Unless the contract or an applicable statute states otherwise, the percentage will be the amount specified by law to be paid on legal judgments. This ranges from 5%–10%, with the most common being 6%. (See Section 26.02(H).)

SECTION 6.09
SPECIAL PROBLEMS OF EMPLOYMENT CONTRACTS

Section 5.11 discussed employment contracts. This section discusses remedies for breach of those contracts.

(A) MONEY AWARDS

The application of two rules discussed in Section 6.06 have particular significance in employment contracts. The first is the rule of avoidable consequences discussed in (D).

That section noted that the courts have not been solicitous toward claims by a breaching party that the loss could have been avoided or reduced by efforts of the claimant.

The law has been even less solicitous toward employers who have wrongfully dismissed an employee and who assert the employee could have reduced or avoided the loss. The employee recovers the full amount of salary that would have been earned for the period of service less the amount the *employer affirmatively proves* the employee has earned or with reasonable effort could have earned from other employment. However, before projected earnings and other employment opportunities not sought or accepted by the employee can be applied to reduce the damage award, the employer must show that the other employment was comparable or substantially similar to the services the employee would have performed under the contract. The employee may reject or fail to seek available employment of a different or inferior status. In addition, as was noted in Section 5.11, punitive damages may be awarded for wrongful dismissal in extreme cases.

The other legal doctrine that has particular importance in employment contracts is the collateral source rule, a topic discussed in Section 6.06(F). Employees who are wrongfully terminated from their employment often receive social benefits, such as unemployment insurance payments either under state or private plans. While no clear answer has emerged in the cases, it is likely that these payments will be considered a collateral source and not applied to reduce the amount of the claim.

(B) EQUITABLE REMEDIES

Direct court orders that an employee work for an employer or that an employer take

---

39. 41 U.S.C.A. § 611 allows interest from the date the claim is received until payment. The rate is set by the Secretary of the Treasury for the Renegotiation Board. From January 1, 1985, through June 30, 1985, the rate is 12⅛%. See 49 Fed.Reg. 50357 (1984).

back an employee are not issued for breach of an employment contract. (Reinstatement *is* common under collective bargaining agreements.) However, based upon a famous English case, *Lumley v. Wagner,*[40] in some states *negative* injunctions, that is, orders barring a breaching employee from working for the plaintiff's *competitor,* are issued. Usually such orders involve employment of a highly unique type where a replacement cannot be easily obtained, such as in contracts for performance of services by professional entertainers or athletes. However, such negative injunctions are sometimes limited by statute.[41] The modern tendency is to avoid them, as they can do by indirection something not directly allowed.[42]

In another area, equitable injunctions play a significant role in remedies for breach of an employment contract. Employees sometimes develop or learn technical or commercial data that an employer wishes to keep secret. Employees, particularly those who work in high technology industries, commonly sign employment agreements under which they promise that if the employment relationship is terminated, they will not compete with the employer or work for a competitor for a designated period of time. These contracts also provide they will not divulge technical or commercially valuable information within a specified period after the relationship terminates. Even without express promises, the law may apply a fiduciary obligation not to divulge trade secrets.

Usually the employer who wishes to enforce such a contract or implied obligation seeks a court order prohibiting an employee from competing or disclosing trade secrets. In deciding whether such an injunction should be granted, the law must balance the legitimate interests of an employer in protecting valuable industrial and commercial information with the often unreasonable burden such restraints place on employees whose only source of livelihood may be using what they have learned while working for their employer.

The law has sought to minimize restraints on the use of all information—commercial and technical. A reconciliation of such policies has resulted in a rule under which covenants not to compete or to disclose industrial or commercial secrets are enforceable by injunction if reasonable. To decide whether a restraint is reasonable, the duration of the restraint; the extent of the restraint geographically and competitively; and the legitimate interests of the employer, the employee, and the public are examined.

SECTION 6.10 SALES OF LAND

Because land is regarded as unique, courts will usually grant a request for "specific performance" if the seller, without legal excuse, refuses to go through with the transaction. (As a corollary to this, the seller may be able to obtain a judgment for the purchase price if the buyer is unjustified in refusing to perform.) The court order will order that the seller give a deed to the buyer. Failure to do so will put the seller in contempt of court. Sellers can be jailed until they purge themselves of contempt by executing the deed.

In some states, the court can execute the deed where the seller refuses to perform. When asked to award specific performance, the law looks at the fairness of the transaction, a factor that is not the subject of inquiry when the court is asked to award a money judgment for damages. In addition to other requirements, specific performance will not be awarded if innocent third parties have purchased the property.

---

40. 42 Eng.Rep. 687 (1852).

41. See West's Ann. Cal. Civ. Code § 3423 (barring negative injunctions except in special cases).

42. See Restatement (Second) of Contracts § 367 (1981).

If the seller refuses to perform and specific performance is either not requested or not granted, the buyer should be put in the position it would have been in had the seller performed. Usually this means that the buyer is entitled to the benefit of the bargain—the difference between the contract price at the time conveyance should have been made and the fair market value of the property. If the contract price had been $20,000 and the buyer can show that the fair market value is $25,000, the buyer receives $5,000. The buyer may be able to recover additional losses caused by the seller's breach if they are proved with reasonable certainty and if the seller could foresee that such losses probably would result.

The same measure of damages will apply if the buyer refuses to perform. The seller recovers the benefit of the bargain—the difference between the contract price and the market price and any other foreseeable expenses or losses resulting from the breach.

Some jurisdictions use the English rule, dividing breaches into ordinary breaches and bad faith breaches. Ordinary breaches are usually caused by sellers' being unable to transfer the title that they thought they had when they agreed to sell. There is no moral blameworthiness attached to a seller's nonperformance in such cases. A bad faith breach is a seller's refusal for no *good* reason, for example, if the seller can sell to someone else for more money. In such bad faith breaches the buyer can recover benefit-of-the-bargain damages. A buyer who cannot show that the breach was in bad faith can recover only out-of-pocket, reliance expenses incurred. In 1983 California, formerly a leading English rule jurisdiction, eliminated the bad faith requirement for recovery of the benefit of the bargain.[43] This may presage a gradual elimination of the English rule.

PROBLEMS

1. Professor Best, an assistant professor at a state university, made a contract with Engineering Books, Incorporated, a publisher of engineering texts. The contract provided that Best would supply a manuscript dealing with a research project he had conducted. The project dealt with an analysis of disputes that resulted in the course of building facilities for waste-water treatment. Best was to receive ten complimentary copies and 15% of all the revenues received by the publisher attributable to the book.

   Best finished the manuscript and submitted it to the publisher. The publisher decided that the manuscript, while of a quality that could be considered publishable, would not be published because it did not believe that the cost could be recovered.

   When Professor Best was notified by the publisher, she was enraged, because failure of the manuscript to be published would have a significantly adverse effect on her chances for tenure in her department. She had also hoped that publication of the manuscript would result in her receiving consultation fees and invitations to conferences and thought that the book would earn substantial royalties. She had not published any other books.

   Professor Best contacted some other potential publishers. When they heard that the book had been rejected despite the contract to publish it, they were cool about publishing the book. She

---

43. West's Ann. Cal. Civ. Code § 3306.

may be able to get the book published and marketed if she pays the publisher for these services.

Professor Best does not believe that she can establish with any certainty that her progress as a professor will be impeded, but she is reasonably certain that her career has suffered a setback. She had asked a number of colleagues if they would review the book. If the book is not published, she feels that she will suffer a diminution of respect in their eyes. She spent approximately one hundred hours lining up potential reviewers, time she could have used in developing her consultation business.

What remedies might she seek? What remedy or remedies do you believe a court would award her?

2. Alice Sullivan was about to take her first job as an architect. On November 28, 1984, she had accepted an invitation to be associated with the firm of Rauch and Burns (R&B), whose offices are located in Denver, Colorado. One reason Alice wanted to work for R&B was their undoubted expertise in the design of prisons and correctional in-

stitutions. R&B was nationally known for its work in this area. Alice had recently seen a study that indicated a big boom ahead in the construction of such projects.

The firm gave Alice an employment contract on November 30, 1984. All terms were agreeable to Alice. But she was puzzled by paragraph 17. It provided the following:

> If the employment relation is terminated for any reason, employee agrees she will not join any other architectural firm as a principal or associate for one year. This provision is agreed upon as necessary to protect R&B proprietary data, including design concepts and prospective clients. If this covenant is breached, R&B can choose either to seek past damages agreed to be $500.00 a day and an injunction *or* agreed damages for each day of violation of $300.00 a day, but not to exceed $50,000.

Would this clause be enforceable if Alice were to sign the contract? Should Alice suggest modifications? (Her bargaining position in this regard is not very good.)

7

Losses, Conduct, and the Tort System: Principles and Trends

SECTION 7.01 RELEVANCE TO THE CONSTRUCTION PROCESS

During the Construction Process events can occur that might harm persons, property, or economic interests. Workers or others who enter upon a construction site might be injured or killed. The owner or adjacent landowners might suffer damage to land or improvements. Participants in the process might incur damage to or destruction of their equipment or machinery. The owner or other participants in the project might incur expenses greater than anticipated. Investors in the project or those who execute bonds on participants might also suffer financial losses.

After completion of a project, persons who enter upon or live in the project might be injured or killed because of defective design, poor workmanship, or improper materials. Those who invest in the project might find investment value reduced for similar reasons.

The construction project is a complex undertaking involving many participants. It is one that has a high risk of physical harm to those actively engaged in it. Sometimes such harm is caused by failure of participants to live up to the conduct required by law. Losses sometimes occur by human error that does not constitute wrongful conduct. Losses sometimes occur

because of unpredictable and unavoidable events for which no one can be held accountable. Because of the varying causes of losses, the many participants in the project and the complex network of laws, regulations, and contracts, placing responsibility is a difficult undertaking.

SECTION 7.02 TORT LAW: SOME BACKGROUND

(A) DEFINITION

"Tort" is derived from the Latin and means "twisted." Tortious conduct can be considered twisted or crooked rather than straight. If "straight" encompasses conduct required by the civil as distinct from the penal laws, the term from which tort is derived has some modern accuracy.

A tort has been defined as a civil wrong, other than a breach of contract, for which the law will grant a remedy, typically a money award. This definition, though not very helpful, mirrors the difficulty of making broad generalizations about tort law in the United States. One reason is the incremental or piecemeal development of tort law necessitated by new activities causing harm. For this reason much of American tort law consists of a collection of wrongs called by particular terms, which were given legal recognition in order to deal with particular problems.

Some basic distinctions are essential. Although tort law and criminal law have features in common—each regulating human conduct—they operate independently. Crimes are offenses against the public for which the state brings legal action in the form of criminal prosecution. Prosecution is designed to protect the public by punishing wrongdoers through fines or imprisonment and to deter criminal conduct. The tort system is essentially private. Only individual victims can use the system. Any sanctions imposed against those who do not live up to the standard of tort conduct are for the benefit of the victim. For example, the automobile driver who violates the criminal law may be fined or imprisoned. Those who are injured because of such criminal conduct are likely to institute a civil action to recover for their losses. This civil action is part of the tort system.

(B) FUNCTION

Tort law has different functions. The particular function most emphasized at any given time depends upon social and economic conditions in which the system operates.

One function is to keep the peace. In a violent period, compensating victims of violence can reduce victim or tribal revenge. In primitive societies this peace keeping was more likely to be a province of the government and its criminal laws. As societies became more developed, with a victim willing to accept compensation and those causing harm able to compensate, the tort system could both reduce revenge *and* deter wrongful conduct.

Because of the precedent system[1] American appellate court decisions not only decide appeals in specific cases but also create law. This adds a *public* dimension to the generally private nature of tort law by giving tort law the potential for social engineering.

Inevitably goals of individuals and groups clash. Choices must be made. One person's desires may come at the expense of another. One person may wish to drive a car at a high rate of speed which exposes others on the road to risks of danger. Property owners may wish the freedom to maintain their property as they wish. But this freedom may come at the expense of those who enter the land and are injured.[2] A manufacturer may wish complete freedom to design a product that will earn the highest profit. But this freedom may come at the expense of buyers of the product who suffer harm from using it.[3] Adjusting these conflicts can reflect conscious decisions to select or favor one competing interest or goal over another.

Tort rules may severely limit a property owner's freedom or the freedom of a manufacturer in order to give greater protection to those who are injured on unsafe property or by defective goods. The historical description in (D) suggests trends in social engineering.

(C) SOME THREEFOLD CLASSIFICATIONS

Two important threshold concepts are threefold. The first describes the interests considered sufficiently important to merit tort protection and often described as follows:

(1) *Personal,* sometimes defined to include psychic or emotional interests.

(2) *Property,* tangible and intangible.

(3) *Economic,* unconnected to harm to a person or damage to property.

---

1. Refer to Section 2.14.

2. See Section 7.08.

3. See Section 7.09.

The second classifies the conduct of the person causing the loss. Such conduct can be as follows:

(1) *Intentional*, including not only desire to cause the harm but the realization that the conduct will almost certainly cause the harm.

(2) *Negligent*, usually defined as failure to live up to the standard prescribed by law.

(3) *Nonculpable*, though in a sense wrongful, in which the actor neither intends harm nor is negligent.

These threefold classifications play important roles in determining which victims will receive reparation from those causing the loss. On the whole, harm to a person is the *most* deserving of protection, with harm to property being considered second in importance. At the conduct end, intentional conduct that causes harm is *least* worthy of protection, followed by negligent conduct, that which does not live up to the standard required by law. These classifications are gross, and many subtle distinctions must be made.

(D) SOME HISTORICAL PATTERNS

Earliest English private law developed during the feudal period in which land dominated society. As a result, property law developed before any significant developments in tort or contract law. The feudal period was dominated by agriculture and a largely illiterate population. There was little need for a developed contract or tort law system.

Although early English legal history saw the development of laws that dealt with finance, banking, and maritime matters, what is known today as tort law, as well as much of what is known today as contract law, did not develop until the Industrial Revolution in the late eighteenth and early nineteenth centuries. The Industrial Revolution moved manufacturing out of

cottages and into factories, necessitating a transportation system and migration of workers from the farms and villages to the towns and cities.

The changes brought significant developments in tort law. The preindustrial agrarian society, with its emphasis upon property, was most concerned with property rights and keeping the peace. The important torts were those that were intentional, as they could invite retribution, and those that invaded property interests. As a result, although persons may not have always acted at their peril in the sense that they would have to account for any damages their activities caused, much liability was "strict." Trespass, an invasion of a property owner's right to exclusive possession of its property, did not require any showing of fault. Any trespass, whether innocent or deliberate, was wrongful. Both the importance of property rights and the inability to deal with subtle concepts such as negligence and fault contributed to the rather simple and often harsh structure of early tort law.

Keeping the peace required protection against serious intentional torts, such as trespass, assault (apprehension of harm), battery (harmful or offensive touching), or false imprisonment (deprivation of freedom of movement). These serious matters could lead to breaches of the peace, and such conduct had to be eliminated or at least minimized through making the actor pay for the harm caused.

Unintentional or negligent conduct did not rate very high on the interest-protection scale both for reasons mentioned and because more important matters required attention. Matters such as harsh words, offensive conduct, or careless jostling were not sufficiently important in such a society to receive protection.

The Industrial Revolution generated factory and transportation accidents as well as migration to population centers. Now the law had to deal with conduct that became serious in crowded towns and cities. Many matters of an earlier day that were too trivial to be dealt with now required attention.

Even more important, difficult choices had to be made when commercial activity caused harm. The law chose to protect new and useful commercial and industrial activities from potentially crushing liability by not making liability as "strict" as it had been in preindustrial times. With some important exceptions,[4] a person who suffered a loss could not transfer the loss to the person causing it unless the injured person was free of negligence and could establish that the person causing the harm did not live up to the negligence standard of conduct. Transferring the loss *only* upon a showing of negligence and the development of other legal doctrines were designed to free useful activities from responsibility. These rules were less concerned with compensating victims of these activities.

Liberalization, as the term was then used, was designed to free economic activity from the shackles of heavy state mercantilistic controls. Rules that protected industrial and commercial activity may also have been generated by the belief that such activities brought long-run social and economic advantages. The nineteenth century emphasized the moral aspects of individual responsibility, and shifting a loss required wrongdoing. Such a system was doomed. As the toll in human misery and economic deprivation rose because of industrial, commercial, and transportation activities and as the automobile replaced the horse, changes were inevitable.

At the beginning of the twentieth century, many industrial countries sought to remove industrial accidents from the tort system, a topic briefly treated in Sections 7.04(C) and 35.02. Workers' compensation brought a more humanitarian approach to industrial accidents. In the 1920s manufacturers began to lose protection, and the consumer revolution of the 1960s has for all practical purposes made manufacturers the insurers of losses caused by their defective products.[5] Similarly, some earlier protection given property owners began to be diluted in the late 1960s,[6] although responsibility was not as extensive as that of manufacturers.

People who furnished services, such as architects and engineers, were in the backwash of these changes. (The effect upon professional liability is better postponed until Chapter 17.) The modern era has sought to give security from certain risks to all members of society, and the tort system has reflected this. Compensation of victims rather than unshackling enterprises became predominant. The shift is sometimes described as enterprise liability. Under it an enterprise can and should bear the normal risks of its activity. These risks can be predicted, computed, and insured. The social costs of the activity can through pricing be spread to all who benefit from the enterprise by using its products.

What of the post-industrial tort law? Some have advocated its abolition. They would replace it with social insurance under which all victims of accidents would be compensated by the state or by private insurers.

Some have advocated and many states have adopted no-fault handling of road accidents. The victim recovers up to a threshold amount of medical expenses (which is quite low) *without* showing any fault. Above this threshold the victim can use tort law. (Low thresholds have meant inflated medical expenses and very little reduction in lawsuits.)

---

4. See Section 7.04.

5. This is a deliberate but only slight exaggeration of the current state of the law. (See Section 7.09.)

6. See Section 7.08.

Some have felt enterprise liability has placed too heavy a burden on manufacturers and professionals. Much has been written about the "malpractice crisis," and recent legislative activity has moderated the harsh treatment given professionals and manufacturers by the courts.

A justification sometimes given for the current tort system is that it will admonish those who can cause harm to be more careful. With the advent of liability insurance, some believe that the deterrent function is no longer accomplished by tort law. They would prefer *direct* control by legislation, similar to the Occupational Safety and Health Act (OSHA) enacted by Congress. Though much criticized mainly for obsessive interest in detailed rules, the Act is a direct attempt to make industrial activities safer as opposed to the indirect method of the tort system.

Although predictions are dangerous, it is likely that the tort system in some form will continue to serve if not the primary, at least an ancillary role in compensating victims and regulating activity.

(E) SOME GENERAL FACTORS IN DETERMINING TORT LIABILITY

Some have examined the unruly and disparate thousands of tort cases and have attempted to articulate factors that affect tort liability. One scholar listed the following items as important: [7]

1. The moral aspect of the defendant's conduct.

2. The burden of recognizing a legal right upon the judicial system.

3. The capacity of each party to bear or spread the loss.

4. The extent to which liability will prevent future harm.

(F) COVERAGE OF CHAPTER

Tort law is simply too diverse and immense to cover in this treatise. By and large, intentional torts [8] such as trespass,[9] assault, battery, false imprisonment, intentional infliction of emotional distress,[10] defamation,[11] invasion of privacy, and interference

---

7. See W. Prosser, *Torts*, pp. 16–23 (4th ed.1971). A subsequent edition added "a recognized need for compensation and historical development." W. Prosser and P. Keeton, *Torts*, pp. 20–26 (5th ed. 1984).

8. Comprehensive treatment of intentional torts can be found in W. Prosser and P. Keeton, supra note 7 at pp. 33–159.

9. In re Catalano, 29 Cal.3d 1, 171 Cal.Rptr. 667, 623 P.2d 228 (1981), held that a union official did *not* violate the *criminal* trespass statute when he refused to leave the site when ordered to do so by the owner. The official was conducting a safety inspection, a power given the union under its contract.

10. Refer to Section 6.06(H).

11. The tort of defamation, usually subdivided into libel and slander, sometimes arises in the construction context. For example, *Diplomat Electric Inc. v. Westinghouse Electric Supply Co.*, 378 F.2d 377 (5th Cir.1967), involved a communication by a supplier of a subcontractor made to the prime contractor and owner which stated the supplier had not been paid by the subcontractor. The court held that if the plaintiff subcontractor could prove that the communication was false, he would be able to recover his losses from the defendant supplier based upon the tort of defamation.

A defamation case involving a design professional was *Priestley v. Hastings & Sons Publishing Co. of Lynn*, 360 Mass. 118, 271 N.E.2d 628 (1971). Here the defendant newspaper published defamatory statements made by a town official regarding the plaintiff architect who had designed and administered construction of a public high school. Earlier decisions of the U.S. Supreme Court had granted privileges to newspapers based on the First Amendment's protection to the press. The plaintiff was required to prove under certain circumstances that the defamatory material was published maliciously. The Massachusetts court concluded that free press protection required malice in this case because the architect, though not a public official, became involved in a matter of public or general concern. This case, and earlier U.S. Supreme Court cases upon which the holding was based, concluded that private persons can become public figures if they inject themselves into public matters. A subsequent decision of the U.S.

with contract or prospective advantage [12] are not discussed.[13] Emphasis is placed upon negligence and strict liability, as those concepts relate to the Construction Process, starting with design and culminating with the finished project.

SECTION 7.03 NEGLIGENCE: THE "FAULT" CONCEPT

(A) EMERGENCE OF NEGLIGENCE CONCEPT

The Industrial Revolution was the precipitating factor for tort law moving from principal emphasis on intentional torts and the need to keep the peace to a system that would deal with increased accidents brought about by industrialization. This movement culminated with the recognition of negligence as the principal basis for liability. To legitimately transfer the plaintiff's loss to the defendant, the plaintiff was required to establish that the defendant did not perform in accordance with the legal standard of conduct which, though somewhat inaccurate, was called the "fault" system.

Today the negligence concept largely governs road accident losses, losses caused by the possessor of land failing to keep the land reasonably safe, losses that occur in the home, losses caused by the activities of professional persons, and, for the most part, losses caused by participants in the Construction Process.

(B) ELEMENTS OF NEGLIGENCE

To justify a conclusion that the defendant was negligent, the plaintiff must establish the following:

1. The defendant owed a duty to the plaintiff to conform to a certain standard of conduct in order to protect the plaintiff against unreasonable risk of harm.[14]

2. The defendant did not conform to the standard required.[15]

3. A reasonably close causal connection existed between the conduct of the defendant and injury to the plaintiff.[16]

4. A legally protected interest was invaded.[17]

(C) STANDARD OF CONDUCT: THE REASONABLE PERSON

Objective Standard and Some Exceptions

Nineteenth century English and American courts rejected a standard based upon the subjective ability of the defendant. It was not sufficient for the defendant to show that it did the best it could. To protect the community, its members are held to a standard that can exceed what they are able to do. In this sense, negligence is not synonymous with fault. The community standard

---

Supreme Court seemed to narrow the protection given the press by concluding that private persons who through no desire of their own find themselves thrust into public matters may be able to recover for defamation without any showing that the defamatory communication was made maliciously. *Gertz v. Robert Welch, Inc.*, 418 U.S. 323, 94 S.Ct. 2997, 41 L.Ed.2d 789 (1974).

12. This intentional tort is noted briefly in Section 17.08(D).

13. Most law relating to torts is called "common law" in that the principal sources of law are reported appellate decisions. Legislative bodies increasingly bar certain conduct, such as improper reasons for refusing to sell, rent or hire.

14. See Section 7.03(E).

15. See Section 7.03(C).

16. See Section 7.03(D).

17. See Section 7.04(F).

requires that the defendant do what the reasonable person of ordinary prudence would have done. Such a standard can hold the defendant liable when it did the best it could. Negligence, then, or much of it, is not congruent with morality. The standard holds persons who live in the community to an average community standard. For example, an inexperienced driver is expected to drive as well as the average driver.

There are exceptions. A person can be held to a lower or higher standard than that of the community. Usually such exceptions are created by designating special subcommunities smaller than the general community and then applying an objective subcommunity standard. For example, children are generally not held to the adult standard but only to the standard of children of similar age and experience. But children who engage in *adult* activities, such as automobile driving, are held to an adult standard. Similarly, persons with physical disabilities such as blindness are not expected to conduct themselves in the same way as the average community member who does not suffer from a disability. Blind persons would be expected to conduct themselves as would average members of the blind community. But persons with mental or emotional handicaps are generally held to the community standard.

Since the application of these standards is generally performed by a jury, some tolerance of human weakness and some exceptions to the objective standard other than those mentioned may find their way into jury decisions.

Another exception applies to those persons who because of special training or innate skill are expected to do better than the average person. Physicians are held to a higher standard in dealing with medical matters than are ordinary members of the community. Architects and engineers are held, as a rule, to the standards of their subcommunity, a topic discussed in greater detail in Chapter 17. Professional truck drivers are expected to drive better than ordinary drivers. The combination of objective and subjective standards in such cases is reflected by statements that defendants are judged by what they knew or should have known or by what they did or should have done.

Unreasonable Risk of Harm: Some Formulas

Courts and commentators seek to refine vague community or reasonable-person standards by articulating factors that should sharpen the inquiry into whether the standard of conduct has been met. One approach is to evaluate the magnitude of the risk, the utility of the conduct, and the burden of eliminating risk.

To determine the magnitude of risk, the relevant factors are the gravity, the frequency, and the imminence of the risk. Clearly the likelihood of harm and the severity of harm that can result are important factors in determining the type of conduct that should be expected to avoid these risks. Driving at an excessive speed on the highway clearly involves a high risk because accidents often result from excessive speed and their consequences are usually serious. Railroad crossings are dangerous because, though the likelihood of a train striking an automobile may be small, the consequences of such an occurrence are serious. Conversely, although throwing a soft rubber ball into a crowd may not cause *serious* harm to *anyone*, it is likely to cause *some* harm to *someone*.

The other factors recognize that imposing liability upon actors restricts human freedom. The social utility of the conduct being regulated is an important criterion in determining whether the legal standard has been met. Suppose a bank robber carelessly jostles a bank patron in the course of robbing the bank. If the law holds the bank robber responsible for the harm caused the patron, one factor is likely to be that bank robbing is not considered a useful activity and can be regulated by tort law as well as by criminal law. Conversely, the same carelessness by a bank security guard in the performance of a socially useful activity, such as organizing the patrons of the bank so that they can make an orderly retreat in the face of a fire, would not expose the bank or guard to the same liability as the bank robber.

The burden of eliminating the risk recognizes that almost all risks can be eliminated or minimized if sufficient resources are mobilized. Undoubtedly the impact of road accidents would be minimized if guardrails were installed on all public highways. Yet doing so would involve an immense expenditure that might not be commensurate with the gain that could be realized. Likewise, the burden of eliminating the risk in this manner would take into account not only cost but the esthetic deprivation caused by universal installation of guardrails.

Where serious harm can be avoided by minimal effort or expenditure, failure to do so will very likely be negligent. For example, if serious burns can be avoided by installation of a five-dollar mixer valve in the bathroom fixtures, failure to do so would very likely be negligent.[18]

Common Practice: Custom

Suppose the defendant conformed to or deviated from common practice or what is sometimes called the custom in the community. Frequently defendants attempt to exculpate themselves by showing that they performed as others do. This arises most frequently in claims against a manufacturer. Often the manufacturer establishes that it performs in accordance with industry practices.

While compliance with customary practices is *evidence* of compliance with the legal standard of care, it is not conclusive.

The customary standard itself may be careless and create unreasonable risk of harm. For example, suppose that most pedestrians jaywalk or that all workers refuse to wear hard hats in a hard-hat area. Similarly, failure to conform to customary practices is not *conclusive* and a defendant may be exonerated despite deviation from customary practices if good reasons existed for deviation and the defendant conducted its activities with reasonable care.

One important exception to this relates to the standard of conduct expected of professional persons. Since this is more appropriately dealt with in examining liability of design professionals, major discussion is postponed until Chapter 17. It is sufficient to state at this point that when defendants are judged by the subcommunity of their profession, customary practices of the profession become the standard.

Violations of or Compliance with Statutes

In the exercise of its responsibility to protect all citizens government frequently prohibits certain conduct and attaches civil or criminal sanctions for violations. The construction process is governed by a multitude of laws dealing with land use, design, construction methods, and worker safety. What effect do violations of those statutes have upon *civil* liability?

It is possible, though uncommon, for the statute to expressly declare that violations of the statute *determine* civil liability. More commonly the statute expressly imposes criminal sanctions only. In such cases, courts can and do look at the statute as a legislative declaration of proper community conduct. To have *any* relevance, however, some preliminary questions must be addressed.

First, the person suffering the harm must be in the class of persons that the legisla-

---

18. *Schipper v. Levitt & Sons, Inc.*, 44 N.J. 70, 207 A.2d 314 (1965) (builder-vendor strictly liable without need to show negligence).

ture intended the statute to protect. Many statutes are intended to protect members of the community at large by achieving public peace and order rather than to protect any particular group or individual. For example, statutes sometimes prohibit certain businesses from operating on Sunday. Such statutes are not designed to protect those who suffer physical harm while a business operates in violation of the Sunday closing laws. Similarly, although statutes require that automobiles be registered, the purpose of such laws is to raise revenue and not to impose liability upon the driver of an unregistered car even though driving properly.

Sometimes the legislation is designed to protect an extremely limited class of persons that may *not* include the injured party. For example, legislation requiring that dangerous machinery be shielded may be designed for the benefit of employees and not of those who enter the plant for other purposes. Similarly, as seen in Section 35.04(F), some states hold that safety regulations imposed upon a contractor-employer are not designed to protect employees of *other* employers on the site.

Second, did the statute deal with the particular risk that caused the injury? For example, suppose a statute limits the time a train may obstruct a street crossing. This is designed to deal with traffic delays and not with the risk of personal harm caused by the delaying train. On the whole, the tendency has been to broadly define the particular risk.

The violation of law can be excused in most instances by showing extraordinary circumstances that made compliance more dangerous than violation. For example, a statute may require that drivers always drive on the right unless they are passing another vehicle or making a left turn. Suppose the driver veers to the left lane to avoid hitting a child. This technical viola-

tion will be excused and have no bearing upon negligence.

Where the policy expressed by the statute is particularly strong, the statute may expressly eliminate any possibility of a violation being excused. Violations of such statutes are conclusive on the question of negligence. Such statutes often impose liability despite assumption of risk or contributory negligence by the injured party which, as shall be seen in Section 7.03(G), often are defenses. This form of strict liability may result if the statute was intended to protect someone from his own immaturity or carelessness. Illustrations are those prohibiting child labor or requiring safety measures in construction work.

Much can depend upon the particular law violated. Some laws seem anachronistic and continue to exist only because the legislature lacks the energy to modernize rules. Violation of such laws may have very little impact.

Suppose it is concluded that the preliminary requirements have been met. The plaintiff is in the class of persons to be protected, the statute was intended to cover the risk in question, and the violation was not excused. Most courts hold the violation to be negligence per se and conclusive on the question of negligence. The trier of fact, whether judge or jury, need not decide whether there had been negligent conduct.

With the exception of special protective statutes of the type described earlier, a per se violation may or may not preclude a defense such as assumption of risk or contributory negligence. It does not preclude the defendant from showing that the violation of the statute did not cause the harm. For example, in *Hazelwood v. Gordon*,[19] an employee fell down a flight of stairs. The stairs were too narrow at the bottom. This, together with an inadequate handrail, violated a city ordinance. However, the court held that the injured party could not re-

---

19. 253 Cal.App.2d 179, 61 Cal.Rptr. 115 (1967).

cover from the property owner because her injury was not *caused* by a violation of the ordinance but by her negligently placing her foot on the top step of the staircase knowing the stairs were dangerous.

Some jurisdictions, however, find that a statutory violation is simply *evidence* of negligence that is given to the jury and weighed along with other evidence to determine whether the defendant lived up to the legal standard of care. Some states that employ negligence per se hold violations of local ordinances, traffic regulations, or administrative regulations to be only *evidence* of negligence. For example, *Bostic v. East Construction Co.*[20] dealt with administrative regulations for fire safety. The court held that administrative regulations can be the basis for negligence per se but are less likely to be. The court also held that the regulations must be understandable and the regulations involved in this case were not sufficiently clear. (In any event the court seemed to believe that the failure to comply with the fire regulations did not cause the injury.)

Compliance with the statutory standard does not necessarily preclude a finding of negligence. The statutory standard is a minimum. Additional precautions can be required. For example, it may not be a statutory violation to park a car on the shoulder of a highway so long as a tail light functions. But under certain circumstances, such as on an extremely foggy night, this conduct may be below the legal standard and be negligent. Such instances are rare.

Res Ipsa Loquitur

Proof of negligence can be by direct testimony of witnesses who testify based upon their own observations of the defendant's conduct. However, sometimes this evidence is not available. Absence of direct evidence does not preclude the plaintiff from establishing indirectly (that is by circumstantial evidence) that the defendant was negligent. This can be accomplished by showing facts relating to the accident that tend to show, in the absence of an explanation by the defendant, that the accident was probably caused by the defendant's negligence. For example, suppose a tool falls from a scaffold and injures a passerby. Once the facts are established and it is known that the contractor's workers were working on the scaffold, it is more likely than not that the accident was caused by the contractor's negligence.

The unfortunate Latin term *res ipsa loquitur* used to describe this process of indirect proof of negligence was employed in an English case in which a passerby was struck by a flour barrel falling from a warehouse window.[21]

Much controversy has developed regarding the *res ipsa* concept, mainly centered around judicial and scholarly statements of *res ipsa* requirements. In reality the supposed requirements are not truly requirements, because the doctrine is sometimes applied despite absence of some of them. However, the stated requirements may provide some assistance in understanding when the concept will be used.

It is usually stated that *res ipsa* requires the following:

1. An event is one that ordinarily does not occur in the absence of someone's negligence.

2. The event must be caused by an agency or instrumentality within the exclusive control of the defendant.

3. The accident must not be due to any voluntary action or contribution by the plaintiff.

---

20. 497 F.2d 712 (6th Cir.1974).

21. *Byrne v. Boadle,* 159 Eng.Rep. 299 (1863).

Some courts have suggested that the evidence must be more readily accessible to the defendant than to the plaintiff. But as stated, these requirements cannot be taken as conclusive, as they are not always found in cases where the doctrine is applied.

The use of circumstantial evidence does not, as a rule, shift the burden of proof from plaintiff to defendant. But if the facts indicate that it was likely that the defendant's negligence caused the accident, the matter will be submitted to the jury. The application of the doctrine does not preclude defendants from introducing evidence that they were not negligent. For example, in the illustration given earlier, the contractor can show that the tool fell because of a strong gust of wind for which it was not responsible. But the doctrine helps plaintiffs get their cases before the jury.

Sometimes the inference of negligence on the part of the defendant is so strong that it may persuade the jury that the plaintiff has met his burden of proof. Similarly, a very strong inference of negligence may justify the judge directing a verdict for the plaintiff.

(D) LEGAL CAUSE: CAUSE IN FACT AND PROXIMATE CAUSE

A reasonably close connection must exist between conduct of the defendant and the harm to the plaintiff. Legal cause is divided into two separate though related questions:

1. Has the defendant's conduct caused the harm to the plaintiff? (This is usually referred to as "cause in fact.")

2. Has the defendant's conduct been the "proximate cause" of the harm to the plaintiff?

The first, considered less complicated, is a factual question decided by the finder of fact, usually the jury. Even this supposedly simple question of causation can raise difficult issues. First, the defendant's conduct need not be the sole cause of the loss.

Many acts and conditions join together to produce a particular event. Suppose an employee of a subcontractor suffers a fatal fall while working on a scaffold high above the ground. Any one of the following events could be considered a cause of the death in the sense that without any of the events the fatal fall would not have occurred:

1. The worker's need to pay medical bills causing the worker to take this risky job.

2. Defective scaffolding supplied by a scaffolding supplier.

3. Failure by the subcontractor or prime contractor to remove the scaffolding when complaints were made about its unsafe condition.

4. Weather conditions that made the scaffold particularly slippery on the day of the accident.

5. A low-flying plane that momentarily distracted the worker.

6. The worker's refusal to wear a safety belt.

7. The subcontractor or prime contractor's failure to enforce safety belt rules.

Although the list can be amplified, remove *any* link in the causation chain and the worker would not have been killed. Yet it would be unfair to relieve any actor whose failure to live up to the legal standard played a significant role in the injury simply because other actors or conditions also played a part in causing the fall.

Liability in such a case would depend upon a conclusion that any of the defendants *substantially* caused the injury. Suppose a claim had been made against the scaffold supplier and the prime contractor. Each could have been a substantial cause of the injury. In the case of an indivisible injury, each would be liable for the entire loss, and whether the party paying the claim would be entitled to recovery from the other party would depend upon whether a right to contribution or indemnity existed.[22]

Cause in fact requires that the harm would not have occurred *without* the defendant's failure to live up to the legal standard. Put another way, the defendant will usually be exonerated if the injury would have happened even if the defendant had lived up to the legal standard of conduct. For example, suppose the prime contractor did not supply a safety belt to the worker as required by law but the worker would have been killed because of a refusal to wear it. Under these conditions it is likely that the prime contractor would not be held liable because the contractor's negligence did not *cause* the harm.

The "but for" defense has one important exception. Suppose two builders are constructing houses on adjacent lots. Each is simultaneously negligent, causing fires to begin on each building site. The fires join together, roar down the street, and burn a number of homes. Either fire would have been sufficient to burn the houses. But neither builder will be able to point to the "but for" rule as a defense. Each will be liable for the entire harm.[23]

Proximate cause, though related to cause in fact, serves a different function. Cause-in-fact judgments are factual and best made by common-sense decisions of juries. Proximate cause, on the other hand, involves a legal policy that draws liability lines to relieve those whose failure to live up to the legal standard of conduct causes harm. Proximate cause serves a similar function as the requirement that there be a duty upon the defendant to act to protect the plaintiff, a concept discussed in Section 7.03(E). Each is designed to minimize crushing liability burdens on those engaged in useful activities. They stop liability from "going too far" as well.

Proximate cause can involve the following:

1. Harm of a different type than reasonably anticipated.
2. Harm caused to an unforeseeable person.
3. Harm caused by the operation of intervening forces.

As an example of the first, suppose a contractor installs a sheltered walkway around a project. It can foresee that defective planking could cause a sprained ankle, but will it be liable if a pedestrian pushing a baby stroller falls on a defective plank causing the baby to tumble from the stroller and fracture its skull? Suppose defective wiring on a high-rise building causes a power failure and shuts down all the elevators. A person who intends to submit a bid on a public project on the top floor cannot reach the awarding authority's office in time to submit the bid. Can he recover the lost profits on the contract from the supplier of the electric wire or the owner of the building?

These freak accidents, though they occur infrequently, are dramatic enough to excite scholarly interest when they reach appellate courts. The most famous illustration of the second type of case was *Palsgraf v. Long Island Railroad Co.*[24] It involved a passenger running to catch one of defendant's trains. Some employees of the defendant sought to help the passenger board the train but did so carelessly, dislodging a package from the passenger's arms which fell upon the rail. The package contained fireworks that exploded with some violence. The concussion overturned some scales many feet down the platform. This was unfortunate

---

22. A discussion of concurrent liability for indivisible loss is found in Section 28.06. A claim by one concurrent wrongdoer against the other is discussed in Sections 36.03 and 36.05.

23. See Section 28.06.

24. 248 N.Y. 339, 162 N.E. 99 (1928).

for the plaintiff, who was struck by a scale, but fortunate for legal scholars and generations of law students who dissected the subsequent appellate court decision. In a four-to-three opinion, the court held that the plaintiff was outside the zone of risk. As an unforeseeable plaintiff, the railroad company did not owe any duty toward her despite its employees being careless toward someone else.

A case illustrating the third type was *Petition of Kinsman Transit Co.*, which involved two claims and two appeals. A negligently moored ship in the Buffalo River was set adrift by floating ice, picking up another ship on the way. Bridge attendants were warned but inexplicably failed to lift a bridge. Both ships collided with the bridge and the bridge collapsed. The ships and ice blocked the river channel by creating a dam. Water and ice backed up damaging factories on the bank as far up the river as the original mooring. One claim involved flood damage to the factories, and the other involved pecuniary loss due to the necessity of transporting ship cargos around the blockage. The court granted a recovery to the factory owners [25] but denied recovery for those ships incurring additional expenses in order to go around the blockage.[26]

This defense was attempted in a construction context in *Diamond Springs Lime Co. v. American River Constructors.*[27] Riverfront land flooded when a dam collapsed because of heavy rains. The contractor did not perform in accordance with the specifications but contended it should be relieved because the heavy rains were a superseding cause of the loss. The court held that the contractor would not be relieved since the heavy rains that caused the flooding were a reasonably foreseeable peril.

It would serve no useful function to explore the many formulas used in these cases, such as direct cause, forseeability, hindsight, and superseding causes. It is sufficient to indicate that freak accidents cause unusual harm, and lines must be drawn. One treatise, after cataloging the various formulas, suggested that those proposing formulas are groping for something that it is difficult if not impossible to put into words. It suggested the need for a method:

> . . . of limiting liability to those consequences which have some reasonably close connection with the defendant's conduct and the harm which it originally threatened, and are in themselves not so remarkable and unusual as to lead one to stop short of them.[28]

While proximate cause analytically is considered an issue of law because it deals with a policy of limiting liability, usually juries are given a vague instruction relating to proximate cause similar to the instruction they are given relating to the standard of conduct. It is hoped that the jury will use common sense in deciding such freak cases.[29]

(E) DUTY

Duty is related to proximate cause. The first duty cases were decided in the mid-

---

25. *Kinsman Transit Co.*, 338 F.2d 708 (2d Cir.1964).

26. *Kinsman Transit Co. v. City of Buffalo*, 388 F.2d 821 (2d Cir.1968).

27. 16 Cal.App.3d 581, 94 Cal.Rptr. 200 (1971).

28. W. Prosser and P. Keeton, *Torts*, 300 (5th ed. 1984).

29. Perhaps a layperson's view based upon a gut reaction of what is going too far is the best solution to these vexatious problems.

In commenting favorably on a hindsight test, one commentator suggested a line be drawn:

short of the remarkable, the preposterous, the highly unlikely, in the language of the street, the cock-eyed and far-fetched, even when we look at the event, as we must, after it has occurred.

Id. at 299.

nineteenth century in a legal climate favorable to new industries developing after the beginning of the Industrial Revolution. Like proximate cause, it was an attempt to draw a line beyond which recovery would not be granted. Unlike proximate cause, it tended to emphasize the relationship between individuals that imposes upon one a legal obligation to watch out for the other. Sometimes in a freak accident of the type described in Section 7.03(D) courts conclude that the defendant did not owe *any* duty to the plaintiff. Even if the defendant did not live up to the standard required by law and even if that failure caused harm to the plaintiff, the plaintiff cannot recover.

The duty concept has found its way into the Construction Process. For example, professional persons such as architects, engineers, and surveyors often supply information as part of their professional services. As seen in Section 7.07, persons other than those who have requested and paid for the information may suffer losses if the information furnished is incorrect. Whether those third persons can recover from the person supplying information is sometimes phrased in terms of duty. Did the surveyor owe a duty to someone who relied upon an inaccurate survey? [30]

Similarly, suppose a manufacturer's defective products harm persons with whom it has no contractual relationship, such as those who have purchased the products from retailers, nonbuyers who use the products, or others injured by the products? As seen in Section 7.09 early American decisions denied recovery against a manufacturer because of the absence of any contractual relationship between the person harmed and the manufacturer. Such decisions sometimes spoke of an absence of duty owed by the manufacturer to the injured party.

The two preceding illustrations involve one corollary of the duty concept frequently described as the "privity rule." Nineteenth century English law denied a third party recovery for damages caused by another's breach of contract, as there was no privity between third party and contract breaker.[31] This decision clearly protected manufacturing and commercial activity.

(F) PROTECTED INTERESTS AND EMOTIONAL DISTRESS

As Section 7.02(C) indicates the particular loss suffered often controls liability. Even if all the requirements for negligence are met, the particular harm that has resulted may not receive judicial protection. Courts speak of whether particular interests are protectable in the process of deciding whether the defendant must respond for certain losses caused the plaintiff.

Harm to the person is most worthy of protection. Death not only ends one's life but also can have a severe financial and emotional impact upon the deceased's survivors. Physical injury often means medical expenses and diminished earnings as well as pain and suffering. Often those who suffer physical harm are low-income persons who may not use insurance to protect themselves and their dependents. As a result for these as well as humanitarian reasons it may be important to extend protection to those who suffer physical harm.

Harm to property such as damage or destruction is also considered worthy of protection because of the importance placed upon property in modern society. But as the harm moves away from personal and property losses, the interest receives less protection. Economic harm such as diminished commercial contractual expectations, lost profits or additional expenses to perform contractual obligations, while often protected, is considered less worthy of

---

30. See *Rozny v. Marnul,* discussed in Section 7.07(G).

31. *Winterbottom v. Wright,* 152 Eng.Rep. 402 (1842). This is no longer the law. See Section 7.09.

protection than harm to person or to property. Even less protection is accorded emotional distress and psychic harm, sometimes called noneconomic losses.

There are a number of reasons for caution when claims are made for economic loss and even more when claims are made for emotional distress. Economic losses result from many causes other than the conduct of the defendant and are difficult to prove. Liability can place crushing burdens on the defendant. Although courts have cautiously extended protection for economic harm, they have been even less willing to extend protection for emotional distress. Part of this is the undoubted difficulty of establishing the genuineness of the claim. Another is the difficulty of placing an economic value upon it. Because of these difficulties, the law has been unsettled in this area and decisions vary from jurisdiction to jurisdiction.

With some unimportant exceptions [32] the plaintiff cannot recover for mental disturbance caused by the defendant's negligence in the absence of accompanying physical injury or consequences. Clearly, however, there can be recovery for mental disturbance if the negligence of the defendant has inflicted an immediate physical injury and the mental disturbance such as pain and suffering was caused by the physical injury. Suppose physical harm *follows* fright or shock of the plaintiff such as a miscarriage suffered after negligent conduct by the defendant had caused a mental disturbance. Many American cases have required the plaintiff to show that there had been some physical impact upon the plaintiff caused by the defendant's negligent acts. Others have not. The recent cases have eliminated the requirement for impact, and the impact rule is likely to disappear soon.

Suppose physical injury resulting from fright occurs when the injured person feared for his own safety? Here most states allow recovery. But fear for someone else's safety has divided the courts. The classic example has involved a mother, though not in danger, seeing her child seriously harmed. Some courts allow recovery. Others do not.[33]

Emotional distress and mental disturbance can arise in the context of construction work. Suppose a worker is on a roof. The roof partially collapses but the worker is not injured. He may have suffered emotional distress from the fear that he was about to fall or that he might have fallen. Also, suppose a worker experiences traumatic shock when he sees someone else fall from the roof of a building.

Emotional distress in a construction context occurred in *Olivas v. United States*.[34] The injured employee was working on a project for the United States. He was ordered to enter a blast valve for cleaning. While in the valve, an Air Force officer negligently ordered the valve activated. Olivas knew that the valve door would be closed in 40 seconds, that it could not be stopped, and that he would be crushed to death unless he could be extricated. Fortunately he was dragged out by a foreman and suffered only minor injuries. However, the trial court concluded that imminent danger of serious injury caused him to sustain an anxiety neurosis of a traumatic type with psychoneurotic reactions to stress. This neurosis caused him permanent total disability.

One reason this judgment was affirmed on appeal was the presence of some physical injury. Where physical harm has been suffered, often emotional distress can be the basis for additional compensation.

---

32. Refer to Section 6.06(H) for claims for emotional distress based upon contract breach. See also W. Prosser and P. Keeton, *Torts*, 361–362 (5th ed. 1984).

33. W. Prosser and P. Keeton, supra note 32 at 365–366.

34. 506 F.2d 1158 (9th Cir.1974).

However, the court also seemed willing to recognize the genuineness of the claim and its pecuniary loss as justification for granting recovery.

(G) DEFENSES

Assumption of Risk

Assumption of risk completely bars recovery even if the defendant has been negligent. Advance consent by the plaintiff relieves the defendant of any obligation toward the consenting party. The plaintiff has chosen to take a chance. Suppose the plaintiff voluntarily entered into a relationship with the defendant with knowledge that the defendant will not protect the plaintiff from the risk. In such cases the plaintiff impliedly assumed the risk. Sometimes the plaintiff is aware of a risk created by the negligence of the defendant but proceeds voluntarily to encounter it.

Express agreements to assume the risk are often given effect if knowingly and freely made by parties of relatively equal bargaining power.[35] But sometimes the relationship is one regulated by law and this freedom is denied. For example, workers are frequently prohibited from signing agreements to assume the risk of physical harm.

Most cases involve implied assumption of risk. Did the plaintiff know and understand the risk? Was the choice free and voluntary? Voluntariness has generated considerable controversy where workers take risks under the threat that they will be discharged if they do not continue working.[36] Under such circumstances, some American courts conclude that the worker even under such pressure has assumed the risk of performing dangerous work.[37] While statutes often preclude assumption of risk in employment relationships, the concept can be applied in third-party actions brought by workers against persons *other* than their employer.[38]

The assumption of risk defense is not favored. As a result the defendant must plead and prove that the plaintiff assumed the risk.

Contributory Negligence

Until quite recently most states would bar the plaintiff if the defendant established that the plaintiff's negligence played a significant role in causing the injury. Juries are thought to use techniques to mitigate the harshness of this defense. Where the plaintiff's negligence was slight and much less than that of the defendant, juries sometimes decide that the plaintiff was not contributorily negligent or that the contributory negligence did not cause the harm. Where both parties were negligent but the defendant much more so, the jury may compromise by finding that there has not been contributory negligence but diminishing the amount of the plaintiff's award to take the plaintiff's negligence roughly into account.

The contributory negligence rule has been criticized because slight negligence bars what would otherwise appear to be a just claim. A number of states, notably Wisconsin and Louisiana, enacted legislation early in the century that modified this rule by requiring that the negligence of the plaintiff and the defendant be compared rather than automatically barring the plain-

---

35. *Delta Air Lines, Inc. v. Douglas Aircraft Co.*, 238 Cal.App.2d 95, 47 Cal.Rptr. 518 (1965). However, a release signed by a person seeking admission to a university hospital was not upheld in *Tunkl v. Regents of the University of California*, 60 Cal.2d 92, 32 Cal.Rptr. 33, 383 P.2d 441 (1963).

36. See Section 35.08(A).

37. A reluctant court concluded that there had been assumption of risk in such a case in *Demarest v. T.C. Bateson Construction Co.*, 370 F.2d 281 (10th Cir.1966). See Section 35.08 for further discussion.

38. Third-party actions are discussed in Chapter 35.

tiff from recovery if the plaintiff is negligent. Within the past twenty years, most states have enacted statutes creating comparative negligence, and a few courts have created comparative negligence without the benefit of legislation.

The details of comparative negligence laws vary considerably. The essential feature is that the plaintiff's negligence does not automatically bar recovery but only diminishes recovery. A plaintiff whose negligence reaches a specific percentage, such as 50%, is barred in some states. In other states the comparison is "pure." An 80% negligent plaintiff would be entitled to recover 20% of the loss from a 20% negligent defendant. Comparative negligence has been used in admiralty law and in claims by employees of certain common carriers subject to the Federal Employers Liability Act.

Independent Contractor Rule

Unlike the two preceding defenses, the defense of the independent contractor rule does not involve conduct or risk taking by the party who suffered the loss. But because of its importance in construction legal problems, it merits brief mention at this point.

Suppose a homeowner hires a handyman contractor to do remodeling work and the contractor negligently drops tools from the roof injuring a passerby. Unless one of the many exceptions to the independent contractor rule applies, the homeowner will have a defense if sued by the passerby based upon the negligence of the *contractor*. Sometimes one person is vicariously liable for the negligence of another.[39] But no vicarious liability exists if the party against whom the action is brought can establish that the negligent party was an *independent contractor*. An independent contractor is asked to achieve a result and is not controlled as to means by which this is accomplished. Vicarious liability requires the right to control the details of the work.

Because many independent contractors are one person or small operations unable to respond for losses caused by their activities, many exceptions to the independent contractor rule exist.[40] Another reason for exceptions is that the independent contractor rule can enable financially solvent employers to insulate themselves from liability for activities that are essentially part of the operation by hiring an independent contractor.

SECTION 7.04
NONINTENTIONAL NONNEGLIGENT WRONGS: STRICT LIABILITY

(A) ABNORMALLY DANGEROUS THINGS AND ACTIVITIES

As indicated earlier,[41] preindustrial law was often strict in the sense that liability was not based upon negligence. Despite the emergence of negligence as the dominant basis for liability that developed in the nineteenth century, some pockets of law found liability "strictly."

One—keeping of animals—is of little importance to construction. (But watch strict liability for keeping a vicious dog on the site to deter vandals.) The other, strict liability for abnormally dangerous things and activities, does find application to modern construction problems and merits some comment.[42]

---

39. See Section 35.05(A).

40. See Section 35.05(C).

41. Refer to Section 7.02.

42. Sometimes the type of activity can determine whether a person who hires an independent contractor is liable for that person's negligence. See Section 35.05(C).

Rylands v. Fletcher,[43] an 1868 English decision, held that a mill owner *not* shown to have been negligent was liable when he built a reservoir whose waters broke through to an abandoned coal mine and flooded parts of the plaintiff's adjoining mine.

Ultimately most American decisions in some form or another found strict liability where the defendant owned dangerous things or engaged in dangerous activities.

The prototype abnormally dangerous activity is often stated to be blasting. The activity may not be negligent because the social utility and the high cost of avoiding the harm makes the conduct reasonable despite the high risk of serious harm. The modern rationale for strict liability for such activities is that such enterprises should pay their way and are better risk bearers than the persons who have been harmed by the activity.

Since tort law emphasizes victim compensation and loss spreading through insurance and pricing goods and services, strict liability—liability without a need to establish negligence—has become more important in construction-related activities. Strict liability for defective products plays an increasingly important role in construction-related litigation, (discussed in Section 7.09). (Also, the professional standard to which design professions are held is increasingly under attack, the assertion being that they should be held to a "stricter" standard.[44]) This trend is reflected in federal laws that make those who generate, transport, or dispose of hazardous waste strictly liable for violation of statutes and regulations that regulate these activities.[45]

For these reasons it will be useful to examine a case involving a hydraulic landfill project. The court relied heavily upon a recent formulation of the doctrine by the American Law Institute.

DOUNDOULAKIS v. TOWN OF HEMPSTEAD

Court of Appeals of New York, 1977.
42 N.Y.2d 440, 398 N.Y.S.2d 401, 368 N.E.2d 24.

BREITEL, Chief Judge.

Plaintiffs, three separate owners of private homes built on filled sandy soil on the south shore of Long Island, brought separate actions, since consolidated, to recover for property damage. The damage, subsidence of their lands, was allegedly caused by a hydraulic landfilling project conducted by defendants, the Town of Hempstead, its contractor, and its design engineer, on 146 acres of swampy meadowland abutting plaintiffs' homes.

Supreme Court, after jury trial, awarded plaintiffs Doundoulakis and D'Angelo judgment against the town, but set aside the verdicts and dismissed their complaints as against the engineer, De Bruin, and the contractor, Gahagan Dredging Corporation. With respect to the complaint of plaintiffs Silver, the verdict in their favor was set aside and their complaint dismissed in its entirety. From an order of the Appellate Division, one Justice dissenting, modifying the judgment to reinstate all of the plaintiffs' verdicts, defendants appeal.

The pivotal issue is whether it was established that hydraulic dredging and landfilling, that is, the introduction by pressure of a

---

43. 3 L.R.–E. & I. App. 330 (1868).

44. See Section 17.07.

45. See also Section 35.04(A).

continuous flood of massive quantities of sand and water, is, under the circumstances, an abnormally dangerous activity giving rise to strict liability. Subsidiary issues involve whether strict liability should be imposed on the contractor and design engineer engaged by the offending landowner and, if so, whether apportionment of relative liability can be had among defendants. . . .

Plaintiffs' pleadings and theory of action were based on negligence in the handling of the landfilling project. The trial court, however, precluded submission to the jury of that issue and instead submitted the case on the theory, never raised by the parties, of "absolute" liability for harm caused by activity in the nature of trespass and dangerous use of abutting or adjacent property. The Appellate Division, in modifying, in effect sustained the trial court's preclusion of the negligence issue and submission of the case on theories of strict liability. It extended liability, however, to all of the defendants. As a consequence, plaintiffs' claims based on negligence have never been determined by a fact finder and, indeed, were expressly rejected by the trial court.

There should be a reversal and a new trial. Although the record suggests that hydraulic dredging and landfilling is not a "normal" activity, in the special sense in which the standard has been developed, the record is unsatisfactory to establish that the activity was in this case abnormally dangerous. A new trial is needed, however, to resolve the issue of negligence, the original basis for liability urged by plaintiffs, and to which the record is largely devoted, an issue never submitted to the jury. On such retrial, plaintiffs are also entitled to establish, if they can, that there is sufficient basis for recovery on a theory of strict liability.

Plaintiffs' homes, constructed around 1961, were part of a housing development built on filled meadowland on Long Island's south shore. Because it was bordered on the west by a navigable stream of water, Parsonage Creek, the property was bulkheaded. Immediately south of plaintiffs' homes, and extending eastward, were 146 acres of marshland owned by the Town of Hempstead.

The town property, on which the town proposed to lay out a public park, was significantly lower in elevation than the filled land on which plaintiffs' homes had been constructed. Building a public park required that one and one-half million cubic yards of sandfill be deposited on the 146 acres. Defendant De Bruin was engaged by the town to prepare the plans and specifications and to act as supervising engineer on the site. Gahagan Dredging Corporation was awarded the landfilling contract.

Starting on September 5, 1966, and continuing 24 hours a day, a dredged mix of 85% water and 15% sand was pumped, under pressure, onto the park site. Over two miles of pipe brought the mixture from the Jones Beach inlet. To impound the waters, dikes had been constructed around most of the landfill site. Settlement of the sand was accomplished and the level of water controlled by a system of exit weirs designed to discharge up to 40,000 gallons of water per minute.

Presumably in part because plaintiffs' land was significantly higher in elevation than the land to be filled, no dike had been installed on the side of the landfill abutting plaintiffs' homes. It also seems that a dike constructed along that border during the development of plaintiffs' homes was believed still in place. In fact, however, part of that dike had been removed when a roadway was put in. And despite the system of exit weirs, a "lake" of over 50 acres in the area south of plaintiffs' homes had accumulated as a result of the hydraulic landfill project.

On September 16, the eleventh day of the operation, the first sign of damage, a partial failure of the Doundoulakis bulkhead, was spotted. A further collapse of that bulkhead occurred five days later, the same day the D'Angelo bulkhead bowed out. Finally, in April, 1968, damage to the Silver bulkhead became manifest. Plaintiffs' actions resulted.

Plaintiffs urge that subterranean percolation or seepage of the water from the landfill site raised the underground water table, thereby increasing the pressure and loosening the bulkhead anchorages. Defendants, on the other hand, attributed the collapse to the assertedly dilapidated and thus unstable condition of the bulkheads as well as an increase in pressure caused by heavy rainfalls.

Despite voluminous testimony on the issue of negligence, the trial court treated the hydraulic landfill project as an abnormal activity giving rise to "absolute" liability. Thus limited to the issue of proximate cause, the jury returned verdicts totaling $51,180 in favor of plaintiffs against all three defendants. For reasons not now material the trial court reduced the total verdicts by $15,007. As noted earlier, however, the verdicts against the dredging contractor and the consulting engineer were set aside so as not to extend responsibility for abnormally dangerous activities beyond offending landowners. In addition, the $6,000 verdict in favor of the Silver plaintiffs against the town was set aside for failure to serve the required notice of claim within the appointed time period (see General Municipal Law, § 50–e). At the Appellate Division a majority agreed that hydraulic dredging was an abnormally hazardous activity. The judgment was modified, however, to reinstate, subject to some reductions in amount, the Doundoulakis and D'Angelo verdicts against the contractor and the engineer and the Silvers' verdict against the town, the contractor, and the engineer.

Imposing strict liability upon landowners who undertake abnormally dangerous activities is not uncommon (see, generally, Prosser, *Torts* [4th ed.], § 78). The policy consideration may be simply put: those who engage in activity of sufficiently high risk of harm to others, especially where there are reasonable even if more costly alternatives, should bear the cost of harm caused the innocent.

*   *   *

Determining whether an activity is abnormally dangerous involves multiple factors. Analysis of no one factor is determinative. Moreover, even an activity abnormally dangerous under one set of circumstances is not necessarily abnormally dangerous for all occasions (see Restatement, Torts 2d, § 520, Comment *f*).

Guidelines, however, are identifiable. The many cases and authorities suggest the numerous factors to be weighed. Particularly useful are the six criteria listed in Restatement of Torts Second (§ 520): "(a) existence of a high degree of risk of some harm to the person, land or chattels of others; (b) likelihood that the harm that results from it will be great; (c) inability to eliminate the risk by the exercise of reasonable care; (d) extent to which the activity is not a matter of common usage; (e) inappropriateness of the activity to the place where it is carried on; and (f) extent to which its value to the community is outweighed by its dangerous attributes".

Thus, what should be considered in determining whether an activity is abnormally dangerous is not the most difficult part of this case. Instead, the difficulty this court has had, in contrast to the courts below, is finding enough in the record to determine how the landfill method employed by defendants measures up to the various applicable factors.

There is little if any information, for example, of the degree to which hydraulic landfilling poses a risk of damage to neighboring properties. Nor is there data on the gravity of any such danger, or the extent to which the danger can be eliminated by reasonable care. Basic to the inquiry, but not to be found in the record, are the availability and relative cost, economic and otherwise, of alternative methods of landfilling. There are other Restatement factors, and perhaps still others, which the parties may develop as relevant, about which there is little or nothing in the record.

Yet, the case strongly suggests that strict liability treatment may be appropriate. . . . Writing for a majority of the Appellate Division, Mr. Justice Marcus G. Christ stated, succinctly and soundly, why strict liability might have been the proper standard for fixing defendants' liability:

"From . . . review of the authorities there emerges a dominant theme, viz., that strict liability will be imposed upon those who engage in an activity which poses a great danger of invasion of the land of others. It matters little whether the force used is dynamite, gunpowder or pressure created by accumulating, massing and diverting large amounts of water by means of hydraulic pumps, pipes and impounding dikes, or whether the invasion is by objects projected by explosion, or water forced or diverted over the surface of the earth or forced underneath and through the earth. Often underlying these invasion-causing activities is a deliberate interference, distortion, wrenching or manipulation of natural forces, resources or equilibrium, frequently on a massive scale.

"At bar, the defendants transported (from a site miles from the plaintiffs' homes), under pressure, continuously and in great quantities, waters from a site near a gate to the Atlantic Ocean, and deposited those waters virtually at the doorsteps of homes previously constructed on a filled-in meadowland. The transported water was impounded between dikes upon its arrival, and although the water was to be discharged through weirs into the bay, it was generally impounded long enough to allow the sand to settle. This operation created a 'lake' of 50 to 70 acres in the immediate vicinity of the plaintiffs' homes. Not surprisingly, the water percolated into the adjoining properties. The operations here were a far cry from the type of incidental changes of grade for which the defendant was held not liable in *Kossoff v. Rathgeb-Walsh,* 3 N.Y.2d 583, 170 N.Y.S.2d 789, 148 N.E.2d 132. There the defendant did not *import* the problem-causing 'diffused surface water' to his property site; nor did he use leader pipes, drains or ditches, much less hydraulic pumps and an impounding dike system" (51 A.D.2d 302, 312–313, 381 N.Y.S.2d 287, 293).

It is not insignificant that alternative methods, perhaps more time-consuming and expensive but less dangerous, might have been available to defendants to accomplish the landfill. It may be significant that the edges of the south shore of Long Island are much like sandbars risen from the sea with the instability appropriate to that kind of land.

That the flood of water under pressure was not treated as abnormally dangerous in *McLoone Metal Graphics v. Robers Dredge,* 58 Wis.2d 704, 711, 207 N.W.2d 616, is not too persuasive. The Wisconsin court expressly stated that it dealt not with the adjoining landowner, but, for what relevance that may have, with the construction firm alone (*id.,* p. 709, 207 N.W.2d 616). And, the corporate plaintiff there was not the owner of a small private home, but of a building "made of concrete block" (58 Wis.2d, p. 706, 207 N.W.2d 616).

* * *

Should it appear on a new trial that, as a matter of law, a prima facie case for strict liability has been made out, the issue who besides the offending landowner may be liable must be resolved. As the Appellate Division noted, responsibility for abnormally dangerous activities is not attributable to landowners alone. In *Spano v. Perini Corp.,* 25 N.Y.2d 11, 302 N.Y.S.2d 527, 250 N.E.2d 31, *supra,* for instance, it was the blasting contractors engaged by the City of New York who were held responsible for injuring neighboring property. Similarly, the central roles of De Bruin, planner of the dredging project, and Gahagan, the contractor, justify imposition of strict liability because they were the actors through whose conduct harm was allegedly suffered by plaintiffs (see Restatement, Torts 2d, §§ 383, 384, 427A, Comment a; see, generally, 41 Am.Jur.2d, Independent Contractors, § 48). Just as the landowner is responsible because for his own benefit he has chosen to engage in an activity of sufficiently high risk of harm to others, so, too, those who intentionally undertake or join in that abnormally dangerous activity must bear the consequences resulting from harm to others (but see *McLoone Metal Graphics v. Robers Dredge*, 58 Wis.2d 704, 709–711, 207 N.W.2d 616, *supra*).

* * *

JASEN, GABRIELLI, JONES, WACHTLER, FUCHSBERG and COOKE, JJ., concur.

Order reversed, etc.

[Ed note: The Court held: 1. Each defendant may claim indemnity from the others even if liability is "strict." 2. Claimants can seek to prove negligence. 3. The Silvers' claim against the municipality is barred because notice was too late.]

(B) VICARIOUS LIABILITY

Sometimes one person is responsible for the negligence of another. The most common illustration is the employment relationship. Despite the absence of negligence by the employer, the employer is liable for harm caused by the negligent conduct of the employee so long as that conduct is within the scope of the employment. Sometimes the employer can be held liable for intentional torts committed by the employee. For example, a subcontractor was held liable for injuries sustained by two employees of the general contractor as a result of an assault committed upon them by employees of the subcontractor.[46] The ruling was made despite the assault having occurred after the assaulting employees had completed their work shift.

One reason sometimes given for vicarious liability or, as it is sometimes called, *respondeat superior*, is the incentive it gives to an employer to choose employees carefully. The real basis for vicarious liability is that enterprises should pay for losses they cause. Also, vicarious liability is premised upon the ability to pay or the deep pocket. The pocket of the employer is likely to be deeper than that of the employee.

The hirer of an independent contractor is not liable for the latter's negligence. Sometimes it is difficult to determine whether the person who has been engaged is an employee or independent contractor. For reasons of enterprise liability and deep-pocket notions, vicarious liability has been expanded and the independent contractor rule has been diminishing in scope.[47]

(C) EMPLOYMENT ACCIDENTS AND WORKERS' COMPENSATION

Although the principal thrust of this chapter is the role of tort law in distributing losses and responsibilities, one exception is the

---

46. *Rodgers v. Kemper Construction Co.*, 50 Cal.App.3d 608, 124 Cal.Rptr. 143 (1975).

47. See Sections 7.03(G), 35.05.

workers' compensation law, which was to supplant tort law in employment accidents. The employer's liability under workers' compensation is strict. Recovery against the employer does not require the latter's negligence. For these reasons it may be useful to briefly describe certain doctrines that deal with employment injuries.

The injured worker did not fare well under English or American common law. Expansion of the Industrial Revolution generated employment injuries but only limited legal protection to those injured. The employer's duties were limited: to furnish the worker a safe place to work and safe appliances, tools, and equipment and to warn the worker of any known danger connected with the work. The injured worker faced, in addition to limited employer duties, other obstacles to compensation. First, the worker had to prove that the employer was negligent. Many accidents occurred because of the nature of the work and not the negligence of the employer.

Even if negligence were established, the worker faced the "unholy trinity" defenses of contributory negligence,[48] assumption of risk,[49] and the fellow servant rule. The first barred a worker whose injury was substantially caused by the worker's negligence unless the employer's conduct had been willful or wanton. The second barred the worker from recovery if taking the job was an assumption of the risk of injuries normally incident to the employment. A worker who stayed on a job under protest after the worker knew or appreciated the danger assumed the risk. With some exceptions, a worker could not recover for injuries caused by negligence of a fellow servant.

These formidable barriers often meant hardship to injured workers and their families. Following German social insurance law, England and the United States began in the early twentieth century to enact what were then called workmen's compensation laws designed to replace tort law with social insurance for industrial accidents. Though some early statutes were held unconstitutional, currently all states have workers' compensation laws.

Although there are considerable variations among the states, certain common issues have arisen. Are employer and worker covered under the workers' compensation law? Many statutes exclude agricultural workers and domestics as well as employers who have only a few employees. The injury must arise out of the employment. Do injuries that occur on company picnics or while parking the worker's car in the company lot arise out of the employment? Doubts generally are resolved in favor of the employee.

Some states cover occupational diseases. Others do not. The worker need not show negligence by the employer. Nor are workers precluded from recovering by their own negligence, by their having assumed the risk, or by the injury having been caused by their fellow workers. Some states deny recovery if the worker was guilty of willful misconduct or intoxication and such misconduct caused the injury.

Most states require that the employer obtain compensation insurance. A few states have set up state funds to pay compensation awards. In some states an employer can be a self-insurer if it makes adequate proof of financial responsibility. All these requirements are designed to insure there will be a financially responsible entity.

The particular problems of the construction industry have been recognized. Many states enacted "subcontractor under" or "statutory employer" statutes under which a prime contractor is the employer of subcontractor employees under certain circum-

---

48. Refer to Section 7.03(G).

49. Ibid.

stances, such as a showing that the subcontractor did not procure the required insurance. This allows recovery against the prime contractor's compensation carrier.[50] Special rules for the construction industry recognize that many subcontractors are not financially sound and may not obtain the requisite insurance. Compensation coverage may be diluted if a prime contractor subcontracted out work that would normally be performed by the prime contractor. This may be done to reduce the number of employees for whom the prime contractor would have to obtain insurance coverage or be excluded as having too few employees.

Typically the award consists of a proportionate amount of the employee's wages as well as reimbursement for medical expenses incurred. Sometimes the employee is also able to receive a specific monetary award for designated injuries. More intangible, noneconomic losses—such as amounts to compensate for the emotional distress caused by disfigurement or for pain and suffering—are generally not recoverable. Workers' compensation awards provide *part* (estimated at from one-third to three-quarters of economic loss) and not *full* compensation. The recovery is intended to insure that the injured worker does not become a burden on others. Compensation recoveries are less and in many states much less than can be recovered in a tort action. This has led to demands for federal minimum standards for compensation awards.

Workers' compensation remedies generally supplant whatever tort remedy the worker may have had against the employer. This is accomplished by statutory immunization of the employer from actions by the workers. Immunity gave the employer something in exchange for giving up existing legal protection in employment accidents.

Elimination of all tort suits would create a total compensation system and keep disputes within the administrative agencies charged with the responsibility for handling such claims. This has not been accomplished. Most states permit injured workers or compensation carriers who have paid them to institute tort actions against *third parties* whose negligence caused the injury.[51] A factory worker may be able to recover in tort against the manufacturer of a defective product. A worker on a construction project may recover in tort against one of the many entities involved in the project other than his own employer. Because of the limited recovery available under workers' compensation, it is becoming increasingly common for injured construction workers to institute third-party actions against anyone they can connect with their injury except their employer (explored in Section 35.02).

Workers' compensation claims are handled by an administrative agency rather than a court. Hearings are informal and usually conducted by a hearing officer or examiner. The employee can represent himself or be represented by a layperson or lawyer. Fees for representation are usually regulated by law. Although awards by the agency can be appealed to a court, judicial review is extremely limited and very few are overturned. Third-party actions, on the other hand, since they involve tort claims, are brought to court and incredibly complicated lawsuits often result.

(D) PRODUCT LIABILITY

Another and perhaps more spectacular illustration of strict liability is imposed upon manufacturers for defective products (discussed in greater detail in Section 7.09).

---

50. See Section 35.02.

51. A few states do not permit third-party actions against those in a common employment (discussed in Section 35.02).

SECTION 7.05
CLAIMS BY THIRD PARTIES

(A) LOST CONSORTIUM

Suppose a spouse is seriously injured. In addition to economic loss, the injury can cause other losses of an intangible nature to the other spouse or a child of the injured person. Such losses are sometimes called *consortium,* a term that encompasses the services and society lost and, in the case of the spouse, sexual relations.

While the husband could recover for loss of consortium when his wife was injured, the law generally did not give corresponding rights to his wife. Slowly the courts have equalized consortium rights of husband and wife. However, a child cannot recover for lost consortium of either parent.[52]

(B) DEATH STATUTES

Before the mid-nineteenth century, the death of the wrongdoer precluded any claim being made against the estate. The wrong died with the wrongdoer. Similarly, death of the party harmed also barred any claim he would have had. However, statutes in England and the United States changed this. The death of the wrongdoer does not preclude a claim against his estate. Similarly, the death of the party harmed gives either his estate or his survivors or a combination of both a claim against the wrongdoer.

The claim is usually measured by any loss the deceased has suffered prior to death resulting from the defendant's negligence and, more important, any loss to the estate or survivors. Most statutes limit recovery to pecuniary losses, although some permit a limited amount of noneconomic losses to be recovered. Recovery for economic losses to the survivors is based upon potential earnings of the deceased during his working life that would have been available to the survivors. The amount recovered for the death of a person with high earnings or high earning capacity is often large. A few states limit the amount of recovery for wrongful death.

SECTION 7.06   IMMUNITY

(A) CHARITABLE ORGANIZATIONS

Initially American law granted immunity from tort liability to charitable organizations. Immunity was originally based upon the charitable and nonprofit characteristics of the organization. The availability of public liability insurance, the recognition that even charitable institutions take on some of the characteristics of commercial enterprises and the injured party's needs are factors that led to virtual abolition of this immunity.

(B) EMPLOYERS AND WORKERS' COMPENSATION

The worker's sole remedy against the employer is under workers' compensation. Employers have immunity, subject to a few exceptions, from tort claims made by their workers. That this has complicated construction accidents is shown in Chapter 35.

(C) PUBLIC OFFICIALS

Judges and high public officials are often granted absolute or qualified immunity from claims against them.

(D) SOVEREIGN IMMUNITY

For various reasons, some metaphysical and some practical, English law immunized the sovereign from being sued in royal courts. This doctrine was adopted early in the nineteenth century by the federal

---

52. A few states do allow recovery. See Note, *Child May Recover for Loss of Parent's Society and Companionship,* 68 Marq.L.Rev. 174 (1984).

courts, and it soon became established that the federal government could not be sued without its consent.

In 1946 the Federal Tort Claims Act was adopted. This legislation gave individuals the right to sue the United States for certain wrongs it committed. The Act had two important exceptions: (1) the federal government could not be sued for certain intentional torts, and (2) certain discretionary functions or duties performed by government officials could not give rise to tort liability. *Dalehite v. United States* clarified this by denying liability when the conduct was a policy or planning decision, also holding that the government could not be held liable unless it was shown to have been negligent.[53] *Feres v. United States* barred a claim by a service person against the United States.[54] This has been extended to a manufacturer who followed government specifications.[55] Strict liability cannot be the basis for any claim against the federal government. Negligence under applicable state law must be established.

States generally adopted the English rule of sovereign immunity, although many states have given consent to be sued for specific claims.[56]

Municipal corporations, such as counties and cities, receive immunity for governmental acts but not for those they have performed in a private or proprietary capacity. The cases employing this distinction are often confused and contradictory.

Immunity has frequently been criticized, and beginning with the mid-twentieth century perhaps slightly more than one-third have abolished it. States that abolished sovereign immunity frequently enacted comprehensive statutes modeled to some degree on the Federal Tort Claims Act. Sovereign immunity as it bears upon the Construction Process is discussed in Section 35.08(B).

Modern justifications are usually based upon either relieving already burdened public entities of serious financial responsibilities or precluding judicial intrusion into the running of government. Those who oppose immunity contend that it is better to spread the loss among all the taxpayers in the public entity than to concentrate it on the person who has suffered it. As to the fear of judicial intrusion, opponents of immunity contend that the current trend is toward increased accountability rather than relieving persons from their negligent acts.

Even when immunity has been eliminated, claims against governmental units require particular attention. Often such claims must be made within a shorter period of time than claims against private persons. In addition, such claims must be presented, as a rule, to legislative bodies of the governmental unit for their review before court action can be begun.

SECTION 7.07
MISREPRESENTATION

(A) SCOPE OF DISCUSSION

While misrepresentation problems can occur in the contract formation process, this section treats the liability of persons whose business it is to make representations. A surveyor makes representations as to boundaries, a soils engineer as to soil conditions, a design professional as to costs and the amount of payment due.

---

53. 346 U.S. 15, 73 S.Ct. 956, 97 L.Ed. 1427 (1953).

54. 340 U.S. 135, 71 S.Ct. 153, 95 L.Ed. 152 (1950).

55. *McKay v. Rockwell International Corp.*, 704 F.2d 444 (9th Cir.1983); *McLaughlin v. Sikorsky Aircraft*, 148 Cal. App.3d 203, 195 Cal.Rptr. 764 (1983). (See Section 7.09(J).)

56. For canvassing of state law, see Comment, *Municipal Tort Liability for Erroneous Issuance of Building Permits: A National Survey*, 58 Wash.L.Rev. 537 (1983).

(B) REPRESENTATION OR OPINION

Representations should be distinguished from opinions. For example, an architect may give his best considered judgment on what a particular project will cost. The prediction, however, may not be intended by him or understood by the client, to be a factual representation that will give the client a legal claim in the event the prediction turns out to be inaccurate. If the statement is merely an opinion and not a representation of fact, it is reasonably clear that the person making the representation will not be liable simply because he is wrong.

(C) CONDUCT CLASSIFIED

The person making the misrepresentation may have had a fraudulent intent. He may have made the representation knowing that it was false, with the intention of deceiving the person to whom the representation was made. The representation may not have been made with the intention to deceive, but may have been made negligently. Finally, the representation may have been made with due care, but turned out to be wrong. This is sometimes referred to as an innocent misrepresentation.

(D) PERSON SUFFERING THE LOSS

Another classification relates to the person who was harmed, the person to whom the representation was made, or a third party. For example, the soils engineer may make a representation of soil conditions to a client. If the representation is incorrect, the harm may be suffered by the client or, in some cases, by third parties such as a contractor or a subsequent purchaser or occupant.

(E) TYPE OF LOSS

Cases can also be classified by the harm that resulted from the misrepresentation. A misrepresentation of soil conditions might result in a cave-in that kills or injures workers. It might also cause damage to property or economic loss unrelated to personal harm or damage to the client's property. The client may have to pay for damage caused to an adjacent landowner's property. A subcontractor may incur additional costs during the excavation because of the misrepresentation.

(F) RELIANCE

In addition to the representation having to be material or serious, it must have been relied upon reasonably by the person suffering the loss. If there is no reliance, or if the reliance is not reasonable, there is no actionable misrepresentation. A principal application of this doctrine is discussed in Chapter 29.

Often the owner makes representations as to soil conditions to the contractor and then attempts to disclaim responsibility for the accuracy of the representation. The disclaimer is an attempt to transfer the risk of loss for any inaccurate representations to the contractor. It is intended to negate the element of reliance, a basic requirement of misrepresentation. Generally, but not invariably, as seen in Section 29.04, such disclaimers are successful in placing the risk of loss upon the contractor. However, they cannot relieve the person making the representation from liability for fraud, and they may not be effective if the representations were negligently made.

(G) SOME GENERALIZATIONS

Generally the more wrongful the conduct by the person making the representation, the greater the likelihood of recovery against the party. For example, a fraudulent misrepresentation will always create liability to the party to whom it was made or to third parties. A negligent one, in addition to providing the basis for a claim by the party who has paid for the representation, may also be the basis for a claim by third parties. An innocent misrepresentation, being least culpable, is the most difficult on which to base a claim. Third parties are rarely able to recover, and even the other party to the recovery will be able to recover damages only if there is a warranty of accuracy.

Liability to third parties often depends on the type of harm suffered. If personal harm such as death or injury results, it is not likely that the absence of a contractual relationship between the person suffering the harm and the person making the misrepresentation will constitute a defense. In a recent Louisiana case, a structure collapsed during construction resulting in a large number of personal injury claims. One party against whom a claim had been made was the architect who, in applying for a building permit, certified that he would inspect the work and verify that certain structural requirements were met. He did neither and was held liable for having made a negligent misrepresentation.[57]

Where harm is economic, lack of privity (contractural relationship) between the claimant and the person who made the misrepresentation causes the greatest difficulty. One reason is the wide range of individuals who are affected by the representations of persons in the business of making them. Such persons can be exposed to enormous liability, often disproportionately high to the remuneration paid for the services. For example, a certified public accountant may make an audit report that causes thousands of investors to buy shares in a particular company. Were the accountant accountable to *all* these investors, he would face enormous risk exposure.

In the construction context, the misrepresentation cases typically involve claims against surveyors and those who provide geotechnical information. Those who provide services often are liable to third parties for economic losses when the representations made were found to be negligent, provided the professional making the representation can reasonably foresee the type of harm that is likely to occur and the persons who may suffer losses. This trend is reflected in *Rozny v. Marnul*[58] in which a surveyor was held liable to a homeowner who had built a house and garage relying upon a survey the surveyor had prepared for a developer. The developer sold the lot to the homeowner and evidently the survey along with it. The house and garage encroached upon a neighbor's lot. After reviewing the legal history of such claims against professionals, the court stated that the factors to be considered are as follows:

1. The express, unrestricted, and wholly voluntary "absolute guarantee for accuracy" appearing on the face of the inaccurate plat.[59]

2. Defendant's knowledge that this plat would be used and relied on by others than the person ordering it, including plaintiffs.

3. The fact that potential liability in this case is restricted to a comparatively small group and that ordinarily only one member of that group will suffer loss.

4. The absence of proof that copies of the corrected plat were delivered to anyone.

5. The undesirability of requiring an innocent reliant party to carry the burden of a surveyor's professional mistakes.

6. That recovery here by a reliant user whose ultimate use was foreseeable will promote cautionary techniques among surveyors.

---

57. *Stewart v. Schmieder*, 376 So.2d 1046 (La.App.1979). See also 386 So.2d 1351 (La.1980), holding the city liable for negligently approving the permit application.

58. 43 Ill.2d 54, 250 N.E.2d 656 (1969).

59. A subsequent California case that applied the rule in the *Rozny* case held that the absence of any guarantee would not affect liability for negligent misrepresentation. *Kent v. Bartlett*, 49 Cal.App.3d 724, 122 Cal.Rptr. 615 (1975).

Bushnell v. Sillitoe[60] held *Rozny v. Marnul* to be "different" in rejecting a claim against a surveyor by a landowner whose land had been encroached upon because of a negligently prepared survey. *Rozny v. Marnul,* according to the court, involved a survey prepared for the use of the buyer of the lot while *Bushnell v. Sillitoe* involved a claim by a person for whose guidance the survey had not been prepared.

(H) DISPUTES BETWEEN CONTRACTING PARTIES

Misrepresentation claims are often made by one contracting party against the other. This creates confusion because of the unclear interrelationship between tort and contract doctrines (discussed in Section 17.11).

SECTION 7.08 DUTY OF THE POSSESSOR OF LAND

(A) RELEVANCE

Tort law determines the duty owed by the possessor of land, that is, the one with operative control, to those persons who pass by the land or enter upon it. Before, during, or after completion of a construction project, members of the public will pass by the land or, with or without permission, enter upon the land with the potential of being injured or killed by a condition on the land or by an activity engaged in by the person "in control" of the land. Workers also might suffer injury or death because of the condition of the land or activities on it. Persons who live in or enter a completed project might suffer injury or death because of something related to the land or the Construction Process. Liability for such harm depends upon the particular nature of the obligation owed to the plaintiff by the possessor of land near which or upon which the physical harm was suffered.

That can depend upon the injured party's permission to be near or on the land and the purpose for being there.

The term *possessor of land* is used in this section without exploring the troublesome question of whether the owner, the prime contractor, or the subcontractors fall into this category during the Construction Process (treated in Section 35.03).

(B) TO PASSERBY

The possessor must use reasonable care to protect those who pass by. Whether the possessor has measured up to the legal standard will depend upon factors discussed in Section 7.03(C). The passerby is entitled to expect that the condition of the land and activities on it will be conducted in such a way so as not to be exposed to unreasonable risk of harm.

(C) TO TRESPASSING ADULTS

The trespasser enters the land of another without permission. In so doing the trespasser is invading the owner's exclusive right to possess the land. Veneration for landowner rights led to a very limited protection for trespassers by English and American law. The possessor was not liable if trespassers were injured by the possessor's failure to keep the land reasonably safe or by the possessor's activities on the land.

Exceptions developed, however, as human rights took precedence over property rights. For example, possessors who know that trespassers use limited areas of their land must conduct their activities in such a way as to discover and protect them from unreasonable risk of harm. Railroads were required to be aware of persons who crossed the tracks at particular places. Another exception was applied frequently to railroads for dangerous activities conducted on the land.

---

60. 550 P.2d 1284 (Utah 1976).

Discovered trespassers are entitled to protection. The landowner must avoid exposing such trespassers to unreasonable risk of harm.

Despite their unfavored position, trespassers are not outlaws. Possessors cannot shoot them or inflict physical harm upon them to protect their property. What steps can possessors take to protect their land from trespassers? Suppose a contractor puts up a barbed wire fence? Suppose a contractor keeps savage dogs on a fenced-in site at night to protect the site and construction work from vandals?

A case that excited controversy was *Katko v. Briney.*[61] The defendant owned an uninhabited farmhouse containing some antiques and old jars. The house had been broken into and the contents removed several times. After requests to law enforcement authorities were unproductive, the defendant installed a spring gun aimed at the legs of anyone who entered the house and sought to enter a particular room. The plaintiff, thinking the house uninhabited and looking for old fruit jars, entered the house and was severely injured by the spring gun's discharge. Charged with a felony, the plaintiff pleaded guilty to a misdemeanor and received a sixty-day suspended jail term.

The plaintiff then sued the defendant for having caused the injury. A jury award against the landowner for compensatory and punitive damages was upheld because the privilege to protect property did not extend to the infliction of serious bodily harm. (For an interesting trespasser case see (G).)

(D) TO TRESPASSING CHILDREN

Special rules have developed for trespassing children. Frequently children do not realize that they are entering the land of another. Sometimes they are not aware of the dangerous characteristics of natural and artificial conditions on the land that they enter. Possessors must conduct their activities in such a way as to avoid unreasonable risk of harm to trespassing children. However, controversy frequently develops regarding the extent to which limited protection given trespassers as to artificial conditions on the land should be applied to trespassing children.

The law's strong protection for landowners and their rights has, for the most part, been qualified by humanitarian concerns for children of tender years injured or killed when they confront or deal with dangerous conditions on someone else's land. The Restatement (Second) of Torts reflects this diminution of landowner protection and states that the possessor of land is liable for injuries to trespassing children caused by artificial conditions upon the land if:

(a) the place where the condition exists is one upon which the possessor knows or has reason to know that children are likely to trespass, and

(b) the condition is one which the possessor knows or has reason to know and which he realizes or should realize will involve an unreasonable risk of death or serious bodily harm to such children, and

(c) the children because of their youth do not discover the condition or realize the risk involved in intermeddling with it or in coming within the area made dangerous by it, and

(d) the utility to the possessor of maintaining the condition and the burden of eliminating the danger are slight as compared with the risk of children involved, and

(e) the possessor fails to exercise reasonable care to eliminate the danger or otherwise protect the children.[62]

---

61. 183 N.W.2d 657 (Iowa 1971).

62. Section 339 (1965).

Children often trespass on construction sites. In the process of doing so, they may engage in an activity that can result in injury or death. This section has presented an overview, although it may be useful to look at this problem more specifically. In so doing, it is not important to focus sharply upon the legal theories, such as attractive nuisance and the playground doctrine, that are sometimes applied by courts to determine the nature of the duty owed trespassing children. What is important is to examine the following:

1. Recurrent fact patterns.

2. The clash of important policies, such as humanitarian protection for children and freedom of landowners that has divided many appellate courts.

As to the first, cases have involved the following:

1. A child who was injured when concrete blocks on which he was climbing collapsed.[63]

2. A child who fell when a scaffold collapsed.[64]

3. A child who was injured when another child threw a clod of dirt found at the construction site.[65]

4. A child who was injured by an exploding cartridge that he had found on a construction site and took home.[66]

5. A child who suffocated in a cave-in of an exposed excavation.[67]

6. A child who drowned while playing in an excavation that had become filled with rainwater.[68]

While generally courts have upheld trial courts that ruled for contractors, the appellate court decisions are often made by divided courts in which the judges have exchanged sharp views on the policies involved in making such decisions. Some judges show solicitude toward children and their propensity to play where they should not. One judge stated:

> It is the instinct of children of the age of appellee to play. Building material, stacked as this was, is peculiarly attractive to them. This is a fact known of everyone. In a populous community this instinct is more than likely to find vent in availing itself of such temptation. Warnings are not enough to make the premises reasonably safe. The material should be stacked so, with the knowledge that the premises will be probably so used in spite of warnings and precautions of the lot owner, that the children playing thereabout will not be subjected to the hazards of falling timbers and material insecurely put up.[69]

In concluding that a particular dispute should have gone to the jury another judge stated:

> Nor does it make a difference that no children actually were present at the very moment of [a visit by the contractor's representative] (which may have been during school hours). In a closely built-up residential neighborhood children are as much a part of the natural scene as grasshoppers. Their intrusive appearance upon

---

63. *Goben v. Sidney Winer Co.*, 342 S.W.2d 706 (Ky.1961) (jury question).

64. *Bloodworth v. Stuart*, 221 Tenn. 567, 428 S.W.2d 786 (1968) (jury question).

65. *Kirven v. Askins*, 253 S.C. 110, 169 S.E.2d 139 (1969) (affirmed trial court judgment for contractor despite jury verdict for child).

66. *Concrete Construction, Inc. of Lake Worth v. Petterson*, 216 So.2d 221 (Fla.1968) (contractor owed duty to child but injury too remote).

67. *Gagnier v. Curran Construction Co.*, 151 Mont. 468, 443 P.2d 894 (1968) (should not have been submitted to jury).

68. *Martinez v. C.R. Davis Contracting Co.*, 73 N.M. 474, 389 P.2d 597 (1964) (jury question).

69. *Louisville Ry. Co. v. Esselman*, 93 S.W. 50, 52 (Ky.1906).

and around the unenclosed premises of such an area is to be expected.[70]

The dissenting judge in the same case took a different approach, stating:

> The majority opinion takes the view that the builder of the structure should have employed certain security measures in order to protect children who might intrude and play on the structure as young Goben did. To require such a practice would make building costs, which are mounting skyward by the hour, well nigh prohibitive for the average person.[71]

The economic burden that can be placed on contractors was emphasized in another case in which the court, after pointing to the contractor's awareness that children played on the site, noted that there were no fences and no signs. The court also pointed to the absence of a security guard which would have cost $609 a week in a $75,000 job.[72]

Judges less sympathetic to trespassing children also emphasize the responsibility of even young children to know what is dangerous and of parents to keep their children away from construction sites. In many cases, part of the responsibility for the child's injury must fall upon parents who do not supervise the child properly. Suppose the child brings an action against the possessor and the latter asserts a cross claim against the parents. The cross claim can be based upon either the possessor having warned the parents that the child was playing without permission or the parents knowing without such a warning that the child was engaged in dangerous activities. Such actions would be barred in those states that give the parent immunity from any action for personal harm brought by the child. However, many jurisdictions allow intrafamily lawsuits. Courts that have faced this new problem have not been unanimous.[73]

A legal rule limiting the duty of the possessor of land makes the dispute one determined by the judge rather than by the jury. Those rulings that appear favorable to children usually conclude that the question of whether the possessor of land has performed in accordance with the legal standard of care should be determined by the jury. Courts that abolished the distinctions among trespassers, licensees, and invitees in favor of a general standard of care (discussed in (G)) decide that juries rather than judges will determine whether the possessor has met the legal standard.

(E) TO LICENSEES

A licensee has a privilege of entering or remaining upon the land of another because of the latter's consent. Licensees come for their own purposes rather than for the interest or purposes of the possessor of the land. Examples of licensees are persons who take shortcuts over property with permission, persons who come into a building to avoid inclement weather or to look for their children, door-to-door salespeople, and social guests.

Some anomalous exceptions exist such as the firefighter who enters a building at night to put out a fire, or the police officer who enters to apprehend a burglar. Logically they should be considered as benefiting the possessor of land, but many cases hold that they are simply licensees.

Early cases held that the only limitation on the possessor's activity was to refrain from intentionally or recklessly injuring a

---

70. *Goben v. Sidney Winer Co.,* supra note 63 at 711.

71. *Id.* at 713.

72. *Bloodworth v. Stuart,* supra note 64.

73. Compare *Cole v. Sears, Roebuck & Co.,* 47 Wis.2d 629, 177 N.W.2d 866 (1970), with *Holodook v. Spencer,* 36 N.Y.2d 35, 364 N.Y.S.2d 859, 324 N.E.2d 338 (1974) as qualified by *Nolenchek v. Gesvale,* 46 N.Y.2d 332, 413 N.Y.S.2d 340, 385 N.E.2d 1268 (1978).

licensee. However, most modern courts require that the possessor of land conduct activities in such a way as to avoid unreasonable risk of harm.

The possessor has a duty to repair known defects or dangerous conditions or to warn licensees of nonobvious dangerous conditions. Licensees cannot demand that the land be made reasonably safe for them. The possessor need not inspect the premises, discover dangers unknown to the possessor, or warn the licensee about conditions that are known or should have been known to the licensee.

(F) TO INVITEES

An invitee receives the greatest protection. The possessor must protect the invitee not only against dangers of which the possessor is aware but also against those that could have been discovered with reasonable care. While not an insurer of the safety of invitees, the possessor is under an affirmative duty to inspect and take reasonable care to see that the premises are safe. Sometimes the possessor can satisfy the obligation by warning the invitees of nonobvious dangers.

Who qualifies for such protection? Some cases have limited invitees to those persons who furnish an economic benefit to the possessor. More jurisdictions, however, seek to determine whether the facts imply an invitation to the entrant. However, the invitation concept does not include those invited as social guests.

The line between licensee and invitee not only is difficult to draw but sometimes seems arbitrary.[74] While most courts consider police officers and firefighters licensees, courts hold building inspectors to be invitees.[75] The tendency is to place increased responsibility on possessors engaged in industrial or commercial activities, based upon charging enterprises with the normal harm their activities cause.

(G) MOVEMENT TOWARD GENERAL STANDARD OF CARE

Undoubtedly the various categories that determine the standard of care are difficult to administer. Exceptions develop within the categories, and the application of the categories is often uneven. For this reason there is some movement toward a rule that would require the possessor of land to avoid unreasonable risk of harm to *all* who enter upon the land. The Supreme Court of California stated:

> Without attempting to labor all of the rules relating to the possessor's liability, it is apparent that the classifications of trespasser, licensee, and invitee, the immunities from liability predicated upon those classifications, and the exceptions to those immunities, often do not reflect the major factors which should determine whether immunity should be conferred upon the possessor of land. Some of those factors, including the closeness of the connection between the injury and the defendant's conduct, the moral blame attached to the defendant's conduct, the policy of preventing future harm, and the prevalence and availability of insurance, bear little, if any, relationship to the classifications of trespasser, licensee and invitee and the existing rules conferring immunity. . . .
>
> We decline to follow and perpetuate such rigid classifications. The proper test to be applied to the liability of the possessor of land in accordance with section 1714 of the Civil Code is whether in the management of his property he has acted as a reasonable man in view of the probability of injury to others, and, although the plaintiff's status as a trespasser, licensee, or invitee may in the light of the facts giving rise to such status have some bearing on the question of liability, the status is not determinative.[76]

The California decision has been influential

---

74. The status of contractor employees are discussed in Sections 35.03 and 35.04.

75. W. Prosser and P. Keeton, *Torts* 428–432 (5th ed. 1984).

76. *Rowland v. Christian*, 69 Cal.2d 108, 70 Cal.Rptr. 97, 103–4, 443 P.2d 561, 567–68 (1968).

in causing some jurisdictions to abolish the classifications.[77]

California's elimination of the threefold common law classification led to the enactment of California Civil Code § 846, designed to encourage those who own property suitable for recreational use to allow the public to enter by limiting the owner's tort liability. This statute was invoked by an owner who was building two homes near a scenic beach. Some persons on their way to a beach picnic decided to explore the houses under construction. The roofs of the two homes were connected by two loose boards. The plaintiff's sister panicked after climbing on one of the roofs. The plaintiff rescued her by guiding her across the boards to the other roof. The sister crossed safely but the plaintiff was injured when he fell to the ground.

The court held that the statute could not be invoked by the owner. Even though the homes were located near scenic beaches suitable for recreational use, the homes themselves were not suitable for recreational use. The case was to be decided based upon general negligence principles.[78] (Testimony at the trial suggested it was negligent to use loose boards to connect the roofs and not to fence the site.)

(H) NONDELEGABILITY OF LEGAL RESPONSIBILITY

Suppose the possessor hires someone to make the land reasonably safe and that person does not do so. Clearly the possessor who fails to use reasonable care to select a contractor will be liable. Similarly, the possessor who does not remove an incompetent contractor or does not inspect the work properly will be liable.

Suppose the possessor has lived up to the legal standard of care. Subject to many exceptions, the employer of an independent contractor is not liable for the latter's negligence. However, as seen in Section 35.05(C), one common exception involves using an independent contractor to comply with an important responsibility such as making the land reasonably safe. Such a responsibility is nondelegable. It is too important to be shifted to an independent contractor.[79]

SECTION 7.09 PRODUCT LIABILITY

(A) RELEVANCE

The historical development of legal rules relating to the liability of a manufacturer for harm caused by its products manifests a shift from protection of commercial ventures toward compensating victims and making enterprises bear the normal enterprise risks. Manufacturer's liability has become important in the Construction Process, since harm can be caused by defective equipment or materials.

(B) SOME HISTORY: FROM NEAR IMMUNITY TO STRICT LIABILITY

In 1842 the English case of *Winterbottom v. Wright*[80] held that an injured party could not recover from the maker of a defective product in the absence of a contractual relationship between them. Without contractual privity, the injured party could not recover even if he could show that his loss

---

77. *Cooper v. Goodwin*, 478 F.2d 653 (D.C.Cir.1973) (but not as to trespassers); *Mile High Fence Co. v. Radovich*, 175 Colo. 537, 489 P.2d 308 (1971); *Mariorenzi v. Joseph DiPonte, Inc.*, 333 A.2d 127 (R.I.1975). However, *Moore v. Denune & Pipic, Inc.*, 26 Ohio St.2d 125, 269 N.E.2d 599 (1971), refused to abolish the classifications.

78. *Potts v. Halsted Financial Corp.*, 142 Cal.App.3d 727, 191 Cal.Rptr. 160 (1983).

79. *Singleton v. Kubiak & Schmitt, Inc.*, 9 Wis.2d 472, 101 N.W.2d 619 (1960).

80. Supra note 31.

had been caused by the defendant's negligently made product.

The privity requirement protected an infant manufacturing industry developing out of the Industrial Revolution. It allowed contracting parties to know their liability exposure and deal with inordinate risks by contract. Privity permitted the manufacturer to be secure in the belief that it would not be held liable to persons other than those with whom it dealt. It could relieve itself from inordinate risks with those it dealt with by contract disclaimers. It could not do so with the many third parties who might be injured by its products or activities.

Protection was no more tolerable than the sad plight of the uncompensated industrial accident victim who ultimately received protection through workers' compensation. By the early twentieth century, the privity rule was no longer acceptable. Injured parties needed a solvent defendant from whom they could recover. Ultimate responsibility should be upon the manufacturer. The latter can insure against predictable losses and pass the cost to those who benefited from the enterprise, such as owners or users of the enterprise's activities.

The major turning point occurred in New York in 1916. Before 1916, New York had held that a negligent manufacturer could be liable despite the absence of privity if there were a latent defect in the goods sold or if the goods sold were inherently dangerous. In 1916 the New York Court of Appeals held in *MacPherson v. Buick Motor Co.*[81] that this exception included goods that were dangerous if made defectively. The *MacPherson* case, one involving an automobile, led to abolition of the privity rule in the United States.

Still the plaintiff had the difficult task of establishing that goods were negligently made by the manufacturer. But the *res ipsa loquitur* doctrine proved of great assistance. It permitted the plaintiff to submit the case to the jury even though the plaintiff introduced no direct evidence of negligent conduct by the manufacturer.

Res ipsa did not, as a rule, deprive the defendant of the opportunity of introducing evidence that it had not been negligent. This was typically done by seeking to establish that the defendant had followed common industry practices and had used a system designed to insure that its products were safe. Yet juries typically ruled for victims of adulterated food or beverages, or defective products. A more efficient way of placing this risk upon the manufacturer was needed. The concept that first accomplished this purpose was implied warranty. This doctrine, borrowed largely from commercial law, held sellers liable when their goods were not merchantable or, under certain circumstances, not fit for the purposes for which buyers bought them. It eliminated the need to establish negligence. Implied warranty would under certain circumstances extend protection to third parties.

Warranty first was used in food and drug cases. It then began to be employed when harm was caused by manufactured goods. This concept accomplished, at least for a time, the purpose of putting normal enterprise risks upon the enterprise.

Early in the use of the implied warranty, some courts recognized that though useful, it was essentially a commercial doctrine that was inappropriate in determining who should bear the risk of physical harm caused by defective products. In addition, warranty carried with it technical rules more appropriate to commercial transactions. As a result, a few courts began to treat claims by injured parties against manufacturers as involving strict liability in

---

81. 217 N.Y. 382, 111 N.E. 1050 (1916).

tort.[82]   Soon other courts fell into line, and the strict liability concept gradually supplanted implied warranty as a risk distribution device.

(C) SECTION 402A OF THE RESTATEMENT OF TORTS

One factor that led to the replacement of implied warranty by strict liability in tort was the decision by the American Law Institute, a private group of judges, scholars, and lawyers, to restate the law of torts by publishing Section 402A in 1965. This section states:

1. One who sells any product in a defective condition unreasonably dangerous to the user or consumer or to his property is subject to liability for physical harm thereby caused to the ultimate user or consumer, or to his property, if
   a. the seller is engaged in the business of selling such a product, and
   b. it is expected to and does reach the user or consumer without substantial change in the condition in which it is sold.
2. The rule stated in Subsection (1) applies although
   a. the seller has exercised all possible care in the preparation and sale of his product, and
   b. the user or consumer has not bought the product from or entered into any contractual relation with the seller.

Section 402A has had significant impact. Increasingly manufacturers of defective products are held liable without any privity between the injured party and the manufacturer and without the injured party having to establish that the manufacturer had been negligent.

Yet these standards, those of defective conditions unreasonably dangerous, mask a number of difficult questions. What is defective? Does the *unreasonably dangerous* requirement in effect put the burden of establishing negligence back upon the plaintiff? Can responsibility for marketing a high-risk product be satisfied by warning the user? What if products are dangerous no matter how much care is taken in their manufacture, such as cigarettes, whiskey, and drugs? What of those who sell blood for transfusions, a necessary activity but one that cannot eliminate the risk of infected blood?

(D) PRODUCT USE

Generally a manufacturer's responsibility extends only to reasonably foreseeable use of its products. For example, a manufacturer of casements to be used as window frames is not liable when workers use them as ladders.[83] However, if product misuse is reasonably foreseeable, a jury can find that the manufacturer had an obligation at least to warn the user or even to design the product with this in mind.[84]

The principal controversy over intended use is centered around so-called second-impact injuries received in automobile accidents. Two lines of authority developed regarding the manufacturer's responsibility to take collisions into account in automobile design.[85] However, the more recent cases indicate that the manufacturer of automobiles must take into account the possi-

---

82. The leading case is *Greenman v. Yuba Power Products, Inc.*, 59 Cal.2d 57, 27 Cal.Rptr. 697, 377 P.2d 897 (1963).

83. *McCready v. United Iron & Steel Co.*, 272 F.2d 700 (10th Cir.1959) (negligence action).

84. *Ford Motor Co. v. Matthews*, 291 So.2d 169 (Miss.1974).

85. *Evans v. General Motors Corp.*, 359 F.2d 822 (7th Cir.), *cert. denied*, 385 U.S. 836 (1966), held that the manufacturer did not have to take collision into account as a possible use. However, *Larsen v. General Motors Corp.*, 391 F.2d 495 (8th Cir.1968), held the contrary.

bility of collision when the automobile is designed.[86]

(E) PARTIES

Many entities play significant roles in the manufacturing and distribution of products. Manufacturers buy component parts and materials from other suppliers. They may obtain independent design and testing services. The product itself may be sold or installed by independent retailers or installers. Sometimes products are distributed through wholesalers who sell to retailers.

It is generally assumed that the manufacturer is best able to spread the loss and avoid harm. But some retailers have this capacity. For example, many products are distributed through large national retail chains that design or set performance standards for products in contracts with smaller manufacturers.

The purchaser of the product is not the only one who may be injured by a defective product. Members of the purchaser's family may use the product. Social guests may be injured if a television set explodes in the living room. A defectively designed car can injure drivers and other vehicles or pedestrians.

Section 402A of the Restatement of Torts took no position as to whether persons other than users or consumers can use strict liability to recover against the manufacturer. Nor did it take a position as to whether the seller of a component part would be liable. Generally, strict liability protection is being given to all those who the defendant could have reasonably anticipated would be injured by the product. Similar-

ly, the trend seems toward holding responsible all those who play significant roles in product manufacture.

(F) DEFENSES

Contributory negligence, where it still exists, is generally not a defense available to a manufacturer. But if the latter establishes that the injured party voluntarily assumed the risk, the injured party cannot recover. Where comparative negligence applies, it has been applied in strict liability claims.[87]

Sometimes product manufacturers expect or require that those to whom they sell will take steps to insure that the product will be used safely. For example, suppose user instructions make it clear that guards are to be used around dangerous machinery. Similarly, suppose a manufacturer of automobiles requires that the car dealer prepare the car in a designated way for the customer. Some courts have held that the manufacturer's duty to make a product that is reasonably safe cannot be delegated to others.[88] However, negligent conduct by the purchaser under some circumstances can be a superseding cause that may relieve the manufacturer. This superseding cause may consist of the purchaser making changes that would affect the way the machine was used.[89]

(G) ECONOMIC LOSSES

Is the manufacturer liable for economic losses such as delay damages or lost profits? Clearly a manufacturer is responsible if there is an express warranty. After some early uncertainty, the trend is to bar recovery for economic losses.[90]

---

86. One of the many cases following the *Larsen* case was *Turner v. General Motors Corp.*, 514 S.W.2d 497 (Tex. Civ.App.1974).

87. *Daly v. General Motors Corp.*, 20 Cal.3d 725, 144 Cal.Rptr. 380, 575 P.2d 1162 (1978).

88. *Vandermark v. Ford Motor Co.*, 61 Cal.2d 256, 37 Cal.Rptr. 896, 391 P.2d 168 (1964); *Bexiga v. Havir Manufacturing Corp.*, 60 N.J. 402, 290 A.2d 281 (1972).

89. *Schreffler v. Birdsboro Corp.*, 490 F.2d 1148 (3d Cir.1974).

90. *Jones & Laughlin Steel Corp. v. Johns-Manville Sales Corp.*, 626 F.2d 280 (3d Cir.1980).

(H) DISCLAIMERS

In the commercial world, sellers frequently seek to limit their risk by disclaiming responsibility for certain losses or by limiting the remedy. One reason for shifting from implied warranty to strict liability was the desire to avoid disclaimers frequently part of the commercial transaction where the party injured was not a real participant in the transaction.

Disclaimers are likely to be given effect between the parties if the parties are business entities of relatively equal bargaining strength.[91] However, disclaimers are not likely to be given effect in consumer purchases.[92]

(I) DESIGN DEFECTS

Early products liability cases involved manufacturing defects. To determine whether a product is defective, the product causing the injury is compared to either the design plans or other products made from the same design. If there is a deviation, the product is defective.

Even in the absence of negligence some products will have unintended manufacturing defects. It is not economically feasible to eliminate all risks of randomly defective products. It is better to predict the likelihood of defects, calculate the liability exposure, insure or self-insure against these risks, and include the cost in the product price. In addition to the other exasperating question of whether there was a manufacturer defect (was it a bad weld or cheap material that made welding difficult), judg-

ing the design has no neat test as does manufacturing.

Recent cases have struggled painfully with design defect definitions. California had earlier rejected the Restatement definition of "unreasonably dangerous" as reinstituting the discarded negligence test.[93] It then held that two tests must be applied. First, did the product meet the expectations of an ordinary consumer as to product safety? Second, even if it did, the manufacturer is liable if it cannot establish (using *negligencelike* criteria) that on balance, benefits of the design outweighed the risks. The court allowed a hindsight look at the design and reserved judgment on whether the state of the art being followed would be a defense. The court did not decide whether products are defective because they lack adequate warnings or directions.[94]

New Jersey expressly permitted hindsight to help the plaintiff prove a defective design. Failure to provide an appropriate warning will make the product defective. The state of the art is not a defense in defective products. The New Jersey Supreme Court held that the manufacturer must reduce risk to the greatest extent possible consistent with the product's utility.[95] Pennsylvania requires liability if a *judge* determines the product is not safe and leaves criteria uncertain.[96]

Even within these jurisdictions there will be confusion. It will be even harder for uncommitted courts to decide which standard to adopt. How can a designer determine whether a proposed design is defective? Even worse, how does an attorney advise his client after a claim has been

---

91. *Keystone Aeronautics Corp. v. R.J. Enstrom Corp.*, 499 F.2d 146 (3d Cir.1974).

92. *Henningsen v. Bloomfield Motors, Inc.*, 32 N.J. 358, 161 A.2d 69 (1960).

93. *Cronin v. J.B.E. Olson Corp.*, 8 Cal.3d 121, 104 Cal.Rptr. 433, 501 P.2d 1153 (1972).

94. *Barker v. Lull Engineering Co.*, 20 Cal.3d 413, 143 Cal.Rptr. 225, 573 P.2d 443 (1978).

95. *Beshada v. Johns-Mansville Products Corp.*, 90 N.J. 191, 447 A.2d 539 (1982).

96. *Azzarello v. Black Brothers Co.*, 480 Pa. 547, 391 A.2d 1020 (1978). For an evaluation of design defect cases, see Diamond, *Eliminating the "Defect" in Design Strict Products Liability Theory*, 34 Hastings L.J. 529 (1983).

made? It is no wonder attempts are being made legislatively to deal with this problem. (See (L).)

(J) GOVERNMENT–FURNISHED DESIGN

Suppose someone suffers physical harm because of a defectively designed product procured by the federal government, the design compelled by the United States in its contract with the manufacturer. The federal government is immune from tort liability except to the extent that it has deprived itself of that immunity in the Federal Tort Claims Act. That Act requires that the government be negligent, and negligence is *not* required to establish the liability of the manufacturer for a defective product. Inasmuch as the federal government is likely to be immune, those who have suffered personal harm or their survivors frequently sue the manufacturer of the product.

The manufacturer is likely to assert two defenses: (1) the *government specification* defense if it followed nonobviously defective government specifications and (2) the *government contract* defense, which gives the manufacturer the same immunity that would be given to the government.

These defenses were relatively uncontroversial until the explosion of the environmental movement in the 1970s. The realization that acts committed many years ago may create seriously harmful risks led to many claims based upon exposure to unsafe chemicals and hazardous wastes. The most

controversial have been those that have related to Agent Orange, a defoliant used by the U.S. Armed Forces in Vietnam manufactured in accordance with government specifications. In a recent trial court opinion a federal judge held that the manufacturer will be given a defense if it can show that it followed government specifications in the manufacture of the defoliant and that the United States knew as much as or more than the manufacturer of the hazards.[97]

The case brings out some of the complexities that involve claims against the United States. Not only is the government given a defense if it has not been shown to be negligent, but it is given a defense if actions are brought against it by members of the Armed Forces. This effectively removes the government as a defendant in claims brought by Vietnam War veterans or their survivors. The Agent Orange case, which gave the manufacturer a government contract defense, would absolve the manufacturer *even* if it were *negligent* if the manufacturer is given the exact immunity given to the government.[98] The law in this area will undoubtedly be cloudy until some of these issues are resolved.[99]

(K) BEYOND PRODUCTS: SELLERS OF SERVICES

Attempts to hold those who perform services, such as architects and engineers, strictly liable or liable based upon implied warranties have not been successful.[100] This creates anomalies and different stan-

---

97. *In re "Agent Orange" Product Liability Litigation,* 506 F.Supp. 762 (E.D.N.Y.1980), 534 F.Supp. 1046 (E.D.N.Y. 1982). After this and a number of other trial court decisions, the claims by veterans against manufacturers were settled. *New York Times,* May 8, 1984, p. B–4.

98. This immunity was granted in *McKay v. Rockwell Int'l Corp.,* 704 F.2d 444 (9th Cir.1983) cert. denied 104 S.Ct. 711 (1984). This involved a claim against the manufacturers by survivors of armed services pilots killed when their ejection mechanism malfunctioned. The manufacturer will be given immunity if it can show it followed specifications required by the United States and it warned the United States of known defects. Similarly, this defense was granted in *McLaughlin v. Sikorsky Aircraft,* supra note 55 (authorities collected).

99. See Note, *Liability of a Manufacturer for Products Defectively Designed by the Government,* 23 B.C.L.Rev. 1025 (1982).

100. *LaRossa v. Scientific Design Co.,* 402 F.2d 937 (3d Cir.1968). (See also Section 17.05.)

dards. Those who mass-produce homes [101] or lots [102] have been held strictly liable in some states. Those who manufacture products are held strictly liable. But suppose a claimant sues an independent designer of a defectively designed product? Or suppose the manufacturer sues the independent designer? In either case the designer would not be held to a standard of strict liability or implied warranty. These anomalies may lead some courts to hold those who are in the business of designing to enlarged, more strict liability.[103]

(L) FUTURE DEVELOPMENTS

Considerable dissatisfaction has been expressed with the evolution of products liability law. Manufacturers have complained that the options available to them are all unsatisfactory. One alternative is simply not to insure when the cost of premiums makes the price uncompetitive or is beyond the financial capacity of the manufacturer. Another is to overdesign a product that will pass even hindsight judicial or jury review evaluation. The product line can be dropped, which can mean diminished competition, fewer consumer choices, and higher prices.

Additionally, the cost of defense is staggering, including attorneys fees, costs of testing, and experts. As if this were not enough, courts are beginning to award punitive damages when a jury decides that design choices did not take safety or public needs into account.[104]

It is no wonder that many legislatures have enacted statutes that have limited common law liability. A recommendation has been made for a uniform federal statute dealing with manufacturer's liability for defective products. Congress currently is considering a number of bills that would reduce manufacturer liability, one of which may be enacted.

SECTION 7.10   REMEDIES

(A) COMPENSATION

The principal function of awarding tort damages is to compensate the plaintiff for the loss. In the ordinary injury case, the plaintiff is entitled to recover economic losses and certain noneconomic losses.[105] Economic losses are, for example, lost earnings and medical expenses. The principal noneconomic loss is pain and suffering.

Recovery for emotional distress has always been given hesitantly. However, where there is physical injury, there has been no difficulty in allowing recovery for pain and suffering. Often the plaintiff's attorney will seek to obtain a large award for pain and suffering by asking the jury to use a per diem or even per hour method to compute the pain and suffering award. Breaking down the period of pain and suffering into small units can generate a large award.

Many have suggested that pain and suffering not be recoverable or that limits be placed upon recovery for pain and suffering. This has been especially attractive to those seeking to minimize malpractice liability of doctors. In the 1970s many states reduced the liability of health care provid-

---

101. *Schipper v. Levitt & Sons, Inc.*, 44 N.J. 70, 207 A.2d 314 (1965); *Kriegler v. Eichler Homes, Inc.*, 269 Cal.App. 2d 224, 74 Cal.Rptr. 749 (1969). But see *Wright v. Creative Corp.*, 30 Colo.App. 575, 498 P.2d 1179 (1972). For a thorough discussion see Comment, *Strict Liability in the Building Industry*, 33 Emory L.J. 175 (1984).

102. *Avner v. Longridge Estates*, 272 Cal.App.2d 607, 77 Cal.Rptr. 633 (1969).

103. See Comment, *Architect Tort Liability in Preparation of Plans and Specifications*, 55 Calif.L.Rev. 1361 (1967); Comment, *Liabilities of California Building Contractors and Construction Professionals*, 15 Cal.W.L.Rev. 305 (1979).

104. *Grimshaw v. Ford Motor Co.*, 119 Cal.App.3d 757, 174 Cal.Rptr. 348 (1981).

105. Refer to Section 6.06(H) for this in the context of a *contract* breach.

ers. One method was to cap noneconomic losses.

Recovery of often open-ended pain and suffering damages has been justified by the large amount of the damage award that usually goes to pay the victim's attorney. Plaintiff advocates emphasize that pain and suffering *is* real and that placing an economic value upon it, though difficult, can give victims a sense that the legal system has taken adequate account of the harm they have suffered.

(B) COLLATERAL SOURCE RULE

Suppose an accident victim's hospitalization costs are paid by an employer, a health insurer, or the government. Suppose the victim recovers lost wages through disability insurance. While the many types of benefits create varied results, on the whole, most sources of compensation are considered collateral and are not taken into account when determining the victim's loss.[106]

Considerable criticism has been made of the collateral source rule, since it can overcompensate. However, in tort cases, three principal justifications have been made for the rule. First, the plaintiff is a wrongdoer who should not receive any credit for benefits provided by third parties. Second, accident victims frequently must use a large share of their award to pay their attorneys. Third, accident victims often receive benefits because they have planned for them or because they are part of payment for their services.

(C) PUNITIVE DAMAGES

Tortious conduct that is intentional and deliberate—close to bordering on the criminal—can be punished by awarding punitive damages. Such damages are not designed to compensate the victim but to punish and make an example of the wrongdoer to deter others from committing similar wrongs. In some areas of tort law, such as defamation, punitive damages play an important role because compensatory damages are often difficult to measure. There has been a tendency to award punitive damages for wrongful refusal to settle claims by insurance companies with their own insureds to insure that there is fair dealing in the claim settlement process.[107] A few courts have awarded punitive damages in claims against manufacturers of defective products where the manufacturer seemed unwilling to place a high value on human life.[108]

(D) ATTORNEY FEES: COST OF LITIGATION

Generally accident victims are not able to recover their attorneys' fees or other costs of litigation from the wrongdoer. This has led to expanded damages through pain and suffering and the collateral source rule as well as the contingency fee contract under which attorneys risk their time if they do not obtain a recovery.[109]

(E) INTEREST

Because tort damages are rarely liquidatable, it is difficult to receive interest from any period of time before award of judgment. However, as indicated earlier,[110] statutes vary considerably, and in some states the trial judge has discretion to award interest from the date legal action was commenced.

---

106. Refer to this rule in the contract claim context in Section 6.06(F).

107. *Gruenberg v. Aetna Insurance Co.*, 9 Cal.3d 566, 108 Cal.Rptr. 480, 510 P.2d 1032 (1973).

108. Refer to note 104, supra.

109. Refer to Section 2.04.

110. Refer to Section 6.08.

PROBLEMS

1. C was constructing a new school in a neighborhood where there was considerable vandalism. To protect materials on the site and to avoid liability for possible injuries, C put up a cyclone fence around the site and a series of locked gates. Inside the fence he kept a fierce watchdog. He placed a sign at various positions along the fence stating that trespassers should beware of the vicious dog.

   a. One night the employee charged with the responsibility of locking the gate did not do so. As a result the dog left the site and attacked a ten-year-old child who was walking by the site. Will the contractor be held liable? If so, what would be the recovery?

   b. Suppose a ten-year-old child climbed the fence, entered the site, and was mauled by the dog. Would the contractor be liable?

2. B purchased a six-step stepladder at a local hardware store. Several days later he decided to use the ladder to trim a hedge. He started up the ladder carrying a connected electric hedge trimmer. The fourth step collapsed, and B fell to the ground. His arm was lacerated by the hedge trimmer and in falling he struck his eye on a sharp thistle on the hedge. What would B have to show in order to recover from the hardware store? What if B weighed 350 pounds? What if the fall caused B to be electrocuted?

8

Restitution: Unjust Enrichment

SECTION 8.01 RELEVANCE

Traditionally, private law unrelated to property divided claims into tort and contract. Tort obligations arise from a failure to perform in accordance with the standard of conduct required by the civil law. Contract obligations are consensual, arising from failure to perform in accordance with a contractual obligation voluntarily undertaken. Restitution is increasingly recognized as a *third* basis for private law claims. The ethical basis behind restitution is *unjust enrichment*.

The three doctrines can overlap. For example, the principal use of restitution in this treatise relates to the making and performing of contracts. Torts and restitution can overlap. Tort claims principally involve demands that the claimant be compensated for harm caused by the defendant. However, some torts, such as conversion (exercising control over another's property), enrich the person who has committed the tort.

Restitution lies on the fringes of those legal rules that determine the rights of those who act under mistake or duress. While mistake may be connected to performance under a contract, it can occur outside the contractual context. Restitution can be "pure," unrelated to contracts, torts, or mistake and based simply on unjust enrichment.

This chapter explores basic restitutionary concepts. Where restitution plays a significant role in particular problems directly related to design or construction, it is discussed in that part of the treatise dealing with those problems.

SECTION 8.02 DEFINITION

Restitution stems from the verb "to restore." Restitution requires defendants to disgorge benefits that the plaintiffs have conferred upon them but only where retention of the benefit would be *unjust*. Not every benefit that one person confers on another requires that restitution be made. A can confer a benefit upon B by making a gift to B. B need not restore the subject of the gift. Similarly, one person cannot as a rule thrust benefits upon another and receive restitution. In neither case is retention of that benefit unjust. While unjust enrichment is a vague standard, a body of law dealing with specific illustrations has been and is still being developed.

Restitution can be accomplished by two different methods, similar to the specific decree and money award discussed earlier.[1]

---

1. Refer to Section 6.03.

Perfect restitution forces the defendant to give the very thing that the plaintiff has conferred upon the defendant and that the defendant is retaining unjustly. Suppose an employee of an architect wrongfully takes a set of drawings from the employer. The purest form of restitution would be a specific court order by the judge exercising the judge's *equitable* powers requiring the employee to return the drawings to the employer. This is accomplished either by a *specific restitution* order or by a remedial remedy called a *constructive trust,* a specific directive to the defendant to convey or transfer *specific* property to the plaintiff when retention by the defendant of that property would be unjust.

In most transactions relevant to this treatise, it is not possible to restore the *very* benefit that has been conferred. Usually the benefit consists of improving another's land. Except for the rare case where the benefit can be itself removed, restitution must be substitutional—a money award representing the benefit.

Often the loss to the plaintiff equals the benefit to the defendant. However, sometimes the loss suffered by the plaintiff is *less* than the gain made by the wrongdoer. For example, suppose the owner gives the architect $10,000 to pay certain bills. Instead of doing this, the architect takes the money and invests in shares of stock which at the time the claim is made are worth $20,000. Breach of trust is a serious wrong, here committed knowingly and consciously. In such a case, the defendant must repay $20,000.[2] This precludes the defendant making a *gain* from the wrongful use of money entrusted to the defendant. But this is done only when the defendant is a *conscious* wrongdoer.

Sometimes the loss to the plaintiff is *greater* than the gain to the defendant. For example, a contractor may spend $50,000 to improve the defendant's property, but the value of the property may be increased only $40,000. Usually, the contractor who has been wrongfully terminated can recover $50,000 and is not limited to $40,000.[3]

SECTION 8.03 "PURE" RESTITUTION UNCONNECTED TO TORT OR CONTRACT

Before looking at restitution as it interrelates with breach of contract and the commission of a tort, it is important to look at restitution unconnected to these two substantive doctrines.

Usually conferring a benefit that is not related to a contract or tort in the context of design or construction results from mistake. Suppose a roofing contractor is hired to replace a roof. By mistake, the roofer reroofs the wrong house. Can the roofer recover from the homeowner whose house she has erroneously reroofed?

Clearly the roofer cannot recover if she simply decided to reroof all the houses in the neighborhood and bill all the owners. Nor is it likely that she can recover if she saw a roof that needed repair and decided to do so without obtaining permission from the homeowner, even if it were difficult to contract the owner. In such cases, the law would classify the roofer as a volunteer or, more sharply, as an intermeddler. She had no business deciding to confer benefits on people who had not asked for them. Any enrichment would not be unjust. But a benefit conferred by mistake creates a better claim for recovery based upon restitution.

The strongly individualistic attributes of the common law generally precluded recovery even in the case of benefits conferred by mistake. There were some exceptions. For example, suppose A mistakenly builds a home on B's land. B seeks an injunction ordering A to get off the land or brings an

2. Restatement of Restitution § 211 (1937).

3. See Section 31.02 dealing with the effect of the contract price.

equitable action called "Quiet Title" to make it clear that A has no title to the land. Since these are requests for an equitable decree, the judge sitting as the equity judge can condition the decree upon B's paying A an amount that would reflect the benefit A conferred on B.

Here, though, the law has expanded restitution beyond such cases. A few courts will allow recovery based upon unjust enrichment. More important, statutes have been enacted in many states that give varying types of relief to the good faith improver of another's land. Some allow the improver to offset the value of the benefit conferred on the other party against any claim the other party makes against the improver, something similar to the equitable exception mentioned. Others allow the improver to take back any improvement made so long as the improver pays for any damage caused by removal of the improvement. A few states allow a direct recovery based on unjust enrichment, with the court instructed to take designated factors into account to determine whether recovery should be granted and the extent of the recovery. Local law must be consulted.

Mistake is not the only basis for a claim of pure restitution unconnected to torts or contracts. An unpaid subcontractor may seek to collect from the owner for improving the owner's property. One defense to such a claim is that the owner *has* paid the prime contractor and has not been unjustly enriched.

SECTION 8.04
ANCILLARY TO CONTRACT LAW

(A) QUASI–CONTRACT

A term sometimes used in connection with restitution based upon unjust enrichment is *quasi-contract*. This term, though used less frequently today, was developed by English common law because of the practical necessity of fitting a restitutionary claim into *either* the tort or contract category. For historical reasons these claims fell within the writ most associated with breach of contract. As a result, English courts and early American writers classified restitutionary claims as involving quasi-contracts.

Unfortunately, even modern law may require that a claim be classified as either tort or contract, often because there is no recognition of an independent restitution claim based upon unjust enrichment. Where the choice is tort or contract, restitutionary claims are classified according to the substantive base to which they are connected directly, such as commission of a tort or a breach of contract. Pure restitution claims are likely to be classified as quasi-contract and placed in the contract category.

(B) CONTRACT FORMATION PROBLEMS

One of the important uses of restitution relates to the conferring of benefits when for a variety of reasons a claim cannot be based upon a valid contract. First, there may be no one with whom a valid contract can be made. For example, suppose a doctor passing a construction site sees a passerby struck by a falling piece of lumber rendering the passerby unconscious. The doctor then administers medical services in an attempt to save the victim. Since the doctor is a person trained in these emergencies, she should be encouraged to act. Her training, the emergency, and the fact that she is not likely to be rendering services gratuitously takes her out of the volunteer category and allows her to recover.[4]

Such cases are rare. More commonly, restitution claims are made where parties intended to make a contract but did not do

---

4. *Cotnam v. Wisdom*, 83 Ark. 601, 104 S.W. 164 (1907). Admiralty law recently allowed a recovery to the owner of a ship that altered its course to help a stricken ship that lacked a medical person. *Peninsula & Oriental Steam Navigation Co. v. Overseas Oil Carriers, Inc.*, 553 F.2d 830 (2d Cir.1977). Reluctance to help is likely to be traceable

so. (Most elements of a valid contract were discussed in Chapter 5.) In the context of design or construction services there are two principal formation defects that require a person who has performed services to look to restitution. First, the contract fails because the terms lack sufficient certainty (usually in contracts for design services), the parties could not agree on terms while work went on,[5] or the parties did not get around to formalizing the contract by executing a formal writing where it was their intention to formalize it that way or there was a misunderstanding over the terms of the contract.[6] In all these instances the parties believed they had or would have a valid contract.[7] Where benefit is conferred under these circumstances, the party who has conferred the benefit will very likely be able to recover in restitution.[8]

The measure of recovery may vary, from the *value* of the expenditures to the *benefit* actually conferred, depending on the equities. It may even, in rare cases, be based upon expenditures that did not benefit the defendant.[9] In a close case as to whether a valid contract has been made, performance by both parties or even one party may tip the scales in favor of a conclusion that a valid contract did exist. Note that in these illustrations neither party had breached (unlike the circumstances to be discussed in the following subsection). As a result, re-

covery here is more likely to be limited to actual benefit conferred.

The second defect requiring restitution that is relevant to design and construction services (more common in construction) relates to contracts made illegally, principally where public contracts require that awards be made competitively. Because of the varying judicial attitude toward the importance of rigorously enforcing those rules and the various levels of illegality (compare a bribed official awarding a contract[10] with a technical irregularity by well-meaning officials[11]), court decisions may not appear consistent. If there is *any* trend, it is toward granting recovery in favor of those who have performed in good faith whose contract was invalid because of technical irregularities.[12] (This is discussed later in greater detail.[13])

(C) PERFORMANCE

Many, perhaps most, restitution claims that relate to design and construction are based upon a serious breach of a valid contract committed by one of the parties. This is a frequent issue in claims by contractors when they have performed under a construction contract but were wrongfully terminated or had legitimate grounds to stop work. The issue also arises when a project has been abandoned and a design professional sues for the value of her services.

---

to fear of a malpractice claim by the victim. Many states have enacted Good Samaritan statutes immunizing health care persons who help in emergencies.

5. *Comm v. Goodman*, 6 Ill.App.3d 847, 286 N.E.2d 758 (1972).

6. *Anderco, Inc. v. Buildex Design, Inc.*, 538 F.Supp. 1139 (D.D.C.1982) (misunderstanding of meaning of terms.)

7. *Dyer Construction Co. v. Ellas Construction Co.*, 153 Ind.App. 304, 287 N.E.2d 262 (1972).

8. Refer to notes 5, 6, and 7 supra, where all recovered.

9. *Minsky's Follies of Florida, Inc. v. Sennes*, 206 F.2d 1 (5th Cir.1953) (failure to have a written memorandum).

10. *Manning Engineering, Inc. v. Hudson County Park Commission*, 74 N.J. 113, 376 A.2d 1194 (1977) (no recovery).

11. *Layne Minnesota Co. v. Town of Stuntz*, 257 N.W.2d 295 (Minn.1977) (no recovery but based upon absence of benefit).

12. *Blum v. City of Hillsboro*, 49 Wis.2d 667, 183 N.W.2d 47 (1971).

13. See Section 22.04(J).

(These cases involve the use of restitution as an *alternative* remedy for breach of contract and are discussed later.[14]

Sometimes restitution is invoked by parties to a contract who, though themselves in default, have conferred a benefit upon the other party. Early American common law denied recovery to defaulting parties despite their having conferred a benefit. However, there is a strong tendency in modern decisions to award even a party who has breached any *net* benefit conferred on the other. In the context of design and construction, this expansion of restitution arises when a contractor who has breached asserts a claim. It is discussed later.[15]

Another illustration arises with some frequency in the marketing of interests of land and merits mention here.[16] In contracts for purchase of land, the buyer may give a down payment at the time of entering into the contract. Usually this amount is to be applied to the purchase price. Suppose the buyer, without legal justification, refuses to complete the transaction. The buyer may be able to show that the seller was able to sell the land for at least as much as the contract price or for more. The buyer may try to compel the seller to repay any down payments made that exceed the loss suffered by the seller. For many years, the law routinely denied such requests for restitution. Parties themselves in default cannot receive judicial assistance.

Increasingly today the buyer can recover the down payments if the seller has not been damaged by the buyer's breach.[17] The seller should not be able to keep the land—and sometimes even make a profit on re-

sale—and still keep the down payment. Other courts continue to permit the seller to retain the deposit—particularly if it is described as liquidated damages and forfeited, or as earnest money—as long as it is reasonable in amount.[18]

Mistake can be the basis for a restitutionary claim between *contracting parties*. An owner who has mistakenly overpaid the prime contractor can recover the overpayment based upon restitution.

(D) CANCELLATION

Restitution as an adjunct of contract law can relate to a decision by the contracting parties to cancel or rescind the contract. If there has been performance, it is likely that the agreement to cancel or rescind the contract will deal with payment for performance that has been received or any other issues raised by cancellation. If it does not, recovery for performance that has benefited either party must be based upon restitution. For example, in *Lindenborg v. M & L Builders and Brokers, Inc.,*[19] the buyer moved into a house in the middle of a dispute. The contractor ejected the buyer and sold the house to a third party. The court held that this was a mutual rescission and granted the buyer restitution—the recovery of his payments.

SECTION 8.05
RESTITUTION AND TORTS

Restitution as an ancillary remedy to tort law is of relatively little importance in design and construction. A few remarks at this point are in order, however.

---

14. See Section 31.02.

15. See Section 26.06(D).

16. As to these transactions, refer also to Section 5.12.

17. *Freedman v. The Rector,* 37 Cal.2d 16, 230 P.2d 629 (1951), a leading case, has been modified by statute. See West's Ann.Cal.Civ.Code § 1675 et seq.

18. *Vines v. Orchard Hills, Inc.,* 181 Conn. 501, 435 A.2d 1022 (1980).

19. 158 Ind.App. 311, 302 N.E.2d 816 (1973).

Construction contracts frequently contain provisions allowing the owner to take over material and equipment belonging to the contractor when the owner has grounds for termination.[20] If the owner takes this action *without* proper grounds for termination or if the provision were held to be unenforceable, the owner's taking of possession of property belonging to the contractor would be a tortious *conversion,* an improper exercise of dominion and control over another's property. The owner can recover the value of the property at the time of conversion or profits made from the wrongful use, and, where appropriate, punitive damages.

As discussed in Section 8.02, suppose an architect embezzles money she holds for her client or a prime contractor embezzles money that belongs to the subcontractor. In addition to the crime of theft, each wrongdoer has converted the money. If either invested the money wisely in stock, the client or the subcontractor by use of the constructive trust based upon unjust enrichment can recover the stock or its value even if it far exceeds the amount of money embezzled.

Restitution can come up in the context of bribery. Suppose a public official is bribed to award a contract to a particular bidder. Was this amount a benefit *diverted* from the public? The public has not been deprived of anything, except perhaps the damage to the integrity of the system. Rather than allow the corrupt employee to keep the bribe, the employee must return the bribe money to the public employer.[21]

The final illustration of restitution relates to theft or misappropriation of intellectual property, such as copyrighted drawings or the manufacture of products that may infringe a valid patent. Conduct of this type is regulated by federal copyright and patent law. Under copyright law, the holder of the copyright can recover profits made by the infringer.[22] A similar remedy is *not* granted for a patent infringement.[23] Where there has been a misappropriation of industrial or commercial property, the wrongdoer must disgorge any benefits that it has acquired as a result of the wrongdoing.[24]

---

20. See Section 38.03(G).

21. *County of Cook v. Barrett,* 36 Ill.App.3d 623, 344 N.E.2d 540 (1975).

22. 17 U.S.C.A. § 504.

23. *Zegers v. Zegers, Inc.,* 458 F.2d 726 (7th Cir.1972).

24. *University Computing Co. v. Lykes-Youngstown Corp.,* 504 F.2d 518 (5th Cir.1974).

Introduction to
Parts II, III, and IV

Part I laid the groundwork for understanding the legal aspects of architecture, engineering, and construction, collectively called the Construction Process. It has provided building blocks for subsequent specialized application of legal doctrines in that process.

The balance of this treatise focuses upon the principal actors and their functions. Part II looks principally at the owner and how it sets into motion the process by which land will be improved. Part III focuses upon the design professional and the functions served by those who prepare the design and as a rule play a role in the administration of the project. Part IV concentrates upon contractors, those entities who execute the design.

This approach recognizes as a basic organizational tool what has been called the traditional construction method. Under this method the owner acquires the right to improve land, commissions an architect or engineer to prepare the design, and awards the contract to execute the completed design to a contractor.

Even in the traditional process, the roles and functions are not always kept tightly within that structure. Before the owner can decide on a site or obtain financial commitments and public approval, it is likely to have engaged a design professional to prepare initial drawings. They usually consist of a site plan, typical floor plans, a set of elevations, and perhaps a color rendering of the finished project. The drawings are only illustrative, representing general concepts, and could not be used to construct a building.

Those who put up the money—the owner and lender—are likely to involve themselves in the design process. They must give the designer a program that outlines their needs, and they will ordinarily approve the design as it proceeds through its various phases.

During construction the design professional plays a monitoring and certifying function, the owner and lender may make changes, and the contractor may be asked or volunteers to involve itself in design.

Despite the blurring of roles, this organizational approach has great value. Much construction is still performed in the traditional manner. Even when new methods are used, the traditional organizational methods are a pattern against which these varieties of systems can be described and evaluated, both administratively and legally. In addition, many legal rules were developed with the traditional process in mind. Again, variations not only may reflect discontent with the traditional system, but also may be developed because the legal rules that applied to the traditional system were proving inadequate. (Applying the traditional legal doctrines to new systems also creates disharmony.)

Another justification, perhaps a negative one, is that the alternative would be to organize this treatise around legal doctrines. This would create an air of unreality and would hinder the development of an understanding of how law interacts with this process. The law is not a self-contained structure powered only by its inner logic and the desires of those who created it and sit at its controls. Law should enable society to accomplish desired objectives, get land improved efficiently and quickly, and deal with the disappointed expectations of contracting parties and the harm caused by the process to others.

To sum up, although the many alternative systems that have been developing cannot be ignored and will be discussed, Parts II, III, and IV are built upon the traditional American contracting system in which:

1. The owner obtains the site, the money, and the right to build.
2. The design professional plans the design.
3. The contractor executes the design.

Before proceeding to Parts II, III, and IV and their focus—the *main* actors—it is important to look at the broad contours of the Construction Process both to see the other "actors" in the process and to appreciate the reasons why this process is prone to disputes. While disputes rarely culminate in a full-fledged trial, they occur with some frequency. This perspective on the construction industry and its unique characteristics is central to the basic theme of this treatise—how law affects the process of construction.

This material is intended not to frighten but to develop a healthy awareness of the complexity of the construction process and the need for all participants to:

1. Create fair, realistic, and effective performance requirements.
2. Communicate these requirements to all concerned parties.
3. Devise and comply with communication systems that inform all participants of relevant events and intended courses of action.
4. Devise and follow methods for resolving disputes fairly and effectively.

THE MAIN ACTORS

Since the variety of owners far exceeds the variety of designers and contractors, Chapter 9 classifies owners in some detail. For now it is sufficient to note that the owner is the person or entity who furnishes the site, the design, and the money for the project. The owner may be a public entity, a private corporation, a business partnership, or an individual building a home.

Whether public entity or private party, the owner may be experienced or unsophisticated in the process of construction. The experienced owner is likely to have developed procedures and contract forms that deal with every aspect of construction and provide skilled contract administration. The unsophisticated owner is often unaware of the customs and practices of construction, the roles of the various participants, and the terminology used.

To all owners, public and private, the amount to be paid for the project and the predictability of the amount is crucial. Some owners have limited financial resources and do not have the financing adequate to cover the risks of unforeseen events or of problems that can increase the ultimate cost of the project.

The contracting industry is highly decentralized with a large number of small and medium-sized firms. Despite concentration in other industries, construction is largely local, with most contractors serving a single metropolitan area. Few construction companies are even regional, let alone national or international.

Half a million companies engage in construction. The average company is family-owned with an average of five to ten permanent employees. Most workers are hired for a particular job through unions or otherwise. Contractors obtain their work by competitive bidding. Profit margins are usually low and bankruptcies high. Because construction requires outside sources of funds, it is often at the mercy of the changing monetary and fiscal policies.

Two out of three contractors are specialty contractors, and one out of two workers is in a specialized trade, such as plumbing, electrical work, masonry, carpentry, plastering, and excavation. This means that in many construction projects, the contractor acts principally as a coordinator rather than as a builder. Its principal function is to select a group of specialty contractors who will do the job, schedule the work, police specialty trades for compliance with the schedule and quality requirements, and act as a conduit for the money flow. Additionally, in a fixed-price contract, the contractors provide security to the owner by giving a fixed price.

The volatility of the construction industry adds to the high probability of construction project disputes. Because the fixed-price or lump-sum contract is so common, a few bad bids can mean financial disaster. Contractors are often underfinanced. They may not have adequate financial capability or equipment when they enter into a project. They spread their money over a number of projects. They expect to construct a project with finances furnished by the owner through progress payments and with loans obtained from lending institutions.

Labor problems, especially jurisdictional disputes, are common. Many of the trade unions have restrictive labor practices that can control construction methods. Some contractors are union; others are nonunion; still others are double-breasted, having different entities for union and nonunion jobs.

Some contractors do not have the technological skill necessary for a successful construction project. Often the technological skill, if there is any, rests with a few key employees or officers. The skill is often spread thinly over a number of projects and can be effectively diminished by the departure of key employees or officers for better paying jobs.

The construction industry has attracted a few contractors of questionable integrity and honesty. These contractors will try to avoid their contractual obligations and conceal inefficient or defective performance. Such contractors are skillful at diverting funds intended for one project to a different project.

The third part of the construction trian-gle, design professionals, find themselves in the often uncomfortable position of working for the owner yet being expected to make impartial decisions during construction. They, like the contractors, have financial problems because there is usually not enough work to go around.

All of the main actors—owners, contractors, and design professionals—suffer because of a chronically sick building industry particularly affected by rapid movements of the economy, changes in public spending policies, and swings of monetary policy, all of which can affect the interest rates, a formidable factor in most construction.

THE SUPPORTING CAST ON A CROWDED STAGE

Along the owner chain, there is usually a lender who is advancing funds for the project (discussed in detail in Chapter 12). The seller may have retained a security interest in the land. The owner may be constructing a commercial structure in which space has been leased in advance to tenants. The owner's creditors may have an interest in the construction project. They may hope to collect their debts from profits made by the project or by having the land seized or sold to pay the owner's debts.

The contractor chain in the single contract system involves, in addition to the prime contractor, a large number of subcontractors and possibly sub-subcontractors. Each one of these contractors, as well as the prime contractor itself, purchases supplies and rents equipment. The material and equipment suppliers also have a substantial stake in the construction project.

The use of surety bonds for prime contractors, and often for subcontractors, brings a number of surety bond companies into the picture. The creditors of the contractors, other than suppliers, are often involved. There may be taxing authorities to whom contractors owe taxes, as well as persons who have lent money to the contractors. The contract chain would not be complete without reference to the trade unions, which have a substantial stake in the construction project.

In addition to the owner and contractor chains, there is a somewhat shorter chain beginning with the design professional, who may hire consultants. Some consultants, such as soil engineers, may have been hired by the owner and be in the owner's chain. An employee of the design professional might be on the site daily in larger projects.

Consultants also have a stake in the construction project. They wish to be paid, and their negligence may cause damage to any number of persons who are also affected by or involved in the Construction Process.

Construction work involves a high probability of harm to persons and to property. Workers and members of the public might be injured. Adjacent landowners and other owners of land in the neighborhood can be affected. Insurance companies insure against injuries and damage to property. The general public will be affected if persons injured do not have the financial resources to take care of themselves and are not able to collect from other parties, such as contractors or their insurance companies. Injured parties who cannot pay their bills and take care of their families are likely to end up as charges upon the public.

The sheer size of the supporting cast generates disputes by exacerbating organizational and accountability complexities.

THE CONSTRUCTION CONTRACT

Building a construction project is a complicated undertaking. It is hoped that the construction documents, particularly plans and specifications, will be clear and complete. At their best, they should give a good indication of the contractor's duties. Unfortunately, even the best design professionals cannot do a perfect job of drafting the construction documents that encompass the complete construction obligation.

This requires contract interpretations, often in American construction, by the person who designed the project and who was selected and is paid by one party. The inevitable interpretation issues can induce corner-cutting contractors to bid low and submit a large bill for extras. The variety of contract documents creates additional dispute possibilities, inconsistencies, and ambiguities.

The competitive bidding process so often used to select a contractor and the frequent use of the fixed-price contract play a significant role in dispute generation. The former emphasizes price rather than quality, and the latter places tensions on the relationship by placing the risk of many unknown and abruptly shifting factors on the contractor.

THE APPLICABLE LAW

The law that regulates the rights of parties along any of the chains should be the contract between those on the chain. For example, the basic law between contractor and owner should be the prime contract. Similarly, the contracts between design professional and owner and prime contractor and subcontractor should be found, explicitly or implicitly, in their contracts.

There is an increasing use of standard, preprepared contracts. It is impossible to anticipate all the problems and deal with them properly in each individual construction contract. Form contracts rely heavily upon the experiences of the past and upon the expertise of persons with wide experience in construction projects.

Good form contracts are planned carefully. However, their existence does not solve the contract problems for lawyers or for design professionals. First, some form contracts acquire the reputation for being heavily slanted in favor of one of the parties. Lawyers, when asked to pass upon these contracts, may reject them completely or make substantial modifications to them. Second, a form contract is often unread or misunderstood by the other party if it is not represented by a lawyer. A possibility exists that the contract will not be *all* the law regulating the relationship between the parties.

A number of other laws regulate the construction project. Building codes and industry standards are often incorporated, expressly or impliedly, in the construction documents. Building codes lack uniformity and consist of complicated and often cumbersome rules that regulate the Construction Process. There are zoning laws and subdivision laws. There are tort doctrines, such as those relating to nuisance and soil support, that affect the use of land. Title and security problems are often difficult. Because of history and the archaic language of surety bonds, interpretation of a surety's obligation is difficult. The rights of injured persons or injured property owners are governed by tort law and the bewildering process of indemnification. If more were needed, ultimate responsibility for some losses seems to require a "slug fest" between insurance carriers, all armed with unreadable policies with hordes of special endorsements. Is it any wonder that disputes are common and litigation time consuming and costly?

THE CONSTRUCTION SITE

The physical site itself contributes to the likelihood of disputes and Construction Process difficulties. No two pieces of land are exactly alike, generating a high probability of subsurface surprises. Testing methods for soil conditions are expensive and often do not give an accurate picture of the entire site. In addition, the physical limitations of the site, together with the large number of persons and contracting parties who must perform within this limited physical area, increase the probability of difficulties.

While the site is usually a restricted physical area, work on the site itself can substantially affect adjacent landowners. Excavation always involves a risk of subsidence on adjacent land. Soil conditions may cause slides and subsidence in the excavation process. Transferring the materials to the site can damage adjacent owners and members of the public.

CONTRACT ADMINISTRATION

Even if the general terms and conditions of the contract documents are well expressed, difficulties often develop because contracting parties often are sloppy in contract administration. Decisions are made on the site, modifications are agreed to, changes are ordered—all without the formal requirements frequently expressed in the general terms and conditions of the written contract documents. Telephone conversations are often used to resolve difficulties and continue the work, but there may be a dispute at a later date as to what was said during the conversations.

In the process of the dispute, one party will often point to the contract clauses requiring that certain directions be given in writing or that certain modifications be expressed in writing. The other party will then state that throughout the entire course of administration these formal requirements were disregarded. These are the seeds from which disputes and lawsuits develop.

UNRESOLVED DISPUTES AND LITIGATION

Disputes between parties may be resolved without litigation when there is a desire to maintain goodwill in the parties' future dealings. This element—the necessity of future relations—may be missing in many construction projects. A dispute may involve a number of parties, such as insurance companies and sureties, that must consent to any settlement. The uncertainty of both the law and the facts and variety of legal issues discourages settlement.

Unless the parties value the goodwill of the owner, not common in much construction, the need to compromise is often absent, another factor discouraging settlement and leading to the courtroom. If there is an arbitration provision, one or both of the parties may not trust the arbitration process. Even if the arbitrator makes an award, the party against whom the award is made may not perform. Such awards must be confirmed by a court.

In summary, a dispute-prone process such as construction will have the propensity to call upon the legal system to enforce contracts or obtain compensation for losses. Participants in the process must be aware of this. They must do all they can to avoid disputes, to seek to settle those that do develop, and to be aware of the role law plays in the process.

Part Two

Initiating the Project:
Focus on Owners

9

Owners Classified: an Infinite Variety

SECTION 9.01 RELEVANCE

This treatise makes many references to the owner. Yet the realities of the Construction Process and a perceptive understanding of how the law interacts with it requires an appreciation that an endless variety of owners exists.

SECTION 9.02 PUBLIC OR PRIVATE

Perhaps the most important differentiation is between owners that are public entities and those that are private. A private owner can select its design professional (by competition, competitive bid, or negotiation), its contractor (by competitive bid or by negotiation), and its contracting system (single contract or multiple prime, separate contracts) in *any manner* it chooses. Public agencies, on the other hand, are limited by statute or regulation. As a rule they must hire their designers principally on the basis of design skill and design reputation rather than on the basis of fee. Construction services generally must be awarded to the lowest responsible bidder through competitive bidding. Often a public entity must use separate or multiple prime contracts because of successful efforts in the legislatures by specialty trade contractors.

Other important differentiations exist between public and private owners. Public contracts have traditionally been used to accomplish goals that go beyond simply getting the best project built at the best price in the optimal period of time. Contracts to build public projects have often been influenced by the desire to improve the status of disadvantaged citizens, to remedy past discrimination, to give preferences to small businesses, to place a floor on labor wage rates, and to improve economic conditions in depressed geographical areas. Considerations exist that are not likely to bear a significant role in the award of private contracts.

A public entity is more likely than are its private counterparts to be required to deal fairly with those from whom they procure design and construction services. Yet public entities, having the responsibility for public monies, usually impose tight controls on how that money is to be spent. As a result, such transactions have often generated intense monitoring by public officials and by the press in order to avoid the possibility that public contracts will be awarded for corrupt motives or favoritism. In addition, public projects are more controversial. To whom the project is awarded, the nature of the project, and the project's location often excite fierce public debate and occasional treks to the courthouse. Public owners are not expected to allow those from whom they procure goods and services to make profits on unperformed work or excessive profits.

Public owners often seek to control dispute resolution, a control that manifests itself in different ways at different times. For example, early in this century when commercial arbitration began to be encouraged, many public entities contended that they had no power to arbitrate and that if they did, the agreements to do so were illegal. With an increased emphasis on the desirability of arbitration, some public owners now *must* arbitrate disputes. In addition, many experienced public owners, such as federal contracting agencies and similar agencies in large states, have developed a specialized dispute resolution mechanism often using specialized regulatory, arbitral, or judicial forums.

While other illustrations can be given, a final one concludes this section. Those who improve the property of a *private* owner are given the benefit of a mechanics' lien against the property if they are not paid for their services. Unpaid designers, if work on the site has begun, unpaid prime contractors, and, most important, unpaid subcontractors and suppliers, can assert a lien against the property they have improved, have the lien foreclosed, and be paid out of the proceeds of the judicial sale.

Those who improve *public* property have no corresponding remedies against property they improve. To remedy this and to encourage work on public projects, alternatives have been developed. Unpaid subcontractors and some unpaid suppliers can recover on bonds that prime contractors must purchase. In some states these unpaid participants can file a "stop notice," which requires that payments from funding sources be halted until unpaid subcontractors and suppliers have been paid. (Some statutes give a similar remedy to those who are engaged in private projects.)

Financing methods, another crucial difference, are discussed in Chapter 12.

SECTION 9.03 EXPERIENCED OR INEXPERIENCED

An experienced owner engages in construction if not on a routine basis at least on a repeated one. This owner is familiar with the common legal problems that arise, with construction legal and technical terminology and with standard construction documents. It may also have a skilled internal infrastructure of attorneys, engineers, risk managers, and accountants.

An inexperienced owner, though often experienced in its business, is likely to find the construction world strange and often bewildering. Such an owner lacks the internal infrastructure and may, even if it has resources to hire such skill, not even know whether it should do so and how it can be done. The prototype, of course, is an owner building a residence for his own use.

Many other owners are "inexperienced." Although as a general rule public owners are more experienced than private ones, care must be taken to differentiate, for instance, between the U.S. Corps of Engineers and a small local school district. The former has experienced contracting officers, contract administrators, and legal counsel and operates through comprehensive agency regulations, standard contracts, and an internal dispute resolution mechanism. The small local school district, on the other hand, may be governed by a school board composed of volunteer citizens, be run by a modest administrative staff, and have a part-time legal counsel who may be unfamiliar with construction or the complicated legislation that regulates the school district.

Similar comparisons can be made between private owners. Compare General Motors building a new plant with a group of doctors building a medical clinic or a limited partnership composed of professional persons seeking tax shelters through building or renting out commercial space. The clinic or limited partnership at least can buy the skill needed to pilot through the shoals of the Construction Process. But private individuals building residences they intend to occupy are often bereft of technical assistance.

This differentiation was recognized in a recent case that involved a claim by an owner against its design professional.[1] The owner paid for modules being manufactured by the contractor based upon certifications made by the architect. The contractor went bankrupt, and the trustee in bankruptcy seized modules for which the owner had paid.[2] It asserted the architect should have warned the owner of the risk of this seizure. In ruling for the architect, the court stated:

> [T]here are no special facts in this case that might lead a court to believe that it would be appropriate to extend the legal liability of professional architects. This case did not involve an unsophisticated or helpless consumer, nor was the omitted information ever solicited. The Authority is a government agency and was represented throughout the relevant negotiations by a Deputy Attorney General of New Jersey, who had extensive experience in construction. . . . Neither the government attorney nor the Authority ever indicated that they expected Ewing, Cole to point out potential legal pitfalls in the proposed arrangement with the contractor, and it is far from clear why an architect would believe that it had a duty to inform the legal counsel of its client about such matters. It seems an inefficient allocation of professional responsibilities to hold architects liable for not alerting lawyers to the legal ramifications of the bankruptcy of a contractor. We can see no reason under the facts of this case to predict that the New Jersey courts would impose such a duty.

Looking next to awarding construction contracts, it is likely that inexperienced private owners will prefer competitive bidding not because they must do so, but because they will not know the construction market well enough to sit across the negotiating table from a contractor. On the other hand, an experienced owner may be able to review the contractor's proposal and be aware of the market and other factors necessary to negotiate. (Ironically, often the public agency, which is in the best position to negotiate, *must* use the competitive process, a reflection of the distrust in public officials.)

An inexperienced owner will need to engage an architect or engineer to design and administer the construction contract much more than will an experienced owner. The latter may have sufficient skill within its own internal organization and not need an outside adviser experienced in construction.

An inexperienced owner will more likely use standard construction contracts such as those published by the American Institute of Architects. It will do so because either it does not wish to spend the money for an individualized contract, it does not have an attorney who can draft such a contract, or it wishes to acquiesce in the suggestions of its architect.

The phenomenon of the inexperienced owner dealing with complex standard contract terms can generate many legal problems. The "form" may not "fit," leading to interpretation problems. In addition, if the persons operating under the contract are not familiar with the contract or do not understand it, provisions are likely to be disregarded, leading to claims that those provisions have been waived.

Performance may also be affected in other ways by this differentiation. Inexperienced owners may make many design changes that mount the cost of construction and increase the likelihood for disputes over additional charges. They may also refuse legitimate contractor requests for additional compensation because they are unaware of those provisions of a contract that may provide the basis for additional compensation. Inexperienced owners may not keep the careful records so crucial when disputes arise.

---

1. *Travelers Indemnity Co. v. Ewing, Cole, Erdman & Eubank,* 711 F.2d 14 (3d Cir.1983).

2. This topic is discussed in Sections 26.02 and 26.03.

Differentiations exist relating to tort law. An inexperienced owner is more likely to be given the defense that it has simply hired an independent contractor and is not responsible for the latter's negligence. This defense would be difficult for an experienced owner to sustain.

Where the language must be interpreted, doubts will very likely be resolved in favor of the inexperienced owner. An experienced owner could have made the contract language clear. A claim made by a contractor that it be excused from default or be given additional compensation because of the occurrence of unforeseen events during performance is *less* likely to be successful against an inexperienced owner.

SECTION 9.04 PROJECT TYPE

Owners can be classified by the two types of projects they seek to accomplish: architectural and engineering. Without discussing the frequently debated differentiation between the two, it is generally assumed that engineering projects will have experienced owners. Such owners are likely to possess an infrastructure that exerts considerable control over the design and construction.

Licensing is another factor that must be taken into account. Licensing laws in many states exempt projects from licensing requirements either as to design or to construction where the public interest is less significant. Illustrations are contracts to build individual private homes or contracts to remodel existing homes.

An owner may seek to build a residence from standardized plans purchased or simply copied from a magazine. When this is done greater difficulties are likely to arise during construction because the parties have not agreed on details and may have to work things out as the project proceeds.

As to project complexity, compare a highrise office building with an individual home or a nuclear plant with road construction. In the complex project, there are more likely to be specialty trades, multiple

primes, technological difficulties, and public regulations.

Even if providing shelter is looked upon as one type of project, compare owners who build large tracts or vast apartment complexes with those who build duplexes, single-family residences, or home improvements. The experience level will be different, customs will be different, and the intervention of the law may vary depending on project type. Consumer protection laws are likely to affect improvement of an existing residence or construction of a single-family home; such regulation will rarely be applicable to a syndicate building a highrise metropolitan office building.

SECTION 9.05 SITE

The owner's choice of the location of the project is important. Designing for a locality where craft unions are strong will be different from localities where nonunion labor is used. If the site is in a city with environmentally sensitive citizens, the project will have characteristics different from those designed for localities where this scrutiny is less intense. The site may affect the likelihood that bribes must be paid to receive contracts or permits. The site may also affect weather conditions, which can influence design and performance. The site may determine the availability of an efficient and honest judicial system to resolve disputes.

SECTION 9.06
FOREIGN OR DOMESTIC

Crucial differences exist between foreign and domestic owners. Contracts made with foreign owners are most likely to involve the sovereign or one of its agencies. Under such circumstances, the persons contracting to provide design or construction services must be aware of the sovereign's power to regulate foreign exchange rates, import or export of goods or money, local labor conditions, the necessity for bribes, the lack of an independent judiciary, and the risk of expropriation.

Even contracting with a private owner in a foreign country involves risks that are not significant when contracting with domestic owners. Problems may still exist of unfamiliar laws, different subcontracting practices, different laws and customs regulating the labor market, and different legal solutions.

SECTION 9.07 SUMMARY

Classifying owners is important to anyone seeking to understand the Construction Process and laws relating to it. Rather than focus simply on "an owner," it is crucial to determine whether the owner is a public or private entity, whether the owner is experienced or inexperienced, the type of project the owner seeks to accomplish, and where the project is to be constructed.

10

Ownership: Legal Power To Improve Property

SECTION 10.01 RELEVANCE TO CONSTRUCTION

Owners will not, as a rule, improve land until they are confident they "own" the land or have exclusive right to possess it. They would not want to improve someone else's land. This would expose them to liability and the risk of the land being taken away by the true owner along with the improvements.

The *design* process can begin, perhaps in a sketchy way, before acquisition of ownership. Design may be planned while the client, a prospective owner, is negotiating a purchase, deciding to exercise an option, or seeking financial commitments.

Sections 10.02 through 10.08 look at forms of ownership. Sections 10.09 through 10.13 examine methods of acquiring ownership. Acquisition by purchase was treated in Section 5.12.

SECTION 10.02 ELEMENTS OF LAND OWNERSHIP

American property law has emphasized interests in land rather than land ownership. (Yet "ownership" is used in this chapter for simplification.) Ownership of land can be divided into its constituent elements. Usually the owners of land have a right to sell the land or to give it away while they are alive or effective upon their death.

Owners have sole and exclusive rights to use their land. Those rights can be divided. The right to sell may be divided among co-owners or owners of successive interests in land. The right to use the land can be divided in a number of ways. The owner of land may permit a tenant to use all or a portion of land. The owner may sell subsurface or air rights to the land.

While owners have exclusive possession of the land (a possessory interest), they may grant an easement (a nonpossessory interest). The easement holder can use all or part of the owner's land for specific purposes. Owners may give neighbors the right to cross their land. The owner of a building may grant an easement to a sign company to use part of the roof to erect an advertising sign. The advertising company has a nonpossessory interest in the land upon which its sign is placed.

The owner of land may wish to borrow money and give a mortgage upon the land to the lender as security, giving the lender a security interest in the land.

This chapter examines some of the more important interests in land most relevant to those involved in the Construction Process.

SECTION 10.03
TOTAL OWNERSHIP:
THE FEE SIMPLE

Owners of a fee simple have the right to exclusive possession of their land. They can exclude all others from entry on the land. They determine how the land is to be used. They may sell the land, give it away during their life, or determine how it is to pass upon their death. These rights are *not* absolute. For example, the right to absolute possession is subject to the right of state officials to enter upon the land for certain purposes if the required legal steps are taken. The right to sell the land may be subject to state-imposed formal requirements and may be taxed by the state. The right to sell to whomever the owner chooses may be affected by laws barring nondiscrimination in the sale or rental of apartments or houses. The right to give away or to leave the land by will is often controlled by state laws designed to protect the owner's creditors and family and may be subject to federal and state estate, inheritance, and gift tax laws. Land ownership can be taken away by the state through its right to exercise eminent domain or by the state through its function of protecting persons who have a security interest in the land such as lenders or mechanics' lien claimants.

SECTION 10.04
SUCCESSIVE OWNERSHIP:
LIFE ESTATES, REMAINDERS,
AND REVERSIONS

Absolute ownership, such as the fee simple, can be carved into pieces that relate to the passage of time or to the occurrences of events. This "carving" often results in *present* and *future* interests. The most common illustration is the life estate. Holders of a life estate are entitled to the possession and use of the property during their lifetime (if the life estate is measured by their life) or for a period of time measured by the life of

someone else. Often owners of the fee simple determine who will get the land when the life tenancy ends. If the land will return to them, the owners have a reversion interest. If they decide the land will go to a third party, that party has a remainder interest. For example, A may leave land to B for B's life. In this case, A has a reversion. If A leaves land to B for the life of B and, upon B's death, to C, C has a remainder interest.

When ownership is divided into successive periods of time, the law must balance the rights of the present and future owners. Rules have evolved for how the land can be used and the apportionment of tax assessments between life tenants and persons with remainder interests.

SECTION 10.05
MULTIPLE OWNERSHIP

(A) JOINT TENANCY

The joint tenancy's most important attribute is the right of survivorship. If A and B own real property jointly, upon the death of either joint tenant the other joint tenant takes the entire interest in the property. The estate of the deceased joint tenant has no interest in the property.

Joint tenancies have retained some popularity as forms of informal will making. No formalities are required to make the survivor the sole owner of the property, other than a simple termination of joint tenancy. However, under tax laws, inheritance taxes and estate taxes will still be due on the transfer of the property effective upon the death of the joint tenant.

Creditors can seize and force a judicial sale of the joint tenant's interest during the joint tenant's lifetime. Obviously joint tenants cannot leave by will their portion of the jointly owned property. If creditors obtain judgments against joint tenants, a court can make their interest in the property liable for payment of their judgment debts and order a public sale of their interest. Older law favored joint tenancies because they tended to preserve land in larger units.

(B) TENANCY IN COMMON

The law today favors tenancies in common. Each tenant in common holds some undivided interest in the same property. One might hold a three-eighths share, another one-quarter, etc. They are tenants in common because the property has not been divided into the individual portions. A tenant in common may sell or divide her interest. It can be transferred by will or be inherited. Tenancies in common can be reached by creditors of the tenant during the life of the tenant or, upon death, by a claim against the estate.

(C) PARTITION

It may be necessary to physically divide land owned in multiple ownership. Suppose the property is seized for debts of one tenant, all the tenants wish to divide the property, or one tenant wishes to take her share. If the parties cannot agree on a fair value and pay the proper share division, a court will order either a physical partition (a division of the property itself) or a sale of the property and apportionment of the proceeds according to the interests of each tenant.

(D) CONDOMINIUMS, COOPERATIVES, AND TIMESHARES

Other forms of multiple ownership have become more popular as society becomes more urbanized. In a condominium, the individual units whether apartments, stores, or offices, are owned by the individuals. The hallways and common facilities are owned by all the individual owners as tenants-in-common.

The cooperative is another form of multiple ownership. Title to the land and improvements are owned by a corporation in which residents are shareholders. Each shareholder has a long-term renewable lease in her own unit.

Condominiums and cooperatives provide the reduced responsibility of a tenant along with the tax advantages and investment potential of an owner.

In the 1980s timeshares were vigorously marketed. They allow purchasers to have a guaranteed period of occupancy of part of a residential unit in a tourist area. The periods could be as short as a week. In addition to complaints of heavy sales pressure, complaints were made that promoters misled buyers as to amenities, upkeep costs (also problems in condominiums), and availability. It is likely that timeshares will be regulated by public authorities.[1]

(E) MARITAL PROPERTY

Community property is a form of co-ownership used in eight western states. Although variations exist, generally all property acquired by either spouse during a marriage, other than property acquired by gift, will, or inheritance, is owned by the spouses equally as community property.

Since the advent of more equal rights between sexes, older rules that gave the husband management and control have given way to rules that make the community more like a partnership. Each must agree.

SECTION 10.06
PHYSICAL DIVISION

The owner of land may sell part of the land itself. For example, the owner of an acre of land may sell one-half an acre. Emphasis in this section, however, is upon transactions that involve the transfer of subsurface or air rights.

Usually the owner of land owns the subsurface of the land and the air rights above the land, subject to the public interest in

---

1. See Peirce & Mann, *Time-Share Interests in Real Estate*, 59 N.D.Lawyer 9 (1983).

permitting flights above the land. The owner can sell or lease subsurface rights or air rights. The owner may sell the entire subsurface to persons interested in extracting minerals, oil, or natural gas from below the surface of the land. The owner may sell the right to extract certain minerals or gases, receiving a royalty based upon the amount extracted. The owner may lease, rather than sell, the surface or mineral rights. Population concentrating in urban areas has made subsurface rights and air rights increasingly valuable.

SECTION 10.07 THE TRUST

Sometimes ownership is divided into legal and beneficial ownership. The deed may name one person as owner, but that person may own the land for the benefit of another. Such a division of legal and beneficial ownership is created by a trust. A trust is often used as a means of holding property given to minors or to persons whom the giver does not wish to entrust with full legal ownership. In some states the trust is a device for hiding the true owner of land. Where trusts are used, the person who will hold legal title is called a trustee, and the person for whom the trustee holds the title is called the beneficiary. The person who creates the trust may be called the trustor, the settlor, or the grantor. The laws relating to trusts are complicated and are not covered in this treatise.

SECTION 10.08 NONPOSSESSORY INTERESTS

(A) EASEMENTS AND LICENSES

The most important nonpossessory interests in land are easements and licenses. An *easement* gives one person the right to use the land of another. Examples of affirmative easements are the right to cross another's land; the right to construct and maintain underground pipes, telephone lines, and utility lines; and the right to use another's land to hold excess water.

A negative easement precludes the owner of land from using the land in a certain way. A landowner may promise not to build on the land in any way that would cut off the light, air, and view of an adjacent landowner. Light, air, and view are important considerations in construction of buildings and residences and are discussed in Section 11.04. Public land use controls, discussed in greater detail in Chapter 11, may protect light, air, and view. Zoning ordinances usually contain height limitations and setback requirements.

Easements may also be created by prescription, or by what is sometimes called adverse use. As seen in Section 10.12, a person may acquire title to the land of another by occupying it under certain conditions for a specified period of time. With regard to possessory interests, this is called acquiring title by adverse possession. Easements may be acquired by using the land of another, without express or implied permission, for a period of time that is roughly parallel to the period of time necessary to acquire title by adverse possession.

The scope of the easement is determined by the language of the deed creating the easement, the facts that give rise to an easement by implication, or the nature of the adverse use. If a public utility has an easement over a farmer's land to erect utility poles and wires, the utility company can enter the land to maintain the utility lines. The time for entry must take into account the farmer's need to use the land.

Suppose there is an easement to cross the land of another. Who or what can use the crossing? If it is a foot path, can it be used by horses? If it is a road, can it be used by heavy trucks? The easement language or the facts surrounding its creation determine such issues.

Easements can expire by their own terms. A may give an easement to B for ten years. An easement may last for the life of the holder of the easement or for the particular purpose of the easement. In such cases, when the life of the holder ends or the purpose for which the easement was created has been accomplished, the easement rights terminate. Most easements have a potentially unlimited duration. Some specific action is required to terminate them.

A *license* is less extensive than an easement. It is generally revocable at *any* time by the person granting or creating the license. Because of its revocability, a license is not considered an interest in land and is looked upon merely as a legal justification for what otherwise would be an unauthorized entry on the land of another.

(B) IMPROVING THE LAND OF ANOTHER: ENCROACHMENTS

An encroachment occurs when one property owner builds on the land of a neighbor. The classic cases are those where one landowner has put a few inches of her house on her neighbor's land.

The encroacher has committed a trespass. In such cases the neighbor usually has the right to recover any damages caused by the trespass. Often they are nominal. The neighbor whose land has been trespassed upon can obtain a court decree ordering that the encroacher remove the encroachment.

An order can cause great hardship. If the encroachment occurred by mistake or in reliance upon a survey, the court may refuse to order removal of the encroachment. In addition to showing that she was acting in good faith, the encroacher would have to show that the cost of removal and the harm removal would cause the encroacher greatly exceeded the harm caused by the encroachment. For example, in *Golden Press, Inc. v. Rylands*,[2] the encroacher built within his lot lines, but underground footings projected two to three inches onto his neighbor's land. Encroachment had been unintentional. In denying

the decree ordering that the encroachment be removed, the court stated:

> Where the encroachment is deliberate and constitutes a willful and intentional taking of another's land, equity may well require its restoration regardless of the expense of removal as compared with damage suffered therefrom; but where the encroachment was in good faith, we think the court should weigh the circumstances so that it shall not act oppressively. . . . [R]elative hardship may properly be considered and the court should not become a party to extortion. . . . Where defendant's encroachment is unintentional and slight, plaintiff's use not affected and his damage small and fairly compensable, while the cost of removal is so great as to cause grave hardship or otherwise make its removal unconscionable, mandatory injunction may properly be denied and plaintiff relegated to compensation in damages.

SECTION 10.09 ACQUISITION BY DEATH OF OWNER

(A) WILLS

Most interests in land can be transferred by the death of the owner of the interest. Usually the dispositive act is a valid will. This requires mental capacity to make the will and compliance with specified formalities for execution of a will. The states have formal rules for insuring that the will is authentic and that the testator, the one making the will, is aware of the seriousness of the testamentary act. Most states require that the will be witnessed by a certain number of people. The testator should read over the will to be certain that it manifests the intention as to the disposition of property.

In some states, holographic wills are valid. They are wills drawn up entirely in the handwriting of the testator. They need not be witnessed but must be dated. Any irregularity in these requirements is likely to result in the will being declared void (of no effect) by a court.

A will can be changed or revoked at any

---

2. 124 Colo. 122, 235 P.2d 592, 595 (1951).

time before death if formal requirements are met.

Laws relating to wills are complex and vary from state to state.

(B) DEATH WITHOUT A WILL: LAWS OF INTESTACY

If a person dies without a valid will, the property passes by the state laws of intestacy. The pattern of distribution attempts to follow a natural distribution. Although there is variation among the states, typically the surviving spouse gets one-third and the children receive the remaining two-thirds of the property. If the decedent is survived by a spouse but no children, the spouse typically receives one-half, with the remaining half going to parents, if any, or to brothers and sisters. If the decedent is not survived by a spouse or children, the estate usually goes first to parents, then to brothers or sisters, and then to other collateral relatives. If there are no relatives and no valid will, the property goes to the state.

(C) MARITAL PROPERTY

The law protects the surviving spouse. Early in Anglo-American history, the wife had a dower interest in the real property owned by her husband. (The husband had a similar, but more limited right called curtesy.) Upon death of the husband and survival by the wife, the widow received a life estate in one-third of the land owned by her husband at any time during the marriage.

The widow's interest in the land attached at the time the land was acquired. A purchaser acquired only the interest of the husband, unless the wife signed the conveyance. If she did not join in the conveyance, she could still assert her life estate if she outlived her husband. Purchasers were reluctant to buy land without consent of the wife where she would have a dower interest.

Today spouses have equal rights. The spouse can elect to take under any will, or can take total ownership of from one-third to one-half in all land and personal property (shares of stock, bank accounts, furnishings, cars, etc.) owned by the deceased spouse at the latter's death. If property is owned by husband and wife as joint tenants, all of the property goes to the surviving spouse because of the survivorship aspect of a joint tenancy. (Refer to Section 10.05(A).)

Eight states classify most marital property as community property. Generally under community property laws, property acquired by the efforts of either spouse during the marriage is owned by the community. Commonly the will of either spouse can dispose of only the half of the community property owned by the decedent. The other half still belongs to the surviving spouse. In the absence of a will, state law determines who gets the deceased spouse's one-half. In some states, if either spouse dies without a will, the other spouse becomes the owner of all the community property. In other states, the property of the deceased spouse who dies without a will passes by the laws of intestacy. As stated, typically this means that the surviving spouse will receive one-third to one-half of the deceased spouse's one-half, depending upon whether there are children.

(D) ADMINISTRATION

A court supervises administration of an estate. The actual work or winding up and distribution of an estate is performed by the executor or the personal representative. Usually the will names an executor. If no one is named, or if the person named is unable or unwilling to serve, the court will name an executor. If the deceased died without a will, the court will name a personal representative to wind up and distribute the estate. Since the executor or personal representative is responsible for the estate, she will come to the court for instructions if in doubt as to what should be done regarding administration.

SECTION 10.10
ACQUISITION BY GIFT

There are three general requirements for a valid gift: donative intent (intent to make a gift), delivery, and acceptance. Acceptance is usually presumed. The difficulties center around intention to make a gift and delivery. The delivery concept originally assumed that the item being given could and would be manually transmitted by the maker of the gift to the receiver of the gift. This requirement demonstrated the intent to make a gift and impressed upon the maker of the gift that she was effectuating a final transfer of ownership in the property. For example, to give a book, the donor (maker) would hand over the book to the donee (recipient).

Suppose the subject matter of the gift could not easily be transmitted manually, for example, a herd of cattle, an intangible right evidenced by a promissory note, or an intangible right not evidenced by a promissory note such as an account receivable or a bank account. Delivery of something closely connected with the subject matter, such as a branding iron, a passbook, or a bond or a note, is a valid delivery.

Manual transmission of land is not possible. For this reason, the gift of land is typically made by the grantor executing and delivering the deed to the grantee, the intended recipient of the gift.

Traditionally the law has been suspicious of gifts. Gift promises are not enforced. Assertions of a gift are frequently made by the donee after the death of the donor. Proof of the donor's intention can be difficult to determine. Today, courts are more likely to venture into the troublesome questions of intention to make a gift.

SECTION 10.11
ACQUISITION BY
FORCED SALE

Sometimes title to real property is acquired by judicial sale, such as tax sales and fore-closure sales. The holder of a security interest (someone who has lent money and has been given an interest in the property as security for the loan) may have the right to have a judicial sale of the property and to be paid from the proceeds if the borrower fails to pay the loan. The purchaser of the property acquires title as a result of the judicial foreclosure sale, subject to a right given the owner to redeem the land within a certain period. (Nonjudicial sales are common in many states. They are quicker and avoid a redemption period. But they bar a deficiency judgment if the sale brings less than the unpaid debt.)

Forced sales may also occur when there is nonpayment of taxes or a court judgment. In such cases public sales will be held, usually by the sheriff, who will issue a title to the high bidder, subject to the owner's right to redeem within a certain period.

SECTION 10.12 ACQUISITION
BY ADVERSE POSSESSION

Adverse possession acquisition of title requires the claimant to occupy the property openly for the period of time established by law. Sometimes the claimant must have a good faith claim of ownership. The true owner of the property must bring a legal action to end the wrongful occupancy within a time specified by law or the owner loses title to the adverse possessor.

Adverse possession has the same rationale as Statutes of Limitations, which require that judicial actions be commenced within a certain period of time. Evidence is lost, witnesses die, and memories become clouded. Judicial resolution of the dispute becomes more difficult with the passage of time. Periods of limitation dealing with ownership of land have additional reasons for requiring that actions be brought within a specified period of time. Often persons who take possession of the land believe that they have a right to be on the land. They may have a document that convinces them that they are the true owners of the land.

During this period they may make substantial improvements to the land. In areas where land titles are uncertain, it would be unjust to permit parties to improve the land and rely upon its being their own and then eject them.

In most states, the statutory period during which the original owner may legally eject the adverse holder is twenty years. Some western states have periods as short as five years, but these states frequently require that the possessor pay taxes as well as occupy the land.

Many technicalities are connected with acquiring title by adverse possession, and the laws vary greatly from state to state.

SECTION 10.13
EMINENT DOMAIN AND DEDICATION: ACQUISITION OF OWNERSHIP BY THE PUBLIC

To function, a government must have the power to appropriate private land for public purposes. Such power, called the "power of eminent domain," is needed to build roads, hospitals, schools, and other public projects. Usually state constitutions or statutes grant this power. Sometimes this power is given to private public service companies, such as public utilities. Some public utilities (those given a monopoly by the state) are private companies (though heavily regulated by the state) e.g. the various spin-offs of American Telephone & Telegraph or the surviving company or Commonwealth Edison, while others are publicly owned, such as most water companies.

State and federal constitutions require that private property not be taken without due process of law. The taking must be for a public purpose, and the person whose land is taken must receive just compensation. The person whose land is being taken is entitled to a fair hearing on these questions. The typical dispute in a condemnation action (the type of action under which the state exercises its rights of eminent domain) is the valuation of the property taken. This is complicated when the property is commercial and where the property is claimed to have a value based upon prospective earnings.

May the state take land from a private owner and turn it over to another private owner who will develop the land with what is conceived to be a greater public interest? This question may come up where land is taken for housing or redevelopment that will be turned over to private parties. While the answer will vary from state to state, there is an increasing tendency to permit the state to make such condemnations of private property to provide housing or economic growth.

A public body may acquire title to property by the process of dedication. An owner may transfer land to be used for a public use. When the public body accepts the property for present or future public use, the dedication is completed. Usually there is public dedication for parks, streets, alleys, and highways. The dedication may consist of a transfer of the entire ownership, or of some lesser interest in the land. Usually dedications are made by deeds or written instruments. Dedications may occur without a written declaration by the owner's acquiescence in the public use. A dedication may occur when a street or road is described in a plat or map relied upon by purchasers of lots in the immediate area.

State statutes deal with the subdividing or platting of land. These statutes often require that the subdivider dedicate portions of land for public use.

Dedication requires that there be an acceptance. If the dedication is accepted, the land itself, or a right to use the land, belongs to the public. This can also make the state liable for injuries received caused by the condition of the dedicated land.

Most land is transferred by sale. Refer to Section 5.12.

11

Limits on Ownership: Land Use Controls

As indicated in Chapter 10, ownership rights, though very broad in a legal system that strongly protects the rights of private property, are not absolute. A variety of controls are examined in this chapter. Sections 11.01 through 11.05 look at *private* land use controls, created by contract or tort law and enforced by courts. Sections 11.06 through 11.15 deal with public land use controls enacted mainly by local entities through powers given them by the state. Administration is handled largely through local regulatory commissions or legislative bodies. Courts, though less important, play mainly a passive, oversight role, determining whether local laws or administrations meet the state legislative and constitutional requirements.

SECTION 11.01 NUISANCE: UNREASONABLE LAND USE

The law protects landowners' rights (these can also include family members or tenants or employees) to enjoy their land free of unreasonable interference caused by the activities of landowners in the same vicinity. This regulates the use of land by prohibiting uses that unreasonably interfere with the use of land by others. The unfortunate term *nuisance* has developed as the means of describing this land use control.

A landowner (landowner as used in this section can include a tenant) must make reasonable use of the land and not deprive landowners in the same vicinity of the reasonable use and enjoyment of their land. To a large degree, the granting of legal protection through nuisance depends upon the activity or use complained of and the effect such activity or use has on other landowners.

Looking first from the perspective of the complaining landowner, the interference can take many forms. At one extreme, there can be actual physical damage to the land. For example, blasting activities by a neighbor or the discharge of solid, liquid, or gaseous matter upon the land may change the physical characteristics and shape of the complaining landowner's property and seriously affect its use.

The offensive activity may consist of disturbing the comfort and convenience of the complaining landowner. For example, activities on the offending land can create loud noise or the omission of noxious odors that may disturb the occupants of adjacent or nearby land.

Moving to even more intangible interference, the neighbor's complaint may be based upon simply knowing the nature of the activities on the adjacent or nearby land. For example, a neighbor's mental tranquility may be disturbed by the knowledge that an adjacent landowner maintains a house of prostitution or a meeting place for people whose activities are bizarre or unconventional.

Moving to the outer extreme of intangible interference, one neighbor may be extremely disturbed by the appearance of a neighbor's house.

An objective standard is used. Would the acts or activities in question have disturbed a reasonable person under such circumstances? Could the interference have been avoided had the complaining landowner taken reasonable measures? For example, would closing the windows have reduced or eliminated excess noise caused by the neighbor? If so, would such protective measures unduly interfere with the use and enjoyment of the land?

Generally the more the activity in question causes an actual physical result, such as changing the characteristics or contours of the land or making a physical impact upon the land, the more likely the law will give relief. As the interference moves toward more intangible matters, such as peace of mind and esthetic judgments, the less likely the law will accord protection. Although such interests may be protected, it would take a significant interference with the use and enjoyment of the complaining landowner's property.

The likelihood of legal protection may depend upon the extent and duration of the interference. Temporary interference or interference that occurs only at long intervals is less likely to justify relief.

Legal protection does not require that the complaining party show that the acts or activities in question were *intended* to interfere with the use and enjoyment of adjacent or nearby landowners. The conduct need not be negligent. However, the *more* intentional and *more* negligent the conduct, the more likely the activity in question is a "nuisance."

The social utility of the activity is important. Certain commercial and industrial activities are necessary and are encouraged by society. The necessity for these activities may overcome minor inconveniences to those in the vicinity. If the activity has little social value, slight interference with the use and enjoyment of the adjacent landowner's land can constitute a nuisance.

Often the biggest difficulty is the remedy to be given the complaining owner when the defendant has made an unreasonable use of the land. The choice is between giving a money award for damages to the complaining party or ordering the offending party to cease the activities, an equitable injunction.

The money damage award must be inadequate before an injunction can be awarded. However, it is often difficult to establish the precise value of the damages caused in these cases. This is especially true if the offensive activity consists of noise or emission of noxious fumes or if activity has caused losses other than physical harm. As a result, it has become relatively easy to obtain injunctive relief in nuisance cases, especially if the nuisance is a continuing one.

Suppose an industrial activity is being conducted on a large scale and has adversely affected the occupancy and use of nearby land. Some early court decisions affirmed orders that required a cessation of the offensive activities. Such an order can close industrial plants upon which the economy of the town in which the plant is located may be dependent. The modern tendency is to award what are called *permanent damages* based upon the diminution in the value of the land affected by the industrial activity rather than ordering that the offending activity cease.[1]

Class actions are increasingly being used, and some landowners maintain the action for the benefit of all landowners similarly situated. Class actions through the aggrega-

---

1. *Boomer v. Atlantic Cement Co.*, 26 N.Y.2d 219, 309 N.Y.S.2d 312, 257 N.E.2d 870 (1970).

tion of small claims against the offending landowner can justify the often large expenses needed to mount a lawsuit against the industrial activity adversely affecting a large number of landowners in the vicinity.

SECTION 11.02 SOIL SUPPORT

The owner of unimproved land has the right to the lateral support by adjacent land. Because this is an *absolute* property right, the owner of unimproved land can receive legal relief if lateral soil support were withdrawn by excavation without the owner having to show that the excavating landowner did not use proper methods.

This "rule of property," an absolute one as contrasted to a flexible one used in nuisance, does not apply if the affected land has been improved, adding weight to the soil. In such a case, the excavator is liable only if the adjacent landowner can show that the excavator did not comply with the legal standard of care. If the adjacent landowner can show that the soil support would have been withdrawn by the neighboring excavation regardless of the added weight caused by the improvement, the adjacent landowner can recover without showing that the excavator did not comply with the legal standard.

Frequently the result in excavation cases is controlled by state legislation. For example, in Illinois[2] the excavator must notify the adjacent landowners of its intent to excavate, the depth of excavation, and when the excavation will begin. Legal rights depend upon the depth of the excavation.

The excavator should comply with any public controls such as the Illinois statute mentioned, give notice of the excavation to the other party, seek to work out an advance agreement on protection and responsibility, and use reasonable care in excavating.

SECTION 11.03 DRAINAGE AND SURFACE WATERS

Construction frequently affects and is affected by drainage at the site and adjacent land. Drainage changes can cause troublesome collection of surface waters.

Three surface-water rules have emerged in the United States.[3] The first is the "common enemy" rule. Each landowner can treat surface water as a common enemy with the right to deal with it as it pleases without regard to neighbors.

The second is the "civil law" rule. A landowner who interferes with the natural flow of water is strictly liable for any damage caused the neighbors.

The preceding rules are absolute. One gave great freedom to deal with surface water. Another created absolute liability. A third rule, that of "reasonable use," is gradually supplanting the absolute rules. Each case in jurisdictions with such a rule is decided upon its own facts. The landowner affecting the water flow must act reasonably. In making this determination the following factors are considered:

1. Is there a reasonable necessity for such drainage?

2. Has reasonable care been taken to avoid unnecessary damage to the land receiving the water?

3. Does the benefit accruing to the land drained reasonably outweigh the resulting harm?

4. When practicable, is the diversion accomplished by reasonably improving the normal and natural system of drainage, or if such a procedure is not practicable, has a

---

2. Ill.Ann.Stat. ch. 111½, § 3301.

3. *Butler v. Bruno*, 115 R.I. 264, 341 A.2d 735 (1975), presents a complete discussion of the history of the three American water law rules.

reasonable and feasible artificial drainage system been installed?

SECTION 11.04 EASEMENTS FOR LIGHT, AIR, AND VIEW

Light, air, and view are important considerations in construction of buildings and residences. *Fontainebleau Hotel Corp. v. Forty-Five Twenty-Five, Inc.*[4] involved an addition built by one luxury hotel in such a way as to cut off sun, light, and view to an adjacent hotel. The hotel upon which an addition was being built, the Fontainebleau, was constructed in 1955; and its adjacent competitor, the Eden Roc, was constructed in 1956. The addition was to be a fourteen story tower. During the winter months from two o'clock in the afternoon and for the remainder of the day the tower would cast a shadow over the cabana, swimming pool, and sunbathing areas of the Eden Roc. The Eden Roc sought to obtain an order restraining the Fontainebleau from proceeding with the construction of the addition at a time when the addition was roughly eight stories high.

The trial court granted the requested order, but the appellate court reversed. The court stated that no American decision, in the absence of some contractual or statutory obligation, had held that a landowner "has a legal right to the free flow of light and air across the adjoining land of his neighbor."[5] The court concluded that the construction of a structure that serves a useful and beneficial purpose could not be restrained by the court even though it cut off light and air and interfered with the view of a neighbor. Despite the partially spiteful motivation for the structure, relief was denied the Eden Roc.

A recent case refused to follow the absolute property rule. It applied the flexible nuisance doctrine to preclude an owner from building in such a way as to reduce the effectiveness of a neighbor's solar heating panels.[6]

An owner who constructs to take advantage of view should consider either obtaining easements from adjacent landowners or buying adjacent land.

Some California communities have adopted tree ordinances. An owner whose view is impeded by a tree on a neighbor's land can request the tree be removed or trimmed. If done, the costs are divided. If the tree owner is unwilling, a local "tree commission" decides the dispute.

Suppose one neighbor puts up a "spite fence," a high, unsightly fence that interferes with a neighbor's view and the entrance of light to the neighbor's land and creates an eyesore.

States have not been uniform in their treatment of spite fences. Some emphasize the right of a landowner to use its land as it wishes and permit spite fences. Other states hold that if the dominant motive in erecting the fence is malicious and if the fence serves no useful purpose, the neighbor can recover damages and be granted a court decree ordering that the fence be removed. In some states, statutes limit the height of any fence erected maliciously or for the purpose of annoying a neighbor. The increasing tendency is to restrict the use of spite fences.

SECTION 11.05 RESTRICTIVE COVENANTS

Generally an individual grantor (one who conveys) of land has no interest in how the land is used after the deed has transferred ownership. But suppose the grantor lives

---

4. 114 So.2d 357 (Fla.App.1959).

5. Id. at 359.

6. *Prah v. Maretti*, 108 Wis.2d 223, 321 N.W.2d 182 (1982). This case is noted in 21 Duquesne L.Rev. 1159 (1983) and 48 Mo.L.Rev. 769 (1983).

or will live in the vicinity of the land transferred? Or, more important, suppose the grantor is a residential developer who is selling lots or houses to a large number of buyers? In either case, the grantor may wish to control land use after the title is transferred by deed. The grantor who is not a developer may wish to control the land use to enhance or maintain the use or value of land it owns in the vicinity of the land transferred. The developer-grantor may wish to assure buyers in the development that the land will retain its residential character to protect the enjoyment of and investment in the land. Although restrictive covenants can be created in agricultural or commercial property, this section emphasizes their use in the development of residential property.

To accomplish the land-planning objectives, the developer-grantor can obtain an express promise relating to land use. There are no particular legal problems between the developer and the original buyer in accomplishing this result by contract, with the exception of racial or religious restrictive covenants (mentioned later in the section). The contract in such cases can and usually does contain promises that limit the buyer's right to use the land. In land law, such promises are called restrictive covenants.

Land ownership in a housing development is likely to be transferred. Since use affects value, the developer wishes to assure *buyers* that restrictions on use will bind future buyers and to assure *buyers or their successors* that they will have enforcement rights if any buyers violate the restrictive covenants. As a marketing device, tightly drawn enforceable restrictive covenants can persuade buyers to buy because of the protection of restrictive covenants. To accomplish this, a developer will include restrictions in the deeds that give buyers of lots or residences in the development at the time the covenants are violated the right to enforce these restrictions.

To be effective over a long period of time, such restrictions must be "tied to" or "run with" the land. All buyers, present and future, must be bound to the restrictive covenants; and all owners, present and future, must be able to obtain judicial enforcement of these restrictive covenants. Early English land law that sought to preserve property values by limiting land use through such restrictive covenants became unsatisfactory. A number of technical requirements often frustrated attempts to enforce these covenants.

The ineffectiveness of the system led to the nineteenth century development of equitable servitudes in England and subsequently in the United States. In the famous case of *Tulk v. Moxhay*,[7] the English court held that a purchaser of land who *knew* of the restrictions on the use of the land would be ordered by the Equity Court not to violate these restrictions even though there was nothing in the deed that restricted the land use. Equitable servitudes avoided technical requirements of covenants "running with the land." They created a viable private system of land use controls that had great impact on the development of cities and that still exist despite the proliferation of public land use controls.

In the United States, an equitable servitude requires an intent to benefit the adjoining land and notice of the servitude or restriction to buyers of the burdened land, that is, the land that is limited in use. Intent to benefit is usually shown by a uniform, common scheme of restrictions. Generally this scheme is shown by a legal document recorded in the land records, although it can be shown in some states by a uniform pattern of use throughout the development.

Generally courts have given considerable freedom to developers and buyers in the

---

7. 41 Eng.Rep. 1143 (Ch.1848).

creation of restrictive covenants. However, contractual freedom is not absolute. In *Shelley v. Kraemer,* the U.S. Supreme Court held that state enforcement of racial restrictive covenants violated the Equal Protection Clause of the Federal Constitution.[8]

Should the constitutional tests that determine the validity of *public* land use controls (as seen in Section 11.15) be applied to *private* restrictive covenants? Some contend that buyers do not have adequate notice of the restrictions, the agreements are contracts of adhesion and not consensual, and *in effect,* broad-scale enforcement of private restrictive covenants is tantamount to private zoning. Private restrictions, especially those that limit construction to single-family residences, can frustrate plans to have integrated housing in urban and suburban areas.

One writer urged that constitutional requirements for zoning ordinances be applied to decisions of those who administer such systems, such as homeowner associations and architectural committees, as they often wield power analogous to local authorities.[9] However, courts generally regard restrictive covenants as a useful private planning device and have not as yet been willing to apply public land use standards to reviewing restrictive covenants and actions of homeowner associations.

Private restrictive covenants can be rigid. Restrictions must be effective for a sufficiently long time to protect buyer investments. Yet the future is difficult to predict. To deal with the need for flexibility, many planned developments create homeowner associations composed of owners in the development.[10] These associations can modify existing restrictions under certain cir-

cumstances and pass upon requests to deviate from the restrictive covenants.

Restrictive covenants have begun to deal with esthetic and architectural aspects of a development. In *Hanson v. Salishan Properties, Inc.,*[11] leases for beachfront lots contained provisions dealing with design aspects that affected the view of others in the development. In addition to being required to comply with specific covenants set forth in the lease relating to these matters, tenants were referred to an architectural checklist drafted by the architectural committee that set forth general concepts relating to esthetics and view for the development. The lease required that plans be approved by the architectural committee.

A tenant submitted preliminary plans for a home that were approved by the architectural committee. Some upland tenants sought an injunction against construction of the home that they claimed violated the restrictive covenants. The Oregon court concluded that the committee acted properly and that its decision would not be disturbed. The court did not have to articulate standards for judicial review of decisions of architectural committees, since the court concluded that the architectural committee's decision was correct.

Generally the association or committee must act reasonably and in good faith. It must act in accordance with predetermined standards.[12] Without regard for whether a predetermined standard is required, such decisions must be consistent with the standards set forth in the original declaration of restrictions and with existing neighborhood conditions.

Sometimes private restrictive covenants prohibit owners from modifying or chang-

---

8. 334 U.S. 1, 68 S.Ct. 836, 92 L.Ed. 1161 (1948).

9. Comment, 21 U.C.L.A.L.Rev. 1655 (1974).

10. These associations are often given the power to seek injunctions for violations even though they do not own land in the affected area.

11. 267 Or. 199, 515 P.2d 1325 (1973).

12. *Carranor Woods Property Owners' Association v. Driscoll,* 106 Ohio App. 95, 153 N.E.2d 681 (1957).

ing the exterior in such a way as to detract from architectural harmony or historic tradition. For example, *Gaskin v. Harris* [13] involved a restrictive covenant that required all exteriors to be Old Santa Fe or Pueblo-Spanish architecture. One homeowner wished to build a modern Oriental swimming pool, and other homeowners objected. New Mexico upheld the restrictive covenant despite the pool builder's claim that the pool had already been constructed and that the restrictive covenant was too vague to be enforceable.

Other techniques, though used less often, can soften some of the rigidity of restrictive covenants. A landowner burdened by a restrictive covenant can seek relief or may resist an injunction sought by other landowners by showing that conditions have so changed that the servitude should be removed or modified. The court will not order an injunction if the conditions in the neighborhood have so changed that enforcement of the restrictive covenant would be unreasonable. Denial of enforcement on the basis of changed conditions requires a showing that enforcement of the covenant would harm the burdened land much more than nonenforcement would harm the other land in the development. Reasons for denial can include a significant increase in noise and traffic, the unsuitability of the burdened land for the restricted purpose, and rezoning of the land for a more permissive use.

Servitudes can be eliminated by agreement of all the affected landowners or occasionally by benefited landowners failing to seek judicial relief for past violations. The servitude can also be eliminated if a public agency takes the land for a use inconsistent with the restriction. This is accomplished by condemnation, and those property owners that would have benefited by continued enforcement generally receive compensation from the agency exercising the power of eminent domain.

Despite the proliferation of public land use controls, private restrictive covenants are still important. They are often the principal land use controls in many existing developments and are used in new developments to some degree as protection from the risks of zoning law changes. Also, some writers have asserted that restrictive covenants are more efficient and less expensive to employ than public land use controls. [14]

SECTION 11.06
DEVELOPMENT OF LAND: EXPANDED PUBLIC ROLE

In the individualistic system preceding restrictive covenants, and after the development and enforcement of restrictive covenants, the law had a minimal and mainly passive role. Individual judgment and the free market predominated. The present dominant system of land use control has expanded the law's role. Today the state itself largely determines the permissible use of land.

It is essential to compare the passive approval of private restrictive covenants and the more active public developmental control to understand why the latter has dominated modern land use controls. The essentially political nature of public control is discussed in Sections 11.07 and 11.08.

Other differentiations exist between private restrictive covenants and public land use controls. The principal objective of restrictive covenants is to preserve property values and maintain the characteristics of the neighborhood. While public land use controls take these factors into account, local authorities must also concern themselves with fiscal matters. Since property

13. 82 N.M. 336, 481 P.2d 698 (1971).

14. See B. Siegan, *Land Use Without Zoning* (1972); Ellickson, *Alternatives to Zoning: Covenants, Nuisance Rules, and Fines as Land Use Controls*, 40 U.Chi.L.Rev. 681 (1973).

taxes have been the principal revenue-raising measure available to local authorities—at least until the taxpayer revolt against property taxes in the late 1970s—permitting industrial, commercial, or expensive residential use can increase the tax base and raise revenue.

Local authorities must furnish fire and police protection, educational facilities, and parks, as well as water and sewerage facilities. Local authorities are also responsible for the overall quality of life within and around the community. They must consider community-wide problems such as housing and crime. Public land use planning is more complicated and difficult than private controls.

Flexibility is another differentiating aspect of public land use controls. While some flexibility is needed in private restrictive covenants, the private system emphasizes stability. On the other hand, public controls must take into account and devise methods for dealing with changes that occur at an astounding pace in urban life.

The developer thinks mainly of profits and creates restrictive covenants to sell lots or homes. The public planning process should marshal the best thinking of many disciplines in order to accomplish legitimate planning of objectives. One judge described the public planning process as bringing to bear upon planning decisions:

> the insights and the learning of the philosopher, the city planner, the economist, the sociologist, the public health expert and all the other professions concerned with urban problems.[15]

The passive role did not seem adequate for these and other reasons. As a result, the government took primary responsibility for urban land development which then became a political process.

Legislatures, both state and local, consist of persons elected by and responsible to voters. The public land use control system that evolved in the twentieth century not only was representatively democratic but often became directly democratic through decisions made by the voters themselves.

SECTION 11.07
LOCAL CONTROL OF
LAND DEVELOPMENT

Although there have been and will be continued attempts to bring the federal government into land use planning, essentially the power over public land use controls belongs to each state. But exercise of state powers is limited because property rights have received special protection in the United States. Owners cannot be deprived of their property without due process of law. If an owner's right to use the land is *too* limited, such a limitation can constitute a "taking" of the property and the state must pay the owner.

To avoid land use control being a taking, most public control over land use is predicated upon the police powers of the state to protect its citizens and the state's concern for the general health, safety, and welfare of the community. To bring legislated land use controls within the police powers of the state, the state legislation stressed objectives such as reduced street congestion, better fire and police protection, and promotion of health and general welfare as the reasons for restricting land use.

Land use control was accomplished by state Enabling Acts, which expressed goals in general terms and allowed local authorities to make specific rules and administer them. Generally, local authorities exercised this power. They enacted and administered ordinances to regulate land development. The traditional method for accomplishing this was to create districts in which only certain activities are permitted.

The original Enabling Acts passed in the 1920s recognized that much of the work needed to create and administer public land

---

15. *Udall v. Haas*, 21 N.Y.2d 463, 469, 288 N.Y.S.2d 888, 893, 235 N.E.2d 897, 900 (1968).

use controls could not be done by the local legislative governing bodies. They generally are composed of unpaid or modestly paid citizens elected to local legislative bodies. The elected officials would not have the time or expertise to actually draft and administer a development system.

The original Enabling Acts allowed two agencies to be created to deal with these matters. A planning commission was authorized to draft a master plan and detailed ordinances. A board of adjustment was authorized to deal with appeals from decisions by the local administrator or public officials. The structures, personnel, and names of these agencies varied considerably from locality to locality.

Generally members were given tenure for a specific number of years to insulate them from improper pressures. Beyond tenure requirement, localities varied in the attributes of such agencies. Compensation, qualifications, staff, and support would largely depend upon the size of the locality. However, even in larger cities, members of planning commissions were frequently interested citizens who served without pay and often with minimal staff.

The final authority locally in the enactment of ordinances and appeals from decisions was usually vested in the local governing body such as the city council.

At this point and in other parts in this chapter reference is made to the Model Land Development Code (MLDC),[16] a Model Enabling Act approved in 1975 by the American Law Institute, a group of scholars, lawyers, and judges devoted to law reform. The MLDC gives greater responsibility to the Local Development Authority (LDA), the agency implementing and operating the development program. The Code removed the local governing body, such as the city council, from decisions relating to specific development plan proposals and day-to-day operation of the system which were given to the LDA.

Development control is given to local authorities because land use is largely a local matter. Local decision makers are expected to be knowledgeable about the community and its needs and represent the persons who will be most affected by the developmental rules.

Should an increasingly interdependent urban society give such controls to often small political units? Insularity and unwillingness to take metropolitan social needs into account have led to increased demands for regional or statewide controls. The MLDC has recognized this, and certain provisions are designed to encourage increased activity by regional and statewide authorities.

On the other hand, criticism has been made of control by governing bodies of large cities that can be insensitive to the needs of local neighborhoods. While local neighborhoods were not given decision-making powers in the MLDC, neighborhood organizations were recognized and given greater stature.

The Copley Place project in Boston, a half-billion-dollar private project demonstrated the increased activity of neighborhood groups in the 1970s. Before the state authorities would grant needed air rights over a freeway, they required the developer to negotiate agreements with surrounding neighborhoods. From 1977 to 1980, fifty public hearings were held with neighborhood groups representing a wide stratum of interests. Issues thrashed out related to the scale of the project and its effect upon neighborhoods, traffic, parking, shadows, wind generation, and jobs. The process is one repeated in many cities where large-scale development is proposed.

Land use legislation and administration engenders controversy. Much of land value depends upon use, and land can involve

---

16. This will also be referred to as "the Code." This code is not law. It can be adopted in whole or in part by state legislatures. Sometimes such model codes have great influence. Sometimes they are ignored.

large economic stakes. These decisions oft-en polarize the community. Land use debates frequently feature clashes between those who favor growth and those who oppose it, and between those who advocate the rights of landowners and those who champion social control through the political process. Land use controls can exclude certain persons and activities, a point discussed in greater detail in Section 11.15. From a business vantage point controls can restrict competition.

Udall v. Haas [17] illustrates some of these features. The municipality had enacted a building zoning ordinance that reclassified a landowner's property from business to residential. The new classification would have permitted only public and religious buildings and residences with a designated minimum size. The event that led to the rezoning appears to have been the submission to local authorities of a preliminary sketch for the development of a bowling alley combined with either a supermarket or discount house to be built on a vacant lot that had been zoned for business use. The powerful economic effect of zoning on property value can be demonstrated by the observation made by the New York Court of Appeals that more than 60% of the value of the land, or over $260,000, was wiped out by the rezoning. Describing the process by which the rezoning was accomplished as a "race to the statute books," the court invalidated the rezoning as not in compliance with the Master Plan.

Another factor that has exacerbated abuse potential has been judicial reluctance to carefully review decisions of local legislative instrumentalities.

SECTION 11.08
REQUIREMENTS
FOR VALIDITY

Generally, local land use control requires an Enabling Act followed by the enactment of a master or local land development plan by the local governing body. In addition to serving planning functions to justify a police power interference with ownership, the master plan is designed to avoid *spot zoning*, different land use for small zones. Spot zoning cannot be justified as a legitimate overall planning device and is susceptible to abuse.

The adoption of local land use controls is generally preceded by notice to property owners and the public hearing at which all interested citizens can express their views. Public hearings in these matters frequently generate lengthy and heated debate.

Although the basic power to regulate land use resides in local legislative bodies, citizens are given greater direct power over these decisions, a power used increasingly to limit growth. Sometimes local voters can use the initiative process. This is direct enactment of law by the voters. The process requires the filing of a designated number of signatures—usually 5% to 25% of the voters—on petitions to place a proposed local ordinance on the ballot. Voters then determine whether the proposed ordinance will be adopted.

Sometimes direct voter power can be exercised *after* a local ordinance is passed by the local legislative body by a referendum that submits the ordinance to the voters for their review. The process can begin with the presentation of a petition with a designated number of voter signatures within a certain number of days after the passage of the challenged ordinance or permit decision. Typically the petition must be filed before the effective date of the ordinance or permit. The petition compels the local governing body to either repeal the ordinance or change its decision on the permit or place the issue on the ballot for voter decision.

Exercise of either initiative or referendum can be a costly process and one not likely to be undertaken unless the stakes are

17. Supra note 15.

high or citizens are greatly distressed with decisions made by their city council or board of supervisors.

Increasingly attacks are made on the land use controls as unconstitutional. On the whole, these attacks have not been successful.

SECTION 11.09 ORIGINAL ENABLING ACTS AND EUCLIDEAN ZONES: THE MLDC

Most zoning enabling statutes passed in the 1920s dealt mainly with the physical characteristics of development. They tended to divide the territory of the municipality into districts of different contemplated uses. Each district was to have uniform regulations. This system was sometimes called Euclidean Zoning after the U.S. Supreme Court case that first validated a zoning plan.[18] Euclidean zones required homogeneous use. Only particular uses were permitted in each district. For example, residential districts permitted only residences, commercial districts only commercial activities, and industrial districts only industrial activity. Major categories were further divided. For example, industry would be divided into heavy industry and light industry. Commercial use was divided into different categories. Residential use was divided into single-family homes, two-family homes, and multiple-family uses. Homogeneous districts were based on the assumption that differing uses within a district would be injurious to property values.

Some cities allow multiuse districts or what is sometimes called cumulative zoning. For example, while residences only might be permitted in the residential zones, commercial zones would also permit residences and industrial zones would permit residential and commercial uses as well.

In addition to creating districts of permitted use, the Enabling Acts of the 1920s

regulated matter such as height, bulk, and setback lines. Rigid patterns of land and development were assumed. Single homes were to be placed on gridlike lots.

The assumption of most early Enabling Acts was that landowners could develop their land as they wished provided they did not contravene specific restrictions expressed in the ordinances. The statutes authorized but did not compel the local authority to control development decisions. There was no exhortation or incentive to owners to undertake *desirable* development. The statutes expressed prohibitions designed to avoid *undesirable* development.

The original Enabling Acts were relatively indifferent to the effect of no or poor local planning upon areas *outside* the municipality. Local public interest dominated.

Much has changed since the 1920s, and the development of the MLDC was a response to such changes. The Code seeks to go beyond simply prohibiting poor planning but require desirable planning. The Code recognizes the need for regional and statewide planning. It seeks to develop flexibility and new planning methods to replace the rigid district techniques of the original enabling statutes. (These are discussed in Section 11.10.) The Code more openly recognized esthetics, environmental problems, and the preservation of historical sites as proper planning and developmental factors (discussed in Sections 11.10 and 11.11).

SECTION 11.10 FLEXIBILITY: OLD TOOLS AND NEW ONES

As stated in the preceding section, the controls of early zoning rarely went beyond height, bulk, and setback regulations. But the ugly, inefficient, and expensive explosion of urban sprawl just before and after World War II compelled planners to find methods to improve the quality of urban life. Planners were asked to enact land use

---

18. *Village of Euclid v. Ambler Realty Co.*, 272 U.S. 365, 47 S.Ct. 114, 71 L.Ed. 303 (1926).

controls that would create and preserve a healthy, esthetic community with proper regard for the finite quality of resources and the historical patrimony of the community. This and the following two sections deal with traditional and new methods to accomplish these more ambitious objectives.

(A) VARIANCES AND SPECIAL USE PERMITS

A variance permits the parcel of land to be used differently than prescribed in the zoning ordinance. Issuance requires the landowner to meet requirements that will appear in the cases discussed later in this subsection. Usually the landowner asserts that the use permitted was not economically feasible and would cause great hardship.

Zoning ordinances often allow permits to be issued for a special use. These permits can be issued by the local land development authority only if the special uses are set forth in the ordinance. Sometimes the zoning ordinances spell out specific uses that can justify the issuance of a special use permit. Sometimes the standards for issuance of these permits are described in general terms.

Ordinances often distinguish use and area variances. To avoid an unconstitutional confiscatory ordinance, use variances are given if the land as zoned cannot yield a reasonable return. But because a use variance is a more drastic disruption of planning, a stronger burden must be sustained than where an area variance is sought.[19]

The line between use and area variances is often difficult to draw. The *Broadway, Laguna* case involved an application submitted by a developer of the luxury apartment building for a variance of the floor area ratio imposed by San Francisco's zoning law. The ratio of lot area to rentable floor space was not to exceed 1:4.8. The

developer's proposal had a floor area ratio of 1:5.51. To obtain a variance, the developer contended that he had recently discovered unusual subsurface conditions that would require excavation costs two and one-half times more than anticipated and that his development would possess attractive architectural features beyond Code requirements. The ordinance stated:

> The Zoning Administrator shall grant the requested variance in whole or in part if, from the facts presented in connection with the application, or at the public hearing, or determined by investigation, it appears and the Zoning Administrator specifies in his findings the facts which establish: (1) that there are exceptional or extraordinary circumstances or conditions applying to the property involved or to the intended use of the property that do not apply generally to other property or uses in the same class of district; (2) that owing to such exceptional or extraordinary circumstances the literal enforcement of specified provisions of the Code would result in practical difficulty or unnecessary hardship; (3) that the variance is necessary for the preservation of a substantial property right of the petitioner possessed by other property in the same class of district; (4) that the granting of the variance will not be materially detrimental to the public welfare or materially injurious to the property or improvements in the vicinity; and (5) that the granting of such variance will be in harmony with the general purpose and intent of this Code and will not adversely affect the Master Plan.

The zoning administrator concluded that none of the five requirements had been met but his decision was reversed by the Board of Permit Appeals. The trial court would not disturb the Board's decision, and the Neighborhood Association that had brought the lawsuit appealed.

The California Supreme Court[20] noted that the floor area ratio was an important density control measure. The court was not persuaded that discovery of the subsurface conditions or the assertion that attrac-

---

19. *Anderson v. Board of Appeals, Town of Chesapeake Beach*, 22 Md.App. 28, 322 A.2d 220 (1974).

20. *Broadway, Laguna, Vallejo Association v. Board of Permit Appeals*, 66 Cal.2d 767, 59 Cal.Rptr. 146, 427 P.2d 810 (1967).

tive features that went beyond the Code would satisfy any of the first three criteria.

The court noted the distinction between:

> those circumstances which prevent a builder from profitably developing a lot within the strictures of the planning code and those conditions which simply render a complying structure *less profitable than anticipated.* If conditions which merely reduce profit margin were deemed sufficiently "exceptional" to warrant relief from the zoning laws, then all but the least imaginative developers could obtain a variety of variances, and the "public interest in the enforcement of a comprehensive zoning plan" [citation] would inevitably yield to the private interest in the maximization of profits.[21]

In addition to demonstrating increasing willingness of some modern courts to review local grants of variance, *Broadway, Laguna* also demonstrates the rigidity of criteria such as the floor area ratio. Perhaps the issuance of a permit conditioned upon certain things being done by the developer would have been a better solution than mechanically applying the floor area ratio.

The purpose for the special exception permit sought may influence whether it should be issued. For example, in *New York Institute of Technology, Inc. v. LeBoutillier,*[22] a college wished to obtain a special exception permit to use a large mansion it had recently acquired for educational activities. The court recognized that it would take less of a showing to grant a permit to educational institutions but concluded that the denial of the permit in this case was justified. The principal reason for not issuing the permit was that the school could have offered the instruction on its original grounds and offering instruction at the recently acquired site would have substantially increased the traffic in the residential part of the town.

For a variance § 2–204 of the MLDC, requires the land not be reasonably capable of economic use under the general development provision and the development not significantly interfere with the enjoyment of other land in the vicinity. In commenting on this section, the Code drafters stated that they had rejected the phrase "unnecessary hardship" and used "economic use" so that the applicant would be required to prove in dollar-and-cents evidence that it cannot realize any reasonable monetary gain from the land as regulated.

(B) NONCONFORMING USES

Though not considered a technique for avoiding the overrigidity of the Euclidean grid, the nonconforming use has had that effect. Most zoning enactments created carefully segregated districts, but the cities upon which these grids were to be imposed did not follow such a neat arrangement. Junkyards existed in residential or shopping areas. Stores were found in residential areas and factories next to retail sales stores. To insure similarity of use within districts, early planners sought to eliminate nonconforming uses. While planners felt that police powers of the state could justify elimination of existing uses as well as prohibition of future uses, the political process dictated that the two be treated differently. To avoid entire zoning plans being turned down by local governing bodies, a system for protecting existing lawful uses was developed.

At first rules were adopted based upon the hope that time would eliminate nonconforming uses. These rules precluded nonconforming uses from being changed. Nonconforming structures could not be altered, repaired, or restored, and a use could not be reinstituted after it had been abandoned. Implementation of these rules was difficult. Defining a nonconforming use was not easy. Things became more com-

---

21. 59 Cal.Rptr. at 151, 427 P.2d at 815.

22. 33 N.Y.2d 125, 350 N.Y.S.2d 623, 305 N.E.2d 754 (1973).

plex when courts sought to differentiate a nonconforming use in a conforming building from a nonconforming use in a nonconforming building.

In any event, nonconforming uses did not wither away. Permission to a nonconforming use created a monopoly, such as a small grocery store in a residential neighborhood. The next attack was to enact amortization ordinances ordering nonconforming uses to cease after a period of time during which the nonconforming user could recover its investment. Amortization periods ranged from one year for billboards to 25 years for gas stations to from 50 to 60 years for substantial buildings. Although courts generally enforced amortization ordinances, such ordinances were not used extensively. A 1971 survey by the American Society of Planning Officials showed that less than one-third of the responding municipalities adopted such ordinances and only 27 of 159 that had adopted them had actually employed them, mainly against billboards.

Though courts generally enforced amortization ordinances, judicial opinions expressed skepticism about their fairness. Research revealed a patchwork quilt created by variances and special use permits rather than homogeneity of uses. Some expressed doubt regarding the desirability of homogeneous zones. As a result of these factors, elimination of nonconforming uses by amortization looked less attractive. Put another way, can there really be nonconforming uses in an urban environment that essentially was nonconformist?

(C) REZONING

The only alternatives for a landowner who cannot meet the often stringent requirements for a variance and who does not wish or is not able to obtain a conditional use permit is to seek rezoning. Rezoning can cause a hardship to the neighboring property owners, and under rigid Euclidean theory, an all-or-nothing choice will have to be made between the landowner and the neighboring property owners.

(D) CONTRACT AND CONDITIONAL ZONING

Some have proposed increased use of contract or conditional zoning. Suppose the developer wishes to build a shopping center. The governing body or administrative agency can rezone the land or issue a permit. Contract zoning allows the governing body to obtain in exchange for its action the applicant's promise to do certain things. In conditional zoning or issuance of a special permit, the effectiveness of rezoning or the permit is conditioned upon the applicant doing certain things required by local authorities. The promised acts or conditions can deal with noise abatement, traffic control, setback lines, erection of fences, or any other devices to enable the center to blend into the surrounding neighborhood. As an alternative to promises by the developer to do these things, the developer can promise to pay the local authority to have these designated things done by the latter.

In the 1980s, conditions for approval of a large-scale project such as a mixed-use development or a high-rise office building may include protection of historic views, employment preferences to local citizens and minorities, and inclusion of low-income housing units or subsidy for them elsewhere.

Although contract or conditional zoning gives flexibility, courts have divided as to their validity. Contract zoning is especially vulnerable to attack, since it appears that the lawmaking power is being traded away. As a result, conditional zoning is more likely to be the method selected.

Section 2–103 of the MLDC allows the imposition of conditions for the issuance of a special development permit. These conditions can include the requirement that the developer set up mechanisms for future maintenance of the property, such as a homeowner association.

(E) FLOATING ZONES

Floating zones provide flexibility within fixed ones. The boundaries of such districts are not determined by ordinance but are fixed by approval of a petition by a property owner to develop a specific tract for a specifically designated use.

(F) BONUS ZONING

The *Broadway, Laguna* case discussed earlier in this section involved a claim by the developer that his apartment had special features that justified an area variance. In some cities, a developer can receive dispensation from normal development requirements by providing bonus features. For example, in 1971, the Bankers Trust Building in New York City was given certain planning dispensations in tower height and floor area in exchange for including a large, elevated open plaza with desirable architectural features and a two-level covered arcade of shops.

The bonus features were attractive to the developer because not only could it get dispensation mainly in floor area but the bonus features provided amenities that could make the project more attractive to tenants and obtain a higher rental.

(G) PLANNED UNIT DEVELOPMENT

The gridlike Euclidean zoning with its emphasis on individual lots is inappropriate for planning large developments on unimproved land. Section 2–211 of the MLDC authorizes the designation of specialized planned areas in which development will be permitted only in accordance with a plan of development for the entire area. The drafters of the Code contemplated relatively undeveloped land where local planners anticipate some development in the future but wish to discourage small scattered uncontrolled developments. The designation can

also be applied to urban areas anticipating major development.

If a specially planned area has been designated, no development can take place until the Land Development Agency adopts a precise plan for the area, which may include street locations, utilities, dimensions and grading of parcels, and siting of structures as well as the location and characteristics of permissible types of development. Developers owning land in the area can obtain a development permit if their development is consistent with the plan specified in the planned area ordinance.

(H) OPEN SPACE

Various techniques have evolved to preserve open space. Greenbelts, developed in England, are buffer zones of open space between developed areas. Cluster zoning allows more than concentrated density, usually in the form of high-rise multiple dwellings, in exchange for commonly used open space. The *overall* density does not exceed density limits.

Localities sometimes seek to preserve open space by creating an open space district. Palo Alto, California, created such a district to protect its undeveloped foothills in 1972. The districts were limited to open space uses, such as public recreation, enjoyment of scenic beauty, and conservation of natural resources. Single-family homes could be built but only on ten-acre minimum lots with no more than 3.5% of impervious area and building coverage. Previously the land had one-acre minimum lot requirements. The ordinance was challenged in California and Federal courts. The ordinance was upheld in *Eldridge v. City of Palo Alto.*[23]

SECTION 11.11
ESTHETICS AND CONTROL

Early in the history of public land use con-

---

23. 57 Cal.App.3d 613, 129 Cal.Rptr. 575 (1976).

trol, doubts existed as to the constitutionality of limits on property rights designed to create a more esthetic and pleasing environment. Police powers could be used to restrict development property rights to protect public health, safety, and welfare. But could property rights be limited in order to accomplish esthetic objectives?

Berman v. Parker, a 1954 decision of the U.S. Supreme Court that upheld an urban redevelopment program, ended any doubt over the validity of regulation designed to create a more esthetically pleasing environment. While esthetics were not directly involved, the opinion included language that was later used to justify esthetics as a land use control objective. It stated:

> It is within the power of the legislature to determine that the community should be beautiful as well as healthy, spacious as well as clean, well-balanced as well as carefully patrolled.[24]

In residential developments, regulations have been enacted to promote uniformity by specifying certain design requirements, such as attached garages, two-level houses, or styles of exteriors. Sometimes ordinances sought to avoid subdivision monotony by prohibiting a house from looking too much like the surrounding houses.

An ordinance passed by the Village of Olympia Fields, Illinois, combined both uniformity and nonuniformity by requiring that a permit would not be issued in the case of a design that was excessively similar, dissimilar, or inappropriate in relation to nearby property.[25]

Sometimes architectural compliance with standards is determined by an Architectural Commission created by the local governing body. The commission passes upon designs, and its approval is required before building permits can be issued.

State ex rel. *Stoyanoff v. Berkeley*[26] involved the denial of a building permit because the architectural board created by the City of Ladue, Missouri, had concluded that the design submitted did not conform to the architectural standards required by the local ordinance. The ordinance required the design to:

> conform to certain minimum architectural standards of appearance and conformity with surrounding structures, and that unsightly, grotesque and unsuitable structures, detrimental to the stability of value and the welfare of surrounding property, structures and residents, and to the general welfare and happiness of the community, be avoided, and that appropriate standards of beauty and conformity be fostered and encouraged.[27]

The design was for a single-family home in the shape of a truncated pyramid with triangular windows. The neighborhood included expensive homes in Colonial, French Provincial, and English styles. The ordinance was sustained and the Review Board's decision upheld.

SECTION 11.12 HISTORIC AND LANDMARK PRESERVATION

National concern has been increasing over the destruction of historically significant buildings. A survey of historic buildings made in 1933 included 12,000 buildings. By 1970 over one-half had been razed. The pressure to demolish or substantially alter historic structures is especially strong in fast-changing urban areas. The response has been activity at every governmental lev-

---

24. *Berman v. Parker*, 348 U.S. 26, 33, 75 S.Ct. 98, 102, 99 L.Ed. 27 (1954).

25. The ordinance was found to be invalid as an improper delegation of decision making to the architectural advisory committee without providing adequate standards. See *Pacesetter Homes, Inc. v. Village of Olympia Fields*, 104 Ill.App.2d 218, 244 N.E.2d 369 (1968).

26. 458 S.W.2d 305 (Mo.1970).

27. Id. at 306–307.

el, but especially at the local level of government.

The principal steps taken have been to survey historic landmarks, create historic preservation districts, acquire ownership of historically significant structures or easements over their facades, and grant tax concessions to owners who will preserve the historic character of their structures. Illustrations of historic district designations are Vieux Carré by the City of New Orleans,[28] a four-block area surrounding the Lincoln house in Springfield, Illinois,[29] and the Old Town District of San Diego, California.[30]

In addition to the creation of historic districts, some local entities have designated specific buildings as historic landmarks despite the possibility that the designation could be considered spot zoning. Illustration of buildings that could justify such a designation would be New York City's Grand Central Station,[31] Lincoln's home in Springfield, Illinois, or Jefferson's home in Monticello, Virginia.

Designation usually means that the landowner cannot demolish or alter the existing structure. To alleviate the possible hardship that such a limitation can cause, the New York Landmark Preservation Ordinance [32] provides that the owner is expected to realize at least a 6% return on the property. If it proves to be an economic hardship because of a lesser return, the commission is given discretion to ease the hardship by effectuating a real estate tax rebate or the commission is afforded the additional right of producing a buyer or lessee who can profitably utilize the premises without the sought-for alteration or demolition. If these remedies prove unrealistic or unob-

tainable, the city is given the power to condemn the property.

The most serious constitutional attack on ordinances creating historical districts or designating landmark buildings has been the claim by the landowner that the use limitation created by the ordinance or designation violates the constitutional prohibition against taking private property without paying just compensation. If the action by local authorities is considered a taking, the owner must be compensated. As compensation can be costly, especially at a time when local government faces continuing fiscal problems, ordinances and designations are usually based upon the police powers. Courts have been hesitant to consider anything short of actual appropriation to be a taking as long as the law can be justified as an exercise of police powers.

A method of relieving against the hardship of placing the entire burden on the owner of the landmark building has been the development in New York City of Transferable Development Rights as a means of preserving historic buildings without incurring the high cost of condemnation. Under this plan, a landmark owner who is prevented from using its property to its full permitted use, such as the maximum floor area ratio permitted by law, is compensated by being given a transferable developmental right for a specific transfer district. The developmental right equals the excess potential the landowner was precluded from using because its building had been designated a historical landmark. It can use this excess potential in the transfer district building even if it exceeds normal density limits for the building. An owner who does

---

28. Upheld in *Maher v. City of New Orleans*, 516 F.2d 1051 (5th Cir.1975).

29. Upheld in *Rebman v. City of Springfield*, 111 Ill.App.2d 430, 250 N.E.2d 282 (1969).

30. Upheld in *Bohannan v. City of San Diego*, 30 Cal.App.3d 416, 106 Cal.Rptr. 333 (1973).

31. Upheld in *Penn Central Transportation Co. v. New York City*, 438 U.S. 104, 98 S.Ct. 2646, 57 L.Ed.2d 631 (1978).

32. N.Y.Admin.Code tit. 8–A, §§ 207.10 et seq. (1965).

not wish to use the development right can transfer it. This device is likely to receive consideration in the future.[33]

SECTION 11.13 THE ENVIRONMENTAL MOVEMENT

During the 1960s the American public began to demand that government make strong efforts to improve the quality of life in America. The goals of this movement can be illustrated by the preamble of the National Environmental Policy Act (NEPA) enacted by Congress in 1969:

> (a) The Congress, recognizing the profound impact of man's activity on the interrelations of all components of the natural environment, particularly the profound influences of population growth, high-density urbanization, industrial expansion, resource exploitation, and new and expanding technological advances and recognizing further the critical importance of restoring and maintaining environmental quality to the overall welfare and development of man, declares that it is the continuing policy of the Federal Government, in cooperation with State and local governments, and other concerned public and private organizations, to use all practicable means and measures, including financial and technical assistance, in a manner calculated to foster and promote the general welfare, to create and maintain conditions under which man and nature can exist in productive harmony, and fulfill the social, economic, and other requirements of present and future generations of Americans.
>
> (b) In order to carry out the policy set forth in this chapter, it is the continuing responsibility of the Federal Government to use all practicable means, consistent with other essential considerations of national policy, to improve and coordinate Federal plans, functions, programs, and resources to the end that the Nation may—
>> (1) fulfill the responsibilities of each generation as trustee of the environment for succeeding generations;
>> (2) assure for all Americans safe,

> healthful, productive, and esthetically and culturally pleasing surroundings;
>> (3) attain the widest range of beneficial uses of the environment without degradation, risk to health or safety, or other undesirable and unintended consequences;
>> (4) preserve important historic, cultural, and natural aspects of our national heritage, and maintain, wherever possible, an environment which supports diversity and variety of individual choice;
>> (5) achieve a balance between population and resource use which will permit high standards of living and a wide sharing of life's amenities; and
>> (6) enhance the quality of renewable resources and approach the maximum attainable recycling of depletable resources.
>
> (c) The Congress recognizes that each person should enjoy a healthful environment and that each person has a responsibility to contribute to the preservation and enhancement of the environment.[34]

To protect the environment, NEPA required federal agencies whose proposed activities would disturb the natural environment to "utilize a systematic, interdisciplinary approach . . . in planning and in decision making which may have an impact on man's environment."[35] Under NEPA, agencies were to develop procedures to "insure that presently unquantified environmental amenities and values" may be given appropriate consideration in decision making along with economic and technical considerations.[36] Agencies were to prepare a detailed statement on "(i) the environmental impact of the proposed action, (ii) any adverse environmental effects which cannot be avoided should the proposal be implemented, (iii) alternatives to the proposed action, (iv) the relationship between local short-term uses of man's environment and the maintenance and enhancement of long-

---

33. See J. Costonis, *Space Adrift* (1974); Costonis, *Development Rights Transfer: An Exploratory Essay*, 83 Yale L.J. 75 (1973).

34. 42 U.S.C.A. § 4331(c).

35. 42 U.S.C.A. § 4332(2)(A).

36. 42 U.S.C.A. § 4332(2)(B).

term productivity, and (v) any irreversible and irretrievable commitments of resources which would be involved in the proposed action should it be implemented." [37] The latter statement was called an Environmental Impact Statement (EIS).

When agency decisions are challenged in court, courts extend considerable deference to agency determinations. Agencies possess scientific and technical skills usually unavailable to judges. The courts must see that the laws are followed and the public protected from overzealous or insensitive public agencies.

Do the EIS requirement and judicial review of whether the agency has complied simply give environmentalists a weapon to delay needed projects? Undoubtedly most projects challenged are built. Nevertheless, the requirement that all factors, including environmental ones, be carefully examined should cause agencies to pause before they rush headlong into projects and may assist those in agencies who argue for greater consideration of environmental factors.

Congress has enacted, in addition to NEPA, other legislation dealing with environmental matters, such legislation requiring that air and water meet certain standards,[38] that solid waste be controlled,[39] and that dangerous hazardous waste sites be cleaned up.[40] States have enacted legislation to protect the environment and preserve places of natural beauty such as coastlines, lakeshores, and scenic canyons.

The environment includes urban areas. The location of a highway through a city can affect those displaced by road construction and have a harmful effect upon urban life. Other projects built in urban areas,

such as large office buildings, schools, hospitals, and housing complexes can also have a serious impact on urban life. Such projects may require an EIS or state equivalent.

Typically projects requiring an EIS involve activity by governmental agencies. However, California held that the California Environmental Quality Act required local entities who issue permits for construction by private parties comply with the statute and furnish an EIS.[41]

SECTION 11.14
JUDICIAL REVIEW

While action or inaction of a local governing body can be challenged politically, such as by initiative, referendum, or the election process, the opponents of an ordinance or those who wish to contest a permit decision often go to court. A preliminary and often serious obstacle is whether the challenger has "standing" to challenge the ordinance judicially.

Generally, judicial remedies are limited to persons who have suffered or are likely to suffer injury to themselves. Persons cannot seek legal relief when the injuries they will suffer are no different from those suffered by the general public. Nor are judicial remedies generally given to those who wish to assert the rights of third parties.

Rules that look at whether the plaintiff has standing to sue have a number of objectives. One is to avoid collusive lawsuits. Such suits are not between true adversaries but are between apparent adversaries who attempt to use the judicial process to establish a precedent both desire.

---

37. 42 U.S.C.A. § 4332(2)(C).

38. 42 U.S.C.A. § 7401 (air); 33 U.S.C.A. § 1251 (water).

39. 42 U.S.C.A. § 6901.

40. 42 U.S.C.A. § 9601.

41. *Friends of Mammoth v. Board of Supervisors of Mono County,* 8 Cal.3d 247, 104 Cal.Rptr. 761, 502 P.2d 1049 (1972) (condominium in rural, scenic area).

Other more important reasons exist for limiting judicial relief. Limitations of this type are designed to sharpen the issues. Sharper issues are more likely to result when the plaintiff can show that it suffered specific injuries than if it broadly claims the defendants have acted illegally.

It is easier for a court to determine whether the denial of a permit for a particular project to be built in a particular place violates the law than to determine whether in general an overall plan that limits development or growth is illegal. The remedy to be awarded will be more difficult to frame and more drastic when there is a broad judicial attack rather than a claim of *specific* injury to the plaintiff. In the background there is a desire to conserve judicial energy and avoid the court's deciding abstract matters of great public importance that should be addressed to other arms of the government.

In land use litigation, a person whose property has been directly affected can always challenge the ordinance or official decision. Increasingly, standing status has been accorded nearby property owners and neighborhood associations.

In *Warth v. Seldin*,[42] the U.S. Supreme Court faced the question of whether certain individuals and groups could question the validity of ordinances claimed to be "exclusionary" (see Section 11.15) and designed to keep out low and moderate income housing from Penfield, New York, a suburb of Rochester. A divided court held that the various groups from Rochester representing low-income persons, Rochester taxpayers, and residential construction firms did not have standing to judicially challenge the Penfield ordinance. Nor did the court grant standing to individuals who claimed they had been denied low-income housing in Penfield. Individuals who could show restrictions to particular proposed projects in which they would have resided or which *they* would have built would have standing. The court concluded that none of the plaintiffs had shown that they would have benefited personally from court intervention.

The case undoubtedly will make it more difficult to use the federal courts to challenge local developmental controls. Persons who wish to challenge local ordinances will have to go farther and seek permits or show in some way that they would have been able to live in the area had it not been for the ordinances.

If there is standing, the principal grounds for legal challenge have been the following:

1. The Enabling Act did not authorize the act in question.

2. The ordinance or permit decision was unconstitutional.

3. The making of decisions by nonelected officials, such as an architectural review commission, was invalid because the local governing body delegated power without proper standards.

Although an occasional attack has succeeded,[43] judicial challenges alleging invalidity of local decisions have rarely succeeded.

In *Village of Belle Terre v. Boraas*,[44] Belle Terre, New York, a village of less than one square mile consisting of 700 people and 220 residents restricted the entire village to single-family dwellings. The village prohibited three or more unrelated persons from living together. The ordinance was challenged by six unrelated students from a nearby university who wished to live together.

42. 422 U.S. 490, 95 S.Ct. 2197, 45 L.Ed.2d 343 (1975).

43. The most successful attacks have been in New Jersey and Pennsylvania. See Section 11.15.

44. 416 U.S. 1, 94 S.Ct. 1536, 39 L.Ed.2d 797 (1974).

In upholding the challenged ordinance, the Supreme Court stated:

> A quiet place where yards are wide, people few, and motor vehicles restricted are legitimate guidelines in a land-use project addressed to family needs. . . . The police power is not confined to elimination of filth, stench, and unhealthy places. It is ample to lay out zones where family values, youth values, and the blessings of quiet seclusion and clean air make the area a sanctuary for people.[45]

The court concluded that the ordinance was a rational means of achieving these objectives. This decision will make constitutional attacks on local development control even more difficult than in the past.

SECTION 11.15 HOUSING AND LAND USE CONTROLS

(A) RESIDENTIAL ZONES

When zoning took center stage in the 1920s, one justification for local control was the need to develop and maintain residential neighborhoods that would keep their value and enable persons to live among their own kind. In the early days of zoning, this was accomplished principally by limiting residences in certain districts to single-family detached homes and severely curtailing the multifamily living unit.

Social aspects of residential zoning were revealed in the first opinion of the U.S. Supreme Court that passed upon the validity of zoning laws. In *Village of Euclid v. Ambler Realty Co.,*[46] the U.S. Supreme Court, after justifying the limitation of property rights by the police power to provide a safe and more pleasant community, expressed the American deification of the single-family home and disdain for apartment living.

Home ownership has always been a prized goal for upward mobile classes and a status symbol of achievement. Before the advent of modern *public* land use controls, private mechanisms existed to develop and maintain quality residential neighborhoods of individually owned homes. Affluent Americans wanted and were able to buy homes in districts exclusively devoted to attractive, well-built homes on spacious lots. Such home buyers generally sought to live with persons of their socioeconomic background. They built or bought in neighborhoods where their children could play and go to school with children of their cultural background and social class. The free market encouraged the development and maintenance of such neighborhoods. Only the affluent could afford fine residential homes because, at least in normal times, the demand for such homes was large and the supply limited. As a result high prices kept out persons of low or moderate income.

Buyers wanted assurance that there would be no change in the residential environment. To induce them to buy in such neighborhoods, developers used restrictive covenants. Buyers would be willing to pay higher prices to live in districts that were carefully restricted to quality single-family homes. Buyers wanted assurance that property values would be maintained, that schools would be kept high quality, and that neighbors shared their interests and values.

Political power of local communities to control land use enabled communities to resist market forces that might otherwise have impinged upon the model of the single-family detached home. Unlike restrictive covenants, public land use controls clash with free-market concepts. This is demonstrated by the frequent alliance of developers and low-income groups who unite to attack zoning laws.

---

45. 416 U.S. at 9, 94 S.Ct. at 1541.

46. Supra note 18.

From its inception, residential zoning excluded certain types of people from designated districts. This was justified as a proper planning control designed to encourage investment and the development of good residential neighborhoods. It was not until the 1960s that strong attacks were made upon zoning laws as *excluding* low- and moderate-income families from districts reserved to the more affluent. This is discussed in greater detail in (C).

(B) SUBDIVISION CONTROLS

Many communities grew when developers subdivided raw land for homes. Subdividing increased population and placed added burdens upon the community to provide streets, fire and police protection, schools, parks, and other municipal services. While many of these costs could have been provided by special property assessments, often many local communities exacted conditions for approval of subdivisions. These conditions could include dedication of part of the raw land for public use for streets, schools, and recreation. Sometimes in lieu payments were conditions for subdivision approval. These payments, sometimes called exactions, were used to reimburse the community for providing facilities and services in the subdivision or services that the community had to furnish to residents of the subdivision.

Some states have gone farther and have permitted local communities to require land or a money exaction for parks and recreation facilities "to bear a reasonable relationship to the use of the park and recreational facilities by the future inhabitants of the subdivision." Such an exaction was upheld with the suggestion that a city could exact a fee to purchase park land some distance from the subdivision but that could be used by subdivision residents.[47]

Subdivision exactions place upon new subdivision residents the cost for additional municipal services they will require. But if exaction substantially increased construction costs and prices of subdivision houses, they can, in addition, be exclusionary, a point developed in greater detail in (C).

(C) EXCLUSIONARY AND INCLUSIONARY ZONING

Zoning has always excluded persons. It was not until the mid-1950s and the 1960s that attacks were made on zoning laws as exclusionary. Undoubtedly changes in demographic patterns brought this to the surface. In *Oakwood at Madison, Inc., v. Township of Madison*, the court recognized this and stated:

> Madison Township, among other municipalities, is encouraging new industry. Industry is moving into the county and region from the central cities. Population continues to expand rapidly. New housing is in short supply. Congestion is worsening under deplorable living conditions in the central cities, both of the county and nearby. The ghetto population to an increasing extent is trapped, unable to find or afford adequate housing in the suburbs because of restrictive zoning.[48]

Focusing upon zoning laws as a cause for social injustice took the form of describing white suburbs surrounding central cities as "tight little islands" that refused to do their fair share of housing the poor.[49]

The decline of central cities and the flight to the suburbs caused fiscal problems for central cities. (By the 1980s, some cities saw a reversal of this trend as the middle classes moved back and gentrified decaying inner-city sections.) Cities were left with the increased costs of social services for the

---

47. *Associated Home Builders v. City of Walnut Creek*, 4 Cal.3d 633, 94 Cal.Rptr. 630, 484 P.2d 606 (1971).

48. 117 N.J.Super. 11, 17, 283 A.2d 353, 356 (1971).

49. Sager, *Tight Little Islands: Exclusionary Zoning, Equal Protection, and the Indigent*, 21 Stan.L.Rev. 767, 791–792 (1969).

poor and a declining tax base because the middle classes, along with many businesses, moved to the suburbs. Central cities looked upon the suburbs as parasite communities composed of persons who earned their livings in central cities and used the central cities' cultural and recreational facilities. These parasite communities, however, did not bear their proper share of the urban costs that were increasing because of the need to care for poor people.

Suburbs sought to attract light industry that paid high property taxes but did not require many employees of low or moderate income. The latter were to be avoided. They lived in modest homes or apartments that brought in less revenue than the expenditures—mainly in social services and education—required to take care of these people. Persons of middle income and above wanted to live with those who shared their social and cultural values. Some techniques used to accomplish these purposes were as follows:

1. large lot requirements
2. minimum house size requirements
3. exclusion of multiple dwellings
4. exclusion of mobile homes
5. unnecessarily high subdivision requirements or in lieu exactions [50]

Judicial attacks have been made on many ordinances by those who wish to build or those who wish to be able to live in sub-urbs. Usually these attacks are based upon claims that such ordinances violated federal and state constitutions.[51]

In 1965 the Pennsylvania Supreme Court in *National Land and Investment Co. v. Kohn*,[52] invalidated a four-acre minimum lot requirement. Noting that evaluating the constitutionality of a local zoning ordinance was not easy and declining to be "a super board of adjustment" or "a planning commission of the last resort," the court nevertheless saw itself as a judicial overseer "drawing the limits beyond which local regulation may not go." [53]

The court stated that controlling density is a legitimate exercise of police power and that each ordinance must be examined individually in light of the surrounding circumstances and the needs of the community. Rejecting the justification for a four-acre minimum, the Pennsylvania Supreme Court focused upon the exclusionary aspects of the ordinance and in oft-quoted language stated:

> The briefs submitted by each appellant in this case are revealing in that they point up the two factors which appear to lie at the heart of their fight for four acre zoning.

> The township's brief raises (but, unfortunately, does not attempt to answer) the interesting issue of the township's responsibility to those who do not yet live in the township but who are part, or may become part, of the population expansion of the suburbs. Four acre zoning represents Easttown's position that it does not

---

50. *Building the American City*, Report of National Commission on Urban Problems, H.R.Doc. No. 91–34, 91st Cong., 1st Sess. 211–216.

51. The first challenge to an exclusionary technique, a minimum dwelling size restriction, failed in *Lionshead Lake, Inc. v. Wayne Township*, 10 N.J. 165, 89 A.2d 693 (1952), appeal dismissed 344 U.S. 919, 73 S.Ct. 386, 97 L.Ed. 708 (1953). New Jersey subsequently invalidated many exclusionary techniques in *Southern Burlington County N.A. A.C.P. v. Township of Mt. Laurel*, 67 N.J. 151, 336 A.2d 713 (1975), a lengthy but interesting opinion. For a later decision, see 92 N.J. 158, 456 A.2d 390 (1983), called Mt. Laurel II, which blasted the inaction of Mt. Laurel. Repeating its earlier decision, the court expanded a "builder's remedy" it created in an earlier case, allowing the builder greater density in exchange for mixing low- and moderate-income housing with expensive housing. As to reinvestment in decaying areas displacing residents see Salsich, *Displacement and Urban Reinvestment: A Mt. Laurel Perspective*, 53 U. of Cinn.L.Rev. 333 (1984).

52. 419 Pa. 504, 215 A.2d 597 (1965). For a more recent case, see Appeal of Elocin, Inc., 501 Pa. 348, 461 A.2d 771 (1983).

53. 215 A.2d at 607.

desire to accommodate those who are pressing for admittance to the township unless such admittance will not create any additional burdens upon governmental functions and services. The question posed is whether the township can stand in the way of the natural forces which send our growing population into hitherto undeveloped areas in search of a comfortable place to live. We have concluded not. A zoning ordinance whose primary purpose is to prevent the entrance of newcomers in order to avoid future burdens, economic and otherwise, upon the administration of public services and facilities can not be held valid. Of course, we do not mean to imply that a governmental body may not utilize its zoning power in order to insure that the municipal services which the community requires are provided in an orderly and rational manner.

The brief of the appellant [homeowners] creates less of a problem but points up the factors which sometime lurk behind the espoused motives for zoning. What basically appears to bother [homeowners] is that a small number of lovely old homes will have to start keeping company with a growing number of smaller, less expensive, more densely located houses. It is clear, however, that the general welfare is not fostered or promoted by a zoning ordinance designed to be exclusive and exclusionary. But this does not mean that individual action is foreclosed. "An owner of land may constitutionally make his property as large and as private or secluded or exclusive as he desires and his purse can afford. He may, for example, singly or with his neighbors, purchase sufficient neighboring land to protect and preserve by restrictions in deeds or by covenants inter se, the privacy, a minimum acreage, the quiet, peaceful atmosphere and the tone and character of the community which existed when he or they moved there." (quoting an earlier opinion) [54]

Pennsylvania and New Jersey have gone far toward what some have called socioeconomic public land use controls. Courts in these states have taken an active role in attempting to compel local communities to confront and deal with housing problems. This activistic approach has been criticized as an undue interference in local affairs and an unwarranted attempt by courts to venture into complicated social and fiscal problems more appropriate for legislative treatment. There has been some question regarding the right of courts and legislative bodies to frustrate the desires of persons to choose the types of people near whom they wish to live.

Other courts have not yet been as willing as Pennsylvania and New Jersey courts to step into local land use decisions. This is demonstrated by cases discussed in (D) that permit local communities to limit growth through the land use control process.[55]

The preceding discussion examined zoning laws that make it legally or economically impossible to build housing for low- and moderate-income persons. Some recent cases have dealt with attempts by local communities to impede the building of such housing by not extending reasonable cooperation needed to get such projects built.[56]

The introduction of the term *exclusionary* has been accompanied by increasing use of what is called inclusionary zoning. New Jersey compelled each community to do its fair share to house persons of low or moderate income.[57] Failure by local communities in New Jersey to do so can result in the communities' zoning laws being invalidated. This, in a sense, is similar to affirmative action employment programs. The court is asking local communities not simply to refrain from excluding certain people but to affirmatively use efforts to *include* them. In 1969 Massachusetts attempted to require municipalities to provide for such housing by an "Anti-Snob" Zoning Law.[58]

---

54. 215 A.2d at 612–13 (the court's footnotes omitted).

55. Also see *Village of Belle Terre v. Boraas*, discussed in Section 11.14.

56. *Village of Arlington Heights v. Metropolitan Housing Development Corp.*, 429 U.S. 252, 97 S.Ct. 555, 50 L.Ed.2d 450 (1977) (violation requires racially discriminatory intent or purpose).

57. *Southern Burlington County N.A.A.C.P. v. Township of Mt. Laurel*, supra note 51.

58. Mass.Gen.Laws Ann. ch. 40B, §§ 20 et seq.

Efforts have been made at local levels of government to include those often excluded. The first ordinance of this type was enacted in Fairfax County, Virginia, in 1971. Similar ordinances were enacted in Montgomery County, Maryland; Los Angeles, California; Berkeley, California; and Palo Alto, California. Typically such ordinances require that developers of more than a certain number of units include a designated percentage for low- and moderate-income tenants. Often they inhibit development.[59]

(D) PHASED GROWTH

The once accepted goal of growth as the proven road to prosperity and social mobility began to be questioned seriously in the 1960s. Some advocated goals that would look less at material measurements such as the gross national product and more at measurements of the quality of life. Different measurement processes and the realization of the finite nature of resources were factors that led to the development of the environmental movement discussed in Section 11.13.

From the standpoint of urban planning, growth became synonymous with urban sprawl, the often uncontrolled development of land at the outskirts of major American cities. What often resulted was an unplanned, market-oriented development of isolated and scattered parcels of land on the fringes of suburbia followed by the gradual urbanization of the intervening undeveloped areas. Urban sprawl has been criticized as unesthetic, wasteful of valuable land resources, and unduly increasing the cost of providing municipal services.

Rethinking of goals and priorities reflected themselves in public land use planning. Communities sought to avoid the disruptive effect of an influx of people into their communities.

Plans attacked as exclusionary and defended as phased growth often have common characteristics. However, a crude comparison can be made between these two classifications. Suburbs whose plans are exclusionary seek to be tight little islands. They wish to keep out low- and moderate-income families for social and fiscal reasons. However, phased growth plans such as those discussed in this subsection, are designed to keep out *all* people, not simply undesirable ones.

Another preliminary caution is that all land use controls tend to exclude because density control is a legitimate and frequent objective. For this reason terms such as *exclusionary* and *phased growth* must be looked at in light of their value-laden characteristics. Exclusionary zoning has been the label critics of suburbs have developed when they attack homogeneous suburbs as tight little islands. Yet phased growth, though frequently a label used to justify an exclusionary plan, carries with it socially desirable goals of protecting nature and wildlife and preserving the values and attributes of small town life.

The first significant case to pass upon a phased growth plan involved a system devised by Ramapo, New York, a suburb of New York City some 25 miles from downtown Manhattan. The master plan adopted in 1966 contemplated an eighteen-year period during which public facilities would be built and after which the town would be fully developed. In 1969 an ordinance was passed to control residential development on any vacant lots or parcels. Applications for permits were to be made to the town board, which was to take into account the availability of major public improvements and services such as sewers or approved substitutes, drainage facilities, parks, or recreational facilities including public school sites, improved roads, and firehouses. The

---

59. *Board of Supervisors v. DeGroff Enterprises, Inc.*, 214 Va. 235, 198 S.E.2d 600 (1973). For a comprehensive discussion, see Kleven, *Inclusionary Ordinances*, 21 U.C.L.A.L.Rev. 1432 (1974).

degree of availability of each facility to the site was to be measured and scored on a scale of 0 to 5, and no permit could be issued unless a minimum of 15 points were obtained. A developer could advance the date for development by providing facilities itself or by obtaining a variance. Defenders of the Ramapo approach pointed to the advantage of each parcel owner's knowing the *eventual* use and density classification of every parcel. The plan only postponed development. Backers of the Ramapo approach pointed to specific criteria governing planning board decision making rather than the fuzziness of most permit-issuing criteria.

The Ramapo approach was upheld by the New York Court of Appeals.[60] The court held that the plan was within the powers given the town by the state enabling laws,

with the effective end to be served by the time controls within the state's police powers and not so unreasonable as to constitute a taking without compensation. (By the 1980s the economic slowdown forced Ramapo to substantially modify the plan in order to *encourage* development.)

A second case involved an attempt by a suburban community to limit growth by controlling the number of units that could be built. The plan was upheld,[61] but again by the 1980s, the plan was *too* effective and changes had to be made to encourage growth.

A third case citing the fragile ecology of a small rural town upheld an ordinance drastically increasing the minimum lot size to block a large development of second homes for urban residents.[62]

---

60. *Golden v. Planning Board of Town of Ramapo*, 30 N.Y.2d 359, 334 N.Y.S.2d 138, 285 N.E.2d 291 (1972), appeal dismissed 409 U.S. 1003, 93 S.Ct. 436, 34 L.Ed.2d 294 (1972). Both majority and dissent called for regional land use planning.

61. *Construction Industry Association v. City of Petaluma*, 522 F.2d 897 (9th Cir.1975).

62. *Steel Hill Development, Inc. v. Town of Sanbornton*, 469 F.2d 956 (1st Cir.1972).

12

Financing the Project: The Lender's Perspective

SECTION 12.01 RELEVANCE

Those who perform design and construction services and those who supply materials and equipment expect to be paid for their services and goods. They also expect to be paid without unreasonable delay. Although the law or the contract may give them rights in the event they are not paid, either against the property or, in the case of subcontractors and suppliers, against a surety, the principal concern is that there be a reliable and steady money flow.

Usually the owner must obtain funding commitments from third parties to pay for the project as it is designed and built. Those third parties, whether public or private, will be furnishing funds for a project and will want a project that will cost what it is expected to cost, will be of the quality called for by the contract, and will be completed and ready for use and occupancy at the promised time.

Section 12.02 touches briefly upon funding from the perspective of the active participants in the process, while Sections 12.03 and 12.04 shift to the concerns of those who put up the funds.

SECTION 12.02 FUNDING PROBLEMS

Those who plan to engage in design or construction should be concerned about the possibility that no funds will be available (causing project abandonment), that the project will be underfunded, or that there will not be a smooth flow of funds as the project proceeds. In addition, those who have made contracts for the performance of design or construction services may find that some funding sources will insist that contracts for design and construction be consistent with the loan or grant terms and not operate in a way that may frustrate the objectives of the lending sources. If so, they may ask for these contracts to be modified as a condition to the loan or grant.

The owner's inability to obtain funds for a project is a particular problem for design professionals. If it appears at the outset of the relationship that funds may not be available, the design professional must make a serious appraisal of the risks before determining whether to perform the requested services. Protective language in a contract may be useful and even effective. However, it is not an ironclad assurance that the owner will not be able to claim, and claim successfully, that there was an "understanding" that the design professional would *not* be paid if the project were abandoned because funds could not be obtained.

The next problem, that of a tight budget for an underfunded project, can have an adverse effect on both design professional and contractor. A tight budget—the amount of money the owner expects to spend of its own and that of an outside funding source—creates performance risks. Such a budget may be too brittle to take the strains of the following:

1. Design adjustments, particularly those that involve corrective work.

2. Escalating labor or materials prices, principally during the design phase but even during construction.

3. Changed conditions during performance that increase the contract price.

4. Claims by participants.

Any of these events can adversely affect the relationship between the participants, lead to a breakdown, generate claims, or culminate in litigation. The problem of the tight budget is even more difficult for design professionals who have the misfortune to undertake or the propensity for undertaking design services for an owner who has quality and quantity expectations *not* commensurate with the budget.

To a modest degree, the American Institute of Architects (AIA) has sought to deal with this problem in its standard documents. For example, B141 ¶ 2.2 (1977) requires the owner to include contingencies for bidding and change in its budget and gives the architect power to request the owner to provide "a statement of funds available for the Project, and their source." A201, ¶ 3.2.1. (1976) gives a contractor the power to request that the owner furnish "reasonable evidence that he has made financial arrangements to fulfill his obligations." Failure to provide this gives the contractor the right to refuse to execute the construction contract or begin work. Although vaguely worded, it is much better than whatever common law rights may allow a contractor to demand assurance before it begins to perform.

Inclusion of these provisions demonstrates the importance of underfunding to

architect and contractor and can provide some assistance to those who perform design and construction services. However, it is likely that requests permitted under these provisions will be made sparingly in order to avoid casting doubt upon the owner's financial capability.

Delayed interim fee payments to the design professional and progress payments to contractors, another financing problem of importance, can result from many causes. Assuming that the money flow is not sporadic or stopped because of the fault of the person to whom payments are made initially, the lack of a smooth and reliable money flow can be traceable to overly rigid or inefficient administration by owner or funding source. If this can be predicted by past experience or reputation, the design professional or contractor concerned may decide not to perform services at all, may build an unusual funding charge into the price, may include, where possible, a stiff late payment charge, or may provide for suspension or termination remedies in the event of recurring late payments.

If a funding source enters the picture, it is likely to insist that its loan agreement or grant conditions take precedence over any of the contracts already made, such as those between the owner and design professional or contractor. It usually has the bargaining power to insist that its language control.

Such a contractual priority system creates the risk of confusion and uncertainty. More important, the funding source can effect drastic changes in the contracts made between owner and designer or contractor by demanding that these changes be made as a condition to the loan agreement or making of the grant. (Owners may be wise to condition their contracts upon lender approval.)

Clauses to which the funding source may object are as follows:

1. Clauses that can increase the price, such as changes in the work or changed conditions.

2. Clauses under which the contractor may be paid too liberally or which create the risk of liens, such as loose payments provisions.

3. Clauses that exculpate performing parties if things go wrong.

4. Clauses that create a contractual dispute mechanism thought *too* prone to awarding the contractor money, such as arbitration.

5. Clauses that make it difficult to insist that a design professional or contractor keep performing if the owner-borrower defaults and the funding source must take over.

If the earlier contracts are binding and not expressly conditioned upon lender approval, design professional and contractor should study the proposed changes, evaluate their effect, and stand firm if changes would expose them to unreasonable risk.

SECTION 12.03
PUBLIC PROJECTS

An extended discussion of public financing of improvements to real property is beyond the scope of this treatise. However, a few general observations are made here.

Federal projects are usually funded by appropriations made by Congress. Federal agencies seek contractual and administrative controls to avoid the cost of a project exceeding the appropriate funds. One method of controlling costs is the labyrinthine system set up by federal agencies under which the only individual with any power to direct changes is the contracting officer, the frequently required contract clauses designed to protect against cost overruns, and the specialized dispute resolution system employed by those agencies. In addition, other branches of the government—the Comptroller-General and Congressional committees—frequently monitor federal procurement to avoid waste and corruption. The result of such tightly controlled systems can be overly rigid, bureau-

cratic rules and seemingly intransigent public officials.

State projects have similar specialized budgetary constraints. Projects are likely to be funded with appropriations by legislatures or by bond issues. Again the importance of avoiding cost overruns that exceed appropriations or bond issues becomes crucial. In dealing with state public contracts, there are likely to be specialized statutes, regulations, and dispute resolution procedures. In addition, it may be important to protect the interests of bondholders to ensure that the funds are used as promised to the bondholders who have lent money for the project. One of the catastrophic events of the early 1980s was "Whoops," the ambitious nuclear power projects in the State of Washington. Because of cost overruns and a judicial decision invalidating commitments of public utilities to pay the bondholders, the projects are in default and bondholders have not been paid. Costly litigation will inevitably result.

Local projects also have similar budgetary constraints. In addition, bonds to fund public improvements frequently require voter approval, often by more than a majority of votes. Projects can die during the design phase. There can be costly trips to the courthouse. Those unhappy with the project often seek judicial decrees to stop the project, usually based upon the inadequacy of an environmental impact statement or irregularities in contract award.

A distinction should be drawn between *general obligation* bonds and *revenue* bonds. Traditionally the former were used to build schools, streets, and other public works. They are backed by the public entity's pledge to tax its citizens and are safe investments. At present, though, they make up only 20% of the bond market.

Revenue bonds are less secure, being backed only by the revenue of the project the bonds are to finance. If the hospital, parking garage, housing project, or power plant does not generate the needed revenue, the bondholders will not be paid.

Revenue bonds have become popular because they pay a higher return, do not need voter approval, and do not burden the credit capacity of the city or state. They can avoid competitive bidding requirements for awarding construction contracts.[1]

Increasingly there are projects that are funded by a mix of public agencies and even those that are funded by a combination of private entities and public agencies. The interstate road system, under which 90% is funded by the federal government and 10% by the states, and waste water projects, under which the federal government pays 75%, the state 12.5%, and the local entity 12.5%, are illustrations of such a mixed funding system.

In waste water projects, money is given in the form of grants to local entities called grantees. The local entities must obtain approval of state officials, the latter often empowered to act for the federal agencies. Many of these projects have failed: the projects have not worked as expected, and contractors have claimed large cost overruns. One difficulty in settling claims has been the mixed nature of the funding. The state and federal governments have been wary of settlements for which they are expected to pay. They have also been concerned that settlements would exonerate participants, such as the public entity, the design professional, and the contractors.

Public agencies that make grants frequently seek to dictate the conditions under which these funds are to be used. The Environmental Protection Agency (EPA) insists upon changed conditions clauses dealing with unforeseen subsurface conditions in waste water treatment contracts it funds. The local entities, however, may be governed by local laws that bar a contractor from receiving funds *beyond* the contract price. This can generate contradictory contract terms and administrative confusion and make settlement of claims difficult.

Another type of combination can be seen in some of the federal legislation that has sought to provide incentives for the construction of housing. Section 236 of the National Housing Act creates a hybrid of public and private funds. The nominal participants are private entities such as contractors, lenders, and owners. However, owners are nonprofit, no-asset corporations especially created for these projects. Contracts are made with them despite their lack of financial resources because lenders advance all of the construction costs and are insured against losses by the Department of Housing and Urban Development (HUD), a federal agency.

Just as EPA and state funding agencies for waste water projects keep a contractual distance from their grantees, HUD avoids direct contact with the participants but keeps a tight control on the money. HUD must approve disbursements, and if a project fails, the lender assigns the debt and mortgage to HUD or forecloses on the mortgage and delivers title to HUD. Either way, HUD is substituted for the lender. Those who have not been paid often seek to recover against HUD.

Public funding agencies often avoid any contractual relationship with active participants in the construction process in part to avoid having claims made against them. Two recent Illinois cases demonstrate that the failure to make contracts with key participants has its risks. The first, *Illinois Housing, etc. v. Sjostrom & Sons,*[2] involved a claim by a state housing development authorized to issue revenue bonds against the contractor and architect and the second, *Illinois Housing, etc. v. M–Z Construction Corp.,*[3] involved a claim by a similar housing authority against the architect. The

---

1. *Willman v. Children's Hospital at Pittsburgh,* 74 Pa.Cmwlth. 67, 459 A.2d 855 (1983).

2. 105 Ill.App.3d 247, 61 Ill.Dec. 22, 433 N.E.2d 1350 (1982).

3. 110 Ill.App.3d 129, 65 Ill.Dec. 665, 441 N.E.2d 1179 (1982).

housing authorities contended that they were not ordinary private lenders for various reasons, among them that the developer puts up very little of its own money and that there is often an identity of interest between the owner and general contractor.

The contracts also gave the housing authorities a variety of powers over the contracts, design, and construction. Nevertheless, the courts in both cases rejected claims by the housing authorities that they were intended beneficiaries of the contracts between the owners, architects, and contractors, as well as the contention that the architects and contractors owed them tort duties.

To a large degree, denial of any right to bring action against the key participants was based upon the desire to avoid complex litigation. The end result, particularly since owners in these publicly assisted projects often are bankrupt, is that the housing authority cannot assert claims of any real value.

Another type of public funding is accomplished by the creation by the legislatures of special agencies authorized to issue revenue bonds. As an example, public building authorities enter into contracts for the construction of facilities to be rented by state agencies. Often building authorities are created because of limitations placed by law on states that bar them from incurring budget deficits. In *People ex rel. Resnik v. Curtis & Davis, Architects & Planners, Inc.*[4] the Illinois Supreme Court permitted an action to be brought by the State of Illinois against architects and contractors who had contracted with the Illinois Building Authority. The two Illinois cases mentioned earlier involving claims by housing authorities distinguished the *Resnik* case because it involved a transaction in which the plaintiff was to occupy the project after it was completed.

These complex funding arrangements, particularly those involved with publicly created agencies and public funds, can generate more than the usual confusion that can be found in public construction.

SECTION 12.04
PRIVATE PROJECTS

Projects of any magnitude usually require that the owners look outside for funds with which to build a project. A developer sometimes uses advance payments from purchasers to finance construction, although more commonly the developer will take out a construction loan from a bank or other lending institution. A commercial developer may use advance payments from tenants to construct tenant improvements. Those who make advance payments run obvious risks. The developer may be dishonest and use the money for improper purposes. The principal focus of this section is upon the professional construction lender.

The era of deregulation has made generalizations hazardous. Traditionally, construction loans for housing have been made by savings and loan associations and for commercial projects by commercial banks. These two sources are used collectively as the "construction lender" for purposes of discussion.

The construction lender lends for a short term, usually only for the time necessary to complete the project. The interest rates are short-term, usually floating, and in any case, generally higher than those charged by the permanent lender, more often an insurance company. (Increasingly, permanent lenders demand equity, preferring to be owners rather than lenders.) The difference between the lending rates and the highly leveraged (building with someone

---

4. 78 Ill.2d 381, 36 Ill.Dec. 338, 400 N.E.2d 918 (1980). A similar result was reached in *Cox v. Fremont County Public Building Authority*, 415 F.2d 882 (10th Cir.1969).

else's money) construction project can lead to many difficulties.

What is described here might not be universal; practices vary. However, it is useful to provide a general model so that the role of the lender during design and construction can be appreciated.

The construction lender, like all lenders, wants to have the loan repaid and a security interest that can be used to obtain repayment in the event the loan is not repaid. To achieve these objectives, the lender makes an economic evaluation in order to determine whether a commercial borrower will derive enough revenue to repay the loan. Although the construction lender will want to be "taken out" by a permanent lender or equity investor as soon as the project is finished, the likelihood that a takeout will occur will depend upon the long-term financial prospects for the project. More important for purposes of this treatise, the lender wants the project to proceed on schedule, be of the anticipated quality and quantity, and stay within the cost budget.

The owner-borrower also seeks these objectives. While lender and borrower have many interests in common, it is the lender's money that is being used. If a project runs into problems, the owner may walk away from the project and avoid the loan obligation by going bankrupt, leaving the lender to look only to the security. While the lender may demand loan guarantees, the principal security will be the improved land. The lender wants to make certain that each dollar it lends goes into the project and enhances the value of the project by at least that one dollar.

Perhaps more important, the lender does not want the project to fail, the owner-borrower to default, and the lender to have to take over. To recapitulate, the lender wants the project completed, the loan repaid, and adequate security if it is not.

Focusing first on project failure, while poor workmanship and delay can cause a project to fail by increasing costs and postponing revenues, the principal danger is that the owner will run out of money and the work will stop. This disaster, at least to the lender, can result from an unrealistic budget, an incomplete or poorly planned design leading to costly design changes and claims for extras by the contractor, unforeseen circumstances such as unforeseen subsurface conditions generating claims, inefficient administration creating contractor claims for delays, and insufficient owner funds to supplement the loan.

Poor quality work can increase costs and cause delay because of correction cost and the delay inherent in redoing the work. Delay—and even worse, a work stoppage—can lead to project failure not only because delay can postpone revenue generation but because it may increase the costs of borrowing, extend overhead costs, and expose the project to losses caused by theft or vandalism.

The other large risk is inadequate security, which can result from various causes. First, the owner may divert loan funds to purposes other than the project. Second, the contractor may divert funds paid that should go to the subcontractors and suppliers which can cause liens to be asserted. Even if the lender's security interest takes priority over these liens (this is a reason for unwillingness to lend if work has begun), filing liens creates uncertainties that may make it difficult to obtain a takeout lender. (Diversion can also cause project failure, since participants may refuse to work or deliver supplies when they are not paid.)

With these points in mind, it is useful to look at the lending and building processes from the lender's vantage point. To obtain a construction loan, the owner presents a loan package to the lender. This package is likely to consist of a site plan, typical floor plans, and perhaps a colored rendering of the finished product. Obviously the development of the design will vary, but it is likely that the lender will reserve its judgment until it is either satisfied that the designer will do an adequate job of design or that the design has been sufficiently developed to be appraised by the lender.

The lender will request information as to who will design the project, who will build it, and the project budget and schedule. If any contracts have been made for design and construction, the lender is likely to insist on seeing them.

Owners typically try to minimize the amount of money they must contribute to the project. From the owner's vantage point, it would be best to build the project entirely with loan funds, as that would mean that the owner risks only profits, not losses. This can tempt an owner to inflate the expected cost of the project, a tactic lenders usually meet by having their own independent appraisals made. From the lender's vantage point, it is important that the owner put up enough of the money to preclude the owner from being tempted to walk away if problems develop. Also, the lender wants sufficient funds available to the owner if additional funds become needed. If the lender is not certain about the financial resources of a corporate owner, it may ask that individuals guarantee corporate obligations.

Construction lenders would like assurance that long-term financing will be available when the project is completed. Often permanent lenders will make an advance loan commitment, but only after they have examined the financial package that the owner has presented to the construction lender. Permanent lenders normally require that a major tenant or suitable purchaser be committed before they will agree to take out the loan.

Avoidance of cost overruns, delays, and work stoppages are prime lender concerns. The lender wants a design complete enough to avoid costly design changes and contractor claims. The lender may also be interested in any language in a construction contract that may increase costs. Section 12.02 looked at the demands a lender may make as to the matters that may adversely affect the interests of design professional and con-

tractor. In this section, the focus is upon the lender's concern about construction contract language.

Lenders will examine a construction contract carefully to avoid clauses that the lenders do not believe are in their best interests. They will look particularly at "loose" changes provisions that may enable a contractor to do large amounts of work on a cost basis. They will also be concerned with clauses that may give the contractor the benefit of the doubt if the design requirements are unclear. They may be concerned with the exact nature of any rights given the contractor to recover additional compensation for differing site conditions. Some lenders may wish to eliminate an arbitration clause if they believe that arbitrators will be too free granting awards to the contractor. Some lenders may change language that exculpates the design professional or contractor from responsibility.

The payment process is crucial. The lender may simply take over the process by using a draw system. In such a system, payments are made directly to subcontractors and suppliers to avoid diversion of funds. Payments are approved through certifications made by a draw inspector, an employer of the lender who will inspect carefully both for progress and quality of work. Even if payments are made through the prime contractor or what is sometimes called the voucher system, the lender may wish to be certain that adequate retainage is preserved and that techniques are developed to avoid diversions and liens. Since the owner also wishes to avoid diversion, this problem is discussed in Section 26.02. (Some lenders insist upon the power to examine the records of all participants, particularly major contractors.)

Also, and quite important, the lender will want to know what is going on at all times. It may insist that it receive copies of contract addenda, modifications, requests for changes, change orders, claims, and even correspondence. It may insist that *it* approve any changes that affect price, quality, or time.

One of the most devastating events as far as the lender is concerned is default by the owner-borrower. If this occurs, lenders, though reluctant to take over the project, may want to have the power to keep other participants on the project with a successor. For this reason, they will frequently eliminate clauses that make it difficult for them to insist upon continued performance by the design professional or contractor.

Lenders should also be aware of the risk of taking too great a role in the Construction Process. If they go beyond traditional lender concerns over its investment, they may find that they have exposed themselves to claims by unpaid subcontractors or suppliers [5] and ultimate users of the improved property who find building defects.[6]

---

5. See Reitz, *Construction Lenders' Liability to Contractors, Subcontractors, and Materialmen,* 130 U. of Pa., L.Rev. 416 (1981).

6. Comment, *Indirect Liabilities of Construction Lenders in a Development Setting,* 127 U. of Pa.L.Rev. 1525 (1979).

Part Three

Designing the Project: Focus Upon Architects and Engineers

Introduction

Following the sequential pattern of Parts II, III, and IV, Part III focuses upon those who design improvements to land. Although the basic model is upon an owner who engages a professional designer, gives the designer the program, and reviews the suggested design solution prepared, it is important to be aware of the different owners described in Chapter 9. Some may have employees who design, with or without the advice and services of an independent designer. Similarly, looking ahead, Part IV focuses upon contractors' other methods of design, such as design furnished by a "Design-Build" contractor or even specialty trades, either as subcontractor or multiple primes. When these variants are discussed, any changes in the legal solutions suggested in Part III may have to be modified.

13

Professional Registration: Evidence of Competence or Needless Entry Barrier?

SECTION 13.01 OVERVIEW: RELATION TO CHAPTER 20

Legal requirements for professional practice is the logical place to begin a study of how law intersects with design and the design professions. This chapter deals with professional registration laws. This chapter also examines state occupational licensing that can and often does include contractors, a latecomer into the field of occupational regulation. This chapter is the basic chapter dealing with *all* these laws, while Chapter 20 deals with special problems that relate to contractor licensing.

SECTION 13.02 PUBLIC REGULATION: A CONTROVERSIAL POLICY

Public regulation of professional activity can take many forms. One form of public regulation, professional liability, is discussed in Chapter 17. This chapter discusses the *direct* legal requirements that determine whether a particular person or business entity may use a particular title or perform design or construction services.

Statutes enacted by legislative bodies and regulations promulgated by state administrative agencies set criteria for professional practice and administer systems that determine who may legally perform these services.

In the United States, the states control registration. The federal government can and usually does determine the qualifications for those who perform professional services for it. However, there is no federal regulation system for registration and licensing of design professionals. Local regulation, where it exists, is designed principally to raise revenue and not to regulate the professions. Occasionally a state statute will allow local regulation, mainly of contractors.[1]

(A) JUSTIFICATION FOR REGULATION

The police and public welfare powers granted to states by their constitutions permit states to regulate who may practice profes-

---

1. New Jersey upheld such regulation in *New Jersey Builders Association v. Mayor & Township Council of East Brunswick,* 60 N.J. 222, 287 A.2d 725 (1972). New York did not in *Bon-Air Estates, Inc. v. Building Inspector of Town of Ramapo,* 31 A.D.2d 502, 298 N.Y.S.2d 763 (1969).

sions and occupations. States exercise this power by setting requirements for those who wish to practice professions or occupations. Such requirements are usually expressed in licensing or registration laws.[2]

Some members of the public who seek to have design or construction work performed are unable to judge whether those who offer to perform design or construction services have the necessary competence and integrity. One purpose of licensing laws, whether they apply to architects, engineers, surveyors, or contractors, is to give members of the public some assurance that those with whom they deal will have at least minimal competence and integrity. The public generally may suffer property damage or personal harm because of poor design or construction. One court, in denying recovery to an architect licensed where he practiced but not where the project was located, refused to allow an exception for an isolated transaction and stated:

> One instance of untrained, unqualified, or unauthorized practice of architecture or professional engineering—be it an isolated transaction or one act in a continuing series of transactions—may be devastating to life, health or property.[3]

In passing upon a contractor licensing law, one court stated that it was designed to

> prevent unscrupulous or financially irresponsible contractors from deceiving and taking advantage of those who engage them to build. . . . It often happens that fly-by-night organizations begin a job and, standing in danger of losing money, leave it unfinished to the owner's detriment. Or they may do unsatisfactory work, failing to comply with the terms of their agreement. The licensing requirement is designed to curb these evils; the license itself is some evidence to the owner that he is dealing with an honest and qualified builder.[4]

These undeniably desirable objectives are

sought to be accomplished by requiring that those who wish to perform certain services have had specified education and experience and are able to demonstrate competence by passing examinations. Persons in business occupations may also have to establish financial responsibility. In addition, *after* entry into the profession, professional misconduct or gross incompetence may justify suspending or revoking a license.

(B) CRITICISM OF LICENSING LAWS

Occupational licensing laws can deny some their only means of livelihood. As a result, imposition or tightening of licensing laws is often accompanied by "grandfathering" those who already are in those occupations. This can relieve the hardship caused by denying persons the right to perform their best and often only skill. It can also dilute standards.

More important, even early in the history of occupational licensing, criticism was made of the state's role of determining who and how many persons would be allowed to enter professions or occupations. Occupational licensing has been described as "fence-me-in" laws designed to limit competition and to insure that undesirable persons (or *any* persons) not be allowed to enter professions or occupations. This criticism became more pronounced when the states went into licensing in a wholesale fashion, regulating occupations in addition to the "learned" professions, such as law, medicine, architecture, and engineering. Licenses are commonly required for barbers, beauticians, auto mechanics, and masseuses, to name a few. Any group with sufficient organization and political strength

---

2. The term *registration* is commonly used for the design professions and is used through this chapter. The term *licensing* is used in Chapter 20 dealing with contractors. In this chapter, *registration* and *licensing* are interchangeable.

3. *Food Management, Inc. v. Blue Ribbon Beef Pack, Inc.,* 413 F.2d 716, 723–24 (8th Cir.1969).

4. *Sobel v. Jones,* 96 Ariz. 297, 394 P.2d 415, 417 (1964).

to demand that standards be set for their group can generate a new licensing law.[5]

The competition-inhibiting aspect of licensing laws was demonstrated in 1973 when 2,149 general contractors took and failed the Florida Construction Industry Licensing Board examination. Either all Florida applicants were incompetent or the Board sought to limit competition by barring new entrants to the field. After indignant protest from builders who had failed, the Board abruptly reversed itself and curved the grades so that 88% were given passing marks and contractors licenses.[6]

The Florida incident also demonstrates the difficulty of determining the proper level of competence necessary to receive state permission to engage in a particular occupation. This issue has triggered protest and lawsuits from minority groups who contend that their low representation in the professions results, among other things, from examination questions and grading either designed to limit their entry or that have that effect. The long experience requirements for designers have been criticized as either inadequate as proper training or solely designed to provide registered designers with cheap labor.

New criticism of licensing and registration laws surfaced during the galloping inflation of the mid-1970s. Many pointed to professional licensing laws, among other protections given to the professions, as causing the high cost of professional services. Educational and experience preexamination requirements mean many unproductive years at great expense. This can lead to fees designed to recoup these losses after entry. Similarly, education, practice,

and testing requirements limit the number of persons who can perform these professional services. This reduces supply and increases fees. Another undesirable byproduct is that high professional fees tend to limit design services to those clients who can afford to pay these fees and to high-cost projects.

Critics frequently state that competence requirements do not accomplish the purpose of insuring that only qualified persons practice in the professions. Critics also claim that the agencies that administer these licensing programs are usually dominated by members of the profession being regulated, self-interest generating practices that keep the number of practitioners low in order to improve the economic status of those already in the profession.

Will the regulatory process, though criticized for the reasons given, continue to proliferate? In 1974 the Federal Trade Commission conducted a study of television repair in three cities, one having a traditional occupational licensing law a second had a modified licensing system and the third had no law. That study, though limited, tended to show that the cost of repairing television sets in the District of Columbia, a *then* unregulated area, was substantially lower than the cost of repairing television sets in the others, with no proven diminution in competence.[7] Yet shortly after this, the District of Columbia enacted a *stringent* licensing law regulating auto mechanics and television repairers.[8]

Some believe that the political process and self-interest of the professions and occupations will lead to *more*, not less, regulation. Yet recent developments in Arizona show the perils of political predictions. In

---

5. See Friedman, *Freedom of Contract and Occupational Licensing, 1890–1910: A Legal and Social Study*, 53 Calif. L.Rev. 487 (1965).

6. See *Wall Street Journal*, Jan. 8, 1975, p. 1.

7. *Economic Report: Regulation of the Television Repair Industry in Louisiana and California* (Federal Trade Commission, November 1974).

8. *Washington Post*, July 23, 1975, p. D–1.

1971 an Arizona intermediate appellate court held that a corporation that did not possess the required contractor's license could recover for work performed since its *manager* did have a license.[9] The legislature responded in 1973 making it clear that there could be *no* recovery in such a case.

In 1977 a different intermediate appellate court held that the 1971 decision was too permissive, cited the immediate legislative response as an indication that the 1971 decision was incorrect, and denied recovery.[10]

The winds of deregulation in the 1980s reached Arizona, however. The legislature in 1981 limited the scope of the contractor licensing laws to residential and up to four-unit apartment construction.[11]

(C) IMPORTANCE OF ATTITUDE TOWARD THE REGULATION PROCESS

Clearly the attitude of lawmakers toward the regulation process will be influential. Legislators who take a beneficent view are likely to enact more licensing laws. Similarly, the attitude of courts who must often pass upon these laws will influence decisions. For example, judges are often called upon to determine the constitutionality and meaning of the legislation. They also decide whether particular conduct has violated the statute, whether substantial compliance with licensing laws is adequate to excuse a violation, and whether a party who has performed work, though unlicensed, will be able to recover compensation.

(D) JUDICIAL ATTITUDES TOWARD REGISTRATION LAWS

Different states have variant attitudes. For example, one court passing upon a particular licensing practicing system noted that it expressed "grave policy."[12] Yet another granted recovery to a contractor who through technical default had allowed his license to lapse, stating:

> It performed in all other respects competently and without injury to any person. . . . We are not involved in aiding an incompetent or dishonest artisan. . . . The defendant received full value under the terms of the contract. The licensing law should not be used as a shield for the avoidance of a just obligation.[13]

Even a particular state over time can change its attitude toward licensing laws. For example, in 1957 California precluded a subcontractor who was unlicensed from collecting from the prime contractor who knew he was unlicensed despite the latter's having been paid by the owner.[14] Yet in 1966 and 1973 California was tolerant toward technical contractor noncompliance.[15] On the whole licensing laws still are considered to express important and desirable policy. Criticism that has been made, however, has begun to be reflected in judicial decisions.

SECTION 13.03 ADMINISTRATION OF LICENSING LAWS

Modern legislatures articulate rules of conduct by statute and create administrative

---

9. *Desert Springs Mobile Home Ranches, Inc. v. John H. Wood Construction Co.*, 15 Ariz.App. 193, 487 P.2d 414 (1971).

10. *B & P Concrete, Inc. v. Turnbow*, 114 Ariz. 408, 561 P.2d 329 (1977).

11. Ariz.Rev.Stat.Ann. §§ 32–1101 et seq.

12. *Hedla v. McCool*, 476 F.2d 1223, 1228 (9th Cir.1973).

13. *Vitek, Inc. v. Alvarado Ice Palace, Inc.*, 34 Cal.App.3d 586, 110 Cal.Rptr. 86, 92 (1973).

14. *Lewis & Queen v. N. M. Ball Sons*, 48 Cal.2d 141, 308 P.2d 713 (1957).

15. *Latipac, Inc. v. Superior Court*, 64 Cal.2d 278, 49 Cal.Rptr. 676, 411 P.2d 564 (1966); *Vitek, Inc. v. Alvarado Ice Palace, Inc.*, supra note 13.

agencies to administer and implement the laws.

Agencies created to regulate the professions can make rules and regulations to fill deliberate gaps left by the statutes and particularize the general concepts articulated by the legislature. For example, the licensing laws state that there must be examinations to determine competence. The details of the examinations, such as the type, duration, and frequency, are determined by the agency.

In addition to having quasi-legislative powers, the agencies have quasi-judicial power. They may, subject to judicial review, decide disputed questions, such as whether a particular school's degree will qualify an applicant to take the examination or whether certain conduct merits disciplinary sanction. They also have power to seek court orders requiring that persons cease violating the licensing laws.

These agencies, called boards or commissions, have great power over the professions they regulate. They are often controlled by the members of the profession they are supposed to regulate.

Increasingly, consumer movements are suggesting or demanding that lay persons be given significant policy-making roles. The principal control upon their quasi-judicial decisions has been the scrutiny given to these decisions by courts when judicial review is sought (discussed in greater detail in Section 13.04(B).)

SECTION 13.04　THE LICENSING PROCESS

(A) ADMISSION TO PRACTICE

Requirements imposed by states or territories vary considerably. Some states and territories require citizenship and residen-

cy. Most have minimum age requirements, usually ranging from 21 to 25. All require a designated number of years of practical experience that can be substantially reduced if the applicant has received professional training in recognized professional schools. All states require at least one examination and some two. Most inquire into character and honesty. Some require interviews. As to interstate practice, see Section 13.06(G).

Most states permit practice through a corporate form [16] but a substantial number do not.

(B) POST–ADMISSION DISCIPLINE

Although the regulatory emphasis has been upon carefully screening those who seek to enter the professions, all states can discipline persons who have been admitted. Discipline can be a reprimand or suspension or revocation of the license.

Grounds for disciplinary action vary considerably from state to state, but they are generally based upon wrongful conduct in the admissions process, such as submitting inaccurate or misleading information,[17] and conduct after admission that can be classified as unprofessional or grossly incompetent.

The Florida regulatory agency brought charges against Markel, a licensed Florida architect. The hearing examiner for the agency found Markel guilty of having placed his name and seal on drawings that he had not prepared and that were not prepared under his responsible supervising control. The hearing officer also found him guilty of having held himself out to the public as being in an architectural partnership when his partner was not a registered architect or engineer. The hearing officer, while apparently preferring a year's suspen-

---

16. Refer to Section 3.06.

17. A board's revocation of a license based upon this reason was not upheld in *State Board of Registration for Professional Engineers v. Antonio*, 159 Colo. 51, 409 P.2d 505 (1966). However, a revocation was upheld in In re: Revocation of Certificate of Registration of Scutt, 46 Dauph. 196 (Pa.1938).

sion, recommended revocation, because he did not want Markel to be reregistered until he had made a showing that he was familiar with Florida law.

The State Board of Architecture revoked Markel's license because he had previously been suspended for placing his seal upon drawings he had neither prepared nor supervised. Markel appealed to the intermediate appeals court, claiming that the statutory language was unconstitutionally vague. However, this contention was rejected.[18]

Markel appealed to the Florida Supreme Court which, while upholding the constitutionality of the statute, felt revocation to be too severe. The court concluded that this was a borderline case of malfeasance. As to Markel having held himself out to the public as a partner with a registered professional, the Supreme Court noted that the unlicensed professional had prepared the sign and the wording had been approved by an attorney. As soon as Markel had been notified that he was violating Florida law, he removed the sign.

The court felt that Markel's signing his name to plans in violation of law presented a close case. Markel had testified that he did not rubber-stamp his draftman's drawings but that he checked them over. However, evidence indicated that Markel played a very small role in overall preparation of architectural drawings. Clients initially would contact and engage the draftsman to prepare their plans. They had no personal contact with Markel. Only after the plans were nearing completion did the draftsman take them to Markel for his inspection and approval.

The court was uncertain whether at that stage Markel could be exercising responsible supervising control. The court concluded that he had not. But the closeness of the matter persuaded the court that revocation was too severe. The court noted that it is the most extreme measure of discipline and should be resorted to only when the conduct is wholly inconsistent with approved professional standards.[19]

The case was remanded and the intermediate appellate court concluded that Markel should be suspended for sixty days.[20] The *Markel* case may simply be an illustration of the traditional unwillingness courts have had to approve the drastic step of revocation. On the other hand, perhaps Markel was out of favor with the registration authorities who were seizing upon a close question to justify a severe sanction.

Coffman v. California State Board of Architectural Examiners [21] is another case in which the agency sanction was judicially moderated. The agency had suspended Coffman's license for thirty days, based upon a finding of dishonest practice. A former client had charged that Coffman had allowed extensive substitutions of substandard materials, causing the building to be unsafe and cheapened in value. The court noted that the evidence was in direct conflict on this question and that the client had not filed her complaint until eighteen months after she had filed a notice of completion. Because of the vagueness of the term *dishonest,* the court felt that any acts complained of would have to be clearly dishonest before a person's rights to practice his or her profession could be even suspended let alone taken away.

Some cases, however, have involved sufficiently flagrant misconduct or incompetence to justify a drastic agency decision. For example, a court affirmed an agency decision revoking the license of a professional engineer who had performed welding without being certified as required by

---

18. *Markel v. Florida State Board of Architecture,* 253 So.2d 914 (Fla.App.1971).

19. *Markel v. Florida State Board of Architecture,* 268 So.2d 374 (Fla.1972).

20. *Markel v. Florida State Board of Architecture,* 274 So.2d 12 (Fla.App.1972).

21. 130 Cal.App. 343, 19 P.2d 1002 (1933).

the state administrative code.[22] In the same case, however, the court would not affirm the agency's revocation where the professional engineer designed and supervised the construction of a garage that collapsed. The court held that the incompetence did not have to consist of continued and repeated acts. However, incompetence must refer to some demonstrated lack of ability to perform professional functions. While recognizing that there was an admitted error in the design of the roof supports for the garage, the court noted that the error was not obvious and that this was the first failure the engineer had experienced in eleven years of practice.

A revocation was affirmed based on serious design errors leading to the failure of the basement wall.[23] Also, there had been conduct causing delay in construction, failure to obtain a building permit, misplacing the building in reference to the lot line, and securing the owner's endorsement of payment without informing him of the facts. Another court affirmed a six-month suspension based upon a deficient ventilation plan, a superficial inspection of the premises, a superficial scanning of the architectural plan, and reliance upon the judgment of two relatively inexperienced employees.[24] In addition, the engineer, after finding that the original certifications were in error and serious defects existed, did not disclose this to appropriate city officials.

Many members of the design professions operate under economically unstable conditions. Swings in the economic cycle can prove devastating. As a result, individual bankruptcies of architects and engineers are not uncommon when the economy turns for the worse. Those in the contracting business face similar risks. However, California held that a contractor's license could not be revoked when he went bankrupt because revocation for this reason would frustrate the federal bankruptcy laws.[25]

The effectiveness of post-admission disciplinary powers has been limited. Attempts to suspend or revoke are almost always challenged by the design professional or contractor whose means of livelihood are being taken away. Challenges often mean costly appeals. Often administrative agencies charged with responsibility for regulating the profession or occupation are underfunded and understaffed. Suspension or revocation are unpleasant tasks. Even when action is taken, courts closely scrutinize decisions of administrative agencies. They *should* extend considerable deference to the agency decision. In matters as important as these, however, courts seem to make a redetermination of what is proper. The combination of agency lethargy and overextensive judicial scrutiny when they do act may be one reason, among many, for increased professional liability. The latter can supplement or even replace the regulatory licensing process as a means by which incompetent practitioners are eliminated.

SECTION 13.05   TYPES OF LICENSING LAWS

Again it must be emphasized that licensing rules are determined by each state. As a result, there is considerable variety in the regulatory controls both in terms of prohibited conduct and sanctions for violations.

---

22. *Vivian v. Examining Board of Architects, etc.*, 61 Wis.2d 627, 213 N.W.2d 359 (1974).

23. *Kuehnel v. Wisconsin Registration Board of Architects & Professional Engineers*, 243 Wis. 188, 9 N.W.2d 630 (1943).

24. *Shapiro v. Board of Regents*, 29 A.D.2d 801, 286 N.Y.S.2d 1001 (1968). Cases involving architect license suspensions are collected in 58 A.L.R.3d 543 (1974) and those involving engineer licenses in 64 A.L.R.3d 509 (1975).

25. *Grimes v. Hoschler*, 12 Cal.3d 305, 115 Cal.Rptr. 625, 525 P.2d 65 (1974). Since this case was decided by a 4–3 opinion, courts in other jurisdictions might reach a different result.

Also, legislation is frequently changed. Case decisions cannot always be relied upon because subsequent legislation may have been enacted to change the result in a particular case. Despite these dangers, some broad patterns have emerged as to the types of controls that have been enacted.

(A) LICENSING OF ARCHITECTS AND ENGINEERS: HOLDING OUT AND PRACTICE STATUTES

Licensing laws fall into two main categories. Some statutes regulate the professional title and are called "holding out" statutes. For example, in 1965 Ohio enacted such a statute stating that

> [n]o person shall use the title "landscape architect" . . . unless he is registered . . . or holds a permit.

A court passing upon the validity of this legislation stated:

> A practitioner, upon qualifying, becomes entitled to use the label, "landscape architect." Only the title, not the practice or profession, is restricted to licensees. Unregistered members may continue to practice, but without employing the title. Thus, appellant's allegation that the law prohibits him from practicing his profession is not well taken; only his use of the title "landscape architect" is proscribed.[26]

Often a holding out statute is the first step toward the second category of licensing laws—the practice statute. For example, Iowa originally had a holding out statute. A court decision in 1962 granted recovery to an out-of-state architect because the statute did not preclude practice but only holding oneself out as licensed.[27] Very likely, reaction to that decision generated legislation in 1965 that changed Iowa's registration laws to a practice statute.[28]

Frequently statutes include both holding out and practice language. Illustrations of how these statutes are interpreted are seen in Section 13.06.

Projects that affect the public generally are more likely to require the services of a licensed design professional. A licensed design professional is more likely to be required where the project is one for human habitation or one in which large numbers of persons will gather. Some states exclude less complex structures, such as single family residences, agricultural buildings, or store fronts. Improvements that cost less than a specified amount—typically $10,000–$30,000—may be exempt. Local laws must be checked.

(B) CONTRACTOR LICENSING

The proliferation of occupational licensing has led to an increased number of state licensing statutes for contractors. Until relatively recently only one-fourth of the states had such statutes, although the number seems to be increasing steadily. The statutes raise different problems and are discussed in Chapter 20.

SECTION 13.06
STATUTORY VIOLATIONS

(A) PRELIMINARY ISSUE: CONSTITUTIONALITY

Because of the controversial nature of professional regulation, constitutional attacks are common. By and large, such legislation is upheld based upon the police powers granted by state constitutions and constitutional provisions permitting the state to legislate in the interest of public welfare.[29] Some successful attacks have been made

---

26. *Garono v. State Board of Landscape Architect Examiners*, 35 Ohio St.2d 44, 298 N.E.2d 565, 567 (1973).

27. *Davis, Brody, Wisniewski v. Barrett*, 253 Iowa 1178, 115 N.W.2d 839 (1962).

28. *Food Management, Inc. v. Blue Ribbon Beef Pack, Inc.*, supra note 3.

29. *Richmond v. Florida State Board of Architecture*, 163 So.2d 262 (Fla.1964); *State v. Beck*, 156 Me. 403, 165 A.2d 433 (1960); *State v. Knutson*, 178 Neb. 375, 133 N.W.2d 577 (1965); *Pine v. Leavitt*, 84 Nev. 507, 445 P.2d 942

that have usually involved language or procedural problems and not the power of the state to regulate the professions and occupations.[30]

(B) HOLDING OUT AND PRACTICE COMPARED

Rodgers v. Kelley [31] involved an action by the plaintiff for design services for a project that was abandoned because of excessive costs. The clients claimed the plaintiff violated the holding out Vermont statute. The plaintiff contended that he had never signed his name or in any way represented that he was an architect. He testified that he had twelve years of experience in the architectural field but contended that so long as he did not label himself, his plans, or his business with the title "architect" he had not violated the statute: However, the court stated:

> But "holding oneself out as" an architect does not limit itself to avoiding the use of the label. The evidence is clear that, in the community of Stowe, this plaintiff was known as a proficient practitioner of all of the architectural arts with respect to homebuilding, at least. It was a business operation from which he received fees. He presented himself to the public as one who does the work of an architect. This constitutes holding oneself out as an architect, and is part of the very activity sought to be regulated through registration.

The court in *Rodgers v. Kelley* seems to have converted a holding out statute into a practice statute by concluding that a person who had designed homes over a period of time was holding out to the community that he was an architect.

Sometimes the licensing statute has language that appears to prohibit both practice and holding out. Such a statute was involved in *Frey v. Kent City Nursing Home, Inc.*,[32] where one plaintiff was a professional engineer and the other was not licensed to practice architecture in Washington. Neither claimed to have represented himself as a licensed architect to the public or the defendant. The defendant retained the plaintiffs to design a nursing home addition. When the defendant became dissatisfied, she terminated the plaintiffs and the plaintiffs sued for the services they had performed.

The Washington statute stated that it was "unlawful for a person to practice architecture unless registered." Other provisions of the statute made clear that a person could perform architectural services "as part of his . . . principal occupation" so long as he did not hold himself out to the public as being an architect. Similarly, those in the practice of engineering could perform architectural work so long as they did not use the title "architect."

Arguably the holding out language was meant to permit some practice of architecture as part of another profession so long as there was no holding out. In any event, the court concluded that the statute read in its entirety required the conclusion that the plaintiffs had not violated the licensing laws since they had not held themselves out to the defendants as architects. This result eliminated the provision making it unlawful to practice architecture, resulting in a holding out statute.

(1968); *Chapdelaine v. Tennessee State Board etc.,* 541 S.W.2d 786 (Tenn.1976) (state required surveyors but not engineers to register).

30. *New Jersey Builders Association v. Mayor, etc.,* supra note 1 (definitions too vague for enforcement); *Jenkins v. Manry,* 216 Ga. 538, 118 S.E.2d 91 (1961) (improper discrimination between self-employed and employed persons and an exemption for public employees); *Nemer v. Michigan State Board of Registration,* supra note 16 (constitution required that the profession being regulated have a majority of the members of the board; three architects out of a seven-person board did not constitute a majority).

31. 128 Vt. 146, 259 A.2d 784, 785 (1969).

32. 62 Wash.2d 953, 385 P.2d 323 (1963).

The *Rodgers* case took what appeared to be a holding out statute and made it a practice statute. The *Frey* case took what appeared to be a practice statute and made it into a holding statute. The motivation to engage in these gymnastics must have been the differing attitudes of each court toward licensing laws as well as the likelihood that absence of a license was an unjust excuse to avoid payment. The latter is an important factor throughout licensing cases.

(C) PRACTICE STATUTES

Which activities fall within architectural practice? Design services can range from simply sketching a floor plan or planning to place a residence upon a designated site all the way to construction documents with sufficient detail to obtain a bid from the contractor. In addition to the varying design activities, a differentiation can be made between those services that are part of the design process and those services performed by a design professional during construction.

Kansas Quality Construction, Inc. v. Chiasson [33] involved services during the design phase. The plaintiff was engaged in the business of building apartments. It represented that it had the necessary "know-how" related to preliminary assistance and advice to assist owners in getting apartments "off the ground." The defendants wished to build an apartment complex and made an agreement with the plaintiff under which the latter would develop a plat layout showing buildings, recreational area, and parking. The plaintiff would also *cause* plans to be drawn for loan submission and assist the owner in obtaining financing. The plaintiff would "[e]rect apartment buildings on owner's ground in accordance with approved plans and specifications" on a turnkey basis. The contract also provided that if the owner contracted with another builder, the plaintiff would receive a fee of

$4,000 for services performed. Plaintiff performed the preliminary services but the defendant selected another builder.

One of the defenses to the plaintiff's claim to the $4,000 fee was that the plaintiff was performing architectural services without being licensed. The Illinois statute defined architecture as including the offering or furnishing of professional services in connection with the construction or erection of any building, structure, or project. The court stated:

> We are of the opinion that offering to develop a plat layout . . . to show placement, total buildings, etc., is not so connected with "construction" that it partakes of doing what architects do. If this is so, then performance likewise is outside of the Act. To be sure, there is a nexus between the layout and construction in the sense that placement must precede construction—but then so must the idea itself. "Construction" at the very least, means getting off the ground by going either up or down, not just thinking about it, and plaintiff's plat layout here, "to show building placement, total buildings, recreation area, parking, etc." is certainly not construction nor connected with it to the extent that it falls within the statutory definition. The layout was developed first to please defendants and upon their approval, to convince the municipal authorities of the project's desirability and to gain their approval. To be sure, a plat layout can at times be a very complicated business calling for the collaboration of a battery of experts—surveyors, engineers, planners, landscapers, even lawyers—but in bringing their varied expertise to bear, none are thereby transmogrified into architects—and vice versa. As is said in 5 Am.Jur.2d 665, § 3, Architects, "[T]he making of a survey of existing conditions, with recommendations and preliminary sketches and layouts, with regard to work needed on a hotel building, constituted nonarchitectural services."

> Likewise . . . causing plans of the proposed building to be drawn for loan submission, is not a statement by plaintiff that it will perform architectural services, rather, it is an agreement that they will cause such plans to be drawn for such purpose—loan submission—which may or may not be eventually connected with the projected construction and which may or may not

---

33. 112 Ill.App.2d 277, 250 N.E.2d 785 (1969).

be drawn by an architect—it all depends on how much detail the banker wants.[34]

Noting that the plaintiff had used a non-Illinois architect to prepare these plans, the court indicated that the plaintiff was not "playing architect." Similarly, the court concluded that obtaining financing "has even less to do with architecture—if anything at all."[35]

Some of the language in the opinion may be traceable to the court's belief that the defendant had put forth a number of non-meritorious defenses to a legitimate obligation. There was little consideration for the public purpose protection so often stated to be the justification for registration laws.

The differentiation between those services performed during the design phase and those performed during construction can be difficult. It has been stated that services performed during the design phase are more likely to be considered within the licensing laws than those performed during the construction phase.[36] Undoubtedly this is based upon the specialized training and professional skill possessed by a design professional being manifested mainly in the design phase.

The design professional does bring skill to the Construction Process by interpreting documents, advising on changes, and judging performance. Some of these skills are intimately connected to other design process skills. The generalization may be true if emphasis is placed upon construction methods and organization, more properly the province of the contractor. Yet the language in the *Kansas City* case described earlier that seems to emphasize professional service "in connection with the construction or erection of any building" can be taken to mean that those professional skills exercised during construction may also fall within the licensing laws.

Perhaps this differentiation is of little practical importance. It is rare for design professionals to perform services connected only with construction. They perform services connected with either design and construction or design alone. However, the problem may come to a head when the law will be asked to determine whether construction managers must have a design professional license, a contractor's license, or neither.[37]

Sometimes the person performs design services for registered design professionals and does not deal directly with the public. For example, a Mississippi case involved an attempt by the state registration board to restrain Rogers from performing engineering services or holding himself out as a mechanical designer.[38] Rogers was not a registered engineer. He operated a small office and did all his work for other architects and engineers and never in any way dealt with members of the public. Mississippi exempted persons who performed design services, such as drafters, under the direct control of a registered design professional. In concluding that Rogers was not violating the registration laws the court stated:

> Inasmuch as Rogers' work was supervised and controlled by the architect or engineer for whom he was doing the work, obviously his status was not that of an independent contractor or practitioner of engineering, but rather a helper or employee.[39]

---

34. 250 N.E.2d at 787. Similarly, a California court held creation of preliminary sketches not to be architectural services. See *Walter M. Ballard Corp. v. Dougherty*, 106 Cal.App.2d 35, 234 P.2d 745 (1951).

35. As to financing, see *Herkert v. Stauber*, 106 Wis.2d 545, 317 N.W.2d 834 (1982), discussed in Section 3.06.

36. Annot., 82 A.L.R.2d 1013 (1962).

37. See Section 21.04(D).

38. *State Board of Registration v. Rogers*, 239 Miss. 35, 120 So.2d 772 (1960).

39. 120 So.2d at 776.

Sometimes the design services are incidental to the selling of equipment or mechanical systems. For example, in *Dick Weatherston's Associated Mechanical Services, Inc. v. Minnesota Mutual Life Insurance Co.*,[40] an insurance company was dissatisfied with the advice it had received from its architects regarding the air-conditioning system. They approached Weatherston, a mechanical contractor with a degree in engineering, for his advice. He made some design changes but made clear that he was not a registered engineer. These changes were ultimately accepted by architects and engineers retained by the insurance company.

Weatherston claimed and proved that he had a valid contract with the insurance company to supply air-conditioning systems. He brought legal action, claiming the insurance company had repudiated the contract. The insurance company claimed Weatherston had been practicing engineering without a license. However, the court concluded after noting that the company knew he was not licensed that his action was brought upon the contract to supply the air-conditioning equipment and not a contract for engineering services. The court seemed to assume that often those who sell or install mechanical equipment provide some engineering advice to buyers. When it appears that the principal basis for the claim is the sale of the equipment, the absence of a license under these circumstances should have no effect.

Does testifying as an expert constitute professional practice? For example, in one case a party challenged a lower court verdict because an unlicensed architect had been permitted to testify.[41] The claim was made that testifying was practicing and violated the licensing laws.

The possession of a license generally does not, however, determine whether a witness will be permitted to testify as an expert. The court permitted the testimony by concluding that testifying was not practicing. However, registration or absence of it goes to the *weight* of the testimony and not to its admissibility.[42]

(D) ARCHITECTURE AND ENGINEERING COMPARED

States generally regulate architecture and engineering separately. Each profession and the agencies regulating them sometimes differ over where one profession begins and the other ends. These conflicts demonstrate the economic importance of registration as well as the secondary role sometimes played by the public interest.

The two professions can be differentiated by project types and their use. One case that peripherally involved these questions stated:

> One prominent architect, in explaining the difference between architecture and engineering, said in effect that the entire structure and all of its component parts is architecture, if such structure is to be utilized by human beings as a place of work or assembly. He pointed out that, if the authorities were going to erect a courthouse as the building in which the [case] was being tried, they would obtain the service of an architect; but, if it was proposed to construct a power plant . . . they should employ an engineering firm. . . .
>
> All of the architects and those who were registered as both architect and engineer agreed that the overall plan of a building and its contents and accessories is that of the architect and that he has full responsibility therefor. As one witness answered it, he is the commander in chief.[43]

---

40. 257 Minn. 184, 100 N.W.2d 819 (1960).

41. *W.W. White Co. v. LeClaire*, 25 Mich.App. 562, 181 N.W.2d 790 (1970). Suppose the licensing authorities claim that the witness had violated licensing laws?

42. Expert testimony is discussed in Section 17.06.

43. *State Board of Registration v. Rogers*, supra note 39 at 774.

In *State v. Beck*,[44] the Maine Supreme Court stated esthetics to be the principal difference between engineering and architecture. To that court an architect was "basically an engineer with training in art." That court also stated:

> While categorically an engineer, the architect—without disparagement toward the professional engineer—is required to demonstrate that he possesses and utilizes a particular talent in his engineering, to wit, art or aesthetics, not only theoretically but practically, also, in coordination with basic engineering.[45]

The court then cited a Louisiana case that stated that an engineer "designs and supervises the construction of bridges and great buildings, tunnels, dams, reservoirs and aqueducts." [46]

The Maine court then described architectural projects:

> Architects are commonly engaged to project and supervise the erection of costly residences, schools, hospitals, factories, office and industrial buildings and to plan and contain urban and suburban development. Health, safety, utility, efficiency, stabilization of property values, sociology and psychology are only some of the integrants involved intimately. Banking quarters, commercial office suites, building lobbies, store merchandising salons and display atmospheres, motels, restaurants and hotels eloquently and universally attest to the decisive importance in competitive business of architectural science, skill and taste. A synthesis of the utilitarian, the efficient, the economical, the healthful, the alluring and the blandished is often the difference between employment and unemployment, thriving commerce and a low standard of existence. Basic engineering no longer suffices to satisfy many demands of American health, wealth or prosperity.[47]

Some statutes permit engineers to perform architectural services incident to engineering work. One case [48] involved an attempt by the state licensing authorities to restrain two officers of a construction company who were licensed engineers from designing a seventy-eight-bed nursing home. The engineers contended that the design work they were performing was incident to their engineering services to build the structure. The court recognized that some design services are required for any structure and that this is the proper function of the exemption for architectural services incident to engineering work. However, the building of a nursing home with the esthetic considerations both in terms of the human beings who would live there and the esthetic aspects of positioning the project was architectural and the defendants could not perform these services.

The difference between architecture and engineering came up in an unusual way in Michigan. An attack was made on the composition of the board that regulated architects and engineers. The Michigan constitution required that a majority of the members of the board be members of the profession being regulated. Michigan subsequently passed a statute that required that only three members of the seven-member board be architects. To deal with the prior constitutional requirement, the legislature simply declared architects and engineers to be members of the same profession.

This legislative declaration was rejected by the court,[49] which stated that the legislature could not declare something that is not so. The court noted that architects and engineers have different educational requirements and different functions.

---

44. Supra note 30.

45. 165 A.2d at 435.

46. *State v. Beck* quoted from *Rabinowitz v. Hurwitz-Mintz Furniture Co.*, 19 La.App. 811, 133 So. 498, 499 (1931).

47. 165 A.2d at 437.

48. *Dahlem Construction Co. v. State Board of Examiners*, 459 S.W.2d 169 (Ky.1970). The result was undoubtedly assisted by a special Kentucky statute requiring that architects design nursing homes.

49. *Nemer v. Michigan State Board of Registration*, supra note 16.

(E) DOLLAR AND SPACE EXEMPTIONS

A substantial number of states exempt projects that do not cost more than a designated amount of money. Obviously, in a rapidly inflating period, fixed limits of this type soon become outmoded.

Such statutes can generate interpretation problems because the cost of a project evolves through various stages. The owner has a budget. The design professional gives cost predictions. The contract price can be higher. Not uncommonly the ultimate *payment* for the project exceeds the contract price because of extras or delay claims.

The statute itself may determine which of these figures determines whether the project is exempt. In one case the unregistered design professional gave a good-faith cost prediction the project could be built for less than the statutory limit. The fact that the ultimate cost for various reasons was in excess of the statutory limit would not deny the person performing design services the right to recover.[50] The person performing design services should be able to know at the time he performs them whether he is violating the law.

A similar standard should apply if the exemption relates to square footage of the structure. However, it should be clear that this exemption should not be abused by the person performing design services. Clearly if that person knows the ultimate cost will go beyond the exempt amount or the ultimate space will exceed the amount allowed, the exemption should not protect the person performing those services.

(F) POSSESSOR OF LICENSE

Traditionally, design professionals performed as sole proprietors or partners. Increasingly, for tax reasons, design professionals are being permitted to perform design services through the corporate structure.[51] Whether the business entity chosen is a partnership or a corporation, it will be necessary to determine who must be licensed. One court held that a partnership itself must be licensed even though the partners were individually licensed.[52] Another court held that a corporation need not be licensed if the managing agent were licensed.[53]

The determination of who actually must hold a license when partnership or corporation forms are used requires an evaluation of the licensing statutes, administrative regulations, and case decisions. For this reason, any design professional who operates through a partnership or corporation should seek legal advice.

(G) OUT–OF–STATE PRACTICE

Increasingly the practice of architecture and engineering as well as the performance of construction services transcends state lines. As a result, it is common for a person licensed in one state to perform services in another.

Sometimes design professionals who perform services on a multistate basis obtain licenses in each state in which they perform services or in which projects for which they perform services are located. This has become easier because of standardized examinations and increased reciprocity.

---

50. *State v. Spann*, 270 Ala. 396, 118 So.2d 740 (1959). But see *Sample v. Morgan*, 66 N.C.App. 338, 311 S.E.2d 47 (1984) petition allowed as to additional issues, 310 N.C. 626, 315 S.E.2d 692 (1984) which limited the contractor's recovery to the amount his license allowed him to build when the ultimate cost exceeded this limit because of owner-directed changes.

51. For a general discussion of the practice of architecture by a corporation, see 56 A.L.R.2d 726 (1957). Refer to Section 3.06 for discussion of the professional corporation.

52. *Nickels v. Walker*, 74 N.M. 545, 395 P.2d 679 (1964).

53. *Hattis Associates, Inc. v. Metro Sports, Inc.*, 34 Ill.App.3d 125, 339 N.E.2d 270 (1975).

Some registration laws do not require professional persons licensed in a foreign state to become registered if they perform work only for an isolated transaction or perform work not to exceed a designated number of days. Often such exemption requires an easily obtained temporary license.

As interstate practice has become more common, an increasing number of cases have dealt with attempts by design professionals licensed in one state to recover for services related to a project in another. For example in *Johnson v. Delane*,[54] an engineer licensed in the State of Washington obtained a commission to prepare plans and specifications for a project to be built in Idaho. The engineer was not to perform any supervisory function. He obtained the commission while visiting in Idaho. He performed the requisite design services in Washington and delivered the plans and specifications to the client in Idaho. The court held that he was not practicing architecture in Idaho, and he recovered for his services.

Is the purpose of Idaho licensing laws to protect Idaho clients from retaining unqualified architects and to protect Idaho citizens from being exposed to risk of harm due to structures that may not comply with local building laws and codes? If so, it would seem that the project being constructed in Idaho for Idaho clients should necessitate the licensing of the architect under Idaho laws.

The result in the *Johnson* case is supportable if the architectural registration requirements for Washington and Idaho are similar and if there are no peculiarities of Idaho building codes that might make them substantially different from those in effect in Washington. As to Idaho's interest in insuring that structures meet Idaho building code standards, the defendant client would have had a defense to any action by the

architect, whether licensed in Idaho or not, that there had been a failure to comply with local building codes.

The decision can be supported on another ground. It is likely that the architect could have registered or been allowed a temporary license to perform the services for one project. If so, granting him recovery despite his not being registered in Idaho would avoid his being uncompensated and prevent unjust enrichment of the client by, in effect, excusing his failure to use this method of complying with Idaho law.

Some state laws do not require a license for performance of services by out-of-state design professionals if they are licensed in the state where they practice principally and so long as there is a licensed design professional in overall charge of the project. However, the licensed local design professional must not be a figurehead but must actually perform the usual design professional functions. Sometimes out-of-state architects or engineers "associate" a local architect as a means of insuring that they will be able to collect their fee.

In *Food Management, Inc. v. Blue Ribbon Beef Pack, Inc.*,[55] plaintiff was an Ohio engineering corporation that had entered into a written turnkey contract with the defendant for the design and construction of a meat-packing plant to be built in Iowa. The plaintiff was not licensed in Iowa but entered into a written contract with an Iowa architect-engineer. This agreement designated the Ohio corporation as the Principal Consultant and referred to the local architect-engineer as the Associate Engineer. The contract stated in part:

C. Consideration of the Associate Engineer's Work: The Principal Consultant shall give thorough consideration to all reports, sketches, drawings, specifications, proposals and other documents presented by the Associate Engineer, and shall inform the Associate Engineer of his

54. 77 Idaho 172, 290 P.2d 213 (1955).

55. 413 F.2d 716 (8th Cir.1969).

decision within a reasonable time so as not to delay the work of the Associate Engineer.

D. Standards: The Principal Consultant shall furnish the Associate Engineer with a copy of any design and construction standards he shall require the Associate Engineer to follow in the preparation of drawings and specifications for This Part of the Project.

The project was ultimately abandoned because of excessive cost. When the plaintiff sued for the services performed, the client claimed the plaintiff was not licensed to practice in Iowa. One of the contentions made by the plaintiff was that the engineering or architecture performed in Iowa had actually been performed by the Iowa architects and engineers. The court held that the Iowa design professionals were not "in charge" and denied recovery to the Ohio corporation.

In *Hedla v. McCool*,[56] a Washington architect performed design services for a project to be built in Alaska and had the plans approved by an Alaskan engineer. The court emphasized the importance of the policy expressed in Alaska's licensing law. The court distinguished *Johnson v. Delane* [57] (discussed earlier) by noting that the Alaska statute was broader, that this was not an isolated transaction, and that conditions in Alaska were different from those in the State of Washington. The court was more persuaded by the holding in the *Food Management* case, which had not looked kindly upon out-of-state architects associating a local architect when the out-of-state architects were principally responsible for design decisions.

One important consideration permeating these cases is whether the unlicensed person seems to have performed properly. If so, denial of recovery, despite the importance of license compliance, may seem unjust. For example, in the *Delane* case, the

work appears to have been performed correctly. But in *Hedla v. McCool*, where recovery was denied, the costs overran considerably and the design was never used—factors that make recovery less attractive.

Perhaps the expansion of reciprocity and the ease with which out-of-state design professionals who are registered in their home states can be allowed to practice in other states will make it more difficult for design professionals to recover if they do *not* use these techniques for legitimating their projects in states where they are not licensed. In any event, any design professionals considering performing design services either in another state or for a project that will be built in another state should receive legal advice on the proper process for insuring that they are not violating the laws of the state where the services are being performed or the project is located.

This section has dealt with whether there *has been* a violation by the out-of-state professional. In Section 13.07(B), two cases that involved attempts to recover for services *illegally* performed by design professionals registered in another state are discussed.

(H) SUBTERFUGES

Sometimes artificial arrangements are made to bypass registration laws. Two cases reflect differing judicial attitudes toward these subterfuges. In *Snodgrass v. Immler*,[58] the owner had approached an unlicensed architect and asked him to design a house. The architect told the owner that he was not licensed and would be violating the licensing statute if he did. But the owner entered into a contract with a licensed architect under which the latter would employ the unlicensed architect to draw plans "under the supervision" of the licensed ar-

---

56. 476 F.2d 1223 (9th Cir.1973). This case is discussed again in Section 13.07(C).

57. Supra note 55.

58. 232 Md. 416, 194 A.2d 103 (1963).

chitect. A contract was also made between the licensed and unlicensed architects. The fee was to be 5% of the construction cost, of which the unlicensed architect would receive 4% and the licensed architect 1%.

Problems developed when costs overran and the design submitted by the unlicensed architect was abandoned. The unlicensed architect sought recovery for his services claiming that he was an intended beneficiary of the contract between the licensed architect and the client. The court, in denying recovery, noted that the licensed architect was merely a front and characterized the two contracts as shams. It did not appear that the licensed architect played any significant part in the relationship between the unlicensed architect and the client. (He "sold" his license.)

Scott-Daniels Properties, Inc. v. Dresser [59] involved a not too dissimilar fact situation. The client wished to build a motel in Minnesota. He had heard of Dresser's reputation and contacted him. After being satisfied that Dresser was an outstanding architect and a former pupil of Frank Lloyd Wright, the client retained Dresser to design and supervise the motel. Evidently the client knew Dresser was not licensed in Minnesota.

The actual identity of the design professional was somewhat cloudy. Dresser had formed a corporation, and the corporation had a licensed architect as an employee. The client contended that his contract was with Dresser *personally*, while Dresser contended that the contract was with the corporation, with Dresser, an employee, to do the work. The trial court concluded that the client contracted with the corporation, and this was affirmed on appeal. However, it appears that the registered architect was not in responsible charge of the work (a requirement of the Minnesota law) and that the corporation was simply a sham through which Dresser, an unlicensed architect, could practice architecture.

The two cases reflect an observation made in the preceding subsection. In the *Scott-Daniels* case it appears that the court was convinced that the architect had performed properly and the motel was actually built. On the other hand, in the *Snodgrass* case the court wondered why the client had not defended on the grounds of the unlicensed architect's design having substantially exceeded the estimated budget. Perhaps the client thought it would be easier to defeat the unlicensed architect's claim by showing failure to comply with the registration laws than to establish that the budget had been substantially exceeded through no fault of the client.

Other devices are sometimes used in an attempt to bypass licensing laws. For example, suppose the unlicensed architect assigns his right to recover under the contract to a third party. Normally, rights under a contract can be transferred by assignment.

A court that believed circumvention would frustrate the purpose of the registration laws barred an assignee (the person to whom the assignment had been made) from recovery.[60] Similarly an attempt to exempt a project based upon a statute exempting projects under a designated dollar amount by dividing one contract into two was not successful.[61] Finally, an arbitration award in favor of an unlicensed contractor was not confirmed by a court.[62]

Circumvention through subterfuges raises ethical and moral questions. But

59. 281 Minn. 179, 160 N.W.2d 675 (1968).

60. *Walker v. Nitzberg*, 13 Cal.App.3d 359, 91 Cal.Rptr. 526 (1970).

61. *Cochran v. Ozark Country Club, Inc.*, 339 So.2d 1023 (Ala.1976).

62. *Loving & Evans v. Blick*, 33 Cal.2d 603, 204 P.2d 23 (1949). But see *Parking Unlimited, Inc. v. Monsour Medical Foundation*, 299 Pa.Super. 289, 445 A.2d 758 (1982) (award upheld where it appeared that the arbitrator held recovery justified).

success will depend upon the troublesome and complex questions of whether *denial* of recovery for work performed will create unjust enrichment.

(I) SUBSTANTIAL PERFORMANCE

Increasingly attempts to recover have been based upon asserted "technical" noncompliance with the law. This is not, strictly speaking, a contention that the law has not been violated. As a result, this doctrine is discussed in Section 13.07(B) and, inasmuch as contractors have been the principals users of this approach, in Chapter 20.

SECTION 13.07
SANCTIONS FOR LICENSING LAW VIOLATIONS

(A) CRIMINAL SANCTIONS

Licensing laws usually carry criminal sanctions. Violations can be punished by fine or imprisonment. However, use of the criminal sanction is relatively rare. Perhaps occasional instances have occurred where fines have been imposed, although it is quite unlikely, though possible, that imprisonment will be ordered. Where a penal violation is found, it is likely that sentence will be suspended or probation granted.[63]

A quasi-criminal sanction can be invoked. If the licensing authorities obtain a judgment from the court ordering that the unlicensed design professional cease practice and the order is disobeyed, failure to comply will be contempt of court and punishable by a fine or imprisonment.

Just as law enforcement officials rarely have the staff or resolve to seek criminal sanctions, the regulatory agencies rarely seek a judicial order that the violator is in contempt.

(B) RECOVERY FOR WORK PERFORMED: COMPLEXITY

The infrequent use of criminal or quasi-criminal sanctions means that the principal sanction for unlicensed practice of architecture, engineering, or construction work has been to deny recovery for work performed. Attempts by unlicensed persons to recover for their work has been the main legal battleground.

To determine whether persons not authorized to perform particular services can recover for work performed, recourse first must be made to the statutes that regulate the conduct. The statute may deny the use of the courts to recover for work performed where there has been a violation of the licensing laws. For example, § 7031 of the California Business and Professions Code provides that contractors cannot bring an action in any court for recovery under a contract unless they establish that they were a duly licensed contractor "at all times during the performance of such act or contract."

Explicit statutory treatment of collection rights is rare. As a result, the law must deal with the effect of making a contract to perform illegal services. This requires an evaluation of the objectives of the statute, the strength of the public evaluation of the objectives of the statute, the strength of public policy expressed in the statute, the intended beneficiary of the statute, and whether denial of recovery for services performed under such a contract will create unjust enrichment.

Generally, neither party can enforce an illegal contract. The law provides no help if the parties are equally guilty. If a citizen bribes a public official, the citizen cannot sue to have the promised performance

---

63. Restitution as a condition to probation was permitted in *State of Washington v. Bedker*, 35 Wn.App. 490, 667 P.2d 1113 (1983), discussed in Section 20.05. Usually this is not done, because the defendant does not have protections available in civil actions such as pleading requirements, discovery, and a jury trial. The court ordered restitution to correct only *dangerous* conditions.

made nor can the public official sue to recover the promised bribe.

Parties protected by the statute are given enforcement rights. For example, suppose an unlicensed architect contracted with the client. The client can enforce the contract against the unlicensed architect and maintain any action for faulty performance.[64]

Returning to the rights of the unlicensed person, suppose the latter has conferred a benefit upon the other party. The party performing may have performed in accordance with the contract and seeks payment, not on the illegal contract but based upon restitution. Leaving the parties in status quo could cause unjust enrichment. Clearly the client is enriched. But is enrichment unjust? For example, if the client *knew* the architect was unlicensed, arguably any enrichment would be unjust. However, if this factor were sufficient to grant recovery based upon unjust enrichment, the legislative purpose expressed through the licensing laws may be frustrated. The unlicensed architect might continue to violate the law hoping that he either would be paid or would be able to recover through restitution based upon unjust enrichment.

Whether the law should not only refuse to enforce such contracts but also deny recovery for work performed is particularly difficult when illegality results from non-compliance with laws licensing contractors. Suppose there is a contract between an unlicensed subcontractor and a prime contractor who knew the subcontractor was unlicensed. If the purpose of the law is to protect homeowners who may not be able to determine the competence of contractors, the unlicensed subcontractor should recover. Yet, as shall be seen, there are different judicial attitudes when the contract is made between contractors rather than between a contractor and an owner.[65]

Despite some recent criticism of licensing laws noted in Section 13.02, the laws generally are still received favorably by the courts. Even in the absence of an express statutory provision dealing with the right to collect, unlicensed persons generally are not able to recover either on the contract or upon unjust enrichment.[66] (Yet, as seen in (C), there is a trend toward *allowing* recovery, undoubtedly a reflection both of expanded notions of restitution based upon unjust enrichment and of greater skepticism toward licensing laws.)

Two cases that involve design professionals practicing outside the state in which they were licensed—an increasing phenomenon—demonstrate the turmoil in resolving these issues.

In *Markus & Nocka v. Julian Goodrich Architects, Inc.,*[67] the plaintiff was an architect licensed in Massachusetts who specialized in hospital design. The defendant was a Vermont architect retained to perform design services for hospital facilities in Vermont. The hospital directed the Vermont architect to hire a specialist in hospital architecture and the Vermont architect hired the plaintiff.

The plaintiff made a study of the design needs, consulted with the Vermont hospital staff, and prepared revisions of preliminary sketches and specifications. The plaintiff's staff made numerous trips to Vermont in performing these services. Ultimately the

---

64. *Hedla v. McCool,* supra note 57; *Domach v. Spencer,* 101 Cal.App.3d 308, 161 Cal.Rptr. 459 (1980) (unlicensed contractor); *Cohen v. Mayflower Corp.,* 196 Va. 1153, 86 S.E.2d 860 (1955).

65. Compare *Enlow & Son, Inc. v. Higgerson,* 201 Va. 780, 113 S.E.2d 855 (1960) (granted recovery) with *Lewis & Queen v. N.M. Ball Sons,* 48 Cal.2d 141, 308 P.2d 713 (1957) (denied recovery). See Section 20.04(A).

66. *Food Management, Inc. v. Blue Ribbon Beef Pack, Inc.,* supra note 56. The case cites many cases from a number of jurisdictions denying recovery. See also *Southern Metal Treating Co. v. Goodner,* 271 Ala. 510, 125 So.2d 268 (1960). See also notes 92, 93 infra.

67. 127 Vt. 404, 250 A.2d 739 (1969).

design recommendations of the plaintiff were not accepted by the hospital staff, and the project was put out to bid based on the design work of the defendant.

The court held that the Massachusetts architects had violated the Vermont registration laws. Since the architectural contract between plaintiff and defendant violated Vermont law, the contract was illegal and the provision for the payment of compensation was unenforceable. The court noted that the construction was to be undertaken in Vermont and that many visits had been made to the Vermont site, along with consultations with Vermont hospital personnel.

Since the consulting contract was illegal, could the plaintiff consulting architects collect for their services? The court held that they could not. In denying recovery the court stated:

> The underlying policy is one of protecting the citizens of the state from untrained, unqualified and unauthorized practitioners.[68]

The court would not apply an exception used by some states that permit consulting architects to recover if their performance was one single, isolated act. The court noted that such an interpretation would weaken the registration laws. The court pointed out that the Vermont registration laws were phrased in broad, positive terms and provided no such exception.

The court also noted that the registration laws for medicine and engineering *specifically* authorized consulting services in Vermont by those properly licensed out of state but that there was no corresponding provision relating to out-of-state architectural consultants. The court noted that Vermont law provided that architects licensed in another state whose standards were not below

those of Vermont could be admitted in Vermont without examination.

The court made it clear that not all consultations across state lines on Vermont projects would necessarily violate Vermont registration laws. But the court concluded:

> . . . when the nonresident architect presumes to consult, advise and service, in some direct measure, a Vermont client relative to Vermont construction, he is putting himself within the scope of the Vermont architectural registration law.[69]

Costello v. Schmidlin[70] involved an action for consultation fees by Costello, who was licensed in New York but not licensed in New Jersey, the state in which the project was being built. Costello was an engineer who specialized in the construction of swimming pools. In addition to being licensed in New York, his principal place of business, he was licensed in Maryland, Illinois, and New Mexico. He agreed to perform consulting services for a New Jersey municipal swimming pool complex through a consulting contract with the New Jersey architect.

Costello's design was used in the bid invitations. However, the initial bids exceeded the cost budget. Costello redesigned, and the project was constructed. There was no assertion that Costello's work did not conform to his contract obligations.

When Costello was not paid, he brought action in federal court. The trial court judge, applying New Jersey law, held that the services were performed under an illegal contract because Costello was not licensed in New Jersey and he could not recover.

The appeals court reviewed some decisions in which design professionals had been denied recovery and noted that in most cases the design professional was not licensed in any state. More important, the

---

68. 250 A.2d at 741.

69. Id. at 742.

70. 404 F.2d 87 (3d Cir.1968).

court emphasized earlier decisions that had noted the distinction between unlicensed persons dealing with members of the public and dealings with persons *within* a profession, such as dealings between design professionals or contractors. Here there was a contract between an unlicensed engineer and a licensed architect, the latter *knowing* that the engineer was not licensed. The court allowed recovery despite the admittedly illegal contract, pointing to the lack of *express* denial of compensation in the licensing laws. (Many cases deny recovery *without* a statute.)

Were recovery based upon unjust enrichment, the court should have determined the reasonable value of Costello's services. However, Costello was allowed to recover an amount based upon the admittedly illegal contract.

The courts in the *Markus* and *Costello* cases reflect genuine differences of opinion regarding licensing laws. The *Markus* opinion emphasized the need to insure that all persons who perform design services for Vermont clients on Vermont projects meet Vermont standards of competence. The federal court in the *Costello* case seemed more concerned with doing justice between the two parties rather than arriving at a result that might frustrate the operation of the licensing laws.

Perhaps the variant holdings can in some way be traceable to an important factual distinction in the two cases. In the *Markus* case the unlicensed professional's work was not used. But in the *Costello* case the work was used and there was no indication that there had been any deviation from Costello's obligations. Just as in the subterfuge cases discussed in Section 13.06(H), courts will be favorably inclined toward the unlicensed person if it appears that that person has performed properly and the other party seems to be using failure to comply with the licensing laws as an excuse for not paying. This observation does not necessarily mean that when these facts occur courts will *always* allow recovery. Obviously, much may depend upon the language of the statute and the court's attitude toward the importance of licensing laws. But where it appears that the work was proper and that it was used, many courts will make a serious attempt to allow recovery.

The differing attitudes expressed in the *Markus* and *Goodrich* cases and the trends that are described in (C) do not change the undeniable fact that, as stated,[71] the general rule is that unlicensed persons cannot use the courts to recover for work they have performed. Yet a number of issues lie beneath the surface and must be explored.

If the contract is not absolutely void, an owner may be permitted to sue for damages for breach of such a contract.[72] More important, denial of the right to use the courts to recover has not led to an obligation requiring the unlicensed person to *return* any money that he *has* received.[73] Payment can mean that the work has been performed properly. Also, the law avoids upsetting completed transactions. If the policy that has made the conduct illegal is very strong, some courts do permit recovery of payment made.[74] Licensing violations have not fallen into that category.

Suppose the owner sends a check to the unlicensed party but stops payment on it. Had *payment* been received, it could not have been recovered. A court held that the

---

71. See note 67 supra.

72. See note 65 supra.

73. *Medak v. Cox*, 12 Cal.App.3d 70, 90 Cal.Rptr. 452 (1970).

74. *Albany Supply & Equipment Co. v. City of Cohoes*, 18 N.Y.2d 968, 278 N.Y.S.2d 207, 224 N.E.2d 716 (1966).

unlicensed contractor cannot recover, since the payment had not been received.[75]

Some courts give partial relief to the unlicensed person, permitting him to set off compensation for work that has been performed against any claim that is made against the unlicensed person. For example, suppose an unlicensed architect had furnished services for which he was still owed $10,000. Suppose that the client on an unrelated matter had $15,000 coming from the unlicensed architect. In such a case, if the client brought a claim for the $15,000, some courts would permit the set-off of $10,000 for work performed under the illegal contract to reduce the claim against the unlicensed architect to $5,000.[76] One court would not.[77]

The claims usually seek a money award. But other uses can be made of the claim.

For example, suppose the unlicensed person files a mechanics' lien based upon a statute giving security to the person who has improved another's property. Although it can be asserted that this remedy is given to avoid unjust enrichment, some courts, pointing to the need for a valid contract, have denied the right to impose a lien.[78] Yet one court allowed a lien despite the illegality of the contract when the obligation was restitutionary based upon unjust enrichment.[79]

Suppose the unlicensed person and the other party make a settlement. One court enforced the settlement, even though it would not have enforced the *original* illegal contract.[80] The court concluded that the settlement had been made in good faith, analogous to payment.

Suppose the parties to the contract arbitrate the dispute and the arbitrator grants an award to the unlicensed person? Will a court confirm it? One court, stressing the importance of the licensing statute (the case was decided in 1949) refused to confirm the award.[81] Another case, decided in 1982, while stressing that the arbitrator could resolve the legal issues, confirmed the arbitration award.[82]

Suppose the claimant presents its claim to the Bankruptcy Court when the other party to the contract has been adjudged a bankrupt? A recent opinion allowed the claim, for whatever it will be worth, based upon the broad discretion given to the Bankruptcy Court.[83] The court was influenced by the bankrupt having been enriched and having acquiesced in continued performance by the claimant *after* the bankrupt knew the performing party was not licensed. (This claim will come at the expense of the other unsecured creditors, *not* the bankrupt.)

The special problems of contractor licensing laws are treated in Chapter 20. At this point the "substantial compliance" doc-

---

75. *Vedder v. Spellman*, 78 Wash.2d 834, 480 P.2d 207 (1971).

76. *Sumner Development Corp. v. Shivers*, 517 P.2d 757 (Alaska 1974); *S & Q Construction Co. v. Palma Ceia Development Organization*, 179 Cal.App.2d 364, 3 Cal.Rptr. 690 (1960).

77. *Hedla v. McCool*, 476 F.2d 1223 (9th Cir.1973). This case, interpreting Alaska law, seems to have been virtually overruled by a subsequent Alaska case that is discussed in Section 13.07(C).

78. *Sumner Development Corp. v. Shivers*, supra note 77; *Chickering v. George R. Ogonowski Construction Co., Inc.*, 18 Ariz.App. 324, 501 P.2d 952 (1972).

79. *Expert Drywall, Inc. v. Brain*, 17 Wash.App. 529, 564 P.2d 803 (1977). Cf. *Bastian v. Gafford*, 98 Idaho 324, 563 P.2d 48 (1977).

80. *Shelton v. Grubbs*, 116 Ariz. 230, 568 P.2d 1128 (1977).

81. *Loving & Evans v. Blick*, 33 Cal.2d 603, 204 P.2d 23 (1949).

82. *Parking Unlimited, Inc. v. Monsour Medical Foundation*, 299 Pa.Super. 289, 445 A.2d 758 (1982).

83. *In re Spanish Trails Lanes, Inc.*, 16 B.R. 304 (Bkrtcy.Ariz.1981).

trine, developed in claims by contractors whose violations were technical (such as failure to renew due to an office manager's mental breakdown [84]), merits mention. The case of *Wilson v. Kealakekua Ranch, Limited*,[85] though not *expressly* discussing substantial compliance, dealt with failure to renew, a staple in claims of substantial compliance.

Wilson sued the ranch for architectural services performed. Though Wilson testified he was licensed, cross-examination revealed he had failed to pay the renewal fee of $15. On that basis, the trial court granted the ranch's motion to dismiss Wilson's claim.

First, as is common when the court is inclined to grant recovery, the appellate court pointed to the absence of any *specific* legislative language precluding recovery. Next, it distinguished statutes to raise revenue from those "for protection of the public against incompetence and fraud." Usually recovery is denied in the latter cases.

According to the court, however, the true issue is one of legislative intent. It concluded that the legislature must not have intended to create a forfeiture out of proportion to the offense that would benefit the other party. Here the penal sanction was a fine not to exceed $500 or imprisonment for one year. Denying recovery would mean a forfeiture of $34,000 by Wilson gained by the ranch.

The court pointed to Wilson's having been licensed and no indication that, had he reapplied, there would have been reinvestigation or reexamination; this according to the court, shows that *renewal* is simply to raise revenue. Renewal is automatic unless a charge has been filed against the architect. Even if this were not true, the court returned to denial of fees as disproportion-

ate and punitive, noting that Wilson is still exposed to the penal sanctions.

True the *Wilson* case is not a frontal attack upon the basic rule *denying* liability. But its determination that the statutory requirement for renewal is to raise revenue, and its highlighting the absence of any direct command from the legislature to *deny* fees must be taken as a sign that claims for recovery by unregistered persons are to be given respectful judicial consideration. The next section reveals a recent trend that may cast doubt upon the general rule of denial.

(C) RECENT DEVELOPMENTS: PORTENT OF FUTURE?

A shift in attitude toward licensing laws may be presaged by recent cases passing upon recoverability for work performed in Alaska that violated the Alaska licensing laws. In 1973 the Federal Circuit Court of Appeals interpreting Alaska law took a strong stand against recovery for work performed by an architect licensed in Washington but not licensed in Alaska.[86]

Later in the year a federal trial court sitting in Alaska handed down a decision precluding an engineer licensed in Washington but not licensed in Alaska from recovering for services performed based upon the federal court decision decided earlier in the year. An appeal was taken to the Federal Ninth Circuit. The appellate court thought the engineer made a persuasive case for recovery. The court stated:

> The equities favor the engineers. Their plans were utilized in construction of the building. The illegality of the contract was a matter of inadvertence and bad timing. One of the engineers' officers applied for the requisite certification less than a month after the contract was made. The illegality caused no loss to the [owners]. The engineers were at all times qualified to perform engineering services in the state of

---

84. *Latipac, Inc. v. Superior Court*, 64 Cal.2d 278, 49 Cal.Rptr. 676, 411 P.2d 564 (1966).

85. 57 Hawaii 124, 551 P.2d 525 (1976).

86. *Hedla v. McCool*, supra note 78.

Washington, and obtained qualification in Alaska before the two final phases of the contract had been performed. Finally, the [owner's] agent appears to have been aware of the possibility that the engineers had not qualified to do business in Alaska.[87]

The court was faced with its decision a year earlier denying recovery in a case that admittedly involved poor facts for the unlicensed person. The court noted that two recent decisions of the Alaska Supreme Court seemed to be more tolerant toward persons who make these contracts. The second decision [88] held that although the contractor licensing law would not allow direct recovery by the unlicensed contractor, he could use that claim as a setoff against the claim being made against him. As a result, the appellate court sent the matter back to the trial court for a further study of Alaska law.

The trial court concluded that the Alaska cases had reflected a distaste for persons going uncompensated for their services and a desire to do justice between the two contracting parties to the illegal contract. Noting that the Alaska licensing law, like the Hawaii statute discussed in the *Wilson* case in the preceding section (and most statutes), does not *specifically* provide that no action can be brought for compensation, the court concluded that the criminal sanctions provided were the sole sanctions, since the contract here was not "evil in itself." [89] As a result the trial court granted recovery to the unlicensed engineer.[90]

The influential Supreme Judicial Court of Massachusetts also seems to be moving in the direction of allowing recovery to unlicensed persons. *Town Planning & Engineering Associates, Inc. v. Amesbury Specialty Co., Inc.* involved a contract for design services made by an unlicensed designer who had been approached by a client. The unlicensed designer retained two registered professional engineers as well as a registered land surveyor to perform services related to the project. However, much of the work was done by the designer and his six drafters. The project was ultimately abandoned when the bids were too high. At first the client told the unlicensed designer to renegotiate the bids or get lower ones, but shortly thereafter the client terminated the contract claiming it was illegal. Subsequently the client negotiated directly with a contractor, obtained a reduced bid for a somewhat scaled-down project, and to some degree used the unlicensed designer's plans.

The court noted that the plaintiff's work had been properly performed. In holding that the plaintiff could recover, the court stated:

> If there was a violation here, it was punishable as a misdemeanor under the statute. Violation of the statute, aimed in part at least at enhancing public safety, should not be condoned. But we have to ask whether a consequence, beyond the one prescribed by statute, should attach, inhibiting recovery of compensation, and we agree with the judge in his negative answer to the question in the present case. To find a proper answer, all the circumstances are to be considered and evaluated: what was the nature of the subject matter of the contract; what was the extent of the illegal behavior; was that behavior a material or only an incidental

---

87. *Food Industries Research & Engineering Inc. v. State of Alaska,* 507 F.2d 865, 865–66 (9th Cir.1974).

88. *Sumner Development Corp. v. Shivers,* supra note 79.

89. *Food Industries Research & Engineering, Inc. v. State of Alaska,* 388 F.Supp. 342 (D.Alaska 1975). The court used a classification sometimes employed in invalidity cases. Some contracts are *malum in se.* Such contracts are so infected with universally accepted immorality, that they should not be the basis for contractual recovery or recovery based upon benefit conferred as creating unjust enrichment. Other contracts, less morally objectionable, are sometimes classified as *malum prohibitum.* These are contracts that are invalid because the statute has declared them so. There is more likelihood of relief in the latter case than in the former.

90. But a contractor was denied recovery because a statute specifically barred unlicensed contractors from using the courts to collect for services. *State for the Use of Smith v. Tyonek Timber, Inc.,* 680 P.2d 1148 (Alaska 1984).

part of the performance of the contract (were "the characteristics which gave the plaintiff's act its value to the defendant . . . the same as those which made it a violation . . . of law"); what was the strength of the public policy underlying the prohibition; how far would effectuation of the policy be defeated by denial of an added sanction; how serious or deserved would be the forfeiture suffered by the plaintiff, how gross or undeserved the defendant's windfall. The vector of considerations here points in the plaintiff's favor.

Our cases warn against the sentimental fallacy of piling on sanctions unthinkingly once an illegality is found. As was said in *Nussenbaum v. Chambers & Chambers, Inc.*, 322 Mass. 419, 422, 77 N.E.2d 780, 782 (1948): "Courts do not go out of their way to discover some illegal element in a contract or to impose hardship upon the parties beyond that which is necessary to uphold the policy of the law." Again the court said in *Buccella v. Schuster*, 340 Mass. 323, 326, 164 N.E.2d 141, 143 (1960), where the plaintiff was allowed to recover his compensation for blasting ledge on the defendant's property although he had not given bond or secured a blasting permit as required by law: "We do not reach the conclusion that blasting without complying with the requirements . . . is so repugnant to public policy that the defendant should receive a gift of the plaintiff's services." Professor Corbin adds: "The statute may be clearly for protection against fraud and incompetence; but in very many cases the statute breaker is neither fraudulent nor incompetent. He may have rendered excellent service or delivered goods of the highest quality, his non-compliance with the statute seems nearly harmless, and the real defrauder seems to be the defendant who is enriching himself at the plaintiff's expense. Although many courts yearn for a mechanically applicable rule, they have not made one in the present instance. Justice requires that the penalty should fit the crime; and justice and sound policy do not always require the enforcement of licensing statutes by large forfeitures going not to the state but to repudiating defendants." 6A

A. Corbin, Contracts § 1512, at 713 (1962) (footnote omitted).[91]

Yet recent cases, like so many that have been referred to in this chapter, come to different conclusions in claims by unlicensed designers [92] and unlicensed contractors.[93] But there have been inroads on the general denial rule.

(D) SUMMARY

Although it must still be generally stated that unlicensed persons are not able to recover based upon either the contract they have made or unjust enrichment, some factors may motivate a court to determine that either the statute was not violated, a subterfuge will be given effect, substantial compliance is sufficient, or recovery will be had despite a licensing violation. Such factors are as follows:

1. The work for which recovery is sought conforms to the contract requirements.
2. The party seeking a defense based upon a violation of the licensing law is in the same business or profession as the unlicensed person.
3. The unlicensed person apparently had the qualifications to receive the license.

SECTION 13.08 THE TRAINED BUT UNREGISTERED DESIGN PROFESSIONAL: MOONLIGHTING

(A) UNLICENSED PERSONS: A DIFFERENTIATION

This chapter has spoken of persons who

---

91. 369 Mass. 737, 342 N.E.2d 706, 711–712 (1976).

92. *Tucker v. Whitehead*, 155 Ga.App. 104, 270 S.E.2d 317 (1980) (dictum denied right to recovery); but see *West Baton Rouge Parish School Board v. T.R. Ray, Inc.*, 367 So.2d 332 (La.1979) (cannot arbitrate but can recover based upon unjust enrichment if contract made in good faith; corporate architect's licensed employee quit after contract made).

93. *Revis Sand & Stone, Inc. v. King*, 49 N.C.App. 168, 270 S.E.2d 580 (1980) (denial); but see *C.B. Jackson & Sons Construction Co. v. Davis*, 365 So.2d 207 (Fla.App.1978) (recovery based upon unjust enrichment).

violate the registration for licensing laws as if all of those who do so can be placed in the same category. Yet differentiation can be made between those who violate these laws who have been educated and trained as design professionals and others, such as contractors, developers, design-builders, or self-styled handypersons.

Ordinarily a long period elapses between the beginning of architecture or engineering training (architecture will be used as an illustration) and registration. As a result, there are many persons with substantial education and training as design professionals but who for various reasons do not work under the supervision of a registered architect yet engage in the full spectrum of design services. To be sure, the law does not differentiate these persons from those without design education and training. Yet that these persons often do perform these services in violation of the law demonstrates not only the financial burden involved in education and training before registration but also that a market exists for such services.

This section is directed toward some of the legal problems faced by what is referred to as the moonlighter, who is likely to be a student, a recent graduate, or a teacher. While some of the discussion anticipates material to be discussed in greater detail in the balance of the treatise, such as professional liability, exculpation, and indemnification, it is useful to focus upon the range of legal problems that such individuals face.

(B) ETHICAL AND LEGAL QUESTIONS

A threshhold issue before the discussion of collection for services performed by moonlighters (C) and liability of moonlighters (D) is whether such work should be performed. A differentiation will be drawn between work by a moonlighter that does not violate the law because the project is exempt and the requirement that a registered designer perform services that do violate the registration laws. The former, of course, raises no ethical or legal questions. The latter does.

The discussion in this section should not be taken as a suggestion that moonlighters should violate the law by engaging in design services prohibited by law. However, two factors necessitate discussion of moonlighting. First, arrangements between clients and moonlighters demonstrate that each group feels it will benefit from such an arrangement or such arrangements would not be made. That there is a good deal of moonlighting demonstrates either that the registration laws only make illegal conduct that should not be prohibited or great disrespect for the registration laws.

Second, the many cases that involve attempts by unlicensed persons to recover for services that they have performed and the periodic success of their claims indicate some dissatisfaction with the registration laws even by those charged with the responsibility of enforcing them indirectly by denying recovery for services performed in violation of them.

Anyone tempted to moonlight and violate registration laws must take into account, in addition to the risk of criminal and quasi-criminal sanctions, other risks discussed in the balance of this section—going unpaid for services rendered or being wiped out financially because of claims by the client or third parties.

(C) RECOVERY FOR SERVICES PERFORMED

Clearly the safest path is to design only those projects that are exempt from the registration laws. Undertaking nonexempt projects runs substantial risks. This subsection looks at these risks and ways of minimizing the likelihood that the moonlighter will not be able to recover for services that have been performed.

Since payments made cannot be recovered by the client, it is best to specifically provide that interim fee payments will be made and to make certain that they are paid. If payment has not been made in full, it is possible at least in some jurisdictions to recover for these services in a restitutionary claim based upon unjust enrichment. Such a claim is more likely to be successful if the client is aware of the moonlighter's nonregistered status, if the moonlighter has had substantial education and training in design, if the work appears to have been done properly, and if it appears that the client has benefited by a lower fee. The stated purpose of registration laws is to deter unqualified persons from performing design services and to protect the public from hiring unqualified professionals. A client who is aware of the status of the designer engaged has given up protection accorded by law, and to allow the client to avoid payment where the services have been properly rendered can create unjust enrichment.

Of course if the project will expose the general public to the risk of physical harm, it is more difficult to justify either entering into such an arrangement or allowing recovery for services performed under one. But as a practical matter, these are not the types of projects that moonlighters will design. The uncertainty of recovery and the high cost of litigation even if there is a recovery should deter moonlighters as a whole from doing nonexempt work. If the deterrent function will be accomplished, assuming that is one of the desirable results of these laws, there appears to be no reason to deny recovery where services have been performed by someone with the requisite education and training.

(D) LIABILITY PROBLEMS

This discussion assumes that the moonlighter will not be carrying professional liability insurance. As a result, any liability that moonlighters incur will be taken out of their pockets if there is anything in their pockets to take.

As shall be seen in Section 17.05, the professional is generally expected to per-

form as others in the profession would have performed. Should this standard be applied when a moonlighter is knowingly engaged by a client?

Differentiation must be made between an express agreement dealing with a standard of performance and one that is implied by law. Suppose the parties agree that the standard of performance will be greater or lesser than that usually required of a *registered* design professional. Would the court enforce any agreement under which the standard would be less than that required by a registered professional?

This depends upon a balancing of laws favoring autonomy, which would enforce such an agreement, and the strength of the policy expressed in the registration laws, which would deny enforcement to such an agreement. Although the resolution of such a question would not be easy, it is likely that the law would not give effect to an express agreement under which someone performing services in violation of law would be judged by some standard less than would be used in the event that the person doing the design were registered.

If no express agreement is made regarding the standard of performing, a case can be made for judging the performance by what the parties are likely to have intended. If the client knowingly engaged a moonlighter who received a lower fee, should the client get what he pays for—a lower level of performance for a lower price? However, again it is unlikely that the court in determining the implied terms of such an illegal agreement would prefer autonomy to a result that might appear to frustrate the registration laws.

Another approach, related but somewhat different from the one discussed in the preceding paragraphs, would be for the moonlighter to seek to persuade the client to exculpate the moonlighter from the responsibility from any performance that would violate the contract or to limit the moonlighter's exposure to a designated portion for the entire fee. Since by definition the moonlighter is not insured, the moonlighter may be able to persuade the client that he would not be in a position to pay for any losses that he caused, and this is taken into account in setting the fee.

Suppose the client will agree to exculpation or liability limitation. The moonlighter must take particular care to express these provisions very clearly, as courts will at the very least construe them against the moonlighter. Even more, there is a risk that a court that feels strongly about the registration laws will not enforce such a clause if it would frustrate those laws. Keep in mind, however, that because the moonlighter is not likely to have either insurance or assets, a claim by the client is likely to be rare.

Third-party liability raises different problems, inasmuch as any arrangement for exculpation or a liability limitation between the moonlighter and the client will not affect the rights of third parties. But again, the moonlighter without insurance or assets is unlikely to be sued by a third party. A moonlighter concerned about the liability to third parties either because of moral considerations or because the moonlighter does have assets should attempt to seek indemnification from the owner. But as shall be seen, such clauses are also narrowly interpreted and in some states are unenforceable.[94]

From the standpoint of liability either to the client or third parties, it does not appear that there is a great risk if moonlighters do not have assets of their own. However, if moonlighters do have assets or if they are concerned about causing loss to someone for which they ought to pay, the liability problems of moonlighters are likely to mean that the moonlighters will be judged by the standard of registered professionals.

PROBLEMS

1. A was registered as a licensed architect in the state in which he had his offices. In that state, licenses must be renewed every five years. Renewals are usually granted unless a large number of complaints have been made with the licensing board.

   Three months before A's license was required to be renewed, A entered into a written contract for the performance of design services for C. A was paid approximately 60% of his fee through interim payments. When the time came for final billing after the services were completed, A realized that he had forgotten to renew his registration. He called the licensing board, and it informed him that he would have been renewed had he applied at the proper time; A then took steps to renew his registration.

   However, C discovered that A had not renewed his registration and now refuses to pay the balance of the fee. He also demands that A return to him the interim fee payments made.

   Should A be entitled to recover the balance of the fee? If your conclusion is that he should not, should he be obligated to repay any interim fee payments made? Would you like to know any additional facts? If so, what and why?

2. You are a graduate architecture student. You have been asked to design a single-family residence by a developer for whom you wish to work after you become registered. What would be the factors you should take into account? Why would they be relevant?

---

94. See Sections 36.05(D) and (E).

14

Contracting for
Design Services:
Pitfalls and Advice

SECTION 14.01
AUTHORITY PROBLEMS

(A) PRIVATE OWNERS

Negotiations to provide design services are
always made with individuals. It is important to determine whether the individual
with whom a designer negotiates has authority to make binding representations and
to enter into a contract. The general agency doctrines that regulate these issues were
discussed in Chapter 4. Reference should
be made to that chapter, particularly the
material in Section 4.06 that deals with apparent authority.

Design professionals will be motivated by
many considerations when they decide
whether to enter into a contract to perform
design services. One factor will be representations made by the prospective client.
For example, a design professional may be
influenced by representations that relate to
the likelihood that the project will be approved by appropriate authorities, that adequate financing can be secured, and that the
design professional will work out the design
with a particular representative of the client
or to other facts that will influence the
design professional to contract to perform
design services for the client.

In the "best of all worlds," the persons at
the "top" of the client's organization would

be contacted to determine the authority of
the person with whom the design professional is directly dealing. However, it may
not be easy to determine who is at the top
of the client's organization, and it may not
always be politic to take this approach.
The realities of negotiation mean that the
design professional will have to rely largely
on appearances and common sense to determine whether the person with whom she
is dealing can make such representations.
This may depend upon the position of the
person in the corporate structure, the importance of the representation, and the size
of the corporation. The more important
the representation and the more important
the contract, the more likely it is that only a
person high up in the organizational structure will have authority to make representations that will bind the corporation. If the
corporation is particularly large, a person
in a relatively lower corporate position may
have authority to make representations.
For example, in a large national corporation, the head of the purchasing department
may be authorized to make representations
that would require a vice-president's approval in a smaller organization.

One useful technique is to request that
important representations be incorporated
in the final agreement. If the person with
whom the design professional is dealing is
unwilling to do so, that person may not

have authority to make the representations that have been made.

This approach has an additional advantage. In the event of a dispute that goes to court the parol evidence rule discussed in Section 14.04(E) may make it difficult to introduce evidence of representations that are not included in the final agreement. Even if admitted into evidence, failure to include it in the written agreement may make it difficult to persuade judge or jury that the representations *were* made if denied by the client.

Next it is important to examine the authority to *contract* for design services. This problem will be approached by looking at the principal forms of private organizations that are likely to commission design services.

Sole Proprietor

The sole proprietor clearly has authority to enter into the contract. Dealings with an agent of the sole proprietor are controlled by concepts of agency and scope of actual or apparent authority. Generally, sole proprietorships are small business operations. For a sole proprietorship, it is likely that a contract for design services or construction is a serious and important transaction. For that reason, it is not likely that agents of a sole proprietor have such authority.

Partnership

Usually partnerships are not large businesses. It is likely that only the partners have the authority to enter into contracts for design services or construction. The partners may have designated certain partners to enter into contracts. Unless this comes to the attention of the design professional, it is unlikely that such a division of authority between the partners will affect her. Although partners have authority to enter into most contracts, it is advisable to get all the partners to sign the contract.

If the partnership is a large organization, agents may have authority to enter into a contract for performance of design services and construction. It would probably be best, however, to have some written authorization from a partner that the agent has this authority.

Corporation

It is common for the articles of incorporation or the bylaws to specify who has the authority to make designated contracts and how such authorization is to be manifested. The more unusual the contract, or the more money involved, the higher the authority needed. Frequently contracts that deal with land, loans, or sales or purchases not in the normal course of business must be authorized by the board of directors. Lesser contracts may require authorization by higher officials, while for contracts involving smaller amounts of money, or those in the usual course of business, subordinate officials may be authorized.

Contracts needing board approval are passed by resolution and entered into the minutes of the board meeting. In addition, bylaws or any chain-of-authority directive will frequently state which corporate officials must actually sign the contract. Again, the importance of the contract will usually determine at what echelon the contract must be signed and how many officials must sign it. The corporate bylaws or state statutes may require that the corporate seal be affixed to certain contracts.

For maximum protection, the design professional should check the articles of incorporation, the bylaws, and any chain-of-authority directive of the corporation. The design professional should see who has the authority to authorize the contract and whether the proper mechanism, such as the appropriate resolution and entry of the resolution in the minutes, was used. Then the design professional should determine whether the person who wants to sign or has already signed for the corporation has authority to do so.

In very important contracts, it may be wise to take all of these steps. It is not unreasonable for the design professional to request that the corporation attach a copy of its articles and bylaws and a copy of the resolution authorizing the particular project in question. The design professional can request that the contract itself be signed by the appropriate officers of the corporation, such as the president and secretary of a smaller corporation or the vice-president and secretary of a larger corporation.

Sometimes such precautions need not be taken. The project may not seem important enough to warrant this extra caution. The design professional may have dealt with this corporation before and is reasonably assured that there will be no difficulty over authority to contract. However, laziness or fear of antagonizing the client is not a justifiable excuse.

For the risks involved in dealing with the promoter of a corporation not yet formed, refer to Section 3.04(C).

Unincorporated Association

Dealing with an unincorporated association seems simple but may involve many legal traps. To hold the members of the association, it is necessary to show that the persons with whom the contract was made were authorized to make the contract. In such cases, it is vital to examine the constitution or bylaws of the unincorporated association and to attach a copy of the resolution of the governing board authorizing that the contract be made. The persons signing the contract should be the authorized officers of the association.

It may be wise to obtain legal advice when dealing with an unincorporated asso-ciation, unless the unincorporated association is a client for whom the design professional has worked in the past and in whom the design professional has confidence.

Spouses and unmarried co-habitants

A design professional may deal with a married or unmarried couple living together. Even with spouses, no presumption exists that one is agent of the other.[1] But one can bind the other if the former acts to further a common purpose or if unjust enrichment would result without a finding of agency.[2] It is advisable to have both spouses or members of an unmarried couple as parties to the contract.

(B) PUBLIC OWNERS

Reference should be made to Section 9.02, which deals with public owners and describes some of the special attributes of dealing with public entities. Authority issues arise with some frequency because of the need to protect public funds and preclude improper activity by public officials.

The law has generally provided little protection to those who deal with public entities, rarely invoking the apparent authority doctrine. Similarly, application of "estoppel" concepts, which would bar the public entity from establishing that representations were unauthorized or that contracts were not made by proper persons or in accordance with law, is also rare.[3] Yet there has been a modest relaxation of the protections accorded public entities. Restitution claims based upon unjust enrichment

---

1. *Oldham & Worth, Inc. v. Bratton*, 263 N.C. 307, 139 S.E.2d 653 (1965).

2. *Capital Plumbing & Heating Supply Co. v. Snyder*, 2 Ill.App.3d 660, 275 N.E.2d 663 (1971) (wife agent for lien purposes).

3. *School District of Philadelphia v. Framlau Corp.*, 15 Pa.Cmwlth. 621, 328 A.2d 866 (1974) (school board president); *County of Stephenson v. Bradley & Bradley, Inc.*, 2 Ill.App.3d 421, 275 N.E.2d 675 (1971) (chairman of board of supervisors). See also *M.A.T.H., Inc. v. Housing Authority of East St. Louis*, 34 Ill.App.3d 884, 341 N.E.2d 51 (1976), which held a promise to renegotiate after completion made by the president of the housing authority to be beyond the latter's authority and unenforceable.

are beginning to have some success,[4] as are assertions that a public entity cannot rely on provisions of a contract where it has misled the contractor.[5]

This should not encourage carelessness in dealing with public entities. There are still substantial risks in relying on representations by government officials and entering into contracts with public entities.

SECTION 14.02
FINANCIAL CAPACITY

(A) IMPORTANCE

A crucial factor in determining whether to undertake design work is the client's financial capacity. The relationship between the design professional and client can deteriorate when the latter cannot or will not pay interim fee payments as required, is unable or unwilling to obtain funds for the project, or does not possess the financial capacity to absorb strains on the budget caused by design changes or market conditions that increase the cost of labor and materials. If the client will not have the financial resources to pay any court judgment that may be awarded, the contract is risky and should be avoided.

Despite constant warnings to design professionals, many enter into contracts and perform work and find that they are unable to collect for their services. Professionals do not like to confront financial responsibility openly. They prefer to assume that clients are honorable people who will pay their obligations. Usually this is the case. However, there are too many instances of uncollected fees to justify cavalier disregard of this problem. Design professionals must

seriously confront the problem of financial capacity.

(B) PRIVATE CLIENTS

Retainers and Interim Fees

Most standard contracts published by the professional associations provide for initial retainers to be paid by the client at the time the contract is signed and for interim fee payments to be made during performance. A major purpose of such provisions is to give the design professional working capital.

Another purpose is to limit the scope of the financial risk taken by the design professional. If services and efforts do not run very far beyond the money paid, risk of nonpayment is substantially reduced. The difficulty is that design professionals frequently do not insist that the client comply with these contract terms. This is *absolutely* essential. If the matter is explained properly to the client, there should be no difficulty. It should alert the design professional to possible danger when the client seems to be insulted when polite requests are made for advance retainers and interim fee payments when due. Clients who react adversely to such requests are often clients who either do not have the money or will not pay even if they do have the money.

Client Resources

AIA Doc. B141, ¶ 2.2.2, permits the architect to request that the client furnish "a statement of the funds available for the Project, and their source." This power can be exercised at any time during the contract period. This is both a promise and a condition to the architect's continued performance.

---

4. *Coffin v. District of Columbia*, 320 A.2d 301 (D.C.App.1974) (dictum: recovery limited to the amount of money that the contracting officer had authority to commit); *Saul Bass & Associates v. United States*, 205 Ct.Cl. 214, 505 F.2d 1386 (1974) (recovery based upon an informal agreement despite failure to execute a formal contract).

5. *Emeco Industries, Inc. v. United States*, 202 Ct.Cl. 1006, 485 F.2d 652 (1973); *Manloading & Management Associates, Inc. v. United States*, 198 Ct.Cl. 628, 461 F.2d 1299 (1972). See also *Hueber Hares Glavin Partnership v. State*, 75 A.D.2d 464, 429 N.Y.S.2d 956 (1980) (dictum).

Failure by the client to comply would permit the architect to suspend performance. If suspension continued or if the client indicated an inability or unwillingness to furnish this information, the architect could treat it as a material breach terminating the obligation to perform and giving remedies for breach of contract. (See Sections 31.02 and 38.04.) The design professional can and should ask for this information *before* deciding whether to perform services. It may be useful to run a credit check if the client is new or if the information furnished does not appear trustworthy.

One-Person and Closely Held Corporations: Individual Liability of Officers

Many small businesses are incorporated, and the shares of the business are held entirely by the proprietor of the business. The proprietor is permitted to do this by law. One purpose is to insulate personal assets from the liabilities of the corporation. Some corporations are small, closely held corporations, with the shares owned by a family or by the persons actually running the business. As noted in Section 3.04(E), the law can pierce the corporate veil and treat the inadequately capitalized one-person or closely held corporation as a sole proprietorship or partnership. If this is done, the design professional can recover from the shareholder or shareholders.

Piercing a corporate veil is rarely done, however. If a credit check reveals that the corporation is merely a shell with very few assets, it may be necessary to demand that the shareholders assume personal liability. To do so, the sole shareholder or shareholders should sign both as representatives of the corporation and as individuals. If those signing as individuals are solvent, this is a reasonably secure method of assuring payment if the corporation is unable to pay the contractual obligations. If the individuals signing are not solvent, individual liability will be of little value. A refusal to sign as

an individual may be a warning that the corporation is in serious financial trouble.

A Surety or Guarantor

Chapter 37 discusses the role of the surety in the construction phase of the project. If the design professional believes the person with whom she is dealing is not financially sound, the design professional may wish to obtain a financially sound person to guarantee the obligation of the corporation or the individual *before* entering into the contract. Subject to some exceptions, promises to pay the debt, default, or miscarriages of another must be in writing. The design professional who requests and obtains a third person to act as surety for the client should make certain that the surety signs the contract or a separate surety contract. Legal advice should be obtained where a surety or guarantor is involved.

Real or Personal Property Security

Another method of securing the design professional against the risk that the client will not pay is to obtain a security interest in real or personal property. This can be done by obtaining a mortgage or a deed of trust upon the land on which the project is to be constructed or upon other assets owned by the client. This may seem like a drastic measure. If that much insecurity is involved, it may be advisable not to deal with the client at all. The law regarding creating and perfecting security interests is beyond the scope of this treatise. Legal advice should be obtained if such protection is needed.

Client Identity

Some clients move quickly in and out of various different but related legal forms. What appears to be a partnership turns out to be a corporation of limited resources. The design professional may find that the corporation with whom she dealt is insolvent while other solvent related corporations are controlled by the client. It can be difficult for the design professional to know for whom she is working or who has legal responsibility. This can be complicated by the different entities described in the preliminary correspondence, the contracts, and the communications during performance. Design professionals should know the exact identity and legal status of the client.

Spouses or Unmarried Couples

As mentioned in Section 14.01(A), dealing with co-habiting couples can raise agency questions. Design professionals may discover that the person with whom they dealt has no assets while the other co-habitant owns all the assets. It is important to have both co-habitants sign the contract.

Mechanics' Liens

Mechanics' liens—more important to contractors and suppliers—are treated in Section 32.07(D). Their occasional utility to professional designers can tempt those who are about to perform design services to avoid some of the techniques previously mentioned for fear of losing a client. For that reason they are discussed here briefly.

State statutes often give persons a security interest in property that they have improved to the extent of any debt owed to the improver by the owner of the property or someone who has authority to bind the owner. The lien holder can demand a judicial foreclosure of the property and satisfy the obligation out of the proceeds.

Some state statutes specifically give design professionals mechanics' lien rights. In some states they have been brought in under general provisions granting liens to those who improve real property. However, a number of stumbling blocks frequently make the lien remedy unavailable. Work usually has to begin upon the land in ques-

tion for the design professional to have a lien.[6] In some states, the lien covers only services that directly benefit the land, such as supervision but not design. Where design services are sufficient, they may have to be tied to supervision. Generally, the plans must be used in the project.

Lien rights cannot be asserted against public improvements. Many technical requirements exist for the creation of such a lien. Notices have to be given, filings have to be made, and foreclosure actions must be taken within specified times. Without strict compliance, there is no lien. Other persons may have equal or prior security rights in the land. The land value may not be enough to pay the lien claims in their entirety.

Mechanics' liens statutes vary considerably from state to state and are frequently changed by legislatures. Design professionals should seek legal advice to see whether they are within the class of persons accorded liens and to ascertain the steps needed to perfect a lien. It is unwise to undertake work for a client who may not be able to pay with the hope that in the event of nonpayment, there will be a right to a mechanics' lien. The possibility of being able to assert a mechanics' lien is never a substitute for a careful consideration of the financial responsibility of the client and collection of interim fee payments.

(C) PUBLIC OWNERS

References should be made to Section 12.03, which deals with methods to fund public projects. Those who deal with public entities run the risk that money will not be appropriated or bonds not issued to pay for the project or design services. An Illinois decision held that an architect could not recover for his services where the evidence showed he was risking compensation on the passage of a bond issue.[7] The court

---

6. *Goebel v. National Exchangors, Inc.*, 88 Wis.2d 596, 277 N.W.2d 755 (1979).

7. *County of Stephenson v. Bradley & Bradley, Inc.*, 2 Ill.App.3d 421, 275 N.E.2d 675 (1971).

pointed to a statement made by the architect during negotiations that he was gambling with the county, which convinced the court that he took this risk.

The best protection against risks is to check carefully on the availability of funds and make certain that work does not begin until it is relatively certain that funds will be available to compensate the design professional.

SECTION 14.03 COMPETITION FOR THE COMMISSION: ETHICAL AND LEGAL CONSIDERATIONS

Historically, associations of design professionals have sought to discipline their members for ungentlemanly competition, both in obtaining a commission for design services and replacing a fellow member who has been performing design services for a client.

As an illustration, AIA Document J330 in effect in 1958 listed among its mandatory standards of ethics:

> 9. An Architect shall not attempt to supplant another Architect after definite steps have been taken by a client toward the latter's employment.
>
> 10. An Architect shall not undertake a commission for which he knows another Architect has been employed until he has notified such other Architect of the fact in writing and has conclusively determined that the original employment has been terminated.

In addition, J330 stated that the architect would not compete on the basis of professional charges.

Note that these restraints were considered ethical. Undoubtedly they are traceable to the desire to preserve these professions *as professions* and to avoid practices that would be accepted in the commercial marketplace. Although the associations must have recognized that there will be competition for work, it was hoped that competition would be conducted in a gentlemanly fashion and would emphasize professional skill rather than price.

Beginning in the 1970s, attacks were made upon these ethical standards by public officials charged with the responsibility of preserving competition. One by one, association activities that were thought to impede competition came under attack, particularly those that dealt with fees. Fee schedules, whether required or suggested, were found to be illegal.[8] Disciplinary actions for competitive bidding were also forbidden.[9]

Standard 9 in J330 attempted to inhibit competition *before* any valid contract had been formed.[10] Standard 10, though less susceptible to the charge that it is anticompetitive, could have had the effect of limiting the client's power to replace one architect with another.

It is clear that design professionals do not like to compete on the basis of price. Although their associations cannot use their disciplinary power to attack this practice, current federal legislation precludes head-to-head fee competition.

In 1972 Congress enacted the Brooks Bill,[11] which determines how contracts for design services can be awarded by federal agencies. The statute declares that the policy of the federal government is to negotiate on the basis of "demonstrated competence and qualification for the type of professional services required and at fair and reasonable prices."

Those who perform design services submit annual statements of qualifications and

---

8. *Goldfarb v. Virginia State Bar*, 421 U.S. 773, 95 S.Ct. 2004, 44 L.Ed.2d 572 (1975).

9. *National Society of Professional Engineers v. United States*, 435 U.S. 679, 98 S.Ct. 1355, 55 L.Ed.2d 637 (1978).

10. *United States v. American Society of Civil Engineers*, 446 F.Supp. 803 (S.D.N.Y.1977) (barred society from disciplining members who supplanted another member).

11. 40 U.S.C.A. §§ 541–544.

performance data. The agency evaluates the statements, together with those submitted by other firms requiring a proposed product, and discusses with no fewer than three firms "anticipated concepts and the relative utility of alternative methods." The agency will then rank the three most qualified to perform the services. The agency will attempt to negotiate a contract with the highest qualified firm and takes into account "the estimated value of the services to be rendered, the scope, complexity, and professional nature thereof."

If a satisfactory contract is not negotiated with the most qualified, the agency will undertake negotiations with the second most qualified and if that fails with the third. The U.S. Justice Department has proposed repeal, arguing that the statute restricts competition. Many states have had comparable legislation regulating the award of contracts for design services by state agencies.

One troubling question that arose as a result of the increased use of Construction Management in the 1970s related to whether contracts for the performance of these services were more like those for design, which did not require competitive bidding, or more like contracting, which did. This is discussed in Section 21.04(D).

The law protects contracts or advantageous relationships from improper interference by third parties. The famous English case of *Lumley v. Gye* [12] allowed the plaintiff impresario to recover from the defendant who enticed an opera singer to breach her contract with the plaintiff and sing for the defendant. The tort granted that recognition—in this case, interference with a contract—must balance protection given a contract with the freedom of parties to seek to persuade others that they can perform services better than someone who has been engaged to perform them.

Although contracts have been protected more vigorously in this century, the interference must be *intentional*. Some states require malice. However, malice may range from *actual* malevolence toward the plaintiff to a mere intent to do the act that will have the effect of interfering with the plaintiff's contracts. For example, in *Blivas & Page, Inc. v. Klein*,[13] an architect successfully maintained a legal action against the client with whom he contracted and a third party who induced the breach. The court defined malice as the commission of an intentional act without justification or excuse. It did not require a malevolent motive or desire to specifically injure the plaintiff. Protection can go beyond an enforceable contract and include relationships that are terminable at will and even prospective economic advantages.

Wrongful interference with a contract is sometimes the basis for a claim against a design professional who suggests to the owner that a contractor be terminated (discussed in Section 17.08(D)).

SECTION 14.04
PROFESSIONAL SERVICE CONTRACTS: SOME REMARKS

(A) PROFITS AND RISK

Ordinarily professional advisers do not have potentially high profit returns. They serve their client and expect to be paid for their services. This does not mean that advisers are always paid for their work. Nonpayment because the client does not have the financial resources is a risk taken by all who perform services.

This section is directed toward defenses by the client to any claim for compensation for services rendered based upon an assertion that no money is due. Usually such defenses are predicated upon an asserted

12. 2 El.&Bl. 216, 118 Eng.Rep. 749 (1853).

13. 5 Ill.App.3d 280, 282 N.E.2d 210 (1972).

understanding that if the project did not go forward, usually because public approval or funds could not be obtained, the professional designer would not be paid for the work.

Although autonomy gives contracting parties the power to make an arrangement under which the professional designer may risk the fee, any conclusion that this risk has been taken should be arrived at only if weighty evidence supports this conclusion. This is based not only upon customary practices but also upon the conclusion that professional designers do not, as a rule, recover profits but recover only payment for services they render.

(B) GOOD FAITH, FAIR DEALING, AND FIDUCIARY RELATIONSHIPS

Increasingly the law finds that all contracting parties owe each other the responsibility of good faith and fair dealing.[14] This is particularly so when objectives sought by the design professional and the client require close cooperation. Each party should help the other achieve its goals under the contract.

The retention of a professional designer by a client—like the relationship between employer and employee and between partners—creates a fiduciary relationship. The core of a fiduciary relationship is trust and confidence. To achieve these, the person to be protected by the relationship must believe in the undivided loyalty of its fiduciary.

The client should be able to trust its professional adviser and receive honest professional advice. Both parties—design professional and client—should be candid and open in their discussions, and each should feel confident that the other will not divulge confidential information.

The fiduciary relationship can be contrasted with the arm's-length relationship. In a commercial setting, parties generally deal at arm's length: each party must look out for itself. On the other hand, a fiduciary relationship is a close one, and the parties must be able to trust each other.

Some aspects of the fiduciary relationship are obvious. Design professionals should not take kickbacks or bribes. They should not profit from professional services other than by receiving compensation from the client.

Funds held by the design professional that belong to the client should be kept separate. Commingling will be a breach of the fiduciary obligation. In such a case, any doubts about to whom the money belongs or for whom profitable investments were made will be resolved in favor of the client.

Financial opportunities that come to the attention of the design professional as a result of the services that she is performing for the client should be disclosed to the client if the services would be an opportunity falling within the client's business.

While the design professional should not disclose confidential information, there may be extreme cases where the design professional has a duty to disclose information even in the face of an *express* contract prohibiting such disclosure. For example, in *Lachman v. Sperry-Sun Well Surveying Co.*,[15] a client brought an action against its engineer who, in violation of an express covenant not to disclose information, notified adjacent landowners of his client's slant drilling. The court held that such a covenant would not be enforced.

One of the most troubling concepts relates to conflict of interest. One cannot serve two masters. The client should be able to trust its design professional to make judgments based solely on the best interests of the client. Advice or decisions by the design professional should be untainted by

---

14. Restatement (Second) of Contracts § 205 (1981): Uniform Commercial Code § 1–203.

15. 457 F.2d 850 (10th Cir.1972).

any real or apparent conflict of interest. The client must believe the design professional serves it and it alone.

For these reasons, design professionals should not have a financial interest in anyone bidding on a project for which they are furnishing professional advice. Likewise, they should not have a financial interest in any contractor or subcontractor who is engaged in a project for which they have been engaged. Design professionals should not have any significant financial interest in manufacturers, suppliers, or distributors whose products might be specified by them. Products should not be endorsed that could affect specification writing, nor should designated products be specified because manufacturers or distributors of those products have furnished free engineering. The purpose of these restrictions is to avoid conflict of interest. Design professionals cannot serve their clients loyally if they might personally profit by their advice.

Generally a client who is fully aware of a potential conflict of interest can nevertheless choose to continue to use the design professional. Consent to a conflict of interest should be binding only if it is clear that the client knows all the facts and has sufficient understanding to make a choice. A design professional who intends to rely upon consent by the client must be certain that such requirements are met.

Some conflicts of interest are clear; others are murky. For example, suppose an architect retained by an airport authority is also retained by airline tenants who plan to rent space in the airport. The city attorney of San Francisco concluded that there was no conflict of interest, principally because the airport authority knew that the architects were working or might work for airline tenants, that such arrangements were common, and that the architects were not "vested with authority to exercise any discretion on behalf of the City." [16]

If the client, after full disclosure, freely chose to proceed, the opinion may be correct. But the *client's* having the power to make decisions—a point emphasized in the opinion—does not eliminate the architect's advisory function. Can the architect advise impartially if she hopes to obtain commissions from airport tenants? The city attorney's conclusion that such dual retentions are efficient and common is questionable.

Clearly the client can avoid responsibility for any act by the design professional that is tainted by a breach of the fiduciary obligation. For example, if a contract is awarded to a bidder with whom the architect or engineer colluded, the contract can be set aside by the client. If the architect or engineer issued a certificate for payment dishonestly or in violation of her fiduciary obligation, the certificate can be set aside if the client so desires.

The breach of a fiduciary obligation can give the client grounds to dismiss the design professional. Any bribes or gifts that have been taken by the design professional can be recovered by the client. Any profit that the design professional has made as a result of breaching the fiduciary obligation can generally be recovered by the client even if the profits were generated principally by the skill of the design professional. Obviously, the design professional must take her fiduciary obligation seriously and make every effort to avoid its breach or the *appearance* of such a breach.

Undoubtedly some of the freedom that courts have felt to develop the fiduciary doctrine occurred because of the frequently brief and incomplete agreements made between clients and design professionals.

Today there is an increasing use of detailed written agreements that deal with almost every aspect of the design professional-client relationship. If there is a detailed written agreement, is there less need for courts to employ the fiduciary concept?

---

16. Letter from Thomas M. O'Connor to Mr. Robert J. Dolan, Clerk, Board of Supervisors, September 7, 1972.

Many aspects of the fiduciary relationship relating to trust and confidence are not likely to be found in standard contracts because the assumption is that they are so obvious that they need not be expressed. In any event, the fiduciary concept has been expanding, and the tendency to have longer and more detailed standard form contracts is not likely to affect such expansion.

(C) VARIETY OF TYPES

Some professional relationships are not accompanied by careful planning in a contractual sense. Such relationships sometimes are created by a handshake without any exploration of important attributes of the relationship. If such attributes are discussed and resolved, the resolution is often not expressed in tangible form.

At the other extreme, the relationship often is cemented by assent to a preprinted standardized form supplied frequently by architect or engineer and occasionally by the client. Often the client will not understand these standardized agreements. On occasion the actual agreement of client and design professional may be different than provisions of the standardized form. Under any of the circumstances mentioned, difficulties can exist if disagreements arise between client and design professional over the services to be performed by the latter.

(D) INTERPRETATION

Contracts between owners and contractors and between contractors and subcontractors generate more interpretation disputes than those between design professional and client. For this reason, the bulk of the discussion in this treatise relating to contract interpretation is found in Chapters 23 and 24. However, a few interpretation guides relating to contracts between design professional and client may be useful here.

One attribute of the design professional client relationship that sets it apart from ordinary commercial contracts is the relative inexperience of many clients in design and design services. Another is the way such agreements are made. Not uncommonly such relationships are created by a vague, informal agreement sometimes followed by assent to a pre-prepared contract form supplied by the design professional. These two attributes frequently generate honest misunderstandings between design professional and client.

The most important lodestar—the common intention of the parties—is determined by an examination of any discussions the parties may have had before entering into the agreement, the language in any written contract to which both parties have assented, the facts and circumstances surrounding the making of the agreement, the conduct of the parties after the relationship has begun, and any custom or usage of the trade of which both parties were aware or should have been aware.

Although discussion between the parties—the relevance of which is determined by the parol evidence rule described in (E)—and surrounding facts and circumstances are important, emphasis here is upon the language of the contract. Although courts still invoke the "plain meaning" rule, the search for evidence outside the writing requiring a preliminary determination that the meaning is not "plain" on its face, by and large courts today will look at any relevant evidence to determine the meaning of language selected by the parties.[17]

As a rule, the most important evidence of the intention of the parties is any written agreement to which the parties have assented. This is true whether the agreement is a carefully negotiated contract between client and design professional of relatively equal bargaining power and experience or

---

17. See Section 24.02.

an agreement between an unsophisticated client and an experienced design professional where the latter has supplied a pre-prepared form contract.

In the former there is likely to be a neutral reading of the language. In the latter, the interpretation guide that contract terms are generally interpreted against the party who has prepared the agreement will be used. This interpretation guide can be justified either because the drafter carelessly or deliberately caused the language ambiguity or because the law assumes that the party who prepares the agreement is in a position to force unfair or onerous terms upon the other party. This doctrine may be an implicit recognition that the party who does not prepare the language may not take the time or have the ability to carefully examine the language selected by the other party.

This discussion assumes that either both or at least one party *had* an intention as to specific language. When lengthy standard contracts published by a third party such as the AIA are used, it is possible that neither party had any idea of the *purpose* of a particular clause, often an important datum invoked to interpret the clause. Since this "absence of intention" can arise in construction contracts, discussion of this topic is postponed until Section 24.02.

The interpretation guide that resolves the dispute against the party who supplied the ambiguous terms will also be applied to contracts published by the professional association of the design professional.[18] Ambiguous language is likely to be interpreted to be consistent with the reasonable expectations of the client rather than the literal interpretation that the language might otherwise bear.

Suppose a design professional is retained by a large institutional client such as a public agency or large private corporation. In such a case, the client often prepares the contract for professional services and presents it to the design professional on a take-it-or-leave-it basis. Ambiguous language in such a case should be interpreted in favor of the design professional.

As to other interpretation guides, handwritten portions of a contract are preferred to typewritten portions or printed portions, and typewritten portions are given more weight than printed provisions in a form contract. Where parties choose language in connection with a particular transaction, the language is more likely to reflect their common intention.

Specific provisions are given more weight than general provisions. For example, suppose a contract between architect and client states that the architect will perform services in accordance with normal architectural professional standards. Suppose further that the same agreement provides that the architect will visit the site at least once a day. If normal professional standards would not require that the architect visit the site this often, a conflict exists between the two provisions. In such a case, it is likely that the provision specifically relating to the number of site visits will control the general provision requiring that the architect perform in accordance with accepted professional standards.

Words are generally used in their normal meanings, unless both parties know or should know of trade usages that have grown up around the use of certain terms. This rule is important in the design professional's relationship to the client. Often a word used in an agreement will have a definite meaning to the design professional but have a different meaning, or at least an indefinable meaning, to the client. For example, suppose that the agreement between

---

18. *Malo v. Gilman,* 177 Ind.App. 365, 379 N.E.2d 554 (1978); *Durand Associates, Inc. v. Guardian Investment Co.,* 186 Neb. 349, 183 N.W.2d 246 (1971).

design professional and client states that the design professional will make periodic visits to the site during construction. Suppose further that the design professional shows evidence of a custom that visits customarily occur at weekly intervals during a certain stage of the project. If the client does not know or have reason to know of this custom, the custom cannot be used to interpret the phrase "periodic visits." Although parties can be said to contract with reference to established usages in a trade, if one party is not or should not have been aware of the usage, it would be unfair to permit evidence of usage to be used to interpret the language in question.

Another interpretation guide, that of practical interpretation, looks at the *practices* of the parties. Their practices may give good evidence of what the parties intended. For example, suppose the language to be interpreted is "periodic visits." Suppose the architect visited the site weekly and this was known by and acquiesced in by the client. A court would consider this persuasive evidence that the parties agreed that the visits would be weekly.

Orput-Orput & Associates, Inc. v. McCarthy [19] involved a disputed oral agreement between architect and client over the existence of a cost limitation (covered in Section 15.03). One issue related to the amount of compensation due the architect. The original construction contract price was approximately $1.2 million. After the work was completed, the client and contractors entered into negotiations that resulted in a final total payout of slightly over one million dollars. The contract between architect and client stated that the fee would be computed on the basis of "the construction cost of the Project." The client contended that the actual expenditure was to govern the fee, while the architect contended that the original contract price controlled.

The appellate court noted that the contract did not assist the court in determining the intention of the parties. The court held that the language should be given a "fair and reasonable interpretation." The trial court had listened to seven witnesses on the question of whether the amount charged was fair and reasonable. The appellate court then noted that though contracts are generally construed against the party who prepares them, the architect's interpretation was fair and reasonable and the trial court's judgment in favor of the architect on this point would be sustained.

Another issue in the *Orput* case involved the question of whether certain services were additional or fell within the basic fee. In addition to looking at the testimony of the architect, the appellate court noted that the architect had billed the client in accordance with the interpretation asserted by the architect and these billings had been received without protest by the client until commencement of the lawsuit.

Although the court was openly skeptical of contentions made after litigation commenced, the holding of the court can also be taken as a recognition of the doctrine of practical (that is, determined by the *practices* of the parties) interpretation. The billings by the architect unobjected to by the client were evidence of what the parties intended when they made the agreement.

(E) THE PAROL EVIDENCE RULE AND CONTRACT COMPLETENESS

One aspect of the parol evidence rule relates to the provability of asserted prior oral agreements when the parties have assented to a written agreement. A client may contend that the design professional agreed to perform certain services that were not specified in the written agreement. The latter's

---

19. 12 Ill.App.3d 88, 298 N.E.2d 225 (1973).

attorney may contend that the writing expressed the entire agreement and that parol evidence or oral evidence is not admissible to "add to, vary or contradict a written document." Sometimes such oral agreements are provable. Sometimes they are not.

Generally courts do not consider that written agreements between client and design professional are complete expressions of the entire agreement.[20] Sometimes a court will refuse to listen to *any* evidence outside the writing if it is convinced that the writing is so clear in meaning that it needs no outside assistance. Most courts will permit either party, especially the client, to show prior oral agreements not included in the writing.

Attorneys for design professionals frequently rely upon the parol evidence rule. However, such reliance is often misplaced. This is especially so if the agreement is sketchy and does not spell out the details adequately and if the client was not represented by an attorney during the negotiations.

Reducing an arrangement to writing does not necessarily protect the design professional from assertions of additional oral agreements. However, the more detail included in the agreement and the greater the likelihood that the client understood the terms or had legal counsel, the greater the probability that the client will *not* be allowed to prove the claimed oral agreement. The parol evidence rule does not apply to agreements made after the written agreement was signed by the parties.

The parol evidence rule relates only to the provability of antecedent or contemporaneous agreements. If such agreements are admitted into evidence, the trial court or the jury—depending upon who makes the determination of fact—must decide whether the evidence shows that such an agreement was made. Very often the attorney for the design professional places heavy reliance upon the parol evidence rule and does not adequately prepare for the more important question of whether the asserted agreement took place.

Most parol evidence cases require a determination of whether the writing was intended to be the complete and final repository of the entire agreement of the parties. To preclude the court from determining that a particular writing was not intended to be complete, contracts prepared by attorneys or by professional associations often contain integration or merger clauses.

Before 1966 the AIA Standard Documents did not contain an integration clause, reflecting the then emphasis upon the close professional relationship between architect and client. Beginning in 1966 the AIA Documents tended to become more concerned with liability and litigation, and this undoubtedly was the reason for inclusion of ¶ 13.1, which currently states:

> This Agreement represents the entire and integrated agreement between the Owner and the Architect and supersedes all prior negotiations, representations or agreements, either written or oral.

In commercial contracts such clauses are generally successful in accomplishing the drafter's objective unless the party attacking the clause asserts that the other party fraudulently induced the agreement.

The effect of such a clause in the context of the design professional client relationship is likely to vary due both to the way in which such relationships are originated and the general disrespect some courts have for the parol evidence rule.

---

20. *Spitz v. Brickhouse,* 3 Ill.App.2d 536, 123 N.E.2d 117 (1954); *Malo v. Gilman,* supra note 18, to be reproduced in Section 15.03.

PROBLEMS

1. X comes to see you at your architectural office. You have been in practice for three months and have not designed any projects. X gives you details regarding the construction of a five-story office building that he wishes you to design. He asks you to begin work immediately. What steps would you take toward protecting the collectibility of your fee?

2. Y approaches you and asks you to perform certain design services. He tells you that he is the vice-president of the Comac Corporation, a company that makes electronic equipment. He gives you his calling card, which identifies him. He tells you that it will take some time to get the formal contracts signed by the appropriate officers of the corporation and that you should begin work immediately. He also tells you that a written, authorized contract will be issued shortly. You note that he is driving a new car with the words "Comac Corporation" neatly embossed upon each door.

You start work. Three weeks later you bill Comac for $1,000 for schematic design work. Comac refuses to pay since it says Y was not authorized to bind the company. Would Comac have a legal defense? Give your reasons. What other facts would you like to know?

15

Professional Design Services: The Sensitive Issues

SECTION 15.01
RANGE OF POSSIBLE PROFESSIONAL SERVICES: FEES AND INSURANCE

The range of *potential* professional services already substantial because of the complexity of design and construction and the centrality of the design professional's position, has been increasing because of the increased public control over design and construction and the greater participation by the design professional in economic and financial aspects of the project. A glance at any pre-prepared standard contract for professional services, especially those labelled as "additional," demonstrates this.

Although compensation is discussed in Chapter 16, one aspect of compensation relates to the potential list of professional services. Although the percentage of construction cost method of compensating the design professional is beginning to lose its domination, it is still common for the fee to be computed in this manner. However, and not without occasional surprise to the client, this amount is called the *basic* fee. In addition, services are frequently desig-

nated as "additional services," which entitle the design professional to compensation *in addition to* the basic fee. In many cases the design professional will be willing to perform any services that the law allows and within his professional skill, provided there is compensation for performance of these services.[1] But disputes may arise over whether the services requested fall under the basic fee or entitle the design professional to additional compensation.

Before 1958 Standard Documents published by the American Institute of Architects that dealt with architectural services contained no description of additional services. From 1958 to 1963 nine additional services were listed. Steady increases culminated in 1977 with twenty-two additional services.

Such a formidable list can be the breeding ground for honest disputes. Although the extensive list may reflect instances where architects were unjustifiably denied compensation for services beyond normal services, its use can cause misunderstanding.

An argument can be made for an extensive list of additional services. If the bulk of the additional services is rarely used,

---

1. But suppose the design professional is precluded by law or does not wish to perform requested services? A literal reading of AIA Doc. B141, ¶ 1.7, would seem to give the client the power to insist the architect do almost anything that relates to the project.

including it within the basic design services can increase the basic design fee when those services are not needed and not performed. Separating the rarely performed services, it is argued, benefits the client by requiring that it pay only for those services used.

The formidable list of additional services includes some that seem performed with relative frequency. If so, and if the client is not made aware of this, the client may complain that it is paying *more* than it expected.

Another factor, professional liability insurance (discussed in greater detail in Section 18.05), must be taken into account in planning services to be performed. From the standpoint of the design professional, it is important to be aware of which services can be covered by professional liability insurance and which are generally excluded.

As a general rule, insurers wish to insure only those services that are routinely performed and that are connected to the professional training and experience of the insured design professional. They include analyzing the client's needs, preparing a design solution, and performing site services whose objective is to see that the design is executed properly. Services that go beyond these may very well be excluded unless there is a *special* endorsement.

Similarly, the client should be aware of which services are included. While it is possible to agree that services be rendered that are *excluded* from insurance coverage, this should be a deliberate choice made by the contracting parties and one not inadvertently made.

At the outset of the professional relationship or as soon as possible thereafter, the client and design professional should determine in advance which of the possible professional services are to be performed by the design professional and whether those services fall within the basic fee. After this is done, the written agreement should reflect the actual agreement, and performance should be in accordance with the con-

tract requirements or any subsequent modifications.

SECTION 15.02
DESIGN SERVICES

Basically, a client retains an architect or engineer to prepare the design. While the *basic* elements of design preparation have not engendered legal problems, except to the extent the law has had to determine the standard of performance (see Section 17.05), some mention should be made of *basic* design services.

Usually the client comes to the designer with a problem that it hopes the designer can solve. The client describes its needs and, as a rule, what it *wishes* to spend. After consultation with the client, a schematic design is developed and possibly a revision of the client's budget.

Next the designer studies the design and prepares drawings and possible models that illustrate the plan, site development, features of construction equipment, and appearance. The designer is also likely to prepare outline specifications and again possibly revise the predicted costs. In small projects, schematic design and design development are still designated as *preliminary* studies.

After the design development has been approved by the client, the designer prepares working drawings and specifications that cover in detail the general construction, the structure, mechanical systems, materials, workmanship, site development, and the responsibility of the parties. Often the designer will supply or draft general conditions and bidding information (discussed in Section 15.07 dealing with services of a "legal" nature).

The final preconstruction phase is generally called the bidding or negotiation phase. The designer helps the client obtain a construction contractor through bidding or negotiation. These phases are discussed again in Section 16.02, as phases often determine timing of interim fee payments.

During construction the designer interprets the contract documents, checks on the progress of the work to issue certificates, participates in the change order process, and resolves disputes. Most, but not all, of those services are part of *basic* design services, to see the project through to *proper* completion. Many of these functions are discussed in greater detail in this chapter and in Part IV.

SECTION 15.03
COST PREDICTIONS

(A) THE INACCURATE COST PREDICTION: A SOURCE OF MISUNDERSTANDING

The relative accuracy of cost predictions is of great importance to public or private clients. Clients must know how much projects will cost. They may be limited to bond issues, to appropriations, to grants, to loans, or to other available capital. Unfortunately, many clients think that cost estimating is a scientific process by which accurate estimates can be ground out mechanically by the design professional. For this reason, design professionals should start out with the assumption that cost predictions are vital to the client and that the client does not realize the difficulty in making accurate cost predictions.

The close relationship between design professional and client often deteriorates or breaks down when the low bid substantially exceeds any cost figures discussed at the beginning of the relationship or even the last cost prediction made by the design professional. Not uncommonly the project is abandoned and the client often feels that it should not have to pay the design professional. The design professional contends that cost predictions are educated guesses, the accuracy of which depend in large part on events beyond the control of the design professional. Clients frequently request cost predictions before many details of the project have been worked out. Clients frequently change the design specifications without realizing the impact such changes can have on earlier cost predictions. Design professionals point to unstable labor and material costs. The amount a contractor is willing to bid often depends upon supply and demand factors that cannot be predicted far in advance. Design professionals contend that they are willing to redesign to try to bring costs down but that unless it can be shown that they have not exercised the professional skill that can be expected of persons situated as they were, they should be paid for their work.

An indication that this is a sensitive area can be demonstrated by the terminology used relating to cost predictions. The most commonly used expression, though used less frequently in contracts made by professional associations, is *cost estimates.* The term *estimate* is itself troublesome, in some contexts meaning a firm proposal intended to be binding and in others only an educated guess. Probably many clients believe that cost estimates will be "in the ballpark," while design professionals may look upon such estimates as educated guesses.

Professional associations seek to minimize the likelihood that their members will go unpaid if the project is abandoned because of excessive costs (explored in greater detail in (C)). Part of this effort is reflected in the terminology chosen. For example, the American Institute of Architects (AIA) seeks to differentiate between "statements of probable construction costs, detailed estimates of construction costs, and fixed limit of construction costs." Only the latter, according to the AIA, creates a risk assumption by the architects that they will go unpaid if the project is abandoned because of excessive costs. The National Society of Professional Engineers (NSPE) refers only to "opinions of probable project costs." The potential for misunderstanding is demonstrated in the two reported appellate cases reproduced in (C).

(B) TWO MODELS OF COST PREDICTIONS

The many reported appellate cases demonstrate not only the frequency of misunderstanding but also the difficulties many design professionals have predicting costs. Recognition of this, along with other factors explored in Chapter 21, has led the sophisticated client to seek more refined methods of controlling and predicting costs. It is important at the outset to distinguish between what can be called the traditional method and these more refined methods.

The traditional method usually involves the design professional's using rough rules of thumb based on projected square or cubic footage, modulated to some degree by a skillful design professional's sense of the types of design choices that will be made by a particular client. Through the development of the design, cost predictions are almost likely to be given, but they are not going to be based on much more than these rough formulas, refined somewhat as the design proceeds toward finalization.

The cost predictions should be more accurate as time for obtaining bids or negotiating with a contractor draws near. In this model, there is a great deal of suspense when bids are opened or when negotiations become serious. Under this system, a greater likelihood exists of substantial if not catastrophic differences between the costs expected by the client and the likely costs of construction as reflected through bids or negotiations. It is this model that gives rise to the bulk of the litigation.

The other model, that of more efficient techniques, is likely to be used by sophisticated clients who are aware of the difficulties design professionals have using the model just described. Clients are likely to engage someone who will be able to give a more accurate cost prediction as the design evolves. Sometimes this is done by hiring a skilled cost estimator—something close to the quantity surveyor used in the United Kingdom—as a separate consultant. Sometimes it is done by engaging a construction manager (CM), who is supposed to have a better understanding of the labor market, the materials and equipment market, and the construction industry, as well as the Construction Process itself. Using a CM is intended not only to free the designer from major responsibility for cost predictions but also to keep an accurate, ongoing cost prediction. Sometimes the CM agrees to give a guaranteed maximum price (GMP) that may vary as the design evolves. To do this, the CM may obtain firm price commitments from the specialty contractors.

This second model—a "fine-tuned model"—is designed to avoid the devastating surprises that are not uncommon under the traditional model. To determine whether the design professional bears the risk of losing the fee, a differentiation between the two models is vital, the risk being greater in the traditional model.

(C) CREATION OF A COST CONDITION: *GRISWOLD & RAUMA v. AESCULAPIUS AND MALO v. GILMAN*

When clients assert that they have no obligation to pay the design professional, they are asserting a cost condition. They are claiming that their obligation to pay was *conditioned* upon the accuracy of a design professional's cost prediction. A cost condition is a gamble by the design professional that the cost prediction will be reasonably accurate. If it is not, the fee is lost unless the client is willing to dispense with the cost condition. In such a case, the client need make no showing that the design professional has not lived up to the professional standard of care.

Clients sometimes contend that the design professional not only has gambled the fee but also has *promised* that the project will be brought in within a designated price. Failure to perform in accordance with this promise makes the design professional responsible for any damage to the client that was reasonably foreseeable at the time the contract was made. In (C), emphasis is upon the question of whether a cost condition has been created, while the question of damages for breach is discussed in (F).

The Parol Evidence Rule and Contract Completeness

The issue that has arisen most frequently in cost cases relates to the parol evidence rule (discussed earlier [2]). This rule determines whether a writing assented to by contracting parties is the sole and final repository of their agreement. If so, testimony relating to agreements made before or at the time of the written contract is not judicially admissible. In the context of a cost condition, the issue is whether the client will be permitted to testify that it had made an earlier oral agreement or had an understanding that it could abandon the project and not pay for design services if the low construction bid substantially exceeded the cost prediction of the design professional.

Generally the client will be permitted to testify that such an agreement had been made if the agreement between design professional and client is oral, if the agreement is written but nothing stated as to the effect of accurate cost predictions, or in some cases even if this problem is dealt with in the agreement. (See *Malo v. Gilman*, reproduced later in this section.) Permitting such testimony is based upon the conclusion that such agreements are not, as a rule, the final and complete repository of the

entire agreement between design professional and client.[3]

Standardized contracts prepared by the AIA or NSPE include language that seeks to protect the design professional from assertions of the existence of a cost condition. Although such language has been useful to design professionals, its presence is not ironclad protection against clients being permitted to testify to their understanding that they could abandon the project if the costs were excessive and not pay the design professional for services. The cases reproduced later in this section involve clauses that sought to make the writing complete. Yet, as shall be seen, testimony seems to have been freely admitted. (The *current* AIA language is discussed later in this section.)

Keep in mind that permitting the testimony does not end the matter. Issues are still likely to exist as to whether the agreement took place, the nature of the agreement, and whether the condition has occurred or been excused.

Existence of Cost Condition

Although cost condition cases generally involve the client claiming an *express* agreement based upon a cost agreed to or set forth in the contract under which it could abandon the project and not pay a fee, a cost condition can be created by implication without any *specific* agreement as to the effect of inaccurate costs.[4]

Certainly design professionals do not operate in the dark. If they know what funds are available and the remoteness of ob-

---

2. Section 14.04(E).

3. *Stevens v. Fanning*, 59 Ill.App.2d 285, 207 N.E.2d 136 (1965). Many cases are collected in 20 A.L.R.3d 778 (1968).

4. *Stanley Consultants, Inc. v. H. Kalicak Construction Co.*, 383 F.Supp. 315 (E.D.Mo.1974) (dictum). In *George Wagschal Associates, Inc. v. West*, 362 Mich. 676, 107 N.W.2d 874 (1961), a cost condition was found in a consultant contract because the consultant knew of the client's budget limit. But if the contract expressly negates a cost condition, it will not be implied. *Kurz v. Quincy Post Number 37, American Legion*, 5 Ill.App.3d 412, 283 N.E.2d 8 (1972).

taining additional funds, any cost specified may be "hard." In such a case, a cost condition may be created despite the absence of an *express* agreement under which this risk is taken.

Evidence that bears upon the softness or hardness of any projected costs discussed by the design professional and client or expressed in their agreement is crucial. This evidence, such as labeling the amounts as merely an estimate or using a cost range, may indicate that the amount or range specified is what is hoped for rather than a fixed-cost limitation. Where the amount is "soft," design professionals are being exhorted to use their professional skill to bring the project in for the amount specified. Where it is "hard," the client may be informing the design professional that the latter is risking the fee upon ability to accomplish this objective.

This distillation of appellate cases appeared in a legal journal dealing with architectural cost predictions:[5]

> Courts have admitted evidence of custom in the profession. Architects have been permitted to introduce evidence that customarily architects do not assume the risk of the accuracy of their cost predictions. Also, courts have been more favorably disposed toward holding for the architect if the project in question has involved remodeling rather than new construction, because estimating costs in remodeling is extremely difficult. The same result should follow if the type of construction involves experimental techniques or materials.
>
> Courts sometimes distinguish between cases and justify varying results on the basis of the amount of detail given to the architect by the client in advance. Generally, the greater the detail, the easier it should be for the architect to predict accurately. However, it is much more difficult for the architect to fulfill the desires of the client within a specified cost figure if the client retains a great deal of control over details, especially if these controls are exercised throughout the architect's performance. For

this reason, some courts have held that a cost condition is not created where the architect is not given much flexibility in designs or materials.

> Some courts have looked at the stage of the architect's performance in which the cost condition was created. If it is created at an early stage, it is more difficult for the architect to be accurate in his cost predictions. Generally, the later the cost limit is imposed in good faith, the more likely it is to be a cost condition. But courts should recognize that if it is imposed later, creation—or, more realistically, imposition—may be an unfair attempt by the client to deprive the architect of his fee.
>
> Occasionally the courts have applied the rule that an ambiguous contract should be interpreted against the person who drew it up and thus created the ambiguity. If the client is a private party, the contract is usually drafted or supplied by the architect. Courts have looked at the building and business experience of the client. If the client is experienced, he should be more aware of the difficulty of making accurate cost estimates. If he has building experience, the client is more likely to be aware of the custom that architects usually do not risk their fee upon the accuracy of their cost estimates.
>
> Courts have sometimes cited provisions for interim payments as an indication that the architect is not assuming the risk of losing his fees on the accuracy of his cost estimates. However, standard printed clauses buried in a contract are not always an accurate reflection of the understanding of the party not familiar with the customs or the forms. If payments have actually been made during the architect's performance, this is a clearer indication that the client is not laboring under the belief that he will not have to pay any fee if the low bid substantially exceeds the final cost estimate. A few cases have looked for good faith on the part of the client. For example, if the client has offered some payment to the architect for his services, this may impress a court as a show of fairness and good faith. (footnotes omitted)

Standard Contracts and Disclaimers: A Look at AIA Standard Contracts

Professional associations have dealt with cost problems by inclusion of language in

---

5. Sweet & Sweet, *Architectural Cost Predictions: A Legal and Institutional Analysis,* 56 Calif.L.Rev. 996, 1006–1007 (1968).

their standard contracts designed to protect users from losing fees when cost problems develop. The protective language has ranged from the brief statement that cost estimates cannot be guaranteed to the currently elaborate contract language contained in AIA Doc. B141 starting with ¶ 3.2 set forth in Appendix A. The length and complexity of the contract language demonstrates the seriousness of the problem.

Paragraph 2.2 requires the client to include in its budget amounts that take into account the possibility that the bids may substantially exceed the budget. Paragraph 3.2.1 requires the architect to use best judgment to evaluate any client-supplied project budget or architect-created cost estimates but states that the architect does not warrant accuracy.

In addition to Art. 13 stating that the writing is complete, ¶ 3.2.2 states that a project budget is not a fixed-cost limit and requires that any such limit be in writing and signed by the parties. While not clear, it is likely that the ordinary signature to the contract as a whole will not be sufficient and that the drafters intended that such a limit be signed or at least initialled separately. If such a fixed-cost limit (not an ordinary budget) is established, the architect is given the right to determine "materials, equipment, component systems, and types of construction" to be included and to make reasonable adjustments in project scope. The architect can include pricing contingencies and alternate bids.

Paragraph 3.2.3 requires an adjustment if bid or negotiation is delayed more than three months after submission of construction documents. Finally, ¶ 3.2.4 is a contractual expression of the obligation of good faith and fair dealing. It states that if the lowest bona fide bid exceeds the budget or fixed-cost limit, the owner must do one of the following:

1. Approve an increase in the fixed limit.

2. Authorize rebidding or renegotiating.

3. Terminate and pay for work performed and termination expenses if the project is abandoned.

4. Cooperate in project revision to reduce cost.

If (4) is done where there is a fixed limit, the architect must bear the cost of redesign, specified as the limit of the architect's responsibility and earning the architect compensation.

Taken as a whole, AIA treatment of cost predictions uses a multifaceted approach. First, there is the attempt to preclude any oral testimony of a fixed limit. Second, there is the requirement that the client give up certain design choices if a fixed limit is employed. Third, additional protection is given the architect if the low bid substantially exceeds the cost limit.

That such clauses can help the architect is shown by the *Griswold & Rauma* case reproduced later in this section.[6] Earlier cases, such as *Stevens v. Fanning*,[7] and some recent cases, such as *Malo v. Gilman*, also reproduced in this section, reflect a different attitude toward such protective clauses. Much will depend upon the facts surrounding the transaction, especially the degree of variance, the extent of client changes in the design, and the way each party behaved before and after the problem surfaced.

Two Illustrative Cases

Two cases, both probably using the same standard form but coming to different conclusions, are reproduced here.

---

6. See also *Torres v. Jarmon*, 501 S.W.2d 369 (Tex.Civ.App.1973).

7. Supra note 3.

GRISWOLD
AND RAUMA,
ARCHITECTS, INC. v.
AESCULAPIUS CORP.

Supreme Court of
Minnesota, 1974.
301 Minn. 121, 221 N.W.2d
556.

PETERSON, Justice
[Ed. note: Footnotes omitted.]

Plaintiff brought this action to enforce and foreclose a lien for $19,438.65 for architectural services provided defendant. Defendant answered by denying that it owed plaintiff anything and filed a counterclaim to recover $17,436.04 already paid plaintiff, defendant's theory being that plaintiff was entitled to no fee because it breached its contract by grossly underestimating the probable cost of construction of the as yet unbuilt building. The trial court found for defendant. We reverse, for reasons requiring an extended recital of the factual setting out of which the litigation arose.

Plaintiff is a corporation engaged in providing architectural services. Defendant is also a corporation, the principal stockholders of which are Drs. James Ponterio, P.J. Adams, and A.A. Spagnolo. Defendant owns a medical building in Shakopee which it rents to the Shakopee Clinic, which in turn is operated by Drs. Ponterio, Adams, and Spagnolo.

In early 1970 defendant decided to expand the Shakopee Clinic to allow for a larger staff of doctors. As a result the members of the corporation and their business manager, Frank Schneider, contacted various architectural firms and selected plaintiff.

At his first meeting with the doctors in February or March 1970 David Griswold, one of plaintiff's senior architects, was shown a rough draft of the proposed addition and given a very general idea of what the doctors wanted. Although the evidence is conflicting, it appears that at this meeting the doctors talked in general terms of a budget of about $300,000 to $325,000.

After a number of subsequent conferences with the doctors and Mr. Schneider, plaintiff prepared and delivered to the doctors on May 8, 1970, a document entitled "Program of Requirements." This document, which outlined and discussed the requirements of the project as then contemplated, contained the following final section:

"BUDGET:

"The design to evolve from this program will indicate a certain construction volume that can be projected to a project cost by the application of unit (per square foot and per cubic foot) costs; and eventually, as the design is developed in detail, by an actual materials take-off. Inevitably the projected cost must be compatible with a budget determined by available funds. It is obvious that adjustment of either the program or the budget may be necessary and that possibility must be recognized.

"The project budget established, as currently understood, is $300,000. It has not been stated if this is intended to include non-building costs such as furnishings, equipment and fees—which may be approximately 25% of the total expenditure—

as well as construction costs. Advice in this respect will eventually be necessary.

"The essential principle to be considered in the design development is as previously stated in the paragraphs of the section titled PROJECT OBJECTIVES.

"The construction shall be as economical as possible within the limitations imposed by the desire to build well and provide all of the facility required for a medical service."

In spite of the suggestion at the end of the second paragraph quoted above, neither party at any time thereafter sought to define more particularly what the budget was intended to include. Mr. Schneider testified, however, that he believed the original budget figure included the cost of construction, architects' fees, and the remodeling of the old building. In contrast, Dr. Ponterio testified that it was his belief that the original budget figure did not include a communications system valued at $12,755, architects' fees, or the $20,000 remodeling of the existing building.

On or about May 22, 1970, plaintiff submitted to defendant two alternate preliminary plans, designated SK–1 and SK–2. Plan SK–1 projected the programmed services to be housed partly in the existing building and partly in the new building. Plan SK–2 projected the programmed services as being housed entirely in the new building. Plan SK–2, as specifically shown on the plans, involved a larger plan in terms of area than SK–1. Defendant indicated its preference for plan SK–2, the larger and more elaborate of the two.

On June 1, 1970, plaintiff provided defendant with a "Cost Analysis" of the plan chosen by defendant. This cost analysis showed the dimensions of the project in square feet as then contemplated, and computed the cost of the project at two different rates per square foot. At the higher rate per square foot, the cost came to $322,140, plus an estimated $20,000 for remodeling the old building, totaling $342,140. At the lesser rate per square foot, the cost came to $284,575, plus $15,000 for the remodeling of the old building, totaling $299,575. The cost analysis memorandum also noted that "the best procedure for projecting costs is by a materials take-off" which was to be done "when sufficient information is available."

It is undisputed that subsequent to the June 1, 1970, cost estimate, no further cost estimates were ever conveyed to defendant. What is disputed is whether in the ensuing months there was any discussion as to whether the project was coming within the budget. According to the testimony of Mr. Schneider, defendant was assured at all times during the preparation of the building plans and in all discussions with Griswold that the construction would come within the budget. Dr. Ponterio also emphasized that Griswold constantly mentioned the budget figure of $300,000 at their meetings. Griswold, however, denied that he had ever assured defendant that the project was coming within the budget.

Although the architectural services began in March and the first billing was May 6, 1970, no written contract was forwarded until June 23, 1970. At that time a standard American Institute of Architects (AIA) contract was forwarded, calling for payment at plaintiff's standard hourly rate and for reimbursement of expenses and recognizing that a lump sum fee for the construction phase would be negotiated prior to its commencement. The following provisions of the contract have relevance to this case:

"SCHEMATIC DESIGN PHASE

"1.1.3   The Architect shall submit to the Owner a Statement of Probable Construction Cost based on current area, volume or other unit costs."

"DESIGN DEVELOPMENT PHASE

"1.1.5   The Architect shall submit to the Owner a further Statement of Probable Construction Cost."

"CONSTRUCTION DOCUMENTS PHASE

"1.1.7   The Architect shall advise the Owner of any adjustments to previous Statements of Probable Construction Cost indicated by changes in requirements or general market conditions."

"THE OWNER'S RESPONSIBILITIES

"2.8   If the Owner observes or otherwise becomes aware of any fault or defect in the Project or non-conformance with the Contract Documents, he shall give prompt written notice thereof to the Architect."

"CONSTRUCTION COST

"3.4   .   .   .   Accordingly, the Architect cannot and does not guarantee that bids will not vary from any Statement of Probable Construction Cost or other cost estimate prepared by him.

"3.5   When a fixed limit of Construction Cost is established as a condition of this Agreement, it shall include a bidding contingency of ten per cent unless another amount is agreed upon in writing.   .   .   .

"3.5.1   If the lowest bona fide bid .   .   . exceeds such fixed limit of Construction Cost (including the bidding contingency) established as a condition of this Agreement, the Owner shall (1) give written approval of an increase in such fixed limit, (2) authorize rebidding the Project within a reasonable time, or (3) cooperate in revising the Project scope and quality as required to reduce the Probable Construction Cost. In the case of (3) the Architect, without additional charge, shall modify the Drawings and Specifications as necessary to bring the Construction Cost within the fixed limit.   The providing of this service shall be the limit of the Architect's responsibility in this regard, and having done so, the Architect shall be entitled to his fees in accordance with this Agreement."

"PAYMENTS TO THE ARCHITECT

"6.3 If the Project is suspended for more than three months or abandoned in whole or in part, the Architect shall be paid his compensation for services performed prior to receipt of written notice from the Owner of such suspension or abandonment, together with Reimbursable Expenses then due and all terminal expenses resulting from such suspension or abandonment."

Between June 1, 1970, and November 25, 1970, when the bids were opened, plaintiff worked actively with defendant both in the design development and construction documents phases of the project, plaintiff continuing to bill defendant on an hourly basis without objection by defendant. Although the project was substantially increased in size and scope during this period, plaintiff did not furnish and defendant did not request any up-to-date cost projections. Significantly important changes in the project made during this period include:

(1) Replacement of offices with examining rooms necessitating additional plumbing;

(2) More extensive X-ray space;

(3) A doubling of the size of the laboratory;

(4) Addition of a sophisticated communications system;

(5) Addition of 2,100 feet of finished space in the basement (to provide facilities originally projected for the remodeled old building);

(6) Enlargement of the structure as follows:

waiting area	1710' to 1724'
first floor	5880' to 6650'
basement	6260' to 6814'

All of the changes were discussed, approved, and understood by defendant. Defendant alleges, however, that it was under the impression that all such changes would be included in the original budget figure.

Bids were opened on November 25, 1970. The low construction bid was Kratochvil Construction Company at $423,380. Deductive alternates agreed to by defendant would bring the total low bid cost, including carpeting, down to $413,037.

Subsequent to the opening of the bids, the doctors called a meeting with Griswold at which the doctors informed Griswold that they could not complete the building according to the cost evidenced by the bids. Thereafter, Griswold met with the doctors and offered suggestions as to how the low bid figure could be reduced. Approximately $42,000 of reductions were projected, so that the final bid as reduced totaled $370,897. This figure included construction, carpeting, and remodeling of the old building and would meet the program of requirements without reducing size in any way. Griswold also pointed out to the doctors that the project cost could be further reduced by eliminating "bays" (series of examination rooms) from the building at a saving of approximately $35,000 per bay. Defendant was willing to accept the $42,000 reduction but was not in favor of eliminating any bays from the proposed project.

From November 25, 1970, when the bids were opened, until October 12, 1971, plaintiff and defendant met and corresponded many times in connection with various possible revisions of the project. During this time defendant never indicated that the project was abandoned and in fact, in January 1971 made a payment of $12,000 on its bill. During this time defendant never asked to be excused from the balance of the bill, and defendant offered to help plaintiff by paying interest on the open account if plaintiff required bank financing by reason of nonpayment. During this time defendant advised plaintiff that ground breaking would be deferred for 6 months, and that request by plaintiff for a further payment was "well taken" but that no further payment would be recommended until construction began.

The building, in fact, was never constructed. Plaintiff filed its lien on April 30, 1971, and commenced this action to enforce it in October 1971.

In analyzing the facts of this case we have considered Minnesota decisions as well as decisions from other jurisdictions. From these decisions we have extracted a number of factors which we believe are relevant to a determination of what the effect on compensation of an architect or building contractor should be when the actual or, as here, probable cost of construction exceeds an agreed maximum cost figure.

One very significant factor is whether the agreed maximum cost figure was expressed in terms of an approximation or estimate rather than a guarantee. Where the figure was merely an approximation or estimate and not a guarantee, courts generally permit the architect to recover compensation provided the actual or probable cost of construction does not substantially exceed the agreed figure.

Another significant factor is whether the excess of the actual or probable cost resulted from orders by the client to change the plans. Where the client ordered changes which increased the actual or probable construction costs, courts are more likely to permit the architect to recover compensation notwithstanding a cost overrun.

A third factor is whether the client has waived his right to object either by accepting the architect's performance without objecting or by failing to make a timely objection to that performance.

A fourth factor, applicable in a case such as this where the planned building was never constructed, is whether the architect, after receiving excessive bids, suggested reasonable revisions in plans which would reduce the probable cost. Courts have held that if the architect made such suggestions and the proposed revisions would not materially alter the agreed general design, then the architect is entitled to his fee, again provided that the then probable cost does not substantially exceed the agreed maximum cost figure.

Considering this case in light of these factors, we conclude that the trial court erred in denying plaintiff's motion for amended findings of fact, conclusions of law, and order for judgment.

First, it does not appear that plaintiff guaranteed the maximum cost figure. The trial court did not expressly state whether the agreed cost figure of $300,000 to $325,000 was an established estimate or a guarantee. However, the intention of the parties, as evidenced by the record, especially by contract provision 3.4, quoted earlier, more reasonably supports an established cost estimate than a guaranteed cost figure.

Secondly, the probable cost of the project in our view did not substantially exceed the cost figure. At trial architect Griswold testified and Mr. Schneider agreed, that the project could be completed for approximately $360,000 to $370,000 without reducing the square footage of the project. A probable construction cost of $370,000 exceeds the agreed cost estimate maximum found by the trial court, $325,000, by only 13 percent. It seems difficult to classify such a degree of cost excess as substantial. A review of cases cited in Annotation, 20 A.L.R.3d 778, 804 to 805, suggests that most courts would not consider such a degree of cost excess substantial.

Thirdly, we think it relevant that defendant approved substantial changes beyond the original plans. Defendant not only adopted and acknowledged these changes but was often active in advocating them (especially in expanding the X-ray facilities and changing the doctors' offices into extra exam rooms). It is true that defendant contends that it was under the impression at all times that such changes were within the original cost estimate. However, this impression seems unreasonable and unjustified in view of the scope of the changes.

Finally, we think it relevant that plaintiff showed defendant how they could reduce the cost of the lowest bid below $370,000. For example, defendant would have reduced the cost drastically by simply agreeing to eliminate one of the "bays" or a fraction of a bay from the project. Provision 3.5.1 of the contract required the parties to revise the project scope and quality if necessary to reduce the probable construction cost. While it might be against public policy to allow an architect, under such a provision, to reduce substantially the area of a proposed project, it seems that, barring a specifically guaranteed area, a reasonable reduction in the size of the project should be allowed when necessary to meet the construction cost.

Reversed and remanded.

SHERAN, C.J., took no part in the consideration or decision of this case.

MALO v. GILMAN
Court of Appeals of
Indiana, 1978.
177 Ind.App. 365,
379 N.E.2d 554.

STATON, Judge

[Ed. note: Footnotes renumbered and some omitted.]

Edward Malo, an architect, completed plans and specifications for Arnold Gilman, who sought to construct an office building. The construction bids received totaled $105,000, 50% more than the preliminary estimated cost of $70,000, which appeared in the contract between the men. Gilman was unable to secure financing and the building was never built. Malo brought an action to recover his fee as architect. Gilman counterclaimed for the $500 he had paid Malo. The trial court found for defendant Gilman and granted his counterclaim. We affirm the judgment.

Malo agreed to provide architectural services to Gilman in the design and construction of an office building. Between May, 1967, and November, 1968, Malo expended considerable time and effort on the project. A verbal agreement was reached on July 28, 1967. Gilman was assured that costs of construction could be kept below $20.00 per square foot. After talking to prospective tenants, Gilman decided he required 3,500 square feet in the building. A standard American Institute of Architects (A.I.A.) form contract was signed on May 14, 1968.[8] Among the terms was the following:

"It is recognized that this written contract ratifies the similar verbal contract entered into July 28, 1967.

"The preliminary estimated cost of this project is Seventy Thousand Dollars, ($70,000.00)."

The contract also contained the following standard clause:

"3.4 Statements of Probable Construction Cost and Detailed Cost Estimates prepared by the Architect represent his best judgment as a design professional familiar with the construction industry. It is recognized, however, that neither the Architect nor the Owner has any control over the cost of labor, materials or equipment, over the contractors' methods of determining bid prices, or over competitive bidding or market conditions. Accordingly, the Architect cannot and does not guarantee that bids will not vary from any State-

---

8. Since the contract was prepared by Malo, it must be strictly construed against him and in favor of Gilman. *Oxford Development Corp. v. Rausauer Builders, Inc.* (1973), 158 Ind.App. 622, 304 N.E.2d 211.

ment of Probable Construction Cost or other cost estimate prepared by him."

Malo completed the plans and specifications for the project in September, 1968. In October, 1968, bids were solicited. The lowest total of bids received was approximately $128,000, which was negotiated down to $105,000. The bids were never accepted. Gilman indicated that the bids were unacceptable to him and that he was unable to secure financing. In mid-December, Gilman sold the land on which the building was to have been erected.

Malo demanded payment of his fee for architectural services in the sum of $9,132.60. Gilman refused to pay. Malo brought an action to collect his fee. Gilman counterclaimed for the $500 he previously had paid Malo. The trial court denied Malo's claim, but granted Gilman's counterclaim.

We hold that the judgment of the trial court can be affirmed on either of two alternate theories: (1) that parol evidence was properly admitted to show a maximum cost limitation of approximately $70,000.00, which was exceeded unreasonably by Malo's plans for construction; or (2) that the estimated cost figure appearing in the contract placed a reasonable limit on the actual cost of the project, which limit was exceeded unreasonably. In either event, architect Malo breached the contract and is not entitled to compensation under the contract.

I.

Parol Evidence to Show a Maximum Cost Limitation

On appeal, Malo argues that no fixed price agreement appeared in the "fully integrated contract" for architectural services. The only figure appearing in the contract, $70,000.00, was merely a preliminary estimated cost figure which was not binding on the architect.[9] Further, even if the trial court properly allowed evidence of a $20 per square foot cost limitation, Malo claims his final design plans contained 5,400 square feet,[10] 50% more space than originally projected. In that case, the bids totaling $105,000 were in the right price range for a building costing $20 per square foot.

Gilman contends that evidence showing the existence of a $20 per square foot cost limitation (or $70,000 to $78,000 total for the project) was properly admitted, since the contract failed to contain a maximum cost limitation. Further, no significant changes in the project occurred to increase its size or cost.

Normally parol evidence may not be considered if it contradicts

---

9. For a discussion of the legal issues raised by cost estimates, see: Sweet & Sweet, *Architectural Cost Predictions: A Legal and Institutional Analysis*, 56 Cal.L.Rev. 996 (1968).

10. Of this space, less than 3,900 square feet was on the main floor, with the remaining square footage comprising a "partially finished" basement.

or supersedes matters intended to be covered by the written agreement. However, parol evidence may be admitted to supply an omission in the terms of the contract. *Caldwell v. United Presbyterian Church* (Common Pleas (1961), 20 Ohio Ops.2d 364, 88 Ohio L.Abst. 323, 180 N.E.2d 638; *Spitz v. Brickhouse* (1954), 3 Ill.App.2d 536, 123 N.E.2d 117. Many contracts for architectural services, as here, fail to include specific requirements such as the size, style, and character of the building, the number of rooms, the quality of the materials to be used, and, finally, the maximum cost. Yet, according to section 1.1.1 of the A.I.A. form contract,

> "The Architect shall consult with the Owner to ascertain the requirements of the Project and shall confirm such requirements to the Owner."

Thus, depending on the specific needs of the owner, these requirements may be integral parts of the contract for architectural services. *Standley v. Egbert* (1970), D.C.App., 267 A.2d 365, 367. A contract that fails to set out the details agreed upon, then, is not a complete and integrated statement of the agreement. *See Levy v. Leaseway System, Inc.* (1959), 190 Pa.Super. 482, 154 A.2d 314. Parol evidence may be considered to determine the agreement with respect to these matters. *See* Annot., 69 A.L.R.3d 1353 (1976).

Ordinarily, the maximum cost of a project is agreed upon prior to commencement of design. The owner who plans to construct a building has in mind a figure for the maximum cost of construction, particularly where, as here, he must secure outside financing. The architect must design the project, keeping in mind this maximum cost limitation. Evidence of the maximum cost limitation should be admissible where the contract fails to show that figure.[11] As noted in an annotation to the *Spitz* case, 49 A.L.R.2d 679, 680 (1956):

> "In the great majority of the cases where the question has been raised the evidence has been held admissible, usually on the ground that the written contract failed to disclose the parties' intention as to the cost of the structure contemplated, and that such contemplated cost was an element which must have entered into the negotiations."

---

11. A line of cases, among them *Wick v. Murphy* (1952), 237 Minn. 447, 54 N.W.2d 805, has admitted parol evidence to show the cost of the project agreed upon by the parties, based upon a different line of reasoning. As the courts interpret the contract, the fee for architectural services depends on the actual construction cost. Where no contract has ever been let, the terms of the contract become ambiguous, that is, the fee cannot be determined based on actual construction cost. Parol evidence is admissible to resolve the ambiguity.

We note that the contract contains a diagram, with percentage plotted against cost of the project, to determine the architect's fees. The contract utilizes the line marked "Group B." We note that the 7.75%, which is marked on the graph, corresponds to a cost figure of $70,000. This is clearly the figure intended to be used to calculate Malo's fee.

Indiana has not yet decided a case on this point. However, the following cases allowed the introduction of parol evidence to show a maximum cost limitation when the contract failed to contain one: *Williams & Assoc., Arch. & Eng. v. Ramsey Prod. Corp.* (1973), 19 N.C.App. 1, 198 S.E.2d 67; *Caldwell v. United Presbyterian Church, supra; Petrus v. Bunnell* (1961), Fla.App., 129 So.2d 702; *Spitz v. Brickhouse, supra; Rosenthal v. Gauthier* (1953), 224 La. 341, 69 So.2d 367; and *Food Management, Inc. v. Blue Ribbon Beef Pack, Inc.,* 413 F.2d 716, 726 (8th Cir.1969), which cited as a general rule "that an oral cost limitation imposed upon an architectural or engineering design, where not contradictory to the express terms of the written contract, may be admitted into evidence. . . ."

We agree that parol evidence of a maximum cost limitation may be introduced where the contract fails to contain such a limitation. The question of fact, whether architect Malo agreed to design a building, the cost of which could not exceed $20 per square foot (or $78,000 for 3,900 square feet), was resolved by the trial court in favor of Gilman. On examining the record, we cannot say that the finding of fact was incorrect as a matter of law.

Gilman testified that, from the beginning, he received repeated assurances that Malo would have "no problem" designing a building costing less than $20 per square foot. The amount of usable space Gilman required was approximately 3,500 square feet. By multiplying the figures, Gilman and Malo arrived at $70,000 as a "firm figure" for maximum cost. Gilman sought to reduce costs, accepting the use of a cost-saving "Uni-Roof" design, and suggesting a cost-cutting relocation of the basement.

Malo testified that he was aware that Gilman was interested in "getting the best price on the market." Yet he claimed that there was no ceiling on the cost of the project, that "it could be as much as a million dollars." Other evidence presented concerning the cost of the project included two bids received on the basis of Malo's preliminary drawings; one was $80,000; the other (which would incorporate the Uni-Roof) was $62,000 to $70,000. Finally, one of plaintiff's witnesses, who was present at a meeting where the bids were tabulated, testified that no specific "cost talk" was discussed, that he did not recall the figure of $70,000, but that $80,000 kept "ringing a bell."

The evidence fails to show that Gilman would have paid *any* sum of money to construct the proposed building. On the contrary, the evidence clearly shows that he wished to construct the building as cheaply as possible. The trial court correctly resolved the factual question of a maximum cost limitation in favor of Gilman. Under the terms of the agreement, then, Malo lost his right to recover compensation when he designed a building impossible of construction within the maximum cost limitation.

II.

Estimated Cost Figure Exceeded Unreasonably

The contract for architectural services contained an estimated cost figure of $70,000. After negotiation, the lowest construction bids totaled $105,000, a figure 50% higher. Appellee Gilman argues that so great a discrepancy should bar Malo from receiving his fee for architectural services.

The court in *Caldwell v. United Presbyterian Church, supra,* supported such a theory. In that case, the building would have cost at least $57,800 to build, or $12,800 (almost 30%) in excess of the $45,000 cost limitation on the project. The court concluded,

> "that plaintiff has not substantially complied with the terms of his written contract and, therefore, is not entitled to recover in this action. . . ."

Id. 180 N.E.2d at 642.

Citing many cases, a federal Court of Appeals decision declared as a general rule

> "That there may be no recovery for engineering or architectural services where the actual cost of the structure substantially or unreasonably exceeds the estimated cost limitation, unless the cost excess is attributable to the owner's action. . . ."

Food Management, Inc. v. Blue Ribbon Beef Pack, Inc., supra, 413 F.2d at 726.

More recently, *Durand Associates, Inc. v. Guardian Investment Co.,* (1971), 186 Neb. 349, 183 N.W.2d 246, involved construction bids which exceeded the estimate by 55%. The court interpreted the section of the contract in which the architect explicitly refused to guarantee the cost estimate, declaring that the figure represented a "reasonable approximation of the cost of the project." *Id.* at 250. This did *not* mean that an architect would never be bound by his estimate. Such a situation would be "contrary to public policy because it would mean that no matter how large the bid for doing the work, defendants would be obligated to pay an architectural fee based on that amount. . . . *Id.* at 250. The court held,

> "that an architect or engineer may breach his contract for architectural services by underestimating the construction costs of a proposed structure. The rule to be applied is that the cost of construction must reasonably approach that stated in the estimate unless the owner orders changes which increase the cost of construction. . . ."

Id. at 251.

III.

Conclusion

Under either theory, the trial court could have found that Malo breached his contract for architectural services and denied Malo compensation under the contract, and found that Gilman was entitled to recover the $500 he paid Malo under the contract. The judgment of the trial court is affirmed.

BUCHANAN, C.J. (by designation), concurs.

HOFFMAN, J., concurs in Part II only.

Negligent Cost Predictions

If a cost condition is created, it is not necessary to determine whether the design professional met the appropriate standards of performance. However, in many cases where the cost estimate is wide of the mark, it is likely that a design professional did not live up to the legal standard of performance. When this occurs, it should not be necessary to determine whether a cost condition has been created. At the very least the design professional represents that he will live up to the professional standard in making cost predictions.

Stanley Consultants, Inc. v. H. Kalicak Construction Co.[12] illustrates a clear case of a failure to estimate properly. The project was a sixty-one-unit housing project to be built in Zaire, Africa. The cost estimate was $8 million and the only bidder submitted a bid of $16 million. The design professional had prepared cost estimates without data from Zaire when such data were available. The driveways contained impassible grades. The sewer lines were twenty feet above the surface with eight-foot supports, and the sewer lines were to be fifty feet above a river without any supports being designated. Structures were located outside the property lines, and the design professional failed to take into account an easement to a religious shrine and created an encroachment. The court found ample evidence that the design professional had not lived up to the reasonable standard of professional skill. Such a finding could be the basis not only for denying the design professional compensation for services rendered but also for holding the design professional responsible for damages. See (F).

(D) INTERPRETATION OF COST CONDITION

Is the design professional allowed some tolerance in determining whether the cost condition has been fulfilled? One court required a reasonable approximation.[13] Roughly 10% seems to be accepted, although this tolerance figure may be reduced if the project is large and the fee justified continual detailed pricing takeoffs. Before 1977, ¶ 3.4.1 of AIA Doc. B141 included a bidding contingency of 10% unless another amount were fixed in writing. Now § 2.2 requires a bidding contingency in the budg-

---

12. 383 F.Supp. 315 (E.D.Mo.1974).

13. *Durand Associates, Inc. v. Guardian Investment Co.*, 186 Neb. 349, 183 N.W.2d 246 (1971).

et.[14]   The degree of tolerance permitted may also depend upon the language used to create the cost limitation.   The more specific the amount, the more likely a small tolerance figure will be applied.

(E) DISPENSING WITH THE COST CONDITION

Generally a cost condition is created for the benefit of the client. If it so chooses, it can dispense with this protection. Courts that conclude that the client has dispensed with the cost condition usually state that the condition has been waived. Where this occurs, the condition is excused and the design professional is entitled to be paid even if the cost condition has not been fulfilled.

Excusing a condition can occur in a number of ways. A condition is excused if its occurrence had been prevented or unreasonably hindered by the client. For example, if the client does not permit bidding by contractors or limits bidding to an unrepresentative group of bidders, the condition is excused. The most common basis for excusing the condition has been excessive changes made by the client during the design phase.[15]

The client proceeding with the project despite the awareness of a marked disparity between cost estimates and the construction contract price can excuse the condition. By proceeding, the client may be indicating that it is willing to dispense with the originally created cost condition.[16] However, proceeding with the project should not *automatically* excuse the condition. It may be economically disadvantageous to abandon the project. Proceeding in such cases may not indicate a willingness to dispense with the condition. In such cases, any recovery of the design professional should be based upon restitution measured by any benefit conferred.[17]

(F) NONPERFORMANCE AS A BREACH: RECOVERY OF DAMAGES

Occasionally the client seeks damages based on the breach of a promise of accuracy or negligence in making the cost prediction. Refer to (C). An understanding of the basis of such claims requires that *promises* be differentiated from *conditions.*

The creation of a cost condition does not necessarily mean that the design professional promises to fulfill it. Design professionals can risk their fees upon the accuracy of the cost prediction, though they may not wish to be responsible for losses caused by nonperformance.

Courts seem to *assume,* however, that a fixed-cost limit constitutes a promise by the design professional that the project would cost no more than the designated amount. Under this assumption, if costs substantially exceed predicted costs, the design professional has breached even though he has lived up to the professional standard in making cost predictions. The breach entitles the nonbreaching party to recover foreseeable losses caused by the breach that could not have been reasonably avoided by the nonbreaching party.

Suppose the project is abandoned. The client can recover any interim fee payments

---

14. Despite a 10% tolerance figure, a court allowed the architect to recover where there was a 13% difference in *Griswold and Rauma, Architects, Inc. v. Aesculapius Corp.,* reproduced in Section 15.03(C). But this factor was taken *with others* in ruling for the architect.

15. *Koerber v. Middlesex College,* 128 Vt. 11, 258 A.2d 572 (1969). See also *Griswold & Rauma, Architects, Inc. v. Aesculapius Corp.,* reproduced at Section 15.03(C).

16. One factor considered in *Kurz v. Quincy Post Number 37, American Legion,* 5 Ill.App.3d 412, 283 N.E.2d 8 (1972).

17. Refer to Section 8.04(C).

based upon restitution.[18]   In spite of this, restitution of fees paid does not occur often. For example in *Stanley Consultants, Inc. v. Kalicak Const. Co.,*[19] (discussed in (C)), the architect had received $18,000 in interim fee payments and the court considered this adequate compensation for the architect's services.   The architect's breach of promise to perform in accordance with the professional standard, something clearly found by the trial court, entitled the client to restitution of any interim fee payments made. Yet the client did not seek restitution of the fees paid.   Even where restitution *is* sought, clients often do not press these claims. (But see *Malo v. Gilman* in Section 15.03(C)).   Denial of a design professional's claim does not necessarily mean services have been performed without any remuneration.

Abandonment of the project may cause other client losses, such as wasted expenditures in reliance on the design professional's promise to bring the project in within a designated cost.[20]

Redesign followed by construction very likely causes delay.   Delay damages are recoverable if they can be proved with reasonable certainty and were reasonably foreseeable at the time the agreement was made.[21]

Suppose the project is constructed and the client seeks to recover the difference between the cost prediction and the actual cost.   The property may be worth what it cost.   If so, the client has suffered no loss. Proceeding with the project knowing the costs would substantially exceed the predictions *may* show that the loss could have been reasonably avoided by abandoning the project.

Kellogg v. Pizza Oven, Inc.,[22] involved a client who rented space for a restaurant. The lease provided that the landlord would pay up to $60,000 for the cost of an improvement to the landlord's building.   The architect had negligently estimated costs at $62,000, but the project cost $92,000.   The tenant-client had to pay the balance of approximately $30,000.   The tenant recovered this amount less a 10% tolerance for errors from the design professional because of the excess cost.[23]

In another case, *Kaufman v. Leard,*[24] the clients, the Kaufmans, purchased a house after Brooks, the architect, indicated that the house could be remodeled and redecorated at a cost of between $12,000 and $15,000.   Brooks drew up plans that were not sufficiently detailed for a fixed-price contract.   After some discussion and design changes, the price of $17,000 was set as a cost ceiling.   Brooks, upon authorization by the clients, engaged two contractors (Leard and Noel) on a time and materials basis. Ultimately the cost soared to $40,000.

---

18. *Durand Associates, Inc. v. Guardian Investment Co.,* supra note 13 (by implication).

19. Supra note 12.

20. The client's claim in *Durand Associates, Inc. v. Guardian Investment Co.,* supra note 13, included excavation costs and losses suffered on a steel prepurchase made to avoid an anticipated price rise.   These losses should have been recovered if they were reasonably foreseeable and not avoidable.   Because they were incurred by an affiliated company of the client and not by the client, they were denied.

21. *Impastato v. Senner,* 190 So.2d 111 (La.App.1966), denied recovery for delay, but *Hedla v. McCool,* 476 F.2d 1223 (9th Cir.1973), allowed recovery.   Under AIA Doc. B141, ¶ 3.2.4 states redesign to be the limit of the architect's responsibility.   This should not relieve the architect from responsibility for delay damages caused by negligence.

22. 157 Colo. 295, 402 P.2d 633 (1965).

23. The landlord ended up with improvements probably worth considerably more than the $60,000 he spent. Perhaps the design professional should have claimed against the landlord based upon unjust enrichment.

24. 356 Mass. 163, 248 N.E.2d 480, 482 (1969).

This computation was the basis for a $24,000 award against the architect based upon the latter's negligence:

Total of Leard's charges to the Kaufmans on time and materials basis contracted for by Brooks		$32,953.86
Total due from the Kaufmans to Noel as fair value of work contracted for by Brooks		7,000.00
		$39,953.86
Limit of cost under the Kaufmans' authorization to Brooks	$17,000.00	
Extras order by the Kaufmans	5,415.00	$22,415.00
Kaufmans' excess liability to the contractors, unauthorized by them.		$17,538.86
Part of architect's fee already paid		785.00
		$18,323.86
Interest		5,900.28
Par. 3 of final decree (Brooks to pay the Kaufmans)		$24,224.14
Plus costs		35.25
		$24,259.39

The court rejected the contention made by the architect that the clients would unjustly be enriched if they recovered the excess costs and retained the house with $40,000 worth of improvements.

Where the project is built, such as in the *Kaufman* case, the architect should not be held for the excess of actual costs over predicted costs. This is not based upon the client's having proceeded despite its knowledge that the costs will be more than anticipated. The client should not be required to give up the project to reduce damages for the design professional. However, the client benefits by ownership of property presumably of a value equal to what it has paid for the improvement. Damages should not be awarded unless the client can prove that the economic utility of the project was reduced in some ascertainable manner because of the excessive costs.

(G) RELATIONSHIP BETWEEN PRINCIPAL DESIGN PROFESSIONAL AND CONSULTANT

Section 15.10(B) discusses whether and to what extent the consultant bears the risk that the principal design professional will not be paid by the client. This problem can arise when a project is abandoned because of excessive costs. *George Wagschal Associates, Inc. v. West* [25] involved an action by the consulting engineer against the principal design professional (the architect) for services he had performed in the design of a school that was never built to these plans because of excessive cost. Apparently the cost overrun was due to engineering overdesign.

Evidently the architect had not been paid by the school district and contended that the engineer should share this loss with him. The court upheld this contention and seemed to hold that the principal design professional and consulting engineer were jointly engaged in the project. The architect risked his fee by his contract with the school district, and the engineer risked his fee because he knew of the budgetary constraints. [26]

The American Institute of Architects (AIA), representing mainly prime design professionals, and the National Society of Professional Engineers (NSPE), represent-

---

25. Supra note 4.

26. If the lost fee resulted from engineering overdesign, should the architect be able to transfer his lost fee to the engineer?

ing consulting engineers, have differed as to the right of a consulting engineer to be paid if the architect has not been paid. NSPE argues that consultants generally do not take this risk. It notes that the prime design professional selects the client and has the best opportunity to evaluate its capacity to pay. It also points to AIA Doc. B141, ¶ 2.2, which allows the architect to request that the client provide a statement of funds available for the project and of the source of funds.

The AIA contends that frequently a long-term relationship exists between prime design professionals and consultants that resembles a partnership even if not cast in legal terms. It believes that under these conditions the risk of nonpayment should be shared.

Obviously the standardized documents published by these associations reflect their positions, as does their unwillingness to endorse the documents of the other. Yet their standard documents do change. For example, in 1974 AIA published C141, a document not endorsed by NSPE. In it, ¶ 6.5 stated that if the architect does not receive full payment from the owner for any cause not the fault of either architect or engineer, the architect shall pay the engineer in the same proportion as payments received bear to the total compensation to the architect. This appears to require that the parties share this loss.

In 1979, AIA published a new C141, also not endorsed by NSPE. Paragraph 6.1.2 was added requiring that the architect disclose to the engineer any contingent or special provisions relating to compensation included in the architect's understanding with the owner or in the agreement between them. No sharing provision comparable to the C141 published in 1974 was included.

AIA Doc. C141a, Instructions to C141, recognizes the possibility of variations and includes prototype clauses that can be included depending upon the arrangements made between prime design professional and consultant. One such clause makes clear that the consultant receives payment whether the prime design professional receives payment or not. Another requires sharing this risk. Another makes clear that the consultant does not get paid if he is the cause of the nonpayment. Another requires the architect to use reasonable and diligent effort to collect from the owner. This is a matter that should be clarified *in advance*.

(H) ADVICE TO DESIGN PROFESSIONALS

Many design professional-client relationships deteriorate or collapse because of excessive costs. Design professionals should try to make the cost prediction process more accurate. Their chief method has been to use the contract to protect their members from losing their fees where their cost predictions are inaccurate. They have included provisions that are supposed to insure that fees will not be lost when cost predictions are inaccurate and that fees will be lost only where the cost predictions are made negligently.

Protective language should be explained to the client. Design professionals who give a reasonable explanation to a client are not likely to incur difficulty over this problem. They should inform the client how cost predictions are made and how difficult it is to achieve accuracy when balancing uncontrollable factors. They should state that best efforts will be made, but that for various specific reasons, the low bids from the contractors may be substantially in excess of the statement of probable construction costs. The suggestion should be made that under such circumstances, the design professional and the client should join to work toward a design solution that will satisfy the needs of the client. In helping the client to be realistic about desires and funds, the design professional should request that the client be as specific as possible as to expectations about the project.

If these steps are taken, some clients may be lost. It may be better to lose them at the outset, rather than spend many hours and either not be paid or be forced to go to court to try to collect. Without an honest discussion with the client at the outset, the design professional takes risks.

Design professionals should also consider greater flexibility in fee arrangements. If the stated percentage of construction costs is used, it may be advisable to reduce or eliminate any fee based upon construction costs that exceed cost predictions.

During performance, the design professional should state what effect any changes made by the client will have upon any existing cost predictions. It is hoped that not every change will require an increase in the cost predictions. If the client approves any design work, the request for approval should state whether any change has occurred in cost predictions.

(I) PROBLEM

A, an architect, had been retained by C to design a small commercial building. The contract was AIA Doc. B141 as set forth in Appendix A. C requested that the building be five stories and contain a specified amount of square feet. He also requested that the building have enough luxury features so that he could attract high-class tenants who would pay high rentals. Since he hoped to attract law firms as tenants, he stated that sufficient space should be segregated for a law library that could be used by all of the lawyer tenants.

In the course of his design performance, A had informed C that he planned to use a very luxurious type of wood paneling in the larger offices and that there would be murals painted in the entrance hall. He also stated that there would be a sauna on each floor. He planned to install piped-in music and a number of other luxury features. All of this was agreeable to C.

The price was not to exceed one million dollars. The construction documents were submitted to five bidders, and the lowest bid was $1.3 million. A then stated that he would replace the luxurious wood panels with a less expensive type of wall construction. He suggested that the murals and saunas be eliminated. He suggested elimination of the music system. Plumbing features would be of cheaper quality. In addition, he wanted to eliminate the library and thus increase the rental area as well as cut costs. He also stated that the size of the windows would be reduced and that a number of other features would be changed to cut the costs.

C was unhappy about all these changes. He stated that he could not charge projected rent unless the luxury features were retained. Does A have the right to make these changes? Are there any limits to A's rights to make deletions or substitutions? Give illustrations of changes that would be permissible and those that would not.

SECTION 15.04 ASSISTANCE IN OBTAINING FINANCING

Frequently a building project requires lender financing. To persuade a lender that a loan should be granted, the client generally submits schematic designs or even design development, economic feasibility studies, cost estimates, and sometimes the contract documents and proposed contractor. (Refer to Chapter 12.) This submission includes material prepared by the design professional.

Must the design professional do more than permit the use of design work for such a purpose? Does the basic fee cover such services as appearing before prospective lenders, advising the client as to who might be willing to lend the client money, or assisting the client in preparing any information that the lender may require? Must this be done only if requested as *additional* service and paid for accordingly?

The professional education and training of a design professional does not include techniques for obtaining financing for a project. Nor is it likely that the design professional will be examined on this activity when seeking to become registered. It should not be considered part of *basic* design services.

Clearly this is true if the design professional has been retained by a large institutional client with personnel experienced in financial matters. The result does not change, however, even if the client does not have the skill within its own organization. Rarely does the client engage a professional designer because the client expects the designer to have and use skills relating to obtaining funds for the project.

An AIA study in 1956 showed that most firms give financing advice to some clients, many furnish financial contacts, and almost all arrange financing occasionally. Yet extra charges for such services are rare. Perhaps the activity is not burdensome, or it merely reflects economic realities of practice. Only if design professionals have *specifically* agreed to perform such services and be compensated for them by the basic design fee should they be denied additional compensation if it is requested and performed.[27]

The AIA does not list such services among those for which the basic fee is paid. The term selected in AIA Doc. B141, ¶ 1.7.2, dealing with additional services—*financial feasibility*—does not clearly cover the services related to obtaining financial backing. Paragraph 1.7.22 describes "services not customarily furnished in accordance with generally accepted architectural practice" as additional services. It would appear to classify "obtaining financing" as "additional services."

Services outside basic design services can involve the risk of increased liability if these services are performed but the client asserts they were not done properly. (See Section 15.05.)

SECTION 15.05 ECONOMIC FEASIBILITY OF PROJECT

Paragraph 1.7.2 of AIA Doc. B141 states that financial feasibility studies are additional services. "Financial feasibility" can mean the availability of financing for the project or the general economic feasibility of the venture. Obviously a lender will take into account the economic feasibility of the project when deciding whether to finance the project. The expanding role of design professional in financial matters can prove dangerous to the design professional. Not only may the design professional be denied compensation beyond the basic fee, but the client may claim it relied upon assurances made by the design professional as to economic feasibility or profitability of the project. These predictions are treacherous, ones for which design professionals are not trained.

In *Martin Bloom Associates, Inc. v. Manzie*, the plaintiff architect sued the defendant clients for design services he rendered to the defendants before the project was abandoned by the clients because they could not obtain financing. The clients owned land in Las Vegas, Nevada, which they wished to develop for investment purposes. The architect produced a written contract under which he was to provide design services and contract administration for a designated compensation fee. Here the issue was not the architect's right to additional compensation but his responsibility for the accuracy of his representations as to financing and profitability.

The clients claimed that the architect had represented that the client would have no difficulty in obtaining financing and that the project would be a profitable one. The architect's version differed, and he contended that the entire agreement was in the letter and that no representation had been made as to profitability.

---

27. Refer to *Herkert v. Stauber*, 106 Wis.2d 545, 317 N.W.2d 834 (1982), discussed in Section 3.06.

First, the trial judge concluded that the written agreement was not the complete contract between the parties. The judge then stated:

> Throughout the two days of trial, the Court had the opportunity to observe the manner of the witnesses while they testified, the consistency of their versions, both internally and compared to those of other witnesses, the probability or improbability of their versions and their interest in the outcome of the lawsuit. Based on these factors, the Court credits the testimony of Manzie [the client] that representations of Bloom [the architect], both express and implied, induced Manzie to enter into the agreement and also constituted part of the agreement itself.[28]

The case illustrates the danger of exceeding one's professional capabilities. If the architect had, as determined by the trial judge, made representations relating to financial feasibility and profitability, he may very well have ventured into an unpredictable area beyond his professional skill. As seen in Section 15.03, architects and engineers often lose fees when their cost predictions are inaccurate. But at least cost predictions, though certainly difficult, should be within the professional competence of a design professional. Economic feasibility goes beyond this and should be avoided.

SECTION 15.06
SECURING APPROVAL OF PUBLIC AUTHORITIES

Greater governmental control and participation in all forms of economic activity have meant that the design professional increasingly deals with federal, state, and local agencies. Must the design professional assist the client or its attorney in preparing a presentation to be made for the planning commission, zoning board, or city council? Must he appear at such a hearing and act as a witness if *requested*? Is the design profes-

sional who does these things entitled to compensation in *addition* to the basic fee?

Reasonable cooperation by the design professional in matters relating to public land use control can reasonably be expected by the client. Design professionals are expected to have expertise in matters that are often at issue in these public hearings. It may be within the client's reasonable expectations that the architect or engineer will render reasonable assistance and advise the client on these matters. Each contracting party should do all that is reasonably necessary to help obtain the objectives of the other.

The increased legal controls and their effect upon design professionals' services are reflected by changes made in AIA Doc. B141. Despite the proliferation of legislation requiring environmental impact statements or reports in the 1970s, the 1974 edition was unclear on treatment of such items. Paragraph 1.1.8 stated that the architect's assistance to the owner "in filing the required documents for the approval of governmental authorities having jurisdiction over the Project" was part of the architect's basic services.

Paragraph 1.3.3 stated that environmental studies were additional services. Failure to specifically include the frequently *required* environmental impact reports or studies made the issue of the architect's right to be paid additional compensation when he performs this service difficult to resolve.

In 1977, ¶ 1.3.4 required the architect to assist the owner in filing *required* documents, while ¶ 1.7.3 states that additional services include:

> . . . preparing special surveys, studies and submissions required for approvals of governmental authorities. . . .

One gap remains. Suppose the design professional spends considerable time as-

---

28. *Martin Bloom Associates, Inc. v. Manzie*, 389 F.Supp. 848, 852 (D.Nev.1975).

sisting the client or its attorney to prepare for any hearings or court proceedings that have become a part of government regulation of land use.[29]

Again, it may be useful or desirable to perform such services without requesting additional compensation. Discussion in this section deals solely with the questions of whether design professionals are obligated to perform the services and whether they are legally entitled to be paid additionally for doing so.

SECTION 15.07 SERVICES OF A LEGAL NATURE

Some design professionals volunteer or are asked to perform services for which legal education, training, and licensure may be required. This may be traceable to the high cost of legal services, the blurred line between those services that can be performed only by a lawyer, and designer belief that repetitive performance of certain services equips the design professional to perform them as well as a lawyer.

Undoubtedly one of the most troublesome activities is drafting or providing the construction contract. It may not be *wise* for lay persons to draw contracts for themselves although the law allows professional designers to supply or draft contracts for their *own* services. Suppose the designer drafts or selects the *construction* contract for a client? Lawyers often contend that design professionals are practicing law when they perform such activities. To meet such complaints, the architects and engineers include language on their standard documents that the document has important legal consequences and that "consultation with an attorney is encouraged with respect to its completion or modification."

AIA Doc. B141, ¶ 1.3.2, requires that the architect assist the owner in the preparation of "bidding information, bidding forms, the Conditions of the Contract, and the form of Agreement between the Owner and the Contractor." In addition, ¶ 2.7 requires the owner to furnish legal services as necessary. Clearly the AIA wants architects not to perform legal services.

Some provisions in construction contracts can be considered "architectural" or "engineering" in that architectural or engineering training and experience are essential to reviewing or even drafting such provisions. This may tempt professional designers to play an aggressive role rather than realize that their *advice* will certainly be useful to the client or its lawyer.

Some cases have involved activities that can be considered "legal." In *Transit Casualty Co. v. Spink Corp.,*[30] a trial court held an engineer liable for not including a hold-harmless clause in the prime contract and failing to have the owner designated as a name insured under the insurance policy. This finding was not challenged on appeal.[31]

Chaplis v. County of Monterey[32] involved a claim that the client had suffered losses because it found that it could not place a laundromat in a building that it had built

---

29. Paragraph 1.7.3 covers preparation of studies, while ¶ 1.7.20 deals with preparing to serve or serving as an expert witness. Under ¶ 1.3.4, assisting the owner in filing required documents is part of basic services. Perhaps such services fall in the catch-all of ¶ 1.7.22.

30. 94 Cal.App.3d 124, 156 Cal.Rptr. 360 (1979), disapproved on other issues 26 Cal.3d 912, 164 Cal.Rptr. 709, 610 P.2d 1038 (1980).

31. See also *Travelers Indemnity Co. v. Ewing, Cole, Erdman & Eubank*, 711 F.2d 14 (3d Cir.1983), discussed in Section 9.03 which involved an unsuccessful claim that services of a legal nature by the architect should have been performed.

32. 97 Cal.App.3d 249, 158 Cal.Rptr. 395 (1979). This case also dealt with the need for expert testimony, (discussed in Section 17.06).

because local authorities refused to issue a use permit. In addition to bringing a claim against the local authorities, the client sought recovery from the designer and contractor, claiming they were negligent in not informing him that a use permit would be required before work began on the laundromat. The court held that professional services include informing the client that a use permit would be required. The court concluded that the use permits are matters about which the designer should advise the client.

This demonstrates the blurred border between legal and nonlegal services in the land use area. Land use matters can and do involve legal skills. Often an architect is knowledgeable about the politics and procedures in land use matters, particularly if he has had experience in dealing with land use agencies.

Another illustration of the overlap in matters relating to land use can be demonstrated by the hearings required before public agencies charged with the responsibility of issuing permits. On the whole, such hearings tend to be political rather than legal, and the orchestration of such a hearing may be best in the hands of someone with political skill and experience. This is the responsibility of the owner, although the architect or engineer can be of great value in advising the owner or even in orchestrating the hearing and the strategy for it.

The line between legal and nonlegal services can also be demonstrated by the not uncommon phenomenon of the design professional being asked to advise the client as to whether a surety bond should be required. The experience of design professionals may put them in the position of being able to determine whether a particular contractor should be bonded. Their experience may also lead them to have strong opinions on the general desirability of surety bonds. Repetitive involvement with such matters may give design professionals confidence that they can handle them.

Here the balance tips strongly in favor of considering these legal services. Sometimes bonds are required for public projects. The use of surety bonds may to a large degree depend upon other legal remedies given subcontractors and suppliers, such as the right to assert a mechanics' lien or to stop payments. The design professional should simply answer specific questions rather than advise generally on such matters. If *asked* to advise on such a matter, the services a design professional provides in response to such a request are additional.

Perhaps the most educational "horror" story involved a project that was abandoned because of excess costs and the unavailability of funds, but not until after a contract had been made with the contractor with no specific language allowing the owner to cancel the construction contract if funds could not be obtained.

The contractor demanded his lost profits. When the owner refused, the contractor demanded arbitration. He pointed to AIA Doc. A201 (incorporated by reference in AIA Doc. A101, which was the only document the owner had signed or seen). Nevertheless, an arbitration was held and the arbitrator awarded lost profits to the contractor. As if this were not bad enough, the architect who had furnished the contract documents sued for services he had performed. The owner contended:

1. When the architect furnished the contract he was acting as a lawyer.

2. A competent lawyer would have included protective language in the contract with the contractor and would certainly have pointed out the arbitration clause to the owner.

3. The architect must pay the arbitration award and be denied any fees for his services—legal services being performed in violation of law and the design services having been performed negligently because of the cost overrun.

The case was settled.

SECTION 15.08 SITE SERVICES

(A) RELATION TO CHAPTER 17 AND PART IV

Section 15.08 concentrates upon the reasonable expectations of the client as to the design professional's role on the site while construction proceeds. It focuses upon the belief that the client *may* have that the design professional has been paid to *insure* that the client receives all it is entitled to receive under the construction contract. To accomplish this, the client may expect the design professional to watch over the job, see to it that the contractor performs properly, and be responsible if the contractor does not. As seen in (B), a design professional may take a different view of his function. The purpose of Section 15.08 is to explore this problem with a view toward seeing what the law has done when faced with resolving the often different expectations of client and design professional.

Site services are but a part of the total professional services performed by the designer. Disputes over design performance may involve determining whether the design professional has performed in accordance with his obligation. Standard of performance is also an issue when claims are made by third parties against the design professional. Because claims, whether made by the client or by others, involve the professional responsibility of the designer, they are treated together in Chapter 17.

The design professional plays an important role during execution of the design that has a significant impact upon not only the owner but also the contractor. As illustrations, the design professional plays a part in the changes process, interprets contract documents, judges performance, condemns defective work, issues certificates for payment and completion, reviews listed subcontractors, passes upon schedules—to name some of the most important activities. Inasmuch as they are all central to the relationship between owner and contractor, they are treated in Part IV.

(B) SUPERVISION TO PERIODIC OBSERVATIONS: *FIRST NAT. BANK OF AKRON v. CANN; SHEPARD v. CITY OF PALATKA; SO. BURLINGTON SCH. DIST. v. CALGAGNI, FRAZIER-ZAJKOWSKI ARCH.*

A review of the standard contracts published by the AIA reveals a drastic shift in the architect's function during construction. Before 1961, the architect had general supervision of the work. In response to the specter of the expanded liability to third parties, this phrase was dropped in favor of language under which the architect observed rather than supervised, made periodic visits rather than conducted exhaustive on-site inspections, and did his best. He was *not* responsible, however, for the failure by the contractor to perform in accordance with the contract documents or for methods of executing the design.

The AIA justified this shift by suggesting that the architect was no longer the masterbuilder exercising almost total domination of the Construction Process from beginning to end.[33] Instead, said the AIA, the architect has turned over the responsibility for executing the design to the contractor who presumably has the necessary skill to accomplish this properly and safely. Even more, in sophisticated construction the architect may simply be part of the management team that may consist of a project manager, a construction manager, and a field representative of owner or lender. This shift in role and function, though principally in response to increased liability of third parties, was also designed to recognize the shift in organization for most construction, the architect no longer being mas-

---

33. See Sweet, *The Architectural Profession Responds to Construction Management and Design-Build: The Spotlight on AIA Documents*, 46 Law & Contemp.Probs. 69 (1983).

terbuilder but being *called in* to interpret or decide disputes and making periodic observations to check on the progress of the work.

It is difficult to determine whether changes in terminology will have an effect on the actual role taken by the architect. Some will undoubtedly still seek to dominate the process for a variety of reasons that cannot be explored here. For example, in *Kleb v. Wendling*,[34] the architect testified that when he used the term *administer the contract* it was, as far as he was concerned, the equivalent of *supervise,* the term architects had previously used. He stated specifically that his role did not change whether he was supervising or administering "and that the change in jargon was to avoid liability under the Structural Work Act."

This testimony cast some doubt on the decision to recast the language of the standard contract to reflect the more diminished role by architects, at least as far as certain architects and projects are concerned. If the architect in the *Kleb* case is typical, the shift in terminology, though perhaps designed to avoid liability to third parties, may generate difficulties between architect and client.

Resolving disputes between clients and design professionals over the overall site responsibilities of the latter usually involves the following issues:

1. General nature of the design professionals' undertaking (supervision v. periodic observation).

2. Compliance with the design professional's contractual requirements.

3. Course of action when problems develop.

Resolving the first issue requires an interpretation of any contract language as well as any expectations that may have been created by conduct. Clearly the standard requirements published by the professional associations are intended to negate any supervisory function and to limit the responsibility of an architect or engineer to periodic observations. Nevertheless, the more extensive "supervision" standard may be applied if changes are made in standard agreements or if changes are contained in contracts negotiated by the parties.

One court held that supervision must be more than superficial, that there is "no real value in supervision unless the same be directed toward securing a workman-like adherence to specifications and adequate performance on the part of the contractor."[35] Such language suggests an outcome standard but does not provide much guidance.

Kleb v. Wendling[36] is instructive. It held the architect to a supervision standard of the contract despite the use of the less comprehensive term *administration.* When he has agreed to supervise or oversee the construction of a building, he must "prevent gross carelessness or imperfect construction." Merely detecting defective workmanship does not relieve him of a duty to prevent it. This appears to require that the architect provide continuous monitoring of the contract or service. The architect's representative had visited the site daily while the architect visited the site at least twice a month. (The legal effect of permanent on-site observation is discussed in (D).) Perhaps this strict standard was traceable to the owner's absence during construction and a representation by the architect that he would keep a "closer tab on the general contractor than he normally would because he doubted the general contractor's competence."

As to compliance with the standard applicable, the *Kleb* case pointed to the reputa-

---

34. 67 Ill.App.3d 1016, 24 Ill.Dec. 434, 385 N.E.2d 346 (1979).

35. *Pancoast v. Russell,* 148 Cal.App.2d 909, 307 P.2d 719 (1957).

36. Supra note 34.

tion of the contractor for corner cutting and the absence of the owner as indications that careful monitoring were expected.

Other factors likely to be examined to determine whether the duration, frequency, and timing of site visits were adequate are as follows:

1. Size of the project.

2. Distance between the site and the design professional's home office.[37]

3. When crucial steps are undertaken, such as pouring concrete or covering work.

4. Type of construction contract (Cost-type requires more monitoring.).

5. Experimental design or unusual materials specified.

6. Extent to which owner has a technical staff that will take over some of these responsibilities.

7. Observation of contractor's performance during visits.[38]

8. Contractor's record of performance on this project.

Some cases involving judicial resolution of these problems are reproduced at this point.

FIRST NATIONAL BANK OF AKRON v. CANN
United States District Court, Northern District of Ohio, 1980.
503 F.Supp. 419, affirmed 669 F.2d 415 (6th Cir.1982).

CONTIE, District Judge
[Ed. note: Footnotes renumbered and some omitted.]

* * *

I. FINDINGS OF FACT

A. The Bank's Remodeling Project

In the mid–1960s the Bank's management decided to refurbish the interior and exterior of its main building located at 106 South Main Street, Akron, Ohio. Up until that time, Harold S. Cassidy had provided all the architectural services required by the Bank since their first association in 1952. Said services included the remodeling of existing structures and the design and erection of new structures, and were performed by Cassidy as an independent contractor.

In 1966 Cassidy was involved in the design and construction of a connecting link and loading dock at the Bank's main building and adjacent to a developing public plaza. The project required the cooperation and approval of the local City government, which Cassidy's long relationship as the Bank's local architect facilitated securing. When the Bank approached Cassidy with the prospect of taking on the more expansive exterior and interior remodeling of the main building, however, he informed them that his office was not in a position to handle the project. This was the first Bank project requiring architectural services that Cassidy was not in charge of since the inception of their association.

B.E.C. is a corporation engaged in the business of assisting

---

37. *Warde v. Davis*, 494 F.2d 655 (10th Cir.1974).

38. *Chiaverini v. Vail*, 61 R.I. 117, 200 A. 462 (1938). (Visits when contractor *not* on site inadequate.)

lending institutions in making improvements to their physical facilities, providing assistance in the selection of an architect, and management of the construction if the proposed improvements are undertaken. In 1966, John T. Huffman was employed by B.E.C. as a senior consultant, and represented that company in its efforts to sell its services. James L. Hilton was the Bank's vice president in charge of property management and security. Huffman initially contacted the Bank through Hilton. The Bank eventually expressed interest in B.E.C.'s services, and the consulting and architectural agreements presented by Huffman were executed. On November 3, 1966, B.E.C. and the Bank entered into the consulting agreement. On November 17, 1966, the Bank entered into a standard architect agreement with William F. Cann, a licensed architect and employee of B.E.C.

At that time Cann was president of B.E.C. All architectural fees paid to Cann were turned over by him to the company. Cann was then paid by the company in his capacity of president; he received no separate compensation as an architect. Cann never personally disclosed this arrangement to the Bank, but held himself out as an individual architect in executing the architect agreement.

In a letter dated April 12, 1967, Cann proposed an association between himself and Cassidy, through which Cassidy would act as a local architect during the construction phase of the architectural agreement. This contact was initiated because it was B.E.C.'s common practice to hire a local architect for intricate, remote jobs such as the Bank's remodeling project. Cassidy signed the proposal and dated it June 6, 1967. This written proposal, approved by Cassidy, was not a fully integrated agreement. By its explicit terms, the proposal covered only the principal items and left open the precise relationship between Cann and B.E.C. to further negotiations.

On June 6, 1967, Cassidy met Cann at his St. Louis office to refine the terms of their relationship. It was agreed that Cassidy would both make periodic visits to the site and assist and supervise the work only when so requested. Cassidy refused to assume any greater responsibility because of the demands of his private practice. Cassidy did agree to turn over his drawings of the connecting link and loading dock. This allowed one architect to coordinate a single project. From January to November, 1967, Cassidy attended a number of meetings between the Bank, Cann, and B.E.C. in order to facilitate coordination of Cann's remodeling plans and Cassidy's work on the connecting link and loading dock.

Thus, retaining Cassidy as the local architect was the logical choice for two reasons of considerable import: first, it achieved the necessary coordination between B.E.C.'s responsibility for the extensive remodeling project and the two minor facets of the Bank's refurbishing already begun by Cassidy; and second, it provided B.E.C. with an experienced liaison who maintained a standing relationship with both the Bank and the City government.

On June 22, 1967, the Bank and B.E.C. entered into a management agreement to achieve the satisfactory completion of the remodeling project. The contract contained a cost incentive clause. Under the agreement B.E.C. became the construction manager for the project and was responsible for all work necessary to complete the project. Thus, B.E.C. was both the consultant and the construction manager, and its president Cann was the architect.

Actual construction commenced on the project in January, 1968, and was completed by November, 1969. The exterior remodeling included the erection of a granite facing on the Bank's south wall. The instant action is solely directed to alleged problems involving the granite facing on the south wall, and no other aspect of the project is in dispute.

B.   The Plans For The South Wall

A curtain wall construction was utilized to erect the granite facing consistent with the following specifications set out in the plans.[39] The type of curtain wall erected on the Bank's south wall is a non-load-bearing design. The granite skin consists of some 500 individual panels two inches thick and otherwise varying in size. Some of the panels weigh up to 800 lbs. The curtain wall's non-load-bearing design provides for intermediate steel supports at each floor level so that the total weight of the granite is not transferred to the bottom.

[Editor's summary: In the fall of 1976 a bank official noticed that caulking had failed and panels on the wall had been dislodged on the ninth floor. Bank employees pushed the panels back and tuck-pointed the joints, cutting out mortar joints and then repointing the joints with mortar, a solution which they thought would remedy the problem.

In the summer of 1977 the bank made a contract under which the contractor would inspect and service the caulking. The contractor determined that the wall had been installed without a bond breaker and the caulking had failed, allowing water to seep in the cavity, freeze and push out the granite panels. It was thought that this could be remedied by simply pushing back the dislodged panels and tuck-pointing. In August 1977 the joints were opened and structural problems with the supporting steel were uncovered. It was determined that the granite panels in the wall had become unstable and presented a dangerous condition. The recommendation was made that they be removed from sections of the wall so that the underlying supporting structure could be observed.

The interior of the curtain wall was examined from the inside.

---

39. Unless more specifically identified, the plans shall collectively refer to the contract documents for job no. 22805, which include: (1) the consulting agreement; (2) the owner-architect agreement, A.I.A. Doc. B131 (ed. 1963); (3) the management agreement; (4) the general conditions of contract, A.I.A. Doc. A201 (ed. 1963); (5) the architectural and structural plans for the project; and (6) the specifications for the project.

Structural problems were uncovered; the contractor was contacted. After some unsuccessful discussions the bank advised the architect that the bank would seek redress for the difficulties in the south wall, the basis being that the granite sections were fixed to the building in a way which deviated from the contract requirements. A similar letter was sent to the contractor. The bank decided to remove the granite and perform corrective work. Investigation determined that there had been substantial deviations in specifications and unorthodox cutting and welding by the employees on the site. When the granite had been removed it appeared that deviations from the plans and specifications created a sponge effect under which moisture entered the curtain wall and accumulated in excess mortar. Also moisture that entered naturally collected in the moist mortar buildup. When inclement weather caused the accumulated moisture to freeze, the resulting thermal pressure dislodged the panels. Also, inspection revealed that there were numerous instances of welding and cutting of the structural steel which caused the ship coat of primer to be damaged. These damaged areas did not receive a field coat of primer. The exposed areas of steel rusted as moisture entered the curtain wall cavity.]

II.   CONCLUSIONS OF LAW

A.   Plaintiff's Contract Claims

The Bank asserts a number of claims against defendants Cann and B.E.C. for breach of contract. It is the duty of the Court to construe the contract documents in their entirety so as to give effect to the intention of the parties. . . . The several parts of a contract must be construed together so that the intent may be properly identified from a consideration of the whole contract. . . .

1.   Cann's Contractual Duties

Plaintiff sets out in the amended complaint three separate claims against Cann for breach of contract. First, the Bank alleges that Cann failed to inspect the project for compliance with the plans and breached article 3.4.3 of the owner-architect agreement. Second, the Bank alleges that Cann breached an express warranty made in article 3.4.3 that the quality of the work is in accordance with the plans. Finally, the Bank alleges that Cann breached the contract documents by failing to provide adequate specifications for the installation of caulking and weep holes.

Article 38 of the general conditions of contract provides in relevant part:

ARCHITECT'S STATUS: ARCHITECT'S SUPERVISION

The Architect shall be the Owner's representative during the construction period. The Architect will make periodic visits to the site to familiarize himself generally with the progress and quality of the work and to determine in general if the work is proceeding in accordance with the Contract Documents. He will not be required to make exhaustive or continuous on-site inspections to check the quality or quantity of the work and he will not be responsible for the Contractor's failure to carry out the construction work in accordance with the Contract Documents. During such visits and on the basis of his observations while at the site, he will keep the Owner informed of the progress of the work, will endeavor to guard the Owner against defects and deficiencies in the work of Contractors, and he may condemn work as failing to conform to the Contract Documents.

Article 3.4.3 of the agreement between the Bank and Cann for architectural services similarly provides in relevant part:

[The Architect] will make periodic visits to the site to familiarize himself generally with the progress and quality of the work and to determine in general if the work is proceeding in accordance with the Contract Documents. He will not be required to make exhaustive or continuous on-site inspections to check the quality or quantity of the work and he will not be responsible for the Contractors' failure to carry out the construction work in accordance with the Contract Documents. During such visits and on the basis of his observations while at the site, he will keep the Owner informed of the progress of the work, will endeavor to guard the Owner against defects and deficiencies in the work of Contractors, and he may condemn work as failing to conform to the Contract Documents. Based on such observations and the Contractors' Applications for Payment, he will determine the amount owing to the Contractor and will issue Certificates for Payment in such amounts. These Certificates will constitute a representation to the Owner, based on such observations and the data comprising the Application for Payment, that the work has progressed to the point indicated. By issuing a Certificate for Payment, the Architect will also represent to the Owner that, to the best of his knowledge, information and belief based on what his observations have revealed, the quality of the work is in accordance with the Contract Documents. He will conduct inspections to determine the dates of substantial and final completion and issue a final Certificate for Payment.

In support of its claim that Cann breached article 3.4.3, the Bank contends that there existed on the part of Cann a contractual duty to conduct on-site job inspections, and that such inspections would have uncovered the variations from the plans that occurred during construction.

Article 38 of the general conditions of contract and article 3.4.3 of the architectural agreement are harmonious on the nature and extent of Cann's duty to observe generally the progress of the work. Under these provisions, Cann is obliged to act as the Bank's representative. The nature of this obligation includes keeping the project under general observation.

One purpose circumscribing the architect's duty to keep the project under general observation is to provide a basis upon which the architect can determine whether to issue the certificate of payment. Thus, periodic visits are required and the architect has the duty to observe generally the progress of work and to insure that the contractor is compensated for the amount of work completed. It is also apparent that the architect represents the owner in protecting the owner's aesthetic interests in the quality of the work and is under a duty to endeavor to guard the owner's interests.

The contract documents further circumscribe the architect's duty to keep the project under general observation. Articles 38 and 3.4.3 specifically provide that the architect is not required to make continuous on-site inspections to check the quality and quantity of the work and is not responsible for the contractor's failure to complete the work in accordance with the plans. That article 3.4.3 provides that the owner may contract for more extensive on-site representation further supports this construction.

Other than reference to the broad guideposts of establishing a basis for issuing certificates for payment and guarding the owner's interests, the scope of the architect's duty generally to observe is not clearly defined. Admittedly, the contract documents are explicit on the provision that exhaustive, continuous on-site inspections are not required. That exhaustive, continuous on-site inspections were not required, however, does not allow the architect to close his eyes on the construction site, refrain from engaging in any inspection procedure whatsoever, and then disclaim liability for construction defects that even the most perfunctory monitoring would have prevented. Even the most general supervision would have placed Cann on notice as to the activities of laborers engaging in unauthorized, extensive welding and cutting of structural steel.

Turning to principles of contract interpretation, the Court construes Cann's duty under article 3.4.3 to include inspections and monitoring of a nature that would have uncovered the vast majority of defective conditions on the south wall.

If their terms are less than clear, contracts are generally construed in favor of the promisee and against the promissor. . . . It is fundamental that if a provision is ambiguous it is construed strictly against the author of the contract. . . . Cann is both the promissor and author of the architect agreement. Thus, the duty of general supervision clause must be construed in favor of the Bank and against Cann.

But it is not solely upon these principles that the Court rests its interpretation of Cann's duty under article 3.4.3. Indeed, elevation of these general principles to the level of categorical rules would work the unequitable result of giving a windfall to the promisee who did not draft the contract but nonetheless executed an ambiguous agreement in an arms length transaction. There are more compelling reasons supporting the Court's interpretation.

Apparently conflicting clauses are reconciled in favor of the one contributing most essentially to the object of the contract. . . . The object of the contract documents taken as a whole was the completion of a remodeling project with good quality workmanship and materials. To the extent that article 3.4.3's general-supervision clause may conflict with the no-exhaustive-or-continuous-on-site-inspection clause, it is the former that was more essential to the satisfactory completion of the project under the facts of the instant case.

Cann represented the Bank's interests on the project. The exact nature of the architect's general supervision duties will turn on the particular facts of a case. *See Pastorelli v. Associated Engineers, Inc.*, 176 F.Supp. 159 (D.R.I.1959). Often, terms of a contract clause are implied in law to accommodate considerations of justice. . . . In identifying whether Cann's duty to supervise would entail inspections sufficiently diligent to detect the defective conditions on the south wall, there are a number of relevant factors that must be considered. . . .

Cann's employment relationship with B.E.C. and their potential conflict of interest as architect and contractor on the Bank job placed a high duty upon the architect to insure that the work was proceeding as planned. The Bank project involved large architectural fees and construction costs. These costs justified a greater supervisory role than would be required on a small, less costly job. The Bank project was remodeling construction, and thus Cann was aware that the approval of variances would be required to meet existing job conditions. With regard to the frequency of visits to the site, at a minimum they should have been made when the crucial step of erecting the supporting steel for the granite non-load-bearing curtain wall was undertaken. Finally, more supervision is required on a cost incentive job like the Bank project because the contractor may be more inclined to cut corners.

For these reasons, the Court construes Cann's duty under article 3.4.3 to include inspections and monitoring of a nature that would have uncovered the defective conditions on the south wall, and finds that Cann's failure so to perform constitutes a material breach of the contract documents.

Plaintiff also contends that article 3.4.3 contains an express warranty by Cann that the quality of the remodeling work would be consistent with the plans. Plaintiff concludes that because the south wall was found to have been constructed contrary to the plans, Cann has breached his express warranty. The provisions of article 3.4.3 support the construction offered by plaintiff.

Article 3.4.3 provides that the architect, upon observation and the data provided, will issue a certificate of payment to the contractor representing to the owner that the work has progressed to the point indicated. By issuing the certificate of payment the architect also represents to the owner that to the best of his knowledge, information, and belief, the quality of the work is consistent with the plans. Thus, Cann warrants only that to the best of his knowledge the quality of the work is consistent with the plans.

In light of the fact that Cann failed to perform his supervision duties adequately, the warranty that the quality of the work was consistent with the plans was not to the best of Cann's knowledge, information and belief, and thus constituted a material breach of article 3.4.3. Even the most cursory discharge of the supervisory duties would at least have given notice of the unauthorized welding and cutting of structural steel on the south wall. . . .

[The Court held Cann and B.E.C. jointly *and* severally liable. The *entire* award can be collected from *either* or both.]

SHEPARD v. CITY OF PALATKA

District Court of Appeal of Florida, Fifth District, 1981. 414 So.2d 1077.

FRANK D. UPCHURCH, Jr., Judge.

Herschel Shepard appeals from a summary final judgment entered against him. Appellee, the City of Palatka, retained Shepard as its architect for the repairing and remodeling of the Bronson-Mulholland House in Palatka. Appellee, Joseph Rothenberg, was the contractor selected to do the work. Appellee, United States Fidelity and Guaranty, was Rothenberg's surety.

After completion of the work, veneer plaster began to peel away from the plaster wallboards because a special type of wallboard necessary to assure adhesion had not been installed. The city contended that the special type of wallboard (one-half-inch Imperial Plaster Base, Type X, paragraph 9F–3C) was mandated by the job specifications. The contractor, in his deposition, acknowledged that he had not used this board but instead used a finishing board, fire resistive, type X, throughout. He contends that the specifications permitted either to be used and that the fire resistant board was approved on the job.

The city sued Shepard claiming he was negligent in the preparation of plans and specifications and also charged that he failed to make timely and proper inspection of Rothenberg's work. The city also sued Rothenberg and his surety for negligent failure to follow the plans and specifications. Shepard cross-claimed against Rothenberg for indemnification. Summary judgment was granted to the city against all three defendants on the issue of liability.

. . . In *Palm Bay Towers Corp. v. Crain & Crouse, Inc.*, 303 So. 2d 380, 383 (Fla. 3d DCA 1974), the court enunciated the standard of care owed by an architect to his employer:

As a general rule, an engineer, like an architect, owes his

employer a duty to exercise and apply his professional skill, ability and judgment in a manner which is reasonable and without neglect.

As to the question of whether there was a failure to make timely and proper inspection of Rothenberg's work, it is necessary to consider the duty of the architect as specified in the contract which provided:

> The Architect shall make periodic visits to the site to familiarize himself generally with the progress and quality of the Work and to determine in general if the Work is proceeding in accordance with the Contract Documents. On the basis of his on-site observations as an architect he shall endeavor to guard the Owner against defects and deficiencies in the Work of the Contractor. The Architect shall not be required to make exhaustive or continuous on-site inspections to check the quality or quantity of the Work. The Architect shall not be responsible for construction means, methods, techniques, sequences or procedures, or for safety precautions and programs in connection with the Work, and he shall not be responsible for the Contractor's failure to carry out the Work in accordance with the Contract Documents.

> The Architect shall not be responsible for the acts or omissions of the Contractor, or any Subcontractors, or any of the Contractor's of the Subcontractors' agents or employees, or any other persons performing any of the Work.

In a proper factual situation, where it is demonstrated that the architect ignored his contractual duty to make periodic visits to the site, liability could possibly lie regardless of such exonerating language. Such was not the case here, however. The architect's deposition reflected that he had made inspections. While he admitted that he had not discovered the misuse of the wallboard, the contract clearly protected him because it imposed no duty upon him to discover the omission of the contractor and clearly absolved him of liability if there were, in fact, such omissions.

The other question was whether the architect was negligent in the preparation of plans and specifications.

Shepard, in his cross-claim against the contractor, Rothenberg, alleged that Rothenberg had failed to follow the plans and specifications. Rothenberg denied and alleged that he had followed the plans and specifications.

The court did not determine liability as between Shepard and Rothenberg. If it is ultimately determined that Rothenberg did follow the plans and specifications, the conclusion is inescapable that Shepard was negligent in their preparation and would be liable to the city. On the other hand, if it is determined that Rothenberg did not follow the plans and specifications then Rothenberg would be liable to the city but Shepard would not. The jury could also find each liable to some extent under a comparative negligence theory. *Hoffman v. Jones*, 280 So.2d 431 (Fla.1973).

We conclude, therefore, that the entry of summary judgment, even though limited to the issue of liability was erroneous. There is an issue of fact as to whether the architect was negligent in the preparation of the plans and specifications. There is also an issue of fact as to whether the contractor complied with the plans and specifications. Whether either the contractor or architect or both are liable to the city cannot be determined as a matter of law at this stage of the case. We must REVERSE.

SHARP, and COWART, JJ., concur.

SOUTH BURLINGTON SCHOOL DISTRICT v. CALCAGNI–FRAZIER–ZAJCHOWSKI ARCHITECTS, INC.
Supreme Court of Vermont, 1980.
138 Vt. 33, 410 A.2d 1359.

HILL, Justice

[Ed. note: The school district brought legal action against architect, prime contractor, subcontractor, and supplier when the roof split and leaking occurred because the roofing insulation was exposed to substantial precipitation. The claim against the architect was based upon negligent design and "supervision." After concluding that summary judgment in favor of the architect based upon design was proper because no expert testimony had been introduced to establish that the architect did not comply with the professional standard of performance, the court dealt with negligent supervision.]

* * *

With regard to South Burlington's negligent supervision claim, the duty owed by CFZ to the plaintiff was set out in their contract. See *Lapoint v. Dumont Construction Co.,* 128 Vt. 8, 10, 258 A.2d 570, 571 (1969). Their agreement provided as follows:

> The Architect shall make periodic visits to the site to familiarize himself generally with the progress and quality of the Work and to determine in general if the Work is proceeding in accordance with the Contract Documents. On the basis of his on-site observations as an Architect, he shall endeavor to guard the Owner against defects and deficiencies in the Work of the Contractor. The Architect shall not be required to make exhaustive or continuous on-site inspections to check the quality or quantity of the Work. The Architect shall not be responsible for construction means, methods, techniques, sequences or procedures, or for safety precautions and programs in connection with the Work, and he shall not be responsible for the Contractor's failure to carry out the Work in accordance with the Contract Documents.

In other sections, the contract provided that CFZ would not be responsible for the acts or omissions of the contractor or subcontractors, and that the contractor alone was responsible for supervision, direction and coordination of the construction.

To hold CFZ liable for breach of contract, South Burlington not only had to establish the existence of an enforceable contract but also had to show that CFZ failed to perform according to their agreement. *Lapoint, supra*, 128 Vt. at 10, 258 A.2d at 571. Viewing the contract in the light most favorable to the plaintiff, it is difficult to see how CFZ failed to perform its contractual obligations. The language of the contract is clear and unambiguous, and this Court's duty is to apply it, not remake it or ignore it. *Simpson v. State Mutual Life Assurance Co.*, 135 Vt. 554, 556, 382 A.2d 198, 199 (1977).

CFZ's duty under the contract in no way related to supervision of the construction project or the duties of the contractor or subcontractors. Indeed, the agreement between South Burlington and the contractor imposed the duty solely on the latter to supervise and direct the job. That the parties understood this division of responsibility is further evidenced by the fact that Kenclif, the contractor, hired a special superintendent to supervise the job.

South Burlington's claim to the effect that CFZ had a duty of care other than that specified in their agreement based on custom or usage in the industry cannot be determined on the record before us because plaintiff introduced no competent evidence that a "reasonable roof designer" would have done anything different. Without testimony by witnesses who could speak authoritatively as to the customs and usages of the roof designing and construction industry, the jury had no benchmark against which to measure CFZ's conduct. South Burlington's claim that the jury was competent to determine whether CFZ failed to meet the standard of care of a professional roof designer misperceives the relationship between the jury and expert witnesses. A jury is competent to draw logical inferences from facts within their knowledge. But where, as here, the facts (*i.e.*, the customs of a profession) are outside their knowledge, competent evidence, usually in the form of expert testimony, must be adduced.

For the above-stated reasons the direction of the verdict in favor of CFZ was proper.

* * *

(C) SUBMITTALS:
WAGGONER v. W&W STEEL CO.

During the course of construction, the contractor is required to submit shop drawings, product data, and samples to the design professional. Shop drawings illustrate how the contractor plans to perform the work. The product data provides information regarding the material, products, and systems. Samples are physical examples that illustrate the material, equipment, or workmanship the contractor intends to supply.

Submittals, particularly shop drawings, are often more detailed than construction documents. Generally, for example, the information given to the steel fabricator must be more detailed than the information contained in the drawings and specifications furnished by the owner.

Some vendors may not accept orders for equipment until they are certain that the materials or equipment has been approved by the design professional. In addition to containing information that is part of or an elaboration of the design, submittals frequently contain information as to how the contractor proposes to do the work, generally an activity within its discretion. However, the design professional's position in the Construction Process—principal professional adviser of the owner, interpreter of the documents, and judge of performance as well as the principal conduit between the owner and prime contractor—can appear to make him responsible if anything goes wrong.

Such an outcome does not take into account the primary responsibility of the prime contractor for selection of subcontractors, the fabrication processes, or construction methods or other activities within its control. Most standard contracts seek to make clear that approval of submittals does not relieve the contractor. Such contracts generally require the contractor to review and approve submittals before submission. If this is not done, the design professional should return them, inasmuch as approval of the submittal is a representation by the contractor that it has verified submitted data and coordinated it with the working contract documents.

The principal legal problem has resulted from information that is included in a submittal that is not part of design but that may indicate that the contractor is choosing a construction method or technique that may cause harm to those working on the site. Can the design professional ignore information that indicates the likelihood of harm to the project or persons?

Disclaimers such as those usually found in standard contracts or those expressed on stamps of approval have generated controversy. Are they effective? Do they abdicate the design professional's professional responsibility?

This highlights the possibility of shared responsibility. The design professional's approval does not relieve the contractor. This does not necessarily mean that the design professional may not be responsible if he has played a role in causing a defect or personal harm by failing to take reasonable steps if aware that harm is likely to occur. Perhaps *ultimate* responsibility will be shared by contractor and design professional through contribution or indemnification.

A case that deals with approval of submittals is reproduced at this point.

WAGGONER v. W&W STEEL CO.

Supreme Court of Oklahoma, 1982. 657 P.2d 147.

BARNES, Vice Chief Justice:

This case presents for resolution on appeal, the following issue:

Is the architect who designed this building responsible for ensuring that the contractor employ safe methods and procedures in performing his work?

During the construction of Presbyterian Hospital in Oklahoma City, Oklahoma, a portion of the steel framework fell, resulting in the death of two workmen and the injury of another. The accident occurred on a Friday, as workers were beginning to secure a portion of the steel in the sixth, seventh and eighth floors which had been erected that day. The three men were on the structure waiting for other workers to bring up guy lines which were to be used in securing the steel before an approaching thunderstorm rolled in over the construction site. Before the task could be completed, a gust of wind hit the unsecured and unbraced steel, causing it to collapse.

In designing this building, the architects had provided for expansion joints which would allow for expansion and contraction with changing weather conditions. The expansion joint was designed so that a shelf, welded to a column, provided a seat for a beam which was held in place by "keeper angles" welded in on either side of the beam. The opposite end of the beam was secured to another column with large bolts. Unfortunately, at the time of the accident, the "keeper angles" had not been installed and the beams were not secured in any other way.

The erector had built portions of three floors rather than completing all the work on one floor at a time. As a result, the east half of the new sections were without interior columns and cross beams which would have provided lateral bracing for the outside columns. These outer columns bore the weight, not only of the steel beams and trusses which had been erected, but also of large bundles of steel decking which were placed on the beams at each floor level. The outside columns were held upright only by temporary "clip angles" provided by the fabricator to facilitate the alignment and welding of the top columns to the columns below.

After the accident, suit was brought against the owners of the hospital, the fabricator and the architect by the injured worker and representatives of the estates of the deceased workers. The case (three cases were consolidated) went to jury trial with the architect as the sole defendant, the other defendants having been released from the lawsuit by dismissals and summary judgments. After plaintiffs and defendant rested, the trial court directed a verdict for the architect. The Court of Appeals, in reversing the lower court, determined that the architect did owe a duty to the workers, in that the architect "undertook to supervise the construction project." With this, we disagree.

Architects are required to exercise ordinary professional skill and diligence and to conform to accepted architectural standards. Because an architect's undertaking does not imply a guarantee of perfect plans or results, he is liable only for failure to exercise reasonable care and professional skill in preparation and execution of plans *according to their contract. Smith v. Goff*, 325 P.2d 1061 (Okla.1958) [emphasis added]. This principal was reaffirmed in *Wills v. Black & West, Architects*, 344 P.2d 581 (Okla.1959) in which we looked to the contract between the parties to determine the responsibilities of the architect. It is stated in 5 Am.Jur.2d Architects § 5 (1962) that:

"(t)he employment of an architect is ordinarily a matter of contract between the parties, and the terms of such employment are governed by the terms of the contract into which they entered. The architect's duties may be limited to the preparation of plans and specifications, or they may include, in addition, the supervision of construction."

At common law, privity of contract was required before an action in tort could arise from a breach of duty created by a contract. However, in *Truitt v. Diggs,* 611 P.2d 633 (Okla.1980), we indicated that in cases involving physical injury to third parties, that restriction has in many cases been eliminated or modified. Therefore, it is possible for an architect to be liable for injuries received by a person with whom he has no privity, but there can be no standard rule. The determination must be made by considering the nature of the architect's undertaking and his conduct pursuant thereto.

To do so in the present case, we must look to the contract which actually consists of several separate documents. Pertinent provisions of the General Conditions of the Contract for Construction are set forth below.

2.2.4 The Architect will make periodic visits to the site to familiarize himself generally with the progress and quality of the Work and to determine in general if the Work is proceeding in accordance with the Contract Documents. On the basis of his on-site observations as an architect, he will keep the Owner informed of the progress of the Work, and will endeavor to guard the Owner against defects and deficiencies in the Work of the Contractor. *The Architect will not be required to make exhaustive or continuous on-site inspections to check the quality or quantity of the Work. The Architect will not be responsible for construction means, methods, techniques, sequences or procedures, or for safety precautions and programs in connection with the Work, and he will not be responsible for the Contractor's failure to carry out the Work in accordance with the Contract Documents* [emphasis added].

4.3 SUPERVISION AND CONSTRUCTION PROCEDURES

4.3.1 The Contractor shall supervise and direct the Work, using his best skill and attention. He shall be solely responsible for all construction means, methods, techniques, and sequences and procedures and for coordinating all portions of the Work under the Contract.

4.13.1 By approving and submitting Shop Drawings and Samples, the Contractor thereby represents that he has determined and verified all field measurements, field construction criteria, materials, catalog numbers and similar data, or will do so, and that he has checked and coordinated each Shop Drawing and Sample with the requirements of the Work and of the Contract Documents.

4.13.5 The Architect will review and approve Shop Drawings and Samples with reasonable promptness so as to cause no delay, but only for conformance with the design concept of the Project and with the information given in the Contract Documents. The Architect's approval of a separate item shall not indicate approval of an assembly in which the item functions.

10.2 SAFETY OF PERSONS AND PROPERTY

10.2.1 The Contractor shall take all reasonable precautions for the safety of, and shall provide all reasonable protection to prevent damage, injury or loss to:

> .1 all employees on the Work and all other persons who may be affected thereby; . . .

10.2.2 The Contractor shall comply with all applicable laws, ordinances, rules, regulations and lawful orders of any public authority having jurisdiction for the safety of persons or property or to protect them from damage, injury or loss. He shall erect and maintain, as required by existing conditions and progress of the Work, all reasonable safeguards for safety and protection, including posting danger signs and other warnings against hazards, promulgating safety regulations and notifying owners and users of adjacent utilities.

Appellants maintain that the trial court was in error when he directed a verdict against them because the architect owed a duty to them to exercise ordinary professional skill and diligence in preparing and approving plans for construction, and therefore the question of the alleged violation of that duty should have been submitted to the jury.

We must determine if, under the contractual provisions set out above, the architect had such a duty.

Section 4.3.1 of the General Conditions specifies that the contractor is to supervise the work, being "solely responsible for all construction means, methods, techniques, and sequences and procedures." Although the architect is to periodically visit the construction site, § 2.2.4 provides that he is not required to make "exhaustive or continuous on-site inspections to check the quality or quantity of the work." It goes on to exclude from the architect's responsibilities those which are outlined in 4.3.1 as belonging solely to the contractor. It should also be noted that the owners of the hospital employed an engineer as the project inspector.

Article 10, entitled "Protection of Persons and Property" contains provisions which require the contractor to protect workmen from injury, comply with safety regulations and laws and designate a superintendent whose job it is to prevent accidents. It is to be noted that the responsibilities for all safety precautions and programs are assigned exclusively to the contractor.

Article 4 sets forth the procedure and purpose concerning the shop drawings which appellants contend should have included specifications for temporary bracing and connections. The shop drawings, prepared by a subcontractor, were submitted to the contractor and the architects, as required by the contract. It is the architects' approval of the shop drawings, without provision for temporary connections on the expansion joints, that appellants maintain was negligence.

However, § 4.13.4 provides that the shop drawings are submitted to the *contractor* for determination and verification of "all field measurements, field construction criteria, materials, catalog number and similar data." By approving them, he represents that he has checked each shop drawing with the requirements of the contract. And, as previously noted, § 4.3.1 of the contract makes the contractor "solely responsible for all construction means, methods, techniques, sequences and procedures."

It is apparent that the shop drawings serve more than one purpose. They are submitted to the contractor for approval regarding aspects of the construction work. But, according to § 4.13.5, *they are submitted to the architects for approval "only for conformance with the design concept of the project and with the information given in the Contract Documents."*

Therefore, it was the duty of the contractor, not the architects, to see that the shop drawings included provisions for temporary connections which fall into the categories of "field construction criteria," "construction means, methods, techniques, sequences and procedures." Since it was not the responsibility of the architects, they obviously would not be negligent in failing to require temporary connections.

Because the contractor, not the architect, was required under the contract to supervise the job and employ all reasonable safety precautions, the architects cannot be held liable for injuries sustained as a result of an unsafe construction procedure. There was no question of fact for the jury. The trial court properly directed a verdict for the architects.

The judgment of the Court of Appeals is vacated and the decision of the trial court is affirmed.

IRWIN, C.J., and LAVENDER, SIMMS, DOOLIN, HARGRAVE, OPALA and WILSON, JJ., concur.

Moving to what the law expects from design professionals when they encounter problems when they *are* on the site, two cases are instructive. *Pastorelli v. Associated Engineers, Inc.,*[40] involved an injury to an employee of a race track caused by a falling heating duct. The accident occurred after the work had been completed and accepted by the injured employee's employer.[41] The injured employee sued the prime contrac-

---

40. 176 F.Supp. 159 (D.R.I.1959).

41. Earlier American law immunized architect and contractor from most accidents that occurred after the project was completed and accepted by the owner. In *Pastorelli v. Associated Engineers, Inc.,* supra note 40, the

tor, the sheet metal subcontractor who installed the duct, and engineers who prepared the plans and who agreed to "supervise the contractor's work throughout the job."

The duct was 20 feet long and weighed 500 pounds. It had not been attached directly to the roof itself or to the joists of the clubhouse but had been suspended from the ceiling of the clubhouse by the attachment of semi-rigid strips of metal called hangers which were then attached to the ceiling. The ceiling was of sheathing and was nailed to the joists, leaving a considerable airspace between sheathing and roof. The specification required that sheet metal work be erected "in a first class and workmanlike manner" and that "the ducts be securely supported from the building construction in an approved manner." [42]

The trial judge concluded that the duct had not been properly installed. The engineer prepared and submitted periodic inspection reports while the work was in progress. The trial judge stated that the engineer's employee who prepared the reports:

> testified that his employer assigned to him the task of supervising the installation of said systems, and that in pursuance of his duties he visited the job site on one, two or three occasions each week to inspect the work of the contractor as it was being done. He also testified, however, that he never observed any of the ducts being hung from the ceiling in said clubhouse, stating that whenever he visited the clubhouse the ducts were either on the floor or already installed. He also admitted that he never climbed a ladder to determine whether the hangers by which they were suspended were attached by nails or lag screws and never tested any of the hangers to see how securely they were attached. [43]

After holding prime and subcontractor negligent, the judge noted that the engineer's employee knew that the safety of persons in the clubhouse required that the ducts be attached to the joists and that he made no attempt to ascertain whether they were so installed. The judge also noted that the employee made no visits at a time when he *could* determine how they were being installed. Holding the engineer negligent, the judge stated:

> In other words, he failed to see that they were properly installed and took no steps after their installation to ascertain how and by what means they were secured. In my opinion he failed to use due care in carrying out his undertaking of general supervision. [44]

The case illustrates three points. First, note that the contract phrase was cast in terms of supervision. This could have heavily influenced the trial court's conclusion that the engineer's employee had not properly checked to see that the duct had been installed safely.

Second, the case illustrates the possibility of precise specifications rather than general ones. If the duct should have been attached to the joists, the specifications should have so required. If so, a more careful inspection would very likely have been made.

Third, suppose the engineer's employee had determined that the ducts were not properly secured. What should he have done? Should he have directed the employees of the sheet metal subcontractor to correct the work? Should he have gone to the superintendent of the prime contractor with a similar request? Should he have ordered the work terminated until proper corrective measures were taken? Should he have gone to a building inspector with a suggestion that the latter order that the work be

---

court rejected this rule and held that completion and acceptance did not furnish a defense to the architect. This is discussed in Section 17.09(C).

42. 176 F.Supp. at 162.

43. Id. at 162–63.

44. Id. at 167.

corrected? These questions show the difficult position in which the design professional can find himself when he does determine that work is not being properly performed, especially when persons may be injured or killed as a result.

If the design professional simply observes, one would think that his principal responsibility would be to call the attention of defective work to the contractor and have the contractor transmit this to the subcontractor in a case such as *Pastorelli*. Yet if it is likely that correction will not come quickly enough and if it is likely that persons may be injured, perhaps registering a complaint indirectly would not be sufficient. Perhaps the owner should be given the information, along with advice as to what should be done. The owner can then determine the course to be followed. If the danger were imminent and the situation urgent, perhaps the building inspector should be called, or if the design professional has the power, he should stop the work.[45]

The second injury case, *Cutlip v. Lucky Stores, Inc.*,[46] involved an action by the wife of a subcontractor's employee who had been killed during construction. Although the case had many legal issues, one part related to a site visit by the architect. The superintendent of the prime contractor called the architect and told him that either certain steel beams were not the proper length or that the concrete piers upon which the beams were to be erected were not in the correct location. Two solutions were proposed. The first was to send the steel back to be refabricated and the second to pour a new concrete base and pillar. The testimony was contradictory but evidently the superintendent, claiming he had been authorized by the architect, poured the new base in order to avoid delay. Two days later the worker was killed when the steel superstructure collapsed.

Although the architect retained a structural engineer to make structural drawings as well as be available for consultation, the architect reserved to *himself* the structural supervising work of the building. An architect who testified as an expert for the plaintiff stated that in his opinion architects such as the architect in the case were not qualified to perform structural engineering inspections. That expert witness also expressed the opinion that when the problem developed the architect should have contacted the structural engineer and authorized him to redesign. In addition to noting this, the court pointed to the fact that the architect could have stopped the work at any time. The appellate court concluded that there had been sufficient evidence to submit the matter of negligence to the jury. (The jury had ruled against the architect.) The court seemed persuaded that it was negligent to agree to perform services for which the architect was not qualified. Note the observation by the expert witness that whether it was proper or not for the architect to have agreed to inspect structural work of this sort, at the very least he should have contacted the consulting structural engineer before making any decision relating to structure.

(D) USE OF PROJECT REPRESENTATIVE

The standard contracts published by the professional associations do not require the

---

45. In 1970 the AIA General Conditions of the Contract A201 eliminated the power of the architect to stop the work and reposed this power solely in the owner. The clause doing this, ¶ 3.3.1, reflects the tendency in AIA Documents to give the architect a more passive role in the hope of eliminating or reducing liability. See Section 17.09(D).

46. 22 Md.App. 673, 325 A.2d 432 (1974).

design professional to have a continuous presence on the site. If this is required, parties can agree to have a permanent representative, such as project representative or a resident engineer, on the site continually. Although the principal problems involved in the use of a project representative relate to his authority,[47] such use can bear upon the responsibility of the design professional.

At the very least, the presence of a full-time project representative means that the design professional is not expected to be on the site continuously. It should also mean that the frequency of visits may be diminished. However, this should not mean that the design professional need not visit the site at times when it would otherwise be appropriate.[48] Also, the design professional must select competent representatives and monitor their work.[49]

(E) USE OF CONSTRUCTION MANAGER

Although newer construction organization methods are discussed in Chapter 21, it may be useful at this point to note the development of Construction Management. The role of the design professional has been substantially changed by standard contracts, particularly in the area of predicting costs, visiting the site, and passing upon submittals. This has left a vacuum that in some types of contracts has been filled by a Construction Manager (CM).

A CM should provide construction experience and skill, particularly at the design stage but also during design execution. There are two models of cost estimating, one that uses rough formulas and the other that does a continuous evaluation of those elements that comprise construction costs. Similarly, during construction the Construction Manager should bring greater skill in monitoring contract compliance. Is work being done in accordance with safety regulations, and will or does the end product meet the contract document requirements? It is likely that the future will see increased use of specialized professionals unconnected with the creation of the design to perform many services mentioned in this chapter.

PROBLEM

Suppose you are the architect who has been retained to design and supervise the construction of a luxury residence that is to cost $250,000. The project is to take about four months, and you are to receive a fee of 10% of the cost. When and how often would you be expected to visit the site? Upon what things would the duration of your visit depend? What would you do while you were at the site? Would the frequency of your visits in any way depend upon whether the contractor has been found to be deliberately skimping on the job or has been late in the scheduled performance?

SECTION 15.09 MEASURING PROFESSIONAL PERFORMANCE

(See Chapter 17)

---

47. See Section 21.05.

48. *Central School District Number 2 v. Flintkote Co.*, 56 A.D.2d 642, 391 N.Y.S.2d 887 (1977).

49. *Town of Winnsboro v. Barnard & Burk, Inc.*, 294 So.2d 867 (La.App.1974), cert. denied 295 So.2d 445 (1974).

SECTION 15.10 WHO ACTUALLY PERFORMS SERVICES: USE OF AND RESPONSIBILITY FOR CONSULTANTS

(A) WITHIN DESIGN PROFESSIONAL'S ORGANIZATION

The design professional may operate through a corporation or a partnership or be a sole proprietor. The actual performance of the work may be done by the sole proprietor; by a principal [50] with whom the client discussed the project; by another principal; by employees of the contracting party (whether the contracting party is a sole proprietor, a partnership, or a corporation); or, as discussed in Section 15.10(B), by a consultant hired by the design professional. Are the client's obligations conditioned upon the performance being rendered by any particular person? Can certain portions of the performance be rendered by persons other than the design professional without affecting the obligation of the client to pay the fee?

Services of design professionals, especially those relating to design, are generally considered personal. A client who retains a design professional usually does so because it is impressed with the professional skill of the person with whom it is dealing or the firm represented by that person. The client is likely to realize that licensing or registration laws may require that certain work be done by or approved by persons possessing designated licenses.[51] Yet the client is also likely to realize that some parts of the performance will be delegated to other principals in the firm with whom the client is dealing, employees of that firm, or consultants retained by the design professional organization.

As to design, unless indicated otherwise in the negotiations or in the contract, the client probably expects the design professional with whom it has dealt to assemble and maintain a design "team," and to control and be responsible for the design. The fleshing out of basic design concepts, such as the construction drawings and specifications, is likely to be actually executed by other employees of the design professional's organization. Although contract administration is probably less personal than design, it is still likely that the client will expect that the principal contract administration decisions will be made, if not by the person with whom it has dealt, at least by another principal in the design professional organization. However, just as in design, the client is likely to realize that the person who has overall responsibility will not actually perform every aspect of contract administration. (As to interpretations and disputes see Section 33.06.)

These conclusions can depend upon the client's knowledge of the size and nature of the design professional's organization. These conclusions may also depend upon whether the client has ever dealt with design professionals before and whether it knew of the division of labor between principals, employees, and consultants.

(B) OUTSIDE DESIGN PROFESSIONAL'S ORGANIZATION: CONSULTANTS

Because of the complexity of modern construction, the high degree of specialization and the proliferation of licensing laws, design professionals frequently retain consultants to perform certain portions of the services they have agreed to provide. Although the principal legal issues have been whether consultant fees are additional ser-

---

50. "Principal" is used to define a partner in a partnership or a person with equivalent training, experience, and managerial control in a corporation.

51. For discussion of the effect of licensing laws on who is permitted to do the work, refer to Section 13.05.

vices not covered by the basic fee [52] and whether the design professionals are responsible for the consultants they hire—a point to be mentioned later in this subsection—inquiry should be directed initially to whether the design professional can use a consultant to fulfill the obligations owed the client.

Generally clients prefer that highly specialized work be performed by highly qualified specialists. Suppose the client insists that all design services be performed within the design professional's organization. The client may believe that it would not be able to hold the design professional accountable if the consultant did not perform properly. In such a case, the client might feel at a serious disadvantage if forced to deal with or institute legal action against a consultant it has not selected and with whom it has no direct contractual relationship. Clearly the client can insist upon this in the negotiations, and the contract could so provide. But suppose nothing is discussed or stated specifically in the agreement.

It is unlikely that any client who is or should be aware of the customary use of consultants for certain types of work can insist upon consultants not being used. Standard contracts published by professional associations contain language relating to professional consultants. Usually such language does not directly relate to the question of whether consultants may be used but deals indirectly by providing that certain consulting services are covered under the basic fee and others are additional services. Inclusion of this language should indicate to the client the likelihood that some consultants will be used and that possibly others will be used if the client consents.

One person can be held for the wrongdoing of another. The most common example is the liability of an employer for the negligent acts of its employee committed in the scope of employment. Although there are many reasons for vicarious liability, one reason sometimes given is that the employer controls or has the right to control the details of the employee's activities.

If the negligent actor is not controlled nor subject to the control of the person who has hired the actor, the actor is an independent contractor and not an employee. Subject to many exceptions, the employer of an independent contractor is generally not liable for the negligence of the latter. For example, the business person who hires an independent garage to service its fleet of trucks generally will not be held liable for the negligence of the garage. An opposite result would follow if the business person had a repair service as part of its organization.

Generally the consultant is an independent contractor. As a rule, the consultant is asked to accomplish a certain result but can control the details of how it is to be accomplished. If an architect retains a structural engineer as a consultant, the latter's negligent conduct that causes injuries to third persons will generally not be chargeable to the architect because the engineer would be considered an independent contractor.

The client who retained the architect is not a third party in the same sense. Permitting the architect to use a consulting engineer as a substitute to perform certain portions of the work should not relieve the architect of responsibility to the client for proper performance of that work. The architect's obligation to the owner is based upon their contract.[53]

A trilogy of consultant cases decided by the Supreme Court of Oregon are instruc-

---

52. See Sections 15.01 and 16.01(G).

53. *Harold A. Newman Co. v. Nero*, 31 Cal.App.3d 490, 107 Cal.Rptr. 464 (1973). A trial court decision coming, incorrectly, to the opposite result was *Whitfield Construction Co. v. Commercial Development Corp.*, 392 F.Supp. 982 (D.V.I.1975).

tive. The first, *Scott & Payne v. Potomac Insurance Co.,*[54] involved a claim by an owner against an architect based upon failure in the heating system that the owner claimed was caused by defective design.

The architect's insurer refused to defend, claiming that the negligent act occurred in a period not covered by the policy. The architect settled the claim and brought an action against his insurance company when it refused to reimburse him. To recover, the architect found himself in the strange position of claiming he was liable. One argument made by the insurer was that the architect was not negligent because he relied upon the advice of a heating engineer. In rejecting this argument, the court stated that the architect should possess skill in all aspects of the building process and cannot shift responsibility to a consultant.

The second case in the trilogy was *Johnson v. Salem Title Co.,*[55] in which an injured third party sued an architect when a wall defectively designed by the consulting engineer collapsed. The court concluded that the engineer was an independent contractor which would normally give the architect a defense. However, the court held the architect liable because of an exception to the independent contractor rule for nondelegable duties created by safety statutes. In this case, the architect was required to comply with building codes, and this duty could not be delegated to the engineer. By nondelegation the court did not mean that the architect could not use the engineer to fulfill code requirements but meant that the architect could not divest himself of ultimate responsibility if the code were not followed.

The action was brought not by the client against the architect but by an injured party. Had the action been brought by the client for its losses, the independent contractor rule would not have been relevant.

The third case in the Oregon trilogy, *Owings v. Rosé,*[56] involved the owner suffering a loss when the floor cracked because of defective design by the consulting engineer. The architect paid the owner's claim and then sought and was given indemnity from the consulting engineer. In passing on the indemnity claim, the court stated that the architect was liable to the owner on the basis of *Scott & Payne v. Potomac Insurance Co.*[57]

Design professionals can exculpate themselves from liability to clients for errors of their consultants. One method—a novation—is a tripartite contract under which the consultant is substituted for the design professional for that part of the work. Another is to seek and obtain from the client exculpation from any responsibility for the conduct of the consultants. This is more likely to be obtainable if the client designates that particular consultants be used. It does not seem unreasonable for the principal design professional in such a case to seek exculpation from the client.

Following this approach to its logical extreme, the principal design professional can suggest or insist that the client contract directly with consultants. Clearly this would relieve the principal design professional unless he had information regarding the consultant that should have been communicated to the owner. Although this approach is commonly used in the retention of soil engineers, principal design professionals frequently prefer to keep overall professional control and are not anxious for clients to contract separately with consultants.

It is likely that the principal design professional will not be relieved from liability to the client for the acts of consultants.

54. 217 Or. 323, 341 P.2d 1083 (1959).

55. 246 Or. 409, 425 P.2d 519 (1967).

56. 262 Or. 247, 497 P.2d 1183 (1972).

57. Supra note 54.

Probably the best the principal design professional can do is to insure that the consultant is obligated to perform in an identical manner to the principal design professional's obligation to the client. It is essential that the principal design professional consider the financial responsibility of consultants. Having a good claim against a consultant may be meaningless if the consultant is not able to pay the claim. The principal design professional should undertake an investigation of the financial capacity of consultants employed. If the consultant is a small corporation, the principal design professional should bind the individual shareholders to the contract so that they are personally liable. The principal professional probably should require that the consultant carry and maintain adequate professional liability insurance.

Does the consulting engineer assume the risk of not being paid if the principal design professional—the architect—is not paid by the client? Refer to Section 15.03(G).

PROBLEM

A has been hired by C to design a project at a cost of about $300,000. A is the senior partner of a five-partner architectural firm. Later C asks what he, A, will do personally. A states that he will not draft anything but will delegate all the design work to draftspersons in his office. A states that he will look at and approve all work that will be done and will offer various suggestions to the draftspersons. As to supervision, he will send out one of his partners from time to time who has good engineering training and knows construction better than A does. C is unhappy and tells A that he wants A to participate more in the work. What are C's legal rights in this regard?

SECTION 15.11
OWNERSHIP OF DRAWINGS AND SPECIFICATIONS

Who owns, more properly, who has use rights of the written material created by the design professional necessary to build the project?

Clients sometimes contend, and many agencies of state and local government insist, that the party who pays for the production of drawings and specifications should have exclusive right to their use.

On the other hand, design professionals contend that they are selling their ideas and not the tangible manifestations of these ideas as reflected in drawings and specifications. This, along with the desire to avoid implied warranties to which sellers of goods are held, is the basis for calling the tangible manifestations "instruments of service."

Design professionals contend that the subsequent use of their drawings and specifications may expose them to liability claims. If another design professional completes the project, the original designer may be denied the opportunity to correct design errors as they surface during construction. Design professionals contend that most projects are "one of a kind" and that in reality design is a trial and error process. Similarly, liability exposure can result if the design is used for an addition to the project or a new project for which it may not be suitable. Even if the design professional is absolved, it may not come until after a lengthy and costly trial.[58] (As seen later in this section, this can be dealt with by indemnification.)

In addition, reuse without adaption may compromise the esthetics or structural integrity of the original design.

Perhaps most important, design professionals contend that the extent of use by the

---

58. See *Karna v. Byron Reed Syndicate*, 374 F.Supp. 687 (D.Neb.1974) (designer absolved when project's use unforeseeably changed).

owner is an important factor in determining the value of design professionals' services and, correspondingly, their fee, a factor recognized by the frequent use of the percentage of construction cost to determine compensation. (See Section 16.01 (B).) Had the design professional known that drawings and specifications would be used again, a larger fee would have been sought.

In the absence of any specific provision in the contract dealing with reuse, the client who has paid for the services has the exclusive right to use the tangible manifestations of the design services performed by the design professional.[59] While cases are rare, this result stems from the analogy to a sale of goods. In ownership terms, the client "owns" the drawings and specifications.

Suppose a design professional establishes a custom that drawings and specifications belong to the person creating them and that the client is allowed their use only for the particular project. The client would not be bound by it unless it knew or should have known of such a custom.[60] If the design professional intends to claim ownership of the plans and specifications, it is safer to include a contractual provision in the agreement with the client. Such provisions are included in standard contracts published by design professional associations. For example, ¶ 8.1 of AIA Doc. B141 states:

> Drawings and Specifications as instruments of service are and shall remain the property of the Architect whether the Project for which they are made is executed or not. . . . The Drawings and Specifications shall not be used by the Owner on other projects, for additions to this Project or for the completion of this Project or for the completion of this Project by others provided the Architect is not in default under this Agreement, except by agreement in writing and with appropriate compensation to the Architect.

Legitimate reasons exist for giving the design professional exclusive right to reuse the drawings and specifications, mainly on the basis of use determining value of the professional services.

Yet, the prohibition against the client using the materials for additions to or completion of the project can be looked upon as a device to discourage the client from retaining a new architect or at least to make it pay compensation if it replaces the original architect. It is as if an implied term of the original retention agreement gave the design professional an option to perform any additional design services required by an addition to the original project. Hiding such "options" in the paragraph dealing with ownership of drawings and specifications can make courts suspicious of the fairness of such standardized contracts.

The National Society of Professional Engineers (NSPE), with the American Consulting Engineers Council, the American Society of Civil Engineers and the Construction Specifications Institute publishes Doc. 1910–1 (1984). Paragraph 7.2 claims ownership for the engineer and states that the documents are not intended or represented "to be suitable for re-use" on extensions or other projects. Their use without written verification or adaptation by the engineer is at the owner's risk, and the owner agrees to indemnify the engineer for such reuse. If verification or adaptation is sought and accomplished, the engineer is entitled to receive additional compensation. Reuse *without* verification or adaptation does not *specifically* entitle the engineer to additional compensation. Protection here would have to come from Copyright Law. See Section 19.04.

Are there any limitations upon the *design professional's* right to reuse the drawings and specifications? Suppose the design professional plans to use the documents to build an identical residence near the completed residence which would diminish the

---

59. 5 Am.Jur.2d Architects § 11 (1962).

60. *Meltzer v. Zoller,* 520 F.Supp. 847 (D.N.J.1981).

exclusivity of the original residence. Although these matters are rarely thought about or expressed in contracts, it is likely that the law would imply a promise by the design professional not to reuse the drawings and specifications in any way that would significantly diminish the value of the original residence.

The hotly contested issue of drawing ownership may justify a contractual compromise that avoids an all-or-nothing solution. One possibility would be *joint* ownership with specific delineation of reuse rights. For example, the owner could be given the right to use the drawings to construct, maintain, repair, or modify the project. The design professional, in the case of owner reuse for another project, could be given an agreed additional fee if the original design and designer were used, and indemnification plus a small fee if another design professional were selected. The design professional could be given reuse rights only if it did not produce a project with similar, distinctive features, which would diminish the uniqueness of the original project.

SECTION 15.12   TIME

If there is no specific provision dealing with time for performance, the parties must perform within a reasonable time. However, the cooperative nature of design, the client providing a program, and the design professional creating a design subject to client approval make it difficult to determine who is responsible for delay. Delay may be caused by failure of public officials to move the administrative process along or by lenders deciding whether to make a loan. Even if a schedule is created, it is likely to require frequent adjustment. For these reasons, claims for a delay during design are difficult to sustain.

For delays in contract administration, claims are more likely to be made by contractors. This is dealt with in Chapter 30.

In 1977 the AIA added ¶ 1.8 to B141 re-

quiring the architect to perform services as "expeditiously as is consistent with the professional skill and care." Upon the owner's request, the architect must submit a schedule adjusted as required as the project proceeds. The language has sufficient looseness to make it quite unlikely that the architect will be held strictly to any schedule that may be submitted and approved.

Yet loose as the provisions are, some architects have expressed reservations about them. They fear the client may justify a refusal to make interim payments or reduce payments by an allegation that the architect has not completed the work according to the schedule. On the other hand, such a schedule, loose as it may be, may be a spur to a slow-performing architect to get the work moving.

SECTION 15.13   CESSATION OF SERVICES: SPECIAL PROBLEMS OF THE CLIENT-DESIGN PROFESSIONAL RELATIONSHIP

(A) COVERAGE

Contracts to perform design or construction services are service contracts, and many of the same legal rules apply. Principles discussed in Chapter 38 dealing with construction contracts also apply to contracts for design services. However, some aspects of the client-design professional relationship have generated special legal rules that are discussed in this section.

(B) SPECIFIC CONTRACT PROVISION AS TO TERM

Often contracts specifically define the duration of the contract relationships. However, this rarely exists in contracts to perform design services because of the high likelihood of delay traceable to the large number of participants in the Construction Process, including not only owners, designers, and contractors, but also lenders and public authorities. (Refer to Section 15.12 dealing with time and schedules.)

At the outset, a number of different issues must be differentiated. First, when do the contractual obligations between client and design professional end? Second, when does delay in the performance of design services entitle the design professional to additional compensation? Third, when has the design professional's dispute resolution power ended? The first is discussed in this section, the second in Section 16.01(G), and the third in Section 33.05.

Usually the design professional's services are divided into phases. Since the construction phase is last, it is the focal point for determining when the client-design professional relationship has terminated. Also, the construction phase is *itself* divided into segments, principally to enable the contractor to be paid as it works. Phases are important for other purposes as well.

Generally the owner would like the design professional to perform until the project is "completed." At this point, the owner takes possession of the work. After that point, except for post-completion services such as furnishing as-built drawings, the owner no longer has a need for the design professional's services.

In most construction contracts executed on standard contracts published by the professional associations, completion has two stages: *substantial* and final. AIA Doc. B141, ¶ 1.5.1, states that the construction phase terminates:

> When final payment to the Contractor is due, or in the absence of a final certificate for Payment or of such due date, sixty days after the Date of Substantial Completion of the Work whichever occurs first.

The first benchmark recognizes that the function of the design professional has been completed when he has certified that final payment is due. The alternative benchmark—sixty days after substantial completion—recognizes that issuance of a final certificate may be delayed because of the contractor's unwillingness to correct punch-list items. Nevertheless, this alternative benchmark may come as a surprise to clients if they believe they can still call upon the architect to perform services until the project is completed. (Perhaps more important, it may come as a surprise to the owner to find that for services performed after whichever event occurs earlier AIA Doc. B141, ¶ 1.7.19, grants the architect additional compensation.)

(C) CONDITIONS

A condition is an event that must occur or be excused before a party is obligated to begin or continue performance. Often contracting parties do not wish to begin or continue performance unless certain events occur or do not occur. For example, an owner may not wish to start construction until it has obtained a loan or permit. Design professionals may wish to condition their obligations upon their ability to rent additional space or hire an adequate staff. Yet each party may wish to make a binding contract in the sense that neither can withdraw at its own discretion.

A condition that is within the *sole* power of one of the parties may prevent a valid contract from being formed. Such a conclusion is avoided by implying an obligation to use good faith to seek occurrence of the condition. However, the creation of a condition does not affect the validity of the contract as long as the condition is described with reasonable certainty.

In contracts for design services, conditions are frequently asserted by the client, the nonoccurrence giving the client the power to terminate the relationship and, depending upon the language of the condition, preclude the design professional from being compensated for services rendered before termination.

Sometimes the client asserts that a condition was created though not expressed in the written contract. (This was discussed in Section 15.03 dealing with costs.) Generally, the client can testify as to an oral condition so long as it does not directly contradict the written contract or unless the written contract is clearly the final and complete repository of the entire agree-

ment. Even if the condition is expressed in the written agreement, problems of reconciling it with other contractual provisions may exist. A case dealing with this question is reproduced at this point.

PARSONS v. BRISTOL DEVELOPMENT CO.

Supreme Court of California, 1965.
62 Cal.2d 861, 44 Cal.Rptr. 767, 402 P.2d 839.

TRAYNOR, Chief Justice. In December 1960 defendant Bristol Development Company entered into a written contract with plaintiff engaging him as an architect to design an office building for a lot in Santa Ana and to assist in supervising construction. Plaintiff's services were to be performed in two phases. He completed phase one, drafting preliminary plans and specifications, on January 20, 1961, and Bristol paid him $600.

The dispute concerns Bristol's obligation to pay plaintiff under phase two of the contract. The contract provided that "a condition precedent to any duty or obligation on the part of the OWNER [Bristol] to commence, continue or complete Phase 2 or to pay ARCHITECT any fee therefor, shall be the obtaining of economically satisfactory financing arrangements which will enable OWNER, in its sole judgment, to construct the project at a cost which in the absolute decision of the OWNER shall be economically feasible." It further provided that when Bristol notified plaintiff to proceed with phase two it should pay him an estimated 25 per cent of his fee, and that it would be obligated to pay the remaining 75 per cent "only from construction loan funds."

Using plaintiff's preliminary plans and specifications, Bristol obtained from a contractor an estimate of $1,020,850 as the cost of construction, including the architect's fee of 6 per cent. On the basis of this estimate, it received an offer from a savings and loan company for a construction loan upon condition that it show clear title to the Santa Ana lot and execute a first trust deed in favor of the loan company.

Shortly after obtaining this offer from the loan company, Bristol wrote plaintiff on March 14, 1961, to proceed under phase two of the contract. In accordance with the contract, Bristol paid plaintiff $12,000, an estimated 25 per cent of his total fee. Thereafter, plaintiff began to draft final plans and specifications for the building.

Bristol, however, was compelled to abandon the project because it was unable to show clear title to the Santa Ana lot and thus meet the requirements for obtaining a construction loan. Bristol's title became subject to dispute on May 23, 1961, when defendant James Freeman filed an action against Bristol claiming an adverse title. On August 15, 1961, Bristol notified plaintiff to stop work on the project.

Plaintiff brought an action against Bristol and Freeman to recover for services performed under the contract and to foreclose a mechanic's lien on the Santa Ana lot. The trial court, sitting without a jury found that Bristol's obligation to make further payment under the contract was conditioned upon the existence of construction loan funds. On the ground that this condition to plaintiff's right to further payment was not satisfied, the court entered judgment for defendants. Plaintiff appeals.

* * *

. . . After providing for payment of an estimated 25 per cent of plaintiff's fee upon written notice to proceed with phase two, paragraph 4 of the contract makes the following provisions for payment:

"4. . . .

"(a) . . .

"(b) Upon completion of final working plans, specifications and engineering, or authorized commencement of construction, whichever is later, a sum equal to SEVENTY–FIVE (75%) PER CENT of the fee for services in Phase 2, less all previous payments made on account of fee; provided, however, that this payment shall be made only from construction loan funds.

"(c) The balance of the fee shall be paid in equal monthly payments commencing with the first day of the month following payments as set forth in Paragraph 4(b); provided, however, that TEN (10%) PER CENT of the fee based upon the reasonable estimated cost of construction shall be withheld until thirty (30) days after the Notice of Completion of the project has been filed.

"(d) If any work designed or specified by the ARCHITECT is abandoned of [sic] suspended in whole or in part, the ARCHITECT is to be paid forthwith to the extent that his services have been rendered under the preceding terms of this paragraph. Should such abandonment or suspension occur before the ARCHITECT has completed any particular phase of the work which entitles him to a partial payment as aforesaid, the ARCHITECT'S fee shall be prorated based upon the percentage of the work completed under that particular phase and shall be payable forthwith."

Invoking the provision that "payment shall be made only from construction loan funds," Bristol contends that since such funds were not obtained it is obligated to pay plaintiff no more than he has already received under the contract.

Plaintiff, on the other hand, contends that he performed 95 per cent of his work on phase two and is entitled to that portion of his fee under subdivision (d) of paragraph 4 less the previous payment he received. He contends that subdivision (d) is a "savings clause" designed to secure partial payment if, for any reason, including the lack of funds, the project was abandoned or suspended. Plaintiff would limit the construction loan condition to subdivision (b), for it provides "that *this payment* shall be made only from construction loan funds" (emphasis added), whereas the other subdivisions are not expressly so conditioned.

The construction loan condition, however, cannot reasonably be limited to subdivision (b), for subdivision (c) and (d) both refer to the terms of subdivision (b) and must therefore be interpreted with reference to those terms. Thus, the "balance of the fee" payable "in equal monthly payments" under subdivision (c) necessarily refers to the preceding subdivisions of paragraph 4. In the absence of evidence to the contrary, subdivision (d), upon which plaintiff relies, must likewise be interpreted to incorporate the construction loan condition (Civ.Code, § 1641), for it makes explicit reference to payment under preceding subdivisions by language such as "under the preceding terms" and "partial payment as aforesaid." Subdivision (d) merely provides for accelerated payment upon the happening of a contingency. It contemplates, however that construction shall have begun, for it provides for prorated payment upon the abandonment or suspension in whole or in part of "any work designed or specified by the Architect." Implicit in the scheme is the purpose to provide, after initial payments, for a series of payments from construction loan funds, with accelerated payment from such funds in the event that construction was abandoned or suspended. Although plaintiff was guaranteed an estimated 25 per cent of his fee if the project was frustrated before construction, further payment was contemplated only upon the commencement of construction. This interpretation is supported by evidence that plaintiff knew that Bristol's ability to undertake construction turned upon the availability of loan funds. Accordingly, the trial court properly determined that payments beyond an estimated 25 per cent of plaintiff's fee for phase two were to be made only from construction loan funds.

* * *

The *Bristol case* demonstrates the importance of clear drafting. Had the parties intended the ultimate result, they should have included the contingency of obtaining construction loan funds in ¶ 4(c) as well as in ¶ 4(b). Also, they should not have included ¶ 4(d).

On the other hand, had the architect not intended to take this risk, the contract language should have clearly indicated that ¶ 4(d) was to control earlier provisions that mentioned the construction loan fund. The court's analysis of the function of ¶ 4(d) seems incorrect. It is likely that its inclusion was designed to protect the architect if he continued to work without being notified that the condition of financing had not occurred.

The case also demonstrates the tendency of courts to construe language against the parties that supplied the language and to protect clients from having to pay for design services when the project is abandoned. Such client protection is based upon the understandable reluctance to force a client to pay for services that it ultimately does not use. But professionals generally expect to be paid for their work. Fee risks *can* be taken. But any conclusion that the risk was taken should be supported by strong evidence that the design professional assumed the risk.

Although the unhappy result for Parsons was that he was not paid for 70% of his work, there are indications in a deleted portion of the opinion that he might have been able to collect for his services had he shown the following:

1. That Bristol did not make reasonable efforts to obtain financing.

2. That Parsons had not been notified immediately when Freeman filed his action, so that he could have suspended work until title matters were cleared up.

3. That he would have stopped work had he been notified that the construction loan had fallen through.

Courts facing claims of finance conditions often come to difficult results depending upon the language, the surrounding circumstances, and judicial attitude toward outcomes that either deny any payment for work performed or force a party to pay for services it cannot use.[61]

(D) SUSPENSION

The power to suspend is sometimes found in construction contracts. However, AIA Doc. B141 does not grant an express power to the architect to suspend performance. Some design professionals include such a power in their contracts mainly to protect them from having to perform services for which they have not been or may not be compensated. Clearly it is advantageous to being able to point to an express provision granting power to suspend performance in the event of nonpayment rather than having to rely upon any common law right to suspend. A client who does not pay usually gives reasons to justify nonpayment. Suspension, if later found to be unjustified, can expose the design professional to a large damage claim.

AIA Doc. B141 *does* give the owner a power to order that the architect suspend his performance. Paragraph 6.4.1 states that if the project is suspended for more than three months, the architect shall be compensated for services performed before receipt of written notice of suspension along with reimbursables and termination expenses. Tacked onto expenses attributable to termination, B141, ¶ 10.4, provides for an additional liquidated amount that depends upon the stage of performance at which suspension occurs.

Paragraph 6.4.1 has three difficulties. First, it can be argued that suspension can apply only if the *project*—meaning after construction begins—is suspended. Very likely the drafters intended to allow the owner to suspend at any time, the use of the word "project" encompassing design as well as construction services.

Second, the project must remain suspended for three months before the architect is paid, despite ¶ 6.1.2 requiring payments monthly. Often this language is changed to give the architect a right to *immediate* payment in the event of suspension. True, the architect may not *wish* to demand payment if the demand may make it difficult financially to resume the project. The inclusion of a power to suspend and be paid immediately does not bar the architect from deciding not to request payment if he believes this may frustrate the possibility of resumption.

Third, the language allowing for resumption after three months seems to leave open the possibility that suspension can continue

---

61. Compare *Campisano v. Phillips*, 26 Ariz.App. 174, 547 P.2d 26 (1976) (architect assumed risk), with *Vrla v. Western Mortgage Co.*, 263 Or. 421, 502 P.2d 593 (1972) (architect recovered).

indefinitely. While the clause does give the architect the right to have compensation equitably adjusted if suspension continues over three months, it would probably be better from the architect's position and fairer as well to make clear that a protracted suspension automatically becomes a termination if the party whose performance becomes suspended elects to treat it so.

(E) ABANDONMENT

Suspension puts services on hold. Abandonment indicates that the client will no longer need the design professional's services. As seen in *Parsons v. Bristol Development Co.* in (C), a financing condition may conflict with language allowing the client to abandon but making it clear that the architect is to be paid for his services.

To make the matter more complicated, B141, ¶ 10.2, speaks of *permanent* abandonment as contrasted to ordinary abandonment set forth in ¶ 6.4.1. What is the difference? Very likely abandonment means that the *client* will no longer pursue the project but hopes to find someone else who will. Permanent abandonment means *no one* can be found willing to continue the project.

Permanent abandonment gives the client the power to terminate the contract by giving a seven-day written notice to the architect. In such cases, as in the case of an over three-month suspension, the architect is entitled to payment for services, reimbursement for expenses due, and termination expenses.

(F) TERMINATION CLAUSES

Contracts frequently contain provisions under which one or both parties can terminate their contractual obligations to perform. Termination does not necessarily—nor does it usually—extinguish any claim either party may have for the other's failure to perform.

Termination clauses vary. Some provide that one party can terminate if the other commits a *serious* breach of the contract. Some allow termination powers for *any* breach, a method intended to foreclose any inquiry into the seriousness of the breach. However, such a power can be abused. For that reason, careful judicial inquiry is likely to be made into the way in which the contract was made. If such a clause would operate unfairly, at the very least it will be interpreted against the stronger party to the contract and perhaps even be found unenforceable.

Some contracts provide that either party can terminate by giving a specified notice without any need to show that the other party has breached the contract. This codifies what the common law called "contracts at will," under which the contract regulates rights and duties of the parties only as long as both wish to continue. For a close, confidential relationship such as one created when a design professional is retained by a client, such a provision may be reasonable.

However, AIA Doc. B141, ¶ 10.2, is a "default" termination clause requiring a substantial failure to perform through *no* fault of the terminating party. As noted, the AIA allows the owner to abandon the contract by giving a seven-day written notice, something resembling construction contract provisions that allow the owner to terminate for its own convenience. AIA Doc. B141, ¶ 10.3, provides a remedy: the architect is entitled to payment for services performed before termination along with reimbursable expenses due and termination expenses.

Termination of a contract can result from a material breach without any specific power to terminate. Although the same result may not be reached in every jurisdiction, it is likely that a contractual termination clause will *not* be exclusive and will not limit any common law right to terminate.[62]

---

62. *North Harris County Junior College Dist. v. Fleetwood Construction Co.*, 604 S.W.2d 247 (Tex.Civ.App.1980 (dictum)). Cf. *Glantz Contracting Co. v. General Electric*, 379 So.2d 912 (Miss.1980).

Default termination clauses often specify that the breaching party be allowed time to cure any default. Since this opportunity to cure arises more frequently in construction contracts, it is discussed in Section 38.03.

Often termination clauses require that a written notice of termination be given. For example, AIA Doc. B141 allows termination "upon seven days' written notice." Although it is clear that termination does not actually become effective until expiration of the notice period, it is not always clear what the rights and duties of the contracting parties are during the period between receipt of notice and the effective date of termination. This may depend upon the purpose of the notice.

The notice period can serve as a cooling-off device. Termination is a serious step for both parties. If it is ultimately determined that there were insufficient grounds for termination, the terminating party has committed a serious and costly breach. A short termination period can enable the party who has terminated to obtain legal advice and to rethink its position.

If the notice period provides a cooling-off period, performance should continue during the notice period but cease when the notice period expires. Only if the termination is retracted *during* the notice period should the parties continue performance after the effective date of termination.

If the right to terminate requires a contract breach, notice can have an additional function. It may be designed to give the breaching party time to cure past defaults and provide assurances that there will be no future defaults. If cure is the function of the notice period, actual termination should occur only if the defaults are not cured by the expiration of the notice period. During the notice period, the parties should continue performance. If it appears that there is no reasonable likelihood that past defaults *can* be cured and reasonable assurances given, performance by the defaulting party should continue only at the option of the party terminating the contract. The latter

should not be forced to receive and perhaps pay for substandard performance.

Probably the principal purpose of a notice is to wind down the work to allow the parties to plan new arrangements made necessary by the termination. A short continuation period can avoid a costly shutdown of the project or the unavoidable expenses that can result if the design professional must stop performance immediately. The notice period can enable the client to obtain a successor design professional while retaining the original professional for a short period. It can also enable the design professional to make work-force adjustments, to get employees back to home base, to cancel arrangements made with third parties, and to allow time to line up work for employees.

If making adjustments is the principal reason for the notice period, each party should be able to continue performing during the notice period. However, if relations have so deteriorated that continued performance would likely mean deliberately poor performance by either or both parties during the notice period, neither should be compelled to perform during the notice period. Certainly work that cannot be finished before the effective date of termination should not be begun.

Contracting parties should decide in advance what function the notice period is to serve and what will be the rights and duties of the parties during this period. Once this determination is made, the contract should reflect the common understanding of the parties, and any standardized contracts should be modified accordingly.

(G) MATERIAL BREACH

Commission of a material (that is, serious) breach empowers the other party to terminate the contract. Whether a breach is material will depend upon a number of factors discussed in Section 38.04(A). Generally a material breach by the client is an unexcused and persistent failure to pay compensation or cooperate in creating the design.

A material breach by the design professional is likely to be negligent performance or excessive delays. As in construction contracts, termination of contracts to perform design services is relatively rare, the principal problem involving client abandonment of the project.

(H) SUBSEQUENT EVENTS

Contract law normally places the risk of performance being more difficult or expensive on the party promising performance. But sometimes events occur after the contract is made that go far beyond the assumptions of the parties at the time of contract making. If so, the law will relieve a party who is affected by these events unless the contract clearly allocates this risk to the performing party. This is more commonly a problem in construction contracts and is dealt with in Section 27.05. However, the highly personal nature of the performance of design services makes it important to discuss one problem that rarely surfaces in the performance of *construction* contracts.

Suppose a key person is no longer available to perform professional services? Is that person's continued availability so important that his inability to perform—because of either disability, death, or an employment change—will terminate the contract? Usually the issue arises if the client wishes to terminate its obligation because a key design person is no longer available. That key person can be the sole proprietor, a partner, or an important employee of a partnership or professional corporation.

Contract obligations generally continue despite the death, disability, or unavailability of persons who are expected to perform. Only in clear cases of highly personal services will performance be excused.

Unavailability of key design persons can frustrate contract expectations. For example, in the absence of a contrary contractual provision, the death of a design professional who is a party to the contract, such as a sole proprietor or partner, will terminate the obligation of each party. The personal performance of that particular design professional was very likely a fundamental assumption upon which the contract was made. A successor to the design professional can, of course, offer to continue performance, and this may be acceptable to the client. However, continuation depends upon the consent of both successor and owner. Without agreement, each party is relieved from further performance obligations.

Suppose the person expected to actually perform design services is an employee of a large partnership or professional corporation. That person's unavailability may still release each party, but it would take stronger showing that the unavailable design professional was *crucial* to the project and that his continued performance was a *fundamental assumption* upon which the contract was made.

The parties should consider the effect of the unavailability of key design personnel and include a provision that states clearly whether the contract continues if that person dies, becomes disabled, or for any other reason becomes unavailable.

AIA Doc. B141, Art. 12, binds owner and architect and their *successors* to the contract. Under this obscure language it appears that the parties contemplate successors stepping in if for some reason a contracting party, such as the architect, can no longer perform. This appears to require that the client continue dealing with the partnership if the partner with whom the client had originally dealt dies, becomes disabled, or leaves the partnership.

Continuity may be desirable. However, the close relationship required between design professional and client may mean that the client does not wish to continue using the partnership if the person in whom it had confidence and with whom it dealt is no longer available. Similarly, a successor may not *want* to work for the client. Specific language should be included dealing with this issue.

PROBLEMS

1. Client and Architect have executed AIA Doc. B141 (found in Appendix A). After Architect worked on the design for two months, Client told him that the project was suspended. Architect would prefer to be paid for what he has done and not be called upon to resume performance. He has found Client difficult to deal with and would rather work elsewhere. Architect claims that the project has been abandoned and not simply suspended. How would it be determined whether the project has been suspended or abandoned, and what would be the result of either under B141?

2. A and C entered into a written contract by which A agreed to perform designated design services for C. The contract contained the following provision: "Either party may terminate the contract by giving the other party seven days' notice in writing." What are the rights and duties of A and C if either gives the other written notice of termination?

SECTION 15.14 JUDICIAL REMEDY FOR BREACH: SPECIAL PROBLEMS OF CLIENT-DESIGN PROFESSIONAL RELATIONSHIP

(A) COVERAGE

Basic judicial remedies for contract breach were discussed in Chapter 6. Part IV in Chapter 31 discusses claims in the context of the construction contract. In this section, basic legal doctrines are applied to special problems found in the relationship between client and design professional.

(B) CLIENT CLAIMS

The principal claims that clients make against professionals relate to defective design. A breach of contract by the design professional entitles the client to protect its restitution, reliance, and expectation interests. (Refer to Section 6.05.)

Although clients occasionally seek to protect their restitutionary interest by demanding return of any payments made, the principal problem relates to the client's expectation interest. If the project is designed defectively, the client may choose or be forced to use the diminished value method of measuring its loss. (This is a measurement used more commonly in claims against contractors and is discussed in Section 31.03(D).)

It may be useful to mention *Bayuk v. Edson,*[63] in which the owner complained about faulty design consisting of, among other things, an improperly designed floor, closets too small, outside doors constructed for a milder climate than where the house was built and of an unusual type that could not be constructed by artisans in the area, unesthetic kitchen tile, sliding doors that did not fit properly in their tracks, and a fireplace that became permanently cracked.

A number of witnesses testified that it would not have made economic sense to repair the defects. One witness testified that tearing out and repairing would cost more than the cost of rebuilding the house in its entirety. The plaintiff produced an expert real estate appraiser who fixed the value of the house without the defects at $50,000 to $60,000 and with the defects at $27,500 to $31,500. The trial court awarded a judgment of $18,500, the least of the possible remainders. This was affirmed by the appellate court.

Suppose, however, that it would not have been economically wasteful to correct the defective work. This would entitle the owner to the cost of correction. In claims

---

63. 236 Cal.App.2d 309, 46 Cal.Rptr. 49 (1965).

against a design professional for improper design, application of this standard involves the Betterment rule, based upon the cost of correction sometimes unjustly enriching the owner. For example, in *St. Joseph Hospital v. Corbetta Construction Co., Inc.,*[64] the hospital sued its architect, the contractor, and the supplier of wall paneling that had been installed when it was disclosed that the wall paneling had a flame spread rating some seventeen times the maximum permitted under the Chicago Building Code.

After the hospital had been substantially completed, it was advised that it could not receive a license because of the improper wall paneling. The city threatened criminal action against the hospital for operating without a license. The hospital removed the paneling and installed paneling that met Code standards. The jury awarded $300,000 for removal of the original paneling and its replacement by Code complying paneling and an additional $20,000 for architectural services performed in connection with removal and replacement.

In reviewing the jury award of $320,000, the appellate court noted that had the architect complied with his obligation to specify wall paneling that would have met Code standards, the construction contract price for both paneling and cost of installation would have been substantially higher. The court stated that the hospital should not receive a windfall of the more expensive paneling for a contract price that assumed less expensive paneling. The paneling that *should* have been specified together with installation would have cost $186,000, while the paneling specified with installation cost $91,000. This, according to the court, should have reduced the judgment by $95,000. The court reduced the award an additional $21,000 for items that were installed when the panels were replaced that

were not called for under the original contract. As a result, the judgment was reduced some $116,000.

Another claim sometimes made by clients is unexcused delay in preparing the design or performing administrative work during construction. Delay can harm the contractor, and most delay disputes are between an owner and contractor. Claims for delay during the design phase are difficult to establish because of the likelihood of multiple causes. But suppose the project is completed 120 days late because of negligent failure by the architect to pass upon submittals of the contractor.

There are two main damage items. First, the owner will lose the use of its project for 120 days. Second, the contractor may make a claim for its delay against the owner based upon the delay wrongfully caused by the design professional. (The second claim is discussed in Section 30.10.)

The first claim by the owner against the design professional was involved in *E.C. Ernst, Inc. v. Manhattan Construction Co. of Texas*[65] The project was delayed 120 days, the court apportioning 60 days of delay to the negligence of the architect in passing upon a request to approve particular equipment for the project. Clearly the measure of recovery should have been the lost use of the project for 60 days. However, some projects do not lend themselves easily to loss-of-use measurements, particularly those not commercial. The court looked at the liquidated damages clause in the *prime* contract and concluded that it was a good indication of the damages suffered by the owner. The court also noted that the architect against whom the claim had been brought had participated in selecting the liquidated damages amount.

The contract between architect and client did *not* agree on damages in advance. But the amount was borrowed from the con-

---

64. 21 Ill.App.3d 925, 316 N.E.2d 51 (1974).

65. 387 F.Supp. 1001 (S.D.Ala.1974), affirmed 551 F.2d 1026, 559 F.2d 268 (5th Cir.1977).

struction contract. Although this may be evidence of the lost-use value, it should have been looked at *with other evidence* of lost use. Note the possible conflict of interest. Architects contemplating the possibility that the liquidated damages clause in the construction contract may be used to measure their liability for delay may be tempted to suggest that a lower amount be included in the construction contract.

Suppose the client claims its design professional exceeded his authority in ordering changes in the work or accepting defective work without authority of the client. Since this usually involves claims by the contractor as well, this topic and the measure of recovery for a valid claim are discussed in Section 25.04(C).

(C) DESIGN PROFESSIONAL CLAIMS

The principal claims made by a design professional against the client relate to the latter's failure to pay for services performed. These claims have not raised difficult valuation questions. Design professionals commonly seek to protect their restitution interests and recover the reasonable value of their services. Occasionally clients have resisted this claim by contending that they did not use the plans and specifications drafted and thereby have not been enriched. Such defenses have been generally unsuccessful.[66] (The more difficult problem—the problem of a contractor who seeks to protect its restitution interest when the owner has breached—is discussed in Section 31.02(E).)

PROBLEM

A has agreed to perform design services for C. A is to be paid 6% of the construction cost. After the professional had been working on schematic designs, the design professional repudiated the contract and stated that he would perform no further. Which of the following items of damages can the client recover:

1. A payment made by the client to the

design professional when the contract was signed.

2. The additional cost that would be incurred in hiring a new design professional.

3. Delay damages such as increased cost of obtaining a construction loan, loss of rentals from prospective tenants, and increase in construction contract costs.

---

66. *Barnes v. Lozoff*, 20 Wis.2d 644, 123 N.W.2d 543 (1963) (measured by rate of pay in community).

16

Compensation for Professional Services

SECTION 16.01 CONTRACTUAL FEE ARRANGEMENTS

(A) LIMITED ROLE OF LAW

In (B) a number of different fee arrangements are discussed. The choice among possible fee arrangements is principally guided by criteria that are professional and not legal. Put another way, design professionals are generally in a better position to determine the type of fee structure than are their attorneys. Nevertheless, the law does play a limited role.

Principally the law interprets any contractual terms that bear upon fee computation when the contracting parties disagree. If no fee arrangement is specified in the contract and the parties cannot determine an agreed-upon fee subsequent to performance, the law may be called upon to make this decision.

Despite this limited role, certain legal principles must be taken into account in choosing a fee arrangement or in predicting the legal result if a dispute arises. For example, faced with the question of whether certain services come within the basic design fee, the law may choose to protect the reasonable expectation of the *client* if the design professional selected the language. Likewise, any fee arrangement that measures compensation by a stated percentage of construction cost may be interpreted to favor the client. This can result from the belief that such a fee formula can be unfair to the client, given the design professional's incentive to run up costs. In rare cases, the contractual method selected will be disregarded because supervening events occur that neither party contemplated.

(B) STATED PERCENTAGE OF CONSTRUCTION COSTS

Though no longer universal, the stated percentage of construction costs is still the most common method of fee computation. In such a method, the fee is determined by multiplying the construction costs by a designated percentage set forth in the contract. Since this formula typically covers only basic design services, this amount is often augmented by payments for additional services and reimbursable expenses. (But see (I) on fee limits.)

This method has been criticized. It can be a disincentive to cut costs and may reward the design professional who is less cost conscious. It can be too rigid, as projects and time spent can vary considerably. It may not reflect time spent. It also tends to subsidize the inefficient client at the expense of the efficient one.

Despite constant criticism, it is still used. Clients seem accustomed to it. It can avoid bargaining over fee. In *normal* projects it may be an accurate reflection of the work performed. Although it may undercompensate on some projects and overcompensate on others, some design professionals feel that these average out. It can avoid extensive record keeping. It also is much less likely to generate a client demand to examine the design professional's records, a not uncommon feature of a cost type of fee formula.

Percentages vary, with the figure selected in major part reflecting the amount of work the design professional must perform and, increasingly these days, the risk of failure. The latter is addressed in greater detail in Chapter 17. The former depends upon a number of factors.

If the design professional has worked for the client before, past experience may be relevant in determining the stated percentage. A client who is inexperienced and inefficient may require more work than an efficient client who has dealt with construction before. Sometimes the percentage is based upon whether the project is residential or commercial, whether the construction contracts are single or separate, and whether the construction contract price is fixed or a cost type. Smaller projects may have a minimum fee.

To avoid criticism that the fee method encourages high costs and discourages cost reduction, some design professionals use a flexible percentage. One method is a percentage that *declines* as the costs increase. The actual percentage for the entire project is determined from a schedule that has variable percentages depending upon the ultimate construction cost. For example, a fee schedule may provide for a 5% fee if the costs do not exceed one million dollars, a fee of 4% if the ultimate cost is between one million dollars and $1.5 million and 3% if the cost is over $1.5 million.

Another method to encourage cost con-

sciousness is to employ a sliding scale under which the highest percentage is applied to a cost up to a specified amount and then the percentage reduces on succeeding amounts. For example, the fee can be 8% on the first million dollars of cost, 7% on the next $4 million, and 6% on all amounts over $5 million.

As to the construction cost multiplied by the percentage, AIA states:

> 3.1.1 The Construction Cost shall be the total cost or estimated cost to the Owner of all Work elements of the Project designed or specified by the Architect.
>
> 3.1.2 The Construction Cost shall include at current market rates, including a reasonable allowance for overhead and profit, the cost of labor and materials furnished by the Owner and any equipment which has been designed, specified, selected or specially provided for by the Architect.
>
> 3.1.3 Construction Cost does not include the compensation of the Architect and the Architect's consultants, the cost of the land, rights-of-way, or other costs which are the responsibility of the Owner . . .[1]

(For equitable adjustments see (J).)

Courts have interpreted fee provisions, but because of the different provisions that can be or are employed, generalizations are perilous. In close cases, courts are likely to favor the position of the client if the design professional selected the contract language. But the client does not always succeed. For example, in *Orput-Orput & Associates, Inc. v. McCarthy*,[2] the court held the construction cost to be the construction contract price and not the amount the client ultimately paid the contractor after a renegotiation. *Simonson v. "U" District Office Building Corp.*[3] held that work done for tenants after completion and after final settlement was made was included in construction cost. Suppose the actual owner

---

1. AIA Doc. B141 (1977).

2. 12 Ill.App.3d 88, 298 N.E.2d 225 (1973).

3. 70 Wash.2d 35, 422 P.2d 1 (1966).

payout to the contractor exceeds the original contract price because of defective design for which the design professional was responsible. Although this is likely to mean increased design service, this increase should not enlarge the design professional's fee.

Suppose responsibility is *shared* by the design professional and the contractor based upon the contractor's *not* having directed attention to obvious design defects. An apportionment of responsibility may be appropriate, with the design professional's fee increased by that portion of the corrected work cost chargeable to the contractor. However, the difficulty of making such allocation and primary responsibility for design being that of the design professional will likely preclude such an apportionment. Yet the major issue is likely to be who bears responsibility for the cost of correction and other losses. If an apportionment has to be made for this purpose, there seems to be no reason why that apportionment cannot be used to determine the design professional's fee.

Techniques for imparting flexibility to what otherwise can be a rigid fee method are mentioned in (J).

(C) MULTIPLE OF DIRECT PERSONNEL EXPENSE: DAILY OR HOURLY RATES

Personnel multipliers determine the fee for basic and additional services by multiplying direct personnel expense by a designated multiple ranging from one to three. The multiple gives the design professional administrative overhead and profits. Article 4 of AIA Doc. B141 defines direct personnel expense as:

> the direct salaries of all the Architect's personnel engaged on the Project, and the portion of the cost of their mandatory and customary contributions and benefits related thereto, such as employment taxes and other statutory employee benefits, insurance, sick leave, holidays, vacations, pensions, and similar contributions and benefits.

Once fringe benefits were truly on the "fringe." Today, they can constitute as much as 25% to 40% of the total employee cost. The contract must *clearly* specify personnel compensation cost beyond the actual salaries or wages.

There are obvious disadvantages to daily or hourly rates. Because a day is a more imprecise measurement than an hour, to the extent either of these methods is used it is likely to be the hourly rate. Such a method requires detailed cost records that set forth the following:

1. the exact amount of time spent
2. the precise project upon which the work was performed
3. the exact nature of the work
4. who did the work

Differential hourly rates may be used for work by personnel of differential skills.

The AIA has recognized the importance of accounting records in B141, Art. 7, which requires that the architect keep records based on generally accepted accounting principles and make them available to the owner. Some clients may prefer a more detailed provision stating with greater particularity the types of records that will be kept, enlarging the right to inspect and make copies, and specifying how long records must be kept.

(D) PROFESSIONAL FEE PLUS EXPENSES

This form of fee arrangement is analogous to the cost type of contracts discussed in greater detail in Section 21.02(B). One advantage of a cost type of contract for *design* services is that the compensation is not tied to actual construction costs and there should be incentive to reduce *construction* costs. It can be a disincentive, however, to reduce the cost of *design* services.

The cost type of contract, or what the AIA calls "professional fee plus expenses," necessitates careful definition of recoverable costs. Costs, direct and indirect, can be an accounting nightmare. Disputes can arise over whether certain costs were excessive or necessary. In cost contracts advance client approval can be required on the size of the design professional staff, salaries, and other important cost factors. Cost contracts require detailed record keeping. (As seen in (I), a ceiling can be placed on costs.)

Suppose the design professional estimates what the costs are likely to be in such a contract. While the client may wish to know approximately how much the design services are likely to cost, an estimate can easily become a cost ceiling.[4] If an estimate is given—*not* intended as a ceiling—it should be accompanied by language that indicates the assumptions upon which the estimate is based and that it is not a fixed ceiling or a promise that design costs will not exceed a designated amount.

(E) FIXED FEE

Design professional and client can agree that compensation will be a fixed fee determined in advance and incorporated in the contract. Before such a method is employed, a design professional should have a clear idea of direct cost, overhead, and profit, as well as appreciate the possibility that contingencies may arise that will affect performance costs. A fixed fee should be used only where the scope of design services is clearly defined and the construction project well-planned. It works best in repetitive work for the same client.

Does the fixed fee cover only *basic* design services? Does it include additional services and reimbursables? Standard contracts published by professional associations usually limit the fixed fee to *basic* design services. A design professional who intends to limit fixed fees to basic services should make this clear to the client. See (I)

for discussion of fee provisions that place an absolute ceiling on compensation.

(F) REASONABLE VALUE OF SERVICES OR A FEE TO BE AGREED UPON

The fee will be the reasonable value of the services where the parties do not agree on a compensation method. If there is no agreed valuation method for *additional* services, compensation is the reasonable value of the services. The reasonable value of a design professional's services will take into account the nature of the work, the degree of risk to the design professional, the novelty of the work, the hours performed, the experience and training of the design professional, and any other factors that bear upon the value of these services, including overhead and a reasonable profit. Proving the reasonable value of services requires detailed cost records. Leaving the fee open is generally inadvisable.

Where this issue does arise, each party usually introduces evidence of customary charges made by other design professionals in the locality as well as evidence that bears upon factors outlined in the preceding paragraph. In many cases, there is a great variation between the testimony of the expert witnesses for each party, and it is not unusual for the court or jury to make a determination that falls somewhere in between.

The parties can agree to jointly determine the fee at the completion of performance. Where the project has gone well, where the parties wish to work with each other again, and where adequate records have been kept, agreement on fees may be reached easily. However, when such fortunate events have not occurred, an agreement on fees can be difficult.

At one time, such agreements were considered unenforceable as simply "agreements to agree."[5] However, if the work

---

4. *Ballinger v. Howell Manufacturing Co.*, 407 Pa. 319, 180 A.2d 555 (1962).

5. Refer to Section 5.06(F) for further discussion.

has been performed, it is likely that the parties must negotiate in good faith to determine a fee or, more likely, to determine what the parties would have agreed to had they bargained in good faith and made an agreement on compensation. The same result can follow if the parties cannot agree and the matter is submitted to arbitration.

(G) ADDITIONAL SERVICES

Whether certain services fall within the basic design fee or are additional services was discussed earlier,[6] with emphasis upon whether particular services were part of *basic* design services or were *additional.* This section treats the following:

1. methods of authorizing additional services

2. compensation for such services

The first depends upon the agreement between design professional and client. In the *absence* of any specific method of authorization designated in the contract, the client must request that these services be performed, and the design professional must perform or agree to perform them or have them performed by a consultant.

Standardized contracts frequently contain provisions that not only define additional services but also specify procedures for their authorization. For example, ¶ 1.7 of AIA Doc. B141 requires services be performed "if authorized or confirmed in writing by the Owner." Confirmation refers to an earlier oral directive to perform the additional services.

Fees to consultants can constitute significant additional services or reimbursables. It may be useful to examine the evolution of AIA standard contracts dealing with consultant fees.

Before 1961 the AIA published different standard contracts that allowed the parties to select either an arrangement where con-

sultant fees were included in the architect's fee or one where such fees were considered reimbursables. For example, in 1958 the AIA published AIA Doc. B101, which required the owner to reimburse the architect for normal consulting services, while AIA Doc. B121 included such consulting fees in the architect's compensation. In 1961 the Institute published AIA Doc. B131, which included normal engineering services as part of the basic fee and not reimbursable to the architect.

In 1967, B131 was amplified. *Normal* structural, mechanical, and electrical services were *not* considered additional and *were* part of the basic design services. But ¶ 5.1.3 stated that fees of *special* consultants when authorized in advance by the owner were reimbursable expenses. In 1970 B131 moved other than normal consultant fees from Reimbursables in Art. 5 to Additional Services under Art. 1. Paragraph 1.3 dealing with Additional Services required authorization by the owner. In 1974 *written* authorization by the owner was required which was extended to include written confirmation in 1977.

The requirement of written authorization for the use of special consultants may have the laudable effect of channeling the sometimes casual procedures employed by design professionals into more regular forms. This requirement also increases the likelihood that legal disputes will arise where the prescribed formalities are not followed. If the prescribed formalities are followed casually or not at all, it is likely that formal requirements have been abandoned. Design professionals should be aware of the formal requirements and follow them.

Paragraph 14.4.2 of AIA Doc. B141 compensates additional consultant services by a designated multiplier "times the amounts billed to the architect for such services." While such a formula avoids or should avoid fee disputes, it can be abused. Other

---

6. See Section 15.01.

additional services are usually compensated by a multiplier of personnel expense.

(H) REIMBURSABLES

Illustrations of reimbursables are as follows:

1. Transportation and living expenses in connection with out-of-town travel.

2. Long-distance communications.

3. Fees paid to secure approval of authorities having jurisdiction.

4. Reproductions, postage, and handling of construction documents.

5. Data processing.

6. Overtime work.

7. Renderings, models, and mockups when requested by owner.

For the current AIA list of reimbursables, see AIA Doc. B141, Art. 5 reproduced in Appendix A.

Incurring obligations for the client and paying for them can impose an administrative burden on the design professional. Sometimes design professionals charge the client a markup for handling reimbursables. For example suppose the design professional incurred expenses of $1,000 for traveling in connection with the project and long-distance calls. Under a markup system, the design professional might bill the client $1,000 plus an additional 10% or $100, making a total of $1,100. The markup percentage can depend upon the number of reimbursables and the administrative overhead incurred in handling them. A design professional who wishes to add an overhead markup should explain this to the client in advance and obtain client approval.

(I) FEE CEILINGS

Owners are often concerned about the total fee, particularly if the fee is cost-based. But even in a compensation plan under which the design professional is paid a fixed fee or a percentage of compensation, the client may be concerned about additional services, reimbursables, or increases based upon an unusual jump in the *construction* cost. Public owners with a specified appropriation for design services may seek to limit the fee to a specified amount. It may be useful to look at a few cases that have dealt with fee limits.

Two New York cases, both involving public entities, are instructive. *Meathe v. State University Construction Fund*[7] held that the fee limit was ambiguous, and testimony of those who negotiated the contract was admitted to determine whether the "upset" price included additional services.

Hueber Hares Glavin Partnership v. State[8] was more complex. The contract limited the fee. It also *excluded* recovery for work to correct design errors. The trial court held the contract severable and that the city could not offset damages for design errors when the architect sued for compensation for design services, an unfortunate use of the divisibility fiction.

The Appellate Division correctly found severability irrelevant because of the fee limit. It held the language unambiguous and the fee limit not an estimate. It could not be exceeded by costs attributed to design errors. By dictum the court stated that the city would have been precluded from asserting the fee limit had it ordered extra services knowing the fee limit had been reached.

Harris County v. Howard[9] involved an AIA Document. The upset price was also held to unambiguously include additional services and reimbursables. The public owner inserted a detailed recital on the fee limit and what it included. The Court re-

---

7. 65 A.D.2d 49, 410 N.Y.S.2d 702 (1978) (3 to 2).

8. 75 A.D.2d 464, 429 N.Y.S.2d 956 (1980).

9. 494 S.W.2d 250 (Tex.Civ.App.1973).

jected the architect's contention that his having been paid $20,000 *over* the limit showed the limit did *not* include additional services or reimbursables.

Substantial changes in project scope should eliminate any fee ceiling that may have been established.[10]

(J) ADJUSTMENT OF FEE

Generally the law places the risk that performance will cost more than planned on the party who has promised to perform. Careful planners with strong bargaining power build a contingency into their contract price that takes this risk into account. Suppose a client directs significant and frequent changes in the design. Suppose for *any* reason the design services must be performed over a substantially longer period than planned. It is unlikely that the law will give the design professional a price adjustment under a fixed-price contract.

However, the AIA has sought to protect architects from these risks in AIA Doc. B141. Most important, § 14.7 grants an equitable adjustment in compensation if the scope of the project or the architect's services are materially changed or the design services have not been completed within a designated number of months from the date the contract is signed. Under ¶ 6.1.3, if the *construction* contract period is extended through no fault of the architect, services performed during this extended period are "additional." Finally, ¶ 6.4.1, as discussed earlier,[11] gives the architect an equitable adjustment if a project is resumed after a three-month suspension.

(K) DEDUCTIONS FROM THE FEE: DEDUCTIVE CHANGES

Acts of the design professional may cause the client to incur expense or liability, and the client may wish to deduct expenses in-curred or likely to be incurred from the fee to be paid to the design professional. Suppose a design professional commits design errors that cause a claim to be made by an adjacent landowner or by the contractor against the client. Suppose the client settles any claims or wishes to deduct an amount to reimburse itself in the event it must pay the claims.

The right to take deductions or offsets in such cases can be created either by the contract or by law. An illustration of the first is the frequent inclusion of provisions in construction contracts that give the owner the right to make deductions and offsets against the contractor. This is usually *not* found in contracts between design professionals and clients. However, even in the absence of such provisions, the client may be able to take deductions for expenses incurred or likely to be incurred by the client as a result of any contractual breach by the design professional. The amount deducted must not be disproportionate to the actual or potential liability of the client.

It is likely that the construction contract price will be increased or decreased during actual performance through issuance of change orders. AIA Doc. B141, ¶ 6.3.1, states that no payments will be withheld from the architect "on account of the cost of changes" unless the architect is "legally" responsible. Presumably this would encompass only *judicially* determined negligent design. Deductive changes, though they may reduce the construction contract price, do not affect the architect's fee even if the fee is based upon a stated percentage of construction costs. Deductive change orders are not only *not* likely to reduce the extent of the architect's services but may *increase* it.

In addition, ¶ 6.3.1 states that there will be no reduction from the architect's compensation if the total contractor payout is

---

10. *Herbert Shaffer Associates, Inc. v. First Bank of Oak Park,* 30 Ill.App.3d 647, 332 N.E.2d 703 (1975).

11. Refer to Section 15.13(D).

reduced because of deductions for liquidated damages or other sums withheld from the contractor, such as deductions for damages suffered because of the contractor's breach. Such deductions do not reduce the contract price; they only reimburse the client for its losses when it does not receive performance it has been promised by the contractor.

(L) THE FEE AS A LIMITATION OF LIABILITY

Looking ahead to professional liability, the fee can serve another function. Some design professionals seek to limit or actually limit their liability exposure to their client to the amount of their fee. This is discussed ahead.[12]

SECTION 16.02 TIME FOR PAYMENT

(A) SERVICE CONTRACTS AND THE RIGHT TO BE PAID AS ONE PERFORMS

In service contracts, the promises exchanged are the payment of money for the performance of services. Unless such contracts specifically deal with this question, the performance of all services must precede the payment of any money. Put another way, the promise to pay compensation is conditioned upon the services being performed.

Such a rule operates adversely for the person performing services. First, if the performance of services spans a lengthy time period, the party performing these services may need a source of financing to perform. Second, the greater the performance without being paid, the greater the risk of being unpaid. For these reasons, the law protects manufacturers and sellers of goods by giving them the right to payment as installments are delivered. However, this

protection was not accorded persons performing services. If the hardships and risks described are to be avoided, contracts for professional services must contain provisions giving design professionals the right to be paid as they perform.

State statutes generally provide that employees are to be paid at designated periodic intervals. However, such statutes do not protect those that perform design services who are not employees of the owner.

(B) INTERIM FEE PAYMENTS

Design professionals commonly include contract clauses giving them the right to interim fee payments. This avoids the problems described in the preceding paragraphs. From the client's standpoint, interim fee payments can ease the financial burden by providing for installment payments. Interim fee payments can be an incentive for the design professional to begin and continue working on the project.

Usually interim payments in design professional contracts become due as certain defined portions of the work are completed. While the 1977 edition of AIA Doc. B141 leaves blanks for interim fee payments, the AIA suggests interim fee payments as follows:

Schematic Design Phase	15%
Design Development Phase	35%
Construction Documents Phase	75%
Bidding or Negotiation Phase	80%
Construction Phase	100%

Any schedule used should depend upon the breakdown of professional services and the predicted work involved in each phase.

Dividing the design services and allocating a designated percentage of the fee to each service can make it appear that the contract is *divisible*. A divisible contract matches specified phases of the work to

---

12. See Section 18.03(D).

specified compensation or a specified portion of the total compensation. In a truly divisible contract, the amount designated is earned at the completion of each phase and the value of the work for each completed phase cannot be revalued.

As an illustration, suppose the contract were considered divisible and the design professional unjustifiably discharged after completion of the construction documents phase. In such a case, the design professional will recover 75% of the fee if this were the amount specified even if the reasonable value of services exceeded this amount. Conversely, the client would not be permitted to show that the reasonable value of the services was less than 75% of the fee. Interim fee payments provisions should *not* make these contracts divisible. The amounts chosen are usually rough approximations and not agreed final valuations for each phase. This is especially true if the standard form contract specifies the phases and allocates a percentage of the fee for each phase. Such payment should be considered only provisional.[13]

(C) MONTHLY BILLINGS

In many projects, months may elapse before a particular phase is completed. To avoid overly long periods between payments, contracts for such projects should provide for monthly billings within the designated phases.[14]

(D) LATE PAYMENTS

Financing costs are an increasingly important part of performing design professional services. Late payments and reduced cash flow can compel design professionals to borrow in order to meet payrolls and pay expenses. The contract should provide that a specified rate tied to the actual cost of money be paid on delayed payments.[15]

In the absence of a rate or formula, interest is the legal rate. This is the amount payable on court judgments. In inflationary periods, this is substantially lower than the market rate for borrowing money.

(E) SUGGESTIONS REGARDING INTERIM FEE PAYMENTS

When clients delay payments, the design professional should make a polite and sometimes strong suggestion that payments should be made when due. If there is a pattern of delayed payments, the design professional should seriously consider suspending performance until the payments are made. Despite the current failure of AIA to grant an *express* power of suspension, the power to suspend exists under the common law.[16] If the suspension continues for a substantial time period, the design professional should consider terminating the contract.

In cases of suspension or termination of performance, it is desirable to notify the client of an intention to either suspend or terminate unless payment is received within a specified period of time. This gives the client an opportunity to make the payment. It also shows the client that failure to make interim fee payments as promised will not be tolerated.

SECTION 16.03 TO WHOM PAYMENTS ARE TO BE MADE

Can the design professional assign the right

---

13. *Herbert Shaffer Associates, Inc. v. First Bank of Oak Park,* supra note 10. But see *May v. Morganelli-Heumann & Associates,* 618 F.2d 1363 (9th Cir.1980) (architect contract divisible).

14. See AIA Doc. B141, ¶ 6.1.2.

15. Paragraph 14.6 of AIA Doc. B141 now provides blank spaces to be filled in for the rate of interest and when payable. Before 1977, ¶ 6.5 specified the "legal rate" payable sixty days from date of billing.

16. Restatement (Second) of Contracts § 237 comment a (1981).

to receive compensation for services rendered for the client? Suppose the design professional wishes to transfer her right to receive compensation to a lender as security for a loan or to placate an impatient creditor.

While early contract law set up restrictions that hampered the assignment of rights to receive money, modern law permits and even encourages such transfers. Permitting assignment enables parties to cash in contract-created rights. The person obligated to pay the money (the obligor) is not unduly burdened if it must pay to the person to whom the right has been transferred (the assignee) rather than the other party to the contract who has transferred the right (the assignor).

Unless special statutes compel a different result, the assignability of contract rights can be prohibited by contract provisions. Contracts commonly contain provisions stating that rights under the contract are not assignable without consent of the party who would have to perform the obligation. Courts have interpreted such nonassignability clauses narrowly. Unless it were quite clear that the obligation to pay money cannot be transferred, general nonassignability clauses prohibit substituted performance but not the payment of money. A carefully drafted clause precludes assignability of money payment rights except where the assignment is made as security for a debt.

Most standard agreements used by design professionals include provisions requiring that consent be given to any assignments. Such provisions are not enforceable where the right assigned is the payment of money and the right is assigned as security for a loan.[17]

SECTION 16.04 PAYMENT DESPITE NONPERFORMANCE

Denying a contracting party recovery for services performed unless it has performed *all* the obligations under the contract can create forfeiture (expenditure by performing party substantially exceeding harm caused by the breach) or unjust enrichment (the other party retaining and using the performance without paying for it). Legal doctrines developed that minimized the likelihood of forfeiture and unjust enrichment.

Where it is not clear from the contract whether *exact* performance is a promise or a condition, the law is likely to classify the nonperformance as a breach creating a right to recover damages rather than a failure of a condition that bars recovery for work performed. The party who has breached can recover if it has *substantially* performed. Failure to *fully* perform does not bar recovery if caused by the other party's prevention or hindrance of performance or if the latter has failed to extend reasonable cooperation necessary to performance.[18] Sometimes the party to whom performance is due may have waived its right to performance by indicating that it was satisfied with less than exact performance.

Repudiation of the contract, denying its validity, or communicating an unwillingness or inability to perform excuses full performance. It would make little sense for the party to complete performance when the other party has indicated it does not wish performance and will not accept it.

Increasingly the law is recognizing the right of a party in default to recover despite failure to perform under the contract.

---

17. U.C.C. § 9–318(4). See *Mississippi Bank v. Nickles & Wells Construction Co.,* 421 So.2d 1056 (Miss.1982) (AIA Doc.); *Aetna Casualty & Surety Co. v. Bedford-Stuyvesant Restoration Construction Corp.,* 90 A.D.2d 474, 455 N.Y.S. 2d 265 (1982).

18. *Carroll Fiscal Court v. McClorey,* 455 S.W.2d 547 (Ky.1970) (architect recovered despite abandonment of project because public entity did not make good-faith effort to obtain matching funds).

These problems arise more commonly in construction contracts, and the legal doctrines have been developed with those contracts in mind. For this reason, a detailed discussion of these doctrines is postponed until Section 26.06.

SECTION 16.05
OTHER CLIENT OBLIGATIONS

AIA Doc. B141, Art. 2, lists other client obligations, including specialized services, information, surveys, and reports. A client engaging in a construction project for the first time may be surprised to discover it must furnish under § 2.5 services of a soil engineer and other subsurface information and under ¶ 2.6 tests, inspections, and reports required by law or the contract documents. The client may be even more surprised to find that under ¶ 2.9 it warrants accuracy of information it has furnished the design professional, a standard that the client will not ordinarily be able to demand from the party from whom it has purchased the information. Finally, ¶ 2.2 requires the client to comply with any request by the architect for a statement of funds available for the project and their source. Although the client's assent to the contract theoretically binds it to all its terms, it is advisable to go over such provisions with a client in advance, particularly one not experienced in construction. (Strong owners often change these provisions).

In addition to express provisions, obligations can be implied into contracts. The client impliedly promises not to interfere with the design professional's performance and to cooperate. For example, the client should not refuse the design professional access to information that is necessary for the performance of the work. Refusal to permit the design professional to inspect the site would be prevention and a breach of the implied obligation owed by the client to the design professional.

Positive duties are owed by the client. The client should exercise good faith and expedition in passing upon the work of the design professional and in approving work at the various stages of the latter's performance. It should request bids from a reasonable number of contractors and should use best efforts to obtain a competent bidder who will agree to do the work at the best possible price. If conditions exist that will require acts of the client, such as obtaining a variance, or obtaining financing, the client impliedly promises to use best efforts to cause the condition to occur.[19]

Although the legal issue was defamation and not implication of terms, *Sharratt v. Housing Innovations, Inc.,*[20] raised an interesting problem. A brochure advertising a housing project incorrectly listed the *associate* architect as the *principal* architect. The principal architect brought legal action against the publisher claiming that he had been defamed because he had informed others in the design profession and construction community that *he* had been named as principal architect. The court held that the architect had been defamed and could recover damages from the publisher of the brochure inasmuch as the publication, though not on its face defamatory, injured him when surrounding facts and circumstances were taken into account.

Suppose the client carelessly or by design made an incorrect announcement similar to the brochure in the *Sharratt* case. Does the client imply that it will make a correct announcement of the engagement of the architect? If an inaccurate announcement is made, can any damages be cured by a retraction and correction? If the mistake cannot be or is not corrected, what would be recoverable damages for such a breach?

---

19. Ibid.

20. 365 Mass. 141, 310 N.E.2d 343 (1974).

The answers will depend upon the culpability of the client and the likelihood that actual damages were suffered. Clearly if this were done in bad faith with an intention to injure the architect, the architect has a valid claim. If the client were careless, the claim should be recognized, but careful inquiry should be made into the type of damages for which recovery should be granted. If the mistake occurred without any fault on the part of the client, it is likely that any claim would be denied. To sum up, the client impliedly promises that it will not be careless or act in bad faith in making such an announcement.

Terms are usually not implied if the contract is directly at variance with the implied term claimed. It is less likely to be implied if the subject of the implication is dealt with in the contract, especially if the contract has been negotiated by parties of relatively equal bargaining strength.

PROBLEMS

1. A contracted to perform design services for the construction of an office building which was to have a contract price of one million dollars. A was to be paid 6% of the construction cost of the project. The construction contract contained a provision for liquidated damages under which the contractor was charged $500 for each day of unexcused delay. After performance was completed, it was determined that there were fifty days of unexcused delay. The owner deducted $25,000 from the final balance owed the contractor. Use AIA Doc. B141 in Appendix A to determine the amount of A's fee.

2. You are about to receive your first commission. Your client has asked you about methods of compensation and the method you would prefer. What would be your answer? Why? What additional facts would be useful?

17

Professional Liability: Process or Product?

SECTION 17.01
CLAIMS AGAINST
DESIGN PROFESSIONALS:
ON THE INCREASE

Undeniably the possibility that a design professional will find himself in court has increased dramatically. One out of every three practicing architects is likely to find himself in litigation. Sometimes litigation is necessitated by the client's failure to pay fees for professional services. This chapter concentrates upon claims made by clients and others that they have suffered losses that should be transferred to the design professional.

Many reasons exist for increased claims against design professionals. This section explores them.

(A) CHANGES IN SUBSTANTIVE LAW

Some defenses that had proved useful when claims were made by parties other than the client (collectively referred to as third parties) have proved of diminished value or have largely disappeared. For example, the requirement of privity between claimant and design professional and acceptance of the project terminating liability of the design professional—both valuable in avoiding third-party claims—have proved *much* less effective. The requirement that expert

testimony be introduced to establish that a professional has not lived up to the standards of his profession has loosened. These substantive changes have resulted from increasing emphasis upon insuring that victims receive compensation rather than protecting socially useful activities.

(B) PROCEDURAL CHANGES

With little difficulty and minimal costs, the claimant can bring legal action against a number of defendants in the same lawsuit. Similarly, those against whom claims have been brought can bring other defendants into that lawsuit with relative ease and without much expense. What has resulted is a complicated lawsuit with a host of parties defending and asserting claims.

Also, statutes of limitations designed to protect defendants from stale claims based upon activities many years before the claims have provided much less protection.

(C) ABILITY OF DESIGN
PROFESSIONALS TO
PAY COURT JUDGMENTS

Usually claimants do not assert legal action against persons who they believe will be unable to pay for a court judgment or are not insured. Although expansion of liability and increased cost of insurance premiums have begun to reduce the percentage of design professionals who carry insurance, most will have professional liability insurance or sufficient resources to respond to court judgments. As a result, claims increase.

(D) ACCESS TO LEGAL SYSTEM

Usually a person who seeks relief through the legal system engages a lawyer. Easier access to legal services will mean more claims and litigation.

Increased accessibility to lawyers began with giving those charged with major crimes a lawyer even if they cannot afford to hire one. Programs were later developed to give legal representation to the poor when they sought to use the legal system. Along with this was increased emphasis upon informing persons of their legal rights. For example, construction trade unions routinely inform their members of their legal rights and encourage them to use the legal process. Increasingly members are provided legal services through prepaid legal insurance plans.

Another reason for easier access to the legal system is the much-maligned contingent fee contract (discussed in Section 2.04). Under such a contract, the client pays the lawyer for his time only out of any recovery obtained, and the client's investment is generally limited to expenses (in injury cases often paid by the attorney).

Another reason for more claims is that the prevailing party does *not* recover *its* costs of defense including attorneys' fees unless it can point to a contract providing for such recovery or to a law granting attorneys' fees to the prevailing party. This can encourage legal claims by insuring the claimant that it does not run the risk of having to pay the *other* party's legal expenses if the claim fails.

(E) SOCIETAL CHANGES

Americans today are less willing to accept their grievances silently. They are more inclined to use the legal system if they feel they have a grievance. The high ratio of lawyers to population in the United States, perhaps the highest in the world, demonstrates this.

American society has become increasingly urban and impersonal. There is much less likelihood that disputants are part of a cohesive social unit. Those units usually provide informal mechanisms for adjustment of rights and discourage resort to outside processes.

A sense of alienation in a large impersonal society makes people feel powerless with no one to protect them or help them. This was undoubtedly one element in the rise of consumerism in the 1960s. Aggressive use of the legal system responded to this phenomenon.

(F) ENTERPRISE LIABILITY: CONSUMERISM

Compensating victims rather than protecting enterprises has been the dominant modern tort motif. This has led to liability rules based upon the belief that it is better that victims recover from the *enterprise* that is in the best position to avoid or spread the losses to those who benefit from the enterprise. Much of this drew unconsciously from the emphasis on security, which became a dominant objective after World War II. Although much was accomplished through social welfare legislation, such as unemployment insurance, public housing, public welfare, and job security, emphasizing compensating victims through expansion of tort rights was a useful adjunct.

Liability expansion is also traceable to the consumer movement of the 1960s. Those who felt consumers were being supplied shoddy goods and services advocated increased liability as a means of bringing home to the business sector the importance of dealing fairly with consumers.

(G) DESIGN CENTRALITY

Increased awareness exists of the centrality of design as a regulator of human conduct and allocator of societal resources. For example, California sought to reduce water consumption by legislation limiting the flushing capacity of toilets.[1] Public officials have stated that much of the responsibility for avoiding fires or minimizing fire losses falls upon those who design structures. A fire chief stated that "good fire protection in high-rise buildings begins on the architect's drawing board." He also stated that it was the builder's responsibility to design "a safe building, not a firetrap."[2]

Law enforcement officials have stated that assaults can be reduced if those who design business and residential areas plan properly. A director of the National Institute of Law Enforcement and Criminal Justice stated:

> Better environmental design can do much more [to reduce crime]. New housing projects, schools, shopping centers and other areas can be designed, for example, with more windows looking out on streets, fewer hidden corridors, and other crime discouraging features.[3]

Two recent cases, though not involving design professionals, demonstrate the centrality of design to human activities. One found a landlord liable when a tenant was robbed because the landlord did not fix a deadbolt lock on the apartment door.[4] The other involved a claim against a bar owner by the survivors of a patron who had been murdered in the bar by a robber because the bar owner had violated a building code that required that egress doors be operable from the inside.[5] It is not inconceivable that the design professional who designs an apartment house without adequate locking facilities or one who remodels a bar with-

out complying with local codes as to egress would be liable as well.

(H) CODES

Liability has also expanded because of a proliferation of detailed building and housing codes. Violation of these codes, while not conclusive on the question of negligence, makes it relatively easy to establish that the design professional is not only liable to the client but to third parties who suffer foreseeable harm because of the violation.

(I) EXPANSION OF PROFESSIONAL SERVICES

Design professionals are expected to either provide or volunteer to provide services that go beyond basic design. These services can relate to availability of funds for the project, the likely profitability of the project, and the likelihood of approval by public officials and agencies who control building and construction. This can increase claims when clients suffer disappointments or losses.

(J) SITE SERVICES

A design professional's varied site services discussed in Chapter 15 and in Part IV make him more vulnerable to a claim traceable not only to design but to the way in which the design was executed, both as to compliance with the design and the methods of accomplishing it.

SECTION 17.02 OVERVIEW OF CHAPTERS 17 AND 18

The preceding section outlines reasons why

---

1. West's Ann. Cal. Health & Safety Code § 17921.3.

2. *New York Times*, August 4, 1974, p. 11.

3. *New York Times*, August 4, 1974, p. 11. See also O. Newman, *Defensible Space: Crime Prevention Through Urban Design* (Collier Books, 1973).

4. *Braitman v. Overlook Terrace Corp.*, 132 N.J.Super. 51, 332 A.2d 212 (1974).

5. *Elliott v. Michael James, Inc.*, 507 F.2d 1179 (D.C.Cir.1974).

professional liability has expanded and the likelihood that more claims will be made against design professionals. This chapter deals with professional liability claims mainly by clients but also by third parties. Chapter 18 discusses techniques to avoid or reduce liability risks through risk management.

(A) APPLICABLE LAW

Professional liability is usually determined by state law. Depending upon various factors, professional liability rules will be those of the state in which the design professional has its principal place of business, where the actual design is created, or where the project is located. Because both legal rules and actual outcomes may vary depending upon the applicable state law, discussion of liability must be general and not directed to a particular state except to the extent that illustrative cases may do so.

(B) TYPES OF HARM

It is useful to divide claims into those that involve personal harm, those that involve harm to property, and those that involve economic loss not connected to personal harm or property damage. As a general rule, the law is more likely to protect claims based upon personal harm than claims based upon harm to property or economic loss and to prefer protecting claims related to property harm to those involving economic loss. Traditionally, though less so today, tort law dealt with harm to person or property.

This hierarchy of protection can help predict the outcome of any lawsuit, particularly where a court is asked to veer from established rules and recognize new legal rights. This is more likely to be done when a claim is based on personal harm and the need for compensation more urgent. Once this inroad has been made, some courts stop at that point. But more often they decide that there is no particular reason to limit it to cases involving person-

al harm. Over time, what was once an established rule may disappear through this process. Disappearance is usually gradual, however. This preserves the appearance of stability while the need for change is accommodated.

SECTION 17.03
CLAIMS AGAINST
DESIGN PROFESSIONALS:
SOME ILLUSTRATIONS

Another way to demonstrate the liability explosion described in Section 17.01 is to note the types of claims that have been made against design professionals. While *most* of the claims to be noted were successful, some cautionary remarks are essential.

Any conclusion that particular conduct gave rise to a valid claim was made in the context of a particular contract and particular facts. Conduct *held* to be below the standard required in a particular case does not mean that this conduct will always fall below the legal standard.

Many appellate opinions simply conclude that the determination made by the finder of the facts—either the trial judge or jury—was within its discretion. An appellate court reviewing a lower court decision may not agree with the decision but will respect the differentiation between the role of trial and appellate courts, the former resolving factual disputes and the latter, as a rule, deciding questions of law.

As to negligence in the design phase, one writer stated:

> Architects might fail to use due care in various ways. The architect may inadequately consider the nature of the soil under the building; he may design an inadequate foundation; he may design a roof too weak to support the weight it will foreseeably have to bear; he may insulate or soundproof the building inadequately. The architect may negligently design a sewer so that waste is carried toward rather than away from the house, he may design windows too small or too large, he may fail to put a handrail on a stairway, or he may specify that nails rather than bolts be used to secure a sundeck.

. . . In addition the architect may negligently fail to notice a defect in the work of a consultant he has hired to help prepare the plans and specifications. The architect would also probably be liable for damage caused by his failure to hire a consultant where a reasonable architect would have done so.

Negligence in design can be based on negligently incomplete specifications as well as upon complete but erroneous ones. The plans and specifications must be complete and unambiguous. For example, specifications are negligently prepared if they are so indefinite that a contractor can bid as if he were going to use first class materials and then build using inferior materials. If measurement of a material is involved, the specification must distinguish between dry and liquid states, or loose or tight packing, where there is any chance of ambiguity.[6]

Since that analysis there have been other illustrations provided in judicial opinions. For example, the following cases involved claims, many of which were successful:

1. Misrepresenting existing topography.[7]

2. Relying on an out-of-date map and building on land not owned by the owner.[8]

3. Specifying material that did not comply with building codes.[9]

4. Positioning the building so as to violate setback requirements.[10]

5. Failing to inform client of potential risks of using certain materials.[11]

6. Drafting ambiguous sketches causing extra work.[12]

7. Designing a house that could not be accomplished by tradespeople in the community where the project was to be built.[13]

8. Designing closets not large enough for the clothing to be contained in them.[14]

9. Designing a project that greatly exceeded the client's budget.[15]

10. Specifying untested material solely because of seller's representations.[16]

11. Designing inadequate solar heating system.[17]

12. Failing to consider energy costs.[18]

13. Failing to disclose an underground high-voltage live wire.[19]

---

6. Comment, *Architect Tort Liability in Preparation of Plans and Specifications*, 55 Calif.L.Rev. 1361, 1370–71 (1967). The author's footnotes to cases are omitted. Cases involving surveyors' mistakes are collected in Annot., 35 A.L.R.3d 504 (1971).

7. *Mississippi Meadows, Inc. v. Hodson*, 13 Ill.App.3d 24, 299 N.E.2d 359 (1973) (dictum).

8. *Jacka v. Ouachita Parish School Board*, 249 La. 223, 186 So.2d 571 (1966). Here the architect was relieved because the client was obligated to and did furnish the out-of-date map.

9. *St. Joseph Hospital v. Corbetta Construction Co.*, 21 Ill.App.3d 925, 316 N.E.2d 51 (1974); *Johnson v. Salem Title Co.*, 246 Or. 409, 425 P.2d 519 (1967).

10. *Armstrong Construction Co. v. Thomson*, 64 Wash.2d 191, 390 P.2d 976 (1964).

11. *Banner v. Town of Dayton*, 474 P.2d 300 (Wyo.1970).

12. *General Trading Corp. v. Burnup & Sims*, 523 F.2d 98 (3d Cir.1975).

13. *Bayuk v. Edson*, 236 Cal.App.2d 309, 46 Cal.Rptr. 49 (1965).

14. Ibid.

15. *Stanley Consultants, Inc. v. H. Kalicak Construction Co.*, 383 F.Supp. 315 (E.D.Mo.1974).

16. *New Orleans Unity Society v. Standard Roofing Co.*, 224 So.2d 60 (La.App.1969) (dictum).

17. *Keel v. Titan Construction Corp.*, 639 P.2d 1228 (Okl.1981).

18. *Board of Education v. Hueber*, 90 A.D.2d 685, 456 N.Y.S.2d 283 (1982) (unsuccessful).

19. *Mallow v. Tucker, Sadler & Bennett, Architects & Engineers, Inc.*, 245 Cal.App.2d 700, 54 Cal.Rptr. 174 (1966).

14. Failing to include owner as named insured and omitting indemnity clause.[20]

15. Failing to advise a need for use permit.[21]

Claims, again mostly successful, relating to the Construction Process phase were as follows:

1. Allowing non-code approved material to be installed.[22]

2. Ordering excess fill to be placed without consulting a soil tester.[23]

3. Failing to make changes needed to comply with codes.[24]

4. Failing to condemn defective work.[25]

5. Performing scheduling and coordination incompetently.[26]

6. Failing to exercise supervisory powers properly.[27]

7. Failing to warn an experienced contractor of general precautions not known in the industry.[28]

8. Failing to engage and check with a consultant.[29]

9. Failing to stop work after discovering contractor using unsafe methods.[30]

10. Issuing payments or certificates negligently.[31]

11. Failing to warn of bankruptcy when paying for materials in contractor's possession.[32]

12. Failing to observe design deviation when checking shop drawings.[33]

SECTION 17.04 SPECIFIC CONTRACT STANDARD

(A) LIKELIHOOD OF SPECIFIC STANDARD

Contracting parties can by agreement determine the standard of performance. Since the client-design professional relationship is created by agreement, a primary source of any agreed-upon standard is the contract itself. Most disputes between clients and design professionals do not involve a *specific* contractually designated standard. Rather, they involve a *general* standard *not* set forth in the contract such as the professional standard to be described in Section 17.05 or an outcome-oriented standard such as implied warranty described in Section 17.07.

Why do most contracts fail to *specifically* state *how* the design professional is to perform? First, many relationships are created without any written agreement—by

20. *Transit Casualty Co. v. Spink*, 94 Cal.App.3d 124, 156 Cal.Rptr. 360 (1979) (trial court finding for client not challenged on appeal) disapproved on other issues in *Commercial Union Assurance Co. v. Safeway Stores*, 26 Cal.3d 912, 164 Cal.Rptr. 709, 610 P.2d 1038 (1980).

21. *Chaplis v. County of Monterey*, 97 Cal.App.3d 249, 158 Cal.Rptr. 395 (1979).

22. *St. Joseph Hospital v. Corbetta Construction Co.*, supra note 9.

23. *First Insurance Co. of Hawaii v. Continental Casualty Co.*, 466 F.2d 807 (9th Cir.1972).

24. *Mississippi Meadows, Inc. v. Hodson*, supra note 7.

25. *Skidmore, Owings & Merrill v. Connecticut General Life Insurance Co.*, 25 Conn.Super. 76, 197 A.2d 83 (1963).

26. *Peter Kiewit Sons' Co. v. Iowa Southern Utilities Co.*, 355 F.Supp. 376 (S.D.Iowa 1973).

27. *Aetna Insurance Co. v. Hellmuth, Obata & Kassabaum, Inc.*, 392 F.2d 472 (8th Cir.1968).

28. *Vonasek v. Hirsch & Stevens, Inc.*, 65 Wis.2d 1, 221 N.W.2d 815 (1974) (dictum).

29. *Cutlip v. Lucky Stores, Inc.*, 22 Md.App. 673, 325 A.2d 432 (1974).

30. *Associated Engineers, Inc. v. Job*, 370 F.2d 633 (8th Cir.1966). See Annot., 59 A.L.R.3d 869 (1974).

31. *Aetna Insurance Co. v. Hellmuth, Obata & Kassabaum, Inc.*, supra note 27.

32. *Travelers Indemnity Co. v. Ewing, Cole, Erdman & Eubank*, 711 F.2d 14 (3d Cir.1983) (unsuccessful).

33. *Jaeger v. Henningson, Durham & Richardson, Inc.*, 714 F.2d 773 (8th Cir.1983).

handshake arrangements. Second, many are made by casual letter agreements drafted by the design professional that are not likely to describe specific standards. Third, those relationships created by assent to standard contracts published by professional associations such as the American Institute of Architects (AIA) do not, despite their completeness as to services and what the design professional is *not* responsible for, specifically describe *how* the work will be done. Even if the standard of performance is discussed in advance—something that is rare—it might not be included in the written contract: the design professional will not want it included or the client may think it unimportant to do so. Both may believe that any assurances as to outcome are simply nonbinding opinions or expectations.

It is, though, common for large organizations of design professionals to include language in their standard agreements specifying the professional standard described in Section 17.05. If challenged, which is more common if clients are strong and experienced such as public entities, the design professional will justify this standard by stating that its professional liability insurance coverage will not include contractual risks that deviate from the professional standard. If the matter cannot be resolved one way or the other, the language may be omitted, leaving the standard to that applied by law. Yet sometimes contractually specified standards of performance are created. Looking at some of them can provide a useful backdrop to the nonspecific standards described in Sections 17.05 through 17.07.

(B) CLIENT SATISFACTION

Sometimes the client's obligation to pay arises only if the client is satisfied with the work of the design professional. Such a contract may be interpreted to be a promise by the design professional to satisfy the client. Though relatively one-sided and one that design professionals generally seek to avoid, if clear evidence exists of such an agreement, the agreement will be enforced.

If satisfaction is a condition to the client's obligation to proceed and to pay, the client need not pay unless it is satisfied or waives this performance measurement. Any legal obligation that may arise must be based upon unjust enrichment created by the client using the work of the design professional.

If satisfaction has been *promised,* failure to perform requires that the design professional compensate the client for any losses the latter may have suffered because of the breach. For example, if a breach caused the project to be delayed or abandoned, the design professional is accountable for any foreseeable losses that can be established with reasonable certainty and that could not have been reasonably avoided.

Although the design professional may, however unwisely, risk the fee, strong evidence of such a risk assumption should be produced before he must respond for losses caused by his failure to satisfy the client.

Two standards of satisfaction exist.[34] If performance can be measured *objectively,* the standard is reasonable satisfaction. Would a reasonable person have been satisfied? Objective standards are more likely to be applied where performance can be measured mechanically. For example, if an engineer agreed with a manufacturer that the manufacturer would pay if satisfied with the performance of a particular machine designed by the engineer, the obligation to pay would require that a reasonable manufacturer be satisfied.

More personal performance invokes a *subjective* standard. Suppose an artist agrees to paint a portrait that will satisfy the person commissioning the portrait. The latter must exercise a good-faith judgment and must be genuinely dissatisfied

---

34. *First National Realty Corp. v. Warren-Ehret Co.,* 247 Md. 652, 233 A.2d 811 (1967) (collecting authorities).

before he is relieved of the obligation to pay. If, for example, the person refused to view the portrait or give it sufficient light to judge its quality, any judgment was not exercised in good faith.

In practice the two standards may not operate differently. If judge or jury thought the performance satisfactory, it will take a strong showing on the part of the person commissioning the portrait that there has been genuine dissatisfaction.

As a rule, the *subjective* satisfaction standard arises in the design phase, particularly in esthetic matters. However, in standard commercial projects, an objective standard may be invoked.

What about a design professional's performance during construction? If the performance in question related to delicate matters, such as how the design professional handled the contractor or public officials or how he dealt with site conflicts, a subjective standard may be applied. Roughly speaking, though, design is more likely to be measured subjectively, while contract administration is more likely to be measured objectively, another illustration of the important difference between design and nondesign site services.[35]

Suppose the client is justified in refusing to pay. This may create a forfeiture. The loss in such a case to the design professional may substantially exceed the loss to the client that would occur if the client, though dissatisfied, were required to accept the design professional's work. Various legal doctrines, among them waiver and estoppel, can be employed to avoid forfeiture. But if this risk was clearly taken, forfeiture will be enforced.

(C) FITNESS STANDARD

A more specific performance standard than satisfaction can be contractually created.

The parties may agree that the completed project will be suitable or fit for those purposes for which the client entered into the project.

The client who plans a luxury residence usually wants a house suitable for a person of his means and taste. In addition to wanting the normal requirements for any residence, such as structural stability, shelter from the elements, and compliance with safety and sanitation standards, the client may want a house that is admired by those who enter it or a residence that can facilitate closing business deals or making business contracts. The client may hope that the opulence of the residence will make social events successful.

The client who plans a commercial office building wishes to make a profit from the rental of space. To accomplish this, suitable tenants at an economically adequate rent must be found. Such a client assumes that the planned use of the structure will be permitted under zoning laws and that the structure will comply with the applicable building codes and zoning regulations relating to materials, safety, density, setback regulations, and other land use controls. In addition, the client who builds an unusually designed office building may hope that the structure will be the subject of national architectural interest.

The client who wishes to build an industrial plant generally assumes that the plant when completed will be adequate to perform anticipated plant activities. The building is expected to comply with applicable laws relating to public health and safety.

Proper design requires that the design professional consider client objectives such as those mentioned in the preceding paragraphs. Some items mentioned in those paragraphs will be discussed and included in the client's program. Some of the matters discussed in the preceding

---

35. *Jaeger v. Henningson, Durham & Richardson, Inc.,* 714 F.2d 773 (8th Cir.1983) (expert testimony not needed for claim based upon site services); *Herkert v. Stauber,* 106 Wis.2d 545, 317 N.W.2d 834 (1982) (obtaining funds not professional design services).

paragraphs would be assumed and probably not discussed. One would not expect client and design professional to discuss the necessity of complying with building codes or regulations dealing with health and safety. Yet beyond these basic objectives discussions may have taken place of economic and social goals less directly connected with basic design objectives.

One *possible* standard to measure the performance of the design professional is whether the project accomplishes the objectives of the client. Put another way, is the structure suitable for the client's anticipated needs, or is it fit for the purpose for which it was built? Is the building an architectural success? Has the client been able to attract good tenants? Has plant production increased? Are the social events successful?

To determine whether suitability or fitness performance standards will be used to measure the design professional's obligation, it is important to look at any antecedent negotiations, discussions, or understandings that may have preceded the client-design professional contract or may have occurred during the course of the design professional's activities.

Suppose there were discussions of client objectives during precontract negotiations or during the design professional's design performance. Were any assurances given by the design professional that related to the fitness or suitability of his design to accomplish particular client objective *promises* that the design would accomplish the objectives or just statements of *opinion* that these objectives would be achieved?

Suppose the design professional made statements relating to such matters. The design professional might have stated that a particular luxury residence would create an artistic stir within the client's social circle. The design professional might have expressed a belief that suitable tenants could be found or that someday a particularly

unusual office building would be considered an architectural landmark. He might have assured the client that the latter could conduct certain activities on the premises of the structure being built.

To determine whether a statement is a promise or an expression of opinion, the law looks at the definiteness with which the statement was made ("I am certain your cost per unit will decrease" vs. "It's my considered opinion that you will improve productivity"), the degree to which the design professional's performance can bring about that objective ("People will like the exterior design" vs. "Your parties will be great successes"), and the degree and reasonableness of any reliance by the client on the statement (using certain types of machinery in a plant vs. redecorating the interior of a house at great expense for the new social season). The more definite the statement, the more within the control or professional expertise of the design professional is the outcome, and the more likely there has been justifiable reliance, the more likely it is the law will find there has been a promise.

If the statement was made before the formation of a contract, a design professional may contend that such statements cannot be proved because of the parol evidence rule.[36] Although results are not always consistent, by and large, the client will be permitted to testify as to these statements.

The design professional should avoid assuring the client that particular objectives will be obtained unless he is willing to risk the possibility of being held accountable if this objective is not achieved. Assurances of certain matters should be given, such as the design meeting public land use controls such as zoning laws and building codes. But the design professional should avoid venturing into areas that are beyond his expertise and that require difficult predictions of the future.

---

36. Refer to Section 14.04(E).

(D) QUANTITATIVE OR QUALITATIVE PERFORMANCE STANDARDS

Sometimes the contract between design professional and client contains a specific performance standard. For example, an engineer may make a contract with a manufacturer under which it was specifically agreed that the machine designed by the engineer would produce a designated number of units of a particular quality within a designated period of time.

Suppose the performing party finds that it is extremely difficult to meet the performance standard or that the performance standard will require an amount of time and money not anticipated by the performing party. In some extreme cases, the performance standard may be impossible to meet.

In such cases, two legal issues may arise. First, is the design professional entitled to be paid for the effort made in trying to accomplish the performance specifications? It may be an onerous contract, but if this is the risk assumed, there will be no recovery. Like satisfaction contracts, interpretation doubts are resolved in favor of the performing party in order to avoid forfeiture. But if this cannot be done, the performing party will be uncompensated unless he can show that efforts expended, although not fulfilling the performance standards, have contributed a benefit to the other party. Although it may be beneficial to the other party to be shown that the performance standards were not possible, in most cases there will be no unjust enrichment and no recovery for the performing party.

Second, has the performing party breached by not accomplishing the objective? Again this is a question of whether the accomplishment of the objective is not merely a condition to the client's obligation to pay but also a promise on the part of the performing party.

Denial of recovery generally is a sufficient burden for the performing party. For this reason, it is not likely that the performing party will be held to a promise to accomplish what turns out to be either an extremely difficult or impossible performance. But in *Gurney Industries, Inc. v. St. Paul Fire & Marine Insurance Co.,*[37] a designer/builder was required to pay damages for failure to meet performance standards. There was no indication that the standards were beyond the state of the art or involved inordinate expense to meet. The contract doctrines that relieve contracting parties where their performance becomes impossible or impracticable may grant relief from any promise to fulfill performance specifications beyond the state of the art or disproportionately expensive.[38]

It is obvious that performance standards place a heavy risk on the performing party. Yet they are attractive to clients because they objectively measure whether the owner is getting what it was promised. Such a standard can be an effective sales device for the designer.

(E) INDEMNIFICATION

The frequency of indemnification in construction contracts necessitates more complete treatment in Part IV. For the purposes of this section, it should be noted that clients increasingly demand that the design professional indemnify *them* against claims that the client believes to be the responsibility of the design professional. Any design professional indemnification creates another specific contractual obligation, one potentially broader than the professional standard.

(F) CONTRACTUAL DIMINUTION OF LEGAL STANDARD

Usually specific contractual standards are higher than the professional standard. Sup-

---

37. 467 F.2d 588 (4th Cir.1972).

38. See Section 27.03(D).

pose, however, that the agreement between the client and design professional specifies a lower standard. Since this more directly involves the extent to which the standard set by law can be varied, it is discussed in Section 17.05(E).

SECTION 17.05
THE PROFESSIONAL STANDARD: WHAT WOULD OTHERS HAVE DONE?

(A) DEFINED AND JUSTIFIED

The Minnesota Supreme Court in *City of Eveleth v. Ruble* [39] stated:

(1) In an action against a design engineer for negligence, the applicable legal principles are held to be:

(a) One who undertakes to render professional services is under a duty to the person for whom the service is to be performed to exercise such care, skill, and diligence as men in that profession ordinarily exercise under like circumstances.

(b) The circumstances to be considered in determining the standard of care, skill, and diligence to be required include the terms of the employment agreement, the nature of the problem which the supplier of the service represented himself as being competent to solve, and the effect reasonably to be anticipated from the proposed remedies upon the balance of the [water] system.

(c) Ordinarily, a determination that the care, skill, and diligence exercised by a professional engaged in furnishing skilled services for compensation was less than that normally possessed and exercised by members of that profession in good standing and that the damage sustained resulted from the variance requires expert testimony to establish the prevailing standard and the consequences of departure from it in the case under consideration. [40]

Four years later that same court was invited to jettison this standard and replace it with the implied warranty standard, an outcome-oriented standard (discussed in Section 17.07). At this point, a portion of that opinion is reproduced.

CITY OF MOUNDS VIEW v. WALIJARVI
Supreme Court of Minnesota, 1978.
263 N.W.2d 420.

TODD, Justice.

[Ed. note: The city became apprehensive because of dampness in the basement of an addition that was being added to a city building. The architect wrote to the city that its design, would, if executed properly, generate a "water-tight and damp-free" basement. But problems grew worse and corrective work was needed.

The city sued the architect based upon claims of negligence, express warranty, and implied warranty.

The trial court held that the language in the letter asserted to constitute a warranty to be merely an expression of opinion and that Minnesota did not recognize implied warranty of a perfect plan or an entirely satisfactory result in an architectural service contract. The trial court granted the architect's motion for summary judgment (no trial needed) on the warranty claims. The Minnesota Supreme Court held that the express warranty claim failed because the contract required that modifications be written and signed by both parties and no evidence had been introduced

---

39. 302 Minn. 249, 225 N.W.2d 521 (1974). See also *Chrischilles v. Griswold*, 260 Iowa 453, 150 N.W.2d 94 (1967); *State, etc. v. Mototan, etc.*, 98 N.M. 740, 653 P.2d 166 (1982).

40. 225 N.W.2d at 522.

of any written agreement by the city. The opinion then dealt with implied warranty. Footnotes have been renumbered and some omitted.]

3. As an alternative basis for recovering damages from the architects, the city urges that we adopt a rule of implied warranty of fitness when architectural services are provided. Under this rule, as articulated in the city's brief, an architect who contracts to design a building of any sort is deemed to impliedly warrant that the structure which is completed in accordance with his plans will be fit for its intended purpose.

As the city candidly observes, the theory of liability which it proposes is clearly contrary to the prevailing rule in a solid majority of jurisdictions. The majority position limits the liability of architects and others rendering "professional" services to those situations in which the professional is negligent in the provision of his or her services. With respect to architects, the rule was stated as early as 1896 by the Supreme Court of Maine (*Coombs v. Beede,* 89 Me. 187, 188, 36 A. 104 [1896]):

> "In an examination of the merits of the controversy between these parties, we must bear in mind that the [architect] was not a contractor who had entered into an agreement to construct a house for the [owner], but was merely an agent of the [owner] to assist him in building one. The responsibility resting on an architect is essentially the same as that which rests upon the lawyer to his client, or upon the physician to his patient, or which rests upon anyone to another where such person pretends to possess some skill and ability in some special employment, and offers his services to the public on account of his fitness to act in the line of business for which he may be employed. The undertaking of an architect implies that he possesses skill and ability, including taste, sufficient to enable him to perform the required services at least ordinarily and reasonably well; and that he will exercise and apply in the given case his skill and ability, his judgment and taste, reasonably and without neglect. But the undertaking does not imply or warrant a satisfactory result." [41]

The reasoning underlying the general rule as it applies both to architects and other vendors of professional services is relatively straightforward. Architects, doctors, engineers, attorneys, and others deal in somewhat inexact sciences and are continually

---

41. Accord: *Gravley v. Providence Partnership,* 549 F.2d 958 (4 Cir.1977); *Ryan v. Morgan Spear Assn. Inc.,* 546 S.W.2d 678 (Tex.Civ.App.1977); *Borman's Inc. v. Lake State Development Co.,* 60 Mich.App. 175, 230 N.W.2d 363 (1975); *Sears, Roebuck & Co. v. Enco Associates,* 83 Misc.2d 552, 370 N.Y.S.2d 338 (Sup.Ct.1975), affirmed, 54 A.D.2d 13, 385 N.Y.S.2d 613 (1976); *Mississippi Meadows, Inc. v. Hodson,* 13 Ill.App.3d 24, 299 N.E.2d 359 (1973); *Scott v. Potomac Insurance Co. of Dist. of Columbia,* 217 Or. 323, 341 P.2d 1083 (1959); *Smith v. Goff,* 325 P.2d 1061 (Okl. 1958); *Ressler v. Nielsen,* 76 N.W.2d 157 (N.D.1956); *Palmer v. Brown,* 127 Cal.App.2d 44, 273 P.2d 306 (1954).

called upon to exercise their skilled judgment in order to anticipate and provide for random factors which are incapable of
precise measurement. The indeterminate nature of these factors
makes it impossible for professional service people to gauge them
with complete accuracy in every instance. Thus, doctors cannot
promise that every operation will be successful; a lawyer can
never be certain that a contract he drafts is without latent ambiguity; and an architect cannot be certain that a structural design will
interact with natural forces as anticipated. Because of the inescapable possibility of error which inheres in these services, the law
has traditionally required, not perfect results, but rather the exercise of that skill and judgment which can be reasonably expected
from similarly situated professionals. As we stated in *City of
Eveleth v. Ruble*, 302 Minn. 249, 253, 225 N.W.2d 521, 524 (1974):

> "One who undertakes to render professional services is
> under a duty to the person for whom the service is to be
> performed to exercise such care, skill, and diligence as men in
> that profession ordinarily exercise under like circumstances."

See, also, *Kostohryz v. McGuire*, 298 Minn. 513, 212 N.W.2d 850
(1973).

We have reexamined our case law on the subject of professional
services and are not persuaded that the time has yet arrived for the
abrogation of the traditional rule. Adoption of the city's implied
warranty theory would in effect impose strict liability on architects for latent defects in the structures they design. That is, once
a court or jury has made the threshold finding that a structure was
somehow unfit for its intended purpose, liability would be imposed on the responsible architect in spite of his diligent application of state-of-the-art design techniques. If every facet of structural design consisted of little more than the mechanical
application of immutable physical principles, we could accept the
rule of strict liability which the city proposes. But even in the
present state of relative technological enlightenment, the keenest
engineering minds can err in their most searching assessment of
the natural factors which determine whether structural components will adequately serve their intended purpose. Until the
random element is eliminated in the application of architectural
sciences, we think it fairer than the purchaser of the architect's
services bear the risk of such unforeseeable difficulties.[42]

---

42. Our decision in *Robertson Lumber Co. v. Stephen Farmers Co-op. Elev. Co.*, 274 Minn. 17, 143 N.W.2d 622
(1966), does not compel a different result. In that case, we affirmed an award of damages for a defectively
constructed grain storage building on a theory of implied warranty. The contract in question, however, was treated
as a construction contract and not an architectural contract. This distinction for warranty purposes between
"professional" services and general contracting services is well established in other jurisdictions. See, e.g., *Kriegler
v. Eichler Homes, Inc.*, 269 Cal.App.2d 224, 74 Cal.Rptr. 749 (1969); *Stuart v. Crestview Mutual Water Co.*, 34 Cal.
App.3d 802, 110 Cal.Rptr. 543 (1973); *Schipper v. Levitt & Sons, Inc.*, 44 N.J. 70, 207 A.2d 314 (1965); *Weeks v.
Slavick Builders, Inc.*, 24 Mich.App. 621, 180 N.W.2d 503, affirmed, 384 Mich. 257, 181 N.W.2d 271 (1970).

The city suggests that many of the design-related tasks performed by modern architects are routine and carry no risk of error if they are performed with professional due care. It is argued that with respect to such tasks, the premise on which the traditional rule rests is inoperative, making the adoption of the implied warranty theory fully proper. We note, however, that architectural errors in relatively simple matters are quite easily handled under the existing cause of action for professional negligence.

Moreover, if implied warranties are held to accompany only uncomplicated architectural endeavors, the finder of fact will be forced in every case to determine, as a preliminary matter, whether the alleged architectural error was made in the performance of a sufficiently simplistic task. Defects which are found to be more esoteric would presumably continue to be tried under the traditional rule. It seems apparent, however, that the making of any such threshold determination would require the taking of expert testimony and necessitate an inquiry strikingly similar to that which is presently made under the prevailing negligence standard. We think the net effect would be the interjection of substantive ambiguity into the law of professional malpractice without a favorable trade-off in procedural expedience.[43]

In addition, we observe that the ills which spurred the creation and expansion of the implied warranty/strict liability doctrine are not really present in this case or in the architect-client relationship generally. The implied warranty of fitness originated primarily as a means of facilitating the legitimate interests of the consuming public and bringing common-law remedies into step with the practicalities of modern industrialism. The outmoded requirement of contractual privity, coupled with manufacturers' sweeping disclaimers of liability, frequently operated to deny effective remedies to those who purchased commercial products at the bottom of a multi-tiered production and distribution network. See, generally, Prosser, Torts (4 ed.) §§ 97, 98. The introduction of the implied warranty doctrine created an effective remedy by allowing plaintiffs to proceed directly against the offending party without reliance on express contractual warranties.

The relationship between architect and client is markedly different. For a client, architectural services are hardly produced by a faceless business entity, insulated by a network of distributors, wholesalers, and retailers. Architects and clients normally enjoy a one-to-one relationship and communicate fairly extensively during the course of the relationship. When a legal dispute arises, the client has no trouble locating the source of his problem, and a remedial device like the implied warranty is largely unnecessary.

---

43. [Ed. note: *Broyles v. Brown Eng'g Co.*, 275 Ala. 35, 151 So.2d 767 (1963), drew this distinction, holding that a civil engineer impliedly warranted the accuracy of his drainage survey.]

Finally, while it is undoubtedly fair to impose strict liability on manufacturers who have ample opportunity to test their products for defects before marketing them, the same cannot be said of architects. Normally, an architect has but a single chance to create a design for a client which will produce a defect-free structure. Accordingly, we do not think it just that architects should be forced to bear the same burden of liability for their products as that which has been imposed on manufacturers generally.

For these reasons, we decline to extend the implied warranty/strict liability doctrine to cover vendors of professional services. Our conclusion does not, of course, preclude the city from pursuing its standard malpractice action against the architects and proving that the basement area of the new addition was negligently designed. That issue remains for the trier of fact in the district court.

Affirmed.

OTIS, J., took no part in the consideration or decision of this case.

That the court felt compelled to justify at some length its decision made four years earlier in *City of Eveleth v. Ruble* demonstrates some dissatisfaction with the professional standard. This is reflected in Section 17.06(A) dealing with exceptions to the requirement of expert testimony, in Section 17.07(D) comparing the professional standard with that of implied warranty, and in Section 17.12(A) analyzing some of the current controversies relating to professional liability.

The court in note 47 differentiated contracts for *design* services from *construction* contracts. Most decisions that have adopted the outcome, implied warranty standard, have involved defendants who designed, built, and sold. Those defendants who design and build are not looked upon as providers of services but as sellers of *products*, who are held to implied warranties of fitness.

(B) EXPERT TESTIMONY

As noted in the preceding section, expert testimony is *usually* required to support a finding that the professional standard has *not* been met. Since this has been a central issue in professional liability claims, it is discussed in Section 17.06.

(C) CONTRACTUAL DIMINUTION OF STANDARD: INFORMED CONSENT

As noted in Sections 17.04 and 17.05, the professional standard is *residual*, applying only where the parties have not contractually agreed to a different standard. Usually any specific standard is more strict; that the design professional did what others would have done will not in itself relieve the design professional. Suppose the design professional can point to a lower specific standard in the contract. Although rare, exploration of this possibility uncovers difficult problems.

Suppose the design professional would have selected X, a material that would have been designated by other professional designers because of a combination of its durability, low maintenance, cost, and appearance. But the client being aware of the tradeoffs orders that Y be used simply because it is less expensive. Suppose the material selected needed replacement earlier than the client expected. The client should not be able to recover any correction costs against the designer. The client and designer agreed to a different standard than the professional standard—in this case, a lower standard. Assuming that this choice does not expose others to unreasonable risk of harm, the law will let the contracting parties decide whether the professional standard or something less will satisfy the designer's contractual obligations.

But suppose an applicable building code requires X, and Y is thought to be unsafe. If the design professional makes the client aware of the code but the client insists upon Y, what are the consequences? Several issues are presented:

1. Will the design professional be denied recovery for services that relate to selection of this material?

2. Will the design professional be given a defense if the client sues him for the cost of corrective work?

3. Will this selection expose the design professional to negligence claims by any third parties who suffer losses because of this choice?

4. Have client and design professional (and any contractor who knowingly violates the code) violated the building laws, exposing them to criminal prosecution?

5. Will this selection be grounds for disciplining the design professional under the registration laws?

The answers to these questions should be yes.[44] Because of the importance of design, the transaction cannot, like the preceding one, be simply regarded as a private one between client and design professional. In matters of safety, public protection takes precedence.

This problem also exposes another issue, an increasingly debated one. Must any client's consent be "informed"? Taking a leaf from the law regulating the relationship between physician and patient, one writer advocated that architects be required to inform their clients of the costs and benefits of design choices.[45] This can, under certain circumstances, be related to the requirement of good faith and fair dealing increasing the duties required of contracting parties.[46]

The uncertain dimensions of such a duty and the increasing concern for expanded liability even under the generally more protective professional standard may be reasons why this obligation has not as yet received overt approval.

SECTION 17.06
EXPERT TESTIMONY

One important component of the professional standard is the need, at least as a general rule, for expert testimony. (A) looks at the exceptions to that requirement, (B) examines expert testimony generally and particularly in the context of claims against design professionals, (C) looks at criticism of the current system, and (D) approaches the problem from the vantage point of the expert witness.

(A) PURPOSE AND EXCEPTIONS TO GENERAL RULE

In the judicial system, decisions are made by judges and juries. To do so they may

---

44. But see *Greenhaven Corporation v. Hutchcraft & Associates, Inc.*, 463 N.E.2d 283 (Ind.App.1984) where the architect recovered despite failure to comply with fire codes at the client's request. The court concluded no public policy problem required resolution as the fire marshal had to approve or grant a variance. The architect did seek a variance but the project was abandoned.

45. Note, *Design Professionals—Recognizing A Duty to Inform*, 30 Hastings L.J. 729 (1979).

46. U.C.C. § 1–203. See also *T.G.I. East Coast Construction Corp. v. Fireman's Fund Insurance Co.*, 534 F.Supp. 780 (S.D.N.Y.1982); *Seaman's Direct Buying Service, Inc. v. Standard Oil Company of California*, 36 Cal.3d 752, 206 Cal.Rptr. 354, 686 P.2d 1158 (1984).

have to hear, evaluate, and judge testimony and exhibits that relate to technical matters unfamiliar to them.

To assist them, the law permits evidence of opinion testimony as an exception to the rules of evidence that *generally* bar opinion testimony. But before such opinion testimony can be admitted, the issue must be one that is too difficult for a judge or jury to decide *without* technical assistance. Also, the person permitted to give opinions must possess the necessary education and experience to be an "expert" on the issue for which the judge or jury needs help.

The professional standard, a *basic* rule, requires expert testimony to support any conclusion that the design professional has not performed in accordance with ordinary professional standards. There are important exceptions to this general rule. For example, in *City of Eveleth v. Ruble,* discussed earlier in this section, the plaintiff city had retained defendant engineer to design a new water treatment plant. After completion, two difficulties developed. First, the intake system that took water from a lake and processed it for distribution into the city's transmission lines and storage reservoir proved inadequate. Second, pressure in the cast-iron distribution lines leading from the water treatment plant to users and storage facilities caused some of the leaded joints in the line to give way. Without any expert testimony, the trial court awarded the city damages against the engineer.

As to the diminished intake capacity, the court noted that the engineer knew that the city decided to build a new water plant in order to increase the intake capacity to a designated number of gallons per minute. The intake capacity was inadequate because there was a failure to anticipate changes in the lake level. The existing intake line that was laid on the bottom of the lake in approximately forty feet of water was not the 18-inch line expected by the engineer but in some places was 16″ and in other places only 12″ in diameter.

The Minnesota Supreme Court concluded that the trial judge was able to assess the validity of these excuses "without the aid of expert testimony." In drawing this conclusion the court stated:

> In our judgment, no expert opinion is needed to demonstrate that a design engineer charged with the responsibility of analyzing the piping and other structural characteristics of an existing plant should be as certain of the dimensions of the intake line as circumstances would possibly permit before recommending a plan the function of which depended upon this critical measurement. It would seem clear that the examination of a photograph of the line would be of little value. Incomplete drawings made available to the Engineer at its request indicate that the intake line was 18 inches in diameter at the point of terminus with the old plant, but we believe that common knowledge would reject this as adequate basis for careful analysis. We find the explanation given for the failure of the Engineer's employees who entered the lake for the purpose of measuring the intake line unsatisfactory when the record shows that others employed for this same purpose by the City were able to obtain the true dimensions of the intake line without difficulties disproportionate to the importance of the task.

> It seems to us that when a professional designs a water plant, in particular a plant intended to deal with increased community needs, that professional should, as a matter of reasonable care, be certain of the size of the piping which provides the plant with raw water and should be equally certain that, ultimately, there will be an adequate supply of water, both in the sense of supply to the plant and in the sense of supply upon which the plant may draw, to meet the operational expectations of the design; or, in avoidance, should more convincingly demonstrate why, under the circumstances, it was good professional practice not to do so.[47]

As to the excess pressures in the distribution lines causing the joints to give way, the trial court found that the high service pumps that distributed the clear water to the storage tanks caused sudden surges

---

47. 225 N.W.2d at 527–28.

"with consequent water hammer" resulting in blowing out the lead-sealed joints. Evidently the cast-iron distribution lines were buried in the ground and had been in use for approximately sixty years. The court concluded that it would not be fair to hold the engineer to any "express or implied commitment that the City's transmission lines would be trouble-free following the installation of the new facility." [48] The court stated that "it would not be reasonable to expect the Engineer to guarantee the performance of these lines under these circumstances." [49] The court relieved the engineer by concluding that expert testimony was required because determining what type of pressure system should be used with pipes of this age was a technical question. In employing the professional standard the court concluded:

> Would a design engineer, in the exercise of that degree of care, skill, and diligence to be expected from this profession, knowing the line pressures which would be created by the operation of the high service pumps and knowing the age of the line and the character of its construction, have reasonably anticipated that it would fail in use? If uncertain, would he have employed tests or techniques of inspection which could have been employed and which would have been employed by a design engineer applying the requisite standards of skill and care which should have revealed the deficiencies in the line? If not, would such an engineer, being unable to ascertain the facts, have recommended the installation of devices such as those now recommended as a precaution against possible

but unpredictable ruptures? We do not think that common knowledge affords answers to these questions. [50]

The wide range of services often performed by the design professional in both design and construction phases will inevitably raise questions as to which services are sufficiently technical to *require* expert testimony and which are sufficiently nontechnical that the lay judge or jury needs no assistance. In many cases the plaintiff is not able or chooses not to offer expert testimony and claims it is not needed, while the defendant contends that the lack of expert testimony bars any judgment of professional malpractice.

Claims based upon design services as a rule will require expert testimony. Choices that involve excavation, design,[51] foundation sufficiency, structural stability, equipment and components,[52] protection against the elements,[53] energy efficiency,[54] and surface water disposal [55] are all matters that require expert testimony. Often they are technical areas for which the prime design professional will retain consultants. Whether such services are performed in-house or by outside consultants, they require decisions that should be made by professionals with specialized education and experience. Often these services must be performed only by persons registered by the state, though this is not determinative. *Seaman Unified School District v. Casson Construction Co.*[56] involved water damage to

---

48. 225 N.W.2d at 529.

49. Ibid.

50. 225 N.W.2d at 530.

51. *Nauman v. Harold K. Beecher & Associates*, 24 Utah 2d 172, 467 P.2d 610 (1970).

52. *Dresco Mechanical Contractors, Inc. v. Todd-CEA, Inc.*, 531 F.2d 1292 (5th Cir.1976), discussed in Section 17.06(B).

53. *South Burlington School District v. Calcagni-Frazier-Zajchowski Architects, Inc.*, 138 Vt. 33, 410 A.2d 1359 (1980).

54. *Board of Education v. Hueber*, 90 A.D.2d 685, 456 N.Y.S.2d 283 (1982).

55. *National Cash Register Co. v. Haak*, 233 Pa.Super. 562, 335 A.2d 407 (1975), reproduced in part in Section 17.06(B). For a recent collection of cases, see 3 A.L.R.4th 1023 (1981).

56. 3 Kan.App.2d 289, 594 P.2d 241 (1979).

a gymnasium following heavy rains. The owner sued the prime contractor, architect, and consulting engineer. The issue was the location of the sidewalk. Disregarding testimony of three experts supporting the design choice made by the architect, the trial court noted that it was common knowledge that water runs downhill. Concluding that this was an error, the appellate court stated:

> The evidence reveals that the plan provided the sidewalk leading to the northwest stairwell was to have a slope of one inch per ten feet to the south and one and one half inches per ten feet to the east. The plan also specified a six tenths of 1 percent overlawn slope to the north for the area immediately east of the walk. This water was not scheduled to run off in a direct line downhill but was to be shunted diagonally by the pitch of the sidewalk and then diverted to the north by the land contours and the plan-specified grade east of the walk. Considering these facts and the size of the surface area to be drained, we do not think it is within the common knowledge of laymen to decide whether such specifications were proper. This conclusion, together with the testimony of three architects that the plan was in accordance with local architectural standards and the absence of any testimony by an *architect* to the contrary, convinces us that the trial court erred in finding directly opposite to these experts' opinion testimony.[57]

As to site services unconnected to design, the issue becomes more cloudy. Does it take expert testimony to assist a judge or jury in deciding whether the design professional should have certified a particular amount to be paid, detected a design deviation in a shop drawing or submittal, noted a deviation from an approved schedule, verified a changed condition, coordinated the work of separate contractors, to name some of the tasks often given to the design professional?

Two recent cases held that no expert tes-timony was needed for work done during the construction phase, the courts concluding that this did not require specialized technical skill possessed only by those with particular education and training.[58]

Unfortunately the courts did not separate those services that take professional skill, such as issuing change orders and interpreting contract documents, but lumped everything into the amorphous "supervision" category. Undoubtedly many of the activities on the site, such as those performed by a project representative, an inspector, or a lender representative, do not take specialized architectural or engineering training. In fact, one case, *Bartak v. Bell-Galyardt & Wells, Inc.*,[59] pointed out that the project representative (who it erroneously stated was supervising the work) was not a registered architect. But certainly many activities that relate to on-site design do involve professional skill.

What about judging performance, inspecting the work, and insuring that the subcontractors are paid? Often persons without specialized architectural and engineering training can perform these functions. But care must be taken to distinguish those activities that are part of the design professional's "judging" role. In attacking a decision made by a design professional, expert testimony may not be needed because the issue is not whether the design professional did as others would have but the finality of his decision.[60] Unfortunately this distinction is not always drawn.

Claims by those who engage one entity to both design and build or who buy from a builder-vendor raise special problems. At the outset, a differentiation must be made between a claim based upon professional malpractice and one based upon implied

---

57. 594 P.2d at 245.

58. *Jaeger v. Henningson, Durham & Richardson, Inc.*, 714 F.2d 773 (8th Cir.1983) (approval of shop drawings); *Bartak v. Bell-Galyardt & Wells, Inc.*, 629 F.2d 523 (8th Cir.1980) (checking on contractor compliance).

59. Supra note 58.

60. See Section 33.09.

warranty, a standard that usually measures the obligation of the builder-vendor or developer. In the latter, negligence is not the standard.[61] But a number of California cases raise the question of whether expert testimony is needed when the claim is based upon negligence.

The first, *Miller v. Los Angeles County Flood Control District*,[62] involved a claim by a homeowner whose house had been destroyed in a flood. His claim was against the defendant developer-builder who had built the home and sold it to him. The home had been built at the base of a 90-degree curve in a road that doubled as a flood control channel. The plaintiffs charged the builder with negligent construction and strict liability for defects in design and construction. The builder knew that the home was to be on a flood channel, but the plaintiffs did not introduce any qualified experts on building practices to testify as to what a reasonable builder would have done about the site.

The California Supreme Court affirmed a judgment for defendant after plaintiff had put in its evidence because the plaintiffs had not proved a violation of the standard of care. The plaintiffs had argued that the claim was based upon ordinary negligence and not professional malpractice. The court held that nonexpert minds could not unaided determine whether the developer had failed to exercise due care in the construction of the home. The court stated:

> Building homes is a complicated activity. The average layman has neither training nor experience in the construction industry and ordinarily cannot determine whether a particular building has been built with the requisite skill and in accordance with the standards prescribed by law

or prevailing in the industry. In the instant case, the issue as to whether or not the Miller home had been negligently constructed involved a multitude of subsidiary questions bearing not only upon the erection of the structure itself but also upon the location of the house on the particular lot, the elevation of the lot, the influence of the surrounding terrain, the possibility of run-offs and floods, and the existence of the debris dam. These were not questions which the jury could have resolved from their common experience . . . .[63]

However, the Court of Appeals in *Chaplis v. County of Monterey*,[64] looked at a similar issue differently. The plaintiff suffered losses because it could not have a laundromat in a building that it had built because the local authorities refused to issue a use permit. In addition to bringing a claim against the local authorities, the plaintiff sought recovery against the designer and contractor, claiming that they were negligent in not informing him that a use permit would be required before work began on the laundromat. The court distinguished the *Miller* case, concluding that in the *Chaplis* case the claim was based on ordinary negligence and not professional negligence. The court noted that the claim was not based on technical or mechanical aspects of design or construction and rejected the arguments of the defendants that expert testimony is needed because construction procedures and zoning ordinances are complex and confusing. The court stated:

> These are simple straightforward acts or failures to act which have nothing to do with the intricacies of the construction industry or the complexities of the county's zoning procedures.[65]

Raven's Cove Townhomes, Inc. v. Knuppe Development Co.[66] held that expert testimony was *not* needed in a claim against a

61. See Section 28.09.

62. 8 Cal.3d 689, 106 Cal.Rptr. 1, 505 P.2d 193 (1973).

63. 106 Cal.Rptr. at 10, 505 P.2d at 202.

64. 97 Cal.App.3d 249, 158 Cal.Rptr. 395 (1979).

65. 158 Cal.Rptr. at 406.

66. 114 Cal.App.3d 783, 171 Cal.Rptr. 334 (1981).

developer for defects in the irrigation system, landscaping, painting, or exterior trim. It distinguished the *Miller* case because the defects in the *Raven's Cove* case were "of a kind which are of such common knowledge that men of ordinary education could easily recognize them." [67]

One argument for the use of the implied warranty standard has been the difficulty determining when expert testimony will be needed in a claim based upon negligence.

(B) ADMISSIBILITY OF TESTIMONY: *NAT. CASH REGISTER CO. v. HAAK; DRESCO MECHANICAL CONTRACTORS, INC. v. TODD–CEA*

Before proceeding to the principal focus of this section—the issue of who can testify and the type of testimony needed—a few preliminary remarks relating to admissibility generally must be made.

First, a party intending to call an expert witness must notify the other party in advance of trial of the identity and qualifications of the expert and the issues about which expert testimony will be elicited. This enables its opponent to investigate or use the deposition process to check on the qualifications of the experts that the other side plans to use.

Second, the expert must give his opinion based upon firsthand knowledge (such as an expert medical witness who has conducted his own examination), upon facts admitted into evidence, or a combination of firsthand knowledge and evidence. If the expert has no firsthand knowledge, the opinion of the expert in many states is based upon a hypothetical set of facts that are included in the question and that are based upon evidence that has been admitted or evidence that the party calling the expert plans to admit. Sometimes the expert can observe the testimony of witnesses and be asked to assume the truth of previous testimony as a basis for his opinion. This can simplify the often bewilderingly complex hypothetical question.

Third, a witness who is not registered in accordance with the registration laws of the state very likely will be permitted to testify.[68] Local law must be consulted.

The qualifications of an expert to give an opinion—an increasingly difficult issue because of overlapping specialization in professions—is discussed in a case reproduced here.

NATIONAL CASH REGISTER CO. v. HAAK

Superior Court of Pennsylvania, 1975. 233 Pa.Super. 562, 335 A.2d 407.

SPAETH, Judge.

[Ed. note: After completion of construction of a manufacturing plant, sinkholes adjacent to dry wells developed, threatening the integrity of the building. The problem was diagnosed by Gannett-Fleming, an engineering firm. Corrective work was accomplished. The owner brought a claim against the architect for negligent design. Footnotes have been omitted.]

The controversy on this appeal centers around the adequacy of the testimony of the four expert witnesses called by appellant.

Charles W. Pickering was admitted as an expert in "civil engineering in hydraulics" (the court's characterization) or as a "civil engineer familiar with hydraulics" (defense counsel's characterization). He was employed by Gannett-Fleming and was the one who

---

67. 171 Cal.Rptr. at 342.

68. *South Burlington School District v. Calcagni-Frazier-Zajchowski Architects, Inc.*, 138 Vt. 33, 410 A.2d 1359 (1980).

had examined the site for that firm and had recommended the removal of the dry wells and the installation of the new system. His opinion on the cause of the sinkhole activity was unequivocal:

> It is my opinion that the Surface Water System as was installed on the NCR site has accelerated the formation of sinkhole activity on the site.
>
> [I]f the present system were to be continued in use, . . . the formation of sinkholes would continue and with the ultimate possibility or ultimate meaning at sometime, some point in time, because it is a natural phenominal [sic] it cannot be predicted, that at sometime there may be serious damage caused to the major facilities on the site.
>
> Q. Would that include the building?
>
> A. Yes.

Timothy Saylor was admitted as an expert in geology. He was also employed by Gannett-Fleming, and his testimony, where relevant to the issues to be considered here, corroborated Mr. Pickering's.

Professor Jacob Freedman of Franklin & Marshall College also testified as an expert in geology. His qualifications indicated extensive experience in his field covering over 25 years, including being frequently called in as a consultant on the geology of Lancaster County, especially with regard to "water problems, foundations problems, [and] studies of quarries." His testimony corroborated Mr. Pickering's and Mr. Saylor's with regard to the causal relationship between the dry wells and the sinkholes. He then added the following at the end of re-direct examination:

> I probably ought to say and I probably haven't said this in my discussion so far along these lines, that I have and I know no geologist who has ever recommended dry wells for construction in an area like this; that, in fact, we urge people not to use this kind of system; and if they have French drains— in fact, as I say, we frequently get calls at school; and anytime anybody mentions a french drain we tell them they are in for trouble, and french drains and dry wells are very similar in their properties. So these are systems that should not have been installed because they lead to trouble.

This brief statement (which was not given in direct response to any question, but which also was not objected to) was immediately explored on recross-examination:

> Q. This is your opinion, Professor, is that correct?
>
> A. Let's say it is not only opinion. It is an observation.
>
> Q. And I understand you to say that no geologist that you know of recommends this system for this type of area?
>
> A. I know all my colleagues invade [sic; "inveigh"?] against them.

Q. These are all of your colleagues at Franklin & Marshall?

A. Everybody I have talked to.

Q. At Franklin & Marshall?

A. No, at other places too.

Q. Pardon me?

A. Other places, other colleagues. We discuss these at meetings and I have been a consultant on a situation where I have seen the result of one of these things.

That was the complete recross-examination of this witness.

Appellant's final witness was F. James Knight, a registered engineer, who was qualified as an "engineering geologist." He was employed by Gannett-Fleming and had supervised the repairs of the sinkholes on appellant's site. He testified as to the extent of the damage and the extent of the repairs necessary, and expressed his professional opinion that the dry wells had "contributed significantly to the accelerated formation of sinkholes . . . on that tract."

At the close of appellant's case an oral motion for compulsory nonsuit was made on the ground that appellant had failed to present sufficient expert testimony to establish the standard of care required of an architect in that locality with respect to the design of surface water disposal systems. The court granted the motion, ruling that "while [appellant] has presented testimony of experts in the field of geology and engineering, they have not presented any testimony in the field of architecture tending to prove that [appellees'] professional services departed from accepted practice in this profession or that [appellees] failed to meet the standards of their professional duties." A motion to take off the nonsuit was denied by the court *en banc* with an opinion by Judge BUCHER, who was also the trial judge. In that opinion the court held that "testimony in the field of architecture" was necessary. It also suggested a second reason for the nonsuit, by asking, "[D]id [appellant] prove any negligence on the part of [appellees'] that justified submitting the case to the jury?" Both of these issues are before us on this appeal.

I

The failure to present an architect as an expert witness was not fatal to appellant's case.

The court below states the general rule as being that "expert testimony is necessary to establish negligent practice in any profession." Wohlert v. Seibert, 23 Pa.Super. 213 (1903), is cited for this proposition (as it has been in other opinions), but in fact its test is more analytical:

The crucial test of the competency of a witness offered as an expert to give testimony as such is the resolution of the question as to whether or not the jury or persons in general who are inexperienced in or unacquainted with the particular subject of inquiry would without the assistance of one who possesses a knowledge be capable of forming a correct judgment upon it.

Id. at 216.

The opinion then goes on to restate the rules for the standard of care to which a physician is held; it does not discuss any other profession or professions in general. The same observation may be made of the section of Wigmore most cited on professional experts:

On any and every topic, only a qualified witness can be received; and where the topic requires special experience, only a person of that special experience will be received [cross reference omitted]. If therefore a topic requiring such special experience happens to form a main issue in the case, the evidence on that issue must contain expert testimony or it will not suffice. Wigmore on Evidence (3d ed. 1940) § 2090(a) at 453.

We have no doubt that expert testimony was required in this case. The "subject of inquiry" was the standard to be applied to one who holds himself out as competent to design and supervise the construction of a surface water disposal system in Lancaster County. This is certainly a subject that "requires special experience." However, there is nothing inherent in the nature of that experience that makes it unique to architects. What the jury needed was not "the assistance of one who possesses a knowledge" of architecture but of surface water disposal systems. Whether the person offering that assistance happened to be an architect, or engineer, or geologist, or something else, was unimportant; what was important was what he knew.

The error committed by the court below was that it literally analogized the instant case to one of medical malpractice. The court reasoned that in medical malpractice cases there must be expert testimony from physicians as to the appropriate standards of medical practice. From this the court reasoned that in a suit against architects, only architects are competent to testify as to the appropriate architectural standards. However, in medical malpractice cases the expert generally must be a physician because only a physician is trained to perform the medical functions that are the subject matter in controversy. In the instant case the subject matter in controversy (the design and installation of a surface water disposal system on the site in question) is not within the exclusive realm of one profession. To the contrary, it is within the realm of at least three professions: architects, engineers, and geologists. Therefore, a member of any of these professions (if otherwise qualified) was competent to state what the appropriate design and installation standards were.

A case similar to this one is Bloomsburg Mills Inc. v. Sordoni Construction Co., 401 Pa. 358, 164 A.2d 201 (1960). There the plaintiff hired some architects to design and supervise the construction of a weaving mill for nylon and rayon. This type of manufacturing requires a constant temperature and humidity. It was thus necessary to build the roof with a "vapor seal" to prevent condensation and leakage of moisture. The plaintiff alleged that the roof was defective in that the vapor seal did not function properly, and that the roof was otherwise inadequately sealed. The defendant appealed a verdict in favor of the plaintiff, claiming insufficient evidence of negligence. As part of this claim the defendant attacked the competency of the plaintiff's expert witness. The expert had had extensive experience with a large roofing manufacturing company and was at the time of the trial a professional roofing consultant. (He had, in fact, been requested by the defendants to submit a bid on the project in question but had declined.) He was not, however, an architect. Nevertheless the Supreme Court held him qualified to testify against the defendant architects. For other decisions in accord, *see* Abbott v. Steel City Piping Co., 437 Pa. 412, 263 A.2d 881 (1970) (witness with extensive experience in masonry competent to testify as to how a certain wall should be built despite lack of formal engineering degree); Willner v. Woodward, 201 Va. 104, 109 S.E.2d 132 (1959) (heating engineer competent to testify against architect regarding a heating and air conditioning duct); Cuttino v. Mimms, 98 Ga. App. 198, 105 S.E.2d 343 (1958) (engineer and contractor both competent to testify against architect in case based on faulty construction); Covil v. Robert & Co. Associates, 112 Ga.App. 163, 144 S.E.2d 450 (1955) (engineer competent to testify against architect who had drawn plans for water works where issue was whether a certain pipe joint was properly secured).[69]

The court below dismissed *Bloomsburg Mills*, saying that the improper design of a roof is a "common problem" that "would hardly require expert testimony." As suggested by our preceding statement, however, in fact the problem was quite complex and involved a special type of structure. This is further apparent from the Supreme Court's quite lengthy and detailed discussion of the construction problems presented, and of the expert witness's qualifications, none of which would have been appropriate had the case presented only a "common problem," requiring no expert testimony at all. We thus find *Bloomsburg Mills* persuasive authority, and conclude that the failure of appellant to present an architect as an expert witness was not fatal to its case.

---

69. [Ed. note: *Perlmutter v. Flickinger*, 520 P.2d 596 (Colo.App.1974), permitted an engineer and contractor to testify about skylight design.]

II

Although, as noted above, the opinion of the court below asks the question, "[D]id [appellant] prove any negligence . . . that justified submitting the case to the jury?", in fact that question is not there addressed. Instead, the entire opinion deals only with whether the trial judge was correct in holding that in an action against an architect acting in his professional capacity, the plaintiff must produce testimony by another architect, which is the issue that we have just disposed of. In these circumstances we have made our own examination of the record, and have concluded that appellant did prove sufficient evidence of negligence to send the case to the jury.

Messrs. Pickering, Saylor, and Knight all testified unequivocally that in their respective professional opinions the sinkholes were aggravated by the dry well system, and that considerable damage to appellant's property resulted. However, none of them testified as to the standards of skill required of one who undertakes to design and supervise the installation of a surface water disposal system in the particular area in question, and, as observed in the preceding section of this opinion, it was essential to appellant's case that there be some testimony by a qualified expert on that point.

Appellees have contended that there was no such testimony. (They do not challenge the testimony of Messrs. Pickering, Saylor, and Knight.) This contention, however, overlooks the testimony of Professor Freedman, which we have already quoted in relevant part, *ante* at 409. It is indeed true that Professor Freedman's testimony was not developed in an orderly manner, and in fact it appears to have emerged almost by chance. It is, nevertheless, in the record, and the professor was cross-examined with respect to it. Summarized, the testimony was that neither the professor, nor any other geologist, nor any one consulted about surface water disposal, "has ever recommended dry wells for construction in an area like this . . . [W]e frequently get calls . . . and . . . we tell them they are in for trouble . . . [T]hese are systems that should not have been installed because they lead to trouble." When we bear in mind that we must give appellant every reasonable benefit from the evidence, Shirley v. Clark, *supra*, this testimony may fairly be read as a statement of opinion, by a qualified expert, that appellants violated the professional standards of care to which they were obliged to conform.

Order reversed.

The setting in which expert testimony is taken is best understood by reference to judicial opinions that look carefully at the testimony and determine whether it meets the legal requirements. A portion of an opinion that does this is reproduced here.

DRESCO MECHANICAL CONTRACTORS, INC. v. TODD–CEA

United States Court of Appeals, Fifth Circuit, 1976. 531 F.2d 1292.

CLARK, Circuit Judge.

[Ed. note: A boiler exploded causing property damage. The owner brought legal action against Todd, who had designed and manufactured the boiler and combustion controls, based upon Todd's negligence. Todd claimed the explosion resulted from design negligence by Austin, a consulting engineer retained by the owner who had specified that a dual timer system be used. The owner prevailed against Todd, and Todd asserted a claim against Austin. The jury verdict against Austin was set aside by the trial judge and an appeal taken.]

Todd's action against Austin rested on a theory of negligent design. To establish a cause of action on such a theory under Georgia law, Todd was required to establish that Austin's acts or omissions breached its duty to observe the standard of care it owed to those with whom it dealt and that this breach was a proximate cause of the occurrence from which damage was suffered. As an engineering firm, Austin's duty of care was that of a responsible *professional* person or organization. It is

> . . . the obligation to exercise a reasonable degree of care, skill, and ability, which generally is taken and considered to be such a degree of care and skill as, under similar conditions and like surrounding circumstances, is ordinarily employed by their respective professions.

* * *

It was Todd's burden to establish the failure to observe this standard by the introduction of expert opinion evidence. *See, e.g., Shea v. Phillips*, 213 Ga. 269, 271(2), 98 S.E.2d 552 (1957).

Todd presented the testimony of several witnesses to show that Austin had established the original requirement for a "completely redundant . . . dual flame safeguard system" and had refused to approve a change to a single timer system when Todd's engineers described it to them as "unsafe." Witness Horton Rucker, an employee of an engineering firm not involved in the suit, explained why he believed the explosion to have been caused by the dual timer system. He stated the dual timer presented a "bad situation" because of the possibility that the timers would get out of synchronization and send improper signals to the system. He said he had no personal experience with such a system in his twenty-five years of experience in the field of boiler and burner-control engineering.

James Warren of Honeywell, Inc., the manufacturer of the timers, testified that he "[didn't] believe [he] would try using these particular programmers in parallel" because "as long as everything [was] going well they would probably work all right. But if things started going wrong, they would get false messages and get confused."

In short, Todd's proof presents a convincing theory of the *cause* of the explosion and for the proposition that Austin participated in—or perhaps even dictated—the decision to install the system whose malfunction lay at the heart of that theory. What is missing is any sort of unequivocal statement by any witness that Austin's specification of the dual timer system or its manner of reviewing Todd's detailed plans failed to conform to the ordinary accepted practices of the engineering profession. An examination of Rucker's testimony demonstrates considerable lack of communication between counsel and witness, and neither the questions nor the answers establish whether Rucker was describing a "bad situation" inherent in the design or a "bad situation" that obviously resulted in this instance. Rucker stated that he was "totally unfamiliar" with dual timer systems prior to his inspection of the system involved in the case *after* the explosion. However, he did not intimate that this lack of familiarity was based upon the fact that such systems were not safe to use or not customarily put to use by others. Thus, he could have no basis for a professional judgment that a dual timer design was inherently unsafe or failed to conform to engineering practice. Likewise, Warren's testimony does not refer to the professional standard. It is true that he said he "didn't think" he would use such a system, but neither his position nor any reasons for it are free from ambiguity.

In marked contrast to these statements is the testimony of Austin's independent expert witness, Alderman. He stated unequivocally that the original Austin specifications were "excellent" and "complied generally and extrapolated on" the generally accepted practice of mechanical engineers in Georgia during the relevant time period. He further stated that the specifications for the dual flame safeguard system in specific met the accepted standard and that it was the standard practice for consulting engineers to review builders' wiring diagrams only for general compliance with the specified concept. He concluded his direct testimony as follows:

Q Mr. Alderman, will you state whether or not the use of dual timers if they are properly wired is a practice which can be carried out safely in a flame safeguard system, if they are properly wired?

A There's no doubt in my mind that an expert wiring control designer and system designer could successfully wire dual flame scanners, relays and timers.

Q Is the practice of not checking the wiring, details of shop drawings for workability, is that the general accepted practice among consulting mechanical engineers in this State and was the practice in 1969, 1970 and '71?

A It is the common practice.

In reviewing the correctness of the trial court's decision to take a case from the jury by directed verdict or judgment notwithstanding the verdict,

> the Court should consider all of the evidence—not just that evidence which supports the non-mover's case—but in light and with all reasonable inference most favorable to the party opposed to the motion. If the facts and inferences point so strongly and overwhelmingly in favor of one party that the Court believes that reasonable men could not arrive at a contrary verdict, granting of the motions is proper. On the other hand, if there is substantial evidence opposed to the motions, that is, evidence of such quality and weight that reasonable and fair-minded men in the exercise of impartial judgment might reach different conclusions, the motions should be denied, and the case submitted to the jury.
>
> . . . [I]t is the function of the jury as the traditional finder of the facts, and not the Court, to weigh conflicting evidence and inferences, and determine the credibility of witnesses.

Boeing Company v. Shipman, 411 F.2d 365, 374–75 (5th Cir.1969) (en banc). When this test is applied here it requires the conclusion that there was no substantial evidence opposed to the motion on the element of Austin's violation of the standard of care imposed on it. Therefore, the trial court properly refused to allow the jury's verdict to stand.

(C) CRITIQUE OF SYSTEM

The expert testimony system has been severely criticized. The complex "hypothetical" question has lead to overtechnical appellate review and confusion of jurors. In addition, complexity has made errors more likely because of the increasing specialization of professions, a point demonstrated by the *Haak* case. When administered too strictly, no experts may be found, a particular difficulty when expert testimony is required. The pool of experts being limited to *local* experts not only has made it more difficult to obtain experts but also has not taken into account the increasingly statewide or national standards of practice.

Even more criticisms have been made of the entire system itself, based largely upon the adversary system discussed in Section 2.11. Each party looks not for the best qualified experts, but for the experts who will best support its case. This, coupled with the high compensation paid experts, has led to skepticism as to the professional honesty of many experts.

Not many years ago there was intense criticism of what was called the "conspiracy of silence," the unwillingness of professionals to testify against one another. This was particularly difficult where local standards were employed. This led to the development of professional expert witnesses (called forensic design professionals because of their ability to persuade judges and juries of the soundness of their professional conclusions). Although this has certainly helped overcome the "conspiracy of silence," it has had some unfortunate results.

Such experts have been looked upon as "hired guns"—too quick to find fault. (To those challenging design professionals, these experts are looked upon as "paladins.") Of course, other experts can be produced by the design professional charged with malpractice. But in the end, judge and jury are often confused. How can experts with such outstanding credentials differ so sharply as to the cause of the harm and the standards of professional practice?

This and the high cost of producing experts has led to more use of expert panels from which trial court judges can call expert witnesses rather than exclusive reliance on the experts called by the parties. The use of an impartial panel conflicts with the adversary system and is not widespread.

(D) ADVICE TO EXPERT WITNESSES

Space does not permit a detailed discussion of all the problems seen from the perspective of a professional designer asked to be an expert witness. Some brief comments can be made, however.

A clear, written understanding should precede any services being performed. Such a writing should include the following:

1. Specific language making clear that the expert will give his best professional opinion.

2. Language that covers all aspects of compensation for time to prepare to testify, travel time, and actual time testifying before a court, board, or commission. (Many experts use an hourly rate for preparation time and a daily rate for travel and testimony time.)

3. Specifying expenses to be reimbursed, using a clear and administratively convenient formula for reimbursing costs of accommodations, meals, and transportation.

4. A minimum fee if the expert is not asked to testify. Some attorneys retain the best experts, use the experts whose opinions best suit their case, and by having retained the others, preclude them from testifying for the *other* parties.

Details as to appearance, description of qualifications, methods of answering questions, explanations for opinions, and defending opinions on cross-examination are usually provided by the attorney calling the expert. It is important to recognize who is being addressed and the reason for seeking expert opinions. The expert should assist the judge or jury, persons often inexpert in evaluating technical material. For that reason, opinions and explanations must be understandable by persons who must evaluate them. The expert should *never* speak *down* to judge and jurors.

SECTION 17.07
IMPLIED WARRANTY:
AN OUTCOME STANDARD

(A) MINORITY DOCTRINE:
SOME ANALOGIES

Most American courts look at the process by which the design professional has performed design services and compare what he did to what other professionals would have done. A few jurisdictions, mainly in cases dealing with defendants who both designed and built, determine liability by comparing the product developed from the design with the communicated or understood expectations of the client.[70] Did the product measure up to the client's purposes of which the design professional was aware or should have been aware? This contrasts to the process-oriented professional stan-

---

70. *J. Ray McDermott & Co. v. Vessel Morning Star*, 431 F.2d 714 (5th Cir.1970) (admiralty); *Hill v. Polar Pantries*, 219 S.C. 263, 64 S.E.2d 885 (1951); cases are collected in 25 A.L.R.2d 1085 (1952). See also note 42 supra. *Broyles v. Brown Engineering Co.*, infra note 71, held that most professionals, including architects, are held to the professional standard, but a civil engineer designing a drainage system for a developer was held to a warranty standard. *Greenhaven Corporation v. Hutchcraft & Associates, Inc.*, supra note 44, used "suitability for purpose" language. But the issue was compliance with codes and ordinances, usually clear cases of professional negligence.

dard with the usual requirement of expert testimony.

Claimants and those against whom claims are directed are usually connected by a contractual relationship. (Some special problems of third-party claims are discussed in Sections 17.08 and 17.09.) Claimants are sometimes allowed to bring a tort claim. Yet essentially the claimant is asserting that it did not get what it was promised. The real issue in these cases is the exact nature of the promised performance.

A promise can be an assurance that the promisor, here the design professional, will do the following:

1. Do what was reasonable (the ordinary negligence standard).

2. Do what other professionals would have done (the professional standard).

3. Accomplish a particular objective sought by the client (a warranty standard).

A warranty can be expressed in the contract. Here the issue is whether such a warranty will be *implied.*

If the parties have made their intention clear—whether the design professional will be judged by the process or by the outcome—the law will respect that intention. In many professional relationships, specificity is absent and the law must look at what the parties must have intended or what is fair and proper. For that reason, those cases in which a warranty has been implied have been transactions where an owner has dealt with someone who both designs and builds. Such an owner expects that the house will be habitable and provide shelter commensurate with the reasonable expectation of the owner. It should not be necessary to establish that the builder was negligent or even professionally negligent. Usually American contract law does not require that a contract claim be based upon

negligence, only that the promisor did not perform in accordance with his promise.

Where a professional is retained to provide *design* services, American law has concluded (not without criticism that is explored in Section 17.12) that the *likely* understanding between the client and the professional designer is not that a *successful outcome* will be achieved when professional services are purchased but that the professional will perform as would other professionals. This was explored in *City of Mounds View v. Walijarvi* (reproduced in part in Section 17.05(A)). The decision to use the process-oriented professional standard may have been based in part upon the convenience of using the standard used in malpractice claims made by third parties.

Although a comparison of the two standards is made in Section 17.07(B) and a critique of the current law in Section 17.12(A), one theory often ignored is the contractual *implied term.* Not all the expectations of contracting parties are *specifically* included in the contract. As a result, inquiry must be made in what the parties must have expected but did not express. As indicated this *can* be used to justify a professional standard. But such an inquiry can reveal a communicated—expressly or impliedly—expectation of a successful outcome.

This concept is expressed in *Board of Education, etc. v. Del Biano & Associates,*[71] which did not receive much attention when it was decided by the Illinois Court of Appeals. The parties had contracted on the standard AIA agreement. After quoting the specific duties provided for in that agreement relating to design and site services, the court stated:

> These expressed duties are not the only obligations which defendant was required to perform. Not all the duties growing out of a contract must have been expressly stipulated for: many duties

---

71. 57 Ill.App.3d 302, 14 Ill.Dec. 674, 372 N.E.2d 953 (1978). See also *Broyles v. Brown Engineering Co.,* 275 Ala. 35, 151 So.2d 767 (1963).

arise *ex lege* out of the relation created by the contract. . . . Consequently, we find that defendant had the implied obligation to specify the use of reasonably good materials, to perform its work in a reasonably workmanlike manner, and in such a way as reasonably to satisfy such requirements as it had notice the work was required to meet.[72]

While not classified as a minority implied warranty case, that court recognized that the client may reasonably expect proper materials to be specified—materials that will accomplish the client's purpose—rather than an assurance by the design professional that he will live up to the professional standard.

(B) COMPARISON WITH PROFESSIONAL STANDARD

A pronouncement of majority and minority rules of law is often misleading. If careful evaluation is made of actual decisions, two supposedly contradictory rules may not appear different.

This is illustrated by *City of Eveleth v. Ruble* [73] (discussed in Sections 17.05 and 17.06). Since the law is more willing to classify the principal issue as one that does *not* require expert testimony or is more flexible in determining who can testify as an expert as in the *Haak* case (reproduced in Section 17.06(B)), the *actual* issue faced in so-called majority and minority jurisdictions may not be different. While the issue in a *majority* jurisdiction still focuses upon the *process* rather than the *product,* as seen in the *Ruble* case, even the professional standard looks at the objectives of the client. This and the often difficult task of determining *which* standard has been ap-

plied is illustrated by *Bloomsburg Mills, Inc. v. Sordoni Construction Co.,*[74] a case involving the design of a weaving plant. The contract between owner and architect specified that the building was to be air-conditioned to maintain a constant temperature of 80 degrees Fahrenheit and a constant humidity of 60% moisture. To effectuate this, a built-up roof with a vapor seal was required that would prevent leakage of moisture from the outside and condensation from the inside.

After a short number of years, the insulation material became saturated, soggy, and inefficient, causing high condensation on the inside ceiling. The roof had to be replaced.

The client based its claim against the defendant architect on professional negligence. The court pointed to the evidence adduced at the trial that supported professional negligence. In addition, an expert witness had testified that the insulation material used was inadequate and not generally accepted by the roofing industry. The expert also pointed to other evidence of negligent design. The court, though explicitly employing the professional standard, stated:

> While an architect is not an absolute insurer of perfect plans, he is called upon to prepare plans and specifications which will give the structure so designed reasonable fitness for its intended purpose, and he impliedly warrants their sufficiency for that purpose.[75]

This startling interjection of implied warranty language and the citation by the court to *Hill v. Polar Pantries,*[76] a well-known minority implied warranty case, blurs the distinction between the professional standard

---

72. 372 N.E.2d at 958.

73. Supra note 39.

74. 401 Pa. 358, 164 A.2d 201 (1960).

75. 164 A.2d at 203.

76. Supra note 75. Similarly, to avoid the owner and architect being joint tortfeasors, one court used the fitness for intended purpose standard despite the law and the owner's complaint applying the professional negligence standard. See *County of Los Angeles v. Superior Court,* 155 Cal.App.3d 798, 202 Cal.Rptr. 444 (1984).

and implied warranty. On the surface, *Bloomsburg* appears to be a professional standard case, but using "reasonable fitness for its intended purpose," along with "impliedly warrants their sufficiency" causes confusion. Obviously, in determining whether the design professional performed properly, the court should take into account the intended use of the structure. However, often a court looks both at what others would have done under similar circumstances *and* at the end result and compares it to the intended purposes of the client.

The murkiness of the formulas is also reflected in *E.C. Ernst, Inc. v. Manhattan Construction Co. of Texas*,[77] which stated that the implied warranty standard that the court claimed it was following did not require a *favorable* result but merely a *reasonable* result. This may not differ much from the professional standard, particularly if expert testimony is not needed.

The line between the two is less distinct if attention is drawn to the exact nature of the implied warranty claimed. For example, in *Allied Properties v. John A. Blume & Associates*,[78] the owner of a luxury hotel located along the ocean in Santa Barbara, California, brought a claim against a naval architect that had designed a pier. The claim was founded upon an implied warranty that the product (the pier) was "reasonably suitable for use by small craft." The court brushed aside the claim of implied warranty by citing earlier California cases accepting the professional standard, noting that the implied warranty cases reflected a minority view.

A careful reading of the opinion shows that the claim was based upon the infrequent use of the pier. But the evidence indicated a number of possible reasons—

rough water, length of the pier, depth of the water, availability of ample moorings nearby, and the formal atmosphere of the hotel not appealing to boating people.

In effect, the client wanted the naval architect to pay for the pier that was not being used much for reasons that had little to do with the naval architect's performance of professional services. Certainly the latter did *not warrant* that patrons would use the pier. Had the pier fallen because of structural instability, it would have been more likely that the court would have implied a warranty of structural soundness by the naval architect.

SECTION 17.08
SPECIAL LEGAL DEFENSES

The principal issue in claims against a design professional relates to the standard of performance and whether there has been compliance. This section briefly outlines a number of defenses that have been used by the design professional when claims are made by the client or third parties. The special problems of third-party claims are discussed in Section 17.09.

(A) APPROVAL BY CLIENT

Suppose the design professional asserts that the design has been approved by the client and this relieves him of any liability, even for negligent design. Ordinarily such a defense is not successful.[79]

The client retains a design professional because of the latter's skill in design, a skill not *usually* possessed by the client. The very purpose of engaging an expert would be defeated if the approval by the nonexpert client relieved the expert design professional from his negligence. Approval *does* au-

---

77. 551 F.2d 1026 (5th Cir.1977), rehearing denied in part, granted in part 559 F.2d 268 (5th Cir.1977).

78. 25 Cal.App.3d 848, 102 Cal.Rptr. 259 (1972).

79. *Eichler Homes, Inc. v. County of Marin*, 208 Cal.App.2d 653, 25 Cal.Rptr. 394 (1962); *Simpson Brothers Corp. v. Merrimac Chemical Co.*, 248 Mass. 346, 142 N.E. 922 (1924); *Bloomsburg Mills, Inc. v. Sordoni Construction Co.*, supra note 74.

thorize the design professional to proceed with the next phase of services, and client-directed changes *after* approval may justify the design professional who is compensated by a method unrelated to costs receiving additional compensation.

There may be an unusual circumstance that could justify approval relieving the design professional. Suppose the design professional points out a design dilemma to the client. The designer may inform the client that particular material may prove unsatisfactory for specified reasons but asks the client's approval of that material because it is less costly. If the client is apprised of all the risks and authorizes that particular material to be used, the client assumes the risk. Even if the design professional has not done what others would have done, approval by the client with *full* knowledge of the risks and with the ability to evaluate them should relieve the design professional.

Claims by third parties raise other problems. Even if approval by the client would bar its claim, this would not bar a third-party claim against the design professional. If approval by the client were *itself* negligent, any third party who suffered harm could recover against the client as well and may give the design professional an indemnification claim against the client.

As a general rule, client approval of the design does not relieve the design professional from liability for negligent design.

(B) PASSAGE OF TIME: STATUTES OF LIMITATIONS

Sometimes the design professional can defend by establishing that legal action was not started within the time required by law. This is usually accomplished by invoking the statute of limitations as a bar to the claim. Since this bar can apply to all claims—not only to those against design professionals—and since most construction-related claims are brought against other participants such as the owner or contractors, full discussion of this defense is postponed until Section 27.03(G). A few observations should be made in this section dealing with claims against design professionals.

Statutes of limitations usually prescribe a designated period of time within which certain claims must be brought. One of the troublesome areas in construction claims relates to the point at which that period *begins.* This can be a formidable problem in all construction-related claims, inasmuch as the defect may be discovered long after the design is created and executed. As a result, even in claims that do not involve the design professional, the law has had difficulty selecting from a number of base points, such as the wrongful act, the occurrence of damage, discovery of a substantial defect, or discovery of any defect.

This troublesome question is exacerbated when claims are made against a professional designer. The design is usually developed by the design professional, approved by the client, given to the contractor to execute, and often changed during construction, either to make design changes or to correct errors. As a result there are additional base points for determining when the period begins. Some jurisdictions, principally New York, look at design as an ongoing, trial-and-error process. With that in mind, New York holds that the period begins when the project has been completed.[80] Completion of the project as a base point is based upon the possibility that design errors can be corrected during construction and, a fortiori, during the design phase.

New York refused to apply this rule in a claim by a third party who brought a tort action based upon professional negligence against a designer, the court holding that

---

80. *Sosnow v. Paul,* 43 A.D.2d 978, 352 N.Y.S.2d 502 (1974), affirmed 36 N.Y.2d 780, 369 N.Y.S.2d 693, 330 N.E.2d 643 (1975).

the period began when the wrongful act was discovered, usually the time of injury.[81] The court held that the completion base point is proper in dealing with claims between client and design professional but not those brought by third parties.

In addition to the complexity resulting from a differentiation between claimants drawn in New York, additional confusion can result because the period for beginning action is usually longer for claims based upon breach of contract than for those based upon other wrongful conduct such as negligence. This becomes a serious issue, since claimants against design professionals are often given the option of bringing actions in tort or in contract. Since the basis for the claim can affect other issues, it is discussed in Section 17.11.

Generally a defense based on the passage of time—though still of value to the design professional—has provided limited protection. This has led to the enactment of completion statutes that seek to cut off liability after a designated period of time following substantial completion of the project. (They are discussed in Section 27.03(G).) Also, the possibility of claims long after performance has been completed has led to contract clauses which seek to bar such claims.

(C) DECISIONS AND IMMUNITY

The design professional is frequently given the power to interpret the contract documents, resolve disputes, and monitor performance. When claims are brought against a design professional by the client or others for activity that can be said to resemble judicial dispute resolution (such as deciding disputes or issuing certificates) design professionals sometimes assert that they should receive quasi-judicial immunity.

This is based upon the analogy sometimes drawn between judges and the design professionals performing judge-like functions. For example, the judge is given absolute immunity from civil action, even for fraudulent or corrupt decisions. The corrupt judge may be removed from office or subject to criminal sanctions. But a disappointed party cannot institute civil action against a judge. Immunity protects judges from being harassed by vexatious litigants and encourages them to decide cases without fear of civil action being brought against them.

Quasi-judicial immunity for design professionals has had a troubled history both in England and in the United States. The English House of Lords reversed an earlier decision and held that the architect can be sued by the owner for a negligently issued certificate.[82]

American decisions have not been consistent. Some have granted quasi-judicial immunity while others have not.[83] Where immunity is given, the design professional cannot be sued for decisions made unless they were made corruptly, dishonestly, or fraudulently. However, immunity does not protect against negligent delay in making a decision.[84]

Again the peculiar position of the design professional—independent contractor as designer, agent of the owner, decider of disputes, and an individual participating in construction projects—has caused difficulty. If immunity is granted, it should be based upon both parties to the contract—owner and contractor—agreeing to give cer-

---

81. *Cubito v. Kreisberg*, 69 A.D.2d 738, 419 N.Y.S.2d 578 (1979), affirmed 51 N.Y.2d 900, 434 N.Y.S.2d 991, 415 N.E.2d 979 (1980).

82. *Sutcliffe v. Thackrah*, 1974 A.C. 727.

83. Authorities are collected in *City of Durham v. Reidsville Engineering Co.*, 255 N.C. 98, 120 S.E.2d 564 (1961). See also *Blecick v. School District No. 18*, 2 Ariz.App. 115, 406 P.2d 750 (1965); Annot., 43 A.L.R.2d 1227 (1955).

84. *E.C. Ernst, Inc. v. Manhattan Construction Co. of Texas*, supra note 77.

tain judging functions to the design professional. Yet the cases that have been quickest to deny immunity have been actions instituted against a design professional by the client—the party who has selected and paid the design professional.[85] Courts that have denied immunity in such cases seem more impressed with the client's selection of and payment to the design professional as indicating the design professional's principal responsibility being to protect the owner. This, in addition to the increasing tendency to hold professional persons accountable, may not mean that immunity will never be available when the claim is made by the client but reduces the likelihood that such a defense will be successful.

One would think that claims by the contractor and certainly third parties should be at least as successful as those of the client. Third parties have not selected the design professional and have been forced in most cases to accept the design professional as judge. Yet a recent case granting immunity was one involving a claim brought by a contractor against the architect.[86]

There has been criticism of immunity even where limited to good-faith decisions. Can the design professional truly be neutral in rendering a decision when he has been selected and paid by the client? Also, as indicated earlier, the contractor rarely has much choice in these matters. The English House of Lords in *Sutcliffe v. Thackrah*[87] dealt with a claim by the client that the architect had negligently overcertified. Lord Reid drew a distinction between a dispute resolver who is a judge or arbitrator and an architect. According to Lord Reid, a true dispute resolver is a passive recipient of information and arguments submitted by the parties. An architect, on the other hand, is a professional engaged to act at his client's instructions and to give his own opinions. The judge or true arbitrator does not investigate but simply decides matters submitted to him. Lord Reid concluded that deciding whether work is defective is not judicial. He noted here there had been no dispute, that the architect was not jointly engaged by the parties, that the parties did not submit evidence to the architect, and that the architect made his own investigation and came to his own decisions.

The reasoning by Lord Reid may not be applicable in the U.S. because of different, though perhaps marginal, American practices. First, the architect is "approved" by both parties although *engaged* by the owner. Although the owner retains the design professional, the contractor knows who the design professional will be when it enters its bid or negotiates to perform the work. Under AIA Doc. A201, the contractor has a limited power of veto over any successor architect.[88] Also, A201 *seems* to contemplate the parties submitting evidence to the architect.[89]

The desirability of immunity was also passed upon recently by an American court in *E.C. Ernst, Inc. v. Manhattan Construction Co.*[90] The case involved a claim against the architect based upon his rejection of certain equipment proposed by the contractor. The court stated:

> The arbitrator's "quasi-judicial" immunity arises from his resemblance to a judge. [The court here is speaking of the architect as arbitrator.] The scope of his immunity should be no broader than this resemblance. The arbitrator serves as a private vehicle for the ordering of economic

---

85. *Newton Investment Co. v. Barnard & Burk, Inc.*, 220 So.2d 822 (Miss.1969).

86. *E.C. Ernst, Inc. v. Manhattan Construction Co. of Texas,* supra note 77.

87. Supra note 82.

88. AIA Doc. A201, ¶ 2.2.19 (1976).

89. Id., ¶ 2.2.12.

90. Supra note 77.

relationships. He is a creature of contract, paid by the parties to perform a duty, and his decision binds the parties because they make a specific, private decision to be bound. His decision is not socially momentous except to those who pay him to decide. The judge, however, is an official governmental instrumentality for resolving societal disputes. The parties submit their disputes to him through the structure of the judicial system, at mostly public expense. His decisions may be glossed with public policy considerations and fraught with the consequences of stare decisis [precedent]. When in discharging his function the arbitrator resembles a judge, we protect the integrity of his decisionmaking by guarding his fear of being mulcted in damages. . . . But he should be immune from liability only to the extent that his action is functionally judge-like. Otherwise we become mesmerized by words.[91]

The court then concluded that such immunity as possessed by the architect as arbitrator did not extend to unexcused delay or *failure* to decide, with immunity limited to "judging."

Determining whether to grant immunity must also take into account the finality of the design professional's decision (taken up in greater detail in Section 33.09). Under many contracts, the initial decision by the design professional can be taken to arbitration. If so, any wrong or even negligent decision can be corrected by the arbitrators. However, if arbitration is not used and the dispute goes to court, the decision by the design professional is likely to have a certain degree of finality. Giving the decision substantial finality and the decision maker immunity may be granting too much power to the design professional—power that can be abused. In Section 33.10 the suggestion is made that the process under which the design professional interprets the contract and decides disputes can be justified principally by expediency—the need to move construction along. This justification will not be adversely affected even if the decision were given very little finality and even if

the design professional is stripped of any immunity.

It is often difficult to determine whether the design professional is acting as *agent* or *judge*. Perhaps it is simply better to jettison immunity as the English have done. Such immunity as exists protects only against the negligent decision and not one made in bad faith. Also, as shall be seen in Section 18.03(E), one way of dealing with this problem is to provide immunity by contract.

(D) INTERFERENCE WITH CONTRACT OR PROSPECTIVE ADVANTAGE: THE ADVISER'S PRIVILEGE

This chapter has concentrated upon claims based upon professional negligence. Another legal doctrine sometimes invoked in claims against a design professional is intentional interference with contracts or prospective advantage. These are usually related to advice given by the design professional to the client or decisions made, such as termination. Claims based upon interference are similar to those based upon an assertion that a decision has been made wrongfully.

The interrelationship of these two types of claims can be illustrated by reference to the leading case of *Lundgren v. Freeman.*[92] The contractor had been terminated upon advice of the architects. The contractor brought a legal action against the school district owner and the architects. The claim against the architects was that they had willfully and maliciously interfered with the contractor's performance of the contract. The contractor claimed that the architects had damaged his reputation by stating that he had failed substantially to perform his contract and damaged his credit standing in the construction industry. The contractor and the school district submitted their dispute to arbitration. The ar-

91. Id. at 1033.

92. 307 F.2d 104 (9th Cir.1962).

bitrators made an award to the contractor; but the dispute between contractor and architects went to court.

The court noted that architects can function in one of three capacities: as agents, as quasi-arbitrators, or on their own. The court held that the contractor could not recover from the architects for acts committed in their capacity as agents because the contractor had elected to pursue the principal, that is, the school district.

As to acts committed in the architect's quasi-judicial capacity, immunity would normally be given, but not *full* immunity. The court noted that the architect is employed and paid by the owner and is often called upon to judge his own work. The court was willing to protect the architect only if he acts in good faith but not if he acts fraudulently or with willful or malicious intent to injure the contractor.

As to the third category, acts committed on their own, the architect is given a qualified privilege to interfere with the contract between contractor and owner. This privilege does not protect actions taken in bad faith or with the intention to injure the contractor. The contractor need not show specific intent to injure the contractor so long as he shows that the acts are willful and intentional. If the contractor can establish malice he can recover punitive damages as well.

The court returned the case to the trial court to determine whether there was a sufficient basis for the claim by the contractor that the advice to terminate was given for a wrongful purpose. (The contractor had alleged that there had been past poor relations between the contractor and the architects, that the architects had cut down the contractor's estimate too much, that the

architects had ordered needless expenditures to complete the work after the contractor had been terminated, and that other facts existed that he claimed showed a wrongful purpose.)

Often the parties have agreed that the very acts complained of can be taken. AIA Doc. A201, ¶ 14.2.1 requires a certification by the architect that reasonable cause exists for termination before the owner can elect to terminate. The contractor is aware of this when it makes the contract. True, if certification is denied for improper purposes, the design professional has wrongfully interfered in the relationship between the contractor and the owner. But in the absence of this, ¶ 2.2.10 bars such a claim.

Increasing claims have been made of *intentional* interference with contracts or prospective advantage, some in construction contexts.[93] It promises to be a new liability exposure area. These claims have been dealt with extensively in the Second Restatement of Torts. The section that applies most directly to claims against a design professional is § 772. It provides that there has not been improper interference if a person against whom the claim has been made has in good faith given honest advice when requested to do so.[94] In many instances, even without a formal power given the design professional to make decisions, the design professional may advise the client to terminate the relationship with a contractor. If the advice is given honestly and in good faith, no claim should be permitted against a design professional.

(E) CAUSATION

Transferring a loss requires the claimant to prove that the person to whom it seeks to

93. *Natco, Inc. v. Williams Brothers Engineering Co.,* 489 F.2d 639 (5th Cir.1974) (willful and malicious interference with performance); *Waldinger Corp. v. Ashbrook-Simon-Hartley, Inc.,* 564 F.Supp. 970 (C.D.Ill.1983) (wrongful rejection of proposed equipment); *Commercial Industrial Construction, Inc. v. Anderson,* 683 P.2d 378 (Colo.App. 1984) (rejected bidder claimed against architect: owner privileged to reject bid).

94. Followed in *Keck Piping Co., Inc. v. Town of Monroe,* 172 Conn. 197, 374 A.2d 179 (1977) (architect advised owner to reject proposed subcontractor).

transfer its loss has caused it. Multiple causation problems are endemic to construction losses because of the many participants who may play a role in causing the harm. Because multiple causation issues involve other participants, this problem is dealt with principally in Section 28.06. A few general observations are made in this chapter.

Two types of claims must be distinguished. First, there may have been wrongful conduct by the *claimant* which played a part in causing the harm. For example, as indicated in Section 17.08(A), the design may have been approved by the client, and in a sense approval can be said to have played a *part* in causing the harm. However, as discussion in that section indicated, often approval by the client is not wrongful.

Suppose the defect or the harm can be traceable to negligence by the design professional and failure to execute design by the contractor. When indivisible loss is caused by more than one participant, the claimant can recover the *entire* loss from *either*, with the ultimate loss between the participants being determined by laws relating to contribution between wrongdoers and indemnification.[95]

Suppose also that the claimant's negligence has played a substantial role in causing the harm. The law must determine the effect of the claimant's *wrongful* conduct being partially responsible. One solution is to decide who is more responsible and place the entire loss on that entity. Another is to share the loss. The modern tendency has been to avoid all or nothing solutions and require that persons who have each substantially contributed to causing the loss bear responsibility on some comparative basis.[96]

(F) SUMMARY

Legal defenses mentioned in this section and in Section 17.09 dealing with claims by third parties have furnished less protection to design professionals, as the law has emphasized compensating victims and placing responsibility upon those who derive economic benefits from activities causing harm. There are exceptions. The privity rule has provided some protection, at least in cases where the claimant has suffered economic losses. Design professionals have been receiving some protection from third-party claims by contract language seeking to make clear that the design professional's *only* function is to do his best to see that the project is built according to the contract documents. But on the whole, diminished effectiveness of these defenses, along with loosening of the need for expert testimony, has expanded liability exposure for design professionals.

SECTION 17.09 THIRD–PARTY CLAIMS: SPECIAL PROBLEMS

Third-party claims raise problems that do not arise when the claimant is connected to the design professional by contract. The proliferation of third-party claims has generated more litigation, varying state rules, and judicial opinions of divided courts than have claims by clients against design professionals. This reflects rules in transition with inevitable strains and contradictions.

(A) POTENTIAL THIRD PARTIES

The centrality of the design professional's position in construction—both designing and monitoring performance—generates a wide range of potential third-party claimants. At the inner core are the other major

---

95. *Northern Petrochemical Co. v. Thorsen & Thorshov, Inc.,* 297 Minn. 118, 211 N.W.2d 159 (1973). But see *Shepard v. City of Palatka,* 414 So.2d 1007 (Fla.App. 1981) (dictum: loss be shared by architect and contractor).

96. *S.J. Groves & Sons Co. v. Warner Co.,* 576 F.2d 524 (3d Cir.1978); *Grow Construction Co. v. State,* 56 A.D.2d 95, 391 N.Y.S.2d 726 (1977); *Environmental Growth Chambers, Inc.,* ASBCA No. 25845, 83–2 BCA ¶ 16,609. But see *Broce-O'Dell Concrete Products, Inc. v. Mel Jarvis Construction Co.,* 6 Kan.App.2d 757, 634 P.2d 1142 (1981).

direct participants in the process who work on the site itself, such as contractors and construction workers. Around the core are those who supply money, materials, or equipment, such as lenders and suppliers. Next are those who "backstop direct participants," such as sureties and insurers. Yet farther from the core are claimants who will ultimately take possession of the project, such as subsequent owners, tenants, and their employees. Farthest from the core are those who may enter or pass by the project during construction or after completion, such as members of the public or patrons.

The wide variety of potential claimants, the varying distance from core participants, the type of harm, the difference between those who have other sources of compensation, such as workers—all combine to insure complexity.

(B) THEORIES: CONTRACT AND TORT

Third-party claims must initially be divided into those based upon contract and those based upon tort. A contract claim must be supported by an allegation that the claimant was an intended beneficiary of the contract. For example, a contractor may base its claim against the design professional upon the assertion that it is an intended beneficiary (judicial resolution of this question varies greatly [97]) of the contract between the owner and the design professional and that the design professional has caused harm by breaching that contract. No negligence need be shown.

A tort claim, though not directly connected with contract breach, requires that the claimant establish that the design professional owes it a duty (see (E)) to avoid exposing the claimant to an unreasonable risk of harm, that the design professional did not live up to the standard of conduct required by law, and that failure to do so caused the claimant to suffer harm. The standard (as seen in Section 17.05) is the professional standard, at least where the claim is based upon professional design services. Although the standard is one of *professional* negligence rather than strict liability, some recent cases have held the engineer to strict liability based on the project having been an ultra-hazardous activity.[98]

The two theories cannot be so easily separated. A tort claim will be unsuccessful unless the claimant can establish that the defendant owes it a duty. That duty in construction cases is usually, though not exclusively, based upon the existence of a contract between the design professional and client, at least in the view of some courts. (See (E).) Those courts believe that the contract can control not only the nature of the duty but also its existence.

As shall be seen in Section 17.11, the classification of the claim, whether in tort or in contract, can have important results.

(C) ACCEPTANCE OF THE PROJECT

Acceptance of the project usually involves the owner taking possession of the completed project. It can, although under standard contracts it usually does not,[99] imply that the owner is satisfied with the work and bar any claim for existing defects or defects that may be discovered in the future.

For purposes of this chapter, acceptance

---

97. Compare *A.R. Moyer, Inc. v. Graham,* 285 So.2d 397 (Fla.1973) (contractor could sue architect in tort but not as intended contract beneficiary) with *John E. Green Plumbing and Heating Co. v. Turner Construction Co.,* 742 F.2d 965 (6th Cir.1984) (contractor could sue CM as intended beneficiary of CM-owner contract). See also, Sections 32.05(B), 32.07(H), and 32.08(D) for subcontractor cases. Third parties usually find tort claims more successful. See Section 17.09(D).

98. *Doundoulakis v. Town of Hempstead,* 42 N.Y.2d 440, 398 N.Y.S.2d 401, 368 N.E.2d 24 (1977), reproduced in part in Section 7.04(A).

99. AIA Doc. A201, ¶¶ 9.5.5, 9.9.4.

can affect any claim third parties have against those who have participated in design and construction. Although some early cases [100]—and occasionally recent ones [101]—have barred some such claims after acceptance, the modern tendency is to hold that acceptance does not bar third-party claims.[102] If acceptance bars claims it is because the intervening act of the owner—that of acceptance—is a *superseding cause* relieving even negligent participants of liability. This rationale is particularly weak when the claim is asserted against the design professional, the person who often decides whether the project has been completed and should be accepted. Even if the owner decides to accept, acceptance may not be negligent. While acceptance by the owner can be looked upon as an intervening cause, it is in most cases a nonnegligent one that should not immunize a design professional from his negligence.

Suppose, though, that acceptance by the owner precluded the design professional from correcting design errors. Should the design professional be immunized from third-party claims? If the owner *knew* that there were defects and *barred* their correction, the owner's intentional acts operate as an intervening cause to immunize the design professional. If acceptance does *not* immunize the design professional from a third party claim, the owner's acceptance should give the design professional a claim for contribution or indemnity against the owner if the latter's acceptance precluded correction of defects.

Immunization because of the acts of others—even negligent acts—are not favored today. Acceptance rarely bars third-party claims against a design professional.

(D) PRIVITY: PERSONAL HARM AND ECONOMIC LOSSES

Historically a party injured by a breach of contract could not sue the breaching party unless there were privity between the party harmed and the contract breacher. Privity was a relationship, usually though not exclusively, based upon a contract between the claimant and the contract breacher.

As suggested in Section 17.09(B), absence of privity usually bars a claim based upon breach of contract unless the claimant was an intended beneficiary of that contract.

As chronicled in Section 7.09, the requirement of privity had been an insurmountable obstacle when the claimant suffered personal harm because of another's negligent breach of contract. Early in this century, the privity requirement was abandoned in cases of personal harm. This elimination of the privity requirement occurred in actions against manufacturers, with negligence ultimately being replaced by strict liability.

Privity is not necessary in claims for personal harm made against a design professional based upon failure on the part of the latter to perform in accordance with the professional standard.

Tort law is not limited to claims for personal harm. In the Construction Process, claimants often suffer economic losses unconnected with personal harm or damage to property. Often they seek to recover these losses from participants with whom they *do not* have a contract. For example, contractors and sureties bring claims against design professionals either as intended beneficiaries or as persons who have suffered economic losses because of the tor-

---

100. *Sherman v. Miller Construction Co.*, 90 Ind.App. 462, 158 N.E. 255 (1927).

101. *Phifer v. T.L. James & Co., Inc.*, 513 F.2d 323 (5th Cir.1975); *Shetter v. Davis Brothers, Inc.*, 163 Ga.App. 230, 293 S.E.2d 397 (1982) (many exceptions); *Fauerso v. Maronick Construction Co.*, ___ Mont. ___, 661 P.2d 20 (1983).

102. *Pastorelli v. Associated Engineers, Inc.*, 176 F.Supp. 159 (D.R.I.1959); *Green Springs, Inc. v. Calvera*, 239 So. 2d 264 (Fla.1970); *Theis v. Heuer*, 264 Ind. 1, 280 N.E.2d 300 (1972); *McDonough v. Whalen*, 365 Mass. 506, 313 N.E.2d 435 (1974); *Totten v. Gruzen*, 52 N.J. 202, 245 A.2d 1 (1968); *Strakos v. Gehring*, 360 S.W.2d 787 (Tex.1962); *Andrews v. Del Guzzi*, 56 Wash.2d 381, 353 P.2d 422 (1960).

tious conduct of the design professional. Claimants have had success in some jurisdictions [103] and have failed in others.[104]

The privity "debate" has taken on a new form similar to the shift from *privity* to *duty* explored in Section 17.09(E). Reformulations are common in the constant struggle between victim compensation and protection against quantum leaps in liability. Stemming from product liability cases courts now examine the propriety of tort law to protect quality expectations usually "purchased" in the commercial world. Some have been unwilling to use tort law to compensate for commercial disappointments unrelated to personal harm or damage to property.[105] Others do not draw this line.[106] See Sections 17.11 and 17.12(C) for further discussion.

(E) DUTY: THE NEW PRIVITY: *KRIEGER v. GREINER; CALDWELL v. BECHTEL, INC.*

Claims against design professionals (mostly for personal harm) have generated great confusion in the past twenty-five years over the hotly debated issue of whether the de-

sign professional owes a duty to other participants, usually construction workers.

A tort claim requires that the defendant owe a duty to the claimant to act in a way that avoids exposing the claimant to unreasonable risk of harm. When the privity requirement was dropped in the late 1950s, design professionals switched to another defense. They asserted that they owed no duty to third party claimants because their project monitoring was directed toward a project fulfilling the contract obligations promised to the *owner*. (Of course, it was more complicated than that, something that is demonstrated in the cases reproduced ahead and the analysis in Section 17.12(B).)

Design professionals complained that courts were unjustifiably placing responsibility upon them if anything went wrong in the Construction Process. Courts did this by focusing upon the construction contract, with its many powers given the design professional, such as to reject work, stop the work, and provide general supervision. These powers, they contended, existed only to implement their "monitoring" function. The design professional, they asserted, had

---

103. *E.C. Ernst, Inc. v. Manhattan Construction Co. of Texas,* supra note 77 (subcontractor v. architect); *Cooper v. Jevne,* 56 Cal.App.3d 860, 128 Cal.Rptr. 724 (1976); *Rozny v. Marnul,* 43 Ill.2d 54, 250 N.E.2d 656 (1969) (surveyor); *Bates & Rogers Construction Corp. v. North Shore Sanitary District,* 92 Ill.App.3d 90, 47 Ill.Dec. 158, 414 N.E.2d 1274 (1981) (contractor v. engineer); *Conforti & Eisele, Inc. v. John C. Morris Associates,* 175 N.J.Super. 341, 418 A.2d 1290 (1980); *Shoffner Industries, Inc. v. W.B. Lloyd Construction Co.,* 42 N.C.App. 259, 257 S.E.2d 50 (1979); *Keel v. Titan Construction Corp.,* 639 P.2d 1228 (Okla.1981). For surety claims, see Section 26.07. See also Annot., 65 A.L.R.3d 249 (1975).

104. *Harbor Mechanical, Inc. v. Arizona Electric Power Cooperative, Inc.,* 496 F.Supp. 681 (D.Ariz.1980); *Peyronnin Construction Co. v. Weiss,* 137 Ind.App. 417, 208 N.E.2d 489 (1965); *Illinois Housing Development Authority v. M–Z Construction Corp.,* 110 Ill.App.3d 129, 65 Ill.Dec. 665, 441 N.E.2d 1179 (1982) (funding agency v. architect); *Delta Construction Co. v. City of Jackson,* 198 So.2d 592 (Miss.1967); *Bernard Johnson, Inc. v. Continental Constructors, Inc.,* 630 S.W.2d 365 (Tex.App.1982).

105. *Redarowicz v. Ohlendorf,* 92 Ill.2d 171, 65 Ill.Dec. 411, 441 N.E.2d 324 (1982) (tort claim to recover cost of repair and replacement of defective chimney, adjoining wall and patio barred). But see *Ferentchak v. Village of Frankfort,* 121 Ill.App.3d 599, 76 Ill.Dec. 950, 459 N.E.2d 1085 reversed on other grounds, Slip Opinion, Jan. 23, 1985. See also *R.H. Macy & Co. v. Williams Tile & Terazzo,* 585 F.Supp. 175 (N.D. Ga. 1984); *Lesmeister v. Dilly,* 330 N.W.2d 95 (Minn. 1983); *R.J. Reagan Co. v. Kent,* 654 S.W. 2d 532 (Tex.Civ.App.) (subcontractor barred from suing architect for negligent design). See Bertschy, *Negligent Performance of Service Contracts and the Economic Loss Doctrine,* 17 J.Mar.L.Rev. 249 (1984).

106. *Berkel & Co. Contractors, Inc. v. Providence Hospital,* 454 So. 2d 496 (Ala. 1984); *J'Aire v. Gregory,* 24 Cal.3d 799, 157 Cal.Rptr. 407, 598 P.2d 60 (1979) discussed in Sections 30.11 and 17.12; *Ferentchak v. Village of Frankfort,* supra note 105. See cases cited infra note 137.

no *right* or *duty* to tell the contractor *how* the work was to be done. Nor was he paid, in the ordinary project, to be a "safety" engineer.

As a result of adverse court decisions, standard agreements made by the professional associations were changed in the 1960s to seek to make clear that the contractor, not the design professional, decided how the work was to be done and that the design professional did not provide continuous on-site observation and certainly did not supervise the work. These changes, designers claimed, simply reflected the true allocation of responsibility for work on the site.

At this point, parts of two cases coming to different conclusions are reproduced.

KRIEGER v. J.E. GREINER CO.

Court of Appeals of Maryland, 1978.
282 Md. 50, 382 A.2d 1069.

SMITH, Judge.

[Ed. note: Footnotes omitted. Krieger was a construction worker injured when a steel column collapsed. He claimed that the erection subcontractor did not support the 780-pound reinforcing bars and the steel column. He asserted that the work was performed under the supervision of defendant Greiner, the prime engineer, and defendant Zollman, a consulting engineer. He asserted that each knew or should have known that the subcontractor was performing in a defective and dangerous manner and that the defendants had stopped work on other occasions when the work did not conform or was performed dangerously. Krieger and his wife brought legal action against Greiner and Zollman. The court noted the split of authority, summarized the leading cases, and stated that the "weight of authority is on the side of nonliability."]

2. This case

The Kriegers in their attempt to recover here have proceeded upon three theories, (1) that under the contracts between the Commission and Greiner and Zollman these engineers are responsible for supervision of the methods of construction, and hence are responsible for safety; (2) that the engineers under their contracts with the Commission are specifically responsible for safety; and (3) that, aside from the contracts, the engineers have assumed responsibility for safety.

a. The contracts with the engineers

The principles relative to construction of contracts are well known and have been enunciated by this Court numerous times: the clear and unambiguous language of an agreement will not give way to what the parties thought the agreement meant or intended it to mean; where a contract is plain and unambiguous, there is no room for construction, and it must be presumed that the parties meant what they expressed; and when the language of a contract is clear, the true test of what is meant is not what the parties to the contract intended it to mean, but what a reasonable person in the position of the parties would have thought it meant. *Board of*

Trustees v. Sherman, 280 Md. 373, 380, 373 A.2d 626 (1977);
Billmyre v. Sacred Heart Hosp., 273 Md. 638, 642, 331 A.2d 313
(1975), and cases there cited.

It will be recalled that in the declaration the Kriegers alleged
that Greiner and Zollman each "expressly agreed to see that the
method of the construction work conformed to all Federal, State,
and local laws and ordinances." The provision of the Greiner
contract from which this is derived states:

"V. GENERAL CONDITIONS

"1. *General Compliance with Laws*

The Consultant will observe and comply with all Federal,
State and Local Laws or Ordinances that affect those employed or engaged by him on the Project, or the materials or
equipment used, or the conduct of the work, and will procure
all necessary licenses, permits and insurance."

There also is an allegation in the declaration that Greiner and
Zollman "expressly agreed, by their respective contracts to be
responsible for all damage to life and property due to their
activities or those of their agents or employees." The relevant
portion of the Greiner contract on this subject appears in the next
succeeding paragraph to that which we have just quoted. It states:

"2. *Responsibility for Claims and Liability*

The Consultant will be responsible for all damage to life and
property due to his activities or those of his agents or employees, in connection with the services required under this
Agreement and will be responsible for all parts of his work,
both temporary and permanent, until the services under this
Agreement are declared accepted by the Commission, it being
expressly understood that the Consultant will indemnify and
save harmless the Commission, its members, officers, agents,
and employees of, from and against all claims, suits, judgments, expense, actions, damages, and costs of every name
and description, arising out of or resulting from the services
of the Consultant under this Agreement."

The contract then contains a covenant against contingent fees, and
goes on to provide for termination, ownership of documents, and
the like. The contract with Zollman contained identical provisions under "General Conditions." To us it is obvious that this
language does not attempt in any manner to place upon the
engineers a duty to see that the contractors obey all laws in
connection with the work. It is simply an agreement on the part
of the engineers that insofar as the work *they* perform on behalf of
the Commission is concerned they will comply with such laws.
Likewise, this quoted language does not purport to hold the
engineers responsible for life and property generally in connection
with the construction of the bridge. It is an agreement on the part

of the engineers that as to the work *they* perform they will be responsible for damage to life and property.

Aside from the two paragraphs of the contract which we have just quoted, obviously taken out of context in the declaration, the recitals of contract provisions in the narr. [the abbreviation of *narratio*, the Maryland name for a portion of the legal pleading in which the plaintiff states the basis of its claim.] are substantially correct.

We have carefully examined each of the contracts in question. We find no provisions in these contracts imposing any duty on the engineers to supervise the *methods* of construction. Some mathematics instructors have been heard to observe that there is more than one solution to a given problem and thus they are unable to say that any given method of solving a problem is the only correct solution, being able only to determine that the correct answer is produced. The same reasoning would apply to methods of construction. One skilled contractor may prefer one method for performing a given task while another such contractor may choose what seems to him a simpler, less expensive way of reaching the same end result, either of which procedures would be a proper method. It could well be, however, that one method might not have occurred to an engineer or another contractor.

We likewise find nothing in the contracts imposing any duty on the engineers to supervise safety in connection with construction.

The duty of the engineers under their contracts is to assure a certain end result, a completed bridge which complies with the plans and specifications previously prepared by Greiner. It will be observed that many of the cases which have held architects and engineers responsible for safety have done so upon the basis of the construction by the courts of the contracts existing between the engineer or architect and the owner. We hold that a fair interpretation of the contracts between the Commission and Greiner and Zollman is that the duties of those engineers do not include supervision of construction methods or supervision of work for compliance with safety laws and regulations. Hence, the Kriegers may not recover from the engineers under the contracts between the owner and its engineers.

b. Assumed responsibilities

Our determination relative to the contractual provisions does not end the matter, however, because in their declaration the Kriegers alleged that Greiner and Zollman had "inspectors and engineers [on the job who] had exercised their right to stop the work on numerous other occasions when said work . . . was being performed in a negligent and dangerous manner which was unsafe for the workmen employed on the project" and that the Commission "required [Greiner and Zollman] to supervise the work as aforesaid in order to insure the safety of . . . the workmen on the job." These alleged facts, if true, might provide a basis for recovery. The declaration, however, refers only to the two contracts with the Commission discussed above, which are treated here under Rule 326 "as if incorporated in the pleading." As we have shown, those contracts negate the allegation that the Commission required the engineers to supervise the work for safety. However, the Kriegers should have an opportunity to amend their declaration to allege—if they have a proper foundation for such an allegation—some other possible hypothesis for recovery, such as an amendment of the contract between the engineers and the Commission, or a separate, supplemental agreement between them.

JUDGMENT REVERSED AND CASE REMANDED FOR FURTHER PROCEEDINGS CONSISTENT WITH THIS OPINION; . . .

LEVINE, J., concurs in the result with opinion in which EL-DRIDGE, J., joins. [Ed. note: concurring opinion omitted]

CALDWELL v. BECHTEL, INC.
United States Court of Appeals, District of Columbia Circuit, 1980. 631 F.2d 989.

MacKINNON, Circuit Judge:

[Ed. note: Footnotes renumbered and some omitted.]

We are here concerned with a claim for damages by a worker who allegedly contracted silicosis while he was mucking in a tunnel under construction as part of the metropolitan subway system. The basic issue is whether a consultant engineering firm owed the worker a duty to protect him against unreasonable risk of harm.

The Shea–S&M–Ball joint venture (hereinafter Shea) entered into a contract with the Washington Metropolitan Area Transit Authority (hereinafter WMATA) in the construction of tunnels for the Washington Metro Subway system and appellant Clem Caldwell was employed by Shea as a heavy equipment operator. Appellee Bechtel, Inc. (hereinafter Bechtel) was also under contract with WMATA, as a consultant to provide, *inter alia*, "safety engineering services" with respect to work to be done by various contractors pursuant to their contracts with WMATA. Among the duties that Bechtel undertook to perform for and on behalf of WMATA under the contract between the parties was the function of overseeing the enforcement of safety provisions in relevant safety codes, and inspecting job sites for violations. . . .

This appeal attacks the district court's grant of defendant Bechtel's motion for summary judgment. . . .

The essence of Caldwell's complaint is that Bechtel "had the function, duty and responsibility, as consultant to Metro [WMATA] to provide, *inter alia*, overall direction and supervision of safety measures and regulations in effect, or needed during the course of construction . . ." (App. 5), and that Bechtel was aware or should have been aware of the danger posed by high levels of silica dust and inadequate ventilation in the Metro tunnels, but failed to take the steps it was duty–bound to take to rectify the situation. This failure, Caldwell alleges, was not only in violation of Bechtel's contract with WMATA and applicable safety codes, but more importantly constituted gross negligence toward Caldwell who was working in the tunnels.

While we must accept Caldwell's allegations of fact, we will closely examine his version of the law, since it stands in contradiction to the district court's conclusion "that the WMATA–Bechtel Contract created no duty owed plaintiff by defendant the breach of which would give rise to plaintiff's action for negligence." (App. 172). The issue in this case, then, is whether the contractual authority vested in Bechtel with respect to job site safety regulations created a special relationship between Bechtel and Caldwell under which Bechtel owed a duty to Caldwell to take reasonable steps to protect him from the foreseeable risk to his health posed by the dust laden Metro tunnels. We find that under applicable tort law principles Bechtel was indeed duty-bound. Accordingly, we reverse the decision of the district court.

* * *

In our view, the analysis of both Bechtel and the district court is overly reliant upon contract theory to the point of losing focus of the nature of the claim made here, which asserts negligence, rather than breach of contract. It has been many years since courts required privity of contract between the plaintiff and defendant before assessing tort liability, yet Bechtel would return us to that distant day. The duties that Bechtel undertook in its contract with WMATA are relevant to this case, not because they illustrate Bechtel's point that a contractual duty was owed only to WMATA, but because by assuming a contractual duty to WMATA, Bechtel placed itself in the position of assuming a duty to appellant in tort. The particular circumstances of this case, including the Bechtel–WMATA contract, Bechtel's superior skills and position, and Bechtel's resultant ability to foresee the harm that might reasonably be expected to befall appellant, created a duty in Bechtel to take reasonable steps to prevent harm to appellant from the hazardous conditions of the subway tunnels.

III

In attempting to convince the court that it owes no duty of reasonable care to protect appellant's safety, Bechtel argues that by its contract with WMATA it assumed duties only to WMATA.[107]

---

107. Bechtel also maintains that it owed appellant no duty to protect him against risk of harm since appellant was an incidental, rather than an intended third party beneficiary under the WMATA–Bechtel contract. Yet, even the question of whether appellant was an intended third party beneficiary is debatable. Certain provisions of the WMATA–Bechtel contract might be construed to indicate an intent to benefit construction workers such as Caldwell. For instance, Bechtel contracted to supervise compliance with the provisions of the Safety Program and the Safety Manual, which are written in terms strongly protective of the workers' right to a safe working environment. It

Appellant has not brought this action, however, for breach of contract but rather seeks damages for an asserted breach of the duty of reasonable care. Unlike contractual duties, which are imposed by agreement of the parties to a contract, a duty of due care under tort law is based primarily upon social policy. The law imposes upon individuals certain expectations of conduct, such as the expectancy that their actions will not cause foreseeable injury to another. These societal expectations, as formed through the common law, comprise the concept of duty.

Society's expectations, and the concomitant duties imposed, vary in response to the activity engaged in by the defendant. If defendant is driving a car, he will be held to exercise the degree of care normally exercised by a reasonable person in like circumstances. Or if defendant is engaged in the practice of his profession, he will be held to exercise a degree of care consistent with his superior knowledge and skill. Hence, when defendant Bechtel engaged in consulting engineering services, the company was required to observe a standard of care ordinarily adhered to by one providing such services, possessing such skill and expertise.

A secondary but equally important principle involved in a determination of duty is to whom the duty is owed. The answer to this question is usually framed in terms of the foreseeable plaintiff, in other words, one who might foreseeably be injured by defendant's conduct. This secondary principle also serves to distinguish tort law from contract law. While in contract law, only one to whom the contract specifies that a duty be rendered will have a cause of action for its breach, in tort law, society, not the contract, specifies to whom the duty is owed, and this has traditionally been the foreseeable plaintiff.

It is important to keep these differences between contract and tort duties in mind when examining whether Bechtel's undertaking of contractual duties to WMATA created a duty of reasonable care toward Caldwell.

[Ed. note: The court quoted scholarly writings and cases.]

These cases illustrate that courts have primarily premised an extension of liability to the site architect upon its contractual undertaking on behalf of the project owner, and upon the resultant foreseeability of injury to workers in the event that the undertaking is negligently performed. We endorse this interpretation of the interrelationship of contractual duties owed to one party upon possible duty in tort owed to another party, and find it applicable to the facts at hand. Before making this application, however, we find it equally helpful to refer to an analogous line of cases in which defendants have been found to owe a duty to a

---

might thus be ventured that Caldwell was an intended third party beneficiary of the WMATA–Bechtel contract. We need not decide this issue, however, since Caldwell sues not to enforce any alleged rights under the contract, but for relief in tort. We accordingly base our opinion upon principles of duty owed in tort law.

plaintiff because of a special relationship between either the defendant and a third party, or defendant and plaintiff.

Analyzing the common law, Prosser[108] noted that courts have found a duty to act for the protection of another when certain relationships exist, such as carrier–passenger, innkeeper–guest, shipper–seaman, employer–employee, shopkeeper–visitor, host–social guest, jailor–prisoner, and school–pupil. These holdings suggest that courts have been eroding the general rule that there is no duty to act to help another in distress, by creating exceptions based upon a relationship between the actors.

* * *

We find that case law provides many such analogous situations from which the principles deserving of application to this case may be culled. The foregoing concepts of duty converge in this case, as the facts include both the WMATA–Bechtel contractual relationship from which it was foreseeable that a negligent undertaking by Bechtel might injure the appellant, and a special relationship established between Bechtel and the appellant because of Bechtel's superior skills, knowledge of the dangerous condition, and ability to protect appellant. We thus look to the relevant facts.

IV

The first component of the duty that Bechtel owed to appellant takes as its point of departure Bechtel's contractual duties to WMATA. In discussion of these contractual provisions we are not unmindful of the limited meanings the contract gives to some words by its "Definitions," *supra.* Under the contract Bechtel was to provide:

> safety engineering services as required to ensure compliance with the provisions of the Metro Construction Safety Manual, the Metro Coordinated Safety Program and Reporting Procedures and other applicable codes, and the contractual obligation of the Authority's contractors, and [Bechtel] shall direct the contractor to correct any unsatisfactory condition which may be detected.

Bechtel was also "to develop and ensure a uniform system of safety and accident prevention procedures including reporting requirements." (App. 168).

Several duties are encompassed in these contractual terms. First, Bechtel was charged to ensure compliance with the Safety Manual, which included among its admonitions that "[n]o man shall be required to work in an unsafe place". Second, Bechtel's

---

108. [Ed. note: Here the court referred to W. Prosser, *Handbook of the Law of Torts* (4th ed.1971), a leading text. A subsequent edition, the fifth, with P. Keeton as co-author, was published in 1984.]

contract directed Bechtel's Resident Engineer to "report on unsafe working conditions" and to receive and monitor copies of the contractor's daily safety inspection reports and atmospheric logs, and gave the Resident Engineer authority to order work stopped "if unsafe conditions exist until such time as the condition is corrected". (App. 142).

Third, Bechtel was to ensure, as defined, that the other contractors on the job obeyed safety requirements and fulfilled their contractual obligations to WMATA.[109] One such duty owed by Shea to WMATA was to ensure "that the methods of performing the work do not involve undue danger to the personnel employed thereon . . . ." (App. 165). And the responsibility fell to Bechtel to "direct [as defined] the contractor to correct any unsatisfactory condition which may be detected." (App. 168).

While the Bechtel–WMATA contract requires only that Bechtel use its "best efforts to persuade [Shea] . . . to comply" with safety regulations (App. 168) and thus Bechtel would not be absolutely liable to WMATA in the event of a safety violation, we are not only concerned with assessing Bechtel's duties to WMATA. Rather, the significance of the Bechtel–WMATA contract is that once Bechtel undertook responsibility for overseeing safety compliance, it assumed a duty of reasonable care in carrying out such duties that extended to the workers on the site.[110]

In the case that forged the limits of the common law concept of duty, *Palsgraf v. Long Island R. Co.*, 248 N.Y. 339, 162 N.E. 99 (1928), Chief Judge Cardozo declared that "the orbit of the danger as disclosed to the eye of reasonable vigilance would be the orbit of duty". Further emphasizing the relationship between the plaintiff and one properly named as defendant in such a case, he added: "[t]he risk reasonably to be perceived defines the duty to be obeyed, and risk imports relation; it is risk to another or to others within the range of apprehension." 248 N.Y. at 343, 344, 162 N.E. at 100.

With these time-honored principles in mind, we have no difficulty in concluding that appellant was "within the orbit of danger" and hence within "the orbit of duty" owed by Bechtel. Among Bechtel's safety engineering services was the monitoring of atmospheric logs, as well as its more generalized duty to ensure that "[n]o man [worked] in an unsafe place". These duties thus focused upon safety in the workplace—appellant's place of employ-

---

109. We reject the district court's conclusion that because primary responsibility for safety fell to Shea, Bechtel could not be liable in tort to appellant. Shared responsibility for safety arising out of contract, as here, poses no barrier to liability of the supervising agent. *See, e.g., Erhart v. Hummonds, supra*, 334 S.W. at 872. While Bechtel's duty in tort to appellant should be considered relative to the circumstances, more than one party can owe a duty in a given set of circumstances, and Bechtel's duty is to be judged here on the basis of *its* relation to the circumstances.

110. Any argument that in fact Bechtel used its best efforts to ensure safe working conditions goes, of course, to the merits and we express no opinion on that point. We do hold, however, that merely notifying WMATA and the contractors of the problems in the tunnel would not necessarily extinguish a duty owed in tort.

ment. The primary workplace of course was the system of Metro tunnels under construction, the source of calculable danger to those employed there. If unsafe conditions were allowed to exist in the tunnels, injury to those engaged in the construction process was foreseeable. This was especially true since the tunnels were shown by atmospheric testing to contain a dangerous level of silica dust. Hence, Bechtel should be held to a duty to perceive and take reasonable steps to rectify the unreasonable risk posed by the hazardous conditions, and the company's duty extends to the protection of the workers, who were, we conclude, foreseeable victims of that danger.

The second component of duty owed by Bechtel is not entirely distinct from the first, since it too is premised in part upon the contractual relationship between Bechtel and WMATA. The contract provides an initiating source of duty, since Bechtel was on the job site, and in charge of safety engineering, because of its contractual duties. Bechtel's presence upon the site signified more than mere fulfillment of contractual duties, however; Bechtel possessed a status that transcended its contractual purposes. Bechtel was the safety engineer on the site, and as such assumed a special relationship to the workers also on the job site. Not only was Bechtel armed with contractual ability to inspect for safety violations *and to stop work,* it also possessed the special skills of one engaged in the profession of safety engineering. Additionally, Bechtel was informed of the high concentration of silica dust and inadequate ventilation in the subway tunnels. It is apparent then that Bechtel was fully possessed of the power to protect appellant, and stood in a superior position from which to do so. To an equal degree, therefore, we rely upon the special relationship [111] formed between Bechtel and appellant as grounds for the imposition of a

---

111. Professor Sweet, one of the nation's leading experts in the field of construction law, proposes that the architect or construction manager be held to a duty to construction workers because of his status, or special relationship.

An architect who has contracted to perform traditional site services owes a duty to construction workers. This duty does not depend upon the architect's having particularly broad powers, such as the power to supervise, direct or stop work, an emphasis mistakenly employed in some of the seminal cases. Nor can the duty be completely negated by the architect's contract. The duty arises from the simple fact that architects and construction workers are coparticipants in a dangerous enterprise. They are both physically on the site, often at the same time. Each would expect the other to act when danger surfaces. They are *not* strangers. [Sweet, *Site Architects and Construction Workers: Brothers and Keepers or Strangers?,* 28 Emory L.J. 291, 327 (1979).]

While there is some intimation in the foregoing passage that Professor Sweet relies less upon the contractual duties of the defendant architect or construction manager than we do, the Professor does define the duty of the architect in his conclusion to be: "to take reasonable steps, *as judged by his contract,* to discover unsafe practices and contract deviations which expose workers to an unreasonable risk of harm." *Id.* at 334 (emphasis added). Hence, while we generally agree with Professor Sweet that the status of an architect or construction manager on the site creates the duty to construction workers, we also believe that the scope of the duty cannot totally be divorced from the architect's contractual undertaking. We note this not to imply that such a duty in tort can be fundamentally changed by denying responsibility when drafting a contract, but rather upon the belief that the actual scope of duties for safety, supervision, etc., assumed in the contract will be apparent, and will also be a reflection of the superior abilities brought by the engineer or architect to the site. This in turn will influence the standard of duty.

duty on the part of Bechtel to take steps reasonable under the circumstances to protect appellant from the foreseeable risk of harm posed by the unsafe level of silica dust.

V

We reverse the summary judgment of the district court, and hold that as a matter of law, on the record as we are required to view it at this time, Bechtel owed Caldwell a duty of due care to take reasonable steps to protect him from the foreseeable risk of harm to his health posed by the excessive concentration of silica dust in the Metro tunnels. We remand so that Caldwell will have an opportunity to prove, if he can, the other elements of his negligence action.

Judgment accordingly

That the issue continues to be hotly debated is demonstrated not only by both the *Krieger* and *Caldwell* cases having generated scholarly comment [112] but also in the varied conclusions in cases since they were decided.[113] But as emphasized elsewhere in this treatise, *factual* differences, such as different contract terms, different actual practices, and whether *obviously* unsafe practices were occurring, may account for different conclusions. Yet even when these *are* taken into account, judicial opinions can still reflect different emphases, such as *Krieger* on the contract language and disclaimers of responsibility and *Caldwell* on tort concepts of compensation, foreseeability, and avoiding harm to persons.

A compromise seems to be emerging. Design professionals who perform the usual functions under terms such as found in AIA Documents will not be liable unless they *know* of unsafe practices and take no steps to advise or warn owner or contractor.[114] Their status as site architects does *not* make them automatically liable. But neither does it automatically *exclude* liability where they do not act reasonably.

(F) PASSAGE OF TIME: STATUTES OF LIMITATIONS

(Refer to Section 17.08(B).)

SECTION 17.10   REMEDIES

The remedies available when the design professional does not perform in accor-

---

112. See Goldberg, *Liability of Architects and Engineers for Construction Site Accidents in Maryland—Krieger v. J.E. Greiner Co.; Background and Unanswered Questions*, 39 Md.L.Rev. 475 (1980); Comment, *Architect's Liability for Construction Site Accidents*, 30 U.Kan.L.Rev. 429 (1982) (*Caldwell* criticized as too broad).

113. For worker, see *Duncan v. Pennington County Housing Authority*, 283 N.W.2d 546 (S.D.1979) (architect knew of OSHA violation and expected to deal with safety). Cases favorable to the architect were *Porter v. Stevens, Thompson & Runyan, Inc.*, 24 Wash.App. 624, 602 P.2d 1192 (1979); *Hortman v. Becker Construction Co.*, 92 Wis.2d 210, 284 N.W.2d 621 (1979); *Hanna v. Huer, Johns, Neel, Rivers & Webb*, 233 Kan. 206, 662 P.2d 243 (1983); and *Welch v. Grant Development Co.*, 120 Misc.2d 493, 466 N.Y.S.2d 112 (1983).

114. *Duncan v. Pennington County Housing Authority*, supra note 113; *Balagna v. Shawnee County*, 233 Kan. 1068, 668 P.2d 157 (1983), citing dictum in *Hanna v. Huer, Johns, Neel, Rivers & Webb*, supra note 113.

dance with the contract or tort obligations were set forth in Sections 7.10 and 15.14.

SECTION 17.11 TORT AND CONTRACT: CLIENT CLAIMS

Sections 17.08(B), 17.09(B), and 17.09(D) noted the possibility that claimants may seek to base their claims upon breach of contract or upon tort, usually—at least in claims against design professionals—requiring a showing of professional negligence. Sometimes classifying a claim as contract or tort can be advantageous to one party and disadvantageous to the other. The two most significant differences for purposes of this chapter are the remedy (more expansive in tort except as noted in Section 17.09(D) in states which bar tort claims for economic losses.) and the statute of limitations (longer for contract). Yet there are others that have been discussed in this treatise such as the parol evidence rule,[115] the Statute of Frauds,[116] punitive damages,[117] prejudgment interest,[118] and governmental immunity.[119] Although there are others, this section concentrates upon a differentiation based upon the passage of time and remedies available.

Before addressing these issues, certain points must be clearly in mind. First, a breach of contract requires proof that the defendant has not done what it has promised to do in its contract, as a rule a nonfault standard. Some limited defenses are provided by law, such as impossibility or frustration.

A tort claim, on the other hand, requires proof that the defendant did not perform in accordance with the requirements the law imposes on everyone. For purposes of this chapter, this means professional negligence, a fault-based standard. Although in theory the tort standard is *more* difficult to show, as has been seen, in contracts for professional services the standard may be in actuality the same because the standard is not expressed specifically in the contract.

Second, clearly the duty owed in a contract-based claim is to the other party to the contract or an intended beneficiary. In tort there is a requirement that the tortfeasor owe a duty to the claimant, usually based upon the foreseeability that acts or failure to act on the part of the tortfeasor will expose the claimant to an unreasonable risk of harm. In claims against a design professional, the underlying duty is almost always created by contract (rarely, the assumption of duty by acting), the engagement or retention of the design professional. Without a *contract* no claim would exist, under the view of even the tort-oriented approach of *Caldwell v. Bechtel, Inc.* (reproduced in Section 17.09(E)).

Tort exposure is broader, as it exposes the design professional not only to a potentially large range of claimants but also to diminished foreseeability limitations (compared to contract) and more expansive remedies again except in those states that bar tort claims for economic losses. Except for a shorter period to bring a legal claim, often ameliorated by a "discovery" rule,[120] tort claims are generally preferable to the claimant.

Third, traditionally the law has been quickest to use tort to protect against harm to persons or property and more reluctant to protect against economic losses that re-

---

115. Refer to Section 14.04(E).

116. Refer to Section 5.10.

117. Refer to Section 6.04.

118. Refer to Section 6.08.

119. Refer to Section 7.06(D).

120. See Section 27.03(G).

late to disappointed consumer quality expectations as noted in Section 17.09(D), economic losses being more appropriate for contract law.

Fourth, it is important to see the difference between framing the issue as one that allows the claimant to elect which theory to employ as opposed to the law seeking the *essence* of the claim, sometimes referred to as the gravamen, when that is relevant. An election given to the claimant to choose which theory to employ assumes that the claim is not in essence one in contract or in tort but one that can be categorized as either.

Some generalizations can be made nationally. It appears that courts in general seem less willing to bar claims because of the passage of time and more willing to grant expansive tortlike remedies. This generalization means that when passage of time is the issue, it is likely that the claimant will be allowed to use breach of contract with its longer statute.[121] When the issue is the remedy, the courts will allow more expansive remedies available through tort, such as less emphasis on foreseeability as a limiting factor and greater willingness to allow damages for emotional distress or to punish. This generalization may not always apply because the language of statutes of limitations varies, the case precedents may be from a period before these trends developed, or judges struggling with these paradoxes may come to different conclusions. Again, like other matters that have been discussed in this treatise, local law must be consulted.

Analysis *can* be useful. It is clear that the underlying basis for claims by a client against the design professional is the contract they have made. As indicated in Section 17.04, that contract may be explicit both as to specified duties and how they are to be performed or it may be explicit as to duties but silent as to the standard. Many retention arrangements are silent as to specific duties and how they are to be performed.

A contract that covers *in detail* the duties of each party and allocates the risks of losses likely to occur is a *plan* under which the parties agree to exchange their performance and apportion risks in a specific manner. To disregard that plan by allowing tort claims exposes design professionals to risks they did not plan to undertake and for which they were not paid. Clients should not have an *election* to bring their actions either in contract or tort. In planned transactions, subject to an exception to be described later, they must base their claims upon breach of the contract.[122]

Emphasis on "the plan" should not ignore the possibility that a contract breach *can* be classified as negligent. For example, suppose the contract specified that the design professional will visit the site twice a week. Failure to visit the site in accordance with the obligation is a breach of contract. The design professional also may be negligent if a second visit in a particular week would have been made by other design professionals because concrete was to be poured on that day. Should a client who can satisfy the theoretical higher tort burden be allowed the tort remedy?

Again the paradox becomes apparent. In most claims there is no difference in the standard of proof. One court drew this distinction, pointing to a detailed contract stating that the design professional would do a number of different things, among them "supervise." The court concluded that this provided a specific standard that would *allow* a claim to be based on breach of contract even though a more general standard, such as the obligation to perform in

---

121. *Board of Education, etc. v. Del Biano & Associates, Inc.*, 57 Ill.App.3d 302, 14 Ill.Dec. 674, 372 N.E.2d 953 (1978); *Sears, Roebuck & Co. v. Enco Associates, Inc.*, 43 N.Y.2d 389, 401 N.Y.S.2d 767, 372 N.E.2d 555 (1977).

122. *Lesmeister v. Dilly*, supra note 105.

accordance with the performance of other professionals, would have classified the claim as a tort claim.[123] However, even if the contract states that the design professional will supervise, the professional standards are likely to be employed to determine *whether* the design professional has supervised properly.

The problem becomes more complex when account is taken of the possibility that design professionals may cause harm in the same manner as any other persons and that as a result they may damage property unconnected with the project itself. For example, suppose the design professional drives to the site by car, backs up the car, and damages the work in process. Must any claim by the client for the property damage be based *solely* upon the contract? Can the claim be based upon tort law?

The latter would be the standard if a mail truck backed up and caused the same damage. It would also be the same standard that would be applied if the design professional injured a worker on the site or the postal employee delivering mail. It would be strained and artificial to hold the design professional to a nonfault contract standard by concluding that each party to the contract, in this case client and design professional, impliedly promised not to damage the other's property, at least in performing collateral activities such as getting to and from the site.

Any analysis must take into account the reason why statutes of limitations are longer for contract claims. (It has already been noted that tort claims for economic losses cannot be justified by a higher *proof* level required for a tort claim.) Either such claims are worthier (the undesirability of having such claims lost solely by the passage of time) or proof of such claims is easier if there is a contract with a set of rules that relieves the court from having to make them. (Again the paradox of the planned transaction versus the unplanned, casual engagement is apparent. In the latter, the court will have to furnish the rules anyway.)

Suppose a line is drawn based upon the completeness of planning. The issue would be, Is this a carefully planned transaction? If so, should design professionals be exposed to claims for *longer* periods of time, penalizing them for having *planned* the transaction carefully? Design professionals who plan the transactions and who can be assured that the law will allow *only* a contract claim have other advantages if clients are not given an *election* to sue in tort. They will be able to rely on the contractual protection they have included, and if the statute of limitations period seems long, they can incorporate a private statute of limitations that can specify either the start of the period, the duration of the period, or both.

The claimant should not be given a choice to use tort unless one of the following occurs:

1. The transaction is essentially unplanned, such as a casual letter or handshake agreement.

2. The claim is based upon harm to person or property that is not dependent upon the underlying contract and that would be equally applicable were the harm caused by anyone else.

This suggestion also is compatible with a desire to encourage persons to plan their transactions by freeing them from the risk of open-ended tort exposure. Contracting parties who have suffered emotional distress can be protected by classifying those transactions, though based upon contract, as torts. Similarly, if there is a need to punish or deter, the claim can be based upon tort law. But other claims based upon failure to perform under the design pro-

---

123. *Securities-Intermountain, Inc. v. Sunset Fuel Co.*, 289 Or. 243, 611 P.2d 1158 (1980).

fessional-client relationship should be based *solely* upon the contract.[124]

SECTION 17.12
CURRENT CONTROVERSIES:
SOME OBSERVATIONS

It would be impossible to comment at length on the issues related to professional liability addressed in this chapter. But this section highlights issues that have generated heated controversies.

(A) THE PROFESSIONAL STANDARD: SHOULD PROFESSIONALS BE TREATED DIFFERENTLY?

The professional standard in essence permits local professional practice to be the legal standard. It has been subjected to intense criticism. Manufacturers of products [125] and those who build and sell homes [126] are held to strict liability. Owners warrant to the contractor that the design they have required will be sufficient.[127] Should professionals be given special dispensation when they have caused harm?

This specialized treatment may have had some justification, according to its attackers, when professionals were practicing in one-person offices or in small firms and in times when the professionals largely learned from those practicing in their localities. Today, professionals such as doctors, lawyers, and many of the design professionals practice in large organizations, make large profits, compete nationally, and receive education and training that uses state or national, not local practices. Those who attack the professional standard point to this as a reason why professionals today do not deserve special treatment. They argue that design professionals must accomplish the objective for which they are retained. Any increased liability that *may* result can be handled by insurance, and the cost spread to those who use the professional services. Even if insurers generally exclude contractual risks that go beyond negligence, those who insure are part of a competitive industry that will respond to market pressures. In any event, critics argue that insurance practices should no more determine the appropriate legal standard than professional practices. Attackers also stress the following:

1. Antitrust laws are increasingly applied to the professions. The law views their members as businesspersons rather than professionals.

2. The commercial nature of the design services being recognized will allow design professionals to contract for a lower standard or limit if not exculpate themselves from liability if they are faced with open-ended rules. (See Section 18.03.)

3. The current standard greatly increases the cost of litigation by requiring a parade of well-paid experts.

4. The line between design, to which the professional standard applies, and site monitoring, where it may not, is so blurred that it adds an added dimension of uncertainty.

5. To obtain a commission, the design professional often stresses it is *better* than others with whom it competes yet asserts it should be measured by the professional standard that looks to *average* local practices.

Indictment of the professional standard

---

124. This can avoid the anomalous result in *Rosos Litho Supply Corporation v. Hansen*, 123 Ill.App. 3d 290, 78 Ill.Dec. 447, 462 N.E.2d 566 (1984) in which an architect was held in tort to his client for economic losses despite a finding that he did not breach his contract.

125. Refer to Section 7.09.

126. Refer to Section 28.09.

127. Refer to Section 27.05(E).

has had an effect on the law such as the loosened requirements for expert witnesses [128] and the beginning of a tendency to hold that some design professionals are engaged in ultrahazardous activities for which they are strictly liable.[129]

Counter arguments must exist that have persuaded courts to hold to the professional standard in the face of more strict standards being applied to others. Even modern cases in an atmosphere in which compensation is emphasized have still applied the professional standard. Why? Perhaps most important is the belief that professionals operate amidst great uncertainty. Will the professional be judged harshly in the event of an unsuccessful outcome when even the *best* professional services would not have been able to create a satisfactory outcome? Does any *outcome* standard run the risk of holding the professional to professional performance that was more appropriate at the time of trial than at the time the professional services were performed?

Defenders contend that the client expects good *professional* service rather than insurance. Those who defend the professional standard fear that any warranty standard will expose design professionals to *unreasonable* expectations of the client that will be resolved in favor of the client by a jury, particularly if the client is unsophisticated in the world of design and construction. Also, what will be the *exact* nature of the warranty if an outcome standard is used? Might it not extend far beyond the function of design services such as seen in *Allied Properties v. John Blume*.[130]

Defenders of the professional standard argue that an outcome standard will generate overdesign at unneeded costs to reduce or avoid the risk of liability. Increased liability also means higher cost of service

resulting in fewer practitioners, higher prices, and more uninsured professionals.

Those who defend the design professions state that the high compensation that *some* professionals receive is rare in the design professions. Only "strong" design professionals will be able to contract out of any outcome standard. The ordinary professional designer lacks the bargaining power to obtain a more limited standard or believes that it is inappropriate to begin a professional relationship by demanding or requesting one.

So it stands. The reason for announcements of the professional standard in clear terms by courts means that it is unlikely that the standard will be abolished in the near future. Dissatisfaction at the privileged position accorded professionals is likely to result in chipping away at its protections as well as sophisticated clients demanding contractual protection beyond the standard.

(B) THE DESIGN PROFESSIONAL'S DUTY TO WORKERS

The *Krieger* and *Caldwell* cases produced in Section 17.09(E) demonstrate the contrary perspectives from which this hotly contested issue can be viewed. It is important to recognize that this is a multilevel problem.

The *Krieger* case reflects what appears to be the emerging rule, that the issue in the ordinary retention should be viewed almost exclusively as a contractual one. Design professionals are retained and paid to use their best efforts to see that the project is built in accordance with the contract documents. The contract makes clear that the contractor, not the design professional, is responsible for how the design is executed. This approach requires an examination of

---

128. Refer to Section 17.06.

129. Refer to Section 7.04(A).

130. Supra note 78.

the contracts, mainly the ones for design services but also for construction. If they do not reveal that the design professional was accepting responsibility for construction methods (Bechtel, the design professional in the *Caldwell* case, was a safety engineer as well), the design professional has *no* duty to workers and there need be no inquiry into his conduct.

The other perspective is to *start* with tort law with its function of compensating victims, deterring wrongful conduct, and avoiding harm. That body of law jettisoned the privity rule to insure that claimants can find a person from whom compensation can be recovered and to avoid technical defenses that bar scrutiny into the conduct of those persons who very likely have caused the harm. Why reinstitute this barrier by looking mainly at the purpose for engaging a design professional and focusing almost exclusively upon the contract?

To be sure the contract for design services is important. Without it, vague or detailed, the design professional would have no business on the site and not be in a position to look out for danger to workers or anyone else. The contract plays a significant role in determining what the design professional should have seen. The design professional cannot be expected to *look* for problems. That is not his function. If he sees unsafe practices or would have seen them had he done what the contract obligated him to do, his conduct will be judged. Did he act reasonably? He may satisfy this obligation by complaining to the contractor's superintendent, by bringing this matter to the client's attention, or even by inviting public officials who deal with safety matters to deal with the question. But if viewed as a tort problem, he has a responsibility to act reasonably.[131]

A second level looks at judicial procedures. Those who support the "no duty to workers" rule seek a defense that will bar a full trial and have the issue decided by looking solely at the pleading or the pleadings and affidavits submitted to support a motion for summary judgment. They wish to escape the high cost of trial even if they believe that the design professional will ultimately prevail if his conduct is judged.

Those who support the tort orientation see no reason why the design professional should escape being judged. They contend that if the facts clearly show that he could not have been expected to know of the unsafe practices or that he did what was clearly adequate he will be able to avoid a full-scale trial. But they object to a rule of law that relieves him. They prefer that this question be treated as a factual question like any other.

A third level relates to workers' compensation law. As noted, one of the principal functions of modern tort law is to see that victims are compensated. Most workers will be compensated under workers' compensation law, a social insurance system that pays a certain portion of the economic losses through an administrative process that does not focus upon wrongful conduct. Workers cannot sue their employers in tort, with workers' compensation being their exclusive remedy. Those who feel that the compensation system is inadequate seek to encourage the development of third-party claims that the injured worker can make to fully recover for the harm suffered; this is usually more than the often inadequate amounts recovered under workers' compensation law.

On the other hand, workers' compensation is sometimes used as an argument by those who oppose the design professional's having a duty to workers.[132] Those opposing such a duty argue that the worker will not be uncompensated, and they see no reason why a worker in a construction-

---

131. A fuller expression of the author's views can be found in Sweet, *supra* note 111.

132. *Balagna v. Shawnee County*, 233 Kan. 1068, 668 P.2d 157, 170–172 (1983) (dissent).

related accident should be placed in a better position than a worker in an industrial accident who may not have as many third parties from whom he can seek tort recovery. They contend that the broadening of third-party claims has hopelessly overcomplicated often simple accident cases and has generated lawsuits with horrendous costs to all participants.

At a fourth level, operations on a construction site are stressed. Those who support the duty state that safety is everyone's business, and the more people concerned the less likely injuries will occur.

Those who oppose a "duty to workers" rule claim that it will induce design professionals to venture into areas where they do not have the expertise and which can only cause blurred lines of responsibility as well as expose the design professional and the client to claims that intervention into these matters is a breach of the contract between the owner and the contractor.

Probably the best solution in an ideal world would be to have a total, enclosed social insurance system under which all workers receive adequate compensation without the necessity of going to court but without having rights against third parties. Until this is achieved (it may be a long way off), it is hard to support a result that would revive the dead privity doctrine and put professional designers in a favored position under which their principal function precludes their conduct being judged.

(C) INJECTION OF TORT LAW INTO THE COMMERCIAL WORLD: A WILD CARD

The commercial world—the world of business dealings between merchants where much of the world's work is accomplished—certainly does not escape from tort law intervention. For example, under certain limited circumstances the tort law will enter when a person has wrongfully induced another to breach a contract or impede a prospective economic advantage. On the whole, tort law interferes here only when there has been intentional wrongdoing or where there has been harm to persons or property.[133] Tort law at least in some states (refer to Sections 17.09(D) and 17.11) has been reluctant to shift economic losses caused by negligent conduct and has left that largely to contract and commercial law.

In the construction world, as in others, tort law has begun to play an increasingly important role in allocating purely economic losses. For example, negligent representation has been the basis of claims made by persons who have relied upon the representations of those in the business of furnishing information, such as surveyors or geotechnical engineers.[134] Even more, there has been the tendency to expose the design professional, among other participants, to claims by other participants, particularly contractors, sureties, and even prospective occupiers of projects, such as buyers or tenants.

Perhaps these tendencies cannot be rolled back in a legal world where tort law and accountability have become dominant factors in private law even in commercial disputes. Those courts faced with these decisions or asked to expand existing rules should consider the effect of injecting this wild card.

As an illustration, some courts allow a contractor to maintain a tort action against

---

133. A subsequent tenant was allowed to bring a tort claim against the architect when the floor settled, walls became damaged and premises became untenantable. See *A.E. Investment Corp. v. Link Builders, Inc.,* 62 Wis.2d 479, 214 N.W.2d 764 (1974).

134. *Rozny v. Marnul,* supra note 105.

a soils engineer for misrepresentation.[135] This was done in the context of a transaction in which it was clear that the construction contract placed the entire risk of unforeseen subsurface conditions on the contractor.[136] Allowing the tort action will induce the soils engineer to request indemnity from the client, increase his contract price to take this risk into account, or price his work to encompass performance designed to insure that his representations are accurate even if "excessive" caution would not be justified. Either way, the system of allocating risks is frustrated, with the client perhaps paying twice for the same risk by increased contractor bids (tort recovery is too uncertain to permit the prudent contractor to reduce the bid because of potential tort recovery) and the higher compensation to the soils engineer.

Perhaps even worse, one court allowed a tenant who lost the use of its premises because of delays caused by the contractor to establish negligence and use that as the basis for a claim against the contractor.[137] What effect does this holding have upon the potential risk of construction contract participants?

Delay is dealt with at length in the construction contract with its time extensions, damage liquidations, or no-damage clauses. Will these clauses apply to tort claims? How can contractors or design professionals faced with this risk deal with it? They can hope the law will protect them. They may decide to build a contingency into their contract price to deal with this risk or

demand indemnification if they have the bargaining power. They may, if they can identify potential claimants, seek exculpation from them or demand that the owner do so. All of these approaches add transaction costs and *must* increase contract prices. It must also be kept in mind that the party who has suffered economic losses usually can transfer such losses to the party with whom it has contracted—the tenant to the landlord or the contractor to the owner.

The tort wild card has caused unnecessary chaos in the construction world. It is bad enough to allow major participants, such as the design professional, owner, and contractors, to sue each other in tort. It is much worse to allow more *remote* participants such as suppliers, potential buyers, tenants, lenders, or even sureties to use tort law as a means of shifting their *economic* losses to the major participants in the Construction Process.

The law should encourage persons to enter into contracts. One way is to limit the exposure for consequential damages suffered by the other *party*. But this protection is diminished in claims by third parties in *tort*. That tort usually in this area requires negligence does not compensate for this added exposure, particularly when claims against design professionals usually use a tortlike standard anyway.[138]

(D) THE EFFECT OF EXPANDED PROFESSIONAL LIABILITY

This section has outlined some of the argu-

135. *M. Miller Co. v. Central Contra Costa Sanitary District*, 198 Cal.App.2d 305, 18 Cal.Rptr. 13 (1961). But see *Texas Tunneling Co. v. City of Chattanooga*, 329 F.2d 402 (6th Cir.1964).

136. See Section 29.04.

137. *J'Aire Corp. v. Gregory*, 24 Cal.3d 799, 157 Cal.Rptr. 407, 598 P.2d 60 (1979). This case sought to put some limits on recovery by barring recovery for losses that involve ordinary business risks; a subsequent decision did not draw this line. *Chameleon Engineering Corp. v. Air Dynamics, Inc.*, 101 Cal.App.3d 418, 161 Cal.Rptr. 463 (1980) (supplier of subcontractor sued by prime for not supplying components). Cases following *J'Aire* are *Keel v. Titan Construction Corp.*, supra note 105, and *Hawthorne v. Kober Construction Co.*, ___ Mont. ___, 640 P.2d 467 (1982). But see *Local Joint Executive Board of Las Vegas, etc. v. Stern*, 98 Nev. 409, 651 P.2d 637 (1982). Refer to notes supra 105 and 106.

138. Refer to Sections 17.05–17.07.

ments for and against expanded professional liability. But the focus of this subsection is to point to the effect this expansion has had.

Expanded liability can be looked upon as a method of eliminating incompetent practitioners from the professions to supplement the unarguably ineffective registration laws or, at least in this area, the inefficient marketplace. But does it have this effect?

Does expanded liability drive out the practitioner who should be removed from the profession? Do incompetents get sued more often than those who are competent? Are they likely to be forced out by increased insurance rates, decisions by insurers not to insure them, or unwillingness of prospective clients to engage them if they cannot be insured? At best these are unprovable. Very likely the answer to all three is no.

Has expanded liability improved professional practice? Undoubtedly designers are more careful, perhaps too careful. The result can be overdesign, an unwillingness to take design risks, and mediocre design. Again it will be difficult to assemble anything beyond anecdotal evidence and polemics to uncover the truth.

Of course, expanded liability can be and has been justified as a process for allocating responsibility to the persons who are responsible and who can best spread the loss. The Construction Process with its wealth of participants, its overlapping functions, and unclear lines of responsibility makes it unlikely that expansion of professional liability will place responsibility on the party upon whom it ought to be placed. All that can be certain is that expanded liability has led to hopelessly complicated and unpredictable lawsuits with the inevitable rise in the overhead of performing professional services.

What is certain is that expanded liability has generated overprotective contract language that may simply drive prospective clients to others, such as those who design and build or manage construction. It, along with other causes of high operational overhead such as taxes and other regulations, has slowly reduced the ranks of sole practitioners and small partnerships and has led to increased specialization. Expanded liability has led to an emphasis upon Risk Management (explored in Chapter 18).

PROBLEMS

1. A was the architect for a large office building. When he selected material to be used for wall construction, he did not realize that the fire insurance rates were dependent upon the type of wall construction material. He specified wall material based upon cost, safety factors, and esthetics. Had he specified different materials of approximately the same cost, there would have been a reduction in the fire insurance rate. When the owner found this out, he claimed that the additional insurance premium should be borne by the architect because the architect should have known that insurance premiums are based upon materials used in construction. Is the owner correct in his statement? What additional facts would be helpful in resolving this problem?

2. A was the architect for a lavish theater to be built in New York. The theater owner wanted a dramatic staircase to be designed that would run from the lobby to the first and second balconies. Two types of floor coverings for the stairway were available. Type A was more pleasing esthetically than Type B and cost 20% less. The principal advantage of Type B was that persons walking on the stair were less likely to slip. The architect chose Type A because it cost less and looked better

even though it was less safe than Type B. Would the architect be liable if a theater patron slipped and suffered injury when ascending the stair from the lobby to the first balcony? What facts would be necessary to answer this question?

3. Suppose an architect designs a building without knowing the federal government recently enacted a statute that gives a tax rebate to those who use particular types of insulation to conserve energy. The insulation that would qualify for the rebate was not specified. Would the architect be responsible to the client for any lost tax rebate? Suppose the contractor knew of this and said nothing? Suppose it was likely that the client would be aware of new legislation of this type?

4. Can the holdings in the *Krieger* and *Caldwell* cases reproduced in Section 17.09(E) be reconciled?

18

Risk Management: Variety of Techniques

Chapter 17 chronicles the expansion of professional liability. An often ignored technique to avoid claims is cultivation of a good client relationship. Honesty in approach, respect for the client's intelligence, appreciation of the proper role of a professional advisor, and common courtesy (answering phone calls and letters) are perhaps the best techniques to avoid claims. These are nonlegal considerations. This chapter suggests legal and planning approaches to deal with this phenomenon.

SECTION 18.01 SOUND ECONOMIC BASIS: BARGAINING POWER

This treatise is not an appropriate place to discuss in detail the economics of the design professions. To implement some if not most of the approaches suggested in this chapter, the design professional must be able to choose which commissions to accept and to request or even demand that particular contract language be excluded or included. Doing this requires sufficient economic strength to pick and choose among projects and contracts. Although this chapter cannot deal with methods to achieve this power, the approaches suggested will be of no value unless economic strength can be attained and mobilized.

SECTION 18.02 EVALUATING THE COMMISSION: PARTICIPANTS AND PROJECT

As a rule there are more design professionals than commissions. Usually this means that most design professionals will take whatever work they can get. But there are design professionals who can pick among projects. This section is directed to them.

It is most important to evaluate the client and its financial resources. A client with limited resources, particularly one with extravagant expectations, may not be able to withstand the shocks of added costs such as those generated by design changes, delays, claims, or other circumstances that will increase the ultimate contract payout. Such a client will be more inclined to abandon a project before construction. If construction does begin, it may be quicker to point at the design professional and other participants if the project does not proceed as planned. The AIA gives architects the power to request information on the financial resources of the owner at any time during their performance.[1] Yet it is likely that

1. AIA Doc. B141, ¶ 2.2.

most architects will not exercise this power, particularly at the start of the professional relationship. Although it is understandable that an architect may wish to make these inquiries at the *beginning* of performance, such inquiries may be crucial *during* performance if financial troubles appear imminent.

Another important client criterion is experience in design and construction. An inexperienced client may be more likely to make claims because it does not realize uncertainties inherent in construction and the likelihood that adjustments will have to be made. Such a client may also be mesmerized by a fixed-price contract and be unduly rigid as to price adjustments. Although construction contract pricing disputes principally affect the relationship between owner and contractor, when that relationship sours there will be more administrative burdens placed on the design professional and the greater likelihood of claims.

The project too must be evaluated. One that involves new materials, untested equipment, and novel construction techniques must be viewed as creating special risks. If disappointments develop, claims (including those against the design professional) are more likely to be made.

It is important to evaluate the other key participants, such as the prime contractor, the principal subcontractors, and consultants, for technical skills, financial capacity, and integrity. The construction contract is another factor that must be taken into áccount. A tight fixed-price contract, a rigid time schedule with stiff liquidation of damage clauses for delay, a multiple prime contractor arrangement, and a fast-track sequence all contain the seeds for controversy and possibly liability exposure.

It may be helpful to develop a point system for evaluating these factors. If the points reach a certain level, the project should not be undertaken without careful executive review. Beyond the next numerical benchmark the commission should be refused unless changes are made. At a point beyond even that the commission should be refused.

SECTION 18.03 CONTRACTUAL RISK CONTROL

Section 18.02 noted the importance of a general appraisal of the contract. This section looks at specific contract clauses that can be useful in risk management.

(A) SCOPE OF SERVICES

The contract should make clear exactly what the design professional is expected to do. Reference was made in Sections 15.01 and 16.01(G) to the difference between basic and additional services. Here the emphasis is upon services the client may expect the design professional to perform that the design professional does not feel are part of her undertaking. Perhaps most important is the design professional's role in determining how the work is being performed and the responsibility of the design professional for the contractor not complying with the contract documents.

(B) STANDARD OF PERFORMANCE

Usually the design professional wishes to be held to the professional standard discussed in Section 17.05. When this is the case, the contract language as well as any other communications should not use words such as *assure, insure, guarantee, achieve, accomplish, fitness,* or *suitability* or any language that appears to promise a specific result or achievement of the client's objectives. It is even better to include language that specifically incorporates the professional standard and, where possible, language that justifies it.

(C) EXCLUSION OF CONSEQUENTIAL DAMAGES

The law does not charge a breaching party with *all* the losses caused by its breach.

Usually it is not chargeable with what are sometimes called consequential or less direct damages. This limits a contracting party's responsibility.

Yet the foreseeability requirement—the standard used most frequently in determining whether consequential damages can be recovered—has been applied by modern courts in such a way as to diminish protection given by earlier courts.

For that reason the design professional should seek to exclude her liability for consequential damages in the contract. This can be done expressly or by limiting the responsibility of the design professional to correction of work caused by defective design or, when that is not economically feasible, to the diminished value of the project.

(D) LIMITING LIABILITY TO CLIENT

Some have advocated that the design professional include a clause in the contract under which the designer's liability to the *client* is limited to a designated amount or a specific portion of the fee, whichever is greater or, in some contracts, whichever is less. Some clauses allow the client to buy out of the limitation by paying a specified amount. Very likely this is done not to encourage a buyout but to help enforce the clause by giving the client an alternative.[2] Although this would not affect third-party claims, client claims still constitute the bulk of liability exposure.

Although the principal debate has centered upon the enforceability of such a clause, the bargaining realities make the *main* problem obtaining client consent. This is difficult to do when dealing with private owners and almost impossible when dealing with public owners where accountability is sensitive.

The issue of enforceability has stirred much debate. Some claim that such clauses are invalid based upon the undesirability of professional persons using a contract to reduce their liability and the assumption of modern courts that those who obtain *any* exculpation *must* be in an oppressive/dominant bargaining position. Those who suggest that such contract clauses would not be enforced also point to the possibility that enforcement would provide an incentive for professionals to perform carelessly.

Those who suggest that such clauses *are* enforceable point to the absence of any strong bargaining power on the part of most design professionals. Also, they note that they are not totally exculpatory clauses but simply limit liability to a particular amount, often more than trifling. They assert that the law is no longer treating professionals differently than commercial enterprises. They point to attacks by law enforcement officials upon the professions through enforcement of antitrust laws. They also point to court decisions that have permitted exculpation between commercial parties who presumably know what is in their best interests.

Not many cases have dealt directly with the enforceability of such clauses. A few implicitly seem to be willing to enforce such clauses.[3]

Contracting parties in this context have sufficient bargaining strength and intelligence to bargain freely over risks. They should be given the freedom to apportion risks as they see fit. Any client who agrees to such a limitation generally has legitimate reasons for doing so. It may decide to go along with such a limitation because those are the only terms under which a particular

---

2. See *Cregg v. Ministor Ventures,* 148 Cal.App.3d 1107, 196 Cal.Rptr. 724 (1983) (consumer renting storage space given opportunity to insure with storage company at a higher rate or self-insure at lower rate: exculpation upheld).

3. *Federal Reserve Bank of Richmond v. Wright,* 392 F.Supp. 1126 (E.D.Va.1975); *Turner, Collie & Braden, Inc. v. Brookhollow, Inc.,* 624 S.W.2d 203 (Tex.Civ.App.1981). In a subsequent opinion by the Texas Supreme Court the issue was not discussed. See 642 S.W.2d 160 (Tex.1982).

design professional will perform. It may also receive a reduced fee generated by lower professional liability insurance premiums. It may be willing to accept such a provision because it feels that it is fair and inclusion of it in a contract will get the professional relationship off on a positive basis. For *whatever* the reason, if the parties have chosen freely to deal on that basis there is no reason for the law to interfere.

A number of legislatures have enacted anti-indemnity statutes that invalidate indemnity clauses that relieve one party from ultimate responsibility for its sole negligence. Some have dealt specifically with attempts by designers to indemnify themselves either generally or for their design negligence.[4]

Such legislation should not affect liability limitations. They are not exculpations. The amount selected usually bears a rational connection to the fee or the project costs. One state with anti-indemnity legislation *appears* to allow liability limitation.[5]

Such clauses, where used, should make clear which expenditures come within the liability limitation, particularly if the design professional has agreed to indemnify the client. The clause should make clear that the liability limitation does not *automatically* entitle the client to recover the specified amount. Damages must be established, with the liability limitation being a ceiling on what can be recovered.

(E) IMMUNITY: DECISION MAKING

Some American courts grant design professionals quasi-judicial immunity when they decide disputes under the terms of the con-struction contract. Many standard agreements published by the professional associations incorporate language that relieves the design professional from any responsibility if decisions are made in good faith.[6] They attempt to incorporate into the contract limited quasi-judicial immunity. If such clauses are to be effective, they must be incorporated in the contracts both for design services and for construction services.

(F) CONTRACTUAL STATUTE OF LIMITATIONS

As noted earlier, judicial claims can be lost simply by the passage of time, accomplished by statutes of limitations. Parties in a contract can include a private statute of limitations as long as it is reasonable and will not deprive one party of any viable judicial remedy. Some design professionals incorporate language in their contracts that states when the period begins and specifies the period itself.[7] These can be useful as risk management tools.

(G) THIRD–PARTY CLAIMS

Third-party claims are sometimes based upon the assertion that the claimant is an intended beneficiary of the contract. This contention can be negated by appropriate contract language.[8] This language may even have some effect on tort claims.[9]

(H) DISPUTE RESOLUTION

Some believe that the most important risk management tool is to control the process by which disputes will be resolved. It shall

---

4. West's Ann.Cal.Civ.Code § 2782.

5. Id. at § 2782.5.

6. AIA Doc. B141, ¶ 1.5.10; A201, ¶ 2.2.10.

7. AIA Doc. B141, ¶ 11.3 (controls *commencement* of period).

8. AIA Doc. A201, ¶ 1.1.2 (bars such claims by contractors against architects).

9. *Harbor Mechanical, Inc. v. Arizona Electric Power Cooperative, Inc.*, 496 F.Supp. 681 (D.Ariz.1980).

be seen that many American standardized construction contracts give first instance dispute resolution power to the design professional and frequently provide for an appeal to arbitration. Because this has become such an important feature of construction contract dispute resolution, it is covered in detail in Chapter 34.

(I) THE RESIDUE

Although a "wish list" can be created,[10] the realities of contract bargaining necessitate that emphasis be placed on those that are most useful in risk management.

(J) SOME SUGGESTIONS

The clauses noted in this section are more likely to be enforced if they are drafted clearly and express specific reasons for their inclusion, if the client's attention is directed to them, and if suggestions are made to the client to seek legal advice if it has doubts or questions about them.

SECTION 18.04 INDEMNITY: RISK SHIFTING OR SHARING

This important risk management device usually involves all key participants in the Construction Process. From the vantage point of the design professional, it seeks to shift any or a part of any loss she suffers related to claims by third parties such as workers, members of the public, or adjacent landowners to another participant, usually the contractor but possibly the client. This process is discussed in Chapter 36.

SECTION 18.05 PROFESSIONAL LIABILITY INSURANCE: RISK SPREADING

(A) COMPARED TO PUBLIC LIABILITY INSURANCE

Public liability insurance must be differentiated from professional liability insurance.[11] For example, public liability insurance will cover a design professional whose employee causes an accident by driving negligently on the way to the site. Professional liability insurance, on the other hand, covers those risks that are peculiarly incident to the performance of professional services.[12]

(B) REQUIREMENT OF PROFESSIONAL LIABILITY INSURANCE

The law does not require design professionals to carry professional liability insurance. However, clients increasingly require that design professionals have and maintain professional liability insurance. The AIA does not specifically require insurance but does state that any excess premium above that usually paid is a reimbursable.[13] Clients may have a claim against the design professional for losses relating to the project or because they have satisfied claims of third parties that are directly traceable to the design professional's failure to perform in accordance with the legal standards. To make any claim collectable, they may require that the design professional carry professional liability insurance. If the design professional has adequate professional liability insurance, third parties injured as a

---

10. Such as exculpation for consultants, a favorable choice of applicable law, waiver of a jury trial, a power to suspend work for nonpayment, insurance premiums as a reimbursable, and a stiff late payment formula, to mention a few.

11. This term will be used instead of the more commonly used errors and omissions insurance.

12. *First Insurance Co. of Hawaii v. Continental Casualty Co.*, 466 F.2d 807 (9th Cir.1972) (decision to dump excess fill: professional); *Bettenburg v. Employers Liability Assurance Corp. Ltd.*, 350 F.Supp. 873 (D.Minn.1972) (design caused building collapse: professional).

13. AIA Doc. B141, ¶ 5.1.6.

result of her conduct may choose to bring legal action against the design professional directly rather than against the owner.

Even if not required, many design professionals carry such insurance. One reason is to protect their nonexempt assets from being seized if a judgment is obtained against them. Another is that many design professionals do not wish to see persons go uncompensated who suffer losses because of the design professional's failure to live up to the legal standard.

(C) PREMIUMS

Insurance premiums continually go up. One reason is expanded liability for professional persons.[14] The amount of the premium itself in any individual case is determined by a number of factors, among which are the type of work performed, the experience of the insured, the locality in which the work or project is located, the gross receipts of the insured, the contracts under which services are performed,[15] and the experience record of the insured.

(D) COVERAGE: OCCURRENCE OR CLAIMS MADE POLICIES

An important coverage problem results from the frequent time lag between the act or omission claimed to be the basis for liability and the making of the claim. If both act or omission and claim occur during the policy period, there is no problem. But where both events do not occur within the policy period, the insurance coverage may depend upon whether the policy is what is called an "occurrence" or what is called a "claims made" policy.[16]

Suppose a claim is made *during* the policy period that is based upon acts or omissions that have occurred *before* the policy. Under an occurrence policy there would be no coverage, but there would very likely be coverage under a claims made policy.[17]

Suppose a claim is made *after* expiration of the policy that is based upon acts or omissions that occurred during the policy period. In a claims made policy there would not be coverage, although coverage would very likely be provided under an occurrence policy. An insured under a claims made policy will be even more exposed if the state in which she practices has a discovery statute of limitations under which the period for making the claim does not begin until the damage is discovered.[18]

Although a few cases have struck down claims made provisions, most have upheld them.[19] California rejected a bill to ban

---

14. Other reasons are poor underwriting by insurers, a decline in the stock market that reduces the value of insurance company investments, a downturn in the economy that can induce insureds to cancel coverage, and the inordinately high costs of defending claims.

15. One insurer gives a reduced premium rate if the insured limits liability in a designated percentage of its contracts with clients. (See Section 18.04(D).)

16. Sometimes a claims made policy is described as a "discovery" policy. To avoid the confusion that can result from statutes of limitations sometimes being described as "discovery" statutes, the "claims made" term will be used.

17. One policy gives retroactive protection when the insured carried some form of insurance at the time the event occurred, the insured had no knowledge of prior occurrences when the policy in effect was obtained, and there is no other valid and collectible insurance available. Other policies can have different conditions. One reason for a court's striking down a claims made clause was that its retroactive coverage applied only if the insured had carried a policy for the earlier period with *that* company. See *Jones v. Continental Casualty Co.*, 123 N.J.Super. 353, 303 A.2d 91 (1973).

18. Refer to Sections 17.08(B) and 27.03(G).

19. *Stine v. Continental Casualty Co.*, 419 Mich. 89, 349 N.W.2d 127 (1984). The most recent collection of cases can be found in Parker, *The Untimely Demise of the "Claims Made" Insurance Form? A Critique of Stine v. Continental Casualty Co.*, 1983 Det.C.L.Rev. 25 (1983). See also Comment, The *"Claims Made" Dilemma* in *Professional Liability Insurance*, 22 U.C.L.A.L.Rev. 925 (1975).

claims made policies. Instead, in 1974 the Legislature enacted legislation requiring that claims made policies conspicuously state the coverage on the policy.[20]

Currently professional liability insurers use claims made policies to avoid the "tail" at the end of the policy. The latter, in an inflationary period, generates large, unpredictable claims, particularly in states that do not start the statute of limitations until discovery. The fiercely competitive nature of the insurance industry can mean that a new entrant or someone seeking to gain a larger share of the market may offer an occurrence policy.

Frequently professional liability insurers provide coverage at reduced rates for design professionals who leave the profession or retire. However, without such coverage, a design professional must bear the cost of defense and any ultimate judgment for claims made after expiration of the policy period.

Design professionals should examine policies carefully to determine the type of coverage provided and discuss with the insurer the possibility of an endorsement for added coverage. The open-endedness and uncertainty of an occurrence policy is the reason for insurers' preference for claims made coverage. If an endorsement changing the policy to occurrence coverage were available, it would very likely be expensive.

(E) COVERAGE AND EXCLUSIONS

Generally, in the absence of a special endorsement, professional liability insurance policies cover liability for performing normal professional services. Tasks commonly undertaken by design professionals are known to the insurer. The insurer is aware of the standard to which the design professional will be held. These known and predictable elements are needed for loss predictions and intelligent rate making.

One policy excluded work not customarily performed by an architect, as well as activities relating to boundary surveys, subsurface conditions, ground testing, tunnels, bridges, and dams. Also excluded were failure to advise or require insurance or surety bonds and failure to complete contract documents or act upon shop drawings in the time promised unless those losses were due to improper design. The policy excluded liability based upon express warranties, guarantees, and estimates of probable construction costs, indemnity liability assumed by contract, and liability for copyright, trademark, or patent infringements.

Excluding contractually created liability can cause problems for design professionals. The contract with the client may require performance that exceeds the legal standard. As to arbitration clauses, see Section 34.15.

Clients sometimes seek contractual indemnification from the design professional. If so, a policy that excludes contractual liability may not, in the absence of a special endorsement, cover indemnity liability.

The formidable list of exclusions in the policy should be a warning to design professionals that risks that they wish to cover must be discussed with the insurer. Sometimes a special endorsement covering these risks can be obtained. Coverage can depend upon the location of the project. Although there may be slight variations in state law relating to the standard of performance required of a professional liability, in foreign countries the law may be so different that insurance companies do not wish to cover these risks without further study.

(F) DEDUCTIBLE POLICIES

Increasingly insurance companies seek to reduce their risk by excluding from coverage claims, settlements, or court awards below a specified amount. Policies that exclude smaller claims are called deductible

---

20. West's Ann.Cal.Ins.Code § 11580.01.

policies. Generally, the higher the deductible, the lower the premium cost. A high deductible means substantial risks are borne by the insured. The recent tendency to raise deductible amounts makes the insured increasingly a self insurer for small claims. Some policies include in the deductible the cost of defense. For example, if the deductible amount is $5,000 and a claim of $3,000 is paid to a claimant, any cost of defense up to the deductible amount is borne by the insured.

Deductible policies can create a conflict of interest between insurer and insured. When small claims are made, the insurer may prefer to settle the claim rather than incur the cost of litigation. The insured may oppose such a settlement because of a belief that it is an admission of negligence and payment will come out of the insured's pocket. One insurer gives the insured the right to veto a settlement recommended by the insurer but provides that if the insured's ultimate liability exceeds that settlement proposal, the insurer's liability will not exceed the amount of the proposed settlement and any expense costs incurred *prior* to the settlement. In effect, such a provision gives the insurer the right to determine settlement.

Usually the deductible amount applies to each occurrence. But where the policy was not clear in this regard, one court applied the deductible amount of $10,000 to *each* claim, and in an accident with eight claimants this made the total deductible amount $80,000.[21]

(G) POLICY LIMITS

American policies generally limit insurance liability.[22] Suppose a claim for $50,000 is made and the policy limits are $100,000. The claimant offers to settle for $40,000. The insurance company exercises a right it has to veto settlement and refuses. The claim is litigated, and the claimant recovers $125,000. The insured may contend that had the insurer settled, it would not have had to pay amounts in excess of the $100,000 policy limit. Some courts hold insurance companies liable for amounts over the policy limit if their refusal to settle was unreasonable in light of all the circumstances.[23]

(H) NOTICE OF CLAIM

Insurance policies usually state that the insured must notify the insurance company when an accident has occurred or when a claim has been made. The notice is to enable the insurer to evaluate the claim and gather evidence for a possible lawsuit.

The insured usually does not hesitate to notify the insurer when anything occurs that may result in liability. But suppose the insured does not know of the accident and, in any event, when it does find out about the claim, believes liability to be quite unlikely. In *Empire City Subway Co. v. Greater New York Mutual Insurance Co.,*[24] the owner hired a contractor to perform excavation, back filling, and pavement replacement in New York City. An insurer issued a liability policy to the owner covering the contractor's work. The work was completed in October 1968. Shortly thereafter a pedestrian fell at the crosswalk five to ten feet from where the excavation work had been performed. The pedestrian brought legal action against the City; and on June 29, 1970, the City brought a cross action against the owner based upon an

---

21. *Lamberton v. Travelers Indemnity Co.,* 325 A.2d 104 (Del.Super.1974).

22. For discussion of whether the surety can be liable for more than the bond limits, see Section 37.10(D).

23. *Comunale v. Traders & General Investment Co.,* 50 Cal.2d 654, 328 P.2d 198 (1958); *Crisci v. Security Insurance Co.,* 66 Cal.2d 425, 58 Cal.Rptr. 13, 426 P.2d 173 (1967).

24. 35 N.Y.2d 8, 358 N.Y.S.2d 691, 315 N.E.2d 755 (1974).

indemnity provision contained in the application to the City for a permit under which the work was to be performed. Shortly before trial and sixteen months after the cross action against the owner, the latter evidently discovered it had insurance and notified the insurer.

The liability policy required a notice be given as soon as practicable after any claim was made or legal action commenced. The court in relieving the insurer from any obligation under the policy stated:

> While a good-faith belief of nonliability may excuse or explain a seeming failure to give timely notice . . . , that belief must be reasonable under all the circumstances. A reasonably prudent person, faced with a complaint alleging injuries sustained because of defects in a highway at a place described only generally but still within "five to ten feet" from where that person had recently completed excavation work, would at least have taken measures to ascertain whether the situs of the accident was within the area where the work was performed before concluding that there was no basis for liability. Where, as here, an accident occurs which may fall within the coverage of an insurance policy the insured may not, without investigation, gratuitously conclude that coverage does not exist.[25]

(I) DEFENSE OF THE ACTION

One important element of insurance protection is the cost of defending the claim. Usually professional liability policies specify that the insurer will furnish a defense or pay for defense costs. Suppose a claim is made or liability determined that is less than the deductible specified in the policy. Some policies provide for the cost of defense in such a case, but if the amount paid on the claim is *less* than the deductible amount, defense costs are considered part of the deductible up to the deductible amount.

Suppose the policy limit is $100,000 and a claim is made for $200,000. Any recovery over $100,000 must be paid by the insured. If the insurer is willing to pay the policy limits, will the insured be obligated to defend the claim?

Professional liability policies generally require the insurer to defend even though the latter is willing to pay the policy limit. The professional liability policy is designed to furnish the *dual* protection of paying the claim and defending the claim, subject to policy limits and deductibles.

In addition to the often formidable attorneys' fees involved in defending claims, there are other expenses. Exhibits must be prepared, and expert witness fees must be paid. Sometimes transcripts must be made of testimony taken in advance of trial or at the trial. Bonds sometimes may have to be provided at stages of the legal action. The insured should determine whether the insurer is obligated to pay for these expenses.

The insured may wish independent legal counsel despite the willingness by the insurer to provide a defense. If a claim is made in excess of the policy limit and it is likely that liability will exceed the policy limit, it may be useful for the insured to have her own attorney to provide settlement and litigation counseling. Costs of such services may have to be borne by the insured.

(J) TERMINATION

Most insurance policies permit the insurer to terminate by giving a designated notice, often as short as thirty or forty-five days. Increasingly legislatures and courts deny the insurer an *absolute* right to terminate.[26]

SECTION 18.06
PREPARING TO FACE CLAIMS

Design professionals should anticipate the likelihood that claims will be made against them. With this in mind, the design professional must be able to document in the clearest and most objective way that a proper job was done. For example, expanded

---

25. 358 N.Y.S.2d at 694–695, 315 N.E.2d at 757–758.

26. *Spindle v. Travelers Insurance Co.,* 66 Cal.App.3d 951, 136 Cal.Rptr. 404 (1977).

liability should not deter design professionals from using new designs, materials, or products. They must, however, prepare for the possibility that if things go wrong they will be asked to explain their choice.

Taking new materials as an example, the design professional should accumulate information directed toward predicting the performance of any contemplated new materials. Information should be obtained from unbiased persons who have used them on comparable projects. A list of such persons can be requested from manufacturers. Manufacturers' representatives should be questioned about instances where bad results were obtained. The manufacturer can be notified as to intended use of the project along with a request for technical data that includes limitations of the materials. Sometimes it is possible to have a manufacturer's representative present when new material is being installed to verify installation procedures. Any representations or warranties obtained should be kept in a readily accessible place.

Design professionals should be able to reconstruct the past quickly and efficiently. A system for efficient making, storing, and retrieving of memoranda, letters, and contracts is essential. Legal advice should determine the proper time to preserve records. If major design decisions have to be made, the design professional should indicate the advantages and disadvantages

and obtain a final written approval from the client. Records should show when all communications are received and responses made. If work is to be rejected, the design professional should support her decision by communications to client and contractor. Similarly, if any previous approvals are to be withdrawn, written notice should be given to all interested parties. Records should be kept of all conferences, telephone calls, and discussions that may later have to be reconstructed in the event disputes develop.

In this regard the instability of many design professional relationships can be troublesome. Design professionals dissolve partnerships frequently. Where dissolution occurs, records that should be kept are often lost or destroyed. When rearrangements occur, those involved should separate records and see that those who may need them have them.

Design professionals need competent legal services at prices they can afford at all stages of their practice. Certainly legal advice only *after* disputes have arisen is insufficient.

Younger groups of design professionals should consider negotiating with those who provide legal services for prepaid legal services plans.

The operations of the design professional, including contracts used, records kept, and compliance with laws regulating employers, should be evaluated periodically.

PROBLEMS

1. A contract between an architect and her client stated that all claims are barred unless they are made within six months after final payment to the contractor is due. Would such a clause be enforced? What added facts would be relevant? If you conclude that it is not enforceable, could it have been modified in such a way as to make it enforceable?

2. An engineer entered into a written

contract with a large manufacturer to design a warehouse. The fee was to be 5% of the cost *estimated* to be one million dollars. The engineer demanded that her liability be limited to 50% of her fee or $10,000, whichever is greater. The client agreed. Would such a clause be enforced? Would it be enforced if it were the full fee *paid* or $10,000, whichever is *less*?

19

Intellectual Property:
Ideas, Copyrights,
Patents, and Trade Secrets

SECTION 19.01 RELEVANCE TO DESIGN PROFESSIONAL

Design professionals use their training, intellect, and experience to solve design, construction, or manufacturing problems of their clients or employers. Usually design professionals reduce the proposed design solution to tangible form. These forms, whether they be sketches, renderings, diagrams, drawings, specifications, or models, communicate the design solution to the client or employer and others concerned. Often the design solution is followed by the completion of the end product, such as the construction project, the industrial process, or product. Additionally, some aspects of the design solution, such as the floor plans, sketches, diagrams, or pictures, may be used to advertise the project or the product.

To sum up, the three steps are as follows:

1. The intellectual effort by which the solution is conceived.
2. Communication of the solution.
3. Development of the end product.

The creator or owner of the tangible manifestation of the design solution or the end product itself may wish that it not be copied or used without permission. The client may not wish another project of an identical design constructed to preserve uniqueness of the project, whether it be a residence or a building. The client might feel wronged if the construction documents were used without payment if the client paid for and received exclusive ownership rights.

Similarly, the manufacturer who invests funds to develop a product or process may not wish others to copy it without permission. The manufacturer may wish to recoup the money which has been invested to develop the process or product either directly through royalties or by retaining a competitive advantage the research investment has given.

The design professional may wish to obtain similar protection. Dealings with the client were discussed in Section 15.11. Likewise, the design professional may wish protection against third parties copying the construction documents, diagrams, or drawings to be used in developing a product or the end products themselves.

The protection accorded those who create or hire others to create tangible manifestations of intellectual effort is the subject of this chapter.

SECTION 19.02 AN OVERVIEW

(A) SPECIFICITY OF DISCUSSION

Certain legal concepts relating to intellectual ideas will be explored in greater detail than others. For example, patents, though of great importance to engineers, will be discussed only briefly. Patent law is a highly technical area. Inventors who wish to obtain or enforce a patent require a patent lawyer. For this reason, only the basic principles and certain salient features of patent law will be mentioned.

Obtaining copyright protection, on the other hand, is a relatively simple process. Persons who wish to *acquire* copyright protection, in contrast to legal *enforcement* of copyright remedies, can generally do so without the assistance of an attorney. For this reason, more detail will be given to copyrights than patents.

(B) PURPOSE OF PROTECTION

Copyrights and patents are given authors and inventors for their writings and discoveries. The primary purpose of granting them is to foster social and industrial development for the public good. This development is accomplished by granting individuals monopoly rights to reward them for their contributions, monopolies that would otherwise be antithetical in a competitive system. One judge stated:

> The economic philosophy behind the clause empowering Congress to grant patents and copyrights is the conviction that encouragement of individual effort by personal gain is the best way to advance public welfare through the talents of authors and inventors. . . . Sacrificial days devoted to such creative activities deserve rewards commensurate with the services rendered.[1]

On the other hand, society can suffer from excessive protection. Much intellectual and industrial progress depends upon free interchange of ideas and free use of the work of others. Commercial and industrial

ventures can be frustrated or impeded if entrepreneurs are compelled to pay tribute to persons who claim that their ideas, designs, or inventions have been used in some way by the entrepreneur. The law attempts to reward truly creative and inventive work, without unduly limiting the free flow of ideas and use of industrial and scientific technology. Patent law, for example gives a seventeen-year monopoly to the inventor of a novel, original, and nonobvious invention in exchange for disclosure to the public. The period was chosen as a compromise that adequately rewards an inventor but does not unduly perpetuate the stagnation that can accompany monopoly.

(C) EXCLUSIONS FROM COVERAGE: TRADEMARKS AND SHOP RIGHTS

The creation of an effective and universally recognized trademark or trade name is an intellectual act. However, design professionals are less concerned with trademarks and trade names. For this reason there will be no discussion of common law or statutory trademarks or trade names.[2]

Shop rights, a doctrine under which an employer under certain circumstances has limited rights in the inventions of employees, will not be discussed. For all practical purposes it has been preempted by near universal use of standard form employment contracts.

SECTION 19.03 COPYRIGHT LAW [3]

(A) COMMON LAW COPYRIGHT ABOLISHED

Section 301 of the federal Copyright Act has

---

1. *Mazer v. Stein*, 347 U.S. 201, 219, 74 S.Ct. 460, 471, 98 L.Ed. 630 (1954).

2. The Lanham Act, 15 U.S.C.A. § 1051 et seq., a federal statute, permits registration of trade names, trademarks, and service marks, as well as provides for remedies. State law also deals with trademarks.

3. In 1976 Congress enacted Pub.L. No. 94–553, 90 Stat. 2541, which replaced the 1909 Copyright Act and went into effect in 1978. References are to sections in the current Act. The statute is also found in 17 U.S.C.A. § 101 et seq.

preempted *state* common law copyright laws that gave the author the power to determine when and if the work would be made available to the public. (See Section 19.04(C).) Preemption was intended to promote uniformity both by replacing state law with federal law and by eliminating the frequently difficult question of when the work had become dedicated to the public, an act that deprived the author of a common law copyright. Common law copyright remains only for works of authorship not fixed in tangible medium of expression that nevertheless can be copyrighted.

Illustrations of works that are not fixed in a tangible medium of expression that can be copyrighted are choreography that has never been filmed or notated, extemporaneous speech, original works of authorship communicated solely through conversations or live broadcasts, and a dramatic sketch or musical composition improvised or developed from memory and without being recorded or written down.

(B) STATUTORY COPYRIGHT

Classification of Copyrightable Works

The classification of works that can be copyrighted was changed from the close-ended thirteen to an open-ended seven. Section 102 permits copyright of the following categories:

1. literary works
2. musical works, including any accompanying words
3. dramatic works, including any accompanying music
4. pantomimes and choreographic works
5. pictorial, graphic, and sculptural works
6. motion pictures and other audiovisual works
7. sound recordings

Drawings or plans would fall under (5), while specifications fall under (1). In that regard, literary works need not be "literary" so long as they express concepts in words, numbers, or other symbols of expression.[4]

Copyright Duration: More Protection

Under the Copyright Act of 1909 the copyright holder had protection for twenty-eight years with the right to renew for an additional twenty-eight years. Much criticism had been made of copyright duration, and longer life expectancy made it inadequate. To bring American law in line with that of most foreign countries, § 302 gives copyright protection for the life of the author and fifty years thereafter. However, if a work has been one made for hire (discussed in greater detail in Section 19.04(D)), the duration of the copyright is seventy-five years after the year of its first publication or one hundred years from the year of its creation, whichever expires first.

Codification of Fair Use Doctrine

Section 107 expressly recognized "fair use." It permits reproduction "for purposes such as criticism, comment, news reporting, teaching (including multiple copies for classroom use), scholarship or research." Factors described as bearing upon whether the use is a fair one include the following:

1. The purpose and character of the use, including whether such use is of a commercial nature or is for nonprofit educational purposes.
2. The nature of the copyrighted work.
3. The amount and substantiality of the portion used in relation to the copyrighted work as a whole.
4. The effect of the use upon the potential market for or value of the copyrighted work.

---

4. *Apple Computer, Inc. v. Franklin Computer Corp.*, 714 F.2d 1240 (3d Cir.1983), held certain computer software copyrightable. An appeal to the U.S. Supreme Court was taken. The case was settled.

Obtaining a Copyright

Although the form of copyright notice was not changed, § 405 gives some relief where the notice has been omitted in certain circumstances, and § 406 gives relief if the notice contains an error in name or date.

The law does not require registration with the Copyright Office before commencement of an infringement action. But § 412 precludes recovery of statutory damages or attorneys' fees if the copyrighted work is not registered within three months after first publication of the work.

Section 407 stiffens the requirement that copyrighted works be deposited with the Library of Congress. The Act requires that two complete copies of the best edition be deposited, although failure to deposit will not affect the validity of the copyright. A person who fails to deposit within three months *after* a demand is subject to a fine of not more than $250 for each work and a fine of $2,500 if refusal is willful or persistent.

Remedies for Infringement: Increase in Statutory Damages

In addition to permitting an injunction to prevent or restrain infringement, § 503 allows a court to order an impounding or destruction of infringing copies and articles by which infringement has been accomplished. Section 504 allows recovery of actual damages and profits made by the infringer attributable to the infringement.[5] In establishing the infringer's profits, the copyright owner is required to present proof only of the gross revenue and the infringer must prove deductible expenses and elements of profit attributable to factors other than the copyrighted work. Statutory damages can now be awarded up to $10,000, with $50,000 for *willful* infringe-

ment. The statutory damages for an *innocent* infringer can be reduced to $100.

Works Commissioned by U.S. Government

Some had advocated that there be no copyright in works commissioned by the U.S. government. However, Congress rejected this position to give procuring agencies discretion to determine whether to give the design professional copyright ownership. Copyright protection will be denied only if the copyrighted work is authored by an employee of the government.

SECTION 19.04 SPECIAL COPYRIGHT PROBLEMS OF DESIGN PROFESSIONALS

(A) ATTITUDE OF DESIGN PROFESSIONALS TOWARD COPYRIGHT PROTECTION

Design professionals vary in their attitude toward the importance of legal protection for their work. Some design professionals want their work imitated. Imitation may manifest professional respect and approval of work. When credit is given to the originator, imitation may also enhance the professional reputation of the person whose work is copied. Some design professionals are messianic about their design ideas and would be distressed if their work were not copied. Many design professionals believe that free exchange and use of architectural and engineering technology are essential.

Even design professionals who want imitation or who do not object to it draw some lines. Some design success is predicated upon exclusivity. Copying the exterior features and layout of a luxury residence or putting up an identical structure in the same neighborhood is not likely to please the architect or client. The same design

---

5. An application of the statute will be found in *Aitken, Hazen, Hoffman, Miller, P.C. v. Empire Construction Co.*, 542 F.Supp. 252 (D.Neb.1982), reproduced in part in Section 19.04(D).

professional who would want his ideas to become known and used might resent someone going to a public agency and without authorization copying construction documents required to be filed there. This same design professional is likely to be equally distressed if a contractor were to copy plans made available for the limited purpose of making a bid. Much depends upon what is copied, who does the copying, and whether appropriate credit is given to the originator.

(B) WHAT MIGHT BE COPIED?

Design professionals may wish protection for ideas, sketches, schematic and design drawings, two-dimensional renderings, three-dimensional models, construction documents sufficiently detailed to enable contractors to bid and build, and the completed project itself. Ideas themselves cannot receive legal protection, and legal protection for the executed project is very unlikely. The principal problems relate to tangible manifestations of design solutions that are a step toward the project. These tangible manifestations vary considerably in the amount of time taken to create them and the amount of time and money saved by the infringer who copies them.

(C) COMMON LAW COPYRIGHT AND PUBLICATION

Traditionally, design professionals sought protection through common law copyright. Perhaps this was traceable to a lack of understanding of statutory copyright. Some design professionals believe that statutory copyright protection against infringement requires the expensive registration of often formidable construction documents after they are created.[6] Whatever the reason, the bulk of the cases involving design works involve claims of common law copyright.

Common law copyright was abolished by the 1976 Federal Copyright Act. For historical reasons, a brief comment on common law copyright is adequate.

The principal difficulty in perfecting a common law copyright had been the frequent claim made by the alleged infringer that the work copied had already been published. If dissemination and the facts and circumstances surrounding it indicated to a reasonable person that the creator had dedicated the work to the public, common law copyright was lost.

Common law copyright did not provide much protection. Design professionals should welcome the abolition of common law copyright. Now they are limited to statutory copyright which, though it too has its weaknesses, is substantially better.

(D) STATUTORY COPYRIGHT: THE *AITKEN* CASE

Looking first at the beginning (creation of ideas) and end (execution of completed projects) of a design professional's services, it is clear that the former is not subject to copyright protection. Before the 1976 Act, the copyright holder of technical drawings had no exclusive right to complete the project. However, drawings or models for an unusual structure that could be classified as a work of art, such as the Lincoln Monument or the Eiffel Tower, gave the copyright holder the exclusive right "to complete, execute and finish it." But the Copyright Office took the position that works of design professionals were copyrightable only as technical drawings and not as works of art. On the other hand, if a structure incorporated features such as artistic sculpture, carving, or pictorial representation that could be identified separately and were capable of existing independently as works of art, such features were eligible for registration. Thus, limited aspects of a

---

6. The prior law did *not* require registration until an infringement action had been commenced. See 17 U.S.C.A. § 411.

building received copyright protection as a work of art. Under some circumstances certain features of design could be sufficiently novel to justify a design patent. On the whole, the pre-1976 law denied the design professional exclusive right to execute the project and freely permitted persons to copy the completed project so long as there was no impermissible copying of writings.[7]

In addition to granting the exclusive right to reproduce and distribute, the 1976 Act gives the copyright holder the right to prepare derivative works based upon the copyrighted work.[8] Although it can be argued that a project is derived from the construction documents,[9] it is not likely that Congress changed the pre-1976 law.[10]

The limited copyright protection accorded design professionals extends to those works such as sketches, renderings, schematics, design development, and construction documents. One court expressed doubt of even their protectibility,[11] based upon an earlier case [12] which had held that even copying would be permitted if it had to be done to employ the system developed by the copyrighted material. However, recent cases seem to give copyright protection to these tangible manifestations of design solutions unless the copier can prove fair use.[13]

Before the 1976 Act, it was assumed that the person commissioning copyrightable works was entitled to the copyright in the absence of an agreement to the contrary. This was one of the reasons for frequent inclusion of clauses in contracts between design professionals and their clients giving the former ownership rights.

Section 201 gives copyright protection to the author of the work. However, if the work is made for hire, the employer or other person for whom the work was prepared is considered the author unless the parties have agreed otherwise in a signed written agreement. Section 101 defines a "work for hire" as prepared by an employee or a work specially ordered or commissioned. The illustrations given do not include a client commissioning a design professional.

Two recent cases have held that the work for hire doctrine does not apply to works commissioned by a client and the author of such a commissioned work has copyright ownership.[14] One, in addition to passing upon the application of the work for hire doctrine to architectural drawings commissioned by a client, provides an illustration of the remedial protection given a copyright owner against an infringer. It is reproduced here in part.

---

7. See Note, *Copyright Protection for Architectural Structures*, 2 U.S.F.L.Rev. 320 (1968).

8. 17 U.S.C.A. § 106(2).

9. This argument was made in Comment, *Copyright Protection for the Architect: Leaks in a Legal Lean-to*, 8 Cal. W.L.Rev. 458, 478 (1972). The preceding article contains a comprehensive study of copyright protection for design professionals before the 1976 Act. For an up-to-date survey see Note, *Innovations and Imitations: Artistic Advance and the Legal Protection of Architectural Works*, 70 Corn.L.Rev. 81 (1984).

10. 17 U.S.C.A. § 113(b).

11. *Scholz Homes, Inc. v. Maddox*, 379 F.2d 84, 85–86 (6th Cir.1967).

12. *Baker v. Selden*, 101 U.S. (11 Otto) 99, 25 L.Ed. 841 (1879).

13. *Imperial Homes Corp. v. Lamont*, 458 F.2d 895 (5th Cir.1972); *Baldwin Cooke Co. v. Keith Clark, Inc.*, 383 F.Supp. 650 (N.D.Ill.1974), affirmed 505 F.2d 1250 (7th Cir.1974). At a subsequent trial the plaintiff was awarded its damages and attorneys' fees as well as the defendant's profits. See 420 F.Supp. 404 (N.D.Ill.1976.)

14. *Meltzer v. Zoller*, 520 F.Supp. 847 (D.N.J.1981); *Aitken, Hazen, Hoffman, Miller, P.C. v. Empire Construction Co.*, 542 F.Supp. 252 (D.Neb.1982). See also *May v. Morganelli-Heumann & Associates*, 618 F.2d 1363 (9th Cir.1980).

AITKEN, HAZEN, HOFFMAN, MILLER, P.C. v. EMPIRE CONSTRUCTION CO.
United States District Court, District of Nebraska, 1982. 542 F.Supp. 252.

URBOM, Chief Judge.

[Ed. note: Footnotes renumbered and some omitted. In 1978 the Aitken architectural firm designed an apartment complex for Belmont. The apartment complex was to be built on land purchased by Empire (for all practical purposes Belmont and Empire were dominated by one individual). Empire had purchased a tract of land and decided to build an apartment on one of the four parcels it had acquired. However, it contemplated building another apartment complex on one of the other parcels in the future. The agreement between Aitken and Belmont was oral, and there was no discussion as to copyright ownership of the plans.

Belmont communicated its ideas to Aitken through sketches and verbal descriptions and indicated some of the design features it wished.

Aitken retained the original of the preliminary plans and prepared eighteen blueprint copies, which it delivered to Belmont. At that time the plans did not contain a copyright notation. The plans were used, and Belmont completed the apartment complex. (1820–22 Knox) in 1979. Empire paid Belmont for the construction and then sold the complex to Amwest. Aitken billed Belmont on an hourly basis and was paid some $13,000 for its services.

A year later, Belmont, without permission of Aitken, copied the plans prepared by Aitken to produce another set of plans from which an apartment complex adjacent to the first would be built. These copies were taken to the Lincoln Lumber Company, which had the plans reviewed and approved by King, a licensed engineer. The apartment complex (1830–32 Knox) was constructed in early 1980 and again sold by Empire to Amwest.

In March 1980 Aitken discovered that its plans had been copied and sent a bill for $36,000 to Belmont and Empire. Each denied liability, and Aitken filed a mechanics' lien against the real estate, but its action in the state court to foreclose the lien was dismissed because it had not established an express or implied contract for the provision of services for the project.

Immediately thereafter, Aitken placed notice of its copyright on the originals of the plans it had prepared. The following day it submitted its application to register the copyright with the U.S. Copyright Office. Shortly thereafter it commenced an action for copyright infringement and sought damages against Belmont, Empire, Lincoln, and King.

The court held that Belmont could not take advantage of the made for hire doctrine, as the 1976 Act limited that doctrine to employers and not to those who commissioned independent contractors. In making this determination, the court stated that Belmont could not exercise the degree of control over the work necessary to make Aitken its employee. The court noted that Aitken was a registered architectural engineering firm governed by professional regulations. Although Belmont could direct the result to be accomplished, it did not have the right to control the detail and means by which the result was to be accomplished.

Next the court rejected Belmont's argument that its participation in the design made it a joint work in which it was a joint author. It also rejected Belmont's claim that its copying was a fair use and that Belmont was an innocent infringer. This assertion was based upon the absence of the copyright notice.

Absence of the copyright notice under Section 405(a)(1) from a relatively small number of copies and under Section 405(a)(2), registration within five years after publication without notice and with reasonable effort to add notice, did not destroy the validity of the copyright. However, an innocent infringer under Section 405(b) who relies upon an authorized copy from which the copyright notice has been omitted incurs no liability for actual or statutory damages for infringements committed before notice of the copyright if that person had been misled by the omission of the notice. However, the court held that Belmont had not been misled by the omission of the notice. In supporting this, the court pointed to expert testimony that it is the custom in the architectural profession that the architect retain ownership of plans unless there is an express agreement to the contrary. Also, Belmont had been informed of Aitken's practice of retaining ownership.

The claims against Lincoln and King were dismissed. The court concluded that they had no reason to know that Belmont had infringed nor were they vicariously liable for Belmont's infringement.

Next the court discussed damages.]

III. Damages

A. *Actual Damages*

Under § 504(b) of the 1976 Act, a copyright owner is entitled to recover the actual damages suffered by him as a result of an infringement. In this case the plaintiff is entitled to the fair market value of its architectural plans as revised for use in constructing the 1830–32 project. See *Nucor Corp. v. Tennessee Forging Steel Service, Inc.,* 513 F.2d 151, 153, n. 3 (C.A. 8th Cir. 1975) (infringers of plaintiff's common law copyright in architectural plans were liable to the plaintiff for the fair market value of the architectural plans).

As established at trial, there was no ready market for architectural plans for apartment complexes. There did exist, however, one potential market source for the 1820–22 architectural plans, and that source was Belmont. Thus, the amount Belmont would reasonably have paid to the plaintiff and the plaintiff would reasonably have expected to receive for the revision and use of the 1820–22 plans in conjunction with the 1830–32 project is the fair market value of those plans at the time of Belmont's infringement.

Belmont paid the plaintiff $13,440.93 for the plaintiff's services in preparing the architectural plans for use in the construction of the 1820–22 Knox Street apartment building. This charge was based on an hourly rate for each of the plaintiff's employees. However, by a letter dated November 24, 1978, the plaintiff notified Belmont and Empire that as of December 1, 1978, payment on any project would be based on a firm percentage rate.

In accordance with this notification, by a letter dated March 28, 1980, the plaintiff billed $35,973.00 for "services rendered in connection with apartments at 1830– and 1832 Knox" based upon an estimated construction cost of $479,643.00 and a percentage fee of 7.5 per cent. Mr. Hoffman, president of the plaintiff firm, testified that a percentage fee of 7.5 per cent was selected, because that was the percentage fee utilized by the plaintiff in connection with similar projects. However, in each of the exemplary contracts introduced by the plaintiff, the architectural services to be provided by the plaintiff for a percentage fee of 7.5 per cent entailed not only the preparation of design drawings and specifications but also the supervision of the contract bidding, supervision of construction, and payment of the contractor. These latter activities were not, of course, performed by the plaintiff in connection with the 1820–22 project and would not have been required in connection with the 1830–32 project, because the plaintiff had been engaged by both the owner (Empire) and the contractor (Belmont) of the projects. Furthermore, with respect to the 1830–32 project, the plaintiff would not have been required to develop the plans from the schematic design stage through the construction design stage, but would only have had to revise the final working drawings it had already prepared. I find, therefore, that a percentage fee of 7.5 per cent is inappropriate as a measure of the value of the use of the plaintiff's plans.

William Speece, the plaintiff's expert witness, estimated the fair market value of the architectural services reflected in the 1830–32 plans to be in the range of $23,750 to $37,500, based on a percentage fee of 4.75 to 7.5 per cent. However, on cross-examination Mr. Speece admitted that he was unaware, in making his estimation, that Belmont had paid only $13,440 for the preparation of the original set of plans. He stated that architects sometimes work on an hourly, rather than percentage, basis, and that the manner of computing fees on an hourly basis varies a great deal.

Even utilizing Mr. Speece's lowest estimated percentage base, the value of the 1830–32 plans is almost twice what Belmont paid for the design and use of the 1820–22 plans. It is inconceivable that Belmont would have been willing to pay or that the plaintiff would reasonably have expected to be paid $23,750 for the revision and use of plans for which Belmont originally paid only $13,440 to have developed from scratch. By preparing the original design plans for a price of $13,440 the plaintiff itself set the value of those plans. There is no evidence before me that the revised set of plans were of any greater value than the original set.

Indeed, there being only one potential buyer for the plans and the 1830–32 plans being a revised, rather than an original, set of plans, the 1830–32 plans may have been of less value. See *Edgar H. Wood Associates, Inc. v. Skene*, 347 Mass. 351, 197 N.E.2d 886, 896 (Mass.1964) ("Any value assigned [to the architectural plans] should reflect the fact that at the time of the conversion and the copying of the plans they had already been used in the Woburn apartment project and hence were not novel."). I find, therefore, that $13,440.93 is the highest reasonable value which can be assigned to the 1830–32 plans or their use.[15] See 3 M. Nimmer, Nimmer on Copyright § 14.02 at 14–8 (1981) (the price previously paid by the parties for the same or similar copyrighted materials may be accepted as evidence of actual damages).

In determining the actual damages suffered by the plaintiff as a result of the infringement of its copyright, there must be, of course, a deduction from the gross amount the plaintiff would have realized if Belmont had paid for use of the plans whatever costs the plaintiff would have incurred in revising those plans. Mr. Hoffman testified that it would have taken two to three eight-manhour days per sheet to trace the 1820–22 architectural plans to produce the 1830–32 architectural plans. He testified that the plaintiff firm would pay its employees from $5.00 to $15.00 an hour for this drafting work. The 1830–32 plans consist of eighteen sheets. Allowing 2.5 eight manhour days per sheet at an average hourly rate of $10.00, it would have cost the plaintiff $3,600.00 to reproduce the 1820–22 plans. Deducting this figure from $13,440.93, I find the actual damages sustained by the plaintiff to have been $9,840.93. Prejudgment interest on this amount is not authorized by the Copyright Act of 1976 and will not be allowed. See *Baldwin Cooke Co. v. Keith Clark, Inc.*, 420 F.Supp. 404, 409 (U.S.D.C.N.D.Ill.1976).

B. *Profits*

In addition to actual damages, the plaintiff is entitled to "any profits of the infringer that are attributable to the infringement and are not taken into account in computing the actual damages." 17 U.S.C. § 504(b). This section further provides that "[i]n establishing the infringer's profits, the copyright owner is required to present proof only of the infringer's gross revenues, and the infringer is required to prove his or her deductible expenses and the elements of profit attributable to factors other than the copyrighted work."

---

15. It is conceivable that if Belmont had had the opportunity to negotiate with the plaintiff concerning the amount to be paid for the revision and use of the 1820–22 plans, Belmont may have paid less than the amount it paid for the design and use of the 1820–22 plans. However, it has been held that the defendant infringer "cannot expect to pay the same price in damages as it might have paid after freely negotiated bargaining, or there would be no reason scrupulously to obey the copyright law." *Iowa State University Research Foundation, Inc. v. American Broadcasting Cos.*, 475 F.Supp. 78, 83 (U.S.D.C.S.D.N.Y.1979).

(1) Belmont

In consideration for the construction of the apartment complex and garage located at 1830–32 Knox Street, Empire paid Belmont $512,569 ($511,250 contract price plus $1,309.00 in reimbursed expenses). Belmont introduced evidence showing that it incurred $451,540.56 in direct, deductible expenses, thus realizing a gross profit of $59,709.44 on the project. Belmont contends that it is entitled to deduct from the gross profit a proportion of its administrative and general overhead expenses by a formula which would reduce its gross profit to a net profit of $12,878.94.

The rule is that the overhead expenses which assist in the production of an infringing work are deductible from the gross profit of the infringer. *Wilke v. Santly Bros.*, 139 F.2d 264, 265 (C.A. 2nd Cir.1943); *Smith v. Little, Brown & Co.*, 273 F.Supp. 870, 874 (U.S.D.C.S.D.N.Y.1967), aff'd, 396 F.2d 150 (C.A. 2nd Cir.1968). The burden is upon the defendant infringer to prove the actual expenditures for ordinary overhead and a fair method of allocating the overhead to the particular infringing activity in question. *Sammons v. Colonial Press*, 126 F.2d 341, 349 (C.A. 1st Cir.1942); *Stearns-Roger Mfg. Co. v. Ruth*, 87 F.2d 35, 41–42 (C.A. 10th Cir. 1936) (patent case). The defendant need not, however, prove that each item of overhead was used in connection with the infringing activity:

> ". . . The law requires no such minutiae, for it would make trials interminable. When appellant proved the actual expenditures for ordinary overhead, and a fair method of allocation, it carried its burden in the first instance. If, on cross-examination or otherwise, it appears that ordinary overhead is not chargeable, in whole or in part, to the infringing business, then a proper charge only should be made. But all allowance should not be denied because stenographers, bookkeepers, janitors, and presidents were not called to testify that they did perform specific tasks on this specific business. Courts and accountants resort to allocation to obviate this particular difficulty." *Stearns-Roger Mfg. Co. v. Ruth*, 87 F.2d at 41–42.

Cf. *Sammons v. Colonial Press*, supra, 126 F.2d at 349 (the burden is upon the defendant to show that each item of general expense or overhead assisted in the production of the infringement).

There is sufficient evidence in this case to establish that Belmont, in copying the plaintiff's architectural plans and constructing the apartment complex from those infringing plans, necessarily utilized the company's administrative personnel and its office facilities.

Belmont's calculated overhead figure of $249,881.00 consists of $217,636.00 in expenses, $3,384.00 in depreciation on office equipment and furniture, and $28,861.00 in interest on a capital funds loan which was not utilized on any particular project. The expenses are itemized by their nature and amount, with the following categories included: officers' salaries, other salaries, payroll taxes, advertising and promotion, auto and travel, rent, employee benefits, telephone, office supplies, general insurance, general taxes, utilities, professional services, bad debts, and other operating expenses. With the exception of the advertising and promotion expense of $4,000.00 and the bad debt expense of $11,621.00, these expenses are justifiable overhead expenses, a proportion of which are attributable to the 1830–32 project.

The advertising and promotion expense is not chargeable to the 1830–32 project, because Karl Witt, as sole owner of Empire, contracted with his own construction company (Karl Witt is majority stockholder of Belmont) for the construction of the 1830–32 project. Presumably, no advertising or promotion was responsible for the consummation of that contract. As to the bad debt expense, Mr. Witt testified on cross-examination that the bad debts were incurred on projects other than the 1830–32 project. The interest on the capital funds loan and the equipment and furniture depreciation are properly allocable overhead expenses. See *Sheldon v. Metro-Goldwyn Pictures Corp.*, 106 F.2d 45 (C.A. 2nd Cir. 1939), aff'd, 309 U.S. 390, 60 S.Ct. 681, 84 L.Ed. 825 (1940). Thus, the total overhead expense, a portion of which is allocable to the 1830–32 project, is $234,260.00.

In determining the proportion of the overhead expense allocable to the 1830–32 project, Belmont calculated what percentage the allocable overhead expenses were of the total net sales of the company, and then applied this percentage to the gross profit of $511,250.00 made on the 1830–32 project. This formula is a reasonably acceptable formula for allocating overhead expenses, except in one respect—Belmont utilized the net sale figure ($2,728,642.00), rather than the total income figure ($2,777,975.00), for the fiscal year in question. The total income figure includes gain on sale of equipment, management income, interest income and miscellaneous income. No explanation was given for the exclusion of this income, and there is nothing in the record to indicate that the company's overhead is not chargeable to it. It will be included, therefore, in the computation of allocable overhead. The $16,109.00 income loss incurred by Belmont's subsidiary will not be included in computing Belmont's total income, because Belmont's overhead expense was not involved in generating that loss. Computed as described, Belmont's total income figure is $2,794,084.00.

Utilizing Belmont's proposed allocation formula, the overhead allocable to the 1830–32 project is $42,863.92.[16] The deduction of

---

16. ($234,260 ÷ $2,794,084) × $511,250 = (.0838) × $511,520 = $42,863.92.

this figure from the gross profit of $59,709.44 realized on the 1830–32 project results in a net profit of $16,845.52.

(2) Empire

On July 14, 1980, Empire entered a written contract with Amwest Properties, Inc. for the sale of the land and buildings located at 1830–32 Knox Street. The total purchase price of $587,250.00 was to be paid as follows: $179,500.00 at the time of closing, $16,250.00 ninety days after closing and $391,500.00 plus interest in 360 monthly amortization installments of $4,407.40 each. Pursuant to paragraph 10 of that agreement, Amwest had the right to pay the monthly installments direct to State Federal Savings and Loan Association on Empire's outstanding real estate mortgage of $275,000.00. Pursuant to the agreement of the parties, this indebtedness was later refinanced for the benefit of Amwest to the amount of $391,000.00.

Empire introduced into evidence an accounting sheet listing the direct costs incurred by Empire in connection with the 1830–32 project. The costs consisted of (1) the amount paid to Belmont for the construction of the apartment complex and garage, (2) amounts paid in arranging the financing for the project, (3) the costs of the 1830–32 real estate, including carrying costs, (4) the amount of real estate taxes paid, and (5) the interest paid on the money borrowed by Empire for payment for the construction of the 1830–32 buildings.

The plaintiff contends that the total gross revenue which will be realized by Empire as a result of the sale of the 1830–32 project is not only the total purchase price of $587,250.00, but also the interest in the amount of $1,195,264.00 to be paid by Amwest on $391,500.00 of the purchase price—a total profit of $1,782,414.00. The evidence in this case established, however, that the interest payments paid by Amwest would not inure to the benefit of Empire, because both the principal and interest payments owing from Amwest would be paid to the savings and loan association as monthly installments on Empire's mortgage indebtedness of $391,000.00 plus interest incurred in connection with the 1830–32 project. Certainly, any interest incurred on amounts borrowed by Empire as working capital for the 1830–32 project was a direct expense deductible from Empire's gross profit on the project. *Sheldon v. Metro-Goldwyn Pictures Corp.*, 106 F.2d at 53 (where the court approved the master's deduction of interest upon a $300,000.00 loan taken for purposes of working capital). Because the interest owing from Amwest is, pursuant to the parties' agreement, being paid to the savings and loan association in satisfaction of the interest accruing on Empire's borrowing of working capital, Empire is not entitled to deduct the accruing interest as a direct cost of the project; because it has not taken such a deduction, it need not recognize the interest paid by Amwest as revenue.

Therefore, the total profit realized by Empire on the project is $587,250.00.

With the exception of the interest payments to Commonwealth Savings Company and the cost of the real estate, I find the expenses listed on Empire's accounting sheet to be deductible expenses. As will be discussed, the cost of parcel No. 2 is an expense which should be deducted from that proportion of the profit realized on the sale of the parcel, rather than from the profit realized on the sale of the constructed buildings. With respect to the interest payments to Commonwealth, the deposition of Karl Witt, introduced by the plaintiff, reveals that the $50,000.00 loan from Commonwealth involved more than just the 1830–32 project. Thus, the interest payments on this loan should not be borne solely by the 1830–32 project. The defendant introduced no evidence as to how much of that loan was utilized for the 1830–32 project, and without such information it is impossible to calculate the amount of interest allocable to the project. Because Empire has failed to satisfy its burden of proof as to this element of direct cost, the deduction of $2,477.54 as costs relating to this loan will not be allowed. There is no evidence that the other direct costs listed on the accounting sheet were not wholly attributable to the 1830–32 project; they will therefore be allowed.

Empire also seeks to deduct from its profit on the 1830–32 project $46,830.50 in overhead expenses allocable to the project. As previously stated, it is Empire's burden to prove the actual expenditures for ordinary overhead and a fair method of allocation. *Stearns-Roger Mfg. Co. v. Ruth*, supra. Merely stating that $46,830.50 was the overhead expense allocable to this project does not satisfy this burden. The court (and the plaintiff, for purposes of cross-examination) is without any knowledge as to what items were included in the overhead figure or the specific amount of each of those items. Furthermore, the method of allocation utilized by Empire in computing this overhead allocation remains a mystery. This amount, therefore, will not be deducted from Empire's gross profit on the 1830–32 project. See *Sammons v. Colonial Press*, supra, 126 F.2d at 349.

Finally, Empire argues that the value of the land on which the 1830–32 apartment complex was built is $50,000.00 and that this part of the total purchase price for the 1830–32 real estate and buildings is, under § 504, an element of profit "attributable to factors other than the copyrighted work." I am inclined to agree. See *Sheldon v. Metro-Goldwyn Pictures Corp.*, 309 U.S. 390, 402, 60 S.Ct. 681, 685, 84 L.Ed. 825 (1940) (apportionment of profits is appropriate where the evidence is "sufficient to provide a fair basis of division so as to give the copyright proprietor all the profits that can be deemed to have resulted from the use of what belonged to him.").

Wayne Kubert, a Nebraska licensed real estate broker and appraiser, testified on behalf of Empire that the fair market value of parcel No. 2 in an undeveloped state as of July, 1980, was $50,000.00. This appraisal is supported by the fact that the contract entered into between Empire and Amwest on July 14, 1980, gave Amwest an option to purchase parcel No. 1, which was slightly smaller than parcel No. 2 and still undeveloped, for $50,050.00. Thus, there is sufficient evidence that $50,000.00 of the purchase price of $587,250.00 was attributable to the real estate, rather than to the buildings constructed thereon. I note, however, that the cost of this land—$14,552.17, is a direct expense which should be deducted from the profit on the real estate and not from the profit attributable to the buildings.

A calculation of the gross profits realized by Empire on the sale of the apartment complex—$587,250.00 minus $50,000.00 (profit attributable to the real estate), minus $560,917.95 (expenses excluding the cost of the land, payments to Commonwealth, and overhead)—reveals that Empire suffered a loss of $23,667.95, not a profit. That such a loss actually occurred is substantiated by the consolidated statement of income and retained earnings for Belmont and Empire, which shows that for the fiscal year in question Empire suffered a loss of $16,109.00. Because Empire realized no profits on the 1830–32 project, no profits can be recovered by the plaintiff.

C. *Costs*

Pursuant to § 505 of the 1976 Act, the full costs of this action will be awarded to the plaintiff against the defendants Belmont and Empire.

D. *Attorney's Fee*

Under § 505 of the 1976 Act, this court may award a reasonable attorney's fee to the prevailing party as a part of costs, except as otherwise provided by the Act. Section 412 prohibits an award of attorney's fee for:

> "(1) any infringement of copyright in an unpublished work commenced before the effective date of its registration; or
>
> "(2) any infringement of copyright commenced after first publication of the work and before the effective date of its registration, unless such registration is made within three months after the first publication of the work."

The defendants argue that under subsection (2) of this section this court is prohibited from awarding the plaintiff attorney's fees. The success of this argument depends upon a finding that the plaintiff's filing of the 1820–22 architectural plans with the city codes administration department on February 10, 1978, for purposes of obtaining a building permit and/or the distribution of the plans to Belmont with knowledge that the plans would be distributed to its subcontractors and suppliers constituted a publication of those plans. Such a finding is contrary to the majority of copyright cases addressing the issue of publication of architectural plans. See *Nucor Corp. v. Tennessee Forging Steel Service, Inc.,* 476 F.2d 386, 390–391 (C.A. 8th Cir.1973) (neither distribution of plans to potential contractors and subcontractors for bidding purposes, permitting persons to view and inspect a building during and after construction, nor distribution of catalogs with photographs of the exterior of a building can be said to be publication of architectural plans); *Masterson v. McCroskie,* 194 Colo. 460, 573 P.2d 547 (1978) (distribution of architectural plans to contractor, subcontractors, building inspector and division developer does not constitute publication, even though no express restrictions were communicated to the recipients); *Seay v. Vialpando,* 567 P.2d 285 (Wyo.1977); *Krahmer v. Luing,* 127 N.J.Super. 270, 317 A.2d 96 (1974); *Shaw v. Williamsville Manor, Inc.,* 38 A.D.2d 442, 330 N.Y.S.2d 623 (App. Div.1972); *Edgar H. Wood Associates, Inc. v. Skene,* 347 Mass. 351, 197 N.E.2d 886 (1964); *Smith v. Paul,* 174 Cal.App.2d 744, 345 P.2d 546 (1959). Contra, *DeSilva Construction Corp. v. Herrald,* 213 F.Supp. 184 (U.S.D.C.M.D.Fla.1962) (filing of plans with building inspector is paramount to publication).

Based upon the facts of this case and the cited copyright cases, I find that there was no general publication of the 1820–22 architectural plans. Section 412(2), therefore, is inapplicable. However, because the 1820–22 plans were an unpublished work and because the defendants Belmont and Empire commenced their infringement of the plaintiff's copyright in these plans before April 29, 1980—the effective date of the plaintiff's registration of copyright—subsection (1) of § 412 prohibits an award of attorney's fee to the plaintiff.

E.   *Treble Damages*

The plaintiff's request for treble damages will be denied. First, there is no statutory authorization in the 1976 Act for such an award. Second, I do not find the defendants Belmont and Empire's infringing activities to be so outrageous or egregious as to justify such an award.

For the reasons stated in this memorandum of decision, the plaintiff's complaint against Lincoln Lumber and William R. King will be dismissed; Belmont and Empire are jointly and severally liable for the actual damages suffered by the plaintiff in the amount of $9,840.93 and for the costs of this action; and Belmont is liable for its profits in the amount of $16,845.52. A separate judgment will be entered.

One reason for design professionals seeking common law rather than statutory copyright protection was the presumed burden of depositing copyrighted works with the Copyright Office. Before and after the 1976 Act, registration could be made at any time

prior to an infringement acti
brought. However, under the 19
failure to register within three month
publication precludes recovery of attorne
fees or statutory damages.[17] But registration need not be difficult or expensive.
Even the pre-1976 copyright regulations
permitted substitutions for the original
materials themselves if they were bulky or
if it would be expensive to require deposit.

The 1976 Act gives authority to the Register of Copyrights to specify by regulation
the nature of copies required to be deposited. These regulations can allow deposit of
identifying material rather than copies.
One rather than two copies can be
permitted.[18]

(E) ADVICE TO DESIGN PROFESSIONALS

If copyright protection is to be sought, design professionals should comply with the
statutory copyright requirements. First,
they should be certain that they have not
assigned their right to copyright ownership.
They no longer need a contract clause giving ownership rights to the plans and specifications to the design professional. But
see Section 15.11.

Second, design professionals should comply with the copyright notice requirements.
The word *Copyright* can be written out or
the notice can be communicated by abbreviation or symbol. The authorized abbreviation is "Copr," and the authorized symbol is
the letter "C" enclosed with a circle (©).
The year of first publication should be giv-
en and the name of the copyright owner or
an abbreviation by which the name can be
recognized or generally known can be used.

ad
and
profess
work wit
Methods a
burden. The
consulted.

SECTION 19.05
SOME OBSERVAT
AND COMPARISON

(A) SCOPE OF COVERAGE

The discussion of patent law is
design professional who wishes to in
legal action for infringement of a paten
a copyright will retain an attorney. A
though the steps for perfecting a copyright
are simple, an inventor who wishes to obtain a patent must secure the services of a
patent attorney. A patent attorney is
needed to guide the inventor through the
maze of patent law and the complexities of
a patent search. Perfection of a copyright,
as a rule, will not require the services of an
attorney.

(B) PATENT AND COPYRIGHT COMPARED

Generally the subjects of patents are products, machines, processes, and designs.[20]
Copyright generally protects writings.

The principal protection accorded by

---

17. 17 U.S.C.A. § 412.

18. 17 U.S.C.A. § 408(c).

19. As indicated, 17 U.S.C.A. §§ 405 and 406 provide relief if the copyright notice is omitted or erroneously made.

20. For an instructional design patent case in an architectural context, see *Blumcraft of Pittsburgh v. Citizens & Southern National Bank of South Carolina*, 407 F.2d 557 (4th Cir.1969).

suits are lengthy,
isive. The plaintiff
id, in some cases, a
or patent infringe-
infringement cases,
treble damages and

s seventeen years.
than copyright pro-
rotection accorded
poly protection ac-
ably the reason for
ig more limited in

ent protection is harder to
copyright but once acquired is
ch more.

414

SECTION 19.06
TRADE SECRETS

(A) DEFINITION

The Restatement of Torts has defined a trade secret as:

> . . . any formula, pattern, device, or compilation of information which is used in one's business, and which gives him an opportunity to obtain an advantage over competitors who do not know or use it. It may be a formula for a chemical compound, a process of manufacturing, treating or preserving materials, a pattern for a machine or other device . . . . A trade secret is a process or device for continuous use in the operation of the business. Generally it relates to the production of goods, as, for example, a machine or formula for the production of an article.

>   *   *   *

> The subject matter of a trade secret must be secret. Matters of public knowledge or of general knowledge in an industry cannot be appropriated by one as his secret. Matters which are completely disclosed by the goods which one markets cannot be his secret. Substantially, a trade secret is known only in the particular business in which it is used. It is not requisite that only the proprietor of the business know it. He may, without losing his protection, commu-

chance of establishing that the patent is not valid.

One study covering patent cases back to 1920 demonstrates the difficulty of sustaining a patent in the courts.[21] From 1920 to 1973 the Supreme Court found 18% of the patents it reviewed to be valid. The Court of Appeals found 35% valid, while the trial courts found validity in 45% of the cases. Even more ominous for patent holders, the recent tendencies have been to find patents invalid, probably reflecting the prevailing hostility toward monopoly. The difficulty of sustaining a patent's validity often induces those who have developed industrial data to keep the information secret rather than publish it and seek patent protection. (Trade secrets are discussed in Section 19.06.)

---

21. Baum, *The Federal Courts and Patent Validity: An Analysis of the Record*, 56 J.Pat.Off.Soc'y 758 (1974).

nicate it to employees involved in its use. He may likewise communicate it to others pledged to secrecy.

The Restatement sets forth the following factors that are considered in determining whether particular information is a trade secret:

(1) the extent to which the information is known outside of his business;

(2) the extent to which it is known by employees and others involved in his business;

(3) the extent of measures taken by him to guard the secrecy of the information;[22]

(4) the value of the information to him and his competitors;

(5) the amount of effort or money expended by him in developing the information;

(6) the ease or difficulty with which the information could be properly acquired or duplicated by others.[23]

(B) CONTEXT OF TRADE SECRET LITIGATION

Trade secret litigation can arise when an employee leaves an employer either to go into business or to work for a new, and frequently competing, employer. If the employee has commercial or technical information, the prior employer may seek a court decree ordering the former employee not to disclose any trade secrets "belonging" to the prior employer and a decree ordering the new employer not to use the secret information. Such a court order can be justified by a confidential relationship between the prior employer and the former employee or the breach of an employment contract between the prior employer and the former employee.

Trade secret litigation can result when the proprietor of a trade secret learns that someone to whom a trade secret has been disclosed on a basis of confidentiality intends to make, or has made, an unauthorized use of the information. For example, the developer of a new product may give technical information relating to the product to the contractor building the plant in which the product is to be manufactured or to the manufacturer who is to build the machinery needed to make the product. Trade secret litigation may result if the person to whom the disclosure is made intends to, or has made, unauthorized use of the information. Similarly, a confidential disclosure of the information may be made to a manufacturer by an inventor who seeks to interest the manufacturer in a process or product developed by the inventor. The unauthorized use of such a pre-contract disclosure can lead to trade secret litigation.

Developers of technology sometimes try to recover research costs by licensing others to use the data. To protect the secrecy of the technology and to enable them to sell the data to others, developers usually obtain a promise from the licensee not to disclose the data to anyone else. Breach, or a threatened breach, of such a nondisclosure promise may cause the proprietor of the trade secret to seek a court decree forbidding any unauthorized use or disclosure.

(C) CONTRAST TO PATENTS: DISCLOSURE vs. SECRECY

Patent law requires public disclosure of the process, design, or product that is the subject of a patent. In exchange for this public disclosure, the patent holder obtains a seventeen-year monopoly. Trade secret protection, on the other hand, requires that the data asserted to be a trade secret be kept relatively private and nonpublic.

A patent requires an invention to be nov-

---

22. Absence of these measures precluded a trade secret in *Jet Spray Cooler, Inc. v. Crampton,* 361 Mass. 835, 282 N.E.2d 921 (1972).

23. Restatement (First) of Torts § 757 comment b. The Second Restatement of Torts omitted this topic. While less authoritative, the First Restatement still provides a useful summary.

el, unique, useful, and not obvious from the prior art. The trade secret need not meet these formidable requirements.[24] Although the courts are not unanimous on the point, it seems clear that the person who asserts ownership of a trade secret must show that he has made some advance on what is generally known. If the information is generally known, or generally available, the information is not a trade secret.

Roughly, trade secret protection has the same relationship to patent protection that common law copyright had to statutory copyright. Both the doctrine of common law copyright and the doctrine of trade secrets are predicated on extending legal protection to creative people by giving them the right to determine when, how, and if the fruits of their intellectual labor should be made generally available. Patent and statutory copyright are predicated upon disclosure. Trade secret protection is accorded by state law and suffers from the same lack of uniformity as common law copyright had. On the other hand, patent law is governed by federal law, resulting in general uniformity throughout the United States.

(D) ADJUSTING COMPETING SOCIAL VALUES

The doctrine of trade secrets, like many other legal doctrines, must consider and adjust various desirable, yet often antithetical, objectives. This can be shown by examining these objectives from the points of view of the various persons affected.

Those who seek trade secret protection—primarily inventors and research-oriented organizations—want to be rewarded economically for their creativity. Restricting others from using the information and technology that they have developed can make their information more valuable. Without adequate economic incentives, perhaps scientific and industrial progress

would be impeded. Protection of trade secrets can discourage industrial espionage and corruption.

Trade secret protection can restrain the freedom of choice and action for research employees. Creative employees can be prevented from making the best economic use of their talents. An employer's failure to consider or develop an employee's research ideas can destroy the employee's creativity. Many space age industries developed when creative people banded together to start new companies. Had they been tied to an older established company unwilling to engage in experimental research, many of these industries might not have developed or might have taken considerably longer to do so.

Overzealous protection of trade secrets can hamper commercial, scientific, and industrial progress. To a great extent, such progress is made possible by the free dissemination of technical and scientific information. Dissemination of such information can avoid costly duplication of research efforts.

Overprotection of trade secrets can also have an anticompetitive effect. Protection of trade secrets can give the developer a virtual monopoly that can hinder the development of competitive products and can result in higher prices to consumers.

Trade secret law has had to consider and adjust all these competing objectives—not an easy task.

(E) AVAILABILITY OF LEGAL PROTECTION

Duty Not To Disclose Or Use: Confidential Relationship and Contract

The circumstances surrounding the disclosure and the nature of the information disclosed are relevant in determining whether a duty exists not to use or disclose the

---

24. *University Computing Co. v. Lykes-Youngstown Corp.,* 504 F.2d 518 (5th Cir.1974); *Raybestos-Manhattan, Inc. v. Rowland,* 460 F.2d 697 (4th Cir.1972).

information. If the disclosure is accompanied by an express promise not to use or disclose, a general duty exists not to disclose. It may still be necessary to interpret the agreement to determine what cannot be disclosed, to whom disclosure is prohibited, and the duration of the restraint.

Suppose a licensing agreement exists by which the licensor permits the licensee to use technological data disclosed by the licensor to the licensee. Does the restraint on disclosure include information that the licensee knew before the disclosure? Does it include information developed by the licensee from the disclosed information? Does the restraint include parts of the technological data disclosed that are known at the time of disclosure or become generally known? Can disclosure be made to an affiliated or successor company? Is there a continuing obligation for either or both parties to communicate new technology? These questions should be and usually are covered in the licensing agreement. If not, courts must interpret the agreement and, if necessary, imply terms.

In some circumstances there is no express provision prohibiting unauthorized use or disclosure. The method by which the information is acquired will often determine whether the disclosure is made in confidence and whether the person to whom it is disclosed obligates himself not to disclose it to others. This is similar to the process by which the law implies certain promises between contracting parties not expressed in the written contract. The communication may be part of a contractual arrangement. For example, the possessor of the information may communicate it to a consulting engineer who has been retained to advise the possessor on the type of machinery to be used in the process. If a written contract exists, the possessor will usually require a promise by the consulting engineer not to divulge certain specified information. Even without such an express promise not to disclose, the law would probably imply such a promise, based upon

surrounding facts and circumstances. The same is true if the disclosure is made to the manufacturer of the machine or to a building contractor.

Circumstances exist where there is no contractual relationship between the possessor of the information and the person to whom it is disclosed. For example, an inventor may disclose information to a manufacturer in order to interest the manufacturer in buying the information. It is possible for the inventor to obtain a promise from the manufacturer not to disclose the information. Even without such a promise, if it is apparent from the surrounding facts and circumstances that the disclosure is made in confidence, any disclosure of the information by the manufacturer would be a breach of the confidence and would give remedies to the inventor.

Nature of Information

If the person to whom the information is disclosed—whether a contractor hired to build a plant, a manufacturer hired to build a machine, or a consulting engineer hired to furnish technical services—knows that the information is not generally known in the industry, this is likely to persuade a court that a confidential relationship was created or that a nondisclosure promise should be implied.

The nature of the information will also determine the legal remedy for a breach of confidence or a breach of contract. Under American law, the normal remedy for a breach of contract is a judgment for money damages. Only if that remedy is inadequate will the law specifically order that a defendant do or not do something. This is of crucial importance in trade secret cases. Typically, if the information is truly valuable and not generally known, the most important remedy is the court decree ordering that the person who has the information not disclose it to anyone else. Violation of such an order is punishable by a fine, or even imprisonment, under the contempt powers of the court. Such a decree puts the plaintiff in a good position to demand a substantial royalty or settlement price if the defendant needs to use the trade secret information.

To obtain such an extraordinary remedy, the plaintiff must show that irreparable injury would occur without such a court order and, as mentioned, that a judgment for money damages would be inadequate. In a trade secret case, the plaintiff seeks to show irreparable economic harm would be suffered if the information claimed to be a trade secret is broadly disseminated. The plaintiff usually asserts that such broad disclosure will enable competitors to draw even despite the plaintiff's research expenditure to turn out a better product or develop a better process. The plaintiff will also claim that it is difficult, if not impossible, to establish the actual damages suffered by general dissemination of the secret information. A court concluding that the information is a trade secret usually gives injunctive relief.

The principal defense in trade secret cases is that the information was not secret.[25] Often defendants point to the existing literature in a given scientific or technical area, with a view towards showing that a person diligently searching for this information could put it together and arrive at the process independently. Courts have not been particularly receptive to this defense. Usually the defendant has not gone through the literature to ferret out the secret. The information is often obtained from an employee of the trade secret possessor, paid for its disclosure by virtue of a licensing agreement, or received through a confidential, limited disclosure. While the defense has occasionally worked, on the whole it has not been successful. Part of the difficulty in arguing for this defense is that sometimes information and data are available, but not in a collected, organized, convenient, and usable form. These factors are the principal advantages of the trade secret. Sometimes the data are collected and organized in readily accessible form but most people in the industry are unaware of this fact or unable to locate the material easily.

Employee Cases

There are special aspects to the cases where the information has been learned or developed by an employee and that employee goes into business himself or herself, joins in a venture with others, or is hired by an existing or potential competitor of the prior employer. In addition to the use of the confidentiality theory, the former employer often points to an employment contract under which the employee agreed not to disclose the information after leaving the employment. Sometimes the limitation on disclosure far exceeds what is reasonable. Often the employee has little bargaining power in deciding whether to sign such an agreement. Some courts have recognized the adhesive (nonbargain) nature of such agreements and have refused to give these agreements literal effect. However, the employer who has an agreement by the employee not to divulge information is in a better position to obtain a court decree ordering the employee not to disclose particular information.

In addition to recognizing the take-it-or-leave-it nature of most employment contracts, some courts feel that agreements under which employees cannot practice their trade or profession or use the information that is their principal means of advancement are unduly oppressive to employees. Such courts are not likely to be sympathetic

---

25. In *ILG Industries, Inc. v. Scott*, 49 Ill.2d 88, 273 N.E.2d 393 (1971), the court held that the possibility of "reverse engineering" (starting from the finished product and working backwards) was not a defense when the process of doing so was quite time-consuming.

to claims for trade secret protection. On the other hand, other courts manifest great concern with immorality and disloyalty on the part of employees and look upon employee attempts to cash in on information of this type as morally indefensible. These courts are likely to deal harshly with employees in trade secret cases.

To sum up, a former employee will be restrained from using confidential information if that restraint is reasonable,[26] taking into account the legitimate needs of the former employer, the former employee, and the public.

(F) SCOPE OF REMEDY

A trade secret claimant can recover damages suffered, profits made by the infringer resulting from the infringement, and a court decree prohibiting him from using or divulging the information.[27] The injunctive relief usually does not exceed the protection needed by the plaintiff. An injunction may be only for a period of time commensurate with the advantage gained through the technological information improperly acquired or used.[28] If the defendant could have ascertained the information within a designated period, the court decree may require that he not use the information for that period of time. Unless the defendant has made the information public, the court order for non-disclosure will apply only until the information is generally known. Some courts take a more punitive attitude and will order that the trade secret not be used even if it becomes generally known.[29] Generally, the

more reprehensible the conduct by the defendant, the broader the injunction.

(G) DURATION OF PROTECTION

The trade secret is protectible as long as it is kept relatively secret. This unlimited time protection has caused some to advocate protecting trade secrets for a limited period of time by according a patentlike monopoly to the developer of a trade secret.

Some trade secrets are patentable. Unlimited duration of protection for a trade secret can frustrate the seventeen-year patent monopoly policy. To the extent that states, through protection of trade secrets, frustrate patent law, such trade secret protection may be unconstitutional.

(H) ADVICE TO DESIGN PROFESSIONALS

Design professionals who invent processes, designs, or products should, wherever possible, use contracts to give them protection against the possibility that persons to whom they divulge the information may disclose the information to others or use it themselves.

Design professionals who occupy managerial positions in companies where trade secrets are important should use all methods possible to keep the information secret. Only those who have an absolute need to use the information should be given access to it, and these persons should expressly agree in writing not to disclose the information. Management should also realize that

---

26. An interesting case examining these factors and restraining the former employee is *B.F. Goodrich Co. v. Wohlgemuth*, 117 Ohio App. 493, 192 N.E.2d 99 (1963).

27. One court held that a plaintiff cannot receive both its losses and profits of the defendant. *Sperry Rand Corp. v. A-T-O, Inc.*, 447 F.2d 1387 (4th Cir.1971). Another stated that losses occur only as a practical matter when the trade secret is destroyed. *University Computing Co. v. Lykes-Youngstown Corp.*, supra note 24. That case explored formulas for determining gain to defendant.

28. *ILG Industries, Inc. v. Scott*, supra note 25 (for 18 months); *Sperry Rand Corp. v. A-T-O, Inc.*, supra note 27 (for two years).

29. A. Turner, *Law of Trade Secrets* 437–438 (1962).

employee loyalty is probably the best protection against the loss of trade secrets. Reasonable treatment of employees is likely to be a better method of preserving trade secrets than litigation.

Design professionals who are technical employees and who wish to take their technological information to start their own business, join in a business venture, or work for a competitor of their present employer should realize that their departure under these circumstances may result in litigation, or at least the threat of litigation. Legal advice should be sought in order to examine the legality of any asserted restraints and to determine the scope of risk involved to the employee who chooses to leave his or her present employment.

Part Four

Building and Projects:
Focus on Contractors

20

Contractor Licensing:
Consumer Protection
or Harmful Cartel?

SECTION 20.01
RELATIONSHIP TO CHAPTER 13

Part IV highlights contractors, although the other main actors—owners and designers—are important figures. In many states before contractors can bid for construction work or execute a design, they must have a contractor's license. Chapter 13, which discussed the registration and licensing laws, included some material dealing with contractor license laws, should be reviewed. This chapter concentrates on the special problems of contractor licensing laws.

SECTION 20.02 SHOULD
CONTRACTORS BE LICENSED?

(A) PURPOSE OF LICENSING LAWS

To protect the public, the state regulates professions, occupations, and businesses. Those who wish to build must retain a contractor and often a designer. Licensing laws should provide a representation to the public that the holder meets a minimal level of competence, honesty, and financial capacity. The state wishes to protect the general public from being harmed by poor construction work and seeks to accomplish this by allowing only those who meet state requirements to design and to build. Yet as seen in Chapter 13, these efforts are not

costless. Educational and experience requirements and the administration of these regulatory systems increase the cost of providing these services. They also reduce the pool of contractors. Are the gains worth the cost? Do contractor licensing laws do the job? Who needs protection, and do the licensing laws provide it?

Professional designers usually must meet certain educational and training requirements before they are allowed to take the examination that will determine whether they will be registered. Although these requirements do raise the cost of entering the profession, it is at least arguable that they improve the quality of designers.

As a rule, contractors need meet only experience requirements and not educational ones. In California, for example, they must have four years of experience in the trade they wish to enter. Under certain circumstances, particular education courses taken can fulfill up to three years of this four-year experience requirement. Applicants in California take a one-day written examination, half of which is devoted to law and the other half to trade practices. As Section 13.02 showed the "pass" line can be drawn arbitrarily. Can examiners determine what is a reasonable level of classroom performance? Can the classroom simulate actual construction?

How will the examiners determine the integrity of the person who applies for a license? Undoubtedly this is accomplished by letters of reference (of dubious value) and perhaps other investigations. Yet will investigations uncover much that will reflect the integrity of the contractor? How will the regulatory agency determine financial responsibility? In California, financial solvency requires a working capital of $2500. Until 1980, an applicant had to post a bond for $2,500 in California. Effective 1980 the applicant must post a bond of $5,000 ($10,000 if a swimming pool contractor). Is this bond, which covers more than complaints by homeowners, adequate to establish financial capacity?

What must a contractor do to keep a license in effect? In California, the license must be renewed every two years. Renewal for an active contractor costs $200 and requires posting of the bond. An inactive contractor can renew for $50. There is no provision for periodic examination or continuing education.

Are the undeniable costs that such a system imposes more than balanced by helping *those who need help* to determine who is competent? This raises a second question. Who needs the protection, and who gets it? Clearly such legislation was intended to protect those who are inexperienced in the world of construction or who cannot without great cost obtain sufficient information to make judgments. Owners that fall into these categories are more likely to make judgments based upon whether a contractor is "licensed and bonded."

Perhaps the inexperienced owner building a residence, a duplex, or a four-unit apartment or the naive owner about to engage in home improvement can be helped by the existence of licensing laws. But does that justify an across-the-board licensing requirement and an expensive licensing apparatus? Particular abuses can be dealt with by already abundant consumer legislation.

In addition to these questions, does this system deter unlicensed persons from building? As seen in Section 20.03, often an unlicensed contractor *can* use the courts to recover for services performed.

(B) HARMFUL EFFECTS

Suppose contractor licensing laws do not screen out the incompetent, nor do they only protect those who need protection. Do they do any harm?

Contractor licensing laws can artificially reduce the pool of contractors, something demonstrated by the Florida experience described in Section 13.02.

Contractor licensing statutes usually draw lines between general contractors and the specialized trades. In addition to the difficulty of drawing these lines, the segregation of specialty trades by licensing tends to reduce the likelihood of better organized, more efficient organizational structures, something so needed in construction.

Other dangers exist. It is not uncommon for licenses to be "bought" and "sold" as any other valuable commodity. An owner who wishes to build a residence or small commercial structure and avoid the cost of a full-time prime contractor sometimes engages an individual with a license, gives that person very little overall control, and limits her activity to periodic visits. This is done to avoid the owner's being liable for violating the licensing laws and gives a licensed contractor a saleable commodity for a relatively small investment.

The issuance of a license can be a false representation by the state that may deceive the unwary consumer. Consumers may believe they are dealing with a competent, honest, and financially reliable builder because that builder is licensed. To obtain business from a gullible public, many builders advertise that they are licensed and bonded. Perhaps elimination of contractor licensing laws would induce consumers to protect themselves.

SECTION 20.03 CONTRACTOR LICENSING LAWS

Along with the general proliferation in occupational licensing there has been an increase of contractor licensing laws. At present about one-half of the states have such laws. Perhaps the deregulation spirit seen in Arizona [1] may signal a slowdown or even rollback. But ideological movements such as deregulation often lack the staying power of organized groups, such as consumer groups and contractors, both prime movers in occupational licensing.

The sanctions imposed upon the unlicensed contractor are discussed in detail in Section 20.04. It should be noted, however, that many states have statutes that specifically bar an unlicensed contractor from recovering for work performed, something usually not found in architectural and engineering registration laws.

Contractor licensing laws have also brought in their wake other regulations. For example, in California a contractor must notify the owner of lien law provisions and the person to whom complaints may be made.[2] Home improvement contracts must contain a notice that the owner or tenant has a right to require the contractor to post a performance and payment bond.[3] The contractor must notify the owner that the contractor's failure to begin work under a home improvement contract within twenty days of the agreed date without excuse violates the state licensing laws.[4]

California licensing laws contain grounds for disciplinary action against the contractor that appear to be no more than simple breach of contract. For example, the following can be grounds for disciplinary action:

1. abandonment of project [5]
2. willful departure from plans [6]
3. material failure to complete project for price stated [7]
4. willful failure to prosecute work diligently [8]

These provisions cause public intervention into relations between owner and contractor (discussed in greater detail in Section 20.05).

SECTION 20.04 THE UNLICENSED CONTRACTOR: CIVIL SANCTIONS

(A) RECOVERY FOR WORK PERFORMED

The chief sanction for violating licensing laws has been to bar the use of the courts to collect for services. Yet for reasons to be noted in (E), denials of recovery, while still common, are less likely when a contractor seeks judicial relief than when such relief is sought by a design professional. This occurs *despite* the specific legislative direction

---

1. Refer to Section 13.02(B). A license is needed *only* for residential and small apartment construction.

2. See West's Ann.Cal.Bus. & Prof.Code § 7018.5. Failure to comply with a predecessor statute did not bar the contractor from recovery. *Gonzales v. Concord Gardens Mobile Home Park, Limited,* 90 Cal.App.3d 871, 153 Cal. Rptr. 559 (1979). The statute *specifically* bars an *unlicensed* contractor.

3. West's Ann.Cal.Bus. & Prof.Code § 7159(f).

4. Id. § 7159(j).

5. Id. § 7107.

6. Id. § 7109.

7. Id. § 7113. Failure need not be willful to constitute grounds for disciplinary action. *Mickelson Concrete Co. v. Contractors' State License Board,* 95 Cal.App.3d 631, 157 Cal.Rptr. 96 (1979).

8. West's Ann.Cal.Bus. & Prof.Code § 7119.

in many contractor licensing laws to bar the state courts to unlicensed contractors.

These specific legislative directives, such as found in California or Washington, do *not* mean contractors *never* recover. Such a harsh sanction has led these states to create a substantial compliance doctrine in order to avoid unjust enrichment (discussed in (C)). Courts have divided on the use of such laws as a defense by a knowledgeable participant in the construction industry.[9] The recent cases show the usual variant results when unlicensed contractors seek recovery.[10]

(B) EXCEPTIONS

Barring the unlicensed contractor from using the courts to recover for services it has performed does not mean that the unlicensed contractor cannot use other avenues. Section 13.07 noted that at least in some states an unlicensed person or entity can make a settlement and have it enforced, can obtain an arbitration award and have it enforced, or present a claim in a bankruptcy proceeding. Most of those cases involve unlicensed *contractors*.

In addition, an unlicensed subcontractor has been allowed to sue a prime contractor for delay damages in states that bar the unlicensed contractor from using the courts to recover under a contract.[11]

The reasoning of these cases, that recovery is *not* sought for *performance*, should allow an unlicensed *prime* contractor to recover from an owner for delay damages.

Other cases demonstrate the ambivalence of courts when asked to deny recovery to unlicensed contractors who have done their work properly. For example, courts have divided as to the effect of noncompliance when the dispute does not involve an ordinary member of the public but involves those knowledgeable about construction.[12]

Another evidence of the greater solicitude shown unlicensed contractors is reflected by *Moore v. Breeden*.[13] The applicable law required that the owner give notice to the contractor that a license is required before the owner can defend any claim made against it based upon the contractor's having been unlicensed. It was held that this notice requirement applies even if the contractor knew he was required to be licensed. This interpretation allowed the court to grant a recovery of almost $300,000 to an unlicensed contractor.

A similar judgment in favor of the claimant was based upon the claimant's not being a contractor at all but simply a foreman who had been engaged by the owner to perform construction work.[14]

These cases should not convey the impression that courts simply disregard contractor licensing laws. They do show the difficulty *some* courts have had in reconciling the legislative requirement for a license with the merits of a particular claim,

---

9. Compare *Lewis & Queen v. N.M. Ball Sons*, 48 Cal.2d 141, 308 P.2d 713 (1957) (subcontractor cannot recover from prime despite latter's knowing former not licensed), with *Dow v. United States*, 154 F.2d 707 (10th Cir.1946) (recovery), and *Enlow & Son, Inc. v. Higgerson*, 201 Va. 780, 113 S.E.2d 855 (1960) (recovery).

10. Compare *C.B. Jackson & Sons Construction Co. v. Davis*, 365 So.2d 207 (Fla.App.1978) (recovery), and *Lignell v. Berg*, infra note 18 (sophisticated owner), with *Cochran v. Ozark Country Club, Inc.*, 339 So.2d 1023 (Ala.1976) (had been paid $35,000, sought $28,000 more: denied); *United Stage Equipment, Inc. v. Charles Carter & Co.*, 342 So.2d 1153 (La.App.1977) (paid all but $1,300: denied); *Revis Sand & Stone, Inc. v. King*, 49 N.C.App. 168, 270 S.E.2d 580 (1980) (need to protect public).

11. *American Sheet Metal, Inc. v. Em-Kay Engineering Co.*, 478 F.Supp. 809 (E.D.Cal.1979); *Gaines v. Eastern Pacific*, 136 Cal.App.3d 679, 186 Cal.Rptr. 421 (1982).

12. Refer to note 9 supra.

13. 209 Va. 111, 161 S.E.2d 729 (1968).

14. *Sobel v. Jones*, 96 Ariz. 297, 394 P.2d 415 (1964).

the denial of which would appear to create unjust enrichment.

(C) SUBSTANTIAL COMPLIANCE

Perhaps most important, a few states, led by California, have adopted a substantial compliance doctrine. Excusable errors by the contractor related to obtaining or renewing a license will not preclude a recovery. In *Latipac, Inc. v. Superior Court*,[15] failure to renew a license was due to the inadvertence of an office manager who had had a mental breakdown. California has struggled with this doctrine but on the whole has used it to allow recovery where it appears that the work was properly performed and failure to obtain a license did not relate to competence but to bureaucratic error.[16]

Washington has also employed this doctrine. Its leading case, *Murphy v. Campbell Investment Co.*,[17] involved a contractor who inadvertently forgot to accompany his application for a license with proof that he had liability insurance. He corrected this mistake, and a license was issued. But at the time the contract was signed, he did not have a valid license. Washington precludes legal action unless the contractor has been licensed at the time the contract is made. The court concluded that there had been substantial compliance and the legislative intent would not be furthered by barring recovery. The dissenting opinion contended that the use of the substantial compliance doctrine seriously reduced the penal-

ties provided by the legislature for failure to obtain a license. Although Utah seems to have recognized the doctrine,[18] North Carolina vehemently rejected it.[19]

(D) INTERSTATE TRANSACTIONS

One difficulty generated by the sporadic adoption of contractor licensing laws—not nearly as universal as professional registration laws—is illustrated by *Conderback, Inc. v. Standard Oil Co. of California*,[20] a case that involved the extraterritoriality of the California licensing statute. The contractor was not licensed in California, the state in which his principal place of business was located. He sought to use the California courts to recover for services performed in Washington, which at that time did not have a licensing law. California requires that a contractor be licensed before maintaining an action for the performance of construction services. But California would not bar its courts to the unlicensed contractor, as that would be applying California law to a Washington transaction. Since Washington—the state with the greatest interest—had no licensing law, California would not impose its policy in this transaction.

Two cases noted in Section 13.07(B) came to contrary results when dealing with the collectibility for design services where a professional was licensed in one state but performed work in another where he was not licensed. Two other cases involving a

---

15. 64 Cal.2d 278, 49 Cal.Rptr. 676, 411 P.2d 564 (1966).

16. Successfully used in *Gaines v. Eastern Pacific*, supra note 11; *Airfloor Co. v. Regents of the University of California*, 84 Cal.App.3d 1004, 149 Cal.Rptr. 130 (1978); and *Vitek, Inc. v. Alvarado Ice Palace, Inc.*, 34 Cal.App.3d 586, 110 Cal.Rptr. 86 (1973), but denied in *General Insurance Co. of America v. Superior Court*, 26 Cal.App.3d 176, 102 Cal.Rptr. 541 (1972), and in *Weeks v. Merritt Building & Construction Co.*, 39 Cal.App.3d 520, 114 Cal.Rptr. 209 (1974) (strong dissent).

17. 79 Wash.2d 417, 486 P.2d 1080 (1971). See also *Expert Drywall, Inc. v. Brain*, 17 Wash.App. 529, 564 P.2d 803 (1977); Comment, 14 Gonz.L.Rev. 647, 659–661 (1979).

18. *Lignell v. Berg*, 593 P.2d 800 (Utah 1979).

19. *Brady v. Fulghum*, 309 N.C. 580, 308 S.E.2d 327 (1983).

20. 239 Cal.App.2d 664, 48 Cal.Rptr. 901 (1966).

similar problem for interstate contractors denied recovery, one over a strong dissent.[21]

(E) OBSERVATIONS

The contractor licensing laws and cases that interpret them demonstrate different attitudes from those that involve architects, engineers, and surveyors. More appears to be expected of design professionals than of contractors. Contractors are often small businesses that may not be aware of the existence of licensing laws and may not have the staff to insure that they are always in compliance. The recent enactment of many contractor licensing laws and the difficulty of formulating standards for determining competence may play a part in some judicial reluctance to bar a contractor. Even in cases where contractors *are* precluded from recovery, all that has been lost was part payment, with contractors often having collected most of their compensation.

SECTION 20.05 INDIRECT EFFECT: FORUM FOR CONSUMER COMPLAINTS

Some make a case for contractor licensing laws by pointing to the power they give the consumer homeowner when a contractor is not responsive to complaints that work has been done poorly. A criminal sanction, though rarely imposed, can be the basis for restitution, at least in aggravated cases.[22] More important, a threat to go to the licensing authorities may provide an incentive for the contractor to take the complaint seriously. If the dispute cannot be resolved amicably, there is likely to be an informal hearing before a representative of the state licensing agency. In effect, these agencies have become small claims courts.

The consumer protection aspect of contractor licensing would be more effective if licensing agencies made available a record of complaints against particular contractors. Yet usually this information is not made available.

This indirect function has its dangers. It can provide too powerful a tool to an owner who uses it in an abusive way to obtain more than promised under the contract or required by law. The system can be costly to operate, particularly where a high level of consumer dissatisfaction exists. This induced California to encourage arbitration of such disputes.[23]

---

21. *Meridian Corp. v. McGlynn/Garmaker Co.*, 567 P.2d 1110 (Utah 1977) (3 to 2); *Jary v. Emmett*, 234 So.2d 530 (La.App.1970), writ refused 256 La. 374, 236 So.2d 502 (1970). But see *Johnson v. Delane*, 77 Idaho 172, 290 P.2d 213 (1955) (engineer licensed in Washington recovered for design of Idaho project).

22. *State of Washington v. Bedker*, 35 Wash.App. 490, 667 P.2d 1113 (1983) (probation conditioned on payment of amount to remedy defects affecting safety). Refer to Section 13.07.

23. See West's Ann.Cal.Bus. & Prof.Code § 7085.

21

Planning the Project: Compensation and Organization Variations

SECTION 21.01 OVERVIEW

(A) SOME ATTRIBUTES OF THE CONSTRUCTION INDUSTRY

Please review the introduction to Parts II, III, and IV with special attention devoted to contractors. Industry characteristics are central to this chapter.

(B) OWNER'S OBJECTIVES

In the design phase, quality, price, and completion date are interrelated. An owner who wishes the highest quality may have to trade off quantity, price, and completion date. If early completion is crucial, the owner may have to sacrifice price and probably quality and quantity.

After these choices have been made, the owner may have to make *additional* choices when it determines *how* it will select a contractor, the contractor or contractors it *will* select, and the type of construction contract or contracts. Again these choices should be made in a way that maximizes the likelihood that the owner will receive quality that complies with the contract documents and on-time completion at the lowest *ultimate* cost. Again compromises may be needed. The contractor who will do the highest quality work is not likely to be cheapest and quickest. The quickest contractor may not be the one who will provide

the best quality. The importance the owner attaches to these objectives will affect the process it uses to select the contractors and how construction contracts are organized. This chapter looks at pricing and organizational variations. Chapter 22 focuses on competitive bidding.

(C) BLENDING BUSINESS AND LEGAL JUDGMENTS

This treatise examines law in the *context* of the construction industry. A sharp line cannot always be drawn between business and legal considerations.

Clearly, choices as to the pricing of a construction contract and how the project is to be organized must seek to achieve the owner's objectives. Choices made in these crucial matters should seek to obtain the best on-time work at the best price. To a significant degree, this will depend upon each participant's knowing what it is supposed to do and being able to do so in the most efficient way. Modern methods designed to bring efficiency to a chaotic process are described in Section 21.04. Although the Introduction to Parts II, III and IV describes some characteristics of the construction industry, one characteristic—operational inefficiency and even chaos—is relevant to this chapter. Section 21.05 looks at internal efficiency, mainly authority and communication. A recent empirical study dealt with "the pervasive and distressing inefficiency" in construction work. One writer stated that the study indicated that only 32% of the total time spent on a construction site involved actual work on the project.[1] He quoted the study as showing that the remainder of the work was divided in the following manner:

1. 7% for equipment transportation delays

2. 13% for travelling on the job site

3. 29% consumed by waiting delays

4. 8% for late starts and early quits

5. 6% for receiving instructions

6. 5% for personnel breaks

Quoting these statistics, the writer concluded that much inefficiency is due to the many contractual relationships and parties all working on the same structure, each "under a different management and each marching to the beat of a different drummer."[2]

Choices of the type described in this chapter that deal with these matters are to a great extent best made by those who know design and construction. What role does the law play? The answer can be divided into two categories: direct and indirect legal controls.

Direct legal controls involve such matters as registration and licensing laws, legal controls dealing with how a construction contract is awarded, the standard to which a contracting party is held, and the way in which the law will treat claims by a party who has suffered losses that it seeks to transfer to someone else.

Less direct legal considerations are often as important. For example, where there are blurred lines of authority and unclear allocations of risk, the law will frequently be called upon to pass upon claims. Similarly, inefficiency and other matters that cause losses are likely to lead to claims that, though usually settled, are done so against the backdrop of what the law would provide in the event the dispute ended up in the courthouse. Increasingly it is being recognized that one of the overhead costs in construction work incurred by all participants is making, avoiding, preparing for, and resolving claims. Choices of the type described in this chapter must be made carefully and intelligently. Failure to do so can only increase the cost of construction.

(D) PUBLIC vs. PRIVATE PROJECTS

Chapter 9 described the various types of owners and indicated that an important criterion is the status of the owner, whether a private party or a public entity. A private party who wishes to build can choose any type of compensation plan it can persuade a contractor to accept. It can award a contract in any way it chooses. It can make one contract with a prime contractor or a number of separate contracts with individual contractors.

Public entities may be limited by laws and regulations in making these choices. Commonly, public entities must award their construction contracts by competitive bidding under which contractors are all given a chance to submit a bid for particular work. Some states require that separate contracts be used for certain types of public work. Often limitations are placed upon cost contracts used in public works. A public entity may be controlled in the way it resolves disputes with its contractors. Early in this century, public entities frequently took the position that they could not arbitrate disputes, as this would be delegating

---

1. Foster, *Construction Management and Design-Build/Fast Track Construction: A Solution Which Uncovers a Problem for the Surety,* 46 Law and Contemp. Probs. 95, 116 n. 119 (1983).

2. Ibid.

power to private arbitrators. Increasingly, by law or regulation, public entities are being required to use arbitration or some other method of resolving disputes.

Those working on public contracts must first examine statutes and regulations applicable to them to determine the restraints placed upon them in awarding or organizing construction contracts. This chapter largely assumes that there are no restraints and that the owner and contractor can make any type of contract they wish and the owner is not limited in the way in which it chooses to organize participants contractually and administratively.

(E) PLANNING
TREATED ELSEWHERE

That this chapter focuses upon planning by no means indicates that planning is not considered elsewhere in this treatise. As illustration, planning can deal with contract interpretation,[3] changes,[4] delay,[5] subsurface problems,[6] and risk management emphasized in Chapters 18, 36, and 37. One of the important principles of this treatise is that an awareness of law and its willingness to allow planning is crucial to a well-managed construction project.

SECTION 21.02
PRICING VARIATIONS

Selecting a compensation system must take into account the responsibility for certain risks. Though there are many variations possible, each major category deals with risk allocation.

(A) FIXED–PRICE OR LUMP–SUM CONTRACTS: SOME VARIATIONS

In American usage, unlike that in Continental Europe,[7] fixed-price and lump-sum contracts are used interchangeably. Under such contracts, the contractor agrees to do the work for a fixed price. Almost all of the performance risks, that is, events that make performance more costly than planned, fall upon the contractor. For example, in *American Casualty Co. v. Memorial Hospital Association*,[8] no relief was given the contractor where the actual costs were twice those anticipated.

Only if the contract itself provides a mechanism for increasing the contract price can the contractor receive more than the contract price. Although not common in *ordinary* American contracts (contracts in the U.K. often use Fluctuations Clauses), some contracts have price escalation clauses under which the contract price is adjusted upward (and sometimes downward), depending upon market or actual costs of labor, equipment, or materials. Such a provision protects the contractor from the risks of any unusual costs that play a major part in its performance. Such a provision usually requires the contractor to use its best efforts to obtain the *best* prices.

Many construction contracts contain changed conditions clauses that allow a price increase if conditions under the ground or in existing structures are discovered that are substantially different than planned by the parties.

Sophisticated procurement systems sometimes provide *variant* fixed-price for-

---

3. Section 24.03(B).

4. Section 25.04.

5. Sections 30.08–30.10.

6. Sections 29.04 and 29.05.

7. There it is much harder to obtain a price increase under a fixed-price contract than under a lump sum. The ameliorating doctrine of *rebus sic stantibus* does not apply to the former.

8. 223 F.Supp. 539 (E.D.Wis.1963).

mulas even in fixed-price contracts. For example, one is the federal fixed-price incentive firm (or FPIF) contract used in *negotiated* contracts. Before finalization, owner and contractor negotiate the following items:

1. Target cost—against which to measure final costs.
2. Target profit—a reasonable profit for the work at target cost.
3. Ceiling price—the total dollar amount for which the owner will be liable.
4. Sharing formula—the arrangement for establishing final profit and price.

After the work is completed, the contractor and the owner negotiate the final costs of the contract, sharing the overruns or underruns according to the agreed formula. To illustrate, suppose the target cost for a contract is one million dollars, the target profit is $100,000, the price ceiling is $1,180,000, and the sharing formula is 75% (owner) and 25% (contractor). Under the formula, the contractor would keep 25% of every dollar saved. To earn a total profit of $120,000, it would have to reduce costs by $80,000 below target cost. Because there is no profit ceiling, profit would continue to increase indefinitely as the amount of underrun increased. Conversely, the contractor would have to overrun the target cost by $80,000 to reduce its profit to $80,000. If it overran by more than $180,000, it would lose money, since there is no minimum profit guaranteed in this contract type. Regardless of the final cost to the contractor, it must meet the contractual specifications, and the owner's liability cannot exceed the ceiling price of $1,180,000.

This form of contract has the advantage of establishing a price ceiling similar to the guaranteed cost to the owner under the fixed-price contract. By penalizing the contractor for cost overruns above the target

estimate (which is always something less than the price ceiling) and by rewarding it for cost savings, this type of contract provides financial motivation to the contractor to perform at the most economical cost.

A contract of this type is appropriate where the owner's plans are not sufficiently detailed to allow fixed-price bidding without excessive provision for contingencies yet are sufficiently advanced that a reasonably accurate target estimate can be made. Such a contract can also, if desired, include monetary incentive provisions to the contractor for early completion.

A fixed-price contract has the obvious advantage of letting the owner and those providing funds for the project know in advance what the project will cost. It works best when clear and complete plans and specifications are drawn. Incomplete contract documents are likely to cause interpretation questions that can lead to cost increases that, under a loose-changes clause, can convert what appears to be a fixed-price contract into a cost contract.[9] The fixed-price contract is used most efficiently when there is a reasonable number of experienced contractors willing to bid for the work. This is less likely to be the case where the design is experimental or where there is an abundance of work.

Another advantage to the fixed-price contract is that the owner need not be particularly concerned with the contractor's record keeping. If changed work or extra work is priced on a cost basis, there may have to be some inquiry into the contractor's cost. On the whole, the fixed-price contract avoids excessive owner concern with cost records of the contractor. Conversely, such a contract is attractive to the contractor, as it need not expose its cost records, something as seen in (B) that occurs in a cost contract.

The fixed-price contract has come under severe attack because of its inherent adver-

---

9. *Rudd v. Anderson,* 153 Ind.App. 11, 285 N.E.2d 836 (1972) (fixed price of $35,000 became time and materials contract for $58,000).

sarial nature. A contractor who reduces its costs increases its profits. So long as the cost reduction does not come at the expense of the owner's right to receive performance specified in the construction contract, the owner cannot object.

In construction, however, the performance required and whether such performance has been rendered is often difficult to establish. In this gray area of compliance, the interests of owner and contractor can clash. Unless the contractor is interested in its reputation for quality work or seeks possible repeat business, it is likely to perform no more than is demanded by the contract. This has led to some of the variations described in Section 21.04, particularly the cost contract with a guaranteed maximum price (GMP), sometimes called a guaranteed maximum cost (GMC).

Another disadvantage to the fixed-price contract is that the risk of almost all performance cost increases falls upon the contractor. A prudent contractor will price these risks and include them in its bid. However, in a highly competitive industry, the prudent contractor may not receive the award because there may be others who are more willing to gamble with a low price and either hope that problems will *not* develop or recoup any losses by asserting claims for extras and delays. Even if a prudent contractor does take these risks into account and does receive the award, if the risks do not materialize, the owner may be paying more than it would have had the risk been taken out of the contractor's bid.

(B) COST CONTRACTS

When prospective contractors cannot be relatively certain of what they will be expected to perform, or where they are uncertain as to the techniques needed to accomplish con-

tractual requirements, they are likely to prefer to contract on a cost basis. For these reasons, it is likely that projects that involve experimental design, new materials, or work at an unusual site or those in which the design has not been thoroughly worked out are likely to be made on a cost basis.

Usually a cost contract allows the contractor to be paid its costs *plus* an additional amount for overhead and profit. This should be distinguished from what is sometimes called a time and materials contract, which at least in one case was held to preclude recovery by the contractor of overhead on direct labor costs.[10]

The cost contract has two principal disadvantages. First, the owner does not know what the work will cost at the time it engages the contractor. Second, as a general rule, a cost contract does not give sufficient incentive to the contractor to reduce costs. These have led to variants on a pure cost contract developed both in public and private contracting systems.

Cost contracts often contain provisions that require the contractor to use its best efforts to perform the work at the lowest reasonable cost. Often provisions are included that require the contractor who has reason to believe that the cost will overrun any projected costs to notify the owner or its representative and give a revised estimate of the total cost. Sometimes these provisions state that failure to give a notice of prospective cost overrun will bar recovery of any amounts over any cost estimates that have been given.[11]

Another method of keeping costs down is to include provisions in the contract stating that a fiduciary relationship has been created between owner and contractor that requires that each use its best efforts to accomplish the objectives of the other and to

---

10. *Colvin v. United States*, 549 F.2d 1338 (9th Cir.1977).

11. In *Jones v. J.H. Hiser Construction Co., Inc.*, 484 A.2d 302 (Md.App. 1984) the court in a cost plus a fixed fee contract implied an obligation by the contractor "to know of, keep track of, and advise the owner that actual costs were substantially exceeding estimates."

disclose any relevant information to the other.[12]

One type of cost contract gives the contractor *cost plus a percentage of cost* for overhead and profit. Obviously, not only is there little incentive to cut costs in such a contract but there is a reward for increasing costs. For this reason, it is not used in federal procurement. However, sometimes it is used to price changed work in private contracts. It provides a readily accepted guideline for determining the percentage in contrast to the less readily definable fixed fee, which is its alternative for compensating overhead and profit. With a contractor of the highest integrity, this type of cost contract is useful.

Another type is the *cost plus a fixed fee* contract. The parties agree that the contractor will be reimbursed for allowable costs and paid a fee that is fixed at the time the contract is made. The fee is normally not affected when actual cost exceeds or is less than the estimated cost. However, if the scope of the work is substantially changed, sometimes the fee is renegotiated. Because the contractor's fee is not affected by cost savings, the contractor has no compensation incentive to reduce costs. For this reason, in federal procurement this type of contract has largely been superseded by cost contracts that create incentives to reduce costs.

One incentive cost contract used in federal procurement is the *cost plus award fee contract,* sometimes referred to as a CPAF contract. Under such a contract, the owner reimburses the contractor for its actual allowable costs and the contractor is paid a base fee that is negotiated before contract award and that usually is a low percentage of the agreed *estimated* total cost of the work.

The contractor is given an opportunity to earn through superior performance an additional specified award fee that may be two or three times the amount of the base fee. The award fee is determined by the owner and is based upon the owner's evaluation of the contractor's performance. The basis for this evaluation is set forth in the contract and focuses upon those goals the owner considers most important. The award fee is designed to give the contractor incentive for high quality performance.

A CPAF contract states the performance factors and the weights assigned to them that will be used in the owner's periodic evaluation of the contractor's performance. If the owner's goals change, the contract can be amended by mutual agreement to revise the weights assigned or add new factors. In an average construction program, evaluation factors might be control and reduction of costs, quality of construction, and maintenance of schedules.

During performance, the owner pays only the *basic* fee each month when that month's portion of the work is completed. Periodically, the owner's designated representatives evaluate the contractor's performance for a given period and give it a numerical grade in those factors set forth in the contract. This overall grade determines the portion of the maximum attainable award fee that the contractor has earned and will be paid for the period being graded. While the owner under such a contract has the sole right to evaluate performance, good owner-contractor relations usually require that the owner discuss evaluations with the contractor so that the contractor knows what the owner expects and how its performance is being rated.

Another type of incentive cost contract pioneered by the federal government is the *cost plus incentive fee* contract, or what is sometimes called the CPIF contract. It is used when firm bids cannot be made in advance of performance but where the

---

12. AIA Doc. A111, Art. 3 cited in *Jones v. J.H. Hiser Construction Co., Inc.,* supra note 11, creates a fiduciary obligation which allows the owner to make informed decisions.

owner wishes to give the contractor profit motivation to reduce costs. In a CPIF contract, owner and contractor before award negotiate the target cost, target fee, minimum and maximum fee, and fee adjustment formula. The formula determines the amount of fee payable to the contractor based upon a comparison between the negotiated target cost and the final total allowable cost.

After the work is completed, contractor and owner negotiate the final fee in accordance with the fee adjustment formula. The formula can provide, for example, that the contractor would be penalized 25% of actual cost overruns above the target estimate and rewarded by 25% of the underruns. In each case, the formula is subject to the previously agreed-upon minimum and maximum fees. Such a contract can have separate incentive provisions for early completion.

Another method of providing cost reduction incentives is value engineering. This is generated by insertion of clauses designed to encourage the contractor *after* award to be efficient in construction and to improve the owner's design. When this is accomplished, the cost savings are determined by the owner and a share paid to the contractor in accordance with a formula established in the contract.

These methods of seeking to keep costs down, although sometimes successful, still do not accomplish the objective of letting the owner or anyone supplying funds for the project know that the costs will not exceed the particular designated amount. To deal with this problem, owners sometimes insist that the contractor give a guaranteed maximum price (GMP) or that there be an "upset" price included in the contract. This is designed to give some insurance that the project will not cost more than a designated amount. This should be differentiated from any cost estimates given by the contractor, although there is always a risk that any cost figures discussed will end up being a guaranteed maximum price.[13]

Construction managers are frequently asked to give a GMP if *they* engage the specialty trade contractors or perform some of the work with their own forces. For that reason, discussion of a GMP is postponed until Section 21.04(C). It should be emphasized at this point, however, that a GMP may not be worth very much if the design is quite incomplete at the time the GMP is given. If costs exceed the GMP, a claim is likely to be made by the contractor that there has been such change in the scope of the work that the GMP no longer applies.

The owner has additional administrative costs in a cost contract. Usually the design professional will seek a higher fee than for a fixed-price contract, because there are many more changes made in a cost contract as the work progresses. The design professional may have additional responsibilities for checking upon the amount of costs incurred by the contractor and for insuring that the costs claimed actually went into the project and were required under the contract. The determination of costs involves not only often exasperating problems of cost accounting but also the creation of record management and management techniques for determining just what costs have been incurred.

Innumerable variations of allowable costs exist. Usually there is no question on certain items, such as material, labor, rental of equipment, transportation, and items of the contractor's overhead directly related to the project. However, sometimes disputes arise over such matters as whether the cost of visits to the project by the contractor's administrative officials, the cost of supervisory personnel employed by the contractor,

---

13. *J.E. Hathman, Inc. v. Sigma Alpha Epsilon Club,* 491 S.W.2d 261 (Mo.1973) (earlier AIA Doc. A111 with *no* blank for a guaranteed maximum price).

preparatory expenses or delay claims by subcontractors are allowable costs.

The drafter must try to anticipate all types of costs that can relate directly or indirectly to the project. A determination should be made as to which will be allowable costs for the purposes of the contract. Some of the troublesome areas can be highlighted by comparing Art. 8 and Art. 9 of AIA Doc. A111 which deal with cost plus arrangements. The first lists reimbursable costs, and the second specifies certain costs that are not to be reimbursed. The years of experience of the federal procurement has generated complicated allowable cost rules. Yet problems still arise in this troublesome area.

(C) UNIT PRICING

One risk of a fixed-price contract relates to the *number* of units of work that will have to be performed. This risk can be removed by the use of unit pricing. The contractor is paid a designated amount for each unit of work performed.

A number of factors must be taken into account in planning unit pricing. First, the unit should be clearly described. The cost of a unit should be capable of accurate estimation. Best unit pricing involves repetitive work in which the contractor has achieved skill in cost predicting.

Second, it must be clearly specified whether the unit prices include preparatory work such as cost of mobilizing and demobilizing apparatus needed to perform the particular work unit.

Third, it is important to decide whether the invitation to bidders will include an "upset" or maximum price or whether the invitation will include a minimum price. The reason for an upset price is obvious.

However, the reason for a minimum price involves understanding the use of an unbalanced bid. Under such a bid, the contractor does not actually base its unit bid price upon its prediction of the cost that will be incurred for performing that unit. Commonly the contractor bids high for unit work that will be performed early and low for work that will be performed later. In a competitive bid evaluation, such a contractor suffers no disadvantage by this distortion. The award is based on the total unit prices multiplied by the total estimated quantities. As a result, such a bidder takes no *apparent* risk when it unbalances the bid. (But suppose units bid high are deleted or reduced.[14])

The reasons for an unbalanced bid, sometimes called "pennying," were revealed in a New Jersey decision, *Boenning v. Brick Township Municipal Utilities Authority*.[15] It involved a bidding invitation that set minimum unit prices for certain portions of work connected with the installation of a municipal sewerage system. The issue was whether the municipality could set *minimum* unit prices.

The court first distinguished *fixed* from *discretionary* items. It described a fixed unit as one that can be measured with certainty by reference to the plans and specifications, such as sewer pipe whose size and length is shown on the plans. (These types form the basis of what the English call measured contracts, with the price based upon so many board feet of lumber, doors, windows, etc.)

Discretionary items (an unfortunate term)[16] are types of work whose quantities cannot be accurately measured or ascertained in advance of the work itself being performed. The contract before the court involved two discretionary items: under-

---

14. *Gregory & Reilly Assoc., Inc.*, FAACAP No. 65–30, 65–2 BCA ¶ 4918 (1965).

15. 150 N.J.Super. 32, 374 A.2d 1214 (1977).

16. Most disputes involve excavation. The units are *not*, as a rule, discretionary or fixed. The issues usually are overruns or underruns (discussed later in this subsection).

ground and restoration work. The underground work required laying sewer pipes in trenches ranging from six to twenty feet in depth. Although test borings had been taken before design, the soil conditions that would be encountered could not be determined with certainty. Some soil conditions would require specified material or a concrete cradle to support the pipe. Timber or steel shoring left in place might be needed to shore up the trench. Similarly, it could not be determined in advance how much restoration work, such as asphalt paving or landscaping, would be required after the sewers had been installed.

In what the court called discretionary work, engineers for the awarding authority usually estimate units that will be required based upon the borings, their knowledge of local conditions, and their experience. (The problem of mistaken estimates is taken up later in this subsection.) An unbalanced bid, or what the court referred to as penny bid, can encourage the contractor who has pennied certain discretionary items to challenge an engineer's orders because the contractor may not want to incur the expenses of doing work for which it made a nominal bid. This can increase the number of disputes and interfere with timely completion of the project. Yet according to the court, it had become common practice in New Jersey bidding for utility work to penny units of discretionary work.

The court then dealt with how a pennying contractor can compete with one who does not penny units. The former may simply gamble that such work will not be needed or may believe that the engineer has overestimated the need for such items. Such a contractor will gain a competitive advantage by "pennying" and can afford to increase its bid on items that are certain which can give it a windfall.

Another, more unsavory, reason for pennying may be the prospect of collusion between the contractor and the engineer, the latter finding the pennying units low and the excessively priced units high. Even without a specific "deal," the contractor may believe that the prospective gains may make it worthwhile to attempt to influence or even bribe the engineer.

The contractor may submit an unbalanced bid to obtain progress payments more quickly and give it added capital for the performance of the work. This can mean a greater chance of late-finishing subcontractors and suppliers not being paid.

The court noted that the minimum prices were fixed below the estimated cost so that a contractor cannot reap a windfall if there is an overrun on any high-priced item. Establishing a minimum at a figure less than the likely cost will encourage competition among bidders. The minimum ensures that a contractor who has bid that price will be paid at least a part of its cost and can minimize gambling by contractors and potentially unnecessary disputes. It can also inhibit front-end loading by use of the unbalanced bid.

The court rejected an argument that if the contractors gamble and lose, their sureties will complete the job and the local authority will receive a cheaper price. The court held that the local authority could use the minimum bid for units inasmuch as it could, if it chose, reserve the right to reject unbalanced bids.

Later that same year, New Jersey faced the unbalanced bid again. In *Riverland Construction Co. v. Lombardo Contracting Co.,*[17] the court held that a nominal unit price does not automatically invalidate the bid unless it can be shown that there are other excessive unit bids or that there has been fraud, collusion, or unfair restriction of competition. However, citing the *Boenning* case, the court noted that an invitation can bar such bids by setting a minimum.

The second legal issue, and perhaps one

---

17. 154 N.J.Super. 42, 380 A.2d 1161 (1977).

that arises most frequently, relates to inaccurate estimates of units to be performed. These usually involve excavation cases in which the actual units substantially overrun or underrun the estimates. Pricing the unit work usually assumes that there will not be a substantial deviation from the estimates. If the actual units substantially underrun, the cost cannot be spread over the number of units planned and will cost on a unit basis more than the contractor expected. If the unit is overrun, the contractor may be expected to perform more unit work in the same period of time, another factor that can increase planned costs.

Often contracts specifically grant price changes if costs overrun or underrun more than a designated amount. For example, the federal procurement system grants an equitable adjustment if costs overrun or underrun more than 15%.[18]

AIA Doc. A201 deals with this problem in a more limited way. Paragraph 12.1.5 grants an equitable adjustment to either party if the quantities are changed *by change order* and application of the unit prices will cause "substantial inequity." This does not grant an automatic adjustment for overruns or underruns.

Many contracts provide equitable adjustments if the subsurface conditions actually encountered vary from the conditions stated to exist or from those that are usually encountered (see Chapter 29).

Some of the more difficult cases have involved claims where the owner simply gives an estimate that turns out to be inaccurate (most commonly grossly inaccurate) without any showing of negligence chargeable to the owner or any contractual adjustment method. As a rule, the mere fact that there is a variation does not entitle the contractor to additional compensation or a change in the unit price. Such a result would be based upon the contractor's having assumed this risk or, occasionally, upon an express provision in the contract stating that estimates cannot be relied upon.[19] However, reasonable reliance on the estimate may justify reforming the contract to grant a price readjustment. For example, in *Peter Kiewit Sons' Co. v. United States*,[20] the contractor estimated its unit prices for three types of work of varying difficulty but then learned that the specifications required it to submit a combined unit price bid. It computed and submitted a unit price bid by averaging the three types of excavation and based its bid upon the units estimated by the government.

When the work was in progress, the government ordered a reduction in one excavation type, a change that, because of the averaging used, the contractor claimed would increase the composite unit cost of the work. The claim was denied based upon the contention that the estimates did not bind the government and could not be relied upon, pointing to language to that effect in the specifications.

One basis for the contractor's claim was that each party had made a mutual mistake of fact as to the quantity of work that would be required. The Claims Court upheld the claim, stating that the contract was intended not to be speculative but to be capable of proper computation and conservative bid. The composite bid being required meant that a great variation in the actual quantities performed either could be ruinous to

---

18. 32 C.F.R. § 7–603.27. This was applied in *Victory Construction Co., Inc. v. United States*, 206 Ct.Cl. 274, 510 F.2d 1379 (1975). The equitable adjustment was based upon the cost differential directly attributable to the volume deviation greater than the "upset" amount and not the average cost of the entire unit work.

19. *Zurn Engineers v. California Department of Water Resources*, 69 Cal.App.3d 798, 138 Cal.Rptr. 478 (1977).

20. 109 Ct.Cl. 517, 74 F.Supp. 165 (1947). See also *Timber Investors, Inc. v. United States*, 218 Ct.Cl. 408, 587 F.2d 472 (1978) (reformation allowed if estimated quantities "grossly inaccurate" and reliance reasonable, such as inability to verify).

the contractor or could cause the government to pay far more than the work was worth. The court concluded that the government did not intend to contract on such an irrational basis. If the parties each believe the estimates to be accurate, a mutual mistake to that effect would be a basis for reforming the contract.

As to language stating that estimates are not guaranteed, the court first pointed to language granting an equitable adjustment in the event of changed conditions, noting that such a provision was in conflict with the one stating that estimates are not guaranteed. Perhaps more important, the court stated that the disclaimer did not mean "that all considerations of equity and justice are to be disregarded, and that a contract to do a useful job for the Government is to be turned into a gambling transaction." [21]

Additionally, this case demonstrates the not uncommon conflict between clauses that grant relief under certain circumstances and those that seek to place risks closely related to the former upon the contractor (explored in greater detail in Chapter 29). The case also reflects a recognition of the contractual doctrine of mutual mistake that is sometimes applied in extreme cases despite the presence of language that appears to place the risk of unexpected quantities upon the contractor.

Bidding mistakes are discussed in Section 22.04(E), but one case that involved a unit price clerical error merits comment here. *Pozar v. Department of Transportation* [22] involved a bid for highway construction. One unit item was "Binder," a dust palliative. The estimated quantity was ninety tons. Pozar's unit price per ton was $20. The total price should have been $1,800, which would have been the low total bid for the work. But by mistake the total binder unit price was $18,000.

The bidding information stated that in the case of discrepancy between per-unit price and unit price totals, the per-unit price prevails unless it is "ambiguous, unintelligible or uncertain."

The awarding authority wanted to use the per-unit price and award the contract to Pozar. But the legal adviser, noting the estimate for binder was $300 a ton (this was, evidently, an unbalanced bid), concluded the bidder must not have intended to bid $20 a ton.

Pozar asked the court to review the award made to another bidder. The court held that the $20 per ton unit bid was not "ambiguous, unintelligible or uncertain" and ordered the award be made to Pozar.

Ironically, as seen in Section 22.04(E), the legal adviser must have been worried about the possibility of a palpable bidding error giving Pozar a *right* to refuse to enter the contract. But here Pozar *wanted* the contract. But suppose its bid had been *much* lower than all others and it wanted to *withdraw* its bid. Very likely, it could have done so unless the unbalanced bid barred relief for a clerical error. This appears to give the bidder the best of both worlds.

(D) CASH ALLOWANCE

Sometimes the owner wishes to select certain items after the prime contract has been awarded. For example, it may wish the right to select particular hardware or fixtures after a contractor is selected. To deal with this, a cash allowance can be specified. This allowance should cover the cost of the items to be selected. If the cost of the items ultimately selected varies from the cash allowance, an appropriate adjustment in the contract sum is made.

Disputes sometimes arise relating to what is encompassed within the allowance. The contract should specify whether the cash

---

21. 74 F.Supp. at 168. For a similar result based upon misrepresentation that induced a particular composite bid, see *Acchione & Canuso, Inc. v. Pennsylvania Department of Transportation*, 501 Pa. 337, 461 A.2d 765 (1983).

22. 145 Cal.App.3d 269, 193 Cal.Rptr. 202 (1983).

allowance encompasses only the net cost of the materials and equipment delivered and unloaded at the site including taxes or whether it also includes handling costs on the site, labor, installation costs, overhead and profits, and other expenses. Commonly, the allowance includes only the direct cost of the items selected.[23]

PROBLEMS

1. Your client is about to secure a contractor for the construction of a ten-story commercial office building. It wishes to know whether it should use a single or separate contract system. It also wants to know whether it should use a fixed-price or cost contract. It asks you how "fixed" is a fixed-price contract and whether it should set aside a reserve above any fixed-price figure in the construction contract. It asks whether it should conduct a competitive bid. What would be your advice? What would be the considerations that would bear upon any advice you would give?

2. A fraternity house was heavily damaged by a fire, a portion of the building having been totally destroyed. Because of the importance of making the house habitable as soon as possible, the fraternity employed an architect who prepared plans and specifications and started negotiations with a particular contractor.

   The work was begun immediately, even though final plans and specifications had not been prepared. Plans and specifications were prepared and submitted during construction, when a written contract was prepared by the fraternity and agreed upon by the contractor. The contract was prepared on a printed form that stated that it was to be used when the cost of the work plus a fixed fee was a basis for payment. It also stated that the contractor was to provide all labor and material and do all things necessary to complete construction according to the plans and specifications. The contract stated that any maximum cost would be adjusted in accordance with change orders. A typed-in provision stated, "Estimated maximum cost of this work is $300,000." The contract also stated that the contractor would be reimbursed for all costs necessarily incurred for the proper execution of the work, with particular items included and others excluded.

a. What sources would be used to determine whether the contractor has a legal right to recover for costs incurred that exceed $300,000?

b. What is likely to have been the intention of the contractor? The fraternity?

c. What if they were different and neither knew or should have known of the other's intention?

SECTION 21.03 TRADITIONAL ORGANIZATION: OWNER'S PERSPECTIVE

This chapter looks at compensation and organizational aspects of the construction process mainly from the owner's perspective. It is generally assumed that contractors who engage in the process are individual legal entities, usually corporations. However, some projects may be beyond the capacity of an individual contractor, who may seek to associate other contractors with it through the use of a joint venture. This organizational method, roughly a partnership for one project, was discussed in Section 3.07.

---

23. AIA Doc. A201, ¶ 4.8.

(A) TRADITIONAL SYSTEM REVIEWED

The traditional system separates design and construction, the former usually performed by an independent design professional and the latter by a contractor or contractors. The sequencing used in this system is creation of the design followed by contract award and execution. (The commonly used method of awarding construction contracts through competitive bidding is discussed in detail in Chapter 22.) The contractor to whom the contract is awarded is usually referred to as the prime or general contractor. It will perform some of the work with its forces but is likely to use specialized trades to perform other portions of the work. The prime contractor is both a manager of those whom it engages to perform work and also a producer in the sense that it is likely to perform much or some of the work with its own forces. The traditional system used in the United States usually gives certain site responsibilities to the design professional who has created the design (discussed in detail in Section 15.08).

The principal advantages of this system are as follows:

1. The owner can select from a wide range of design professionals.

2. For inexperienced owners, an *independent* professional monitoring the work with the owner's interest in mind can protect the owner's contractual rights.

3. Not awarding the contract until the design is complete should enable the contractor to bid more accurately, making a fixed-price contract less likely to be adjusted upwards except for design changes.

4. Subcontracting should produce highly skilled workers with its specialization of labor and should create a more competitive market because it takes less capital to enter.

(B) WEAKNESSES

The modern variations to the traditional system described in Section 21.04 developed because, despite its strengths, the traditional system developed weaknesses. It is important to see these weaknesses as a backdrop to Section 21.04.

Separation of design and construction deprive the owner of contractor skill *during* the design process, such as sensitivity to the labor and material markets, knowledge of construction techniques, and their advantages, disadvantages, and costs. A contractor would also have the ability to evaluate the coherence and completeness of the design and, most important, the likely costs of any design proposed.

Sequencing the work in the traditional system not only precludes work from being performed while the design is being worked out but also deprives the contractor the opportunity of making forward purchases in a favorable market. Although the leisurely pace that often accompanies the traditional Construction Process at its best can produce high-quality work, it is almost inevitable that such a process will take more time than one that allows construction to begin while the design is still being completed.

The frequent use of subcontractors selected and managed by the prime contractor causes difficult problems. Subcontractors complain that their profit margins are unjustifiably squeezed by contractors demanding they reduce their bids and, more important, reducing the price for which they will do the work after the prime contractor has been awarded the contract. Subcontractors also complain that contractually they are not connected to the owner—the source of authority and money. They also complain that prime contractors do not make sufficient effort to move along the money flow from owner to those who performed work and that they withhold excessive amounts of money through retainage.

The traditional system tends to keep down the number of prospective prime contractors who could bid for work and thereby reduces the pool of competitors. The traditional system, with its emphasis upon a fixed-price contract and competitive bidding, also can create an adversarial relationship between owner and contractor (described in Section 21.04(C)).

Another weakness of the traditional system relates to the role of the design professional during construction. For the reasons outlined earlier,[24] modern design professionals seek to exculpate themselves from responsibility for the contractor's work and to limit their liability exposure. Perhaps even more important, many design professionals lack the skill necessary to perform these services properly.

Under the traditional method, the managerial functions of the prime contractor may not be performed properly. The advent of increasing, pervasive, and complex governmental controls over safety often found many contractors unable to perform in accordance with legal requirements. In addition, the managerial function of scheduling, coordinating, and policing took on greater significance as pressure mounted to complete construction as early as possible and as claims for delay by those who participate in the project proliferated. A good prime contractor should be able to manage these functions, as its fee is paid to a large degree for performance of these services. Not all prime contractors were able to do this managerial work efficiently. Some owners believed the managerial fee included in the cost of the prime contract could be reduced.

The division between design and construction, while at least in theory creating better design and more efficient construction, had the unfortunate result of dividing responsibility. When defects develop, as shall be seen in Chapter 28, the design professional frequently contends that such defects were caused by the contractor's failure to execute the design properly, while the contractor asserts that the design was defective. This led to bewildered owners not being certain who was responsible for the defect as well as to complex litigation.

SECTION 21.04
MODERN VARIATIONS

(A) INTRODUCTORY REMARKS

Generalizations are avoided in this section. Although separate contracts have been used for a substantial period of time, construction management (CM) and design/build (D/B) as well as phased construction (fast track) have only recently been the center of attention. As a result, crystallized and accepted practices do not exist.

Much of the standardization in the traditional system is attributable to general acceptance of contract forms, published by the American Institute of Architects (AIA) and endorsed by the Associated General Contractors (AGC), that relate to that system, such as A101/201 and B141 (B141 is *not* endorsed by AGC). These associations have developed well-understood and accepted practices. However, in CM and D/B, the failure of the AIA and AGC to jointly develop standard agreements has precluded generally accepted industry practices. Perhaps this was inevitable, not only because the two organizations seek different objectives but because drastic changes in a well-established system inevitably lead to experimentation. Use of these modern variations is most likely to occur in large-scale projects. Owners who commission such projects of widely varying types often have different ideas of the most efficient way of building a project. As a result, a wide variety of contracts is used. Descriptions must be general and tentative.

As to legal conclusions, this section raises more questions than it can answer. Any

---

24. Refer to section 15.08.

observations will have to be tentative, not only because of the lack of crystallized practices but because the law must deal with a series of new problems that cannot be easily solved by use of simple reference to earlier precedents or statutes. Law functions to a large degree through crude categories aided by analogies, largely for administrative convenience. Law often lags behind organizational and functional shifts in the real world. Inevitably variations of the type to be described in this section will create a period of temporary disharmony. It is hoped that predictable solutions will emerge.

Many legal rules are premised on the traditional system, with owners engaging independent professionals such as architects and engineers to design the project. These professionals are retained because their skill is needed to solve technical problems for their clients. They are engaged based principally upon their predicted or reputed *professional* skill rather than on competitive pricing. They expect to be paid for their services. They do not make entrepreneurial profits.

After the design is completed, the design, along with the site, is turned over to the contractor for execution. While contractors increasingly are licensed, they are still principally thought of as *businesspersons* who have managerial and technical skill and who are in construction to make a profit. During construction they control the site. The owner, acting through its design professional, makes changes and payments and monitors performance.

Modern variations change this system. The use of separate contracts (multiple primes) (to be discussed in (C)) shifts responsibility for coordination from a prime contractor to someone else, often an independent professional adviser retained by the owner or a principal (not a prime) contractor who has no direct contracts with those contractors whose work it must, as a manager, coordinate.

Phased construction (fast tracking dis-

cussed in (B)) allows construction to proceed during design. When the spotlight is placed upon construction management (in (D)), a new adviser or coordinating contractor is engaged (not, as a rule, after the design is completed but during design) by the owner, who also plays a significant monitoring role during construction. Here the owner, through at least some of the permutations of construction management, has a more active role in managing the project and is not simply turning over the design to the contractor, a businessperson engaged in the venture to earn a profit.

In turnkey and design/build (discussed in (E) and (F)), the owner has not engaged an independent adviser to prepare the design but has turned over everything to a businessperson who represents that it has paid or can pay for the skill to both design and build. Even more difficult, the D/B owner can, as noted in Section 21.04(F), span a wide range of owners.

(B) PHASED CONSTRUCTION (FAST TRACKING)

Phased construction, or what has come to be known as fast tracking, is not an organizational variation. It differs from the traditional method in that construction can begin before design is completed. There is no reason why it cannot be used in a traditional single contracting system, although it is likely to mean that the contract price will be tied to costs. However, it has received greatest attention since emphasis has been placed upon modern variations on the traditional method discussed in this section.

Construction can begin while design is still being worked out. Ideally this means that the project should be complete earlier. If a contractor is engaged during the design phase and knows which material and equipment it will have to use, it can make purchases or obtain future commitments earlier and often cut costs.

Its disadvantages center principally around design changes. Clearly a contractor which does not know exactly what it will have to do must protect itself through its pricing. If not, claims will result.[25] Since these problems principally relate to construction managers or design/builders who guarantee a maximum price, this problem is discussed in (D) and (F).

Two other potential disadvantages need to be mentioned here. First, there is greater likelihood that there will be design omissions—items falling "between the cracks." Needed work is not incorporated in the design given *any* of the specialty trade contractors or subcontractors.[26] Although omissions of this type can occur in any design, they are more likely to occur when the design is being created piecemeal rather than prepared in its entirety for submission to a prime contractor.

Second, in fast tracking there is a greater likelihood that one participant may not do what it has promised and affect adversely the work of *many* other participants. For example, an opinion of a federal contracts appeal board discussing fast tracking in the context of construction management and multiple primes stated:

> "Phased" construction has been analogized to a procession of vehicles moving along a highway. Each vehicle represents a prime contractor whose place in the procession has been predetermined. The progress of each vehicle, except that of the lead vehicle, is dependent on the progress of the vehicle ahead. The milestone dates have been likened to mileage markers posted along the highway. Each vehicle is required to pass the mileage markers at designated times in order to insure steady progress.[27]

When any of the vehicles do not pass the mileage markers assigned to them, claims for delays and complex causation problems are likely to result.

(C) SEPARATE CONTRACTS (MULTIPLE PRIMES): *THE RUTGERS CASE*

Separate contracts developed because of owner concern over the subcontracting process briefly described in Section 21.03(B). Separate contractors give the owner greater control in specialty contracting and avoid the subcontractor complaints that they were a contract away from the source of power and funds. Separate contracts were used if owners were not confident of prime contractor management skill. Their use was also spurred by successful legislative efforts by subcontractor trade associations to require that state construction procurement use separate contracts.

Another factor, though of less significance, was the hope that separate contracts could develop lower contract prices. Breaking up the project into smaller bidding units allows more contractors to bid. Owners hoped the managerial fees could be reduced by taking this function from the prime contractor. These hoped-for pricing gains can compensate for the additional expense of conducting a number of competitive bids.

In the traditional contracting system, the linked set of contracts determines communication and responsibility. If subcontractors have complaints, they can look to the prime contractor. If the prime has prob-

---

25. *Armour & Co. v. Scott,* 360 F.Supp. 319 (W.D.Pa.1972) (changes made it a cost contract) affirmed 480 F.2d 611 (3d Cir.1971); *E.C. Ernst, Inc. v. Koppers Co.,* 476 F.Supp. 729 (W.D.Pa.1979) (100% overrun in manhours), modified in 626 F.2d 324 (3d Cir.1980); *City Stores Co. v. Gervais F. Favrot Co.,* 359 So.2d 1031 (La.App.1978) (completed drawings must be consistent with incomplete drawings).

26. *Grinnell Fire Protection Systems Co. v. W.C. Ealy & Associates, Inc.,* 552 S.W.2d 747 (Tenn.App.1977).

27. *Pierce Associates, Inc.,* GSBCA No. 4163, 77–2 BCA ¶ 12,746, citing *Paccon, Inc. v. United States,* 185 Ct.Cl. 24, 399 F.2d 162 (1968) at 61, 942.

lems, it can look to the subcontractors or the owner.[28]

Separate contractors are required to work in sequence or side-by-side on the site. But they do *not* have contracts with one another. Disputes between them must be worked out by the participant performing the coordinating function, something that is not so neat in the separate contract system as in the single contract system. To be sure, even in the traditional system, this problem can arise if there are disputes between subcontractors as they have no contracts with each other. The traditional system handled this with relative clarity by requiring that these matters be dealt with by the prime contractor as part of its mana-

gerial function and the owner remove itself from these problems.

Who performs the managerial function? Who has legal responsibility?

Under the construction management system, this often falls to the construction manager (CM). But the managerial function can be performed not only by a CM but by the design professional if he has the skill and willingness to do so, by a staff representative of the owner, or by a managing or principal separate contractor. If it is not *clear* who will coordinate, police, and be responsible in *any* system, the project will be delayed and claims made. A case that generated all these problems is reproduced here.

BROADWAY MAINTENANCE CORP. v. RUTGERS

Supreme Court of New Jersey, 1982.
90 N.J. 253, 447 A.2d 906.

SCHREIBER, J.

[Ed. note: Footnotes renumbered and some omitted.]

Two contractors engaged in the construction of the Rutgers Medical School in Piscataway Township sought to recover damages from Rutgers, The State University (Rutgers), for its failure to coordinate the project and to compel timely performance by a third contractor. Rutgers asserts that it allocated the sole duty to coordinate the work of the prime contractors on the project and ensure their timely performance to Frank Briscoe Co., Inc. (Briscoe), its general contractor. Plaintiffs deny their right to sue Briscoe as third party beneficiaries of Briscoe's contract with Rutgers. We must determine first, whether plaintiffs were intended to be beneficiaries of that contract. Even if they were, we must determine whether Rutgers, as owner, retained any duty to coordinate or supervise which could give rise to a cause of action in plaintiffs. Finally, we must determine whether Rutgers is excused from any such liability by an exculpatory clause in the contract.

On October 31, 1966, Rutgers signed contracts for general construction work with Briscoe for $7,392,000, electrical work with plaintiff, Broadway Maintenance Corp. (Broadway), for $2,508,650, and plumbing and fire protection with plaintiff, Edwin J. Dobson, Jr., Inc. (Dobson), for $998,413. Six other contracts, also entered into, covered precast concrete; structural steel; elevators; heating, ventilating and air conditioning; laboratory furni-

---

28. Even in the traditional system, lines were becoming blurred as owners and subcontractors tried to sue each other. See Sections 32.05(B), 32.07(H), and 32.08(B). AIA prime contracts gave the owner power to intervene to a degree in the prime-subcontractor relationship. See AIA Doc. A201, ¶¶ 5.2.2, 5.3.1, 9.5.2.

ture and independent inspection and testing.[29] In contrast with construction projects in which the owner contracts with a general contractor who undertakes the entire project and coordinates operations of subcontractors, Rutgers entered into contracts with each of several prime contractors. In Rutgers' agreement with Briscoe, Briscoe agreed to act as the supervisor on the job and coordinator of all the contractors.

After the work was finished, at a date well beyond the scheduled time for completion, Dobson and Broadway filed separate complaints against Rutgers in the Superior Court, Law Division, asserting a variety of claims, including damages due to delays and disruptions caused by Rutgers' failure to coordinate the activities of the various contractors on the site. Rutgers filed third party complaints seeking indemnification from Briscoe and its surety. The two actions were consolidated for trial with a pending third suit brought by Briscoe against Rutgers for money due under the Rutgers-Briscoe contract. Dobson and Broadway never added Briscoe as a party defendant in their actions. The suit between Rutgers and Briscoe was settled before trial, except for two claims for indemnification. Briscoe is not a party to this appeal.

The non-jury trial proceeded for 43 days. Rutgers produced no evidence and rested at the end of plaintiffs' case. The trial court in an extensive written opinion granted plaintiffs judgments for some of their claims, but denied recovery against Rutgers for failure to coordinate the activities of the prime contractors, including Briscoe. 157 *N.J.Super.* 357, 384 *A.2d* 1121 (Law Div.1978). Dobson and Broadway appealed to the Appellate Division, which affirmed. 180 *N.J.Super.* 350, 434 *A.2d* 1125 (App.Div.1981).

Each plaintiff petitioned for certification. We granted both petitions to consider three questions: (1) in a multi-prime contract, is each prime contractor liable to the other, (2) in such a contract, does the owner have a duty to coordinate the work of the contractors, and (3) does the exculpatory clause in the prime contracts at issue here shield Rutgers from liability for damages due to delay. The Mechanical Contractors Association of New Jersey, Inc. and the Building Contractors Association of New Jersey were granted leave to file briefs as *amici curiae.*

[Ed. note: *Amici Curiae* are "Friends of the Court" allowed to present their views by court permission.]

[The court noted that jurisdictions had reached different results when faced with the question of whether one separate contractor could sue another. It concluded that the record supported the findings of the trial court that held that all parties had agreed that

---

29. Rutgers apparently believed that it was subject to *N.J.S.A.* 52:32–2, which requires the letting of multiple contracts in connection with the construction of public buildings. This Court has since determined, however, that the statute does not apply to Rutgers. *Rutgers v. Kugler,* 58 *N.J.* 113, 275 *A.2d* 441 (1971), aff'g 110 *N.J.Super.* 424, 265 *A.2d* 847 (Law Div.1970).

the separate contractors could maintain legal action against each other for damage due to unjustifiable delay. In determining this question, the court held that the terms and conditions of the contract should be examined to determine the intent of the parties.]

The existence of a third party claim does not necessarily extinguish all claims between the parties to the contract. Here plaintiffs argue that Rutgers breached its agreement with them and is liable irrespective of any third party claims they may have against the general contractor, Briscoe. The plaintiff's contention is sound at least with respect to those matters that Rutgers had contractually obligated itself to do and for which it would be responsible. The trial court did award damages to the plaintiffs against Rutgers for certain contractual breaches. Rutgers has not appealed from those determinations.

The narrow questions before us on this appeal are whether Rutgers had agreed to synthesize the operations of the prime contractors, including Briscoe, and, if so, whether Rutgers breached that duty and would be liable for the delay flowing from that breach. The order granting the petitions for certification was limited to the subject matter of coordination of the operations of all the prime contractors including Briscoe. Damages flowing therefrom involve only delay on the job and its consequential costs.

Plaintiffs urge various claims that they relate to delay such as additional expenses incurred because of lack of elevators and stairs. These particular claims were disallowed by the trial court, which pointed out, among other things, that Briscoe's delay did not cause the asserted damages. Though these items are not before us on this appeal, we have reviewed the record and are satisfied that the trial court's findings have adequate factual support.

Plaintiffs also continue to press before us matters such as Rutgers' alleged default in not withholding funds due Briscoe in order to satisfy plaintiff's claims for delay against Briscoe, and Rutgers' non-fulfillment of its supposed duty to place the site in condition so that plaintiffs could proceed. On this appeal we are concerned solely with the general supervision over the contractors. What duties Rutgers had in this respect depends upon what obligations were imposed upon it by the contract.

In the absence of any compelling public policy, an owner has the privilege to eliminate a general contractor and enter into several prime contracts governing the construction project.[30] In that event the owner could engage some third party or one of the

---

30. Use of multiple prime contracts assumes the owner will benefit from savings that will accrue from elimination of the overhead and profits of the general contractor. Provisions for construction of any public buildings by the State provide for separate bids for different major aspects of the job and bids for all work in one contract, the award to be made to whichever method results in a lower cost. *N.J.S.A.* 52:32–2.

contractors to perform all the coordinating functions. Where all the parties enter into such an arrangement the owner would have no supervisory function. The situation would be analogous to one where a general contractor had been engaged to construct the project on a turnkey basis. Surely the subcontractors would have no claim against the owner for failure to coordinate.

If no one were designated to carry on the overall supervision, the reasonable implication would be that the owner would perform those duties. In so doing, the owner impliedly assumes the duty to coordinate the various contractors to prevent unreasonable delays on the project. See, *e.g.*, *Born v. Malloy*, 64 *Ill.App.*3d 181, 184, 21 *Ill.Dec.* 117, 120, 381 *N.E.*2d 52, 55 (1978); *Carlstrom v. Independent School District No. 77, Minn.*, 256 *N.W.*2d 479 (1977). That is a reasonable assumption because the contracting authority has the power to use its superior position and to invoke its contractual rights to compel cooperation among contractors. *Shea-S & M Ball v. Massman-Kiewit-Early*, 606 *F.*2d 1245, 1251 (D.C.Cir.1979). The owner is impliedly obligated to act in good faith and to do that which it reasonably can to ensure that the other contractors adhere to the time schedules established for the project. *Paccon, Inc. v. United States*, 399 *F.*2d 162, 169–70 (Ct.Cl. 1968). An owner's failure to take action in the face of unnecessary and unreasonable delays by one of the contracting parties would ordinarily evidence bad faith and constitute a breach of its implied duty to coordinate.

It is of course also possible that the owner might fractionalize those supervisory functions. Where the owner has chosen to engage several contractors directly, the bottom line is to ascertain who the parties agreed would orchestrate and harmonize the work. The answers may be found in the contract language as illuminated by surrounding circumstances.

The complications that arise in this case are due in part to the unclear nature of who was to perform what supervisory function. Briscoe had agreed to supervise the job generally. It was entrusted with the "oversight, management, supervision, control and general direction" of the project. It was to control "the production and assembly management of the building construction process." Briscoe was obligated to have sufficient executive and supervisory staff in the field so as to handle these matters efficiently and expeditiously. The General Conditions also recited that Rutgers relied upon Briscoe's management and skill to supervise, direct and manage Briscoe's own work and the efforts of the other contractors; indeed, the agreements stated that the contractors also relied upon Briscoe's supervisory powers to deliver the building within the scheduled time.

However, Rutgers also designated its Department of New Facilities to represent it in technical and administrative negotiations with the contractors and the Department could stop the work if necessary. Rutgers also selected a critical path method consultant whose progress chart was to be followed, but could be amended with the consultant's approval. Further, Rutgers engaged an architect. The architect agreed to "[s]upervise the construction of work by periodic inspections sufficient to verify the quality of the construction and the conformity of the construction to the plans and specifications . . . and by which supervision the architect shall coordinate the work of the various contractors and expedite the construction of the Project." If a contractor were delayed in completing the work whether due to the owner, any other contractor, the architect, or other causes beyond the contractor's control, the architect alone was authorized to determine extensions of time to which a prime contractor would be entitled. In this respect the trial court found that the architect acted as an impartial and independent umpire, and as such was not the agent of either Rutgers or the contractor.

Plaintiffs contend that Rutgers retained supervisory control because Briscoe had not been given the power to enforce its coordinating authority. They argue that Rutgers had the economic weapons of terminating the contracts and withholding payments, so that only Rutgers could effectively cause the prime contractors, particularly Briscoe, to keep up to schedule. Though this contention has some surface appeal, it fails to account for the entire contractual structure governing the project. Rutgers' power to terminate a contract or withhold funds did not alter its expressed intent to have someone else supervise the work. Rutgers had delegated that overall responsibility to Briscoe. Rutgers never intervened to coordinate the operations. It never assumed control.

The plaintiffs also claim that Rutgers retained coordination and supervision over the job because of the roles of the Department of New Facilities and critical path method consultant. These contentions are misconceived. The Department's functions were primarily quality control and only incidentally affected time of performance. The consultant plotted actual progress as against the scheduled performance, but did not supervise and coordinate the work. Nor does the architect's supervisory role support plaintiffs' position. First, the architect's agreement was not incorporated in the contractors' contracts. Second, Rutgers' delegation of coordinating functions to the architect confirms, if anything, that Rutgers itself was not engaged in any such undertaking. Lastly, plaintiffs rely on an indemnity clause in each contract that provided that if a contractor or subcontractor sued Rutgers, the contractor would defend, indemnify and save Rutgers harmless. However, that provision simply confirms the intent that Rutgers was not to be responsible for defaults of other prime contractors, including Briscoe.

When viewed in its entirety, the contractual scheme contemplated that if a contractor were adversely affected by delays, it could maintain an action for costs and expenses against the fellow contractor who was a wrongdoer. Furthermore, a contractor had a right to obtain extensions of time to complete the work when delayed by "any act or neglect" of any other contractor. Other than those remedies, the contractor could not look to Rutgers for recourse because of its failure to coordinate the work.[31]

[The court reviewed the enforceability of a "no-damage" (no pay for delay) clause and concluded that it would be given effect in this case. The court rejected an argument that the clause be given effect only when the delays are reasonable and concluded that Rutgers did not actively interfere with the progress of the work nor did it act in bad faith.]

In summary then, we hold that a third party beneficiary may sue on a contract when it is an intended and not an incidental beneficiary. Resolution of that issue depends upon examination of the contractual provisions and the attendant circumstances. In a construction project where the owner directly hires the various contractors, the owner may engage a separate contractor to coordinate and supervise the project and agree with the several contractors that the general supervision will be carried out in that manner. Lastly, the owner may exculpate itself from liability for damages to the extent delineated in the contract, in the absence of any public policy reasons to the contrary.

The judgment is affirmed.

For affirmance—Justices PASHMAN, CLIFFORD, SCHREIBER, HANDLER and O'HERN—5.

For reversal—None.

The *Rutgers* case exposes some of the administrative and legal difficulties inherent in the separate contracting system. Injection of a new "entity" into the process, in that case a CPM consultant (perhaps a limited CM), without a well-defined role or clearly delineated responsibility can create problems. The case demonstrates the administrative difficulties of giving one entity managerial responsibilities without the tools to effectively manage, one reason the *Rutgers* case has been criticized.[32]

If managerial functions are given the managing contractor, what effect does this have on any owner responsibility? For example, had that power been given to the architect or even a CM, it is likely that the owner would be responsible if either had failed to coordinate properly.

The AIA in its frequently used general conditions, A201,[33] under ¶ 6.1.3. gives the *owner* coordination responsibility. Each separate contractor agrees to cooperate. A separate contractor may claim another has

---

31. [Ed. note: On *less* specific language, *Hanberry Corp. v. State Building Commission,* 390 So.2d 277 (Miss.1980), exonerated the owner, the court stating, somewhat simplistically, that the purpose of separate contracts is to exculpate the owner from coordination responsibility.]

32. Bynum, *Construction Management and Design-Build/Fast Track Construction from the Perspective of a General Contractor,* 46 Law & Contemp.Probs. 25, 33 (1983).

33. This is not substantially different in A201 CM (1980) and B801 (1980), its CM Documents.

damaged or destroyed its work or property. Most important, a separate contractor can claim that another has failed to perform in accordance with the latter's contractual obligation and adversely affected the former's scheduling and performance.

Under ¶ 6.2.5., if a dispute exists between separate contractors regarding a claim by one that its property has been damaged or destroyed, the separate contractors must attempt to promptly settle the dispute. If they do not, and one brings legal action or initiates arbitration against the *owner*, the owner notifies the contractor who must defend such proceedings at the *owner's* expense. If there is any judgment or award against the owner, the contractor must pay it or satisfy it and reimburse the owner for its attorneys fees and other costs.

Very little in A201 deals with economic loss claims by one separate contractor against another, such as those involved in the *Rutgers* case. A201 only notes that the owner has overall coordination responsibility [34] and that each separate contractor agrees to connect and coordinate with others.

Unlike the system in the *Rutgers* case, the AIA does *not* place sole responsibility upon the coordinating architect or construction manager. The AIA places responsibility on the owner who may have a claim against the party who has promised to coordinate, such as the architect or construction manager.

If responsibility for the loss is traceable to one of the separate contractors, under A201, the owner, though having indemnification rights, is *initially* responsible, probably *strictly*, for any damage incurred by a separate contractor because of another separate contractor.

AIA's indemnification process is cumbersome. The owner must fund the cost of the defense. (Can the indemnitor bill the owner as those defense costs accrue?) If there is any award against the owner, the contractor pays it and reimburses the owner for its costs, including attorneys' fees. If the contractor defends successfully, the owner absorbs the cost of defense. However, suppose there is a large claim made by one separate contractor and a very modest award. It appears that under A201 all of the costs of defense will have to be borne by the separate contractor whose conduct has been challenged. It would be better to have the separate contractor pay the costs if the claimant separate contractor recovers an award in excess of any amount tendered by the challenged separate contractor. Alternatively, the costs of defense can be shared.

Should the owner have residual responsibility, that is, responsibility for the failures of its professional advisers entrusted with managerial responsibilities? While case decisions vary on this point,[35] it is suggested that a promise be implied by the owner (warranty) that each separate contractor will be allowed to perform its work in accordance with normal industry standards. It is hoped that such an implied promise will not tempt owners to intervene and generate administrative confusion.

Owners contemplating a separate contract system should consider these problems. Clearly, it ought not to be used unless the managerial function can be better performed and at a lower cost than if it is performed by a prime contractor under a

---

34. Under the CM Documents this is done by the CM.

35. The Court of Claims was not able to resolve this issue squarely. Compare *Fruehauf Corp. v. United States*, 218 Ct.Cl. 456, 587 F.2d 486 (1978) (warranty based on suspension of work clause), with *Paccon, Inc. v. United States*, 185 Ct.Cl. 24, 399 F.2d 162 (1968) (not liable if it takes reasonable steps). In *Blinderman Construction Co. v. United States*, 695 F.2d 552 (Fed.Cir.1982), only best efforts were required. Similarly, *Shea–S & M Ball v. Massman-Kiewit-Early*, 606 F.2d 1245 (D.C.Cir.1979), required the owner to invoke its authority to compel cooperation. See generally Goldberg, *The Owner's Duty To Coordinate Multi-Prime Construction Contractors, A Condition of Cooperation*, 28 Emory L.J. 377 (1979) for an exhaustive treatment.

single contract system. If the separate contract system is selected, all contracts should make clear who has the administrative responsibility and who has the legal responsibility if a contractor has been unjustifiably delayed. More particularly, if this responsibility is to rest solely in the hands of the persons who are given the responsibility for managing the contract, all contracts should make clear whether the owner is exculpated.

(D) CONSTRUCTION MANAGEMENT

Before seeking to define construction management it is important to outline the reasons for its development. Principally it developed because of perceived weaknesses and inefficiencies in the existing Construction Process with special reference to the inability of designers and contractors to use efficient management skills. Designers were faulted because of their casual attitude toward costs, their inability to predict costs, and their ignorance of the labor and materials market, as well as costs of employing construction techniques. Owners were also concerned about the tendency of design professionals to take less responsibility for quality control, policing schedules, and monitoring payments. Contractors also came in for their share of blame. Some lacked skills in construction techniques and the ability to work with new materials. Others did not have the infrastructure to comply with the increasingly onerous and detailed workplace safety regulations.

Construction management instead would be an *efficient* tool for obtaining higher quality construction at the lowest possible price and in the quickest possible time.[36] In addition it can avoid competitive bidding requirements for public construction.[37]

One reason for emphasizing construction management rather than construction managers is the great variety of approaches taken to achieve these objectives. All that can be done in this treatise is to briefly describe some of these techniques. The advantages and disadvantages have been covered elsewhere.[38]

Sometimes a CM is part of the design professional's organization, and the design professional in his contract with the owner agrees to provide CM services. (As seen in *First National Bank of Akron v. Cann* [reproduced in Section 15.08(B)] an arrangement under which the architect and CM are one entity can turn into a design-build contract.) More commonly the CM is separately retained by the owner in addition to a design professional. (There is rarely a contract between the design professional and the CM.) The design professional still furnishes a design and interprets the contract documents. But the CM advises the owner and the design professional as to design and takes over many of the site services performed by the design professional in the traditional system. The method of compensating the CM when acting as professional adviser can be any of the various methods used to compensate a design professional, such as a percentage of construction costs, a personnel multiplier, cost plus a fee, or a fixed price.[39]

One of the problems that has made construction management so difficult to fit into traditional legal concepts relates to those

---

36. For an admirable analysis of bonding in CM, see Foster, *Construction Management and Design-Build Fast Track Construction: A Solution Which Uncovers a Problem for the Surety,* 46 Law & Contemp.Probs. 95 (1983).

37. See *Willman v. Children's Hospital of Pittsburgh,* 459 A.2d 855 (Pa.Commw.1983) (trade contracts awarded by CM need not be competitively bid).

38. For a series of articles on Construction Management, see 46 Law & Contemp.Probs., Winter 1983. See also C. Thomsen, Developing, Marketing, and Delivery Construction Management Services (1982).

39. For a case describing functions and compensation of a CM, see *Gibson v. Heiman,* 261 Ark. 236, 547 S.W.2d 111 (1977).

arrangements under which CMs themselves engage specialty contractors. Sometimes they do this as agent for the owner, with the contractual relationship then being between the owner and the specialty trades. Here CMs look like professional advisers. Alternatively, CMs may contract on their own with specialty trades, a system that makes them look very much like prime contractors. CMs who contract on their own usually guarantee a maximum price (GMP)—a cost contract with a "cap." In such arrangements, CMs may also perform some construction with their own employees.

When CMs prepare a GMP, they usually agree to give a GMP after the design has been worked out to a designated portion of finality, say 75%. They protect themselves by getting as many fixed-price contracts from specialty trades as they can and usually build enough into their GMP to take into account not only their fees but also contingencies that may arise from both design changes and unforeseen circumstances.[40]

Despite this, one of the risks inherent in a GMP is that the specialty trades or the CM who gives a GMP will claim that there have been drastic scope changes that eliminate any contractual price commitments they have made. Usually, though not invariably, a construction management system will use fast track, a principal justification for construction management, to complete the project in a shorter time. Yet it is *possible* to have CM without having a fast track system. The dangers of fast tracking—sloppier work and uncontrolled costs—may persuade an owner who wishes to use the construction management system not to fast track the job.

The legal issues have at their core whether the CM is more like a design professional or an entrepreneurial contractor. For example, must the CM be registered or licensed by the state, and if so, which type of registration or license is needed?[41] This may depend upon which form of CM is used, whether the CM is engaged solely as a professional adviser, or whether the CM undertakes to perform some construction himself.

Must awards for CM services be made in the same manner as other professional services?[42] The role of the CM in awarding construction contracts also raises the issue of analogies. For example, one case held that the CM represented the owner when he awarded a contract, thereby allowing the trade contractor to demand arbitration with the owner.[43] If he is simply a professional adviser who, unlike most design professionals, has been given the authority to award contracts, the contractor has made a contract with the owner *through* the CM. Here he looks more like a design professional than a contractor. However, if he contracts on his own with specialty trade contractors, he looks like a prime contractor.

Liability problems also involve analogies. For example, it has been held that a contractor can bring a negligence claim against the CM just as a negligence claim can be instituted against a design professional.[44] If the CM is analogized to a design professional or performs services usually performed by a design professional,[45] will the CM be

---

40. For a case involving a D/B contract and the permissible design changes, see *City Stores Co. v. Gervais F. Favrot Co.*, 359 So.2d 1031 (La.App.1978).

41. See Bynum, supra note 32 at 27–28; Lunch, infra note 50 at 86–87.

42. Compare *Attlin Construction, Inc. v. Muncie Community Schools*, ___ Ind.App. ___, 413 N.E.2d 281 (1980), with *City of Inglewood—L.A. County Civic Center Authority v. Superior Court*, 7 Cal.3d 861, 103 Cal.Rptr. 689, 500 P.2d 601 (1972).

43. *Seither & Cherry Co. v. Illinois Bank Building Corp.*, 95 Ill.App.3d 191, 50 Ill.Dec. 672, 419 N.E.2d 940 (1981).

44. *Gateway Erectors Division v. Lutheran General Hospital*, 102 Ill.App.3d 300, 58 Ill.Dec. 78, 430 N.E.2d 20 (1981).

45. *Pierce Associates, Inc.*, supra note 27 (CM interprets contract and judges performance).

given the benefit of the professional standard described in Section 17.05 or quasi-judicial immunity noted in Section 17.08(E)? The professional standard has its greatest application when the design professional is performing design rather than administrative services.[46] Inasmuch as the services performed by a CM are connected more to administration than to design, it is possible that the *ordinary* but not *professional* negligence standard will be applied, meaning that no expert testimony will be required. But if the CM is simply a professional adviser performing administrative tasks for the owner, he looks *less* like a design professional. As a result, the CM may be able to defend himself more successfully if sued by other participants by asserting that he was simply advising the owner and not acting as an independent professional.

In some states, persons in charge of the construction work are strictly liable for violations that injure workers. Those in charge are more commonly prime contractors, although it is possible to include design professionals or CMs if they have broad powers over the process.[47] If CMs look like prime contractors, either because they are managing the work of the specialty trades and have overall safety responsibility *or* because they are performing some of the construction with their own forces, they may be found to be contractors with contractorlike liability.[48] If they have broad powers, they may expose the *owner* to greater liability.[49]

Those planning to engage a CM or those about to engage in CM work must also determine whether the insurance coverage of the CM will be adequate. The CM who is principally a contractor will find that his comprehensive general liability coverage will not include design, while a CM who is essentially a design professional may find that his professional liability insurance excludes any coverage that relates to the Construction Process or the work of the contractor.[50]

These legal uncertainties demonstrate the difficulty of placing the CM in traditional categories in the Construction Process, a task made even more difficult because there has yet been no agreement on the proper function of the CM. Perhaps this is inevitable during a shakedown period while the industry struggles to come to some consensus based upon a sufficient experimentation time. The future may see lawmakers, such as legislators and judges, revise existing laws to take this new actor into account if he cannot be fit within the existing categories.

(E) TURNKEY CONTRACTS

The discussion in this subsection must be taken together with the material in (F) dealing with design/build. Both subsections involve the owner contracting with an entity who agrees to both design and build. However, because turnkey contracts have other functions, they should be taken separately, both on their own and as an introduction to design/build.

Although a variety of definitions is employed, the core element to a turnkey contract is, at least in theory, that the owner simply tells the contractor what it wishes

---

46. Refer to Section 17.05.

47. *Kenny v. George A. Fuller Co.*, 87 A.D.2d 183, 450 N.Y.S.2d 551 (1982) (also liable as agent of owner).

48. *Caldwell v. Bechtel, Inc.*, 631 F.2d 989 (D.C.Cir.1980) (liable to worker); *Bechtel Power Corp. v. Secretary of Labor*, 548 F.2d 248 (8th Cir.1977) (OSHA liability for injury to CM employee); *Lemmer v. IDS Properties, Inc.*, 304 N.W.2d 864 (Minn.1980) (liability as "possessor of land"). As to OSHA, see *Skidmore, Owings & Merrill*, 5 O.S.H.C. (BNA) 1762 (1977).

49. *Everette v. Alyeska Pipeline Service Co.*, 614 P.2d 1341 (Alaska 1980).

50. See Lunch, *New Construction Methods and New Roles for Engineers*, 46 Law & Contemp.Probs. 83, 93–94 (1983).

and does not appear on the scene again until the contractor says the project is completed and hands the owner a key and the owner "turns it." [51]

Sometimes turnkey contracts are used to build public housing projects.[52] Developers are invited to submit bids to build housing. Each developer proposes to construct housing on a site it selects according to its plans and specifications. If its proposal is accepted, upon completion the building is turned over to the local housing authority. At that time the developer is paid. Such a system avoids competitive bidding requirements, as the sites and plans and specifications are different. Also, at least in theory, the owner plays no role during construction and puts the entire responsibility on the developer/builder.

Some turnkey contracts require the contractor not only to build the building but also to provide interior equipment and furnishings.

A turnkey contract looks more like a sale than a contract for services. As a result, one court held that a turnkey contract created warranties that made the seller/contractor responsible for any defects.[53] This is something that may also be found in any design/build contract.

(F) COMBINING DESIGNING AND BUILDING (D/B)

The principal advantage of D/B (called in the U.K. a "package" job), usually associated with fast tracking, is speed. One entity replaces *different* entities who design and build with the inevitable delay that using two entities whose work intersects will cause. Those who design and build fre-

quently do repetitive work and acquire specialized expertise. Perhaps most important from the standpoint of an unsophisticated owner, the system concentrates responsibility on the designer-builder. Owners are often frustrated when they look to the designer who claims that the *contractor* did not follow the design, with the latter claiming that the problem was poor design.

D/B has weaknesses. The absence of an *independent* design professional selected by the owner can deprive the owner of the widest opportunities for good design. An unsophisticated owner often lacks the skill to determine whether the contractor is doing the job well or as promised. This can reflect itself not only in substandard work but in excessive payments being made early in the project or slow payment or nonpayment of subcontractors.

Design flexibility, though costly in a traditional contract system, is more difficult to achieve in the design/build system. Excessive changes are likely to lead to a cost-plus contract. The system works *best* in repetitive work of a simple type, such as warehouses and uncomplicated residential construction, although it is also used in high technology work.[54]

Care must be taken not to lock D/B to fast tracking. One advantage of combining designing and building is the ability to fast track the project. However, the two need not go together. One entity who *both* designs and builds may create the complete design before starting construction.[55]

Anyone contemplating entering into a contract under which it will *both* design and build must check local registration laws. It is likely that a registered design professional will have to design the project

51. *Glassman Construction Co., Inc. v. Maryland City Plaza, Inc.*, 371 F.Supp. 1154 (D.Md.1974).

52. *United States Constructors & Consultants, Inc. v. Cuyahoga Metropolitan Housing Authority*, 32 Ohio Misc. 243, 291 N.E.2d 790 (1972), affirmed 35 Ohio App.2d 159, 300 N.E.2d 452 (1973).

53. *Mobile Housing Environments v. Barton & Barton*, 432 F.Supp. 1343 (D.Colo.1977).

54. Advantages and disadvantages are explored in materials cited supra note 38.

55. See Dibner, *Construction Management and Design-Build: An Owner's Experience in the Public Sector*, 46 Law & Contemp.Probs. 137, 143–144 (1983).

and often states bar a corporation from practicing architecture or engineering. If so, the entity that will both design and build will have to engage an independent design professional or associate the design professional in a joint venture.

If the design professional must be independent and retained by the D/B contractor, legal problems may develop. For example, the owner, with whom the design professional has no contract, may wish to bring a claim against the design professional based upon negligence.[56] Similarly, the D/B contractor may be giving sellerlike warranties to the purchaser-owner but may be able to hold an independent design professional only to the professional standard unless a more specific standard is set forth in a contract between builder and design professional.

On the other hand, suppose the designer is an *employee* of the D/B contractor. The owner may believe that an independent design professional will perform the traditional services that such a person performs when retained by the owner. Employees who work for a design/build entity need not look out for the interest of the owner as they would were they retained by the owner. This may raise problems if the employee is a registered architect or engineer who may be expected to look out for the public interest rather than simply being an employee of a contractor.

Chapter 9 noted the great variety of owners. Although this system can operate competitively,[57] it will not be permitted in some public contracts.[58] Owners who hire one entity to both design and build range at one extreme from a homeowner building a single-family home "patterned" on a house that the builder has already built to a manufac-

turing company with its own infrastructure, such as engineers, accountants, or lawyers, that wishes to use D/B for highly technical work because it has confidence in the technical skill of a particular design/build entity. (In between are simple projects like warehouses that can be done quickly and efficiently by a D/B who has acquired skill through repetition of work.)

An owner building a single-residence home, a warehouse, or simple commercial building relies entirely upon the contractor for design and construction. Here warranties of fitness are appropriate. The owner is buying not services but a finished product.[59] The sophisticated owner who employs D/B and has within its own organization the services of skilled professionals to control design or monitor performance has not bought a product but rather expert services to prepare and execute the design *it* chooses. Such an owner should not receive a sale-of-goods-like warranty.

The differentiation between the two types also reflects itself in pricing. It is likely that the homeowner commissioning a builder to design and build a house will contract on a fixed price. The sophisticated owner building a petrochemical plant will contract on a cost basis with a GMP, at least until the design is put in final form.[60] These projects create immense exposure for consequential damages. It is likely that the D/B will limit its obligation to correction of defects.

(G) SUMMARY

The modern variations described in this section developed because of weaknesses in the traditional system, particularly leisurely performance, divided responsibility, and

---

56. This was permitted in *Keel v. Titan Construction Corp.*, 639 P.2d 1228 (Okl.1981).

57. For an example, see *Ogden Development Corp. v. Federal Insurance Co.*, 508 F.2d 583 (2d Cir.1974).

58. *Negley v. Lebanon Community School Corp.*, 173 Ind.App. 17, 362 N.E.2d 178 (1977) (separate building district prohibited from using D/B as bids based upon different designs).

59. Refer to Section 15.07 and note 53, supra.

60. See Foster, supra note 36 at 118–119.

complaints of subcontractors. However, as these new systems become used, they will become time tested and court tested. In the process, deficiencies will be revealed. This does not in any way diminish the utility of these variations, although it does impose a serious burden upon those who wish to engage in them to anticipate the problems in deciding whether to use a variant from the traditional process and to plan in such a way so as to minimize the difficulties, both administrative and legal.

SECTION 21.05
ADMINISTRATIVE PROBLEMS

(A) OVERVIEW

This chapter has examined how a construction project can be organized, mainly the problems of selecting the type of contract (fixed-price or cost), the type of contract system to use (single, separate, D/B), and methods to bring efficiency to the management of the process (prime contractor, design professional, or construction manager). Throughout this chapter reference has been made to the need for efficient organization of any project, with emphasis principally upon the coordination of the activities of the various participants.

Yet another factor that is essential to construction efficiency must be examined. A successful construction project recognizes the importance of clear and efficient lines of communication among the main participants. It is particularly important to know who has authority to bind the owner (with the hope that such a person can be identified and found), how communications (and there are many) are to be made, to whom they should be directed, and what the rules are for determining their effectiveness (effective when posted or received?).

Communications are essential elements in the responsibility for defects and delays, and equitable adjustments for unanticipated subsurface conditions. When those topics are discussed, the problems of communications will be addressed. After a contract has been awarded and before performance begins, the participants must address these administrative issues openly and seek to make clear at the outset who speaks for whom, to whom and how communications are to be directed, and what the rules are for their implementation. Usually they are addressed at a preconstruction meeting attended by the representatives of the major participants in the process. This section deals with some of these problems.

(B) AUTHORITY: SPECIAL PROBLEMS OF CONSTRUCTION CONTRACTS

Were the construction contract an ordinary one, it would be sufficient simply to refer to Chapter 4 dealing with agency and authority. Special characteristics of the construction project make it essential to devote additional discussion to that topic. One relates to the multifaceted roles of an independent design professional engaged to design and monitor performance, to act as agent of the owner, to interpret and judge performance, and in general to be the central hub around which the process revolves. When a construction manager is used the problems do not become easier, with the process revolving about two participants of often unclear responsibilities.[61]

In construction projects that will justify it, the owner may have a full-time site representative who observes, records, and reports the progress of the work. The representative (called a clerk of the works, project representative, resident engineer, or otherwise) can be a regular employee of the owner, an employee of the design professional or CM and paid for either by the employer or by the owner, or an independent entity retained to perform this service. The representative's permanent presence invites difficulties, with the contractor often

---

61. Even more complex was an arrangement with two CMs, one called an "Overall CM." See *DiSalvatore v. United States,* 456 F.Supp. 1079 (E.D.Pa.1978), 472 F.Supp. 816 (E.D.Pa.1979).

contending that he directed the work, knew of deviations, accepted defective work, or knew of events that would be the basis for a contractor claim for a time extension or additional compensation.[62]

Communication is vital to construction. Problems often develop that should be brought to the attention of the appropriate party. For example, if there are design difficulties, they should be reported to the design professional. If there is intervention by public authorities, clashes between participants, material shortages, or strikes—to name only some—the owner and design professional or construction manager should be made aware of them. Early discussion of problems or potential difficulties should effect prompt corrective action, efficient readjustments, and gathering of necessary information for the ever-present "claim." A successful construction project requires clear lines of communication and their use.

Looking first at the traditional organization (CMs are discussed later), the design professional, though an agent of the owner, has limited authority;[63] the authority of a project representative is even *more* limited.[64] Which acts of the design professional or project representative will be chargeable to the owner? Before looking at that question, it is important to differentiate between *acts* of the design professional that bind the owner by principles of agency from those *decisions* made by the design professional as interpreter of the contract and judge of its performance. Although courts sometimes blur this distinction, this section concentrates on the first. Chapter 33 examines the second.

Generally the owner retains a design professional to provide professional advice and services, *not* to make or modify contracts or issue change orders. Although design professionals may enter into discussions with prospective contractors or even assist in the administration of a competitive bid, participants in the construction industry recognize that design professionals do not have authority to make contracts, modify them, or issue change orders on behalf of owners.

This does not mean that a design professional will never bind the owner. First, a design professional may be given express authority to perform any of these functions. For example, the design professional is frequently given authority to make *minor* changes in the work.[65] In addition, the design professional may be vested with certain authority that by implication includes other authority. For example, an architect who had the power to require that a contractor post a payment bond was authorized to represent to a subcontractor that such a bond would be required.[66]

Similarly, the owner may, by its acts, cloak the architect with apparent authority. For example, suppose the architect directs major changes for which he does not have express authority. If the owner stands by and does not intervene to deny such authority or pays for the changes, either may

---

62. See Section 25.04.

63. *Crown Construction Co. v. Opelika Manufacturing Corp.*, 480 F.2d 149 (5th Cir.1973); *Metropolitan Sanitary District v. Anthony Pontarelli & Sons, Inc.*, 7 Ill.App.3d 829, 288 N.E.2d 905 (1972); *Kirk Reid Co. v. Fine*, 205 Va. 778, 139 S.E.2d 829 (1965). But an actual employee may have more authority. *Grand Trunk Western R. Co. v. H.W. Nelson Co.*, 116 F.2d 823 (6th Cir.1941) (employee design professional authorized to make important contract).

64. *Lemley v. United States*, 317 F.Supp. 350 (N.D.W.Va.1970) affirmed 455 F.2d 522 (1971); *Missouri Portland Cement Co. v. J.A. Jones Construction Co.*, 323 F.Supp. 242 (M.D.Tenn.1970) affirmed 438 F.2d 3 (6th Cir.1971); *Samuel J. Creswell Iron Works, Inc. v. Housing Authority of the City of Camden*, 449 F.2d 557 (3d Cir.1971); *Acoustics, Inc. v. Trepte Construction Co.*, 14 Cal.App.3d 887, 92 Cal.Rptr. 723 (1971); *Savignano v. Gloucester Housing Authority*, 344 Mass. 668, 183 N.E.2d 862 (1962); *Hunt v. Owen Building & Investment Co.*, 219 S.W. 138 (Mo.App.1920).

65. AIA Doc. A201 ¶ 12.4.

66. *Bethlehem Fabricators, Inc. v. British Overseas Airways Corp.*, 434 F.2d 840 (2d Cir.1970).

represent to the contractor that the architect has apparent authority to order that work be changed.[67]

Care must be taken to determine whether the principal, that is, the owner, has authority that can be exercised through its agent, the design professional. For example, suppose the engineer directs the contractor as to *how* the work is to be performed where this choice by contract is given the contractor. The *owner* would not have the authority to direct the methods. Any directions given by the engineer would not have to be followed any more than any direction by the owner. This is not an agency issue but one of contract interpretation.

Accepting defective work raises problems that are discussed in Section 27.03(F). A few points need to be made in this chapter, however. First, design professionals, as interpreters and judges of performance, decide whether work complies. If they decide that the work is proper, they have not accepted defective work. Of course, their determination may be reversed by arbitrators or a court. However, if they determine that the work has not conformed, any acceptance by them would be changing the contract or waiving the owner's right to proper performance, neither of which they have authority, at least in the ordinary case, to do.

Because of the failure to make this distinction, it is more likely that a design professional's acceptance of the work, even defective work, will be considered to bind the owner. To preclude this result, many contracts provide that acts of the design professional will not be construed as accepting defective work.[68]

As indicated earlier in this section, the Construction Process requires many communications among the participants. Clearly a contract can provide that a notice can or must be given to the design professional.[69] If this is the case, the notice requirement has been met. The notice need not come to the owner's attention. In the absence of specific contract language of this type, will a notice that should go to the owner be effective if it is delivered to the design professional?

Because of the close relationship between owner and design professional and the tendency of the law generally to be impatient with notice requirements if they would appear to bar a meritorious claim, there is a strong likelihood that a notice given to the design professional will be *as if* given to the owner. For example, in *Lindbrook Construction, Inc. v. Mukilteo School District Number 6*,[70] the contractor discovered unanticipated subsurface conditions substantially at variance with the contract documents. He notified the architect and claimed an equitable adjustment in the price and additional time. The architect denied that the contractor was entitled additional compensation and directed him to proceed. The contractor completed the work and sued the owner.

The issue before the appellate court was whether the contract requirement that notice be given to the *owner* had been satisfied by notice being given to the *architect*. The court held that the architect was the only representative of the owner with whom the contractor had contact. It noted that the architect had complete knowledge of the changed condition and that the contractor was going to claim additional compensation. There was no evidence of a breakdown in communication between the architect and the owner, and the court noted that it would be unbelievable to assume that the architect did not notify the owner. The court held that the notice to the architect was imputed to the school district.

---

67. See Section 25.04.

68. AIA Doc. A201 ¶¶ 9.5.5, 9.9.4.

69. Id. at ¶ 8.3.2 (time extensions).

70. 76 Wash.2d 539, 458 P.2d 1 (1969).

The dissenting judges argued that the architect's knowledge of the extra work could not be imputed to the owner because he had no authority to modify the contract that required that notice be given to the *owner*. He was therefore, according to the dissenters, not operating within the scope of his authority, and the school district should not be bound. The dissenters also pointed to specific language stating that the architect could not commit the district to cost allowances and noted the importance of protecting public funds from unlawful expenditure.

Sometimes the notice problem is dealt with specifically by statute, such as provisions in mechanics' lien laws that state that notices can be given to the architect.

Are facts known to the design professional *as if* they were known by the owner? For example, one case held that the notice of a limited warranty given to a mechanical engineer consultant bound the owner-buyer.[71] Similarly, a design professional's knowledge of the construction industry custom was held to be chargeable to the owner.[72] A more bizarre use of the imputation doctrine was unsuccessful in a claim by an owner against its engineer based on conspiracy with the contractor to prepare and submit false estimates and vouchers. One contention made by the defendant engineer was that his knowledge of the conspiracy was imputed to the *owner*. This unusual use of the imputation doctrine was rejected, with the court noting that imputation applies only when the agent had knowledge that he has a duty to communicate to his principal. Where he had the motive to con-

ceal, knowledge cannot be imputed to the principal.[73]

Clearly if the design professional has limited authority, the project representative has even less. Usually the project representative is simply authorized to observe, keep records, and report.[74] Use of a project representative is often accompanied by a document that describes these functions and limits the authority of any project representative.[75] There is always a risk that a project representative will become overactive and seek to perform unauthorized activities, such as directing or accepting work. Although these are clearly unauthorized and the contractor should realize it, the contractor may submit and then later claim that these acts bind the owner. Usually such assertions are not successful.[76]

This section has assumed the traditional construction delivery system. However, the problem of authority may have to take into account the way in which an architect or engineer, in a traditional system, may differ from a CM. Again, great caution must be exercised in making generalizations because of the fluidity and imprecision of construction management.

Nevertheless, it is likely that the difference in professional status between a design professional and a CM may reflect itself in the extent to which acts of the CM may be held to bind the owner. In many instances it may be difficult to distinguish the activities of regular employees of a sophisticated owner, such as members of an engineering department, from the activities of a CM. Even if the CM is an independent entity, much of what the CM does, though resembling what design professionals were ex-

---

71. *Trane Co. v. Gilbert,* 267 Cal.App.2d 720, 73 Cal.Rptr. 279 (1968).

72. *Fifteenth Ave. Christian Church v. Moline Heating and Construction Co.,* 131 Ill.App.2d 766, 265 N.E.2d 405 (1970).

73. *Metropolitan Sanitary District v. Anthony Pontarelli & Sons, Inc.,* supra note 63.

74. But in *Town of Winnsboro v. Barnard & Burk, Inc.,* 294 So.2d 867 (La.App.1974) certiorari denied 295 So.2d 445 (1974), knowledge of a project representative was chargeable to the architect.

75. AIA Doc. B352 (1979).

76. See note 64, supra.

pected to do in the traditional system, is more geared toward the efficiency expected from the owner's own employees. As a result, it may be easier to establish apparent authority of a CM than of a design professional.[77]

To sum up, independent advisers such as design professionals, CMs, and project representatives, though agents of the owner, do not have apparent authority to make contracts, modify existing contracts, accept defective work, or waive contract requirements on behalf of the owner. Yet doctrines such as apparent authority or ratification may, in a proper case, justify a conclusion that acts of these professional advisers will be chargeable to the owner.[78]

(C) COMMUNICATIONS

A communication can be made personally, by telephone, or by a written communication. Generally, written communications should be required where possible. If it is not possible for a written communication to be made, at the very least a person making the oral communication should give a written confirmation as soon as possible. If the contract provision that deals with communications of notices does not state how communications are to be made, the communication can be made in any reasonable manner.

Section 21.04(B) discussed the problems of authority and noted that it is essential to have designated persons with authority for each party to make decisions and to take responsibility. With regard to communications, each party should know to whom it should direct any particular communication. The authorized person should be designated by name in the contract and address given. It is important to notify the other party if there has been a change in personnel and notices are to be sent to someone else.

If the contract does not designate time requirements for notices, any notice or communication required must be given within a reasonable time. Commonly, the construction contract will specify time requirements. For example, ¶ 8.3.2 of AIA Doc. A201 requires that a notice be given twenty days after start of the delay. The absence of such a base point, such as simply stating that a twenty-day notice must be given, can create difficulties. Does the notice period begin to run when the event occurred, when the contractor found out about its occurrence, when it could have found out about its occurrence, or when the event had sufficient impact upon the work to cause the delay? Such a dispute would be best resolved by concluding that the period of time for giving notice began when it was reasonably clear that a delay would occur.

Suppose the owner justifiably notifies the contractor that the contract will be terminated unless the contractor pays certain subcontractors within ten days. When does the ten-day period begin? Suppose the letter were dated on September 1, mailed on September 2, and received on September 4. Under such circumstances, when does the time period begin?

It is likely that the time would begin when the letter was received. Doubts will be resolved against the party sending the letter because the latter could have clarified any possible uncertainty. The receiver would probably expect the time period to begin to run when the letter was received unless the receiver knew that there had been an inordinate delay in the mail.

At the other end of the notice, suppose a contractor must respond to a communication within ten days. He posts the letter on the tenth day but it is not received until the twelfth. Has he complied? American law generally follows what is

---

77. *Everette v. Alyeska Pipeline Service Co.*, 614 P.2d 1341 (Alaska 1980) (owner denied independent contractor defense because CM could direct work).

78. See Section 25.04.

called the mailbox, or dispatch, rule. Posting a reply within the time period will be effective unless it is made clear that actual receipt must be had within the time period. The mailbox rule will apply if the communication is sent by a reasonable means of communication. The mailbox concept, though developed in cases involving formation of contract, would very likely be applied to construction project communications.[79]

Many of the difficulties discussed can be avoided by careful contract drafting. It is best to make absolutely clear when periods begin, when they conclude, and whether receipt will be effective when placed in the means of communication or whether a communication must be actually received.

Systems should be developed that make it likely that communications will be received and that proof exists that communications were sent and received. Although false claims of dispatch and receipt are probably rare, a carefully conceived and administered construction project should consider the importance not only of proving that notices were sent or received but of when these events occurred.

Communications received and copies of communications dispatched should be logged and kept in readily accessible files. These records should be kept for a substantial time after completion of the project to deal with the possibility of long-delayed claims being asserted.

Contracts frequently set forth formalities to avoid difficult proof problems. During the course of contract administration, parties often dispense with formal requirements. If they do, there is a serious risk that the formal requirements have been eliminated by the conduct of the parties. For this reason, formal requirements should be complied with throughout contract administration. If the formal requirements must be dispensed with, it is important for the party who wishes to rely on the formal requirements at a later date to notify the other party that the dispensation of the requirements on this occasion will not operate as an elimination of the formal requirements for the balance of the contract. Sometimes general conditions contain provisions stating that waiver of formal requirements in one or more instances will not operate to eliminate formal requirements in the future. These clauses may not mean much if the parties, by their conduct, show an intention to generally dispense with formal requirements.

Sometimes legal rights, such as the right to a mechanics' lien, may require compliance with statutory provisions for notices. Many of the problems that have been discussed in this section can arise when statutory notices are required. Particularly in the area of mechanics' liens, it is necessary to obtain and follow legal advice regarding notice requirements.

PROBLEMS

1. The statute in a state requires that all contracts for over a designated amount of money be awarded by competitive bidding, "except those that involve professional services." A public agency in that state is planning a massive public building program. It intends to hire independent architects, technical engineers, construction managers, and critical path method (CPM) consultants. Which, if any, of these persons can be selected by the public agency without competitive bidding?

2. O owned unimproved land upon which

---

79. *Palo Alto Town & Country Village, Inc. v. BBTC Co.*, 11 Cal.3d 494, 113 Cal.Rptr. 705, 521 P.2d 1097 (1974) (mailbox rule applied to the exercise of an option).

she wished to build a large suburban office building. She was inexperienced in construction and decided to engage PM as project manager. The contract between O and PM authorized, among other things, PM to enter negotiations with specialty trade contractors based upon contract documents that would be prepared by an architect O would engage. The drawings and specifications were partially completed, and PM started negotiations with a number of specialty trade contractors. His negotiations with E, an electrical contractor, were successful, and E agreed to do the work for $5 million as long as the completed contract documents did not differ materially from those that had been the basis of their negotiations. PM wanted to close the deal, as he believed he had made a very good contract for O. O, however, was out of the country and could not be reached. PM decided to make a binding contract with E. PM prepared AIA Document A101 CM and attached to it an A201 CM. He inserted O's name as owner in the first part of the form, typed it in at the end, and under that signed his name as project manager.

When O returned from her trip, she conferred with PM regarding contract negotiations. PM told her that he had made a contract with E. O became infuriated at PM's having exceeded his authority. Is O bound to this agreement? If she is bound, does she have a claim against PM? What additional facts would you need to answer her questions? Assuming you develop

these facts, what would be your advice to her? What if PM had been designated as a construction manager with the "usual" construction manager functions?

3. A construction contract between O and C stated that completion of the project would be no later than December 1. A provision of the contract stated that O could accelerate the completion date, not to exceed sixty days, if he notified C of this in writing no later than April 1.

C's main office was in Chicago; O's main office was in New York; and the construction project was in Toledo, Ohio. On April 1, O drafted a letter requesting a forty-five-day time acceleration. The letter was mailed on April 1 and was directed to C's office in Chicago. The normal course of post between New York and Chicago is one day by airmail and two days by regular mail. By mistake the clerk placed regular mail postage on the letter, and the letter did not arrive in Chicago until April 3. The employee receiving the letter at C's Chicago office did not know that it pertained to the Toledo job and, for that reason, actual notice to the Toledo employees of C was not received until April 5. Does O have the legal right to accelerate the construction date forty-five days? What would your answer be if, on April 1, an employee of O called the project supervisor of C at Toledo and told him to accelerate the performance forty-five days and then sent the letter?

22

Competitive Bidding: Theory, Realities, and Legal Pitfalls

SECTION 22.01 BASIC OBJECTIVES RECONSIDERED

The owner wishes to obtain a completed project that complies with the contract requirements as to quality, quantity, and timeliness at the lowest possible cost. Just as the organizational method and pricing discussed in Chapter 21 must take this into account, so must the method selected to designate the contractor.

In looking at cost—the factor most commonly emphasized in competitive bidding—attention must be directed not only to the bid price but also to other costs. For example, it is likely to cost more to conduct a competitive bid than to negotiate a contract. Similarly, the ultimate cost must take into account any administrative costs (increasingly these days called "claims overhead") incurred to obtain the promised performance and resolve disputes. These are likely to be greater under competitive bidding because the low bidder's performance and administration can increase the cost of monitoring performance and resolving disputes,

as well as create a greater likelihood of claims increasing the ultimate contract price because of extra work.

The ultimate cost of a project should take into account the cost of maintenance and the durability of the project. There are methods that take these factors into account.[1]

Although many of these costs are difficult to quantify, any owner who can *choose* to have competitive bidding should take them into account. Techniques within the competitive bidding system can reduce some of these risks.[2]

SECTION 22.02 COMPETITIVE BIDDING: THEORIES AND SOME PITFALLS

Competitive bidding, according to an Ohio court:

> gives everyone an equal chance to bid, eliminates collusion, and saves taxpayers money. . . . It fosters honest competition in order to obtain the best work and supplies at the lowest

---

1. Vickrey and Nicol, *Total-Cost Bidding—A Revolution in Public Contracts?* 58 Iowa L.Rev. 1 (1972) (bidders bid initial cost, guarantee maintenance costs, and give a price to repurchase goods: variables totaled and bid made to lowest *total* cost bidder).

2. Comment, *Requests for Proposals in State Government Procurement*, 130 U.Pa.L.Rev. 179 (1981) (competitive negotiation).

possible price because taxpayers' money is being used. It is also necessary to guard against favoritism, imprudence, extravagance, fraud and corruption.[3]

This is a large order. This section examines the central assumptions which underlie competitive bidding.

First, competitive bidding assumes that goods or services requested can be objectively evaluated or compared, preferably before award, or at least after. An award of pencils of a standardized type is an illustration. All pencils would be comparable, and all that would be needed would be to compare prices. If a number of competing sellers will bid, the price should be as low as can be obtained.

Even here there are performance uncertainties. Will the party awarded the pencil contract deliver as promised? But if it did not, replacement pencils could be obtained and the excess procurement cost charged to the contractor or its surety. For example, California held that a public award for a construction manager must be made competitively[4] while Massachusetts held that competitive bidding was not required in the purchase of an existing vessel for the use of a ferry boat.[5] The Massachusetts court noted that competitive bidding does not work well where the various properties offered for purchase are not identical, such as in the sale of real property. This noncomparable factor is a reason given for holding that turnkey contracts for public housing are not subject to the requirement for competitive bidding.[6] However, the same argument can be made in favor of permitting a negotiated contract for a construction manager. Yet the California court was more

impressed with the need for competition as a method of reducing costs.

A procurement for the award of police cars that would meet designated standardized performance specifications would fit the competitive bidding model. Either before or after award, testing can be done to determine whether the bidders have prototype vehicles that will or have complied with the specifications. Again there can still be the problem of predicting whether all the cars will conform and be delivered on time.

The principal objective of the awarding agency will be frustrated if the performance standards are not met. Time has been lost and the objective not yet accomplished. Even if the successful bidder is financially solvent or has furnished a surety bond, the objective has still not been achieved. Collecting for nonperformance certainly was not the objective of the procurement.

The difficulty in predicting whether a prospective bidder will be able to meet the performance standards is an important reason why competitive bidding is rarely used in research and development contracts. In these procurements the agency must find a contractor who has the technological skill to perform properly. Competitive bidding, unless it is accompanied by some preselection process, may not be the best method to select a contractor in such a procurement.

Although most construction work is not as sophisticated or experimental as building nuclear submarines or space capsules, there are still elements in construction work that do not make it perfectly suitable to competitive bidding. First, what is the likelihood that prospective bidders have the willing-

3. *United States Constructors & Consultants, Inc. v. Cuyahoga Metropolitan Housing Authority,* 35 Ohio App.2d 159, 163, 300 N.E.2d 452, 454 (1973).

4. *City of Inglewood—L.A. County Civic Center Authority v. Superior Court,* 7 Cal.3d 861, 103 Cal.Rptr. 689, 500 P.2d 601 (1972). But refer to Section 21.04(D).

5. *Douglas v. Woods Hole, Martha's Vineyard, & Nantucket S.S. Authority,* 66 Mass. 459, 319 N.E.2d 892 (1974).

6. *United States Constructors & Consultants, Inc. v. Cuyahoga Metropolitan Housing Authority,* supra note 3. Refer to Section 21.04(E).

ness and the ability to do the work they promise? Second, how difficult will it be to determine whether they have done the work they have promised? Third, what is the likelihood that the actual price will substantially exceed the contract price for reasons other than owner-directed changes or circumstances over which neither party had control?

The competitive bidding process can be structured to minimize some of these risks. The invitation may state that the contract will be awarded to the lowest *responsible* or lowest and *best* bidder. Preselection can screen out bidders who do not have the requisite competence or capacity. Nevertheless, despite these provisions, the likelihood of success depends upon the integrity and ability of the contractor which are often difficult to measure in competitive bidding where the tendency is to look solely at price.

Competitive bidding also assumes that there are free bids and true competition. If there is collusion among the bidders to "take turns" or submit fictitious bids, not only are antitrust laws violated but competitive bidding cannot accomplish its objective of obtaining the lowest price. Although construction is, on the whole, a fiercely competitive business, collusion between competitors in the competitive bidding process is not unknown.

Anticompetitive devices can be found in competitive bidding. If product specifications do not provide for alternative products and a viable method for substitutes, competitive pricing may be unduly restricted. New Jersey prohibits specifications in local public contracts that limit free and open bidding, such as specifying brand names without allowing equivalent products to be substituted.[7]

To sum up, the competitive bidding system is presently, and is likely to continue to

be, the major method of obtaining construction contractors. However, there are pitfalls in competitive bidding, and, in the proper circumstances, serious thought should be given to another method of obtaining a contractor.

SECTION 22.03 THE COMPETITIVE BIDDING PROCESS

(A) OBJECTIVES

Competitive bidding should result in contract awards made impartially at the lowest price. Nonconforming bids are usually disregarded because conformity is needed for a proper comparison of bids and to give each bidder an equal opportunity. The competitive bidding system cannot function properly unless honest and capable bidders have enough confidence in the fairness of the system to submit bids. Submitting a bid proposal is an expensive and time-consuming operation. The bidders are entitled to reasonable assurance that they will be treated fairly and that the owner will follow its own rules.

(B) INVITATION TO BIDDERS

No effort will be made to cover every aspect of the competitive bidding process. Most public agencies have standard forms for the competitive bidding process, and they are often regulated by statutes, regulations, and ordinances. Private owners often have engaged in substantial construction work and have also developed their own forms and methods. Most design professions have developed standard or recommended forms for bidding documents.[8] They are usually available to the design professional. The principal objective of this subsection and others following is to present an overview of the process.

---

7. N.J.Stat.Ann. 40A:11–13.

8. AIA Doc. A701 (Instruction to Bidders, 1978).

The initial step in conducting a competitive bid is the Invitation to Bidders. In federal procurement, this is known as the Requests for Bids or the RFB. Frequently, public agencies are required to invite bids by public advertising. By using the broadest dissemination, such as trade newspapers, professional journals, and government publications, the largest number of bidders can participate. This gives all bidders a chance and should obtain the lowest price.

Although inviting the maximum number of competitors should result in a lower price, dangers exist in having too many bidders. Good bidders may be discouraged if the large number of bidders make their chances of winning quite remote.

Devices are available to avoid the competence risks inherent in the price-oriented competitive bid. In addition to stating that the award will be made, if at all, to the lowest responsible or the lowest and best bidder, a prequalification system can be used. The owner selects a group of bidders, all of whom are likely to have the capability of doing a competent job. The design professional or construction manager may request information from specific contractors on prior jobs completed, capital structure, machinery and equipment, and personnel (including supervisory personnel). After evaluation of this information, the owner's professional adviser usually determines which contractors should be permitted to receive an Invitation to Bid.[9]

The advantage to prequalification is that there is a better chance of finding a good contractor. However, there will be administrative time and expense expended in making this preselection or prequalification, and the competitive aspect is likely to be diminished. Such a method should be used where the project is of great magnitude or involves new construction techniques and where cost is less important than insuring the quality of performance.

The federal procurement regulations sometimes permit a "two-step" formal advertising procedure.[10] This is used where definite specifications cannot be prepared and offered for fixed-price competitive bids. As a first step, unpriced technical proposals are solicited to meet certain specified requirements that are set forth in the solicitation for proposals. Typically these requirements are for an end-result product. All proposals received are evaluated, and those that are considered to be within the range of acceptability are discussed with the proposers to obtain clarification and more detailed definition.

The second step is a request for price bids from all those whose first-step proposals met the criteria specified in the original solicitation for proposals. Award is then made in accordance with the procedures for award of a fixed-price contract, with each bidder pricing its own technical proposal previously approved as having met the specified criteria.

In the procurement of products or equipment, the federal government can protect itself from having to wait for a product that may not be adequate by inserting a "first article" clause that requires that the contractors deliver for government testing a preproduction model of what the contractor has agreed to furnish.

Some states have prequalification systems. For example, New Jersey requires that all school construction be done by persons certified by the state board of education.[11]

---

9. For a discussion of competitive negotiated contracts in state procurement, see note 2 supra. See AIA Doc. A305 (Contractor's Qualification Statement, 2d printing 1983).

10. See *Wheelabrator Corp. v. Chafee*, 455 F.2d 1306 (D.C.Cir.1971).

11. The method is discussed in *Donald F. Begraft, Inc. v. Borough of Franklin Board of Education*, 133 N.J.Super. 415, 337 A.2d 52 (1975).

Bidders should be given adequate opportunity to study the bidding information, to make tests, to inspect the site, and to obtain bids from subcontractors and suppliers.[12] Even when there is adequate time, bidders often wait to complete their bids until the bid closing deadline is imminent. Generally this is due to reluctance on the part of subcontractors to give sub-bids to bidders until shortly before the deadline for bid submissions. The reasons for this reluctance are explored in Section 32.02. Even if bidders do not make proper use of the time available to them, having allowed reasonable time for bid preparation can be helpful to the owner if a dispute arises over claimed computation errors or unforeseen subsurface conditions.

Invitations can preclude bids from being changed, corrected, or withdrawn after submission. However, there is a tendency toward more flexibility. More commonly, Invitations permit changes, corrections, and withdrawals of submitted bids before bid opening.

This may be due to the tendency of modern courts to release a low bidder who has made computation errors even *after* opening of bids but while there is still an opportunity to award the contract to the next lowest bidder. Permitting changes can help the owner by permitting and encouraging reduction of bids in a changing market. However, overliberality in permitting changes, corrections, and withdrawals may cause bidders to lose confidence in the honesty of the bidding competition.

Where the right to change is given, it may be limited to reducing the bid. If changes, corrections, or withdrawals are to be permitted, the invitation can limit such a right to a designated time period and require that any change, correction, or withdrawal be expressed in writing and received by the owner by a designated time.

In public procurement and occasionally in private procurement the invitation requires the bidder to deposit a bid bond, a cashier's check, or a certified check to provide security in the event that the bidder to whom the award is made does not enter into the contract. The amount is either a stated *percentage* of the bid or a *fixed* amount determined by a designated percentage of the estimated costs made by the design professional.

The Invitation should also state how long the securities will be held, as this is a matter of importance to bidders. After award, the owner should release the securities of all but the lowest three or four bidders. Perhaps the bidder to whom the award is made will not enter into the contract. It still may be possible under some circumstances to hold the next lowest bidder. However, there is no justification for holding the securities of bidders who are not likely to be awarded the contract. After the successful bidder *signs* the contract, all securities should be released.

Usually the bidders must make a small monetary deposit when they request information in order to study a possible proposal. Sometimes this has the effect of discouraging persons who are not serious about making a bid proposal from obtaining the plans and specifications merely out of curiosity. The deposit is usually refunded when the bidding documents are returned.

The owner should keep all the records connected with the bidding process. Time logs should be kept that show exactly when the bidder obtained the bidding information. Changes in bidding information should be sent to all persons who have picked up bidding information, with copies of the changes kept for future reference. Changes after the bidding information is disseminated should be kept to a minimum.

Bidding documents sometimes provide that bidding technicalities can be waived by

---

12. Under AIA Doc. A201, ¶ 1.1.2 the contractor warrants that it has visited the site and checked local conditions.

the owner. This allows the owner to accept the lowest bid despite minor irregularities. However, advance notice that technicalities can be waived can encourage careless proposals and may also create the impression that favoritism may be shown to certain bidders.

Sometimes larger projects can be divided into designated stages. The owner may decide not to build the entire project if it does not have the money or if bids are too high for certain portions of the work. For this reason the Invitation to Bidders can be divided into the stages or project alternates. But alternates can mean that favoritism can be accomplished by the award. For example, there may be an Invitation to Bidders involving a hospital, a dormitory for nurses, and a parking structure. One alternate could be the hospital alone. A second could be the hospital with the parking structure, and a third the hospital with the nurses' dormitory. One bidder may be low on the total bid. Another bidder may be low on the first alternate, another on the second, and another on the third. The determination of which alternate is to be awarded could be based upon favoritism to one of the bidders. To avoid this difficulty, the Invitation to Bidders should state the preferred alternate choices.

The same type of award manipulation is possible if the alternates consist of different methods of construction or materials. A list of preferences within price limits should avoid suspicion of possible favoritism.

(C) INFORMATION TO BIDDERS

Information to Bidders, called in federal procurement the Information for Bids (IFB), usually consists of drawings, specifications, basic contract terms, general and supplementary conditions, and any other documents that will be part of the contract. Sometimes soils test reports are included.

Alternatively the information to bidders may state that designated reports are available in the office of a particular soils engineer or the design professional for examination by the bidder. Soil information is discussed in Section 29.03.

The drawings and specifications should be detailed and complete so that bidders can make an intelligent bid proposal. Imprecise contract documents and much discretion to the design professional may discourage honest bidders from submitting bids and encourage bidders of doubtful integrity who make low proposals in the hope that they will later be able to point to ambiguities and make large claims for extras.

The Information to Bidders can specify that any uncertainties observed by the bidder must be resolved by a written request for a clarification to the design professional before bid award. Any clarification issued should be in writing and sent to all persons who have been invited to bid.

Increasingly the law requires that the party conducting the competitive bid must disclose information in the bidding documents under certain circumstances (discussed in Section 22.04(B)).

(D) BID PROPOSALS

Bid proposals should be submitted on forms provided by the owner. To properly compare bids, all bidders must be proposing to do the same work under the same terms and conditions. The bid proposal form should include or make reference to any important disclaimers contained in the bidding information. Such disclaimers include denying responsibility for subsurface information and specifying any rules that seek to govern the rights of the bidder to withdraw the bid, such as limiting withdrawal to clerical errors or setting forth other requirements such as a deadline for claiming mistake.[13]

The bid proposal should be signed by an

---

13. See also Section 22.04(H) dealing with bidding documents generally.

authorized person. If the bidder is a partnership, the entire name and address of the partnership should be given. If the bidder is a corporation, the bid should be signed by appropriate officers and the corporate seal should be attached. It may be desirable to attach to the bid a resolution of the board of directors approving the bid.

(E) BID OPENING

Bid proposals are sent in a sealed envelope to the owner or the person designated to administer the competitive bidding, such as a design professional or construction manager. In public contracts, they are opened publicly at the time and place specified in the Invitation to Bidders. Usually the person administering the process announces the amount of the bids and the bidders. The Invitation specifies a designated period of time in which the owner can evaluate the bids. At bid opening, care must be taken to avoid any impression that the low bidder has been awarded the contract. Often the person administering the process states that a particular bidder is the "apparent low bidder." [14]

In private competitive bids, the Invitation can state that bids need *not* be opened in public. When the bid opening is private the bidders, at least at that time, are not able to compare their bids with the others. This minimizes the likelihood that the low bidder may begin to suspect a bidding error if its bid is much lower than the others. One case involved the successful bidder not finding out that its bid had been much lower until it was well into the project. At that time, it threatened to walk off the project unless it was given a price adjustment,

claiming that the owner must have known its bid was erroneous. Although the court did not grant relief to the contractor, it is likely that its determination to grant the contractor relief on another theory was affected by the belief that the owner should have notified the contractor of the wide discrepancy between its bid and the others. [15] This may reflect the law beginning to impose upon contracting parties, or even those in the process of making a contract, an obligation of good faith and fair dealing, such as drawing attention to a possible mistake. [16] A private bid opening may tempt the owner not to notify a bidder when it is apparent that a mistake has been made.

(F) EVALUATION OF BIDS

Some legal problems relating to bid evaluations and awards are discussed in Sections 22.04(E) and (F). This section outlines some steps that should be taken when evaluating bids.

First, the owner or its representative should follow the procedures set forth in the Invitation to Bidders. The bids should be checked to see whether they conform.

A nonconforming bid is a proposal based upon performance not called for in the Invitation or Information or that offers performance different from that specified. Nonconformity may relate to the work, the time of completion, the bonds to be submitted, or any requirements of the contract documents. If a bid is not in conformity and the defect cannot be waived, the bid must be rejected.

Second, the owner [17] should examine the Invitation to Bidders to determine whether

---

14. In *McCarty Corp. v. United States*, 204 Ct.Cl. 768, 499 F.2d 633 (1974), the contracting officer stated that the plaintiff was the "apparent low bidder." This, among other facts, precluded a finding that the plaintiff's bid had been accepted.

15. *Paul Hardeman, Inc. v. Arkansas Power & Light Co.*, 380 F.Supp. 298 (E.D.Ark.1974).

16. See Section 22.04(E).

17. "Owner" includes design professional or construction manager advising the owner.

any bid must be accepted. Typically, the Invitation reserves the right to reject all bids. The Invitation must be reviewed to determine the *basis* for making an award if one is made. Commonly, the award is to be made to the lowest responsible or lowest and best bidder.

To determine the lowest responsible bidder, the owner can take into account the following factors, as well as any others that bear upon who would be the bidder most likely to do the job properly:

1. expertise in type of work proposed
2. financial capability
3. organization, including key supervisory personnel
4. reputation for integrity
5. past performance

These illustrations relate to the basic objectives of obtaining a contractor who is likely to do the job properly at the lowest price and with the least administrative cost to the owner.

In contracts for the purchase of machinery, bid evaluation should take into account the following factors:

1. length and extent of warranty
2. availability of spare parts
3. service and maintenance
4. cost of replacement parts
5. cost of installation
6. durability

Gary Aircraft Corp. v. United States,[18] illustrates the methods of evaluating a bidder used by the Department of the Air Force.

The contract contemplated a three-year fixed-price contract. Two bids were received, but the awarding authority doubted the capability of the low bidder. The contracting officer first screened information regarding the two bidders. This information included credit reports; Defense Department records; reports from customers, suppliers, and bankers; financial data, and current and past production records.

Because the information created doubts regarding the low bidder, an advisory preaward survey was made. The survey of the low bidder took four days, and the report following it recommended that the award *not* be made to the low bidder because of deficiencies in the following areas:

1. quality control system
2. past performance
3. meeting required schedules
4. property control system
5. plant facilities and equipment

The award was *not* made to the low bidder.

Although the owner should evaluate criteria other than price in selecting a contractor, there is a substantial risk of legal action if the bid is awarded to someone other than the low bidder. Sometimes bid awards in public contracts can be judicially challenged. This should not deter even a public owner from awarding the contract to someone other than the low bidder. However, compelling reasons for not awarding it to the low bidder must exist and be documented.

Third, if bidders were asked to bid separately on project alternates, the owner should determine whether any bidders had made "all-or-nothing" bids. Such a bid indicates an unwillingness on the part of the bidder to perform any part of the project except the entire project. An all-or-nothing bid is permitted unless the Invitation specifically precludes it.

(G) NOTIFICATION TO BIDDERS

When the successful bidder is selected, the successful and unsuccessful bidders should be officially notified. However, sometimes with or without legal justification the successful bidder will not enter into the contract. For this reason the notification to

---

18. 342 F.Supp. 473 (W.D.Tex.1972).

unsuccessful bidders should state that their bids still remain available for acceptance by the owner for the period of time specified in the Invitation to Bid. This is done to preclude any later contention that awarding the contract released the unsuccessful bidders.

(H) POST AWARD CHANGES

If the owner changes the contract terms after the bid has been awarded and the change makes the obligation less burdensome to the successful bidder, this can be unfair to the other bidders. For this reason, changes of this type should be done with caution.

(I) SIGNING THE FORMAL CONTRACT

The successful bidder is sent the formal contract for signature or requested to meet at a specified time and place to sign the formal documents. Records of all correspondence should be kept in the event that subsequent disputes arise between the owner and the contractor relating to the bidding process and awarding of the contract.

The successful bidder must enter into the formal contract awarded unless legal grounds exist for refusal (discussed in Section 22.04(E)). The effect of signing the formal contract, as well as its effect upon bidding documents and the function of the formal contract, is discussed in Section 22.04(G). The legally effective date of the contract (when awarded or formal contract signed) is discussed in Section 30.02.

(J) READVERTISING

Sometimes the bids are all too high and the owner decides to readvertise the project. In public contracts there are procedures for readvertising. Readvertising requires time and administrative expense. Some feel that readvertising without reducing the scope or quality of the project is unfair, because it is an attempt to beat down the bids of the

contractors, often resulting in deficient workmanship and substandard materials. Frequently readvertising means higher bids from all bidders, because the less skillful bidders have seen what the better bidders have bid.

(K) SPECIAL RULES FOR PUBLIC CONTRACTS

Owners generally seek a construction contract that will give them the balance they choose among quality, timely completion, and cost. On the basis of these criteria, the owner looks for a contractor who will do the best work at the best price.

Looking only at the project and these goals, the owner will not be concerned with the racial characteristics of the contractor's labor force, the wages the contractor pays its employees, and the source of the supplies. Obviously, this unconcern is not absolute. The owner could be concerned with the wages if it felt that workers who were not paid the prevailing rate would not perform properly. The source of the supplies could be important if it felt that supplies from certain sources were higher quality. But on the whole, the owner who wishes to obtain the best price through competitive bidding or through negotiation must give broad latitude to the contractor in these matters.

But public contracts involve billions of taxpayer dollars. Elected officials frequently look to the procurement process and public spending as ways of accomplishing goals that go beyond the best quality at the best price. In doing so, they often respond to the pressure of interest groups who also see procurement as a means of obtaining their objectives. The federal government has been the pioneer in using the procurement process to achieve social and economic objectives. State and local public agencies and even some private owners have also begun to see procurement in this light.

This treatise cannot deal in detail with the many rules that seek to achieve a variety of objectives through the public contract process. However, it should be noted that some goals sought to be achieved by the public contract process are as follows:

1. Providing employment opportunities to disadvantaged minorities, women, handicapped persons, or disabled war veterans.
2. Setting aside certain procurement awards for small businesses or disadvantaged minorities.
3. Favoring contractors, suppliers, or workers who reside in a particular state or city.
4. Awarding contracts to bidders located in economically depressed areas.
5. Insuring that workers are paid at the local prevailing wage rate.
6. Avoiding corruption in procurement.

Clearly such rules have their costs. In addition to higher bid prices, there are likely to be administrative costs incurred by all participants to see that these rules are followed. These costs may be worth incurring if they accomplish the objectives sought. However, attempts to use the public contract process for these objectives have generated intense controversy, both as to the fairness of the programs and whether the objectives can be accomplished in a different way at a lower cost.

Every aspect of awarding public contracts, especially competitively bid awards, must avoid even the appearance of impropriety. For example, public officials who make procurement decisions are expected to have the interests of their agency in mind and avoid conflict of interest.

As a result of public concern over contract awards, public officials often are rigid and unbending in such matters as bidding irregularities and withdrawal of bids.

SECTION 22.04
SOME LEGAL ASPECTS
OF COMPETITIVE BIDDING

(A) INVITATION TO BIDDERS

Ordinarily the Invitation is not an offer and does not create a power of acceptance in the bidders. It is a request that bidders make offers to the owner that can be accepted or rejected. But the Invitation is an important document. It sets up the ground rules for the competitive bid. If a contract is formed with one of the bidders or if the successful bidder refuses to enter into the contract, the Invitation may have legal significance. Of course, the Invitation can be modified by the formal agreement between bidder and owner, but often the formal contract simply memorializes what has been agreed to in the Invitation and bid proposal.[19]

(B) DUTY TO DISCLOSE

The early common law rarely obligated a contracting party to disclose information that the other party would want to know. Contracting parties were generally expected to look out for themselves. They could not make deliberate misrepresentations but could conceal matters within their knowledge, even if they knew that the other party would want to know them.[20] But the common law has been finding a duty to disclose vital information that the other party is not likely to discover.[21]

This has been reflected in an increasing tendency to require that the party conducting a competitive bid disclose certain information to bidders. The Court of Claims has held that a federal procurement agency must disclose procurement plans of other federal agencies of which it knows

---

19. See Section 22.04(G).

20. *Swinton v. Whitinsville Savings Bank*, 311 Mass. 677, 42 N.E.2d 808 (1942).

21. *Obde v. Schlemeyer*, 56 Wash.2d 449, 353 P.2d 672 (1960).

and which may affect the pricing assumption of bidders.[22] Similarly, it has held that the procuring agency must disclose technical information that it possessed relating to the manufacturing process that it knew the contractor intended to use.[23] The court noted that on many occasions each party has an equal opportunity to uncover the facts. But where one party knew much more than the other and that the latter was proceeding in the wrong direction, it could not betray the contractor into a "ruinous course of action by silence." State courts have also recognized a duty to disclose.[24]

Yet contractors cannot rely too heavily on the possibility of legal protection. Generally parties who *can* protect themselves *must* do so.[25] Contracting parties, although expected to cooperate, do not owe each other fiduciary obligations to look out for each other.[26]

(C) BID PROPOSAL

The bid proposal is an offer and creates a power of acceptance in the owner. The owner has a right to close a deal and bind a bidder without a further act of the bidder. Subject to many exceptions,[27] offers are generally revocable, even if stated to be irrevocable for a specified period of time.

This raises an issue that does not usually surface in construction litigation. Suppose either *before* bid opening or *after*, but *before* the formal contract is signed, the bidder revokes its bid. It is generally assumed that despite the common law rule of revocability the bid is irrevocable for the period stated in the Invitation unless the Invitation *permits* withdrawal before bid opening. What is the justification for such an assumption?

Sometimes public contract competitive bidding is authorized by statutes or regulations that state bids to be irrevocable. This may supersede the common law rule of revocability.[28] Sometimes the bid is revocable, but any deposit made with it is forfeited.[29] One approach suggested is that the bidder receives the benefit of having the bid considered by the owner. This "fiction" has been rejected.[30] The bid may be irrevocable if the owner has justifiably relied upon the bid. The owner may have given up the opportunity of negotiating a contract having relied on the bidders' stating that their bids would be irrevocable or, at the very least, expended considerable time and money in conducting the competitive bid process.[31] If the transaction is one covered by the Uniform Commercial Code, the bid may

---

22. *J.A. Jones Construction Co. v. United States,* 182 Ct.Cl. 615, 390 F.2d 886 (1968) (Corps of Engineers knew of Air Force plans); *Bateson-Stolte, Inc. v. United States,* 145 Ct.Cl. 387, 172 F.Supp. 454 (1959) (must prove Corps of Engineers knew of AEC plans). See also *Hardeman-Monier-Hutcherson v. United States,* 198 Ct.Cl. 472, 458 F.2d 1364 (1972) (duty to disclose weather and sea conditions after contractor request).

23. *Helene Curtis Industries, Inc. v. United States,* 160 Ct.Cl. 437, 312 F.2d 774 (1963).

24. *Welch v. California,* 139 Cal.App.3d 546, 188 Cal.Rptr. 726 (1983) (records of earlier attempt to repair that documented tidal difficulties); *Pennsylvania Department of Highways v. S.J. Groves & Sons Co.,* 20 Pa.Cmwlth. 526, 343 A.2d 72 (1975) (another contractor would occupy site after access given).

25. *H.N. Bailey & Associates v. United States,* 449 F.2d 376 (Ct.Cl.1971) (information obtainable elsewhere).

26. See Section 23.02.

27. Refer to Section 5.06(B).

28. 1 A. Corbin, Contracts § 46 (1963).

29. Id. at § 47.

30. *Sooy v. Winter,* 188 Mo.App. 150, 175 S.W. 132 (1915).

31. *Drennan v. Star Paving Co.,* 51 Cal.2d 409, 333 P.2d 757 (1958) (subcontractor bid irrevocable).

be irrevocable as a "firm offer." [32] If the procurement involved the sale of goods with installation incidental, the Code would govern.[33] Even if it does not, it may be possible to use the Code by analogy.[34]

Once the bid has been opened, it is generally assumed that it has become irrevocable and can be withdrawn only if legal grounds exist for doing so, usually a mistake in preparing the bid.[35]

Can the owner hold bidders to whom the bid has *not* been awarded for the period specified in the Invitation to Bidders? The not uncommon refusal of the bidder to whom the contract has been awarded to enter into the contract makes it imperative that no expectation be created that the unsuccessful bids are no longer binding when an award has been made. Of course, the period of time cannot be extended, but it should not be shortened simply because the award appears to have been made to someone else.

(D) AWARD: WAIVER OF IRREGULARITIES

Usually the owner promises to award the contract to the lowest responsible or lowest and best bidder, reserving the right to reject *all* bids. To make this decision, the owner must evaluate the bids. This is done between bid opening and award, if any. Yet this evaluation period is not open-ended. Bids are irrevocable for the period specified in the Invitation to Bidders. This is illus-

trated in *Hennepin Public Water District v. Petersen Construction Co.*,[36] where the Invitation had stated that the awarding authority would have to obtain financing within sixty days from bid opening. There was a tentative acceptance before the awarding authority obtained financing. Then sixty-seven days after bid opening, the awarding authority forwarded the formal contract to the successful bidder. The court held the acceptance too late and the bidder released.

Suppose the contractor had sued the awarding authorities for not submitting a formal contract within the sixty-day period. Such an action would not have been successful, because the sixty-day period was very likely a condition to the award but not a promise by the awarding authority to obtain financing within the period.

The U.S. Court of Claims has held that a bidder is entitled to honest consideration of its bid. To show that the bid was not properly considered, the bidder must establish that the government acted arbitrarily and capriciously and that there was no reasonable basis for the government's decision.[37]

On the other hand, California rejected a bidder's contention that the awarding authority must consider bids in good faith. The Invitation permitted the awarding authority to reject all bids. The court held that the awarding authority had complete discretion, could act capriciously or arbitrarily, and would not be liable in tort or contract. But the awarding authority can-

---

32. U.C.C. § 2–205.

33. *Bonebrake v. Cox,* 499 F.2d 951 (8th Cir.1974).

34. *Transatlantic Financing Corp. v. United States,* 363 F.2d 312 (D.C.Cir.1966). Analogous use of the U.C.C. was rejected in *Joseph W. O'Brien Co. v. Highland Lake Construction Co.,* 17 Ill.App.3d 237, 307 N.E.2d 761 (1974) (construction contract).

35. *Elsinore Union Elementary School Dist. v. Kastorff,* reproduced in Section 22.04(E); *A.J. Colella, Inc. v. City of Allegheny,* 391 Pa. 103, 137 A.2d 265 (1958).

36. 54 Ill.2d 327, 297 N.E.2d 131 (1973).

37. *Keco Industries, Inc. v. United States,* 428 F.2d 1233 (Ct.Cl.1970). Absence of fraud or palpable abuse of discretion is required in Mississippi. *Warren G. Kleban Engineering Corp. v. Caldwell,* 361 F.Supp. 805 (N.D.Miss. 1973), vacated on other grounds, 490 F.2d 800 (5th Cir.1974). New Jersey requires a bona fide judgment. *Mendez v. City of Newark,* 132 N.J.Super. 261, 333 A.2d 307 (1975).

not solicit a bid without any intention of considering it.[38]

Although the California rule gives maximum discretion to the awarding authority, there is an undeniable trend toward holding government agencies accountable for their acts.

Procedures where the low bidder is rejected vary. For example, New Jersey precludes the local awarding authorities from rejecting the highest bid for property or low bid for work without giving a hearing to the disappointed bidders.[39]

But California does not require a full-fledged courtlike hearing if the low bid is rejected. Rather, the awarding authority must give the low bidder access to any evidence that has reflected upon its responsibility received from others or produced as a result of an independent investigation. The bidder must be afforded an opportunity to rebut such adverse evidence and present evidence that it is qualified to perform the contract.[40]

States undoubtedly vary as to the requirement for a hearing and its nature. But again, with the emphasis upon greater accountability, it is likely that low bidders who are not awarded the contract can at least demand that reasons be given and that they be given an opportunity to rebut adverse evidence.

As for specific reasons for rejecting a bidder, New Jersey held that disputes over a previous job were insufficient to deny award to the low bidder.[41]

But Mississippi upheld a provision in the Invitation that allowed the awarding authority to reject a bidder for being in arrears on an existing contract or in litigation with the awarding authority for having defaulted on a previous contract.[42] The bidder's wholly owned subsidiary was suing the awarding authority on another matter, and on this basis the bidder's bid was rejected. The court recognized the provision as possibly coercing contractors from asserting their legal rights but held that the standard of rejecting bidders was one within the discretion of the authority. This allows an awarding authority to use the procurement process for an improper purpose.

Massachusetts upheld rejection of a subcontractor who had had a dispute with the prime contractor on an earlier job and had made misstatements of his previous work experience.[43] A federal court applying Mississippi law upheld rejection of a bidder because of his poor reputation for quality work.[44] Courts have varied on rejection of bidders who intended to use nonunion workers.[45]

An awarding authority or owner conducting a competitive bid should be able to reject a bidder if a good-faith determination has been made that the bidder is not likely

---

38. *Universal By-Products, Inc. v. City of Modesto,* 43 Cal.App.3d 145, 117 Cal.Rptr. 525 (1974); *Pacific Architects Collaborative v. California,* 100 Cal.App.3d 110, 166 Cal.Rptr. 184 (1979); *Commercial Industrial Construction, Inc. v. Anderson,* 683 P.2d 378 (Colo.App. 1984).

39. *Mendez v. City of Newark, supra,* note 37; *D. Stamato & Co. v. Township of Vernon,* 131 N.J.Super. 151, 329 A.2d 65 (1974).

40. *City of Inglewood—L.A. County Civic Center Authority v. Superior Court, supra* note 4.

41. *D. Stamato & Co. v. Township of Vernon, supra* note 39.

42. *M.T. Reed Construction Co. v. Jackson Municipal Airport Authority,* 227 So.2d 466 (Miss.1969).

43. *Kopelman v. University of Massachusetts Building Authority,* 363 Mass. 463, 295 N.E.2d 161 (1973).

44. *Warren G. Kleban Engineering Corp. v. Caldwell, supra* note 37.

45. A Massachusetts court upheld an awarding authority's rejection of a bidder who intended to use nonunion workers in *Modern Continental Construction Co. v. Massachusetts Port Authority,* 369 Mass. 825, 343 N.E.2d 362 (1976). The court held that the award could have caused a labor dispute and was not in the public interest. See also *Image Carrier Corp. v. Beame,* 567 F.2d 1197 (2d Cir.1977). But see *Wittie Electric Co. v. New Jersey,* 139 N.J.

to be able to complete the required performance. As the reason for rejection departs from evaluation of the bidder's experience, financial ability, integrity, and availability of facilities necessary to perform the contract, it is more likely that the rejection has not been made in good faith.[46]

Statutes and invitations frequently require that the award be made, if at all, to the lowest or best bidder. The California Supreme Court held that the low bidder cannot be rejected unless it is found to be not responsible and the awarding authority cannot simply award to the best bidder.[47]

Suppose there are irregularities in the proposal or process. The awarding authority may believe them minor and wish to waive them and accept the bid. The legal issues have been whether the irregularity in question is minor and whether waiver would encourage carelessness, create opportunity for favoritism, and operate unfairly to other bidders.

The awarding authority's discretion to waive a nonresponsive bid depends upon the importance of the deviation and possible prejudice to other bidders.[48] The awarding authority's past conduct may have led a bidder to reasonably believe nonconformity would be disregarded. In such cases, the bid may not be rejected where the awarding authority had authority to waive the irregularity.[49]

Many cases involve bid submission deadlines. *H.R. Johnson Construction Co. v. Board of Education*[50] held that an awarding authority could not accept a bid made one minute late. The court was not persuaded that one minute could not give a bidder a competitive advantage over other bidders. The court felt that if the awarding authority is given the discretion to allow a one-minute deviation, it could stretch this power to even fifteen minutes or an hour.

In *William F. Wilke, Inc. v. Department of Army*,[51] the bids were to be opened at 3:00 p.m. At that time a representative of the low bidder was in the room, but he neglected to put the sealed bid in the receptacle designated for that purpose. At 3:04 the contracting officer gathered the receptacle and started to sort out the bids. During the sorting the low bidder's representative added his sealed bid to the box of yet unopened bids. The awarding authority awarded the bid to the low bidder, and the next low bidder complained.

The court concluded that the bid should not have been accepted, as it had been deposited four minutes late. But because it did not appear that the low bidder had obtained any competitive advantage, the court would not order that the bid be awarded to the next low bidder but limited the disappointed bidder to recovery of its bidding expenses. Other cases have allowed the awarding agency discretion to

---

Super. 529, 354 A.2d 659 (1976). Much can depend upon the language of the statute or local ordinance that expresses the power of the agency to reject bids as well as the judicial attitude toward the trade union movement.

46. *D. Stamato & Co. v. Township of Vernon*, supra note 39.

47. *City of Inglewood—L.A. County Civic Center Authority v. Superior Court*, supra note 4.

48. *Albano Cleaners, Inc. v. United States*, 455 F.2d 556 (Ct.Cl.1972). The court held that a substantial deviation affects price, quality, or quantity. In *Rossetti Contracting Co. v. Brennan*, 508 F.2d 1039 (7th Cir.1975), the court held that nonconformity as to affirmative action hiring was not correctable. Here the court was facing a substantial deviation but one that the bidder offered to correct. The court was fearful that allowing "correction" would give the bidder an "option" exercisable after seeing the other bids.

49. *Albano Cleaners, Inc. v. United States*, supra note 48.

50. 16 Ohio Misc. 99, 241 N.E.2d 403 (1968).

51. 485 F.2d 180 (4th Cir.1973).

waive such irregularity, especially if the bidding information permitted this.[52]

(E) WITHDRAWAL OR CORRECTION OF MISTAKEN BIDS: *ELSINORE UNION v. KASTORFF*

Sometimes a bidder will seek to withdraw or correct a bid. Such a request usually occurs just after bid opening or, more rarely, after award. Usually the basis for the request is computation errors, such as omitting a large item, making a mathematical miscalculation in determining an item price or making an error in adding bid price components.[53]

Early cases would not relieve a bidder for these mistakes.[54] A reason given for denying relief was that the mistake was "unilateral," one made only by the bidder and not shared in by the owner. There was, as a rule, negligence in computing the bid. The fear of false claims and the integrity of the bidding system were other reasons for denying relief.

Some courts began to moderate the strictness of this doctrine. They pointed to the rule that a person to whom an offer has been made cannot accept the offer if she knows or should know that the offer was made by mistake. For example, if the owner or design professional knew or should have known that an entire item had been left out of the bid or that there had been a mistake in adding the total, it would be unfair to accept the proposal and seek to bind the bidder. For example, in *Santucci Construction Co. v. County of Cook*,[55] the awarding authority had estimated that the cost of drain work would be $1.9 million.[56] The cost of the drainpipe, including labor, was expected to cost $1.4 million. Santucci submitted a total bid of $1.1 million, with $775,000 for the drainpipe, including labor. Three other contractors submitted total bids between $1.7 million and $1.8 million with their bids for the drainpipe, including labor, being between $1.2 million and $1.4 million. The engineer for the awarding authority had thought that Santucci's bid was "cheap, low." A day after bid opening, Santucci claimed a clerical error and sought to withdraw his bid. The request was refused, and Santucci would not enter the contract. The awarding authority retained his bid deposit. Santucci brought legal action to recover it.

Noting that ultimately the work was let for $1.6 million, the Appellate Court affirmed the finding of the trial court that Santucci had made a mistake and that the awarding authority should have known of this mistake. It rescinded Santucci's bid and ordered that the deposit be returned to him. The Appellate Court emphasized that Santucci's bid was $600,000 less than the next lowest bidder and over $800,000 less than the awarding authority's estimate for the project.

The "snap-up" concept used by the court in the *Santucci* case has been the vehicle for bidder relief where courts have thought it appropriate. Yet an honest appraisal of the cases and an awareness of the immense variation of the bids for many types of construction work leads inescapably to the

---

52. *William M. Young & Co. v. West Orange Redeveloping Agency*, 125 N.J.Super. 440, 311 A.2d 390 (1973) (two minutes late preceded by a telephone call from the bidder stating he would be a few minutes late because of inclement weather); *Gostovich v. City of West Richland*, 75 Wash.2d 583, 452 P.2d 737 (1969) (three *days* late due to mail mixup).

53. Occasionally claims for relief are based upon a sub-bidder's refusal to contract at the price proposed or for an error of judgment relating to performance cost.

54. *Steinmeyer v. Schroeppel*, 226 Ill. 9, 80 N.E. 564 (1907).

55. 21 Ill.App.3d 527, 315 N.E.2d 565 (1974).

56. Amounts are approximations.

conclusion that it is not the fact that the awarding authorities *should* have known of the mistake, but the unfairness of holding a bidder who has made an honest mistake when the next bid can still be accepted.[57]

Some jurisdictions deny relief.[58] These jurisdictions emphasize the need to protect the process from possible favoritism that may accompany the power to allow withdrawal. Such decisions reflect skepticism that the fact-finding process can determine whether honest mistakes have been made. Even in these jurisdictions, relief can be granted if the facts appear to make it inequitable to hold the bidder to its bid.[59]

Most jurisdictions will relieve the bidder if the mistake is clerical and involves a substantial portion of the total bid or a large amount of money and if the owner has not relied to its detriment on the mistaken bid.[60] An illustrative case is reproduced here.

ELSINORE UNION ELEMENTARY SCHOOL DISTRICT v. KASTORFF

Supreme Court of California, 1960
54 Cal.2d 380, 353 P.2d 713, 6 Cal.Rptr. 1

SCHAUER, Justice.

[Ed. note: Footnotes omitted.] Defendants, who are a building contractor and his surety, appeal from an adverse judgment in this action by plaintiff school district to recover damages allegedly resulting when defendant Kastorff, the contractor, refused to execute a building contract pursuant to his previously submitted bid to make certain additions to plaintiff's school buildings. We have concluded that because of an honest clerical error in the bid and defendant's subsequent prompt rescission he was not obliged to execute the contract, and that the judgment should therefore be reversed.

Pursuant to plaintiff's call for bids, defendant Kastorff secured a copy of the plans and specifications of the proposed additions to plaintiff's school buildings and proceeded to prepare a bid to be submitted by the deadline hour of 8 p.m., August 12, 1952, at Elsinore, California. Kastorff testified that in preparing his bid he employed worksheets upon which he entered bids of various subcontractors for such portions of the work as they were to do, and that to reach the final total of his own bid for the work he

---

57. Cases are collected in Annot. 2 A.L.R.4th 991 (1980). See Jones, *The Law of Mistaken Bids*, 48 U.Cin.L.Rev. 43 (1979).

58. *Anco Construction Co. v. City of Wichita*, 233 Kan. 132, 660 P.2d 560 (1983); *Nelson Inc. of Wisconsin v. Sewerage Commission*, 72 Wis.2d 400, 241 N.W.2d 390 (1976) (despite statute). Care must be taken to distinguish cases that refuse relief, because the elements are *not* established from those that hold that no relief *can* be granted in any case.

59. Pennsylvania holds that a bidder cannot revoke after bid opening and must forfeit the bond. But to forfeit the bond, the bidder must give *formal* notice of withdrawal of the bid or actually refuse to execute the contract documents presented. In *Travelers Indemnity Co. v. Susquehanna County Commissioners*, 17 Pa.Commw. 209, 331 A.2d 918 (1975), the court refused to forfeit the bid bond because the bidder had not actually withdrawn his bid but had simply requested to do so and the awarding authority had failed to present the bidder with contract papers for execution. On this technicality, the bidder obtained relief.

Likewise, a federal district court, though stating that the bidder probably should have been relieved, felt precluded because Arkansas law denied relief because the contractor was negligent. Yet the court granted a large recovery to the plaintiff contractor on the grounds that the owner's termination was not made in good faith. *Paul Hardeman, Inc. v. Arkansas Power & Light Co.*, 380 F.Supp. 298 (E.D.Ark.1974) (a thorough discussion).

60. Regulations allowing relief in federal contracts are found in 32 C.F.R. §§ 2–406.2 and 2–406.3.

carried into the right-hand column of the worksheets the amounts of the respective sub bids which he intended to accept and then added those amounts to the cost of the work which he would do himself rather than through a subcontractor; that there is "a custom among subcontractors, in bidding on jobs such as this, to delay giving . . . their bids until the very last moment"; that the first sub bid for plumbing was in the amount of $9,285 and he had received it "the afternoon of the bid-opening," but later that afternoon when "the time was drawing close for me to get by (sic) bids together and get over to Elsinore" (from his home in San Juan Capistrano) he received a $6,500 bid for the plumbing. Erroneously thinking he had entered the $9,285 plumbing bid in his total column and had included that sum in his total bid and realizing that the second plumbing bid was nearly $3,000 less than the first, Kastorff then deducted $3,000 from the total amount of his bid and entered the resulting total of $89,994 on the bid form as his bid for the school construction. Thus the total included no allowance whatsoever for the plumbing work.

Kastorff then proceeded to Elsinore and deposited his bid with plaintiff. When the bids were opened shortly after 8 p.m. that evening, it was discovered that of the five bids submitted that of Kastorff was some $11,306 less than the next lowest bid. The school superintendent and the four school board members present thereupon asked Kastorff whether he was sure his figures were correct, Kastorff stepped out into the hall to check with the person who had assisted in doing the clerical work on the bid, and a few minutes later returned and stated that the figures were correct. He testified that he did not have his worksheets or other papers with him to check against at the time. The board thereupon, on August 12, 1952, voted to award Kastorff the contract.[61]

The next morning Kastorff checked his worksheets and promptly discovered his error. He immediately drove to the Los Angeles office of the firm of architects which had prepared the plans and specifications for plaintiff, and there saw Mr. Rendon. Mr. Rendon testified that Kastorff "had his maps and estimate worksheets of the project, and indicated to me that he had failed to carry across the amount of dollars for the plumbing work. It was on the sheet, but not in the total sheet. We examined that evidence, and in our opinion we felt that he had made a clerical error in compiling his bill. . . . In other words, he had put down a figure, but didn't carry it out to the 'total' column when he totaled his column to make up his bid. . . . He exhibited . . . at that time . . . his worksheets from which he had made up his bid." That same morning (August 13) Rendon telephoned the school superintendent and informed him of the error and of its

61. [Ed. note: Should the "snap-up" concept be used here? When the city "accepted" it did not know of Kastorff's mistake. Evidently the court concluded that the contract was not made until the formal contract was signed. See Section 22.04(G).]

nature and that Kastorff asked to be released from his bid. On
August 14 Kastorff wrote a letter to the school board explaining
his error and again requesting that he be permitted to withdraw
his bid. On August 15, after receiving Kastorff's letter, the board
held a special meeting and voted not to grant his request. Thereafter, on August 28, *written notification* was given to Kastorff of
award of the contract to him. Subsequently plaintiff submitted to
Kastorff a contract to be signed in accordance with his bid, and on
September 8, 1952, Kastorff returned the contract to plaintiff with
a letter again explaining his error and asking the board to recon-
sider his request for withdrawal of his bid.

Plaintiff thereafter received additional bids to do the subject
construction; let the contract to the lowest bidder, in the amount
of $102,900; and brought this action seeking to recover from
Kastorff the $12,906 difference between that amount and the
amount Kastorff had bid. Recovery of $4,499.60 is also sought
against Kastorff's surety under the terms of the bond posted with
his bid.

. . . Judgment was given for plaintiff in the amounts sought,
and this appeal by defendants followed.

In reliance upon M.F. Kemper Construction Co. v. City of Los
Angeles (1951) and Lemoge Electric v. County of San Mateo
(1956), . . . defendants urged that where, as defendants claim is
the situation here, a contractor makes a clerical error in computing a bid on a public work he is entitled to rescind.

In the Kemper case one item on a worksheet in the amount of
$301,769 was inadvertently omitted by the contractor from the
final tabulation sheet and was overlooked in computing the total
amount of a bid to do certain construction work for the defendant
city. The error was caused by the fact that the men preparing the
bid were exhausted after working long hours under pressure.
When the bids were opened it was found that plaintiff's bid was
$780,305, and the next lowest bid was $1,049,592. Plaintiff discovered its error several hours later and immediately notified a member of defendant's board of public works of its mistake in omitting
one item while preparing the final accumulation of figures for its
bid. Two days later it explained its mistake to the board and
withdrew its bid. A few days later it submitted to the board
evidence which showed the unintentional omission of the $301,769
item. The board nevertheless passed a resolution accepting plaintiff's erroneous bid of $780,305, and plaintiff refused to enter into
a written contract at that figure. The board then awarded the
contract to the next lowest bidder, the city demanded forfeiture of
plaintiff's bid bond, and plaintiff brought action to cancel its bid
and obtain discharge of the bond. The trial court found that the
bid had been submitted as the result of an excusable and honest
mistake of a material and fundamental character, that plaintiff
company had not been negligent in preparing the proposal, that it
had acted promptly to notify the board of the mistake and to

rescind the bid, and that the board had accepted the bid with knowledge of the error. The court further found and concluded that it would be unconscionable to require the company to perform for the amount of the bid, that no intervening rights had accrued, and that the city had suffered no damage or prejudice.

On appeal by the city this court affirmed, stating the following applicable rules . . . :

"[1] Once opened and declared, the company's bid was in the nature of an irrevocable option, a contract right of which the city could not be deprived without its consent unless the requirements for rescission were satisfied. [Citations.] . . . [2] . . . the city had actual notice of the error in the estimates before it attempted to accept the bid, and knowledge by one party that the other is acting under mistake is treated as equivalent to mutual mistake for purposes of rescission. [Citations.] [3] Relief from mistaken bids is consistently allowed where one party knows or has reason to know of the other's error and the requirements for rescission are fulfilled. [Citations.]

"[4] Rescission may be had for mistake of fact if the mistake is material to the contract and was not the result of neglect of a legal duty, if enforcement of the contract as made would be unconscionable, and if the other party can be placed in statu quo. [Citations.] In addition, the party seeking relief must give prompt notice of his election to rescind and must restore or offer to restore to the other party everything of value which he has received under the contract. [Citations.]

"[5] Omission of the $301,769 item from the company's bid was, of course, a material mistake. . . . [E]ven if we assume that the error was due to some carelessness, it does not follow that the company is without remedy. Civil Code section 1577, which defines mistake of facts for which relief may be allowed, describes it as one not caused by 'the neglect of a legal duty' on the part of the person making the mistake. [6] It has been recognized numerous times that not all carelessness constitutes 'neglect of legal duty' within the meaning of the section. [Citations.] On facts very similar to those in the present case, courts of other jurisdictions have stated that there was no culpable negligence and have granted relief from erroneous bids. [Citations.] [7] The type of error here involved is one which will sometimes occur in the conduct of reasonable and cautious businessmen, and, under all the circumstances, we cannot say as a matter of law that it constituted a neglect of legal duty such as would bar the right to equitable relief.

"[8] The evidence clearly supports the conclusion that it would be unconscionable to hold the company to its bid at the mistaken figure. The city had knowledge before the bid was accepted that the company had made a clerical error which resulted in the omission of an item amounting to nearly one third of the amount intended to be bid, and, under all the circumstances, it appears that it would be unjust and unfair to permit the city to take advantage of the company's mistake. [9, 10] There is no reason for denying relief on the ground that the city cannot be restored to status quo. It had ample time in which to award the contract without readvertising, the contract was actually awarded to the next lowest bidder, and the city will not be heard to complain that it cannot be placed in statu quo because it will not have the benefit of an inequitable bargain. [Citations.] [11] Finally, the company gave notice promptly upon discovering the facts entitling it to rescind, and no offer of restoration was necessary because it had received nothing of value which it could restore. [Citation.] We are satisfied that all the requirements for rescission have been met."

In the *Lemoge* case . . . the facts were similar to those in Kemper, except that plaintiff Lemoge did not attempt to rescind but instead, after discovering and informing defendant of inadvertent clerical error in the bid, entered into a formal contract with defendant on the terms specified in the erroneous bid, performed the required work, and then sued for reformation. Although this court affirmed the trial court's determination that plaintiff was not, under the circumstances, entitled to have the contract reformed, we also reaffirmed the rule that "Once opened and declared, plaintiff's bid was in the nature of an irrevocable option, a contract right of which defendant could not be deprived without its consent unless the requirements for rescission were satisfied. [Citation.] Plaintiff then had the right to rescind, and it could have done so without incurring any liability on its bond." . . .

The rules stated in the *Kemper* and *Lemoge* cases . . . appear to entitle defendant to relief here, ...

Further, we are persuaded that the trial court's view . . . that "Kastorff had ample time and opportunity after receiving his last subcontractor's bid" to complete and check his final bid, does not convict Kastorff of that "neglect of legal duty" which would preclude his being relieved from the inadvertent clerical error of omitting from his bid the cost of the plumbing. . . . Neither should he be denied relief from an unfair, inequitable, and unintended bargain simply because, in response to inquiry from the board when his bid was discovered to be much the lowest submitted, he informed the board, after checking with his clerical assistant, that the bid was correct. He did not have his worksheets present to inspect at that time, he did thereafter inspect them at what would appear to have been the earliest practicable moment, and thereupon promptly notified plaintiff and rescinded his bid. Further . . . Kastorff's bid agreement, as provided by plaintiff's own bid form, was to execute a formal written contract only after receiving written notification of acceptance of his bid, and such notice was not given to him until some two weeks following his rescission.

If the situations of the parties were reversed and plaintiff and Kastorff had even executed a formal written contract (by contrast with the preliminary bid offer and acceptance) calling for a fixed sum payment to Kastorff large enough to include a reasonable charge for plumbing but inadvertently through the *district's* clerical error omitting a mutually intended provision requiring Kastorff to furnish and install plumbing, we have no doubt but that the district would demand and expect reformation or rescission. In the case before us the district expected Kastorff to furnish and install plumbing; surely it must also have understood that he intended to, and that his bid did, include a charge for such plumbing. The omission of any such charge was as unexpected by the board as it was unintended by Kastorff. Under the circumstances the "bargain" for which the board presses (which action we, of course, assume to be impelled by advice of counsel and a strict concept of official duty) appears too sharp for law and equity to sustain.

Plaintiff suggests that in any event the amount of the plumbing bid omitted from the total was immaterial. The bid as submitted was in the sum of $89,994, and whether the sum for the omitted plumbing was $6,500 or $9,285 (the two sub bids), the omission of such a sum is plainly material to the total. In Lemoge . . . the error which it was declared would have entitled plaintiff to rescind was the listing of the cost of certain materials as $104.52, rather than $10,452, in a total bid of $172,421. Thus the percentage of error here was larger than in Lemoge, and was plainly material.

The judgment is reversed.

State legislation increasingly regulates attempts by *public* bidders to withdraw bids because of mistake after bid opening. In 1945 California adopted legislation that allowed a bidder to be relieved from a bid if he established to the satisfaction of the court that:

1. A mistake was made.
2. He gave the department written notice within five days after the opening of the bids of the mistake, specifying in the notice in detail how the mistake occurred.

3. The mistake made the bid materially different than he intended it to be.

4. The mistake was made in filling out the bid and not due to error in judgment or to carelessness in inspecting the site of the work or in reading the plans or specifications.[62]

---

62. Cal.Stats.1945, ch. 118, p. 501, now West's Ann.Cal.Pub.Contract Code § 10202. A bidder relieved cannot bid on the same job. For an interpretation of this provision see *Colombo Construction Co. v. Panama Union School District*, 136 Cal.App.3d 868, 186 Cal.Rptr. 463 (1982) (bidder barred).

This statute as originally enacted applied only to certain state public agencies.

In 1982 California incorporated this approach in its Public Contract Code, which applies to all public entities—state, county, and local as well as other public entities.[63]

California held the exclusive method of a public bidder obtaining relief is under the state statute, and the flexible "fairness" rationale seen in the *Kastorff* case has been displaced by the standards set forth in the statute.[64] The statute also reflects a more mechanical method of dealing with the problem, particularly in limiting relief where notice is given within five days after bid opening and specifying in detail how the mistake occurred. There is no requirement under the statute that holding the bidder would be unconscionable or that the city have knowledge before the bid was accepted that there had been a clerical mistake making it unjust and unfair for the city to take advantage of the bidder's error. Nor does the statute require any evaluation of the degree of negligence.

The evolution of legal rules dealing with bidding mistakes reflects a slow process starting first in those transactions where it is determined that the awarding authority knew or should have known of the mistake but limiting mistake to clerical errors such as the *Kastorff* case.[65] Yet suppose a mistake has been made that is not a *simple* clerical arithmetic error. Recent cases have differed as to whether relief is confined to *clerical* arithmetic mistakes. It may be useful to look at cases refusing to draw this distinction.

For example, *Balaban-Gordon Co. v. Brighton Sewer District Number 2*[66] in-volved separate general construction and specialty trade contracts. By mistake the bidder included equipment in its *plumbing* bid, which should have been in its *general* construction bid. It was awarded only the *general* construction contract. It was relieved from its bid. The court refused to draw a line between clerical and judgmental errors, noting the mistake to have been objectively discoverable.

In an astounding case, *White v. Berenda Mesa Water District*,[67] the court granted relief when the mistake related to the amount of hard rock that would have to be excavated. The bids ranged from $427,000 to $721,000, and the bid next lowest to that of White was $494,000. When the bids were opened, White reviewed his bid. He learned that the soil report upon which he relied was in error but that the project specifications were accurate. Despite the fact that White *knew* that soil reports are prepared before detailed specifications and that the specifications and not the soil report control, and despite White's having a copy of the specifications that indicated that the owner did not warrant the accuracy of the soil report, White was granted relief. The court felt that a line between clerical and judgmental errors was too fine a line to draw.

The *White* case is instructive. First, there would have been no problem had the work been bid on a unit price basis. Second, it seems as if there was no justification for the error. Third, the variations in bids were not so gross as to infer that the District should have realized that White made a mistake. Fourth, were the *White* case governed by the California statute described earlier, White

---

63. See West's Ann.Cal.Pub.Contract Code § 5103.

64. *A & A Electric, Inc. v. City of King*, 54 Cal.App.3d 457, 126 Cal.Rptr. 585 (1976).

65. See also *Osberg Construction Co. v. City of The Dalles*, 300 F.Supp. 442 (D.Or.1969); *Missouri State Highway Commission v. Hensel Phelps Construction Co.*, 634 S.W.2d 168 (Mo.1982).

66. 41 A.D.2d 246, 342 N.Y.S.2d 435 (1973). See also *Wil-Fred's Inc. v. Metropolitan Sanitary District*, 57 Ill.App. 3d 16, 14 Ill.Dec. 667, 372 N.E.2d 946 (1978) (relief despite error as to how work was to be done).

67. 7 Cal.App.3d 894, 87 Cal.Rptr. 338 (1970).

would have been barred from relief because the error was one of judgment.

Courts refusing to bar relief for judgment errors seem persuaded that if proof *exists* of the mistake and the owner can simply accept the next low bid, it has not been harmed. To these courts it seems unconscionable to force the bidder to pay the difference between its bid and the next low bid, often a substantial amount. Yet another method of relieving from large forfeitures exists, though it is often ignored. It looks at the *security* bidders are usually requested to deposit, such as a certified check, a cashier's check, or bid bond, based upon a designated percentage of the bid. Suppose a bidder who has posted security seeks to withdraw its bid based upon an asserted clerical or mathematical error of the type discussed in this section. Usually one of the reasons to allow withdrawal is a great variation between the bid submitted and the bid intended to be submitted or the other bids. But suppose the owner sought only to keep the deposit rather than attempt to hold the bidder to its bid and seek damages based upon the difference between the low bid and next low bid?

The owner would contend that the bidder has simply breached the obligation to *enter into* the construction contract and not the construction contract itself. The stipulated damage for such a breach is the amount of the deposit. In such a case, the owner would claim the right to retain the deposit but not damages based upon the difference between the bid and the next low bidder.

This would seem a fair solution to the vexatious bidding mistake cases. The contractor's exposure would be the deposit. However, courts have not drawn this distinction and have generally held that the contractor is relieved from its performance *entirely* if the requirements for relief are established. In such a case, it is entitled to recover the deposit. While the deposit may be only 5% to 10% of the bid, in large jobs this amount is substantial. Where the facts are sufficient to allow the mistake doctrine to be applied, most courts would prefer to relieve the bidder entirely.

Generally courts have been reluctant to allow a bidder to *correct* a mistake even where the facts would have permitted the withdrawal.[68] This is particularly true in public awards and the emphasis upon avoiding favoritism and obtaining the work at the lowest price. However, a federal court held that a bidder would be entitled to correct his bid even after the work had been completed in a private project.[69]

Relief after award, whether withdrawal or correction, is not permitted in public contracts. However, in private contracts this may be possible, especially where the bid opening is *not* done in public. In such cases, the bidders do not know what other bidders have proposed, and they may not find out that there was great discrepancy until problems develop. Two cases[70] seemed willing to give relief, including, in one case, correction,[71] but the facts in each case did not persuade the court that relief was proper.

Can an *owner reduce* the contract price because of a claimed mistake made by the

---

68. *Lemoge Electric v. County of San Mateo,* 46 Cal.2d 659, 297 P.2d 638 (1956); *Anco Construction Co. Ltd. v. City of Wichita,* supra note 58. But it was allowed where the U.S. "overreached" and rescission was no longer possible in *Bromley Contracting Co. v. United States,* 219 Ct.Cl. 517, 596 F.2d 448 (1979).

69. *Nat Harrison Associates, Inc. v. Louisville Gas & Electric Co.,* 512 F.2d 511 (6th Cir.1975). But the appellate court reversed a finding for the plaintiff contractor since it was not convinced that there had been a unilateral mistake or inequitable conduct by the owner in accepting the bid. The bid was low but not much lower than other bids and the owner's estimate.

70. *Paul Hardeman, Inc. v. Arkansas Power & Light Co.,* supra note 15; *Nat Harrison Associates, Inc. v. Louisville Gas & Electric Co.,* supra note 69.

71. *Nat Harrison Associates, Inc. v. Louisville Gas & Electric Co.,* supra note 69.

bidders? In *Edward D. Lord, Inc. v. Munic-ipal Utilities Authority*,[72] two sub-bidders agreed to cross-excessive bids with each bidding to prime bidders at agreed excessive prices. Ultimately this was discovered, and the bidder to whom the contract had been awarded cancelled the subcontract and rebid the job for much less than the cross-excessive bids.

New Jersey held that the awarding authority could reform the contract because of mutual mistake. Both bidder and awarding authority assumed that the bidding was fair and not rigged and the price agreed upon reflected this basic assumption. However, the assumption turned out to be incorrect because of the collusive conduct of the subcontractors. Since public funds were involved, the court concluded that the awarding authority would be granted reformation reducing the contract price to take into account the excessive cross bids.

Suppose the Invitation to Bidders states that bidders will not be released for errors. One court interpreted this language to cover errors of judgment and not clerical errors.[73] Another, though, refused to employ this interpretation technique to give relief for a clerical error.[74] The former approach is preferable. If the Invitation *clearly* covers clerical errors, the risk of even clerical errors should be placed on the contractor. But it is more likely that such language will not tie the hands of courts to grant relief for such mistakes if the enforcement of the mistaken bid would be unconscionable.

(F) BID DEPOSIT

The bidders are usually requested to submit deposits with their proposals. Is the deposit a security deposit out of which the owner can take whatever damages it has incurred?

Does the payment limit the damages to a specified figure that still obligates the owner to prove damages up to that figure? Is the deposit submitted in an attempt to set damages in advance by agreement of the parties?

Suppose Bidder A submits a proposal for one million dollars and the invitation to bidders requires it to submit a bid bond for 5% of its bid. A bid bond for $50,000 is deposited. The bids are opened and A is low. The next lowest bidder has submitted a bid of $1.1 million. A is offered the contract but without any legal justification declines to enter into it. The contract is offered to the next low bidder for $1,100,000. Is the owner entitled to $100,000 damages, with $50,000 being a security deposit out of which it can assure itself that it will be able to collect at least a part of its damages? Or is the owner limited to $50,000 because this is what the parties have agreed will be the actual damage amount whether the actual damages are higher or lower?

Suppose the next low bidder had been $1,025,000 instead of $1.1 million. In such a case, can the owner keep the entire $50,000 or be limited to $25,000?

To a certain extent, the parties are free by their contract to determine whether the amount submitted or the bid bond deposited liquidates (agrees in advance on the amount) damages, is a security deposit, or is a limitation of liability. A security deposit is an amount of money deposited with one party, out of which the latter can satisfy whatever damages to which it is entitled. If the 5% deposit is merely a security deposit, the owner can retain this amount and sue for any balance to which it is entitled or it must return the excess of the deposit over damages.

If the deposit is a valid liquidated dam-

---

72. 133 N.J.Super. 503, 337 A.2d 621 (1975).

73. *M.F. Kemper Construction Co. v. City of Los Angeles*, 37 Cal.2d 696, 235 P.2d 7 (1951).

74. *City of Newport News v. Doyle & Russell, Inc.*, 211 Va. 603, 179 S.E.2d 493 (1971).

ages clause, the parties have agreed in advance that whatever the amount of the actual damages, the breaching party will pay the amount stipulated in the clause. The owner could retain the $50,000 whether the damages were $200,000 or one dollar.

A valid liquidated damages clause requires that it be difficult to ascertain the damages at the time the contract is made and that the amount agreed be a genuine preestimate of the potential damages.

An amount disproportionate to the actual or anticipated damage chosen merely to coerce performance is a penalty and unenforceable. For example, if the deposit were 50% of the bid and if it were most unlikely that there would be damages approaching this amount, the clause would be a penalty and unenforceable. The owner would be entitled only to actual damages without regard for the amount of the deposit, and it would have to refund the excess of the deposit over its actual damages.

Legislation, state or local, may give the public agency damages based upon the difference between the low bid and the next lowest bid if the low bidder does not enter into the contract awarded to it. In such a case, the deposit is for security and is not an attempt to liquidate damages. A liquidated damages clause *establishes* the damages.

Legislation sometimes provides that the public agency may retain the amount deposited but only to the extent of the difference between the defaulting bidder's bid and the amount for which the contract is ultimately awarded. For example, suppose the amount deposited was $50,000 or 5% of the one-million-dollar bid and the next bidder was awarded the contract at $1,025,000. In such a case, the owner would be entitled

to retain $25,000. Such statutes set up a liquidation of damages that will apply only if actual damages are greater than the amount deposited. If actual damages are less, the deposit is simply security.

Is it desirable to liquidate damages? From the owner's standpoint, the chances of collecting an amount in excess of the deposit from the contractor are remote. On the other hand, the owner would like to retain the amount deposited without having to show actual damages. For that reason, it is preferable to liquidate damages rather than use the deposit solely as security. The pure security deposit does allow the owner to seek to recover an amount beyond the deposit. This would occur if the discrepancy between defaulting bidder's bid and the next bidder is *more* than the deposit. In such a case there is a strong likelihood of a mistake that would permit the bidder to withdraw his bid. Many contractors would not be able to satisfy a large court judgment. This is a reason for providing the security deposit. Liquidated damages protect the contractor from the risk of excessive damages and guarantee the owner a reasonable amount of collectable damages.

A properly drafted liquidated damages clause is likely to be enforced if created by legislation that specifically permits the awarding authority to forfeit the deposit.[75] In the absence of such legislation, courts divide. Some enforce such a clause; [76] others do not.[77]

The uncertainty of the amount of damages must exist at the time the contract is made and at the time of the deposit the amount of damages is not yet known. Some courts ignore this and seem to look at whether the amount of damages can be easily ascertained at the time of breach.

---

75. *A & A Electric, Inc. v. City of King*, 54 Cal.App.3d 457, 126 Cal.Rptr. 585 (1976). *Petrovich v. City of Arcadia*, 36 Cal.2d 78, 222 P.2d 231 (1950), decided earlier did not seem willing to enforce such a clause.

76. *Bellefonte Borough Authority v. Gateway Equipment & Supply Co.*, 442 Pa. 492, 277 A.2d 347 (1971).

77. *Petrovich v. City of Arcadia*, supra note 75. *Ogden Development Corp. v. Federal Insurance Co.*, 508 F.2d 583 (2d Cir.1974), held that the clause forfeiting the deposit was a penalty.

These courts seem unwilling to forfeit an amount in excess of actual damages. This approach does not take note of the "long haul" aspects of denominating the forfeiture clause as stipulated damages. Over the long haul of many competitive bids, the losses to the public agency probably average per competition the amount stipulated in each competitive bid. Though the long haul may seem unfair to the *particular* bidder who must lose more than what it *appears* the agency has been damaged in *this* competitive bid, the particular bidder is relieved from any risk *beyond* the deposit amount.

There are administrative costs when the next low bidder is selected. Admittedly, that loss often seems much less than the amount forfeited. But again, the parties *expect* the amount to be deposited to be forfeited, and the bidder is relieved from the risk of loss beyond the deposit. As long as the amount selected is reasonable, the forfeiture clause should be considered a valid liquidated damages clause.

The clause should be *clearly* enforceable if it is necessary to rebid the entire project. Rebidding entails substantial additional administrative expense.

Suppose both low bidder *and* the next low bidder unjustifiably refuse to enter into the contract? Can the owner retain the deposit by both bidders? While it may seem unfair to retain both bidders' deposits, it is not logically indefensible. Each bidder has breached, and each has been to some degree relieved from the risk of excessive damages by the use of an agreed damage provision. However, goodwill and the avoidance of litigation may necessitate some solution such as retaining one-half of each, rather than trying to retain both deposits.[78]

(G) THE FORMAL CONTRACT

The culmination of a successful competitive bidding process is the award by the awarding authority to the successful bidder. Usually a formal contract is forwarded or given to the successful bidder for its execution.

Suppose the award is made *before* the expiration of the period during which the bid is irrevocable but the formal contract is *not* executed within that period. Although one case discussed earlier held that a tentative acceptance within the period was not a sufficient acceptance,[79] two Wisconsin cases held that the validity of the contract did not require execution of the contract. In one,[80] the awarding authority had voted to accept the bid in the presence of the bidder. In the other,[81] approval by a federal regulatory agency that conditioned the award was not received until the morning of the final day of the period during which the bid was irrevocable. On that morning, a Friday, the engineer notified the contractor that the formal contracts would be in the mail that day. The following day, Saturday, the contractor wrote that it was withdrawing its bid as the forty-five-day period during which its bid was irrevocable had expired. The contract was received on the following Monday. The court could have held that the acceptance took place on the forty-fifth day when the contracts were put in the mail.[82] The court held in favor of the awarding authority. The contract had been formed and the formal contract merely memorialized the agreement that had been already made.

78. West's Ann.Cal.Pub.Contract Code §§ 10181–10182 permit forfeiture of the security of lowest, second lowest, *and* third lowest if none will enter into the contract.

79. *Hennepin Public Water District v. Petersen Construction Co.,* supra note 36, discussed at § 22.04(D).

80. *Nelson Inc. of Wisconsin v. Sewerage Commission,* supra note 58.

81. *City of Merrill v. Wentzel Brothers, Inc.,* 88 Wis.2d 676, 277 N.W.2d 799 (1979).

82. Refer to Section 5.06(C).

Suppose the formal contract is not consistent with earlier communications exchanged between the successful bidder and the awarding authority. The Court of Claims held that the formal contract, while typically superseding all previous negotiations, documents, etc., is merely a reduction to form of the actual agreement made by the advertisement, bid, and its acceptance.[83] This is another recognition of the formal contract often being simply a memorial of the agreement that has been made. However, the date of the formal contract *may* set into motion any time commitment of the contractor.

(H) BIDDING DOCUMENTS

The culmination of the complex competitive process usually is assent by both parties to the construction contract. This process has generated a series of many, long, and complex writings, such as the Invitation to Bid, Information to Bidders, and the Bid Proposal. Included among these are other material, such as plans, specifications, general conditions, supplemental conditions, addenda, and the agreement forms. Chapters 23 and 24 deal with the problems generated by this wealth of written material. But what about the material generated by the bid process itself?

AIA Doc. A201, ¶ 1.1.1, seeks to deny them *any* legal effect by excluding them as contract documents. They have been superseded by execution of the construction contract. Although it is important to avoid contradiction in the voluminous contract documents, a provision can state that in the event of conflict, the construction contract takes precedence over the bidding documents. Although implementation of provisions of this type is not as simple as it appears, the solution AIA has selected—*excluding* bidding documents—is undesirable.

Most participants in the process believe that the bidding documents do have legal efficacy. Information is given in the bidding documents that is relied upon by the bidders. For example, soil reports are frequently included in the Information to Bidders. Under AIA Doc. A201, ¶ 12.2, unless this information is found in the specifications, it cannot be the basis of any claim by the contractor for an equitable adjustment for subsurface conditions different from those usually encountered or disclosed by the contract documents. Similarly, AIA Doc. A701 (Instruction to Bidders) imposes many requirements on the contractor, such as liquidating damages for failure to enter into the construction contract and requiring that a particular type of surety bond be used. If the award is made and the construction contract documents signed *without* these provisions, if taken literally the A201 would discharge any obligations the contractor may have that are expressed in the Instruction to Bidders. There may be factual material in the bidding instructions or information that the owner may wish to point to at some later date if a claim has been made.[84]

Attempts to deny any legal effectiveness to the bidding documents may not be in the owner's best interest and may frustrate the reasonable expectations of both parties.

(I) JUDICIAL REVIEW OF AGENCY ACTION

Public procurement encompasses a vast number of social and economic goals. The result has been a complicated set of statutes and regulations with the increasing likelihood of irregularities.

To challenge an agency decision, the

---

83. *Dana Corp. v. United States*, 200 Ct.Cl. 200, 470 F.2d 1032 (1972).

84. *D.A. Collins Construction Co. v. New York*, 88 A.D.2d 698, 451 N.Y.S.2d 314 (1982) (information avoided contractor delay claim).

challenger must have "standing." [85] Many states allow taxpayer suits to challenge official actions sometimes done in the name of the disappointed bidder or someone acting on its behalf. It may also be done by a taxpayer who simply believes that the agency action will cost the taxpayers money.[86] Massachusetts grants a disappointed bidder the right to challenge the agency action.[87] Pennsylvania allows *only* the low bidder to challenge the agency decision.[88] Much depends upon state statute, but at the state and local level, standing does not seem to have been the problem it has been at federal level.

Before 1970 disappointed bidders were not granted standing to challenge federal agency decisions based upon *Perkins v. Lukens Steel Co.,*[89] which held that the government has the sole right to choose with whom and under what terms it will contract. A disappointed bidder has only a privilege and not a right to do business with the government. In addition to this wooden logic, the court was concerned that judicial interference with government procurement would cause delay and involve the courts in decision making beyond their competence.

But in *Scanwell Laboratories, Inc. v. Shaffer,*[90] a federal appeals court cited a statute passed after *Perkins v. Lukens Steel Co.* and held that a disappointed bidder had standing to seek a judicial order stopping an allegedly invalid procurement award.

The *Scanwell* case seemed to open some federal courthouse doors to disappointed bidders seeking federal contracts. This would involve courts in difficult procurement problems. A year later the same court had second thoughts. It held that judicial interference with a procurement award required a demonstration by the challenger that there was no "rational basis" for the award. In addition, trial courts were given broad discretion to refuse to interfere with the procurement.[91] It is easier for a disappointed bidder to recover damages that are usually limited to bidding expenses *if* it can establish a defect in procurement procedures.[92]

(J) ILLEGAL CONTRACTS

The imposing number of legal controls on public contracts, the interest generated by public projects, staff inadequacies in small public entities, all create a substantial risk that an award may be made illegally. Suppose the party to whom the award had been made partly or fully performed.

Clearly, recovery cannot be made under the illegal *contract.* But two other issues can arise. First, and most frequently, can the contractor who has performed under an illegal contract recover for the work it has performed based upon restitution? Second,

---

85. Standing was also discussed in Section 11.15.

86. But in *Lynch v. Devine,* 45 Ill.App.3d 743, 359 N.E.2d 1137 (1977), plaintiff had no standing as bidder *or* taxpayer.

87. *Grant Construction Co. of R.I. v. City of New Bedford,* 1 Mass.App. 843, 301 N.E.2d 463 (1973).

88. *Pullman Inc. v. Volpe,* 337 F.Supp. 432 (E.D.Pa.1971).

89. 310 U.S. 113, 60 S.Ct. 869, 84 L.Ed. 1108 (1940).

90. 424 F.2d 859 (D.C.Cir.1970).
Not all federal circuit courts give the disappointed bidder standing.

91. *M. Steinthal & Co. v. Seamans,* 455 F.2d 1289 (D.C.Cir.1971).

92. *Keco Industries, Inc. v. U.S.,* 428 F.2d 1233 (Ct.Cl.1970). Lost profits were not allowed in *Armstrong & Armstrong, Inc. v. United States,* 514 F.2d 402 (9th Cir.1975); *Swinerton & Walberg Co. v. City of Inglewood,* 40 Cal. App.3d 98, 114 Cal.Rptr. 834 (1974); *Paul Sardella Construction Co. v. Braintree Housing Authority,* 329 N.E.2d 762 (Mass.App.1975).

can any payments that have been made to a contractor be recovered by the awarding authority?

These issues of unjust enrichment, like attempts by bidders to withdraw their bids, generate sharp differences of opinion. Those who would deny recovery or even require repayment stress the importance of an honest competitive bidding system and the need to protect public funds. Although a recognition exists of the occasional unfairness of denying a contractor recovery for work it has performed because of technical irregularities, those who take a hard line cite the difficulty of making these judgments and the importance of not allowing any loopholes in the laws regulating public contracts.

Those who take a softer approach are willing to concede that there should not be recovery or that there should even be repayment where there is venality or corruption, but they draw a distinction between those cases and ones that do not involve serious criminal misconduct. Where corruption is pervasive they would concede that only harsh and unremitting punishment has a chance of deterring such corruption. On the other hand, in some jurisdictions inefficiency is common and corruption rare. In those jurisdictions there should be greater willingness to allow payment to a contractor where the award was not tainted with bad faith, fraud, or corruption.

Generally, illegally awarded contracts cannot be the basis for restitution. But in 1971 restitution *was* allowed in *Blum v. City of Hillsboro.*[93] Immediately after acceptance of the $47,000 contract, council members of the awarding authority asked the contractor if it would be willing to do additional work for a designated price. The contractor stated he would, and the award-ing authority, through its mayor and city council, specified the work to be done and drew up an amendment to the original contract.

The contractor performed the additional work which increased the amount of the contract to $154,000. The awarding authority paid $82,000 but refused to pay the balance of $72,000. The contractor sought to recover the balance, and the awarding authority counterclaimed for the amount it paid in excess of the original contract price. This claim was based upon the failure to follow state law, which required that the additional work be competitively bid.

The court held that recovery of restitution based upon benefit conferred (though a minority view) was justified in this case. According to the court, failure to allow profits would be sufficient deterrence. If recovery is not granted, a claim will very likely be made upon the municipality to use its discretionary power to pay moral claims. The court limited recovery to actual costs including overhead not exceeding actual benefit but denied profit. Nor could recovery exceed the unit cost of the original contract that had been properly awarded.

In 1977 the Minnesota Supreme Court followed the *Blum* case in a well-drilling contract that was illegally awarded because it violated the competitive bidding statute. Whether there had been a violation was a close question, inasmuch as the awarding authority thought there had been a sufficient emergency that granted it an exception from the competitive bidding requirements. Yet because the contractor had not actually found water, the court denied recovery because there had been no benefit to the awarding authority.[94]

Sloppy administration surfaced in *Mc-*

---

93. 49 Wis.2d 667, 183 N.W.2d 47 (1971). Relief was also granted in *Board of Commissioners v. Dedelow, Inc.,* 159 Ind.App. 563, 308 N.E.2d 420 (1974) (need to start work despite challenge and contractor's lack of culpable misconduct).

94. *Layne Minnesota Co. v. Town of Stuntz,* 257 N.W.2d 295 (Minn.1977).

Cuistion v. City of Siloam Springs.[95] The contract was illegally awarded, as it had never been formally authorized and approved by the city council. Additionally the contractor faced a difficult problem of establishing that there had *ever* been a written contract made. The contractor testified he signed the contract in the office of the city attorney and that he either saw executed copies or was told by the city attorney that they had been executed. He had never been furnished an executed copy. All the parties seemed to *assume* that a written contract had been made, but *no one* was ever able to find the contract or a record of it. The court held that the contractor could recover.

Yet other jurisdictions have taken a much harder line. *Manning Engineering, Inc. v. Hudson County Park Commission*[96] involved pervasive corruption in the awarding of contracts in Jersey City, New Jersey. The court not only denied the engineering company recovery for work that it had performed but indicated that the awarding authority would have had a good claim had it sought repayment of funds that had been paid. It cited a New York case[97] that had involved a contractor who had been convicted of conspiring to violate state bribery laws through a kickback system. When the contractor sued for the unpaid balance, the city successfully defended the claim and recovered payments that it had paid.

Gerzof v. Sweeney[98] demonstrated not only the difficulty of fashioning an appropriate remedy but how the "tough" New York court can be persuaded to relax its harsh rules.

The Village of Freeport had advertised for bids for a 3500-watt generator. Enterprise bid $615,000, and Nordberg bid $674,000. After an advisory committee had recommended acceptance of Enterprise's bid, a new village election was held at which a new mayor and two new trustees were elected. Shortly thereafter, Nordberg's *higher* bid was accepted.

Enterprise obtained a court order setting the award aside. The board of trustees then drew up *new* specifications for a 5000-watt generator with the active participation of Nordberg. The specifications were so rigged that only Nordberg could comply. As expected, Nordberg was the only bidder and its bid of $757,000 was accepted. Nordberg installed the generator and was paid.

After a court declared the *second* award invalid, the trial court held that the village should retain the generator and recover the $757,000 from Nordberg. The intermediate appellate court modified that judgment by providing that Nordberg could retake the machine upon posting a bond for $357,000 to secure the village against damages from removal and replacement of equipment. New York's highest court, the Court of Appeals, first emphasized the importance of protecting the public against corruption and collusion between public officials and bidders. In the *normal* case it would make no difference whether the Village was defending a claim brought by Nordberg or was seeking to recover the money paid Nordberg. But this was *not* a normal case. Granting recovery of payments made would cost Nordberg three-quarters of a million dollars, and the Village would have its generator. Motivated by the enormity of the forfeiture, the court awarded a remedy different from that awarded by the trial court *or* the intermediate appellate court. The Court of Appeals stated that the award should have been made to Enterprise. Had

---

95. 268 Ark. 148, 594 S.W.2d 233 (1980).

96. 74 N.J. 113, 376 A.2d 1194 (1977).

97. *S.T. Grand, Inc. v. City of New York,* 32 N.Y.2d 300, 344 N.Y.S.2d 938, 298 N.E.2d 105 (1973).

98. 22 N.Y.2d 297, 292 N.Y.S.2d 640, 239 N.E.2d 521 (1968).

this been done, the Village would have had a 3,500-watt generator for $615,000. The court awarded judgment against Nordberg based upon the difference between the $757,000 paid Nordberg and the $615,000 that the Village would have paid Enterprise. To this was added $37,000, the difference between what it cost the Village to install the Nordberg generator and what it would have cost to install the one offered by Enterprise. In addition, the Village was awarded interest.

Suppose the contract should have been awarded to X but was awarded illegally to Y. Suppose X seeks the profits from Y that Y made on the contract. Although some cases have denied recovery,[99] a federal court decision applying Iowa law employed unjust enrichment to award a bidder who *should* have been awarded the contract

the profit of the contractor who had been awarded the contract improperly.[100]

SECTION 22.05
SUBCONTRACTOR BIDS

The relationship between prime and subcontractor will be discussed in greater detail in Chapter 32. However, one aspect of that relationship should be mentioned briefly in this chapter. A legal problem that has surfaced frequently relates to the right of a prime contractor to hold a subcontractor to its bid after the former has used that bid in computing its own bid and submitting it to the owner. While the cases are by no means unanimous, the clear trend is toward holding the subcontractor's bid irrevocable after it has been used by the prime contractor.[101]

PROBLEMS

1. O decided to construct a building and to obtain competitive bids. He selected five contractors and sent them Invitations to Bid, as well as bidding information. The Invitation stated that bids would be irrevocable until thirty days after bid opening. It also stated that the bid would be awarded to the lowest responsible bidder. Bids were to be submitted no later than 2:00 p.m. on June 10. Each bidder was to accompany the bid with a certified check or bid bond for 10% of the bid. This amount would be forfeited if the bidder were awarded the contract but did not sign it.

   When the bids were opened, they disclosed that four of the bids ranged

   from one million dollars to $1.4 million. The fifth bid by Acme Construction was for $840,000. The owner asked Acme if there had been some mistake. Acme said that he was certain that this was the correct bid. The owner then told Acme that the job was his and that he did not have to evaluate Acme's responsibility because he knew that Acme had an excellent reputation.

   The next day, Acme came in and stated that his bid was underbid by $120,000. There was a computation error of $80,000 caused by his bookkeeper who had not added the columns correctly. Also, the electrical subcontractor had claimed that he left a

---

99. *Savini Construction Co. v. Crooks Brothers Construction Co.*, 540 F.2d 1355 (9th Cir.1974); *Royal Services, Inc. v. Maintenance, Inc.*, 361 F.2d 86 (5th Cir.1966).

100. *Iconco v. Jensen Construction Co.*, 622 F.2d 1291 (8th Cir.1980).

101. The leading case is *Drennan v. Star Paving Co.*, 51 Cal.2d 409, 333 P.2d 757 (1958). See Section 32.02.

$40,000 item out of his bid figures, and he canceled his bid.

a. Would you release Acme?

b. Would you let him "correct" his bid?

c. Would a court permit Acme to cancel? To correct?

d. Can the owner accept the next lowest bid if Acme refuses to perform?

e. If Acme unjustifiably refuses to perform, can the owner retain his damages for the difference in bids or is he limited to the 10% deposit?

2. O wanted to build a residence on a lot she owned. She asked three contractors to give her a fixed bid price on the project. None would do so, because experimental methods of construction were called for and because the architect was considered difficult to get along with. O asked C to make a fixed-price bid. She did not tell C that three other contractors had refused to do so. C gave a written bid of $100,000, which O accepted. Three days later, C found out that the three contractors had refused to bid on the job. He claims O should have told him of this. Does C have a legal defense if he refuses to go through with the contract? Explain.

23

Sources of Construction Contract Rights and Duties: Contract Documents and Legal Rules

SECTION 23.01
CONTRACT DOCUMENTS

(A) BIDDING DOCUMENTS

Bidding documents were discussed in Section 22.03.

(B) BASIC AGREEMENT

The basic agreement culminates competitive bidding or negotiation. As an illustration, AIA Doc. A101 set forth in Appendix B identifies the parties and the architect and contains provisions dealing with the work to be performed, time of commencement and completion, the contract sum, and provisions for progress payments and final payment. In essence the basic agreement sets forth the principal incentives of each contracting party. The owner seeks a project completed on schedule, and the contractor seeks agreed compensation for its performance.

The construction project is a complex undertaking. The basic agreement forms but one part of the total package of construction documents. Frequently these other documents are incorporated by reference in the basic agreement. For example, A101 includes as contract documents the general and supplementary conditions, drawings, specifications, addenda issued prior to execution of the basic agreement, and all modifications issued subsequently. Article 7 provides space to list the documents incorporated by reference.

Frequently contracts incorporate industry standards. The parties can even incorporate a document not yet in existence. For example, in *Randolph Construction Co. v. Kings East Corp.,*[1] the basic agreement incorporated plans that had not yet been completed. But according to the court, if the completed plans are substantially different than anticipated, no contract exists unless the parties agree to the completed plans.

Incorporation by *reference*—a technique for giving legal effectiveness to writings not physically attached to the contract—should be differentiated from *referring* to other writings. Often reference is made to other writings for informational purposes without any intention of making them part of the contract obligations. Classifying a writing referred to in the contract as simply providing information can be a technique to avoid binding a party to a writing of which it was unaware. Although this can occur when an owner signs an A201, it is more of a prob-

---

1. 165 Conn. 269, 334 A.2d 464 (1973).

lem for subcontractors. Frequently reference is made in subcontracts to the prime contract, and there can be serious questions as to whether the prime contract provisions are incorporated into the subcontract as well as to reconciling contrary provisions in prime and subcontracts. For that reason, this topic is discussed ahead.[2]

(C) DRAWINGS (PLANS)

The drawings graphically depict the contractor's obligations. Together with the other contract documents, particularly the specifications, they define and measure the contract obligation.

Drawings are of great importance to the Construction Process. Compliance with them usually relieves the contractor if the project is unsuccessful or does not accord with the owner's expectations.[3] Incomplete drawings and those that are inconsistent with the specifications almost always generate increased construction costs.[4] Defective drawings or drawings that do not fit with the specifications usually result in liability to the design professional.[5]

A common Construction Process problem is inconsistency between the drawings and specifications (dealt with in Section 24.03).

(D) SPECIFICATIONS

Specifications use words to describe the required quantity and quality of the project. They also provide information that will help the contractor plan its price and performance, such as subsurface conditions, site access for heavy equipment, and availability of temporary power. They should be clear and complete and should "fit togeth-

er". If not, there are likely to be defects, disputes, increased costs, and litigation.

Specifications are classified by type. Most important are *design, performance,* and *purchase description.*

Design Specifications (sometimes called Materials and Methods or Detail Specifications) state precise measurements, tolerances, materials, construction methods, sequences, quality control, inspection requirements, and other information. They tell the contractor *in detail* the material it must furnish and how to perform the work.

Performance Specifications state the performance characteristics required, such as the pump will deliver fifty units per minute, a heating system will heat to 70°C within a designated time, or a wall will resist flames for a designated period. Design and measurements are not stated or considered important so long as the performance requirements are met. Under a pure performance specification, the contractor accepts responsibility for design, engineering, and performance requirements, with general discretion as to how to accomplish the goal. Sometimes the contract documents give the contractor suggestions for foundation work. They may be clearly labeled as indicative of "general requirements" or accompanied by a statement that the foundation shall be redesigned to suit soil conditions. Performance specifications are more common in large-scale industrial work where the contractor agrees to design and build a plant that will turn out a designated number of units of a particular quality in a specified period of time. However, performance specifications may also be used in residential or commercial work.

---

2. See Section 32.04.

3. See Sections 28.01 and 28.02, which deal with performance specifications and the risk of defective design, respectively. See also Note, *Liability of a Manufacturer for Products Defectively Designed by the Government,* 23 B.C.L.Rev. 1025 (1982).

4. See Section 25.01, which deals with changes.

5. Refer to Section 17.03, which treats professional liability.

Purchase Description Specifications designate the product or equipment required by manufacturer, trade name, and type. Sometimes the contractor can select from an approved list. These specifications usually increase contract cost as they limit the contractor's ability to use material or equipment that may be just as good as those specified and cost less. For this reason, public contracts and many private contracts frequently create a method by which the contractor can seek approval to use alternative products or material. When such systems are used, a number of legal problems can arise. First, does use of substitute material or product transfer the risk of design failure?[6] Second, does unreasonable delay in passing upon such a request expose the design professional to liability to the owner or contractor?[7] Third, will the contractor have a claim for intentional interference with its contract if the specifications and methods for approving alternates are rigged by the design professional in favor of a particular manufacturer?[8]

The proposal to use an alternate is made most commonly before bid opening but occasionally during performance. When such proposals are made, the design professional should be prompt and fair in passing upon whether the alternate product or material is the equal or equivalent of the brand specified. If the design professional has a reputation for intransigence or unreasonably delayed decisions, a contractor is likely to assume that alternates will not be available and bid accordingly. Unreasonable delay in passing upon requests for alternates can result in liability to the design professional.[9]

Specifications should state that the determination of whether a proposed alternate is the equivalent of that specified can take into account not only function and performance but also esthetics, manufacturer's warranty, and the reputation of the manufacturer for servicing the product and supplying spare parts.

Specification writing has many legal ramifications. Design professionals can find themselves enmeshed in antitrust litigation if a manufacturer contends that the design professional, through specification writing, participated in a scheme to illegally restrain trade. The design professional's role in the specification drafting process can make the design professional a target for high pressure tactics or bribery by a manufacturer who wishes to limit competition by the specification process.[10]

Specification writing can play an important part in relations between contractors and any labor union representing employees. Traditionally, most of the labor that goes into a construction project is performed at the site. The relatively high wages paid to unionized craft workers, the frequent disruption caused by labor disputes at the site,[11] and delays caused by weather conditions have encouraged systems under which more work would be performed in factories.

Construction specifications increasingly require integrated units or prefabricated materials designed to reduce the amount of

---

6. See Section 28.03.

7. *E.C. Ernst, Inc. v. Manhattan Construction Co. of Texas*, 551 F.2d 1026, 559 F.2d 268 (5th Cir.1977) (claim by owner). As to contractors, refer to Section 17.09.

8. *Waldinger Corp. v. Ashbrook-Simon-Hartley, Inc.*, 564 F.Supp. 970 (C.D.Ill.1983).

9. *E.C. Ernst, Inc. v. Manhattan Construction Co. of Texas*, supra note 7.

10. *George R. Whitten, Jr., Inc. v. Paddock Pool Builders, Inc.*, 508 F.2d 547 (1st Cir.1974), involves an antitrust action between two competitors where design professionals were courted with gifts to draft specifications in a certain way.

11. See Section 27.05(B).

labor performed at the site.[12] But reduction of work performed at the site affects job security of those in the construction craft trades. As a result, many craft unions have won what are called "work preservation" clauses in their collective bargaining agreements with employers under which the employers, mainly subcontractors, agree not to handle products upon which labor is performed at the factory when that labor traditionally was performed at the site. Under these work preservation clauses, unions are given the right to strike if there is a violation.

Enterprise Association of Steam, etc., Local Union Number 638 v. NLRB involved the construction of the Norwegian Home for the Aged in New York City. The prime contractor prepared the specifications, which required that climate control units manufactured by a designated supplier be installed as integrated units. These units contained factory-installed internal piping. If the units were installed complete with factory prepiping, the supplier guaranteed all units for a year.

The subcontractor who had agreed to install the heating, ventilating, and air conditioning had had a collective bargaining agreement for many years with the plumbers local that contained a provision requiring the subcontractor employer to preserve certain cutting and threading work for performance at the job site by its own employees. That agreement would have required the internal piping already installed in the climate control units be cut and threaded at the job site.

When the units arrived at the job site, the union's business agent inspected them and informed both prime contractor and subcontractor that the union employees would not install the units. The dispute delayed completion, and the prime contractor filed an unfair labor practice charge with the National Labor Relations Board, alleging that the union could not instruct members to refuse to handle the units, as their object would have been to force a "neutral"—the prime contractor—to cease using the supplier's products.

In dealing with work preservation cases, the National Labor Relations Board and some of the federal circuit courts of appeals had adopted a "right to control" test. If the struck employer (here the subcontractor) had the right to control work assignments, the strike is legal. The pressure must be on him in such a case as he can control work assignment. But if he does not have the right to control, the pressure must be on a "neutral"—the prime contractor. Then the strike is illegal.

Applying the right to control test to the *Local 638* case, the National Labor Relations Board concluded that the subcontractor never had the power to assign the disputed piping work to its union employees. This, according to the NLRB, made the subcontractor a neutral. The union's principal target was the prime contractor, inasmuch as the subcontractor had no right to control the work assignment. Pressure was being put on the subcontractor to force the prime contractor to stop buying the units and restore the work. Such pressure was a secondary boycott and an unfair labor practice.

However, the Federal Circuit Court of Appeals for the District of Columbia,[13] following opinions in other federal circuit courts, refused to follow the "right to con-

---

12. The recent subject matter of work preservation clauses illustrates some of the technologically motivated decisions being made by those drafting specifications. For example, cases have involved precast walls, *NLRB v. Carpenters District Council of Kansas City and Vicinity*, 439 F.2d 225 (8th Cir.1971); a prepiped sink unit, *Associated General Contractors of California, Inc. v. NLRB*, 514 F.2d 433 (9th Cir.1975); prefabricated fireplaces, *Western Monolithics Concrete Products, Inc. v. NLRB*, 446 F.2d 522 (9th Cir.1971); and prepiped heating and ventilating controls discussed in the text of this subsection.

13. 521 F.2d 885 (D.C.Cir.1975).

trol" test and concluded that the subcontractor was not a neutral, inasmuch as the union was simply attempting to enforce its lawful work preservation clause with the employer subcontractor. The court also suggested that the pressure could have been placed on the subcontractor to negotiate a compromise or to terminate the contract with the prime contractor. Because the core of the union's grievances was with the subcontractor, the court held the strike lawful. However, the U.S. Supreme Court reversed the circuit court and concluded that this was an illegal secondary boycott.[14]

Design professionals drafting specifications must take work preservation clauses into account. Craft unions will oppose specifications that reduce their work. Various steps, some legal and some probably illegal, are likely to be taken by the craft union if substantial work traditionally performed by their members will be performed by others. This does not mean that the best and cheapest technology should never be specified. However, in taking into account the advantages of specifications that involve prefabricated products or sealed-at-the-factory components, the disadvantages of potential work stoppages and legal battles must be considered.

(E) CONDITIONS: GENERAL AND SUPPLEMENTARY

Construction is a complex undertaking and a dispute-prone activity. Guidelines are needed to spell out clearly and completely the rights and duties of the parties. For these reasons most construction documents include general conditions (often supplemented by supplementary conditions) of the contract. They are the ground rules under which the project will be constructed.

They are often lengthy and deal with subjects of the following type:

1. Scope of contract documents and resolution of conflicts between them.
2. Roles and responsibilities of the principal participants in the project.
3. Subcontractors and separate contractors.
4. Time.
5. Payments and completion.
6. Protection from and risk of loss to persons and property.
7. Changes.
8. Corrections.
9. Termination.
10. Disputes.

A differentiation should be made between a legal condition, general conditions, and supplementary conditions. The first is a legal classification. It is an event that must occur or be excused before an obligation to perform arises. Whether or not a particular event—sometimes all or part of a promised performance and sometimes an event not within the control of a contracting party—is a condition depends upon any contractual language manifesting this conclusion, the probable intentions of the parties, and elements of fairness. However, this is not the sense in which the term *conditions* is used in this subsection.

General conditions are usually expressed in standardized pre-prepared printed contract form often published by professional associations such as the American Institute of Architects (AIA) but sometimes prepared by a sophisticated contract maker to be used in all of its contracts. Although frequently called general conditions, the provisions within it are almost never automatically considered legal conditions although there is nothing that prevents language within the general condition creating a legal condition. General conditions are prepared

---

14. *NLRB v. Enterprise Association of Steam, etc., Local Union Number 638,* 429 U.S. 507, 97 S.Ct. 891, 51 L.Ed.2d 1 (1977).

in a way that allows them to be used in many types of transactions, either by different contract makers or by a single contract maker.

Yet any individual construction contract may have attributes that make it necessary to have supplementary conditions. For example, the indemnity provisions in general conditions prepared for national use may not be enforceable or desirable in states that have specific statutes regulating indemnification. Similarly, the frequent existence of specialized statutes dealing with arbitration may make it essential to add supplementary conditions if those requirements are to be met. Not all contract makers wish to use general conditions drafted by others, such as the AIA. As a result these contract makers may use an AIA Document A201 but either modify it or attach their own supplementary conditions and make those conditions take precedence over the general conditions.

Two further observations must be made. First, some owners or design professionals may lull contractors into a false sense of security by prescribing *general* conditions with which the contractors are familiar and comfortable but may make many changes in the *supplementary* conditions that destroy some of the protections accorded contractors by the general conditions.

Second, the law will frequently be called upon to sort out inconsistencies between general conditions and supplementary conditions. While it is best to delete from general conditions any provisions that are supplemented by the supplementary conditions, often this requires more work than the attorney wishes to expend. As a result the attorney may simply state that the supplementary conditions take precedence over the general conditions. This often requires

a judge or arbitrator to seek to reconcile apparently conflicting language. This problem is an endemic one to construction with its wealth of contract documents.

The role of the design professional in drafting or suggesting that particular general conditions be used was discussed earlier.[15]

(F) SOIL AND SITE TEST REPORTS

Frequently Invitations to Bidders contain soil test reports or site data or a reference that such data are available in the office of a soil consultant engaged by the owner. This information and data are of great importance in the construction project. The legal effect of furnishing this information is discussed in Chapter 29.

(G) PRIOR NEGOTIATIONS AND THE PAROL EVIDENCE RULE

The parol evidence rule relates to the provability of oral agreements made before or at the same time as a written contract. This rule is based upon the concept of "completeness of writings." If the written agreement and other written documents incorporated by reference or attached to it are the complete agreement of the parties, prior or contemporary oral agreements are not binding on the parties. Even if they were made, they were integrated, or merged, into the written, complete document.[16]

Most construction contracts are complete. Nevertheless, a possibility exists that oral agreements have been made before or at the same time as the final written agreement. If the written agreement contains an integration or merger clause, one that specifies that the written agreement is the complete and final agreement, it will be difficult for either party to prove a prior oral agree-

---

15. Refer to Section 15.07.

16. Refer to Sections 14.04(E) and 15.03(C) for further discussion.

ment.[17] It will not be impossible, however, because doctrines exist that permit proof of an oral agreement even if there are provisions specifying that the written agreement is the complete document. If no contract provision deals with the question of completeness, most courts will admit the testimony claiming an oral agreement.

It is important to incorporate the entire agreement into the writing. Even if the writing is prepared and ready for execution, the design professional, and certainly the owner's attorney, should insist on incorporating any changes or additions into the writing itself. The entire document need not be retyped if time does not permit. If the oral agreement is not included in the writing, a substantial risk exists that it cannot be proved.

Suppose there is a dispute over meaning of terms used in the writing. If the words chosen by the parties are ambiguous, the court can look at the surrounding facts and circumstances to determine how the parties used those particular words. These circumstances include the setting of the transaction, the contracting parties' objectives in making the contract, and any conversations they may have had. Even if a writing is considered complete, antecedent or subsequent conversations may be admissible to interpret the writing.

This doctrine was applied in a case, *Marbury-Pattillo Construction Co. v. Bayside Warehouse Co.,*[18] which involved an integration clause and demonstrates the not uncommon phenomenon of substantial work being performed before the formal contract is made. The owner sought bids for storage tanks. The first bids were too high, and the owner decided to use an alternative foundation method to cut costs. The owner had been advised that the cheaper foundation would settle sooner or later.

The contractor made a letter proposal accompanied by specifications and outline drawings that provided for the alternative foundation method. A few days later the owner sent a letter authorizing the contractor to proceed and stated that a "formal" contract would follow. Two to three months elapsed before the owner's attorney prepared and submitted a formal contract to the contractor which he signed. During the delay much of the foundation work was performed.

When the tanks settled, the owner refused to pay the $60,000 retainage, claiming defective design. When the contractor was forced to go to court to obtain the retainage, the owner claimed one million dollars for delay due to defective design. To substantiate this "claim," the owner relied upon a clause in the formal contract stating that the formal contract expressed the entire agreement and then pointed to language in the formal contract that placed design and construction risks upon the contractor. The owner sought to exclude evidence of what had transpired prior to the execution of the formal contract, such as the change in design and assumptions as to design risks.

The court listened to the evidence of surrounding facts and circumstances and ruled for the contractor. On appeal, the court concluded that the trial court had been correct, noting that the activities and contentions of the owner were perilously close to fraud. The basis for affirming the trial court's decision was that evidence of facts and circumstances is admissible not to vary but to determine the true intention of the parties.

(H) MODIFICATIONS

Modifications should be differentiated from changes. Changes occur frequently during

---

17. *Dixie Belle Mills, Inc. v. Specialty Machine Co.,* 217 Ga. 104, 120 S.E.2d 771 (1961). Such a clause will not be effective if fraud is alleged.

18. 490 F.2d 155 (5th Cir.1974).

a construction project and are discussed in detail in Chapter 25. They are generally governed by carefully drawn and complete provisions under which the owner has the right to order changes and the contractor must perform them, with appropriate contract language dealing with compensation for deletions or additions caused by the changes. A modification is a change agreed upon by both parties in the basic obligation not based upon any contractual provision giving the owner right to order changes.

In addition to agreement, a valid modification requires consideration.[19] To demonstrate, suppose there is a construction project for $100,000. The obligations of the contractor are expressed in the contract documents. Suppose during the term of the agreement that the parties mutually agree that the contract price will be increased to $110,000. The owner must receive something in exchange for the additional compensation. If, in exchange for this, the contractor agrees to an increase in the quality or quantity of the building or to shorten the time for completion, there is an appropriate exchange and the modification agreement is binding. Difficulties arise when the modification agreement encompasses an increase in price without any change in the contractor's obligation. Such an agreement can be invalidated by the preexisting duty rule. Unless one of the many exceptions applies, the agreement is not binding if the party—in this case the contractor—is obligating itself to do no more than it was previously obligated to perform under the original contract.

This preexisting duty rule is criticized. It limits the autonomy of the parties by denying enforceability of agreements voluntarily made. Implicit in the rule is an assumption that an increased price for the same amount of work is likely to be the result of expressed or implied coercion on the part of the contractor, as if the contractor is saying "pay me more money or I will quit and you will have to whistle for the damages." However, suppose the parties have arrived at a modification of this type voluntarily. There is no reason for not giving effect to their agreement. There are a number of exceptions that can relieve against the sometimes harsh effect of the preexisting duty rule.

In the construction contract, minor changes in the contractor's obligation have been held sufficient to avoid the rule even where the increase in price was not commensurate with the change in obligation on the part of the contractor. This approach permits contract modification to be enforced if the parties have had enough foresight or legal knowledge to provide for some minor and relatively insignificant change in the contractor's obligation as a means of enforcing the increased price.

An exception that developed in construction contracts enforces the modification if it is fair and equitable in view of the circumstances not anticipated by the parties when the contract was made.[20] Since this exception (frequently referred to as the unforeseen circumstances exception) has occurred mainly in the area of subsurface conditions, it is discussed in greater detail in Section 29.01(D).

Another indication of dissatisfaction with the requirement of consideration for a modification is manifested by Sec. 2–209 of the Uniform Commercial Code under which modifications are valid without consideration. However, a modification obtained through bad faith is not valid.

Curiously, despite the movement toward abolition of the consideration requirement for a modification, the rule is applied vigorously in federal public contracts. Unless the government receives something for an increase in price or a reduced price for a

---

19. *Mainland v. Alfred Brown Co.,* 85 Nev. 654, 461 P.2d 862 (1969).

20. *Angel v. Murray,* 113 R.I. 482, 322 A.2d 630 (1974).

deletion, the modification is not valid. This rigid adherence to the consideration requirement may be justified as a means of insuring that there be no gifts of public funds and preventing corruption and collusion between contractor and public official. Giving some advantage to a contractor without the government getting anything in return may be unfair to the other bidders.

Frequently contracts state that modifications must be in writing. Such provisions generally are not valid.[21] In enforcing oral modifications despite such clauses, the courts have stated that by making a subsequent oral modification, the parties have changed the agreement requiring that the modification be in writing. However, under Sec. 2–209 of the Uniform Commercial Code, oral agreements modifying the written agreement are not effective if the written agreement contains a provision requiring that modifications be in writing. There are some exceptions to this, but the Code expresses a policy that such contractual provisions requiring a writing as a condition to enforcement of an asserted modification agreement should be given more effect than the courts have given them in the past. It remains to be seen whether this change in the law relating to the sale of goods will have any impact upon judicial thinking in other types of cases.

SECTION 23.02 JUDICIALLY DETERMINED TERMS

(A) NECESSITY TO IMPLY TERMS

When making a contract, parties frequently do not consider all problems that may arise. Some matters may have been considered,

but the parties believed the resolution of the matters to be so obvious that contract coverage would be unnecessary. Matters have been discussed by the parties during negotiations, but the parties could not agree on a contract solution to the problem. Yet these parties may intend to have a binding contract despite their inability to resolve all problems during negotiation. In such cases, the parties may state in the contract that they will agree in the future on certain less important contract matters or omit the matter from the contract entirely.

Courts may be asked to fill in the gaps not covered by the contract or decide matters left for future agreement where the parties cannot agree. Courts will be more likely to perform these functions if convinced that the parties intended to make an enforceable agreement, especially where performance has commenced.

Until recently, common law judges hesitated to imply terms. These judges saw themselves as enforcing contracts made by the parties rather than making contracts for the parties.

One court stated:

1. The implication must arise from the language used or it must be indispensable to effectuate the intention of the parties.

2. It must appear from the language used that it was so clearly within the contemplation of the parties that they deemed it unnecessary to express it.

3. Implied covenants can be justified only on the grounds of legal necessity.[22]

4. A promise can be implied only where it can be rightfully assumed that it would have been made if attention had been called to it.

5. There can be no implied covenant where the subject is completely covered by the contract.[23]

---

21. *United States v. Klefstad Engineering Co.*, 324 F.Supp. 972 (W.D.Pa.1971). A similar provision in an AIA Document was not binding in *Fishel & Taylor v. Grifton United Methodist Church*, 9 N.C.App. 224, 175 S.E.2d 785 (1970). Four states, most notably California, require that modifications of written contracts must be in writing. There are many exceptions.

22. This probably means courts will imply a promise if it is necessary in order to have a valid contract and the parties have so intended.

23. *Stockton Dry Goods Co. v. Girsh*, 36 Cal.2d 677, 681, 227 P.2d 1, 3–4 (1951).

These requirements reflect judicial reluctance to imply terms. In addition, courts during this period hesitated to provide a term where the parties had agreed to agree but did not.[24]

Modern judges seem less insecure and more realistic about their role in contract disputes. They are beginning to recognize that they "make" contracts for the parties. Their recognition of the proliferation of standard form contracts often made in an adhesion context makes them more willing to imply terms to redress unequal bargaining than were courts a quarter century ago.

The law should exercise restraint in implying terms where the contract has been negotiated. In such cases, it is likely that most of the major problems were considered and the absence of a promise may be deliberate. Implying a term in such cases could frustrate the bargain.

Even in negotiated contracts, it still may be necessary to imply terms that were so obvious that the parties did not think it necessary to express them. A court should "complete" the deal by filling in gaps where parties intended to make a contract, left minor terms for future agreement, but have not been able to agree.

Again, a word of caution. The existence of express contract terms generally preclude terms on that subject being implied.[25] Also, custom takes precedence over implied terms. First, the contract must be examined. If the contract deals with subject matter that relates to the proposed implication, a court is less likely to imply terms. However, sometimes the effect of implication is achieved by interpreting particular contract terms in a way consistent with the implication.

One court discussing the owner's duty to furnish a site stated:

> Each party to a contract is under an implied obligation to restrain from doing any act that would delay or prevent the other party's performance of the contract. . . . A party who is engaged to do work has a right to proceed free of let or hindrance of the other party, and if such other party interferes, hinders or prevents the doing of the work to such an extent as to render the performance difficult and largely diminish the profits, the first may treat the contract as broken and is not bound to proceed under the added burdens and increased expense.[26]

The court here speaks of the conduct of the owner that may delay or prevent the contractor from performing the work. Implied obligations can go further and, under certain circumstances, require that the owner perform positive acts to assist the contractor in performance. However, the law will be more reluctant to impliedly require affirmative acts of cooperation than to preclude negative acts of hindrance or prevention. Positive acts are more likely to have been thought about if they were important, and failure to express them in the contract may indicate that neither expected them to be done. For example, *C.A. Davis, Inc. v. City of Miami*[27] involved a claim by a defaulting contractor that the city had failed to cooperate and thereby hindered the contractor's performance. Evidently the contractor contended that the city had a duty to secure utility company cooperation with regard to underground obstructions encountered on the project. The trial court judge refused the contractor's request that the jury be instructed that the city had a duty to cooperate. The appellate court concluded that the trial judge had been correct in

---

24. Refer to Section 5.06(F).

25. *Weber v. Milpitas County Water District,* 201 Cal.App.2d 666, 20 Cal.Rptr. 45 (1962).

26. *United States v. Guy H. James Construction Co.,* 390 F.Supp. 1193, 1206 (M.D.Tenn.1972), affirmed without opinion, 489 F.2d 756 (6th Cir.1974); *Lewis-Nicholson, Inc. v. United States,* 550 F.2d 26 (Ct.Cl.1977). As to subcontracts, see *Elte, Inc. v. S.S. Mullen, Inc.,* 469 F.2d 1127, 1132–1133 (9th Cir.1972).

27. 400 So.2d 536 (Fla.App.1981).

refusing to give this instruction, as there had been no evidence that the city had failed to cooperate. The court stated:

> The evidence before the court demonstrated that the City repeatedly and routinely cooperated in an effort to keep the project going. In fact, the City had no implied duty to do [the contractor's] work, but only a duty not to hinder or impede that work.[28]

This language indicates that implied terms of the type under discussion must often be correlated with contract interpretation. Had the contractor been obligated to secure utility company cooperation, it would have been difficult to imply a term obligating the city to do so.

Any duty to cooperate should not require undertaking heavy burdens of cooperation that would very likely frustrate the contractual allocations of responsibility made by the parties. However, if one party can assist the other party's performance at a minimal cost, this cooperation should be required.

Increasingly the law is requiring that the parties to a contract show good faith and fair dealing toward each other during performance. Although American law has been somewhat hesitant to imply such an imprecise standard, there are signs that courts are beginning to do so in the construction context. For example, one court held that a termination of the contract by the owner that would otherwise have been justified was wrongful because it had been done in bad faith.[29] Similarly, another court held that a prime contractor negotiat-

ing an impact delay damage claim on behalf of a subcontractor must do so in good faith.[30] However, the elasticity of this requirement may mean that some courts will refuse to venture into such claims or, though willing to recognize the duty exists, will find its violation only in extreme cases.

Some specific illustrations of implied terms in the construction contract have been mentioned earlier, such as the standard of the design professional's performance[31] or the obligation of the owner to coordinate work of separate contractors or stand behind the obligation of another entity given that responsibility.[32]

There are other illustrations. The owner impliedly promises that the site will be ready for the contractor to commence performance.[33] (Often, as seen in Section 30.02, this is dealt with by not starting the obligation to commence performance until a notice to proceed has been given.) The owner impliedly promises to obtain the necessary easements or rights to enter the land or land of another necessary for the contractor's performance.

The prime contractor impliedly promises the owner to use proper workmanship and materials and to complete the project free and clear of liens. A contractor in a cost-type contract promises to inform the owner of prospective overruns of cost estimates.[34] One court held that the prime contractor promised the subcontractors that the work would be coordinated and that the prime contractor would process subcontractor

---

28. Id. at 539.

29. *Paul Hardeman, Inc. v. Arkansas Power & Light Co.,* 380 F.Supp. 298 (E.D.Ark.1974).

30. *T.G.I. East Coast Construction Corp. v. Fireman's Fund Insurance Co.,* 534 F.Supp. 780 (S.D.N.Y.1982). See also *Seaman's Direct Buying Service v. Standard Oil Company of California,* 36 Cal.3d 752, 206 Cal.Rptr. 354, 686 P.2d 1158 (1984).

31. Refer to Section 17.05.

32. Refer to Section 21.04(C).

33. *North Harris County Junior College District v. Fleetwood Construction Co.,* 604 S.W.2d 247 (Tex.Civ.App.1980).

34. *Jones v. J.H. Hiser Construction Co., Inc.,* 484 A.2d 302 (Md.App. 1984).

claims to the owner.[35]   But another court refused to imply a promise that the prime contractor would create and maintain an efficient schedule when this was neither customary nor bargained for.[36]

The owner impliedly promises that it and its designated representative, usually the design professional, will perform in such a way as to reasonably expedite the contractor's performance. For example, the owner impliedly promises that the design professional will give contract interpretations and pass on sufficiency of shop drawings within a reasonable time.

Usually the law implies a promise by the owner to supply adequate drawings and specifications (discussed in greater detail in Section 27.05(E)). However, that such terms will be implied does not eliminate the necessity of focusing more closely on the nature of the implication. For example, suppose the owner designates specified material. That will very likely mean that if the contractor uses that material, the contractor will not be responsible if that material proves to be unsuitable.[37] However, does specification of a material imply that it is available? Does it imply that the material is in stock of local suppliers? Questions such as these expose the basic function of implying terms, that of allocating risk and responsibility.

A federal appeals board held that the owner does not warrant that a product specified will be in stock inasmuch as parties do not usually guarantee the performance of third parties.[38] Such a result may also be based upon the likelihood that owners know no more about the stock of suppliers than does the contractor.

Usually the owner reserves certain powers during the contractor's performance to protect its interests. This should not be automatically converted into an affirmative duty for the benefit of the contractor or subcontractors. For example, it has been held that the power to monitor contractor performance and stop the work if necessary did not imply that the owner had *promised* to direct the contractor's workers.[39]

Courts are sometimes asked to imply completion times to construction contracts. Usually a specified time for completion is given in the contract. If there is no express provision dealing with this question, courts hold that the contractor must complete performance within a reasonable time. Reasonableness will be determined in light of all the surrounding facts and circumstances.

Courts should not imply a reasonable time for performance if performance has not yet commenced and if the failure to agree on a time for performance indicates that the parties have not yet intended to conclude a contract. The absence of agreement on such an important question may mean that the parties are still in a bargaining stage.

(B) CUSTOM

Customary practices are important in determining rights and duties of contracting parties, particularly in complex transactions such as construction, where everything cannot be stipulated in the contract. Mention has been made in Section 15.11 of the frequent claim by design professionals that customarily they retain ownership of drawings and specifications. Custom plays other significant roles in construction. In placing such a heavy emphasis on customary prac-

---

35. *Citizens National Bank of Orlando v. Vitt*, 367 F.2d 541 (5th Cir.1966).

36. *Drew Brown Limited v. Joseph Rugo, Inc.*, 436 F.2d 632 (1st Cir.1971).

37. See Section 28.02.

38. *James Walford Construction Co.*, GSBCA No. 6498, 83–1 BCA ¶ 16,277.

39. *Nat Harrison Associates, Inc. v. Louisville Gas & Electric Co.*, 512 F.2d 511 (6th Cir.1975).

tices when *interpreting* contract terms, courts often state that parties contract with reference to existing customs. In that sense, courts can be said to be simply giving effect to the actual intention of the parties. However, in many cases parties have not thought about the problem, and courts find that an established custom can be a more convenient and proper method of filling gaps than a judicial determination of reasonableness.

For example, if customarily the contractor obtains a building permit, the court is likely to place this responsibility upon the contractor, but only where there is a contract gap on this point.[40] If the contract does not deal with the matter, custom would be preferable to a judge's determination of who should obtain the permit.

Custom can assist the court when the terms must be interpreted or conflicts exist in the contracts documents. For example, in *Fifteenth Avenue Christian Church v. Moline Heating & Construction Co.,*[41] the owner brought an action against the heating contractor claiming that the contractor was obligated to install a "one hour fire resistive" ceiling in the boiler room of the church because provisions of the local building code required it. The specifications made no reference to the walls or ceiling of the boiler room. After completion and final payment, the city fire marshal called the owner's attention to the fact that the ceiling did not meet code requirements. About a year later, the owner demanded that the ceiling be replaced but the heating contractor refused.

The general conditions required the contractor to comply with all local laws and to notify the engineer if the specifications were at variance with local laws. Technical

provisions included in the contract documents also gave precedence to code regulations over specifications.

Witnesses called by the contractor stated that provisions and specifications requiring that the contractor comply with all local laws were standard provisions and referred to the work *specifically* included in the contract. They testified that if the plans and specifications did not specifically provide for the installation of a particular type of ceiling in the boiler room, the *heating* contractor would not be obligated to install such a ceiling. They further testified that the local codes referred to in the specification related to the code rules relating to the type of work to be done in the specifications and that the installation of ceilings in a boiler room fell within the general category of work done by the general construction trades and not the plumbing and heating trades.

The appellate court held that it was proper to receive evidence of custom to interpret the language in the contract as long as the custom is known or should have been known to the parties and there is no specific stipulation in the contract to the contrary. In affirming the decision for the contractor, the court noted that the engineers who represented the owner were professionals and must be presumed to know the customs and usages of the trade and that this knowledge was imputed to the owner.

The *Fifteenth Avenue* case is instructive. Custom *is* important.[42] Also, the owner will be held responsible for what the design professional knew or should have known, which of course can ultimately mean enlarged responsibility for the design professional. Blanket recitals that the work will be in compliance with local codes should

---

40. In *Weber v. Milpitas County Water District*, supra note 25, custom that the *owner* procured these permits was not relevant since there was an *express* provision in the contract requiring the *contractor* to obtain such permits.

41. 131 Ill.App.2d 766, 265 N.E.2d 405 (1970).

42. In *Chicago Bridge & Iron Co. v. Reliance Insurance Co.*, 46 Ill.2d 522, 264 N.E.2d 134 (1970), the court admitted custom that subcontractors execute lien waivers despite not having received payment.

not be taken to enlarge the contractor's obligation when it has bid upon specifications that did not include disputed work and when that disputed work is customarily performed by a different contractor.

Custom is an important part of construction contracts. However, parties should never rely upon custom when there are express provisions to the contrary. Custom can be a trap for a contractor who performs a project in a locality where customs may be different from those where it customarily works. Custom can often be difficult to establish. In the absence of an express provision controlling the dispute, the party seeking to establish custom will be given an opportunity to do so.

(C) BUILDING LAWS

Building codes and land use controls play a pervasive role in construction. Frequently it is implied or expressed in the contract that the contractor will comply with applicable laws. Further analysis is needed to separate several related but distinct problems.

Where design and construction are separate, the former is the responsibility of the owner and is usually accomplished by the design professional. Suppose the design violates building code requirements. The contractor may contend that its job is to build the design, but the contractor cannot ignore violations of law. It should direct the design professional's attention to any *obvious* code violations in the design. Similarly, any obvious land use violations should be brought to the design professional's attention. This duty does not divest the owner of ultimate responsibility.

This problem is dealt with by AIA Document A201, ¶ 4.7.3, which states that it is not the contractor's responsibility to make

certain that the design accords with legal building requirements but that *if* the contractor observes variances, it must notify the architect in writing. Under ¶ 4.7.4, the contractor's performance of work *knowing* it to be contrary to such laws without notifying the architect places full responsibility upon the contractor.[43]

This solution effectuates a sound middle ground between putting an unreasonable responsibility on the contractor and allowing the contractor to close its eyes in the face of danger. A contractor who does not comply with the requirements of ¶ 4.7.3 is exposed to liability, not only for the cost of correction and consequential damages that may be suffered by the owner but also to losses suffered by third parties.

As noted, the contractor is not usually responsible for design when that design is furnished to it by the owner. Suppose the contractor both designs and builds. In *Quedding v. Arisumi Brothers, Inc.*,[44] the contractor supplied the design. It clearly specified only *vertical* rebars. Local building codes required *both vertical* and *horizontal* rebars. The court held that the existing law was part of the contract unless stipulated to the contrary. As a result, the contractor breached when he violated the code even if he followed the specifications. Here the court properly protected the reasonable expectations of the owner that the contractor would build in accordance with the local building laws. The owner would not be expected in such a transaction to know that the design violated the building codes.

In this case there was no provision requiring the contractor to comply with local building law, yet the court implied this obligation. Had the contract provided that the contractor would comply with local build-

---

43. In *Greenhaven Corporation v. Hutchcraft & Associates, Inc.*, 463 N.E.2d 283 (Ind.App. 1984) the architect recovered his fee despite an agreement with the owner to violate the fire code. They hoped to obtain a variance from the fire marshal.

44. 661 P.2d 706 (Hawaii 1983).

ing codes, there would have been a conflict between that provision and the specifications. Although it is usually held that the specific provision in a contract takes precedence over a general provision, in such a case an unsophisticated owner unaware of the conflict should have its expectations protected by giving preference to the provision requiring that the contractor comply with building laws.

Yet in many instances no *direct* clash exists between building codes and the design. In such instances, the owner and the design professional expect the contractor to follow building codes. It is in this sense that the contractor must build in conformity with legal requirements. Essentially, this is a gap-filling instrumentality. Additionally, such code compliance provisions are intended to obtain compliance with worker safety rules.

Other legal requirements can consist of permits issued by public officials at various stages of the performance. These permits are not limited to building and occupancy permits. Frequently permits must be obtained from public utilities to connect the utilities of the project to lines of the public utility. Drainage rights of way may be needed as well as permits from state highway officials for certain types of construction. Who must obtain permits required by law?

Construction documents should cover these matters. If the contract does not, custom may allocate responsibility for obtaining permits. In the absence of an express contract provision or accepted custom, the law will be likely to imply that the owner will obtain the more important permanent permits such as land use control permits,[45] but the contractor should get operational permits such as building permits [46] and occupancy permits. The contractor is likely to be required, by implication of law, to obtain permits from public officials that are associated with facilities and equipment. The contractor should obtain utility hookup permits and permits for temporary construction. The determination of who should obtain particular construction permits will depend upon the extent of experience the particular owner has had in construction projects.

Suppose the contractor is required to obtain certain permits but does not. In *Lew Bonn Co. v. Herman*,[47] the electrical subcontractor sued the prime contractor for work that had been performed. The prime contractor contended that the contract was unenforceable because the plaintiff subcontractor had failed to file a copy of the plans with the city building inspector as required by law. The court held that this slight violation of the statute would not make the contract unenforceable and allowed the subcontractor to recover for his work. The court emphasized that the public was not placed in jeopardy, and the subcontractor's failure should not entitle the prime contractor to a windfall. Generally the subcontractor's failure to comply with any requirement to obtain a work permit or submit the plan does not affect its right to recover compensation.[48] However, it is dangerous to rely upon a court's subsequently determining that the violation was technical.

---

45. *COAC, Inc. v. Kennedy Engineers*, 67 Cal.App.3d 916, 136 Cal.Rptr. 890 (1977) (Environmental Impact Report).

46. In *Drost v. Professional Building Service Corp.*, 153 Ind.App. 273, 286 N.E.2d 846 (1972), the contract documents required the contractor to furnish "all necessary . . . drawings for construction of this facility as described for the use in obtaining the necessary building permits." The court held that this language placed a duty upon the contractor to obtain these permits.

47. 271 Minn. 105, 135 N.W.2d 222 (1965).

48. See Annot., 26 A.L.R.3d 1395 (1969).

24

Contract Interpretation: Chronic Confusion

SECTION 24.01
BASIC OBJECTIVES

The basic objective in contract interpretation is to determine the intention of the parties. However, many problems lurk within this relatively simple standard, some of which follow:

1. What can be examined to ascertain the intention of the parties?

2. Once the relevant sources are examined, how is the intention of the parties determined?

3. What if each of the parties had different intentions?

4. What if one party knew of the other party's intention?

5. What if the parties had no particular intention about the matter in question?

6. Can the court go beyond these presumed intentions of the parties and interpret in accordance with what the court thinks they would have intended had they thought about it?

7. Can a court disregard the intention of the parties test and base determination of the rights and duties of the parties upon judicial notions of proper allocation of risk?

The problems can be even more complicated when the language has been selected not by the parties or by one of the parties but by a third party such as the American Institute of Architects (AIA). Should the intention of AIA personnel who selected the language be examined if it can be ascertained? If it cannot, what should the law do when neither party had any intention whatsoever when it agreed to use AIA documents, not an uncommon phenomenon as to certain types of contract clauses.

The preceding list is given merely to indicate that phrasing the test as the process of ascertaining the intention of the parties is deceptively simple, often hiding difficult interpretation problems.

SECTION 24.02
LANGUAGE INTERPRETATION:
NEWSOM v. U.S.

Words have no inherent meaning. Yet words develop meanings because people who use them as tools of communication attach meanings to them. A judge asked to interpret contract terms could decide all interpretation questions simply by using a dictionary and choosing the dictionary meaning that seemed most appropriate. However, even dictionary meanings are not exclusive. The choice of which dictionary meaning to be selected can be a formidable task. For these reasons, interpretation of contract terms should take into account the setting and function of the transaction and other matters not found in the contract or the dictionary. Courts seek to put themselves in the position of the contracting parties and determine what the contracting parties must have meant or intended when they used the language in question.

Yet judges or juries do not have absolute freedom to determine the meaning of words. This constraint may be traceable to the fear that juries, and sometimes trial judges, will be unduly sympathetic to a hard luck story and too inclined to protect the party in the weaker bargaining position. Underlying this reluctance is the skepticism the law has toward the ability of the trial process to separate truth from falsehood, especially when parties differ in their testimony as to what transpired during the negotiations.

Reluctance to give absolute discretion to the fact finders may also be traceable in part to the fear that juries, and sometimes trial judges, may not realize the importance to the commercial world of attaching consistent, commercially accepted, meanings to terms.

The "plain meaning rule" has been the method employed to limit interpretation powers of trial judges and juries. A judge must first determine whether the words used by the parties had a plain meaning. If so, the judge cannot look beyond the document itself to determine what the parties meant when they used those particular words.

If the judge determines that language cannot be interpreted solely by looking at the writing, other evidence can be examined. A judge will look at the surrounding facts and circumstances that led to the making of the contract, the preliminary negotiation and statements of the parties, any written codes the parties may have adopted, custom and usage, and any interpretation the parties may have placed upon the words by their own acts.

Perhaps even judges who state the meaning is plain on its face have already formally or informally reviewed evidence outside the writing and determined that those outside sources were not helpful.

Despite the plain meaning rule, courts will look at evidence of custom and usage even though the custom and usage seems at variance with the apparently plain meaning of the terms in the contract. For example, suppose a contract specified that meat to be delivered must not have a fat content exceeding 50%. Yet custom in the industry could be introduced that sellers are allowed to deliver meat with a fat content of 50.5%.[1]

The plain meaning rule applies only to litigation. For example, an arbitrator could examine evidence outside the writing without first having to determine that the language was susceptible to more than one meaning.

Appellate courts have stressed that rarely is language sufficiently plain in meaning to preclude extrinsic evidence from being introduced. As a result, modern trial judges seem hesitant to invoke the plain meaning rule and preclude extrinsic evidence. Often the result is a lengthy, expensive trial that does not seem to have advanced the inquiry beyond the language.[2]

Surrounding facts and circumstances often determine how language is interpreted. Typically the relevant surrounding facts and circumstances are those that existed at the time the contract was made. However, sometimes courts look at circumstances that existed at some time during *performance* to interpret contract language.[3]

---

1. *Hurst v. W.J. Lake & Co.*, 141 Or. 306, 16 P.2d 627 (1932).

2. *Metropolitan Paving Co. v. City of Aurora, Colorado*, 449 F.2d 177 (10th Cir.1971) is an illustration.

3. *Contracting & Material Co. v. City of Chicago*, 20 Ill.App.3d 684, 314 N.E.2d 598 (1974). The holding was reversed, not on the admissibility of the evidence, by finding the evidence irrelevant because of a strict interpretation given to the contract clause in question. See 64 Ill.2d 21, 349 N.E.2d 389 (1976).

Evidence of the surrounding facts and circumstances that courts will *not* examine are any *undisclosed* intentions of the parties they claim existed at the time they made the contract. If these intentions are made known to the other party, they may be relevant.

Disclosed intention was relevant in *United States v. F.D. Rich Co.*,[4] a dispute between a subcontractor and prime contractor. The contract documents upon which the subcontractor prepared his bid were inconsistent as to certain work. The specifications appeared to require that the work be done although the drawings did not. The subcontractor met with the prime contractor before submitting the bid. The subcontractor testified that at the meeting the contract documents were examined and discussed. The subcontractor testified that he made it clear that his bid was based upon not doing the work in question.

During performance the public agency insisted that the disputed performance was required. The prime contractor ordered the subcontractor to do the work, which was done under protest. Later, the subcontractor sued for the additional costs of doing the disputed work.

In sustaining a finding by the trial court that the work did not fall within the subcontract, the appellate court stressed that the prime contractor *knew* that the subcontractor's bid did not include the disputed work.

Statements and, more important, acts of the parties *before* the dispute arose may indicate how the parties interpreted the language. Courts often invoke and give considerable weight to those acts under what is called practical interpretation. The prac-

tices of the parties often indicate their intentions at the time the contract was made. For example, making two progress payments without a showing that the work complied with the contract manifested an intention that the contractor was entitled to progress payments despite noncompliance.[5] Likewise, the prime contractor periodically billing the owner in accordance with certain unit prices and the owner paying these billings indicated that the unit prices were correct.[6] The contractor doing what the owner had directed without complaint indicated the contractor's acquiescence in the owner's interpretation.[7] But the doctrine of practical interpretation applies only to language susceptible to more than one interpretation.[8] This rarely is a difficult obstacle, however.

Not infrequently the surrounding circumstances and the predispute conduct of the parties provide little assistance. When this occurs courts sometimes resort to secondary assistance called "canons of interpretation." These canons are interpretation guides. Some are used frequently. One, *expressio unius est exclusio alterius,* excludes an item from relevance when there is a list of items and the item in question was not expressed. For example, suppose a party were excused from performance in the event of strikes, fire, explosion, storms, or war. Under the *expressio* guide, the occurrence of an event not mentioned, such as a drought, would not excuse performance. Where the parties have expressed five justifiable excuses, they must have intended to exclude all other excuses for failure to perform. This guide is sometimes harshly applied and does not give realistic

4. 434 F.2d 855 (9th Cir.1970).

5. *Giem v. Searles,* 470 S.W.2d 327 (Ky.1971).

6. *Berry v. Blackard Construction Co.,* 13 Ill.App.3d 768, 300 N.E.2d 627 (1973).

7. *Bulley & Andrews, Inc. v. Symons Corp.,* 25 Ill.App.3d 696, 323 N.E.2d 806 (1975).

8. *Dana Corp. v. United States,* 470 F.2d 1032 (Ct.Cl.1972).

recognition of the difficulty of drafting a complete list of events.

Another guide, *ejusdem generis,* states that the meaning of a general term in a contract is limited by the specific illustrations that accompany it. Anything not specifically mentioned must be similar in meaning to those things that are. For example, damages payable for harm to crops, trees, fences, and premises probably would not include depreciation in market value of the land. The particular item in question is too unrelated to those items listed to be covered by the contract terms.

As to other guides, one court stated:

> . . . where one interpretation makes a contract unreasonable or such that a prudent person would not normally contract under such circumstances, but another interpretation equally consistent with the language would make it reasonable, fair and just, the latter interpretation would apply.[9]

This does not give the court the power to rewrite the language. Nor does it preclude the contracting parties from agreeing to language that a court might consider unreasonable. However, for an unreasonable interpretation to be selected, the language must make clear that this is what the parties intended.

Another court required that the language be interpreted "so as not to put one side at the mere will or mercy of the other."[10] Another stated that a particular interpretation should be rejected because it "would certainly be both unconscionable and inequitable," and the recognized rule is that the interpretation that makes a contract fair and reasonable will be preferred to one leading to a harsh and unreasonable result.[11]

Another important interpretation guide, *contra proferentem,* interprets ambiguous language against the party who selected the language or supplied the contract. Usually this guide is not applied to negotiated contracts but to those mainly prepared in advance by one party and presented to the other on a take-it-or-leave-it basis.

One basis for this guide is to penalize the party who created the ambiguity. Another, and perhaps more important, rationale is the necessity of protecting the reasonable expectations of the party who had no choice in preparing the contract or choosing the language. This rationale had its genesis in the interpretation of insurance contracts. Frequently insureds were given protection despite what appeared to be language precluding insurance coverage. This preference recognizes that insurance policies are difficult to read and understand and the insured's expectations as to protection are derived principally from advertising, sales literature, and salespeople's representations. Giving preference to the insured's expectations may also rest upon the judicial conclusion that insurance companies frequently exclude risks that should be covered.

The *contra proferentem* guide can be a tie breaker when all other evidence is either inconclusive or unpersuasive. It can be mentioned by the court to bolster an interpretation that has already been determined by other evidence.

This guide has been used in the construction context, such as disputes between contractor and owner[12] and between prime and subcontractor.[13]

Frequently design and construction work are performed after parties have assented to a standard prepared contract form created

---

9. *Elte, Inc. v. S.S. Mullen, Inc.,* 469 F.2d 1127, 1131 (9th Cir.1972).

10. *Contra Costa County Flood Control & Water Conservation District v. United States,* 206 Ct.Cl. 413, 512 F.2d 1094, 1098 (1975).

11. *Glassman Const. Co. v. Maryland City Plaza, Inc.,* 371 F.Supp. 1154, 1159 n. 3 (D.Md.1974).

12. Ibid.

13. *United States v. Klefstad Engineering Co.,* 324 F.Supp. 972 (W.D.Pa.1971).

by associations such as the American Institute of Architects (AIA) or the National Society of Professional Engineers (NSPE). Sometimes such printed contracts are agreed to without careful consideration of the language. On other occasions, all or some of the language is considered carefully by the parties before entering into the contract. Sometimes the parties have dealt with and understood the language, while on other occasions one party may be unfamiliar with the terminology and concepts employed in the standard contract.

A look at a few cases in the construction context may be instructive. *Durand Associates, Inc. v. Guardian Investment Co.*[14] construed an AIA standard contract against the engineer who had supplied it despite the fact that the owner was an investment company about to build a medical clinic that later was changed to an apartment complex. Although the owners in the case seemed to be experienced business persons, evidently the court felt that their knowledge and experience did not equip them to carefully appraise the language of the AIA document. Yet they *read* the document and made some revisions that were accepted by the engineer. Perhaps the court construing the document "strictly" against the engineer was based more upon its reluctance to compel the owner to pay a fee when the project was abandoned because of excessive cost.[15]

Two other cases that involved disputes between owners and contractors would not interpret the document against the owner despite the latter having furnished the AIA document. One noted that the contractors were sophisticated businesspersons and that the owner was legally represented as indications that the document should not be con-

strued against either party.[16] The other pointed to the contract having been the result of arm's length bargaining.[17]

Although the contract involved was not one published by a professional society, *W.C. James, Inc. v. Phillips Petroleum Co.*[18] is instructive. The plaintiff, a large pipeline contractor contracted with Phillips, a large gasoline supplier. The contractor contended that the contract had been pre-prepared or "boiler plated" and presented to bidders on a take-it-or-leave-it basis. The trial court noted that the clause in question—one waiving damages for delay—was common in construction work. The trial court stated that the contractor entered into the contract voluntarily, intelligently, and knowingly. The trial court upheld the clause, stating that it was not unfair and that the contract was not one of adhesion.

The Federal Circuit Court of Appeals affirmed the trial court, noting that the pipeline contractor was of sufficient size, even in relationship to Phillips, so that it could not seek relief on the grounds of an adhesion contract.

It is likely that the pipe contractor in the *James* case had no choice. Yet if the contractor was aware that it was taking the risk of delay damages, it could adjust its price and take this risk into account. It is not in the same position as a consumer who is usually not in the position to adjust to the risk that the contract language seeks to place upon her.

The preceding discussion has demonstrated the need for an analysis that will recognize some of the particular problems of dealing with standard contracts published by the AIA or NSPE. First, it is important to look at the surrounding facts

---

14. 186 Neb. 349, 183 N.W.2d 246 (1971).

15. Refer to Section 15.03 for more discussion on this subject.

16. *Robinhorne Construction Corp. v. Snyder*, 113 Ill.App.2d 288, 251 N.E.2d 641 (1969) aff'd 47 Ill.2d 349, 265 N.E.2d 670 (1970).

17. *Cree Coaches, Inc. v. Panel Suppliers, Inc.*, 384 Mich. 646, 186 N.W.2d 335 (1971).

18. 485 F.2d 22 (10th Cir.1973).

and circumstances that led to the use of the standard agreement. It should not always be assumed that the architect dictates that an AIA document be used for either design or construction services. Nor should it be assumed that an owner dictates the use of an AIA document for construction services. It is not inconceivable that an AIA document is selected by each party to the contract because of the reputation of the document for fairness or familiarity. This is more likely to be the case when AIA construction documents are used, unless the owner has dealt with design services before.

Second, changes are frequently made in AIA documents, principally in the area of payments, changes, indemnification, arbitration, and the responsibilities of the design professional. It would be administratively inconvenient to use different standards of interpretation when substantial changes have taken place. Where there have been substantial changes, it is likely that the parties have considered the entire document, in some cases even jointly negotiating the agreement. Where the latter is the case, the agreement should be interpreted neutrally. Where one party has used some clauses from a standard agreement that favor it and then incorporated other clauses more favorable than contained in the standard agreement, clearly the language should be interpreted against that party.

Third, it is possible to interpret any B-series agreements—those that deal with design services—in favor of the client, as the AIA receives no input from other organizations for these documents and is likely to be drafting with the best interests of architects in mind.[19] However, the AIA obtains endorsement and approval for some of its construction services documents, such as A201, from the Associated General Contractors (AGC). This can lead to an assumption

that the agreement has been jointly negotiated by representatives of owners and contractors and should be interpreted neutrally. However, although in some respects the AIA can be said to be thinking of the interests of the owner, there is too much in A201 that is done with the interests of the architect in mind and there is no direct owner representation. It would be a mistake to treat the agreement as if it had been jointly worked out by representatives of owners and contractors.

Fourth, courts usually seek to find the intention of the contracting parties as the "lodestar" of contract interpretation. However, it is not uncommon for neither party to an AIA document to have any intention whatsoever as to certain clauses and their function when they used an AIA document as the basis for its agreement. Does that mean that parties should be allowed to introduce any evidence of the intention of AIA representatives, if any can be obtained? This is analogous to the use of legislative history by courts in interpreting statutes. Inasmuch as there is currently no official AIA publication that seeks to explain the reasons for language it has selected, any such evidence would have to consist of informal letters or affidavits by AIA officials. Such evidence is untrustworthy at best, and the parties would have no opportunity to question those who have sought to describe the intention of the Institute. It is probably best for courts to ignore this evidence and decide intention questions based upon other criteria.

Courts often look at the particular clause in question to determine whether to enforce it or how to interpret it. Courts give indemnification clauses careful scrutiny, either invalidating them entirely or construing them strongly against anyone seeking to be indemnified for its own negligence. Such scrutiny is based on a desire to redress an

---

19. Some cases in which this has been done are *Malo v. Gilman*, 177 Ind.App. 365, 379 N.E.2d 554 (1978), reproduced in Section 15.03, and *Kostohryz v. McGuire*, 298 Minn. 513, 212 N.W.2d 850 (1973).

unequal bargaining position and to avoid an interpretation that could encourage carelessness.

The competitive bidding process has developed special rules. Prospective bidders are given a complex set of construction documents that have been prepared over a lengthy period of time. Bidders are asked to review these materials within the short period of time available to them before submitting a bid. They are also asked to report any errors or inconsistencies that they have observed. This laudable attempt to catch errors and invoke the knowledge and skill of the contractor often generates problems because many errors are not actually detected until after the work is in progress, mainly through the review of shop drawings.

One way of dealing with interpreting construction contracts in this context is to accept the interpretation that would have been given by a reasonably prudent bidder.[20] However, courts have had to struggle, sometimes painfully, with the question of whether the contractor should have sought clarification before bidding because the ambiguity was patent and glaring.[21] A case exploring this problem is reproduced at this point.

NEWSOM v.
UNITED STATES
United States Court of
Claims, 1982.
676 F.2d 647.

SMITH, Judge:

[Ed. note: Footnotes renumbered and some omitted.]

This case is an appeal by petitioner, George E. Newsom, of a decision of the Veterans Administration Board of Contract Appeals (board). The board found that certain parts of the contract for hospital improvements were patently ambiguous and that, having failed to consult with the contracting officer about the ambiguities, petitioner was barred from recovering for work done beyond that required under petitioner's interpretation of the contract. We affirm the decision of the board.

On August 28, 1978, the Veterans Administration (VA) issued an invitation for bids for building medi-prep and janitor rooms in the VA hospital at Knoxville, Iowa. Drawings and specifications for the work to be done were supplied to the prospective bidders.

Paragraphs 4, 5, and 6 of the specifications described, respectively, buildings 81, 82, and 85. Each paragraph had two parts: the first described the first floor of the building and referenced page 7 of the drawings; the second described the second floor of the building and referenced page 8 of the drawings. Conversely, the caption block on page 7 of the drawings indicated that it described work for all three buildings, 81, 82, and 85. However, page 8 of the drawings indicated only building 85.[22] Petitioner at no time inquired about this discrepancy.

As a consequence, petitioner included in his bid the costs of the

---

20. *Corbetta Construction Co. v. United States,* 198 Ct.Cl. 712, 461 F.2d 1330 (1972).

21. *Brezina Construction Co. v. United States,* 196 Ct.Cl. 29, 449 F.2d 372 (1971). See also *Wickham Contracting Co. v. United States,* 212 Ct.Cl. 318, 546 F.2d 395 (1976) (scale of drawings).

22. It is not entirely clear whether a drawing of building 85 on page 8 of the drawings would have also described buildings 81 and 82, or whether separate drawings of buildings 81 and 82 were omitted from page 8. We do not believe that there is a difference of legal significance between these two possibilities.

second floor of building 85 only. He was the low bidder and the contract was awarded to him on October 13, 1978. It was not until March 29, 1979, that the parties realized that there was a discrepancy between what the VA had intended and what petitioner had understood. Petitioner then did the work as intended by the VA at an additional cost of $14,600, and he appealed the decision of the contracting officer denying relief to the Veterans Administration Board of Contract Appeals. The board held against petitioner on the ground that the error on page 8 of the drawings was a patent ambiguity which imposed upon the contractor a duty to inquire about it. Petitioner now appeals that finding to this court under the Contract Disputes Act.

The doctrine of patent ambiguity is an exception to the general rule of *contra proferentem* which requires that a contract be construed against the party who wrote it. If a patent ambiguity is found in a contract, the contractor has a duty to inquire of the contracting officer the true meaning of the contract before submitting a bid.[23] This prevents contractors from taking advantage of the Government; it protects other bidders by ensuring that all bidders bid on the same specifications; and it materially aids the administration of Government contracts by requiring that ambiguities be raised before the contract is bid on, thus avoiding costly litigation after the fact. It is therefore important that we give effect to the patent ambiguity doctrine in appropriate situations.

The existence of a patent ambiguity is a question of contractual interpretation which must be decided de novo by the court. This determination cannot be made upon the basis of a single general rule, however. Rather, it is a case-by-case judgment based upon an objective standard. In coming to our decision, we are bound neither by the legal conclusions of the board, nor by the subjective beliefs of the contractor, subcontractors, or resident engineer as to the obviousness of the ambiguity.

The analytical framework for cases like the instant one was set out authoritatively in *Mountain Home Contractors v. United States*.[24] It mandated a two-step analysis. First, the court must ask whether the ambiguity was patent. This is not a simple yes-no proposition but involves placing the contractual language at a point along a spectrum: Is it *so* glaring as to raise a duty to inquire? Only if the court decides that the ambiguity was not patent does it reach the question whether a plaintiff's interpretation was reasonable. The existence of a patent ambiguity *in itself* raises the duty of inquiry, regardless of the reasonableness . . . of the contractor's interpretation. It is crucial to bear in mind this analytical framework. The court may not consider the reasona-

---

23. *Beacon Constr. Co. v. United States*, 161 Ct.Cl. 1, 6, 314 F.2d 501, 504 (1963); *Blount Bros. Constr. Co. v. United States*, 171 Ct.Cl. 478, 495–96, 346 F.2d 962, 971–972 (1965).

24. *Mountain Home Contractors v. United States*, 192 Ct.Cl. 16, 20–21, 425 F.2d 1260, 1263 (1970).

bleness of the contractor's interpretation, if at all, until it has determined that a patent ambiguity did not exist.[25]

Examining the contract itself, we find that a patent ambiguity existed. Two parts of the contract said very different things: the specifications required construction on the second floors of buildings 81, 82, and 85, whereas the drawings required construction on the second floor of only building 85. It is impossible from the words of the contract to determine what was really meant. The contractor speculated that it meant that part of the project had been dropped along the way. Looking at the same language, the Government can insist that it was clearly a drafting error. We do not consider which interpretation is correct; at this stage we determine only whether there was an ambiguity. What is significant about the differing interpretations is that neither does away with the contract's ambiguity or internal contradiction. There is simply no way to decide what to do on the second floors of buildings 81 and 82 without recognizing that the contract also indicates otherwise.

Mountain Home, discussed above, involved a very similar ambiguity. The specifications ordered inclusion of kitchen fans in certain housing units, but the drawings appeared to indicate that kitchen fans were not to be installed. There is a crucial difference between that case and this, however. In *Mountain Home* the indication on the drawing was that the fans were to be under an alternate bid. Thus, the drawings indicated that the Government was reserving the option of either including the fans in the main contract or ordering them separately. The drawings did not state that the fans were simply not to be included. The *Mountain Home* contract, therefore, was susceptible of an interpretation which did not leave significant ambiguities or internal contradictions. Here, even petitioner's interpretation acknowledges that the contract is not internally consistent. Petitioner's interpretation explains the reason for the inconsistency but does not eliminate it.

We recognize that the instant case does not represent a difference in kind from the *Mountain Home* facts, but this area of the law involves a case-by-case determination of placement along a spectrum. In our opinion, this case is closer to *Beacon Construction*[26] than to *Mountain Home*. In *Beacon Construction*, the specifications stated only that "weatherstrip shall be provided for all doors," while the drawings describing the weatherstripping clearly indicated weatherstripping around the windows as well. The conflict between the specifications and the drawings was direct, as in the instant case. And the court was not swayed by the mere fact that the contractor was able to come up with a highly plausible interpretation of the ambiguity. No interpretation could

---

25. If the court finds that a patent ambiguity did not exist, then the reasonableness of the contractor's interpretation becomes crucial in deciding whether the normal *contra proferentem* rule applies.

26. *Beacon Constr. Co. v. United States, supra* note 3, 161 Ct.Cl. at 4–5, 314 F.2d at 502–03.

in *Beacon Construction,* or can in the instant case, eliminate the substantial, obvious conflict between the drawings and the specifications.

Finally, we emphasize the negligible time and the ease of effort required to make inquiry of the contracting officer compared with the costs of erroneous interpretation, including protracted litigation. While the court by no means wishes to condone sloppy drafting by the Government, it must recognize the value and importance of a duty of inquiry in achieving fair and expeditious administration of Government contracts.

Accordingly, upon consideration of the submissions, and after hearing oral argument, the decision of the Veterans Administration Board of Contract Appeals is

AFFIRMED.

Note the difference between the drawings and the specifications in the *Newsom* case. Often contracts seek to deal with this problem by incorporating a precedence of documents clause, which gives more weight or precedent to one document than to others. This method of dealing with ambiguity in the construction context is discussed in Section 24.03.

Inasmuch as many of the disputes that have dealt with the duty of the contractor to draw attention to defective design have, such as in the *Newsom* case, come before a federal contracting agency appeal board, it may be useful to look at criteria that have been developed by those boards to determine whether a defect should have been brought to the attention of the owner. Some of them are as follows:

1. Did other bidders discover the error and seek clarification?

2. Did skilled professionals for the agency detect the error?

3. Were the construction documents so complicated that a small detail could have been easily missed?

4. Was the cost of correcting the defect a relatively small part of the contract price?

5. Did the defect occur in one item out of many items involved in the bid?

6. Was there a single prime contractor or many (a "multi-prime" job)?

7. Will the contractor make a profit from a failure to inquire? [27]

Interpretation guides are only guides. Courts look principally at the language, the surrounding facts and circumstances, and the acts of the parties. Nevertheless, interpretation guides are useful in close cases and provide the court with a rationalization for a result already achieved.

SECTION 24.03
RESOLVING CONFLICTS AND INCONSISTENCIES

(A) WITHIN THE WRITTEN AGREEMENT

The difficult process of contract interpretation requires a court to put itself in the position of the parties and determine their

---

27. For agency decisions dealing with this issue, see *Pathman Construction Co.,* ASBCA No. 22343, 81–1 BCA ¶ 15,010; *George Hyman Construction Co.,* ENG BCA No. 4506, 81–2 BCA ¶ 15,363; *Sam Bonk Uniform & Civilian Cap Co.,* ASBCA No. 23592, 81–1 BCA ¶ 14,840; *R&C Corp.,* GSBCA No. 6041, 81–2 BCA ¶ 15,369; *Linde Construction Co.,* GSBCA No. 5840, slip op. (1981); *S&W Contracting Co.,* IBCA 1370–10–79, 81–1 BCA ¶ 15,133.

intentions at the time of the agreement. This process is complicated by the modern tendency toward longer agreements. Often long agreements are not carefully examined to avoid conflicts and inconsistencies. This is especially true where a set of standardized general conditions or contract terms is appended to a letter agreement that expresses the basic elements of the contract.

Generally courts interpret the contract as a whole and seek, wherever possible, to reconcile what may appear to be conflicting provisions. It is assumed that every provision was intended to have some effect. Yet this process of reconciliation may not accomplish its objective, and it may be necessary to prefer one provision over another.

Where one clause deals generally with a problem and another deals more specifically with the same problem, the specific takes precedence over the general. The specific clause is likely to better indicate the intention of the parties. For example, suppose a building contract specified that *all* disputes were subject to arbitration but also stated in a different paragraph that disputes as to artistic effect were to be resolved by the design professional whose decision was to be final. It is likely that the parties intended the specific clause dealing with nonreviewability of specific decisions to control the general clause.

If parties have expressed themselves *specifically*, their failure to change the general clause is likely due to a desire to avoid cluttering up the contract with exceptions and provisos. Also, the parties may not have noticed the discrepancy.

Later clauses take precedence over earlier ones. While this is rarely used, it is premised upon the assumption that parties sharpen their intentions as they proceed, much like an agreement made Tuesday displaces one made the preceding Monday.

Operative clauses take precedence over "whereas" clauses that seek to give the background of the transaction. As to inconsistency between printed, typed, and handwritten provisions, handwritten will be preferred over typed and typed preferred to printed. These preferences are premised upon the assumption that the parties express themselves more accurately when they take the trouble to express themselves by hand. Likewise, specially typed provisions on a printed form are more likely to be an indication of the intention of the parties than the printed provisions.

Some of these rules may have to be reevaluated in light of the frequent use of contracts printed by word processors. Such agreements may appear to have been typed particularly for this transaction but may have been prepared for a variety of transactions just as a printed contract form. The test in seeking to reconcile conflicting language within a written document should be whether some language was specially prepared for this transaction and indicated with greater accuracy the intention of the parties.

These priorities are simply guides to assist the court in resolving these questions. Guides to the meaning of language *within* a document are similar to the canons of interpretation that were explored in Section 24.02. Usually each party can point to guides which support its position. What must be done is to focus upon Section 24.01 and its emphasis upon the intention of the parties. Which meaning is most likely to accomplish those objectives without grossly distorting the language?

(B) BETWEEN DOCUMENTS: *UNICON MANAGEMENT CORP. v. U.S.; FRANCHI v. U.S.*

Construction contracts present particularly difficult interpretation questions because of the number and complexity of contract documents and the frequent incorporation of bulky specifications by reference. Suppose work is called for by one document but not specifically required in another. Suppose one document sets up procedures for changes different from that set forth in another document. At the outset it may be instructive to reproduce a case dealing with this problem.

UNICON MANAGEMENT CORP. v. UNITED STATES

United States Court of Claims, 1967.
375 F.2d 804.

DAVIS, Judge. In March 1959 the contractor agreed with the Corps of Engineers to construct, for a fixed price, two phases of the Missile Master Facilities near Pittsburgh. The current claim is that the Government required plaintiff to install in one room a steel-plate flooring which was not called for by the plans and specifications. The contracting officer and the Armed Services Board of Contract Appeals (65–1 BCA para. 4775) refused the demand for an equitable adjustment and this suit was brought. The problem arises because the most pertinent specification, if read alone, could be said to contemplate a wholly concrete rather than a partial steel-plate floor, while the most pertinent drawings, if read alone, direct the steel-plate covering. The contractor resolves the difficulty by relying, mainly, on the contractual clause that "in case of difference between drawings and specifications, the specifications shall govern." The Board and the defendant invoke another provision that "anything mentioned in the specifications and not shown on the drawings, or shown on the drawings and not mentioned in the specifications, shall be of like effect as if shown or mentioned in both." [28] Since the question is one of contract interpretation, we are free to decide the matter for ourselves.

The room was Equipment Room No. 1 (also designated as Room 117) in one of the buildings being erected by plaintiff. To secure heavy equipment, this space was to have pallets along the floor with parallel cable trenches for bringing electric current to the machines. Steel beams were to be used in and along the floor in connection with these pallets. It is agreed that the top cover of the floor, apart from the portions devoted to the pallets and the trenches, was to be a resilient floor tile. There was also to be a concrete base. The dispute is whether the tile was to be laid directly on this concrete or over a quarter-inch steel plate above the concrete. The specifications, in a section of the Technical Provisions on miscellaneous metalwork, contain a paragraph dealing with the floor of this room and with the pallets (TP 17–23).[29]

---

28. [Ed. note: Footnotes renumbered and some omitted.]
These sentences are from Article 2 of the General Provisions which reads as follows:

"The Contractor shall keep on the work a copy of the drawings and specifications and shall at all times give the Contracting Officer access thereto. Anything mentioned in the specifications and not shown on the drawings, or shown on the drawings and not mentioned in the specifications, shall be of like effect as if shown or mentioned in both. In case of difference between drawings and specifications, the specifications shall govern. In any case of discrepancy either in the figures, in the drawings, or in the specifications, the matter shall be promptly submitted to the Contracting Officer, who shall promptly make a determination in writing. Any adjustment by the Contractor without this determination shall be at his own risk and expense. The Contracting Officer shall furnish from time to time such detail drawings and other information as he may consider necessary, unless otherwise provided."

29. "17–23 *Equipment Room No. 1:* AAOC Main Building. Floor shall have steel beams embedded in the concrete floor so that they run transversely to the equipment pallet lengths. The flange surfaces shall be flush with the finished cement. Two beam runs shall be used, one at each end of the pallets, as shown on the drawings. These beams will be used to anchor the equipment pallets, to level the equipment pallets and can be used as references for cement finishing tools during pouring of the floor. These beams shall be connected to the building ground system

By itself, this provision can be interpreted as implicitly envisioning an all-concrete floor; there is no reference to steel plate but there are references to "the concrete floor", "a concrete floor", "pouring of the floor", "pouring of the floors", and steel beams being "embedded" in the concrete floor.

The relevant drawings, by themselves, give a very different impression. The most significant sketch—a cross-section of the floor of Equipment Room No. 1—shows the trench covered by a removable steel cover plate, with a depression into which the plate's handle is fitted flush with the floor level; the open-floor portion of the drawing shows the top of two steel beams covered by ¼″ steel plate and the resilient floor tile on top of this plate. This is a specific directive to use steel plate immediately beneath the tile. To the same effect, another detail drawing of the trenches and beams in the room shows steel plate on top of an "I" beam and the resilient tile above the steel plate. The other drawings are uninformative or unclear on the point, but they do not contradict the sketches explicitly showing the steel plate. The result is that the drawings as a whole affirmatively support the Government's understanding as to the nature of the floor of Equipment Room No. 1.

We agree with the Board that these plans need not, and should not, be construed as in conflict with specification TP 17–23 footnote [22], supra but, instead, as supplementing the latter. The parties directed in the contract that "anything . . . shown on the drawings and not mentioned in the specifications, shall be of like effect as if shown or mentioned in both" (footnote [23] supra). This rule is peculiarly applicable here because TP 17–23 obviously does not cover the entire subject of floor construction. For instance, it does not mention the important item of resilient floor tile, or the metal plates for the trenches, nor does it even describe fully the construction of the steel beams running transversely to the pallets. When the contract is viewed as a whole, the function of the paragraph is seen as calling attention to the steel beams set in the floor to support the pallets—the references to concrete are in that connection and in that special light—not as drawing together all of the requirements on the make-up of the floor. Certainly, as the Board pointed out, the specification does not provide in terms or by necessary implication that the resilient tile is to rest directly on concrete; nothing is said about the tile or its placing. It is therefore proper to read TP 17–23 as open to complementation by the drawings (or by other specifications) insofar as these cover items or aspects other than the features of the steel beams discussed in the former provision. Plaintiff seems to insist that, if

---

so that they will serve as a grounding means for the equipment pallets. The steel beams shall be on one continuous length, with portions removed to provide clearance for the cable troughs. The cross section area of the beams or jumpers, at the points where metal is removed shall provide at least as much conductivity as the ground cables used to connect the beams to the ground system. The grounding system must be installed prior to the pouring of the floors, as specified in Section 47, 'Electrical Work'."

the specification *can* be read as conflicting with the drawings, that reading must be adopted even though a more harmonious interpretation is also reasonably available. The rule, however, which the courts have always preferred is, where possible, to interpret the provisions of a contract as coordinate not contradictory. . . . Contractors, too, have long been on notice that in reading contract documents they should seek to find concord, rather than discord, if they properly can.

This brings us to another reason why plaintiff's position is weak. There is no evidence as to the view actually taken by the company's estimators before it submitted its bid (plaintiff did not call them to the stand in the administrative proceeding). But assuming . . . that they read TP 17–23 the way plaintiff now does, if they examined the plans and specifications carefully they could not have helped notice the drawings which specifically embodied the contrary requirement. If they were not aware of this fact they should have been. The contract provided that "in any case of discrepancy either in the figures, in the drawings, or in the specifications, the matter shall be promptly submitted to the Contracting Officer, who shall promptly make a determination in writing. Any adjustment by the Contractor without this determination shall be at his own risk and expense" (footnote [22] supra). A warning of this kind calls upon the bidder to bring to the Government's attention any serious or patent discrepancy of significance, of which he is or should be cognizant. . . . The discrepancy here—if TP 17–23 was then thought to mean what plaintiff now contends—surely met that standard of importance. Yet there is no suggestion that plaintiff brought it to the contracting officer or sought the required guidance before bidding. If, on the other hand, plaintiff did not study the plans and specifications before bidding, it cannot complain that the Board and this court strive, in accordance with the established canon, to read the relevant contract provisions together rather than at odds.

The plaintiff is not entitled to recover. Its motion for summary judgment is denied and the defendant's is granted. The petition is dismissed.

Other cases have dealt with conflicts in documents. First, like the *Unicon* case, the court attempts to reconcile the apparently conflicting language.[30] When this cannot be done, the court will often base its decision upon contract language that seeks to resolve potential conflicts.[31]

A case that applied the reconciliation

---

30. *Warren G. Kleban Engineering Corp. v. Caldwell,* 361 F.Supp. 805 (N.D.Miss.1973), vacated on other grounds, 490 F.2d 800 (5th Cir.1974); *Contracting & Material Co. v. City of Chicago,* supra note 3.

31. *Graham v. Virginia,* 206 Va. 431, 143 S.E.2d 831 (1965) (specifications over general conditions); *Dunlap v. Warmack-Fitts Steel Co.,* 370 F.2d 876 (8th Cir.1967) (specifications over bid proposal). AIA Doc. B141, ¶ 11.2, requires terms be given same meaning as A201.

concept of *Unicon,* provides an instructive exercise in reading contract documents, and illustrates the use of the federal procurement system precedence of documents clause is reproduced here.

FRANCHI
CONSTRUCTION CO.
v. UNITED STATES
United States Court of
Claims, 1979.
609 F.2d 984.

PER CURIAM.

[Ed. note: A "per curiam" opinion is one by the entire court and not written by a particular judge. Footnotes renumbered and some omitted.]

* * *

Plaintiff criticizes the trial judge's calling the discrepancy between the specifications and the drawings "patent." It says if we agree with defendant that the precedence clause should be confined to instances of latent discrepancies, we should consider whether the discrepancy here was not rather a latent one. In the present case the conflict was not in fact noticed before plaintiff bid, and probably never would have been noticed except by one who was seeking light on the specific problem the discrepant provisions dealt with, and in the course of his investigation, placed them side by side. The trial judge did not define what he meant by a patent discrepancy. It is obvious that no careful writer of contracts would deliberately create a discrepancy, patent or latent, and then leave it to be resolved by the precedence clause. A discrepancy even with the precedence clause therefore indicates a probable mistake, which may be more or less serious. We would assume arguendo that a bidder, who noticed or should have noticed a serious mistake in the invitation or other of the contract documents, must divulge what he has or should have noticed to the government, and will not in equity be allowed to profit by not doing so, as it would be an instance of overreaching. That is not this case, whether the discrepancy be patent or latent. Defendant does not accuse plaintiff of overreaching nor could it do so. The standard clause prescribes a precedence only in case of discrepancies between specifications and drawings, while one in figures, drawings, or specifications, each by themselves, must be promptly reported to the contracting officer. This no doubt reflects defendant's experience that the latter class of discrepancies is much more likely to be discovered early. We cannot in the circumstances say in face of the precedence clause, our characterization of a discrepancy as patent automatically triggers an obligation to report. The clause itself seems designed to excuse such reporting, instances where equity would intervene aside. Accordingly, we do not deem it necessary to address in a critical manner the characterization of the discrepancy as patent.

Since the court agrees with the trial judge's recommended decision and conclusion, as hereinafter set forth, it hereby affirms and adopts the decision and conclusion, together with the paragraph inserted above by the court, as the basis for its judgment in this case.

OPINION OF TRIAL JUDGE

WILLI, Trial Judge:

By this suit plaintiff, a construction contractor, seeks reversal of a decision of the Armed Services Board of Contract Appeals (the Board) affirming the contracting officer's denial of its claim for an equitable adjustment to compensate it for an alleged change in the terms of its fixed-price construction contract to build a three-story, 116-bed hospital for the Army at Fort Devens, Massachusetts.

* * *

The controversy concerns only the sequence in which vinyl-asbestos floor tile was to be installed on the ground floor of the building in relation to the Gypsum Wallboard (GWB) partitions to be constructed there; the question being whether the tile was to abut such partitions after they were erected on the concrete subfloor or was to be laid first, with the partitions placed on top of it. Incongruity as between the specifications and drawings is at the heart of the parties' disagreement over priority of installation.

The dispute did not surface until approximately a year into performance when plaintiff had the building under roof and in a weather-tight condition. At that point it began erecting GWB partitions on the concrete surface of the first floor. When it had installed about 5 or 10 percent of them in that manner, the contracting officer rejected the work contending that the contract required that such partitions be placed on top of the vinyl tile with which the floor area was to be finished. Though disagreeing with that interpretation of the contract, plaintiff removed the partitions complained of and complied with the directive. It thereafter filed a claim under the Changes article. It is the Board's affirmance of the contracting officer's denial of that claim that precipitated this suit.

In all significant respects, the facts of this case, as found by the Board, are beyond legitimate dispute. The result that follows from them depends upon whether the duty of inquiry that devolves on a construction contract bidder confronted with patently conflicting specifications and drawings persists even though the contract on which he is bidding contains a so-called order of precedence clause. If it does, as held by the Board, plaintiff cannot recover. If the clause displaces that duty, by eliminating the conflict, plaintiff prevails. For reasons that will appear, it is held that on the facts of this case the clause predominates.

As required by the Armed Services Procurement Regulations,

Article 2 of the General Provisions of the subject contract reads as follows:

SPECIFICATIONS AND DRAWINGS
(JUNE 1964)

The Contractor shall keep on the work a copy of the drawings and specifications and shall at all times give the Contracting Officer access thereto. Anything mentioned in the specifications and not shown on the drawings, or shown on the drawings and not mentioned in the specifications, shall be of like effect as if shown or mentioned in both. *In case of difference between drawings and specifications, the specifications shall govern.* In case of discrepancy either in the figures, in the drawings, or in the specifications, the matter shall be promptly submitted to the Contracting Officer, who shall promptly make a determination in writing. Any adjustment by the Contractor without such a determination shall be at his own risk and expense. The Contracting Officer shall furnish from time to time such detail drawings and other information as he may consider necessary, unless otherwise provided. [Emphasis added.]

There are specifications dealing specially with GWB and with vinyl floor tile.

As the Board found, section 6C of the specifications, "Gypsum Wallboard (Dry Wall)," is inconclusive on the question at issue. Without elucidation, it directs simply that the base members (runners) of the partitions be "securely attached to floors." The vinyl tile specification is not similarly equivocal, however. Paragraph 6 of section 9E of the specifications, "Flooring; Vinyl-Asbestos Tile," is captioned "Installation of Tile." Subparagraph 6.1 directs that: "Installation of tile shall be deferred until all other work that might cause damage to the flooring has been completed." Subparagraph 6.4 specifies: "Tile shall be laid in accordance with the pattern layout starting from axes that will produce tile against opposite walls of equal width and not less than half the tile width." The Board read these provisions as being consistent *only* with a sequence of construction that provides for the walls or partitions to be installed before the tile is laid.

Pointing squarely in the opposite direction is a note (note 3) on one of the contract drawings,[32] No. 32–02–01, sheet 41, titled "Door Elevations & Interior Finish Details." It directs that: "All partitions with GWB on both sides are to set on top of finish floor." Referring to the meaning and interrelationship of the drawing note and the tile specifications previously quoted, defendant says:

---

32. Though there is a reference on another drawing, No. 32–02–21, sheet 25, titled "Color & Finish Schedule No. 1," under the caption "Color Schedule," to laying vinyl-asbestos floor tile "wall-to-wall," it is clear on the face of the drawing that the reference is only to color configuration; the directive being that the field color pattern is to extend throughout the rooms in which installed, without a perimeter border of tile of a different color.

"Both requirements are explicit and they are inconsistent." Defendant's Opposition to Plaintiff's Motion For Summary Judgment and Defendant's Cross-Motion For Summary Judgment, at 19.

The notation is not, in terms, related to any of the detail drawings showing cross-sections of the base portion of five different types of interior walls and partitions that are shown above it and none of the partitions that are illustrated is of GWB construction on both sides. All of the cross-sections depicted, however, are oriented at their respective bases to a horizontal plane comprised of two closely aligned parallel lines, the upper one of which is labeled "Finish Floor." One of them shows the base of a partition finished with GWB on one side and plaster on the other positioned atop the upper horizontal line identified as "Finish Floor."

Relying on a paragraph [33] of section 3a of the specifications dealing with concrete, plaintiff's principal contention before the Board was that the "Finish Floor" appearing in the drawing note referred to the concrete surface to which tile was to be applied, rather than to the exposed surface of the installed tile. It therefore contended that as used in the note "Finish" meant texture (*i.e.,* monolithic finish), not state of completion of the work. The Board properly rejected this contention as opposed to both the obvious purport of the detail drawings and to the specific acknowledgement of plaintiff's project engineer in a written submission made by him in the early stages of the controversy. . . .

The configuration of the drawings affirmatively discredits plaintiff's interpretation. It is undisputed that in no instance was the completed interior floor surface abutting the partitions shown on the drawings to consist of exposed concrete.[34] Those surfaces were to be of either ceramic or vinyl tile. Accordingly, if the line (labeled "Finish Floor") shown on the drawings as passing under the base of a partition represented tile surface on both sides of it, the line (being straight and uninterrupted) necessarily denoted the same material under the partition.

In sum, the Board correctly concluded that "Finish Floor" as used in note 3 meant vinyl tile. Although, with the drawing note so construed, it recognized that the specifications and drawings

---

33. That paragraph provided in part:

"22. CONCRETE FLOOR AND ROOF SLAB FINISHES. . . .

"22.1 *Monolithic finish.* Except where otherwise specified, the concrete for floor and roof slabs shall be screeded and floated with straightedges to bring the surface to the required finish level with no coarse aggregate visible. The concrete, while still green but sufficiently hardened to bear a man's weight without deep imprint, shall be wood-floated to a true, even plane with no coarse aggregate visible. Sufficient pressure shall be used on the wood floats to bring moisture to the surface. After the surface moisture has disappeared, surfaces shall be steel-troweled to a smooth, even, dense finish, free from blemishes including trowel marks."

34. Paragraph 22.1 of the concrete specifications, note 7, *supra,* does not apply to the surface texture of concrete that is not to be covered. Such surfaces are governed by paragraph 21.1 of the same specifications. That paragraph, titled "Smooth Finish" directs that: ". . . smooth finish shall be given to interior and exterior concrete surfaces that are to be painted or exposed to view as finished work. . . ."

were in irreconcilable conflict as to those areas of the building where both GWB partitions and vinyl tile were to be installed, it went on to deny the claim by construing the contract in a way that it felt satisfactorily removed the contradiction. It invoked two familiar and related principles of interpretation, *viz*, that potentially conflicting provisions should, if their language reasonably permits, be assigned meanings that will place them in harmony rather than discord. *Unicon Management Corp. v. United States*, 375 F.2d 804, 806, 179 Ct.Cl. 534, 537–38 (1967), and that a provision directed to a particular matter prevails over one which is general in its terms. *Hol-Gar Manufacturing Corp. v. United States*, 351 F.2d 972, 980, 169 Ct.Cl. 384, 396 (1965).

The Board reasoned that congruity resulted if the tile specification's scope were confined to only the second and third floors of the building, where no GWB partitions were to be found, leaving the drawing note free to govern the areas (primarily the ground floor) calling for such partitions. The difficulty with this approach is that it relieves conflict by means of segregation rather than integration. Thus, the provisions no longer collide only because their respective coverages have been made mutually exclusive. The coordination of which *Unicon* speaks occurs where competing provisions can reasonably be read to operate compatibly on the *same* aspect of the contract work. There the court found a flooring specification that was silent on the includability of a steel plate to be open to complementation by a drawing calling for such a plate. The language of the tile specification here involved affords no such latitude. That it does not serves both to distinguish the situation at hand from that presented in *Unicon*, *supra*, and to render the specific versus general principle inapplicable. The generality comprehended by that principle must emanate from the contract language. In the case of a provision explicit in its terms, generality is not created by a mere showing that the provision can be subordinated in the definition of one aspect of the contract work and still have room to function viably elsewhere. As the defendant forthrightly confirms, the tile specification is no less complete and specific in terms than is the drawing note. Accordingly, the textual prerequisite to application of the principle is notably lacking in this instance.

Ultimately, the Board alluded to the real crux of this case—the impact of the order of precedence clause. Properly characterizing the interrelationship of the tile specification and drawing note as one of patent conflict, the Board declared: "A contractor may not resolve such a patent conflict between specifications and drawings by resort to the General Provision Article 2 admonition that specifications control over drawings, since that pertains only to such differences as do not require clarification at the bidding stage. See *Peter Kiewit Sons' Company*, ASBCA No. 15855, 71–2 BCA ¶ 8959." 75–1 BCA at 53,160. Though the Board was faithful to the prior holding, albeit *obiter*, on which it relied to support the limitation that it imposed on the scope of the order of precedence clause, it failed to acknowledge earlier Board pronouncements to the contrary. For example, in *Sides Construction Co.*, ASBCA No. 9539, 1964 BCA ¶ 4342 the Board, dealing with a standard order of precedence clause, said: "We have no difficulty in affirming the principle that in such cases of *patent and irreconcilable discrepancy* between the specifications and drawings, the specifications govern." (Emphasis added.) In *Mapp Contracting Co.*, ASBCA No. 12205, 68–1 BCA ¶ 6754, the Board declared that the standard clause "would require that the specifications be given precedence over an antagonistic indication on the drawings" where there was "essential repugnancy" between the two.

It distinctly appears that in order to constitute the "difference" contemplated by the order of precedence clause, specifications and drawings must be in the same degree of affirmative conflict that the present Board deemed to disqualify the clause and impose a duty of inquiry on the contractor. With due deference, that conclusion cannot be permitted to stand. The Government authored the order of precedence clause as a mechanism to automatically remove conflict between specifications and drawings by assigning preeminence to the former. As an additional initiative, the clause imposes an affirmative duty of inquiry on the contractor who is confronted with a particular dilemma, *viz*, discrepancies within, as opposed to between the separate categories of, *inter alia*, drawings and specifications. *Glen N. Allen*, ASBCA No. 12728, 68–1 BCA ¶ 6807, at 31,467. In this case there is no conflict among either the drawings, as a whole, or the constituent provisions of the specifications.

The plaintiff is entitled to take the Government sponsored order of precedence clause at face value. *WPC Enterprises Inc. v. United States*, 323 F.2d 874, 876–77, 163 Ct.Cl. 1, 6–7 (1963). Once its right to do so in the present situation is recognized, no conflict sufficient to occasion inquiry remains for, as the Board acknowledged, the specifications in issue are consistent *only* with the sequence of work from which the plaintiff was compelled to depart. The Board found that this enforced change significantly increased the cost of construction. Plaintiff is therefore entitled to a compensatory equitable adjustment under the Changes article.

* * *

Another technique for dealing with potential conflict is seen in AIA Doc. A201 ¶ 1.2.3, which states:

> The intent of the Contract Documents is to include all items necessary for the proper execution and completion of the Work. The Contract Documents are complementary, and what is required by any one shall be as binding as if required by all. Work not covered in the Contract Documents will not be required unless it is consistent therewith and is reasonably inferable therefrom as being necessary to produce the intended results.

This paragraph warns the contractor to assume that work called for under *any* document will be required even if apparently omitted from other documents. The paragraph requires work not specifically covered if that work is necessary to produce the intended results. Undoubtedly such a provision gives considerable power to the architect inasmuch as she may be called upon to determine whether particular work is reasonably inferred though not specifically required. A cautious contractor must take this into account when submitting a bid. Even though there are limits to the architect's power, the contractor may find itself at the mercy of an architect who wishes to cover up for poor document drafting. Understandably, a provision is needed to take into account that it is not possible to specifically include *every* aspect of work that should be done. However, such a provision can generate performance difficulties if the architect does not exercise this power fairly.

Note that the AIA has chosen not to include a precedence-of-documents clause such as those discussed in the *Unicon* and *Franchi* cases. It believes that such a clause can be a disincentive for contractors to report errors, omissions, or inconsistencies in the design documents. Certainly if such a clause is read to eliminate the need to make inquiry at the bidding or negotiations stage, the AIA's position can be justified. The AIA also believes that inclusion of such a clause can lead to "wooden" interpretation decisions that ignore the documents as a whole. As seen in the *Unicon* and *Franchi* cases, the reconciliation process is not always an easy one. The AIA also believes that there is no standardized precedence system that can be the basis for a provision in a nationally used contract form. It suggests that any parties that wish to include such a provision can simply add it.

Other arguments against a precedence-of-documents clause can be made. As the *Franchi* case shows, it may be difficult to know what to do with notes that appear on the drawings. The common law has developed a method of reconciling apparently inconsistent clauses within a writing and among writings, which may do as good a job as a precedence-of-documents clause, particularly where the language can be easily supplemented by evidence of industry custom and a course of dealing between the parties. As can be seen in the Court of Claims decisions reproduced in this section, not only is it often difficult to determine whether a precedence clause diminishes or eliminates the obligation to inquire, but it also requires mental gymnastics to distinguish differences within a document from those differences that appear between documents. A precedence-of-documents clause does not always deal well with problems where the evidence of a lesser ranked document is strong while the evidence of a more highly ranked document is weak.

Yet the frequent inclusion of precedence-of-documents clauses must indicate that they have their value. In the often insoluble interpretation problem, they may provide some method of arriving at a solution. It is likely that this is the principal reason why they are used despite their weaknesses.

To conclude this subsection it may be instructive to examine a case that demonstrates the complexity that can be generated by the wealth of contract documents. *William F. Klingensmith, Inc. v. United States* [35] involved a contract dispute between the contractor and the U.S. General Services Agency over the base layers of certain roadways and paved areas. The contractor asserted that it was required only to use "bank run gravel" for a particular base layer. However, the GSA took the position that macadam was required.

The specifications supported the GSA's position, although the drawings contained

---

35. 205 Ct.Cl. 651, 505 F.2d 1257 (1974).

details that supported the contractor's position. However, the general conditions expressly provided that "in case of difference between drawings and specifications, the specifications shall govern." The court concluded that because there was no ambiguity in the specifications, there was no need to invoke the rule that ambiguities are resolved against the drafter.

Yet the contractor had another arrow in its quiver. The contract had incorporated certain provisions of the Maryland State Roads Commission's specifications dealing with materials, highways, and bridges. The Maryland specifications stated that in the event of discrepancy between plans and specifications, the *plans* controlled. However, this arrow did not reach the target. The court concluded that the *specific* precedence of documents language in the contract *itself* took precedence over precedence language of specifications *incorporated* by reference. The specific controls the general.

A number of other contractor arrows did not reach the mark. The contractor's contention that macadam had not been used as a base course in the locality in recent years was rejected. Likewise, its showing that the government's estimator also assumed gravel could be used was not persuasive. (Evidently the government estimator also relied upon the drawings.)

The contractor's estimator, having noticed the conflict between drawings and specifications, called the architect-engineer who had been employed by GSA but was told that the matter was no longer in its hands and suggested that the estimator seek clarification from GSA. Upon requesting such a clarification from GSA, the estimator was advised that GSA could not respond to an inquiry on the telephone "since such clarifications were required to be made through a formal addendum issued to all prospective bidders." This was impossible, as bid opening was only three hours away.

The contractor claimed that this conversation showed that the government *knew*

the contractor's interpretation and was bound by it. However, the court rejected this, pointing to provisions of the Invitation for Bids stating that requests for clarifications or interpretations must be submitted within a specified period of days from bid opening and the request was too late. The court concluded by stating that resolving ambiguities against the drafter cannot be done in federal government contracts where the contractor does not seek a clarification of potential conflicts.

The *Klingensmith* case demonstrates the immense complexity in construction document interpretations. It also shows the importance of reading specifications carefully before preparing bids and allowing enough time to seek clarifications.

SECTION 24.04
REFORMATION OF CONTRACTS

Sometimes parties to a contract reduce their agreement to a writing, and for various reasons, the writing does not correctly express the intention of the parties. The "equitable" remedy of "reformation" rewrites the contract to make the writing conform to the actual agreement of the parties. Because there were no juries in equity cases, courts felt freer to go into the difficult questions of intention and mistake.

Most cases have involved an improper description of land in a deed. These descriptions were complicated and often taken from old deeds and tax bills. It was not unusual for the person copying the old description to make a mistake. If it could be shown by reliable evidence that there had been a mistake, the court would reform the contract. This would be a judicial declaration that the contract covered the particular land that the parties actually intended to buy and sell rather than the land covered in the description.

In addition to the cases of mutual mistake in description, other types of mistake would justify reformation. Usually these mistakes would involve the process by which the actual agreement was reduced to writing.

The reformation doctrine has been given an interesting application by the U.S. Court of Claims in federal procurement contracting. In *National Presto Industries, Inc. v. United States,*[36] the contractor under a fixed-price contract sought more money when its costs were substantially increased because a method that it had intended to use was not feasible for mass production. The court found mutual mistake in that the contractor and the government each assumed that the particular process in which the U.S. had interest could be used. Usually the remedy for mutually mistaken fundamental assumptions is to relieve the performing party—in this case the contractor—from the obligation to perform. But the Court of Claims extended reformation beyond its normal correction of mistakes function to change the fixed-price contract to one of a joint enterprise in which each joint enterpriser would share the unforeseen expenses.

Such a doctrine can impair the certainty and risk assumption features of procurement. As a result, two years later the Court of Claims seemed to have had second thoughts about its decision in *National Presto* and sought to make clear that the reformation remedy awarded in *National Presto* could be justified only where there is a joint enterprise experimental situation in which neither party has assumed the particular risk, a great concern on the part of the government in the process and not merely the end product and a distinct benefit to the government from the contractor's period of trial and error.[37] Nevertheless, in exceptional circumstances, reformation may be the vehicle to redistribute risks at least in federal procurement.[38]

Reformation can correct writings that do not reflect the actual agreement of the parties,[39] but the requirements for invoking this equitable remedy are formidable. Parties obviously cannot rely upon the possibility that the law will correct their mistakes. They must make every effort to make the writing conform to their actual agreement.

PROBLEMS

1. O owned a large house. He entered into negotiations with C regarding the painting of the house. They signed a written contract under which it was agreed that C would receive $2,000. C drafted the contract. In part the contract stated:

   > C will do a first-class job of painting O's house.
   >
   > (Garage not included.)

   A dispute arose over the number of coats of paint to be used and whether C was to paint the cement steps leading to the house. C claimed two coats were adequate, while O stated that three coats were needed for a first-class job. C claimed the contract did not include the cement steps. O states that he pointed to the steps when they inspected the house and that he said that the steps needed a coat of paint. C admitted O said this, but C contends that when it was not included in the contract, he assumed he would not

---

36. 167 Ct.Cl. 749, 338 F.2d 99 (1964), certiorari denied 380 U.S. 962, 85 S.Ct. 1105, 14 L.Ed.2d 153 (1965).

37. *Natus Corp. v. United States,* 178 Ct.Cl. 1, 371 F.2d 450 (1967).

38. *Dynalectron Corp. v. United States,* 207 Ct.Cl. 349, 518 F.2d 594 (1975).

39. *Timber Investors, Inc. v. United States,* 218 Ct.Cl. 408, 587 F.2d 472 (1978) (mistaken estimates in quantities). Refer also to Section 21.02(C).

have to do it.   How should a judge decide this dispute?

2. A contract for the construction of a major water delivery pipeline required the contractor to excavate a trench, place bedding material in the bottom of the trench, install the pipe, and place backfill material around and above the pipe.   The contract specified four zones of bedding and backfill material—Zones 1, 2, 3, and 4.   Zone 3 backfill material was to be placed at certain points depending upon whether steel or concrete pipe was used.   Also, depending upon the type used, all or a portion of the backfill was to be compacted.

The contract provided that backfill used in Zones 1 and 2 should contain no material larger than three-quarters of an inch.   The dispute involves the size limitation, if any, of Zone 3 material.   The specific provision dealing with Zone 3 backfill reads:

> 12.2.44 *Zone Backfill Material.* Zone 3 backfill material shall consist of selected material from the trench excavation, free from frozen material and lumps or balls of clay, organic or other objectionable material. When compaction of

Zone 3 backfill is called for the material shall be well graded and easily compacted throughout a wide range of moisture content.   Alternatively, if flooding, jetting and vibration are to be used for placing and compaction, the material shall meet the additional requirements specified in paragraph *Zone 1 and Zone 2 Bedding Material* for material to be placed and compacted by flooding, jetting and vibration.   The maximum size shall pass a 2-inch U.S. Standard Series sieve.

The contractor contended that no size limitation existed whatsoever on Zone 3 backfill material, unless compaction was by flooding, jetting, or vibration, in which case a two-inch limitation existed.   The owner contended that all Zone 3 backfill material must meet the two-inch test.

a. If this dispute came before a court, should it look at evidence outside the writing?

b. What evidence would be relevant if the court were willing to look beyond the writing?

c. Assuming that all the relevant evidence is examined, how should the clause be interpreted?

25

Changes: Complex Construction Centerpiece

SECTION 25.01 FUNCTIONS OF A CHANGES CLAUSE: *WATSON LUMBER CO. v. GUENNEWIG*

After award of a construction project, the owner may find it necessary to order changes in the work. The contract documents are at best an imperfect expression of what the design professional and owner intend to be performed by the contractor. Circumstances during the Construction Process develop that may make it necessary or advisable to revise the drawings and specifications.

Design may prove to be inadequate. Methods specified become undesirable. Materials designated become scarce or excessively costly. From the owner's planning standpoint, program or budget may change. Natural events may occur that necessitate changes. For any of many reasons it often becomes necessary to direct changes after the contract has been awarded to the contractor.

"Changes" questions can arise in at least two contexts. Most cases dealing with changes involve contentions by contractors that they have performed work not included in the contract requirements for which they should be paid extra. This is a principal issue in *Watson Lumber Co. v. Guennewig*, reproduced in this section.

Sometimes contractors defend an action brought against them by the owner by asserting that the original contract requirements were changed and the contractor has complied with the obligations as changed.

Preliminarily a differentiation should be made between *changes*, *extras*, and *deletions*, though the AIA encompasses all of them in AIA Doc. A201, Art. 12, dealing with changes. However, reference shall be made in this chapter to "extras," that, for example, being the way the court classified the problem involved in the *Watson* case. This *type* of change highlights the difficulties that a long post-completion list of extras can present to the owner or those furnishing funds for the project. Extras usually involve additional or more expensive items than those in the original contract documents. Conversely, a deletion, depending upon how it is priced, can adversely affect planning. Although technically both are changes, additions or deletions that affect contract price or time are sensitive areas.

A slight variant of this differentiation involves a comparison between *additional* work and changed work. This can relate to whether or not the owner *must* order changes from the contractor (discussed in Section 25.02).

Changes must be contrasted with modifications and waivers. Modifications are two-party agreements in which owner and contractor mutually agree to change portions of the work. They have been discussed in various contexts in Sections 5.11(C), 23.01(H), and 29.01(D). (One way to avoid formal requirements in a changes clause is to conclude that the work in question was a modification agreed upon by the parties, *not* a change. See Sections 25.03(A) and 25.04(H). A change, on the other hand, is the term used in construction contracts that allows the owner to unilaterally direct that changes be made *without* obtaining consent to perform the work by the contractor.

A waiver is generally based upon the owner's acts that either manifest an intention to dispense with some of the contractual requirements or lead the contractor to reasonably believe that the owner is giving up its right to have required work performed. In such a case, the contractor can recover the full contract price despite not complying with the contract documents (discussed in Section 23.06(E)).

What have been described are changes in their classic form. An understanding of how changes work in construction contracts requires some appreciation of the changes process. An appreciation of that process demonstrates the multifaceted function of the changes clause.

This can be demonstrated by looking at Article 12, the changes clause in AIA Doc. A201. Although ¶ 12.1 recognizes the classic use of the changes clause by giving the owner the authority to change the work, ¶ 12.1.1 states that contract price and time "may be changed only by Change Order." Sprinkled throughout A201 are provisions stating that under certain circumstances (as illustrations ¶ 5.2.3 authorizes a price change if there is a substitution of subcontractors and ¶ 12.2 grants the contractor an equitable adjustment if certain unforeseen subsurface conditions are encountered) "an appropriate Change Order shall be issued." This recognizes that under certain circumstances the contract price and the completion date may be changed without a design

change, the classic use of the changes clause. In that sense, the changes clause operates as a bookkeeping system, with the owner knowing *what* it owes, the contractor knowing what it is entitled to receive, and each knowing the required date of completion.

In addition, ¶ 12.1.1 states that a change order signed by the contractor "indicates his agreement therewith, including the adjustment in the Contract Sum, or the Contract Time." This recognizes that it is useful for the parties to agree on any matters upon which there is disagreement and which may be the basis for subsequent claims. The change order process can function as a mechanism for resolving disputes between owner and contractor.

The changes process is an important element of cost control. Owners and, perhaps more important, lenders, fear a "loose" changes mechanism. They are fearful that bidders of questionable honesty and competence will bid low on a project with the hope that clever and skillful post-award scrutiny of the drawings and specifications will be rewarded by assertions that requested work is not *required* under the contract and generate claims for additional compensation. A changes clause, which either gives too much negotiating power to the contractor, or too much discretion to the design professional or arbitrator, will convert a fixed price contract into an open-ended cost type tied to a generous allowance for overhead and profit. This is the reason owners and lenders want tight, complete specifications, a mechanical pricing provision, such as unit pricing, and limits on overhead and profit. Their horror is the prospect that the end of the project will witness a long list of claimed extras which, if paid, will substantially increase the ultimate construction contract pay out.

A case exploring the changes mechanism mainly from the latter perspective is reproduced here.

WATSON LUMBER
COMPANY v.
GUENNEWIG
Appellate Court of Illinois,
1967.
79 Ill.App.2d 377, 226
N.E.2d 270.

EBERSPACHER, Justice. The corporate plaintiff, Watson Lumber Company, the building contractor, obtained a judgment for $22,500.00 in a suit to recover the unpaid balance due under the terms of a written building contract, and additional compensation for extras, against the defendants William and Mary Guennewig. Plaintiff is engaged in the retail lumber business, and is managed by its president and principal stockholder, Leeds Watson. It has been building several houses each year in the course of its lumber business.

* * *

[Ed. note: The project was a four bedroom, two bath house with air conditioning for a contract price of $28,206. The total amount claimed as extras and awarded by the trial court was $3840.09.]

The contractor claimed a right to extra compensation with respect to no less than 48 different and varied items of labor and/or materials. These items range all the way from $1.06 for extra plumbing pieces to $429.00 for an air-conditioner larger than plaintiff's evidence showed to be necessary, and $630.00 for extra brick work. The evidence, in support of each of these items and circumstances surrounding each being added, is pertinent to the items individually, and the evidence supporting recovery for one, does not necessarily support recovery for another.

* * *

Most of the extras claimed by the contractor were not stipulated in writing as required by the contract. The contractor claims that the requirement was waived. Prior to considering whether the parties, by agreement or conduct dispensed with the requirement that extras must be agreed to in writing, it should first be determined whether the extras claimed are genuine "extras". We believe this is an important area of dispute between these parties. Once it is determined that the work is an "extra" and its performance is justified, the cases frequently state that a presumption arises that it is to be paid for. . . .

No such presumption arises, however, where the contractor proceeds voluntarily; nor does such a presumption arise in cases like this one, where the contract makes requirements which any claim for extras must meet. . . .

* * *

The law assigns to the contractor, seeking to recover for "extras", the burden of proving the essential elements. . . . That is, he must establish by the evidence that (a) the work was outside the scope of his contract promises; (b) the extra items were ordered by the owner, . . . (c) the owner agreed to pay extra, either by his words or conduct, . . . (d) the extras were not furnished by the contractor as his voluntary act, and (e) the extra items were not rendered necessary by any fault of the contractor. . . .

The proof that the items are extra, that the defendant ordered it as such, agreed to pay for it, and waived the necessity of a written stipulation, must be by clear and convincing evidence. The burden of establishing these matters is properly the plaintiff's. Evidence of general discussion cannot be said to supply all of these elements.

The evidence is clear that many of the items claimed as extras were not claimed as extras in advance of their being supplied. Indeed, there is little to refute the evidence that many of the extras were not the subject of any claim until after the contractor requested the balance of the contract price, and claimed the house was complete. This makes the evidence even less susceptible to the view that the owner knew ahead of time that he had ordered these as extra items and less likely that any general conversation resulted in the contractor rightly believing extras had been ordered.

In a building and construction situation, both the owner and the contractor have interests that must be kept in mind and protected. The contractor should not be required to furnish items that were clearly beyond and outside of what the parties originally agreed that he would furnish. The owner has a right to full and good faith performance of the contractor's promise, but has no right to expand the nature and extent of the contractor's obligation. On the other hand, the owner has a right to know the nature and extent of his promise, and a right to know the extent of his liabilities before they are incurred. Thus, he has a right to be protected against the contractor voluntarily going ahead with extra work at his expense. He also has a right to control his own liabilities. Therefore, the law required his consent be evidenced before he can be charged for an extra . . . and here the contract provided his consent be evidenced in writing.

The amount of the judgment forces us to conclude that the plaintiff contractor was awarded most of the extra compensation he claims. We have examined the record concerning the evidence in support of each of these many items and are unable to find support for any "extras" approaching the $3,840.09 which plaintiff claims to have been awarded. In many instances the character of the item as an "extra" is assumed rather than established.[1] In order to recover for items as "extras", they must be shown to be items not required to be furnished under plaintiff's original promise as stated in the contract, including the items that the plans and

---

1. [Ed. note: Footnotes renumbered and some omitted.]

We cite as some examples: An extra charge was made for kitchen and bathroom ceilings, concerning which Watson testified that he was going to give these as gifts "if she had paid her bill." According to the testimony, the ceilings were lowered to cover the duct work. We consider it unlikely that the parties intended to build a house without duct work or with duct work exposed. Likewise an "extra" charge was made for grading, although the contract clearly specifies that grading is the contractor's duty. An "extra" charge is sought for enclosing the basement stairs, although the plans show the basement stairs enclosed. An extra charge is sought for painting, apparently on the basis that more coats than were provided in the contract were necessary.

specifications reasonably implied even though not mentioned. A promise to do or furnish that which the promisor is already bound to do or furnish, is not consideration for even an implied promise to pay additional for such performance or the furnishing of materials. The character of the item is one of the basic circumstances under which the owner's conduct and the contractor's conduct must be judged in determining whether or not that conduct amounts to an order for the extra.

The award obviously includes items which Watson plainly admits "there was no specific conversation". In other instances, the only evidence to supply, even by inference, the essential element that the item was furnished pursuant to the owner's request and agreement to pay is Mr. Watson's statement that Mrs. Guennewig "wanted that". No specific conversation is testified to, or fixed in time or place. Thus it cannot be said from such testimony whether she expressed this desire before or after the particular item was furnished. If she said so afterward, the item wasn't furnished on her orders. Nor can such an expression of desire imply an agreement to pay extra. The fact that Mrs. Guennewig may have "wanted" an item and said so to the contractor falls far short of proving that the contractor has a right to extra compensation.

* * *

Many items seem to be included as "extras" merely because plaintiff had not figured them in the original cost figures.

It is clear that the contractor does not have the right to extra compensation for every deviation from the original specification on items that may cost more than originally estimated. The written contract fixes the scope of his undertaking. It fixed the price he is to be paid for carrying it out. The hazards of the undertaking are ordinarily his. . . .

* * *

"If the construction of an entire work is called for at a fixed compensation, the hazards of the undertaking are assumed by the builder, and he cannot recover for increased cost, as extra work, on discovering that he has made a mistake on his estimate of the cost, or that the work is more difficult and expensive than he anticipated." 17A C.J.S. Contracts § 371(6), p. 413.

Some so called "extras" were furnished, and thereafter the owner's agreement was sought.[2] Such an agreement has been held to be too late. . . .

---

2. The drain tile around the foundation of the house, according to the evidence, was already in place when it was disclosed to the owner that it was more expensive material. It was only then that the contractor secured the owner's consent to pay for one-half the cost of the more expensive material.

The judge, by his remarks at the time of awarding judgment, shows that the definition of extras applied in this case was, indeed, broad. He said,

> "substantial deviation from the drawings or specifications were made—some deviations in writing signed by the parties, some in writing delivered but not signed, but nevertheless utilized and accepted, some delivered and not signed, utilized and not accepted, some made orally and accepted, some made orally but not accepted, and some in the trade practice accepted or not accepted".

While the court does not state that he grants recovery for all extras claimed, he does not tell us which ones were and were not allowed. The amount of the judgment requires us to assume that most were part of the recovery. It can be said with certainty that the extras allowed exceeded those for which there is evidence in the record to establish the requirements pointed out.

Mere acceptance of the work by the owner as referred to by the court does not create liability for an extra. . . . In 13 Am.Jur. 2d 60, "Building & Cont." § 56, it is stated, that: "The position taken by most courts considering the question is that the mere occupancy and use do not constitute an acceptance of the work as complying with the contract or amount to a waiver of defects therein". Conversation and conduct showing agreement for extra work or acquiescence in its performance after it has been furnished will not create liability. . . . More than mere acceptance is required even in cases where there is no doubt that the item is an "extra". . . .

The contractor must make his position clear at the time the owner has to decide whether or not he shall incur extra liability. Fairness requires that the owner should have the chance to make such a decision. He was not given that chance in this case in connection with all of these extras. Liability for extras, like all contract liability, is essentially a matter of consent; of promise based on consideration. . . .

The Illinois cases allow recovery for extra compensation only when the contractor has made his claim for an extra, clear and certain, before furnishing the item, not after. They are in accord with the comments to be found in 31 Ill.Law Rev. 791 (1937). There the author, after reviewing the cases, makes the following analysis:

> "The real issue in these cases is whether or not the contractor has, at the time the question of extra work arises, made his position clear to the owner or his agents and that would seem to be the true test in situations where a written order clause is sought to be disregarded. If he does expressly contend that work demanded is extra, the owner certainly cannot be said to be taken unawares, and if orders are given

to go ahead it is with full knowledge of the possible consequences."

The contractor claims that the requirement of written stipulation covering extras was waived by the owner's conduct. The defendants quite agree that such a waiver is possible and common but claim this evidence fails to support a waiver of the requirement. There are many cases in which the owner's conduct has waived such a requirement. . . . In all the cases finding that such a provision had been waived thus allowing a contractor to collect for extras, the nature and character of the item clearly showed it to be extra. Also, in most cases the owner's verbal consent of request for the item was clear beyond question and was proven to have been made at the time the question first arose while the work was still to be finished. The defendants' refusal to give a written order has in itself been held to negative the idea of a waiver of the contract requirement for a written order. . . . We think the waiver of such a provision must be proved by clear and convincing evidence and the task of so proving rests upon the party relying upon the waiver. . . .

[Ed. note: The court ordered a new trial, stating that the contractor could recover only for those extras he could prove were ordered as such by the owner in the proper form, unless he could show that the owner waived the requirements of a writing by clear and convincing evidence.]

Undoubtedly, the court is correct in emphasizing the owner's right to know whether particular work will be asserted as extra. However, in a small construction project, the Illinois court's requirements would place an inordinate administrative burden on the contractor. It would not only have to make clear its position that particular work was extra but also obtain the written change order executed by the owner. If it did not obtain a written change order, it *might* be able to assert the doctrine of waiver if it could persuade the court that the work was extra and that the owner was made aware of this and allowed the work to proceed.

A contractor in a small project, such as that involved in the *Watson* case, might better add a contingency in the contract price to cover small extras that are very likely to be requested rather than comply with the excessively formal requirements set forth by the court in the *Watson* case.

The *Watson* case shows that courts often ignore the cost of complying with rules of law. Undoubtedly, larger projects will bear the administrative costs of doing things correctly and "according to the book." But smaller jobs may not permit "to the book" contract administration.

That Illinois has not retreated from the formidable requirements set forth in the *Watson* case is demonstrated by *Watson Lumber Co. v. Mouser.*[3] In rejecting a claim

3. 30 Ill.App.3d 100, 333 N.E.2d 19, 29 (1975).

for numerous extras claimed by the contractor, the court stated:

> The contractor does not offer any evidence of the specific time, place or nature of any conversation with, or conduct of, purchasers concerning these alleged extras. To the contrary, the evidence fails to disclose that the purchasers made any response at all to the floor plan as it was being implemented. If the floor plan, as implemented, was as radically different from the original contract as contractor would have us believe, we consider it inherently improbable that some specific acts or conversations by purchasers thereto appertaining would not have resulted. Such specific evidence is required for the recovery of extras, particularly where no formal bill for such alleged extras is ever submitted. In short, it was the contractor's burden to prove each and every element of each and every alleged extra by clear and convincing evidence. This he has here failed to do.

Returning to the various functions served by the changes process designed principally for classic design changes, that process, at least under AIA Doc. A201, involves the following:

1. The ordering of a change or a direction the contractor contends is a change in the work.

2. Methods for the contract or the parties to price the change and its effect on time requirements.

3. A residuary provision that controls the price in the event the parties do not agree.

This residual provision consists of the architect determining the price adjustment based upon "the reasonable expenditures and savings" and a reasonable allowance for overhead and profit. This will be looked at in greater detail later. What is important to note at this point is that a changes clause that states that the contract sum can be changed only by change order and that particular circumstances that entitle the contractor to additional compensation must be determined by change order may invoke a changes mechanism that is more suitable to classic changes.

SECTION 25.02 SHIFTS IN BARGAINING POWER

To appreciate the centrality of the changes process to construction, the shifts in bargaining power because of changes must be appreciated. When an owner is preparing to engage a contractor, as a rule the owner has superior bargaining power. The hotly competitive construction industry and the frequent use of competitive bidding usually allow the owner to control many aspects of the construction contract terms.

A few contractors will deliberately bid low and drive out more prudent and experienced contractors with the hope that they will be able to demand and receive additional compensation by pointing to design ambiguities and amassing large claims toward the end of the project.

The contractor who is performing moves into a much stronger bargaining position. This is clearly so if the owner *must*, for either practical or legal reasons, order any additional work from the contractor. It would be in an even stronger bargaining position if it could refuse to execute the change unless there were a mutually satisfactory agreement on the effect of the change on price or time. Yet any bargaining advantages to the contractor by being in the position to refuse to do the work until there is an agreement are usually avoided by contract provisions that require the contractor to do the work even if there is no agreement on the price or time. (To counter this, a contractor can assert that the direction is a cardinal change.) This can be made even worse by the dominance some owners have over the changes process through their control over the purse strings. This power, exercised either directly or through the design professional, to withhold payment until the contractor agrees on price and time can exert immense pressure upon the contractor to accept whatever the owner or design professional is willing to pay.

The changes mechanism can operate adversely to the contractor if the owner makes many small changes but is niggardly in its proposals for adjusting the price. (This can backfire, however, generating claims by the contractor, particularly in a losing contract, that the cumulative effect of any changes has created a "cardinal" change.)

To sum up, the changes mechanism on the whole favors the owner except if it is dealing with a clever claims-conscious contractor who uses the changes mechanism to extract large amounts of money at the end of the job. Judicial resolution of disputes that involve changes may take into account this factor, particularly if the owner seems to have abused its power.

SECTION 25.03
CHANGES CLASSIFIED

In addition to the classifications noted in Section 25.01, two others, mainly developed in federal procurement law, merit special attention.

(A) CARDINAL CHANGE

Though obviously essential to construction, the changes clause, as suggested earlier in this chapter, is amenable to abuse. Although a contractor can expect *some* changes, the owner should not be given a blank check to order the contractor to do *anything* it wishes, compel it to perform, and take care of compensation for the changed work and *nothing else* later. As a result, changes clauses generally limit the power to the general scope of the work. For example, an owner might be able to order a 10% increase in floorspace of a residential home, a carport, or even a swimming pool. But it could not order a beach house be built twenty miles away.

The cardinal change—the current method of designating a direction that goes beyond the power granted by a normal changes clause—developed not primarily to deal with an abuse of the changes clause. It developed because of jurisdictional aspects of federal procurement. Before 1978, a dispute between a contractor and a federal procurement agency would have to first go before an agency board of contract appeals if it arose *under the contract*. To bypass the Board and bring the dispute before the U.S. Court of Claims the contractor had to show a *breach of the contract*. To accomplish this, the contractor would seek to show a course of conduct—usually a large number of changes, drastic changes, or other wrongful agency conduct that it could establish as a breach of contract. (Conversely, this jurisdictional "divide" generated the constructive change fiction to prevent or avoid excessive claims being brought before the Court of Claims.)

Despite legislation enacted in 1978 that allows the contractor to choose which route to take,[4] the concept still has utility in both federal procurement and private disputes.

Clearly any changes clause must be interpreted. Not *any* change can be ordered. A direction that goes beyond the scope of the work, if that is the standard, need not be obeyed. Such an order shall be called a "one-shot" scope change, to distinguish it from the more common "nibbling" or "aggregate of changes," currently the most common claim.

Suppose the contractor complies. Now there has been an agreement. But does it fall within the jurisdiction of the changes clause with its procedural and pricing mechanisms? Of course, the facts may indicate that the contractor, knowing it could not be compelled to perform because the order went beyond the scope of the work, agreed to waive any "beyond the scope" defense and allow the work to be governed by the changes clause. In the absence of such evidence, the parties have simply made a new agreement. If they cannot

---

4. 41 U.S.C.A. §§ 601 et seq.

agree upon price, the contractor is entitled to reasonable compensation.[5]

A cardinal change is considered a *breach*, largely because of its origin in federal procurement jurisdictional issues. Can the contractor choose to treat it as a serious material breach and terminate any further contractual obligations it owes the owner?[6]

Now the real nature of the direction or order can emerge. It is simply a proposal, an order, that the contractor can *choose* to accept. It is *not* a breach of contract unless the owner asserts that it will *not* proceed further unless it *is* accepted. If so, this is a contract repudiation, and the contractor can terminate and recover any damages it may have suffered.

The other type of cardinal change does not occur "all at once," as in the one-shot direction. It occurs over the performance of the contract. It consists of many changes,[7] drastic changes,[8] or other conduct that has gone beyond the reasonable expectations of the contractor and made the transaction different than what the parties had in mind when they made the contract.[9] This *is* a breach. But what are the remedies?

Here two scenarios emerge. One involves the contractor deciding in the *middle* of the project that it has "had enough," that it feels that it can walk off the job because

the total effect of the owner's conduct constitutes a material breach.[10] If upheld, the contractor can recover the reasonable value of its work based upon restitution[11] or damages.[12] (The former is selected in losing contracts.)[13]

The second, and more common, involves *completion* of the work by the contractor and a claim for the reasonable value of the services and materials without regard for the contract price or damages.[14]

Whether the owner can direct an "acceleration" is discussed in Chapter 30 dealing with time.[15]

(B) CONSTRUCTIVE CHANGE

Again, reference must be made to federal procurement law. Claims for breach of contract could be taken *only* to the Court of Claims before 1978. A claim had to go before any procuring agency appeals board if it arose "under the contract." Relief had to be provided by statute or by a clause in the contract. To keep claims within the appeals board, the constructive change developed. Clearly, had a change order been issued, any claim came *under* the contract.

In many cases the contractor claimed that a direction or order was a change. The contracting officer acting for the procuring agency refused to issue a change order be-

---

5. *Nat Harrison Associates, Inc. v. Gulf States Utilities Co.,* 491 F.2d 578 (5th Cir.1974). Refer also to Section 8.04(B).

6. See Section 38.04(A).

7. *Wunderlich Contracting Co. v. United States,* 240 F.2d 201 (10th Cir.1957) (6000 changes).

8. *Saddler v. United States,* 152 Ct.Cl. 557, 287 F.2d 411 (1961) (doubling excavation in small contract).

9. *Allied Materials & Equipment Co. v. United States,* 215 Ct.Cl. 406, 569 F.2d 562 (1978). *Wunderlich Contracting Co. v. United States,* 173 Ct.Cl. 180, 351 F.2d 956 (1965) (cumulative effect of magnitude and quality of changes).

10. See Section 38.04(A).

11. See Section 31.02(E).

12. See Section 31.02(D).

13. *Rudd v. Anderson,* 153 Ind.App. 11, 285 N.E.2d 836 (1972). See also Section 31.02(D).

14. *Luria Brothers & Co. v. United States,* 177 Ct.Cl. 676, 369 F.2d 701 (1966) (home office overhead and lost productivity). See also Section 31.02(F).

15. See Section 30.03.

cause he asserted that the work ordered was *within* the contract. The fiction of the constructive change allowed the boards of appeal to take jurisdiction by concluding that a change *should* have been issued if it were to later agree with the contractor. It is then *as if* a change order *had* been issued. The doctrine was also used if specifications were defective requiring the contractor to do additional work. Even though the doctrine has no *jurisdictional* significance because of 1978 federal legislation,[16] it still merits comment.

Federal procurement disputes clauses require the contractor to keep working pending resolution of a dispute. If a contractor wishes to stop work, it may assert that its claim is based upon breach and not arising *under* the contract. Here the agency may assert that there has been a constructive change and the contractor must continue *under* the contract. The changes clause precludes claims after final payment. Post-final payment claims may be asserted to be based upon *breach* not "under the contract." In private contracts the problem can still arise although not in a jurisdictional sense.

Suppose that there is a dispute between design professional and contractor over disputed work. The design professional contends that the work is called for under the contract, while the contractor claims it is not. Suppose the contractor performs the work but makes clear that it considered the position of the design professional unjustified and that it intended to claim additional compensation.[17]

Here, as in so many other aspects of the Construction Process, it is important to recognize the design professional's power to interpret the document and his quasi-judicial role. If the design professional determines that the work falls within the contract, many construction contracts make his decision final unless it is overturned by arbitration or litigation. If the contractor later requests additional compensation for the work and is met with the contention that no change order has been issued, the absence of a written change order should not bar the contractor's claim as long as it made clear that it intended to claim additional compensation. However, the claim should be denied *if* the design professional's decision is binding and has *not* been overturned. A contractor dissatisfied with the design professional's decision should invoke any process under which that decision can be appealed. If it later is determined that the design professional had been incorrect and the decision is overturned, the absence of a written change order should not bar the claim. Although no change order had been issued, there was a *constructive* change order—one *should* have been issued.[18] (Some problems of this type are handled under "waiver," largely because the party directing the change had no authority to do so.[19])

Another type of constructive change also developed in federal procurement law—the constructive acceleration (discussed in Chapter 30 dealing with time).

(C) MINOR CHANGES

Sometimes it becomes necessary to make minor changes in the drawings and specifications that are not intended to affect the contract price or completion date. AIA Doc. A201, ¶ 12.4.1, gives the architect au-

---

16. See note 4 supra. Before 1968, constructive changes were also used to get around the *Rice* doctrine, *United States v. Rice,* 317 U.S. 61, 63 S.Ct. 120, 87 L.Ed. 53 (1942), having barred recovery for additional expense of *unchanged* work.

17. See 32 C.F.R. § 7–602.3(b); AIA Doc. A201, ¶ 12.3.1.

18. *Chris Berg, Inc. v. United States,* 197 Ct.Cl. 503, 455 F.2d 1037 (1972).

19. *Weeshoff Construction Co. v. Los Angeles County Flood Control District,* 88 Cal.App.3d 579, 152 Cal.Rptr. 19 (1979).

thority to execute a written change order for such changes so long as they are consistent with the intent of the contract documents. (Even if the architect is given authority to make minor changes, it is best not to make changes when the owner is available.)

Suppose the contractor believes the change is *not* minor. It demands additional time or compensation. Paragraph 12.4 requires that the contractor carry out the order and dispute its propriety later. It should make its intention to claim additional compensation clear at the time it performs the work.[20] A dispute of this type should not be submitted to the architect, since his decision is being questioned.

Although A201 does not define a minor change, one court held that the strength of the walls, some of them load-bearing, "must be considered nothing short of major."[21]

SECTION 25.04
CHANGE–ORDER MECHANISM

(A) JUDICIAL ATTITUDE TOWARD CHANGES MECHANISM

From a planning standpoint, a changes mechanism is essential. Design flexibility and cost control cannot be accomplished without a system for changing work.

Yet judicial attitude toward the changes mechanism often determines how courts will interpret contract language and how quickly courts will find that the changes mechanism has been waived.

Judicial attitude is reflected by language in opinions that must pass upon these questions. For example, Massachusetts held that the contractor could not recover for extra work and concluded:

> Although it seems a hardship for the [contractor] not to be able to recover for the extra work which apparently it performed in good faith, yet such failure results from its not obtaining from the architect or his agents written authority to perform the work . . . .[22]

A federal court granted recovery to a subcontractor despite the absence of formal requirements by pointing to the fact that "[f]rom the beginning of the contract work the parties to the subcontract ignored the provisions as to written orders and proceeded with the work with little or no regard for them."[23] It was obvious that the court thought the formal requirements simply technicalities that should not preclude a contractor from recovering for work undoubtedly beyond the contract requirements.

In a decision denying recovery for extras ordered orally by the city engineer, the Pennsylvania Supreme Court seemed unwilling to force the City of Philadelphia to pay for work not properly ordered and which could not be returned by the city.[24] Here control over public funds predominated. Likewise, the Supreme Court of West Virginia denied recovery for work performed at the direction and with the knowledge of the county court president and other board members because of statutory requirements that all proceedings of the county court be entered into the record books of such court. In justifying its decision the court stated:

> So the requirements of the statute and the decision that county courts must enter of record its orders for the expenditure of the funds must not be construed as meaningless, but must be

---

20. AIA Doc. A201, ¶ 12.3.1.

21. *Whitfield Construction Co. v. Commercial Development Corp.*, 392 F.Supp. 982 (D.V.I.1975).

22. *Crane Construction Co. v. Commonwealth*, 290 Mass. 249, 195 N.E. 110, 112 (1935).

23. *Ross Engineering Co. v. Pace*, 153 F.2d 35, 49 (4th Cir.1946).

24. *Montgomery v. City of Philadelphia*, 391 Pa. 607, 139 A.2d 347 (1958).

enforced for the benefit of the whole public and not for the benefit of any particular individual who may suffer on account of a mistaken reliance upon invalid acts of individual officers, however unfortunate, harmful or deceptive any such acts may have been, unless there is a predominant reason not to do so. To extend the doctrine of liability in every instance because of unjust enrichment is to open the door to all such claims as have not been properly authorized. . . . Too often is it apparent that the expenditures first authorized under a contract are enlarged by so-called "extras" without proper authorization, either intentionally or otherwise, and any such practice should certainly not be given judicial sanction. Nor should sloven management of county affairs be approved. In the absence of special reason, based on competent evidence, why the paramount consideration inherent in the legislative act requiring formal orders by county courts should not be made effective, there should be no digression therefrom as to its enforcement.[25]

It is more difficult to grant recovery to a contractor where there has not been compliance with formal requirements in contracts for public works. However, as seen in the *Watson Lumber* case (reproduced in § 25.01), courts are also protective of inexperienced owners who are building their first home. They must be made aware of price increases for proposed changed work.

The judicial attitude toward formal requirements must take into account the contract language, the experience of the parties in construction, the need to protect public funds, the reasonable expectations of the owner, and the potential unjust enrichment that can result if the contractor is unable to recover for admittedly extra work.

(B) AUTHORITY TO ORDER CHANGES

In the absence of any contract clause dealing with the question of authority to make changes, the question of who can order a change is controlled by doctrines of agency. The owner can order changes. Members of the owner's organization may also have the authority to order changes expressly, impliedly, or by the doctrine of apparent authority. Usually carefully thought-out construction contracts specify which members of the owner's organization have the power to order changes.

Neither the design professional nor, surely, the project representative has inherent authority by virtue of his position to direct changes in the work. Yet sometimes the ultimate outcome is based upon a contractor's claim that a *direction* had been given by someone clearly *without* the authority to direct the change when this unauthorized act was known by authorized officials. For example, *Chris Berg, Inc. v. United States*[26] involved a contract awarded under federal procurement rules. The contractor painted a stairway clearly not required under the contract to be painted. A field memo had been executed by a project engineer and a resident engineer, neither of whom had authority to issue a change order, which impliedly directed the contractor to paint the stairway in question.

Responsible officials in the agency *knew* that the plaintiff was doing work not called for under the contract. The plaintiff had requested a ruling on whether the stairway should be painted and that it be advised of the government's interpretation of the field memo. The court stated that the agency had ample opportunity to warn the plaintiff that he was painting a stairway not called for under the contract. The court concluded that this was a *constructive change* and the contractor was entitled to an equitable adjustment. In addition, the constructive change was based upon the contractor's having painted certain areas after having received the tacit and undoubted oral approval of the project engineer.

Similarly, in *Weeshoff Construction Co. v.*

25. *Earl T. Browder, Inc. v. County Court*, 143 W.Va. 406, 102 S.E.2d 425, 432–433 (1958).

26. Supra note 18.

Los Angeles County Flood Control District,[27] the contractor performed additional work in a road construction contract largely under the direction of the agency's site inspector. The court noted the problem of the site inspector's lack of authority to waive a written change-order requirement. However, it concluded that the change order had already been issued, inasmuch as the procuring agency had threatened to do the work with its own forces if the contractor did not do it. The court concluded that the contractor was justified in relying on the site inspector's statements.

These two cases demonstrate that directions by project representatives coupled with other facts that would make denial of recovery unjust may be the basis for the contractor's receiving additional compensation. (Excusing formal requirements is discussed in Section 25.04(E).)

The construction contract frequently specifies who has the authority to change the work. AIA Doc. A201, ¶ 12.1.1, requires that change orders be signed jointly by the owner and architect. Until 1976, A201 gave the architect authority to issue a written change order without the owner's signature if he had written authority from the owner to do so, the authorization furnished the contractor upon its request. In 1976 this was eliminated to reduce the architect's liability despite contentions by some that there were still situations whereby the owner's absence or unavailability might make it necessary to give the architect this authority. In the latter case, the form should be changed.

One of the chronic problems in construction administration is the need for a mechanism that can move the project along when interpretation problems develop. In earlier AIA documents, the architect was authorized to issue field orders that interpret the documents or that order minor changes

that do not affect price or time. If such a power were abused, at least in theory, the contractor could demand arbitration under the AIA Document or reserve its right to contest the decision at some later date. Nevertheless, this power was deleted from AIA Documents in 1976.

The need for quick decisions "on the spot" motivated the AIA to publish AIA Document G713, the Construction Change Authorization, in 1979. It went beyond the old field order by contemplating an increase in the price and extension of time. The AIA wanted a method that would process change orders expeditiously and avoid delaying the project.

The form is initiated by the architect. Based on conversations between the architect and the contractor, the architect fills in blanks based upon information provided by the contractor that relate to the change in price or time, including a blank for *firm, estimated,* or *maximum* changes in price or time. The form is then sent to the owner for authorization, inasmuch as the architect cannot authorize additional expenditures or time. (Does the contractor do the work in the meantime? If not, how does this expedite the process?) After owner authorization, the form is sent to the contractor for its confirmation. When this is done, a change order should be processed promptly.

Suppose the owner will *not* authorize the change that *presumably* has been performed at the direction of the architect. There is likely to be recovery by the contractor. As a result, some owners may not use such a method, as it can enlarge the authority of the architect. Use of G713 depends upon the urgency of changes in the field and the owner's confidence in the architect.

Just as the owner can give the design professional express authority, the owner may cloak the design professional with ap-

---

27. Supra note 19.

parent authority.[28] The owner's acts can lead the contractor reasonably to believe that the design professional has the authority to change the work. Suppose the owner knows that the design professional is executing change orders and makes no objection or, even further, pays based upon change orders issued by the design professional. This activity could reasonably lead the contractor to believe that the design professional has authority to change the work. The apparent authority doctrine is less likely to be successfully invoked by the contractor if the project involves public work.[29]

Fletcher v. Laguna Vista Corp.[30] illustrates enlarged authority. Changes were required to be ordered in writing by owner and architect. However, in concluding that the architect had authority to execute a written change order the court stated:

> . . . the manner in which the parties themselves have interpreted the contract through their course of dealings is of utmost importance. The record in this case is filled with testimony to the accord that both [owners] and [contractor] had relied on architect Frye to make adjustments in the contract sum and had abided by his decision. [Owners] knew that there would be at least a slight overage in the sums spent by [contractor] for overhead but had never objected. [Owners] accepted decreases in the cost of millwork which were incorporated into a change order signed only by architect Frye and the contractor. The parties themselves have interpreted the contract to allow an increase and a decrease in the contract sum with only the written signature of architect Frye. Even if the contract does not grant this authority to architect Frye, the parties through their course of dealings have interpreted and modified the document so as to place in the hands of architect Frye the final authority to authorize increases and decreases in the contract sum.[31]

It is possible for the owner to give the architect this authority. However, if the basis for enlarging the architect's normal authority is the architect's power to interpret the contract documents, the case is incorrect. The court failed to distinguish the normal power given to the design professional to interpret the contract documents from the power to order changes.

Sometimes there is insufficient time to obtain authorization from the owner for work needed immediately because of impending danger to person or property. The contractor should have authority to do such emergency work. Frequently contracts provide that extra work in emergencies can be performed without authorization. Even without express or implied authority, the contractor should be able to recover for emergency work based upon the principle of unjust enrichment.

(C) MISREPRESENTATION OF AUTHORITY

If the contractor performed the additional work at the order of the design professional and this order was beyond the latter's actual authority, what recourse does the contractor have?

First, the contractor would seek to establish that the design professional had *apparent* authority to order the work. But in the absence of the owner having led the contractor to believe that the design professional had this authority, the contractor will not be successful. Likewise any claim that the owner has been unjustly enriched by the work would not be successful. The contractor would be considered a volunteer and denied recovery.

Next, the contractor would look to the design professional. The threshold question would be whether the contractor rea-

---

28. Refer to Section 4.06.

29. This is discussed in greater detail in Section 25.04(G).

30. 275 So.2d 579 (Fla.App.1973).

31. Id. at 580–581 (the court's footnotes omitted).

sonably relied upon the misrepresentation of authority. In many cases, reliance would not be reasonable because typically design professionals are not given this authority and because the contractor should have checked with the owner. But the reliance element is often downgraded, and in any event suppose compliance with the order had been reasonable under the circumstances.

In a case of misrepresentation of authority the contractor can recover from the design professional.[32] The recovery would be the cost of performing the unauthorized work and any cost of correction made necessary to make the work conform to the contract documents. It would be unlikely that the design professional held to have misrepresented his authority could recover from the owner on the theory of unjust enrichment.

Suppose the contractor *had* recovered from the owner because the latter's acts cloaked the design professional with apparent authority. It must be remembered that the design professional acted without authority even though the acts of the owner may have created apparent authority. Suppose the owner seeks to transfer this loss to the design professional because of the latter's misrepresentation of authority.

Passing by the question of whether the owner had suffered any real loss other than having work it did not choose, the loss was caused *both* by the design professional's misrepresentation of authority and by the owner's acts making it appear that the design professional *had* the authority. Since the more culpable act appears to be the misrepresentation of authority, this loss should be borne by the design professional.

Yet if the misrepresentation resulted in the value of the property having been increased, the owner has received a benefit. Although the law does not explicitly divide losses in such cases, it would be fair to do

so here. Another method of sharing the loss would be to give the owner any difference between what the owner had to pay the contractor and the enhanced value of the project because of the unauthorized work.

At this point, a distinction should be made between the contractor's recovery having been based upon implied authority or upon apparent authority. In the former case, there has been no wrongdoing by the agent, who was authorized to act even though authority was not expressly given. For example, the architect's representation to the subcontractor in the *Bethlehem* case noted in Section 32.08(B) was authorized implicitly though not explicitly. Had the architect's statement been *beyond* his authority—express or implied—recovery against the owner would have had to have been based upon apparent authority. If so, the owner would have had a claim against the architect. In the *Bethlehem* case, there was no unjust enrichment of the type present in the preceding discussion. As a result, if the subcontractor's recovery against the owner had been based upon an unauthorized representation by the architect, the owner should be reimbursed for what it paid the subcontractor.

(D) DUTY TO ORDER CHANGES FROM CONTRACTOR

Clearly the changes clause gives the owner the power to order that work within the general scope of the contract be performed by the contractor. Suppose the owner asserts the right to award work within the general scope of the contract to a third party or perform that work with its own forces. The power to do this would give the owner some bargaining advantage in negotiating the price for the changed work with the original contractor.

Most changes clauses do not deal with this question specifically. The contractor

32. *Brown v. Maryland Casualty Co.*, 246 Ark. 1074, 442 S.W.2d 187 (1969) (dictum).

could contend that giving the owner this power would be unfair, putting the contractor at the mercy of a changes clause without the compensating advantage of being able to perform all work within the general scope of the contract. Giving the contractor the right to do the changed work, it could assert, would not put the owner at a disadvantage. Usually there are pricing formulas that apply to the changed work if the parties do not agree on a price. Although as a practical matter the owner may find it inexpedient to have the original contractor working side-by-side with a substitute or its own employees, the express language of most changes clauses does *not* give the contractor the power to perform changed work.

In *Hunkin Conkey Construction Co. v. United States*,[33] the contractor had entered into a contract with the Corps of Engineers for the construction of a dam. During performance, subsurface problems developed that necessitated a significant change in design. The contractor and the Corps discussed alternative designs but were unable to agree upon a price for an alternative design desired by the Corps. As a result, the Corps negotiated a contract with a third party to perform the alternative design work at a cost of $200,000 *less* than that proposed by the original contractor.

After completion of the project, the contractor contended that the work should have been awarded to it. The Court of Claims was not persuaded that the power to direct changes meant that the government had a duty to order those changes from the original contractor. But the court directed its attention to another provision of the contract. This provision allowed the government to undertake or award other contracts for "additional work" and required the original contractor to cooperate with other contractors or government employees. This, according to the Court of Claims, had to be read together with the changes clause. When so done, it was clear that the government was not obligated to award the additional work to the contractor.

Although the conclusion that the work in question was additional can be debated, the line of demarcation may be useful both from a practical, administrative basis (it would be difficult for the *two* contractors in such a dispute to work side-by-side) and from the normal expectations of the owner. The owner may not want to be "locked in" to the contractor for "extra" work. Conversely, though, the contractor may assert an "expectation" that its price was predicated upon a "monopoly" on "extras."

(E) FORMAL REQUIREMENTS

Generally construction contracts require that change orders be written and signed by the person or persons authorized to execute change orders. Obviously it is *best* to have the change order issued and price agreed upon *before* the work is started. If price cannot be agreed upon before the work is begun, issuance of the change order does assure the owner that the changed work will be compensated in accordance with a changes clause formula if no agreement is reached and assures the contractor that the owner will not contend that the work in question falls within the contract requirements.

Is it best to require that the change order be issued before the work is begun? If the change order *is* issued after the work is begun or completed, issuance generally forecloses any question of whether the work was within the contract requirements. However, *not* requiring that the change order be issued *before* the work is begun invites issuance of oral change orders with the assurance by the owner that a written change order will follow. Although this sometimes may be necessary under certain circumstances, suppose the owner *denies* issuing an oral change order. In such cases,

---

33. 198 Ct.Cl. 638, 461 F.2d 1270 (1972).

the owner will insist that no additional compensation should be paid and the contractor will contend that a requirement for a writing has been dispensed with by the issuance of an oral change order.

It is probably advisable to require that the change order be issued *before* the start of work even if this means some delay while the written change order is issued.[34]

Under AIA Doc. A201, Art. 12, the sequence is the issuance of a written change order, an *attempt* to agree on price, the issuance of a *subsequent* written order by the owner directing that the work be done even if the price cannot be agreed upon, and the residual power in the architect to decide price or time changes.

Some contracts do *not* require a written change order in *advance* of the work being done. The contractor can recover if a *subsequent* written change order is issued. Here the obvious advantage is speed, but the disadvantage is the possibility of subsequent disputes over whether the work was extra. At the very least, the issuance of a written change order eliminates this difficulty. Perhaps the impetus for "looser" change mechanisms is traceable to the reluctance courts have to deny compensation to the contractor despite the absence of a written change order if it appears that the owner has directed that work that the court decides was beyond the contract documents be performed.[35]

Another problem that surfaces with regularity relates to the interaction between a changes clause with *its* formal requirements and other provisions of the contract that may have *different* formal requirements. This arises most frequently when the contractor encounters subsurface conditions different from those specified or normally expected and seeks to recover an equitable adjustment. (Inasmuch as this is a special problem, it is discussed in Chapter 29.)

(F) INTENTION TO CLAIM A CHANGE

A lengthy list of extra work submitted after completion is a sad, though not infrequent, part of construction work. Section 25.03(B), which treated constructive changes, noted that a vehicle for such claims is the contention that particular directions or instructions given by owner or design professional were changes entitling the contractor to extra compensation. One method of minimizing this problem is to require the contractor to give a written notice within a designated number of days after the occurrence of any event that the contractor will claim justifies an increase in the contract price. Paragraph 12.3.1 of AIA Doc. A201 requires such a notice twenty days after the occurrence of the event. Except in an emergency, the notice must be given before beginning the work.

A similar provision is used by the California Department of Public Works. A general contractor doing work for the state must submit a written protest to the state architect within thirty days after receiving a *written* order from the state architect to perform any disputed work. Failure to do so precludes compensation for the work.[36] The California provision also requires that the protest notice specify in detail how requirements were exceeded and the appropriate change in cost resulting. This content requirement goes considerably beyond ¶ 12.3.1, which simply requires written no-

---

34. In *Uhlhorn v. Reid*, 398 S.W.2d 169, 175 (Tex.Civ.App.1965), the clause stated that "No extra work or charges under this contract will be recognized or paid for unless agreed to in writing before the work is done or the charges made."

35. As examples, see *Universal Builders, Inc. v. Moon Motor Lodge, Inc.*, 430 Pa. 550, 244 A.2d 10 (1968) and *Ecko Enterprises, Inc. v. Remi Fortin Construction, Inc.*, 118 N.H. 37, 382 A.2d 368 (1978). See also § 25.04(F).

36. This provision was interpreted in *Acoustics, Inc. v. Trepte Construction Co.* and *Weeshoff Construction Co. v. Los Angeles County Flood Control District*, discussed in Section 25.04(G).

tice of an intention to make a claim in the contract price.

(G) EXCUSING FORMAL REQUIREMENTS

Not uncommonly the change order mechanism is disregarded by the parties. This may be due to unrealistically high expectations by the drafters, time pressures, or unwillingness by the parties to make and keep records. When contractor claims for extras are denied by the owner because of the absence of a written change order or intention to claim an extra, contractors frequently assert that the requirements have been waived. Waiver questions can be divided into three subissues:

1. Is the requirement waivable?
2. Who has the authority to waive the requirement?
3. Did the facts claimed to create waiver lead the contractor to reasonably believe that the requirements have been eliminated or indicate that the owner intended to eliminate the requirements?

Except where the waiver concept cannot be applied to public contracts,[37] formal requirements can be waived. They are not considered to be an important element of the exchange and are often viewed as simply technical requirements.[38]

As a rule, only parties who have authority to order changes have authority to waive the formal requirements.[39] Usually only the owner or its authorized agent can waive the formal requirements, although a design professional with authority to order changes should have authority to waive the writing requirement.

Acts that can be the basis for waiver generally are oral orders by the owner or its authorized representative. Generally if it appears that the owner orally ordered the changed work, it will be assumed that the writing requirement has been waived.[40]

Unfortunately, courts do not differentiate between cases where the owner *admits* ordering the work but claims it was within the contract requirements from owner admissions that the work was extra and was ordered. One purpose for having a requirement for a written change order is to obviate the question of whether the work was extra. The absence of a writing indicates at the very least that the owner did not consider the work extra. However, courts seem to disregard this factor. If they determine that the work was extra and the contractor was ordered to perform it, absence of a written change order is not likely to prevent the contractor from recovering. But in a public contract, recovery could be denied even in such a case, unless a *constructive* change is found.

Owner payment based upon oral change orders by the design professional or conceivably by a resident engineer or project representative can be a waiver. Payment can lead the contractor to reasonably believe that the design professional has the authority to order changes orally.[41] If payment is accompanied by a notice that makes clear to the contractor that the for-

---

37. In *Delta Const. Co. of Jackson v. City of Jackson*, 198 So.2d 592 (Miss.1967), the court held that waiver could operate against a municipality but not where the formality in question was a supplemental agreement on price. In *Metropolitan Sanitary Dist. of Greater Chicago v. Anthony Pontarelli & Sons, Inc.*, 7 Ill.App.3d 829, 288 N.E.2d 905 (1972), the court held that a provision giving the public agency the right to recover illegal and excessive payments meant that there was no way that the required approval by the board of trustees could be waived.

38. See Section 25.04(A).

39. See Section 25.04(B).

40. *T. Lippia & Son, Inc. v. Jorson*, 32 Conn.Super. 529, 342 A.2d 910 (1975). See note 35, supra.

41. *Oxford Dev. Corp. v. Rausauer Builders, Inc.*, 304 N.E.2d 211 (Ind.App.1973).

mal requirements are *not* being waived, no waiver should be found.

It may be useful to examine language in some waiver cases.

In *Rivercliff Co. v. Linebarger*,[42] the court found a waiver of the writing requirement, stating:

> For a second ground, appellant contends that . . . the trial court should not have made any allowance to the contractor because the extra work was not authorized in accordance with the terms of the contract. This contention appears to be supported by the terms of the contract, which provides that extras must be approved in writing prior to execution. This provision was not complied with but it does not constitute a defense available to appellant, because, as we hold, a strict compliance with this provision of the contract was waived by appellant in this instance. It is not disputed that the extra excavation was done with the knowledge and at the direction of Smith who was not only the architect supervising the work for Rivercliff but was also a part owner of the appellant corporation. From his testimony we gather that he refused to approve an allowance for extras mainly because he did not think the contractor was entitled to anything as a result of the changed method of constructing the foundation. It appears that other changes in construction had been made and paid for where no written change order had been previously issued. Although it was shown that several such changes had been made and paid for during the construction of the four buildings, yet Mr. Smith testified that only one written change order had been made.[43]

Another court stated:

> Several situations may form the basis for waiver: (1) when the extra work was necessary and had not been foreseen; (2) when the changes were of such magnitude that they could not be supposed to have been made without the knowledge of the owner; (3) when the owner was aware of the additional work and made no objection to it; and (4) when there was a subsequent verbal agreement authorizing the work.[44]

Finally, in a subcontractor context, a court stated:

> . . . the court finds that plaintiff is entitled to recover for several of the jobs which it performed that were not incidental to the building of a cofferdam and that were required to be done due to the insistence or inaction of James. Where a party is aware that extra work is being done without proper authorization but stands by without protest while extra work is being incorporated into the project, there is an implied promise that he will pay for the extra work. See *United States v. Klefstad Engineering Co.*, 324 F.Supp. 972 (W.D.Pa.1971). In *Klefstad*, the prime contractor was given the right to recover for work it performed which was the responsibility of the subcontractor; whereas in the present case the subcontractor, E & R, is entitled to recover for work it performed which was the responsibility of the prime contractor, James. Whether the theory of recovery is considered as quasi-contract, implied-in-fact contract, or promissory estoppel, a subcontractor is entitled to be compensated for extra work it performed as the result of inducing statements and conduct by the prime contractor.[45]

The difficulty in predicting when a court will find that formal requirements have been excused is demonstrated by two California cases. In *Acoustics, Inc. v. Trepte Construction Co.*,[46] the court held that compliance with contractual provisions for written orders is indispensable and denied recovery to a contractor who had been *verbally* ordered to perform changed work, with the court noting that the state inspector who had ordered the changes had no authority to waive the formal requirement and that the contractor erred in expecting payment without a written change order

---

42. 223 Ark. 105, 264 S.W.2d 842 (1954).

43. 264 S.W.2d at 846.

44. *Nat Harrison Associates, Inc. v. Gulf States Utilities Co.*, 491 F.2d 578, 583 (5th Cir.1974).

45. *United States v. Guy H. James Construction Co.*, 390 F.Supp. 1193, 1223 (M.D.Tenn.1972), affirmed 489 F.2d 756 (6th Cir.1974).

46. 14 Cal.App.3d 887, 92 Cal.Rptr. 723 (1971).

from the state architect as required in the contract.

Eight years later in *Weeshoff Construction Co. v. Los Angeles County Flood Control District*,[47] which involved a similar claim with the only differentiation being the apparent knowledge by the awarding agency that a site inspector was directing that changes be made, the court came to the opposite result based upon, though not explicitly, the *constructive* change rationale.

The formal mechanism in this contract for road construction work required not only that change orders be issued in writing but also that the change orders contain price or payment information. If the price is not negotiated, the contractor may proceed and be paid on a cost basis provided it submits "daily written expenditure reports which are reconciled and signed by the parties."

After noting the *Acoustics* case and other California cases requiring a writing, the court noted that particular circumstances may waive these requirements. Even in this *public* contract the court noted that if the parties by *their* conduct clearly assent to a change or an addition to their required performance they have *waived* the change order requirement.

To back up its conclusion that there had been a waiver, the court pointed to pressure being placed by the agency and its representatives upon the contractor to do work a particular way that the court later determined was *not* required by the contract. (The agency always took the position that the work was required by the contract.) The court noted that the agency stated that *it* would and did do the work if the contractor did not, and it would charge the contractor. This induced the contractor to do the rest of the work in accordance with the demand of the agency.

The failure to issue a change order was consistent with the agency's position that the work fell within the contract requirements. Once the court determined that the work did not, however, it pointed to the conduct of the agency as indicating a waiver of the written change order. Clearly this was not a case where the parties *agreed* that the work *was* extra but agreed that it should be done without the requirement of the written change order. Yet the court found a waiver, probably because the contractor had made its position clear that it considered the direction to be a change, with the ultimate issue to be resolved later.

The court also had to face the contractor's failure to submit daily expenditure reports. The court referred to the earlier *Acoustics* case, which had found that failure of conditions precedent barred recovery. But the court in the *Weeshoff* case pointed to the contract provision in the *Acoustics* case stating that failure to perform these conditions would be a waiver of any right to demand compensation and the *absence* of such a provision in the *Weeshoff* contract. The court noted that without an express contractual stipulation barring the claim, the court could determine whether the contractor's failure to supply these daily reports would bar his recovery.

The court decided the issue in favor of the contractor by pointing to the language in the contract stating that the agency could direct the contractor to proceed if they could make an agreement on price. This, according to the court, waived any right to demand the daily expenditure reports. In addition, the court noted that the daily expenditure requirement was linked with the issuance of a change order, and since the district had never issued the change order, it was reasonable for the contractor to believe that this requirement was no longer applicable and that he needed only to record his position for a future determination. The court also noted that detailed cost records *were* produced at the trial by

---

47. Supra note 19.

the contractor and that the agency had provided constant supervision on the construction site and "had ample notice of the extra work as it was performed."

Three observations can be made. First, cases of this type are strong motivations for drafters to "overdraft" and make sure that everything is stated as clearly as possible and as often as possible. Second, even the best drafting is not likely to bar a contractor from recovery if the court believes the work to have been extra and the owner not misled as to the contractor's position. Third, there was no reference in the case to the dispute process mechanism under which a representative of the owner is given the power, at least in the first instance, to interpret the contract. Assuming this existed in the *Weeshoff* case, the resolution of whether the work was extra should have been made under the dispute resolution process. Had his issue been resolved in the contractor's favor, the absence of a written change order surely should not have barred a recovery.

Another way of avoiding the formalities of the changes clause is to conclude that the work was not a "change." Section 25.03(A) noted that work may be considered outside the changes clause. Likewise, work made necessary because of errors of the design professional must be paid for despite the absence of the formal requirements set forth in the changes clause.[48] Such work can be regarded as a remedy for defective work given to the contractor.

(H) PRICING CHANGED WORK

Pricing changed work is an important part of the changes mechanism. A tightly drawn pricing formula, along with clear and complete contract documents, can discourage a deliberately low bid made with the intention of asserting a long list of claims for extra work.

Where possible, work should be compensated by any unit prices specified in the agreement. It is important, however, to take into account the possibility of great variations in units of work requested and the effect upon contractor costs. (Refer to Section 21.02(C).)

As to other work, ¶ 12.1.3 of AIA Doc. A201 specifies two methods of pricing changed work:

.1 by mutual acceptance of a lump sum properly itemized and supported by sufficient substantiating data to permit evaluation; . . .

.3 by cost to be determined in a manner agreed upon by the parties and a mutually acceptable fixed or percentage fee.

The first method presupposes that the contractor will make a proposal and substantiate it for consideration by the owner. Paragraph 12.1.3.3 assumes that the parties will negotiate and agree to compensation.

Paragraph 12.1.4 states that if none of the methods in 12.1.3 results in agreement, the contractor shall, after another order signed by the owner, proceed with the work. In such a case, the architect will determine compensation on the basis of "reasonable expenditures and savings" including "in the case of an increase in the Contract Sum, a reasonable allowance for overhead and profit." The contractor must keep and present itemized accounting and supporting data.

Some contracts provide that if the parties cannot agree, the contractor will be paid cost plus a designated percentage of cost in lieu of overhead and profit. Pricing extra work in this fashion can discourage the contractor from reducing costs. However, it may be difficult to arrive at a fixed fee for overhead and profit when the nature of the extra work cannot be determined until it is ordered. For this reason, it is likely that

---

48. *Watson Lumber Co. v. Mouser*, supra note 3.

cost plus a percentage of cost will continue to be used to price changed work.[49]

When a cost formula is used, the design professional must examine the cost items to see if they are reasonable and required under the change order. The changes clause can specify that only certain material, equipment, and labor costs or direct overhead costs are to be included as cost items.

Public contracts—federal and state—frequently provide that if the parties cannot agree there will be an "equitable" adjustment of the price.[50] This raises a troublesome question noted in Section 25.01 that relates to the difference between contract relief expressed in a provision granting the contractor an *equitable adjustment* and relief granted, whether or not described as an equitable adjustment, effectuated by *change order*. If the latter, must the entire change-order mechanism be employed? For example, in AIA Doc. A201, this would involve the architect's residual powers to decide the dispute with costs as a guideline, along with overhead and profit.

But suppose the changes mechanism is *not* used to resolve any disputes over an equitable adjustment. In such a case the contract must be examined to determine whether there is any contractual dispute mechanism established. For example, AIA Doc. A201, ¶ 2.2.9, requires that claims and disputes that relate "to the execution or progress of the Work or the interpretation of the Contract Documents shall be referred initially to the Architect". If the dispute over the equitable adjustment falls within ¶ 2.2.9, the architect, at least initially, resolves the dispute but *not* through the changes mechanism of Art. 12. Any guides expressed in Art. 12 need not be employed. In addition, the interrelation between the architect's residual powers under Art. 12 and his power to resolve disputes under Art. 2 may cause difficulty if legal action is brought against the architect because of his decision. Article 2 contains language protecting the architect if his decisions are made in good faith, something absent in Art. 12. At the very least, contracts should seek to make clear whether adjustments of the contract price by change order invoke the entire changes mechanism or whether failure to agree on matters such as an equitable adjustment is like all other disputes. Since this pricing problem arises most frequently when subsurface conditions are encountered that are different than described or expected, it is treated separately in Chapter 29.

Suppose the change *reduces* the work? Does deductive change require that overhead and profit also be deducted for deleted work? Paragraph 12.1.4 of AIA Doc. A201 seems to permit the architect to add a reasonable allowance for overhead and profit *only* for an increase in the work but appears to preclude reduction if work is deleted. On the other hand, a federal procurement decision held that overhead and profit would be deducted when work is deleted.[51]

Should the contract specify a mechanism for arriving at an agreement? For example, the contract can state that the contractor will propose and support a proposed price to the owner. The owner can accept this pricing proposal or propose its own price based upon other supportive evidence. The contract can provide that a failure by one party to object within a specified period to a

---

49. Federal procurement regulations have developed weighted guidelines for determining profit. The factors taken into account are degree of risk, relative difficulty of the work, size of job, period of performance, contractor's investment, assistance by government, and subcontracting. Each of these factors is weighted depending upon the particular procurement. An illustration of how these guidelines are used can be found in *Norair Eng'g Corp.*, ASBCA No. 10856, 67–2 BCA ¶ 6619.

50. 32 C.F.R. § 7–602.3(d). For a municipal contract containing a similar clause, see *Fattore Co. v. Metropolitan Sewerage Commission*, 454 F.2d 537 (7th Cir.1971).

51. *Algernon Blair, Inc.*, ASBCA No. 10738, 65–2 BCA ¶ 5127.

price proposal by the other party is an acquiescence to that price proposal. This technique has advantages. It forces the party to whom a proposal has been made to set forth its own proposal within a specified period of time. This accelerates the agreement process. Sometimes such a provision is included with the hope that the other party will be dilatory and forget to object to a proposal, thereby adopting it by acquiescence.

Suppose a changes clause specifies that the parties will agree on compensation for changed work and the parties cannot agree. Early cases concluded that such an agreement was not enforceable as simply "an agreement to agree." [52] Modern courts would very likely determine a reasonable price for the work where the parties do not agree.[53] One case involving changed work on a subcontract held that where the parties could not agree on a price, the subcontractor could receive cost plus overhead and profit.[54]

A changes clause that specifies that the parties will agree should also provide an alternative if the parties do not agree, such as a pricing formula or a broadly drawn arbitration clause.

Earlier federal procurement cases seemed to prohibit contractors from recovering for added costs of doing *unchanged* work caused by the change order.[55] However, federal procurement regulations have been changed to permit the contractor to recover additional costs of performing unchanged work.[56]

In addition to claims involving the cost of performing unchanged work, an excessive amount of change orders of the type that can be considered "cardinal" changes are, along with other acts of the owner, often the basis for the now frequent delay damage claim (discussed in Section 30.10).

(I) CHANGE ORDERS AND SETTLEMENTS

Change-order provisions sometimes seek to resolve any disputes incident to the change by allowing the disputes to be resolved in the change order itself if the contractor agrees to accept any proposals made in the change order. This raises the problem of disputes that arise at the time change orders are issued.

Disputes as to whether there *is* a change are referred to in this chapter. However, the more difficult settlement problems arise when the contractor, though satisfied with the amount the owner proposes to pay for the changed work, asserts that the cost of doing unchanged work has increased and that the entire project has been delayed. This problem arose in *Merritt-Chapman & Scott Corp. v. United States,*[57] in which during the contractor's performance the United States made a drastic change in the contract requirements that consisted of requiring that the entire building be air-conditioned. Subsequently the contractor proposed a price adjustment to compensate it not only for the changed work but for suspension of the work and delays resulting from the change and stop orders. The proposal was turned down by the United States. A second proposal by the contractor was followed by a change order for the precise amount specified in the final proposal made

---

52. Refer to Section 5.06(F).

53. *Purvis v. United States,* 344 F.2d 867 (9th Cir.1965).

54. *Hensel Phelps Construction Co. v. United States,* 413 F.2d 701 (10th Cir.1969).

55. *United States v. Rice,* supra note 16.

56. 32 C.F.R. § 7–602.3(d).

57. 198 Ct.Cl. 223, 458 F.2d 42 (1972). A similar result in a private contract occurred in *J.A. Jones Construction Co. v. Greenbriar Shopping Center,* 332 F.Supp. 1336 (N.D.Ga.1971), affirmed 461 F.2d 1269 (5th Cir.1972).

by the contractor. The Claims Court concluded that the change order accepted the proposal of the contractor and discharged any claims related to the change, including a subcontractor claim for damages because of the changed work. The decision by the court did not affect any claims that the subcontractor might have against the prime contractor.

Sometimes owners seek to take advantage of the contractor's need for payments by insisting that payment for the changed work to which there is *no* dispute is conditioned upon the contractor's giving up any claims it may have for delay damages. This can be accomplished by language in the change order itself or a restrictive endorsement on the check tendered to the contractor. The common law would in such cases usually bar the contractor from asserting any additional claims if it has accepted tender of the money knowing the conditions under which it has been tendered.[58] This, while expediting the resolution of disputes, this rule can operate harshly because of the inordinate pressure it may place upon a contractor.[59] This has been recognized by recent legislation in California that has barred state agencies from using this approach.[60] Some cases have used the Uniform Commercial Code to keep a claim alive if the claimant makes clear at the time it accepts the payment that it is not giving

up any claims it may have.[61] Here, as in other areas, case decisions are likely to have different outcomes.[62]

Some contractors accept the amount tendered conditionally, reserving the right to submit a delay damage claim later. Suppose the parties cannot agree upon delay damages at the end of the job. Can the contractor contest the amount accepted? This depends upon whether the facts show that accepting conditionally was designed *simply* to protect its delay damage claim, or whether *everything* connected to the change can be reopened.

SECTION 25.05
EFFECT OF CHANGES
ON PERFORMANCE BONDS

When sureties were usually uncompensated individuals, courts held that any changes in the contract between the principal and the obligee would discharge the surety. This would be unjust where a professional surety bond company is used, especially since changes and modifications are common in construction contracts. Most surety bonds provide that modifications made in the basic construction contract will not discharge the surety. Some changes clauses permit changes up to a designated percentage of the contract price without notifying the bonding company.

PROBLEMS

1. A was the architect for a construction project being built by C for O. The specifications called for soundproof tile for certain room ceilings. C wanted to put in tile of a certain brand that was represented by the manufacturer to be

---

58. E.A. Farnsworth, Contracts 282 (1982).

59. Compare *Huber, Hunt & Nichols, Inc. v. Moore,* 67 Cal.App.3d 278, 136 Cal.Rptr. 603 (1977) (claim barred), with *North Harris County Junior College District v. Fleetwood Construction Co.,* 604 S.W.2d 247 (Tex.Civ.App.1980) (pressure by owner unconscionable).

60. West's Ann.Cal.Pub. Contract Code § 7100.

61. *Ruble Forest Products, Inc. v. Lancer Mobile Homes,* 269 Or. 315, 524 P.2d 1204 (1974).

62. *Connecticut Printers, Inc. v. Gus Kroesen, Inc.,* 134 Cal.App.3d 54, 184 Cal.Rptr. 436 (1982) (claim barred).

soundproof. A insisted that another brand, which was more expensive, be used. C claimed that this would be an extra and demanded a change order. A claimed that as judge of performance, he was ruling that the tile that the contractor wanted to use did not meet the contract requirements. C used the tile specified by A and, after completion of performance, demanded that he be paid extra for the tile he had used. A instructed O not to pay, because the tile used was required under the contract and because the contract required that a written change order be issued before there could be extra payment for any work. Is A correct in his position? (Assume that no

tiles are actually soundproof but that the tile demanded by A resisted sound better, and cost more, than the tile that C wanted to use.)

2. C contracted with O to build a five-story commercial building under a changes clause allowing changes in "the general scope of the work." Must C perform an order to:

a. Add an underground garage?
b. Build a sauna bath on the roof?
c. Substitute toilets with a lower water tank capacity for those designated?

Must C delete a library taking up one-half of one floor if ordered to do so? If the library *were* deleted, how should the price be reduced?

26

Payment: Money Flow as Lifeline

SECTION 26.01 THE DOCTRINE OF CONDITIONS

This chapter deals principally with the process by which the contractor is paid for performing work required by the construction documents. The rules that control this process are derived principally from the contract documents, most noticeably the basic agreement and the general conditions. An important backdrop to these contract provisions is the legal doctrine of conditions.

This important legal doctrine seeks to protect the actual exchange of performance specified in the contract. A party should not have to perform its promise without obtaining the other party's promised performance. For example, suppose a contract is made under which a supplier agrees to deliver supplies to a small manufacturer. The supplies arrive by truck at the warehouse. However, the seller's truck driver refuses to unload the supplies until payment is made. The buyer's employee refuses to pay until the supplies are unloaded and placed on the buyer's receiving dock. Obviously, such a dispute could have been dealt with in the first instance by appropriate contract language dealing with the question of whether delivery precedes payment. When the contract does not deal with this

question, the law must determine the sequence of performance.

Before proceeding to the legal resolution of this question, it should be noted that had the seller unloaded the supplies and not been paid, the seller would have had a valid claim for payment. Conversely, had the buyer paid but the supplies not been unloaded, the buyer would have had a valid claim for the value of the supplies or return of the money. However, neither party wishes to exchange its actual performance for a legal claim. Each would prefer to receive the other's performance before rendering its own.

Likewise, buyer and seller of real estate wish to avoid performance without obtaining the other's performance. The seller does not wish to transfer ownership by deed without obtaining the money. Conversely, the buyer does not wish to pay without receiving the deed. Protecting the desire of each is usually accomplished in a real estate purchase by use of a third party or escrow holder. The seller will transfer the deed, and the buyer will deliver the money to the third party. Each believes that the third party will effectuate the exchange.

In the absence of a third-party system or specific contract clause dealing with sequence of performance, the law must determine whether performance or payment must come first. The common law required that the performance of services *precede* the payment for those services.[1] This protected the party receiving the services by permitting that party to withhold payment until *all* the services were performed. Not only did this rule allow the paying party to avoid the risk of paying and then not receiving performance, but it also allowed the paying party to dangle payment before the performing party, a powerful incentive for rendering performance.

However, the "work first and then be paid" rule is disadvantageous to the performing party. The latter must finance the entire cost of performance. The party performing services must take the risk that *any* deviation, however trivial, would enable the paying party to withhold money greatly in excess of the damages caused by the deviation. The performing party must assume the risk that the paying party might *not* pay after complete performance, leaving the performing party with a legal claim.

The doctrine of conditions is central to the discussion of progress payments (discussed in Section 26.02) and the right of the contractor to recover despite noncompliance (discussed in Section 26.06). Though less important, it is also relevant to retainage and final payment, discussed in Sections 26.03 and 26.05, respectively.

SECTION 26.02
PROGRESS PAYMENTS

(A) FUNCTION

As indicated in the preceding section, the doctrine of conditions requires that work be performed before any obligation to pay arises. Such a rule in a construction context places severe financial obligations on the contractor and creates a substantial risk of nonpayment. To avoid these problems, construction contracts generally provide for periodic progress payments made monthly or at designated phases of the work. This section examines the process by which progress payments are made and some common problems involved in this process.

(B) SCHEDULE OF VALUES

To facilitate the computation of the amount to be paid under progress payments in a fixed-price contract, the contractor is generally required to submit to the design professional a schedule of values before the first application for payment. This schedule, when approved, constitutes an agreed valuation of designated portions of the work. The aggregate of the schedule should be the contract price.

(C) APPLICATION FOR PAYMENT CERTIFICATE

The contractor submits an application for a progress payment. This application is generally submitted a designated number of days before payment is due. The application is usually accompanied by documentation that supports the contractor's right to be paid the amount requested. If payment is requested for materials or equipment stored on or off the site, AIA Doc. A201, ¶ 9.3.2 requires the contractor to submit bills of sale or comply with other procedures that establish the owner's title to such property or equipment. If the material is stored off-site, ¶ 9.3.2 requires the contractor to show that the material has been properly insured and will be transported to the site.

Some contracts require the contractor to give assurance or proof that it has paid its subcontractors and suppliers when progress payments applications are made. Alternatively, such contracts allow the prime contractor to submit documents executed by subcontractors or suppliers that give up their rights to any mechanics' liens.

Neither the standard contract published

---

1. E.A. Farnsworth, Contracts 588 (1982).

by the American Institute of Architects (AIA) nor the one published by the National Society of Professional Engineers (NSPE) requires such assurance or proof of payment or submission of lien waivers at the time *progress* payments are made. This is justified in part by the formidable administrative requirement that can result in large projects if partial lien waivers or evidence of partial payment is required. These steps may not be necessary if other techniques exist that protect against such risks. One function of a retainage (discussed in Section 26.03) is to protect against liens. Also, ¶ 9.3.3 in AIA Doc. A201 requires the contractor to warrant that all work and material will be free of liens. This, of course, will not preclude liens but will create a right against the contractor or surety under a payment bond. A similar warranty approach is used by the NSPE in ¶ 14.2.

Perhaps allowing design professionals to withhold progress payments when they learn that subcontractors are not being paid or that liens have been or are likely to be filed is the best owner protection. Usually such information is communicated quickly to the design professional.

AIA Doc. A201, ¶ 9.4.1, requires that the architect either issue a certificate or notify the contractor in writing with reasons for withholding a certificate within seven days after receipt of the contractor's application for payment. Under ¶ 9.6.1, the architect can decline to certify or withhold the certificate "in whole or in part." The seven-day requirement will be met if there is partial certification and written reasons given for a failure to certify the entire amount requested. Since ¶ 8.1.4 defines days as calendar days, the seven-day period is short. Proof of payment and lien waivers required for final payment are discussed in Section 26.05.

(D) OBSERVATIONS AND INSPECTIONS

Before issuing a payment certificate, the design professional visits the site to determine how far the work has progressed. This is the basis upon which progress payments are made.

The inspection should uncover whatever an inspection principally designed to determine the *progress* of the work would have uncovered. The AIA has sought to limit the architect's responsibility by defining a certificate as a representation by the architect based solely upon her "observations at the site." This attempt to inject a subjective standard is not always successful. The principal focus should be upon the inspection *promised* by the architect. The AIA makes clear that the architect is not promising to make "exhaustive or continuous on-site inspection to check quality or quantity of the work." Although this means that the architect is *not* expected to be on the site *full time*, it does not diminish the importance of the inspection to determine whether and to what extent a payment certificate should be issued.

(E) AMOUNTS CERTIFIED FOR PAYMENT

The amount certified for payment depends upon the pricing provisions in the construction contract. In fixed-price contracts, the pricing benchmark is the contract price. In cost contracts, the principal reference point for determining payments is the allowable costs incurred by the contractor. In unit priced contracts, the progress payments are based upon the number of units of designated work performed.

Each pricing provision commonly uses a retainage system under which a designated amount is withheld from progress payments to provide security to the owner. (Retainage is discussed in Section 26.03.)

The computation of the amount to be paid is facilitated by the use of a schedule of values. Such an agreed schedule determines the extent to which the work has progressed. Smaller contracts without a schedule of values sometimes provide that specified payments are to be made at designated phases of the work with an appropriate allowance for retainage where used.

Despite a properly prepared schedule of values, measurement problems can develop. A significant amount of time elapses between the contractor ordering material and equipment and incorporating that material and equipment into the work. The contractor prefers payment as early as possible. The owner prefers not to have to pay until the material and equipment have been incorporated into the project. Paragraph 9.3.2 of AIA Doc. A201 generally requires that the work be incorporated before payment is to be made. However, as noted in (C), it allows for payment for materials and equipment not incorporated but stored on and off the site. Also as noted in (C), payment prior to incorporation into the project raises legal and insurance questions.

Another problem relates to the possibility that material or equipment left in the possession of the contractor may be seized by its creditors or by its trustee in bankruptcy. Seizure would be based upon leaving the material or equipment in the possession of the contractor *manifesting* to the outside world that it belongs to the contractor. A recent case involved a claim by a public entity against the architect when the trustee in bankruptcy seized material left in the contractor's possession for which the owner had paid. The claim against the architect was based upon the assertion that the architect should have warned the owner of the legal risks in paying for materials and allowing them to be in the contractor's possession. The claim was not successful.[2]

Construction contracts allow the design professional to make *partial* certification for payment. Partial certificates authorize payments of an amount less than requested by the contractor. Usually such certificates are based upon the design professional's determination that the work has not progressed to the extent claimed by the contractor or upon a determination that the work does not meet the requirements in the contract documents.[3] If an architect intends to withhold the certificate either in whole or in part, ¶ 9.6.1 of AIA Doc. A201 requires the architect to notify the contractor of such a decision and to give reasons for this action. Under ¶ 9.6.1, failure by the contractor and architect to agree on a revised schedule allows the architect to issue a certificate for payment for the amount "for which he is able to make such representations to the Owner."

Construction contracts frequently give the design professional the power to revoke a previously issued certificate. Revocation can be effectuated by a partial certificate or by withholding a certificate for work that has been performed. One reason for revocation is the discovery of defective work that had been the basis of a previously issued certificate.

Paragraph 9.6.1 of AIA Doc. A201 gives the architect power to make payment adjustments. These adjustments are designed to protect the owner from losses that have occurred or may occur in the future. Losses can relate to nonconforming work, nonpayment of subcontractors and suppliers, or claims made by other contractors or other third parties against the owner for which the contractor may be responsible.

Although the law would give the owner certain offset rights under these circumstances without a contractual right of offset, it is advisable to expressly recognize this right. Doing so avoids the necessity of es-

---

2. *Travelers Indemnity Co. v. Ewing, Cole, Erdman & Eubank*, 711 F.2d 14 (3rd Cir.1983). (This was discussed in Section 15.07.)

3. *Berry v. Blackard Construction Co.*, 13 Ill.App.3d 768, 300 N.E.2d 627 (1973), held that compliance with contract documents did not condition contractor's right to recover progress payments. This is justified only if there had been a course of conduct of payment despite noncompliance. Normally the architect should not certify for payment work that the architect knows is not in compliance with the contract documents.

tablishing a right to offset and can exceed the protection given by law. However, the design professional should not be unreasonable or arbitrary in determining when and how much to withhold from certificates otherwise earned by the contractor.[4] Issuing partial certificates or, more important, withholding certificates can result in contractor default. For this reason, language that requires design professional and contractor to discuss and negotiate on these matters is useful.[5]

Federal procurement policy allows reduction or suspension of progress payments despite the absence of any present contract breach by the contractor where the latter's financial position makes a future nonperformance likely.[6]

(F) TO WHOM PAYMENT SHOULD BE MADE

Contractors or subcontractors often must borrow funds to operate their businesses. Lenders commonly require collateral to secure them against the possibility that the borrower will not repay the loan. Sometimes collateral consists of funds to be earned under specific construction contracts. Lenders often seek information from owners regarding the construction contracts whose payments are to be used as security and assurances by the owner that payments will be made to the lender or that checks will be issued jointly to lender and contractor.[7]

Rather than rely on a promise, more commonly lenders demand that assignments be made to them of the payments. An assignment transfers the right to receive payment and effectuates a change of ownership in the rights transferred. It is a more substantial security than a promise. The party making the transfer is the *assignor*. The party to whom ownership is transferred is the *assignee*. The party owing the obligation being transferred is the *obligor*. Using the fact pattern of the prime contractor seeking a loan, the prime is the assignor, the bank the assignee, and the owner the obligor.

Such assignments were difficult if not impossible to accomplish early in English legal history. But modern law not only makes such assignments possible but encourages them. However, these assignments should not put the obligor in a substantially worse position.

Encouragement of assignments has gone to the extreme of invalidating contract clauses that prohibit assignment of such rights where the assignments are given as collateral to obtain loans.[8] Terms in contracts precluding assignment of payments for such purposes are invalid.

Although the validity of such assignments no longer raises serious questions, other legal problems exist. They center principal-

---

4. *S.I.E.M.E., S.r.1.*, ASBCA No. 25642, 81–2 BCA ¶ 15,377 (improper to withhold $10,000 when $1,000 would have been sufficient to complete the few remaining items).

5. This is required in federal procurement. See 32 C.F.R.App. E ¶ 524.

6. This power was exercised in *National Eastern Corp. v. United States*, 477 F.2d 1347 (Ct.Cl.1973).

7. Sometimes subcontractors seek to borrow funds, and information or assurances are sought from the prime contractor.

Whether giving information and assurances is legally enforceable by the lender is often difficult to determine. *Central National Bank & Trust Co. of Rockford v. Consumers Construction Co.*, 5 Ill.App.3d 274, 282 N.E.2d 158 (1972), held a prime contractor liable to a lender who lent to a subcontractor in reliance on letters sent by the prime contractor. Yet another Illinois decision held that a lender could *not* recover from the owner. *Bank of Marion v. Robert "Chick" Fritz, Inc.*, 9 Ill.App.3d 102, 291 N.E.2d 836 (1973).

8. *Mississippi Bank v. Nickles & Wells Construction Co.*, 421 So.2d 1056 (Miss.1982); *Aetna Casualty & Surety Co. v. Bedford-Stuyvesant Restoration Construction Corp.*, 90 A.D.2d 474, 455 N.Y.S.2d 265 (1982). See U.C.C. § 9–318(4).

ly upon the extent to which the original contract obligation can be changed because a third party (the assignee) now owns contract rights.

An obligor prime contractor was held liable to the assignee bank who had lent to an assignor subcontractor when the prime contractor did not see to it that the funds were paid to the bank *despite* the prime's assent to the assignment being conditioned upon its being given no greater burden.[9]

Normally the obligor must pay the assignee after it has received notice of the assignment. But sometimes the obligor, such as the owner, can pay to the assignor contractor to enable it to complete the contract.[10]

Modification of existing contracts whose rights have been assigned can be made "in good faith and in accordance with reasonable commercial standards" as long as payments assigned have not as yet been earned.[11] The obligor (owner) can offset against the assignee (lender) amounts owed it by the assignor (prime contractor).[12]

Owners who wish to be certain that subcontractors and suppliers are paid sometimes issue a joint check made out to the prime contractor jointly with the subcontractor for its work or with the supplier for its materials. This method is discussed in greater detail in Section 26.05.

(G) SURETY REQUESTS THAT PAYMENT BE WITHHELD

Payment bonds require the surety to pay unpaid subcontractors and suppliers. To recover any losses caused by having to make such payments, sureties often demand that the owner stop paying the prime contractor when the latter has defaulted in its payments to subcontractors or suppliers.

In public work, unpaid subcontractors and suppliers do not have lien rights. As a result, the owner may decide to pay the prime contractor to enable the latter to continue performance and complete the work. In doing so, the surety's right to reimbursement can be affected adversely. Cases decided in the U.S. Court of Claims have given the contracting officer broad discretion to pay earned progress payments to the contractor despite some minor defaults.[13] The court has recognized the government's interest in obtaining a completed project by giving discretion to make payments even though payment can harm the surety.

Suppose the owner accedes to the surety's requests or demands. In such a case, the unpaid prime contractor may, in addition to claiming a right to the payment withheld, assert a claim against the surety for wrongful interference with the contract that the prime contractor has made with the owner.[14]

(H) REMEDIES FOR NONPAYMENT

For convenience, the discussion will assume an unpaid prime contractor. However, any conclusions expressed in that context are likely to apply to unpaid first- or second-tier subcontractors. The law recognizes the im-

---

9. *Bank of Yuma v. Arrow Construction Co.*, 106 Ariz. 582, 480 P.2d 338 (1971).

10. *Fricker v. Uddo & Taormina Co.*, 48 Cal.2d 696, 312 P.2d 1085 (1957).

11. U.C.C. § 9–318(2).

12. Id. at § 9–318(1).

13. *Argonaut Insurance Co. v. United States*, 434 F.2d 1362 (Ct.Cl.1970). Although the court seemed to give the contracting officer considerable discretion, a subsequent opinion by the Court of Claims noted that the *Argonaut* case required that discretion be exercised "responsibly" and that the surety's interest be considered. *United States Fidelity & Guaranty Co. v. United States*, 475 F.2d 1377, 1384 (Ct.Cl.1973).

14. Such a claim was unsuccessful in *Gerstner Electric, Inc. v. American Insurance Co.*, 520 F.2d 790 (8th Cir. 1975), the court failing to find the requisite malice or bad faith. Refer also to Section 17.08(D).

portance of prompt payment. Yet under certain circumstances, drastic remedies for nonpayment may not be appropriate. Sometimes delay in payment is unavoidable. Delay may not harm the contractor. Care must be taken to avoid an unpaid party using minor delay as an excuse to terminate an unprofitable contract and cause economic dislocation problems. Despite these possibilities, on the whole nonpayment is—and should be—considered a serious matter.

At the outset, there must be a determination that failure to make payment is a breach of contract. Because construction contracts are detailed and contract procedures often ignored, those accused of not paying in accordance with the contract often assert that prompt payment has been waived.[15]

If payment is not made as promised, the contractor is entitled to interest on the payment. Since, as a rule, the payments are liquidated or relatively certain in amount, the interest runs from the date payment was due.

In the absence of a contract specified interest rate, the rate at which interest is computed is the legal rate of interest that is the amount added to court-ordered judgments when payment is not made at the time the judgment is entered. A survey in 1979 showed that the legal rate varied from 5% to 8%, with the most common rate being 6%.[16] During the highly inflationary 1970s and early 1980s, the legal rate was substantially below the market rate. This encouraged owners to delay progress payments.

Generalizations as to the differential between the legal rate and the market rate can be dangerous. For example, effective 1983, California increased the legal rate to 10%.[17] Shortly thereafter, inflation sharply declined. Perhaps it is not impossible that in the near future the legal rate may be more than the market rate for borrowing money. For this reason, the construction contract should stipulate a realistic rate for late payment rather than leave the rate unspecified, invoking the legal rate. Any market rate selected should be based upon the interest the contractor will have to pay in the market where it borrows and should recognize the possibility that a lower-than-market rate may discourage prompt payment.[18]

Contractors *unable* to borrow money may suffer other losses, such as the ability to bid on other projects, loss of key personnel who leave when they are not paid, or eviction caused by nonpayment of rent. If such losses are reasonably foreseeable at the time the contract is made and could not have been reasonably avoided by the contractor, they can be recovered by the contractor if they can be proven with reasonable certainty. The obstacles to recovering for losses of this sort are formidable, and as a rule, interest is about as much as can be recovered for delayed payment or nonpayment.

A formidable weapon in the event of nonpayment is the power to suspend work. In addition to placing heavy pressure upon the owner, suspension avoids the risk of further uncompensated work. The availability of a mechanics' lien is a pale substitute when there is a substantial risk of nonpayment. Until recently the law was not willing to recognize a remedy short of termination for such breaches. Recent cases hold that the

---

15. *Bart Arconti & Sons, Inc. v. Ames-Ennis, Inc.*, 275 Md. 295, 340 A.2d 225 (1975) (waiver by accepting late payments 50% of the time); *Silliman Co. v. S. Ippolito & Sons*, infra note 20 (no waiver).

16. *National Law Journal*, June 11, 1979, pp. 18–19.

17. West's Ann.Cal.Code Civ.Proc. § 685.010.

18. AIA Doc. A201, ¶ 7.8.1, specified the legal rate at the *place* of the project.

contractor can suspend work if it is not paid.[19]

Certainly nonpayment should not automatically give the right to suspend performance. Shutting down and starting up a construction project is costly. It would be unfair to allow the contractor to shut down the job simply because payment is not made absolutely on schedule.

Under AIA Doc. A201, ¶ 9.7.1, seven days after a progress payment should have been made, the contractor can give a seven-day notice of an intention to stop work unless payment is made. Failure to pay by expiration of the notice period permits suspension. Paragraph 9.7.1 also states that the contractor shall receive a price increase for its reasonable costs of shutdown, delay, and startup. Undoubtedly the contractor should be able to suspend work when it is not paid after a reasonable period of time. However, the two seven-day periods may be too short.

Can an unpaid contractor terminate its obligation to perform? Suspension is temporary. Termination relieves the contractor from the legal obligation of having to perform in the future. In the absence of any contract provision dealing with this question, termination is proper if the breach is classified as "material."

Although materiality of breach is discussed in greater detail in Section 38.04, it may be useful to look at some factors that relate to nonpayment as a material breach. The most important are effect on the contractor's ability to perform and the likelihood of future nonpayment. Persistent nonpayment may indicate that the problem is a serious one. A clear statement that performance will *not* be made is a repudiation and clearly gives the right—and probably the obligation—to stop performance. Termination is an important decision. The law looks at *many* factors before deciding whether there has been a material breach. Yet many cases have concluded that a failure to pay gives the contractor the right to terminate its obligation to perform under the contract.[20]

Emphasis in the discussion and in the cases has been upon the effect of nonpayment upon the contractor's ability to perform. This can operate to the disadvantage of a financially sound contractor. Another approach, and perhaps a better one, is to permit termination if a changed cash flow will cause the contractor to finance the project to a larger degree than anticipated. If, however, the facts clearly demonstrate that a particular contractor was selected for its capacity to absorb payment delay, among other factors, this *may* show an intention by

---

19. *Watson v. Auburn Iron Works, Inc.*, 23 Ill.App.3d 265, 318 N.E.2d 508 (1974) (facts indicated more than simple nonpayment); *Zulla Steel, Inc. v. A & M Gregos*, 174 N.J.Super. 124, 415 A.2d 1183 (1980); *Aiello Construction, Inc. v. Nationwide Tractor Trailer, etc.*, 413 A.2d 85 (R.I.1980). See also Restatement (Second) of Contracts § 237 (1981).

20. *Guerini Stone Co. v. P.J. Carlin Construction Co.*, 248 U.S. 334, 39 S.Ct. 102, 63 L.Ed. 275 (1919); *United States v. Western Casualty & Surety Co.*, 498 F.2d 335 (9th Cir.1974); *St. Paul-Mercury Indemnity Co. v. United States*, 238 F.2d 917 (10th Cir.1956) (subcontractor); *United States v. Premier Contractors, Inc.*, 283 F.Supp. 343 (N.D.Me.1968) (supplier); *Integrated, Inc. v. Alec Fergusson Electric Contractors*, 250 Cal.App.2d 287, 58 Cal.Rptr. 503 (1967) (many factors in addition to nonpayment); *Silliman Co. v. S. Ippolito & Sons, Inc.*, 1 Conn.App. 72, 467 A.2d 1249 (1983) (subcontractor); *Berry v. Blackard Construction Co.*, supra note 3 (owner's refusal based upon contention that work did not conform); *Leto v. Cypress Builders, Inc.*, 428 So.2d 819 (La.App.1983) (nonpayment made work precarious); *Zulla Steel, Inc. v. A & M Gregos, Inc.*, supra note 19 (protracted suspension converted to termination); *Shapiro Engineering Corp. v. Francis O. Day Co.*, 215 Md. 373, 137 A.2d 695 (1958); *Aiello Construction, Inc. v. Nationwide Tractor Trailer, etc.*, supra note 19 (substantial underpayment for prolonged period); *Tennessee Asphalt Co. v. Purcell Enterprises, Inc.*, 631 S.W.2d 439 (Tenn.App.1982) (subcontract: substantial delay required); *Darrell J. Didericksen & Sons, Inc. v. Magna Water & Sewer Improvement District*, 613 P.2d 1116 (Utah 1980); *H.E. & C.F. Blinne Contracting Co.*, ENG BCA No. 4174, 83–1 BCA ¶ 16,388.

the contractor to accept the risk of substantial alteration of financing the work.

Suppose the contractor continues performance. Continued performance may indicate that nonpayment was not sufficiently serious to constitute termination. Continued performance may manifest an intention to continue performance that is relied upon by the owner. However, continued performance should not invariably preclude nonpayment from justifying termination. An unpaid contractor may *choose* to continue work for a short period while awaiting performance.[21] This would be especially true if the contractor clearly indicated that if payment were not forthcoming, work would cease.

AIA Doc. A201, ¶ 14.1.1, permits the contractor to terminate if the work has been suspended for thirty days for nonpayment. The contractor must first give a seven-day written notice that it intends to terminate. Does this express termination provision affect any common law power to terminate for nonpayment? That depends upon whether the express power to terminate is exclusive. Generally the law hesitates to make specified remedies in a contract exclusive unless there is a clear indication that this is the intention of the parties.[22] This has been codified in A201, ¶ 7.6.1, which states that remedies specified are in addition to any remedies otherwise imposed or available by law.

Can the owner preclude termination by paying all unpaid progress payments with interest during the seven-day period? Is the notice period designed to allow the owner to cure past defaults?[23] Since termination is a drastic remedy, common sense and the obligation of good faith and fair dealing support the conclusion that the owner can cure past defaults during the notice period. Remedies to a contractor who has justifiably ceased performance or who has been wrongfully removed from the site is discussed in Section 31.02. Some courts allow the contractor to recover only for work performed but not for lost profits.[24] However, the general tendency is to treat this breach as no different than any other.[25]

The effect of owner nonpayment upon the right of a subcontractor to recover for its performance is treated in Section 32.06.

(I) PAYMENT AS WAIVER OF DEFECTS

Sometimes contractors contend that making progress payments waives any deviations from contract document requirements. Generally payment in and of itself does not waive defects.[26] Suppose the owner knows or should know of the defect and makes payment. This may lead the contractor reasonably to believe that the owner intends to pay despite the defect. If so, minor defects are waived.

To preclude waiver in such cases, construction contracts frequently contain provisions stating that progress payments do not waive defects. For example, ¶ 9.5.5 of AIA Doc. A201 states that progress payments are not an acceptance of work that is not in compliance with the contract documents. The payment certificate should make clear that payment does not constitute waiver.

---

21. *Darrell J. Didericksen & Sons, Inc. v. Magna Water & Sewer Improvement District,* supra note 20. But see *Drew Brown Limited v. Joseph Rugo, Inc.,* 436 F.2d 632 (1st Cir.1971).

22. *Glantz Contracting Co. v. General Electric,* 379 So.2d 912 (Miss. 1980); *Bender-Miller Co. v. Thomwood Farms, Inc.,* 211 Va. 585, 179 S.E.2d 636 (1971).

23. See Section 38.03(F).

24. *Palmer v. Watson Construction Co.,* 265 Minn. 195, 121 N.W.2d 62 (1963).

25. *Leto v. Cypress Builders, Inc.,* supra note 20.

26. See Annot., 66 A.L.R.2d 570 (1959). Waiver generally is discussed in Section 26.06(E).

Acts that can constitute waiver of defects more commonly arise when the project is accepted or completed and final payment is made. Detailed discussion of this problem is found in Section 28.05.

(J) PROGRESS PAYMENTS AND THE CONCEPT OF DIVISIBILITY

Divisibility, sometimes called severability, is a multipurpose legal concept sometimes applied to a contract where performance is made in installments or is divided into designated items. As an illustration of the first, the contract for the sale of goods could allow twelve monthly deliveries and payments at the time of each delivery. An illustration of the second would be a contract for the sale of goods that consisted of five different items to be delivered with different prices for each item.

If a contract is considered divisible, it would be as if there were twelve separate contracts in the first illustration and five in the second. Each partial performance, whether a monthly delivery or delivery of less than all the five items, would be considered the equivalent of the money promised for each installment or item. Without describing the many legal issues that can sometimes depend upon the divisibility classification, the progress payment mechanism should not result in the construction contract being considered divisible. The amounts certified are approximations and not agreed valuations for the work as it proceeds. The amounts can be—and frequently are—adjusted at the time of final payment.[27]

SECTION 26.03   RETAINAGE

Retainage is a contractually created security system under which the owner retains a specified portion of earned progress payments to secure itself against certain risks. For example, suppose the schedule of values establishes that the contractor is entitled to a progress payment of $50,000. Construction contracts commonly provide that a portion of this $50,000 will be retained by the owner and paid at the end of the project or at some later date. The purpose of retainage is to provide money out of which claims that the owner has against the contractor can be collected without the necessity of a lawsuit. It is *not* an agreed damage or damage limit.[28]

Retainage has become controversial. Contractors and subcontractors frequently contend that retention of money that they have earned is unfair to them and costly and unnecessary to the owner. They assert that the reasons for creating the security, such as defective work or the prime contractor not paying subcontractors or suppliers, are dealt with by the contractor furnishing performance and payment bonds. Contractors and subcontractors also contend that financing costs that are inevitably incurred because of retention are ultimately transferred to the owner through the contract price. They then argue that the owner is paying twice for the same risk—once through financing costs included in the contract price and once by the cost of surety bond premiums.

This argument equates the efficacy of retainage with bonds. Obviously it is better for the owner to actually have funds within its control that can be used to secure the owner against these risks rather than having to deal with or send unpaid subcontractors or suppliers to the surety. Prudent owners can earn interest on the funds retained.

There has been some tendency to reduce retainage, and some contracts eliminate it entirely. Sometimes retainage is limited to

---

27. *Dravo Corp. v. Litton Systems, Inc.,* 379 F.Supp. 37 (S.D.Miss.1974); *Shapiro Engineering Corp. v. Francis O. Day Co.,* supra note 20; *Kirkland v. Archbold,* 113 N.E.2d 496 (Ohio App.1953).

28. *Gurney Industries, Inc. v. St. Paul Fire & Marine Insurance Co.,* 467 F.2d 588 (4th Cir.1972).

the first 50% of the work. The frequent modification of California statutes that control retainage in state contracts illustrates the retainage controversy.[29]

Complaints of subcontractors have resulted in a recent decision by the Office of Federal Procurement Policy [30] to effectuate a uniform governmentwide retainage policy. Retainage must not be used as a substitute for good contract management. The agency cannot withhold funds without good cause. Determinations concerning the use of retainage should be based upon an assessment of the contractor's past performance and the likelihood that such performance will continue. It is suggested that retainage not exceed 10% and that it be adjusted downward as the contract approaches completion, particularly if there is better than expected performance or alternate safeguards. Once all contract requirements have been completed, all retained amounts should be paid promptly to the contractor.

Who should receive interest on retained funds? Usually interest goes to the owner. High interest can lead to payment delay. Standard forms in some countries provide that an interest be shared or paid to the contractor.[31]

As to other areas of law, it is becoming more common for tenants to receive interest on security deposits they leave with landlords. Similarly, there has been a movement toward giving home loan borrowers the interest earned on tax impounds (monthly payments they make to lenders out of which lenders pay for property taxes and insurance). If the retainage is the *contractor's* money that is simply being held as a security by the owner, the contractor can contend that the owner, holding this money as a trustee, must pay any interest earned to the contractor.

But does the money belong to the contractor? Unlike deposits made by tenants or payments made by debtors, this is money that has never been in the possession of the contractor. It is not likely that contractors have the bargaining power to obtain this in their contracts. (Perhaps many would not want it. They would have to pay much of the interest to subcontractors.) This can embarrass a contractor if the money were attached by its creditors or its trustee in bankruptcy.

One problem incident to retainage can develop at the end of the project. Suppose the retained amount is $50,000 and the owner is entitled to take $20,000 from it to remedy defects in the work. A solvent contractor will be paid the balance.

However, in the volatile construction industry, it is not uncommon for a contractor to have run into financial problems. In such a case, a horde of claimants descends upon the owner and demands the money. The claimants can be lenders who have received assignments of contract payments as security for a loan, sureties who have had to discharge the obligations of the contractor, taxing authorities who claim the funds when the contractor has not paid its

---

29. The statute that prescribes retention for contracts with California State agencies is West's Ann.Cal.Pub. Contract Code § 10261. Before 1981, it was West's Ann.Cal.Gov't Code § 14402. In 1945, 10% was required (1945 Cal.Stat. Ch. 118). In 1957, the state was permitted to eliminate retainage after 50% of the work (1957 Cal.Stat. Ch. 481). In 1970, the state could reduce retainage after 95% of the work was completed to 125% of the work needed to complete if the surety approved (1970 Cal.Stat. Ch. 1080). In 1971, retainage was reduced to 5% (1971 Cal.Stat. Ch. 1286). See *Economy Forms Corp. v. City of Cedar Rapids*, 340 N.W.2d 259 (Iowa 1983), which upheld an Iowa statute requiring a public agency to retain double any claim made by third parties.

30. OFPP Policy Letter 83–1, 48 Fed.Reg. 22,832 (1983). This has been implemented by the GSA. See 48 Fed. Reg. 37,997 (1983).

31. A recent ad by a commercial bank stated that it had developed a system under which the owner retained control but the contractor received interest. *Wall Street Journal,* Oct. 5, 1983, p. 22.

taxes, and, in the event the contractor has gone bankrupt, the trustee in bankruptcy.

The scramble for funds usually is dealt with by the owner's beginning what is called an "interpleader" action. The owner pays the disputed funds into court, starts a lawsuit in which it names as parties all claimants to the funds, and withdraws from the fray. The court must unscramble the claims.[32]

Although the possible multiple claimants' confusion at the end of the job may not in itself be a reason to limit or eliminate retention, the troublesome disputes that relate to retainage ownership should be taken into account in determining whether and to what extent retainage should be used.

SECTION 26.04
SUBSTANTIAL COMPLETION

Sometimes construction contracts provide that a payment will be made to the contractor upon substantial completion. Perhaps more important, the "end of the job" process, which begins with the contractor's representation that it has substantially completed the project and concludes with a certificate of final completion, is one about which a number of legal issues cluster (described in this section and in Section 26.05).

Although time problems are discussed in Chapter 30, it should be noted that under AIA Doc. A201, ¶ 8.1.1, compliance with the contractor's time obligations is measured by *substantial* completion. The extent to which the contractual mechanism for determining substantial completion has analogies in the legal doctrine of "substantial performance" is discussed in Section 26.06(B).

The process for determining substantial completion under A201, ¶ 9.8.1, begins with a submission to the architect by the contractor of a list of items to be completed or corrected, followed by an inspection by the architect to determine whether she should issue a certificate of substantial completion. Such inspections are more carefully made than are inspections for ordinary progress payments. More important issues are involved.

SECTION 26.05
COMPLETION AND FINAL PAYMENT: SOME LEGAL ISSUES

The AIA process begins with an act by the contractor, a notice that the work is ready for final inspection and acceptance and an application for final payment. The architect inspects the work to determine whether the certificate should be issued—an inspection that should be undertaken with great care.

This stage in the Construction Process is a benchmark that has serious implications for major participants in that process. If the owner has not taken possession of the project at the time of substantial completion, it will certainly do so at this point. Possession by the owner involves legal responsibilities that rest upon the possessor of land.[33] Perhaps more important, the end of the job often has an effect upon allocation of risks, such as those that relate to defects, and the continued existence of any claim. Because completion is sometimes equated with acceptance—an important legal doctrine that relates to defects—it is discussed in Sections 17.09(C) and 28.05. Similarly, the effects of completion, final payment, and acceptance can also bear upon claims the contractor has against the owner.

Mechanics' liens can surface at the end of the job. Although they are discussed in greater detail in Chapter 32 dealing with

---

32. For one of the many articles dealing with this issue, see Mungall, *The Buffeting of the Subrogation Rights of the Construction Contract Bond Surety by United States v. Munsey Trust Co.*, 46 Ins. Counsel J. 607 (1979).

33. Refer to Section 7.08.

subcontracts, this section looks briefly at liens from the vantage point of the owner, with particular reference to the end of the job.

Owners employ a variety of techniques to avoid liens. Some relate to the payment process. In some contracts, though not those of the AIA, owners seek to avoid liens by requiring evidence that subcontractors and suppliers who have lien rights either have been paid or have executed lien waivers before the contractor receives *progress* payments. Even if this is not required for a progress payment, it is almost certainly required for final payment. For example, AIA Doc. A201, ¶ 9.9.2, conditions final payment upon the contractor's submitting an affidavit that it has paid all its bills and gives the owner the right to require other data "such as receipts, releases and waivers of liens arising out of the Contract, to the extent and in such form as may be designated by the owner." Because of concern that lien waivers or—even worse—"no lien" contracts [34] (under which the prime contractor waives all liens in advance for itself, its subcontractors, and suppliers) may be unfair to subcontractors and suppliers, legal controls have been enacted by the states that regulate these methods.[35] Where this is the case, owners may choose to issue final payments as well as progress payments through joint checks.

Although these legislative activities have been designed to protect subcontractors and suppliers, other legislative activity protects owners from liens. For example, an Iowa statute provides that the owner need not make final payment until sixty days after completion of the project if the project would be subject to a lien.[36] This does not apply if the prime contractor has filed a payment bond or supplied signed receipts or lien waivers. Generally, liens must be filed a designated number of days after substantial or final completion.[37] One way of avoiding liens is to delay final payment until the time for filing liens has expired.

Owner protection does not eliminate the requirement that contracting parties be fair to one another. For example, a recent case involved a contract with lien avoidance conditions before final payment. Yet the contractor was awarded final payment *without* complying because the owner had insisted that the contractor waive its delay damage claim before final payment would be made. The court found this tactic to be unconscionable.[38]

SECTION 26.06
PAYMENT FOR WORK
DESPITE NONCOMPLIANCE

(A) THE DOCTRINE OF CONDITIONS

The promise by the owner to make any payments is conditioned upon the contractor's complying with the contract documents. During the course of the work, the contractor is not entitled to be paid unless, as discussed in Section 26.02, there are provisions for progress payments. At the end of the project, the doctrine of conditions in its strictest sense requires absolute compliance with the contract documents before the contractor is entitled to any unpaid part of the contract price.

Although the owner would be permitted to hold back all the money until it receives what has been promised, in construction contracts strict application of the doctrine

---

34. Upheld in *Beacon Construction Co. v. Matco Electric Co.*, 521 F.2d 392 (2d Cir.1975) (New York law).

35. West's Ann.Cal.Civ. Code § 3262.

36. Iowa Code Ann. § 572.13.

37. West's Ann.Cal.Civ. Code § 3116 (ninety days after completion or cessation of work or thirty days after notice of cessation or completion filed by owner).

38. *North Harris County Junior College District v. Fleetwood Construction Co.*, 604 S.W.2d 247 (Tex.Civ.App.1980).

can cause hardship to the contractor. During performance, progress payments are essential to finance the job and avoid the risk of going unpaid for work. At the end of the job, a strict application of the doctrine of conditions can cause a loss to the contractor disproportionate to the loss caused by nonperformance. Such loss can result in unjust enrichment where the owner is occupying the project. This section outlines some of the legal doctrines that have developed to relieve the contractor when performance has not been in *exact* compliance with contract documents.

The doctrine of conditions can apply to a contractor who fails to complete the project on time. A subcontractor's promise to complete by a designated time was held to be a promise and not a condition to the prime contractor's obligation to pay the contract price.[39] The court noted that conditions are not favored because they can create forfeiture and unjust enrichment. The contractor can recover the unpaid balance, but it must pay any damages caused by delay. The absence of a valid liquidation of damages clause usually means that damages are very difficult to prove. The ultimate outcome is likely to be that the contractor will collect for the work despite its late performance. However, were timely performance a condition to final payment—unlike defects of quality that can be cured—the prime contractor would have the benefit of the subcontractor's work without paying fully for it.

(B) SUBSTANTIAL PERFORMANCE: *PLANTE v. JACOBS*

Most substantial performance cases involve construction contracts for a number of reasons. Construction is a complex undertak-ing with detailed contract requirements. A strong likelihood exists of minor deviations that surface at the "end of the job." Work performed by prime contractor employees may escape the scrutiny of the prime's superintendent. Much of the work is performed by subcontractors, and the exact quantity and quality may be difficult to determine while the work is being performed.

The complexity of contract documents frequently generates interpretation questions, and it is often difficult to get an authorized interpretation by the design professional while work is being performed. Likewise, it may be difficult to find someone with authority to direct a change or approve substituted materials or products claimed by the contractor to be equal or equivalent to those specified.

Unlike other types of contracts, the breaching party—that is, the contractor—cannot take back its performance. It has become incorporated into the owner's land, and very frequently the owner has taken possession of the project.

With some limited exceptions,[40] substantial performance issues usually arise at the end of the job. The prime contractor contends that the owner has *essentially* what it bargained for and often points to the owner's occupying and using the project. One court defined substantial performance as:

> . . . when construction has progressed to the point that the building can be put to the use for which it was intended, even though comparatively minor items remain to be furnished or performed in order to conform to the plans and specifications of the completed building.[41]

Another case added that the defects must not "so pervade the whole work that a deduction in damage will not be fair compen-

---

39. *Landscape Design & Construction, Inc. v. Harold Thomas Excavating, Inc.*, 604 S.W.2d 374 (Tex.Civ.App.1980).

40. *Nordin Construction Co. v. City of Nome*, 489 P.2d 455 (Alaska 1971) (owner sued to recover payments).

41. *Southwest Engineering Co. v. Reorganized School District R–9*, 434 S.W.2d 743, 751 (Mo.App.1968).

sation."[42] Yet another case concluded that there had not been substantial performance when there was an accumulation of small defects, even though the owner was occupying the premises.[43]

The ratio of defect correction costs to contract price and the determination of whether the building has met its essential purpose are relevant.[44] But one court affirmed a judgment of substantial performance despite cost of correction of incomplete and defective work of 31% of the contract price.[45]

Jacob & Youngs, Inc. v. Kent,[46] a leading case, found that there had been substantial performance when it was discovered *after completion* that one brand of pipe had been substituted for the one specified. The court noted that the pipe substituted was of equal quality to that designated.

Kirk Reid Co. v. Fine,[47] involved the installation of an air-conditioning system. The owner proved that the following deviations existed:

1. The system was 46–50 tons short in capacity.

2. The primary air unit was of a lower rating by 7850 cubic feet per minute.

3. The condenser water pipe was 120 gallons per minute short of capacity.

Yet the court held that the contractor had substantially performed stating:

> It appears from the . . . finding that "there was a workable air conditioning plant installed and operating by May 1, 1959," the time called for by the contract. It further appears from the record that the system so installed and operating has been used, since that time, by the defendant without complaint by him that it is insufficient to perform its task.[48]

A case dealing with construction of a church held that the contractor had substantially performed when the deviation consisted of a two-foot lower ceiling, windows shorter and narrower than called for, and seats narrower than designated by the specifications.[49] Another court held that the contractor substantially performed when it was discovered at the completion of the project that the roof shingles were discolored.[50] The court noted that the shingles could not be seen from the street and did not constitute a functional defect.

Where the contractor leaves the project *before* completion, it is more difficult, though not impossible, to establish substantial performance.[51]

Early cases required the contractor to show that it was free from fault and that the breach was not willful, such as by establishing that the deviation was caused by a

---

42. *Jardine Estates, Inc. v. Donna Brook Corp.*, 42 N.J.Super. 332, 126 A.2d 372, 375 (1956).

43. *Tolstoy Construction Co. v. Minter*, 78 Cal.App.3d 665, 143 Cal.Rptr. 570 (1978).

44. *Stevens Construction Corp. v. Carolina Corp.*, 63 Wis.2d 342, 217 N.W.2d 291 (1974). See also 3A A. Corbin, Contracts § 706 (1960).

45. *Jardine Estates, Inc. v. Donna Brook Corp.*, supra note 42.

46. 230 N.Y. 239, 129 N.E. 889 (1921).

47. 205 Va. 778, 139 S.E.2d 829 (1965).

48. 139 S.E.2d at 837.

49. *Pinches v. Swedish Evangelical Lutheran Church*, 55 Conn. 183, 10 A. 264 (1887).

50. *Salem Towne Apartments, Inc. v. McDaniel & Sons Roofing Co.*, 330 F.Supp. 906 (E.D.N.C.1970). But in *O.W. Grun Roofing & Construction Co. v. Cope*, 529 S.W.2d 258 (Tex.Civ.App.1975), the court held that streaks in a structurally sound roof precluded substantial performance and required the roof to be replaced.

51. In *Watson Lumber Co. v. Mouser*, 30 Ill.App.3d 100, 333 N.E.2d 19 (1975), though, a trial court conclusion of substantial performance was affirmed.

subcontractor or was not done knowingly.[52] However, recent cases have reflected ambivalence on this issue. One case held that substantial performance could be used despite a finding that the contractor knowingly deviated from the contract.[53] Other cases have held that the reason for the breach must be taken into account with other factors to determine whether there has been substantial performance.[54]

Although substantial performance is designed to avoid forfeiture and unjust enrichment, it would be difficult to conclude there was unjust enrichment if a contractor testified that it had substituted one brand of pipe for another because "it was just as good" or because it did not feel like going back to the warehouse.[55]

It has been stated that explicit contract language precludes substantial performance from determining the contractor's rights to the contract price.[56] Where such statements are made, they are likely to be in cases where the contractor has been found to have substantially performed.[57] The dilemma posed by this question requires that the essential nature of the substantial performance doctrine be determined. If the doctrine merely reflects a common intention that minor deviations will not preclude recovery, a contract clause negating this intention should be controlling. However,

if the doctrine is based upon the avoidance of forfeiture by the contractor and corresponding unjust enrichment of the owner, an express contract clause should not bar use of this doctrine. Resolving this difficult question will be aided by an evaluation of *Plante v. Jacobs* (reproduced ahead).

Following this case is a further critique of the substantial performance doctrine and its relationship to contract language dealing with strict compliance.

Usually the contractor must prove it has substantially performed.[58] However, once substantial performance has been established, as a rule the owner must prove its damages.[59] Substantial performance, though entitling the contractor to recover the balance of the contract price, is still a breach. Since damages reduce the amount to which the contractor is entitled, the law places the burden of proving damages upon the owner once substantial performance has been established. It is likely that the *extent* of damages will be an important factor in the owner's attempt to establish that there has not been substantial performance.

A few states place the burden of establishing damages upon the *contractor*, an anomalous result in light of the first line of attack by the contractor usually being that it has *fully* performed.[60]

Damages are sometimes measured by the

---

52. *Jacobs & Youngs, Inc. v. Kent,* supra note 46.

53. *Kirk Reid Co. v. Fine,* supra note 47.

54. *Hadden v. Consolidated Edison Co. of N.Y., Inc.,* 34 N.Y.2d 88, 356 N.Y.S. 249, 312 N.E.2d 445 (1974); *Nordin Construction Co. v. City of Nome,* supra note 40; *Watson Lumber Co. v. Mouser,* supra note 51 (requiring performance be in good faith and breach not be willful).

55. *O.W. Grun Roofing & Construction Co. v. Cope,* supra note 50.

56. *Jacob & Youngs, Inc. v. Kent,* supra note 46.

57. Ibid. But see *Winn v. Aleda Construction Co., Inc.,* 315 S.E.2d 193 (Va. 1984) (language barred substantial performance).

58. *A.W. Therrien Co. v. H.K. Ferguson Co.,* 470 F.2d 912 (1st Cir.1972).

59. *Maloney v. Oak Builders, Inc.,* 224 So.2d 161 (La.App.1969); *Hopkins Construction Co. v. Reliance Insurance Co.,* 475 P.2d 223 (Alaska 1970).

60. *Vance v. My Apartment Steak House of San Antonio, Inc.,* 677 S.W.2d 480 (1984).

cost of correction of the defect, sometimes by the diminished value of the project, and sometimes by a combination of the two.[61]

The interrelation of various damage measurements can be seen in *Plante v. Jacobs*, reproduced here.

PLANTE v. JACOBS
Supreme Court of
Wisconsin, 1960.
10 Wis.2d 567, 103 N.W.2d
296.

Suit to establish a lien to recover the unpaid balance of the contract price plus extras of building a house for the defendants, Frank M. and Carol H. Jacobs, who in their answer allege no substantial performance and breach of the contract by the plaintiff and counterclaim for damages due to faulty workmanship and incomplete construction. . . . After a trial to the court, judgment was entered for the plaintiff in the amount of $4,152.90 [62] plus interest and costs, from which the defendants, Jacobs, appealed and the plaintiff petitioned for a review. . . .

The Jacobs, on or about January 6, 1956, entered into a written contract with the plaintiff to furnish the materials and construct a house upon their lot in Brookfield, Waukesha county, in accordance with plans and specifications, for the sum of $26,765. During the course of construction the plaintiff was paid $20,000. Disputes arose between the parties, the defendants refused to continue payment, and the plaintiff did not complete the house. On January 12, 1957, the plaintiff duly filed his lien.

The trial court found the contract was substantially performed and was modified in respect to lengthening the house two feet and the reasonable value of this extra was $960. The court disallowed extras amounting to $1,748.92 claimed by the plaintiff because they were not agreed upon a writing in accordance with the terms of the agreement. In respect to defective workmanship the court allowed the cost of repairing the following items: $1,550 for the patio wall; $100 for the patio floor; $300 for cracks in the ceiling

---

61. These damage measurements are discussed in Section 30.03. In *Jacob & Youngs, Inc. v. Kent*, supra note 46, the measure of recovery for substitution of pipe was the diminished value of the project that the court concluded was nominal. But in *Salem Towne Apartments, Inc. v. McDaniel & Sons Roofing Co.*, supra note 50, the diminished value of the project was found to have been about one-half the cost of correction.

62. Ed. note: The computation of the judgment must have been:

Original contract price	26,765.00	
Extras	960.00	
Contract price as adjusted		27,725.00
Payments received by plaintiff	20,000.00	
Balance unpaid		7,725.00
Less: 1) Amount conceded by plaintiff for omissions (kitchen cabinets, gutters and downspout, sidewalk, closet clothes poles, and entrance seat)	1,601.95	
2) Cost of correction for patio wall ($1550), patio-floor ($100), cracks in living room and kitchen ceiling ($300), and credit balance for hardware ($20.15).	1,970.15	
		3,572.10
		$4,152.90

of the living room and kitchen; and $20.15 credit balance for hardware. The court also found the defendants were not damaged by the misplacement of a wall between the kitchen and the living room, and the other items of defective workmanship and incompleteness were not proven. The amount of these credits allowed the defendants was deducted from the gross amount found owing the plaintiff, and the judgment was entered for the difference and made a lien on the premises. . . .

HALLOWS, Justice. The defendants argue the plaintiff cannot recover any amount because he has failed to substantially perform the contract. The plaintiff conceded he failed to furnish the kitchen cabinets, gutters and downspouts, sidewalk, closet clothes poles, and entrance seat amounting to $1,601.95. This amount was allowed to the defendants. The defendants claim some 20 other items of incomplete or faulty performance by the plaintiff and no substantial performance because the cost of completing the house in strict compliance with the plans and specifications would amount to 25 or 30 per cent of the contract price. The defendants especially stress the misplacing of the wall between the living room and the kitchen, which narrowed the living room in excess of one foot. The cost of tearing down this wall and rebuilding it would be approximately $4,000. The record is not clear why and when this wall was misplaced, but the wall is completely built and the house decorated and the defendants are living therein. Real estate experts testified that the smaller width of the living room would not affect the market price of the house.

The defendants rely on Manitowoc Steam Boiler Works v. Manitowoc Glue Co., . . . for the proposition there can be no recovery on the contract . . . unless there is substantial performance. This is undoubtedly the correct rule at common law. . . . The question here is whether there has been substantial performance. The test of what amounts to substantial performance seems to be whether the performance meets the essential purpose of the contract. In the Manitowoc case the contract called for a boiler having a capacity of 150 per cent of the existing boiler. The court held there was no substantial performance because the boiler furnished had a capacity of only 82 per cent of the old boiler and only approximately one-half of the boiler capacity contemplated by the contract. In Houlahan v. Clark, . . . the contract provided the plaintiff was to drive pilings in the lake and place a boat house thereon parallel and in line with a neighbor's dock. This was not done and the contractor so positioned the boat house that it was practically useless to the owner. Manthey v. Stock, . . . involved a contract to paint a house and to do a good job, including the removal of the old paint where necessary. The plaintiff did not remove the old paint, and blistering and roughness of the new paint resulted. The court held that the plaintiff failed to show substantial performance. The defendants also cite Manning v. School District No. 6, . . . However, this case involved a contract to install a heating and ventilating plant in the school building which would meet certain tests which the heating apparatus failed to do. The heating plant was practically a total failure to accomplish the purposes of the contract. See also Nees v. Weaver, . . . (roof on a garage).

Substantial performance as applied to construction of a house does not mean that every detail must be in strict compliance with the specifications and the plans. Something less than perfection is the test of specific performance unless all details are made the essence of the contract. This was not done here. There may be situations in which features or details of construction of special or of great personal importance, which if not performed, would prevent a finding of substantial performance of the contract. In this case the plan was a stock floor plan. No detailed construction of the house was shown on the plan. There were no blueprints. The specifications were standard printed forms with some modifications and additions written in by the parties. Many of the problems that arose during the construction had to be solved on the basis of practical experience. No mathematical rule relating to the percentage of the price, of cost of completion or of completeness can be laid down to determine substantial performance of a building contract. Although the defendants received a house with which they are dissatisfied in many respects, the trial court was not in error in finding the contract was substantially performed.

The next question is what is the amount of recovery when the plaintiff has substantially, but incompletely, performed. For substantial performance the plaintiff should recover the contract price less the damages caused the defendant by the incomplete performance. Both parties agree. Venzke v. Magdanz, . . . states the correct rule for damages due to faulty construction amounting to such incomplete performance, which is the difference between the value of the house as it stands with faulty and incomplete construction and the value of the house if it had been constructed in strict accordance with the plans and specifications. This is the diminished-value rule. The cost of replacement or repair is not the measure of such damage, but is an element to take into consideration in arriving at value under some circumstances. The cost of replacement or the cost to make whole the omissions may equal or be less than the difference in value in some cases and, likewise, the cost to rectify a defect may greatly exceed the added value to the structure as corrected. The defendants argue that under the Venzke rule their damages are $10,000. The plaintiff on review argues the defendants' damages are only $650. Both parties agree the trial court applied the wrong rule to the facts.

The trial court applied the cost-of-repair or replacement rule as to several items, relying on Stern v. Schlafer, . . . wherein it was stated that when there are a number of small items of defect or omission which can be remedied without the reconstruction of a substantial part of the building or a great sacrifice of work or material already wrought in the building, the reasonable cost of correcting the defect should be allowed. However, in Mohs v. Quarton, . . . the court held when the separation of defects would lead to confusion, the rule of diminished value could apply to all defects.

In this case no such confusion arises in separating the defects. The trial court disallowed certain claimed defects because they were not proven. This finding was not against the great weight and clear preponderance of the evidence and will not be disturbed on appeal. Of the remaining defects claimed by the defendants, the court allowed the cost of replacement or repair except as to the misplacement of the living-room wall. Whether a defect should fall under the cost-of-replacement rule or be considered under the diminished-value rule depends upon the nature and magnitude of the defect. This court has not allowed items of such magnitude under the cost-of-repair rule as the trial court did. Viewing the construction of the house as a whole and its cost we cannot say, however, that the trial court was in error in allowing the cost of repairing the plaster cracks in the ceilings, the cost of mud jacking and repairing the patio floor, and the cost of reconstructing the non-weight-bearing and nonstructural patio wall. Such reconstruction did not involve an unreasonable economic waste.

The item of misplacing the living room wall under the facts of this case was clearly under the diminished-value rule. There is no evidence that defendants requested or demanded the replacement of the wall in the place called for by the specifications during the course of construction. To tear down the wall now and rebuild it in its proper place would involve a substantial destruction of the work, if not all of it, which was put into the wall and would cause additional damage to other parts of the house and require replastering and redecorating the walls and ceilings of at least two rooms. Such economic waste is unreasonable and unjustified. The rule of diminished value contemplates the wall is not going to be moved. Expert witnesses for both parties, testifying as to the value of the house, agreed that the misplacement of the wall had no effect on the market price. The trial court properly found that the defendants suffered no legal damage, although the defendants' particular desire for specified room size was not satisfied. . . .

On review the plaintiff raises two questions: Whether he should have been allowed compensation for the disallowed extras, and whether the cost of reconstructing the patio wall was proper. The trial court was not in error in disallowing the claimed extras. None of them was agreed to in writing as provided by the contract, and the evidence is conflicting whether some were in fact extras or that the defendants waived the applicable requirements of the contract. The plaintiff had the burden of proof on these items.

The second question raised by the plaintiff has already been disposed of in considering the cost-of-replacement rule.

It would unduly prolong this opinion to detail and discuss all the disputed items of defects of workmanship or omissions. We have reviewed the entire record and considered the points of law raised and believe the findings are supported by the great weight and clear preponderance of the evidence and the law properly applied to the facts.

Judgment affirmed.

Returning to the question posed before *Plante v. Jacobs*, to what extent can contract language preclude use of the substantial performance doctrine? *Plante v. Jacobs* stated:

> Substantial performance as applied to construction of a house does not mean that every detail must be in strict compliance with the specifications and the plans. Something less than perfection is the test of specific performance *unless all details are made the essence of the contract*. [Editor's emphasis.] This was not done here. There may be situations in which features or details of construction of special or of great personal importance, which if not performed, would prevent a finding of substantial performance of the contract.

A distinction must be drawn between different methods by which the contract itself might preclude use of the doctrine. First, a clause could simply state that strict compliance with all requirements of the contract documents is required. Such a clause is likely to be found among the many clauses included in the general conditions of the contract.

Second, the contract could create, as does AIA in A201, a system dealing with minor defects discovered when an inspection is made to determine whether there has been substantial completion and a mechanism for dealing with these defects between substantial and final completion.

Third, interpretation of specific language in the contract documents and the surrounding facts and circumstances may indicate that exact compliance with certain contract requirements was expected. For example, suppose in *Plante v. Jacobs* that the contractor knew or should have known that the Jacobses had purchased expensive furniture that could not be used because the living room wall had been misplaced. Alternatively, suppose the Jacobses had informed Plante that for esthetic reasons the living room *must* comply in all respects to the floor plan dimensions and *no* deviation would be allowed. In either case, it could be argued that the substantial performance should not be used, as the owners expected strict performance and this was communicated to the contractor at the time the contract was made.

If the doctrine of substantial performance is based upon the implication that "close to perfect" compliance is all the owner could reasonably expect, the doctrine should not be applied to contracts using any of the methods previously outlined. Any method could indicate that the owner expected full compliance and that the contractor was aware of this responsibility. Perhaps a clause of the first type would be less persuasive, since it could be part of boilerplate, preprinted clauses and would not sufficiently bring to the contractor's attention the need for strict compliance. Even here the contractor should be aware of the importance of strict compliance.

A misplaced wall could be troublesome even if the contract appeared to negate use of the substantial performance doctrine. This is the very type of defect for which the doctrine is designed. If so, another objective of the doctrine is to avoid economic waste, a goal that originally helped generate the doctrine.

In *Plante v. Jacobs,* correction of the misplaced wall would have cost $4,000, but the misplaced wall was found not to have diminished the value of the house. Although the court did not face a contract method for avoiding substantial performance, it is doubtful that the court would have either ordered the contractor to correct the mistake or permitted the owner to take a $4,000 deduction from the final payment for damages. Actual replacement of the wall either by court order or by allowing the Jacobses to reimburse themselves out of the final payment after they corrected the work themselves would cause or endorse an uneconomic expenditure of $4,000 that would not increase the value of the house.

Suppose the Jacobses were allowed to deduct the $4,000 *without* the need to show they were sufficiently concerned to correct the mistake. Would the Jacobses simply take the money and not make the correction? In the likely event they would not make the correction, they would be given a windfall. But this windfall possibility assumes that the Jacobses built their house principally as an investment and not as a place to live. Even assuming that the Jacobses would not invest the $4,000 to correct the mistake, some recognition should be given to the possibility that the Jacobses might have been damaged in some difficult-to-measure way by having to live in a house with a living room one foot shorter than they had planned even though the misplaced wall gave them an extra foot in the kitchen. Although some might doubt that a mistake of this kind could cause emotional distress, as was seen in Section 6.06(H), the law is slowly moving toward the recognition that certain contracts are made not simply for commercial reasons but for reasons of personal solicitude. Were this the case here, such a loss could have been taken into account in determining the damages caused by the misplaced wall.

The substantial performance doctrine straddles concepts of contract and unjust enrichment. There *can* be risk assumption by contract, though economic waste and windfall *cannot* be excluded. As evidence becomes clear that the risk was specifically assumed by the contractor, there may be less room for the doctrine. But as it becomes apparent that there will be economic waste or windfall, the doctrine has more scope despite contract language that might seem to preclude its use.

Plante v. Jacobs illustrates yet another aspect of the interrelationship between the contract and substantial performance. The court stated:

> In this case the plan was a stock floor plan, no detailed construction of the house was shown on the plan. There were no blueprints. The specifications were standard printed forms with some modifications and additions written in by the parties. Many of the problems that arose during construction had to be solved on the basis of practical experience.

In this type of transaction, the contract documents, such as they were, may have only been starting-out points from which the parties work together toward a particular solution. In such a case, it would be unfair to hold the contractor to the original requirements of the contract because contractor and owners were essentially designing the house as it was being built.

Plante v. Jacobs also illustrates possible abuse of the doctrine. The defects at the end of the job consisted of the contractor's failure to furnish the kitchen cabinets, gutters and downspouts, sidewalk, closet clothes poles, and entrance seat. The estimated cost of furnishing these items was $1,610.95. The amount retained was $7,725. Superficially it would seem as if Plante had substantially performed and should be entitled to the outstanding balance less the cost of correction. In theory the deduction from the outstanding balance would compensate the owner for the omissions. In actuality the legal measure of recovery does not accomplish this. There are additional costs, such as obtaining a substitute contractor, the delay in effectuating the necessary corrections, and the risk of a substitute contractor not performing, which would be difficult to deduct from the outstanding balance.

Even if the contractor *had* removed his workers from the project, it would not impose any serious burden upon him to return and complete the job as promised. Arguably, it would not be unreasonable for the owner to withhold the entire $7,725 as an unashamedly coercive device to get the contractor to finish the work if the owner chose this route rather than engaging a substitute contractor. If the amount withheld appears excessive, perhaps it could be reduced to two or three times the projected cost of correction. But the amount retained must be enough to make it worthwhile for the contractor to finish the job properly. If not, some contractors will refuse to perform and invoke the substantial performance doctrine to recover the outstanding balance less the amount deducted as damages. In such a case, it is likely that a court would conclude that there had been substantial performance. The court noted that the house had been decorated and occupied by the owners even though the owners were dissatisfied with the contractor's work. Certainly the house, with the omission, was fit for its primary purpose, and close to strict completion had been accomplished.

The above criteria were generated from cases that clearly involved economic waste or windfall because the defects were discovered at the end of the project and would have involved substantial redoing of the project at expenditures greatly in excess of the diminished value caused by the deviation. However, where these elements are not present, the substantial performance

doctrine should not be a device by which contractors can walk away without having completed the project and expect to be paid the balance less the damages allowed by law. In this regard one court expressed the view:

> The doctrine of substantial performance is a necessary inroad on the pure concept of freedom of contracts. The doctrine recognizes countervailing interests of private individuals and society; and, to some extent, it sacrifices the preciseness of the individual's contractual expectations to society's need for facilitating economic exchange. This is not to say that the rule of substantial performance constitutes a moral or ethical compromise; rather, the wisdom of its application adds legal efficacy to promises by enforcing the essential purposes of contracts and by eliminating trivial excuses for nonperformance.[63]

This court views the doctrine as one to protect contractors from owners who seek an excuse to avoid paying. This attitude can lead a contractor to walk away from a commitment and be able to force the court to perform an "accounting job," compute the balance owed by deducting the cost of correction or completion, or use the diminished value, a result *not* bargained for by the owner. It can promote a "just as good" philosophy and encourage sloppy, incomplete work. Whether it promotes promise making is debatable.

AIA Doc. A201, ¶ 8.1.3, defines substantial completion as that state of completion that allows the owner to "occupy or utilize" the work, a definition much like that used for substantial performance.

Does the contractual system used in A201 *displace* the substantial performance doctrine? Earlier in this section it was noted that courts frequently state that the substantial performance doctrine is subject to express provisions stating that 100% compliance is required. Has the contract *itself* generated a substantial performance mechanism that makes the common law substantial performance doctrine superfluous?

---

63. *Bruner v. Hines*, 295 Ala. 111, 324 So.2d 265, 269–270 (1975).

This rarely discussed issue requires a description of the process set forth in A201. As noted in Section 26.04, the contractor who considers its work substantially complete submits a punch list of items to be completed to the owner. The architect, upon determining that the work has been substantially completed, issues a certificate of substantial completion. As noted in Section 26.05, the process for final completion begins with a notification by the contractor that the work has been completed. If after inspection the architect determines this to be the case, the architect issues a final payment certificate and the owner must pay.

A comparison between this system and the common law substantial performance doctrine can be made at both substantial and final completion stages. Suppose the architect does *not* believe the work sufficiently complete for the owner to occupy or utilize it. In the role as issuer of certificates and judge of performance, the architect's decision is final unless overturned by arbitration or litigation. Were the contractor to go to court because of a disagreement over the resolution of this question, the issue would be the finality of the architect's decision (discussed ahead).[64] At *this* point the issue should be not substantial performance but the finality of the architect's decision. Similarly, were the owner dissatisfied with the architect's determination that there had been substantial performance, that decision would stand unless overturned by arbitrators or by courts.

The next stage, that of final completion, does present a conflict between the AIA system and the common law doctrine of substantial performance. Suppose the architect refuses to issue a certificate of final completion. The architect's having issued the certificate of substantial completion is a determination that there has been substantial performance, the assumption being that substantial completion and substantial performance are roughly the same. Invocation of the substantial performance doctrine would entitle the contractor to the balance of the contract price less any damages caused by its breach. Under the AIA system, it would not be entitled to the *balance* until the architect issues the final payment certificate. In theory, the entire retainage can be withheld, creating the type of forfeiture and unjust enrichment that the common law substantial performance doctrine sought to avoid. However, if obligations of good faith and fair dealing or specific contract language created an obligation to retain only that which is reasonable, there would appear to be no need for the substantial performance doctrine.

Suspicion of the architect's impartiality and the deeply ingrained nature of the substantial performance doctrine and its underlying premises would combine to make it difficult to oust the substantial performance doctrine at either substantial or final completion stage.[65] Yet a study of the cases shows that most involve transactions between unsophisticated owners often *without* a design professional and detailed contract documents, the very type of transaction where the AIA Documents are *not* likely to be used.

(C) DIVISIBLE CONTRACT

For various reasons the law sometimes artificially treats one contract as if it were composed of a number of separate contracts. Section 26.02(J) discussed whether progress payments in construction contracts make the contract divisible.

One function that can be served by the divisibility fiction is to enable a contractor to

---

64. Section 33.09.

65. Two cases that applied substantial performance in a contract that provided for architect certification are *Standard Millwork & Supply Co. v. Mississippi Steel & Iron Co.,* 205 Miss. 96, 38 So.2d 448 (1949), and *Southwest Engineering Co. v. Reorganized School District R–9, supra* note 41.

recover despite not having fully performed the contract. For example, suppose a construction project has five well-defined phases and designated payments for each phase. Suppose the contractor completes three phases but omits the remaining two. In such a case, classifying this essentially single contract into a series of five contracts would enable the contractor to recover for the three phases completed, with the owner being left with a deduction for the damages caused by the failure to complete the final two.

The presence of progress payments does not mean that the parties have agreed that each segment of work is an agreed equivalent for the amount of the progress payment being made. The progress payments are approximations and not intended as agreed figures. Although it is not inconceivable that a court would classify a construction contract as divisible in order to permit the contractor to recover despite default, the doctrine of substantial performance discussed in Section 26.06(B) and the doctrine of restitution, to be discussed in Section 26.06(D), are more appropriate vehicles for giving relief to a contracting party who has not fully performed yet has benefited the other party.

(D) RESTITUTION: UNJUST ENRICHMENT

The defaulting building contractor may not have performed sufficiently to take advantage of the substantial performance doctrine. Another concept sometimes available that can enable the contractor to recover for any net benefit that its performance has conferred upon the owner is restitution, a concept sometimes called quasi-contract, or *quantum meruit*.

Restitution, unlike its substantial performance counterpart, has not received uniform acceptance. Early cases were not sympathetic to defaulting parties, and many held that a party in default could not recover for work performed.[66] However, there has been some relaxation of this attitude, and in many jurisdictions contractors are able to recover restitution despite their own breach in order to avoid unjust enrichment of the owner.[67]

Suppose there is a construction contract in which a promise to pay $100,000 is exchanged for the promise to construct a building. Suppose the contractor unjustifiably leaves the project during the middle of performance. The defaulting contractor may have conferred a net benefit despite its having breached by abandoning the project. For example, suppose the defaulting contractor has received $20,000 in progress payments but the work has been sufficiently advanced so that it can be completed by a successor contractor for $75,000. Absent any delay damages, the defaulting contractor has conferred a net benefit of $5,000 upon the owner and should be entitled to this amount.

Some of the same problems that have been discussed in relation to substantial performance apply when the contractor seeks to use restitution to recover for the benefit that has been conferred. Such recovery may be precluded if the contractor's breach is considered "willful" and similar problems of measuring the damages caused by the breach arise when restitution rather than substantial performance is the basis of recovery.

Restitution cases do not arise as frequently as do "end of the job" substantial performance disputes. Part of this may be due

66. *Kelley v. Hance,* 108 Conn. 186, 142 A. 683 (1928); *Steel Storage & Elevator Construction Co. v. Stock,* 225 N.Y. 173, 121 N.E. 786 (1919).

67. *American Surety Co. of New York v. United States,* 368 F.2d 475 (9th Cir.1966); *United States v. Premier Contractors, Inc.,* supra note 20; *Starling v. Housing Authority of City of Atlanta,* 162 Ga.App. 852, 293 S.E.2d 392 (1982); *R.J. Berke & Co. v. J.P. Griffin, Inc.,* 116 N.H. 760, 367 A.2d 583 (1976); *Nelson v. Hazel,* 91 Idaho 850, 433 P.2d 120 (1967); *Kreyer v. Driscoll,* 39 Wis.2d 540, 159 N.W.2d 680 (1968). See Annot., 76 A.L.R.2d 805 (1961).

to the unlikeliness of a contractor's walking off a project that would be profitable. Only in these cases is there likely to be a net benefit to the owner that exceeds progress payments that have been made.

(E) WAIVER

During performance or at the end of the project, the owner may manifest a willingness to pay the full contract price despite the contractor's not having fully complied with the contract documents. This manifestation is often described as a waiver. If the deviation is an important aspect of the contract, any premise by the owner to pay despite noncompliance would have to be supported by what the law would regard as consideration. Consideration, as discussed in Section 5.08, usually consists of something given in exchange for accepting the deviation, such as a price reduction. Alternatively, the waiver of even important matters would be enforced if the contractor had reasonably relied upon the statement. Examples are stating "Do the extra work and I won't insist on a written change order" or misleading the contractor by a course of conduct such as paying despite noncompliance.[68]

Waivers of less important matters are effective despite the absence of any consideration or reliance. All that is needed in such cases is evidence that an intention to waive or give up the requirement was communicated to the other party.[69]

A waiver can be directly expressed (You told me you would pay even though I omitted the doorstops). It can be expressed indirectly by implication (When you made the final payment knowing I omitted the doorstops, I assumed you were waiving the defect). Waiver is more likely to be found when the defect is small (doorstops) or technical (failure to obtain a written change order). It is more likely to be found where it is directly *expressed* (I'll pay anyway, or Change orders are a bother and I know you did the extra work). It is less likely to be found when expressed by implication. (When you paid, I thought you were waiving those defects of which you were aware, or When you didn't complain, I assumed you waived the defects). These factors recognize that small things can be given up if the evidence is clear that the party entitled to them was willing to do so but that acts are often ambiguous.

The act frequently contended to indicate waiver of defects is the owner accepting by using or occupying the project; particularly those defects apparent at the time the owner used or occupied the project will be asserted to have been waived. Although this topic is treated in greater detail in Section 28.05, it should be noted that court decisions are not consistent, though the trend is against acceptance as waiver of a claim.[70]

General conditions of construction contracts sometimes seek to eliminate the possibility of waiver. For example, ¶ 9.5.5 of AIA Doc. A201 states that payment is not an acceptance of nonconforming work. Sometimes clauses more generally state that when one party does not insist upon its full contract rights, it shall not be precluded from insisting upon these rights in the future. Clauses of this type, though supportive of a conclusion that there has been no

---

68. Refer to Section 25.04(G).

69. *Texana Oil Co. v. Stephenson*, 521 S.W.2d 104 (Tex.Civ.App.1975).

70. Although moving into the house and signing a form of acceptance for a federal agency was held to waive defects in *Cantrell v. Woodhill Enterprises Inc.*, 273 N.C. 490, 160 S.E.2d 476 (1968), *Honolulu Roofing Co. v. Felix*, 49 Hawaii 578, 426 P.2d 298 (1967), held that mere occupancy is not a waiver, particularly when the owners had to move in because they had vacated their own home. See also *Aubrey v. Helton*, 276 Ala. 134, 159 So.2d 837 (1964), which held that occupancy alone would not waive defects: only where there is accompanying conduct clearly indicating that there has been acceptance will occupancy and use constitute acceptance.

waiver,[71] are not absolutely waiver proof. Where it appears that there has been a course of conduct that deviates from contract requirements or clear evidence that a defect has been given up, waiver can be found. The clause is removed by concluding that the nonwaiver clause itself has been waived.

Waiver, unlike substantial performance or restitution, allows recovery of the entire contract balance without any deduction for damages. It must be remembered that waiver usually deals with technical requirements such as written change orders[72] or minor defects in the work.

The party asserted to have waived contract rights must have been authorized to do so. Authority is determined by agency concepts and is often dealt with in the contract.[73] However, it is likely that it would take less authority to waive a minor contract defect than to make a contract in the first instance or to modify an existing contract. Waiver is less likely to be applied in public contracts.[74]

SECTION 26.07
THE CERTIFICATION
AND PAYMENT PROCESS:
SOME LIABILITY PROBLEMS

The certification and payment process can generate professional liability problems for design professionals. Their conduct in this process can cause losses to various participants in the construction project that can be caused by:

1. Certifying an incorrect amount for payment.

2. Not reasonably discovering defects during inspections conducted incident to certificate issuance.

3. Issuing certificates without determining whether the contractor has been paying subcontractors or suppliers.

4. Not requiring that lien waivers be filed when they should have been.

5. Unreasonable delay in issuing certificates.

The amount of the certificate raises a separate legal problem that must precede discussion of the other claims against the design professional. Issuing certificates, like other functions often performed by the design professional, can be considered quasi-judicial in nature. The contract that creates this power may accord certain finality to the decisions that culminate in a payment certificate. Likewise, the law may grant the design professional some immunity when this quasi-judicial function has been performed. These matters were discussed elsewhere.[75]

Claims by the owner for not conducting inspections properly, for not avoiding liens, and for delay will be regulated largely by the contract between owner and design professional (discussed in Section 15.08). Contracts published by professional associations of design professionals, such as the AIA or NSPE, seek to limit the extent of and responsibility for the inspection that precedes issuance of certificates for payment.

More difficult problems arise when the claim is made by persons not a party to the owner-design professional contract, such as the prime contractor, subcontractors, suppliers, or sureties. A claim by the contrac-

---

71. *Armour & Co. v. Nard*, 463 F.2d 8 (8th Cir.1972).

72. Refer to Section 25.04(G), and see Section 30.08 dealing with waiving formalities to obtain time extension.

73. On the question of authority, refer to Section 21.05(B).

74. Refer to Section 14.04(C). But in *Kenny Construction Co. v. Metropolitan Sanitary District of Greater Chicago*, 52 Ill.2d 187, 288 N.E.2d 1 (1972), the court treated waiver in a public contract much as if it had been a private contract.

75. Refer to Section 17.08(C).

tor is likely to be based upon under-certification or delay in certification. As for the first, the prime contract may determine the finality of the decision.[76] As for the second, the prime may be met with the argument that there is no "privity" between prime and design professional—that there is no contractual relationship between the two that creates a duty owed to the contractor by the design professional. Cases are rare in this area because the standards for determining unreasonable delay are sometimes unclear. Even when they are clear, courses of conduct frequently develop that may constitute abandonment of any clearly drawn time requirement. Delay may be caused by persons other than the design professional. More likely, claims for delay will be made against the owner rather than the design professional.

While privity as a defense is weakening in claims of personal harm, it still has some vitality where the claim is an economic loss, such as delay damages.[77] Although the answer is not settled, it is likely that an absence of privity in most jurisdictions would not preclude the contractor from suing the design professional for unreasonable delay.

The most difficult cases are those involving claims by unpaid subcontractors, unpaid suppliers, or sureties who pay them in accordance with payment bonds. Over-certification usually shrinks the retainage, an important source of surety reimbursement. Paying the prime contractor without determining whether subcontractors or suppliers are or have been paid can permit diversion, which causes losses to these subcontractors and suppliers when the prime contractor does not pay them and becomes insolvent.

Liability for over-certification depends upon the finality of the certificate and any immunity given to design professionals. (These problems are discussed elsewhere.) [78] Actions brought by third parties based upon the design professional's conduct causing or not preventing prime contractor diversion of funds must surmount the absence of privity discussed earlier. Although the cases are not unanimous, absence of privity is not likely to be a bar to recovery.[79]

PROBLEMS

1. A construction contract called for oak panelling in certain rooms. Because oak was unavailable, a subcontractor installed birch panel. The prime did not notice this deviation. Oak and birch panelling cost the same. When the owner saw the birch panelling, he told the prime contractor he liked it.

---

76. See Section 33.09.

77. Refer to Section 17.09(D).

78. See Section 33.09. In *Palmer v. Brown,* 127 Cal.App.2d 44, 273 P.2d 306 (1954), an owner recovered from the architect for negligently issued certificates without discussing finality or immunity. The case also involved a claim that certificates were issued fraudulently because the architect was working for the contractor on another project at the time certificates were issued. The court held that this raised a conflict of interest and could justify avoidance of the certificate on that basis.

In *Newton Investment Co. v. Barnard & Burk, Inc.,* 220 So.2d 822 (Miss.1969), a claim by the owner against the engineer was denied because the owner "took over" the payment process and there was no showing of negligence. Again, neither finality nor immunity was discussed.

79. Allowing recovery were *State v. Malvaney,* 221 Miss. 190, 72 So.2d 424 (1954); *Westerhold v. Carroll,* 419 S.W.2d 73 (Mo.1967); *Aetna Insurance Co. v. Hellmuth, Obata & Kassabaum, Inc.,* 392 F.2d 472 (8th Cir.1968). Denying recovery were *Peerless Insurance Co. v. Cerny & Associates, Inc.,* 199 F.Supp. 951 (D.Minn.1961), and *Engle Acoustic & Tile, Inc. v. Grenfell,* 223 So.2d 613 (Miss.1969) (sharply limiting *State v. Malvaney,* supra).

Finality is discussed in Section 33.09, and design professional immunity was discussed in Section 17.08(C).

The next day the owner told the prime contractor he wanted oak panelling installed. Is the contractor obligated to replace the panelling? If he does not, what, if anything, can the owner recover?

2. On February 1, C was entitled to a progress payment of $10,000 based upon a certificate issued by the architect. The owner informed the contractor on February 2 that payment would be delayed for a week because the proceeds of the construction loan had run out and the owner was in the process of getting a new loan. What are the contractor's rights at this point? What additional facts would be helpful in resolving this question?

27

Expectations and Disappointments: Some Performance Problems

SECTION 27.01
INTRODUCTION TO
CHAPTERS 27 THROUGH 32

Each party to a construction contract (the contract between owner and contractor will be illustrative) has expectations when the contract is created. The owner expects to receive the project specified in the construction contract at the time promised. It also expects to pay the contract price as adjusted for change orders.

The contractor's expectations are more complex. Its performance entails a variety of tasks dependent upon cooperation by many entities as well as upon weather conditions. The contractor expects that it will be able to purchase labor and materials at or within certain prices and that work will proceed without excessive slowdowns or stoppages due to weather, labor difficulties, acts of public authorities, or failure by the owner to cooperate. The contractor does not anticipate that construction of the work will be impeded or destroyed by fire, earthquake, or vandalism. It expects to be paid as it performs.

Yet either or both may be disappointed. The owner may not receive the project it expected due to a variety of causes explored in Chapter 28 dealing with defects and

Chapter 29 dealing with subsurface problems. The owner may not receive the completed project at the time promised, a disappointment discussed in Chapter 30. The methods of measuring any claims the owner may have against the contractor for not performing are explored in Chapter 31.

The contractor's disappointments that relate to failure to receive progress payments treated in Chapter 26 are explored in greater detail in Chapter 31. The disappointments that relate to obstacles to its performance or insecurity as to the performance by the owner are dealt with in this chapter. The special problems of increases in its cost of performance due to unexpected subsurface conditions are treated in Chapter 29. Its disappointed expectations as to the pace with which it will work are treated in Chapter 30.

Disappointments can encompass losses caused by subcontractors whose special problems are discussed in Chapter 32.

This chapter introduces performance problems discussed in Chapters 28 through 32. Initially, the essential attributes of legal doctrines central to performance problems are described and briefly illustrated. In addition, some performance problems that do not justify full chapter treatment are noted in this chapter.

SECTION 27.02 AFFIRMATIVE LEGAL DOCTRINES: THE BASES FOR CLAIMS

(A) INTRODUCTION: THE SHOPPING LIST

This section summarizes in mostly nontechnical terms the legal theories that can be the basis for construction-related claims between owner and contractor. Unfortunately, it has become common for claimants to assert a variety of claims based upon assorted legal theories, with the law generously permitting, at least up to a point in a lawsuit, the use of such a "shotgun" approach. This chapter deals only with claims by contract-connected parties. As will be seen, the variety of doctrines is astounding.

(B) FRAUD

Fraud is the intentional deception of one contracting party by the other. The deception induces the former to make a contract, one that it would not have made had it not been deceived. Usually the deception consists of giving false information or half-truths or occasionally concealing information. The deception must relate to important matters. The party defrauded relies by making the contract. For example, an owner may defraud a contractor who is concerned about the owner's resources by falsely telling the contractor it has received a loan commitment. The contractor may defraud the owner by falsely representing it has access to equipment needed to perform particular work. In most states, fraud can consist of making a promise to induce the making of a contract without intending to perform it.

The defrauded party can rescind the contract and recover any benefit it has conferred on the other party based upon restitution. It can affirm the contract, cease or continue performance, and recover damages. Punitive damages are generally available.

Fraud should not play a significant role in construction performance disputes. Although it can on rare occasions be committed during performance, such as deceiving an owner by telling it that certain work has been performed or has been approved by a city inspector, it usually consists of preperformance conduct. Performance disputes over unanticipated subsurface conditions sometimes involve allegations that an owner deliberately gave false information that was relied upon by the contractor.

Fraud is easy to allege, particularly in a free-swinging construction dispute. Although fraud claims have a nuisance and settlement value, they are rarely successful.

(C) CONCEALMENT OF INFORMATION

Concealing information, though not always fraudulent, can be the basis for a claim. The early common law placed no duty upon a contracting party to disclose information to the other party during negotiation. But as noted in Section 22.04(B), there is an increasing tendency to require a party to disclose information to the other party that the latter would want to have known and that it is not likely to discover on its own. This, like fraud and misrepresentation to be discussed in (D), relates mainly to subsurface problems, as discussed in Chapter 29. Failure to disclose when disclosure should have been made permits the innocent party to rescind the contract and recover for any benefit it may have conferred upon the other party based upon restitution. Alternatively, it can be the basis for a claim for damages, based upon either the additional expenses incurred or the difference between the expenses that should have been incurred and those that were incurred.

(D) MISREPRESENTATION

Misrepresentation, discussed in Section 7.07, is intentionally, negligently, or innocently furnishing inaccurate information that is relied upon by the other party in deciding to make the contract and its terms—principally price and time. Although, like fraud and concealment, it is principally a preperformance claim, it can be based on conduct during the performance of the contract such as misrepresenting the value of work done or its quality. But like fraud and concealment, misrepresentations occurring before making the contract are usually discovered during performance. Again, this arises mainly, though not exclusively, in subsurface disputes.

(E) NEGLIGENCE

Negligence claims generally involve parties not connected by contract, such as a person delivering materials to a site against the contractor based upon the latter's negligent conduct. But as seen in Section 17.11 it can be the basis of a claim by one party to a contract against the other. A negligence claim requires a duty owed to the claimant, a violation of that duty, and a suffering of harm caused factually and proximately by the negligent party.[1]

Negligence claims have become more common in construction *contract* disputes that relate to defects or delays, usually with the hope of involving the contractor's public liability carrier, avoiding exculpatory clauses in the contract, or obtaining a more expansive remedy than one that could be obtained for breach of contract.[2]

Negligence claims require that the claimant establish that the defendant's conduct did not live up to the standard required by law, something not required in a simple breach-of-contract claim. Negligence claims are barred after a shorter period than contract claims.

The remedy for negligent conduct related to construction is usually the additional expense incurred. Although it is *roughly* similar to damages for breach of contract, certain doctrines used in contract law, such as foreseeability,[3] offer less protection in a negligence claim. Negligence as a tort can be more easily the basis for a claim for emotional distress.[4] If wrongdoing goes beyond simple negligence and involves intentional misconduct or recklessness, punitive damages may be recovered.

(F) STRICT LIABILITY

As noted in Section 7.09, strict liability has its greatest modern use in claims against manufacturers who market defective products. It does not require a showing of negligence. In that sense, it is like a breach of contract or its near relative, breach of an implied warranty. Although it is often the basis of claims by those who buy homes or lots built by developers,[5] it is rarely used as the basis for a claim between owner and contractor in the ordinary construction contract.

(G) BREACH OF CONTRACT

To determine whether a contract breach has occurred, the following must be interpreted:

1. The written contract, if any.
2. Antecedent negotiations not superseded by a written contract.
3. Terms implied by law.

The contract may prescribe conditions that relate to claims, set forth exculpatory provisions, or control remedies. Breach does not require fault, although doctrines described in Section 27.04 may provide a defense. Remedies were discussed general-

---

1. Refer to Section 7.03.

2. See Sections 27.10, 28.07.

3. Refer to Section 6.06(C).

4. Refer to Section 6.06(H).

5. See Section 28.09.

ly in Chapter 6 and are discussed more particularly in Chapter 31. Basically the claimant can recover gains prevented and losses incurred that are established with reasonable certainty, were caused substantially by the breach, could not have been reasonably avoided, and were reasonably foreseeable at the time the contract was made.

(H) EXPRESS WARRANTY

A warranty is an assurance by one party relied upon by the other party to the contract that a particular outcome will be achieved by the warrantor. For example, a manufacturer may expressly warrant that a tank it promises to build will travel one hundred miles on one hundred gallons of fuel, will not be inoperable more than ten days a month, and will hit a moving target ten miles away while the tank is travelling at twenty-five miles per hour.

Failure is a breach. Fault is not required. (Defenses, though rare, are discussed in Section 27.03.) Frequently the *method* in which the promised outcome is to be achieved is determined by the warrantor. Failure to give the warrantor freedom to determine the method is something that courts have been reluctant to do.

Express warranty for various purposes has been considered a *special* form of contract breach. It *can* be considered simply an outcome promise by the warrantor. Perhaps express warranty continues to be used to emphasize its nonfault nature. It is less susceptible to contract doctrines that can *excuse* nonperformance by a contracting party. In some jurisdictions, breach of a contract to perform services must be based upon negligence.[6] Warranty, express or implied, in such jurisdictions

may simply bring the law back to the normal nonfault contract standard. (Express warranty can create rights for a third party. It was used before the modern law of products liability allowed a user to sue a manufacturer directly without privity and based upon nonfault conduct in manufacturing the product.)[7] Its use in construction relates mainly to warranty or guarantee clauses discussed in Section 28.08. Its remedy is the same as that for breach of contract.

(I) IMPLIED WARRANTY

This is another "subspecie" of contract breach that developed historically as a separate basis for a claim. It is like express warranty in that it must be relied upon and an outcome is warranted. Liability is strict. No negligence need be shown. Remedy is the same as for breach of express warranty. In sale of goods, it usually relates to merchantability or fitness for the buyer's known purposes. Implied warranty developed as an offshoot of modern tort law mainly to allow a claimant who had suffered harm, usually personal, to maintain an action against a manufacturer with whom it was not in privity. It was also used to avoid real property doctrines that often barred an occupier of a home from maintaining an action against a builder who had sold the home to it or to a predecessor owner, again, with no need to show negligence.

In construction contract disputes, as shall be seen,[8] a contractor often claims that the owner *impliedly* warrants the sufficiency of the design. This could have been described as an implied term, that the owner "promises" that executing the design will achieve the expected outcome of both owner (a suc-

---

6. *Samuelson v. Chutich*, 187 Colo. 155, 529 P.2d 631 (1974).

7. See *Baxter v. Ford Motor Co.*, 168 Wash. 456, 12 P.2d 409 (1932). It is still used when the product was *not* defective. *Collins v. Uniroyal, Inc.*, 64 N.J. 260, 315 A.2d 16 (1974).

8. See Sections 27.05(E) and 28.02.

cessful project) and contractor (performance in a reasonably efficient manner).

Although sometimes implied terms are justified as simply reflecting the intention of the parties,[9] warranties are often implied because of "fairness," a conscious judicial allocation of risk. Since it is implied, it can be controlled by an express warranty, just as any express contract term takes precedence over or may bar an implied term. This is based not only upon preemption but also upon the absence of any reliance upon any asserted implied warranty. It is an issue in Chapter 29, which deals with subsurface claims, with the owner often giving information but disclaiming responsibility for its accuracy.

(J) CONTRACTUAL

Sometimes a claim is based on an *express* provision of the contract, seen most vividly in Chapter 29 dealing with subsurface problems.

(K) A SUGGESTION

The wide assortment of theories that can be used to justify claims in the construction context unnecessarily causes complex litigation. Claims should be divided into those based upon preperformance conduct and those based upon performance. Remedies should be classified by the type of misconduct—intentional, negligent, or innocent.

Normal performance claims should be based *solely* upon breach of contract. There is no need for express warranty, implied warranty, or negligence. The reasons for using these doctrines do not exist in normal performance disputes between owner and contractor. Although continued use of warranties *can* be justified as an indication of more *absolute* liability, this can be handled within normal contract rules. The performing party can be found to have assumed the risk of events that may make its

performance more difficult or expensive as a basis for denying any defenses based upon impossibility of performance.

Only intentional wrongdoing—rare in construction performance disputes—may need supplementation by tort law. Supplementation need not encompass recovery for emotional distress. Where appropriate, it can be dealt with by expanding remedies for breach of contract.

SECTION 27.03
DEFENSES TO CLAIMS

(A) INTRODUCTION:
CAUSATION AND FAULT

On the whole, the doctrines described in this section relate to the occurrence or discovery of supervening events that one contracting party asserts should relieve it from its obligation to perform. A party asserting such a defense must show that the event has seriously disrupted performance planning or performance itself. If the occurrence of the event could have been prevented or if the effect of the event upon performance could have been minimized or avoided, the occurrence of the event does not provide a defense. For example, suppose laborers struck because of illegal conduct by the contractor who employed them. The contractor will not receive relief, because its conduct has caused the disruptive event. Likewise, if work were shut down because of a court order issued because the owner did not comply with land use controls, the owner will not be given relief if a contractor makes a claim.

Similarly, a party who does not appear to be "innocent" is less likely to be given relief, either because its conduct has caused the event or because its fault denies it the right to obtain relief from its contractual obligations. On the whole, the doctrines described in this section deal with events that

---

9. Refer to Section 23.02.

affect performance that cannot be said to have been the fault of either party.

(B) CONTRACTUAL RISK ASSUMPTION

Most of the doctrines discussed in this section are common law. They involve claims for relief based upon doctrines developed by courts to deal with *drastic* changes of circumstances. The willingness or even power of courts to use these common law doctrines is often affected negatively by *specific* contracts clauses. For example, an owner may defend against a contractor's claim for additional compensation for delays caused by wrongful acts of the owner by pointing to a clause under which the contractor is limited to a time extension.

Similarly, the contractual doctrine of impossibility is not likely to be employed if there is a *force majeure* provision, which grants relief under specified circumstances to the performing party. For example, contract clauses frequently allow for a time extension if certain events occur. The contractor may defend a claim by the owner that the contractor has not performed on time by pointing to contract language that justified a time extension.

A fixed-price contract itself is a contractual assumption of risk that in many instances bars the contractor from receiving additional compensation in the event its performance costs have increased because of events over which it may have had no control.

Contractual risk assumption is not limited to express contract terms. Courts often hold that the owner warrants the "sufficiency" of its design. The law will protect a contractor who follows the design if the owner asserts a claim for a defect by concluding that the owner impliedly warranted the design.

As contracts become more detailed, the potential for judicial intervention may narrow. Yet the more open recognition of the adhesive nature of many construction contracts may tempt a court to judge the *fairness* of contract clauses dealing with these claims for relief.[10]

(C) MUTUAL MISTAKE

Each contracting party, particularly the party agreeing to perform services, has fundamental assumptions often not expressed in the contract. For example, often the contracting parties assume that the subsurface conditions that will be encountered will not vary greatly from those expected by the design professional or the contractor. Although some deviation may be expected, drastic deviation that has a tremendous effect upon the contractor's performance may be beyond any mutual assumption by the parties. In such a case, the contractor may be relieved from further performance (discussed in Chapter 29).

For example, *Gevyn Construction Corp. v. United States* [11] involved a contract with the Post Office Department. The specifications referred to a letter sent by Michigan state officials to the post office that stated that the contractor could tap a drain into a designated storm sewer of the State of Michigan. Relying upon this, the contractor made its bid. But the State of Michigan refused permission, and the contractor was forced to connect the drain to a more distant outlet. Although the court to some degree emphasized that the Post Office Department invited the contractor to rely upon the letter, the main basis for relief was the parties having based their agreement upon the false assumption that the drain could be tapped into the state's storm sewer.

Another illustration of the occasional use of mutual mistake as a basis for relief is

---

10. Refer to Section 5.04(C).

11. 357 F.Supp. 18 (S.D.N.Y.1972).

seen in Section 27.05(B) dealing with disruptive labor activities.

(D) IMPOSSIBILITY

Despite the generally harsh attitude taken by nineteenth century common law courts toward claims of impossibility, in that century contracting parties could be relieved if it were concluded that their performance had become impossible.[12] However, it soon became recognized that true impossibility is rare, and what is usually asserted as a defense by a performing party is that performance cannot be accomplished without excessive and unreasonable cost. This recognition emphasized "commercial impracticability" rather than actual impossibility.[13] Despite this apparently increased willingness to grant relief to a contracting party, the need for commercial certainty meant that relief would be granted sparingly. First, something unexpected must occur. Second, the risk of this unexpected occurrence must not have been allocated to the performing party by agreement or by custom. The occurrence of the event must have rendered performance commercially impracticable.[14] These are formidable requirements, and relief requires extraordinary circumstances.

In addition to the differentiation between impossibility and impracticability, other differentiations are useful. First, *objective* impossibility is sometimes differentiated from *subjective* impossibility. To grant relief, it must usually be shown not simply that the contracting party could not perform (subjective impossibility) but that other contractors similarly situated (objective impossibility) would not have been able to do so.

Temporary impossibility should be differentiated from *permanent* impossibility. For example, suppose there is a severe material shortage that makes it impossible for the contractor to continue performance. This shortage may be relieved by elimination of the event that caused the shortage, such as a transportation strike. In such a case, impossibility is only temporary. Unless the contractor is held to assume these risks, the only relief it should obtain is a time extension and not termination. However, if it appears that the event will continue an indefinite time or for so long a time that it will drastically affect the costs of the contractor, termination may be appropriate.

Partial impossibility must be differentiated from *total* impossibility. Suppose a material shortage is caused by a strike. As a result, a supplier who has contract and customer commitments of 1,000 units has only 500 in stock and cannot obtain others. It will be allowed relief if it makes a good-faith allocation of its available supply to all its customers.[15] Suppose it has a contractual commitment to deliver ten units to a contractor. If it would be an exercise of commercial good faith to supply all customers—those with contracts and those with commercial commitments—alike, the contractual obligation would be discharged if the supplier delivered five units to the contractor.

(E) FRUSTRATION

Frustration of purpose developed early in the twentieth century. It looked at the effect of subsequent events not on performance but on desirability of performance.

---

12. *Taylor v. Caldwell,* 122 Eng.Rep. 309, 3 Best & S. 826 (Q.B.1863) (theatre owner exonerated when theatre burned.)

13. U.C.C. § 2–615(a).

14. *Transatlantic Financing Corp. v. United States,* 363 F.2d 312 (D.C.Cir.1966) (charterer not relieved when Suez Canal closed in 1956.)

15. U.C.C. § 2–615(b).

For example, the leading English case [16] giving rise to this doctrine involved a contract under which the plaintiff let rooms to the defendant from which the defendant would have a good view of the coronation of King Edward VII. The coronation was called off because the King became ill. The defendant was relieved from his obligation to pay for the rooms. Clearly this did not involve impossibility. The coronation parade cancellation made the contract much less *attractive* than it was originally. This doctrine has great similarity to mutual mistake. Each party probably assumed that the coronation would take place, and each was equally mistaken.

It can be seen that relief under this doctrine must be awarded sparingly or contracts would lose much of their effectiveness. The leading American case, *Lloyd v. Murphy*,[17] involved a claim by a commercial tenant at the onset of World War II that it should be relieved of its obligation to pay rent on premises from which it intended to sell new cars that became unavailable because of the war. The court held that relief under this doctrine required that the value of the contract be almost totally destroyed by an event that was not reasonably foreseeable by the party seeking relief. The court held that the tenant did not meet these requirements.

This doctrine is rarely applied in construction transactions. But suppose a law were passed during construction of a racetrack that made horse racing illegal in the state. Suppose the owner sought to relieve itself of its contractual obligation by pointing to this event. If the project can be used only for an activity that is now illegal, it can be argued that the owner should be freed from its construction contract. However, much would depend upon whether the legislative action were reasonably foreseeable and whether the project could be used for other purposes such as a go-cart track, tennis court, commercial exhibitions, a nine-hole golf course, or auto racing.

(F) ACCEPTANCE

Sometimes the contractor claims a defense when a claim is asserted against it for defective work based upon the owner's having accepted the project. Inasmuch as this defense is principally related to defects, it is discussed in Section 28.05.

(G) PASSAGE OF TIME: STATUTES OF LIMITATIONS

Sometimes a claimant is met with a defense that it did not begin legal action on the claim within the time required by law. Though this is not strictly speaking a defense based upon the contract or common law doctrines, it has generated immense complexity and must be looked at in any treatment of claims generally.

Statutes of limitations (discussed elsewhere [18]) deal with the effect of the passage of time upon the maintainability of claims. They are designed to protect defendants from false or fraudulent claims that may be difficult to disprove if not brought until relevant evidence has been lost or destroyed and witnesses become unavailable. In addition, entirely apart from the merits of particular claim, the law seeks to promote certainty and finality in transactions, especially commercial transactions, by terminating contingent liabilities at specific points in time. This second function is sometimes expressed as providing a "statute of repose." [19]

Although most agree that these statutes are necessary, their implementation has

---

16. *Krell v. Henry*, K.B. 740 (C.A.1903).

17. 25 Cal.2d 48, 153 P.2d 47 (1944).

18. Sections 2.03 and 17.08(B).

19. *Gates Rubber Co. v. United States Marine Corp.*, 508 F.2d 603 (7th Cir.1975).

generated great difficulty in construction performance claims, claims often discovered long after the breach by the contractor or completion of the work. Such statutes frequently are unclear as to the time the statutory period begins. In construction, the period can begin when the wrongful act occurred, when damage occurred, or when the claimant knew or should have known of its claim. For example, suppose a contract is made to build a commercial building in 1980. Work proceeds during that year. The contractor does not use proper workmanship in the application of adhesive materials in the construction of the roof. The building is accepted and occupied in 1980. In 1985, severe roof leaks occur that are traceable to improper workmanship by the contractor.

Suppose the applicable statute of limitations states that a claim for breach of contract must be begun within four years from the time the "cause of action accrued." If the cause of action accrued at the time of the improper workmanship or the acceptance of the building, the claim has been barred by the passage of time. If the claim accrued at the time of *damage,* should the period begin at the time of completion (at that time the improper workmanship could not have been corrected) or at the time the poor workmanship was discovered? If the cause of action accrued in 1985, legal action can be brought until 1989. If so, the contractor will be forced to defend a claim based upon events that occurred nine years earlier. Its defense may be hampered by its inability to produce witnesses and documentary evidence. Yet if the earlier period is selected, the owner has lost a claim before it becomes aware it had one unless an objective standard is used and a reasonable inspection would have discovered the poor workmanship during the time of performance.

Participants in the construction industry persuaded most state legislatures to enact Completion Statutes that sought to cut off liability a designated number of years after substantial completion. The statutes varied as to details. Some dealt only with claims based upon breach of contract while others included tort claims. Most differentiated between latent and patent defects, the former being defects not discoverable by reasonable inspection and the latter being defects that were reasonably discoverable. The statutes varied as to the persons who could take advantage of these statutes. Typically, those protected were design professionals and contractors, with protection not accorded owners or suppliers. The period selected by the legislatures also varied.

The statutes were attacked in many states, principally based upon the unfairness of depriving a party of the claim that it did not realize it had. More technically, statutes were attacked based on a variety of constitutional doctrines, mainly as violating constitutional equal protection of the law. Some persons were given legislative protection while others were not. Courts were divided in their willingness to uphold these statutes.[20] In addition, they generated a large amount of litigation, not only passing upon their constitutionality but also seeking to integrate them with the normal statutes of limitations of the state. Since this treatise cannot provide a detailed description

---

20. The leading case refusing to uphold the statute is *Skinner v. Anderson,* 38 Ill.2d 455, 231 N.E.2d 588 (1967) (statute subsequently revised). The leading case upholding such a statute is *Freezer Storage, Inc. v. Armstrong Cork Co.,* 234 Pa.Super. 441, 341 A.2d 184 (1975). The cases are canvassed in Note, *People Who Live in Glass Houses Should Not Build in Vermont: The Need For A Statute of Limitations For Architects,* 9 Vt.L.Rev. 101 (1984). Hawaii recently redrafted its statute after it was found unconstitutional and the redrafted statute was found to have still violated equal protection of laws. *Shibuya v. Architects Hawaii Limited,* 65 Hawaii 26, 647 P.2d 276 (1982). The most recent case, *State Farm Fire & Casualty Co. v. All Electric, Inc.,* 660 P.2d 995 (Nev.1983) held, over a strong dissent, the statute invalid.

and analysis of these statutes, local law must be consulted.

One observation that relates to risk management (discussed in greater detail in Chapters 36 and 37), merits mention at this point. Racing to the legislatures may have distracted from another approach to the long-delayed claim. It is possible for the contracting parties to regulate this problem. Although contractual regulation will not affect claims of third parties, such as those who may be injured if a building collapses years after it is completed, the bulk of the exposure in this area usually involves claims by the owner against the contractor or design professional.

Parties in a construction contract can regulate both the period of time in which the claim can be made and when that period begins. AIA Doc. B141, ¶ 11.3 seeks to control the latter but A201, the General Conditions of the Contract, does neither. By way of analogy, § 2–725(1) of the Uniform Commercial Code which creates a four-year period of limitation allows a contractually created period of not less than one year and in no event longer than the four-year statutory period.

Sometimes contractors contend that any warranty or guarantee clause is a private period of limitations (discussed in Section 28.08 dealing with these clauses).

SECTION 27.04   RESTITUTION

Some of the doctrines described in Section 27.03 must be looked at in connection with restitution (treated generally in Chapter 8). If a contractor is relieved because of some of the common law doctrines described in Section 27.03, it may seek affirmatively to recover for the benefit that it has conferred before the time that its obligation to perform was terminated. One aspect of this, destruction of the work in progress, is discussed in Section 27.05(C). For two reasons, this problem has not generated much difficulty in construction disputes. First, the contractor is usually being paid as it performs, and any restitutionary claim it might have would be reduced by these progress payments. This means that the amount at stake is likely to be small. Second, courts have used these relief doctrines sparingly in construction cases. However, in a series of Massachusetts cases, subcontractors were granted broad restitution remedies when the prime contract was found to have been illegally awarded.[21]

Restitution, however, has another use. Section 27.02 discussed claims and their bases. Sometimes a party wishes to measure its recovery for its claim not by the other party's promised performance but by the benefit it has given the other party (discussed in greater detail in contractor claims analyzed in Section 31.02).

SECTION 27.05
SPECIFIC APPLICATIONS
OF GENERAL PRINCIPLES

(A) INCREASED COST
OF PERFORMANCE

In the course of performance, the contractor may find that its cost of performance has risen dramatically. Illustrations can be drastic increases in the cost of materials, equipment, supplies, utilities, labor, or transportation. These risks are generally assumed by the contractor under a fixed-price contract.

There are exceptions, most relating to specific contract provisions or terms implied into the contract. For example, sometimes the owner impliedly warrants the design. Such a warranty may justify a contractor claim for additional compensa-

---

21. *Albre Marble & Tile Co. v. John Bowen Co.*, 338 Mass. 394, 155 N.E.2d 437 (1959); *Boston Plate & Window Glass Co. v. John Bowen Co.*, 335 Mass. 697, 141 N.E.2d 715 (1957); *M. Ahern Co. v. John Bowen Co.*, 334 Mass. 36, 133 N.E.2d 484 (1956).

tion. Section 29.05 deals with claims for additional compensation resulting from unanticipated costs and indicates that express contract provisions and sometimes those implied can be the basis for a contractual claim for additional compensation. Section 30.10 notes that unjustifiable acts of the owner may result in additional performance expense because of delay. Section 30.08 states that time extensions are sometimes available when certain events impede performance. Sometimes construction contracts contain escalation clauses under which the contract price will automatically increase and sometimes decrease in the event that certain costs change from a base formula expressed in the contract.

The common law provided little protection to a contractor who performs under a fixed price. To be sure, extraordinary and unanticipated events may occur that drastically affect the cost of the contract of performing and may be the basis for a claim for relief. Here again the remedy sought may be influential if not determinative. Additional time is easiest to obtain, although this depends upon the language of the clause that grants time extensions. Additional compensation and termination, if sought, are more difficult to obtain.

The shock waves generated by steep rises in the cost of energy sources in the 1970s generated litigation. Sellers who had agreed to deliver energy-related items for a fixed price over a long contract period and even those that had included escalation provisions sought relief from such contracts. Although most decisions refused to grant any relief,[22] one important trial court decision "rewrote" the price term to give the seller its anticipated profit margin.[23] It is unlikely that this will be done in a construction contract, even one that requires performance over a long period.

(B) LABOR DISRUPTIONS: THE PICKET LINE

The labor dispute is one of the most disruptive events in a construction project. Such disputes can cause frequent and lengthy work stoppages. Craft unions sometimes dispute who has the right to perform certain work. Not uncommonly employees of unionized contractors refuse to work on a site with nonunion employees of another contractor. Workers sometimes refuse to work on a project because prefabricated units are introduced on the site in violation of a "work preservation" clause in a subcontract or collective bargaining agreement. Any of these situations can result in a strike, a picket line, or both. Workers frequently refuse to cross a picket line. Such refusals can shut down a job directly because no workers are willing to work on the site or indirectly through the refusal of workers to deliver materials to a picketed site.

Pressure and even coercion are common tactics in the struggles between employers and those groups attempting to organize the workers as well as represent them in negotiating wages, hours, job security and working conditions. Each of the disputes—employer and union—not only seek to persuade but also use economic weapons to obtain a favorable outcome.

Disputants direct pressure at each other by use of economic weapons such as strikes or lockouts. Even these direct primary pressures affect neutrals not involved in the labor dispute. A strike affects nonstriking workers, the families of the strikers, and those who deal with the struck employer such as those who supply it goods or purchase its products or services. Likewise, a lockout by the employer affects neutrals.

Economic warfare often expands to include pressure and coercion upon third par-

22. *Louisiana Power & Light Co. v. Allegheny Ludlum Ind.*, 517 F.Supp. 1319 (E.D.La.1981); *Publicker Industries, Inc. v. Union Carbide Corp.*, 17 U.C.C.Rep. 989 (E.D.Pa.1975).

23. *Aluminum Co. of America v. Essex Group, Inc.*, 499 F.Supp. 53 (W.D.Pa.1980).

ties important to the employer, such as suppliers or customers. The pressure can be direct through communicated coercive threats or indirect by the use of a picket line that briefly describes the union's reason for its grievance against the employer and may ask certain activity or nonactivity be taken. A simple illustration may be helpful in understanding how economic warfare in labor relations can spread.

Suppose there is a strike between a manufacturer of kitchenware and a union seeking to organize its plant workers or obtain better wages, hours, or working conditions. Initially, the economic warfare might consist of a strike at the manufacturer's plant that would be accompanied by a picket line at the entrance designed to discourage union drivers or drivers sympathetic to the union's cause from delivering supplies to the manufacturer.

In addition, the union might broaden its attack by picketing places that sell the kitchenware made by the manufacturer, such as hardware stores or supermarkets. The picket signs might appeal to union drivers not to deliver goods to the hardware store. The picket signs might also be directed to those who might buy in the hardware store. Such signs might ask patrons not to buy kitchenware products made by the manufacturer and might even go to the extent of asking that patrons not purchase *anything* at the hardware store because the store sells the manufacturer's kitchenware.

The purpose for such picket lines would be to reduce the sales of kitchenware items by the manufacturer, either directly by encouraging people not to buy them or indirectly by coercing hardware store outlets for the manufacturer's products into canceling orders for the manufacturer's goods.

As these tactics broaden the field of eco-nomic warfare, they may begin to seriously harm neutrals and become what are called secondary boycotts. Such impermissible activities are unfair labor practices giving the party injured by such activity the right to damages and more important the right to obtain an injunction ordering that such activities cease or be modified.

The U.S. Supreme Court has stated that the law reflects:

> the dual congressional objectives of preserving the right of labor organizations to bring pressure to bear upon offending employers in primary labor disputes and of shielding unoffending employers and others from pressures and controversies not their own.[24]

In addition, the court recognized that the law was concerned not only with pressure brought to bear upon the other disputants but also with pressure brought to force third parties "to bring pressure on the employer to agree to the union's demands."[25]

The line between permitted primary and prohibited secondary boycotts is difficult to draw in industrial collective bargaining warfare. It is even more difficult in construction work because of the transient nature of the workers, the seasonal nature of the work, the proliferation of craft unions, the frequent occupation of the construction site by employees of many bargaining entities such as contractors and subcontractors, and the special rules that govern collective bargaining in the construction industry.

One of the above-described factors merits additional comment. As indicated, the construction project usually involves work by a number of contractors sometimes at different stages of the work but often at the same time. Some of the employers may have collective bargaining agreements with a union while others may not. Some employers may have collective bargaining agree-

---

24. *NLRB v. Denver Building & Construction Trades Council*, 341 U.S. 675, 692, 71 S.Ct. 943, 953, 95 L.Ed. 1284, 1297 (1951).

25. *NLRB v. Local 825, International Union of Operating Engineers*, 400 U.S. 297, 303, 91 S.Ct. 402, 406, 27 L.Ed. 2d 398, 404 (1971).

ments with one union while others may have collective bargaining agreements with a different one. For example, there may be a nonunion prime contractor working alongside union subcontractors. Sometimes there are union subcontractors working on a site with employees of a subcontractor who does not have a collective bargaining agreement with any union. A union engaged in an economic struggle with an employer can maximize its bargaining power if it can shut down the entire project by putting up a picket sign that no union workers will cross. The very effectiveness of "common situs" picketing is the reason it can also enmesh many neutrals and cause frequent and costly work stoppages.

The law limited common situs picketing by allowing the prime contractor to set up separate gates, one for the employees involved in the labor dispute and the other for those not involved in the dispute. This "two gate" system prevents the project from being totally shut down. Picketing can be done at the first gate but not at the second.

Congress passed a Common Situs Picketing Bill in 1975 that would have allowed this type of picketing for a limited period of time and that also set up machinery for more rapid settlement of these disputes at a national level. The bill died because of a presidential veto.

But suppose a labor dispute does shut down or curtail a project. What effect will this have on the construction contract obligations of participants in the project?

Labor disputes, like damage to a partially completed project, can involve three separate but related questions:

1. Is the contractor entitled to a time extension for the period that work is shut down or curtailed?

2. Can the contractor collect additional compensation for expenses caused by the shutdowns or curtailment of work?

3. Can the contractor whose work is severely disrupted by a labor dispute to which it is not a party terminate its obligation to perform under the contract?

As for the first, construction projects typically provide time extensions for work disrupted by labor difficulties. For example, ¶ 8.3.1 of AIA Doc. A201 awards the contractor a time extension for delays caused "by labor disputes." [26] Without such a provision granting relief, a contractor will have a difficult time obtaining a time extension in the event of labor difficulties. Generally the contractor assumes the risk of events that would increase the cost and time of performance.

Moving to the second question, it is generally difficult for the contractor to recover additional compensation when its costs are markedly increased because of unusual labor disruptions. *McNamara Construction of Manitoba, Limited v. United States,*[27] denied a claim for additional compensation by a contractor who completed performance on a $21 million project in the midst of labor problems that caused 303 days of work stoppages in a contract with a completion date of three years. The work stoppage resulted from seemingly never ending jurisdictional disputes, featherbedding practices of the unions, and what appeared to be the gross abuse of union power in matters relating to employee discipline. The contractor was given time extensions but sought an additional $6 million that it claimed were costs caused by the unusual labor situation.

To recover additional compensation, the contractor contended that both contractor and government labored under a mutual

---

26. New Jersey recently held a strike clause used in all elevator installer contracts not unconscionable. *Curtis Elevator Co. v. Hampshire House Inc.,* 142 N.J.Super. 537, 362 A.2d 73 (1976).

27. 509 F.2d 1166 (Ct.Cl.1975).

mistake of fact when they made the contract, each "assuming" that labor unions would behave in a responsible manner. In view of how things turned out, the contractor contended that this was a mutually mistaken assumption. To justify the recovery of additional compensation, the contractor claimed the contract should be "reformed"[28] and that it should be awarded additional expenses.

Over a strenuous dissenting opinion arguing that the contractor faced not ordinary but abnormally difficult labor difficulties, the court concluded that there was no mutual mistake. Labor problems, according to the majority of the court, are inherent in the vast majority of construction contracts and are a fact of modern commercial life. Rejecting the contention that the parties assume the risk of only normal strikes, the court emphasized that contractors in a fixed-price contract assume the risks of unexpected costs, a matter that a contractor should have taken into account when pricing the work.

The third question, that of termination, will depend principally upon the construction contract. Interestingly, the AIA standard documents, though providing for time extensions, do not allow termination. The absence of a clause specifically allowing termination should not necessarily preclude termination from being a proper remedy under certain circumstances. Of course, some courts are likely to conclude that a contract that specifically mentioned labor problems and did not make them the basis for termination indicated that the parties contemplated continued performance with time extensions being the only remedy.

It may be useful to examine briefly a case that involved a defense of impossibility asserted to be grounds for terminating performance obligations where the contract did not treat labor difficulties. In *Mishara Construction Co. v. Transit-Mixed Concrete*,[29] a prime contractor brought legal action against a supplier who claimed he could not deliver because of a picket line at the site. The court rejected the prime contractor's contention that a labor dispute that makes performance more difficult *never* constitutes an excuse for nonperformance. The court instead held that whether a labor dispute excuses the supplier's obligation to perform would be a fact question that must be determined by a jury. The court noted that a picket line might constitute a mere inconvenience and not make performance impracticable. Also, according to the court, a contract made in the context of an industry with a long record of labor difficulties shows that the parties assumed the risk of labor disputes. For example, if the supplier knew he were agreeing to deliver concrete to an employer who had had chronic and bitter labor difficulties, his making a contract *without* providing for contractual protection could indicate that he was assuming this risk. The court stated:

> Where the probability of a labor dispute appears to be practically nil, and where the occurrence of such a dispute provides unusual difficulty, the excuse of impracticability might well be applicable.[30]

The court concluded by noting that the tendency has been to recognize strikes as an excuse for nonperformance. Although the court indicated that under certain circumstances absence of protective language might be considered an assumption of risk, the Uniform Commercial Code, with its emphasis upon commercial impracticability, might not find the absence of a clause *con-*

---

28. Refer to Section 24.04 dealing with reformation as a method adopted by the old Court of Claims to deal with unanticipated expenditures made in the performance of a fixed-price contract.

29. 365 Mass. 122, 310 N.E.2d 363 (1974).

30. 365 Mass. at 130, 310 N.E.2d at 368.

clusive on the question of whether a particular risk was assumed.[31]

(C) PARTIAL OR TOTAL DESTRUCTION OF PROJECT

During project construction the project may be partially or totally destroyed by circumstances for which neither party is chargeable. The work may be destroyed by fire, unstable soil conditions, or violent natural acts such as earthquakes or hurricanes. Although destruction is often discussed in terms of impossibility or assumption of risk, it is essential to recognize that at least two separate issues can arise when such events occur:

1. Are the parties relieved from any further obligation to perform?
2. Is the contractor entitled to be paid for work incorporated into the structure prior to destruction?

As for the first question, the performance obligation is not terminated in the absence of a clause relieving the contractor from its performance obligation. Courts have stated that a party who in unqualified terms promises to perform is itself at fault when it does not expressly protect itself from these contingencies.[32] If the contract is for repairs or addition to an existing structure or for *part* of a new structure and the work is destroyed in whole or in part without the fault of either party, each party is relieved from further contract obligations.[33]

As to the second question, that of recovery for work performed *prior* to destruction, the result depends upon, as a rule, the termination issue. If the contractor is not discharged from its obligation to perform, it cannot recover for work performed. If it is discharged, such as where it had agreed to build only *part* of the structure or to repair an existing structure, generally the contractor can recover for work performed prior to destruction.[34]

Some cases look at insurance and insurability, concluding that where there is or ought to have been insurance the contractor should recover for what has been performed prior to destruction. More commonly, cases proceed as if progress payments and insurance were not common components of the Construction Process.

Usually each participant in the Construction Process insures *its* property from loss of the type discussed in this subsection. Typically, the owner insures the work in progress while the contractor insures its equipment and other property that will not go into the project. When the project is destroyed during construction, the owner receives proceeds from the insurance company. It holds these proceeds as trustee for those who have suffered property damage covered by the proceeds. Since much of the work has been paid for by progress payments, the proceeds that are held for the contractor are typically work for which progress payments have not yet been received and work for which progress payments were received but from which the retainage has been deducted. A well-planned property insurance system should reimburse the parties for the losses suffered. If this occurs, an argument can be made for the rule that does *not* discharge the parties from their obligation to perform further. This

---

31. New Jersey recently stated that even without a clause excusing performance hindered by labor disputes, the doctrine of commercial impracticability might give the performing party relief. *Curtis Elevator Co. v. Hampshire House, Inc.*, supra note 26.

32. *Stees v. Leonard*, 20 Minn. 494 (1874). Although the principal cases articulating this rule were decided in the nineteenth century, the rule was recently applied in *Dravo Corp. v. Litton Systems, Inc.*, 379 F.Supp. 37 (S.D.Miss. 1974).

33. *Fowler v. Insurance Co. of North America*, 155 Ga.App. 439, 270 S.E.2d 845 (1980).

34. Annot., 28 A.L.R.3d 788 (1969).

result follows whether the construction was for an entire structure or part of one or for the repair of an existing structure.

If a substantial period of time has elapsed from the making of the contract until the destruction of the project, it would be unfair to require each party to begin over. Even if property losses have been reimbursed through insurance, the original bid was based upon prices and conditions existing at the time bids were made. The prime contractor might not be able to hold subcontractors to the original subcontracts in states where those who build only part of a whole structure are relieved from further performance obligations. In these states new subcontracts would have to be negotiated. Where it would be an essentially different contract than originally made, holding that the contractor must still perform would be unfair. Likewise the owner's perspective may be different after the project has been destroyed while in the process of construction.

New information may bear upon the economic viability of the project, and it might be unreasonable to compel the owner to continue as if nothing has happened.

It is advisable to incorporate a provision in the construction contract that would give either party the right to terminate if the project is totally or nearly destroyed after a designated period of time has expired from start of performance.

Although discussion has emphasized destruction for which neither party can be held accountable, a third issue often arises over responsibility. The most frequent cause of destruction—destruction by fire—is usually an insured risk. Fire losses are often traceable to human failings. When the insurer pays, it often looks for reimbursement from a party other than the insured who it can claim negligently caused the fire.

It is common for construction contracts to require that owner and design professional be designated as additional insureds on the contractor's insurance policies, or that each party waive any subrogation claims it has against the other for certain risks, mainly destruction of the property.[35] These are intended to eliminate any claim by the insurer who pays for the loss that construction participants negligently caused the fire. However, some insurers do not look kindly upon such waivers. They see them as their insured "giving away" any claims they may have, similar to a motorist admitting he was wrong at the scene of an auto accident. One court held that such a waiver of subrogation eliminated coverage.[36] Before this waiver of subrogation is included, legal advice should be obtained. Often the insurer will consent.

(D) GOVERNMENTAL ACTS

Governmental interference can take different forms. A law may be passed after the contract has been made that makes performance of the contract illegal. For example, suppose a law is passed prohibiting the construction of any nuclear facility. Certainly enactment of such legislation would terminate any contract to build a nuclear plant. Performance would require an illegal act. The contractor's right would be limited to recovery for any work performed before the enactment of the legislation. Such recovery would be restitutionary and based upon unjust enrichment.

Another form of governmental interference would be the issuance of a judicial or administrative order shutting the project down. For example, a project might be shut down because of improper construction methods, an invalid building permit, or a design or use that violated land use controls. If the project is shut down for an appreciable period of time, the party not

---

35. AIA Doc. A201, ¶ 11.3.6. A201, ¶ 11.3, sets up a mechanism for distribution of insurance proceeds.

36. *Liberty Mutual Insurance Co. v. Altfillisch Construction Co.*, 70 Cal.App.3d 789, 139 Cal.Rptr. 91 (1977).

responsible for the shutdown should be relieved from further performance obligations and may have a cause of action against the other party. For example, if a shutdown is due to improper design or failure to comply with land use controls, the contractor may have a valid claim for damages as well as be relieved from further performance obligations.[37] If the shutdown is due to poor construction methods or failure to comply with the contract documents, the owner should have an action against the contractor for breach as well as be able to terminate any further obligations owed the contractor. The mere fact that performance was stopped by a government official does not necessarily absolve a party from contract breach even if the work is shut down. If the shutdown was due to the unexcused nonperformance by one of the parties, there can be breach as well as a possible termination.

Despite the generalization made in the preceding paragraph, that government intervention may be traceable to the fault of one of the parties or some one for whose acts they are responsible, the difficulty of placing clear responsibility on one of the parties for acts of government intervention make it unlikely that government intervention in the ordinary case will be chargeable to one of the contracting parties.

Suppose the governmental acts that stop performance are not the responsibility of *either* contracting party. The site may have been condemned by the state's exercise of its power of eminent domain. In wartime the state may have drafted workers, requisitioned essential materials, or commandeered all transportation facilities. These acts would relieve the contractor from further contractual obligations. (They might also discharge the owner's obligations through frustration of purpose. See Section 27.03(E).) Any claims for work performed by the contractor prior to shutdown would be restitutionary.

Suppose a work stoppage results because important equipment being used by the contractor has been repossessed by the owner of the equipment or someone with a security interest in it. Repossession is often accomplished by court order. However, such a court order is a risk clearly assumed by the contractor. That the actual act that interferes with performance is an order by a public official should not relieve the contractor from its obligation to perform.

Suppose the job was shut-down because a vital piece of *subcontractor* equipment was seized by court order. The result would be the same as if the equipment were owned by the prime contractor. It would not be entitled to a time extension either because this was an assumed risk, as in the preceding paragraph, or because the job was not really shut down by the state.

Suppose legislation is enacted that would bar a method of performance contemplated at the time the contract was made. It is likely that the performing party will not receive relief if the legislation was reasonably foreseeable and did not increase costs astronomically.[38]

(E) DEFECTIVE SPECIFICATIONS

Contractor allegations that the specifications were defective are common in construction performance disputes. Unfortunately, such allegations often produce more heat than light and obscure the real issues in the dispute. Before looking at these issues, it is useful to reproduce the fountainhead case that relates to such an allegation.

---

37. In *Gordon v. Indusco Management Corp.*, 164 Conn. 262, 320 A.2d 811 (1973), the failure of the owner to obtain a building permit did not give the owner a defense when sued by the contractor since the defendant owner had assured the contractor that the building permit could be obtained.

38. *Levine v. Rendler*, 272 Md. 1, 320 A.2d 258 (1974) (change in requirements for composition of road in a residential development).

UNITED STATES v.
SPEARIN

Supreme Court of the
United States, 1918.
248 U.S. 132, 39 S.Ct. 59, 63
L.Ed. 166.

MR. JUSTICE BRANDEIS delivered the opinion of the court.
[Ed. note: Footnotes renumbered.]

Spearin brought this suit in the Court of Claims, demanding a balance alleged to be due for work done under a contract to construct a dry-dock and also damages for its annulment. Judgment was entered for him . . . .

First. The decision to be made on the Government's appeal depends upon whether or not it was entitled to annul the contract. The facts essential to a determination of the question are these:

Spearin contracted to build for $757,800 a dry-dock at the Brooklyn Navy Yard in accordance with plans and specifications which had been prepared by the Government. The site selected by it was intersected by a 6-foot brick sewer; and it was necessary to divert and relocate a section thereof before the work of constructing the dry-dock could begin. The plans and specifications provided that the contractor should do the work and prescribed the dimensions, material, and location of the section to be substituted. All the prescribed requirements were fully complied with by Spearin; and the substituted section was accepted by the Government as satisfactory. It was located about 37 to 50 feet from the proposed excavation for the dry-dock; but a large part of the new section was within the area set aside as space within which the contractor's operations were to be carried on. Both before and after the diversion of the 6-foot sewer, it connected, within the Navy Yard but outside the space reserved for work on the dry-dock, with a 7-foot sewer which emptied into Wallabout Basin.

About a year after this relocation of the 6-foot sewer there occurred a sudden and heavy downpour of rain coincident with a high tide. This forced the water up the sewer for a considerable distance to a depth of 2 feet or more. Internal pressure broke the 6-foot sewer as so relocated, at several places; and the excavation of the dry-dock was flooded. Upon investigation, it was discovered that there was a dam from 5 to 5½ feet high in the 7-foot sewer; and that dam, by diverting to the 6-foot sewer the greater part of the water, had caused the internal pressure which broke it. Both sewers were a part of the city sewerage system; but the dam was not shown either on the city's plan, nor on the Government's plans and blue-prints, which were submitted to Spearin. On them the 7-foot sewer appeared as unobstructed. The Government officials concerned with the letting of the contract and construction of the dry-dock did not know of the existence of the dam. The site selected for the dry-dock was low ground; and during some years prior to making the contract sued on, the sewers had, from time to time, overflowed to the knowledge of these Government officials and others. But the fact had not been communicated to Spearin by anyone. He had, before entering into the contract, made a superficial examination of the premises and sought from the civil engineer's office at the Navy Yard information concerning the conditions and probable cost of the work; but he had made no special examination of the sewers nor special enquiry into the possibility of the work being flooded thereby; and had no information on the subject.

Promptly after the breaking of the sewer Spearin notified the Government that he considered the sewers under existing plans a menace to the work and that he would not resume operations unless the Government either made good or assumed responsibility for the damage that had already occurred and either made such changes in the sewer system as would remove the danger or assumed responsibility for the damage which might thereafter be occasioned by the insufficient capacity and the location and design of the existing sewers. The estimated cost of restoring the sewer was $3,875. But it was unsafe to both Spearin and the Government's property to proceed with the work with the 6-foot sewer in its then condition. The Government insisted that the responsibility for remedying existing conditions rested with the contractor. After fifteen months spent in investigation and fruitless correspondence, the Secretary of the Navy annulled the contract and took possession of the plant and materials on the site. Later the dry-dock, under radically changed and enlarged plans, was completed by other contractors, the Government having first discontinued the use of the 6-foot intersecting sewer and then reconstructed it by modifying size, shape and material so as to remove all danger of its breaking from internal pressure. . . .

The general rules of law applicable to these facts are well settled. Where one agrees to do, for a fixed sum, a thing possible to be performed, he will not be excused or become entitled to additional compensation, because unforeseen difficulties are encountered. *Day v. United States,* 245 U.S. 159; *Phoenix Bridge Co. v. United States,* 211 U.S. 188. Thus one who undertakes to erect a structure upon a particular site, assumes ordinarily the risk of subsidence of the soil. *Simpson v. United States,* 172 U.S. 372; *Dermott v. Jones,* 2 Wall. 1. But if the contractor is bound to build according to plans and specifications prepared by the owner, the contractor will not be responsible for the consequences of defects in the plans and specifications. *MacKnight Flintic Stone Co. v. The Mayor,* 160 N.Y. 72; *Filbert v. Philadelphia,* 181 Pa.St. 530; *Bentley v. State,* 73 Wisconsin, 416. See *Sundstrom v. New York,* 213 N.Y. 68. This responsibility of the owner is not overcome by the usual clauses requiring builders to visit the site, to check the plans, and to inform themselves of the requirements of the work, as is shown by *Christie v. United States,* 237 U.S. 234; *Hollerbach v. United States,* 233 U.S. 165, and *United States v. Utah &c. Stage Co.,* 199 U.S. 414, 424, where it was held that the contractor should be relieved, if he was misled by erroneous statements in the specifications.

In the case at bar, the sewer, as well as the other structures, was to be built in accordance with the plans and specifications fur-

nished by the Government. The construction of the sewer constituted as much an integral part of the contract as did the construction of any part of the dry-dock proper. It was as necessary as
any other work in the preparation for the foundation. It involved
no separate contract and no separate consideration. The contention of the Government that the present case is to be distinguished
from the *Bentley Case, supra,* and other similar cases, on the
ground that the contract with reference to the sewer is purely
collateral, is clearly without merit. The risk of the existing system
proving adequate might have rested upon Spearin, if the contract
for the dry-dock had not contained the provision for relocation of
the 6-foot sewer. But the insertion of the articles prescribing the
character, dimensions and location of the sewer imported a warranty that, if the specifications were complied with, the sewer
would be adequate. This implied warranty is not overcome by the
general clauses requiring the contractor, to examine the site,[39] to
check up the plans,[40] and to assume responsibility for the work
until completion and acceptance.[41] The obligation to examine the
site did not impose upon him the duty of making a diligent
enquiry into the history of the locality with a view to determining,
at his peril, whether the sewer specifically prescribed by the
Government would prove adequate. The duty to check plans did
not impose the obligation to pass upon their adequacy to accomplish the purpose in view. And the provision concerning contractor's responsibility cannot be construed as abridging rights arising
under specific provisions of the contract.

* * *

The judgment of the Court of Claims is, therefore,
Affirmed.

Before attempting to break down the imprecise term *defective specification* into
more workable categories, it is important to
note the different legal issues affected by
the quality of specifications. The risk of building defects may depend on whether
the defect was caused by the design. If so,
the entity who supplied design specifications (materials and methods), usually the
owner, will not be able to transfer the cost

---

39. "271. *Examination of site.*—Intending bidders are expected to examine the site of the proposed dry-dock and
inform themselves thoroughly of the actual conditions and requirements before submitting proposals."

40. "25. *Checking plans and dimensions; lines and levels.*—The contractor shall check all plans furnished him
immediately upon their receipt and promptly notify the civil engineer in charge of any discrepancies discovered
therein. . . . The contractor will be held responsible for the lines and levels of his work, and he must combine all
materials properly, so that the completed structure shall conform to the true intent and meaning of the plans and
specifications."

41. "21. *Contractor's responsibility.*—The contractor shall be responsible for the entire work and every part
thereof, until completion and final acceptance by the Chief of Bureau of Yards and Docks, and for all tools,
appliances, and property of every description used in connection therewith. . . . "

of correction to the contractor who has executed the design. (Whether it can transfer this loss to the design professional was discussed in Chapter 17.) Put another way, the contractor who executes the required design is not liable for the cost of correction, as seen in *United States v. Spearin.*

Suppose the contractor claims that it has expended more than it anticipated because of defective specifications. Chapter 29 deals with this in the context of subsurface conditions. This subsection looks briefly at this issue in other contexts.

A contractor who cannot comply with performance specifications, particularly when the contractor has been required to follow detailed specifications, sometimes asserts that the specifications were "impossible" [42] (discussed in (F)).

It may be helpful to start with the attributes of good or high-quality specifications. This requires an understanding of what specifications include and what they are designed to accomplish.

In the sense that specifications measure the contractor's contract obligations, the specifications should make clear to the contractor what it will be expected to do. A design specification tells the contractor what it is to do and how it is to do it. A purchase description specification is even more limiting, telling the contractor that the materials and equipment it *must* furnish will be made by a particular manufacturer and be of a designated model or type. A performance specification describes a specific outcome. When it is combined with a design specification, the owner has sought to tell the contractor what it must do, how it must do it, and the result it must achieve. As shall be seen in Section 28.03, this can cause problems.

In addition, specifications as seen in *United States v. Spearin* can provide relevant information that the contractor needs and uses to determine whether it will enter into the contract, what will be its price, and how it expects to perform. In that case, the United States did not tell the contractor that there was a dam in the seven-foot sewer. Put another way, it failed to provide complete information. It described the sewer but did not inform the contractor that there was a dam in it or that there had been prior flooding problems. Information can relate to the site, its subsurface characteristics, its accesses, and any conditions that would be helpful in planning performance.

High-quality specifications clearly inform the contractor what it will be expected to do and the conditions under which it will perform. They enable the contractor to plan its performance and price with the expectation of predictable and efficient work sequences. From the owner's standpoint, high-quality specifications will advance the owner's anticipated objectives, particularly those objectives expressed in or implied by the contract documents.

Defective specifications do not accomplish these objectives. More detailed classification helps understand some of the legal issues that surround this term, one that hides a multitude of sins.

Erroneous specifications contain factual errors of the type described in *United States v. Spearin* and subsurface data errors to be discussed in Chapter 29. They also include *legal* errors, such as noncompliance with applicable laws such as building and housing codes and environmental regulations. [43] Usually legal errors are the responsibility of the owner, with ultimate responsibility be-

---

42. Sometimes specifications are impossible in the sense that they are not coherent, that they do not "fit." Compliance with one part makes compliance with another part an "impossibility." In (F), impossibility means beyond the state of the art.

43. *St. Joseph Hospital v. Corbetta Construction Co.*, 21 Ill.App.3d 925, 316 N.E.2d 51 (1974) (panelling violated code); *Atlantic National Bank of Jacksonville v. Modular Age, Inc.*, 363 So.2d 1152 (Fla.App.1978) (wall violated code).

longing to the design professional.[44] Only if the contractor *knew* of the errors may this risk have been shifted or shared.

Another erroneous specification is mechanical. Illustrations are physical problems with the project, such as water intrusion, foundation settling, insufficient heating or cooling, or a partial or total collapse of the structure. Although such problems are often classified as defects (the subject of Chapter 28), it can be noted here that the party responsible for defects of this type that are caused by the design is the party who created the design or had the design created for it, usually the party with a preponderance of expertise in the element of design that failed. Typically this is the owner, unless the owner has transferred this risk to the contractor.

Errors can be functional, such as a project failing to accomplish the owner's desired objectives. This can result from bad business judgment or changed circumstances—both owner risks. The owner may be able to transfer this risk to the design professional if the latter warranted a successful outcome or if the design professional's failure to perform as other design professionals would have caused the project to fail.

Defective specifications can "fail" as a method of communication. For example, specifications that do not describe what will be demanded or that fail to give sufficient design detail to enable the contractor to accomplish the desired objective are "incomplete." Sometimes the specifications are confusing or contradictory, again a problem of failing to communicate the performance that will be demanded.

Jasper Construction, Inc. v. Foothill Junior College District[45] is a case that excited interest in 1979. It is instructive, and an analysis of it can assist in pinning down the elusive defective specifications charge so often made in construction disputes.

At the outset, it is important to note what was *not* involved. Jasper was not claiming that it incurred additional expense because of inaccurate information as did the contractor in *United States v. Spearin* or the contractors making claims in Chapter 29 dealing with subsurface conditions. Nor was Jasper defending any claim made against him because of a defect by maintaining he had "followed the design," an issue discussed in Chapter 28.

Jasper's claim related to the specifications as a tool of communication as a method of informing him what he would be *obligated* by his *contract* to do. This contention requires an examination of the specification. It stated:

11. CONSTRUCTION JOINTS

A) Locations and details of construction joints shall be as indicated on the structural drawings, or as approved by the Architect. Relate required vertical joints in walls to joints in finish. In general, approved joints shall be located to least impair the strength of the structure.[46]

The locations of the construction joints were not shown on the drawings. However, Jasper contended that certain structural drawings indicated *to him* that the steel was from "floor to floor" and therefore the concrete would be poured in the same manner. He began to pour the basement "floor to floor" but the architect informed him that the joints would have to be "wall to wall." The contractor complained, contending that this would leave him with many wooden concrete forms he could not use and a change of techniques that would cost him

---

44. But see *Green v. City of New York*, 283 A.D. 485, 128 N.Y.S.2d 715 (1954), and *Quedding v. Arisumi Brothers Inc.*, 661 P.2d 706 (Hawaii 1983), where contractors who designed and built were liable where specified material violated code because of express or implied promise to comply with the law.

45. 91 Cal.App.3d 1, 153 Cal.Rptr. 767 (1979).

46. 153 Cal.Rptr. at 769.

an additional $500,000. Jasper did the work as directed and made a claim.

The jury was instructed that a public entity that issues plans and specifications *impliedly warrants* that they are *free of defects*, are *complete*, and will, if followed, result in the project intended. [Emphasis added.]

The appellate court found this jury instruction erroneous. The court recognized that warranties do attach to owner-supplied specifications. However, the court limited the implied warranty doctrine to *affirmative misrepresentations* or concealment of material facts that misled the contractor. The court concluded that the contractor had not been misled. It could have cleared up any ambiguities or incompleteness in advance by seeking a clarification from the architect.

The case was criticized for failing to recognize the difference between misrepresentation and implied warranty.[47] Although there is some justification for this criticism, the muddled thinking that often accompanies analysis of this issue is exposed if the focus is upon the specifications as tools of communication.

Assuming that Jasper had adequate time to review the specifications, several possibilities exist. First, Jasper as a reasonable bidder[48] might have believed that he *would be allowed* to pour floor-to-floor. If so, any direction to pour wall-to-wall was a change for which Jasper should have been compensated. If he had not been reasonable, he might have been misled, but his reliance was not reasonable. *Were* Jasper reasonable, his contention need *not* be premised upon any defective specifications. Of course, this term could be plugged in by

finding that the school district impliedly warranted that Jasper would be allowed to pour floor-to-floor. This does not advance analysis; it only muddles it.

Alternatively, the specifications could have been unclear. But were they unclear? It is likely that Jasper's attack on the specifications was their incompleteness, at least this is what it looked like to the trial judge. If not indicated on the drawings, the concrete joints would have to be approved by the architect, with the standard to be applied being structural stability. *If* unclear to *Jasper*, Jasper could have sought a pre-award clarification to determine *in advance* what the architect *would approve*. If Jasper chose *not* to do so, he took his chances upon the architect's decision. Jasper could either build in a contingency for an unfavorable decision or hope that the architect's decision would be one in accord with the technique that Jasper expected to use. If so, Jasper's complaint is with the architect's decision. If the architect was wrong, Jasper should have attacked the decision. But if Jasper believed the specifications unclear, his failure to seek a clarification that could have been communicated to *all* bidders should bar his later contention that the architect was wrong. The issue is whether Jasper was misled, not whether the owner impliedly warrants the completeness of its specifications.

The result in the *Jasper* case in no way undermines the importance or existence of implied warranties. Warranties are implied either where the issue is who bears the risk of inaccurate information, such as in the subsurface cases, or in disputes that involve the outcome of compliance with specifications. In either case, an implied warranty

---

47. Note, *How Does A Contractor Prove Breach Of Implied Warranty Of Plans And Specifications?* 18 Cal.W.L. Rev. 499 (1982).

48. Jasper's Reply Brief contended that none of the other bidders sought clarification. This could have been based upon a belief by other bidders that they were agreeing to perform to the architect's determination of what was least likely to impair the strength of the structure. Suppose Jasper's bid had been much *lower* than the others, a common phenomenon in these disputes. This would have supported a conclusion that his expectation was unreasonable.

may be found if it is more equitable to make the owner responsible for inaccurate information relied upon reasonably by the contractor or to make the owner pay for defects caused by the design or unanticipated expenses of complying with that design. Implied warranties do not depend upon fault. They are found if this is what the parties are very likely to have intended or, more commonly, if this is the fairest way of allocating the risk for particular losses.

The result in the *Jasper* case was correct if it can be assumed at best that the specifications were unclear and Jasper should have sought a determination. But its limiting of the implied warranty doctrine to affirmative misrepresentation unduly limited the utility of the implied warranty doctrine.

Most implied warranty cases involve owners who were public entities. There is no reason to differentiate public from private owners. Private owners who have the same resources and expertise of public entities should be held to similar implied warranties of quality specifications. If a private entity does *not* have these resources and expertise (and there may be cases where public entities do not either) and relies upon the contractor, the owner may be the beneficiary of an implied warranty by the contractor.

(F) "IMPOSSIBLE" SPECIFICATIONS

Performance specifications require the performing party to accomplish a designated objective.[49] For example, an aircraft company might agree to manufacture an airplane that will fly twice the speed of sound or a machinery manufacturer might agree to build a system for a plant that would turn out 1,000 units of a particular quality per hour. Suppose the airplane manufacturer or machinery maker fails. Ordinarily the failure to comply with performance specifications is a breach of contract.[50] But suppose the party promising to meet these specifications asserts that it was "impossible" to do so. Does proof of this provide a defense? [51] Does such proof entitle the performing party to reimbursement for the expenses incurred while seeking to meet the performance specification?

The answers to these questions in the first instance depends upon the contract. The contract can place such risks upon one or both parties. Often contracts are not clear on this point and the law must determine the answers to these difficult questions.

Unfortunately such problems are classified as "impossible" specification cases since the performing party, who will be referred to as the contractor, claims that it was "impossible" to meet these performance specifications. But the word *impossible* has a number of subtle shadings that often complicate cases and make prediction uncertain.

Another difficulty with the "impossible" label is that it can obscure the crucial issue

---

49. For a cavalier judicial treatment of performance specifications, see *Kurland v. United Pacific Insurance Co.,* discussed in Section 28.03.

50. *Gurney Industries, Inc. v. St. Paul Fire and Marine Insurance Co.,* 467 F.2d 588 (4th Cir.1972). One court intimated that the doctrine of substantial performance (discussed in Section 26.06(B)) might be applicable to performance specifications. *Votaw Precision Tool Co. v. Air Canada,* 60 Cal.App.3d 52, 131 Cal.Rptr. 335 (1976). Under this approach, coming "close" might be sufficient to justify receiving the unpaid balance of the contract price, less damages caused by the breach.

51. The discussion in this subsection emphasizes attempts by the performing party to be reimbursed for its efforts, since this has been the principal issue in those cases raising this problem. If there were a sufficient degree of impossibility to justify reimbursement, it seems obvious that the performing party would have a defense if sued by the other party for not complying with the performance specifications. The contractor was given a defense when the plans were considered "impossible" in *City of Littleton v. Employers Fire Insurance Co.,* 169 Colo. 104, 453 P.2d 810 (1969).

of risk assumption. Clearly a party can promise to do the impossible, although evidence that such a foolish risk was taken should be clear.[52] Conversely, the fact that it is not physically possible, that is, beyond the state of the art, should not invariably resolve the matter against the contractor.

Foster Wheeler Corp. v. United States,[53] a Court of Claims opinion, dealt at length with an impossible specifications problem. Foster Wheeler Corporation (FWC) entered into a fixed-price supply contract under which it agreed to design, fabricate, and deliver within thirteen months two boilers and perform a "dynamic shock analysis" study called a DDAM (dynamic design-analysis method) that would demonstrate that the boilers could withstand shock up to certain designated intensities set forth in the contract specifications. The total contract price was $280,000. The boilers were ultimately to be installed in naval ships. Because the boilers could not be subjected to actual shock testing, the DDAM, a mathematical model to represent a piece of equipment and the use of dynamic inputs to substitute for physical stresses and failure criteria, was to substitute.

After many months of design work and creation of mathematical models, the contractor ceased performance and sought an equitable adjustment of $192,000, claiming "impossibility" specifications. This claim was based upon the impossibility of meeting the performance specifications of "shock hardness." Given other design requirements, this could not be demonstrated by the DDAM.

The court recognized that the term *impossibility* does not require *absolute* impossibility but encompasses impracticability, a type of commercial impossibility caused by extremely unreasonable expense to perform. *Absolute* impossibility, in the sense

of requiring performance beyond the state of the art, would entitle the contractor to recover its cost in attempting to perform unless it assumed this risk. The court concluded that demonstrating the boiler to be shock hard by the DDAM method was *both* "commercially and absolutely impossible."

After giving facts that supported a conclusion of absolute impossibility, the court went on to the more controversial "commercial impossibility," stating:

> Under this theory, it is contended that the construction of shock-hard boiler, even if ultimately possible, could not be accomplished without commercially unacceptable costs and time input far beyond that contemplated in the contract. To design a shock-hard boiler by means of a mathematical model and dynamic analysis could . . . take an infinite amount of time. . . .
>
> The evidence shows . . . that the . . . contract contained specifications which were impossible to meet, either commercially or within the state of the art.[54]

This did not end the matter. The government argued that FWC assumed the contractual responsibility for performing the impossible. To determine who should assume this risk, the court examined which party had the greater expertise in the subject matter of the contract and which party took the initiative in drawing up specifications and promoting a particular method or design. The court ruled for FWC.

(G) WEATHER

Sometimes weather conditions both increase the cost of performance and cause delay. As a rule, the owner does not warrant that weather will not interfere with a contractor's performance. Unusual weather will rarely if ever be a sufficient basis for a claim for additional compensation. Only if the owner breached a duty to disclose information or if the weather is a "changed

52. *J.C. Penney Co. v. Davis & Davis, Inc.,* 158 Ga.App. 169, 279 S.E.2d 461 (1981).

53. 206 Ct.Cl. 533, 513 F.2d 588 (1975).

54. 513 F.2d at 598.

condition" would there be a basis for a claim for additional compensation.[55] However, it can frequently be the basis for a request for a time extension (treated in greater detail in Section 30.08).

(H) FINANCIAL PROBLEMS

After the construction contract is made and before completion of each party's performance, either party can suffer severe financial reverses. These reverses may manifest themselves in difficulties that range from short delay in paying bills to insolvency and even bankruptcy.

Financial reverses of this sort can raise two related but separate legal questions. First, the party in financial difficulty may contend that it should be relieved from further performance because it does not have the financial capacity to continue performance. Second, one party may be unwilling to continue performance if it appears that the other party's financial difficulties will make it unlikely that the latter will perform as promised. Under certain circumstances, the law allows a party who has legitimate concern over the other party's ability to perform to refuse to continue performance until it is reassured that the other party will perform its promise.[56]

AIA, in its documents relating to both design and construction services, gives the performing parties—the architect in B141 and the contractor in A201—the right to demand certain information regarding the owner's financial resources. The architect is given the power to ask for evidence of financial capacity at *any time* during his performance.[57] The contractor, on the other hand, can request this information as to the owner's financial arrangements *only* before it enters into the contract.[58]

Limiting the contractor's power to make this demand to an interval between award and execution of the formal contract provides an automatic remedy—that of not having to enter into the contract.

Suppose the contractor asserts that the owner has made a *promise* to supply adequate information and that its failure to do so constitutes a breach. This assertion would be based upon the furnishing of information not only being a *condition* to the contractor's obligation to execute the contract but also being a *promise* by the owner that it will furnish this information if requested. If so, the contractor is entitled to damages. At the very least, this should encompass its expenses incurred in negotiating or submitting a competitive bid. However, it should not recover lost profits. To avoid this, it is likely that such a provision would be held to create a *condition* rather than a *promise*.

Suppose this power to demand this information *can* be exercised at *any* time during performance, an approach AIA took in earlier drafts of A201. If this information is not provided, can the contractor *suspend* work until it is? (The power to suspend work has been expressly given to contractor if it is not paid.)[59] If the information is not furnished for a period of time, this can be a material breach allowing the contractor to *terminate* its obligation to perform. Again, it must be kept in mind that the contractor has an *express* power to terminate under A201 if it stops work for thirty days because of nonpayment.[60] If failure to furnish this evidence constitutes a material breach, it

---

55. *Hardeman-Monier-Hutcherson v. United States*, 198 Ct.Cl. 472, 458 F.2d 1364 (1972).

56. U.C.C. § 2–609. See Restatement (Second) of Contracts, §§ 251, 252 (1981).

57. AIA Doc. B141, ¶ 2.2.

58. AIA Doc. A201, ¶ 3.2.1.

59. Id. at § 9.7.1.

60. Id. at ¶ 14.1.

would also be the basis for the contractor recovering the reasonable value of its services, a convenient remedy for a contractor in a losing contract.

Some owners may object to such a provision. (Such a clause can give the contractor an opportunity to get out of the contract if it finds that it bid too low.) They may justify their refusal by noting that contractors in *private* construction have a right to mechanics' liens in the event they are not paid. However, the contractor would prefer a battery of weapons to deal with financial insecurity and nonpayment.

Suppose an A201 is used, but that portion that gives the contractor this power is deleted. Would a court use common law doctrines that might otherwise be available to the contractor in the event it has reasonable insecurity as to the owner's ability to pay? Deletion would probably indicate that the parties did not intend that this "gap filler" be part of this transaction.

Suppose A201 is used with the contractor's power limited to that period between award and execution of the formal contract. Would this preclude a court from using any common law financial insecurity doctrines that might otherwise be available during performance? [61] Does a contract that deals with a particular problem signal a court that the law should not imply terms? The detailed treatment of this problem would make legal intervention unlikely.

Financial problems encountered by the contractor are not likely to provide it with any justification with refusal to continue performance. However, construction contracts usually grant the owner certain remedies in the event the contractor runs into financial difficulties.[62] This should make it unnecessary for the law to employ any common law doctrines dealing with financial insecurity. This is particularly likely in light of the frequent use of surety bonds to provide financial security for the owner in the event the contractor does not perform as promised.

PROBLEMS

1. P was the prime contractor for the construction of a five-story office building. The project was destroyed by fire after approximately 50% was completed. All participants were paid for the work they had performed either by earlier progress payments or insurance proceeds turned over to them by the owner. Are the participants, the owner, the prime contractor, and subcontractors obligated to begin performance again? Should they be?

2. A prime contractor failed to comply with safety rules. This was reported to a union representing the employees. The union shop steward ordered the workers to leave the job until the safety rules were corrected. Should the contractor be given a time extension under a contract that includes a clause granting time extensions "for labor disputes"? Additional compensation?

3. O is a highly paid executive for a computer company. He made a construction contract with C under which C agreed to build a luxurious home and O promised to pay $350,000. During construction, O was fired because of a financial scandal in which he was implicated. C is concerned that O will not be able to make progress payments. What are his legal rights under AIA Doc. A201? Under the common law?

---

61. See Section 38.04(B).

62. As to bankruptcy see Section 38.04(C).

28

Defects: Design, Execution, and Blurred Roles

SECTION 28.01
INTRODUCTION:
THE PARTNERSHIP

The sad but not uncommon discovery that the project has defects often generates a claim by the owner against parties it holds responsible for having caused the defect. Claims against the design professional were discussed in Chapter 17. This chapter concentrates upon owner claims against the contractor.[1]

Defects in a house can include a leaky roof, a sagging floor, structural instability, and an inadequate heating or plumbing system. A commercial structure can include these defects as well as an escalator that is unsafe or inefficient or that requires excessive repairs. An industrial plant can include the preceding defects as well as inadequate space to install machinery or the inability of the computer system to operate the assembly line.

A differentiation must be made between temporary and permanent work. Usually defects deal with *permanent* work—the finished product that the owner intends to use. Under AIA Doc. A201, ¶ 4.3.1, *how* the permanent work is to be accomplished, such as means, methods, and sequences including

methods for accomplishing the *temporary* work, such as false work or coffer dams, is usually the responsibility of the contractor.

It is important to differentiate the different types of owners discussed in Chapter 9. The most important is a differentiation between an owner who supplies a design—usually by an independent design professional—and one who hires a contractor to both design and build. (Design-build (DB) was discussed in Section 21.04(F)).

It is also important to review the principal types of specifications noted in Chapters 23 and 27. They are *design* specifications, *performance* specifications, and *purchase description* specifications. The first requires the contractor to use designated methods and materials. The second gives the contractor particular goals, with the contractor often but not always able to determine how they are to be achieved. The third designates material and equipment by name of manufacturer, trade name, and type. Sometimes purchase description specifications allow the contractor to request authorization to substitute something that is equal to or the equivalent of the designated product. Specifications can combine types, such as design and perform-

---

1. For claims against public liability insurers see Section 28.07.

ance. Also, they vary as to completeness of material and methods requirements.

Often it is difficult to determine what causes a defect. It can be caused by the design, by workmanship, by extraneous factors such as weather, or a combination of factors. Perhaps the greatest difficulty involves the defects traceable to material (discussed in Section 28.02).

Looking ahead, one *rough* classification is to charge the owner with defects due to design while holding the contractor accountable for defects caused by failure to follow the design or by poor workmanship. This can be deceptive unless account is taken of the increasingly *blurred* roles related both to design and its execution.

Although design is principally the responsibility of the owner, usually acting through the design professional, the contractor plays a role in design, particularly a contractor retained because of its specialized skill in certain work or because it retains subcontractors with those skills. Sometimes bidders are asked to provide design alternates or do so voluntarily. Even more important, at the bidding stage, contractors are frequently required to examine the contract documents and report any errors they observe. Before or after award or during performance, the contractor may request to substitute different equipment or material than that specified. Although approval is usually required by the design professional, often approval is based upon representations or even warranties by the contractor that the proposed substitution will be at least as good as that specified or will accomplish the desired result. The contractor, though clearly subordinate to the design professional, plays an important role in design.

Similarly, the design professional frequently monitors the contractor's performance during the work and at the end of the job. Some design professionals may even direct *how* the work is to be done, although this is typically not within their power or responsibility.

There is a rough "partnership" between owner and contractor that makes it difficult to neatly divide responsibility into design and its execution.

SECTION 28.02 BASIC PRINCIPLE: RESPONSIBILITY FOLLOWS CONTROL

As a basic principle, responsibility for a defect rests upon the party to the construction contract who essentially controls and represents that it possesses skill in that phase of the overall Construction Process that substantially caused the defect. Usually defects caused by design are the responsibility of the owner in a traditional construction project and the contractor who both designs and builds. Control does not mean simply the *power* to make design choices. Usually every owner has the power to determine design choices. For example, an owner may require a particular tile to be used, a power within its contract rights. But the control needed to invoke the basic principle means a skilled choice, either one made by an owner who has professional skill in tile selection or an adviser such as an architect with those skills.

This principle recognizes that the owner who supplies the design is responsible for design that does not accomplish the owner's objective and yet may not have a claim against the design professional. Usually the standard to which the design professional is held is whether she would have performed as would have other design professionals similarly situated.

Under this principle, the owner, though faultless, bears the cost of correcting defects. The owner has the principal economic stake in the project and will benefit from a successful project. There is no reason why it cannot be responsible for project failures even though it is blameless and cannot transfer the loss.

Many cases have held that the contractor who follows the design is not responsible for a defect unless it warrants the design or was negligent.[2] This establishes the principle that the owner is responsible for any design it has furnished. Similarly, the owner impliedly warrants the accuracy of specific information it furnishes that is reasonably relied upon by the contractor.[3] It also warrants that any required materials, design features, or construction methods will create a satisfactory end product within the completed time [4] and without extraordinary unanticipated expense. Cases supporting this principle usually involve traditional construction in which the owner furnishes and monitors the design but is not primarily responsible for its execution. Risk allocation is based upon the probable intention of the parties, the greater skill possessed or supplied by the owner, the contractor's lack of discretion, and the owner's being in the best position to avoid the harm, as well as its ability to spread, absorb, or shift the risk to the design professional.

Some cases have barred a contractor who has failed to execute the design from even pointing to a design defect.[5] This is too punitive and too exculpatory of bad design and denies the possibility of shared responsibility.[6] Similarly, contractors are generally held to have impliedly warranted the quality of their workmanship.[7]

Contractors who *both* design and build usually warrant that the finished product will be fit for the purposes of the owner of which the contractor knew or should have known for the same reasons that were given to place the risk on the owner in a traditional method. Also, these warranties are similar to those placed upon *sellers* of goods. For example, one court found an implied warranty [8] where:

1. The contractor holds itself out, expressly or by implication, as competent to undertake the contract; and the owner;

2. has no particular expertise in the kind of work contemplated;

3. furnishes no plans, designs, specifications, details or blueprints; and

4. tacitly or specifically indicates its reliance on the experience and skill of the contractor, after making known to him the specific purposes for which the building is intended.

Defective material generates the most difficult problems. The huge variety of available materials, much often untested, the

---

2. For a few of the cases see *United States v. Spearin* (reproduced in Section 27.05(e)); *Northern Pac. Ry. Co. v. Goss*, 203 F. 904 (8th Cir.1913); *Lewis v. Anchorage Asphalt Paving Co.*, 535 P.2d 1188 (Alaska 1975); *Kurland v. United Pacific Insurance Co.*, 251 Cal.App.2d 112, 59 Cal.Rptr. 258 (1967); *Home Furniture Inc. v. Brunzell Construction Co.*, 84 Nev. 309, 440 P.2d 398 (1968); *Burke County Public Schools Board of Education v. Juno Construction Corp.*, 50 N.C.App. 238, 273 S.E.2d 504 (1981) affirmed 304 N.C. 159, 282 S.E.2d 779 (1981); *Teufel v. Wienir*, 68 Wash.2d 31, 411 P.2d 151 (1966). More are collected in Sweet, *Defects: A Summary and Analysis of American Law*, in *Selected Problems of Construction Law* 97, note 20 (Univ. Press, Fribourg, Switzerland, 1982).

3. See Section 29.03.

4. Refer to Section 27.05(E).

5. *Valley Construction Co. v. Lake Hills Sewer District*, 67 Wash.2d 910, 410 P.2d 796 (1965); *Robert G. Regan Co. v. Fiocchi*, 44 Ill.App.2d 336, 194 N.E.2d 665 (1963), cert. denied 379 U.S. 828, 85 S.Ct. 56, 13 L.Ed.2d 37 (1964).

6. *Northern Petrochemical Co. v. Thorsen & Thorshov, Inc.*, 297 Minn. 118, 211 N.W.2d 159 (1973). See also Section 28.06.

7. *Northern Pac. Ry. Co. v. Goss*, supra note 2; *Trahan v. Broussard*, 399 So.2d 782 (La.App.1981); *Smith v. Erftmier*, 210 Neb. 486, 315 N.W.2d 445 (1982). But see *Samuelson v. Chutich*, 187 Colo. 155, 529 P.2d 631 (1974) (fault required).

8. *Dobler v. Malloy*, 214 N.W.2d 510, 516 (N.D.1973). See also *Rosell v. Silver Crest Enterprises*, 7 Ariz.App. 137, 436 P.2d 915 (1968); *Barraque v. Neff*, 202 La. 360, 11 So.2d 697 (1942); *Robertson Lumber Co. v. Stephen Farmers Cooperative Elevator Co.*, 274 Minn. 17, 143 N.W.2d 622 (1966). But see *Milau Associates v. North Avenue Development Corp.*, 42 N.Y.2d 482, 398 N.Y.S.2d 882, 368 N.E.2d 1247 (1977) (no implied warranty of fitness in service contract).

pressure to cut costs or weight, and the inability to test or rely on manufacturers all combine to make this a prime cause of defects.

Material specified may be *unsuitable* and will *never* accomplish the purpose for which it has been specified. As this is part of design, responsibility for such material falls upon the person who controls the design or to whom the risk is transferred.[9]

Suppose the materials were suitable but were faulty goods that came off the assembly line on a particular day that did not measure up to the quality requirement of the manufacturer. This problem is addressed in greater detail in Section 28.08(C) dealing with the AIA warranty clauses. A few observations can be made at this point.

Arguments can support placing the risk of faulty materials on either owner or contractor. As to putting the risk on the owner, it is often difficult to determine whether the material was unsuitable or faulty. It is more efficient to place responsibility for *any* specified material upon the party in control of the design—the owner in the case of traditional construction and the contractor who both designs and builds. It is unfair to use purchase description specifications and then seek to hold the contractor accountable. When a purchase description specification is used, the contractor is not a seller but simply a procurer of goods ordered by someone else. It should not be held to the warranties of sellers or manufacturers.

Cogent arguments can be made for holding the contractor responsible for faulty materials. Although it may be difficult to determine whether a defect is caused by unsuitable or faulty material, it is also difficult to determine whether the defect is caused by material *at all* or by the contractor's failure to install it properly. Very likely a contractor is responsible for faulty material if it knew or should have known it

was faulty. If so, it is administratively more efficient to make the contractor responsible for faulty material, particularly since the contractor is likely to have a better claim against the supplier or manufacturer than the owner.

This deceptively simple problem is one that *should* be handled specifically in any contract with design or purchase description specifications.

SECTION 28.03 DISPLACING THE BASIC PRINCIPLE: UNCONSCIONABILITY

Autonomy (freedom of contract) generally allows contracting parties to determine how particular risks will be borne. As a rule, the owner is the party who usually seeks to take advantage of autonomy.

Displacement should be differentiated from clauses that clarify or augment the basic principle. A warranty clause (discussed in Section 28.08) can transfer risks for design to the contractor. However, such a clause may make *clear* that the contractor is responsible for poor workmanship and specify a remedy. The latter warranty clause seeks not to displace but to augment the basic principle.

Any clauses that seek to displace the basic principle will be scrutinized carefully to determine both whether the contracting party upon whom the risk is placed was *aware* of the risk allocation and the *fairness* of displacement of the basic principle. Although cases do not always openly recognize this, the courts are more willing to determine the fairness of such clauses.

W.H. Lyman Construction Co. v. Village of Gurnee [10] involved a specification that required that the contractor use a particular manhole base and seal. The specifications also stated that the contractor assumed the risk of complying with infiltration limits. If the contractor thought the design would

---

9. *Teufel v. Wienir*, supra, note 2.

10. 84 Ill.App.3d 28, 38 Ill.Dec. 721, 403 N.E.2d 1325 (1980).

be inadequate, the contractor was to direct attention to this in writing at the time it submitted its bid.

The court held that this was an "impermissible attempt on the part of the Village to shift the responsibility for the sufficiency and adequacy of the plans to the contractor, without providing the contractor the corresponding benefit of something to say about the plans that he is strictly bound to follow." [11] After noting the "possible" unconscionability of such a clause, the court concluded that the clause would not shield the Village from its negligence because of public policy. Shielding the Village would discourage bidders, and the public interest would suffer in the long run.

Kurland v. United Pacific Insurance Co.[12] demonstrates the unwillingness of courts to simply *apply* language that seeks to displace this basic principle. It involved a claim by an owner against a subcontractor's surety. The subcontractor had undertaken to install an air-conditioning system in an apartment building. The plans and specifications designated equipment to be used and required the contractor to meet performance standards for cooling and heating. The subcontractor did as required, but the air-conditioning system did not function as required.

In affirming a judgment for the surety, the court noted that the plans and specifications had been prepared by the architect. This was an *owner* design choice, perhaps a negligent one. The court held that it could not be *reasonably* concluded that the subcontractor would assume the responsibility for the adequacy of the plans and specifications. The court concluded that warrantylike language was simply an undertaking by the subcontractor that it would work

as effectively as possible to achieve the desired result. But it was *not* a warranty of a successful outcome. Nevertheless, courts do not *always* relieve the contractor if it follows the plans and specifications dictated by the owner. In *United States Fidelity & Guaranty Co. v. Jacksonville State University*[13] the contractor did as it was required by the specifications. But the court held the contractor responsible for wall leaks caused by unsuitable sealing material, as language in the contract was found to be a guarantee by the contractor.

The *Kurland* and *Jacksonville* cases, along with *Teufel v. Wienir*,[14] raise a problem that relates to subcontracting. Often the prime contract contains language that requires the prime contractor to obtain a warranty from the subcontractor of its work. As seen in the *Kurland* case, such warranties may not mean much if purchase description specifications are used.

Subcontractor warranties raise another issue reflected in the *Jacksonville* and *Teufel* cases. Do these warranties affect any *prime* contractor obligation? Commonly, contract language makes the prime contractor responsible for the subcontractor's work. On the other hand, the contract may indicate that the owner was exclusively relying upon the *subcontractor's* warranty. This exonerates the prime contractor.

That a subcontractor can be or has been sued should not change the legal obligation of owner and prime contractor unless the owner, by either demanding or accepting the subcontractor's warranty, manifests an intent to release the prime. For example, the trial court judge gave the prime contractor a defense in *Teufel v. Wienir* because the

---

11. 403 N.E.2d at 1332.

12. Supra note 2. Also see *Fanning & Doorley Construction Co. v. Geigy Chem. Corp.,* 305 F.Supp. 650 (D.R.I.1969) and *Wood-Hopkins Contracting Co. v. Masonry Contractors, Inc.,* 235 So.2d 548 (Fla.App.1970), which relieved contractors who simply installed what they were told.

13. 357 So.2d 952 (Ala.1978).

14. Supra note 2.

specifications indicated that the owner agreed to look only to the subcontractor.[15]

Suppose the design is created by the *subcontractor*. As between owner and prime, responsibility very likely falls upon the contractor.[16]

Another problem of risk shifting results from proposals by the prime contractor for substitutions of material or equipment for that specified. Although substitutions must be approved by the design professional, to protect herself and to recognize that she may be relying upon representations of the contractor, the design professional often obtains a guarantee by the contractor. Courts do not seem to be willing to interpret warranty language in this context neutrally. For example, a recent unpublished opinion held that the contractor warranted only its workmanship.[17]

Varying results show that language will not *automatically* displace the basic principle of control outlined in Section 28.02. Yet the outcome can often be predicted. Which party really made the design choice? Which party had the greater experience and skill? Which party relied on the other? The answers do not always implicate the owner, even in a traditional system. Even with these questions answered, account must be taken of the different attitudes toward the use of contract language to achieve what appears to be an unconscionable result.

SECTION 28.04 GOOD FAITH AND FAIR DEALING: A SUPPLEMENTAL PRINCIPLE

Parties who *plan* to enter into a contract are not, as a general rule, expected to look out for each other. Although there are exceptions, this is still a strong principle in American contract law.

Once a contract has been made, the law, led by the Uniform Commercial Code (UCC) § 1–203 dealing with certain commercial transactions, increasingly expects parties to act in good faith and to deal fairly with one another. In construction contracts either party, though most commonly the contractor, must warn the other when the other is proceeding in a way that will cause failure. Sometimes this is reflected in contract clauses that require that the contractor bring to the attention of the owner or design professional design or other problems that can adversely affect the project.[18]

Suppose there is no specific contract obligation? What does the law demand of the contractor? Must it:

1. Examine the contract documents, but only to prepare its bid, with design errors being none of its business?

2. Examine the contract documents principally to prepare its bid but also to note and report any errors observed?

3. Examine the contract documents, both to prepare its bid and to check for errors, it being held for errors it should have observed?

Even making allowance for factual variations, the cases display diverse results. This is common when a new doctrine of a vague nature limits the powerful principle of autonomy. A few appear to permit the contractor to ignore any design errors it may even observe, with its responsibility simply

---

15. The appellate court affirmed a judgment for the contractor by finding that the prime contractor simply followed the design. *Teufel v. Wienir*, supra, note 2.

16. *Stevens Construction Corp. v. Carolina Corp.*, 63 Wis.2d 342, 217 N.W.2d 291 (1974).

17. *Habenicht & Howlett v. Jones-Allen-Dillingham*, Cal. Court of Appeals, 1 Civ. 46449 (1981). Such an opinion cannot be cited in California. Yet a warranty was upheld in *New Orleans Unity Society v. Standard Roofing Co.*, 224 So.2d 60 (La.App.) cert. denied 254 La. 811, 227 So.2d 146 (1969).

18. AIA Doc. A201, ¶¶ 4.2.1, 4.7.3.

to build and not design.[19] The better reasoned cases and, as indicated, the AIA require the contractor to warn the owner if the contractor believes a suitable result cannot be obtained from the design.[20]

Is the contractor's conduct measured objectively? Is it responsible for an examination being what it *should* have discovered and what it *should* have reported?

Cases that have given the contractor who has followed the design a defense have stated that this defense will be lost if the contractor has been negligent, an apparently objective standard.[21] However, the AIA has used a subjective standard, perhaps a fairer result. The owner is paying for whatever design skill the contractor possesses and uses. The principal purpose for the contractor's reviewing the contract documents is to prepare its bid. The owner should expect attention to be drawn only to those errors that the contractor *does* discover.

The owner often has years to prepare the design contrasted to the thirty days or so given the contractor to review the design and prepare its bid.[22] An objective approach may operate in a way as to place design risks unfairly on the contractor.

The contractor who *does* notify the owner of errors is not charged for defects that result despite its warnings and is entitled to be paid for what it has done.[23] The contractor who does *not* report obvious errors should not be given advantage of the "following the plans" defense. If it performs work knowing that it violates legal requirements or technical competence, it will not be able to recover for the work it performed.

If failure to comply with its requirement causes a loss, as it will undoubtedly do, should the contractor be responsible for the *entire* loss? Sharing responsibility, rather than placing it all on the contractor, is fairer, particularly if the contractor did not have actual knowledge of the error. This will be discussed in Section 28.06.

SECTION 28.05
ACCEPTANCE OF PROJECT

If the owner communicates a *clear* intention to relinquish a claim for obvious defects, the claim is barred.[24] Acts frequently *asserted* to have indicated such an intention are payment, particularly the final payment and taking possession of the project.

Standard forms increasingly make clear that these acts do *not* constitute a waiver of any claim for defective work.[25] The universal presence of warranty (guarantee) clauses makes it quite unlikely that such a claim will be lost because of the occurrence of these acts.[26]

---

19. *Lewis v. Anchorage Asphalt Paving Co.*, supra note 2; *Luxurious Swimming Pools, Inc. v. Tepe*, 177 Ind.App. 384, 379 N.E.2d 992 (1978); *Rubin v. Coles*, 142 Misc. 139, 253 N.Y.S. 808 (Cty.Ct.1931).

20. *Lebreton v. Brown*, 260 So.2d 767 (La.App.1972); *Hutchinson v. Bohnsack School District*, 51 N.D. 165, 199 N.W. 484 (1924); *Home Furniture, Inc. v. Brunzell Construction Co.*, supra note 2.

21. See supra note 2.

22. *Southern New England Contracting Co. v. State*, 165 Conn. 644, 345 A.2d 550 (1974); *Pittman Construction Co. v. Housing Authority of New Orleans*, 169 So.2d 122 (La.App.1964).

23. Architects who pointed out design difficulties can recover if the owner still proceeds. *Greenhaven Corporation v. Hutchcraft & Associates*, 463 N.E.2d 283 (Ind.App. 1984) (design violated fire code); *Bowman v. Coursey*, 433 So.2d 251 (La.App.) certiorari denied 440 So.2d 151 (1983) (design below acceptable professional practice).

24. *A.H. Sollinger Construction Co. v. Illinois Building Authority*, 5 Ill.App.3d 554, 283 N.E.2d 508 (1972); *Maloney v. Oak Builders, Inc.*, 224 So.2d 161 (La.App.1969); *Salem Realty Co. v. Batson*, 256 N.C. 298, 123 S.E.2d 744 (1962); *Stevens Construction Corp. v. Carolina Corp.*, supra note 16.

25. AIA Doc. A201, ¶¶ 9.5.5., 9.9.4.

26. AIA Doc. A201, ¶¶ 4.5.1. and 13.2.2.

Generally the owner by payment [27] or by taking possession of the project [28] does not waive its claim for defective work. These acts alone do not *unambiguously* indicate the owner's intention to give up a claim for work to which it was entitled and for which it paid. Possession may be taken for reasons other than satisfaction with the work. Such holdings are frequently supported by contract clauses denying that these acts have waived claims for defects. Cases that have *found* waiver have been ones that have involved disputes over particular work followed by some act of the owner, such as payment or taking possession, which communicated satisfaction with the work and an intention not to assert any claim. [29]

SECTION 28.06
DIVIDED RESPONSIBILITY

In a venture as complicated as a construction project, a defect can often be traced to multiple causes. Where proof can establish which defect or part of a defect is attributable to the owner and which to the contractor, the result is clear. [30] Suppose there is one defect, one indivisible loss, and responsibility cannot be traced exclusively to either party. It is clear that *each* party did not perform as promised, with each having played a significant role in causing the defect.

Similarly, design for which the owner is generally responsible and execution of the design for which the contractor is responsible are not watertight compartments. Under certain circumstances, the contractor plays a role in design and the owner, either directly or through the design professional, has a limited role in execution. Each may play a role in the other's sphere of responsibility which can mean shared responsibility. Legal responsibility does not require that *all* causes, except that of the party against whom the claim has been made, be eliminated. Contract law, borrowing from tort law, requires only a showing that the defendant's breach was a substantial factor in causing the harm. [31] For example, suppose a defect is caused principally by the failure of the contractor to use proper workmanship but also by factors for which the contractor was not responsible, such as abnormal weather conditions, or third parties for whom the contractor is not responsible. If the other factors played a minor or trivial part, the contractor is responsible for the entire loss.

Suppose the owner played a significant role in causing the harm. How will responsibility be apportioned? Here again, that law has followed paths laid out by tort law, with its greater exposure to multiple causes and multiple wrongdoers. Three-fourths of the American states now use comparative negligence, negligence no longer barring a claimant under the mostly discarded contributory negligence doctrine. Instead negligence is *compared*. Subject to variations

27. *Metropolitan Sanitary District of Greater Chicago v. Anthony Pontarelli & Sons, Inc.*, 7 Ill.App.3d 829, 288 N.E.2d 905 (1972); *Parsons v. Beaulieu*, 429 A.2d 214 (Me.1981); *Handy v. Bliss*, 204 Mass. 513, 90 N.E. 864 (1910); *Burke County Public Schools Board of Education v. Juno Construction*, supra note 2.

28. *Coastal Modular Corp. v. Laminators, Inc.*, 635 F.2d 1102 (4th Cir.1980) (by implication); *Aubrey v. Helton*, 276 Ala. 134, 159 So.2d 837 (1964); *Honolulu Roofing Co. v. Felix*, 49 Hawaii 578, 426 P.2d 298 (1967); *Kangas v. Trust*, 110 Ill.App.3d 876, 65 Ill.Dec. 757, 441 N.E.2d 1271 (1982); *Hemenway Co., Inc. v. Bartex, Inc. of Texas*, 373 So.2d 1356 (La.App.1979); *Bismarck Baptist Church v. Wiedemann Industries*, 201 N.W.2d 434 (N.D.1972); *Hurley v. Kiona-Benton School District No. 27*, 124 Wash. 537, 215 P. 21 (1923).

29. *Saldal v. Jacobsen*, 154 Iowa 630, 135 N.W. 18 (1912); *Grass Range High School District v. Wallace Diteman, Inc.*, 155 Mont. 10, 465 P.2d 814 (1970); *Cantrell v. Woodhill Enterprises, Inc.*, 273 N.C. 490, 160 S.E.2d 476 (1968). See generally, Sweet, *Completion, Acceptance and Waiver of Claims: Back to Basics*, 17 Forum 1312 (1982).

30. *S.J. Groves & Sons Co. v. Warner Co.*, 576 F.2d 524 (3d Cir.1978) (prime v. subcontractor).

31. *Krauss v. Greenbarg*, 137 F.2d 569 (3d Cir.) cert. denied 320 U.S. 791, 64 S.Ct. 207, 88 L.Ed. 477 (1943).

in state laws, negligence by the claimant simply *reduces* the damages. Similarly, the *ultimate* responsibility between co-wrong-doers increasingly is being shared on a comparative basis.

It is, of course, easier to apply tort doctrines when the owner, or someone for whom the owner is responsible such as the design professional, and the contractor, or anyone for whom it is responsible such as subcontractor, have been negligent.[32] Although the law is not yet clear, it is likely that when *each* is legally responsible, the loss will be shared, with the owner's claim being reduced by use of a rough formula.[33] For example, suppose an indivisible loss occurred with no way to trace it solely to either party. If each party seems equally blameworthy, the owner should be able to recover 50% of the loss from the contractor.

Suppose there was a design error that the contractor should have discovered. The owner may be held 80% accountable and the contractor 20%. (If the contractor saw it and said nothing, it might be 60/40%.)

Contract provisions often provide a remedy. AIA clauses provide a partial remedy when a contractor does not report design errors of which it was aware.[34] Similarly, contract clauses usually deny the contractor a defense where it performs defective work even if it shows that the architect stood by

and did not complain or condemn the work.[35] Such clauses take precedence over any tortlike comparison.[36]

One reason for the scarcity of cases is the tendency for the owner to bring legal action against both its independent design professional *and* the contractor who usually seeks a *total* defense by claiming it has followed the plans.[37] As a result, few attempts have been made by the contractor to reduce its liability because of acts of the design professional.

The tendency to sue the design professional has generated these two additional, difficult issues:

1. How is an indivisible loss borne when there are two or more "wrongdoers."? (Each is responsible for the entire loss.)

2. How is the *ultimate* responsibility determined when two or more are legally responsible? (Either can transfer all or a part of the loss if it can find a rational apportionment formula or employ indemnity.) [38]

SECTION 28.07 CLAIMS AGAINST LIABILITY INSURER

Commonly the contractor is required to carry comprehensive general liability (CGL) insurance. This indemnifies the insured contractor against claims by third parties who assert that they have suffered losses because

---

32. *Northern Petrochemical Co. v. Thorsen & Thorshov, Inc.,* supra note 6 (owner can recover from either architect or contractor for an indivisible loss with those responsible, as between themselves, permitted to establish an apportionment formula). But see note 37 infra.

33. *Grow Construction Co. v. State,* 56 A.D.2d 95, 391 N.Y.S.2d 726 (1977); *Circle Electrical Contractors, Inc.,* DOT CAB No. 76–27, 77–1 BCA, ¶ 12,339. But one court refused to use comparative negligence in a contract claim for economic loss. *Broce-O'Dell Concrete Products Co. v. Mel Jarvis Construction Co.,* 6 Kan.App.2d 757, 634 P.2d 1142 (1981).

34. AIA Doc. A201, ¶ 4.2.1 (contractor exculpated if it reports errors it discovered), ¶ 4.7.4 (contractor responsible for work that it knew violated legal requirements).

35. AIA Doc. A201, ¶ 4.3.3.

36. Cf. *Rogers v. Dorchester Associates,* 32 N.Y.2d 553, 347 N.Y.S.2d 22, 300 N.E.2d 403 (1973).

37. *Shepard v. City of Palatka,* 414 So.2d 1077 (Fla.App. 1981) reproduced in Section 15.08(B) by dictum stated responsibility for owner's loss would be shared by architect and contractor where loss caused by both design and execution.

38. See note 32 supra. See also Section 36.03.

the insured contractor has not acted in accordance with tort law. Usually claimants are workers on the job who are not employees of the contractor (the latter are usually covered by workers' compensation) or members of the public. Claims covered are those that involve personal harm or property damage.

In the 1970s, contractors began to assert claims against their CGL insurers based upon "public" liability for defects as "property damage." (This was and often is the reason for an owner claim to be based upon negligence.) Such an approach is advantageous to the contractor. The insurer, unlike a surety, cannot claim against its insured.

After a few successful claims, the preponderance of the cases held that defects negligently caused were not covered by CGL policies.[39]

call upon them to perform further work or to respond to a legal claim. For that reason, they frequently contend that acceptance of the project manifests owner satisfaction and the owner cannot make any further complaints about the work.

The law has been reluctant to find that there has been sufficient "acceptance" to bar any future claims against the contractor. Undoubtedly this reluctance is traceable to the difficulty of discovering defects at the time the project is turned over. Many defects will not be apparent until the owner has taken over the project and used it. One method of dealing with the risk that acceptance will be found and claims barred is to insert a provision making clear that liability does not end upon the project's being turned over to the owner. The exact nature of that liability is discussed in (B).

SECTION 28.08 WARRANTY (GUARANTEE) CLAUSES

(A) RELATIONSHIP TO ACCEPTANCE

Acceptance of the project is an important benchmark in the relationship between owner and contractor. Contractors would like to know when the law can no longer

(B) PURPOSES: *ST. ANDREW'S EPISCOPAL DAY SCHOOL v. WALSH PLUMBING CO.*

The deceptively simply warranty clause obscures a variety of possible purposes. Before discussing the many purposes, a case involving such a clause is reproduced.

ST. ANDREW'S EPISCOPAL DAY SCHOOL v. WALSH PLUMBING CO.

Supreme Court of
Mississippi, 1970.
234 So.2d 922.

ROBERTSON, Justice.

The appellant, St. Andrew's Episcopal Day School, a charitable corporation, brought suit in the Chancery Court of the First Judicial District of Hinds County, Mississippi, against Appellee Walsh Plumbing Company, the contractor of the mechanical work, and Appellee The Trane Company, the manufacturer of the major portion of the air conditioning system installed in the new Day School building. The suit was one for breach of warranty or guaranty. The chancellor, after a full trial, dismissed the bill of complaint, and the Day School appeals from this judgment.

On April 8, 1965, appellant entered into a contract with Walsh

---

39. *Western Employers Insurance Co. v. Arciero & Sons, Inc.*, 146 Cal.App.3d 1027, 194 Cal.Rptr. 688 (1983), *Vernon Williams & Son Construction, Inc. v. Continental Insurance Co.*, 591 S.W.2d 760 (Tenn.1979).

Plumbing Company whereby, in consideration of $149,420, Walsh agreed to:

> furnish all of the materials and perform all of the work shown on the Drawings and described in the Specifications entitled: Item II, Mechanical Construction, St. Andrews Episcopal Day School, Old Canton Road, Jackson, Mississippi.

The General Conditions of the Contract provided in Article 20:

CORRECTION OF THE WORK AFTER SUBSTANTIAL COMPLETION

> The Contractor shall remedy any defects due to faulty materials or workmanship and pay for any damage to other work resulting therefrom, which shall appear within a period of one year from the date of Substantial Completion as defined in these General Conditions, and in accordance with the terms of any special guarantees provided in the Contract. The Owner shall give notice of observed defects with reasonable promptness. All questions arising under this Article shall be decided by the Architect subject to arbitration, notwithstanding final payment.

The Construction Specifications, in Paragraph 22 of Mechanical Construction, required:

> This contractor shall guarantee each and every part of all apparatus entering into this work to be *the best of its respective kind and he shall replace within one year from date of completion all parts which during that time prove to be defective and he must replace these parts at his own expense.*

> He shall guarantee to install each and every portion of the work in strict accordance with the plans and specifications and to the satisfaction of the owner.

> Guarantee to include replacement of refrigerant loss from air conditioning and refrigerant system. [Emphasis added.]

On August 19, 1966, Lomax, North and Beasley, consulting engineers, by letter, advised Biggs, Weir, Neal & Chastain, architects, that the mechanical contractor should furnish "as built drawings which locate all underground piping and clean-outs, framed operating instructions in the boiler rooms, CFM figures for all air units, and certification that all safety valves and devices have been tested." The engineers ended their letter with this comment:

> Other than the above, we feel that the mechanical work is ready for final certification, subject to contract guarantee provisions.

On August 30, 1966, the engineer wrote the architects:

> Final inspections of the subject project have been completed and we recommend final certification of the mechanical contract, subject to contract guarantee provisions.

In accordance with the requirements of Paragraph 22 of the Construction Specifications, on August 16, 1966, Walsh Plumbing Company wrote St. Andrew's Episcopal Day School:

> "We hereby guarantee all work performed by us on the above captioned project to be free from defective materials and workmanship for a period of one (1) year, unless called for in the specifications to be a longer period of time."

Between August 16, 1966, and July 17, 1967, Appellee Walsh was called on several times to remedy defects in the air conditioning system, and Walsh always responded promptly. On July 17, 1967, the air conditioning system broke down completely and ceased to function. The headmaster, the Reverend James, immediately tried to contact Ray Walsh, only to find that he was out of town. Mrs. Walsh suggested that the School call somebody else.

During the first two weeks of July, 1967, James E. Davis, Jr., operator of Davis-Trane Service Agency and also a salesman and representative of The Trane Company, was contacted by John B. Walsh and together with Mr. Walsh attempted to determine why the air conditioning system was not cooling. Mr. Davis described the meeting that took place on July 18, 1967, in these words:

> "[T]he meeting that we had on a particular day at the school, in which Mr. Nicholson was there, and Mr. Ray Walsh was there, and Mr. Nicholson said, 'We want to get this thing fixed,' and *Mr. Walsh told me to fix it.* He said, 'You have the people, the personnel and the know-how, *you go ahead and do it* and I will just stay out of it,' so Mr. Nicholson said, 'Okay, Davis-Trane Service Agency, go ahead and fix this machine.' " [Emphasis added.]

Forrest G. North, the mechanical engineer, testified that the failure of two safety devices, the flow control switch and the freeze protection thermostat caused the copper tubes inside the chiller shell and the shell itself to rupture. The purpose of the flow control switch was to prevent the operation of the machine when there was no circulation of water in the chiller shell. The flow control switch was installed by Walsh outside the chiller and was not Trane equipment. The freeze protection thermostat was a Trane part, and was installed inside the chiller unit by Trane at its factory.

The repairs made by Davis-Trane Service Agency pursuant to Ray Walsh's instructions to Davis to go ahead and fix it amounted to $6,813.05. One of the major items of expense was a new chiller unit purchased from the Trane Company. The Trane Company and Davis-Trane billed the appellant, and the appellant paid Trane separately for the new chiller unit, and Davis-Trane for all the repairs.

The appellant is in the business of running a Christian day school; it is not in the air conditioning business. Appellant does not profess to have any knowledge or expertise about air conditioning systems or equipment. That was the main reason for the provision in the Construction Specifications that the mechanical contractor "shall guarantee each and every part of all apparatus entering into this work to be *the best of its respective kind,* and he shall replace within one year from date of completion all parts which during that time prove to be defective and he must replace these parts at his own expense."

Walsh was a reputable, responsible and knowledgeable contractor of mechanical work; and when Lomax, North and Beasley, consulting engineers for the Day School, recommended to the appellant that the bid of Walsh Plumbing Company be accepted, the duty and responsibility was placed squarely on Walsh's shoulders to purchase and properly install the best air conditioning system on the market. Not only was Walsh to purchase and install, he was to guarantee the system and its installation for one year. This was what Walsh contracted to do, and this was what the appellant paid Walsh to do. Not knowing anything about air conditioning systems, the appellant employed experts in this field and reposed full confidence in these experts to look after its interests.

The chancellor was correct in finding that the breakdown of the air conditioning system occurred "within the time of the warranty by Walsh," and that the repairs were made necessary to properly repair the system.

The chancellor was in error in holding that:

> "St. Andrews never gave any written notice or made any demand on Walsh to comply with his warranty to fix the machine, which would be as I hold a condition precedent to hiring someone else to do the work."

The sole purpose of notice is to give the contractor who selected, purchased and installed the system the first opportunity to remedy the defects at the least possible expense to him. Appellee Walsh was given this opportunity.

The evidence is undisputed that Walsh was at the Day School building on July 18, 1967, the day after the breakdown, with James E. Davis, Jr., of Davis-Trane Service Agency and John W. Nicholson of the Day School. Davis, called as an adverse witness by the appellant, testified that Ray Walsh said at that time:

> " *'You have the people, the personnel and the know-how, you go ahead and do it and I will just stay out of it'* . . . ."
> [Emphasis added.]

Walsh was afforded the opportunity of doing the work himself or employing somebody else to do it. With full knowledge and full notice he chose to employ Davis to go ahead and remedy the defects and make the necessary repairs.

Appellee Trane was the major supplier of items and equipment going into the air conditioning system and Trane was well paid for these. It is unfortunate that Trane's one-year guaranty to Walsh had run out at the time of the complete breakdown of the air conditioning system. Trane guaranteed the items and equipment furnished by it for one year from the date of shipment; these parts and equipment were shipped in March, 1966. The complete breakdown did not occur until July, 1967. The chancellor was correct in holding that Trane's warranty had expired.

The chancellor found that $6,813.05 was "a reasonable amount to make the repairs" and put the air conditioning system back in operation.

The judgment of the chancery court is affirmed as to Appellee The Trane Company, but the judgment is reversed as to Appellee Walsh Plumbing Company, and judgment is rendered against Walsh Plumbing Company on its warranty for $6,813.05, together with 6% interest from July 17, 1967.

Judgment affirmed as to the Trane Company, but reversed and rendered as to Walsh Plumbing Company.

GILLESPIE, P.J., and JONES, BRADY and INZER, JJ., concur.

From the owner's vantage point, one of the problems in the traditional contracting system is the possibility of being "whip-sawed" between the design professional and the contractor when a defect is discovered. The design professional contends that the defect resulted from improper execution of the design. The contractor contends that the defect existed because of design inadequacy. To make matters worse, each or both may contend that the defect was caused by conditions over which neither had control, such as unusual weather, misuse, or poor maintenance by the owner. One way of dealing with divided responsibility is to hire one entity to both design and build.

Another method of dealing with this problem is to incorporate a provision under which one entity is responsible for defects *however* and by *whomever* caused. The entity upon whom the owner is likely to place this risk is the contractor. For example, the owner could require that the contractor give a full warranty on the roof which would cover defects *however* they may be caused.

Despite the undoubted advantage in placing the risks of defects *however caused* upon the contractor, this is not done often. Contractors may respond to such a roof guaranty by drastically increasing the contract price to take into account the possibility of design errors, the additional cost of checking upon the quality of the design, the cost of any special endorsement that will have to be obtained to cover design risks, or the cost of a roofing bond. Nevertheless, the owner may decide that these additional costs are worth the advantage of centralizing responsibility. Although enforcement requires very clear language (even *that* may not be sufficient), such clauses have been enforced.[40]

---

40. *Potler v. MCP Facilities Corp.*, 471 F.Supp. 1344 (E.D.N.Y.1979); *Bryson v. McCone*, 121 Cal. 153, 53 P. 637 (1898) (construction of industrial plant); *New Orleans Unity Society v. Standard Roofing Co.*, supra note 17; *St. Andrews Episcopal Day School v. Walsh Plumbing Co.*, reproduced in Section 28.08(B); *Burke County Public Schools*

Another use of a warranty clause relates to the nonfault nature of express warranty. In some states, one who contracts to perform services need only perform nonnegligently.[41] In those states, a warranty clause as to workmanship can avoid the burden of having to establish that the contractor performed negligently.

Many issues lurk in a simple warranty clause. Earlier discussion emphasized the *owner's* purpose in having such a clause. The *contractor* is also concerned with how such a clause will function. Although it would prefer to be relieved from responsibility upon acceptance, it may be willing to accept a warranty clause if it believes that expiration of the period will terminate its obligation. Similarly, the contractor may be concerned with exposure created by such a clause even within the warranty period. Failure to take immediate steps to deal with a defect may increase the loss. The cost of correcting a defect is likely to be greater if correction is done by someone other than the contractor. For that reason, contractors sometimes contend that the clause is for their benefit, that its function is to give the contractor notice as quickly as possible to enable the contractor to cut the losses and repair the defect as inexpensively as possible.[42] On the other hand, the owner may contend that the clause is for its bene-fit. If so, it need *not* call back the contractor (discussed in (C)).

Another possible purpose is to bar the contractor from contesting *how* the owner has chosen to correct the defect if the contractor refuses to attempt to correct the work when requested to do so. The clause may place the burden on the warrantor to establish that *other* causes, such as abnormal weather or owner misuse, caused the defect.

Moving back to the warranty clause as putting a time limit on liability, although a few decisions have held the clause a private period of limitation,[43] most have held that such a provision is *not* intended to cut off liability at the end of the warranty period.[44] Clearly AIA has *not* made its warranty clause a period of limitation. The law will not bar a claim by a contractual provision of this type unless the clause *clearly* shows this intention.[45]

Does the clause affect the remedy? Usually the clause states that the contractor will correct any defect within the warranty period. As discussed in Section 28.08(C), will this affect the issuance of a court decree *specifically ordering* the contractor to come and correct the work? If the contractor fails to correct the work, the owner can do so and charge the contractor. This recognizes the adequacy of the money award,

---

etc. v. Juno Construction, supra note 2 (agreement to maintain roof for five years); *Shuster v. Sion*, 86 R.I. 431, 136 A.2d 611 (1957); *Shopping Center Management Co. v. Rupp*, 54 Wash.2d 624, 343 P.2d 877 (1959); *cf. Pinellas County v. Lee Construction Co. of Sanford*, 375 So.2d 293 (Fla.App.1979).

41. *Samuelson v. Chutich*, supra note 7.

42. See *St. Andrew's Episcopal Day School v. Walsh Plumbing Co.*, reproduced in this section. If it is clear that the contractor will not correct the defect, giving the notice is excused. *Orto v. Jackson*, 413 N.E.2d 273 (Ind.App. 1980).

43. *Cree Coaches, Inc. v. Panel Suppliers, Inc.*, 384 Mich. 646, 186 N.W.2d 335 (1971); *Independent Consolidated School District No. 24 v. Carlstrom*, 277 Minn. 117, 151 N.W.2d 784 (1967).

44. *First National Bank of Akron v. Cann*, 503 F.Supp. 419 (N.D.Ohio 1980), aff'd 669 F.2d 414 (6th Cir.1982); *Norair Engineering Corp. v. St. Joseph's Hospital, Inc.*, 147 Ga.App. 595, 249 S.E.2d 642 (1978); *Board of Regents v. Wilson*, 27 Ill.App.3d 26, 326 N.E.2d 216 (1975); *Michel v. Efferson*, 223 La. 136, 65 So.2d 115 (1953); *Newton Housing Authority v. Cumberland Construction Co.*, 5 Mass.App.Ct. 1, 358 N.E.2d 474 (1977); *City of Midland v. Waller*, 430 S.W.2d 473 (Tex.1968).

45. *Glantz Contracting Co. v. General Electric*, 379 So.2d 912 (Miss. 1980); *Bender-Miller Co. v. Thornwood Farms, Inc.*, 211 Va. 585, 179 S.E.2d 636 (1971) (remedy not exclusive).

usually a bar to judicially ordered specific performance. A warranty clause should not affect specific performance.

Remedy can be approached another way. Suppose there is *no* warranty clause. The owner can recover the cost of correction unless it would be disproportionately high in relation to the diminished value caused by the defect. If the latter, the measure is the difference in value between the project as built and as it *should* have been built. In addition, the owner may be able to recover consequential damages, such as lost use or profits if reasonably foreseeable and proved with reasonable certainty.

A warranty clause is a promise by the contractor to correct defects within the warranty period. Does this make correction cost the only remedy? Is diminished value barred? Does the clause limit liability to cost of correction?

Very likely the clause was not intended to deal with any of these subtle issues. The law should not give the clause *any* remedial effect.[46]

Although historically such clauses may have been justified as a method to avoid acceptance barring claims, is there a current justification for inclusion of a warranty clause?

An unfortunate problem that often plagues the owner is the unwillingness of the contractor to return and correct defects for which it is responsible. Including a specific provision under which the contractor promises to return and correct defects if notified to do so within a designated period may persuade a contractor that it must do as it has promised.

Exhortation, of course, may not be the *sole* function of such a clause or even *a* function of a particular clause. A warranty clause may be designed to shift design risks to a contractor or to affect the legal standard conduct. Nevertheless, on average *modern* warranty clauses are principally exhortations to get defects corrected quickly before they generate greater losses. This does not mean that the drafter may not want a *different* purpose. If so, that objective should be expressed clearly in the contract. Even so, clauses will always need interpretation as to coverage.[47]

(C) AIA WARRANTY CLAUSE

AIA Doc. A201 has two provisions that can be said to be warranty provisions. Paragraph 4.5.1, designated "Warranty," states that the contractor warrants to the owner and *architect* that "materials and equipment" and all "Work will be of good quality, free from faults and defects." Since ¶ 1.1.3 defines "Work" as including all labor, the essence of ¶ 4.5.1 is that material and equipment will be new and free of faults and defects and that all workmanship will be of good quality. Clearly this warranty as to workmanship avoids any state rules that will not imply that labor will be of good workmanship.[48] More important, the warranty will *appear* to place the responsibility on the contractor for *all* materials or equipment at least if the materials were defective and possibly even if they were unsuitable.[49]

Complications develop because AIA has dealt extensively with post-completion de-

---

46. *United States v. Franklin Steel Products, Inc.*, 482 F.2d 400 (9th Cir.1973) (consequential damages recoverable); *Oliver B. Cannon & Son, Inc. v. Dorr-Oliver, Inc.*, 336 A.2d 211 (Del.1975). But see *Leggette v. Pittman*, 268 N.C. 292, 150 S.E.2d 420 (1966). This holding was, though, limited in *Salem Towne Apartments, Inc. v. McDaniel & Sons Roofing Co.*, 330 F.Supp. 906 (E.D.N.C.1970).

47. *Sheldon v. Ramey Builders, Inc.*, 156 Ga.App. 670, 275 S.E.2d 743 (1980) (warranty as structural flaws included flooding caused by failure to waterproof).

48. Refer to note 39 supra.

49. But see *Teufel v. Wienir*, supra note 2, which refused to hold the contractor when it used material specified. Similarly, *Wood-Hopkins Contracting Co. v. Masonry Contractors Inc.*, 235 So.2d 548 (Fla.App.1970).

fects usually known as warranty problems in Art. 13. This lengthy and complex article deals with the process by which defects are discovered and corrected *during* construction and *after* completion. Section 13.2.2 provides that upon receipt of written notification the contractor will correct defective work or work not in accordance with the contract documents.

A few things are clear. AIA in ¶¶ 13.2.2 and 13.2.7 makes clear that the one-year period is *not* a period of limitation. Other issues that have been discussed in this subsection are not so clearly resolved.

Does the contractor have "first crack" at defective work that is discovered within the first year? The clause does state that the contractor shall correct it promptly after receipt of written notice and that the owner shall give notice promptly after discovery of the condition. Also, ¶ 13.2.4 gives the owner the right to correct *if* the contractor does not.

Clearly the contractor cannot be expected to correct the work unless it is notified that a defect has been discovered. It is not so obvious that the requirement of a notice indicates that the clause gives the contractor an absolute right to make the correction.

If this clause is not intended for the contractor's benefit, the owner is not obligated to give the contractor the first chance to make the corrections.[50] Applying the obligation of good faith and fair dealing, it would not be fair for the owner to have the absolute right to determine whether to give the contractor the first chance to correct the work. It can refuse *only* if the owner has had good reason to lose faith in the contractor, such as prior assurances by the contractor having proved to be unreliable. However, it would be equally unfair to give

the contractor the right to do so under *all* circumstances. Where the language is not clear, good faith and fair dealing are the best solution.

The AIA approach highlights the possibility that a defect can be discovered during performance, within the one-year period, or after expiration of the warranty period but before the claim would be barred by law. Article 13 and the sections to which it refers dealing with remedy contemplate the discovery of defective work *during* performance. As a result, some of the mechanisms, such as those that involve the architect, would not be applicable for defects discovered during the one-year period. The architect is no longer "on board," and any role she might have can no longer be fulfilled during the one-year period.[51]

What if the defect is discovered *after* the expiration of warranty period? Must the owner notify the contractor? Notification provisions should apply *only* to defects discovered within one year. Yet ¶ 13.2.2 covers *either* defects discovered during the year period or those not barred by the statute of limitations. One court troubled by this problem pointed to ¶ 7.4.1, which requires either party who suffers injury or damage to make a claim in writing to the other party within a reasonable time.[52] This was very likely intended to apply only to harm suffered during performance and not to place a notice requirement as a condition precedent for *any* claim.

Does the warranty clause control the remedy? It is true that ¶ 13.2.4 allows the owner to *correct* the work if the contractor fails to do so, apparently whether failure occurs during the contractor's performance and after the project has been turned over to the owner, whether within the year period or not. Yet it is unlikely that this was

---

50. *Baker Pool Co. v. Bennett,* 411 S.W.2d 335 (Ky.1967). But note *St. Andrew's Episcopal Day School v. Walsh Plumbing Co.,* reproduced in Section 28.08(B).

51. AIA Doc. B141, ¶ 1.5.1, A201, ¶ 2.2.2.

52. *First National Bank of Akron v. Cann,* 669 F.2d 415 (6th Cir.1982).

intended to bar any diminished value measure of damages or limit the owner to correction cost, barring other losses resulting from the defect.

AIA takes the position that the purpose of the warranty clause is *remedial,* to give the owner a right to specific performance (judicial order to the contractor to correct the defect).[53] It supports this by pointing to its language in ¶ 13.2.1 stating that the contractor has a "specific" obligation to correct the work. It is not likely that a court would accord the language any control over the remedy, the money award, or the specific performance. (Note that the clause is *not* limited to one year.)

What of the exhortation function suggested in (B)? The power to demand the contractor return and correct under ¶ 13.2.2 applies to *any* defect discovered that is not barred by the statute of limitations, not simply the one-year period. This would seem to negate any "exhortation" function. Yet it is more likely that the AIA's inclusion in ¶ 13.2.2 of *all* defects resulted from its desire to be certain that the clause did not create a private period of limitations— something it had made clear in ¶ 13.2.7. Very likely this was "overdrafting." If taken seriously, the clause is left without any legitimate function. Perhaps all that is left is putting the burden on the warrantor contractor to establish exonerating causes such as unusual weather or owner misuse. But should that apply beyond the one-year period?

SECTION 28.09
IMPLIED WARRANTIES
IN THE SALE OF HOMES

(A) HOME BUYERS AND
THEIR LEGAL PROBLEMS

Project success in the preceding section assumes construction documents that set forth the contractor's requirements and are the basis for judging the design professional's performance. Yet persons often buy partially or totally completed homes without the protection of plans that must be met before final payment is made. They purchase homes from developers or builder-vendors, each of whom builds homes for immediate sale to home buyers. Suppose after the purchase the basement leaks, the walls crack, or the heating system malfunctions. What recourse does the buyer have?

Before the mid-1960s, home buyers who faced such problems had little recourse against the sellers, or for that matter against anyone. Part of this was traceable to legal rules designed to avoid uncertainties in real property transfers and to transfer certain risks by ownership change. In addition, the common law expected buyers to protect themselves, a concept expressed in the maxim "caveat emptor" or "let the buyer beware." Even if these barriers were surmounted, the homeowner discovering these defects could only, as a rule, pursue the party who had sold her the house. Recovery could not be obtained from those more responsible, such as the builder, the designer, or the developer who had not sold the house to the person discovering the defect. In a fast turnover market, such requirements of privity often left the discoverer of the defect with a claim only against the person who had sold the owner the house.

(B) THE IMPLIED WARRANTY
EXPLOSION OF THE 1960s

Dissatisfaction with the traditional denial of protection meshed with the general consumerism movement that began in the 1960s. Consumers of goods began to demand products that worked, and home buyers demanded similar protection. In response to this demand, courts recognized an

---

53. AIA Handbook of Architectural Practice, D–3 at 18 (1981). But *Gerety v. Poitras,* 126 Vt. 153, 224 A.2d 919 (1966) refused to order specific performance of a guaranty clause.

implied warranty of habitability in the sale of new homes.[54]

(C) CURRENT PROBLEMS

The breakthrough consisted of overthrowing the old rules that had denied any form of recovery. With the exception of a few jurisdictions,[55] most courts that have recently faced the question have done this. But the courts and legislatures must still work out details. Courts have articulated similar but somewhat variant formulas for describing the nature of the implied warranty.[56]

The exact nature of the warranty will depend upon the price of the house, the customary standards of the community, and the reasonable expectations of the buyer. Perhaps it would be useful to examine a few cases to see what specific facts have constituted breach of the implied warranty.

A number of cases have involved sewerage, water, and moisture damage. One involved an overflowing septic tank,[57] another, a leaky basement,[58] still another, water and fill seepage,[59] and yet another basement seepage.[60] One unfortunate home buyer found water seepage in the basement, surface waters flowing in from the outside, and a septic tank that backed up.[61]

Yet not all water disappointments fall within the implied warranty. For example, Connecticut refused to imply a warranty in a contract to sell a house and dig a well that the water supplied would be potable, quoting the trial judge as having stated that only the Lord can guarantee the quantity or quality of water from a well.[62]

Other defects that have justified a claim that the implied warranty has been breached were structural defects, such as a collapsing stairway[63] or cracked walls.[64] Lead-based paint violated the implied warranty of habitability.[65] Finally, claims have been based upon a defective heating system,[66] a defective air-conditioning system,[67] and a malfunctioning fireplace.[68]

The preceding list of defects seems to require that the developer or vendor-builder furnish basic shelter from the elements and reasonable comfort. Obviously, whether these general requirements have been met will depend upon the contract of purchase

---

54. One of the most interesting of the many opinions is *Humber v. Morton,* 426 S.W.2d 554 (Tex.1968).

55. *Druid Homes, Inc. v. Cooper,* 272 Ala. 415, 131 So.2d 884 (1961), held that there was no implied warranty for improvements on real property in Alabama.

56. *Hartley v. Ballou,* 20 N.C.App. 493, 201 S.E.2d 712 (1974) (suitable for habitation); *Padula v. Deb-Cin Homes, Inc.,* 111 R.I. 29, 298 A.2d 529 (1973) (fit for human habitation); *Gable v. Silver,* 258 So.2d 11 (Fla.App.1972) (fit and merchantable); *Crawley v. Terhune,* 437 S.W.2d 743 (Ky.1969) (major structural features constructed in a workmanlike manner with suitable materials) Conn.Gen.Stat.Ann. § 47–121 (built in accordance with building codes).

57. *Rutledge v. Dodenhoff,* 254 S.C. 407, 175 S.E.2d 792 (1970).

58. *Crawley v. Terhune,* supra note 56.

59. *Wawak v. Stewart,* 247 Ark. 1093, 449 S.W.2d 922 (1970).

60. *Elmore v. Blume,* 31 Ill.App.3d 643, 334 N.E.2d 431 (1975).

61. *Norton v. Burleaud,* 115 N.H. 435, 342 A.2d 629 (1975).

62. *Bertozzi v. McCarthy,* 164 Conn. 463, 323 A.2d 553 (1973).

63. *Rogers v. Scyphers,* 251 S.C. 128, 161 S.E.2d 81 (1968).

64. *Oliver v. City Builders, Inc.,* 303 So.2d 466 (Miss.1974).

65. *City of Philadelphia v. Page,* 363 F.Supp. 148 (E.D.Pa.1973).

66. *Kriegler v. Eichler Homes, Inc.,* 269 Cal.App.2d 224, 74 Cal.Rptr. 749 (1969).

67. *Gable v. Silver,* supra note 56.

68. *Humber v. Morton,* supra note 54.

and the circumstances surrounding the particular transaction.

For whose protection does the implied warranty doctrine exist, and who is subject to its requirements? Clearly a buyer can sue the seller. Beyond that the law varies in the different jurisdictions. Most courts allow a subsequent purchaser to maintain an action [69] while some do not.[70] One case held that the infant son of a tenant could bring an action against the developer.[71]

As to other potential defendants, successful actions have been brought against a lender who took more than a normal lender's role in creating the development [72] and an engineer who conducted soil studies.[73]

What types of property carry with it implied warranty protection? It has been held to include, at one extreme, a condominium,[74] and at the other, a sale of a lot.[75] Courts seem hesitant to imply a warranty in the sale of an existing home not built for immediate sale.[76] Here it is more difficult to justify warranty implication as part of the normal enterprise risk of developers or builders-vendors. The caveat emptor doctrine, with the modern qualification that sellers disclose serious defects of which they know and which the buyer would not likely be able to discover, seems proper. In this regard, the California Supreme Court, in holding builders and vendors to an implied warranty that the completed structure was designed and constructed in a reasonably workmanlike manner, stated:

> In the setting of the marketplace, the builder or seller of new construction—not unlike the manufacturer or merchandiser of personalty—makes implied representations, ordinarily indispensable to the sale, that the builder has used reasonable skill and judgment in constructing the building. On the other hand, the purchaser does not usually possess the knowledge of the builder and is unable to fully examine a completed house and its components without disturbing the finished product. Further, unlike the purchaser of an older building, he has no opportunity to observe how the building has withstood the passage of time. Thus he generally relies on those in a position to know the quality of the work to be sold, and his reliance is surely evident to the construction industry.[77]

Most plaintiffs seek damages based upon either the cost of correction of the defect or the diminished value of the property.[78] A host of structural defects and incurable water problems should enable the buyer to call the deal off and receive any money paid less any benefit received by occupying the premises. Like cars, houses can be lemons.

Can the seller disclaim the implied warranty by a contract clause? Usually express contract language takes precedence over

---

69. *Kriegler v. Eichler Homes, Inc.* supra note 66; *McMillan v. Brune-Harpenau-Turbeck Builders*, 8 Ohio St.3d 3, 455 N.E.2d 1276 (1983) (negligence needed); *Keyes v. Guy Bailey Homes*, 439 So.2d 670 (Miss.1983).

70. *Barnes v. MacBrown & Co.*, 264 Ind. 227, 342 N.E.2d 619 (1976). See Note, *Gupta v. Ritter Homes, Inc.; Extending the Implied Warranty of Habitability to Subsequent Purchasers*, 35 Baylor L.Rev. 670 (1983).

71. *Schipper v. Levitt & Sons, Inc.*, 44 N.J. 70, 207 A.3d 314 (1965). See Comment, *Strict Tort Liability to the Builder-Vendor of Homes: Schipper and Beyond?* 10 Ohio No.U.L.Rev. 103 (1983).

72. *Connor v. Great Western Savings & Loan Association*, 69 Cal.2d 850, 73 Cal.Rptr. 369, 447 P.2d 609 (1968). See Ferguson, *Lender's Liability for Construction Defects*, 11 Real Est.L.J. 310 (1983).

73. *Avner v. Longridge Estates*, 272 Cal.App.2d 607, 77 Cal.Rptr. 633 (1969).

74. *Gable v. Silver*, supra note 56.

75. *Avner v. Longridge Estates*, supra note 73.

76. However, a reconditioner seller, the FHA, was held in an implied warranty case in *City of Philadelphia v. Page*, supra note 65.

77. *Pollard v. Saxe & Yolles Development Co.*, 12 Cal.3d 374, 115 Cal.Rptr. 648, 651, 525 P.2d 88, 91 (1974).

78. These measures of recovery are discussed in more detail in Section 31.03.

implied warranties. It is likely that exculpatory clauses will not be effective where the plaintiff has suffered personal harm. Exculpation may be effective where there is damage to property or other economic loss if the exculpatory language was brought to the buyer's attention in such a way as to *clearly* indicate that the buyer assumed this risk.[79] Marketing methods of most developers and vendor-builders are not likely to employ this open approach.

(D) INSURANCE PROTECTION

The preceding discussion may have overemphasized the value of the implied warranty to buyers. Although the 1960s did see an explosion of protection for home buyers, the law in any given jurisdiction may still be unclear, especially as to the particular nature of the warranty. Often the defects complained of are not large, and instituting individual legal action to enforce the warranty may cost more than what can be recovered in court. This is a reason for increased use of the class action in which a number of buyers "similarly situated" join together in one lawsuit. In any event, there are often long delays in litigating these cases.

Developers and vendor builders are engaged in a hazardous business and often are either out of business or unable to pay for damages by the time a claim is brought. As a result, private industry has been offering homeowner's warranties to cover the types of defects that have been discussed in this section.

(E) DECEPTIVE PRACTICES STATUTES

Many states have enacted legislation designed to protect consumers from deceptive trade practices by those who sell consumers goods or services. Like much consumer legislation, such consumer protection laws usually provide harsh remedies that may not only give the consumer its attorney's fees but also provide for punitive or treble damages. For example, the Texas Deceptive Trade Practices Act has been applied in construction defect cases.[80]

SECTION 28.10 A SUGGESTION: DEFECT RESPONSE AGREEMENTS

This section is directed to owners who are about to engage in projects of a substantial nature, particularly those with the likelihood of defects developing. It is an attempt to outline a skeletal proposal for dealing with defects.

Dealing with defects involves three stages. Stage I, the *diagnostic* stage, involves correcting the defect. Stage II is the *preparation* to resolve the allocation of legal responsibility for the losses caused by the defect. Stage III, the actual trial, involves *judicial* resolution of legal responsibility.

Although Stage III is most dramatic (and costly), Stages I and II are the most crucial. The defect must be fixed for the owner to receive the project for which it has paid and to avoid large losses. Stage II, the gathering of information that can enable the parties to try to voluntarily settle a dispute, involves a large expenditure of time and money before the parties are in the position to settle the dispute. Stage III is rarely reached. Regardless of how expensive Stages I and II are, the expenses of Stage III with a trial of from one to six months add immense direct and indirect cost and often generate unsatisfactory outcomes. As a result, most disputes and large projects never go beyond Stage II, with the dispute being settled.

Before the parties are in the position to seriously discuss settlement, they must have

---

79. See Anderson, *Disclaiming the Implied Warranties of Habitability and Good Workmanship in the Sale of New Homes,* 15 Tex.Tech.L.Rev. 517 (1984).

80. *Norwood Builders, Inc. v. Toler,* 609 S.W.2d 861 (Tex.Civ.App.1980).

some idea of what caused the dispute, who is legally responsible, and how the losses should be apportioned. To prepare for serious negotiations, much expense must be incurred, such as expenses of lawyers, experts, testers, and the often ignored indirect expense of officers and key employees of the major participants having to spend time that does not earn their employers any revenue. In disputes of this nature, the stakes are sufficiently high so that no participant feels that it can "cut costs."

When all the major participants in the dispute—and there are many—have sufficient information or feel under sufficient pressure to discuss negotiations seriously, it is likely that there is enough blame to spread around, enough technical uncertainty and disagreements among experts, and uncertainty as to the legal outcome. As a result, the major participants—the owner, contractors, design professionals, major suppliers, manufacturers, sureties, insurers, and funding agencies, to name the most visible—each decides it will pitch some money into a settlement "pot."

How do these participants determine how much they are willing to pitch into the pot? It will very likely result from hard bargaining with all the variable factors that make negotiation an art, such as willingness or reluctance to continue the fight, financial strength and weakness, appraisal of the strength and weakness of each participant's position, and a prediction of how the matter will be resolved, whether decided by arbitration or litigation.

All of this takes place after a large expenditure of time and money needed to amass the information that each party feels it needs before it can enter into meaningful negotiations. Is there any way that this can be avoided? One method is to simulate in a rough way the type of negotiations that result at Stage II but to do it *before* the project begins. This can be accomplished by a Defect Response Agreement (DRA) made between *all* the major participants. To accomplish this, the major participants enter into the negotiation as they "sign on" the project, all under the direction and supervision of the owner. The DRA deals solely with the response to defects, how to correct the defects, and who is responsible for the "immediate" response as well as with the formula for determining the ultimate loss distribution and possibly the security to back up any such formula.

Since all the major parties are signatories, there would be none of the increasingly complexed third-party problems. The formula itself would be determined by negotiation, influenced very likely by the same matters that affect any negotiation—a calculation of risk and profits. Fear of the open-ended consequential damages would lead to each participant's giving up claims it would have for such damages.

A DRA would have to be imposed by a strong private owner [81] that recognizes that *it* bears the large burden of the wasteful cost incurred in the present system. It can "persuade" other participants that they will benefit in the long run by a system that will avoid the worst aspect of the current dispute resolution process. [82]

PROBLEM

The City Museum engaged the services of an architect to prepare plans and specifications for the construction of a one-story, brick and block warehouse building for the storage of paintings and other works of art. The owner furnished the

---

81. Public owners would be too fearful of accountability requirements to try such a no-fault approach.

82. Another suggestion is to use a property damage policy that would cover defects, however caused, with *all* participants waiving subrogation claims and consequential damages. Defects would then be an insurance problem.

architect with a topographic survey and test boring reports previously prepared by a soils engineer employed by the owner. The parties then executed AIA Document B 141, "STANDARD FORMS OF AGREEMENT BETWEEN OWNER AND ARCHITECT."

The architect delivered the plans and specifications to the owner with an estimated cost of one million dollars. Upon approval by the owner, Invitations to Bid were advertised, and the lowest acceptable bid submitted by a prime contractor resulted in the execution of AIA Document A 201, "GENERAL CONDITIONS OF THE CONTRACT FOR CONSTRUCTION." See Appendix C.

The prime contractor engaged the services of several subcontractors for the various specialties of the project. Contractual arrangements were created by the use of AIA Document A 401, "STANDARD FORM OF AGREEMENT BETWEEN CONTRACTOR AND SUBCONTRACTOR." See Appendix E.

After construction began, the prime contractor notified the architect that there was disturbed earth where the concrete foundation footings were to be placed at the northeast corner of the lot. Rather than undergoing the expense of a change in the foundation as originally designed, the architect revised her plans to provide for reinforced poured concrete grade beams and additional steel at the point where the earth was disturbed. The boring report provided by the owner did not show any water at this location.

The architect further revised the plans to provide for a specific brand of waterproofing materials to be applied from the bottom of the grade beam to a point well above the grade.

The waterproofing material was a new product and had not previously been used by the architect or contractor, but the manufacturer was a reputable company that had a long history of producing acceptable building material and had tested the product and recommended its use to the architect. The architect did not make any independent test of the material before specifying its use after the job condition developed.

During one of the periodic visits with her consulting engineer, the architect determined that the work was not being performed according to the plans as revised with regard to the installation of steel at the northeast corner by the subcontractor who had been employed by the contractor to perform this work. The contractor and the subcontractor complained that the objection was frivolous and that the work was being performed according to the revised plans and specifications.

Following an exchange of recriminations between the parties, the architect rejected the work and then directed the contractor to remove a portion of the beam already in place and to use more and a different type of steel in accordance with the plans. The contractor finally agreed to these demands, stating that it did so under protest, and instructed the subcontractor to make the changes.

Upon completion of the building, the architect issued a certificate of final payment to the contractor and the contractor was paid in full.

A month later the Museum found it necessary to move a number of valuable paintings into the warehouse and placed them on the floor at the northeast corner of the building. The following night a torrential downpour caused an accumulation of water sufficiently deep to destroy and damage a number of valuable paintings.

Subsequent investigation revealed that the beam in the northeast corner of the warehouse had cracked and had settled to a point where a gap was exposed between the beam and the concrete floor. It was further found that the subcontractor had failed to install the additional concrete beams and steel as demanded by the architect and instructed by the contractor. The investigation also revealed that the waterproofing material had been torn as a result of the settlement and the flooding, but testing showed that the waterproofing material may not have repelled all of the moisture in any event.

QUESTIONS

1. Does the owner have a claim against the architect for the water damage to the building based on design error? (Review Chapter 17.)

2. Is the architect liable for selecting defective waterproofing material? (Review Chapter 17.)

3. Would the architect be liable had she not specified the type of waterproofing material and it was selected by the contractor?

4. Is the architect liable for failure to "monitor" the construction work? (Review Chapter 17.)

5. Is the contractor liable to the owner for the subcontractor's failure to comply with the architect's instructions? (Consult Chapter 32.)

6. May the contractor defend against the owner's claim for damages on the grounds that the architect's plans were defective? (Review Chapter 27.)

7. Does the owner have a claim against the architect based on the issuance of the final certificate of payment? (Review Chapters 17 and 26.)

8. Does the owner's claim for damages extend to the damaged and destroyed paintings resulting from the flooding? (Review Chapter 6.)

9. How should the owner's claim for damages be apportioned between architect and contractor?

10. Had the contractor not made the discovery of the soil condition, would it have been liable for damages resulting from defective plans? (Consult Chapter 29.)

29

Subsurface Problems: Predictable Uncertainty

SECTION 29.01 DISCOVERY OF UNFORESEEN CONDITIONS

(A) EFFECT ON PERFORMANCE

For reasons to be explored in (B), the discovery of unforeseen subsurface conditions is not unusual in the Construction Process. When such conditions are discovered, they usually have an adverse effect on the contractor's planned performance and its prediction of performance costs. Sometimes subsurface materials encountered are more difficult to excavate or extract, with many cases involving the discovery of hardpan or more rock than expected. When this occurs, performance is likely to take more time and cost more.

Sometimes the subsurface conditions encountered generate a great increase or decrease in the quantities to be excavated. This also can affect the time and costs, particularly if the contractor has bid a composite unit price that may be adversely affected by finding that certain work overruns and other work underruns. (This was discussed in Section 21.02(C) dealing with unit pricing.) Material (borrow pits) that the contractor expects to use for fill or compaction may turn out to be unsuitable. If so, the contractor must obtain material from a site more distant than planned or more costly to extract. When these condi-

tions are discovered, the contractor may request more time and more money.

(B) CAUSES

Why are unexpected subsurface conditions frequently encountered? Soil testing is expensive, and only a limited number of borings are usually made. As a result, the data reported may not reflect subsurface conditions throughout the entire area. Subsurface conditions even within small areas may vary greatly.

Geotechnical engineers may be exposed to great liability if the information they report is incorrect or their suggestions do not accomplish the owner's expectations. Although this chapter concentrates mainly upon subsurface conditions encountered *during* construction, problems such as structural instability, settling, or cracking can develop later. They are often traceable to the subsurface, often leading to claims against geotechnical engineers. As a result, geotechnical engineers may not perform services where liability exposure is greatly in excess of anticipated profit, or they may sharply increase their fees. This can eliminate testing in smaller projects or make less extensive testing more likely.

Information, however reliable, must be gathered and used by designers and contractors. When what is encountered varies from what is anticipated, the risk will have to be borne by someone. This chapter examines contractor claims for additional compensation when contractors' costs are more than expected because of unforeseen subsurface conditions.

(C) TWO MODELS

Again it is important to recognize the different methods by which construction is accomplished. First consider construction done by a contractor who is given a site and asked to both design and build a particular project. *Stees v. Leonard*,[1] cited many times for the proposition that the contractor bears all the risks of subsurface conditions, involved a contract under which the contractor both designed and built a house for the owner. In such a case, it was easy to place the risk of success upon the contractor.[2]

However, the traditional American Construction Process *divides* design and construction, with the owner engaging a design professional to design, a geotechnical engineer to gather data, and a contractor to build. This makes it more difficult to determine who will bear the risk of unforeseen subsurface conditions.

Although there are abundant factual variations, typically the owner—private or public—calls for bids on a construction project that will involve subsurface excavation. Building specifications may be furnished, but methods of excavation and construction will probably be left to the contractor.

Depending on owner identity and project size, soil tests are taken by an independent geotechnical engineer for purposes of cost estimation, design, and scheduling. The reports will be made available to the design professional and the owner. Although the information is likely to be available to the

contractor, owners take different routes to whether they will stand behind the accuracy of the information.

Whichever approach is taken, the bidder will usually be warned that it must inspect the site and, under some contracts, conduct its own soil testing. However, often the bidder will not make independent soil tests because the profit potential may be too small to justify such expenditure or because the bids are due in a relatively short period of time. As a result, contractors frequently bid without knowledge of actual conditions that will be encountered. Bids are calculated on the basis of expected conditions, and if the unexpected is encountered, the actual cost will vary widely from that anticipated. The focus of this chapter is upon who will bear this risk.

(D) ENFORCEABILITY OF A PROMISE TO PAY MORE MONEY

When unforeseen subsurface conditions are discovered, the owner may promise to pay additional compensation. The enforceability of such a promise depends on the application of the preexisting duty rule.[3] The traditional application of this rule would deny enforceability of such a promise because the owner is getting for its promise nothing more than it was entitled to get under the contract. However, special rules have developed in subsurface cases that can in some jurisdictions justify enforcement of the promise. Sometimes a promise is enforced because the contractor gave up its right to rescind the contract for a mutual mistake as consideration for the promise.[4] More commonly, the law enforces such a promise so long as "the modification is fair and equitable in view of circumstances not

---

1. 20 Minn. 494 (1874), cited in *Dravo Corp. v. Litton Systems, Inc.*, 379 F.Supp. 37 (S.D.Miss.1974).

2. Refer to Section 28.02.

3. Refer to Section 5.11(C).

4. *Healy v. Brewster*, 251 Cal.App.2d 541, 59 Cal.Rptr. 752 (1967).

anticipated by the parties when the contract was made." [5]

(E) SUPERVENING GEOTECHNICAL CONDITIONS AND MISTAKE CLAIMS

Most problems involve encountering unexpected *preexisting* subsurface conditions. Additional costs can result from a change in geological conditions because of unexpected weather (unseasonable rains or frost) or third-party interference (flooding from adjacent lands). A distinction can be drawn between the types of risks involved. The risk of a preexisting geological condition can be eliminated if the parties wish to spend enough time and money. On the other hand, the likelihood of a future occurrence, whether from third-party conduct or atmospheric conditions, can rarely be ascertained with any degree of accuracy.

Supervening changes of geological conditions are not discussed in this chapter. Usually when such events occur, the contractor seeks a time extension but not additional compensation. The differing site conditions clause used by the federal government and discussed in Section 29.05(A) generally does not apply to atmospheric difficulties or third-party interference.[6]

Claims based upon mutual mistake, though peripherally relevant to the discussion in Section 29.02 dealing with basic risks, are not emphasized in this chapter. Mutual mistake may be relevant to the enforceability of a promise to pay additional compensation. Since most subsurface disputes involve claims by the contractor for additional compensation—the topic focused upon principally in Sections 29.03 and 29.04—mistake does not play an important part in this chapter.

SECTION 29.02
COMMON LAW RULE

Unless the owner has furnished site or subsurface information upon which the contractor can rely, the risk of added costs of performance falls upon the contractor. In *United States v. Spearin* (reproduced in Section 27.05(E)), the court stated that a contracting party who agrees to perform something possible for a fixed sum will not be excused or entitled to additional compensation because unforeseen difficulties make performance more costly than anticipated.

Similarly, though not the focus of this chapter, if the completed project slides, settles, or is destroyed because of subsurface conditions, generally the contractor is responsible for redoing the work or for the diminished value of the premises unless the owner has designated the excavation method to be used or the contractor has relied upon misrepresentations made by the owner.[7]

As a general proposition, then, the contractor will bear the risk of unforeseen subsurface condition unless 1) it can establish that it has relied upon information furnished by the owner (discussed in Sections 29.03 and 29.04), 2) the contract itself provided protection (discussed in Section 29.05), 3) the owner did not disclose information that it should have disclosed (discussed in Section 22.04(B)) or 4) the cost of performance was extraordinarily higher than could have been anticipated, a fact that would justify the application of mutual mistake or impossibility as discussed in Section 27.03.

The number of exceptions to the basic rule should not convey the impression that

---

5. *Angel v. Murray*, 113 R.I. 482, 322 A.2d 630, 636 (1974), quoting what is now Restatement (Second) of Contracts § 89(a). See also *Linz v. Schuck*, 106 Md. 220, 67 A. 286 (1907).

6. *Hardeman-Monier-Hutcherson v. United States*, 198 Ct.Cl. 472, 458 F.2d 1364 (1972). It can apply if representations are made to the contractor regarding sea or climate conditions.

7. *Stees v. Leonard*, supra note 1.

the policy of placing these risks on the performing party no longer exists or does not reflect, at least to some courts, an important legal principle in fixed-price contracts. For example, in *W.H. Lyman Construction Co. v. Village of Gurnee*,[8] a contractor had been engaged to perform a sanitary sewer project. One basis for a claim against the public entity was that the sewer had to be constructed through subsurface soil that was for the most part waterbearing sand and silt rather than clay as indicated by the soil-boring log shown on the plans. A high groundwater table was also discovered. This required the contractor to install numerous dewatering wells. Responsibility of the public entity for inaccurate information if furnished is discussed in greater detail in Sections 29.03 and 29.04.

It is important to note the judicial attitude in the *Lyman* case. It dealt with the claim by the contractor that the public entity impliedly warranted that the plans and specifications would enable it to accomplish its promised performance in the manner anticipated. In rejecting the contractor's claim, the court stated:

> It is well settled that a contractor cannot claim it is entitled to additional compensation simply because the task it has undertaken turns out to be more difficult due to weather conditions, the subsidence of the soil, etc. To find otherwise would be contrary to public policy and detrimental to the public interest.[9]

The court looked upon the common law rule as expressing a principle of great importance, one needed to protect public entities and public funds. Whether the court would have felt as strongly as it did were the contract a private one is problematical. Owners and those who supply funds for the project are also greatly concerned with the ultimate cost of the project and rely heavily upon the contract price in their planning.

SECTION 29.03 INFORMATION FURNISHED BY OWNER

Owners who intend to build substantial projects often make subsurface and soil reports available to prospective contractors. One exception to the general allocation of unexpected costs to the contractor relates to the owner furnishing information that is relied upon by the contractor. The owner may make this information available but state that it is *not* responsible for its accuracy. This section emphasizes the effect of providing this information with the full realization that the contractor is likely to rely upon it and without any technique to shield the owner from responsibility.

At the outset, differentiation must be made between the types of information that may be made available to the contractor. Usually reports of any tests that have been taken are among the information made available. The reports may also contain opinions or inferences that the geotechnical engineer may have drawn from observation and tests taken. The information may include estimates as to the type and amount of material to be excavated or needed for fill or compaction. (This information may also be included in the specifications.) Though less common, the report can recommend particular subsurface operational techniques.

Misrepresentation is the basic theory upon which the contractor bases its claim. A threshold question involves what constitutes a misrepresentation. Differentiation must be made between facts and opinions. Reporting the result of tests is clearly a factual representation, while professional judgments that seek to draw inferences from this information may be simply opinions.

A misrepresentation claim may be based

---

8. 84 Ill.App.3d 28, 38 Ill.Dec. 721, 403 N.E.2d 1325 (1980).

9. 403 N.E.2d at 1328.

upon improperly selected test sites, the inference being that a contractor may believe that the test sites selected will generally represent the site. Misrepresentation can consist of a combination of providing some information but not disclosing all the information that qualifies the information given. Half-truths can be just as misleading as complete falsehoods. This is distinguished from simply failing to disclose *any* information or information that the owner knows would be valuable to the contractor and that the contractor is not likely to be able to discover for itself.

Representations may be fraudulent, that is, made with the intention of deceiving the contractor. Fraud claims, though having a heavy burden of proof, are most valuable to a contractor. The contractor's negligence in failing to check the data generally does not bar recovery.[10] Recovery would be barred only if the contractor relied upon information *it* gathered. Fraud gives the contractor a variety of remedies. It can rescind the contract and refuse to perform further, raise fraud as a defense if sued for nonperformance, or—most important—complete the contract and recover additional compensation in an action for damages.

More commonly, though, misrepresentations are not made with the intention of deceiving the contractor. If the misrepresentations were made negligently, some added complexities develop. Very likely the negligence is that of the geotechnical engineer, and owner recovery against the geotechnical engineer may not always be available.[11] Clearly if the negligence was that of the owner, the contractor can re-cover any losses it suffers that could not have been reasonably avoided.

What is the responsibility of the owner for any *negligent* representations in the soil information generated by the geotechnical engineer? If the geotechnical engineer is an independent contractor, which is likely to be the case, the owner would not be chargeable with the latter's negligence. However, it is very likely that the owner will be chargeable either through one of the exceptions to the independent contractor rule,[12] for breach of contract (that the breach is caused by a person whom it engages to perform services does not relieve it of responsibility),[13] innocent misrepresentation, or implied warranty of the accuracy of the information by the owner that is relied upon by the contractor.[14]

The least culpable conduct is that of innocent misrepresentation. In such cases, there is neither intention to deceive nor negligence. Generally, innocent misrepresentation allows the contracting party misled to rescind the contract. Some states allow a restitutionary damage remedy—the difference between what was exchanged [15]—an ineffective remedy in subsurface cases.

Actual rescission in these subsurface cases is rare. Many legal and factual issues may make it difficult to predict whether the right to rescind is available. Walking off the job under these conditions exposes the contractor to liability. Even if its rescission were determined to be justified, it can recover only the reasonable value of its services. This should, at least in theory, put the contractor in the position it was in at the time it made the contract.

---

10. *Seeger v. Odell*, 18 Cal.2d 409, 115 P.2d 977 (1941).

11. Refer to Section 15.07 and see Section 29.04 text at note 23.

12. See Section 35.05(C).

13. *Harold A. Newman Co. v. Nero*, 31 Cal.App.3d 490, 107 Cal.Rptr. 464 (1973).

14. Restatement (Second) of Torts § 552 (1977). The remedy is the difference in value between what was received and what was paid out, a "restitutionlike" remedy.

15. Id.

It may be difficult to recover lost overhead and profit using this restitutionary remedy.[16] More likely the contractor will continue performance and later claim additional compensation. Since innocent misrepresentation, though, is not a contract *breach*, contractors will shift to implied warranty to justify their claims for additional compensation. The principal problem with the implied warranty theory is the owner's attempt to exonerate itself by the use of contract language negating the necessary reliance.[17]

SECTION 29.04
DISCLAIMER: PUTTING
RISK ON CONTRACTOR

(A) ILLUSTRATIONS: *WIECHMANN ENGINEERS v. STATE DEPARTMENT OF PUBLIC WORKS* AND *STENERSON v. CITY OF KALISPELL*

Owners use a variety of techniques to relieve themselves of responsibility for the accuracy of subsurface information they have obtained and made available. Although emphasis is upon making the information available and then seeking to disclaim responsibility for it, attention should preliminarily be directed to the possibility that the owner will choose simply not to make this information available. Owners may seek to do this if they believe that no sure technique exists that will relieve them of the responsibility for the information's accuracy. Owners are increasingly expected to disclose information that would be valuable to the contractor if it is likely that the contractor would not be able to discover this information for itself. Yet an opinion of the Pennsylvania Commonwealth Court [18] held that the owner's failure

to include information on subsurface conditions was not constructive fraud. Even more, an opinion of the Post Office Board of Contract Appeals held that the public entity was not required to make a subsurface investigation if it did not wish to do so.[19]

In these cases, other factors made it difficult for the contractor to recover. For example, in the first case, the court noted that the contractor was experienced and had access to public documents describing the mine water levels in the project area. In the second case, the court noted that the contractor was aware of the existence of subsurface rock, that before bidding it anticipated that some rock would be encountered in excavation, and that the amount of rock actually encountered was not unusual for the area. These factors might have been sufficient to enforce any disclaimer were the information given and responsibility for it disclaimed.

More commonly, the owner needs the information, commissions it, and makes it available to the contractor. A number of techniques are employed by some owners to avoid responsibility for the accuracy of the information. Sometimes owners place the responsibility on the contractor to check the site and make its own tests. As seen in *United States v. Spearin* (reproduced in Section 27.05(E)), these generalized disclaimers are not always successful.

Some owners use a different approach. They do not include subsurface data in the information given bidders but state where such information can be inspected by the contractor. An illustration will be seen in the *Wiechmann* case reproduced in this section.

Another approach is to give the data but state that it is information and is not intend-

---

16. See Section 31.02.

17. See Section 29.04.

18. *Tri-County Excavating, Inc. v. Borough of Kingston,* 46 Pa.Commw. 315, 407 A.2d 462 (1979).

19. *Wyman Construction, Inc.,* PSBCA No. 611, 80–1 BCA ¶ 14,215.

ed to be part of the contract. This is another way of stating that the contractor cannot rely on the accuracy of the information, with the hope being that the owner is shielded from any claim based upon misrepresentation or warranty.[20] Attempts to use these techniques generate varied outcomes, a reflection of the difficulties courts have in deciding whether a party can use the contract to shift a risk to the other party. Two cases reflecting different approaches to this problem are reproduced here.

WIECHMANN ENGINEERS v. STATE DEPARTMENT OF PUBLIC WORKS

California District Court of Appeal, 1973.
31 Cal.App.3d 741, 107 Cal. Rptr. 529.

CARTER, Associate Justice. Plaintiff appeals from a judgment in favor of the State of California following a court trial. This action arose out of a contract between plaintiff corporation and the State of California for the clearing, excavation and grading of approximately 10 miles of road in Modoc County. Plaintiff claims it is not liable for liquidated damages,[21] and that it is entitled to recover additional money for work performed and damages for breach of contract by reason of defendant's fraudulent concealment.

We find the appeal to be without merit and will accordingly affirm the judgment of the trial court.

On February 14, 1964, defendant State of California published a notice to contractors calling for bids for road construction work in Modoc County. The work to be performed generally involved the grading and surfacing with aggregate base of approximately nine and a half miles of county road. The notice provided sealed bids would be received by the state until 2:00 p.m. on March 18, 1964. Plaintiff received a bid package from the state which incorporated by reference Standard Specifications—January 1960, a thick document containing contractual provisions applicable to all projects bid for the Division of Highways. The bid package also included Special Provisions, detailed diagrams and materials information. The road, except for the most northerly end, was to follow an existing unimproved road generally referred to as the Lookout-Hackamore Road. After obtaining the bid information, plaintiff's vice-president and general manager, James Barkley, visited the proposed job site on two successive weekends prior to the submission of plaintiff's bid. During these pre-bid inspections Barkley admitted he traversed the roadway, it was clear of snow, and he noticed boulders in and along the road construction site.

Thereafter, Barkley telephoned the highway district office in Redding and inquired as to the nature of a borrow pit located about 10 miles from the job site. Barkley was a graduate mining engineer with considerable road construction experience. He conceded that since the project necessarily involved cuts and fills, he assumed that subsoil tests had been made in order to design the

---

20. *City of Columbia v. Paul N. Howard Co.*, 707 F.2d 338 (8th Cir.1983). This case is discussed in Section 29.05.

21. [Ed. note: Liquidated damages are discussed in Section 30.09.]

road project. Thereafter, without making any inquiry as to either the existence, nature or contents of such possible subsurface tests, Barkley proceeded to prepare the bid. An itemized bid was required with such items as clearing and grubbing, developing a water supply and finishing the roadway calling for lump sum amounts. Other items such as cubic yards of roadway excavation, aggregate subbase and various other items called for a unit price with a stated estimate of quantity. Plaintiff's bid was admittedly unbalanced in the sense that certain items were deliberately bid below cost. Prior to the bid submission, Barkley conferred with Mr. Wiechmann, president of plaintiff corporation, as a result of which a $15,000 reduction of Barkley's calculations was made to insure that plaintiff would be the low bidder. Plaintiff was awarded the contract with a low bid of $333,156. The next lowest bids were $361,841 and $374,800.25, respectively. The award date was April 14, 1964, with an original completion date of September 2, 1964, an allowance of 100 working days from the authorized date of commencement to completion.

By November 6, 1964, plaintiff had been given an extension of 24 working days, and on this date was granted a winter suspension of performance until May 17, 1965. Plaintiff overran the allowed contract time by 157 days, and pursuant to section 4 of the contract's Special Provisions, was assessed liquidated damages at the rate of $175 per calendar day in the total sum of $27,475, which was withheld from final payment by the state.

Subsequent to completion of the job, plaintiff filed a claim for $435,298.63 for additional costs due to alleged latent conditions, namely, subsurface boulders uncovered along the length of the project, and for release of the retained assessment for liquidated damages. After a hearing by the board of review, the claims were denied by the state engineer and thereafter rejected by the State Board of Control. On August 3, 1967, plaintiff instituted the action now before this court.

In 1962 and 1963, prior to the call for bids, the state caused to be drilled 18 test holes along the roadway. The reports disclosed the presence of subsurface boulders at various stations. Thirteen of the 18 tests contained such comments as, "generally boulderous," "area is very boulderous," and "hole stopped on nested boulders." This information was not in the bid package given to the prospective bidders. The reports were on file in the office of the district engineer in Redding.

Section 5–105 of the Special Provisions of the contract provides in part: "The bidder shall examine carefully the site of the work contemplated, the plans and specifications, and the proposal and contract forms therefor. The submission of a bid shall be conclusive evidence that the bidder has investigated and is satisfied as to the conditions to be encountered, as to the character, quality, and quantities of work to be performed and materials to be furnished, and as to the requirements of the proposal, plans, specifications, and the contract.

"Where the Department has made investigations of subsurface conditions in areas where work is to be performed under the contract, or in other areas, some of which may constitute possible local materials sources, such investigations are made only for the purpose of study and design. Where such investigations have been made, bidders or Contractors may, upon written request, inspect the records of the Department as to such investigations subject to and upon the conditions hereinafter set forth. Such inspection of records may be made at the office of the district in which the work is situated. . . .

"The records of such investigations are not a part of the contract and are shown solely for the convenience of the bidder or Contractor. It is expressly understood and agreed that the Department assumes no responsibility whatsoever in respect to the sufficiency or accuracy of the investigations thus made, the records thereof, or of the interpretation set forth therein or made by the Department in its use thereof and there is no warranty or guaranty, either express or implied, that the conditions indicated by such investigations or records thereof are representative of those existing throughout such areas, or any part thereof, or that unlooked for developments may not occur, or that materials other than, or in proportions different from those indicated, may not be encountered.

"When a log of test borings showing a record of the data obtained by the Department's investigation of subsurface conditions is included with the contract plans, it is expressly understood and agreed that said log of test borings does not constitute a part of the contract, represents only the opinion of the Department as to the character of the materials encountered by it in its test borings, is included in the plans only for the convenience of bidders and its use is subject to all of the conditions and limitations set forth in the provisions in this section.

"In some instances, the information from such subsurface investigations considered by the Department to be of possible interest to bidders or Contractors has been compiled as 'Materials Information.' Said 'Materials Information' is not a part of the contract and is furnished solely for the convenience of bidders or Contractors. It is understood and agreed that the fact that the Department has compiled the information from such investigations as 'Materials Information' and has exhibited or furnished to the bidders or Contractors such 'Materials Information' shall not be construed as a warranty or guaranty, express or implied, as to the completeness or accuracy of such compilations and the use of such 'Materials Information' shall be subject to all of the conditions and limitations set forth in this section and Section, 'Materials', of these special provisions."

As previously noted, plaintiff's vice-president, Mr. Barkley, telephoned the district office in Redding to inquire about the nature of the borrow pit referred to in the materials information report given to prospective bidders. The borrow pit was snowed in at the time and impossible to reach. Barkley did not inquire about the results of any subsurface test that may have been made, although he assumed the state had that information when the cuts and fills were designed, and he knew further that if there were reports on the subsurface conditions, they could be inspected at the district office in Redding. Barkley agreed at the trial that it would have been reasonable to have made an inquiry as to subsurface tests in light of the fact that he could see boulders in and along the roadway area; he was aware of volcanic activity in the area; the presence of snow prevented a clear view of the area surrounding the project, and he had not made any subsurface investigation himself.

When plaintiff commenced excavation work it uncovered larger boulders than contemplated. Plaintiff's scraping equipment could not readily move boulders of that size, and it was necessary for the plaintiff to purchase or rent heavier equipment including a front-end loader and rock type trucks. In addition, it was Barkley's opinion that many of the boulders were too large to be used in the fill portions of the road because they would have extended above the grade level. Some of the boulders were blasted but most were hauled away and borrow material from an outside source was brought in to replace areas where the boulders could not be used as fill.

* * *

Plaintiff urges there was an affirmative duty on the part of the state to refer specifically to the existence of the subsurface tests or to attach a copy of the test results as a part of the bid package and that failure to do so was fraudulent concealment, this on the premise that since subsurface tests had been made which disclosed boulderous conditions, such information was naturally of interest to bidders. We note an onsite inspection made the same disclosure. No specific reference to the test borings was required to apprise the bidder of the reasonable probability that larger or more numerous boulders than appeared on the surface might be found.

Generally, fraudulent concealment gives rise to an action in tort; . . . .

The California Supreme Court recently discussed fraudulent concealment in Warner Constr. Corp. v. City of Los Angeles . . .: "'It is the general rule that by failing to impart its knowledge of difficulties to be encountered in a project, the owner will be liable for misrepresentation if the contractor is unable to perform according to the contract provisions.' [Citations.] As explained in Souza & McCue Constr. Co. v. Superior Court . . .: 'This rule is mainly based on the theory that the furnishing of misleading plans and specifications by the public body constitutes a breach of an implied warranty of their correctness. . . .' [Citations.]"

Here, the trial court made detailed findings (V through XXV) which essentially negated plaintiff's contentions that it was misled, subject to fraud, bad faith, gross error or arbitrariness. The lower court specifically found: "That defendant, neither in the plans, specifications, maps, or drawings furnished plaintiff, nor in any other manner, made any representation as to subsurface conditions that might be encountered at the project site or as to the nature of the material to be excavated." (Finding IX.)

"That defendant misrepresented no fact concerning subsurface conditions, either by an affirmative representation or nondisclosure, and breached no duty owed by it to plaintiff." (Finding XXIV.)

"That plaintiff's Vice President, Mr. James Barkley, made two pre-bid inspections of the project site and observed numerous partially exposed boulders in the existing roadway, and based upon his observation of these boulders, made a judgment as to their size, the means by which they could be excavated, and an estimate of the cost of such excavation." (Finding XIII.)

We have set out the trial court's findings in some detail because there is virtually no conflict in the evidence supporting them. The conclusions of law are clear. There was no partial disclosure, no half-truths which purported to be the whole truth or which were likely to be mislead. The sum total of all the facts known or to be known about the project were readily discoverable by plaintiff; they were not known and accessible only to the state. There was no active or passive concealment of facts.

This court observed in A. Teichert & Son, Inc. v. State of Cal. . . .: "If the contracting agency furnishes inaccurate project information, such as soil reports, *as a basis for bids*, it may be liable for damages on a breach of warranty theory. . . . If the agency makes geological data available under a disclaimer of responsibility, the contractor bears any loss occasioned by unexpected conditions. . . . The contracting agency's disclaimer does not protect it from liability for deliberate misrepresentation or concealment.

* * *

Plaintiff's reliance on *City of Salinas* [v. Souza & McCue Const. Co.] is misplaced. In that case the city's chief engineer in charge of the project not only knew of the highly unstable conditions existing in the subsoil along the proposed construction line; that particularly difficult conditions were likely to be encountered in an extensive slough area, but had directed an independent testing firm to make testings at preselected spacings and locations which avoided the area of greatest unsettled conditions; that the method of taking the tests was misleading; that these boring tests were sent to bidders only a few days before the opening of bids, and while it would have been proper practice to warn bidders of anticipated difficult conditions, the city officials did not do so.

As previously noted, in the case at bench there is no evidence of a deliberate or calculated attempt by the state to create false or misleading information as to subsurface conditions; secondly, no false or misleading information respecting the subsoil was placed in the hands of prospective bidders which it reasonably might be expected they would rely on or use in the preparation of their bid; third, this is not a case in which it reasonably can be contended the state had a duty to warn prospective bidders of boulderous conditions, since the hazard and risk of such a condition was readily apparent as the result of an onsite inspection.

Courts uniformly distinguish between the misleading half-truth, or partial disclosure, and the case in which defendant says nothing at all. The general rule is that silence alone is not actionable. . . .

Plaintiff forcefully urges the applicability of Warner Construction Corp. v. City of Los Angeles . . . a case involving the installation of reinforced concrete pilings in drill holes in sandy unstable soil. The City of Los Angeles had dug two test holes prior to calling for bids and furnished a log as summary of the test hole findings to all bidders. This information erroneously reported the nature of the material to be encountered by the contractor and further the city failed to inform bidders that caveins had occurred during the drilling requiring special drilling techniques. In affirming a verdict against the city, the court stated . . . : "In transactions which do not involve fiduciary or confidential relations, a cause of action for nondisclosure of material facts may arise in at least three instances: (1) the defendant makes representations but does not disclose facts which materially qualify the facts disclosed, or which render his disclosure likely to mislead; (2) the facts are known or accessible only to defendant, and defendant knows they are not known to or reasonably discoverable by the plaintiff; (3) the defendant actively conceals discovery from the plaintiff." . . .

We conclude that *Warner* is distinguishable for several reasons. Here, there was no representation of any kind as to subsurface conditions. Absent such a representation, there was no disclosed fact which was likely to mislead plaintiff.

Secondly, knowledge of the boulderous condition was not known or accessible only to the state, nor did the state have such facts as were not known or reasonably discoverable by plaintiff, if plaintiff had made what would have been admittedly a reasonable and prudent inquiry.

As previously pointed out, section 5–1.05 of the Special Provisions of the contract provided in part: "Where such investigations [of subsurface conditions in areas where work is to be performed] have been made, bidders or Contractors may, upon written request, inspect the records of the Department as to such investigations. . . . "

Nothing in this language in any way limited accessibility or precluded plaintiff from obtaining all information available if it desired to inquire. Thus, there was no concealment of the boulderous condition on the job site or the test hole surveys. The record clearly shows actual visibility of boulders in the job area, a fact readily apparent and *known* to plaintiff *before* the bid was submitted, as evidenced by Barkley's detailed testimony.

Finally, Barkley, the very person entrusted with the responsibility to investigate and prepare the contractor's bid, not only assumed the state had test information as to the road subsurface, but testified he simply decided not to inquire about the same, fully mindful of the fact that the movement of rocks and boulders necessarily would be involved in the performance of the contract. We observe that had plaintiff elected to examine the available test hole surveys, it merely would have confirmed what onsite observations disclosed; namely, that the work of construction was to be undertaken in a boulderous area and the degree and nature of the condition would be something to consider when submitting a bid. Plaintiff elected to make its decision in this regard based on its own expertise in performing the work and its own judgment that further inquiry as to subsoil conditions was not required.

A public entity is not liable for an imprudent or careless investigation on the part of a contractor. Nor can a public entity invade the contractor's judgment arena and actively assist in the evaluation process undertaken by competitive bidders for public projects. It must treat all bidders alike and with equal fairness.

We hold there was no misrepresentation of factual matters within the state's knowledge or concealment of material information.

A careless contractor cannot convert his own lack of diligence into a case of fraudulent concealment against a public entity. As said in Wunderlich v. State of California . . . : " '. . . All the information the State had concerning the soil conditions was available to claimant and claimant had been invited to make an investigation of its own. Under these circumstances the State is not chargeable for claimant's loss. . . . ' "

The State of California is not the guardian of every contractor who seeks to perform services for the public and at public expense. Such a concept would be grossly unfair to the prudent and careful contractor who is frequently underbid by a careless competitor. A contractor who submits a bid for public work which proves unprofitable because of his negligence in failing to ascertain all the facts concerning it from sources readily available, cannot thereafter throw the burden of his negligence upon the shoulders of the state by asserting that the latter was guilty of fraudulent concealment in not furnishing him with information which he made no effort to secure for himself. This is particularly true when the careless contractor is possessed of the kind of knowledge which would concededly prompt a prudent bidder to make further inquiry.

* * *

Judgment is affirmed.

STENERSON v. CITY OF KALISPELL

Supreme Court of Montana, 1981.
629 P.2d 773.

HASWELL, Chief Justice.

Plaintiffs Stenerson and Schmidt, contractors, brought this action in District Court, Flathead County, seeking to recover $28,301.31 from the City of Kalispell. The district judge entered judgment for the contractors for the total cost overrun on a rough grading job on the Buffalo Hills Golf Course. The City appeals.

In March 1975, the City of Kalispell asked for bids for rough grading and related work on an addition to the Buffalo Hills Golf Course. Respondent contractors obtained copies of the plans and specifications for the proposed work, including a booklet entitled "Rough Grading Specifications" and a map entitled "Rough Grading Plans". Based on these documents, respondents submitted a bid of $94,991.50.

On April 18, 1975, respondents entered into a contract with the City for the lump sum of $94,991.50, with a unit cost of $1.10 per cubic yard for any additional work. Plaintiffs began working shortly thereafter and completed the work in the allotted 60 days. During this period, respondents performed additional work which was negotiated separately and was paid for by the City. They also entered into a contract for additional construction, called Phase Two, which was terminated after partial completion because of lack of funds.

While engaged in Phase One, the contractors apparently advised the City on several occasions that they were moving more material than was contemplated in the contract. They did not request a change order, nor demand extra compensation before contracting for Phase Two, nor did they ever refuse to continue working.

Final payment for Phases One and Two was completed prior to January 1976. On January 7, 1976, contractors demanded by letter that the City compensate them for the extra 27,477 cubic yards of earth which they had moved which had not been computed into the bid price. The City refused to pay and the contractors filed suit.

The City moved to dismiss, and following plaintiffs' filing of an amended complaint, moved for summary judgment. The City argued that by reason of certain exculpatory clauses in the grading contract, the contractors had assumed the risk of making an erroneous bid and could not hold the City responsible. Because of the provisions in the contract documents indicating that the calculations on the "Rough Grading Plan" were "approximate," and that the contractors "shall make [their] own determination as to the amount of topsoil and grading work to be done before submitting a bid," the City asserts that the contractors can have no claim for the additional work done.

The district judge found that plaintiff contractors justifiably relied on the information provided in the specifications and plans in making their bid, and that the specifications and plans were in error in setting forth the amount of material which would have to be moved. He awarded $28,301.31 to plaintiffs to compensate for the overrun of 27,477 cubic yards moved. The City appeals, raising several issues which we frame as follows:

. . . [first issue omitted]

2. Does the evidence support the findings of the District Court?

3. Is the City entitled to judgment under the language of the contract?

After the contractors filed their amended complaint, the City moved for summary judgment. The City argued that the contract documents plainly advised all bidders to make their own determination as to the amount of material to be moved and not to rely on the figures on the plans and specifications. By entering into the contract, the bidders bound themselves to do the rough grading at the price bid and can get no additional compensation. Thus, the City argues, the plaintiffs' complaint fails to state a claim, there can be no issues of material fact before the court, and summary judgment should have been granted.

* * *

Appellant next attacks the findings of fact and conclusions of law, arguing that the evidence does not support the findings, and that the findings and conclusions do not support the judgment. In making this argument, appellant again urges this Court to ignore the long line of cases from this Court which give a contractor the right to rely on plans and specifications furnished to him in bidding on and contracting a job. The district judge here considered the evidence in light of these decisions and based his findings and conclusions on the evidence presented. We find no error.

In 1965, in the case of *Sandkay Const. Co. v. State Highway Comm'n* (1965), 145 Mont. 180, 399 P.2d 1002, this Court addressed the following issue:

"[W]here plans and estimates or specifications are used as a basis for bids, is a contractor who has been led to believe that the conditions indicated in such plans exist, able to rely on them and recover for expenses necessary by conditions being other than as represented by such plans?" *Sandkay*, supra, 145 Mont. at 184, 399 P.2d at 1005.

In *Sandkay*, we answered that question in the affirmative, finding there that the conditions actually encountered by the contractors in performing the contract could not have been reasonably anticipated, which, in effect, put the contractors into the position of having to " 'perform an entirely different contract than [sic] the one upon which they bid.' " *Sandkay,* supra, 145 Mont. at 185, 399 P.2d at 1005. See also *Hash v. Sundling and Son, Inc.* (1967), 150 Mont. 388, 436 P.2d 83.

Several other cases have presented the issue of the effect of exculpatory clauses on a contractor's right to recover in situations in which reliance is alleged by the contractor. In *Haggart Const. Co. v. State Highway Comm'n* (1967), 149 Mont. 422, 427 P.2d 686, Haggart bid on a highway construction job and was told that he could use gravel in State-optioned pits as described in the "Available Surfacing Materials Reports." The gravel was later found to be unsuitable and the plaintiff incurred additional expense in obtaining gravel elsewhere. The defendant did not deny that the materials reports were misleading, but rather relied on exculpatory provisions in the contract as a defense to the suit. Those provisions indicated that the Commission made no guaranty as to the quality and quantity of the materials available, and further that if the contractor chose to furnish his own materials, he would be responsible to produce satisfactory material.

The district judge found that the exculpatory provisions were not enforceable. He noted that Haggart received the materials reports only 14 days before bid letting, giving him little time to investigate independently. He further found that few contractors bidding on such projects have sufficient time or test facilities to make an intelligent appraisal of materials. Despite the exculpatory clauses, the district judge concluded:

> "[T]here is nothing to show that appellant expect[ed] less than complete reliance on its materials reports.

> "If the State Highway Commission were allowed to rely on the exculpatory provisions of the contract, the purpose for which such reports are offered would be sadly frustrated, if not totally destroyed. No prudent contractor would proceed in reliance on such reports at his absolute peril; the necessity to guard against unforeseen deficiencies would result in much higher bids than conditions would normally warrant." *Haggart,* supra, 149 Mont. at 425, 427 P.2d at 687–688.

We affirmed the district judge in that decision but noted:

> "We are not here holding that such exculpatory clauses may not be enforced in other situations, that detrimental reliance may be assumed in all cases, or that parties to such contracts are bound to exercise anything less than reasonable and prudent judgment. In other words we will look to 'justifiable reliance.' " *Haggart,* supra, 149 Mont. at 428, 427 P.2d at 689.

The district judge in the instant case noted factors supporting his determination of justifiable reliance, similar to those we approved in *Haggart*. We found that the defendant knew or should have known that the information on the plans and specifications would be used by prospective bidders. Finding No. 6 notes that "[a]lthough the rough grading specifications stated that such figures were approximate only, such figures did not appear to be round, ball park figures but irregular amounts that appear to be precisely calculated." Additionally, he found that the on-site inspections would not have revealed any information to the bidders that was not on the plans, and that "to have learned that the plans failed to show truthfully the amount of material that would actually have to be moved . . . the bidder would have had to have a resurvey of the whole site, at considerable expense and delay."

Additionally, the judge found that there was no information in the specifications indicating that the plans were based on an aerial map with a contour interval of five feet, which was then depicted on the plans with a contour interval of one foot. We also note that as in the *Haggart* case, the bidders had less than three weeks to submit bids, and all on-site inspection was conducted during the spring break-up.

In sum the district judge found that despite the exculpatory clauses in the contract signed by Stenerson and Schmidt, the contractors were justified in relying on the representations made by defendant in furnishing the plans and specifications. Both Schmidt and Stenerson testified that they bid this job in the usual manner, which included relying on the plans and specifications. They testified to making on-site inspections prior to bidding, and could see no discrepancy between the ground elevations as seen on the ground and the ground elevations as shown on the plans. Mr. Schmidt indicated on cross-examination that he and his partner used the figures on the plans because "in my business we have to rely on the expertise of the people who make these." He testified that he had no reason to believe that the figures were inaccurate, and had no way of so determining without surveying. Mr. Stenerson testified that in the course of his work career, he had bid on jobs perhaps 200 times, using plans and specifications such as these as the basis of his bid. His testimony also indicated that in his business of earthmoving it was not customary to double-check the architect or engineer by resurveying.

We find this evidence to be sufficient to support the judge's finding that the contractors justifiably relied on the erroneous information provided to them by defendant. We will not overturn the district judge where there is substantial evidence to support his findings and conclusions. *Marta v. Smith* (1980), Mont., 622 P.2d 1011, 1015, 38 St.Rep. 28, 32.

The order of the District Court awarding the contractors the total cost of overrun is affirmed.

MORRISON, DALY, SHEEHY and WEBER, JJ., concur.

The *Wiechmann* and *Stenerson* cases can be reconciled. The contractor in the *Wiechmann* case had the opportunity to examine the information but did not. It also observed the site and knew that boulders were present. It would be difficult to justify a claim for additional expenses under these conditions. The public entity made available all it knew, and the information was readily accessible to the contractor. The contractor lost its claim because it did not *rely* on the information. This invoked the common law rule that places the risk of performance that proves to be more expensive than anticipated on the contractor.

In the *Stenerson* case, the court concluded that the public entity knew or should have known that the information would be relied upon and that any on-site inspection would not have revealed any information that was not on the plans. The *Wiechmann* case condemned the contractor because it did not perform as the court felt contractors *should* perform. The court in the *Stenerson* case concluded that the contractors did as other contractors would have done. They relied upon the expertise of the geotechnical engineers.

The fact remains that the *Wiechmann* case reflects an unwillingness to deprive the public entity of the opportunity of disclaiming responsibility and placing the risk on the contractor. The *Stenerson* case, on the other hand, was unwilling to allow the public entity to exonerate itself when it knew that the contractor did rely on the information, with reliance in the long run being beneficial to the public entity because it generated a lower price.

That two courts in different jurisdictions came to different results and expressed different attitudes toward disclaiming responsibility may not frustrate contract planning and counseling. Usually planners and counselors must deal only with the law of one jurisdiction. If they know that law and develop the facts, they should be able to plan and counsel efficiently.

It is more difficult when apparently conflicting decisions exist in the *same* jurisdiction, not an uncommon phenomenon.[22] Here great burdens are placed upon planners and counselors. Even when those results can to some degree be reconciled, the different attitude toward the process of disclaiming responsibility for information furnished is often reflected in the judicial opinions.

Although all judges are likely to give great autonomy to contracting parties to apportion risks as they choose, they often differ when one party has dictated a risk allocation plan. To be sure, a contractor can adjust its bid price to take into account the risks that it is being asked to bear.

The law operates through judges who may view the disclaimer process differently. Those judges who are unwilling to allow owners to furnish information and disclaim responsibility for its accuracy (the issue here is not fraud or negligent misrepresentation) seem more influenced by the apparent unfairness of allowing the owner to place the risk upon a party often less in a position to distribute that risk or shift it, particularly when the owner knows that the information will be relied upon and derives a benefit from reliance through lower bid prices. They are also influenced by the belief that the contracting parties owe each other the duties of good faith and fair dealing. They believe autonomy should not be used as a means of placing subsurface risks upon the contractor when the contractor

---

22. Compare *Wunderlich v. California*, 65 Cal.2d 777, 56 Cal.Rptr. 473, 423 P.2d 545 (1967) (enforced), with *E.H. Morrill Co. v. California*, 65 Cal.2d 787, 56 Cal.Rptr. 479, 423 P.2d 551 (1967) (not enforced). Compare *Golomore Associates v. New Jersey State Highway Authority*, 173 N.J.Super. 55, 413 A.2d 361 (1980) (not enforced), with *Sasso Contracting Co. v. State of New Jersey*, 173 N.J.Super. 486, 414 A.2d 603 (1980) (enforced). See generally Parvin & Araps, *Highway Construction Claims*, 12 Pub. Contract L.J. 255 (1982).

did as others and relied. Judges favoring broad autonomy are less influenced by these considerations. They seek only to determine *how* the risk was apportioned by the contract.

Planners seek to have a construction project accomplished efficiently and at the lowest ultimate cost. Before they choose to use the disclaimer system, they must look carefully at several factors. If they place the risks on the contractor, they must anticipate that bidders will add a contingency into their bids to take these risks into account. To be sure, some contractors may not. This may generate a low bid but administrative difficulties and costs later.

Choosing the disclaimer system must take into account the likelihood of its success. A jurisdiction that frowns upon disclaimers in this context *may* not enforce such a clause. Uncertainty may make it inadvisable to use a disclaimer system.

Similarly, the applicable laws must be examined to determine whether a contractor who suffers these losses in reliance on inaccurate information will have a claim against the geotechnical engineer.[23] If so, a geotechnical engineer faced with the possibility of a claim may increase its price for this contingency factor if it cannot use other methods of risk management, such as liability limitation or indemnification. Faced with a disclaimer system, the prudent contractor must raise its bid to take this risk into account, but the contractor cannot eliminate this contingency by assuming that it will have a claim against the geotechnical engineer. Even in jurisdictions that will allow such claims, the claim may not be successful because the contractor cannot establish negligence. But the risk exposure may induce the design professional to raise its bid to take this factor into account. This

puts the owner in the uncomfortable position of paying twice for the same risk. (A way of avoiding such a claim is to exculpate the geotechnical engineer, when possible, in the owner-contractor agreement. The language must be clear.)[24]

These factors have led some contracting parties to use a system under which the contractor can transfer its unanticipated expenses to the owner (discussed in Section 29.05).

SECTION 29.05 CONTRACTUAL PROTECTION TO CONTRACTOR

(A) PUBLIC CONTRACTS: THE FEDERAL APPROACH

In the absence of contract protection, misrepresentation, or breach of warranty, the contractor must bear the risk of unforeseen site and subsurface conditions. If disclaimer language is chosen carefully and the bidder given a reasonable opportunity to observe and test, it is likely that the disclaimer will be effective. Even where this is not so, at best, the owner will be held responsible for factual representations, not opinions or inferences.

The owner may find that this risk shift to the contractor is not in the owner's best interest. In such a case, the owner may choose to accept the risk of unforeseen site and subsurface conditions. This is accomplished in federal construction through what was formerly called the Changed Conditions Clause and is now known as the Differing Site Conditions (DSC) Clause. It states:

> (a) The Contractor shall promptly, and before such conditions are disturbed, notify the Contracting Officer in writing of: (1) subsurface or latent physical conditions at the site differing materially from those indicated in this contract,

---

23. Compare *Texas Tunneling Co. v. City of Chattanooga*, 329 F.2d 402 (6th Cir.1964) (claim denied), with *M. Miller Co. v. Central Contra Costa Sanitary District*, 198 Cal.App.2d 305, 18 Cal.Rptr. 13 (1961) (claim allowed). The latter has been followed in the Restatement (Second) of Torts, § 525 (1977).

24. See Section 36.05(A).

or (2) unknown physical conditions at the site, of an unusual nature, differing materially from those ordinarily encountered and generally recognized as inhering in work of the character provided for in this contract. The Contracting Officer shall promptly investigate the conditions, and if he finds that such conditions do materially so differ and cause an increase or decrease in the Contractor's cost of, or the time required for, performance of any part of the work under this contract, whether or not changed as a result of such conditions, an equitable adjustment shall be made and the contract modified in writing accordingly.

(b) No claim of the Contractor under this clause shall be allowed unless the Contractor has given the notice required in (a) above; *provided,* however, the time prescribed therefor may be extended by the Government.

(c) No claim by the Contractor for an equitable adjustment hereunder shall be allowed if asserted after final payment under this contract.[25]

The following explanation for these clauses was given by the Court of Claims:

The starting point of the policy expressed in the changed conditions clause is the great risk, for bidders on construction projects, of adverse subsurface conditions: "no one can ever know with certainty what will be found during subsurface operations." Kaiser Industries Corp. v. United States, *supra*, 340 F.2d at 329, 169 Ct.Cl. at 323. Whenever dependable information on the subsurface is unavailable, bidders will make their own borings or, more likely, include in their bids a contingency element to cover the risk. Either alternative inflates the costs to the Government. The Government therefore often makes such borings and provides them for the use of the bidders, as part of a contract containing the standard changed conditions clause.

Bidders are thereby given information on which they may rely in making their bids, and are at the same time promised an equitable adjustment under the changed conditions clause, if subsurface conditions turn out to be materially different than those indicated in the logs. The two elements work together; the presence of the changed conditions clause works to reassure bidder that they may confidently rely on the logs and need not include a contingency element in their bids. Reliance is affirmatively desired by the Government, for if bidders feel they cannot rely, they will revert to the practice of increasing their bids.

The purpose of the changed conditions clause is thus to take at least some of the gamble on subsurface conditions out of bidding. Bidders need not weigh the cost and ease of making their own borings against the risk of encountering an adverse subsurface, and they need not consider how large a contingency should be added to the bid to cover the risk. They will have no windfalls and no disasters. The Government benefits from more accurate bidding, without inflation for risks which may not eventuate. It pays for difficult subsurface work only when it is encountered and was not indicated in the logs.

All this is long-standing, deliberately adopted procurement policy, expressed in the standard mandatory changed conditions clause and enforced by the courts and the administrative authorities on many occasions. [Citations omitted.] Faithful execution of the policy requires that the promise in the changed conditions clause not be frustrated by an expansive concept of the duty of bidders to investigate the site. That duty, if not carefully limited, could force bidders to rely on their own investigations, lessen their reliance on logs in the contract and reintroduce the practice sought to be eradicated—the computation of bids on the basis of the bidders' own investigations, with contingency elements often substituting for investigation.[26]

A DSC creates two methods of obtaining an equitable adjustment, conditions different from those represented (Type I) and unanticipated conditions (Type II).

A Type I claim requires a material variation between the actual condition encountered and that indicated. The actual condition encountered need not be contrary to the express representations in the plans and specifications. A representation may be inferred if from a reading of the contract document as a whole the contractor would reasonably be led to believe that it would not encounter the conditions that were in fact encountered. For example, one contractor obtained relief where the specifica-

---

25. 32 CFR § 7–602.4.

26. *Foster Construction C.A. & Williams Brothers Co. v. United States,* 193 Ct.Cl. 587, 435 F.2d 873, 887 (1970).

tions described a dry excavation procedure but the contractor encountered water.[27]

Type II requires a variance between the site condition actually encountered and that to be reasonably expected. Not only should the contractor have had no information from the owner, but it must not have had knowledge of such conditions from any other source. It must not be able to have reasonably anticipated such conditions at the time the contract is made.

The contractor will be judged as a prudent and responsible bidder who normally makes a reasonable site investigation, studies the contract documents, and makes an intelligent assessment of the requirement of the job based on sound construction experience. If after prebid investigation a reasonable bidder would have reasonably anticipated the conditions encountered, the contractor will be denied recovery. It should have known what was likely to be encountered. The condition must be unusual and not one ordinarily encountered in works of a similar nature. But the condition need not be unique.

As to material difference, suppose the boring logs locate and the soil report correctly identifies a certain type of organic silt but upon excavation the soil behaves differently from what is normally expected of that kind of subsurface condition. The issue is whether the public entity made a representation as to quality or what would normally be expected of a certain type of subsurface condition.

Quantity variation is not insignificant simply because the equitable adjustment is relatively minor. There is a distinction between the test to determine whether the difference is material and the test to measure the amount of equitable adjustment. A minute variation may be considered immaterial and may not be considered to justify an equitable adjustment. Yet the fact that the final figure for the equitable adjustment for a differing site condition is a relatively minor sum as a result of bargaining and offsets does not preclude a contractor from being entitled to an adjustment.[28]

The location at which the differing site condition is encountered need not necessarily be at the construction site itself. If the particular area in which the condition was encountered was designated in the contract documents, the condition will be considered to have occurred "at the site." For example, where the contract documents specified a quarry as a source of construction material and the quarry was composed of unsuitable material, the contractor was entitled to an equitable adjustment.[29]

Normally a DSC is a condition occurring below the surface or otherwise not ascertainable by normal inspection. In addition, the condition must be physical. Examples of conditions that fit within these requirements include excessive groundwater, tougher than anticipated ground composition, underground obstacles, uncharted utilities, and unsuitable material in borrow pits. This chapter deals principally with existing subsurface conditions and not events that subsequently occur and make subsurface work more expensive.

To obtain an equitable adjustment, the contractor must notify the contracting officer promptly in writing. This generates the inevitable difficulty over substantial compliance and waiver.[30] Courts seem

---

27. *Foster Construction Co., etc.,* supra note 26; *Metropolitan Sewerage Commission v. R.W. Construction, Inc.,* 72 Wis.2d 365, 241 N.W.2d 371 (1976).

28. *Roscoe-Ajax Construction Co. v. United States,* 198 Ct.Cl. 133, 458 F.2d 55 (1972).

29. *Stock & Grove, Inc. v. United States,* 204 Ct.Cl. 103, 493 F.2d 629 (1974).

30. *Moorhead Construction Co. v. City of Grand Forks,* 508 F.2d 1008 (8th Cir.1975); *Metropolitan Paving Co. v. City of Aurora, Colorado,* 449 F.2d 177 (10th Cir.1971) (waiver: owner denied claim as condition anticipatable); *Acchione & Canuso, Inc. v. Commonwealth Department of Transportation,* 501 Pa. 337, 461 A.2d 765 (1983).

more willing to waive strict notice requirements if it appears that the owner knew that the contractor had encountered unforeseen subsurface conditions and that it was likely that a claim would be made. In *Kenny Construction Co. v. Metropolitan Sanitary District of Greater Chicago,*[31] the contractor notified the engineer, who told the contractor to go ahead and they would figure costs later. Despite the absence of a written order by the engineer, the court granted recovery, since the contractor relied on the engineer's statement and the public entity was estopped to assert the condition.

In *Centex Construction Co.,*[32] the written notice requirement was waived because the project representative was aware of the oil dump initially at the prebid site investigation and again when the contractor orally advised him of the condition when work commenced. The Board of Appeals also noted that failure to provide written notice did not prejudice the government's position. But as in other cases dealing with notices, outcomes will vary.

References have been made to cases outside the federal procurement system. The use of a DSC is becoming more common in public contracts made by entities other than the federal government. This can create difficulties. For example, one case involved a clause that incorporated a Type I DSC but not a Type II DSC.[33] Similarly confusion can develop because a local public entity may be compelled to use a DSC because the project is being partially funded with federal money granted by the Environmental Protection Agency (EPA).

But the local entity making the contract may have had a practice of using disclaimers, and these disclaimers are sometimes still included in the contract documents. One case faced with this question concluded that the DSC provision took precedence.[34]

The contractor's remedy is an equitable adjustment that can involve a variety of measurement techniques, all of which are approximations rather than precise. Formulas can be placed in the following categories:

1. *Additional cost* of work affected plus *overhead* and *profit,* measured either objectively or subjectively.

2. The *total cost* differential, the difference between the actual cost and what it would have cost had the conditions been as represented or expected.

3. A *jury verdict,* an approximate amount determined in a "rough" way and similarly to that which would be done by a jury.

Measures are discussed in Section 31.02. Discovery of a changed condition can *decrease* the contract price.

Does the right to receive an equitable adjustment preclude the contractor from terminating its obligation if the condition discovered would have granted it that power? In addition to taking this potentially disasterous risk away from the contractor, the clause is designed to keep the job going. It should substitute for any common law rights that the contractor might be accorded.[35]

---

31. 52 Ill.2d 187, 288 N.E.2d 1 (1972). A subsequent opinion is found in 56 Ill.2d 516, 309 N.E.2d 221 (1974).

32. ASBCA Nos. 26830 through 26849, 83–1 BCA ¶ 16,525.

33. *Metropolitan Sewerage Commission v. R.W. Construction, Inc.,* supra note 27.

34. *Fattore Co. v. Metropolitan Sewerage Commission,* 505 F.2d 1 (7th Cir.1974). See also *Andrew Catapano Co., Inc. v. City of New York,* 116 Misc.2d 163, 455 N.Y.S.2d 145 (S.Ct. 1980).

35. *North Harris County Junior College District v. Fleetwood Construction Co.,* 604 S.W.2d 247 (Tex.Civ.App.1980) appears to allow termination. Actually the court held that a refusal by the owner to acknowledge that there was a changed condition could have justified the contractor's terminating.

(B) AIA APPROACH: CONCEALED CONDITIONS

In 1937 the AIA adopted a similar approach to unknown subsurface conditions. AIA Doc. A201, ¶ 12.2.1, currently provides the following:

> 12.2 CONCEALED CONDITIONS
>
> 12.2.1 Should concealed conditions encountered in the performance of the Work below the surface of the ground or should concealed or unknown conditions in an existing structure be at variance with the conditions indicated by the Contract Documents, or should unknown physical conditions below the surface of the ground or should concealed or unknown conditions in an existing structure of an unusual nature, differing materially from those ordinarily encountered and generally recognized as inherent in work of the character provided for in this Contract, be encountered, the Contract Sum shall be equitably adjusted by Change Order upon claim by either party made within twenty days after the first observance of the conditions.

Whether the occasional owner who builds once in a lifetime is better off with such a provision is debatable. Similarly, it is questionable whether lenders will want a provision that can substantially expand the ultimate cost of the project.

In AIA Doc. A201, ¶ 1.2.2, the contractor represents that it has visited the site and familiarized itself with local conditions under which the work is to be performed. This can justify denial of a claim. As in the *Wiechmann* case (reproduced in Section 21.04), when conditions could have been observed under a normal site visit, the contractor cannot recover additional compensation.

This raises an issue often ignored. Suppose the contractor *should* have seen certain site conditions or actually *saw* site conditions that varied from the conditions *indicated* by the contract documents. If relief should be accorded only when the conditions are concealed, a contractor who knows or should know what it will encounter because of access to the site should not receive a Type I equitable adjustment under ¶ 12.2.1.

A comparison of ¶ 12.2.1 with the federal DSC reveals gaps in the AIA's clause, traceable perhaps to the different type of projects that use AIA Documents. The contractor need only make a claim within twenty days after first observance. (The same requirement applies to any claim by the owner for a decrease.) On the other hand, the federal DSC requires that a written notice be given *promptly* and specifies information that must be furnished.

Perhaps more important, the federal DSC requires that the notice be given *before* the conditions are disturbed, a requirement not found in ¶ 12.2.1. Giving the notice as soon as possible allows the design professional and geotechnical engineer to study the conditions to determine whether relief is justified as well as the need for design changes. Strictly speaking, under ¶ 12.2.1, the contractor could continue the work so long as it gave the notice within twenty days. Although the obligation of good faith and fair dealing may cure the deficiency, leaving such an important matter to implication is unwise. It may be useful to require that the notice contain any projected impact on the contractor's schedule caused by the discovery of differing site conditions in order to prepare for the inevitable "impact" delay claim.

A question endemic to standard construction agreements relates to the variety of notices required and the not uncommon phenomenon of different formal and substantive requirements. Suppose the contractor discovers that the actual conditions vary from those represented or expected under ¶ 12.2.1. As noted, a notice—not necessarily written—must be given within twenty days. Perhaps more formal notice is required under ¶ 12.3.1 if the contractor will claim additional compensation. Yet ¶ 12.2.1 deals specifically with this problem. The notice given in accordance with the requirements of *that* section should be sufficient.

Suppose the contractor intends to ask for a time extension. Is the notice under ¶ 12.2.1 adequate, or must it also comply with ¶ 8.3.2, which deals with time extensions? The period in ¶ 12.2.1 begins with the *observance* of the conditions, as in ¶ 12.3.1, but in ¶ 8.3.2 it begins with the start of the delay. Suppose the contractor asks for an amount to compensate for its delay. Must it then comply with ¶ 12.3.1 dealing with claims for additional cost?

Another possibility is that the contractor claims that its excavating equipment has been damaged because it encountered conditions for which the equipment was not appropriate. Must it give notice in accordance with ¶ 7.4.1, which requires a notice "in writing" within a reasonable time after first observance of such damage when there is damage to property because of an act of the other party? This complexity bewildered a Texas appellate court, which seemed to require a notice to be given under the predecessor of ¶ 12.3 rather than the notice requirement under the predecessor of ¶ 12.2.1.[36]

Paragraph 12.2.1 can generate another issue. As noted, the federal DSC simply provides for an equitable adjustment, first by the contracting officer, subject to appeal to a Board of Appeals. Under ¶ 12.2.1, the price is to be equitably adjusted by change order, which creates a residuary power to resolve time and price in the architect and provides a cost plus overhead and profit guide.

Suppose the paragraph simply provided for an equitable adjustment. If the change order mechanism does *not* apply, failure by the parties to agree would be a dispute that under ¶ 2.2.9 must first be referred to an architect whose decision is subject to arbitration. However, the specific guides provided for in Art. 12, the changes clause, need not be applied. For example, the architect could divide the additional costs in some manner that he thought equitable rather than simply

award the additional cost to the contractor, the approach prevalent under the federal system.

There is another important feature of ¶ 12.2.1. A Type I equitable adjustment requires a comparison between what is encountered and "the conditions indicated by the Contract Documents." Under AIA Doc. A201, ¶ 1.1.1, the bidding documents are excluded from the contract documents.

If the subsurface information is included in the specifications, there will be no difficulty. Suppose the subsurface information is included in the instruction to bidders or simply is made available at the geotechnical engineer's office. It would be very difficult for the contractor to be able to claim an equitable adjustment.[37]

A variety of techniques are used to place a risk upon the contractor, one of them labelling the subsurface data as solely for information or in some way indicating that they are not part of the contract. This is an indirect and often uncertain method of relieving the owner from responsibility. This technique and a number of other points are revealed in *City of Columbia v. Paul N. Howard Co.*[38] The contract involved the construction of a public sewer improvement and contained a differing site condition clause, with the contractor being entitled to reimbursement of cost resulting from conditions differing materially from those indicated in this contract.

The contractor received a document entitled "Specifications and Documents" with an appendix that contained test boring logs. Its claim was based upon subsurface conditions differing from those indicated in the logs. The city claimed the logs were not part of the contract and pointed to a disclaimer on

---

36. Id.

37. But see *North Harris County Junior College District v. Fleetwood Construction Co.*, ibid, in which the soil report was referred to in the specifications and thereby a part of the contract document.

38. Supra note 20.

the cover page of the test boring logs that stated:

> These reports are for reference only and are not part of the contract documents.

The trial court ruled for the city.

The Court of Appeals looked at other parts of the contract and not simply at the disclaimer on the test boring logs. It noted that the contract documents were defined as items listed in the table of contents. The logs were listed in the table of contents under "Appendix." In addition, a supplementary condition stated:

> Test hole information represents subsurface characteristics to the extent indicated, and only for the point location of the test hole. Each Bidder shall make his own interpretation of the character and condition of the materials which will be encountered between test hole locations.

This supported, according to the court, the city's argument that it was not responsible for unanticipated conditions between the *test holes*, but it also supported the contractor's argument that it could rely on logs at least as to the *actual* conditions at the location of the test holes.

The court stated that the trial court should have reconciled *all* these provisions and concluded that the various parts could be interpreted to mean that the contractor could rely upon the logs only for their accuracy and not for conditions between the test hole locations.

The court conceded the city's argument that the appendix was not an item listed in the table of contents and therefore was not part of the contract. But the court, using a maneuver that some might find adroit and others bizarre, held that the equitable adjustment depended not upon the logs being part of the contract but solely upon conditions differing materially from those indicated in this contract. The court concluded by noting that though the logs were not part of the contract, they were *indicated* in the supplementary conditions, at least enough to make the differing site conditions provision available.

The second part of the *Columbia* case reveals how courts can make law complex. The contractor also contended that it had a claim based upon misrepresentation exclusive of the changed condition clause. This injection of a common law theory into a risk allocation system in which the parties fully covered this problem creates uncertainty. The court looked at this second claim with some sympathy, concluding that the contractor should have the opportunity to prove the elements needed for a tort claim.

It is likely that the disclaimer on the log and its being described as not part of the contract were designed to preclude any claim based upon misrepresentation by negating the reliance element.

The *Columbia* case demonstrates a point that has been made earlier, the confusion that can be created when a DSC is incorporated and disclaimers as to the responsibility of the information are also employed.

Section 29.04 discussed the advisability of using the disclaimer system that puts the risks on the contractor. What about a DSC? There is much debate over the likelihood that incorporation of a DSC actually induces lower bids. Some contend that bids are based upon work load, the desirability of keeping the work force together, and the prospects of a well-administered construction project. If lower bids are not actually induced, much of the reason for using a DSC disappears.

Even if a prudent contractor who thinks it will have a chance of obtaining a contract *does* reduce its bid, it will do so only if it believes that the clause will be fairly administered. Suppose it believes that obtaining an equitable adjustment will be difficult or administratively expensive or that the amount of the equitable adjustment will not provide adequate compensation. If so, a contractor that might otherwise reduce its bid will not do so. There are administrative costs connected with a DSC. The owner must pay the price of uncertainty as to actual pay out, something that may be difficult for a small private owner or even a small public entity to deal with effectively.

Another consideration often ignored is the effect that the choice of system has upon the likelihood that subsurface information will be as accurate as can be reasonably expected. A justification sometimes given for the disclaimer system is the likelihood that testing will be relatively complete and information will be given to the contractor if the owner is not held responsible for the accuracy of the information. On the other hand, it can be argued that the owner who will be charged with an equitable adjustment if the information is inaccurate will hire the best soil testers and give them the best opportunity to develop accurate information.

SECTION 29.06
SOME ADVICE TO COURTS

Undoubtedly arguments can be made for a disclaimer system or a system that uses a differing site conditions clause. Courts should effectuate *whichever* choice has been made. If it appears clear that the contractor has, happily or not, accepted the risk of unforeseen subsurface conditions, the court should effectuate that risk allocation and not seek to destroy it by tortured interpretation generated by a belief that it is unconscionable for the owner to place these risks upon the contractor. Only if it is clear that the contractor was not made aware of the risk or if the imposition of the risk violates the obligation of good faith and fair dealing should the court entertain a result that does violence to the loss distribution selected and expressed in the contract.[39]

Similarly, courts should *not* allow a contractor to maintain a tort action against a geotechnical engineer if it is clear that the contractor was expected to assume this risk. Allowing the contractor to transfer this risk to the geotechnical engineer through a tort action frustrates the efficiency of the system selected.

PROBLEMS

1. O hired A to design a house and prepare construction drawings and specifications, but not to supervise or inspect. A did his work and was paid. O then hired C to build a residence for $35,000.

   O thought the soil might be spongy in the area of his lot, since a neighbor encountered some minor slide problems. However, he made no soil tests and did not call in a soils engineer for an inspection. Shortly after the house was completed, the exterior steps began to crack. The house settled, cracks developed, and the living room tilted. Does O have a valid claim against C? If so, how would his damages be computed? (To avoid these settling problems, it would be necessary to rebuild the house.)

2. During the winter of 1982–1983, county engineers became concerned about earth slides occurring on Edgewood Avenue in Mill Valley. The county decided that it would be necessary to perform slide repair work to prevent the recurring of these incidents. The H–L consulting firm was retained to perform a soil investigation and make a report of its findings in preparation for the correction work. The H–L report contained a description of the slide area, specific recommendations, and complete specifications for performing the correc-

---

39. Illustrations would be deliberately misleading the contractor or not giving the contractor sufficient time to examine the data, visit the site, or take its own tests. Whether this would encompass using a disclaimer system knowing that the contractor does rely is a close case. See *Stenerson v. City of Kalispell,* reproduced in Section 29.04.

tive work. It also contained twelve boring logs or diagrams of test holes that were drilled in and around the slide area in order to ascertain the nature of the subsoil materials encountered at various depths.

The county thereafter put the project up for competitive bidding. Included within the contract documents provided to prospective bidders was a set of plans and specifications for the project and a section entitled "Special Provisions" describing the materials and procedures to be used in performing the work. As an appendix, the contract document packet reproduced four of the boring logs from the H–L report. The report itself was not included in the documents provided to prospective bidders but was referred to in the following manner: "A soil investigation has been conducted at the project location and a copy of the Report is available for inspection at the office of the Department of Public Works, Room 304, Civic Center."

The plans and specifications prepared by the county called for the construction of a tieback retaining wall composed of concrete-encased steel piles, which were to be held in place by rock anchor bolts. The bolts would be constructed by drilling a series of twenty-six holes horizontally into the slope, placing PVC (plastic) pipe into them, and inserting high-strength steel strand cable tiebacks. The steel cables were then to be grouted into thirty feet of "competent rock." After the grout was injected, the cables were to be tensioned and secured to I-beams in the center of each concrete pile.

Valentine's bid was accepted and it entered into a written contract with the county to perform the slide correction work. The total contract price was $51,409, with the price for the rock anchor bolt tiebacks listed at $17,160.

On November 7, 1983, Valentine entered into a written subcontract with Judd Drilling to install five hydraugers and to drill the twenty-six three-inch-diameter holes needed for the tieback. The subcontract specified that this was a drill only job, meaning that if a hole had to be redrilled for any reason, Judd would be entitled to compensation for extra work.

Judd installed the hydraugers without much difficulty in late November 1983 and began working on the tieback system on December 4 or 6. From the very beginning, the holes drilled by Judd would not stay open: as soon as a hole was drilled, the overburden (material overlying the rock formation) would collapse and the hole would cave-in. PVC piping, as called for in the specifications, can be used only in an open hole, since plastic is not strong enough to be driven or twisted into the earth. Judd was therefore forced to switch to steel casing, which could be driven into the slope at the same time the hole was being created. Delays and standstill time were incurred while Valentine went out in search of steel casing. Judd also incurred extra expenses for redrilling holes and bringing in additional equipment.

Problems were also encountered when the drillers attempted to find competent rock, which is rock stable enough to maintain its structure so as to achieve the design purposes—in this case to hold the cables inside the holes when grout (cementlike filler material) is injected into the holes. In fact, the rock was not competent but contained numerous pockets, voids, and fissures. This would sometimes cause the hole to collapse and make it impossible to withdraw the drill. Therefore, the workers had to inject the slope with large quantities of grout to make the rock competent. Often, grout would be injected into a hole and wind up coming out somewhere else.

The work was being performed during a period of heavy rains, and Judd and Valentine requested to pull off the job because of the bad weather and drilling difficulties. At a meeting among all three parties, the county informed them that it was essential that they stay on the job and implied that a refusal to do so would result in lawsuits. In a letter dated December 11, 1983, the county's chief engineer, wrote Valentine, "We will give thorough consideration to any further data or information you wish to submit in writing concerning alleged changed conditions from the plans and specifications." Davidson told Valentine that the county would pay for the steel

casing and the additional grout but not for the extra labor.

The job was finally completed on February 17, 1984.

On November 15, 1984, Judd filed a complaint against Valentine for breach of contract, breach of implied warranty, and a restitution claim for reasonable value of work performed.

Judd's claim consisted of costs of labor to install casing and for redrilling, standby time, and equipment rental. Valentine in turn filed a cross-complaint against the county, seeking indemnity for Judd's claim as well as for its own additional expenses as the result of alleged breach of contract and implied warranty. How should these claims be decided?

30

Time: A Different but Important Dimension

SECTION 30.01 AN OVERVIEW

The law has *not* looked at time as part of the *basic* construction contract exchange—that is money in exchange for the project. This may be due to the frequency of delayed construction projects.[1] Timely completion depends upon proper performance by the many participants as well as optimal conditions for performance, such as weather and subsurface.

Delayed performance is less likely to be a valid ground for termination.[2] Delayed performance is also less likely to create legal justification for the owner's refusal to pay the promised compensation, with the owner left to the often inadequate damage remedy.[3] Similarly, delayed payment by the owner is less likely to automatically give the contractor a right to stop the work or terminate its performance.[4]

Computation of damages is *also* different. If the owner does not pay or the contractor does not build properly, measur-

ing the value of the claim is, relatively speaking, simple. If the owner does not pay, at the very least contractors are entitled to interest. If the contractor does not build properly, the owner is entitled to cost of correction or diminished value of the project.[5]

Delay raises more serious claims measurement. The owner's *basic* measurement for unexcused contractor delay is lost use of the project. The contractor's basic measurement for owner-caused delay is added expense. Lost use is difficult to establish in noncommercial projects. Added expense is even more difficult to measure. Because of measurement problems, each contracting party—whether it pictures itself the potential claimant or the party against whom a claim will be made—would like a contractual method to deal with delay claims, either to limit them or to agree in advance on amount. This does not mean that time is not important in construction. The desire to speed up completion to minimize financ-

---

1. *Johnson v. Fenestra, Inc.,* 305 F.2d 179 (3d Cir.1962), rejected a contention that delays are calculated risks and not breaches of contract.

2. *Hartford Electric Applicators of Thermalux, Inc. v. Alden,* 169 Conn. 177, 363 A.2d 135 (1975) (forty-day delay in furnishing site not material breach).

3. *Landscape Design & Construction Inc. v. Harold Thomas Excavating, Inc.,* 604 S.W.2d 374 (Tex.Civ.App.1980).

4. Refer to Section 26.02(H).

5. See Section 31.03.

ing costs and accelerate revenue-producing activities was a large factor in leading to techniques intended to generate efficient methods of organizing construction (discussed in detail in Section 21.04). Time is another dimension, however.

SECTION 30.02
COMMENCEMENT

The very nature of construction sometimes makes complicated what in other contracts is simple. In an ordinary contract, such as an employment contract, or in a long-term contract for the sale of goods, as a rule the commencement dates are simple to establish. In construction, the date when the performance can begin *in earnest* (procurement can precede site access) is usually the date when site access is given. This cannot always be precisely forecast. The owner may need to obtain permits, easements, and financing before the contractor can be given site access. To avoid responsibility for site access delay and to measure the contractor's time obligation fairly, the commencement of the time commitment period, when measured in days, is often triggered by the contractor's being given access to the site, usually by a Notice To Proceed (NTP). When an NTP is used, the contractor assumes the risk of ordinary delays in site access but not those that go beyond the reasonable expectations of the contracting parties.

Commencement raises other problems. For example, must the contractor actually begin work at the site when an NTP is given? Is the actual commencement date of the NTP a date specified in the NTP or the date the NTP is received? The contract should deal with these issues but frequently does not. What are reasonable expectations of the parties? The custom in the industry? What is fair?

If the NTP does not expressly specify the commencement date, the date should begin when the NTP is received. Any date specified in the NTP should be effective only if it meets the standard of good faith and fair dealing. For example, the notice should not specify a date that the owner knows the contractor cannot meet or that fails to take into account realistic commencement requirements.

Suppose there is a delay between bid opening, bid award, and execution of a formal contract. These issues were involved in two instructive cases. *Quin Blair Enterprises v. Julien Construction Co.*[6] involved a competitively bid contract to build a motel. Julien's bid stated that he would complete 240 days from the date the contract was signed. Julien was awarded the contract and was asked to prepare the agreement. He used AIA Doc. A101 and filled in the blank with 240 days, without anything specific as to when the period began. But A101 stated that the contract was made on October 8, 1971.

AIA Doc. A201, ¶ 8.1.2, stated that the date of commencement, if there is no notice to proceed, begins on the date of the Agreement "or such other date as established therein."

Blair signed on October 22 and Julien on October 25. Julien could not start until Blair cleared the site. That was not completed until November 18. When did the 240-day period commence? On October 8? On October 25? On November 18?[7] The trial court chose October 8, but the Supreme Court disagreed. Recognizing the ambiguities and seeking to harmonize all the writings, the court concluded that the time began on October 25.

Because no date was specified in A101, the court referred to A201. Since there was no NTP, the commencement date should have been the date of the agreement. But

---

6. 597 P.2d 945 (Wyo.1979).

7. Julien's failure to submit the required notice of an intention to ask for an extension barred a time extension.

was that date October 8, the date on the agreement, or October 25, when the agreement was signed by the owner? Since A201 states the date of the agreement, one would think it was the date on the agreement, not the date when it was made. According to the court, each party agreed that the bid was part of the contract. The bid stated that time would begin upon *signing* the contract, and although parties *can* designate a retroactive date, there was nothing to indicate that "either party ever intended . . . [a] retroactive date." [8]

Bloomfield Reorganized School District No. R–14, Stoddard County v. Stites [9] did find a retroactive date. The contract was dated August 8, 1955, and provided that the contract was to be substantially completed in 395 calendar days. The architect mailed the contract to the contractor on August 17, 1955. The contractor signed the contract and returned it to the architect, who delivered it to the school superintendent for execution by the school board. School board officials signed the contract, and the superintendent mailed it to the architect on September 14. The latter forwarded the signed copy to the contractor on September 22, *six weeks* after the contract date. Yet the court looked *solely* at the language of the contract, which stated that the agreement had been made on August 8, 1955.

SECTION 30.03   ACCELERATION

(A) SPECIFIC: THE CHANGES CLAUSE

One method to accelerate the completion date is a specific directive by the owner that the contractor must complete in a time shorter than that originally agreed. Power to accelerate is usually determined by the changes clause.

AIA Doc. A201, ¶ 12.1.2, defines a change order as an order that can authorize a change in contract time. Yet the power to order changes is limited to "additions, deletions or other revisions." "Other revisions" can encompass acceleration, although this stretches the language. While ¶ 12.1.2 also states that ordering changes can require an adjustment in contract price *or* time, references to time adjustment, though probably contemplating time changes incident to quality or quantity changes in the work, *can* support the conclusion that "revisions" includes acceleration. Acceleration's drastic effect on performance should require a *specific* power.

Any power to accelerate is limited by the usual "scope of the work" aspects of a changes clause. Such power can be specifically limited to a designated percentage of the construction time. Acceleration drastically disrupts the contractor's schedule. Any compensation adjustment is rarely adequate.

(B) CONSTRUCTIVE ACCELERATION

Constructive acceleration originated in federal procurement law. Although the original jurisdictional basis for its development is no longer applicable,[10] it can be applied to all construction contracts.

Constructive acceleration is based upon the owner's *unjustified* refusal to grant a time extension. It requires that a cause exists that would have justified a time extension, a request for a time extension, denial of that request, demand, (express or

---

8. Id. at 951.

9. 336 S.W.2d 95 (Mo.1960).

10. Like constructive changes, constructive acceleration was a claim based on the contract and not its breach. This gave jurisdiction to the agency appeals board. Since 1978 a claimant can choose to bring a claim before either an appeals board or the Claims Court. Refer to Section 25.03(C).

implied) that performance be completed on time, and an actual acceleration.[11]

The modern justification for constructive acceleration is that denial of a deserved time extension *can* force additional expenses when work is not performed in the order planned. Suppose, though, that the contractor continues to perform as it would have performed had an extension been granted. This will very likely lead to untimely completion. If the time extension *should* have been granted and it is done so *later* (by agreement, by an arbitrator, or by a court), any attempt by the owner to recover actual or liquidated damages would be unsuccessful. The constructive acceleration doctrine gives the contractor the option of speeding up its performance and recovering any additional expenses it can establish or using the wrongful denial of the time extension as a defense to any claim that might be brought against the contractor by the owner for late completion.

(C) VOLUNTARY: EARLY COMPLETION

Delays are so common in construction that attention is rarely paid to the legal effect of the contractor's completing early or claiming that it would have completed early had it not been delayed by the owner.

Early completion may be desirable to some owners. This can be evidenced by a penalty-bonus clause.[12] On the other hand it may, if unexpected, frustrate owner plans. For example, suppose a contractor building a factory finishes substantially earlier than planned. The owner may have to take possession before it can install its machinery. Early completion can require payments in advance of resource capabilities. It can be as disruptive as late completion. AIA Doc. A101, Art. 3, requires the con-

tractor to substantially complete the project "not later than" a specified date. This appears to give the contractor the freedom to complete early even if this were to disrupt owner plans.

Construction contracts of any magnitude usually have schedules. It is unlikely that the owner will be greatly surprised by early completion. Yet even awareness during construction that performance will be completed earlier than required may not enable the owner to make the adjustments needed to avoid economic losses.

The obligations of good faith and fair dealing require that a contractor notify an owner if it intends to finish much earlier than expected or when it appears that this is likely to be the case. If this is done or the owner is aware of that prospect, the contractor should receive additional compensation if the owner interferes with any realistic schedule under which the contractor would have completed earlier than required by the contract.[13]

SECTION 30.04 COMPLETION

In the absence of any specified completion date, performance must be completed within a reasonable time. AIA Documents divide completion into *substantial* and *final*. Completion, whether substantial or final, is an important legal and practical benchmark. For example, it can determine the validity of a mechanics' lien or a claim against a surety. It can affect whether a legal action has been barred by the passage of time. Completion can affect existing claims. Completion and final payment are often asserted to bar claims for defects subsequently discovered, as noted in Section 28.05. Just as contractors sometimes contend that this benchmark terminates any owner claims, the owner sometimes con-

---

11. *M.S.I. Corp.*, GSBCA No. 2429, 68–2 BCA, ¶ 7377.

12. See Section 30.09.

13. *BECO Corp.*, ASBCA 27090 (1983).

tends that this benchmark terminates any claims that the contractor may have against the owner.

Although mechanisms for ending disputes are useful, claims should not be barred unless it appears that a claimant knew or led the other party to believe that reaching these Construction Process benchmarks extinguished claims. Yet these benchmarks provide an opportunity for parties to resolve any claims that each may have against the other. In the *ordinary* case, however, their arrival should *not automatically* extinguish claims.

Although the AIA does not use completion as a benchmark, final *payment* can affect claims. While making it clear that most major *owner* claims are not waived by final payment,[14] AIA's General Conditions [15] state that acceptance of final payment waives all *contractor* claims "except those previously made in writing and identified by the Contractor as unsettled at the time of the final Application for Payment."

Although useful, these benchmarks can "trap" a contractor who does not pay careful attention to the general conditions, particularly in view of the changes AIA makes from edition to edition. For example, in 1970 contractor claims were waived unless they had previously been made in writing and were "still unsettled." Since 1976 the contractor must identify claims as unsettled *in writing* at the time of its application for final payment.

SECTION 30.05
SCHEDULES: SIMPLE AND CRITICAL PATH METHOD

A project schedule is a formal summary of the planned activities, their sequence, the time required, and the conditions necessary for their performance. A schedule alerts the major participants of the tasks they must accomplish to keep the project on schedule. It can reduce project cost by increasing productivity and efficiency, facilitates monitoring of the project, and can support or disprove delay claims.

The schedule for a *very* simple project, such as the construction of a garage, may simply be starting and completion dates. A somewhat more complex project, such as a residence, may add designated stages of completion, mainly as benchmarks for progress payments. When construction moves upscale, for example, from a simple commercial structure to a nuclear energy plant, the schedule will take on more complex characteristics. Until the past twenty-five years, schedules in anything but the simplest project would be a bar chart, sometimes referred to as a Gantt Chart, after its inventor. One such bar chart [16] is shown in figure 30.05a.

Bar charts have deficiencies. They provide no logical relationship between work packages. There are limits to the number of work packages that can be represented in a bar chart—perhaps thirty to fifty—until the level of detail becomes unwieldy. Rates of progress within a package may not be uniform. According to two authors, however:

> . . . the visual clarity of the bar chart makes it a very valuable medium for displaying job schedule information. It is immediately intelligible to people who have no knowledge of CPM [17] or network diagrams. Its familiarity breeds confidence. It affords an easy and convenient way in which to present information developed from a CPM study. It is far less expensive than CPM. For these reasons, bar charts undoubtedly will continue to be widely used in the construction industry.[18]

---

14. Refer to Section 28.07(A).

15. AIA Doc. A201, ¶ 9.9.5.

16. M. Callahan & M. Hohns, Construction Schedules 38 (1983).

17. Critical path method (CPM) is discussed in (B).

18. M. Callahan & M. Hohns, Construction Schedules 39 (1983).

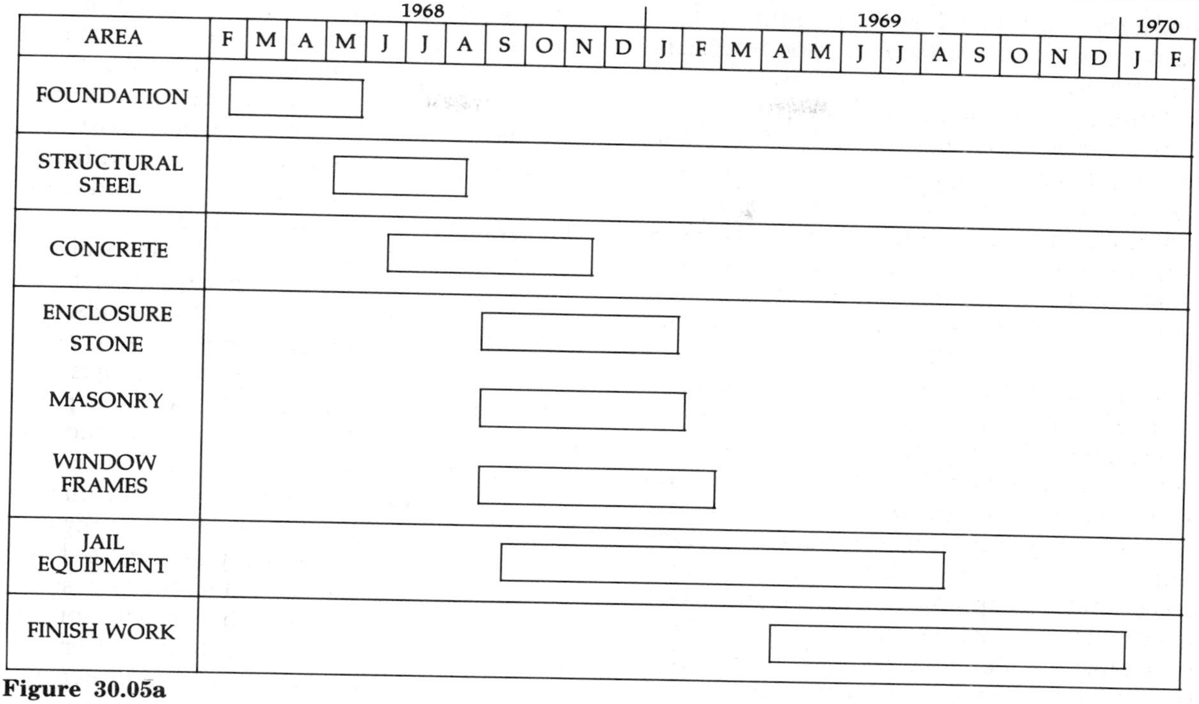

AREA	1968 F M A M J J A S O N D	1969 J F M A M J J A S O N D	1970 J F
FOUNDATION			
STRUCTURAL STEEL			
CONCRETE			
ENCLOSURE STONE			
MASONRY			
WINDOW FRAMES			
JAIL EQUIPMENT			
FINISH WORK			

Figure 30.05a

BAR CHART   PROGRESS SCHEDULE

Construction contract scheduling received great attention in the 1970s traceable to the following:

1. Greater use of management techniques and computers.

2. Roaring inflation and high cost of money.

3. Increased willingness of the law to compensate the contractor when the owner disrupted the construction schedule.

Some of these factors led to the development of new construction specialties, such as construction management and CPM consultants, as well as construction techniques such as fast tracking.

A variety of legal issues have surfaced. Of course, they are not new issues, but the now common delay damage claim, sometimes known as an impact or ripple claim, necessitates greater emphasis on the schedule, from both an operational and a legal

standpoint. This section first notes the AIA's handling of scheduling, describes the CPM method, and concludes with some description of legal issues involved.

(A) AIA

The approach taken by AIA is reflected in AIA Docs. B141 and A201. B141, ¶ 1.5.5, disclaims any architect responsibility for sequences or submissions of contractors. Before 1976, A201 required that the contractor's schedule be approved by the architect. However, beginning in 1976 ¶ 4.10.1 provided the following:

> The Contractor, immediately after being awarded the Contract, shall prepare and submit for the Owner's and Architect's information an estimated progress schedule for the Work. The progress schedule . . . shall provide for expeditious and practicable execution of the Work.

Additionally, ¶ 8.2.2 requires the contractor to "carry the Work forward expeditiously with adequate forces."

The change from requiring approval of the schedule to one that looks upon furnishing the schedule simply for *information* purposes reflects AIA's policy to limit the liability of the architect, avoid any implication that the contractor's schedule was reasonable, and help defend against any potential contractor delay claims.

Neither details nor schedule type is required. All is left to the contractor. Of course, a contractor is also interested in completing the project as promised. It *should* develop and meet a schedule that will accomplish that objective. But should responsibility be put *solely* in the hands of the contractor? Failure to meet the completion date, though giving a claim against the contractor, is not getting an ontime project. Of course, payments are keyed to work progress [19] and can be an incentive to move the work along. Failure to supply enough skilled workers may be grounds for termination,[20] though rarely will this be sought. Taken as a whole, A201 seems to ignore delay claims and contains few levers to obtain timely completion.[21]

(B) CRITICAL PATH METHOD (CPM) [22]

This system has generated burgeoning literature.[23] The description here must be simple, its goal mainly to point out the essential characteristics of the process and note the effect of float or slack time. The CPM process also relates to measuring claims (discussed in Section 31.03(E)).

To show how a CPM schedule operates, a very simple construction project will be used as illustration, without all the complexities of arrow diagrams, precedence diagrams, and nodes.

First, the contractor divides the total project into different activities or work packages. A major project may have thousands of activities, with each subcontractor generally performing a different activity.

Next, the contractor determines the activities that must be completed before other activities can be started. These constraints are the key to the CPM schedule. For example, usually excavation must be completed before foundation work can be begun. Conversely, plumbing and electrical work can usually be performed at the same time, since neither is dependent upon the other. Subcontractors performing this work can work side-by-side.

Finally, the contractor estimates how long it will take subcontractors to complete their activities. This estimate is made after the contractor consults with its subcontractors and analyzes the design drawings. This data influences the number of days allocated to each activity.

In the sample project, the following are activities and their respective durations and constraints:

Activity	Duration	Constraint
1. Excavation	7 days	None
2. Form work	5 days	Excavation
3. Concrete pour	5 days	Formwork and plumbing
4. Plumbing	4 days	Excavation
5. Electrical	2 days	Excavation
6. Roof	4 days	All

Constraints dictate the form of the CPM schedule. Since excavation has no constraints, it can be performed first. Once it

---

19. Section 26.02(E).

20. AIA Doc. A201, ¶ 14.2.

21. Much more is required in the Construction Management Documents, B801 and A201–CM.

22. This discussion owes much to student research done by James K. Graves and Jeffrey M. Chu.

23. M. Callahan & M. Hohns, Construction Schedules (1983); K. Cushman, Construction Litigation 153 (1981); Wickwire, *The Use of Critical Path Method Techniques in Contract Claims*, 7 Pub.Cont.L.J. 1 (1974).

is completed, the formwork, plumbing, and electrical activities can be performed. The concrete pour activity cannot be performed until the formwork and plumbing are completed. The roof cannot be installed until the concrete pour and electrical have been completed. The CPM schedule for this project is illustrated in figure 30.05b. The total project under this schedule should be completed in twenty one days.

The critical path, the longest path on this simple schedule, consists of those activities that will cause a delay to the *total* project if *they* are delayed. In the above example, excavation, formwork, concrete pour, and roofwork are on the critical path. A delay to any of these activities will delay the entire project.

In contrast, plumbing and electrical activities are not on the critical path. Their delay, *up to a point,* will not delay the total project. If electrical work is delayed seven days, the total project will not be delayed. The number of days each noncritical path activity can be delayed before the total project is affected is called float or slack time. In the illustration, plumbing and electrical work have one day and eight days of float, respectively.

If a noncritical path activity is delayed beyond its float period, it becomes part of the critical path. Moreover, some activities that were previously on the critical path will no longer be there. Suppose there is a three-day delay to plumbing. Originally, plumbing had one day of float. Now the CPM must be adjusted as shown in figure 30.05c.

The total project has now been delayed two days. Plumbing has become part of the critical path and formwork has moved off the path.

The use of this method can be illustrated by *Morris Mechanical Enterprises v. United States.*[24] The contractor was to deliver and install a chiller within 120 days. The contractor delivered the chiller 231 days late. As a result, the government withheld $23,100 as liquidated damages from the final payment.

In ruling for the contractor, the court pointed to the CPM. The schedule showed that delivery and installation of the chiller was originally on the critical path. However, it was later taken off because of delays to other activities for which the contractor was not responsible. The chiller was to be installed in an equipment room. Another contractor who was responsible for completing the equipment room had difficulty procuring materials. When the chiller was actually delivered, the room had not yet been completed. Even though the chiller was delivered 231 days late, the contractor's

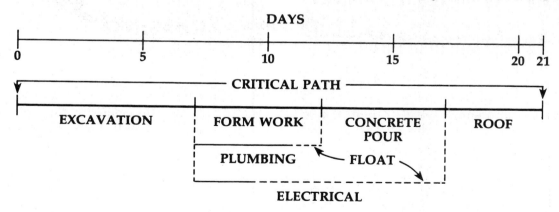

Figure 30.05b

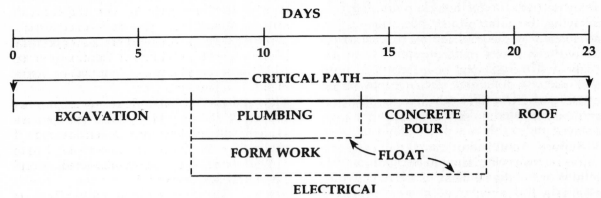

Figure 30.05c

breach did not delay the total project inasmuch as its performance was no longer on the critical path. The court relieved the contractor by concluding that it should have been given a time extension, precluding the agency from deducting from the unpaid contract balance.

Requiring the contractor to *construct* and *maintain* a CPM schedule has at least three advantages.[25] First, it should require the contractor to work more efficiently. Second, it gives the owner notice of the actual progress of the work. Third, from a litigation standpoint, requiring the contractor to maintain a CPM schedule helps prove or disprove the impact of the owner caused delay. (Schedules are more persuasive evidence if they are actually used during the construction of the project.)

There are disadvantages to a CPM schedule. First, it will increase the total contract price. Such schedules are expensive to create and maintain. Second, the contractor may believe such a requirement an unnecessary intrusion into its work. In such a case, the contractor's creation and maintenance of the schedule during construction may be haphazard.

As discussed earlier, float is the number of days a noncritical path activity can be delayed before it becomes part of the critical path. Both owner and contractor prefer that the project have float periods. The float periods reflect that every project has some flexibility. A noncritical path activity can be started a few days after it theoretically can begin without delaying the project. By contrast, a critical path activity must start immediately once the preceding critical path activity has been completed. Furthermore, a noncritical path activity does not have to be completed, as shown in the *Morris* case, by the date it was scheduled for completion. Again, by contrast, the total project will be delayed only if a critical path activity is delayed. The only constraint on a noncritical path activity is that it cannot be delayed longer than its float period.

A number of cases have explored the problem of "who owns the float."[26] Suppose activity A has a float period of thirty days and is the only activity delayed on the project. If either the owner or the contractor causes activity A to be delayed for fewer than thirty days, neither is responsible for

---

25. It is misleading to speak of *a* CPM schedule. In any project there can be up to four: as planned, as properly adjusted, as built and as would have been built. They can be vital in delay disputes. Refer to note 22 supra.

26. *Compare Brooks Towers Corp. v. Hunkin-Conkey Construction Co.*, 454 F.2d 1203 (10th Cir.1972), with *Natkin & Co. v. George A. Fuller Co.*, 347 F.Supp. 17 (W.D.Mo.1972). See generally, M. Callahan & M. Hohns, Construction Schedules, § 4–6 (1983).

delay damages. The delay did not result in delaying the total project. It can still be completed by the contract completion date. Conversely, if one of the parties causes activity A to be delayed beyond thirty days, that party is liable for delay damages because activity A has become part of the critical path and the project has been delayed.

Suppose contractor and owner each cause a twenty-day delay to activity A. The total project will be delayed ten days, since originally this activity had a thirty-day float. But who is liable for the ten-day delay? The project would not have been delayed at all if either owner or contractor had not delayed the activity.

If the contractor owned the float, the owner could not charge the contractor for project delay and would be liable for any contractor delay expenses. The contractor's argument that it owns the float can stress that *it* is responsible for scheduling the activities of the project. In a sense, it created the float because it created the CPM schedule.

But what if the owner requires a CPM schedule or demands ownership of the float? The contractor could change the float period by changing the critical path. Also, owner and contractor rarely bargain for the ownership of the float. Usually the contract provides that the contractor must complete by a certain date and the owner must pay a designated sum of money for the completed project. Implicit is that the owner will not interfere with the contractor's performance. If there were interference, the contractor would not be able to complete the project by the completion date (this seems to be the AIA hands-off approach). If the owner takes part of the float period, it interferes with the contractor's performance. If interference causes the contractor to incur delay damages, the owner should be liable for the time-related added costs. These are contractor arguments to claim the float.

Some owners take float ownership by contract.[27] Where this is done, the contractor does not receive a time extension even if delayed by the owner in a noncritical path activity if the delay can be absorbed. Completion is not delayed.[28]

A responsible contractor will inflate its bid price to cover its potential liability for a delayed project. Also, it must absorb its delay costs. At the bidding stage, the contractor cannot estimate the number of days of float the owner will use. Depending upon the complexity of the project, the total increase in the bid can be substantial. In addition, initial enforcement of float ownership clauses may be uncertain, principally because of the owner's superior bargaining position.

The *project* can own the float using one of two methods. First, the party actually causing the delay to the project is liable. Suppose, in the above example, the owner caused the first twenty-day delay. Then the contractor caused another twenty-day delay. The contractor is liable for the ten-day delay of the total project. The owner's delay had used up twenty days of the total thirty-day float, leaving ten days of float remaining. The contractor's delay used up the balance of that period but then caused the activity to be delayed beyond its float period.

This method works fairest if both parties innocently or negligently caused delays to the same activity. However, if the owner's delay had been caused intentionally or will-

---

27. As an example, ". . . . Contractor shall be entitled to an extension of time only with regard to the completion requirements affected by such delay and only by the amount of time he is actually delayed thereby in the performance of the work . . . " is language used by a strong industrial owner.

28. *E.C. Ernst, Inc. v. Manhattan Construction Co. of Texas,* 387 F.Supp. 1001 (S.D.Ala.1974) affirmed 551 F.2d 1026, 559 F.2d 268 (5th Cir.1977).

fully, the owner rather than the contractor should be liable.

The second method is borrowed from tort law. Liability for the delayed project depends upon the comparative fault of the parties. In the above example, since both parties had caused an equal delay to activity A, each is liable for one-half of the other's time-related losses. This method can become complicated if liquidated damages are used (see Section 30.09).

Each of these solutions has advantages and disadvantages. Giving *either* the contractor or the owner the float provides some certainty. A party is exonerated for delays that use up the float of a noncritical path activity if *it* exclusively owns the float.

In complex projects, it may be best to agree that neither owns the float. In such projects, it is equally likely that both parties will cause delay to the project. The owner will anticipate that it is likely to delay the project but plan on using some of the float. Nevertheless, the contract price may increase if the *project* owns the float. A responsible contractor will increase its bid when some of its previously recognized rights are taken away, but not as much as if the *owner* exclusively owns the float.

Where the project is simple, it may be best that the contractor own the float. This should develop the lowest contract price. Fewer delays by owner and contractor are likely to occur. If the owner does cause a delay and uses part of the float, the contractor's delay damages may be less than the addition to the contract price that would result if the owner took the entire float for itself.

Generally the CPM system is practical for only major construction projects. Contractors that bid for these projects already appreciate the usefulness of the schedule. The amount of money spent to construct and maintain the schedule is nominal in comparison to the total cost of the project. Liability exposure in such projects can be vast.

(C) SOME LEGAL ISSUES

Sometimes contractors, mainly subcontractors, contend that they are entitled to have their work scheduled properly. One court would not imply an obligation by the prime contractor that it would supply a schedule for subcontractor work because this was not customarily done and the subcontractor did not ask for it in negotiations.[29] The court also observed that the subcontractor had not made a showing that scheduling was important in that particular project. However, the increasing use of obligations of good faith and fair dealing may be the basis for implying that there would be a schedule, particularly if this were customarily done.

Although schedules are important and the contractor's performance can be terminated for consistent failure to meet the progress schedule, schedule dates must be more flexible than completion dates. The contractor must be given some latitude both in creating and in maintaining the schedule. Yet changing the schedule, not complying with it, and even abandoning it can be serious enough to be a breach, perhaps even a material one.

SECTION 30.06   CAUSATION

Delays can be caused by a variety of participants and events. Delay can be caused by acts of the owner or someone for whose acts the owner is responsible, such as the design professional. Sometimes delays are caused by the contractor or someone for whose acts the contractor is responsible, such as a subcontractor. Sometimes delays are caused by events not chargeable to either owner or contractor, such as nonnegligent fires, unpreventable labor difficulties, or unforeseeably extreme weather condi-

---

29. *Drew Brown Limited v. Joseph Rugo, Inc.*, 436 F.2d 632 (1st Cir.1971).

tions. Delays can be caused by third parties, such as a union shutting down the project over a labor dispute.

The multiple causes for delay create a number of legal problems. For example, delays caused by events *not* chargeable to either party will not justify additional compensation unless there is a warranty by one party that those events will not occur. If a delay is caused by both owner and contractor, it may be extremely difficult to determine which portion of the delay is caused by either party. This can mean that neither party will be able to recover losses it suffers because of the delay [30] or that any clause liquidating damages will not be upheld.[31]

Frequently the cause or causes of delay are not easy to trace. This, plus the not uncommon multiple causes of delay, makes delay a particularly complicated legal problem.

SECTION 30.07
ALLOCATION OF DELAY RISKS

At the outset, it is important to divide risk allocation relating to delay into two categories. First, the party whose performance has been delayed may seek relief from its obligation to perform by a particular time. For example, a contractor who has agreed to complete the project by a given date may justify its failure to do so by pointing to the causes that it claims excuses its obligation to complete by the time specified. Second, this contractor may claim additional compensation based upon an increase in its anticipated spending because events occurred that delayed its performance. As a general rule, the law has been more willing to excuse performance than to grant addi-

tional compensation. Because of this differentiation, speaking simply of risk allocation of events that impede performance can be misleading. This shall be demonstrated in this section.

Generally a party who has agreed to perform by a specific time assumes the risks of most events that may delay its performance. In the absence of any common law defenses, such as impossibility or mutual mistake, a contractor will not be relieved of its obligation to perform as promised, let alone receive any additional compensation. Relief, if any, must come from the contract.

Normally the contractor does not assume the risk that it will be unreasonably delayed by the owner or someone for whom the owner is responsible. However, under certain special circumstances, a contractor may have assumed even this impediment to its performance. For example, suppose an owner constructs an addition to a functioning plant. The contract provides that the owner can order the contractor off the site if the owner's manufacturing operation requires that be done. The contractor has assumed the risk of these delays, though not delays that are beyond those normally contemplated, such as constant or excessive delays or those caused by the owner's bad faith.

Weber Construction Co. v. State [32] involved a railroad construction contract that required work by the contractor to be coordinated with that of the railroad. The contract provided that the contractor should take into account possible delay or interruption caused by the operation or maintenance of the railroad at the time it determined its contract price. Although the contractor received time extensions when

---

30. *Westinghouse Electric Corp. v. Garrett Corp.,* 601 F.2d 155 (4th Cir.1979) (discretion in trial judge to deny *both* damages); *Armour & Co. v. Scott,* 360 F.Supp. 319 (W.D.Pa.1972) (owner received restitution for overpayments), affirmed 480 F.2d 611 (3d Cir.1973); *Hartford Electric Applicators of Thermalux, Inc. v. Alden,* supra note 2, (breach by *both* converts completion date to reasonable time).

31. *Hartford Electric Applicators of Thermalux, Inc. v. Alden,* supra note 2. But see Section 30.09.

32. 37 A.D.2d 232, 323 N.Y.S.2d 492 (1971) aff'd 30 N.Y.2d 631, 331 N.Y.S.2d 443, 282 N.E.2d 331 (1972).

the railroad delayed its performance, it was not given additional compensation.

The law can take into account the blameworthiness of the party causing the delay. For example, in *Broome Construction, Inc. v. United States*,[33] the Court of Claims held the government not liable for delay in making a worksite available where it sought to do so in good faith. The contractor assumed the risk of *this* delay but would not have assumed the risk of *negligent* delay. The contractor's request for additional compensation requires a clause expressly warranting that the government was making itself *strictly* liable for its failure to furnish the site.

Fault can also play a role when a party who has been delayed seeks to be relieved from its responsibility. For example, in *J.D. Hedin Construction Co. v. United States*,[34] the Court of Claims held that a contractor was entitled to a time extension because of a cement shortage. The court criticized findings by the Agency Appeals Board that the delays were foreseeable and were the fault of the contractor. The court stated that the contractor could not be expected to show prophetic insight but need only resort "to the usual and long-established methods employed by the commercial world in general."[35]

Many events can occur that cause delay in construction. The common law placing almost all these risks upon the contractor has led to the frequent use of *force majeure* clauses, which single out specific events as justifying relief to the contractor. As an illustration AIA Doc. A201 provides the following:

> 8.3.1 If the Contractor is delayed at any time in the progress of the Work by any act or neglect of the Owner or the Architect, or by any employee of either, or by any separate contractor employed by the Owner, or by changes ordered in the Work, or by labor disputes, fire, unusual delay in transportation, adverse weather conditions not reasonably anticipatable, unavoidable casualties, or any causes beyond the Contractor's control, or by delay authorized by the Owner pending arbitration, or by any other cause which the Architect determines may justify the delay, then the Contract Time shall be extended by Change Order for such reasonable time as the Architect may determine.

Note that "adverse weather conditions" must not be "reasonably anticipatable," an attempt to tighten up weather time extensions.[36]

The first catchall, "any causes beyond the Contractor's control," would be subject to the *ejusdem generis* guide.[37] Events claimed to fall within the general clause must be *similar* to the specific events. For example, it may not encompass an extreme rise in labor wages due to the outbreak of war but may encompass delay in transportation of workers to the site caused by a bankruptcy of a transportation company upon which the contractor relied.

This catchall can invoke the *expressio unius est exclusio alterius* rule.[38] This interpretation guide asserts that anything not included in the catalog of events must have been intended to be excluded. Suppose there is an abrupt and sharp rise in the cost of material. This has not been included.

---

33. 203 Ct.Cl. 521, 492 F.2d 829 (1974). But in *Hartford Electric Applicators of Thermalux, Inc. v. Alden,* supra note 2, a 40-day delay in furnishing the site was found to be an owner breach but not a material breach allowing termination.

34. 187 Ct.Cl. 45, 408 F.2d 424 (1969).

35. 408 F.2d at 429.

36. See *Shea-S&M Ball v. Massman-Kiewit-Early,* 606 F.2d 1245 (D.C.Cir.1979) (flood must be of unprecedented occurrence of unusual proportions and not foreseeable).

37. Refer to Section 24.02.

38. Refer to Section 24.02.

This would normally exclude it, except for the catchall. But *ejusdem generis* must be applied. Is this event "like" the events listed?

Suppose the prime contractor seeks to excuse its responsibility for delay by pointing to the delay having been caused by a subcontractor. Unlike some contracts, ¶ 8.3.1 does *not* expressly include subcontractor-caused delay among those that justify a time extension. It does give the architect the right to grant a time extension for any cause that justifies the delay. However, failure to specifically include a common cause of delay, such as subcontractor-caused delay, should mean that the subcontractor-caused delay should not be encompassed within the catchall or within the broad grant of power given the architect.

Another reason for not granting a time extension is the single contract system's objective of centralizing administration and responsibility in the prime contractor. Only if subcontractor-caused delay is *specifically* included should it excuse the prime contractor.

The independent contractor rule, though subject to many exceptions, relieves the employer of an independent contractor for the losses wrongfully caused by the latter.[39] Contractors sometimes assert that the subcontractor is an independent contractor inasmuch as the subcontractor is usually an independent business entity and can, to a large extent, control the details of how the work is performed. Even so, the independent contractor rule does *not* relieve the employer of an independent contractor when the independent contractor has been hired to perform a contract obligation and the party who suffers the loss caused by the independent contractor is the party to whom the contract obligation was owed.[40]

In the construction contract context, the owner usually permits the prime contractor to perform obligations through subcontractors. This does not usually mean that the prime contractor is relieved of its obligation to the owner unless the owner specifically agrees to exonerate the prime contractor.

The residuary power granted to the architect to grant time extensions has been criticized as leading to a deterioration of any fixed completion date. It does have the virtue of not forcing the drafter to think of every possible event and include it in the "catalog of events" justifying a time extension.

Delays caused by a separate contractor by another separate contractor were discussed in Section 21.04(C).

SECTION 30.08
TIME EXTENSIONS

Construction contracts usually provide a mechanism under which the contractor will receive a time extension if it is delayed by the owner or by other designated events such as those described in Section 30.07. This section examines the time extension process.

Ideally, owner and contractor should agree upon the issuance and extent of a time extension. In many construction contracts, the resolution of these issues, at least in the first instance, is given to the design professional or construction manager. The finality of her decision in the event of subsequent arbitration or litigation is discussed ahead.[41] Whether a time extension should be granted usually requires that the *force majeure* clause be applied to the facts asserted to constitute justification for time extension.

Suppose a time extension is justified.

---

39. See Section 35.05(B).

40. *Harold A. Newman Co. v. Nero*, 31 Cal.App.3d 490, 107 Cal.Rptr. 464 (1973). See also Section 35.05.

41. See Section 33.09.

How is the amount of time extension determined? To some degree, this was discussed in Section 30.05 dealing with scheduling. Depending upon the existence of float and who can take the benefit of it, delay in the performance of a particular activity may not justify any time extension.

Suppose abnormal weather conditions not normally anticipatable precluded work from October 1 to October 14. Suppose that period contained ten working days. Unless the contractor can show that it would have worked on other than normal working days, the time extension should be ten days.[42]

However, the extent of time extension need not necessarily be the same as the number of days of delay. Suppose a fourteen-day delay caused the contractor to work during a period when the weather was more rigorous than during the period of delay. It may be fair under these circumstances to grant a time extension of fifteen days when the contractor was actually precluded from working for ten days.

One court faced the imprecision of measuring the exact impact of delay. That court arbitrarily granted a time extension of 65 days for a 131-day delay, because this amount was "as accurate an estimate as can be made from the actual resulting delay."[43] Whoever determines the amount of delay will be given considerable latitude. Mathematical precision is rarely possible. However, it is vital for both owner and contractor to keep careful and detailed records. See Section 30.10(E).

Usually time extension mechanisms provide that the contractor must give notice of the occurrence of an event that is to be the basis for a time extension claim. For example, ¶ 8.3.2 of AIA Doc. A201 provides the following:

> Any claim for extension of time shall be made in writing to the Architect not more than twenty days after the commencement of the delay; otherwise it shall be waived. In the case of a continuing delay only one claim is necessary. The Contractor shall provide an estimate of the probable effect of such delay on the progress of Work.

Some courts seem to consider such notices as technicalities. For this reason, these courts seem quick to find that the notice condition has been waived if it appears that the owner, or someone with actual or apparent authority, knew of the delay-causing event and was not harmed by failure to give the notice. Likewise, the requirement will be waived if in any way the owner has misled the contractor into believing that the notice will not be required.[44]

However, notice conditions serve a useful function. In dealing with the requirement that a homebuyer give notice of a breach of warranty within a reasonable time after discovery of the breach, a court stated:

> The requirement of notice of breach is based on a sound commercial rule designed to allow the defendant opportunity for repairing the defective item, reducing damages, avoiding defective products in the future, and negotiating settlements. The notice requirement also protects against stale claims.[45]

Likewise, in the context of a time extension mechanism, the notice of an intention to claim a time extension serves a number of useful functions. The notice informs the design professional or owner that persons for whom they are responsible, such as other separate contractors, consultants, or the

---

42. *Missouri Roofing Co. v. United States*, 357 F.Supp. 918 (E.D.Mo.1973), appears to support this conclusion.

43. *E.C. Ernst, Inc. v. Manhattan Construction Co. of Texas*, supra note 28.

44. *Travelers Indemnity Co. v. West Georgia National Bank*, 387 F.Supp. 1090 (N.D.Ga.1974) (provision that HUD approve time extension waived). For additional discussion on waiver of technical requirements refer to Section 25.04(G).

45. *Pollard v. Saxe & Yolles Development Co.*, 12 Cal.3d 374, 115 Cal.Rptr. 648, 652, 525 P.2d 88, 92 (1974).

design professional, are delaying the contractor. This can enable owner or design professional to eliminate the cause of the delay and minimize future delays or damage claims. A timely notice should permit the design professional to determine what has occurred while the evidence is still fresh and witnesses remember what actually transpired.

The notice shows that the contractor has been adversely affected and can eliminate long-delayed, sometimes spurious contractor claims made after completion of the work. If the owner or design professional knew of the event causing the delay, the impact of the event upon the contractor's performance, and that it was quite likely that the contractor would make a time extension request, waiver may be proper.[46] However, the value of the notice requirement is substantially reduced if the requirement is too frequently ignored by courts.[47]

SECTION 30.09 UNEXCUSED CONTRACTOR DELAY AND DAMAGE LIQUIDATION: *BETHLEHEM STEEL CORP. v. CITY OF CHICAGO*

Delay as justification for termination is discussed in Section 38.04(A). This section deals with the recovery of damages for delay. Although measurement of claims owners and contractors have against each other is discussed in greater detail in Chapter 31, this section deals with damage liquidation, the contractual method used most frequently to deal with contractor delay.

As a backdrop to damage liquidation, it should be noted that the damage formula applied most frequently by the common law to delayed contractor performance is the value of the lost use of the project caused by the delay. For example, delayed completion of a residence to be occupied by the owner would very likely be measured by the lost rental value.[48] Although the owner may have suffered other losses, such as the inconvenience of living in a motel or with relatives or having to transport a child to a more distant school, such consequential damages would generally be difficult to recover.

In commercial construction, the owner will very likely be able to recover the lost use value in the event of unexcused delay by the contractor.[49] Even in projects that have readily ascertainable commercial value, losses are often suffered that may be difficult to recover. This is even a greater problem when the project is built for a public entity that intends to use it as a school, an office building, or a freeway. Although some public projects have a readily ascertainable use value, many do not.

Because proof of delay damages is very difficult, particularly in public projects, construction contracts commonly include provisions under which the parties agree that certain types of unexcused delay will result in damages of a specific amount. They are usually known as liquidated damages clauses. A court recently stated:

> There was a time when the courts were quite strong in their view that almost every contract clause containing a liquidated damage provision was, in fact, a forfeiture provision which equity abhorred, and therefore, nothing but actual damages sustained by the aggrieved party could be recovered in case of contract breach caused by delay past the proposed completion date.

---

46. *Southwest Engineering Co. v. Reorganized School District R–9,* 434 S.W.2d 743 (Mo.App.1968).

47. Strictly enforced in *Herbert & Brooner Construction Co. v. Golden,* 499 S.W.2d 541 (Mo.App.1973), and *Quin Blair Enterprises, Inc. v. Julien Construction Co.,* supra note 6. Narrowly construed in *Hartford Electric Applicators of Thermalux, Inc. v. Alden,* supra note 2.

48. *Muller v. Light,* 538 S.W.2d 487 (Tex.Civ.App.1976) ($100 a day clause held a penalty).

49. *Ryan v. Thurmond,* 481 S.W.2d 199 (Tex.Civ.App.1972).

But, in modern times, the courts have become more tolerant of such provisions, probably because of the Anglo-Saxon reliance on the importance of keeping one's word, and have become more strongly inclined to allow parties to make their own contracts and to carry out their own intentions, free of judicial interference, even when such non-intervention would result in the recovery of a prestated amount as liquidated damages, upon proof of a violation of the contract, and without proof of actual damages.[50]

The early common law also felt that contractually stipulated damages could be used unconscionably by parties possessing strong bargaining power. The common law saw its role in awarding contract damages as compensating losses and not punishing or effectuating abuse of power.

The earliest use of such clauses were penal bonds, a precursor of surety bonds explored in Chapter 37. Penal bonds were *absolute* promises by the maker to pay a designated sum (the penal sum) followed by a condition that this promise would be null and void *if* certain things occurred, such as the maker, usually a debtor or a surety, paying the full amount of the debt. Frequently the penal sum greatly exceeded the harm caused. Equity courts would not enforce penalties. (This is the historical basis for differentiating liquidated damages clauses from penalties.) In addition, English courts felt that damages were the exclusive province of the courts. They would not enforce clauses which would "oust the court of jurisdiction."

Despite the 19th century common law courts belief in liberty and autonomy, they continued what equity courts had started by placing strict limits on enforcement of these clauses. Courts were willing under *limited* circumstances to enforce liquidated damages clauses. This would be done if there were a showing by the party seeking to enforce the clause that the damages were extremely difficult to ascertain at the time the contract was made and that the amount selected was a genuine preestimate of the damages likely to occur from the breach. In addition, some courts would not enforce these clauses unless there were actual damages and unless it were shown that the parties intended the clause to compensate and not to punish. The amount selected must be the sole money award. The proponent should not have the option of choosing liquidated or actual damages.

Modern courts, particularly in public contracts, recognize the difficulty of proving damages generally, particularly those relating to delay and the certainty that these clauses can provide *both* parties. As indicated, the law is more willing to be relieved of the burden of measuring damages.[51] Although the results are by no means unanimous, clauses liquidating damages for construction delay are generally enforced if they are reasonable as judged by the circumstances existing at the time the contract was made.[52] Often a comparison is made between the amount stipulated and the contract price.[53] Some states even require that such clauses be inserted in public construction contracts.[54] A case that typifies the modern judicial attitude toward such clauses is reproduced here.

50. *Sides Construction Co. v. City of Scott City*, 581 S.W.2d 443, 446 (Mo.App.1979) annotated in 12 A.L.R.4th 891 (1982) collecting many cases.

51. *Sides Construction Co. v. City of Scott City*, supra note 50 (avoids laborious item-by-item damage recitations.)

52. *Dahlstrom Corp. v. State Highway Commission of State of Mississippi*, 590 F.2d 614 (5th Cir.1979); *Pembroke v. Gulf Oil Corp.*, 454 F.2d 606 (5th Cir.1971); *Dave Gustafson & Co. v. State*, 83 S.D. 160, 156 N.W.2d 185 (1968). Cases collected in 12 A.L.R.4th 891 (1982). The Uniform Commercial Code (UCC) enforces them if reasonable in light of *actual* or *anticipated* damages. See § 2–718.

53. *Dahlstrom Corp. v. State Highway Commission of State of Mississippi*, supra note 52.

54. *Westinghouse Electric Corp. v. County of Los Angeles*, 129 Cal.App.3d 771, 181 Cal.Rptr. 332 (1982).

BETHLEHEM STEEL CORP. v. CHICAGO

United States Court of Appeals, Seventh Circuit, 1965.

350 F.2d 649.

GRANT, District Judge.

[Ed. note: Footnotes omitted.] Plaintiff-Appellant (Bethlehem) brought this action to recover an item of $52,000.00 together with certain items of interest, etc., withheld by the Defendant (City), as liquidated damages for delay in furnishing, erecting, and painting of the structural steel for a portion of the South Route Superhighway, now the "Dan Ryan Expressway", in the City of Chicago. . . . [T]he District Court concluded that Plaintiff's claims on the items in controversy should be denied and entered judgment accordingly. We agree and we affirm.

The trial court's findings included the following uncontroverted facts:

*   *   *

"The work which Bethlehem undertook was the erection in Chicago of structural steel for a 22-span steel stringer elevated highway structure, approximately 1,815 feet long, to carry the South Route Superhighway from South Canal Street to the South Branch of the Chicago River. Bethlehem's work was preceded and followed by the work of other contractors on the same section.

"The 'Proposal and Acceptance' in the instructions to bidders required the bidders to '. . . complete . . . within the specified time the work required. . . .' Time was expressly stated to be the essence of the contract and specified provisions were made for delivery of the steel within 105 days thereafter, or a total of 195 days after commencement of work, which was to be not later than 15 days from notification. The successful bidder was to submit to the Commissioner of Public Works a 'Time Schedule' for his work and if 'less than the amount . . . specified to be completed' were accomplished 'the City may declare this contract forfeited. . . .' The work had to be completed irrespective of weather conditions.

"The all important provision specifying $1,000 a day 'liquidated damages' for delay is as follows:

'The work under this contract covers a very important section of the South Route Superhighway, and any delay in the completion of this work will materially delay the completion of and opening of the South Route Superhighway thereby causing great inconvenience to the public, added cost of engineering and supervision, maintenance of detours, and other tangible and intangible losses. Therefore, if any work shall remain uncompleted after the time specified in the Contract Documents for the completion of the work or after any authorized extension of such stipulated time, the Contractor shall pay to the City the sum listed in the following schedule for each and every day that such work remains uncompleted, and such moneys shall be paid as liquidated damages, not a penalty, to partially covered losses and expenses to the City.

'Amount of Liquidated Damages per Day . . . $1,000.00.

'The City shall recover said liquidated damages by deducting the amount thereof out of any moneys due or that may become due the Contractor. . . . '

"Provision was made to cover delay in a contractor's starting due to preceding contractor's delay. Unavoidable delays by the contractor were also covered, and extensions therefor accordingly granted."

Bethlehem's work on this project followed the construction of the foundation and piers of the superhighway by another contractor. Bethlehem, in turn, was followed by still another contractor who constructed the deck and the roadway.

Following successive requests for extensions of its own agreed completion date, Bethlehem was granted a total of 63 days' additional time within which to perform its contract. Actual completion by Bethlehem, however, was 52 days after the extended date, which delay the City assessed at $1,000.00 per day, or a total of $52,000.00 as liquidated damages.

Bethlehem contends it is entitled to the $52,000.00 on the ground that the City actually sustained no damages. Bethlehem contends that the above-quoted provision for liquidated damages is, in fact, an invalid penalty provision. It points out that notwithstanding the fact that it admittedly was responsible for 52 days of unexcused delay in the completion of its contract, the superhighway was actually opened to the public on the date scheduled.

In other words, Bethlehem now seeks to re-write the contract and to relieve itself from the stipulated delivery dates for the purposes of liquidated damages, and to substitute therefor the City's target date for the scheduled opening of the superhighway. This the Plaintiff cannot do.

In Wise v. United States, . . . the Supreme Court said:

". . . [T]he result of the modern decisions was determined to be that . . . courts will endeavor, by a construction of the agreement which the parties have made, to ascertain what their intention was when they inserted such a stipulation for payment, of a designated sum or upon a designated basis, for a breach of a covenant of their contract . . . . When that intention is clearly ascertainable from the writing, effect will be given to the provision, as freely as to any other, where the damages are uncertain in nature or amount or are difficult of ascertainment or where the amount stipulated for is not so extravagant, or disproportionate to the amount of property loss, as to show that compensation was not the object aimed at or as to imply fraud, mistake, circumvention or oppression. *There is no sound reason why persons competent and free to contract may not agree upon this subject as fully as upon any other, or why their agreement, when fairly and understandingly entered into with a view to just compensation for the anticipated loss, should not be enforced.*

". . . *The later rule, however, is to look with candor, if not with favor, upon such provisions in contracts when deliberately entered into between parties who have equality of opportunity for understanding and insisting upon their rights, as promoting prompt performance of contracts and because adjusting in advance, and amicably, matters the settlement of which through courts would often involve difficulty, uncertainty, delay and expense. . . .*

". . . It is obvious that the extent of the loss which would result to the Government from delay in performance must be uncertain and difficult to determine and it is clear that the amount stipulated for is not excessive . . . .

"The parties . . . were much more competent to justly determine what the amount of damage would be, an amount necessarily largely conjectural and resting in estimate, than a court or jury would be, directed to a conclusion, as either must be, after the event, by views and testimony derived from witnesses who would be unusual to a degree if their conclusions were not, in a measure, colored and partisan." (Italics supplied.)

* * *

Affirmed.

The *Bethlehem* case illustrates increased judicial cordiality toward liquidated damages clauses in the construction context, particularly public contracts. The law is more willing to enforce these contracts if they have been bargained for and if actual damages would be difficult to establish in court. Although the amount stipulated is rarely the result of bargaining in a competitive bid contract, the contractor can adjust its bid to take this risk into account.[55]

Despite greater willingness to enforce such clauses, additional legal issues reflect some judicial ambivalence. For example, courts generally will not enforce liquidated damages clauses unless there is a rational apportionment of that portion of the damage caused by the contractor and the owner. Usually this is accomplished by a clause granting time extensions for certain delay.[56] In addition, jurisdictions vary as to whether the proponent of a liquidation clause must

---

55. See Sweet, *Liquidated Damages in California*, 60 Calif.L.Rev. 84, 118–123 (1972).

56. *E.C. Ernst v. Manhattan Construction Co. of Texas*, supra note 28 (enforced); *Jasper Construction Inc. v. Foothill Junior College District*, 91 Cal.App.3d 1, 153 Cal.Rptr. 767 (1979) (enforced); *Hartford Electric Applicants of Thermalux, Inc.*, supra note 2 (not enforced).

bear the burden of establishing its enforceability. Jurisdictions hostile to liquidation place the burden upon the proponent,[57] but the modern tendency has been to enforce such clauses unless the opponent can show invalidity.[58]

Despite increased enforceability of reasonable liquidated damages clauses, traditional suspicion of these clauses makes it important to consider particular issues that surface when such clauses, though clearly enforceable, must be interpreted.

Should a liquidation clause be applied if the contractor *abandons* the project? It can be contended that parties drafting such clauses are thinking principally of *delayed* completion by the contractor and not *abandonment*, particularly if followed by abandonment of the project by the *owner*.[59] While application can appear to create open-ended liability, the clause can be applied for a reasonable period of time, not for infinity. Another way of dealing with this problem is to contractually "cap" the liquidation amount and thereby avoid the open-ended liability that can generate a forfeiture vastly disproportionate to the actual damages.[60]

Unjustified abandonment by the contractor or proper termination by the owner can cause the owner *two* harms: additional cost to complete the project and additional cost caused by delayed completion. Damage liquidation can apply to the second, delay in completion determined by the actual completion by a successor or when the project

could have been reasonably completed by a successor contractor.[61]

Another problem relates to the difference between substantial and final completion of the project. AIA Doc. A201, ¶ 8.1.1, defines completion for liquidation purposes as substantial completion. This issue was treated in *Hungerford Construction Co. v. Florida Citrus Exposition, Inc.*[62] The contract involved an exhibition center for the citrus industry. One important feature of the project was a concrete dome 170 feet in diameter that was to operate as its roof. The roof was to be waterproof without independent waterproof covering. Completion time was 180 calendar days, and the liquidated damages were specified to be $200 per calendar day of delay.

The project was completed within 180 days, and the owner moved into the project. However, from the beginning, the roof leaked and it was necessary to do corrective work. The corrective work did not require that the owner leave the premises, but it did preclude the owner from making the premises available on a rental basis to those who might want to use it for exhibition purposes. The court referred to this use as a secondary use. Even after the corrective work, the secondary use was diminished because there was unsightly discolored plaster across the roof caused by the leaking and correction.

The court held that the liquidated damages clause would not be applied in this case. The court stated that the building was available to the owner for its primary

---

57. *Utica Mutual Insurance Co. v. DiDonato,* 187 N.J.Super. 30, 453 A.2d 559 (1982).

58. A modern California statutory revision puts the burden on the party attacking the clause. West's Ann.Cal.Civ. Code § 1671(b).

59. 5 S. Williston, Contracts, § 785 (3d ed.1961).

60. *Reed & Martin, Inc. v. Westinghouse Electric Corp.,* 439 F.2d 1268 (2d Cir.1971).

61. Compare *City of Boston v. New England Sales & Manufacturing Corp.,* 386 Mass. 820, 438 N.E.2d 68 (1982), and *Austin Griffith, Inc. v. Goldberg,* 224 S.C. 372, 79 S.E.2d 447 (1953) (limited to abandonment after completion date), with *Continental Realty Corp. v. Andrew Crevolin Co.,* 380 F.Supp. 246 (S.D.W.Va.1974) (refused to apply clause).

62. 410 F.2d 1229 (5th Cir.1969).

use and that the loss of secondary use was entirely speculative. The court stated:

> One auto show may have been lost but there is no evidence as to the amount of rent which would have been realized out of this transaction or whether the loss occurred during the period in suit. The only other loss of use claimed was in the form of a daily admission charge to the public to see the building and its contents. The proof demonstrated only that this use did not rise above a suggested use. In any event, such loss of secondary use as there may have been is capable of proof. The proof that was offered was entirely disproportionate to the sum of $200.00 . . . per day. It is thus clear that the claim for liquidated damages was far in excess of such compensatory damages as would be indicated from the slight deprivation of secondary use claimed by the owner.[63]

To deal with this problem, the contract can create a two-tier liquidation system, with one tier dealing with substantial completion and the other with final completion. This would avoid the *Hungerford* case result in which it is likely that the owner will be unable to establish losses with sufficient certainty for the reduced use value of the structure while corrective work was being performed.

Applicability of the clause also was an issue in *Northern Petrochemical Co. v. Thorsen & Thorshov, Inc.*[64] After completion, serious structural defects were discovered. This necessitated large-scale redesign and reconstruction. Correction of the defects took eight months. The court affirmed an award based upon lost profits because of delayed occupancy and excess operating costs while awaiting occupancy. The court rejected application of the contractual liquidated damages clause, noting that it was intended to apply to *normal* delay in the "rate of construction," not an extraordinary

eight months delay due to redesign and reconstruction.

The court's conclusion would limit liquidation to normal, expectable delays where there is *completion*. But suppose the proof of actual damages was so difficult and the liquidation amount high. The owner would then have sought to enforce the clause. Theoretically, such an attempt would fail.

Suppose the defects took one month to correct. Delays caused by defects discovered and corrected during performance and before completion are within the clause. There is no reason to draw a line at completion. But a line *can* be drawn, as did the court, at *extraordinary* delay whether caused by correction of defects during or after completion. To be sure, this does inject an element of uncertainty. But the law should respect the intention of the parties, and the intention was not to deal with extraordinarily long delays.

The use of fast tracking[65] can generate particular problems, especially if the project is being built by a number of separate contractors or multiple prime contractors. In *Casson Construction Co.,*[66] the contractor, though finishing on time, failed to complete a phase of its work that delayed a follow-on contractor. The follow-on contractor was ordered to accelerate. It did so by the use of overtime and double shifts for several weeks. This acceleration cost $174,000.

The breaching contractor claimed that its liability was limited to the liquidation of damage amount of $240 a day contained in *its* contract. The cost of acceleration was $644 a day. The Board pointed to another provision in the contract that stated that the contractor would indemnify the public agency for acceleration payments made to other contractors. The court held that the latter clause controlled and that the same

---

63. Id. at 1232.

64. 297 Minn. 118, 211 N.W.2d 159 (1973).

65. Refer to Section 21.04(B).

66. GSBCA No. 4884, 78–1 BCA. ¶ 13,032.

breach can invoke different clauses. The Board would not apply the liquidated damages clause to milestone date delays, only delay in completing the *entire* contract performance. The Board noted that the standardized language used in the contract was adopted long before the advent of fast track construction.

Subcontracting can raise special problems. In *P & C Thompson Brothers Construction Co. v. Rowe* the subcontract liquidated damages for delay at $250 a day.[67] The prime contractor deducted from the subcontractor's final payment its *actual* damages, the liquidated sum and a pro rata portion of the liquidation of damages it had to pay the owner. The court would not enforce the clause, as it would be more than the prime contractor would have to pay the owner. The court found this to be unconscionable.

Actual damages suffered by the prime contractor itself can be separated from amounts that the prime contractor must pay the owner.[68] There is no reason why the liquidated damages clause cannot encompass *both*. Although sometimes the amount that the prime contractor may have to pay the owner can be ascertained at the time the contract is made (the stipulated amount in the owner-prime contract), the *other* damages that the prime may suffer *in addition* may be difficult to establish. Two provisions *can* be included, one making clear that the subcontractor will reimburse the prime for any amounts it must pay the owner attributable to the subcontractor's breach and another to liquidate damages for other harm suffered by the prime. But the parties should be able to include *both* in one clause.

How important is enforceability? If the clause is unenforceable, the owner can still prove actual damages. For this reason, some drafters insert a stiff liquidation amount, with actual damages still recoverable if the clause is found to be a penalty. This is not desirable. It gives up the possibility of a *sure* liquidation. Actual delay damages are extremely difficult to prove. (But an unforceable penalty may still get on-time completion.)

Project completion dates can be "hard" or "soft," depending upon the importance attached to timely completion. When a high liquidation amount is selected, the contractor should increase its contract price either to insure timely completion (double time, expedited deliveries, more workers) or to pay damage liquidation. Coupling a soft completion date with a stiff damage liquidation amount will generate timely completion at a high cost when timely completion is not crucial.

When the completion date is hard, the amount selected must be sufficiently high (but not too high so as to risk finding that it is a penalty) to make it more profitable for the contractor to finish on time than to delay and pay liquidated damages. Selecting an amount to accomplish this objective requires a sophisticated understanding of contractor costs. It involves awareness of the increased administrative costs of such a clause. Invariably there will be more requests for time extensions. Any stiff clause should be accompanied with a contractual justification for the amount, as in the *Bethlehem* case.

This advice is inconsistent with the stated requirement that liquidated damages be based upon a genuine preestimate of actual damages that will result in the event of delay. In many construction contracts, mainly public but also private, the amount selected does not in reality represent an

---

67. 433 So.2d 1388 (Fla.App.1983). See AIA Doc. A401. ¶ 12.5.1. See also *Industrial Indemnity Co. v. Wick Construction Co.,* 680 P.2d 1100 (Alaska 1984) (subcontractor could take advantage of liquidation clause in prime contract because of "flow through" clause). See Section 32.04.

68. *United States v. Foster Construction (Panama) S.A.,* 456 F.2d 250 (5th Cir.1972).

attempt to estimate damages. In such contracts, it is almost impossible to estimate the economic loss caused by delay. For example, it is likely that the $1,000 a day selected in the *Bethlehem* case did not reflect the City of Chicago's judgment as to the economic loss that it or its citizens would suffer in the event the project were delayed. It is more likely that the amount was selected to make it more economical for the contractor to perform on time than to delay and pay damages. Although courts do not *overtly* concede that they are doing so, their enforcement of liquidated damages clauses for a delay in many construction contracts amounts to enforcing reasonable penalties.

There are possible variations to a per diem formula. For example, a recent contract made by the San Francisco Golden Gate Bridge District for repair of the Golden Gate Bridge specified that for each ten minutes of delay, the liquidation amount would be $1,000. This was done to obtain the promised performance by the contractor of finishing before the morning rush hours.[69]

Another variant was observed in *Grenier v. Compratt Construction Co.*,[70] which involved a blasting contract. The work had to be done before constructing the roads in a subdivision. The contract stated that damages were liquidated at $1,500 at the end of the first week, an additional $2,000 at the end of the second week, an additional $2,500 at the end of the third week, and $3,000 for each additional week of unexcused delay. Despite a "penalty" label, the court enforced it.

Why did the amount go up for each week's delay? Perhaps the developer would incur escalating liability or lost income as the delay continued. It is more likely that the developer was seeking to coerce performance by making it increasingly expensive for the contractor not to complete on time.

Drafting also must take into account the difference between a damage liquidation and a limitation of liability. As indicated earlier, a valid liquidated damages clause establishes the damages in advance. The stipulated amount is recoverable whether actual damages, if they are established, are higher or lower. A clause liquidating damages requires that actual damages be proved but limits the amount recoverable. In *Burns v. Hanover Insurance Co.*,[71] the clause read:

> The Owner will suffer financial loss if the Project is not completed by the above date. The Contractor shall be liable for and paid to the Owner on an actual expense basis as established by receipts, not more than $1,000 for packing and storage of furnishings and $30 per day for temporary accommodation.

The majority held this to be a valid liquidated damages clause. The dissent correctly stated that the clause simply put a limitation of liability but required that actual losses be established.

Sometimes liquidation clauses are joined with bonus provisions. Under such clauses, the contractor forfeits a designated amount for each day of unexcused delay but gains a designated daily bonus if it completes the project in advance of the completion date. Although enforceability does not require that a bonus be attached to a damage liquidation, it may be tactically desirable to do so. It may help to enforce the liquidation clause, with its "mutuality" attractiveness. Its use may also make it appear that the amount had actually been bargained by the

---

69. San Francisco Sunday *Examiner & Chronicle,* Feb. 19, 1984 p. B–2. The English Department of Transportation charges repair contractors who are late a "lane rental" fee of up to $14,000 a day. The Economist, 1–7 Sept. 1984, p. 27.

70. 189 Conn. 144, 454 A.2d 1289 (1983).

71. 454 A.2d 325 (D.C.App.1982).

contractor and owner. However, a bonus clause should not be used unless it is very important to obtain performance in advance of the completion date.

Liquidated damage clauses should specifically give the owner the right to deduct the amount from the final payment or any retainage even if it may be possible to do so without such a provision.

Damage liquidation for unexcused contractor delay is usually desirable, but it must be drafted carefully. First, the applicable law must be determined. Second, the clause must be tailored to the particular type of delay to which it is expected to apply. Third, the amount selected should take into account the importance of timely completion, the likely lost use value, and the likelihood the amount selected will actually achieve the objective.

SECTION 30.10
OWNER–CAUSED DELAY

(A) ASSUMED CONTRACTUAL RISK

Increasingly contractors make large claims for delay damages. As a result, it is becoming even more common for clauses in the contracts—particularly public contracts and private ones drafted largely with the interests of the owner in mind—to attempt to make the contractor assume the risk of owner-caused delay.

This process can take various forms. The contract may specify that the owner has the right to delay the contractor and that the interference is not a contract breach. Another technique is to contend that any time-extension mechanism is the *exclusive* remedy. Generally the availability and use of the time-extension mechanism does not impliedly preclude delay damages.[72] The AIA's time-extension mechanism specifically states that it does *not* bar the contractor from recovering delay damages.

An indirect technique for limiting delay claims is to require written notice by the contractor to the owner or design professional if events have occurred that later will be asserted as justifying delay damages. Often this notice is stated to be a condition precedent to any right to delay damages. Although such clauses are not looked upon favorably by courts and are frequently found to have been waived,[73] noncompliance with a notice provision can be the basis for barring a claim for delay damages.[74]

Strangely, AIA Doc. 201, ¶ 8.3.2, requires a notice as a condition for a time extension but does not apparently require a notice as a condition for delay damages. The claims provision ¶ 7.4.1 requires a notice only for claims relating to "injury or damage to person or property." This anomaly can result in failure to give notice barring a time extension but not a delay damage claim.

Many public and some private construction contracts meet the delay damage problem head on. Rather than setting up notice conditions or specifying what can be recovered, these contracts contain no-damage or no-pay-for-delay clauses. Such clauses attempt to place the entire risk for delay damages upon the contractor and to limit the contractor to time extensions. Generally such clauses are enforced.[75] However,

---

72. *Selden Breck Construction Co. v. Regents of University of Michigan,* 274 F. 982 (E.D.Mich.1921). See *Glantz Contracting Co. v. General Electric,* 379 So.2d 912 (Miss. 1980); *Bender-Miller Co. v. Thomwood Farms, Inc.,* 211 Va. 585, 179 S.E.2d 636 (1971) (remedy not exclusive).

73. *Montgomery Ward & Co. v. Robert Cagle Building Co.,* 265 F.Supp. 469 (S.D.Tex.1967).

74. *Tuttle/White Constructors, Inc. v. State,* 371 So.2d 1096 (Fla.App.1979).

75. *Kalisch-Jarco, Inc. v. City of New York,* 58 N.Y.2d 377, 461 N.Y.S.2d 746, 448 N.E.2d 413 (1983) (public contract described as "arm's length" made "by sophisticated contracting parties"); *W.C. James, Inc. v. Phillips Petroleum Co.,* 485 F.2d 22 (10th Cir.1973) (private contract).

exculpation clauses of this type are rarely attractive to courts, and under some limited situations they will not be given effect. Like all contract clauses, they must be interpreted to determine what they are intended to cover. Using the process of interpretation, no-damage clauses do not preclude recovery for delay damages if the delay was not the type contemplated, if the delay was an abandonment by the owner, or if the delay was caused by bad faith or active interference.[76]

It may be instructive to look at some recent cases that have dealt with no-damage clauses. *Broadway Maintenance Corp. v. Rutgers* [77] was reproduced in part in Section 21.04(C). One basis for denying the contractor's claim was a no-damage provision. In applying the clause the court stated:

> No damage provisions are a part of the economic package upon which the parties agree. The contractor who chooses to accept these risks will reflect the accompanying responsibility in his price.[78]

It is important to look at the intention of the parties. This inquiry determines whether the type of delay that occurred was the type the parties intended to exclude as a basis for a delay claim.

The contractor had complained of backfill problems, slow-pouring concrete, and lack of temporary heat. The court affirmed a finding by the intermediate court that they were *not* sufficiently exceptional situations to fall *outside* the no-damage clause. They are ordinary and usual types of damage that most contractors frequently encounter.

In *Bates & Rodgers Construction Corp. v. North Shore Sanitary District*,[79] the court had to face a similar problem. The contractor's claim was based upon the owner's impliedly warranting that the plans and specifications would enable the contractor to successfully do the work. Breach of such a warranty can be the basis for the contractor's recovering any additional costs.

Interestingly, the contractor sought to persuade the court that it was not seeking damages for delay inasmuch as the no-damage clause appeared to bar such a claim. The court noted that the claim was based upon "cost overrun, excessive labor costs, labor 'add-ons', excessive supervision, winter protection of the work, increased overhead, bond and insurance costs, interest or money borrowed and expended and the loss of anticipated profit." The court noted that these losses were incurred because the contractor was forced to perform its work out of sequence, with its performance becoming less productive. The court barred the damages because of the no-damage clause. Another court [80] held such a clause would bar *delay* damages, which it defined as time lost when work cannot be performed because materials have not been delivered or preliminary work done. But it does *not* bar a claim for *hindering* the work, such as failure to coordinate or supply temporary heat. This may only illustrate *active* interference. But distinctions such as this make the clause of doubtful utility.

---

76. *Peter Kiewit Sons Co. v. Iowa Southern Utilities Co.*, 355 F.Supp. 376 (S.D.Iowa 1973) (citing many authorities). In 1984 California added § 7102 to its Public Contract Code, stating that no-damage clauses do not apply to unreasonable delays or those not contemplated by the parties. See Cal.Stats.1984, ch. 42. See Parvin and Araps, *Highway Construction Claims*, 12 Pub.Cont.L.J. 255, 277–281 (1982); Vance, *Fully Compensating the Contractor for Delay Damages in Washington Public Works Contracts*, 13 Gonz.L.Rev. 410 (1978); Comment, *The Enforceability of "No Damage for Delay" Clauses in Construction Contracts*, 28 Loy.L.Rev. 129 (1982).

77. 90 N.J. 253, 447 A.2d 906 (1982).

78. 447 A.2d at 914.

79. 92 Ill.App.3d 90, 47 Ill.Dec. 158, 414 N.E.2d 1274 (1981).

80. *John E. Green Plumbing & Heating Company, Inc. v. Turner Construction Co.*, 742 F.2d 965 (6th Cir. 1984).

One of the frequently mentioned exceptions is bad faith or active interference. This exception was applied in *United States Steel Corp. v. Missouri Pacific Railroad Co.*[81] The claim of active interference was based upon the owner's directing the contractor to proceed before requisite work of an earlier contractor had been completed. Access to the work was denied during completion of the earlier contractor's work, causing a delay of 175 days. The owner granted a time extension for this delay but, based upon a no-damage clause, refused to pay any delay damages.

The court noted that the owner had an implied obligation to refrain from anything that would reasonably interfere with the contractor's opportunity to proceed with its work and to allow the contractor to carry on that work "with reasonable economy and dispatch." The court concluded that issuance of the notice to proceed was an affirmative, willful act and that the owner's bad faith was demonstrated by its knowledge of circumstances that would prevent the contractor from proceeding timely with its work.

Finally, a well-publicized New York case [82] required more than *active* interference to avoid the "no damage" clause. The majority, in a four-to-three opinion, required fraud, malice, bad faith, or a reckless indifference to the rights of others. Unlike the *John Green* case described earlier, in New York, at least, such a clause will bar almost *any* delay damage claim.

No-damage clauses have two principal difficulties. One is the uncertain enforceability of such clauses—not knowing whether the contractor will be able to invoke an exception. (Some clauses apply for a designated number of days, with damages beyond that period being recoverable.) Second, such clauses are likely to gener-

ate higher bids. Like provisions requiring the contractor to bear the risk of unforeseen subsurface conditions, requiring the contractor to bear the risk of its delay costs should motivate prudent contractors to add bidding contingencies to take this anticipated cost into account. Those that do not may submit low bids, receive awards, and look to the claims-filing technique to make up these costs.

(B) SUBCONTRACTOR CLAIMS

It is readily apparent that delays caused by the owner not only harm the prime contractor but also may harm subcontractors. The absence of a contract between subcontractors and the owner generally precludes direct legal action and, as a rule, direct negotiations between subcontractors and owners over delay claims. Often the subcontractor's claim is processed by the prime contractor. This processing was dealt with in *T.G.I. East Coast Construction Corp. v. Fireman's Fund Insurance Co.*[83]

The owner issued a stop order during performance that suspended work for six months. The contractor submitted a joint claim to the owner on behalf of itself and its subcontractors. At that time, the prime contractor and one of its major subcontractors entered into a memorandum of understanding in which the subcontractor agreed to accept the owner's decision as to the value of the subcontractor's pass-through claim. A settlement was reached and a portion of that settlement allocated to the subcontractor.

Subsequently the subcontractor defaulted, and the contractor sued the subcontractor for damages. The subcontractor counterclaimed, alleging that the contractor had breached the memorandum of understanding by failing to make a good-faith effort to recover the subcontractor's original claim.

---

81. 668 F.2d 435 (8th Cir.1982).

82. *Kalisch-Jarco, Inc. v. City of New York,* supra note 75.

83. 534 F.Supp. 780 (S.D.N.Y.1982).

The court stated that all contracts contain a covenant of good faith and fair dealing. The agreement between prime contractor and subcontractor imposed upon the former a duty to make a good-faith effort to present the subcontractor's claim in a fair and serious manner. The court concluded that a factual issue had been raised as to whether the contractor had carried out this duty.

(C) LIQUIDATED DAMAGES

Until recently it was rare for owners to liquidate damages for delay they caused. Public owners usually protected themselves, or at least hoped to, by the use of no-damage clauses. However, it is becoming more common to liquidate damages as an alternative to exposure to the often open-ended and difficult to establish or to disprove delay damage claims. Although no judicial analysis of these clauses has been found, very likely courts will enforce them. The process of establishing the additional costs incurred by the contractor for delay is even *more* complex and *more* difficult than establishing lost use by owners, even public owners. This should encourage enforcement, provided the amount bears some reasonable relationship to the anticipated or actual damages.

(D) MEASUREMENT

The measurement of the value of the contractor's claim when without excuse the owner delays the contractor's performance is dealt with in Section 31.03(E).

(E) RECORDS

In delay disputes, the party with the best records has a great advantage. Each party should keep job records such as the site representative's daily field reports, corre- spondence, memoranda, photographs, or video recordings, and change orders. These records should include data on labor, equipment, and material used for each activity and should document the cause and impact of every delay.

SECTION 30.11
THIRD–PARTY CLAIMS

Delayed completion can harm third parties, such as tenants who expect to move into the completed project, tenants who are displaced for an unreasonable period during construction, or even others, such as restaurants, who expect to earn revenue from tenants of the completed project or their employees or customers. Suppose these third parties seek to transfer their losses to those they claim responsible for the delay, such as owner, design professional, or contractor.

Claims against a contract-connected party, such as a claim by a tenant against its landlord-owner would be decided by reference to the contract, contract law, and contract remedies. Claims by a noncontract-connected party, such as a tenant against the contractor or design professional, are more complex. The issue arose in *J'Aire Corp. v. Gregory,*[84] (discussed in Sections 17.09(e) and 17.12(c)). The court allowed a tenant to assert a claim against a contractor for negligent delay in completing a renovation project that delayed the tenant's reopening of its restaurant business. In essence, the tenant contended that the contractor *negligently* interfered with its restaurant business. Usually the law requires that interference with contractual relations or prospective advantages be *intentional.*[85] But the court, using an "easy to satisfy" forseeability test, held that the tenant should be allowed to show that the

---

84. 24 Cal.3d 799, 157 Cal.Rptr. 407, 598 P.2d 60 (1979).

85. Restatement (Second) of Torts §§ 766(B)(C) (1979).

contractor's negligence caused it economic harm.

As noted earlier,[86] injection into a series of planned transactions—the contract between owner and contractor and owner and tenant—of tort concepts can destroy contract planning. Yet claims against major construction participants based upon tort or assertions that the claimant is an intended beneficiary are growing. Risk management planning, such as realistic pricing, contract language, and other devices, is becoming essential. It was discussed in Chapter 18 in the context of the design professions. It is also the subject of Chapters 36 and 37 dealing with indemnification and suretyship.

PROBLEMS

1. C and O contract for C's construction of a five-story office building for $500,000. The completion date is December 1. The contractor provides that for every day of unexcused delay, C shall be chargeable with liquidated damages of $3000 a day. The building was completed twenty days late, and the delay was not excused by any events set forth in the *force majeure* clause. What can O recover for the delay? What added facts would be helpful in answering this question?

   Suppose the agreed damage figure had been $1000 a day and C abandoned the project on December 30. O obtained a replacement who completed the project twenty-five days later. The substitute contractor cost $60,000 over the contract price. What can O recover from C?

2. You have been asked to design and monitor construction of a luxury single-family home for O. The contract price is $500,000. You know that O is a particularly demanding client. He wants things "just so." Also, he is likely to make many design changes and he likes to visit the job site. Should you advise his attorney to include a no-damage clause? What would be your reasons?

3. Your client wants to use an AIA Doc. A 201, but she is concerned about the completion point for determining time commitments being substantial completion. She asks if this is fair. How would you answer her question?

---

86. Refer to Sections 17.11 and 17.12(C).

31

Claims: By-Product of Construction Process

SECTION 31.01
INTRODUCTION

This chapter completes the material dealing with performance problems. Sections 31.02 and 31.03 deal with measuring claims that contracting parties have against each other, using as an illustration the construction contract between owner and contractor. Section 31.04 treats claims against more than one participant.

Please review Chapter 6, the basic building block dealing with remedies for breach of contract. Special reference should be made to Section 6.06(F), which deals with attempts by contractors to recover consequential damages not directly associated with the construction contract, such as losses incurred by being unable to receive or complete *other* contracts.

The law seeks to compensate the contracting party who has suffered losses because of the other party's breach of contract. Generally it does not punish the party that has breached the contract.[1] The court judgment seeks to put the injured party in the position it would have reached had there been *no* breach (expectation) or seeks to restore the party to the position it occupied before performance began (restitution).

The law developed conventional formulas to implement the compensation objective. At times, a choice can be made between different formulas. Judicial attempts to determine what *would* have happened can lead to unsupported speculation and guesswork. The complexity of construction performance, especially in large projects, makes it difficult to determine what *has* happened.

As a result, claims measurement uses rough approximations that may be all that is available. This chapter reveals a constant tension between demanding specific and solid proof and the willingness to use formulas or approximations.

It is important to recognize that appellate courts rarely interfere with what trial courts have concluded on these issues, many of which are factual in nature and not legal. The trial court is in the best position to evaluate the evidence.

SECTION 31.02
MEASUREMENT:
CONTRACTOR vs. OWNER

(A) ILLUSTRATIONS

Claims by contractors against owners can

---

1. Refer to Section 6.04.

arise in many contexts. The principal claims are based upon the following:

1. Refusal of the owner to permit the contractor to commence performance after the contract has been awarded.
2. Wrongfully terminating the contractor's right to perform during performance.
3. Committing acts that justify the contractor in ceasing performance.
4. Failing to pay the contractor for work performed under the contract.
5. Committing acts that increase the contractor's cost of performance.

(B) COST–TYPE CONTRACTS

This section does not deal with claims by the contractor for additional compensation under cost contracts. Usually, these claims are based upon assertions that costs were incurred and that the owner's payment did not cover these costs. Those questions generally involve factual determinations of whether particular work was done or whether it was called for under the contract and the value of the work performed. Although the contractor will very likely have the burden of establishing the particular work that was done and its cost, the owner has the burden of showing that the cost of particular work was unreasonable.[2] The purpose of the cost contract is to place the risk of costs incurred upon the owner.

(C) PROJECT NEVER COMMENCED

Although there may be consequential damages asserted when the contractor is not given the opportunity to commence performance, general damages in such cases are the lost profits on the contract. Because lost profits arise when the contractor has commenced performance and is then wrongfully terminated or has justified grounds for ceasing performance, discussion of lost profits is in the next subsection.

(D) PROJECT PARTIALLY COMPLETED: DAMAGES

The damages award should place the contractor in the position it *would* have been in had the owner performed in accordance with the contract. Three possible formulas, which for convenience are referred to as Formulas I, II, and III, can determine the amount of recovery. Formula I determines what *would* have happened by awarding the contractor the contract price. This is the amount it *would* have had upon completion. But from this, the cost of completion must be deducted. This is an expense saved by the breach. Progress payments received must be deducted. Completion cost is what a reasonably prudent contractor in the contractor's position would have had to spend to complete the work.[3]

A contractor who finds it difficult to prove what its costs *would have* been can use the costs of a successor, provided those costs were reasonable.[4] If the owner can establish that the contractor would have sustained a loss had it completed the project, this loss will be deducted from the recovery.[5] Such a deduction may not be permitted if the claim brought by the contractor is based upon restitution.

Formula II is the contractor's expenditures in part performance, including preparation, and, if the contractor can establish

2. *Sloane v. Malcolm Price, Inc.,* 339 A.2d 43 (D.C.App.1975).

3. *Watson v. Auburn Iron Works, Inc.,* 23 Ill.App.3d 265, 318 N.E.2d 508 (1974).

4. *Carchia v. United States,* 485 F.2d 622 (Ct.Cl.1973). But in *Edward Electric Co. v. Metropolitan Sanitary District,* 16 Ill.App.3d 521, 306 N.E.2d 733 (1973), the contractor's attempt to base the cost of completion upon the completed work was denied.

5. *Watson v. Auburn Iron Works, Inc.,* supra note 3.

them, profits on the entire project.[6] To illustrate Formulas I and II, assume that a project has the following figures:

1. Contract price—$100,000
2. Expenditures in part performance—$60,000
3. Cost of completion—$30,000
4. Progress payments received—$50,000

Under Formula I the contractor would receive:

$100,000	(contract price)
−30,000	(cost of completion)
$ 70,000	
−50,000	(progress payments received)
$ 20,000	(net recovery)

Under Formula II, the recovery would be:

$ 60,000	(expenditures in part performance)
+ 10,000	(profits—$100,000 contract price less $90,000, total cost of project)
$ 70,000	
−50,000	(progress payments received)
$ 20,000	(net recovery)

The result using Formulas I and II is identical.

But suppose expenditures in part performance were $90,000. Had the contract been completely performed, the contractor would have lost $20,000. Using Formula I, the contractor would still receive $20,000, inasmuch as the two principal items of the formula are not changed. However, the contractor has already expended $90,000 and has received $70,000 through progress payments ($50,000) and the money award ($20,000). This would create the same loss of $20,000 that it would have suffered had it completed the contract.

Formula II, that is, expenditures in part performance coupled with profits or less losses, would compute recovery as follows:

$ 90,000	(expenditures in part performance)
−20,000	(loss that would have been suffered)
70,000	
−50,000	(progress payments received)
$ 20,000	(net recovery)

Under Formula II, the contractor would be in the same position as under Formula I. The contractor expended $90,000 and received $70,000, leaving the same $20,000 loss.

Formula III entitles the contractor to such proportion of the contract price as the cost of the work done bears to the entire cost of completing performance and, for the remaining portion of the work, the profit that would have been made as to that work.[7] Formula III should produce the same recovery on profitable contracts but will produce a different result in losing contracts. However, since Formula III is rarely used, detailed illustrations of its application will not be given.

Compensation can involve the difficult question of establishing lost profits. At the outset it should be noted that some contracting systems make a sharp differentiation between profits on performed and unperformed work. For example, Federal Procurement policies rarely give recognition to the profit on unperformed work.[8]

However, the ordinary contract measure-

---

6. Restatement (Second) of Contracts § 347, Comment d (1979).

7. *Kehoe v. Borough of Rutherford*, 56 N.J.L. 23, 27 A. 912 (1893).

8. *General Builders Supply Co. v. United States*, 409 F.2d 246 (Ct.Cl.1969). A weighting formula developed by the Chief of Engineers for determining profit on performed work was cited and applied in *Norair Eng'g Corp.*, ASBCA No. 10856, 67–2 BCA ¶ 6619.

ment formulas that have been set forth do not make this differentiation. In such cases, the principal problem has been the standard of certainty required to recover lost profits.

Early case decisions took a negative view toward lost profits as a measure of contract damages. Many cases held that profits could not be established for a new business.[9] Even lost profits by someone in an existing business were closely scrutinized. Was the profit reasonably foreseeable by the contract parties at the time the contract was made? In contracts for the sale of goods, this often excluded recovery of unusual resale profits.

In construction contracts, however, the principal problem has been certainty. How does the contractor establish not only the profits on work performed but also the profits on unperformed work? If it uses Formula I—that of contract price less cost of completion—lost profits need not be established directly. However, what it *would* have cost to complete the project can often be difficult to show. As a result, more commonly contractors use Formula II—expenditures in part performance coupled with profits. Although the former can be difficult to establish if the contractor has not kept good records, it is generally easier to prove than cost of completion. Howev-

er, the contractor who seeks complete damages will have to establish profits by some method other than deducting the cost of completion from the contract price.

Although a court refused to award profits in a construction context because of the lack of experience, personnel, equipment, and background of the contractor,[10] the principal questions have been the type of proof that will support a claim for lost profits. One court allowed a subcontractor to establish general profitability of the job.[11] Another submitted lost profits to the jury based upon testimony of the president of the plaintiff contractor as to profits on other similar work, his method of estimating profits, and his expectation as to profits.[12] In the latter case, another witness had testified as to a range of profits. However, the appellate court reduced the judgment to the smallest profit that could be supported by the testimony.

For lost profits on other contracts, refer to Section 6.06(F).

(E) PROJECT PARTIALLY COMPLETED: RESTITUTION [13]

The law generally permits a contracting party who has *not* fully performed to use an alternative measure of recovery where the other party has committed a serious

---

9. D. Dobbs, Remedies, 153–155 (1973).

10. *Electric Service Co. of Duluth v. Lakehead Electric Co.,* 291 Minn. 22, 189 N.W.2d 489 (1971).

11. *Construction Limited v. Brooks-Skinner Building Co.,* 488 F.2d 427 (3d Cir.1973).

12. *Natco, Inc. v. Williams Brothers Engineering Co.,* 489 F.2d 639 (5th Cir.1974).

13. One confusing aspect of restitution is the frequent mention of the term *quantum meruit.* Often courts state that a particular action is brought upon a *quantum meruit.* Early in English legal history, a party seeking relief had to bring itself within the language of writs issued by public officials commanding persons to appear before one of the King's courts. One frequently used writ was *Assumpsit.* This writ was broken down into a number of subwrits, one of which was called *General Assumpsit.* Because of common usage of certain subwrits such as *General Assumpsit,* many of them became known as Common Counts. One of the Common Counts was given the name of *quantum meruit,* which usually dealt with a method of recovering for the reasonable value of services furnished the defendant at his or her request. It was used as a method of recovering for breach of contract and could be used when the claim was based upon unjust enrichment. This treatise refers to restitution rather than *quantum meruit* even though some courts still use that term in describing the process by which the plaintiff recovers based upon benefit conferred rather than damages.

breach.[14] Breaches sufficiently serious to justify restitution have been failure to make progress payments,[15] excessive changes,[16] and failure to perform those acts during performance that would allow the contractor to perform in the most expeditious way.[17]

Expectation (discussed in (D)) looks forward and seeks to determine what would have happened had the parties completed performance. Restitution looks backward to the position the parties were in at the time they entered into the contract. In some contracts, this is accomplished by a party's returning any performance the other party may have conferred upon it. In construction, actual restitution is ordinarily impracticable, as the work has been attached to the owner's land. Instead, the contractor is entitled to receive the value that its performance has benefited the owner, generally the reasonable value of the material and labor it contributed to improving the owner's land.

Restitution should be measured objectively, with the contractor recovering an amount that equals what the owner would have had to pay to purchase the material and services from one in the contractor's position at the time and place the services were rendered.[18] Ordinarily the contractor seeks to introduce evidence of its actual costs incurred in performing its contractual obligations. Although this is not conclusive, it is likely to be accepted by the court unless the owner can establish that they were not market rate costs or that they were costs incurred to correct deficiencies chargeable to the contractor. One case affirmed a judgment for a subcontractor in its claim against the prime contractor by pointing to the opinion testimony of the subcontractor as to the direct costs and the court's examination of the site.[19] Another court looked at the bids submitted by unsuccessful bidders in determining the reasonable value of the services.[20]

The contractor should be able to recover its overhead costs, as they are incurred by the contractor in improving the owner's land and should be considered as benefiting the owner.[21]

Clearly the contractor who seeks restitution cannot recover profit on unperformed work. It has been allowed to recover profit on work that has been performed.[22] This would be included in what other contractors would have charged.

The most difficult problem relates to the effect of the contract price on the contractor's recovery. Should the law ignore the contract price, use it as evidence of the value of the benefit conferred, use the pro rata contract price to actually measure the recovery, or use the contract price to put a limit on the contractor's restitutionary claim? Although the general tendency has been to use the contract price only as evi-

---

14. *Kass v. Todd*, 362 Mass. 169, 284 N.E.2d 590 (1972) (dictum). The seriousness is similar to that needed for a power to terminate. See Section 38.04(A). But see *United States v. Mountain States Construction Co.*, 588 F.2d 259 (9th Cir.1978) (Washington law), which required a tortious breach. This is not in the mainstream.

15. *United States v. Algernon Blair, Inc.*, 479 F.2d 638 (4th Cir.1973).

16. *Glassman Construction Co. v. Maryland City Plaza, Inc.*, 371 F.Supp. 1154 (D.Md.1974); *Rudd v. Anderson*, 153 Ind.App. 11, 285 N.E.2d 836 (1972).

17. *Leo Spear Construction Co. v. Fidelity & Casualty Co. of New York*, 446 F.2d 439 (2d Cir.1971).

18. *United States v. Algernon Blair, Inc.*, supra note 15.

19. *Leo Spear Construction Co. v. Fidelity & Casualty Co. of New York*, supra note 17.

20. *Paul Hardeman, Inc. v. Arkansas Power & Light Co.*, 380 F.Supp. 298 (E.D.Ark.1974).

21. *Leo Spear Construction Co. v. Fidelity & Casualty Co. of New York*, supra note 17.

22. Id.

dence of value,[23] some cases have limited the contractor to the contract price.[24]

The effect of contract price on recovery has stirred great debate. The majority rule—that of giving the contract price only evidentiary effect—can allow a contractor who has entered into a losing contract to at least break even. One argument given to justify this result has been that the restitutionary principle, that is, looking backward and avoiding unjust enrichment, has as much validity as expectation, that is, looking forward to what would have happened had the contract been performed properly. Another justification for the majority rule is that an owner who has breached a contract should not be able to take advantage of the contract price. A reason frequently given is that the expenses most likely exceed the contract price because of breaches by the owner. Perhaps—most persuasive—restitution not limited to the contract price is relatively easy to administer. If expectation (discussed in (D)) were used, losses would be deducted. This computation would require a determination of cost of completion, often at best a guess. If the contract price is not controlling, all that is needed is to determine the reasonable value of the work performed.

Those that support using the contract price either to measure the value or to limit recovery point to the anomaly that can result if the contractor *does* complete performance. Restitution *cannot* be used if the contractor has completed performance and all that is left is for the owner to pay the balance of the contract price. If the owner's breach has caused the contractor to lose money, the contractor must establish such losses. (As shall be seen, various crude formulas are sometimes allowed,

making proof easier. See Section 31.02(F).) Barring the owner from using the contract price because it has breached does not take into account the possibility and even the likelihood that the question of breach has been a close one, with much fault attributable to both parties. Why take away the benefit of the bargain made by the owner because it is ultimately determined that the owner has breached?

The difficulty with the majority rule is that it can encourage a contractor in a losing contract to claim that it has adequate grounds for termination and to stop work. If it continues to perform, it will almost certainly lose money and face difficult problems if it seeks to recover compensation based upon breaches by the owner. If it stops and claims the right to terminate, it runs the risk that it will be found to be in default and have to pay heavy damages. These damages may be no greater, however, than the contractor will suffer if it continues to perform a losing contract when there is very little hope of turning things around. In the confusion of a construction dispute in court, a good chance exists that the contractor may be exonerated.

(F) PROJECT COMPLETED: *LURIA BROTHERS & CO., INC. v. UNITED STATES; BERLEY INDUSTRIES, INC. v. CITY OF NEW YORK*

A contractor who has fully performed but has not been paid the balance of the contract price cannot measure its recovery by the reasonable value of its services.[25] It recovers only the unpaid balance of the contract price and any other losses it can prove.

Limitation on the restitutionary remedy is sometimes justified on historic grounds. However, the modern justification is the

---

23. *United States v. Algernon Blair, Inc.,* supra note 15; *Paul Hardeman, Inc. v. Arkansas Power & Light Co.,* supra note 20; *Kass v. Todd,* supra note 14.

24. *United States v. Mountain States Construction Co.,* supra note 14; *Johnson v. Bovee,* 40 Colo.App. 317, 574 P.2d 513 (1978).

25. *Kass v. Todd,* supra note 14.

ease of using the contract price as a measure compared to making the more complicated factual determination of the reasonable value of the services. Here, unlike the past performance discussed in (E), there is no need to determine value.

Most complex construction contract disputes involve claims by the contractor who has completed the project and not, as a rule, for the unpaid balance of the contract price (it has probably been paid), except for the additional costs it incurred for which it claims compensation. The contractor asserts that defective specifications required it to perform in an inefficient and costly manner, that it encountered unforeseen subsurface conditions that required that it perform its work in a more costly manner, and, most important, that unexcused delays caused by the owner increased its cost of performance.

The problems of defective specifications and unforeseen subsurface conditions, to the extent that they do *not* involve delay, cause less difficulty. It is not difficult, comparatively speaking, to establish the value of additional services and material that were necessary because of defective specifications or unforeseen subsurface conditions. Even more, in the case of claims based on unforeseen subsurface conditions, the justification for additional compensation is usually based upon a Differing Site Conditions Clause, which either may contain a formula for measuring the additional compensation or has developed legal rules that provide a solution for measuring compensation. However, the greatest difficulty arises in those cases where the contractor claims that it had to perform work in an inefficient manner for which the owner is responsible. To set the scene, a court facing such a claim stated:

We note parenthetically and at the outset that, except in the middle of a battlefield, nowhere must men coordinate the movement of other men and all materials in the midst of such chaos and with such limited certainty of present facts and future occurrences as in a huge construction project such as the building of this 100 million dollar hospital. Even the most painstaking planning frequently turns out to be mere conjecture and accommodation to changes must necessarily be of the rough, quick and *ad hoc* sort, analogous to ever-changing commands on the battlefield. Further, it is a difficult task for a court to be able to examine testimony and evidence in the quiet of a courtroom several years later concerning such confusion and then extract from them a determination of precisely when the disorder and constant readjustment, which is to be expected by any subcontractor on a job site, become so extreme, so debilitating and so unreasonable as to constitute a breach of contract between a contractor and a subcontractor. This was the formidable undertaking faced by the trial judge in the instant case and which we now review on the record made by the parties before him.[26]

Walter Kidde Constructors, Inc. v. Connecticut[27] involved a contractor (Kidde-Briscoe) claim in a dispute over a project that finished almost 900 days later than planned. Much of the responsibility fell upon the public owner. It failed to make certain parts of the site available as promised. It issued many hold orders that substantially interfered with and disrupted construction while plans were being redesigned. It was late in processing change orders and approving shop drawings while still insisting that the contractors accelerate their work under threat of imposing a $250-per-day penalty for not completing within the time requirements of the contract. In describing the effect of these delays the court stated:

The long delay in completing construction had a devastating impact on Kidde-Briscoe. They were not only required to provide labor and materials far in excess of and different from that called for in the original contract but to perform work out of sequence at an accelerated rate and in a manner not planned and not uti-

---

26. *Blake Construction Co. v. C.J. Coakley Co.*, 431 A.2d 569, 575 (D.C.App.1981).

27. 37 Conn.Sup. 50, 434 A.2d 962, 977 (1981).

lized under normal conditions. The unanticipated almost two and a half years of delayed
construction extended through a period of increasing inflation and escalation in labor rates
and material costs, through two winters with
adverse building conditions and into a time
when Kidde-Briscoe was faced with an 89 day
regional strike by the sheet metal workers and
137 day regional strike by the plumbers union.

A computation of the consequent damages to
Kidde-Briscoe involves damages of two sorts:
(a) delay damages due to the extended periods of
field and home office overhead and (b) damages
due to disruption, loss of productivity, inefficiency, acceleration and escalation.

In addition to the items noted in the *Kidde*
case, contractors often claim and are sometimes awarded the following: [28]

1. Idleness and underemployment of facilities, equipment, and labor.

2. Increased cost and scarcity of labor and
materials.

3. Utilization of more expensive modes of
operation.

4. Stopgap work needed to prevent
deterioration.

5. Shutdown and restarting costs.

6. Maintenance.

7. Supervision.

8. Equipment and machinery rentals and
cost of handling and moving.

9. Travel.

10. Bond and insurance premiums.

11. Interest.

A portion of a delay damages case is
reproduced here.

**LURIA BROTHERS &
CO., INC. v. UNITED
STATES**
United States Court of
Claims, 1966.
177 Ct.Cl. 676, 369 F.2d
701.

WHITAKER, Senior Judge

* * *

Loss of Productivity. The second item of damage, not allowed
by the trial commissioner, claimed by plaintiff to have been
caused by defendant's delay was loss of productivity of its labor
force. First, it says the delay required it to work during severe
winter weather between December 1, 1953, and March 10, 1954,
and again between November 25, 1954, and January 19, 1955;
second, from March 11 to August 11, 1954, it had to work under
adverse water conditions; and third, the constant revisions in the
contract drawings resulted in confusion and interruption of the
orderly progress of the work.

For this loss of productivity of its labor, it claims damages in
the aggregate amount of $131,116.66. . . .

The testimony on loss of productivity is far from satisfactory.
Plaintiff's sole witness on this point was John Crawford, who had
been in plaintiff's employ for about 10 years and who was its chief
of construction at the time the work on this job was being done.
However, at the time of his testimony he had left plaintiff's employ
and was then chief engineer in the New York area for the Frouge
Corporation of New York City and Bridgeport, Connecticut, a
company with which plaintiff had no connection. It was a large

---

28. For an opinion that illustrates a claim involving a large number of these items, see *Contracting & Material
Co. v. City of Chicago,* 20 Ill.App.3d 684, 314 N.E.2d 598 (1974). The opinion was reversed by the Supreme Court
based upon failure by the contractor to comply with a contractual condition precedent of working double shifts, 64
Ill.2d 21, 349 N.E.2d 389 (1976) but the facts provide a good illustration.

company, with $50 million worth of building construction in that area, consisting of housing developments, apartments, and office buildings. Mr. Crawford had graduated as a civil engineer from Columbia University in 1924, since which time he had been engaged in both heavy and building construction work.

He was a competent witness, well-qualified to express an opinion on the loss of productivity of the labor. But, strangely, plaintiff offered no corroboration of his testimony. But, stranger still, defendant did not cross-examine him on this point and offered no testimony in rebuttal. His testimony stands unchallenged.

The defendant's sole reply to plaintiff's request that the trial commissioner find the facts as testified to by this witness was in its exceptions to plaintiff's proposed findings of fact. In this document, defendant made only this reply:

> Plaintiff's productivity losses, so-called, were based on estimates, entirely unverifiable.

> Defendant considered this item unproved. It entirely neglects allowances for time consumed by strikes, plaintiff's own delays, and other nongovernment delays.

That loss of productivity of labor resulting from improper delays caused by defendant is an item of damage for which plaintiff is entitled to recover admits of no doubt, . . . ; nor does the impossibility of proving the amount with exactitude bar recovery for the item.

* * *

It is a rare case where loss of productivity can be proven by books and records; almost always it has to be proven by the opinions of expert witnesses. However, the mere expression of an estimate as to the amount of productivity loss by an expert witness with nothing to support it will not establish the fundamental fact of resultant injury nor provide a sufficient basis for making a reasonably correct approximation of damages. . . .

Crawford's testimony is unrebutted. Defendant, out of whose pocket the money must come to pay the large sum testified to by him, did not undertake to discredit his testimony, and so, while it is the sacred duty of this court to protect the Government from unrighteous demands as well as to protect the citizen from imposition by the Government, we cannot wholly reject this witness's testimony on the question of amount of damage.

However, we cannot ignore the fact that he was plaintiff's former employee and had been over a period of 10 years. While he was not in plaintiff's employment at the time he testified, he quite properly had a certain predilection for his old employer and wanted to "help them out" all he could. His sympathy was naturally with his former employer rather than with the Government. We do not mean the witness was dishonest, but we do think he made his estimates as high as he could to the extent his conscience would permit.

Crawford testified that between December 1, 1953, and March 10, 1954, the productivity of plaintiff's labor force was reduced inasmuch as the men had to work outside on trench excavations and foundation construction in winter weather. This required them to wear gloves and warmer clothing, and to work on ground which was frozen and/or extremely wet because of the rising ground water during that time of year. Based on his over-all experience in construction work and his observation of this particular job, he estimated that the average loss of productivity of labor during this period was 33⅓ percent. With regard to the period between March 11, 1954, and August 11, 1954, Crawford testified that the labor productivity was reduced 25 percent because of adverse water conditions encountered at the construction site. During the period from August 12, 1954, to November 24, 1954, Crawford estimated that the average loss of productivity of plaintiff's labor force was 20 percent because of confusion and interruption of normal job progress as a result of several revisions by defendant of the contract drawings for the leanto building. Crawford also estimated that the loss during the period from November 25, 1954, to January 19, 1955, was 20 percent due to cold weather conditions when the men had to work in an only partially completed structure.

That winter weather and adverse water conditions reduce the efficiency of a labor force in the performance of construction work only stands to reason. It has been held by this court that when loss of productivity brought about by these conditions results from defendant's breach of contract, the plaintiff is entitled to recover its additional costs occasioned thereby as damages. . . . However, with respect to the loss of efficiency caused by the revisions to the leanto plans during the period from August 12, 1954, to November 24, 1954, there is no testimony that there was a loss of efficiency on this account other than Crawford's estimate of the amount thereof. Proof of damage is essential before estimates can be received of the amount thereof. Plaintiff's claim for loss of productivity during that period must therefore be rejected. . . .

Notwithstanding the fact that Crawford's estimates regarding the other three periods are unrebutted, we cannot ignore the fact that the percentages testified to were merely estimates based upon his observation and experience. Furthermore, his estimates are much higher than those testified to in other cases in which the conditions were not materially different from those present here. Taking these things into consideration and in view of the fact that no comparative data, no standards, and no corroboration support his testimony, we are constrained to reduce his estimates based on the record as a whole and the court's knowledge and experience in such cases to 20 percent, or $11,091.02 for the period from December 1, 1953, to March 10, 1954; to 10 percent, or $13,014.87 for the period from March 11 to August 11, 1954; and to 10 percent, or $925.50 for the period from November 25, 1954, to January 19, 1955.

The total additional cost to plaintiff during the 518-day overrun period because of loss of productivity of its labor force was thus $25,031.39. . . . Because of a duplication of costs included in another of plaintiff's claims, $4,550.63 must be deducted from this total figure. This results in a net cost to plaintiff of $20,480.76 because of loss of productivity of its labor force. Since the defendant was responsible for 420 days of the 518-day overrun, plaintiff is entitled to recover 81 percent of this amount, or $16,589.42.

We have found that the plaintiff is entitled to recover $62,948.33 for excess home office overhead. Adding these two items to the $85,544.92 found by the trial commissioner makes a total of $165,082.67. A judgment for this amount is entered in favor of plaintiff against defendant.

As the *Luria* case shows, claims of lost productivity are best supported by a careful marshaling of documentary evidence. Where this cannot be done, expert testimony will be needed. In any event, it can be seen that about the best that courts can do in many of these cases is to make rough approximations.

Proof of lost productivity and home office overhead discussed in the *Berley* case reproduced ahead in this section should motivate participants in the construction process not only to recognize the central significance of an efficient record keeping system but to relate this problem to dispute resolution. While explored in greater detail in Chapter 34, it is important to note here one difference between judicial and non-judicial modes of resolving these claims.

While generalizations are perilous, it is likely that non-judicial resolution of such disputes will prove more satisfactory to the contractor than to the owner. Judges may feel more constrained by legal rules and possible appellate court reversal. They will be less willing to accept expert testimony or rough approximation formulas, such as total costs and jury verdicts (explored later in this section) than would their non-judicial counterparts, particularly in claims against *public* entities. Non-judicial dispute resolution involves practical people aware of the realities of record keeping and more willing to guess, compromise, or use rough formulas in these intractable proof problems.

Clearly, contractors can recover for extended home office overhead. Illustrations of aspects of overhead usually included in the construction contract price are home-office salaries, supplies, utilities, rent, taxes, insurance costs, depreciation, bad debts, estimating expenses, and fees for other professional services. Such overhead is sometimes referred to as indirect, as it is not identifiable with any particular project but is incurred for the mutual benefit of all contracts that a contractor performs.

However, great difficulty has developed in measuring the actual amount of home-office overhead that can be chargeable to the party causing the delay. One formula that was developed in a federal procurement agency appeals board case, the Eichleay formula, is discussed and demolished in a case reproduced here.

BERLEY
INDUSTRIES, INC. v.
NEW YORK

Court of Appeals of New
York, 1978.
45 N.Y.2d 683, 412 N.Y.S.2d
589, 385 N.E.2d 281.

OPINION OF THE COURT

FUCHSBERG, Judge.

[Ed. note: Footnotes renumbered.] This action for breach of a construction contract addresses the circumstances in which a mathematically stated rule, such as the so-called "Eichleay formula",[29] may be applied to determine the amount of home office overhead, if any, which may be included in arriving at a contractor's delay damages.

Plaintiff, Berley Industries, was the heating, ventilating, and air conditioning contractor for a combined 48th precinct police station and firehouse then being constructed by the City of New York. The contract price was in excess of $472,000. It is conceded that, for causes attributable to the city, plaintiff was unable to start on this project for many months after it received notice to proceed. As a result, some of its work was not completed until 355 days after the two years stipulated in the agreement had gone by. At the two-year mark, as measured in monetary terms, Berley had completed 87% of its undertaking, leaving only approximately $60,000 worth left to be done. It should also be noted that this was but one of 11 construction contracts aggregating over $5,800,000 which plaintiff was engaged in performing at the same time.

Berley's suit, *inter alia* [Ed. note: among other things], sought damages for *increased* home office overhead expenses during the period of the overrun; presumably, to the extent of all but that represented by the final $60,000, it had already been compensated for its *regular* overhead expenses by the very terms of its contract with the city. At trial, the sole proof it introduced to establish the increase took the form of testimony by its Comptroller. This witness, who had not been in Berley's employ at the time the contract was fulfilled and had no personal knowledge of the facts, calculated the increase in home office overhead at $19,262. He did so solely by applying an arithmetical procedure which he claimed he had developed, but which counsel later referred to as the "Eichleay formula". It consisted of no more than the use of Berley's total billings during the period as a denominator of a fraction of which the billings for the 48th Street precinct was the numerator, the multiplication of that fraction by Berley's total home office expenses for the same period, the division of the sum so produced by the number of days involved in the period and, finally, treating the result as though it were the daily additional overhead for 48th Street, multiplication of it by the 355-day overrun.[30] He provided no substantiation that even a single penny's

---

29. From *Eichleay Corp.* ([CCH] 60–2 BCA, par. 2688).

30. The steps in this process may be restated as follows:

1. $\dfrac{\text{Billings for this Contract}}{\text{Total Billings for Period}} \times \text{Total Overhead for Period} = \text{Allocable Overhead}$

increase in overhead beyond what the rest of the plaintiff's business would in any event have required had in fact been incurred.

Over objection, Trial Term submitted the Comptroller's formula and calculations to the jury, which returned a verdict in Berley's favor. The judgment entered thereon was affirmed by a divided Appellate Division, but only on constraint of its prior decision in *Mars Assoc. v. Board of Educ.*, 53 A.D.2d 532, 383 N.Y.S.2d 889. The ensuing analysis persuades us that its order should be reversed.

It is fundamental to the law of damages that one complaining of injury has the burden of proving the extent of the harm suffered . . .. Delay damages, including ones in overhead, are no exception. A contractor wrongfully delayed by its employer must establish the extent to which its costs were increased by the improper acts because its recovery will be limited to damages actually sustained . . .. Particularly in actions *ex contractu,* however, when it is clear that some injury has been occasioned, recovery will not necessarily be denied a plaintiff when it is apparent that the quantum of damage is unavoidably uncertain, beset by complexity or difficult to ascertain . . .. The law is realistic enough to bend to necessity in such cases. A jury then may draw reasonable inferences from the other, though lesser, proofs actually presented in order to arrive at an estimate of the amount of extra costs which are the natural and probable result of the delay . . .. Even then, there must be a definite and logical connection between what is proved and the damages a jury is asked to find . . ..

It is relevant to the application of these principles that, in the case before us now, any increase in home office overhead was not merely hard to measure, but was lacking altogether. True, unlike job site overhead increases whose relationship to a particular job will usually be capable of direct proof, the connection between home office overhead increases and delay in a particular project will more often be indirect. But because proof is indirect does not mean it does not exist. Here there was no showing that the delay caused an increase in home office activity or expense of any kind. So far as the record shows, all expenditures for office overhead may have been completed when the 87% level was reached. Nor was there any showing that this was a case where the delay precipitated, or was precipitated by, engineering or design problems that call for central staff consideration. Above all, there was

---

2. $\dfrac{\text{Allocable Overhead}}{\text{Days of Performance}}$ = Daily Contract Overhead

3. Daily Overhead × No. of Days Delay = Overhead Claimed

The applicable figures were:

1. $\dfrac{\$473,071}{5,801,952}$ × $568,352 = $46,342   2. $\dfrac{\$46,342}{806}$ = $57.50   3. $57.50 × 335 = $19,262

no claim by the plaintiff that proof was not available. It fell back, instead, on the formula it put forth through the Comptroller.

The mathematical formula did not fill the void. At not a single point in the question which it set up was there a component which represented an actual item of increased costs, whether attributable to the delay on the city's job or not. The computation it essayed was therefore no less speculative because it was cast in a mathematical milieu. And, insofar as it was offered as a substitute for direct evidence of overhead damage, there was no accompanying foundation from which it could be found that, because of the character of Berley's business, increased overhead attributable to delay was impossible of proof without the aid of the formula. Nor was there any attempt to prove that the formula was logically calculated to produce a fair estimate of actual damages. Absent these preconditions, it was but an unsupported opinion. It did not serve to expand Berley's ability to recover any more than the reasonable value of any additional home office expenses it might have been able to prove by other means.

In rejecting the formula we, of course, are aware of the fact that, under the *Eichleay* nomenclature, it has been applied in a series of home office cases, which for the most part accept the formula largely without analysis and almost as a matter of administrative convenience (e.g., *Luria Bros. & Co. v. United States,* 369 F.2d 701, 177 Ct.Cl. 676 [FBI audit in advance of trial accepted pursuant to stipulation between parties]; *Hedin Constr. Co. v. United States,* 347 F.2d 235, 171 Ct.Cl. 70 [Government did not challenge commissioner's figure]).

The case before us readily reveals how the mechanical imposition of a formula akin to the one advanced by the plaintiff can all too easily bring a harsh daily penalty when only compensatory damages are warranted or even when the doctrine of *damnum absque injuria* [Ed. note: loss without violation of a legal right] is in order. For all practical purposes, it would completely ignore the safeguards against overreaching and arbitrariness to which the law of evidence has long been committed. As Justice (now Presiding Justice) Murphy pointed out in his dissent below, "The damages computed under the 'Eichleay formula' would be the same in this case whether the plaintiff had completed only 1% or 99% of the job on the scheduled completion date of May 7, 1971. This rather bizarre result is caused by the fact that the 'Eichleay formula' focuses on the length of the delay to the exclusion of many other important factors bearing on actual damages. If, on May 7, 1971, the plaintiff was merely required to spend $100 to complete the job, the 'Eichleay formula' would still require that the defendant pay $19,262 for the 335-day delay . . . I can only conclude that the mathematical computations under the 'Eichleay formula' produce a figure with, at best, a chance relationship to actual damages, and at worst, no relationship at all". Neither at argument nor in its brief did plaintiff deny the implications of this critique.

Nor is this all. Assuming there had been proof of increased home office expenses in this case, the formula here would have distorted them. Only $60,000 of work was to be performed during the overrun time. Yet, in the fixing of its proportion of any assumed increase, the formula would be asking the city to bear not just a share based on the mathematical relationship between that sum and the plaintiff's total billings during the period of protraction. It would also be burdened with a sum based on the 87% of the $472,000 contract which had not been delayed and, consequently, did not reflect work to be performed during this period. The failure to recognize the distinction between the 87% already performed and the 13% remaining to be performed would multiply the amount of the damages.

The reach of the hypothetical damages which the formula would generate is also demonstrated by the present case. The $19,262 which the Comptroller arrived at as the city's share of "increased" home office overhead here was more than all other damages found for the delay, including the costs of materials and labor at the job site where the balance of the contract was actually fulfilled.[31]

Accordingly, since it was error to submit the Comptroller's formula to the jury, the order of the Appellate Division should be reversed and the case remitted to the trial court for a new trial limited to the issue of delay damages.

JONES, Judge (concurring).

I am in agreement with the majority as to the disposition of this appeal. I cannot concur, however, in as broad an apparent condemnation of the so-called Eichleay formula as the majority opinion might be read to express.

In my analysis the Eichleay formula may be a useful and acceptable tool in some cases for measuring damages when there is an unavoidable inability precisely to allocate items of head-office expense to a particular construction project. Indeed on the new trial that could prove to be true in this case.

The difficulty with the determination of damages made by the jury here, which mandates reversal, is that the claimant, which had the burden of proof, wholly failed in this instance to introduce evidentiary proof which might have served as a predicate for application of the Eichleay formula (or indeed of any other formula). As the majority opinion points out, there is no direct proof in this record that the delay caused any additional office overhead expense. Absent evidence that there was some such damage, there

---

31. The $19,262 was part of a larger claim for total delay damages in the amount of $45,499 (inclusive of any job site overhead damage). Two nonoverhead items in the amount of $5,649 and $2,223 having been withdrawn from that total, the net claim was for $37,627, leaving only $18,365 for all but the $19,262 home office overhead claim. (The total verdict was for $34,696.55.)

is no occasion to apply any rule to measure its extent. Because the jury here did not separately identify the component of office overhead in its general verdict as to delay damages, there must be a new trial on the issue of damages.

If it can be shown that the maintenance of head-office functions, in whole or more likely in part, was required to complete the balance of the work to be performed, a determination may then be in order as to the proper amount of the total office expense properly to be allocated to the conceded delay. There must be some evidence, however, of the causal relationship between delay and additional expense incurred. Depending on the proof, the Eichleay formula might be one acceptable means of measuring this component of the damages sustained by claimant in consequence of the delay.

I would also suggest that it is irrelevant that in the abstract the Eichleay formula, related as it directly is to the period of delay, would produce the same result whether there remained 1% or 99% of the job to be completed. The determinative consideration must be the extent of head-office services required for the completion of the job. That in turn would depend, not on any ratio of the work to be completed to total work to be performed, but on the nature of the work that remained to be completed and the extent to which head-office services and facilities were reasonably required for such completion.

In short, in my analysis, the difficulty in this instance is not with the Eichleay formula itself but arises from the total absence of proof to warrant its application.

GABRIELLI, WACHTLER and COOKE, JJ., concur with FUCHSBERG, J.

JONES, J., concurs in a separate opinion in which BREITEL, C.J., and JASEN, J., concur.

Order reversed, with costs, and the case remitted to Supreme Court, New York County, for a new trial limited to the issue of delay damages.

It is beyond the scope of this treatise to discuss extensively problems such as the proper method of establishing an amount for extended home-office overhead.[32] Some courts seem to be strict,[33] as was the court in the *Berley* case, while others are more willing to give the contractor the benefit of the doubt.[34] What is important to recognize

---

32. For an extensive discussion, see Note, *Home Office Overhead As Damages For Construction Delays*, 17 Ga.L. Rev. 761 (1983). See also McGeehin, *A Farewell to Eichleay*, 14 Pub.Contr.L.J. 276 (1984).

33. *W.G. Cornell Co. v. Ceramic Coating Co.*, 626 F.2d 990 (D.C.Cir.1980).

34. *Southern New England Contracting Co. v. State*, 165 Conn. 644, 345 A.2d 550 (1974), followed in *Walter Kidde Constructors, Inc. v. Connecticut*, supra note 27. For the latest decision in Federal Procurement law, see *Capital Electric Co. v. United States*, 729 F.2d 743 (Fed.Cir.1984) (approved *modified* Eichleay formula).

is the almost insurmountable difficulty that a contractor faces if it must be put to a high degree of proof that the conduct of the owner cost a specified amount of money. For that reason, contractors and their attorneys frequently seek to receive a judicial stamp of approval on formulas such as the Eichleay formula or some other method that will bypass the necessity for actually establishing delay losses. On the other hand, whatever the outcome, these cases generate huge costs in terms of both attorneys' fees and expert witnesses.

In some disputes, the contractors seek to use an even rougher formula than those that have been noted in this section. A contractor faced with the almost insurmountable obstacle of establishing actual losses knows that it has lost money and feels that its losses were attributable not to its own poor estimating or inefficient performance but to acts of the owner or to those for whom the owner is responsible. As a result, the contractor will sometimes seek to use what is called the total cost method, a comparison of the actual costs of performance with what the contractor contends should have been the cost of the project. The Court of Claims has allowed this method but only where the following occurs:

1. The nature of the particular loss is such that it would be impossible or at least highly improbable that the court could ascertain the amount of damages with reasonable certainty based upon available records.

2. The contractor's bid or estimate was reasonable.

3. The contractor's actual costs were reasonable and not inflated.[35]

Judicial attitude toward the total cost method varies. It will be used if the court is convinced that a loss did occur and that this is the best method available of establishing the amount.[36] But it will not, if it is fearful that use of such a method will reward a contractor who either has estimated too low or has performed inefficiently.[37]

Another rough measurement sometimes employed is the jury verdict, best described as an educated guess by the fact finder, whether judge or jury. Although originally developed in the Court of Claims, the jury verdict was recently used in *Fattore Co. v. Metropolitan Sewerage Commission*,[38] in which the court rejected the total cost method but still made a sizable award based upon what it called a jury verdict. The court was sympathetic to the contractor's contention that damages in these cases cannot be proved with mathematical precision, stating:

> Given the inherently imprecise nature of an equitable adjustment, the existence of sufficient evidence for a court to come to a fair and reasonable conclusion, and the quantum issue presented in this case, the District Court's "jury verdict" must stand.[39]

The court announced limitations, however. There must be clear proof that the contractor suffered a loss and there was no more reliable method for computing damages.

---

35. *WRB Corp. v. United States*, 183 Ct.Cl. 409 (1968). The leading case is *F.H. McGraw & Co. v. United States*, 131 Ct.Cl. 501, 130 F.Supp. 394 (1955).

36. *Moorhead Construction Co. v. City of Grand Forks*, 508 F.2d 1008 (8th Cir.1975); *E.C. Ernst, Inc. v. Koppers Co.*, 626 F.2d 324 (3d Cir.1980) (labor costs); *Department of Transportation v. Hawkins Bridge Co.*, 457 So.2d 525 (Fla.App. 1984); *State Highway Commission of Wyoming v. Brasel & Sims Construction Co.*, 688 P.2d 871 (Wyo. 1984).

37. *Namekagon Development Co. v. Bois Forte Reservation Housing Authority*, 395 F.Supp. 23 (D.Minn.1974); *Huber, Hunt & Nichols, Inc. v. Moore*, 67 Cal.App.3d 278, 136 Cal.Rptr. 603 (1977).

38. 505 F.2d 1 (7th Cir.1974). Similarly, see *Meva v. United States*, 206 Ct.Cl. 203, 511 F.2d 548 (1975); *State Highway Commission of Wyoming v. Brasel & Sims Construction Co.*, supra note 36.

39. 505 F.2d at 5.

Use of a jury verdict requires sufficient evidence for a fair and reasonable approximation.

Can the contractor recover profits if its claim is based upon an equitable adjustment granted by the contract? One court held that the contractor could recover a designated percentage of cost as allowance for overhead and profit on performed work as the contract stated that the equitable adjustment could be computed under the changes clause.[40] Where there is no specific provision tying the equitable adjustment to the changes clause, it is not clear whether the contractor can recover profit on performed or unperformed work.[41]

SECTION 31.03
MEASUREMENT:
OWNER vs. CONTRACTOR

(A) ILLUSTRATIONS

The principal measurement problems relate to the contractor's unjustified failure to start or complete the project, to complete the project as specified, or to complete the project on time.

(B) PROJECT NEVER BEGUN

Damages are determined by subtracting the contract price from the market price of the work. This is usually based upon the best competitive price that can be obtained from a successor contractor for the same work. Suppose the contract price were $100,000 and the successor cost $120,000. The owner would be entitled to $20,000. This would protect its contract bargain.

However, claims for lost bargains do not receive the protection that claims based upon out-of-pocket losses do. Very likely in a case of this type it would be concluded that there was no difference between contract and market price, no lost bargain, and no recovery for that element of the breach.

In addition to the expectation interest (loss of bargain), the owner would be entitled to any losses caused by the delay (discussed in (E)).

(C) PROJECT PARTIALLY COMPLETED

Suppose the contractor ceases performance unjustifiably or is properly terminated by the owner. Restitution is one way of measuring the owner's claim.

This measure seeks to restore the status quo as it existed at the time the contract was made. This can be accomplished by restoring to the owner any progress payments it has made to the contractor. From this is deducted the benefit to the owner of the partially completed project. However, the difficulty of measuring the benefit conferred may lead a court to conclude that restitution is not proper. For example, one case denied the owner restitution of its payments in a well-drilling contract in which the well produced 80% of the required amount of water.[42] In such a case, expectation would be more appropriate. The owner would be entitled to the cost of performance that would generate the promised performance by the contractor less the payments it had made.

The owner was allowed to recover progress payments when a fallout shelter leaked and finally caved in.[43] To restore the status quo, the contractor would be entitled to deduct the extent to which its work added value to the land. The more likely event is that it added no value, with the owner being entitled to recover the cost of

---

40. *Kenny Construction Co. v. Metropolitan Sanitary District of Greater Chicago*, 52 Ill.2d 187, 288 N.E.2d 1 (1971). See also AIA Doc. A201, ¶ 12.1.4.

41. See *Fattore Co. v. Metropolitan Sewerage Commission*, supra note 38 (profit on unperformed work denied).

42. *Village of Wells v. Layne-Minnesota Co.*, 240 Minn. 132, 60 N.W.2d 621 (1953).

43. *Economy Swimming Pool Co., Inc. v. Freeling*, 236 Ark. 888, 370 S.W.2d 438 (1963).

removing the shelter and restoring the land to its precontract condition.[44]

Restitution is rarely used to measure owner claims in construction contract disputes. More commonly, failure to complete generates two types of damages. First, the owner very likely will have to pay more than the balance of the contract price to have the work completed by a successor. In many cases, the work will have been performed defectively, and this too may justify additional compensation (discussed in (D)). There is likely to be delayed completion, with delay damages being discussed in (E).

(D) DEFECTIVE PERFORMANCE: CORRECTION COST OR DIMINISHED VALUE?

Claims relating to defective performance may arise when a contractor has performed improperly, never having reached the level of substantial performance.[45] More commonly, defective performance cases involve attempts by the contractor to recover the balance of the contract price by alleging either substantial or full performance, the secondary issue being the deduction available to the owner if the work was only substantially performed.

The law seeks to place the owner in the position that it would have been in had the contractor performed properly. One way of doing this is by giving the owner the *cost of correction*, the amount necessary to correct the defective work and bring it to the state required under the contract. This is the preferred remedy.

Another method of giving the owner what it was promised is to award the *diminished value*, the difference between the value of the project as defectively built and what it would have been worth had it been completed as promised. This measure gives the owner's balance sheet what it would have had by full performance. It combines the value of the project as it sits and adds to it the diminished value because of less than complete performance.

Most commonly, cost of correction is awarded unless it would constitute economic waste.[46] The classic case in which cost of correction was refused for this reason involved a contract for a residence requiring a specified type of pipe. The contractor used another brand of pipe, presumably just as good as that specified.[47] Suppose the owner were awarded the cost of replacing the pipe. One possibility is that the owner will not correct the defect. Its cost would be much greater than the value added to the house. This would give the owner a windfall. Another possibility is that the owner makes the correction, creating a wasteful expenditure of societal resources.

The most difficult cases are those that involve a residence constructed for the use of the owner. In one case, the contractor had installed roofing shingles that were discolored. The cost to correct this properly functioning roof by replacing the roofing material was estimated at between $14,000 and $25,000. The contractor contended that the weathering of the shingles would ultimately lead to a uniform color and that the roof was not visible to passersby. The

---

44. Compare *Mayfield v. Swafford*, 106 Ill.App.3d 610, 62 Ill.Dec. 155, 435 N.E.2d 953 (1982), to be discussed in (D).

45. Refer to Section 26.06(B).

46. *Wentworth v. Airline Pilots Association*, 336 A.2d 542 (D.C.App.1975); *C.A. Davis Inc. v. City of Miami*, 400 So. 2d 536 (Fla.App.1981); *Dickerson Construction Co. v. Process Engineering*, 341 So.2d 646 (Miss.1977) (one and one half to three times original cost); *Estate of H.C. Jessee v. White*, 633 S.W.2d 767 (Tenn.App.1982); *Baldwin v. Alberti*, 58 Wash.2d 243, 362 P.2d 258 (1961); *Maryland Casualty Co. v. Rittiner*, 133 So.2d 172 (La.App.1961), seems to hold that cost of correction is always used.

47. *Jacob & Youngs, Inc. v. Kent*, 230 N.Y. 239, 129 N.E. 889 (1921).

court concluded the measure of recovery to be $7,500 based upon diminution in value.[48] However, another case involving the installation of a shingled roof concluded that the contractor had *not* substantially performed because the roof had yellow streaks though the roof was structurally sound and would protect the owner from the elements. The court concluded that in matters relating to homes and their decoration, taste or preference "almost approaching whimsy" may be controlling with the homeowner so that variations that under other circumstances might be considered trifling would preclude a finding that there had been substantial performance.[49] This gave the owner a defense to any contractor claim for the balance and be the basis for an owner claim measured by the cost of replacing the roof.

Other criteria are sometimes used to determine which measurement should be applied. Some courts look at the willfulness of the breach.[50] Others are less influenced.[51] While not always openly articulated, it is likely that if the breach were in good faith, the contractor would be given the benefit of whichever measure were lower.[52]

Sometimes a combination of the two measures is appropriate. For example, some deviations can be measured by diminution in value, while others can employ cost of correction.[53] The two formulas can be applied together if the cost of correction will not achieve the outcome promised under the contract. For example, in one case there had been settling and slanting due to a failure to sink piles far enough. The court held that the owner could recover not only the cost of doing as well as could be done to correct the problem but also the diminished value, with the cost of correction never accomplishing the contractual requirements.[54]

When cost of correction is used, several other factors must be considered. First, it should be clear that the cost of correction cannot be recovered if the corrective work includes work not called for under the original contract. In one case, replacement of leaky gutters and downspouts with galvanized gutters was not the proper measure of recovery when aluminum had originally been called for.[55] The measure should have been the cost of installing aluminum gutters that would have stopped the leaking, assuming this was the contractor's responsibility.

This generalization may not apply if the basis for the claim is an implied warranty of habitability. For example, in *Parsons v. Beaulieu,*[56] the defendant had contracted to build a septic tank for the plaintiff. The completed work did not comply with the legal requirements. The plaintiff replaced the tank with a larger one that did comply. The court held that the defendant contrac-

---

48. *Salem Towne Apartments, Inc. v. McDaniel & Sons Roofing Co.,* 330 F.Supp. 906 (E.D.N.C.1970).

49. *O.W. Grun Roofing & Construction Co. v. Cope,* 529 S.W.2d 258 (Tex.Civ.App.1975).

50. *Shell v. Schmidt,* 164 Cal.App.2d 350, 330 P.2d 817 (1958), cert. denied 359 U.S. 959, 79 S.Ct. 799, 3 L.Ed.2d 766 (1959). *Kangas v. Trust,* infra note 56 (willful based upon contractor's failure to pay attention to plans and building the way he usually did).

51. *Grossman Holdings Limited v. Hourihan,* 414 So.2d 1037 (Fla.1982).

52. Cf. *Shell v. Schmidt,* supra note 50.

53. *Plante v. Jacobs,* reproduced in Section 26.06(B).

54. *Kahn v. Prahl,* 414 S.W.2d 269 (Mo.1967). See also *Northern Petrochemical Co. v. Thorsen & Thorshov, Inc.,* infra note 70.

55. *St. Joseph Hospital v. Corbetta Construction Co.,* 21 Ill.App.3d 925, 316 N.E.2d 51 (1974); *Steinbrecher v. Jones,* 151 W.Va. 462, 153 S.E.2d 295 (1967).

56. 429 A.2d 214 (Me.1981). See also *Kangas v. Trust,* 110 Ill.App.3d 876, 65 Ill.Dec. 757, 441 N.E.2d 1271 (1982) (better material needed because contractor's breach made it more costly to repair).

tor had to pay for the cost of the larger septic tank, since the original one would not function in a way that would meet the contractor's implied warranty. The court noted that the owner had installed the least expensive one that would meet the legal requirements. In the *Parsons* case, the owner was not being put in a better position, as it had contracted for one that met the legal requirements.

Suppose the correction takes place a year after completion. The installation of new materials and equipment can be said to put the owner in a better position than that bargained for. The owner now has new material and equipment to replace material and equipment that had used up some of its useful life and had become depreciated. Generally the law takes into account any extended life that the replacement may give the owner.[57] But in *Dickerson Construction Co. v. Process Engineering Co.*,[58] the court refused to deduct any amount for extending the life of the building, inasmuch as the defects had been discovered only six months after completion.

At what point is the cost of correction determined? Usually it is the time of the breach. This assumes that the owner knows of the defect and is in the position to have the defective work corrected. However, some recent cases have used the time of *trial* as the benchmark for determining when the cost of correction should be computed. These cases had looked upon the highly inflationary period of the 1970s and noted that an award based upon time of breach rendered years after the work has been completed would not put the owner in the position it would have been had the work been done right in the first place.[59]

A similar problem can arise when the issue is diminution in value. A recent Florida case is instructive on this and other points. *Grossman Holdings Limited v. Hourihan*[60] involved a claim by the owners against a contractor for failure to build a house as promised. The owners contracted with the developer to purchase a house to be built in a planned development. Both the model and the office drawings showed the house with a southeast exposure, and the contract stated that the developer would construct the house "substantially the same" as in the plans and specifications or as in the model. A short time later, a new model and map went on display that showed the owners' lot and to-be-built house facing the opposite direction from that which they expected and wanted. The owners brought this discrepancy to the attention of the developer and demanded the construction of the house that they had contracted for. (They wanted the house facing a particular direction for esthetic and energy-saving reasons.) The developer refused to change the plans and began constructing the house facing in the direction opposite of that expected by the owner.

The trial court refused to award any damages, finding that the cost of turning the house around would have been economically wasteful and out of proportion to the good to be attained. It also noted that the value of the house had risen *above* the contract price since the date of the contract.

The intermediate court of appeals applied a different measure of damages. That court held that the economic waste doctrine does not apply to residential construction. It also found that the developer's willful and intentional failure to perform as prom-

---

57. *Freeport Sulphur Co. v. S.S. Hermosa*, 526 F.2d 300 (5th Cir.1976).

58. Supra note 46.

59. *Anchorage Asphalt Paving Co. v. Lewis*, 629 P.2d 65 (Alaska 1981); *Corbetta Construction Co. v. Lake County Public Building Commission*, 64 Ill.App.3d 313, 21 Ill.Dec. 431, 381 N.E.2d 758 (1978) (contractor denied breach and refused to correct).

60. Supra note 51.

ised barred it from using the substantial compliance doctrine. The court held that the proper damages would have been an amount necessary to reconstruct the dwelling to make it conform to the plans and specifications.

The Florida Supreme Court did not agree. First, it noted that the proper remedy for breach of a construction contract is usually the reasonable cost of correction if this is possible and does not involve unreasonable economic waste. Where it does, the measure is the difference in value between the product contracted for and the value of the performance that has been received. The court saw no reason to separate residential buildings from other construction. The court concluded that repositioning the house would result in economic waste but held that the trial court had incorrectly applied the proper formula—diminution in value. The court stated that the measurement is to be determined at the date of breach. Fluctuations in value after breach do not affect the measure of recovery. (The house would have been worth even more had it been built properly.)

The *Grossman* case also reflects the difference between objective and subjective measurements of value. The impersonal marketplace may not have found any difference in the house facing in one direction or the other. But the Hourihans did.

There is something to be said for the intermediate appeals court decision that refused to use the economic waste argument in residential construction. Balance sheet gains or losses are appropriate for certain types of commercial ventures but hardly make sense when a couple is buying a home in which they expect to live.

The case also rejects, or at least seems to reject, the statement made in other cases that allows the economic waste measurement to be used only if the breach is not willful. Here the owners complained as

soon as they noticed that the developer was starting to turn the house around, and the developer simply refused to make any change. To be sure, the developer might have wanted all the houses facing in a particular way and might have thought it had sufficient discretion to make that sort of change inasmuch as the contract stated that the house that was built would be *substantially* the same as the model and house as shown on the maps. Yet this could be classified as a wilful and deliberate breach.

Another illustration of the difficulty of applying the diminished value rule can be seen in *Mayfield v. Swafford*.[61] The contract in question involved the construction of a swimming pool for $7,000. Within a short period of time, substantial defects were discovered that made the pool useless. The cost of correcting the defects would have been $11,000, or about 50% more than the contract price. As a result the court first held that cost of correction cannot be used because its cost was clearly disproportionate to the probable loss in value to the owner.

How should diminution in value be measured? The pool can be looked at as part of the real estate, and a measurement can be taken of the diminished value of the entire real estate, pool included. Alternatively, the diminished value can be what the *pool* is currently worth, compared to what it should have been worth had it been built properly.

The pool in its present condition, however, has no use, function, or utility to the owner. Its presence detracts from the value of the property. The diminished value of either property or pool should reflect the cost of removal of the pool and restoration of the site to its original condition.

The measurement first looks at the value that a properly constructed pool would have added to the *property*, an amount the court determined to be the $7,000 contract price.

61. 106 Ill.App.3d 610, 62 Ill.Dec. 155, 435 N.E.2d 953 (1982).

This is compared to the value of the property with no pool at all, which must also take into account the cost of removing the pool and restoration of the site, less any salvage obtained in the removal process. Suppose the cost of removal and restoration less salvage is $3,000. There is a negative enhancement that would make the diminution in value a net $10,000. The court held that the owners are entitled to the judgment for the lesser of the two amounts to be compared: the cost of repair or the diminution in the value of the owner's property.

(E) DELAY

Chapter 30 discussed the problems of delay, and Section 30.09 looked at contractual attempts to agree on damages in advance, liquidating damages. Delay was noted in Section 31.02, dealing with a contractor's claims that it was not allowed to perform its work in the most economic and efficient manner. This section looks briefly at owner claims for unexcused contractor delays that were not liquidated in the contract.

The basic measurement formula used for delay is lost use, measured as a rule by rental value.[62] The difficulty arises when the owner makes claims that go beyond lost use value—usually added finance charges[63] and lost rentals. Clearly the owner cannot recover both the interest it had to pay on a loan used to build the project and the lost rental value. One court stated:

This is not a proper element of damage where damages are sought for the reasonable market rental value of the building in question. The rental income the owner received for his building is the interest he receives from his investment. If he has to borrow money to build the building, the interest he pays is part of his expense in furnishing the building. It is the loss of the rental income from the building that makes up the element of his damage, if any.[64]

In another case, the owner was able to be compensated for a loan extension fee that had to be paid to the lender after the contractor had been forty-five days in default.[65]

Suppose the owners made leases with tenants who could not be put into possession because of delayed construction. One case allowed the owner to recover for such rentals that were not paid.[66] Other cases have concluded that no damages were suffered when the leases were *extended* to the full lease period as long as the tenants made no claims against the owner for the delay.[67]

The uncertainties over the lost rental value and other related expenses provide an incentive to liquidate damages for contractor-caused delay.

(F) CONSEQUENTIAL DAMAGES

Sometimes owners seek to recover amounts to compensate them for losses they have suffered that are not directly related to the defective or untimely completion of the project. Such damages are recoverable if they were reasonably foreseeable at the time the contract was made and the loss was proven with reasonable certainty.

As a practical matter, these claims do not play an important part in construction disputes.[68] On occasion, a case is decided in which the owner seeks unusual damages.

---

62. *Ryan v. Thurmond,* 481 S.W.2d 199 (Tex.Civ.App.1972).

63. *Ralph D. Nelson Co. v. Beil,* infra note 67.

64. *Ryan v. Thurmond,* supra note 62.

65. *Herbert & Brooner Construction Co. v. Golden,* 499 S.W.2d 541 (Mo.App.1973).

66. Id.

67. *Ryan v. Thurmond,* supra note 62; *Ralph D. Nelson Co. v. Beil,* 671 P.2d 85 (Okl.App.1983) (owner may not *keep* property for full term of twenty-five-year lease.)

68. But see *Northern Petrochemical Co. v. Thorsen & Thorshov, Inc.,* infra note 70, where lost operating profits were recovered. See Section 31.04.

R.E.T. Corp. v. Frank Paxton Co.,[69] in addition to reviewing some material covered earlier in this treatise, was such a case. The owner built two apartment complexes. The second complex had problems because tenants complained of cold apartments and high heating bills. An investigation revealed that the contractor had not complied with the insulation requirements of the contract. Corrections were attempted that after some initial difficulties seemed to cure the problems, and tenant complaints dropped. But by the time the corrections were made, almost three years after completion, the complex was in serious financial trouble. The tenants forced management to offer rent inducements and adjust the rents. Vacancies increased, and rents could not be increased sufficiently to cover increasing costs. The second apartment complex dragged down the first, the business went into receivership, and the property ultimately was sold at a substantial loss.

The claim against the builder was, as is customary, based upon a number of theories. The trial court found that the builder had breached its contract, breached express and implied warranties, and negligently installed the insulation. It awarded damages of $105,000 for repair costs, $237,000 for lost rents, and $650,000 as the difference between the fair market value of the property (the amount the sale should have realized) and the reduced price (that finally obtained). The total damages were almost one million dollars.

The Iowa Supreme Court held that the trial court had been correct in using both cost of correction and diminished value, since no single formula would sufficiently compensate the plaintiff. However, the reviewing court found that the trial court had erroneously concluded that the damages are the same for breach of contract and for negligence. The court held that foreseeability is a limitation if the claim is based upon breach of contract but not if based upon tort. The court held that the cost of repairs and lost rentals could well have been contemplated by the parties when the contract was made, as they are routine elements of a damage recovery for breach of a construction contract. However, the court concluded that parties would not normally contemplate diminution in the value of the complex and its subsequent distress sale. This would be recoverable in a claim based upon tort but not based upon contract. Yet the court concluded that the error was harmless and affirmed the judgment.

The court dealt with another issue raised in (D). The trial court had determined the diminution in value as of 1978 at the time the building was sold, rather than in 1974–1975, the time the complex was built. The owner argued that the injury was a continuing one that extended until the final distress sale. The court concluded that these were special facts that militated against using the usual rule that diminution is determined as of the time of completion. The court concluded that the injury had no impact upon the plaintiff so as to become final until the complex went into receivership and was sold.

SECTION 31.04 CLAIMS AGAINST MULTIPLE PARTIES

Chapter 28 noted the not uncommon phenomenon of defects that can be traceable to design and to failure by the contractor to execute the design properly. That chapter looked at disputes between owner and contractor. In this section, emphasis is upon a claim made by the owner against both its independent design professional and the contractor. How is recovery measured when there are multiple causes?

The problem is best illustrated by *Northern Petrochemical Co. v. Thorsen & Thor-*

---

69. 329 N.W.2d 416 (Iowa 1983).

shov, Inc.[70] The owner contracted with an architect to design its new headquarters, which contained offices, a warehouse, and a manufacturing plant. It also contracted with a general contractor under a fixed-price contract. The construction began in the fall of 1967 and was virtually completed in April of 1968. At that time, it became clear that there were major structural flaws in the building. Large cracks were developing. Walls and columns were out of plumb. An investigation led to remedial action, but the deterioration continued. It became apparent that the building was moving in fits and starts. An agreement was made by the major participants to perform corrective work.

Until actual reconstruction began, it was believed that the sole reason for the building's failure was a structural defect in a support wall. When corrective work began, it was determined that the reinforcing steel for the concrete flooring had not been imbedded in the concrete and the underlying fill had not been compacted as required. These omissions left large voids under the floor that caused cracking and uneven settling. Correcting these faults required a massive reconstruction process that took approximately eight months. The trial court concluded that the owner's damages were approximately $750,000. It determined that some of the problem was traceable to design, some to faulty construction, and the balance to both.

Interestingly the court treated this as a tort case, finding that both the architect and the contractor had been negligent. It then shifted to tort principles applicable when there are co-wrongdoers and there is no difference in the culpability of wrongdoing. It concluded that where it is not reasonably possible to make a division of the damage caused by the separate acts of negligence closely related in point of time, the negligent parties, even though they acted independently, are jointly and severally liable. Each was responsible for the entire indivisible loss, even though they did not act together. The court adopted a rule that puts the burden on either architect or contractor to limit its liability by providing a method of apportionment. In the absence of a rational method of apportionment, each is responsible for the entire loss. (The rights and duties between the co-wrongdoers are discussed in Section 36.03.)

The increased use of comparative negligence and shared responsibility can lead to another approach. The loss can be divided by comparing negligence, or in the absence of a tort standard being appropriate, the level or intensity of "wrong-doing" in a nontort sense.[71]

Suppose the architect *were looked upon* as having caused 60% of the loss and the contractor 40%. This would place 60% of the loss on the architect and 40% on the contractor. Here cause would be replaced by "fault." This could also by-pass complicated contribution and indemnity issues. But it would expose the owner to a greater likelihood of being uncompensated if either defendant cannot pay the judgment.

PROBLEMS

1. O and C entered into a construction contract under which C was to construct a commercial office building and receive payment of $100,000 from O. During performance O wrongfully terminated the contract and ordered C to leave the project. C's costs before termination were $60,000. It would have

---

70. 297 Minn. 118, 211 N.W.2d 159 (1973).

71. *Shepard v. City of Palatka*, 414 So.2d 1077 (Fla.App. 1981) (dictum) reproduced in Section 15.08.

cost C an additional $60,000 to complete the work. C has been paid progress payments of $50,000. How much should C be able to recover from O?

2. O and C entered into a construction contract under which C was to build a commercial building in accordance with plans and specifications drafted by A, an architect retained by O. The contract stated that for every day of unexcused delay O would deduct $2000 as a "penalty." The contract price was $250,000. C was twenty-five days late. Can O recover $50,000? (Review Section 30.09.) If the clause is invalid, what are O's damages?

3. O hired DB to design and build a warehouse. Six months after completion and final payment, the building developed cracks and leaks in the roof. How would O's claim against DB be measured? (Assume that DB is responsible for the defects.)

32

The Subcontracting Process: an "Achilles Heel"

SECTION 32.01
AN OVERVIEW OF THE PROCESS

At the risk of oversimplification, the subcontracting process, as used in this chapter, is the method of construction organization under which the prime contractor is allowed to perform some or even much of its contract obligations through other contracting entities. The latter contracting entities are first-tier subcontractors. Likewise, the process in a large construction project can involve first-tier subcontractors performing their contract obligations through other contracting entities called second-tier subcontractors, or sub-subcontractors.

Frequently other business entities furnish equipment, machinery, products, supplies, or materials incorporated into the project or used to construct the project. These entities collectively called suppliers, usually make contracts with prime contractors or subcontractors. Although the line between subcontractors and suppliers is sometimes a difficult one to draw, for purposes of the discussion, subcontractors are defined as persons who perform significant services at the site.

What results in the subcontracting process is a chain of contracts that runs from owner to prime contractor or separate contractors, from prime or separate contractors to subcontractors, and from subcontractors to sub-subcontractors. Likewise, there are direct contract lines between contractors that for purposes of discussion include prime contractors and subcontractors and their suppliers.

Some legal problems generated by the subcontracting process have been discussed, and others are discussed in subsequent portions of the treatise. But subcontracting is the legal Achilles' heel of the Construction Process. It generates many legal problems and merits a separate chapter.

The principal advantage of a subcontracting system is improved efficiency, accomplished by breaking down work into categories that require a small number of related skills and the development of those skills by repetition. Those who perform services, whether they be laborers working on the site, cost estimators making bid proposals, or managers making procurement decisions, should become more skilled as they repeat performance of these specified services.

The subcontracting process, if working properly, can reduce costs not only by more efficient work but also by the creation of many competitive prime contractors and subcontractors. Entry into the prime contract field is facilitated by allowing contractors to conserve capital by relieving them of investment and financial burdens to the extent that subcontractors are used. The subcontracting system enables prime contractors to shift over much of the contract risks to subcontractors.

As for subcontractors, the subcontracting system should encourage many smaller, highly competitive subcontractors to enter the construction field. The subcontracting system may be one of the reasons why the construction industry, unlike the automobile industry, is made up of many contractors with specialized talents who operate mainly in limited localities. This has meant vigorous competition for work that, though it has the disadvantage of economic instability, frequent financial failures, and disputes, should, through competition, reduce construction cost.

An often ignored advantage of the subcontracting system is that many subcontracting entities can create social mobility and avoid the rigidity of more closed class structures. This is accomplished by permitting individuals or small business units to enter the field with a minimum amount of capital.

These undoubted potential and actual advantages have their cost. The subcontracting system generates a large share of construction legal problems.

The principal subcontract problem deals with payments. As shall be seen in Section 32.06 the subcontracting system creates risk of delayed payment and nonpayment to subcontractors. Typically the prime contractor on any substantial project is paid monthly as work progresses, with a customary retainage of from 5% to 10%. This creates a delay of cash flow from prime contractor to those to whom it owes payment. Even in no retainage contracts, the prime contractor often has cash flow problems caused by the lag between payments and obligations. When there are retainages, the cash flow problem is more serious. To the extent of payment delay, the prime contractor is providing financing

services for the owner that it seeks to transfer to subcontractors.

Prime contractors generally use subcontract payment provisions to minimize cash flow problems. These provisions frequently permit the prime contractor to delay paying subcontractors until the former have been paid by the owner. Such delayed payment provisions may squeeze subcontractors who have to pay for their labor and supplies. Even those with credit face financial hardship when credit is withdrawn or limited.

Cash flow problems are increased as a greater percentage of work is performed by subcontractors. Increased subcontracting means a greater likelihood that nonperformance by one subcontractor will delay payment to others who have performed properly.

The construction industry is composed of many small businesses with limited financial capacity and credit. A short delay in the flowthrough of funds from the owner to those contractors performing services on the site can cause financial hardship that may deprive a contractor, especially one at the lowest tier, of the funds needed to continue performance. The subcontracting system heightens the financial stress inherent in the flowthrough process because it increases the distance of the money flow.

The extent of work subcontracted also plays a role in creating legal problems. Subcontractors frequently assert that prime contractors are merely assemblers or brokers for the services of others. Prime contractors deny this and claim that they supply much of the material and services themselves. Prime contractors and subcontractors agree that the *amount* of work subcontracted depends upon the *type* of construction. Whether prime contractors or subcontractors are correct on this controversial question, it is clear that the greater percentage of subcontracted work, the greater the likelihood of severe cash flow problems. A prime contractor who does not have much money tied up in the project is less concerned about the swiftness of the cash flow. This is especially true if the prime contractor's obligation to pay subcontractors is conditioned upon receiving payments from the owner.

Moving from cash flow delay to nonpayment problems, the volatility of the construction industry must again be emphasized. Business failures and bankruptcies frequently occur in the construction industry. The higher the proportion of work subcontracted, the greater the risk that there will be unpaid subcontractors who will seek some type of legal relief when their work has benefited the owner. One type of relief often sought by unpaid contractors and suppliers is a mechanics' lien. Valid liens usually result in owners paying lien claimants to remove the liens. To avoid liens, owners create payment structures to minimize payment diversion by the prime contractors. Owners—private and especially public—frequently require payment bonds to give an effective remedy to unpaid subcontractors and suppliers. Bond requirements inject the complexity of surety bonds and additional parties into the already complicated construction structure.

Problems of delayed payment and nonpayment highlight another significant feature of the subcontracting process. The subcontractor is "a contract away" from the major source of power and control over the project—the owner. As it has no direct contractual relationship with the owner, it must look to the prime contractor for payment, for dealing with disputes over performance, and to process claims against the owner. This can create a sense of powerlessness in subcontractors, leading to friction and lawsuits at worst and poor communication at best.

The subcontracting process compounds the already difficult problems caused by the multitude of construction documents that regulate the relationships in the Construction Process. One principal cause of legal problems is the wealth of potentially conflicting documents that regulate the rela-tionship between owner and contractor. Add first- and second-tier subcontracts, which frequently refer *generally* to contract provisions in contracts above them on the contract chain, and additional ingredients for disputes result.

The subcontracting process generates a potentially large number of construction contractors—all working on the same site often at the same time. For purposes of this discussion, contractors include prime contractors, separate contractors, and the various tiers of subcontractors. Contractors may disagree on who will have access to a particular part of the site at a particular time. One contractor's performance often depends upon another contractor's work being completed or at least in a certain state of readiness. One contractor's work may be disturbed or ruined by another contractor's workers, and disputes may develop over which contractor is responsible. One contractor may employ workers who belong to a construction trade union, while another uses nonunion workers. One contractor's employee may be injured or killed, and the responsibility may be asserted against or shared by a number of other contractors. The sheer number of different contractors at work can create immense administrative and, ultimately, legal problems.

The illustrations given do not exhaust the legal problems incident to subcontracting. Additionally, there is frequent bargaining disparity present in construction contracts and especially in subcontracts. Generally, the dominant bargaining strengths parallel the money flow. Lenders can exact terms from owners because the total number of borrowers seek more money than lenders have to lend. The owner who has funds or can borrow them seeks a contractor from among the many willing to perform the work. This gives the owner substantial bargaining advantage over the prime contractor. Although the prime contractor awarded the contract does not as yet have funds, it has contract rights. These contract rights generally give the prime contractor bargaining advantage over subcontractors, many of whom are looking for work. As a result, at this stage the prime contractor usually has the bargaining power to demand, and often obtain, favorable terms from subcontractors, sometimes terms that are more favorable than those in the prime contract. Likewise, subcontractors frequently exert parallel bargaining advantage over sub-subcontractors.

The bargaining position of suppliers and contractors cannot be so easily generalized. Suppliers range from large manufacturing companies with strong financial positions to small distributors with limited financial capacity. As a result, some suppliers are in the position to dictate terms to subcontractors and even to prime contractors. Yet even suppliers in this advantageous position must sell and must often extend credit to do so. Often their extension of credit is predicated upon statutory lien rights or the existence of payment bonds.

In the typical construction project, one finds the lowest tier subcontractor with the weakest bargaining power. This bargaining pattern is illustrated by the frequent passing down of increasingly harsh and one-sided indemnity agreements. In addition to the strong bargaining pressures from prime contractors or higher tier subcontractors, lower tier subcontractors also face strong bargaining power of large suppliers and construction trade unions. Subcontractors, especially those at the lowest tier, often find themselves squeezed on all sides because of their poor bargaining position.

Institutional bargaining disparities generate legal problems. Harsh terms exacted at the bargaining table may be resisted by the weaker party when disputes develop. The law looks with disfavor upon harsh terms, even though the legal system generally allows parties to write their own contracts. As a result, the uncertainty of enforcement of harsh terms exacted by the dominant party generates legal disputes that often require judicial resolution.

Those in vulnerable bargaining positions may seek other avenues of relief. For example, industry associations sometimes obtain legislative enactments such as anti-indemnity legislation to overturn contract clauses exacted from them by parties in a stronger bargaining position. Injection of legislative rules in an area regulated largely by private contracts adds an additional complicating feature to construction problems.

Associations composed of members who are often in a weak bargaining position often seek to remedy this weakness by participation in the process by which standard construction contracts such as those published by the American Institute of Architects (AIA) are created. For example, prime contractors, through the Associated General Contractors (AGC), seek to persuade the AIA to incorporate or eliminate language in AIA Standard Documents that they might not otherwise be able to do in dealing with an owner. Correspondingly, associations representing subcontractors frequently request the AIA to incorporate provisions in the AIA prime contract that require that certain rights be accorded subcontractors when these rights might not be obtainable in normal bargaining between prime contractors and subcontractors. When the prime contract deals with subcontracting and subcontractors, an increased likelihood exists of conflicting contract documents and a more generally complicated prime contract.

Subcontractor associations have sought protective language not only in prime contract documents published by the AIA but also in the AIA's Standard Subcontract, A401. Just as prime contractors prefer an AIA Document over a construction contract drafted by the owner, subcontractors are likely to prefer a subcontract drafted by the AIA to one prepared by the prime contractor.

One final aspect of the subcontracting process adds additional complexity. Contracts for the performance of design and construction services are basically regulated by common law rules, that is, rules that have evolved through court decisions. On the other hand, contracts for the sale of goods are governed by the Uniform Commercial Code (UCC), a comprehensive statutory regulation. When a subcontractor, as is usually the case, provides both goods and services, which rules apply—the common law or UCC? Generally the test is whether goods or services are the predominant aspect of the transaction.[1]

To sum up, although the subcontracting system has undoubted advantages, it generates a host of legal problems. With this in mind, some particular subcontract problems are addressed.

SECTION 32.02 SUBCONTRACTOR BIDDING PROCESS

(A) USE OF SUB–BIDS

In technical projects requiring specialized skills, the prime contractor may do very little of the work itself. In such projects, the cost to the prime contractor of work to be done by subcontractors is not likely to be known until subcontractors submit subbids. The prime contractor cannot bid until hearing from all the prospective subcontractors, something that does not usually occur until close to time for submitting the bid to the owner. The prime contractor relies on the subcontractor's bid in submitting its own.

The prime contractor's reliance on subcontractors is most acute in the mechanical specialty trades (e.g., electrical, plumbing, air-conditioning). Prime contractors utilize a large number of subcontractors from the nonmechanical specialty trades (e.g., masons, roofers). This activity can be bid with relative accuracy by the prime contractor, with the sub-bids providing downside protection. These subcontractors primarily provide the prime contractor with a skilled workforce, one that the prime contractor could not possibly afford to keep employed between projects.

Prime contractors generally use the bids given by subcontractors in computing their bids. Suppose a subcontractor withdraws its bid (usually because it contends that its bid had been inaccurately computed or communicated). Can the prime contractor hold the subcontractor to its bid by contending that it had relied on the sub-bid in making its bid?

(B) IRREVOCABLE SUB–BIDS

The early cases dealing with this problem used traditional contract analysis to allow the subcontractor to revoke its bid. Generally the prime contractor does not make a contract with the subcontractor, conditioned on the prime contractor's being awarded the bid. The prime contractor wishes to preserve maximum freedom to renegotiate with the low bidder or other bidders. Ignoring the prime contractor having relied upon the sub-bid in making its own bid, *James Baird Co. v. Gimbel Brothers*[2] did not hold the subcontractor to its bid. However, contract law has expended reliance as a basis for making an offer irrevocable. This carried over into subcontractor bid cases, the leading case being *Drennan v. Star Paving Co.*[3] The subcontractor had submitted the lowest sub-bid for

---

1. *Bonebrake v. Cox*, 499 F.2d 951 (8th Cir.1974), followed in *Pittsburgh-Des Moines Steel Co. v. Brookhaven Manor Water Co.*, 532 F.2d 572 (7th Cir.1976).

2. 64 F.2d 344 (2d Cir.1933).

3. 51 Cal.2d 409, 333 P.2d 757 (1958), followed in the Restatement (Second) of Contracts, § 87(2) (1981). Recent cases following the *Drennan* case are *Preload Technology, Inc. v. A.B. & J. Construction Co., Inc.*, 696 F.2d 1080 (5th Cir.1983); *Alaska Bussell Electric Co. v. Vern Hickel Construction Co.*, 688 P.2d 576 (Alaska 1984); and *Mead Associates, Inc. v. Antonsen*, 677 P.2d 434 (Colo.App. 1984).

the paving portion of the work, a sub-bid the prime contractor used in computing its overall bid. The prime contractor listed the defendant on the owner's bid form as required by statute.[4] The prime contractor was awarded the contract and stopped by the subcontractor's office the next day to firm up the "subcontract." Upon arrival, the subcontractor immediately informed the prime contractor that it had made a mistake in the preparation of its bid and would not honor it. The prime contractor brought suit and was awarded the difference in cost between the subcontractor's sub-bid and the cost of a replacement.

The court concluded that the use of the sub-bid did not create a bilateral—or two-sided—contract between the plaintiff and the defendant.[5] But the court held the subcontractor to its bid because its bid was a promise relied upon reasonably by the prime contractor when it submitted its bid. The possible uncertainty of subcontract terms was brushed aside.[6]

Reliance making the bid irrevocable created a one-sided contract despite submission of the sub-bid inviting acceptance by the prime contractor.[7] Subcontractors do not wish to subsidize the prime contractor's costs of bid preparation. They want the subcontract.

(C) BARGAINING SITUATION: SHOPPING AND PEDDLING

The *Drennan* rule improves the prime contractor's already powerful bargaining position. Although the prime contractor under the *Drennan* rule is not free to delay its acceptance or to reopen bargaining with the subcontractor and still claim a right to accept the original bid, it can at least for a short period seek or receive lower bid proposals by other subcontractors.

Under the *Drennan* rule, until the prime contractor is ready to sign a contract with the subcontractor whose bid it has used, the subcontractor is not assured of getting the job. Before the award of the prime contract, the plurality of competing prime contractors' bidding on a project tends to diffuse their bargaining power over subcontractors. This competition before the award of the contract should result in lower sub-bids and consequently lower overall bids by the prime contractors, a definite benefit to the owner. Although subcontractors often wait until the last minute to submit their sub-bids in an effort to minimize the prime contractor's superior bargaining position, a substantial amount of competition still exists among the subcontractors themselves.

After award of the contract, the relative bargaining strengths of the successful prime contractor and the competing subcontractors change drastically. The prime contractor now has a "monopoly" and a substantially superior bargaining position over the subcontractors under the *Drennan* rule. The sub-bids used provide the prime contractor with a protective ceiling on the cost of the work with no obligation to use the subcontractors. The prime contractor is therefore free to look elsewhere for yet a

---

4. As to Listing Laws, see § 32.03. In *Southern California Acoustics Co. v. C.V. Holder, Inc.*, 71 Cal.2d 719, 79 Cal. Rptr. 319, 456 P.2d 975 (1969), the court held that an improper substitution created a claim based upon the statutory violation against the prime contractor in favor of the improperly substituted subcontractor. This approximates binding *both* parties. But generally listing the subcontractor to comply with the statute does not create a bilateral contract. *Holman Erection Co. v. Orville E. Madsen & Sons, Inc.*, 330 N.W.2d 693 (Minn. 1983).

5. For a recent case holding use of the bid is not an acceptance see *Mitchell v. Siqueiros*, 99 Idaho 396, 582 P.2d 1074 (1978).

6. Refer to Sections 32.03(F) and (G).

7. For a justification of a one-sided contract see *Holman Erection Co. v. Orville E. Madsen & Sons, Inc.*, supra note 4.

better price and is able to increase its profits by engaging in post-award negotiations.

Post-award negotiations have become controversial. Sometimes they are called "bid shopping"; the prime contractor uses the lowest sub-bid received as a bargaining tool to negotiate still lower sub-bids. "Bid peddling" is the converse, with other subcontractors attempting to undercut the sub-bid to the prime, in essence engaging in a second round of bidding. Subcontractors often refer to these post-award negotiations as "bid chopping" and "bid chiseling." (These tactics can also be used *before* prime bids are submitted.)

Both subcontractors and owners have reasons to condemn post-award competition. The subcontractors assert that the preparation of a bid involves considerable expense. Subcontractors who "bid peddle" may not even prepare their own bids, saving overhead expenses. The subcontractor who went to the expense of calculating a bid subsidizes the bid-peddling subcontractor's overhead costs as well as the prime contractor's costs of bid preparation.

Subcontractors who fear bid shopping often wait until the last minute to submit their sub-bids to the prime contractor to give the prime contractor as little time as possible to bid shop. This last minute rush is the cause for many mistakes by both subcontractors and prime contractors. Some subcontractors simply refrain from bidding on jobs where bid shopping is anticipated to save the expense of preparing a bid. To that extent, competition among subcontractors is diminished and higher prices can result.

Subcontractors feel that they must pad their bids to make allowance for the eventual post-award negotiations. This "puffing"

raises the cost to the owner, as the inflated bid is the bid the prime contractor uses to compute its overall bid. Any subsequent negotiations that result in a reduction of the price inure to the prime contractor alone.

The superior bargaining position of a successful prime contractor spurs cutthroat competition among subcontractors, resulting in lost profits that can upset industry stability. Prime contractors respond by noting that sub-bids are often unresponsive to the specifications and require further clarification and negotiation. This may be especially true when prime contractors are dealing with subcontractors with whom they have never dealt. They must investigate the subcontractor's reputation and work experience before making a firm contract.

Prime contractors state that bids are often requested for alternative proposals and they lack the time to evaluate all the alternatives in the short time available between receipt of the sub-bids and bid closing. They assert that negotiations are sometimes required to decide upon the specific alternative to be chosen. Prime contractors also justify post-award negotiation by stating that estimating a job is a normal cost of overhead in the construction industry.

(D) AVOIDING DRENNAN

Subcontractors can avoid their sub-bids being firm offers that bind them and not the contractor. They can denominate their bids as "requests for the prime to make offers" to them or call them "quotations" given only for the prime contractor's convenience.[8] They may also try to annex to their sub-bids language that states that the use of the bid constitutes an acceptance that ties the prime contractor to them.[9] The

8. *Leo F. Piazza Paving Co. v. Bebek & Brkich*, 141 Cal.App.2d 226, 296 P.2d 368 (1956); *Cannavino & Shea, Inc. v. Water Works Supply Corp.*, 361 Mass. 363, 280 N.E.2d 147 (1972).

9. *Sharp Brothers Contracting Co. v. Commercial Restoration, Inc.*, 334 S.W.2d 248 (Mo.App.1960) (bid must be accepted in ten days). But see *S.M. Wilson & Co. v. Prepakt Concrete Co.*, 23 Ill.App.3d 137, 318 N.E.2d 722 (1974) (protection lost by continuing to deal after expiration of deadline).

difficulty with these methods of avoiding the *Drennan* rule is that either the subcontractors do not have the bargaining power to implement them or the process does not make it convenient for them to use them.

The *Drennan* rule can also be avoided if too many crucial areas have been left for further negotiation.[10] Finally, though unsuccessful in the *Drennan* case, a recent case upheld the subcontractor's claim of mistake.[11]

(E) BID DEPOSITORIES

Whenever individuals are weak at the bargaining table they seek alternative approaches. Subcontractors have organized themselves and created bid depositories (sometimes called bid registries). A typical bid depository is a facility established and operated by a trade association that receives bids from the subcontractors for the supplying of construction services for large construction projects. The depository collects the bids from the subcontractors and delivers them to the prime contractors who intend to bid on that particular project. Bid depositories are used most frequently in the mechanical specialty trades, these subcontractors most vulnerable to the detrimental effects of bid shopping and bid peddling.

The subcontractors must submit their bids to the depository before the cutoff time, typically four hours before bid closing. The bids are submitted in sealed envelopes, one copy addressed to the prime contractor and the other to the depository. To facilitate compilation of the bids, special bid forms are often provided. After the cutoff time, no bids may be received and none received may be revoked or amended.

At closing time, the bids addressed to the prime contractors are distributed to whom they are addressed. The prime contractors have four hours before bid closing to prepare their own bids. The other copy of each bid is retained by the depository. After bid closing, the depository may publish either all the sub-bids, only the lowest sub-bids, or none at all. In any event, at least the depository knows the identity and price of the lowest sub-bidder who used the depository. Further renegotiations with the subcontractor who submitted the low bid becomes highly suspect and may subject the prime contractor to a disciplinary action by the depository.

Depositories are generally open to any subcontractor in the applicable trade. Membership is usually required before the subcontractor may use its facilities. Prime contractors can be members, but normally an agreement to abide by depository rules is all that is required of the prime contractor.

Most depository rules state the following:

1. Subcontractors must use the depository exclusively, if at all.

2. Prime contractors may not accept bids from subcontractors except through the depository.

3. The prime contractor must agree to accept the lowest bid from the depository or not use the depository at all.

4. Bid splitting (combining different parts of different sub-bids together to get a yet lower aggregate bid) is prohibited.

5. Violation of the depository rules results in a forfeiture of a filing fee, imposition of a fine, or both.

Some bid depositories with tight controls have been found to violate laws against price fixing.[12] They eliminate bid shopping, a form of free market competition.

---

10. *Preload Technology, Inc. v. A.B. & J. Construction Co., Inc.* supra note 3 (dictum).

11. *Tolboe Construction Co. v. Staker Paving & Construction Co.*, 682 P.2d 843 (Utah 1984). See Section 22.04(E).

12. *Oakland-Alameda County Builders' Exchange v. F.P. Lathrop Construction Co.*, 4 Cal.3d 354, 93 Cal.Rptr. 602, 482 P.2d 226 (1971) (invalid). But for a more permissive attitude, see *Cullum Electric & Mechanical, Inc. v. Mechanical Contractors Association of South Carolina*, 436 F.Supp. 418 (D.S.C.1976) (used peaceable persuasion and

Those who feel that free competition will achieve the best result see all negotiation as not only useful but also desirable. To subcontractors, negotiation from weakness means harsh terms.

Subcontractors also look to the owner—public or private—to assist them, contending that the owner is also disadvantaged by this negotiation. Involving the owner through an approval process is discussed in Section 32.03.

(F) UNIFORM COMMERCIAL CODE

Offers generally at common law are revocable even if stated to be irrevocable. Dissatisfaction with this rule has generated solutions taking many forms, including the *Drennan* rule making the bids irrevocable. One method adopted was § 2–205 of the Uniform Commercial Code (UCC). It recognizes firm offers and enforces them if certain requirements are met.

The UCC does not apply to service contracts or those which involve interests in land. The UCC has also had limited use in making sub-bids by suppliers irrevocable. Section 2–205 requires a written offer stating that it will be held open. Bids are often communicated over the telephone.

The UCC is not a viable solution to the problem of sub-bids for another reason. Rarely will all the needed terms be expressed in a telephone conversation. Even if the bid is communicated by a writing with all the terms, rarely do prime contractors intend to be bound to those terms, choosing to either use terms used before or dictate terms later. The firm offer of § 2–205 does not fit the sub-bid, another

reason why the *Drennan* rule, a substitute for the UCC firm offer, was incorrectly decided.

California revised its version of the UCC in 1980. Under § 2205(b) of its Commercial Code, a written or oral sub-bid for goods made to a licensed contractor that the bidder knows or should know will be relied upon is irrevocable for twenty days after the prime contractor is awarded the contract but not later than ninety days after the bid. Oral bids of over $2500 must be confirmed in writing within forty-eight hours. The bid can limit the duration of the offer.

Again the problem of terms can arise. The statute does not preclude the sub-bidder from asserting that the terms needed to cure any completeness requirement were not included in the offer, particularly if it were oral.[13] Although the statute as revised expands the enforcement of oral sub-bids for goods, it has the same defects as the *Drennan* rule (noted in (G)).

(G) A SUGGESTION

The *Drennan* rule, now firmly in the saddle in most American jurisdictions, was a laudable attempt to avoid the common law rule of revocability and expand reliance as a method of enforcing *promises that should be enforced*. It is singularly inappropriate for construction sub-bids. First, it gives advantage to the prime contractors that already have a strong bargaining position. Second, it creates a binding offer before there has been agreement on important terms such as bonds, payment, indemnification, and dispute resolution.[14] Third, its

---

no intent to monopolize; bid peddling not essential to pure price competition), affirmed 569 F.2d 821 (4th Cir.1978), cert. denied 439 U.S. 910, 99 S.Ct. 277, 58 L.Ed.2d 255 (1978).

13. Refer to *Preload Technology, Inc. v. A.B. & J. Construction Co., Inc.,* supra note 3.

14. This problem was recognized in *Saliba-Kringlen Corp. v. Allen Engineering Co.,* 15 Cal.App.3d 95, 92 Cal.Rptr. 799 (1971). But too much emphasis on this would frustrate the *Drennan* rule, an outcome the court was not willing to endorse. As to certainty, compare *Debron Corp. v. National Homes Construction Corp.,* 493 F.2d 352 (8th Cir. 1974), with *C.H. Leavell & Co. v. Grafe & Associates, Inc.,* 90 Idaho 502, 414 P.2d 873 (1966). The former was

nature has raised technical difficulties.[15] The traditional common law rule expressed in the *Baird* case [16] is more appropriate. One-sided contracts should be found only when this is clearly intended or where absolutely necessary. Neither occurs here.

Use of the bid should be an acceptance of the offer and a contract concluded *if* there are sufficient terms agreed upon by *both* parties. This can be based upon well-established custom or a course of dealings between the parties supplying the ancillary terms in addition to the work to be performed and payment. When this does not exist, neither party should be bound until they work out the terms needed to make a binding contract.[17]

SECTION 32.03
SUBCONTRACTOR SELECTION AND APPROVAL: THE OWNER'S PERSPECTIVE

Owners wish to have competent contractors building their projects. They want subcontractors treated fairly and given an incentive to perform the work expeditiously. Some owners contract directly with the major specialty trades to avoid the prime contractor as an "intermediary" between owner and specialized trades. Some contract directly with the specialized trades and assign those contracts to a main or prime contractor. Some owners dictate to the prime contractor which subcontractors will be used,

known in the United Kingdom as the nominated subcontractor system.

These methods of direct intervention are not common in the traditional contracting system. Some owners leave subcontracting exclusively to the prime contractor. Others take a role that gives them *some* control but that does not involve the owner *directly* with the subcontractor. Using the prime contractor as a buffer is done for administrative and legal reasons. The subcontracting system requires a well-defined organizational and communication structure under which each participant knows what it must do and with whom it must deal. From a legal standpoint, the owner does not want to be responsible for subcontract work, does want the prime contractor to be responsible for defective subcontract work, and does not want to be responsible if subcontractors are not paid.

One "part-way" control is to require that prime contractors list their subcontractors at the time they make their bid. Statutes in some states, called Listing Laws,[18] impose this upon prime contractors in public projects. Although undoubtedly some of the impetus for such laws came from subcontractor trade associations, one reason for Listing Laws is to assure the awarding authority that only competent subcontractors will perform on the project.

Legislatures frequently justify listing statutes by condemning bid shopping and bid peddling that they assert cause poor quality

---

unimpressed with the lack of certainty argument, while the latter was receptive. Refer to *Preload Technology, Inc. v. A.B. & J. Construction Co., Inc.*, supra note 3.

15. Is it a contract for Statute of Frauds purposes? See *N. Litterio & Co. v. Glassman Construction Co.*, 319 F.2d 736 (D.C.Cir.1963). Is the claim "at law" or "in equity" an issue that controls the requirement of a jury? See *C&K Engineering Contractors v. Amber Steel Co.*, 23 Cal.3d 1, 151 Cal.Rptr. 323, 587 P.2d 1136 (1978) (equitable). Does breach by anticipatory repudiation apply to the prime before it makes a subcontract? See *Alaska Bussell Electric Co. v. Vern Hickel Construction Co.*, supra note 3 (dictum stating it does not).

16. See note 1, supra.

17. This approach has been suggested in *Loranger Construction Corp. v. E.F. Hauserman Co.*, 376 Mass. 757, 384 N.E.2d 176 (1978). See generally Closen & Weiland, *The Construction Industry Bidding Cases*, 13 J.Mar.L.Rev. 565 (1980).

18. West's Ann.Cal.Gov't Code §§ 4100 et seq.

of material and workmanship to the detriment of the public and also deny:

> the public . . . the full benefits of fair competition among prime contractors and subcontractors, and lead to insolvencies, loss of wages to employees, and other evils.[19]

Usually Listing Laws regulate substitution of listed subcontractors by providing specific justifications for substitutions and a procedure for determining the grounds for replacing one subcontractor with another. Complications have developed not only as to grounds for substitution but also as to who has a claim for violation of the statute.[20] From the subcontractor's viewpoint, Listing Laws dampen any attempt by the prime contractor to reopen negotiations.

Listing is used by the American Institute of Architects (AIA) in a more limited way. AIA Doc. A201, ¶ 5.2.1, requires the contractor "as soon as practical after the award" to furnish the owner and architect the name of subcontractors and those who will furnish materials and equipment fabricated to a special design. Before 1976 the architect had to *approve* subcontractors. To make the architect's role more passive and to minimize the likelihood of liability, A201 currently states that the architect "will promptly reply to the Contractor in writing stating whether or not the Owner or the Architect, after due investigation, has reasonable objections to any proposed" subcontractor. Failure to "reply promptly shall constitute notice of no reasonable objection." If the architect has reasonable objection, ¶ 5.2.3 requires a substitution and an increase or decrease in the contract price. This has been criticized as encouraging a prime contractor to submit subcontractors who may not be satisfactory to get

a price increase when the substitution is made. Before 1976, A201 allowed the architect or owner to revoke approval at any time. Currently A201 bars rejection once performance begins.

Some control over subcontractors, though desirable, has negative features. Requiring that a subcontractor not be removed unless there is "due investigation" can invite a subcontractor who has been removed to assert a claim that the architect or owner has intentionally interfered with its actual or prospective contract. Although there may be defenses,[21] this does increase potential liability exposure.

To sum up, owner intervention into the relationship between prime contractor and subcontractor may be essential, but it carries risks. If the risks are so great but the need for intervention so strong, the owner should consider a method other than the traditional contracting system.

SECTION 32.04
SOURCES OF SUBCONTRACT RIGHTS AND DUTIES

The principal source of contract rights and duties between prime contractor and first-tier subcontractor is the subcontract itself. Similarly, the principal source of contract rights and duties between first- and second-tier subcontractors is the sub-subcontract, and so on down the subcontract chain. However, this source is not exclusive. As with any contract, the express terms must be interpreted by language often outside the writing. Express terms will be supplemented by terms implied judicially into the subcontract relationship.

Like other aspects of the subcontract sys-

---

19. Id. at § 4101.

20. See *Southern California Acoustics Co. v. C.V. Holder, Inc.*, supra note 4, giving an improperly delisted subcontractor a statutory claim only against the prime contractor. Generally, listing in accordance with bidding requirement, does not constitute an acceptance. *Mitchell v. Siqueiros*, supra note 5.

21. *Commercial Industrial Construction Co. v. Anderson*, 683 P.2d 378 (Colo.App. 1984) (architect given advisor's privilege). See Section 17.08(D).

tem, each contract on the subcontract chain can be and frequently is affected by contracts higher up the chain. The subcontract relationship is usually affected and perhaps controlled by terms in the prime contract. Correspondingly, second-tier subcontract relationships are affected and perhaps controlled by both first-tier subcontract and prime contract provisions. For convenience, discussion focuses upon the relationship between prime contractor and first-tier subcontractor and the effect of the prime contract to which the subcontractor is not a party upon that relationship.

The discussion in this section centers principally upon the subcontractor's being bound to provisions of the prime contract because the latter is referred to or incorporated in the subcontract. However, it has been contended that a series of interrelated contracts can create a single contract binding all parties to the entire series even though each party may not have signed each contract in the series. Although such a contention was rejected in the context of an architect-owner contract followed by an owner-prime contractor contract,[22] a California decision intimated that such a conclusion might be sustained in a case involving a subcontract followed shortly by a prime contract.[23]

There is a superficial attractiveness in *directly* binding all participants in the series of linked contracts to the terms of each contract in the series. It would certainly avoid some of the problems discussed later in this section. Such a fusion of parties would frustrate the intention of the parties when they create the linked series of con-

tracts such as that found in subcontracting. The contracts, though clearly interdependent for many purposes, are *not* intended to join all the participants in one contractual arrangement. The very purpose of setting up a linked though separate set of contracts is to create a hierarchical structure that establishes lines of authority and responsibility. That this attempt is not entirely successful for all purposes, especially where personal harm is involved [24] or a participant is not paid,[25] does not mean that the system should be scrapped entirely and all the participants involuntarily joined together in one contract arrangement.[26]

In addition, binding all parties to one contract, with the prime contract taking precedence, takes away the autonomy of prime contractor and subcontractor to make a contract with *different* terms. Although often the prime contractor can dictate subcontract terms, this is not inevitably the case. Autonomy provides flexibility that would be lost or diminished if the linked set of contracts were considered one contract. This approach provides many opportunities for conflicting language. Perhaps if there were one contract that would be signed on by each party as it comes onto the project, the system might work. However, considering the linked set as if it is one consolidated contract probably raises more problems than it solves.[27]

Returning to the series of linked individual contracts, which contract—prime or subcontract—should be considered the dominant contract in dealing with disputes between prime contractor and subcontractor when the contracts do not make this

---

22. *C.H. Leavell & Co. v. Glantz Contracting Corp.*, 322 F.Supp. 779 (E.D.La.1971).

23. *Varco-Pruden, Inc. v. Hampshire Construction Co.*, 50 Cal.App.3d 654, 123 Cal.Rptr. 606 (1975). The language expressing this opinion was not necessary to the holding of the case and must be considered as "dictum." A subsequent court would not be required to follow such a conclusion.

24. Refer to Section 17.09(D).

25. See Section 32.07.

26. Refer to Section 28.10 for use of such a system to deal with defects.

27. A similar argument can be made for AIA Docs. B141, A201, and A401.

clear? There are arguments to prefer the subcontract. First, this is the writing directly assented to by prime contractor and subcontractor. Second, the references to or incorporation of prime contract documents may often be vague, and the subcontractor rarely has the time or ability to compare the two for inconsistency. Third, the law does not compel the subcontract and prime contract to "track together."

There are arguments for a contrary conclusion. Often the subcontractor sees only the prime contract bidding documents at the time a sub-bid is submitted. Suppose the subcontractor refuses to enter into the subcontract because the subcontract contains terms different from those in the prime contract. If its bid is irrevocable, it can contend that it should be able to perform under the terms of its firm offer that include the prime contract provisions that apply to the subcontractor. This is yet another reason why the *Drennan* rule does not fit well with subcontractor bidding.

But returning to precedence, the bargaining pressures possessed by the prime contractor can result in harsh subcontract provisions that go beyond the provisions of the prime contract and often seem unnecessarily one-sided. In addition, the owner wants satisfied, competent subcontractors who will do good work and stay on the job. It permits subcontracting for reasons of efficiency, but the owner would not be in favor of the prime contractor's taking advantage of its strong bargaining position to take more from the subcontractor than is taken from the owner and give less to the subcontractor than is given by the owner. This is the reason for the increasing attempts by owners to require that prime contractors include certain provisions in subcontracts.

Pioneer Industries v. Gevyn Construction Corp.,[28] which illustrates these difficulties, involved legal action by a subcontractor against the prime contractor and its surety for work performed before the prime contract was terminated by the public awarding authority. The prime contractor disputed the termination and sought arbitration in accordance with the prime contract arbitration provision. However, the subcontractor sought to have the dispute decided by the courts rather than by arbitrators. The prime contractor pointed to a provision in the subcontract that incorporated the arbitration clause of the prime contract into the subcontract. The court pointed to subcontract provisions, stating that any conflict between prime contract and subcontract would be controlled by the subcontract and that disputes arising out of termination would be arbitrated only if the dispute involved *less* than $3,500. Since the dispute involved in the lawsuit involved *over* $3,500, the court concluded that the dispute did not fall within the arbitration clause of the subcontract. This holding illustrates that the *specific* subcontract arbitration clause takes precedence over the *incorporation by reference* of the prime contract arbitration clause. The probably unintended result was an arbitration between prime contractor and owner and litigation between prime contractor and subcontractor.[29]

---

28. 458 F.2d 582 (1st Cir.1972).

29. Additionally, the subcontract limited the subcontractor's rights to the amount received by the prime contractor from the owner in the event of termination. The anomalous result is that the amount the subcontractor will recover will depend upon the outcome of the arbitration between owner and prime contractor in which the subcontractor is not likely to participate.

During 1975, tentative drafts of what became the 13th edition of AIA Doc. A201 (1976) contained provisions that permitted the subcontractor to be present and submit evidence in arbitrations affecting its rights. These earlier drafts gave procedural choices to the subcontractor if the subcontractor's rights were solely involved in the arbitration. As a result of strenuous objection by the Associated General Contractors, these provisions were deleted. A similar problem can confront subcontractors in federal procurement (treated in Section 24.08(c)).

Sometimes rather than using a *specific* incorporation by reference, the subcontract simply *refers* to the prime contract. In such a case, conflicts between the two contracts are likely to be resolved in favor of the subcontract.[30]

Terms *properly* incorporated by reference into a contract become part of that contract.[31] The antecedent negotiations[32] or other evidence[33] may indicate that particular material incorporated was not to be effective.

The degree to which material incorporated by reference is given effect may also depend upon the particular provision claimed to be incorporated. As the *Pioneer* case illustrates, reluctance to incorporate arbitration clauses by reference may be traceable to occasional judicial antipathy toward arbitration or a feeling that the subcontractor rarely has a choice.[34]

Sometimes provisions in the prime contract are said to flow through into the subcontract. The principal reason for "flowthrough," "passthrough," or "conduit" clauses is to insure that subcontractors commit themselves to the performance and administrative requirements of the prime contract. This is essential from the viewpoints of both the owner and the prime contractor. The owner seeks performance required in the prime contract and wants the subcontractor to commit itself in the same manner as the prime contractor. Similarly, the prime contractor has under-

taken specific obligations to the owner and wishes to be certain that the subcontractor agrees to perform those obligations. As an illustration, AIA Doc. A201, ¶ 5.3.1, requires that the contractor:

> require each Subcontractor, to the extent of the Work to be performed by the Subcontractor, to be bound to the Contractor by the terms of the Contract Documents and to assume toward the Contractor all the obligations and responsibilities which the Contractor . . . assumes toward the Owner and Architect.

Such flowthrough provisions sometimes deal with benefits given the prime contractor and state that they should flow down to the subcontractor. For example, ¶ 5.3.1 gives the subcontractor all the "rights, remedies and redress" against the contractor that the contractor has against the owner *unless* the subcontract specifically provides otherwise. For example, prime contracts will usually contain *force majeure* provisions that excuse delayed performance if certain events occur. Such a flowthrough of benefits provision would give identical rights to the subcontractor if a claim is made against it by the prime contractor for delay.

A201's flowthrough of benefits applies only if the subcontract does not specify otherwise. The flowthrough provision may not be of much value to the subcontractor, as subcontracts frequently contain provisions less favorable to the subcontractor than the prime contractor has in its con-

---

30. *Oxford Development Corp. v. Rausauer Builders, Inc.,* 158 Ind.App. 622, 304 N.E.2d 211 (1973) (conflict over work to be performed); *United States v. Foster Construction (Panama) S.A.,* 456 F.2d 250 (5th Cir.1972) (conflict over terms and conditions).

31. *West Bank Steel Erectors Corp. v. Charles Carter & Co.,* 248 So.2d 52 (La.App.1971), held that a subcontractor was required to perform in accordance with specifications in the prime contract. For further discussion of incorporation by reference, refer to Section 18.01(b).

32. *J.T. Majors & Son, Inc. v. Lippert Brothers Inc.,* 263 F.2d 650 (10th Cir.1958).

33. *United States v. Klefstad Engineering Co.,* 324 F.Supp. 972 (W.D.Pa.1971).

34. *Beacon Construction Co. v. Prepakt Concrete Co.,* 375 F.2d 977 (1st Cir.1967), refused to compel arbitration, while *J.S. & H. Construction Co. v. Richmond County Hospital Authority,* 473 F.2d 212 (5th Cir.1973), describing the arbitration process in laudatory terms, came to a different conclusion.

In *Pearl Street Development Corp. v. Conduit & Foundation Corp.,* 41 N.Y.2d 167, 391 N.Y.S.2d 98, 359 N.E.2d 693 (1976), conflicting clauses in prime contract and subcontract relating to arbitration were resolved by the arbitrator.

tract with the owner. Note that ¶ 5.3.1 requires the contractor to make available to proposed subcontractors, *before* execution of the contract, copies of the general conditions to which the subcontractor will be bound and "identify to the Subcontractor any terms and conditions of the proposed Subcontract which may be at variance with the Contract Documents." The AIA seems to be giving the subcontractor some bargaining muscle when the prime contractor seeks to force provisions upon the subcontractor that are harsher than those in the prime contract. Whether this will have any effect on the bargaining power in the actual subcontracts is debatable.

Another aspect of flowthrough provisions is often ignored. For example, suppose the prime contractor owns the float.[35] As the result of a subcontractor delay in noncritical path activities, the prime contractor is not liable to the owner. Can the prime contractor enforce any liquidated damages provisions in the subcontract? Similarly, suppose the liquidated damages clause in the prime contract is $100 a day while the liquidation clause in the subcontract is $200 a day. Can the prime contractor collect an amount in excess of what it has to pay the owner? Can the subcontractor take advantage of the liquidated damages clause in the prime contract? [36]

Flowthrough provisions raise other problems. For example, AIA Doc. A201, ¶ 3.2.1, gives the prime contractor a right to inquire into the financial arrangements made by the owner. Does the flowthrough of benefits provision give a similar right to the subcontractor to inquire into the financial sources of the prime contractor or the owner? The prime contractor may with justification contend that this power constitutes a trap for unwary prime contractors who may not realize how the use of a boiler-plated flowthrough provision can compel it to disclose sensitive information to subcontractors.

The attempt to give negotiation power to the subcontractor to augment its weak position at the bargaining table has other difficulties. For example, A201 sets up a dispute resolution system under which initially disputes are decided by the architect subject to arbitration. If this is a *burden* under the prime contract, the subcontractor must accept this method of dispute resolution. If, however, it is a *benefit,* the subcontractor is entitled to it only if there is nothing specific in the subcontract to the contrary.

Structuring the flowthrough of benefits provision to apply only if nothing to the contrary is found in the subcontract raises the inevitable problem of seeking to determine when a benefit has been specifically excluded by the subcontract. Not only that, this attempt to give bargaining power to the subcontractor seems to be dependent upon a requirement in ¶ 5.3.1 that the contractor make available to proposed subcontractors copies of the prime contract and identify to the subcontractor "any terms or conditions" of the subcontract that may be "at variance" with the prime contract. Is there a difference between "at variance" and "specifically provided otherwise"?

This flowthrough of benefits provision demonstrates some of the difficulties in construction contract drafting. It seems attractive to include a clause that helps the subcontractor in its often difficult negotiations, if there are any at all, with the prime contractor. However, often drafters do not think about the possible applications of a general clause both as to substance and procedures. This well-meaning attempt to aid the subcontractors may only create more problems in construction contract administration and in dispute resolution.

---

35. Section 30.05.

36. *Industrial Indemnity Co. v. Wick Construction Co.,* 680 P.2d 1100 (Alaska 1984) concluded it could. See also *P&C Thompson Brothers Construction Co. v. Rowe,* 433 So.2d 1388 (Fla.App.1983) (discussed in Section 30.09).

SECTION 32.05
SUBCONTRACTOR DEFAULTS

(A) CLAIMS BY CONTRACT–CONNECTED PARTIES

A subcontractor's unexcused refusal or failure to perform in accordance with its contract obligations can damage those with whom the subcontractor has contracted, such as the prime contractor or a lower tier subcontractor. As shall be seen in Section 32.05(C) a prime contractor may be accountable to the owner for defaults of the subcontractor. This can mean the extent of the claim and the method of resolving any dispute over it may be affected by provisions in contracts other than the subcontract.[37]

(B) CLAIMS BY THIRD PARTIES

Tort claims against a subcontractor by those who suffer personal harm are discussed in Chapter 35. This section looks briefly at claims against a subcontractor for economic losses of third parties, principally the owner.

Owners employ three theories to recover losses they have suffered that were caused by a subcontractor breach. First, an owner may contend that the subcontractor has breached its contract with the prime contractor, the contract intended for the benefit of the owner.[38] Second, the owner may

assert that the subcontractor's breach was tortious in that it failed to live up to the legal standard of care.[39] Third, the owner may contend that the prime contractor was merely a conduit between owner and subcontractor, essentially a contention based upon the prime contractor's contracting as an agent of the owner.[40] Because of the unsettled state of the law, the third party's *principal* claim is against someone with whom it has a contractual relationship and that party institutes action against the subcontractor. For example, if the owner has suffered a loss, it is likely that the owner's action will be against the prime contractor and the prime contractor will bring an action against the subcontractor.

The third party claim, that is, the direct claim against the subcontractor, may also be asserted. But the uncertainty of its success makes it ancillary to the principal claim unless the claim against the contract-connected party would be uncollectable.

(C) RESPONSIBILITY OF PRIME CONTRACTOR

The subcontracting process permits the prime contractor to discharge its obligations by the use of a subcontractor. However, this power to perform through another does not relieve the prime contractor of responsibility for subcontractor nonperformance.[41] Most standard construction

---

37. *United States v. Foster Construction (Panama) S.A.*, supra note 30 (sub liable for prime's lost overhead and liquidated damages); *Reed & Martin, Inc. v. Westinghouse Electric Corp.*, 439 F.2d 1268 (2d Cir.1971) (sub pays liquidated damages only if prime does). See also *P&C Thompson Brothers Construction Co. v. Rowe*, supra note 36, discussed in Section 30.09.

38. Compare *United States v. Ogden Technology Laboratories, Inc.*, 406 F.Supp. 1090 (E.D.N.Y.1973), and *Gilbert Financial Corp. v. Steelform Contracting Co.*, 82 Cal.App.3d 65, 145 Cal.Rptr. 448 (1978), both of which allowed recovery, with *Vogel v. Reed Supply Co.*, 277 N.C. 119, 177 S.E.2d 273 (1970), which did not.

39. Cf. cases cited in Section 17.09(D).

40. *National Cash Register Co. v. UNARCO Industries, Inc.*, 490 F.2d 285 (7th Cir.1974); *Oliver B. Cannon & Son, Inc. v. Dorr-Oliver, Inc.*, 336 A.2d 211 (Del.1975).

41. *Kahn v. Prahl*, 414 S.W.2d 269 (Mo.1967); *Waterway Terminals Co. v. P.S. Lord Mechanical Contractors*, 242 Or. 1, 406 P.2d 556, 13 A.L.R.3d 1 (1965) (prime responsible for subcontractors).

contracts make the prime contractor responsible for subcontractor defaults.[42]

In *Norair Engineering Corp. v. St. Joseph's Hospital, Inc.,*[43] one issue involved the responsibility for defective work performed by a subcontractor who was hired at the owner's demand by the contractor who had been brought in by the bonding company to replace the original prime. The court concluded that the contractor was well aware that it was taking the risk of the subcontractor's default when it entered into the prime contract, that it was not without remedies against the subcontractor, and that it would never have been awarded the contract to complete the work unless it had agreed to the provisions under which the owner could select and approve subcontractors.

Owners can exercise a variety of controls over the selection of subcontractors. That the owner approves a proposed subcontractor should not relieve the prime contractor of responsibility. Clearly this is a matter for negotiation, and the law should enforce the choice that the contracting parties have made. Where there is no clear evidence one way or the other, a strong argument can be made that the owner *dictating* a particular subcontractor should relieve the prime contractor of responsibility.[44] In such a case, the prime contractor may be acting as an agent of the owner to engage a particular subcontractor and should not be responsible unless it knew or should have known that the subcontractor was incompe-

tent.[45] If the owner steps in and takes away the prime contractor's power to manage the subcontractors, the prime contractor cannot be held accountable for the work of the subcontractors. Its obligation is conditioned upon its right to manage the job without reasonable interference.[46]

Sometimes the prime contractor supports a contention that it should not be held responsible by pointing to the independent contractor rule. This rule, though subject to many exceptions, relieves the employer of an independent contractor from responsibility from the latter's improper performance.[47] However, this defense, where available, does not relieve the employer of the independent contractor when the independent contractor's failure to perform properly has damaged a party to whom the employer of the independent contractor owed a contract right of proper performance.[48]

SECTION 32.06
PAYMENT CLAIMS
AGAINST PRIME CONTRACTOR

One of the particularly sensitive areas in the subcontract relationship relates to money flow, with subcontractors frequently contending that they invest substantial funds in their performance and are entitled to be paid as they work and to be completely paid when their work is completed. These issues surface around two concepts: line item retention and payment conditions. The

---

42. AIA Doc. A201, ¶ 5.3.1. This is accomplished by requiring that the prime contractor obtain agreements from subcontractors that preserve and protect the rights of the owner "so that the subcontracting thereof will not prejudice such rights."

43. 147 Ga.App. 595, 249 S.E.2d 642 (1978).

44. Cf. *National Cash Register Co. v. UNARCO Industries, Inc.,* supra note 40. Here the owner was allowed to sue the subcontractor that it ordered the prime contractor to use. Very likely the prime would be relieved of responsibility.

45. *Seither & Cherry Co. v. Illinois Bank Building Corp.,* 95 Ill.App.3d 191, 50 Ill.Dec. 672, 419 N.E.2d 940 (1981) (CM agent of owner in hiring contractor).

46. Refer to Section 23.02(G).

47. See Section 35.05.

48. *Harold A. Newman Co. v. Nero,* 31 Cal.App.3d 490, 107 Cal.Rptr. 464 (1973).

subcontractors wish to divorce their work and payment for it from the prime contract. They argue *for* line item retention and *against* payment conditions.

Line item retention gives the subcontractor the right to be paid after it has fully performed. Delay usually occurs because the owner holds back retainage—a designated amount of the contract price for the entire performance, that of prime contractor and *all* subcontractors. When all the work is completed and the project accepted, the owner will pay the retainage. However, early finishing subcontractors may have to wait a substantial period of time after they have completed their work for the retention allocated to their contract because the entire project is not yet completed. They would like to disassociate their contract from the rest of the subcontracts and the prime contract.

A payment condition makes payment to the prime contractor a condition to the prime contractor's obligation to pay the subcontractor. Prime contractors seek to create such a condition by including language in the subcontract stating that the prime contractor will pay "if paid by the owner," "when paid by the owner," or "as paid by the owner." An endless number of cases have interpreted this language. Does it create a condition to payment or simply indicate that the payment flow contemplated some delay, with payment in any event being required after the expiration of a reasonable time?

All courts agree, or at least so they state, that the parties can make a payment condition under which the subcontractor assumes the risk that it will not be paid for its work. The legal issue has centered around the requisite degree of specificity needed to

create such a condition. On the surface the courts simply examine the language and seek to determine the intention of the parties. Other factors are operating, however. This section first looks at what courts have done, discusses some reasons for such clauses, and concludes by looking briefly at the system used in the AIA Standard Documents.

In 1933 New York held that language of the type that appears to tie the subcontractor's right to payment to the contractor receiving payment created a payment condition precluding the subcontractor from recovering when the prime contractor had not been paid.[49] In 1962 the leading case of *Thomas J. Dyer Co. v. Bishop International Engineering Co.*[50] came to a different conclusion. The court stated that performing parties usually expect to be paid for their work. Any result under which a party would suffer a forfeiture (that is, performing the work and not being paid) must be supported by clear and specific language that this risk has been taken. The court stated that the subcontractor, in addition to its mechanics' lien protection, contracts mainly on the basis of the solvency of the prime contractor. To change this normal credit risk, the contract should "contain an express condition clearly showing that to be the intention of the parties."[51] The court concluded that the language dealing with linkage of payment to payment to the prime contractor was designed "to postpone payment for a reasonable period of time after the work was completed, during which the general contractor would be afforded the opportunity of procuring from the owner the funds necessary to pay the subcontractor."[52]

Most recent cases have followed the rea-

49. *Mascioni v. I.B. Miller, Inc.*, 261 N.Y. 1, 184 N.E. 473 (1933).

50. 303 F.2d 655 (6th Cir.1962).

51. 303 F.2d at 661.

52. Ibid.

soning in the *Dyer* case.[53] But some decisions have concluded that a payment condition has been created.[54] Courts here are not simply construing contract language. The language used indicates that the prime contractor need not pay *until* it is paid. Courts have tilted the balance toward the subcontractor because of the feeling that the subcontractor has little choice as to the contract language, that the subcontractor has no opportunity to evaluate the solvency of the owner, and that it would be unfair for subcontractors to have to bear this risk when the prime contractor is in the best position to evaluate the credit of the owner.

The *reason* for nonpayment is important. Two clear cases should be set aside. First, if the nonpayment to the prime contractor results from the *subcontractor's* failure to perform properly, the subcontractor is not entitled to recovery. Second, if the reason for nonpayment is improper performance by the prime contractor or its unwillingness to make reasonable efforts to obtain payment, the subcontractor is entitled to be paid. Even if a condition to payment *has* been created, prevention of its occurrence or not taking reasonable efforts to make it occur excuses the condition allowing the subcontractor to recover.

The more difficult questions involve the owner's nonpayment under circumstances for which neither the prime contractor nor the subcontractor can be held directly responsible. Suppose the nonpayment results from the design professional's refusal to issue a payment certificate or revocation of a previously issued certificate. For example, suppose work by *another* subcontractor had been previously accepted but was discovered to be defective after payment to the subcontractor. In such a case, the latter's correction of the work will start the money flow again. Under these circumstances, the prime contractor should be required to pay the unpaid subcontractor who *had* performed properly. The prime contractor hired the subcontractor whose performance was improper and is in the best position to put pressure on *that* subcontractor to correct the defective work. Additionally, the prime contractor could have required the subcontractor whose work had been found to be defective to post a performance bond, and this would have entitled the prime contractor to look to the surety for the nonperformance. Allowing the prime contractor to refuse to pay the unpaid subcontractor whose work was properly performed would also strengthen the already strong bargaining position the prime contractor is likely to have. This can give the latter additional leverage to force settlement of disputes between prime and unpaid contractor even

---

53. *Nicholas Acoustics & Specialty Co. v. H & M Construction Co.*, 695 F.2d 839 (5th Cir.1983) (clause gives prime contractor opportunity to see if subcontractor had completed its work); *Culligan Corp. v. Transamerica Insurance Co.*, 580 F.2d 251 (7th Cir.1978); *Midland Engineering Co. v. John A. Hall Construction Co.*, 398 F.Supp. 981 (N.D.Ind. 1975); *Yamanishi v. Bleily & Collishaw, Inc.*, 29 Cal.App.3d 457, 105 Cal.Rptr. 580 (1972); *Sasser & Co. v. Griffin*, 133 Ga.App. 83, 210 S.E.2d 34 (1974); *Chartres Corp. v. Charles Carter & Co.*, 346 So.2d 796 (La.App.1977); *D.K. Meyer Corp. v. Bevco, Inc.*, 206 Neb. 318, 292 N.W.2d 773 (1980); *Schuler-Haas Electric Co. v. Aetna Casualty & Surety Co.*, 40 N.Y.2d 883, 389 N.Y.S.2d 348, 357 N.E.2d 1003 (1976); *Grossman Steel & Aluminum Corp. v. Samson Window Corp.*, 78 A.D.2d 871, 433 N.Y.S.2d 31 (1980); *Elk & Jacobs Drywall v. Town Contractors, Inc.*, 267 S.C. 412, 229 S.E.2d 260 (1976).

In addition, courts have found other ways to protect subcontractors. See *Excavators & Erectors, Inc. v. Bullard Engineers, Inc.*, 489 F.2d 318 (5th Cir.1973) (separate oral contract not governed by subcontract payment condition clause); *Pioneer Indus. v. Gevyn Construction Corp.*, supra note 28 (payment clause violates Massachusetts payment bond statute).

54. *Brown v. Maryland Casualty Co.*, 246 Ark. 1074, 442 S.W.2d 187 (1969); *D.I. Corbett Electric, Inc. v. Venture Construction Co.*, 140 Ga.App. 586, 231 S.E.2d 536 (1976); *Miller v. Housing Authority of New Orleans*, 175 So.2d 326 (La.App.1965). Note that in this and the preceding footnote cases are cited for opposite conclusions in the same jurisdiction. The result can depend on the specificity of the language and/or the cause for the delay. For example, in the *Chartres* and *Midland* cases there was extremely long delay, in each case almost two years.

for projects unaffected by the payment dispute.

Another difficult problem involves non-payment due to owner insolvency or bankruptcy. Generally, such risks should be borne by the prime contractor. It has entered into the contract with the owner and is in the best position to make financial capacity judgments.[55] Although the subcontractor's willingness to perform work may have rested to some degree upon its evaluation of the financial position of the owner, it would not be sufficient to conclude that the prime contractor has transferred the risk to its subcontractors in the absence of a clear showing that this was intended.

Other factors exist that may determine whether a payment condition has been created. If the prime contractor has invested substantial amounts of money into the project itself, it can more persuasively contend that it and the unpaid subcontractor should share the loss. If the prime contractor has invested very little of its own money, a stronger argument can be made for refusing to find a payment condition, with the prime contractor in such a case acting as a broker who should take the risk of the owner's insolvency.

Sometimes it is contended that the unpaid subcontractor has a right to a mechanics' lien if it has improved the owner's property. Although the laws relating to mechanics' liens are varied and often difficult to decipher, the right to a lien should be based upon a legal right to payment from the prime contractor. If a payment condition has been created, the unpaid subcontractor should not be able to obtain a mechanics' lien. This strengthens its argument that a payment condition should not be found unless it is extremely clear that the subcontractor took this risk. To be sure

some court decisions look solely at the enrichment of owners through the improvement on their property as the basis for a mechanics' lien. At best, the uncertainty of the mechanics' lien points toward a conclusion that a payment condition generally has *not* been created.

The AIA has been buffeted by both sides of this controversy. The AGC has argued that matters between the prime contractor and subcontractor should be left to them and should not be regulated in A201, the AIA's standard general conditions. On the other hand, subcontractor associations have contended that the AIA should use A201 to insure that the subcontractors are properly treated.

As a result AIA Doc. in A201, ¶ 9.5.2, is vague on this question, the AIA leaving the ultimate determination to court decisions. Yet A401, the AIA's standard construction subcontract, not only seeks to make clear there is *no* payment condition but also creates line item retention.[56]

SECTION 32.07 PAYMENT CLAIMS AGAINST PROPERTY FUNDS OR PERSONS OTHER THAN PRIME CONTRACTOR

(A) COURT JUDGMENTS AND SPECIFIC REMEDIES

Legal relief generally comes in the form of court-issued judgments. Most judgments simply state that the judgment creditor—the person who seeks and obtains the judgment, is entitled to a specific amount of money from the judgment debtor—the person against whom the judgment is issued. The judgment creditor is given methods of collecting on these judgments, but as mentioned in Section 2.12, such remedies are often cumbersome and ineffective. If the

---

55. This conclusion would be strengthened under AIA Doc. A201, ¶ 3.2.1. A contractor can require the owner at the time the contract is executed to furnish the contractor reasonable evidence that the owner has made financial arrangements to fulfill its obligation.

56. Paragraphs 5.3, 12.4.

judgment debtor does not pay voluntarily, collection problems sometimes make the judgment worthless.

Specific remedies, on the other hand, are remedies that either command the defendant to do or not do a particular act (specific performance decrees or injunctions) or operate against specific property (liens against particular funds, goods, or property). Specific remedies are equitable decrees backed up by the contempt power of the court. They are more effective, as a rule, than the ordinary money award court judgment.

One reason unpaid subcontractors seek mechanics' liens [57] or other specific remedies is that they operate against particular property and are more effective than ordinary judgments, whether obtained against the prime contractor or third parties such as the owner. Sometimes even specific remedies are ineffective, but on the whole, they are preferred to ordinary court judgments.

(B) STATUTORY AND NONSTATUTORY REMEDIES

Most of the material discussed in this section, such as mechanics' liens, stop notices, and trust fund protection, have been created by the state legislatures. Many bonds are required on public work because of legislative enactments. The statutory nature of these remedies has added complications.

Reported appellate cases that seem contradictory are often traceable to the variant statutes before the courts. Statutes change frequently. Compliance is difficult and the

law becomes murky. Although state statutes follow broad patterns of similarity, variance in details places a heavy burden upon those contractors who operate in different states and may in part account for the local character of the construction industry.

Much of the law in this area involves interpretation of these frequently complex state statutes. Such state statutes create many requirements for the creation of lien rights or stop-notice protection. Who will be accorded statutory protection, how such protection is achieved, and the nature of the protection are often resolved by reference to the complicated, almost unreadable statutes. Frequently such interpretation questions are resolved by holding that the statutes are designed to prevent unjust enrichment by protecting unpaid subcontractors, the intended beneficiaries of the legislative protection. Matters can become even more complicated in jurisdictions that state that the standards for perfecting a mechanics' lien will be *strictly* required, but once the lien has been perfected, the remedy will be administered *liberally*.[58] Strained interpretations and language distortion often result, yet construing these statutes to protect subcontractors does not invariably result in lien protection. The principal legacy of this approach is legal uncertainty and unpredictability.

The extensiveness of the statutory system can make it more difficult for subcontractors to assert nonstatutory claims. Where the subcontractor has been given statutory protection but did not take the necessary steps to obtain it, courts sometimes deny the *nonstatutory* claim.[59] Although the stat-

---

57. The many mechanics' lien laws make generalizations perilous. For example, generally a mechanics' lien creates only a *security* interest in the property improved by the lien claimant. However, Florida permits a lien claimant a *personal* judgment against the owner even if there is no contract relationship between them. West's Fla. Stat.Ann. § 713.75(1). Similarly, what are called "liens" against public buildings may only be liens against funds. See note 100 infra.

58. *Talco Capital Corp. v. State Underground Parking Commission,* 41 Ohio App.2d 171, 324 N.E.2d 762 (1974).

59. *Engle Acoustic & Tile, Inc. v. Grenfell,* 223 So.2d 613 (Miss.1969). See generally, Comment, *Mississippi Law Governing Private Construction Projects: Some Problems and Proposals,* 47 Miss.L.J. 437 (1976).

utory systems are not exclusive, their existence can persuade courts that they offer sufficient protection to justify denial of a nonstatutory claim.

(C) PUBLIC AND PRIVATE WORK

The nature of the remedy available may depend upon whether the work is public or private. Mechanics' liens, to the extent that they permit foreclosure rights against public buildings, are not available. As a result, the federal government enacted the Miller Act, a compulsory prime bonding system. Many states have enacted comparable legislation for state construction projects.

Sometimes systems developed in public work spill over into private construction. For example, stop notices [60] were originally developed to compensate contractors for public work because they were not accorded lien rights. However, in those states that have stop notices, the stop notice is often available for private work.

(D) MECHANICS' LIENS

Mechanics' lien laws are complicated and vary considerably from state to state. For that reason, it would be inadvisable to attempt a summary of all aspects of these statutory protections accorded certain participants in the Construction Process. Instead, the discussion focuses upon rationales for such protection, salient features of lien laws, and current criticisms of lien laws.

Participants in the Construction Process who can in various ways trace their labor and materials into property improvements of another are given lien rights against the

property in the event they are not paid by the party who has promised to pay them. The most important lien recipients are prime contractors, subcontractors, suppliers, laborers, and design professionals. The remedy accorded a lien holder is the right to demand a judicial foreclosure or sale of the property and be paid out of the proceeds, including, in some states, the legal costs of perfecting the lien, such as attorneys' fees.[61]

Lien claimants are divided into two principal categories, those who have direct contract relations with the owner and those who do not. Typical illustrations of the first are design professionals and prime contractors. Illustrations of the second are subcontractors, laborers, and suppliers to prime contractors and subcontractors. Owners can avoid liens by paying their design professionals and prime contractors. But since the majority of the difficult problems are generated by lien claimants not connected by contract with the owner and because this chapter focuses upon subcontractor problems, the discussion centers around the second class of lien claimants. The usual justification given for granting these liens is unjust enrichment. Those whose labor or material has gone into the property of another should have lien rights in the property when they are not paid as promised.[62]

The first mechanics' lien law was enacted in Maryland in 1791. One reason was to provide a quick and effective remedy for unpaid workers who cannot wait until a full trial to collect their wages. Quick and certain remedies can induce workers to work on construction by assuring them they will

---

60. See Section 32.07(E).

61. However, Delaware held that the state statute allowing recovery of attorneys' fees by successful plaintiffs in mechanics' liens actions but not by successful defendants violates the constitutional guarantee of equal protection. *Gaster v. Coldiron*, 297 A.2d 384 (Del.1972).

62. The unjust enrichment rationale loses some of its attractiveness when "double payment" is considered. An owner who pays the prime contractor may have to pay *again* to an unpaid subcontractor if it wants to remove the lien.

be paid. As an illustration, the Maryland lien laws were enacted at the urging of Thomas Jefferson and James Madison to stimulate and encourage the rapid building of the City of Washington.

Amplification of this inducement so vital to a developing country could and did lead to expansion of lien beneficiaries to include not only laborers but also all those who participate directly in the Construction Process. The state gives credit to prime contractors by granting subcontractors lien rights and encourages persons to furnish labor and material for construction. This state credit was especially needed to bolster an unstable construction industry composed of many contractors unwilling or unable to pay subcontractors and suppliers. This is probably the principal reason for giving lien rights today.

Expansion of lien laws is undoubtedly traceable to the realities of the political process. Once some participants in the Construction Process received lien rights on a frequently asserted unjust enrichment theory, it was relatively easy to expand the list of lien beneficiaries. Those who might oppose lien expansion, such as owners, are often unrepresented as an organized group in the legislatures. This too may have accounted for expansion of lien beneficiaries and lien rights.

Some salient characteristics of lien laws have been mentioned. A lien is a security interest in the property and can be foreclosed upon by the lien claimant. Some states known as having New York type of statutes limit the amount of the lien to the unpaid balance owed the prime contractor by the owner. Some known as Pennsylvania type of statutes do not place a limit of this type on the lien. States with an open-ended lien permit the owner to limit the lien to the contract price by filing a copy of the contract and posting a bond—something rarely done.

Although owners typically set the lien-creating events into motion by contracting with prime contractors, tenants can create liens by hiring contractors. Property owners can avoid such liens by posting a notice of nonresponsibility within a designated time after the owners learn the improvement is being made.

Those entitled to liens are usually set forth in the statute, and typically the list is lengthy and expanding. Whether particular work qualifies for a lien is often unclear because of lien statutes. Some statutes use generic terms such as improvement, building, or structure. Others that attempt to be detailed do not always keep up with changes in the Construction Process. As an example of this problem, liens have been denied where the lien claimant had placed engineering stakes and markers,[63] where a claimant had graded and installed storm and sanitary sewers, paving, curbing, and seating,[64] where a claimant had performed demolition work,[65] where a claimant had installed a swimming pool,[66] and where a claimant had performed electrical work on

---

63. *South Bay Engineering Corp. v. Citizens Sav. & Loan Association,* 51 Cal.App.3d 453, 124 Cal.Rptr. 221 (1975).

64. *Sampson-Miller Associates Companies v. Landmark Realty Co.,* 224 Pa.Super. 25, 303 A.2d 43 (1973). The court reached its result reluctantly and urged that the legislature liberalize the statute to permit liens for work similar to that done by the claimant. The court noted a number of states where liens are available for preliminary work, such as California, Hawaii, Texas, and Illinois.

65. *John F. Bushelman Co. v. Troxell,* 44 Ohio App.2d 365, 338 N.E.2d 780 (1975). The court noted that it could not extend the right to demolition work but left it for legislative action.

66. *Freeform Pools, Inc. v. Strawbridge Home for Boys, Inc.,* 228 Md. 297, 179 A.2d 683 (1962). The statute gave liens for buildings, and the court concluded that a swimming pool was not a building.

a modular home erected at a factory.[67] Most states deny liens to lessors of equipment used in construction unless the items are consumed in the process of use.[68] Not all work that would be considered part of the Construction Process can qualify for a lien.

Lien laws typically create an obstacle course of technical requirements for claimants. For example, a California subcontractor lien claimant must file a preliminary notice with the owner, the general contractor, and construction lender within twenty days after furnishing the materials.[69] The claimant must also record the claim of lien within ninety days of completion of the work of improvement or, if a notice of completion or a notice of cessation of the work is recorded, within thirty days of such notice.[70] The lien terminates unless an action to enforce it is begun within ninety days of the completion of the improvement [71] and such action is subject to discretionary dismissal if not brought to trial within two years.[72] Generally any failure to comply with these requirements will invalidate the lien. Substantial performance is insufficient.[73]

If the lien amounts exceed the value of the property after those with security interests that take priority are paid, all claimants are treated equally. In some states, liens of prime contractors are subordinated to other lien claimants, while in other states, laborers are sometimes given preference.

As between lien claimants and others with security interests in the property, such as the seller of the property who retains a security interest or a construction lender, the party who perfects its interest first takes priority. For this reason, lenders will not make construction loans if work has begun on the project for fear that their security interest will not take priority over those who have already begun work. As a rule, lenders and other security holders in the land perfect their security interest *before* work begins. As a result, lien claims can become valueless if trouble develops and prior security holders foreclose on the property. This occurs frequently because of market imperfections. The lender is typically able to buy in at less than the amount owing on the construction loan because the liens of other claimants are not extensive enough to justify bidding in or they may not have sufficient funds to be able to compete with the lender.

To deal with this problem, New Hampshire enacted a statute in 1971 that gives mechanics' lien claimants priority over construction lenders.[74] Additionally, one court found a way to grant lien claimants priority over construction lenders.[75]

---

67. *C & W Electric, Inc. v. Casa Dorado Corp.*, 34 Colo.App. 117, 523 P.2d 137 (1974). The court noted that no owner of real property had requested that the work be performed, and this was a requirement of the statute.

68. 3 A.L.R.3d 573, 578 (1965). In California the seller of equipment who retains a security interest cannot obtain a mechanics' lien. *Davies Machinery Co. v. Pine Mountain Club, Inc.*, 39 Cal.App.3d 18, 113 Cal.Rptr. 784 (1974). But Louisiana granted a lien for nails, lumber, and plyform consumed in temporary work. *Slagle-Johnson Lumber Co. v. Landis Construction Co.*, 379 So.2d 479 (La.1979).

69. West's Ann.Cal.Civ.Code 3114.

70. Id. at 3116.

71. Id. at 3144.

72. Id. at 3147.

73. *IGA Aluminum Products, Inc. v. Manufacturers Bank*, 130 Cal.App.3d 699, 181 Cal.Rptr. 859 (1982).

74. N.H.Rev.Stat.Ann. 447:12–a.

75. *Security Bank & Trust Co. v. Pocono Web Press, Inc.*, 295 Pa.Super. 455, 441 A.2d 1321 (1982) (trade contractors given lien even if they did not begin work before lender lien filed despite their notice of lender's mortgage).

Lien claimants must establish that they have performed work under the terms of a valid contract. As has been emphasized throughout this treatise, there has been a proliferation of legal controls on the Construction Process. Licensing laws, land use controls, building and housing codes, and the controls imposed on projects built in part with public funds are illustrations. As a result a lien may be denied because of a technical violation of law or regulation.[76]

Mechanics' lien laws have been attacked as violating the U.S. Constitution. Attacks were based upon decisions of the U.S. Supreme Court that had invalidated certain remedies granted before a full trial as depriving the defendants against whom the remedy was allowed the use of their property without due process of law.[77]

Generally such attacks have been unsuccessful.[78] Courts have held that the owner's deprivation of its property is not substantial. Although marketability may be impaired to a degree, there is no interference with the owner's possession. Courts that have not sustained these attacks have emphasized that the work of the claimants has improved the value of the property and that the owner can force an expeditious adjudication of the lien claim within a short period.

The brief examination of some of the salient characteristics of mechanics' lien protection has revealed the weaknesses of the remedy. Most important, the statutes are complex and change frequently. Care-lessness in compliance can result in the lien's being lost. On the other hand, strict compliance is costly, perhaps, in smaller jobs, more than the value of the lien. The lien is most important in construction projects that fail. It is in these situations that it is most likely that lien claimants will find their claims wiped out because prior security holders have foreclosed and the funds left over for lien claimants are nonexistent. These deficiencies have led subcontractors to seek other forms of legislative and judicial relief, the focal point of the balance of Section 32.07.

Discussion has centered upon the weaknesses of mechanics' lien laws as protection for unpaid subcontractors and suppliers. However, another frequently made criticism of such laws is that they can compel an inexperienced owner to pay twice for the same work. The owner may pay the prime contractor and then have to pay an unpaid subcontractor to remove a lien. The possibility of double payment has led to some legislative change to protect homeowners.[79]

(E) STOP NOTICES

The ineffectiveness of mechanics' liens in private work and the unavailability in public work has led some states to supplement mechanics' lien protection by enactment of stop-notice laws.[80] Like mechanics' lien laws, these statutes vary from state to state, but for simplicity, reference is made to the California stop-notice law.

Those entitled to mechanics' liens can

---

76. Liens were upheld in *Excellent Builders, Inc. v. Pioneer Trust & Savings Bank*, 15 Ill.App.3d 832, 305 N.E.2d 273 (1973) (design violated setback, sideline, and sideyard requirements) and *M. Arthur Gensler, Jr. & Associates Inc. v. Larry Barrett, Inc.*, 7 Cal.3d 695, 103 Cal.Rptr. 247, 499 P.2d 503 (1972) (failure to obtain a new building permit when costs increased).

77. The first of this long series of cases was *Sniadach v. Family Finance Corp.*, 395 U.S. 337, 89 S.Ct. 1820, 23 L.Ed.2d 349 (1969).

78. *Connolly Development Inc. v. Superior Court of Merced County*, 17 Cal.3d 803, 132 Cal.Rptr. 477, 553 P.2d 637 (1976), upheld the mechanics' lien and stop-notice laws. See Note, *Constitutionality of Mechanics' Liens Statutes*, 34 Wash. & Lee L.Rev. 1067 (1977).

79. *Grier Lumber Co. v. Tryon*, 337 A.2d 323 (Del.Super.1975).

80. Note, *Mechanics' Liens: The "Stop Notice" Comes to Washington*, 49 Wash.L.Rev. 685 (1974).

take advantage of stop-notice laws. Using a subcontractor as an illustration, in private projects a subcontractor must file with owner and lender a preliminary notice twenty days after beginning work that describes the work to be done, the parties, the site, and a statement that if bills are not paid, the property being improved may be subject to a mechanics' lien.[81] In public work, a somewhat simplified preliminary notice must be filed twenty days after the work is begun.[82] In private work, a stop notice sent to a lender may be accompanied by a bond in the amount of one and one quarter times the claim to protect the lender from damages if the claim is not ultimately established as valid. If a *nonbonded* stop notice is filed with a lender, the latter *may* withhold construction funds to pay the claim.[83] If a *bonded* stop notice is filed with a lender the latter *must*—unless a payment bond has been filed—set aside construction loan funds to pay the claim.[84]

A stop notice sent to an owner, public or private, need not be accompanied by a bond.[85] Owners, public or private, must withhold funds from money owed to the prime contractor to answer the stop-notice claims unless a payment bond has been filed.[86]

In California, the claim can include only materials furnished or work performed.[87] The notice need not state that the claimant has not yet been paid.[88]

If the funds withheld are insufficient to pay all stop-notice claimants, each claimant shares pro rata in the funds without regard to when the stop notices were filed.[89] An owner, lender, or prime contractor who questions the amount of the notice can file a bond for one and one fourth times the amount stated with the party served with a stop notice. When such bond is filed, the funds withheld must be released.[90] California statutes authorizing stop notices on public works specify a detailed, speedy procedure to deal with disputed stop notices.[91]

The stop notice is more effective than a mechanics' lien in obtaining payment. It can stop the flow of funds, which can jeopardize the project. Effectiveness for a subcontractor does not mean that the possible unfairness to the prime contractor should be ignored. An effective subcontractor remedy may not always be fair to prime contractors, especially where there is an honest dispute over the subcontractor's performance.

(F) TRUST FUND LEGISLATION: CRIMINAL PENALTIES

Some states, such as Michigan, Minnesota, New York, New Jersey, Oklahoma, and Texas, have enacted legislation designed to avoid prime contractor diversion by designating funds received for work performed by others as trust funds that the trustee prime contractor must pay to those

---

81. West's Ann.Cal.Civ.Code § 3097.

82. Id. at § 3098.

83. Id. at §§ 3083, 3162.

84. Ibid.

85. Id. at § 3103.

86. Id. at §§ 3161, 3186.

87. Id. at § 3161.

88. Id. at § 3103.

89. Id. at § 3167.

90. Id. at § 3171.

91. Id. at §§ 3197–3205.

whose work generated the funds. Some trust fund statutes apply to public works, some to private works and some to both.[92]

Sanctions for violation of the trust vary. Frequently the breach of trust caused by diversion is a crime. For example, in New York a diverting prime contractor is guilty of larceny [93] and faces a maximum sentence of seven years in prison. In New Jersey the contractor can be convicted of theft [94] and be imprisoned up to three years; the punishment can include a $1,000 fine. Although the penal sanction is rarely used,[95] its existence should deter diversion and result in payments to subcontractors and suppliers.

New York law gives civil remedies to trust fund beneficiaries.[96] Oklahoma allows the additional possibility of pursuing a civil action directly against certain officers of any diverting corporation.[97] Some states allow a civil remedy without express language in the statute creating such a remedy.[98]

Rather than employ the trust fund designation, some states directly designate diversion as a crime. For example, under certain circumstances the California Penal Code makes willful diversion of amounts over $10,000 punishable by up to five years' imprisonment and a fine of up to $10,000.

Diversion of amounts under $10,000 are misdemeanors punishable by imprisonment not to exceed six months or a fine not exceeding $500 or both. The California Penal Code makes it a crime to submit a false voucher to obtain construction loan funds.[99]

The injection of trust fund and penal legislation into subcontractor collection problems demonstrates the serious risk that construction payment cash flow will not be rapid or continuous.

(G) COMPULSORY BONDING LEGISLATION

Reference has been made to the general unavailability of lien remedies against public structures.[100] To compensate for this and to encourage work on public projects, most public work—federal and state—requires that prime contractors post payment bonds to give protection to subcontractors and suppliers. Statutes and case decisions vary as to how far down the line such protection exists.

Like mechanics' lien laws, the compulsory bonding legislation has its pitfalls. Again, rules determine who must file notices,[101] to whom they must be sent, and what they must state as well as requirements relating to when the lawsuit must be filed. On the whole, with their deficiencies,

---

92. One ignored aspect of this theory is whether the subcontractor is entitled to interest for the period the funds are held in trust. Refer to Section 26.03.

93. N.Y.—McKinney's Lien Law § 79–a.

94. N.J.Stat.Ann. 2C:20–9.

95. *People v. Van Keuren,* 31 A.D.2d 711, 295 N.Y.S.2d 892 (1968).

96. N.Y.—McKinney's Lien Law § 77(1).

97. 42 Okl.Stat.Ann. § 153.

98. *B.F. Farnell Co. v. Monahan,* 377 Mich. 552, 141 N.W.2d 58 (1966); *Hiller & Skoglund, Inc. v. Atlantic Creosoting Co.,* 40 N.J. 6, 190 A.2d 380 (1963).

99. West's Ann.Cal.Penal Code §§ 484b, 484c.

100. Some states have what are called mechanics' lien statutes that apply to public work. (See Ky.Rev.Stat. § 376–210.) These statutes do not give the lienholder the right to foreclose on the public work. The remedy is against the funds for the project in the hands of the public agency similar to a stop notice.

101. Generally, Miller Act notice requirements need not be complied with strictly. *United States v. Merle A. Patnode Co.,* 457 F.2d 116 (7th Cir.1972).

payment bonds on public work are a more effective remedy for subcontractors and suppliers than are mechanics' liens.[102]

(H) NONSTATUTORY CLAIMS

Subcontractors and suppliers who do not receive payment for work performed sometimes seek remedies against third parties not based upon statutes of the type described earlier in the section. Instead or alternatively, they seek recovery based upon a variety of theories against a number of parties other than the party with whom they have made the construction contract. To illustrate, an unpaid subcontractor may seek recovery for work performed from the owner, the lender, or a design professional. Similarly, a claim may be made by another party farther down on the subcontract chain, such as a second-tier subcontractor against all parties on the contract chain other than the first-tier subcontractor with whom the contract was made.

Nonstatutory claims can be further divided into specific remedies and ordinary court judgments. Illustrations of specific remedies are claims against funds in the hands of the owner or lender earmarked for the project. Illustrations of other remedies are claims generally against the owner, lender, or design professional that would result in an ordinary judgment and not a claim against specific property.

Claims to specific funds such as funds in the hands of a lender or owner are usually based upon the assertion of an equitable lien, an equitable remedy based upon unjust enrichment. Such claims generally point to the claimant's having improved the property of the defendant owner or having improved the land in which the lender has a security interest.

Despite the availability of the statutory stop-notice remedy in California, the equitable lien was successfully asserted by claimants against California construction lenders based upon transactions made before 1967. Claimants were allowed to recover against the unexpended construction loan funds if the claimant, typically a subcontractor, had relied upon the funds when it entered into the subcontract.[103]

The California decisions granting this remedy relied upon earlier cases that had required the claimant to show that the fund contained *surplus* in the sense that the value of the improvement equalled the amount of the total loan. Under these circumstances, unexpended loan funds resulted in *unjust enrichment*.

However, the California decisions in the 1960s awarding equitable liens against lenders simply required that claimants show *reliance* upon the construction loan funds. This reliance was shown typically by evidence indicating that the prime contractor or owner lacked resources and that the land was heavily encumbered. California courts noted that the construction lender was in the best position to prevent misappropriation by the owner or prime contractor.

In 1967 the California legislature enacted California Civil Code § 3264, which shut the door on equitable liens by making the stop-notice statutory remedy exclusive against construction loan funds.

Although California made the statutory remedies exclusive, some courts grant equitable liens on construction loan funds held by owners [104] or by lenders.[105]

---

102. See Note, *Rights and Remedies Under Mississippi's New Public Construction Bond Statute*, 51 Miss.L.J. 351 (1980–81).

103. *McBain v. Santa Clara Savings & Loan Association*, 241 Cal.App.2d 829, 51 Cal.Rptr. 78 (1966).

104. *Avco Delta Corp. v. United States*, 484 F.2d 692 (7th Cir.1973). Another case which found an equitable lien was *Kennedy Electric Co. v. United States Postal Service*, 508 F.2d 954 (10th Cir.1974).

105. *Trans-Bay Engineers & Builders, Inc. v. Hills*, 551 F.2d 370 (D.C.Cir.1976) (prime contractor recovered from lender and HUD); *Bankers Trust Savings & Loan Association v. Cooley*, 362 F.Supp. 328 (N.D.Miss.1973) (sub

Although the law is somewhat unsettled, it seems likely that claimants such as unpaid subcontractors or suppliers or those sureties that have had to pay such parties will be able to recover from the design professional if the latter did not perform in accordance with the professional standards established by the law when issuing payments to the prime contractor.[106] Alternatively, if the claimant can show that the design professional breached the contract with the owner and that this contract was in part for the claimant's benefit, recovery is possible. These burdens have been difficult to sustain.[107] Claims against owners are based upon similar theories and have had some, but limited success.[108]

One differentiation between claims brought against design professional and those brought against owner is that claims against the latter can be based upon unjust enrichment. Claimants assert that their work or materials have benefited the owner and that it would be unjust for the owner to retain this benefit without paying the claimants.

As a rule, such claims have not been successful because the owner can show that it has paid someone, usually the prime contractor, or that the retention of benefit was not unjust because the claimant could have protected itself by using the statutory remedies.[109] However, one case granted recovery where the owner had not paid anyone, with the prime contractor's having left town before being paid.[110]

The subcontractor's reliance upon a promise made by the owner's authorized

---

recovered from lender). *Gee v. Eberle*, 279 Pa.Super. 101, 420 A.2d 1050 (1980), granted recovery to a subcontractor against a lender on an equitable lien theory despite the recognition that this was a minority view. See generally Reitz, *Construction Lenders' Liability to Contractors, Subcontractors, and Materialmen*, 130 U.Pa.L.Rev. 416 (1981).

106. Section 26.07. But refer to Sections 17.08(C), 17.09(D), and 17.12(C).

107. *Engle Acoustic & Tile, Inc. v. Grenfell*, supra note 59, denied recovery against owner and architect. Similarly, a contractor sued an architect in contract and in tort and was unsuccessful. *C.H. Leavell & Co. v. Glantz Contracting Corp.*, supra note 22. On the other hand, an architect was held liable to a contractor for economic loss in *United States v. Rogers & Rogers*, 161 F.Supp. 132 (S.D.Cal.1958). The confusion in this area can be shown by *A.R. Moyer, Inc. v. Graham*, 285 So.2d 397 (Fla.1973), which held that a contractor could sue the architect in tort but not as the intended beneficiary of the owner-architect contract.

108. Subcontractors were unsuccessful in *Helash v. Ballard*, 638 F.2d 74 (9th Cir.1980) (against owner); *Urban Systems Development Corp. v. NCNB Mortgage Corp.*, 513 F.2d 1304 (4th Cir.1975) (against lender); *United States Fidelity & Guaranty Co. v. United States*, 201 Ct.Cl. 1, 475 F.2d 1377 (1973) (against owner); *Stratton v. Inspiration Consolidated Copper Co.*, 683 P.2d 327 (Ariz.App. 1984) (against owner); *Rochelle Vault Co. v. First National Bank*, 5 Ill.App.3d 354, 283 N.E.2d 336 (1972) (against owner); *Engle Acoustics & Tile Co. v. Grenfell*, supra note 59 (against owner); *Haggard Drilling, Inc. v. Greene*, 195 Neb. 136, 236 N.W.2d 841 (1975) (against owner); and *Seegers v. Sprague*, 70 Wis.2d 997, 236 N.W.2d 227 (1975) (against owner).

In addition to the equitable lien cases cited supra at notes 104 and 105, *Votaw Precision Tool Co. v. Air Canada*, 60 Cal.App.3d 52, 131 Cal.Rptr. 335 (1976), seemed willing to grant recovery on the third-party beneficiary theory, but the facts did not justify a recovery in the case. Subcontractors were successful in a claim against a public corporation lender that did not require the prime contractor to post a required bond that would have protected the subcontractor. *New England Concrete Pipe Corp. v. D/C Systems of New England, Inc.*, 495 F.Supp. 1334 (D.Mass. 1980) vacated on jurisdictional grounds, 658 F.2d 867 (1st Cir.1981). But see *Haskell Lemon Construction Co. v. Independent School District.*, 589 P.2d 677 (Okl.1979). A special statute justified a recovery by the subcontractor against the owner in *Sweetman Construction Co., Inc. v. South Dakota*, 293 N.W.2d 457 (S.D.1980). *Seither & Cherry Co. v. Illinois Bank Building Corp.*, 95 Ill.App.3d 191, 50 Ill.Dec. 672, 419 N.E.2d 940 (1981), allowed a subcontractor to recover against the owner by concluding that the construction manager acted as an agent of the owner. A subcontractor was successful in *Port Chester Electric Construction Corp. v. Atlas*, 40 N.Y.2d 652, 389 N.Y.S.2d 327, 357 N.E.2d 983 (1976), where owner and prime contractor were all controlled by one person.

109. *Seegers v. Sprague*, supra note 108. See generally Annot., 62 A.L.R.3d 288 (1975).

110. *Costanzo v. Stewart*, 9 Ariz.App. 430, 453 P.2d 526 (1969).

agent that the prime contractor would be required to file a surety bond was the basis for a successful subcontractor claim against the owner.[111]

With the exception of an occasional recovery based upon reliance on a direct promise or unjust enrichment, claimants such as subcontractors and suppliers have had only occasional success against owners. Subcontractors must use statutory remedies or bonds or they will be left to whatever claim they have against the party with whom they dealt directly.

Second-tier subcontractors, sometimes, like first-tier subcontractors, look up the subcontract chain for a solvent party to pay for work when the party who has promised to do so does not. In *Friendly Ice Cream Corp. v. Andrew E. Mitchell & Sons, Inc.,*[112] an unpaid second-tier subcontractor asserted claims based upon a number of theories against a prime contractor but was unsuccessful. But a third-tier subcontractor was allowed to recover against a first-tier sub based upon the latter's promise to the prime contractor to pay all suppliers.[113]

The preceding discussion should indicate the importance of dealing with a party who has the desire and financial capability to pay for work that is ordered. Obviously, a subcontractor's principal concern should be the financial responsibility of the prime contractor with whom it has dealt. Yet the panoply of legislative protection and the occasional nonstatutory relief given indicates that such parties frequently are not paid and seek other methods to collect.

SECTION 32.08 OTHER SUBCONTRACTOR CLAIMS

(A) AGAINST PRIME CONTRACTOR

Suppose a subcontractor asserts that its cost of performance was wrongfully increased because of acts or omissions by the prime contractor or someone for whom the prime contractor is responsible. Generally the law implies an obligation on the part of the prime contractor to take reasonable measures to see to it that the subcontractor can perform expeditiously and is not unreasonably delayed.[114] However, terms will not be implied if express provisions in the subcontract directly deal with this matter. Commonly subcontracts, amplified by relevant provisions in prime contracts incorporated into the subcontract, deal in some way with this problem. Express provisions can place such a responsibility on the prime contractor, such as the contract's requiring the site to be ready by a particular date or that the work be in a sufficient state of readiness by a designated time to enable the subcontractor to perform specific work.

Conversely, contract provisions may indicate that the subcontractor has assumed certain risks regarding the sequence of performance. For example, the subcontract might require that the work be performed "as directed by the prime contractor," and such a provision would give the prime contractor wide latitude to determine when the subcontractor will be permitted to work.

Additional clauses in the prime contract to which the subcontract refers or that are

---

111. *Bethlehem Fabricators, Inc. v. British Overseas Airways Corp.,* 434 F.2d 840 (2d Cir.1970) (promise made by architect).

112. 340 A.2d 168 (Del.Super.1975).

113. *Aetna Casualty & Surety Co. v. Kemp Smith Co.,* 208 A.2d 737 (D.C.App.1965). Similarly, a second-tier subcontractor successfully sued a prime contractor in *Merco Manufacturing, Inc. v. J.P. McMichael Construction Co.,* 372 F.Supp. 967 (W.D.La.1974).

114. Annot., 16 A.L.R.3d 1252, 1254 (1967). Refer also to Section 18.02(A).

incorporated into the subcontract are also relevant. The contract can specify that certain delay-causing risks were contractually assumed risks or grant only a time extension.[115]

In addition, contract clauses may control by denying recoverability of delay damages. For example, a no-damage clause that limited the subcontractor to time extension in the subcontract, especially if tracked with a no-damage clause in the prime contract, would very likely preclude recovery of delay damages by the subcontractor against the prime contractor for owner or prime contractor caused delays.[116]

The problems of tracking or parallelism in prime contracts or subcontracts are a complicating factor in these claims. As a general rule, prime contracts and subcontracts are parallel in terms of rights and responsibilities. For example, should a prime contractor be held liable for delay to the subcontractor for acts caused by the owner when the prime contractor is precluded from recovering from the owner because of a no-damage clause?

The bargaining situation sometimes permits the prime contractor to better its position in the subcontract. For example, it might be able to include a no-damage clause in the subcontract when the prime contract allows the prime contractor its delay damages against the owner. Although a careful subcontractor might be able to preclude this, often time or realities of the process frustrate parallel rights. An abuse of prime contractor bargaining power can result in discontented subcontractors and poor performance. It is becoming increasingly common for owners to insist that prime contracts contain provisions requiring the prime contractor to give the subcontractor benefits parallel to those given the prime contractor in its contract with the owner. Refer to Section 32.04.

(B) AGAINST THIRD PARTIES

Suppose the costs of performance are increased because of acts of third parties such as the owner or design professional. In addition to any claim that might be available against the prime contractor, suppose a subcontractor seeks legal relief directly against those third parties who it asserts caused the loss. For example, Section 32.07 stated that subcontractors sometimes seek payment against third parties such as owners or design professionals when the prime contractor does not pay them. Claims for additional cost of performance are not treated as favorably as claims to be paid for work performed. Even the latter claims are difficult to sustain against third parties. Very likely claims against third parties for increased cost of performance will meet more severe resistance. However, a subcontractor who can persuade a court that the third party did not live up to the legal obligation imposed by law and thereby committed a tort or that the third party promised to perform in a certain way to the prime contractor and that this promise was for the benefit of the subcontractor can recover against third parties.[117]

---

115. *Loughman Cabinet Co. v. C. Iber & Sons, Inc.*, 46 Ill.App.3d 873, 361 N.E.2d 379 (1977) (subcontractor denied delay damages where prime contractor granted only time extension).

116. *McDaniel v. Ashton-Mardian Co.*, 357 F.2d 511 (9th Cir.1966) (subcontractor assumed the risk of government-caused delays). But in *J.J. Brown Co. v. J.L. Simmons Co.*, 2 Ill.App.2d 132, 118 N.E.2d 781 (1954), a subcontractor recovered delay damages from the prime contractor when the former was delayed by another subcontractor.

117. The owner who had negotiated a contract with subcontractors and then assigned the contract to a prime contractor was held liable to a subcontractor for the owner's change of the critical path method. See *Natkin & Co. v. George A. Fuller Co.*, 347 F.Supp. 17 (W.D.Mo.1972). See also cases cited in notes 107–113 supra and Section 17.09(D).

(C) FEDERAL PROCUREMENT AND THE SEVERIN DOCTRINE

A significant but not insurmountable barrier to subcontract recovery in federal procurement is the Severin doctrine. Subcontractors cannot bring direct actions against the U.S. agency because of an absence of a contractual relationship between subcontractor and the U.S. government.[118] As a rule, subcontractor claims against federal agencies, usually for extra work, are brought by the subcontractor in the name of and with consent of the prime contractor. The Severin doctrine [119] precludes the prime contractor from bringing an action against the government for damages suffered by the subcontractor if the prime contractor is not liable to the subcontractor because the subcontract had absolved the prime contractor. However, the courts have not been comfortable with this limitation, and many exceptions have been developed that have enabled the subcontractor, through the prime contractor, to process claims against the government even though the prime contractor would not have to respond to the subcontractor unless the claim were meritorious.[120]

(D) AN ILLUSTRATIVE CASE

International Erectors, Inc. v. Wilhoit Steel Erectors & Rental Service [121] illustrates the complexity of the multiparty construction contract. The project involved construction of a plant for Sunbeam Electronics Company. Sunbeam contracted with a prime contractor, who subcontracted the work of fabricating and erecting the structural steel portion of the project to Southern Engineering. Southern, as subcontractor, sub-subcontracted the labor part of its undertaking to Wilhoit. Wilhoit sub-sub-subcontracted

the labor work to International Erectors. International incurred delay damages because the steel had not been delivered as scheduled. International sued Wilhoit, the party with whom it made the contract, and sued Southern, one step removed on the contractual chain, claiming that it, International Erectors, was a third-party beneficiary of the promise Southern had made to Wilhoit.

Southern was to supply the steel for the project. The contract between Wilhoit and International stated that International would supply the labor and that Wilhoit's obligation was specified as "none." The court held that this indicated that Wilhoit was not responsible for the delivery of the steel. Although it is true that Wilhoit itself was not going to supply the materials, Wilhoit could have been held to have impliedly promised that the materials would be available when scheduled.

The court rejected the contention that Wilhoit, by virtue of a general clause under which it assumed Southern's obligations, had undertaken to be responsible in the event Southern did not deliver the structural steel as scheduled. The court also rejected the contention that it was inconceivable that an experienced steel erector would enter into an agreement to erect the steel for a large complicated building within thirty-five days without obtaining an assurance from the party *with whom it dealt* that the steel would be ready for erection as scheduled. In rejecting this, the court stated:

> Prudence and perhaps foresight might have insisted that a provision creating such an obligation be included in the written contract, but the written memorial of the parties' intention expressly and unequivocally negated any such obligation. This was an arms-length transaction be-

---

118. *Clifton D. Mayhew, Inc. v. Blake Construction Co.,* 482 F.2d 1260 (4th Cir.1973).

119. *Severin v. United States,* 99 Ct.Cl. 435 (1943) cert. denied 322 U.S. 733, 64 S.Ct. 1045, 88 L.Ed. 1567 (1944).

120. *Keydata Corp. v. United States,* 205 Ct.Cl. 467, 504 F.2d 1115 (1974).

121. 400 F.2d 465 (5th Cir.1968).

tween contractors of considerable experience in such matters and we cannot rewrite the contract just because one of the parties would in retrospection have written it differently.[122]

After rejecting International's claim against Wilhoit, the court considered International's claim against Southern. Southern had promised to have the steel available by a certain time. Yet Southern's promise had been to Wilhoit and not to International. The court rejected International's claim against Southern, since it held that this promise was not made to International nor was it for International's benefit.

Judicial rigidity and careless contract making forced International to assume the risk of delay damages caused by the steel not being at the site on schedule.

PROBLEMS

1. Prime contractor and subcontractor enter into a contract under which the prime contractor would pay "as payments are received from the owner." The subcontractor completed his work, but the prime contractor was not paid because the owner went bankrupt. Should the prime contractor be obligated to pay the subcontractor?

2. Prime contractor and subcontractor enter into a subcontract that incorporated all provisions of the prime contract not directly inconsistent with the subcontract. Prime contractor and subcontractor disputed whether particular performance by the subcontractor was in conformity with the plans and specifications. The prime contractor pointed to a provision in the prime contract stating that the architect would be the judge of performance and her decisions would be final. No provision relating to judging performance could be found in the subcontract. However, the subcontractor claimed that the architect should not judge, as she was biased, since the subcontractor's claim was that the specifications were unclear and this was the responsibility of the architect.

When the subcontractor had bid on the job, the prime contract general conditions were part of the bidding package, but the subcontractor did not read them. His reason for not reading them at that time was that he expected to be given a copy before he entered into the subcontract but was told at the contract signing that it was unavailable. Should the prime contract provision making the architect the judge of performance control the dispute between prime contractor and subcontractor?

---

122. Id. at 470.

33

Design Professional as Judge: Tradition Under Attack

SECTION 33.01 RELEVANCE

In many construction projects, the design professional makes decisions affecting the rights and duties of owners, contractors, and subcontractors. Usually the power to make these decisions is created by the contract between owner and contractor. For example, the design professional interprets the contract documents and determines whether performance complies. To do this the design professional is given the right to make inspections and to uncover work. The design professional has the right to grant time extensions, approve substitution of subcontractors, and pass upon the sufficiency of shop drawings submitted by the contractor. The design professional plays an important role in terminating the contractor's right to perform, having either the power to terminate or, evcn where the owner is given that power, the power to issue a certificate that cause exists for termination.[1]

Not all the illustrations given in the preceding paragraph apply to all construction contracts. Because of increased liability, there has been a slight tendency to reduce these powers in contract administration. For example, standard contracts published by the American Institute of Architects (AIA) and the National Society of Professional Engineers (NSPE) no longer give the design professional the power to order that the work be stopped. Generally the design professional has a "judging" role in most construction projects. This chapter deals with the legal aspect of this role.

SECTION 33.02 DOCTRINE OF CONDITIONS

Parties to a contract can condition specific obligations upon the occurrence of designated events. Normally, unless these events occur or are excused, the duty to perform does not arise or ceases to exist.

Commonly parties to a construction contract give the design professional power to make certain decisions. Such third-party decisions can be expressed as conditions. For example, if a certificate issued by the design professional conditions payment, issuance of that certificate or the existence of facts that excuse issuance is necessary before the owner is obligated to pay.

Although a decision of a third party can be a valid condition, there are two aspects of the circumstances under which this is created in construction contracts that will bear upon legal treatment of such condi-

---

1. AIA Doc. A201, ¶ 14.2.1.

tions. First, the owner pays, and for all practical purposes selects, the design professional. Under these circumstances, one basic element of a decision-making process—that of an impartial judge—may not be present in the construction contract dispute process. Second, in many construction contracts the contractor, and even more so the subcontractor, has no real choice but to accept the decision-making process under which the design professional is given broad powers.

As shall be seen,[2] the law accords a degree of finality to decisions made by the design professional. But the degree to which these two factors cast doubt on the impartiality of the design professional may cause courts to scrutinize these decisions carefully.

SECTION 33.03
EXCUSING THE CONDITION

Under certain circumstances, the condition can be excused. If so, any disputes between owner and contractor would have to be negotiated by the parties, submitted to arbitrators if the parties have agreed to use this process, or decided by a court.

The condition is excused if the design professional becomes unavailable or unable[3] to interpret the contract or judge its performance. The construction contract can provide for a successor design professional when the one originally designated cannot or will not perform this function. For example, AIA Doc. A201, ¶ 2.2.19, allows the owner to appoint a successor

"against whom the Contractor makes no reasonable objection." Even if no successor mechanism is specified in the contract, the parties can agree to use a replacement design professional. But if there were neither an agreement nor a mechanism for a successor design professional, any obligation conditioned upon the issuance of a certificate or the resolution of a dispute by the design professional would be excused and no longer be part of the contract.

Other acts can excuse the condition. For example, one court excused the condition when the architect examined the work, found it satisfactory, but still refused to issue the certificate.[4] Similarly, another court held that the issuance of a payment certificate had been waived because the architect failed to reinspect the work and specify which items remained to be corrected.[5] The condition was excused in a private contract for development of a road when the city engineer, whose approval was a condition, simply refused to act.[6]

Clearly collusion between the design professional and either owner or contractor excuses the condition.[7] A condition can be excused if the parties agree to eliminate it. Sometimes this is described as a waiver. For example, one court stated:

> Upon all the facts and circumstances presented by the record herein, it may reasonably be said the appellees [owner] waived the condition in the building contract which provided for final payment upon inspection and approval by the architect, evidenced by the issuance of a final certificate of the architect. The waiver is implied from the conduct of the appellees throughout the construction of the building in making

---

2. See Section 33.09.

3. *Grenier v. Compratt Construction Co.,* 189 Conn. 144, 454 A.2d 1289 (1983); *United States v. Klefstad Engineering Co.,* 324 F.Supp. 972 (W.D.Pa.1971) (surveyor lost records).

4. *Anderson-Ross Floors, Inc. v. Scherrer Construction Co.,* 62 Ill.App.3d 713, 19 Ill.Dec. 914, 379 N.E.2d 786 (1978).

5. *Hartford Electric Applicators of Thermalux, Inc. v. Alden,* 169 Conn. 177, 363 A.2d 135 (1975).

6. *Grenier v. Compratt Construction Co.,* supra note 3.

7. *Metropolitan Sanitary District of Greater Chicago v. Anthony Pontarelli & Sons, Inc.,* 7 Ill.App.3d 829, 288 N.E.2d 905 (1972).

changes and additions without resort to, or consultation with, the architect and in complete disregard of the terms of the contract; their conduct in refusing to pay the balance due under certificate No. G–8 issued by the architect; their conduct in moving into and occupying the building; and their delay in calling for final inspection by the architect and in having submitted, through the architect, the items which were to be either corrected or finished before the project would be considered complete.[8]

Waiver here is based upon the owner's leading the contractor to believe that the condition will not be asserted, an illustration of estoppel based upon reliance. (Sometimes evidence of an intention to give up the condition is sufficient.) The court seemed influenced by the owner praising the work.

The condition can be excused if the party for whom the condition is principally inserted manifests an intention to perform its obligations despite nonoccurrence of the condition. Although one case held that the condition of a certificate is for the owner's benefit and the owner can waive it,[9] there is no reason why such a condition cannot be considered for the benefit of *both* parties, particularly where the contractor establishes that it entered into the contract on the assumption that a *particular* design professional would make these decisions.

SECTION 33.04
DESIGN PROFESSIONAL
AS JUDGE: REASONS

To a continental European, the design professional as judge seems incongruous. With the exception of the United Kingdom, European construction administration does not give the design professional such responsibility. The close association between owner and design professional based upon the former's selection and payment of the latter precludes this role being given to the latter.

Despite the close association between owner and design professional, most American construction contracts, both public and private, give the design professional broad decision-making powers.[10] A number of reasons can be given for the development and continuation of this system.

First, the stature and integrity of the design professions may give both parties to the construction contract confidence that the decisions will reflect technical skill and basic elements of fairness.[11]

Second, the design professional's role in design before construction equips the design professional with the skill to make decisions that will successfully implement the project objectives of the owner. In a sense, the role as interpreter and judge is a continuation of design.

Third, owners are often unsophisticated in matters of construction and need the protection of a design professional to obtain what they have been promised in the construction documents. Implicit is the assumption that without a champion and protector, the owner might be taken advantage of by the contractor. Couple this with the owner's bargaining strength and the present system results.

Fourth, even assuming that complete objectivity is lacking and that the contractor rarely has much choice, the alternative

---

8. *Steffek v. Wichers*, 211 Kan. 342, 507 P.2d 274, 281–82 (1973).

9. *Halvorson v. Blue Mountain Prune Growers Co-op*, 188 Or. 661, 214 P.2d 986 (1950).

10. The history is described in Dreifus, *The "Engineer Decision" in California Public Contract Law*, 11 Pub. Contract L.J. 1 (1979).

11. See *Zurn Engineers v. State Department of Water Resources*, 69 Cal.App.3d 798, 138 Cal.Rptr. 478 (1977). One reason for enforcing such a mechanism was the court's belief that the parties had mutually agreed to accept the expertise of the state engineer rather than the lay opinion of a judge or jury. This controversial opinion led to California legislation that eliminated the decision-making powers of the state engineer and substituted for it a judicialized arbitration. (See Section 34.20(B)).

would be worse. Suppose the design professional did not act as interpreter and judge. Such matters would have to be resolved by owner and contractor which for many owners would require professional advice. If owner and contractor cannot agree, the complexity of construction documents and performance will necessitate many costly delays, because the alternative forums for owners and contractors who cannot agree—litigation or arbitration—still involve time and expense. The present system, with its deficiencies, is probably more efficient than any alternative.

Fifth, despite the dangers of partiality and conflict of interest (the design professional may overlook defective workmanship to induce the contractor not to press a delay claim or a claim for extras based upon the design professional's negligence, or the design professional may find a defect due to poor workmanship rather than expose himself to a claim for defective design), the system seems to have worked well. Arbitrations in construction disputes are rare, and litigation is even more rare. This can mean genuine satisfaction with the system or the recognition that the alternatives are worse. Perhaps a quick decision that may at times be unfair is better than a more costly, cumbersome system that *might* give better and more impartial decisions.[12]

The potential for bias toward the owner [13] and decision making often involving the design professional's own prior design work are bound to reflect themselves in the judicial treatment of the design professional's decisions and whether he will be held personally responsible for decisions (discussed in Section 17.08(C)).

SECTION 33.05 JURISDICTION OF DECISION–MAKING POWERS

The design professional's interpreting and judging function is created by the contract between owner and contractor. As a result, interpretation of that contract determines whether the design professional has the jurisdiction to resolve the dispute. The contract will also govern the finality of the design professional decision (discussed in greater detail in Section 33.09).

Generally, jurisdictional grants to decide disputes are likely to be broader in matters that involve the experience and expertise of the design professional.[14] Jurisdictional grants are likely to be broad where an on-the-spot decision during the operational phases, as opposed to post-completion disputes, needs to be resolved.[15] Despite these basic guidelines, decisions are likely to vary, depending upon the particular language of the contract and judicial attitude toward this decision-making process.

A look at AIA Documents is instructive. Can the architect decide the increasingly acrimonious disputes that relate to contractor claims that the owner had unjustifiably forced a change in the sequencing of the work? AIA Doc. A201, ¶¶ 2.2.8 and 2.2.9, empower the architect to interpret and resolve disputes relating to "proper execution or progress of the Work." Does this encompass delay claims?

---

12. *Cofell's Plumbing & Heating, Inc. v. Stumpf,* 290 N.W.2d 230 (N.D.1980).

13. To avoid creating the false impression that the cases generally involve the design professional's refusing to award a certificate, it should be noted that many cases involve the owner's refusal to pay despite the design professional's having issued a certificate.

14. But see *Paschen Contractors, Inc. v. John J. Calnan Co.,* 13 Ill.App.3d 485, 300 N.E.2d 795 (1973) (architect had *no* jurisdiction, as the subcontractor's claim involved architect's work); *Edward Electric Co. v. Metropolitan Sanitary District of Greater Chicago,* 16 Ill.App.3d 521, 306 N.E.2d 733 (1973) (jurisdiction based on "engineering nature" included performance costs).

15. *Cofell's Plumbing & Heating, Inc. v. Stumpf,* supra note 12. See also *County of Rockland v. Primiano Construction Co.,* 51 N.Y.2d 1, 431 N.Y.S.2d 478, 409 N.E.2d 951 (1980).

Very likely the language contemplated progress of the *work itself* and not with the way in which the work is sequenced. Yet if the language is read broadly and if emphasis is placed upon the architect's expertise in deciding these matters, a court may conclude that this language gives the architect jurisdiction, at least initially, to resolve contractor delay claims.

When does jurisdiction expire? AIA Doc. A201, ¶ 2.2.2, states that the architect will be the owner's representative "during construction and until final payment is due." This would appear to terminate the architect's powers to interpret the contract, judge performance, and resolve disputes when final payment becomes *due*.[16]

AIA Doc. B141 (the standard agreement between owner and architect) uses a slightly different terminal benchmark. Paragraph 1.5.1 states that the construction phase—the final phase—terminates not just when final payment is due but sixty days after substantial completion, "whichever occurs first." The AIA recognized that final payment may be delayed after the final certificate of payment is issued. It makes clear that in no event is the architect "on board" more than sixty days after substantial completion.

The two provisions can be reconciled. A201 can still govern the duration of the dispute resolution powers while, as noted in Section 16.01(G), B141, ¶ 1.7.19, gives the architect *additional compensation* if he must perform work more than sixty days after substantial completion.

The conflict of interest noted briefly in Section 33.04 must be addressed. Architects should not be given the power to interpret, judge performance, or resolve disputes if the competence of their own work is a serious issue.[17] It is better practice for architects to withdraw from such judging roles if they would be placed under a serious conflict of interest. Where this is a *serious* issue, they should seek permission from *both* parties before proceeding to interpret, judge performance, or decide disputes. Since the design professional's design and administrative functions are central, particularly in the traditional contracting system, it would not be wise to push this conflict of interest problem too far. If it were pushed to its extreme, it would almost eliminate the design professional's judging function. Since that function is deeply ingrained in American Construction Process practices and seems to have worked tolerably well, design professionals should remove themselves only if it is quite clear that their work is *directly* being challenged.

SECTION 33.06 WHO CAN MAKE THE DECISION?

The interpreting and judging powers given the design professional by the contract are important, and the parties—particularly the owner—can expect those powers to be exercised by the particular person or persons in whom the design professional has confidence. If a design professional partnership is named, any principal partner should be able to make such interpretations and judgments. However, the owner may expect that the particular partner with whom it dealt or the partner with principal responsibility for the design would be the appropriate and suitable person to interpret and judge. An individual design professional who is empowered to make such determina-

---

16. It can be contended that this affects only the power of the architect to bind the owner through *agency*. But if so, there is no *specific* provision ending dispute resolution jurisdiction. Very likely, the differentiation was not appreciated.

17. *Paschen Contractors, Inc. v. John J. Calnan Co.,* supra note 14.

tions cannot delegate this power without the permission of the contracting parties.[18]

SECTION 33.07 THE CONTRACT AS A CONTROL ON DECISION–MAKING POWERS

In addition to jurisdiction, there are other limits to the decision-making powers of design professionals. For example, a contractual power to grant time extensions usually provides standards for determining whether a time extension should be granted. The clause may set forth the events that will justify a time extension and the impact that these events must have upon performance. Although without specific authorization, such as AIA Doc. A201, ¶ 8.3.1, design professionals would not be able to arbitrarily enlarge the list of events, they will have to determine whether the event specified occurred, what effect the occurrence of the event had upon the performance of the contractor, and the amount of time extension.

Similarly, the power to decide disputes does not give the design professional authority to change the specifications. For example, in *Northwestern Marble & Tile Co. v. Megrath,*[19] the power to interpret the contract did not permit the design professional to require a subcontractor to install galvanized iron when the specifications permitted the installation of either galvanized iron or mild steel pipe.

SECTION 33.08 PROCEDURAL MATTERS

(A) REQUIREMENTS OF ELEMENTAL FAIRNESS

The design professional as interpreter or judge invites comparison with arbitration and litigation. Should the design professional conduct a hearing similar to that used in arbitration or litigation? Clearly, the formalities of the courtroom would be inappropriate and unnecessary. Even the informal hearings conducted by an arbitrator would not be required. Unless a clause that confers jurisdiction requires a hearing, no hearing at all is necessary.

Cogent reasons exist for some semblance of elemental fairness to both parties. First, continued good relations on the project necessitate a feeling on the part of the participants that they have been treated fairly. Each party should feel it has been given a fair chance to state its case and be informed of the other party's position. A fair chance need not necessarily include even an *informal* hearing. The design professional should listen to the positions of each party where feasible before making a decision.

The exact nature of what is fair will depend upon the facts and circumstances existing at the time the matter is submitted to the design professional. Small matters and those that require quick decisions may not justify the procedural caution that would be necessary where large amounts of money are at stake or where an urgent decision is less important.

The second reason for elemental fairness is the likelihood that a decision made without it will not be accorded much finality. For example, *John W. Johnson, Inc. v. Basic Construction Co.,*[20] involved a dispute in which the architect had ordered a prime contractor to terminate a particular subcontractor. The trial judge stated:

> This amazing directive was issued by the architect's office without notice to the plaintiff and

---

18. *Huggins v. Atlanta Tile & Marble Co.,* 98 Ga.App. 597, 106 S.E.2d 191 (1958). But see *Atlantic National Bank of Jacksonville v. Modular Age, Inc.,* 363 So.2d 1152 (Fla.App.1978), where a different architect judged the quality of modular units, the delegation issue not having been raised. An A201 was used.

19. 72 Wash. 441, 130 P. 484 (1913).

20. *John W. Johnson, Inc. v. Basic Construction Co.,* 292 F.Supp. 300 (D.D.C.1968), affirmed 429 F.2d 764 (D.C.Cir. 1970).

without giving the plaintiff any opportunity to be heard, orally or in writing, formally or informally. This action on the part of the architect's office was contrary to the fundamental ideas of justice and fair play. The suggestion belatedly made at the trial that it was not appropriate for the architect to maintain any contacts with subcontractors is fallacious in this connection. Any such principle as that did not bar the architect's representative from according a hearing to the subcontractor before directing that his subcontract be cancelled.[21]

Termination is a serious matter, and more process fairness in such matters can be expected. But the case also reflects judicial unwillingness to uphold rash, impetuous decisions.

The third reason relates to the immunity sometimes given to design professionals when they act in a quasi-judicial role (explored in Section 17.08(C)). The more design professionals act like judges or arbitrators, the more likely they will be given judicial protection from suit by someone dissatisfied with the decision.

A look at AIA Doc. A201 is instructive. Paragraph 2.2.12 states the design professional's decision is subject to arbitration. In determining when a demand for arbitration *can* be made, one base point is ten days "after the parties have presented their evidence to the Architect or have been given a reasonable opportunity to do so." The parties must be given the opportunity of presenting their evidence to the architect, although there is no requirement of a hearing. But suppose the arbitration clause is deleted, something that is not uncommon in construction contracts. Is there still a requirement that the parties be given an opportunity of presenting their evidence to the architect? The language applies only where there is arbitration. These procedural requirements should apply even if arbitration is eliminated.

As to the design professional's power to inspect the work, suppose a subcontractor is required to furnish modules manufac-

tured in the subcontractor's plant. Suppose the architect is concerned over the quality control and demands that the modules be inspected in the plant. The subcontractor refuses, contending that its methods are trade secrets that it does not wish to expose.

AIA Doc. A201, ¶ 2.2.5, gives the architect access to the work "wherever it is in preparation and progress." Also, ¶ 7.7.2 empowers the architect to order special inspection or testing, while ¶ 7.7.4 states that the architect will observe inspections, *"where practicable,* at the source of supply."

Interpreting contract language, particularly language of a standard contract form, should, where a serious dispute arises, balance the *legitimate interests* of *each* contracting party. The owner, through the architect, is entitled to check upon the *likelihood* that the performance will meet the contract requirements, preferably as early as possible. On the other hand, if the process is truly a trade secret, or at least the subcontractor honestly believes that the process is proprietary data, an effort must be made to accommodate the legitimate interests of the subcontractor in protecting its proprietary data. Probably the best solution is to allow a carefully controlled inspection both as to time and as to the personnel, in addition to imposing reasonable contractual restraints dealing with nondisclosure. In this way, the legitimate interests of both parties can be accommodated.

(B) STANDARD OF INTERPRETATION

Section 33.09 deals with the finality to be accorded decisions by the design professional. This section emphasizes the process from the vantage point of the design professional. How should relevant contract language be interpreted?

This treatise has dealt extensively with methods by which contracts—especially construction contracts—are to be inter-

---

21. Id. at 304.

preted.[22] Were it not for the fact that design professionals are often called upon to interpret construction documents that they either have prepared or have had a role in preparing, earlier discussion would suffice. However, the complexity of interpreting and judging one's own work requires some additional comments in this section.

In addition to the limits imposed by the contract language being interpreted, contracts that give interpreting and judging powers to the design professional often specify general standards of interpretation to be used by the design professional. For example, AIA Doc. A201, ¶ 2.2.10, requires that all "interpretations and decisions of the Architect . . . be consistent with the intent of and reasonably inferable from the Contract Documents." Despite the limits and guides, the design professional is often faced with the formidable task of deciding disputes over the meaning of contract document language. The prior participation in the development of the contract documents makes this difficult task even more difficult.

N.E. Redlon Co. v. Franklin Square Corp.,[23] held that the architect must not take into account the intention of the owner or contractor or his intention at the time he drafted the documents in question. According to the court, the architect should first look at the contract documents to see whether they have provided guidance in interpretation matters. In the *Redlon* case, the contract specifications stated that terms were to be used in their trade or technical sense. After applying any contractual guides given, the architect was to use an objective standard to determine the meaning of contract language.

If an objective standard is desired, design professionals should not consider their or the owner's intention. Perhaps the court went too far when it directed the architect to ignore any intention the *contractor* may have had when it made the contract.

The design professional should judge the contract documents from the perspective of an honest contractor examining them before bid or negotiation. The preparation of the contract documents by the owner through the design professional takes a long time, much longer than given to the contractor to examine and bid. Any ambiguities that the contractor should not have been expected to notice and to which attention should not have been directed prior to bid submission or negotiation should be resolved in favor of the contractor. Conversely, unclear language to which the contractor should have directed attention should be resolved against it.

This process should not be eliminated by a printed clause in the general conditions or specifications that seeks to put the risk of all unclear language in the contract documents upon the contractor unless it brings this to the design professional's attention. For reasons described elsewhere,[24] this is an unfair burden to place on the contractor.

Admittedly, a standard that looks at the honest contractor can place design professionals in a difficult position where the language in question was derived from the drawings and specifications they prepared.[25] Openly acknowledging that the specifications were unclear can be a confession of professional failure. The standard of interpretation suggested—that of favoring the contractor under certain circumstances—can inhibit the design professional from ever finding language unclear for fear that it would reflect upon his work. An honest design pro-

---

22. Refer to Section 24.02.

23. 89 N.H. 137, 195 A. 348 (1937).

24. Refer to Section 28.03.

25. The discussion assumes that the language in question was not part of the basic contract or general conditions, writings that should have been drafted or supplied by an attorney. Refer to Sections 6.09, 18.01(B), and 28.01(E).

fessional should be fair to both owner and contractor despite this possibility.

Perhaps it will be expecting too much for the design professional to step back from his own work and judge it objectively. This possibility may be a reason to accord less finality to the decision. In any event, if unwilling to construe unclear language in favor of the contractor, he will likely use or claim to use the objective standard required in the *Redlon* case.

(C) FORM OF DECISION

The contract clause giving the design professional the power to interpret documents can require that a particular form be followed when a decision is made. AIA Doc. A201, ¶ 2.2.10, requires that interpretations be in writing or in the form of drawings. For this and other reasons it is generally advisable for decisions to be made in writing and communicated as soon as possible to each party. Where it is not feasible to make a decision in writing on the spot, any oral decision should be confirmed in writing and sent by a reliable means of communication to each party. A written communication giving the design professional's interpretation of his decision need not give reasons to support the decision. The essential requirement is that the parties know that a decision has been made and that they know the nature of the decision. However, the process will work more smoothly if the participants are given a reasoned explanation for the decision. The decision need not be elaborate or detailed but should specify the relevant contract language and facts and the process by which the decision has been made.

AIA Doc. A201, ¶ 2.2.12, does not require that the written decision state that it is final and can be appealed to arbitration within thirty days. The paragraph states that *if* such a statement is made, arbitration must be demanded within thirty days or the architect's decision becomes final and binding.

Probably the refusal to require such a statement was based upon a desire to discourage arbitration of decisions and judicializing the design professional decision-making process. On the other hand, the present language of ¶ 2.2.12 would appear to allow a demand for arbitration be made at *any* reasonable time, since the thirty-day provision applies only when the decision *states* that it is final but subject to appeal. This can lead to delayed claims, one of the most troublesome features of construction contract disputes.

(D) APPEAL

Some construction contracts allow decisions made by the design professional to be taken to arbitration. Often, as indicated in the preceding subsection, the time limit for arbitration may begin when one party receives notice of an adverse final decision. For example, AIA Doc. A201, ¶ 2.2.12, requires a written decision by the architect that *can* set into motion the time for a demand for arbitration. Likewise in federal public contracts, final decisions of the federal contracting officer must be appealed within ninety days from the date of receipt of the contracting officer's decision.

(E) COSTS

Although decisions made by design professionals require time and effort, the design professional is not compensated for this work. It is a part of the service compensated by the fee. The costs incurred by the parties, such as transporting witnesses to any informal hearing, obtaining any expert testimony, or attending any informal hearings, will be borne by the parties who incur them unless the contract provides otherwise. Since most dispute resolution done by the design professional is informal, costs of this sort are not likely to be comparable to those that will be incurred in an arbitration or litigation.

Sometimes judging performance requires the design professional to order that work be uncovered. Uncovering work is costly. Often contracts specify who will pay for the cost of uncovering and recovering. AIA Doc. A201 is an illustration.

Under ¶ 13.1.1, if work had been *improperly* covered by the contractor, such as covering work despite a request by the architect that it *not* be covered, the cost must be borne by the contractor even if the work had been properly performed.

Paragraph 13.1.2, deals with work *properly* covered. If the work is found "in accordance with the Contract Documents", costs are borne by the owner. If the work did *not* comply, the contractor must pay the cost of uncovering and recovering unless the deviation was caused by the owner or a separate contractor. Placing the entire cost upon the contractor even if the deviation is slight is also the solution if the architect exercises his power under ¶ 2.2.13 to require special inspection or testing as set forth in ¶ 7.7.2.

Should the contractor be required to pay the entire cost if *any* deviation is discovered? One purpose of having the architect visit the site periodically is to observe work before it is covered. The assumption under ¶ 13.1.2 is that the architect did not see the defective work before it was covered, nor request the contractor not to cover it until he had a chance to inspect it. Under such circumstances he may be fearful of a claim being made against *him* unless *some* defective work is found.

The "all or nothing" solution is justified only if there were major deviations or if the work were covered to hide defective work. It is better to share the cost of covering and uncovering if the defect is slight or inadvertent.

SECTION 33.09 FINALITY OF DESIGN PROFESSIONAL DECISION: *LAUREL RACE COURSE, INC. v. REGAL CONSTRUCTION CO., INC.*

Before examining a case that dealt with this question, a preliminary observation must be made. Parties to a contract can create a mechanism by which disputed matters or other matters that require judgment can be submitted to a third party for a decision. The finality of that decision—that is, whether it can be challenged and the extent of the challenge—can range from 0 to 100%. It can be *absolutely* unchallengeable. At the other extreme, a decision by a third party can be simply advisory.

The American design professional's decision falls between these extremes. Most decisions have *some* finality but not the finality of an arbitral award [26] or a court decision. They are clearly more than advisory. Unless they are shown either to have been dishonestly made or to be clearly wrong, they are likely to control.

The preceding discussion has dealt with decisions by third parties. A design professional cannot be said to be a disinterested third party. In discussing the reasons for limiting the effect of the refusal by an architect to issue a certificate for payment, a New York judge stated:

> The rule is based upon the fact that the architect, in contracts of this sort, rarely a disinterested arbiter, is usually the representative of the party, often the owner, who must ultimately bear the cost of the work.[27]

This factor—the dubious impartiality of the decision maker—makes the law uncertain and apparently contradictory. A case exploring these problems is reproduced at this point.

---

26. See Section 34.14.

27. *Arc Electric Construction Co. v. George A. Fuller Co.,* 24 N.Y.2d 99, 299 N.Y.S.2d 129, 133 n. 2, 247 N.E.2d 111, 113–114 n. 2 (1969).

LAUREL RACE
COURSE, INC.
v. REGAL
CONSTRUCTION CO.,
INC.
Court of Appeals of
Maryland, 1975.
274 Md. 142, 333 A.2d 319.

LEVINE, Judge.

[Editor's summary of facts: Laurel Race Course hired Regal to build an all-weather track to be designed by Watkins, an engineering firm. Watkins also was to perform certain contract administration services during Regal's performance. Watkins was given the power to interpret the documents and decide all disputes.

The specifications contained detailed requirements for the base, subbase and cushion of the main track. Oversize rock or other deleterious material which could harm a running horse were to be removed at the contractor's expense.

Regal claimed to have substantially completed performance but Watkins recommended withholding the final payment because of oversized materials, compaction problems and drainage difficulty. Regal claimed it had performed the work and that the work had been informally approved by Watkins. After the dispute arose, Laurel and Regal each performed corrective work claiming the other was responsible.

After hearing the disputed evidence as to performance, the trial judge stated that he held the owner, contractor and engineer all accountable for the oversized stones and the presence of those stones was not a contract deviation. Also the lack of a final certificate, he stated, did not preclude Regal's recovery as Regal had substantially performed and was entitled to the balance.]

Almost a century ago, our predecessors held . . . where work was "to be done . . . to the satisfaction of the City Commissioner [of Baltimore]," and payments during the progress of the work were to be made only in accordance with his "monthly estimates," that those estimates were a condition precedent to recovery of such payments, absent bad faith or collusion.

From that holding has emerged the general rule, followed uniformly by decisions of this Court, that where payments under a contract are due only when the certificate of an architect or engineer is issued, production of the certificate becomes a condition precedent to liability of the owner for materials and labor in the absence of fraud or bad faith, ... Apart from fraud or bad faith, the only other exceptions to this rule are waiver or estoppel, ...

The durability of this rule may be more readily appreciated when one considers the emphasis with which it was enunciated by our predecessors. For example, in Lynn v. B. & O. R.R. Co., 60 Md. 404 (1883), Judge Miller said for this Court:

". . . So, in the case before us, it was not enough that the jury might believe from the evidence that Legge *unreasonably* rejected the ice, or that he was *grossly* wrong in his judgment . . .; they must go further, and actually infer and find fraud or bad faith. By this contract, which is perfectly lawful, the parties expressly agreed to submit the question whether the ice to be supplied was 'good, clear, and solid,' to the judgment of this third party, and his judgment, *no matter how erroneous or mistaken it may be, or how unreasonable it may appear to others, is conclusive* between the parties, unless it be tainted with fraud or bad faith. To substitute for it the opinions and judgments of other persons, whether judge, jury or witnesses, would be to annul the contract, and make another in its place." 60 Md. at 415 [emphasis added].

There is no question but that under subsection 24 of the General Conditions, payment of the "balance due . . . including the percentage retained during the construction period" is expressly conditioned upon production of the engineer's " 'Final Certificate.' " [28] It is equally clear that the amount awarded by the trial court under count I of the declaration was the alleged "balance due . . . including the percentage retained." Nor is there any contention advanced by Regal that any of the exceptions to the general rule—fraud, bad faith, waiver or estoppel—were established here.

The argument interposed by Regal to the applicability of the general rule rests exclusively on the second paragraph of subsection 2 of the General Conditions.[29] It draws upon this language for the contention that the condition precedent in subsection 24 is dispensed with once "either party resorts to legal action." Otherwise stated, the argument is that the engineer's certificate is controlling—a condition precedent to payment—only until the parties reach the courthouse; but once they do so, the engineer's decision that the specifications have not been met is reviewable by a judge or jury. Thus, under this view, if the trier of fact were to determine that the specifications had been met, the absence of the

---

28. [Editor's note: Court footnotes renumbered.]

In pertinent part, subsection 24 provides: "Upon notice that the work is ready for final inspection and acceptance, the Engineer shall make such inspection; and *when he finds the work acceptable* under the Contract and *the Contract fully performed,* he shall promptly issue a 'Final Certificate' over his signature stating in effect that the work provided for in the Contract has been satisfactorily completed and recommending its acceptance by the Owner.

"The balance due the Contractor, including the percentage retained during the construction period, *will then be paid* to the Contractor by the Owner. This final payment will be made within sixty days after date of the Engineer's 'Final Certificate', and said final payment shall evidence the Owner's acceptance of work unless it is accepted in writing prior to said final payment." [Emphasis added.]

29. Subsection 2 in its entirety states:

"The Engineer shall have general inspection and direction of the work as the authorized representative of the Owner. He shall have authority to stop the work whenever such action may be necessary to insure the proper execution of the contract. He shall also have authority *to reject work and materials which do not conform* to the plans, specifications and contract documents, to direct the place or places where work shall be prosecuted, and to have the Contractor's force increased or decreased as in his judgment is required. *He shall decide all engineering questions* which arise in the *execution* of the work.

"The Engineer shall *also interpret* the meaning and requirements of the plans, specifications and contract documents, and *decide all disputes* that arise. The Engineer's decisions on *these* matters shall be final and binding on both the Contractor and the Owner unless both parties agree to submit the dispute to arbitration or either party resorts to *legal action* for settlement." [Emphasis added.]

engineer's certificate would not bar recovery by the contractor under the written contract.

Although Regal's argument purports to rest on the last several words of subsection 2, "either party resorts to legal action," it is necessarily keyed to the final phrase in the immediately preceding sentence of that paragraph, "and decide all disputes that arise." Under the construction advocated by Regal, this specific phrase alone not only refers to disputes under the first part of that sentence—concerning the engineer's interpretation—but applies to all disputes arising under the contract. Hence, Regal would say, the binding effect of the engineer's decisions on all disputes— whether arising out of interpretation or performance—is subject to the "legal action" clause. In sum, the engineer's decisions on all disputes of whatever nature, arising under the contract, would be binding only until there is a "legal action."

The difficulty we encounter with this argument is that it completely ignores paragraph 1 of subsection 2 and the manifest intention of the parties. While this intent is readily discernible from the clear and unambiguous language of these provisions, we nevertheless think it appropriate to stress that the intention of the parties to an agreement must be garnered from the terms considered as a whole, and not from the clauses considered separately, ...

The engineer, pursuant to paragraph 1 of subsection 2, possesses the "authority to reject work and materials *which do not conform* to the plans, specifications and contract documents," and to "*decide* all engineering questions which arise in the *execution* of the work." (emphasis added). In accordance with this paragraph, decisions of the engineer on questions pertaining to performance and execution of the work are controlling and unqualified. Paragraph 2, however, is confined to disputes arising out of the engineer's role as an interpreter of the technical provisions contained in the various documents. The words "these matters," to which the "legal action" exception applies, pertain solely to such disputes. In this limited respect only are the engineer's decisions, though otherwise final, subject to the "legal action" exception.

At first blush, perhaps, one might question whether the positioning of the comma immediately preceding the words, "and decide all disputes that arise," was intended to mean that this phrase should refer to disputes under the first paragraph as well, and hence, whether the "legal action" exception should not apply to decisions under both paragraphs. This construction, however, would permit a simple comma to alter what we regard as the clear intent of the agreement. "The authorities make it plain that punctuation cannot control or alter the effect of language that is plain in its meaning."

It is uncontroverted that the disputes under the written contract all relate to the rejection of "work and materials which do not conform to the plans, specifications and contract documents . . ."

within the contemplation of the first paragraph of subsection 2. Regal claimed at the trial that its performance did conform and Laurel insisted that it did not. Nowhere in the testimony is there a conflict over an interpretation of the "meaning and requirements of the plans, specifications and contract documents." Therefore, the "legal action" clause is not applicable.

As we see it, therefore, the supremacy of the engineer's certificate on all matters pertaining to conformance and execution survived the resort to "legal action," and should not have been ignored, absent a finding of bad faith, fraud, waiver or estoppel. No such finding was made here. Hence, production of the engineer's certificate was a condition precedent to the liability of Laurel under count I of the declaration. It is fundamental that where a contractual duty is subject to a condition precedent whether express or implied, there is no duty of performance and there can be no breach by nonperformance until the condition precedent is either performed or excused. . . . Here, the condition precedent was neither performed nor excused; therefore, no judgment should have been rendered against Laurel under the written contract.

* * *

It is questionable that the parties intended to draw a line like that drawn by the court which differentiated judging performance from interpreting contracts. It was convenient for the court that the parties agreed that their dispute concerned performance matters. But often the functions cannot be so easily separated. To judge whether there has been proper performance it is often necessary to determine what was required, something which cannot be determined without contract interpretation.

Also, disputes can involve legal concepts. For example, Regal contended it had substantially performed.[30] Suppose this argument persuaded the engineer to award the certificate but Laurel refused to pay, claiming that the substantial performance doctrine should not have been applied. What degree of finality would an arbitrator or judge have given the engineer's decision?

Perhaps very little in view of the legal nature of the dispute.

Disputes submitted to the design professional can range from purely factual (Did the work meet certain specific standards?) to matters which though sometimes called legal are really factual (How should this clause be interpreted?) to legal questions (Is the substantial performance doctrine applicable where the design professional is to judge performance?).

The wide range of issues, along with the ambivalent status of the design professional, has led to uncertainty over finality of the design professional's decision.

Finality in the first instance depends upon the language of the contract giving the design professional the power to make decisions. If it says *nothing* about conclusiveness or finality, the decision would be purely advisory. But generally clauses giving

---

30. Refer to Section 26.06(B).

this power state that the decision shall be, as in the *Laurel* case, "final and binding."

Again it is instructive to look at AIA Documents. A201, ¶ 2.2.12, states that matters referred to the architect are subject to arbitration. That paragraph specifies time limits to invoke arbitration. One benchmark— that of thirty days—is tied to the architect's written decision stating that its decision is final but subject to appeal. Failure to demand arbitration within the thirty-day period results in a decision "becoming final and binding."

Suppose the parties delete ¶ 7.9, the clause dealing specifically with arbitration. They are likely to delete ¶ 2.2.12 also, inasmuch as much of that paragraph deals with arbitration. If those two paragraphs are deleted, there is nothing in A201 that gives *any* finality to the architect's decision.

Suppose the architect does not state in the written decision that it is final and binding, an alternative given under ¶ 2.2.12. Does this strip the architect's decision of any finality, making the decision purely advisory? Although undoubtedly the AIA intended to give some finality to the architect's decision, tying finality to arbitration may mean that none will be granted unless the *decision* states it is final but subject to appeal.

Suppose, though, as was probably intended, the architect's decision is final and binding unless arbitration is invoked. If the arbitration clause has been deleted, the *court* will extend a considerable amount of deference to the architect's decision. (If arbitration is sought, the arbitrators need pay no attention to the architect's decision.

This can encourage arbitration by the party that is dissatisfied with the architect's decision.)

As noted earlier in this section, however, language of finality does not actually mean that the decision is final. The *Laurel* case held the decision to be binding unless there were fraud or bad faith. The Restatement (Second) of Contracts makes the decision binding so long as it is made honestly and not on the basis of gross mistake as to the facts.[31]

New York gives greater finality to a certificate that has been issued than to the absence of a certificate. In the absence of a certificate, the contractor can recover if it can show substantial performance, with refusal to issue the certificate being unreasonable.[32]

These varying, though related, degrees of finality make the decision conclusive if honestly made unless it is clear that the design professional made a serious mistake. But the standards do not tell the entire story. Other relevant factors determine the degree of finality.

The particular nature of the dispute is important. If the dispute is more technical and less legal, the decision will be given more finality.[33]

AIA Doc. A201, ¶ 2.2.11, states that architect decisions as to artistic effect "will be final if consistent with the intent of the Contract Documents." In addition, ¶ 2.2.12 makes clear that such decisions cannot be arbitrated. The creation of an apparently unreviewable decision can lead to abuse. Certainly the architect should not use this power to force the contractor to perform in

---

31. Restatement (Second) of Contracts § 227 comment c, illustrations 7 and 8 (1981). The First Restatement was applied in *James I. Barnes Construction Co. v. Washington Township,* 134 Ind.App. 461, 184 N.E.2d 763 (1962). Other somewhat variant standards are expressed in *Perini Corp. v. Massachusetts Port Authority,* 2 Mass.App. 34, 308 N.E.2d 562 (1974) (binding unless arbitrary or in bad faith), and *E.C. Ernst, Inc. v. Manhattan Construction Co. of Texas,* 387 F.Supp. 1001 (S.D.Ala.1974) affirmed 551 F.2d 1026, 559 F.2d 268 (5th Cir.1977) (engineer must use good faith).

32. *Arc Electric Construction Co. v. George A. Fuller Co.,* supra note 27.

33. *Yonkers Contracting Co. v. New York State Thruway Authority,* 25 N.Y.2d 1, 302 N.Y.S.2d 521, 250 N.E.2d 27 (1969); *John W. Johnson, Inc. v. J.A. Jones Construction Co.,* 369 F.Supp. 484 (E.D.Va.1973).

a way not required by the contract, and the language of ¶ 2.2.11 appears to recognize this. But the very subjectivity of artistic effect and the power for abuse that such a clause can create will inevitably lead to a difference of opinion among courts as to the enforceability of such a clause.[34]

If a certificate is issued and the *owner* refuses to pay, it is likely that the certificate will be considered more final.[35] Undoubtedly this is a recognition that the owner for all practical purposes has selected the design professional and should be given less opportunity to challenge the decision. If a subcontractor's rights are at stake, perhaps less finality will be given.[36] The subcontractor had even less of a role in selecting the design professional. Power to the design professional is often accomplished by incorporating prime contract general terms by reference into the subcontract. The subcontractor may not have had much opportunity to present its case to the design professional.[37]

The availability of arbitration may also play a role. If the decision can be appealed to arbitration, arguably more finality can be given.[38] Another factor that may bear upon the degree of finality to be given design professional decisions is the process by which the decision is made. If it appears to have been made precipitously without elemental notions of fairness, less finality if any, will be accorded.[39]

Another problem relates to the interaction between language giving finality to a design professional decision and language that bars acceptance of the project from waiving claims for defective work subsequently discovered.[40] One case held that the architect's power to judge performance did not give his decisions finality because of a clause stating that neither the issuance of a final certificate nor final payment relieved the contractor from responsibility for faulty materials or workmanship.[41] Yet another decision more sensibly reconciled these two clauses by concluding that the issuance of a certificate is conclusive where defects are patent or obvious, but not where defects are latent, that is, not reasonably discoverable.[42]

SECTION 33.10
FINALITY: A COMMENT

If the parties to a contract *voluntarily* designate a third party to make certain determinations or decide certain disputes and agree that these determinations or decisions shall be binding on both parties, the law should give effect to such an agreement. Parties who genuinely agree to abide by such third-party decisions and who are satisfied that the decision was honestly made are likely to perform in accordance with the decision. The third party may be better equipped in the view of the contracting parties to give a fair and quick decision. Failure to make such agreements final may discourage parties from agreeing to submit determinations and disputes to third parties or encourage

---

34. Compare *Baker v. Keller Construction Corp.*, 219 So.2d 569 (La.App.1969) (reviewed decision), with *Mississippi Coast Coliseum Commission v. Stuart Construction Co.*, 417 So.2d 541 (Miss.1982) (refused to review).

35. *Hines v. Farr*, 235 S.C. 436, 112 S.E.2d 33 (1960).

36. *Walnut Creek Electric v. Reynolds Construction Co.*, 263 Cal.App.2d 511, 69 Cal.Rptr. 667 (1968).

37. *John W. Johnson, Inc. v. Basic Construction Co.*, supra note 20.

38. *Roosevelt University v. Mayfair Construction Co.*, 28 Ill.App.3d 1045, 331 N.E.2d 835 (1975).

39. *John W. Johnson, Inc. v. Basic Construction Co.*, supra note 20.

40. Section 28.05.

41. *Flour Mills of America Inc. v. American Steel Building Co.*, 449 P.2d 861 (Okl.1969).

42. *City of Midland v. Waller*, 430 S.W.2d 473 (Tex.1968).

them not to live up to such agreements. The result can be an increasing burden on an already burdened judicial system.

Contract language and courts often state that in making certain determinations, the design professional is acting in a quasi-judicial function and not as representative of one of the contracting parties. This does not change the realities. The design professional is *not* a neutral judge, and the contractor has little choice but to accept the current system.

Generally determinations and decisions by the design professional are followed by the parties. This may be because the parties are satisfied, because it is too costly to arbitrate or litigate, or because of the need to retain the goodwill of the design professional. In most cases, this will mean that giving the design professional decision-making powers, *in the first instance*, will provide a quick method of handling construction disputes. According such decisions some degree of finality simply gives effect to the superior bargaining position the owner frequently enjoys at the time a construction contract is made.

In *Cofell's Plumbing & Heating, Inc. v. Stumpf*,[43] one issue related to the finality of an engineer's decision under a contract that gave the engineer "binding authority to determine all questions concerning specification interpretations in the execution of the contract." After completion, a dispute arose that related to how particular work should be priced. The project engineer had concluded that the work should be compensated under a particular provision in the contract. The owner refused to pay, but the trial court concluded that the contractor was entitled to be paid in accordance with the engineer's decision.

The appellate court held that a line must be drawn between "disputes or specification interpretations which deal with work performance and require immediate resolution and those which deal with other contractual disputes, such as rates, method, or time of payment, and do not require prompt on-the-site determination."[44] The court held that the contract should be interpreted to give finality to the former but not to the latter. The project will continue whatever the determination by the engineer, at least in the view of the majority.

The dissenting judge wished to expand the finality provision to *both* types of decisions to prevent unnecessary delays in the Construction Process while disputes are settled. He contended that the majority's decision will invite delays by contractors who will be forced to stop work while resolving other types of disputes that the engineer, under the interpretation of the majority, cannot resolve with any degree of finality.

It is important to keep the project moving. Decisions needed to accomplish this may be given finality to accomplish this goal. Yet it is likely that the project can be expedited by getting a decision but still allowing a dissatisfied party to challenge it later.

Some American jurisdictions give the design professional immunity when acting as a judge.[45] In those jurisdictions, to both give the design professional immunity and accord substantial finality to his decisions concentrates too much power in the hands of the design professional. Because this power will occasionally be abused, in those jurisdictions very little, if any, finality should be given to the design professional's decision. Even where the design professional has *no* immunity, it is better to accord no greater finality to his decisions. The system will work without it, and a needless issue can be removed from construction litigation.

---

43. Supra note 12.

44. 290 N.W.2d at 234.

45. Refer to Section 17.08(c)

PROBLEM

Under the terms of a construction contract between O and C, A, the architect, was given the power to terminate C's performance if C was not making reasonable progress. A informed C that his progress was not satisfactory and ordered him to leave the job. C tried to show A records of weather conditions to prove that unseasonably cold weather cut down his efficiency, that there had been a jurisdictional strike between the unions of two subcontractors, and that A had taken too long to approve shop drawings and render interpretations. A refused to look at the evidence or to consider these contentions. He also refused to give any time to the contractor to catch up. C left the site. There was no arbitration provision.

C sued O and A claiming that he had been terminated without just cause. Does C have a valid claim against O? Against A? If he has a valid claim against either, what should be his measure of recovery?

34

Construction Disputes: Arbitration and Other Methods to Avoid the Courthouse

SECTION 34.01
INTRODUCTION

This chapter deals principally with a voluntary method of resolving disputes by submitting disputes to a third party and agreeing to be bound by that party's decision. Third-party resolution can be used to determine narrow issues such as the strength of concrete or the value of property. The former would be classified as third-party testing and the latter as third-party appraisal. The parties can go further and authorize a third party to decide *any* dispute that might arise between them in the performance of a contract. Such a general referral to the third party [1] can encompass specific disputes narrow in character or broad disputes that can encompass any matter that might be resolved by a court. This latter system, frequently called arbitration, is the focal point of this chapter.

Although it is valuable to compare arbitration and litigation,[2] arbitration does *not* operate completely apart from the judicial system. The court may be the place that resolves whether particular disputes are to be arbitrated. Court assistance may also be sought if one of the parties seeks to frustrate the arbitration process. The judicial system with its enforcement machinery is likely to be the method by which arbitral awards are enforced if the parties do not comply voluntarily.

Despite the importance of the law's auxiliary role in arbitration, it is useful to look at arbitration as a substitute for litigation. This is particularly important in the Construction Process with the high likelihood of technical disputes that seem inappropriate for resolution in a courtroom.

SECTION 34.02
THE LAW AND ARBITRATION

Until the 1920s the law was openly hostile to arbitration. Although it would enforce an arbitral award *after* it was made, it frustrated agreements to arbitrate disputes that might arise in the future. Some courts found such agreements invalid, some allowed a party to revoke such an agreement prior to award, and some would give only nominal damages for breaching a contract to arbitrate future disputes. Arbitration could not thrive in such a legal system.

---

1. Third parties as used in this chapter, with the exception of Section 34.20 dealing with public contracts, do not encompass decisions being made by one of the parties to the contract or its employee or decisions made by persons closely connected with financing the transaction, such as a lender or governmental agency.

2. This comparison is made in Section 34.17.

In the 1920s commercial arbitration was greatly encouraged by statutory enactments in a substantial number of states, including commercially important states, which sought to remedy some of the deficiencies in the legal treatment of arbitration. Principally, these arbitration statutes accomplished the following:

1. Made agreements to submit future disputes to arbitration irrevocable.

2. Gave the party seeking arbitration the power to obtain a court order compelling the other party to arbitrate.

3. Required courts to stop any litigation where there had been a valid agreement to arbitrate a pending arbitration.

4. Authorized courts to appoint arbitrators and fill vacancies when one party would not designate the arbitrator or arbitrators withdrew or were unable to serve.

5. Limited the court's power to review findings of fact by the arbitrator and her application of the law.

6. Set forth specific procedural defects that could invalidate arbitral awards and gave time limits for challenges.

Forty-three states and the federal government currently have modern arbitration statutes. Clearly this has greatly encouraged the use of arbitration.

The more favorable attitude toward arbitration has been tempered by the modern judicial recognition that many agreements to arbitrate are forced upon the weaker party in an adhesion contract. This can be demonstrated by two cases.

Spence v. Omnibus Industries[3] involved a dispute between a homeowner and a remodeling contractor over a remodeling contract. The contract, a standardized contract, included a clause requiring arbitration in accordance with the rules of the American Arbitration Association.

After the dispute arose, the homeowner brought an action against the remodeling contractor seeking damages of $37,000. The remodeling contractor filed a petition for arbitration. The petition was granted, and the court ordered that the homeowner pay the arbitration filing fee of $720. The homeowner was willing to arbitrate but appealed the court's decision requiring him to pay the filing fee.

The court reversed the decision regarding the filing fee and in doing so compared the cost of beginning an action in court and in arbitration. The court noted that the filing fee for commencing an action in court would have been $50.50, but it would cost the homeowner $720 to submit the matter to arbitration. The court stated:

> The reason for this disparity is obvious. Courts are established and supported by the State in order to afford forums to which all, rich and poor alike, may present controversies at minimum cost to the parties. Arbitration is supported by the parties. If the parties are equal in bargaining power, arbitration is good. If the parties are not equal, arbitration may deny a forum to the weaker.[4]

After characterizing the contract as one of adhesion imposed by the party of greater bargaining power upon the weaker,[5] the court noted that the contract was over 2,000 words jammed into a tightly printed jumble of "terms and conditions." The court noted that it was quite unlikely that one homeowner in one hundred would ever read the massive information on the reverse page. Despite the judicial policy favoring arbitration, the court pointed to the strong judicial policy to protect the weaker party to the bargain. The court felt that a $720 fee could discourage a homeowner from presenting a claim against a builder.

Concluding that the homeowner waived

---

3. 44 Cal.App.3d 970, 119 Cal.Rptr. 171 (1975).

4. 119 Cal.Rptr. at 172.

5. Refer to Section 5.03(C).

her right to arbitrate by filing an action in court,[6] the court held that the contractor seeking arbitration became the initiating party and required the latter to pay the filing fee.[7] Had the homeowner wanted arbitration, according to the court, she would have had to pay the filing fee.

The holding in the *Spence* case points to some of the inadequacies of the adhesion contract analysis. It is just as likely as not that the homeowner had stronger bargaining power. However, when faced with a pre-prepared, complicated form contract that includes matters beyond the understanding of the homeowner, the homeowner may become the weaker party, not because she lacks bargaining power but because she does not have the skill or ability to understand the terms agreed to.[8] This problem becomes even more difficult as legislatures begin to bar arbitration clauses in adhesion contracts.[9]

The second case, *Player v. George M. Brewster & Son, Inc.,*[10] involved an arbitration clause in a subcontract. The work was to be performed in California, and the arbitration clause required that arbitration be governed by New Jersey law and be held in New Jersey. The arbitration clause specified that each party would select one arbitrator and the third would be picked by a New Jersey trial judge.[11] The prime contractor, though having its home office in New Jersey, performed much of its work in the western part of the United States.

The court interpreted the clause to determine whether it covered the particular dispute. Like the court in *Spence v. Omnibus Industries,* the court in this case noted that this was a contract of adhesion and should be construed in favor of the subcontractor. The court stated:

> As a whole it appears to be a "house attorney" prepared form intended by Brewster to be submitted to all of its subcontractors on a take-it-or-leave-it basis.[12]

The court then paid some tribute to arbitration stating:

> The law favors contracts for arbitration of disputes between parties. They are binding when they are openly and fairly entered into and when they accomplish the purpose for which they are intended.
>
> *   *   *
>
> Our trial courts are clogged with cases, many of them involving disputes between contracting parties. One of the principal purposes which arbitration proceedings accomplish is to relieve that congestion and to obviate the delays of litigation.[13]

The court was concerned that a strong prime contractor that made all of its subcontractors agree to such clauses can deprive subcontractors access to the courts in their states. The prime contractor would possess a powerful weapon enabling it to force the subcontractors to arbitrate thousands of miles away from the place of business of the subcontractor and have the third arbitrator, the neutral, be selected by

---

6. See Section 34.05.

7. A court less impressed with the adhesive aspects of the contract held that the party who commenced litigation would have to pay the initiating fee when the other party demanded arbitration. See *A.P. Brown Co. v. Superior Court, County of Pima,* 16 Ariz.App. 38, 490 P.2d 867 (1971).

8. Suppose a client signs an AIA Doc. B141 with its arbitration clause?

9. Iowa Code Ann. § 679A.1.

10. 18 Cal.App.3d 526, 96 Cal.Rptr. 149 (1971).

11. By the time the matter went to court, the prime contractor was willing to have the arbitration heard in California but insisted that the third arbitrator be picked by the New Jersey trial court judge.

12. 96 Cal.Rptr. at 154.

13. Ibid.

a hometown judge where the prime contractor had its offices. The court concluded:

> A skepticism is born when we read paragraph 13 as to whether it was written as Brewster wrote it for the purpose of expeditious disposition of controversies with its subcontractors. Its plan, it is suggested, may have been designed to effectuate a more unilateral benefit to itself.

> * * *

> We think the courts would and should scan closely contracts which bear facial resemblance to contracts of adhesion and which contain cross-country arbitration clauses before giving them approval.[14]

Modern courts are beginning to look carefully at arbitration clauses in an adhesion context. Is the clause designed to frustrate the weaker party's right to make claims rather than provide a fair and efficient resolution of disputes? Is the clause one that is not brought to the attention of the weaker party and phrased in a way so as *not* to be understood? Is the clause, though expected and understood, one which essentially rigs the process in favor of one of the parties?[15] If the answer to any of these questions is yes, despite the favorable attitude toward arbitration[16] as an alternative dispute resolution system, the law may not enforce the clause.

SECTION 34.03
AGREEMENTS TO ARBITRATE AND THEIR VALIDITY

Arbitration is a voluntary system based upon a valid contract to arbitrate. For that reason, analysis of the validity of a general arbitration clause involves the requisite elements for a valid contract as well as an awareness of the special treatment accorded agreements to arbitrate by the courts.

(A) MANIFESTATIONS OF MUTUAL ASSENT: AN AGREEMENT TO ARBITRATE

An agreement to arbitrate is usually manifested by each party signing a written contract that contains the arbitration clause or makes reference to arbitration in another document. However, in consumer transactions, even a signature may be insufficient to create a valid contract to arbitrate. In such transactions, it may be necessary to show that the consumer had her attention drawn to the arbitration clause and understood its import.[17]

In construction contracts, the principal problems have been incorporation by reference of arbitration clauses into subcontracts. Some courts have not been willing to hold subcontractors to arbitration clauses incorporated in this manner.[18] Those that see arbitration as a useful and normal method of resolving disputes permit this method to be used.[19]

Even in prime contracts, incorporation by reference can be a problem for an unsophisticated owner. An owner who uses AIA Documents will sign the Basic Agreement, A101, which states nothing about arbitration but includes A201, the General Conditions, among the Contract Documents and states that they are part of the contract. To complicate the process, A201's arbitration

---

14. 96 Cal.Rptr. at 156.

15. *Graham v. Scissor-Tail, Inc.*, 28 Cal.3d 807, 171 Cal.Rptr. 604, 623 P.2d 165 (1981) (arbitration clause in a contract between an experienced rock promoter and a performer that required arbitration by a committee of the performer's union invalidated). See Section 34.03(B).

16. *La Stella v. Garcia Estates Inc.*, 66 N.J. 297, 331 A.2d 1 (1975).

17. *Spence v. Omnibus Industries*, supra note 3, discussed in Section 34.02; *Deutsch v. Long Island Carpet Cleaning Co.*, 5 Misc.2d 684, 158 N.Y.S.2d 876 (1956) (carpet cleaning contract).

18. *Vespe Contracting Co. v. Anvan Corp.*, 399 F.Supp. 516 (E.D.Pa.1975).

19. *Bigge Crane & Rigging Co. v. Docutel Corp.*, 371 F.Supp. 240 (S.D.N.Y.1973).

clause [20] is among a number of provisions labeled "Miscellaneous." Even more difficult for the unsophisticated owner, ¶ 7.9.1 requires arbitration in accordance with Construction Industry Arbitration Rules of the American Arbitration Association (CI Rules).[21] These formidable rules are not part of the Contract Documents.

Although such a double reference system may be needed to save space, an owner may justifiably complain when it discovers that its signature on A101 means that all disputes must be arbitrated under rules that it is not likely to have seen.

Modern arbitration statutes generally require simply that the agreement be written. However, states sometimes enact special formal requirements for agreements to arbitrate. For example, Texas requires that the arbitration clause be typed in underlined capital letters or rubber-stamped prominently on the first page of the contract.[22] Local statutes must be consulted.

(B) VALIDITY ATTACKS ON ARBITRATION CLAUSES

Agreements to arbitrate future disputes were generally not enforceable at common law. This has been cured in most states by the enactment of modern arbitration statutes. The U.S. Arbitration Act, a modern arbitration statute, applies to maritime contracts and those that affect interstate commerce. Expansion of the federal statute in many states may create concurrent jurisdiction, with the arbitration clause being enforced under either the state or the federal statutes. This has become an additional complication. It makes it easier to enforce arbitration clauses in states without modern arbitration statutes. It can also lead to increasing complexity and trips to the courthouse.[23]

In addition to the now decreasingly successful claims based upon the invalidity of arbitration clauses, other attacks are being made. One frequently used attack by the party who does not wish to arbitrate is that the contract has been fraudulently procured. Unquestionably a contract procured by fraud is invalid. But who decides the disputed issue of fraud—the arbitrators or the court?

The clause should be no more enforceable than the contract that creates it. The claim of fraud is easily made and difficult to sustain. A judicial determination of fraud can take time and would suspend the arbitration hearing while lengthy hearings are held in court. As a result, a claim of fraud can have a tactical advantage for the party wishing to delay. In addition, dividing the dispute between those that should be resolved by the court and those by the arbitrator causes other problems. Suppose the fraudulent claim is *not* sustained by the court but is upheld by the arbitrator.

To encourage arbitration and discourage broad assertions of fraud that did not *specifically* attack the arbitration clause itself, a number of courts employ a legal fiction that artificially severs the arbitration clause from the rest of the contract. This divisibility concept permits the fraud issue to go to the arbitrator unless the arbitration clause

---

20. AIA Doc. A201, ¶ 7.9.

21. A number of references to those rules are in this chapter and are referred to as the CI Rules (reproduced in Appendix F).

22. Vernon's Ann.Tex.Civ.Stat. art. 224–1.

23. *Moses H. Cone Memorial Hospital v. Mercury Construction Corp.,* 460 U.S. 1, 103 S.Ct. 927, 74 L.Ed.2d 765 (1983) (federal court applying federal statute can order arbitration despite challenge to arbitration in pending state court proceeding); *Burke County Public School, Board of Education v. Shaver Partnership,* 303 N.C. 408, 279 S.E.2d 816 (1981) (federal statute applied to interstate architectural services despite contract clause stating state law controlled); *Huber, Hunt & Nichols, Inc. v. Architectural Stone Co.,* 625 F.2d 22 (5th Cir.1980) (federal statute appeared to cover construction contract).

itself was attacked as having been fraudulently obtained.[24] Other courts have refused to use this fiction and submitted issues of fraud to the court.[25]

In 1967 the U.S. Supreme Court adopted the divisibility concept in *Prima Paint Corp. v. Flood & Conklin Manufacturing Co.*[26] This made attacks of fraud less effective and strengthened arbitration. Although state courts deciding cases under state arbitration laws were not bound to follow this decision, most have.[27] Yet the court in its desire to remove impediments to arbitration may have encouraged contracting parties not to agree to general arbitration clauses for fear that matters as important as fraud will not go to a court but will go to an arbitrator.

Another attack is that the arbitration clause is unconscionable. California held that a contract between a rock promoter and entertainers that included an arbitration clause giving the power to arbitrate disputes to a committee of the American Federation of Musicians was unconscionable.[28] Despite the court's conclusion that the promoter knew what it was doing, the court found the clause unconscionable. The cards were stacked in favor of the performers. Similarly, Texas recently enacted a statute precluding arbitration if the clause is unconscionable.[29]

Another attack on an arbitration clause is that it lacks mutuality, when one party is required to arbitrate while the other is not. For example, a Florida appellate court passed upon an agreement that contained an arbitration provision requiring the subcontractor to submit its claims to arbitration while allowing the contractor to pursue its claims through litigation. The court held that the arbitration clause was unenforceable for lack of mutuality.[30] Although traditional contract analysis would find mutuality in the other promises in the contract, this court, much like *Prima Paint*, looked upon the arbitration clause as a collateral contract and would not base enforcement on other contractual promises. Yet Alaska came to the opposite conclusion when it was presented with a subcontract arbitration clause under which the contractor could demand arbitration but the subcontractor could not. The arbitrators had found the clause binding, and their determination was upheld in court, based upon looking at the arbitration clause as part of the contract as a whole. The court found that the clause was not unconscionable and that there was no evidence of coercion.[31]

Section 34.04 deals with the varied types of arbitration clauses that can be found. Note that in the two preceding cases, both of which involved subcontracts, the prime contractor in the first case forced the subcontractor to arbitrate, while in the second case it demanded the right to arbitrate only for itself. Obviously the difference, assum-

---

24. *Robert Lawrence Co. v. Devonshire Fabrics, Inc.*, 271 F.2d 402 (2d Cir.1959).

25. *Lummus Co. v. Commonwealth Oil Refining Co.*, 280 F.2d 915 (1st Cir.1960).

26. 388 U.S. 395, 87 S.Ct. 1801, 18 L.Ed.2d 1270 (1967). The same concept was applied to a dispute that arose after a mutual termination of the contract in *Clifton D. Mayhew, Inc. v. Mabro Construction, Inc.*, 383 F.Supp. 192 (D.D.C.1974) (noting split in federal courts).

27. For example, see *Security Construction Co. v. Maietta*, 25 Md.App. 303, 334 A.2d 133 (1975); *Weinrott v. Carp*, 32 N.Y.2d 190, 344 N.Y.S.2d 848, 298 N.E.2d 42 (1973); *Ericksen, Arbuthnot, etc., Inc. v. 100 Oak Street*, 35 Cal.3d 312, 197 Cal.Rptr. 581, 673 P.2d 251 (1983).

28. *Graham v. Scissor-Tail, Inc.* supra note 15.

29. Vernon's Ann.Tex.Civ.Stat. art. 224.

30. *R.W. Roberts Construction Co., Inc. v. St. Johns River Water Management District*, 423 So.2d 630 (Fla.App. 1982).

31. *Willis Flooring, Inc. v. Howard S. Lease Construction Co. & Associates*, 656 P.2d 1184 (Alaska 1983).

ing the prime contractors had the superior bargaining power, was based upon different attitudes toward the arbitration process.

Suppose one party terminates the contract for an asserted material breach by the other. This does not eliminate any jurisdiction that had been created by an arbitration clause in the original but now terminated agreement.[32] The terminating party or party being terminated can still seek arbitration. Termination is not an attack on the validity of the contract.

SECTION 34.04
SPECIFIC ARBITRATION CLAUSES: JURISDICTION OF ARBITRATOR AND TIMELINESS OF ARBITRATION REQUESTS

(A) JURISDICTION CONFERRED BY CLAUSE

A frequently disputed issue relates to whether the arbitration clause covers the particular matter in dispute. Obviously, much depends upon the language of the clause, and a broadly drafted arbitration clause can cover almost anything that relates to the contract between the parties.

As a general rule, close questions as to jurisdiction are likely to be resolved against the party that prepared the clause.[33] However, courts favorably disposed toward arbitration are unwilling to narrowly interpret clauses to deny arbitrability of particular disputes. This issue can arise at the time

arbitration is challenged or after an arbitration award has been taken to court for confirmation. Often, one ground for refusing to confirm an arbitration award is that it exceeded the arbitrator's jurisdiction. One court noted that overtechnical judicial review of arbitration awards on the basis of the scope of the arbitrator's authority can frustrate the basic purposes of arbitration.[34] It is more difficult to attack jurisdiction *after* the arbitration has taken place. The clause can specify that it applies *only* to claims that do not exceed a certain amount or claims that involve *factual* as opposed to *legal* disputes.[35]

Courts have often come to apparently different conclusions on interpretation issues. The results often depend upon the language or judicial attitude toward arbitration. A few examples illustrate this. Courts have differed as to whether delay damages fall within an arbitration clause.[36] A court seeking to encourage arbitration would not limit arbitration to damages solely to person or property.[37] Another court less favorable to arbitration held that a general arbitration clause did not cover the owner's possible liability for water damage.[38] That court required that the clause be "crystal clear" before it would confer jurisdiction on the arbitrator.

Another court held that a general arbitration clause in a subcontract did not confer jurisdiction on the arbitrator to determine which portion of the funds the prime contractor received to train minority workers

---

32. *County of Middlesex v. Gevyn Construction Corp.*, 450 F.2d 53 (1st Cir.1971); *Riess v. Murchison*, 384 F.2d 727 (9th Cir.1967).

33. *Player v. George M. Brewster & Son, Inc.*, supra note 10.

34. *Federal Commerce & Navigation Co. v. Kanematsu-Gosho, Limited*, 457 F.2d 387 (2d Cir.1972); *Muhlenberg Township School District Authority v. Pennsylvania Fortunato Construction Co.*, 460 Pa. 260, 333 A.2d 184 (1975).

35. *Doyle & Russell, Inc. v. Roanoke Hospital Association*, 213 Va. 489, 193 S.E.2d 662 (1973).

36. *Harrison F. Blades, Inc. v. Jarman Memorial Hospital Building Fund, Inc.*, infra note 40, held a delay damage claim beyond the scope of arbitration, while *Aberthaw Construction Co. v. Centre County Hospital*, 366 F.Supp. 513 (M.D.Pa.1973), held that arbitration was required.

37. *Muhlenberg Township School District Authority v. Pennsylvania Fortunato Construction Co.*, supra note 34.

38. *Silver Cross Hospital v. S.N. Nielsen Co.*, 8 Ill.App.3d 1000, 291 N.E.2d 247 (1972).

should go to the subcontractor.[39] That same court refused to permit the architect to arbitrate a dispute between prime contractor and subcontractor when the subcontractor's principal claim was that the architect had committed design errors.

Courts have differed in their willingness to encompass implied terms under the arbitration clause.[40] Most courts hold that the arbitration clause will be applied to disputes that arise after the work is completed despite a provision in the arbitration clause stating that work will continue while the dispute is being arbitrated.[41]

The AIA has denied jurisdiction to an arbitrator to consider disputes over artistic effect. Another jurisdictional issue relates to the arbitrability of tort claims.[42] Some arbitration clauses allow one party to arbitrate but not the other. As noted earlier, judicial resolution of the jurisdictional question is likely to be influenced by the court's attitude toward arbitration, the relative bargaining power of the parties, and the apparent appropriateness of arbitration for a particular dispute.

Expiration of any time limit specified in the contract or statute for making the award generally ends jurisdiction of the arbitrator, and any awards made after expiration are invalid. This rule has been eroded where an award made after expiration of the time limit is enforced on the basis of waiver, such as proceeding with the arbitration without protest after expiration or failure to protest after a late award has been rendered. Waiver can also be predicated upon a course of conduct that shows that the parties did not consider time to be of the essence, especially where no prejudice is shown. If the expiration period for making the award has been waived, the award must be made within a reasonable time.[43] One of the more difficult jurisdictional questions relates to the application of an arbitration clause to a disputed termination (discussed in Section 38.03(I)).

Jurisdiction problems highlight the variety of possibilities between *no* arbitration and a *general* arbitration clause. Drafting variations are discussed in Section 34.17.

(B) TIMELINESS OF ARBITRATION DEMAND

AIA Doc. A201 sets two standards for timeliness. Decisions by the architect that state that they are final and subject to appeal must be appealed to arbitration within thirty days.[44] Other disputes require that arbitration be requested within a reasonable time.[45] Two cases that have involved prior

---

39. *Paschen Contractors, Inc. v. John J. Calnan Co.*, 13 Ill.App.3d 485, 300 N.E.2d 795 (1973).

40. *Roosevelt University v. Mayfair Construction Co.*, 28 Ill.App.3d 1045, 331 N.E.2d 835 (1975), and *Allentown Supply Corp. v. Hamburg Municipal Authority*, 463 Pa. 167, 344 A.2d 477 (1975), held that arbitration encompassed implied terms, while *Harrison F. Blades, Inc. v. Jarman Memorial Fund Hospital Building Fund Inc.*, 109 Ill.App.2d 224, 248 N.E.2d 289 (1969), would not. Implied and express warranty came within the arbitration clause in *Harman Electrical Construction Co. v. Consolidated Engineering Co.*, 347 F.Supp. 392 (D.Del.1972).

41. *Warren Brothers Co. v. Cardi Corp.*, 471 F.2d 1304 (1st Cir.1973); *Hudik-Ross, Inc. v. 1530 Palisade Ave. Corp.*, 131 N.J.Super. 159, 329 A.2d 70 (1974). A case to the contrary was *Hussey Metal Division v. Lectromelt Furnace Division*, 471 F.2d 556 (3d Cir.1972) (clause stating that no demand for arbitration could be made after final payment).

42. *Harman Electrical Construction Co. v. Consolidated Engineering Co.*, supra note 40; *Morton Levine & Associates, Chartered v. Van Deree*, 334 So.2d 287 (Fla.App.1976).

43. Annot., 56 A.L.R.3d 815 (1974).

44. ¶¶ 7.9.1, 2.2.12. Without an *express* statement of the right to appeal, the thirty-day period does not apply. *Roosevelt University v. Mayfair Construction Co.*, supra note 40; *Niagara Mohawk Power Corp. v. Perfetto & Whalen Construction Co.*, 52 A.D.2d 1081, 384 N.Y.S.2d 299 (1976).

45. ¶ 7.9.1.

or current AIA language illustrate some of the problems.

An earlier edition of AIA Doc. A201 came before the court in *Harman Electrical Construction Co. v. Consolidated Engineering Co.*[46] The contract required that a written notice of claim be given within a reasonable time after its first observation but not later than final payment. The arbitration demand was required to be filed within a reasonable time after the dispute had arisen.

The first observance of this problem occurred in July of 1969. Formal claims were submitted on July 16, 1969. For over two years the parties negotiated, but one party finally rejected the claim on October 26, 1971. Two and a half months later the claimant requested arbitration.[47] The court held that the latter demand had been made within a reasonable time, pointing to the absence of any prejudice having been shown by the party seeking to avoid arbitration.

Milton Schwartz & Associates, Architects v. Magness Corp.[48] involved a dispute in September 1972 between the architect and a land developer over excess costs. Nonpayment was followed by a formal demand for payment on April 4, 1973, by the architect's lawyer in which the latter advised the developer that suit would be instituted if the defendant did not act promptly to insure payment. On June 11, 1973, the developer's lawyer informed the architect's lawyer that the contract was terminated because of the architect's breach and that the developer was considering a counterclaim if the architect instituted legal action. On August 8, 1973, the architect brought legal action against the developer, and the latter re-

sponded immediately by demanding arbitration in accordance with the arbitration clause that required that the demand be made within a reasonable time after the dispute had arisen. The architect contended that eleven months had elapsed from the time the dispute developed in September of 1972 until the demand for arbitration on August 8, 1973, and this exceeded a reasonable period of time.

The court noted that Pennsylvania law, which governed the contract, looked with favor upon arbitration and concluded that the reasonable time standard depended upon who sought arbitration—the claimant or the party opposing the claim. The court concluded that the opponent of the claim—in this case the developer—should not have the burden of considering an arbitration remedy *prior* to the time his potential adversary makes it clear that some form of enforcement proceedings will be forthcoming.

The formal demand was not made until April 4, 1973, and suit not brought until August 8, 1973. Whether the time for determining whether arbitration would be sought began at the time of the formal demand or at the time of the commencement of the lawsuit was irrelevant in light of the court's conclusion that the defendant had sought arbitration within a reasonable time judged from either point.

The choice to arbitrate can be made even *after* the court process has been begun by the other party as long as the delay in requesting arbitration has not prejudiced the other party. Similar problems arise when the litigation has proceeded through preliminary phases before one party invokes the arbitration process (discussed in

---

46. Supra note 40.

47. Two months after filing a mechanics' lien and one month after an action to foreclose it was a reasonable time to demand arbitration in *Frederick Contractors, Inc. v. Bel Pre Medical Center, Inc.*, 274 Md. 307, 334 A.2d 526 (1975). See also *Chase Architectural Associates v. Moyers Corners*, 99 A.D.2d 646, 472 N.Y.S.2d 219 (1984) (demand made five years after completion reasonable, as final payment certificate never issued).

48. 368 F.Supp. 749 (D.Del.1974).

greater detail in Section 34.05 dealing with waiver).

(C) WHO DECIDES JURISDICTION AND TIMELINESS

One of the most troublesome questions relates to whether the court or arbitrator will decide certain threshold issues. Section 34.03 dealt with attacks on the arbitration process. Suppose one of the parties attacks arbitration by claiming that the arbitrators have no jurisdiction to decide the particular dispute or that the process has not been requested within the time period specified in the arbitration clause. Both questions can involve contract interpretation, but the second in particular involves a factual determination that often involves applying the standard of reasonableness.

Some decisions have held that the threshold questions should be decided by the judge, while others hold they can be decided by the arbitrator.[49] The uneasy partnership between arbitrator and judge and the complexity that can be generated by these preliminary procedural questions can be demonstrated by the *County of Rockland v. Primiano Construction Co.*[50] After performance had been completed, the prime contractor claimed that it had suffered delay damages because of acts of the public owner. The owner claimed that the demand for arbitration had come too late.

The New York Court of Appeals differentiated the role of arbitrator and judge in deciding these preliminary questions. The court held that the judge decides whether there was a valid agreement to arbitrate, including the frequently made contention that the dispute is one not arbitrable under the terms of the contract.

Next the court referred to what it called a second threshold issue, which also requires judicial determination, such as whether any preliminary requirements or conditions precedent have been met. As illustrations, the court gave clauses that require that disputes be first submitted to a third party for determination and that set specific contractual limitations to arbitration itself. It sharply distinguished them from what it called "procedural stipulations" that must be observed in the conduct of the arbitration proceedings itself, such as requirements that the demand for arbitration be made within a designated period of time. These, according to the court, are to be resolved by the arbitrator.

Although the court recognized that both can be classified as conditions precedent to the arbitration, it distinguished requirements that are "a prerequisite to entry into the arbitration process" from "a procedural prescription for the management of that process."[51] Putting time limits into the second category, the court concluded that the issue of whether the arbitration demand had been made within the requisite time was to be decided by the arbitrators. Maryland held that timeliness should be resolved by the judge.[52]

SECTION 34.05 WAIVER OF ARBITRATION

Not infrequently one of the parties to the contract contends that the other has waived the right to arbitrate. Such a contention is usually based upon activity that appears to be inconsistent with an intention to arbitrate. For example, one party may institute

---

49. See Annot., 26 A.L.R.3d 604 (1969).

50. 51 N.Y.2d 1, 431 N.Y.S.2d 478, 409 N.E.2d 951 (1980).

51. 431 N.Y.S.2d at 482, 409 N.E.2d at 954.

52. *Frederick Contractors, Inc. v. Bel Pre Medical Center, Inc.*, supra note 47.

legal action or file a mechanics' lien[53] or even go beyond these points and take preliminary steps before trial and *then* seek to arbitrate. In such a case, the other party may contend that the prior activities were inconsistent with the desire to arbitrate and constitute a waiver.[54]

Two lines of authority have developed.[55] One examines the activities of the party alleged to have waived the arbitration clause to see if they manifest an intention not to arbitrate. The other requires that the party claiming waiver show that it has been prejudiced by the activities of the other party. For example, one court found prejudice when the party claiming waiver expended a substantial amount of money or substantial preparation for trial.[56] Another court held that prejudice could consist of the party asserted to have waived the arbitration clause having discovered information through early litigation activity that would not have been available in arbitration.[57]

As has been seen in preceding sections, the inevitable question of who will decide disputed preliminary issues has played an important role in waiver cases. Will it be the arbitrator or the judge? Just as court decisions are not consistent as to the *existence* of waiver, they are inconsistent on the question of who will decide whether there has been a waiver.[58]

Although cases will continue to seem contradictory because of the many factual situations, the general trend may be indicated by a recent case that stated that arbitration was favored and that waiver would not be lightly inferred. The court noted the necessity for showing prejudice rather than simply showing inconsistent acts and indicated that recent waiver cases had involved demands for arbitration long after suits had been started and after discovery proceedings had taken place.[59]

The party wishing to protect its right to arbitrate takes risks if it engages in activity that can indicate an intention to forgo the arbitration process. It may be necessary to take such steps, but they should be taken only upon advice of counsel and with the understanding that the risk is present that arbitration will not be available.

SECTION 34.06 PREHEARING ACTIVITIES

The party initiating arbitration usually files a notice of an intention to arbitrate and, under the CI Rules, must pay a fee based upon the amount of the claim. The notice usually contains a statement setting forth the nature of the dispute, the amount involved, and any remedies sought.[60]

Once it is clear that arbitration will take place—and sometimes even before an arbitrator has been selected—one or both of the parties may wish to examine persons who may be witnesses for the other party and examine documents in the other's possession in order to evaluate the other party's case, to prepare for trial, and possibly to

---

53. Some statutes deal specifically with whether filing a mechanics' lien waives arbitration. Local law should be consulted.

54. In *People ex rel. Delisi Construction Co. v. Board of Education*, 26 Ill.App.3d 893, 326 N.E.2d 55 (1975).

55. *Burton-Dixie Corp. v. Timothy McCarthy Construction Co.*, 436 F.2d 405 (5th Cir.1971), stressed inconsistent conduct, while *Hudik-Ross, Inc. v. 1530 Palisade Ave. Corp.*, supra note 41, stressed the necessity for showing prejudice. See Note, *Enforceability of Arbitration Agreements*, 13 Ariz.L.Rev. 479 (1971).

56. *Commercial Metals Co. v. International Union Marine Corp.*, 294 F.Supp. 570 (S.D.N.Y.1968).

57. *Vespe Contracting Co. v. Anvan Corp.*, supra note 18.

58. See note 49, supra.

59. *Gavlik Construction Co. v. H.F. Campbell Co.*, 526 F.2d 777 (3d Cir.1975).

60. CI Rules § 9.

settle. In judicial proceedings, these activities are classified as "discovery."

An argument frequently made against arbitration is the absence of discovery. Although discovery is still unavailable in many states, some statutes permit it,[61] and it is becoming more common for courts to order discovery under certain circumstances to aid the arbitration process.[62] In complicated arbitrations that involve stakes high enough to justify prehearing activities, it may be useful to permit or even suggest that the arbitrating parties submit in advance a written statement giving *in detail* the facts and legal justification for the contentions of the arbitrating parties. Such an advance submission can eliminate much irrelevant testimony at the hearing. In addition, advance statements can give each party some idea of what the other party will seek to assert. Like discovery, this can aid the parties to prepare for the hearing, can make the hearing more expeditious, and might lead to settlement.

SECTION 34.07
SELECTION OF ARBITRATORS

Generally the method for selecting arbitrators is specified in the arbitration clause. Sometimes a particular arbitrator or specific panel of arbitrators is designated by the parties in advance and incorporated in the arbitration clause. Advance agreement upon the arbitrators or a panel of arbitrators should build confidence in the arbitration process. However, in construction contracts, such advance agreement is not common.

Generally a procedure to select arbitrators rather than a designation of particular individuals is used. For example, the procedure can require each party to name an arbitrator, with the two-party appointed arbitrators designating a third or neutral arbitrator. Some arbitration clauses provide that each party will appoint an arbitrator and only if they cannot agree upon the disposition of the dispute do they appoint a third arbitrator who makes the decision.

Some procedures permit the neutral administrator, such as the American Arbitration Association (AAA), to designate the arbitrator if the parties cannot agree. The AAA seeks to match the particular panel of arbitrators they propose to the type of dispute. Parties are given an opportunity to make choices within the panels, and sometimes this can result in party agreement. Where they cannot agree, as stated, the AAA chooses the neutral arbitrators.

In a three-person arbitration generally each arbitrator is expected to be neutral. Sometimes persons appointed by the parties are expected to be partisans for the party selecting them, and they advocate the position of that party. With the exception of arbitrators who are expected to be partisans, the process usually assumes that the arbitrators, especially the neutral, will render an impartial decision based upon the evidence submitted as well as the expertise they bring to the dispute.

In *Commonwealth Coatings Corp. v. Continental Casualty Co.*,[63] the U.S. Supreme Court dealt with the extent to which the neutral arbitrator must disclose facts that may bear upon the arbitrator's impartiality. It has stirred controversy and has had an impact on arbitration law.

The arbitration had involved a dispute between a subcontractor and the surety on a prime contractor's bond. The neutral member performed services as an engineering consultant for owners and building contractors. One regular customer was the

---

61. M. Domke, *Commercial Arbitration* § 27.01 (G. Wilner ed. 1984).

62. *Burton v. Bush*, 614 F.2d 389 (4th Cir.1980). See also *Bigge Crane & Rigging Co. v. Docutel Corp.*, supra note 19, noted in 44 Univ.Cin.L.Rev. 151 (1975).

63. 393 U.S. 145, 89 S.Ct. 337, 21 L.Ed.2d 301 (1968).

prime contractor with whom the subcontractor had the dispute in question. That relationship was sporadic. The arbitrator's services were used only from time to time, and there had been no dealings between the arbitrator and the prime contractor for about a year before the arbitration. The prime contractor had paid fees of about $12,000 over a four-or five-year period. These facts were not revealed by the arbitrator until *after* the award had been made. When this was disclosed, the subcontractor sought to invalidate the award.

The court noted the resemblance between a judge and an arbitrator exercising quasi-judicial powers. Clearly the judge must avoid any appearance of partiality. But the court felt it even more important that an arbitrator avoid any appearance of impartiality because of the limited review of the arbitrator's decision.

The court refused to confirm the arbitration award because the arbitrator had not disclosed this relationship. Although the court recognized that arbitrators cannot sever all their ties from the business world, the court held that failure to disclose prior activities of this type would create the impression of possible bias.

The concurring justices emphasized that arbitrators are people of affairs and part of the marketplace and cannot be compared strictly to judges. The concurring justices were concerned that too great a burden of disclosure would disqualify the best informed and most capable arbitrators. These justices felt that the arbitrator cannot be expected to provide the parties with a complete and unexpurgated business biography. Nevertheless, the concurring justices felt that in this case the relationship was *more* than trivial and for that reason agreed that the arbitration award should be set aside.

The dissenting justices felt that the losing party was simply grasping at straws and such a requirement of disclosure would discourage arbitrators from serving and would render arbitration less effective because it would be too easily challengeable.

Note that *all* judges wished to encourage arbitration but looked upon disclosure differently in terms of achieving that objective. The majority felt that parties will be more willing to arbitrate and abide by the arbitrator's decision if they are *absolutely* certain that the arbitrator will be impartial. The concurring and dissenting judges felt that arbitration will not work effectively if capable arbitrators are discouraged from serving and if the arbitrator's award is too easily challengeable on the basis of the arbitrator's failure to disclose information.

Subsequent cases have reflected ambivalence toward the *Commonwealth Coatings* requirement of disclosure, especially in the context of a trade association or closely knit industry.[64] Yet the spirit of *Commonwealth* has been followed in other decisions.[65]

SECTION 34.08
PLACE OF ARBITRATION

The arbitration clause need not specify where the arbitration is to take place. Any contractual designation of locale will be given effect so long as the selection is rea-

---

64. *Garfield & Co. v. Wiest,* 432 F.2d 849 (2d Cir.1970), cert. denied, 401 U.S. 940 (1971) (stock exchange arbitration: waiver); *Baar & Beards, Inc. v. Oleg Cassini, Inc.,* 30 N.Y.2d 649, 331 N.Y.S.2d 670, 282 N.E.2d 624 (1972) (need for specialized knowledge and skill); *Reed & Martin, Inc. v. Westinghouse Electric Corp.,* 439 F.2d 1268 (2d Cir.1971) (arbitrator need not provide complete and unexpurgated business biography); *William B. Lucke, Inc. v. Spiegel,* 131 Ill.App.2d 532, 266 N.E.2d 504 (1970) (bias too remote).

65. *Sanko S.S. Co. v. Cook Industries, Inc.,* 495 F.2d 1260 (2d Cir.1973); *J. P. Stevens & Co. v. Rytex Corp.,* 41 A.D.2d 15, 340 N.Y.S.2d 933, affirmed 34 N.Y.2d 123, 356 N.Y.S.2d 278, 312 N.E.2d 466 (1974) (stating major burden of disclosure properly falls upon arbitrator). The cases are discussed generally in Annot., 56 A.L.R.3d 697 (1974). For CI Rules, see Appendix F.

sonable.[66] If hearing site selection is based upon factors that will expedite the process and make it less expensive, it will certainly be considered reasonable. On the other hand, if the place is designated by the stronger party to frustrate the weaker party's right to have disputes heard, such a clause will not be given effect.[67] Section 11 of the CI Rules provides that if the parties cannot agree on the hearing location, the AAA shall determine the location and the decision shall be final and binding.[68]

SECTION 34.09
MULTIPLE PARTY
ARBITRATIONS: JOINDER
AND CONSOLIDATION

Frequently a linked set of construction contracts contains identical arbitration provisions. For example, there are identical arbitration clauses in AIA standard contracts between owner and contractor and between owner and architect. Likewise, identical arbitration clauses are contained in the AIA standard contracts between architect and engineer and between prime contractor and subcontractor.

Suppose there is a building collapse and the owner wishes to assert a claim. It may not be certain whether the collapse resulted from poor design or poor workmanship. The first would be chargeable primarily to the design professional and the second to the contractor. The owner may wish to arbitrate. Suppose there are identical arbitration clauses in its contracts with the design professional and contractor. The owner *can* arbitrate separately with each, but there are disadvantages. Two arbitrations will very likely take longer and cost more than one. The owner may lose *both* arbitrations. (Although this may be unpalatable to the owner, the result may be correct if the design professional performed in accordance with the standards required of design professionals and the contractor executed the design properly.)

Inconsistent findings may result. For example, one arbitrator may conclude that the design professional performed in accordance with the professional standard while the other did not. One may conclude that the contractor followed the design while the other did not. To avoid this and save time and costs, the owner may wish to consolidate the two arbitrations.

Suppose the owner demands arbitration only with the contractor. The contractor may believe that the principal responsibility for the collapse was defective design. In such a case, it might wish to add the design professional as a party to the arbitration or, as stated in legal terminology, "join" the design professional in the arbitration proceedings.[69]

A contractor can be caught in a similar dilemma if the contractor felt that the responsibility for the owner's claim against it was work of a subcontractor. In such a case, the prime contractor may wish to seek arbitration with the subcontractor and consolidate the two arbitrations. Similarly, the architect may find it advisable to consolidate any arbitration she might have with her consulting engineer and any arbitration proceedings with the owner.

---

66. *Central Contracting Co. v. Maryland Casualty Co.*, 367 F.2d 341 (3d Cir.1966). Similarly in an international context, see *Republic International Corp. v. Amco Engineers, Inc.*, 516 F.2d 161 (9th Cir.1975).

67. *Player v. George M. Brewster & Son, Inc.*, supra note 10.

68. This provision was upheld in *Reed & Martin, Inc. v. Westinghouse Electric Corp.*, supra note 64. See also *Aerojet-General Corp. v. American Arbitration Association*, 478 F.2d 248 (9th Cir.1973).

69. This assumes that contractor and design professional have agreed to arbitrate under the same rules as owner and contractor. Ordinarily contractor and design professional do not have a contract, but they can agree to arbitrate their disputes. "Joining" the design professional as a party to the original arbitration is similar but procedurally slightly different from consolidation. The latter merges two existing arbitrations into one.

Similar problems surfaced in early American and English law when commercial transactions became more complex. The common law, or King's Courts, refused to permit joinder and consolidation. The equity courts did so if claims resulted from the same transaction and the dispute could be handled efficiently in one lawsuit. Modern American procedural law ultimately adopted the view of the equity courts.

Modern courts have had to face this problem in the arbitration context. The decisions passing upon the power of a court to order or permit the arbitrator to consolidate separate jurisdictions or permit joinder of additional parties have not been consistent.[70] Where court decisions have not been favorable, some legislatures have sought to encourage arbitration consolidation and joinder by statute.[71] Again, local laws must be consulted.

Current AIA Standard Documents do not permit consolidation of arbitrations between architect and owner and between owner and contractor without consent of all parties.[72] Consolidation of separate arbitrations between owner and prime contractor and prime contractor and subcontractor are permitted if a common question of fact or law requires resolution of the dispute.[73]

Why treat the two consolidation questions differently? One reason given is that an arbitration between architect and owner will involve a determination of whether the architect has lived up to the professional standard while the issue between owner and contractor is whether the contractor has performed in accordance with the contract documents. Although there may be some *slight* difference in the legal standards, this is not sufficient justification to differentiate the two consolidations.

For reasons mentioned, consolidation and joinder are generally desirable. The possibility of confusion because of the potentially large number of parties in a consolidated arbitration can be handled by according the arbitrator the power to decide the number of parties and issues that would make consolidation or joinder too confusing. If so, a request to do so could be denied.

Another method by which participants in the linked set of contracts can be involved in arbitration is what is called the "vouching in" system. This is a method by which a person involved in the arbitration—though not a party—is given the opportunity to participate and to seek to persuade the

---

70. Compare *Grover-Dimond Associates v. American Arbitration Association*, 297 Minn. 324, 211 N.W.2d 787 (1973) (power to consolidate), with *Stop & Shop Co. Inc. v. Gilbane Building Co.*, 364 Mass. 325, 304 N.E.2d 429 (1973) (no power to consolidate in the absence of statute or specific clause permitting consolidation). Massachusetts subsequently did pass a statute. Mass.Gen. Laws Ann. c. 251 § 2A. See also *Episcopal Housing Corp. v. Federal Insurance Co.*, 273 S.C. 181, 255 S.E.2d 451 (1979), which affirmed the trial decision to consolidate separate arbitrations since the party objecting could not show that it would be prejudiced by consolidation. For an interesting case dealing with this issue, see *Jefferson County v. Barton-Douglas Contractors Inc.*, 282 N.W.2d 155 (Iowa 1979).

71. Unhappy with *Atlas Plastering, Inc. v. Superior Court*, 72 Cal.App.3d 63, 140 Cal.Rptr. 59 (1977), which barred consolidation, the California Legislature enacted § 1281.3 of the California Code of Civil Procedure giving broad authority to consolidate.

72. AIA Doc. A201, ¶ 7.9.1. The language was tortured to *permit* consolidation in *Garden Grove Community Church v. Pittsburgh-Des Moines Steel Co.*, 140 Cal.App.3d 251, 191 Cal.Rptr. 15 (1983).

73. Some cynics have observed that the reason for the differentiation is that architects do not wish to participate in the same arbitration with contractors for tactical reasons. Their interests are not so directly at stake when there is a consolidation between the arbitrations of owner and contractor and of contractor and subcontractor.

arbitrators to make an award in accordance with its contentions.[74]

SECTION 34.10   THE HEARING

(A) A DIFFERENTIATION OF ISSUES: DESIRABLE vs. REQUIRED

Section 34.14 discusses attempts by the party satisfied with the award to have it confirmed in court or by the party dissatisfied with the award to have it vacated—that is, upset. That section refers principally to defects in the process that are of sufficient importance to justify not confirming or vacating the award. Even though complaints are made regarding the arbitration hearing that will not justify upsetting the award, examination of complaints might provide a blueprint for conducting a fair hearing. The arbitrator should not consider only what is compelled, which, as shall be seen, is relatively minimal. But the arbitrator should consider the type of hearing that will persuade the parties that they have been treated fairly.

The exact type of hearing will to a large degree depend upon the seriousness of the dispute in terms of both the intensity of feelings of the parties and the amount at stake as well as the need for an expeditious decision. The discussion in this section assumes that a serious matter is brought before the arbitrators, one with sufficient economic importance to justify a careful and fair hearing. Because attorneys are frequently present in these arbitrations, unless otherwise indicated, it is assumed that the parties will be represented by legal counsel. This assumption does not negate the possibility or even likelihood that many arbitrations do not justify some of the steps suggested because of matters adverted to earlier in this paragraph. Nor does it ignore the important differentiation between arbitration and litigation as to speed. Overjudicialization of the arbitration process by giving to it those attributes of the legal process that led to arbitration in the first place is clearly undesirable.

(B) WAIVER

Parties can agree to submit the dispute to arbitration based solely upon written statements, the contract documents, or any other written data the parties feel relevant. Such a paper submission can save time and could be valuable in minor disputes. However, such submissions are often deceptive and may provide insufficient information to make an award. Even if the parties have agreed that no hearing is necessary, an arbitrator can request a hearing in the presence of both parties if resolving the dispute requires it.

(C) TIME

The arbitrators should attempt to schedule the hearing as early as possible, but the arbitrators must take into account the time needed to prepare for the hearing. A complicated dispute may require the opinions or services of engineers, architects, accountants, attorneys, and photographers, among others. If adequate time is not allowed for preparation, continual requests will be made for recesses after the hearing process has begun.

Frequently one or both of the parties request postponement of or recesses in the hearings which should be granted when reasonable. However, arbitrators must take into account that requests for delays and recesses are sometimes bargaining tactics used by the party who feels the other party's financial position will not tolerate delay.

---

74. Such a system is described in detail in *Perkins & Will v. Syska & Hennessy & Garfinkel, etc.*, 50 A.D.2d 226, 376 N.Y.S.2d 533 (1975).

(D) PROCEEDING WITHOUT THE PRESENCE OF ONE OF THE PARTIES

Suppose one of the parties to the arbitration indicates that it will not participate in the hearings. Under such conditions, should the arbitrator proceed with the hearings or can the party who *does* attend be given the amount of the claim? It is better practice for the arbitrators to hear the evidence submitted by the party attending before making the award. Obviously, doubts will be resolved against the party who has chosen not to attend the hearings.

(E) THE ARBITRATORS

At the outset the parties should be permitted to question the arbitrators relating to any matters that could affect their impartiality. The arbitrators must disclose matters that could affect their impartiality.

The arbitrators should comply with any state arbitration laws requiring arbitrators to take an oath at the beginning of a hearing. Even if not required, it is good practice for arbitrators to take an oath that they will conduct the hearing and render their award impartially and to the best of their ability.

(F) RULES FOR CONDUCTING THE HEARING

Generally arbitration clauses do not prescribe detailed rules relating to the method of conducting the hearing. Certain arbitration statutes give general directives, such as requiring that the arbitrator permit each party to present its case and to cross-examine witnesses for the other party. Arbitration associations or trade groups that conduct arbitration often have simple rules relating to the conduct of the hearing. In the absence of rules, the arbitrator determines how the hearing is to be conducted.

(G) OPENING STATEMENTS

In a complicated arbitration, or even in matters that may not appear to be compli-

cated, it is often helpful to permit the parties or their attorneys to make a brief opening statement. The statement can help the arbitrator determine which evidence is relevant to the dispute and can serve the function of reducing the number of issues by having the parties agree to certain facts and issues.

(H) PRODUCTION OF EVIDENCE: SUBPOENA POWERS

Most states give arbitrators the power to issue a *subpoena* that compels witnesses to appear and testify and that requires persons to bring in relevant records. Without such power the arbitrator cannot compel witnesses to appear or documents to be produced.

Usually the arbitrating parties produce witnesses and supply whatever records are advantageous to their position. If issues can be resolved more easily if certain witnesses are produced or certain documents are presented, the arbitrator can resolve those questions against the party who refuses to produce these witnesses or records within its control. If the arbitrator indicates this to a reluctant party, the latter is likely to produce the witness or the records.

(I) LEGAL RULES OF EVIDENCE

The rules of evidence applied in courts need not be followed by the arbitrator. Principally, the rules of evidence that can be dispensed with are those that relate to the *form* that evidence must take to be admissible. However, certain principles that are part of the legal rules of evidence should guide the arbitrator when determining whether material submitted by the parties should be considered. These principles relate to *relevance* and *administrative expediency*.

The arbitrator should not go into matters not germane to the dispute. This is often difficult to determine. At the outset, the arbitrator should not cut off a line of testimony that may not appear to be relevant at the moment it is presented. Perhaps this testimony will become relevant as the hearing proceeds. However, the arbitrator should ask the party presenting the evidence what it intends to establish. If it is then determined that what the party intends to establish is not germane to the dispute, the evidence should be disregarded and testimony cut off. The evidence presented should be relevant and should not be cumulative. Once a fact has been established, it is usually not necessary to establish it again.

(J) DOCUMENTARY EVIDENCE

As stated, the many technical rules of evidence that relate to the admissibility of documentary evidence need not be followed by the arbitrator. Any documentary evidence submitted by the parties can be examined provided that it is relevant and not cumulative. The authenticity of the document can be taken into account. An excessive preoccupation with form, notarial seals, and witnesses to the document can slow down the hearing. Unless a party questions the authenticity of a document, the arbitrator should consider it.

(K) QUESTIONING WITNESSES

One of the important constituents of a hearing is the testimony of witnesses. Failure to hear testimony of witnesses is likely to be procedural misconduct that can vitiate any award made by the arbitrator. Arbitrators should follow any state laws that require that witnesses be placed under oath. Even where the law does not require this, a simple oath adds to the dignity of the hearing and may induce truthful testimony.

Generally, legal rules of evidence prohibit an attorney from asking witnesses leading questions (questions that suggest the answer in the question). However, leading questions are permitted by the opposing attorney. This process, called cross-examination, is designed to test the credibility of the

witness and expose dishonest or inaccurate statements.

Arbitrators need not follow these courtroom rules. However, cross-examination, whether by the opposing attorney or the arbitrators, can be a useful device to test the veracity of the witness. Sometimes the narrative method permits the witness to testify in a logical and understandable fashion. However, the form of examining witnesses lies largely within the discretion of the arbitrators.

(L) VISITING THE SITE

Generally the hearing will be conducted in a hearing room or in an office. It is possible, and often helpful, to conduct a hearing at the site where the evidence is available to the arbitrator or arbitrators. Even when the hearing is not conducted at the site, the arbitrator—either on her own motion or when requested by a party or the parties—may view the premises. Preferably, the viewing of the premises should not be done without the presence of the arbitrating parties or their attorneys.

(M) EX PARTE COMMUNICATIONS

Ex parte communications are information or arguments communicated by one party to a dispute or by a third party to the person deciding the dispute without the knowledge of the other party. For example, it would be improper for a judge to receive privately communicated information relative to a pending case from one of the parties, one of the attorneys for the parties, or a third party not connected with the case. The attorneys who represent the litigating parties should know what communications are being made to the judge in order to respond to them or to point out inaccuracy. This is one reason why arbitrators should notify the parties if they plan to view the premises and set a time that will enable the parties to be present.

Section 31 of the CI Rules requires that all evidence "be taken in the presence of all of the arbitrators and all of the parties, except where any of the parties is absent in default or has waived the right to be present." Section 40 prohibits any communication between the parties and the arbitrator except at the oral hearing.

(N) TRANSCRIPT

Normally there is no requirement that testimony be transcribed or that a written transcript be made. Some arbitrators prefer to have the testimony transcribed and reproduced for their own use as well as to settle any disputes that may arise between the arbitrators and the parties over the exact testimony of witnesses.

(O) REOPENING HEARING

After the hearing has been closed, either party may request to reopen the hearing in order to introduce additional evidence. The arbitrator can determine whether to reopen the hearing. Usually, newly discovered evidence will be a sufficient basis to reopen the hearing as long as the arbitrators have not yet made their award. If the evidence that is proposed to be introduced at an additional hearing could have been available for the original hearing or is merely cumulative, the arbitrator should not reopen the hearing. However, such matters are largely within the discretion of the arbitrator.

SECTION 34.12
SUBSTANTIVE STANDARDS

Sometimes the arbitration clause or the rules under which the arbitration is to be held give general guidelines to the arbitra-

tor regarding standards by which to decide the dispute.

Some arbitration clauses or rules permit the arbitrator to do almost anything regardless of the language of the contract as long as it accords with justice or fairness. As a rule, however, arbitrators decide the dispute based upon the evidence, and in a contract dispute, the contract terms play a central role.

Arbitrators need not follow case precedents. One of the reasons frequently given to arbitrate is that the arbitrator is free to make a proper decision *without* the constraint of earlier precedent or rules of law. However, in complicated arbitrations attorneys may present legal precedents in an attempt to persuade the arbitrators. Although the arbitrators would be free to consider these precedents, clearly they would not be required to follow them.

Must the arbitrator apply legal rules that would provide a defense to the claim were it litigated? For example, suppose the claim would be barred because of the statute of limitations [75] or because the claimant is not licensed in accordance with state law.[76] Usually such defenses are asserted when the demand for arbitration is made. The party opposing arbitration goes to court contending that the contract including the arbitration clause no longer has any legal validity because the claim has been made beyond the period allowed by law. Similarly, the party may seek to oppose arbitration by contending that the party demanding arbitration cannot be allowed to recover because it does not have the requisite license.

If the matter has proceeded to arbitration and the claims have not been asserted as a

---

75. As to statutes of limitations, see Annot., 94 A.L.R.3d, 533 (1979).

76. As licensing see *Merkle v. Rice Construction Co.*, 271 So.2d 220 (Fla.App.1973) (issue for the arbitrator); *Parking Unlimited, Inc. v. Monsour Medical Foundation*, 299 Pa.Super. 289, 445 A.2d 758 (1982) (arbitrator upheld). See also *Starr v. J. Abrams Construction Co., Inc.*, 16 Mass.App. 74, 448 N.E.2d 1311 (1983). The court affirmed an award despite the parties having misled the FHA by failing to inform FHA of a special agreement as to costs. But see *Loving & Evans v. Blick*, 33 Cal.2d 603, 204 P.2d 23 (1949) (refused to confirm award, as party not licensed). (For a discussion of the rights of unlicensed parties refer to Section 13.07(B).)

bar to arbitration, the claims may have been waived. If asserted in the arbitration, arbitrators generally have discretion to apply them. Awards are generally confirmed even if based upon mistake of law. If the defenses appear technical rather than substantive, it is likely that the arbitrators will not bar the claim once the arbitration has proceeded.

SECTION 34.12 REMEDIES

Commonly the arbitration clause or rules under which the arbitration is being held states remedies that can be awarded. For example, § 43 of the CI Rules allows the arbitrator to grant any remedy for relief that is "just and equitable and within the terms of the agreement between the parties." The arbitrator may also assess arbitration fees and expenses equally or in favor of any party and any administrative expenses in favor of AAA.[77] No specific provision allows the arbitrator to award attorneys' fees. However, *Harris v. Dyer*,[78] which involved AIA Doc. A201, awarded the winning party who had filed a mechanics' lien its attorneys' fees incurred in arbitration. State law granted the prevailing party attorneys' fees connected to filing and foreclosing a mechanics' lien.

Arbitrators generally award money damages based upon direct losses occurring or gains prevented.[79] Suppose the arbitrator specifically *orders* a party to perform in accordance with a contractual promise. Such an order cannot be legally enforced by the *arbitrator,* since the arbitrator does not have the power possessed by a judge to fine or jail a person who does not perform in accordance with a coercive court order. This absence of enforcement power on the part of the arbitrator is the reason why arbitration

awards are brought to court for enforcement when the parties do not comply.

Grayson-Robinson Stores, Inc. v. Iris Construction Corp.[80] involved a specific order made by an arbitrator in a dispute over a lease between a department store and shopping center developer. The developer did not build a shopping center, claiming he could not obtain funds. The arbitrator did not find this to be a defense and *ordered* the developer to *build* the shopping center. When the developer refused to comply, the department store sought confirmation of the award in court. Generally courts do not specifically order contracts to be performed (refer to Section 6.03). Yet the court enforced the arbitrator's award, indicating that the arbitrator may have a remedial power not possessed by the judge.

Sometimes arbitrators order that the party perform and state in the award that failure to perform in accordance with the order will entitle the other party to a specified amount of damages. This technique, where enforced, can give a method of enforcement without seeking court confirmation. However, such a technique is not likely to be effective unless the alternative damage claim is high enough to coerce performance. Although this may seem to be justified as giving the arbitrator enforcement power, selecting an abnormally high amount of damages might only lead to a trip to court.

The arbitrator's determination of whether damages or a specific order is appropriate should take into account whether the parties will continue to be able to work together after the arbitration. One reason that courts are hesitant to order parties to perform in accordance with promises is that by the time the matter has reached court, the parties will no longer cooperate

---

77. Annot., 57 A.L.R.3d 633 (1974).

78. 292 Or. 233, 637 P.2d 918 (1981).

79. See Note, *Punitive Damages In Arbitration: The Search For A Workable Rule*, 63 Corn.L.Rev. 272 (1978).

80. 8 N.Y.2d 133, 202 N.Y.S.2d 303, 168 N.E.2d 377 (1960).

with each other. The same can be true in arbitration. However, if there is a reasonably cooperative attitude between the parties and if the work continues to be performed during arbitration, it may be useful to specifically order that work be performed rather than to award damages.

SECTION 34.13 AWARD

Before making the award, the arbitrators review any documents submitted and listen to or read any transcription of the hearings. They consider any briefs that may have been submitted by the parties or by their attorneys. Submission of briefs is uncommon except for disputes involving large amounts of money. The decision need not be unanimous unless the arbitration clause or the rules under which the arbitration is being held so requires.

The form of the award can be simple. The arbitrator need not give reasons for the award. There are arguments for and against a reasoned explanation of the decision accompanying the award. An explanation may persuade the parties that the arbitrators have considered the case carefully. This may lead to voluntary compliance, which is obviously better than costly court confirmation.

Many arbitrators do not accompany the award with reasons because of the additional time it would entail and the fear that it is more easily challengeable. Yet the scope of judicial review may be a reason *for* explaining an award. For example, one court refused to confirm an award that appeared to have ignored a contract clause even though the arbitrator had the *power* to refuse to enforce an unconscionable clause.[81] Had the arbitrator in his award stated that the clause had been disregarded because it was unconscionable, the award would have been sustained.

A disadvantage of giving an explanation is the possibility that the dissatisfied party or parties may refuse to comply and seek to reopen the matter by objecting to the reasons given. Perhaps an even greater argument against reasons accompanying the award is the additional time and expense entailed. Making the arbitration too much *like* a court trial can lose some of the advantages of arbitration.

Taking all of this into account, a short, reasoned explanation accompanying the award is advisable though not required.

The rules under which an arbitration is conducted often specify when the award must be made. For example, § 41 of the CI Rules requires an award be promptly made and, unless otherwise agreed or required by law, "not later than 30 days from the date of closing the hearings, or if the oral hearings have been waived, from the date of transmitting the final statements and proofs to the arbitrator." Failure to make the award by the designated time may terminate the jurisdiction of the arbitrators, although more commonly the parties agree to waive any time deadlines. More important, failure to make the award in the time required may deprive the arbitrators of any quasi-judicial immunity (discussed in Section 34.16). If there are no specific time requirements, the award must be made within a reasonable time.

SECTION 34.14
ENFORCEMENT AND LIMITED JUDICIAL REVIEW: *CHILLUM–ADELPHI VOLUNTEER FIRE DEPARTMENT, INC. v. BUTTON & GOODE, INC.*

Failure to comply with an arbitration award may necessitate judicial involvement and review. A party wishing enforcement may have to go to court to obtain confirma-

---

81. *Granite Worsted Mills, Inc. v. Aaronson Cowen, Limited,* 25 N.Y.2d 451, 306 N.Y.S.2d 934, 255 N.E.2d 168 (1969).

tion. The party seeking to challenge the award may go to court and ask that the award be vacated.

Most state arbitration statutes specify grounds for reviewing an arbitrator's award. The tendency to consider these grounds *exclusive* has limited the scope of judicial review. As to grounds, § 12 of the Uniform Arbitration Act, enacted in whole or with minor variations in a substantial number of states, permits an award to be vacated where any of the following exists:

(1) The award was procured by corruption, fraud or other undue means;

(2) There was evident partiality by an arbitrator appointed as a neutral or corruption in any of the arbitrators or misconduct prejudicing the rights of any party;

(3) The arbitrators exceeded their powers;

(4) The arbitrators refused to postpone the hearing upon sufficient cause being shown therefor or refused to hear evidence material to the controversy or otherwise so conducted the hearing, contrary to the provisions of Section 5, as to prejudice substantially the rights of a party; or

(5) There was no arbitration agreement . . . and the party did not participate in the arbitration hearing without raising the objection;

But the fact that the relief was such that it could not or would not be granted by a court of law or equity is not grounds for vacating or refusing to confirm the award.

Section 13 of the Act allows modification or correction of an award within ninety days after delivery of a copy of the award where any of the following exists:

(1) There was an evident miscalculation of figures or an evident mistake in the description of any person, thing or property referred to in the award;

(2) The arbitrators have awarded upon a matter not submitted to them and the award may be corrected without affecting the merits of the decision upon the issues submitted; or

(3) The award is imperfect in a matter of form, not affecting the merits of the controversy.

Similar language is contained in the U.S. Arbitration Act.[82] Grounds for vacating are limited and principally look to serious procedural misconduct on the part of the arbitrators. Similarly, case decisions have employed language indicating a very limited judicial review of arbitration awards, mainly the manner of holding the arbitration. One court stated that the court will not inquire whether the determination was right or wrong.[83] Another stated that errors of fact or law are not sufficient to set aside the award.[84] Another stated that an error of law was not reviewable unless the arbitrator gave a completely irrational construction to the provision in dispute.[85] Another stated that honest errors were not reviewable.[86]

The arbitrator's power on evidentiary matters was demonstrated in *Norwich Roman Catholic Diocesan Corp. v. Southern New England Contracting Co.*[87] The arbitrator did not examine the specifications that

---

82. Grounds for vacating awards under this statute are contained in 20 A.L.R.Fed. 295 (1974).

83. *Drake-O'Meara & Associates v. American Testing & Engineering Corp.*, 459 S.W.2d 362 (Mo.1970). *Seither & Cherry Co. v. Illinois Bank Building Corp.*, 95 Ill.App.3d 191, 50 Ill.Dec. 672, 419 N.E.2d 940 (1981) noted in 22 A.L.R. 4th 356 (1983).

84. *Mars Constructors, Inc. v. Tropical Enterprises Limited,* 51 Hawaii 332, 460 P.2d 317 (1969). See also *Clinton Water Association v. Farmers Construction Co.*, 163 W.Va. 85, 254 S.E.2d 692 (1979); *Lawrence v. Falzarano*, 380 Mass. 18, 402 N.E.2d 1017 (1980).

85. *Jones v. Kvistad*, 19 Cal.App.3d 836, 97 Cal.Rptr. 100 (1971). See also *Firmin v. Garber*, 353 So.2d 975 (La. 1977) (award upheld, not grossly irrational but simply debatable).

86. *Reith v. Wynhoff*, 28 Wis.2d 336, 137 N.W.2d 33 (1965).

87. 164 Conn. 472, 325 A.2d 274 (1973).

defined excavation materials that were designed to compute the unit prices for the work. The court held that since the arbitrator had jurisdiction, it would not go into the merits of the dispute or evidentiary issues. The court pointed to CI Rules § 30 under which the arbitration was being administered, which stated that the arbitrator would be judge of the admissibility of the evidence.

The application of the limited scope of judicial review to the context of a construction dispute is illustrated by reproducing a judicial opinion dealing with this problem.

CHILLUM–ADELPHI VOLUNTEER FIRE DEPARTMENT, INC. v. BUTTON & GOODE, INC.

Court of Appeals of Maryland, 1966.
242 Md. 509, 219 A.2d 801.

BARNES, Judge.

[Ed. note: Court's footnotes omitted.] This suit was brought by Button & Goode, Inc. (appellee) to enforce an arbitration award entered after Button & Goode and Chillum-Adelphi Volunteer Fire Dept. Co. (appellant) had submitted to arbitration proceedings in regard to a dispute which arose concerning whether Chillum-Adelphi could keep certain sums due Button & Goode under a contract for the erection of a fire house. This money was retained by Chillum-Adelphi as liquidated damages occasioned because of Button & Goode's delay in completing construction of the building. Button & Goode was granted summary judgment in its suit to enforce the arbitration award. This appeal followed.

On April 30, 1962 Button & Goode (contractor) and Chillum-Adelphi (owner) entered into a construction contract whereby Button & Goode agreed to erect two buildings for Chillum-Adelphi. Plans and specifications had been drafted by the owner's architect, Philip W. Mason. The arbitration proceedings and this suit are concerned only with one of the two buildings, the other having been fully completed as required by the contract.

Article 2 of the construction agreement provided that work to be performed under the contract was to commence upon written notice; and the building was to be substantially completed 180 calendar days from the date of such notice. Article 45 of the American Institute of Architects' General Conditions of Contracts, made part of the construction agreement in this case by Article 1 of that agreement, provided that the time in which the contractor agreed to complete the work was of the essence of the contract, and failure to complete the work within the time specified would entitle the owner to deduct as liquidated damages out of any money which may be due the contractor under the contract, the sum of $50.00 for each calendar day in excess of the 180 days until the building should be substantially completed.

The owner's architect specified that one of the buildings was to be constructed of pre-cast concrete framing. Button & Goode could not commence work until that material was delivered to the building site, and the long and protracted delay of Nitterhouse Concrete Products, Inc. (Nitterhouse) in delivering the concrete frames caused a delay in completing the building beyond the 180 days agreed upon as the time within which construction was to be substantially completed. Chillum-Adelphi retained $21,426.48 of the contract price as damages occasioned because of Button & Goode's delay in substantially completing the building.

Article 40 of the General Conditions of Contracts provided that the owner and contractor would submit all disputes, claims or questions arising under the contract to arbitration under the procedure then obtaining in the Standard Form of Arbitration Procedure of the American Institute of Architects (AIA). Button & Goode filed a demand for arbitration with the American Arbitration Association (AAA). Chillum-Adelphi objected to the arbitration procedure provided by the AAA; however, the parties agreed to submit their dispute to arbitration by the AAA provided that the procedure complied with that of the AIA whereby the parties would be given the opportunity to examine and cross-examine all witnesses and introduce exhibits at any time during the hearing.

It was agreed between Button & Goode and Chillum-Adelphi that the issues to be decided by the board of arbitrators would be: (1) What damages, if any, should be assessed against the contractor in this case, and (2) Was the building completed at the time of arbitration?

A hearing was held by the board of arbitrators on August 26, 1964. The arbitrators found that the owner's architect had specified that pre-cast concrete materials of Nitterhouse's manufacture be used in construction of the building, that the contractor had made repeated attempts to have some other company substituted for Nitterhouse to supply the pre-cast concrete frames, but the architect refused to authorize a change because he expected delivery from Nitterhouse sooner than from another company since the order had been pending there for such a long time. Furthermore, a change of suppliers would have necessitated a change in the plans of the building.

Article 18 of the General Conditions provided that the owner's architect should extend the time for the completion of the building if the contractor be delayed in the progress of the work "for any cause beyond the contractor's control". The arbitrators found that Chillum-Adelphi was bound by the decision of its agent, its architect Mr. Mason, to use a product in the construction of the building which proved to be unavailable. The contractor was therefore not responsible for any delay in construction until January 11, 1963, the date Nitterhouse delivered the concrete frames. Under the circumstances, the delay was "beyond the contractor's control" and the architect should have extended the time for completion of the job.

After the pre-cast framing was delivered, Button & Goode proceeded promptly to resume work on the job. The building was substantially completed on August 10, 1963, 211 days after the framing was received from Nitterhouse.

The arbitrators found that Button & Goode was entitled to 180 days from January 11, 1963 for the completion of the job. Since the contractor required 211 days to substantially complete the building from the date the pre-cast frames were delivered, Chillum-Adelphi was entitled to $1,550.00 as liquidated damages, or $50.00 per day for 31 days. Chillum-Adelphi had retained $21,426.48 from the amount due the contractor under the construction agreement. The board of arbitrators therefore awarded Button & Goode $19,876.48 and divided the costs equally between the parties.

Button & Goode filed a petition for judgment on the arbitration award . . . .

An arbitration award is the decision of an extra-judicial tribunal "which the parties themselves have created, and by whose judgment they have mutually agreed to abide." . . . When suit is brought to enforce the award, a court will not review the findings of law and fact of the arbitrators, but only whether the proceedings were free from fraud, the decision was within the limits of the issues submitted to arbitration, and the arbitration proceedings provided adequate procedural safeguards to assure to all the parties a full and fair hearing on the merits of the controversy. . . .

In City of Baltimore v. Allied Contractors, Inc., . . . Judge Hammond, for the Court, said:

> "Mistakes by an arbitrator in drawing incorrect inferences or forming erroneous judgments or conclusions from the facts will not vitiate his award. (citations omitted)
>
> . . . the decisive primary question is not whether the judgment was right or wrong but whether impropriety, to a significant extent, brought about its obtention." . . .

Although a court may modify an arbitration award for a mistake of form such as an evident miscalculation of figures . . . an arbitrator's honest decision will not be vacated or modified for a mistake going to the merits of the controversy and resulting in an erroneous arbitration award, unless the mistake is so gross as to evidence misconduct or fraud on his part. . . .

In short, where parties have voluntarily and unconditionally agreed to submit issues to arbitration and to be bound by the arbitration award, a court will enter a money judgment on that award and enforce their contract to be so bound unless, notwithstanding that the arbitrator's decision may have been erroneous, the facts show that he acted fraudulently, or beyond the scope of the issues submitted to him for decision, or that the proceedings lacked procedural fairness. A court does not act in an appellate capacity in reviewing the arbitration award, but enters judgment on what may be considered a contract of the parties, after it has made an independent determination that the contract should be enforced.

* * *

We hold on this appeal that the trial court properly granted Button & Goode's motion for summary judgment on the arbitration award.

There is no merit in Chillum-Adelphi's contention that the arbitrators went beyond the issues submitted to them for determination. . . .

Chillum-Adelphi's second contention is likewise without merit. The fact that arbitrators may fail to follow strict legal rules of procedure and evidence is not a ground for vacating their award. . . . The procedure followed at the arbitration hearing was fair and in full compliance with the AIA procedural rules which the parties agreed would govern the determination of their dispute. The record in the arbitration proceedings remained open for a full six months before the final award was entered. Additional evidence could have been presented to the arbitration board at any time during that six month period, and upon good cause shown the hearing could have been reopened.

Finally, we must discount Chillum-Adelphi's bald assertion that the determination of the arbitration board was unsupported by the evidence. There is no showing of lack of good faith or fraud on the part of the arbitration board, and we will not review the award on the merits. . . .

The preceding case emphasizes the finality of an arbitrator's decision.[88] The concept of limited judicial review was undoubtedly designed to encourage arbitration by limiting the likelihood that an award can be overturned in court. However, limited judicial review almost to the point of no review can make contracting parties reluctant to use the arbitration process. As shall be seen,[89] one reason sometimes given for reluctance to enter into arbitration agreements is the absence of any meaningful review of the arbitrator's decision.

The limited review does not mean that arbitration awards are *never* upset. For example, a court that articulated the standard of complete irrationality nevertheless upset an arbitrator's decision by concluding that the words of the contract were so clear that there was nothing left to interpret.[90] For all practical purposes, however, an arbitrator's decision is final.

SECTION 34.15
INSURERS AND SURETIES

Generally a public liability insurance policy carried by a contractor or a professional liability insurance policy carried by a design professional indemnifies the insured if it incurs liability to a third party. Insurers in fixing their rates must be able to predict their losses. The professional liability insurer assumes that its liability will be based

---

88. Review is even more difficult because an arbitrator's testimony, whether voluntary or compelled, cannot generally be used to impeach the award. *Grudem Brothers Co. v. Great Western Piping Corp.*, 297 Minn. 313, 213 N.W.2d 920 (1973).

89. See Section 34.17.

90. *O–S Corp. v. Samuel A. Kroll, Inc.*, 29 Md.App. 406, 348 A.2d 870 (1975).

upon the normal professional activity performed by its insured. Similarly, an insurer who issues public liability insurance to a contractor expects to indemnify the insured only if accidents occur that arise out of its normal construction activities.

To standardize risks, insurers usually exclude liability assumed by contract. An insured who wishes to do so, such as a contractor who has contractually agreed to indemnify the owner or architect, must obtain a special endorsement. This gives the insurer the opportunity to examine the risk and decide whether to accept it, refuse it, or accept it with a premium adjustment.

Arbitration, though not *imposing* liability by contract, substitutes one form of dispute resolution for another. This is a more serious problem for professional liability insurers. Unlike claims against a contractor, claims against a design professional are likely to be made by parties with contracts (often containing arbitration clauses) against the insured design professional.

Generally insurers are at least *wary* of arbitration. Although they do not specifically *exclude* arbitration, they may counsel that it not be used, suggest that only a certain clause be used, or, in rare cases, deny coverage for a design professional who intends to use or uses a general arbitration clause. If insurers are aware of such a clause at the time they insure or make no objection if they are asked their opinion, they have consented. Design professionals should check their policies, bring them to the insurers if they are in doubt as to the advisability of arbitration or its effect on coverage, and notify their insurer if they plan to submit an existing dispute to arbitration.

Sureties are discussed in greater detail in Chapter 37. The surety issues a bond to the prime contractor to protect the obligee—usually the owner—from the risk that any claim the owner will have against the prime

contractor cannot be collected. Usually the surety bond either incorporates the construction contract by reference or refers to it. For this reason, it would normally be assumed that if a valid arbitration award is made against the contractor, the owner can recover from the surety. However, the law is not that simple.

Preliminarily, it is important to note that *two* steps in the dispute may involve the surety. First, a question may arise as to whether the surety can or must participate in the arbitration process. Second, a question may arise regarding whether the surety must pay any valid award that has been made against its principal.

Usually the obligee prefers that the surety participate in the arbitration. This insures that the surety will be liable for the award. The owner *may* feel that the surety's participation will make it face *two* lawyers instead of one. Most often, though, the surety does not participate and then seeks to deny its responsibility to pay the award.

The surety is responsible for the award if it has involved itself in some way in the process, such as taking an assignment of the construction contract, being subrogated (placed in the contractor's position) to the contractor's rights, completing the project, or participating in the arbitration. It must pay the award if it has expressly promised to participate or be bound by the award. The difficulties arise if the construction contract is *simply* incorporated or referred to in the bond.

The majority of states hold that the award is admissible only as prima facie evidence in any claim against the surety, very much like a judicial award made against the principal debtor, the contractor.[91] New York holds that a surety cannot be compelled to participate but is conclusively bound to pay the award if the contract is incorporated by reference in the

---

91. *P.R. Post Corp. v. Maryland Casualty Co.*, 403 Mich. 543, 271 N.W.2d 521 (1978).

bond.[92] Some states make the arbitration award inadmissable against a nonparticipating surety.[93]

The claim against a surety is crucial where the principal—the contractor—is insolvent. If the surety will not arbitrate, must the owner bring a separate legal action against the surety? This is contrary to the intention of the owner at the time the contract was made and the bond issued.

Usually the owner-obligee expects that the surety will stand behind the contractor-principal and that if it must arbitrate with the contractor and receives an award, the surety will pay. Notice to the surety that there is an arbitration clause should manifest the surety's consent to be bound by any arbitration award. The owner should contact the surety if it wishes to litigate despite an arbitration clause, since failure to obtain its consent may give the surety a chance to claim that there has been a material alteration discharging its liability. Similarly, the surety's consent should be obtained if the owner and contractor wish to submit an existing dispute to arbitration. Where the law is unclear or would allow the surety to deny responsibility for an arbitration award, the surety's *express* consent should be obtained at the time the bond is issued.

SECTION 34.16
ARBITRATOR IMMUNITY

Some American states grant the design professional quasi-judicial immunity when acting as judge. More clearly, arbitrators are granted quasi-judicial immunity.[94] Arbitrators are even more like judges than are design professionals. However, *Baar v.*

Tigerman [95] stripped an arbitrator of his quasi-judicial immunity when he did not make an award in accordance with the time requirements of the arbitration rules. In addition, the court refused to grant immunity to the American Arbitration Association (AAA).

In the *Baar* case, a dispute among partners was submitted to arbitration. During a four-year period, the arbitrator held forty-three days of hearing and allowed ten days of closing arguments. Both sides submitted final briefs on July 17, 1980, and on the following day the AAA considered the arbitration "submitted." The AAA set the arbitrator's deadline at August 30, 1980, thirty days after final submission, as required by AAA rules. The arbitrator requested and received an extension until November 30, 1980, but did not meet the deadline. This failure, if not waived, would strip the arbitrator of the authority to make any arbitration award.

One party brought legal action against the arbitrator and AAA, alleging breach of contract and negligence. Although the trial court held that both defendants were immune, the Court of Appeals disagreed, concluding that immunity covered only the *actions* of the arbitrator and not *failure* to render an award. The court was not willing to grant the immunity to the arbitrator that the law grants to a judge, noting the differences between their powers and responsibilities. The opinion did not have to face the difficult question of how the damages would be measured.

The case is a warning to arbitrators and those who run arbitral systems to set up reasonable deadlines for making the award

---

92. *Fidelity & Deposit Co. v. Parsons & Whittemore Contractors Corp.*, 48 N.Y.2d 127, 421 N.Y.S.2d 869, 397 N.E.2d 380 (1979). See also *Kearsarge Metallurgical Corp. v. Peerless Insurance Co.*, 383 Mass. 162, 418 N.E.2d 580 (1981).

93. See, e.g., West's Ann.Cal.Civ.Code § 2855. See generally Ruck, *Can A Contract Bond Surety Be Compelled To Arbitrate Claims Against It?* 16 Forum 765 (1981).

94. *Baar v. Tigerman*, 140 Cal.App.3d 979, 189 Cal.Rptr. 834 (1983). The case is noted in 67 Marq.L.Rev. 147 (1983).

95. Supra note 93.

and then to comply with them or obtain extensions.

SECTION 34.17 ARBITRATION AND LITIGATION COMPARED

Arbitration is voluntary. Parties can *choose* to employ it. Choice involves a number of considerations, some of which are apparent and some not.

A lawsuit begins with the filing of pleadings by a lawyer. An arbitration process can start by the filing of a claim, which does not *require* a lawyer. However, the claimant may wish legal advice to decide whether arbitration or litigation is the best method to handle the dispute. Generally it is quicker and less expensive to begin arbitration than litigation.

On the other hand, the filing fees for beginning litigation are modest compared with those required to initiate arbitration (refer to Section 34.02). The CI Rules employ a sliding scale of filing fees. For example, there is a $200 minimum filing fee. A $20,000 claim requires a filing fee of $600. A claim for $160,000 requires a filing fee of $1,800. Claims between $160,000 and $5 million require $1,800 plus one-quarter of one percent of the excess over $160,000. Under the rules, claims that exceed $5 million require a fee to be determined by AAA. These fees apply to both claims and counterclaims, something common in construction disputes.

Arbitration and litigation differ in the ease with which basically one dispute involving a large number of claimants and claims resistors can be handled. Generally joinder of parties and consolidation of disputes is more easily accomplished in a lawsuit than in arbitration (discussed in Section 34.09).

Another major difference is that in many states arbitration does not allow for an effective method of discovery, the process by which attorneys in litigation can obtain evidence from the other party to prepare for the hearing. Although some states require

it, an arbitration system can *contractually* compel it, and although the arbitrator can frequently compel the production of evidence at the hearing, none is as effective as discovery in a lawsuit.

Discovery availability demonstrates the balance between process fairness and a quick resolution of disputes. Its availability can make arbitration more desirable to attorneys. Yet its use can be costly, and its availability can slow down the process and allow one party to delay for bargaining purposes.

One of the advantages frequently given for arbitration is the informality and speed of the process. Although lawyers are generally required in litigation, arbitration does not require that the parties be legally represented. However, parties arbitrating important matters usually have lawyers. This can hamper the arbitration, since there is no requirement that arbitrators be lawyers. Having nonlawyer arbitrators deciding legal questions argued by attorneys can be undesirable. In addition, difficult legal questions may have to be decided by nonlawyer arbitrators.

The skill with which a judge or arbitrator conducts a hearing not only determines how quickly the hearing can be completed but also affects a party's perception of the fairness of the process. Because of the greater hearing experience possessed by judges generally, the hearing process in court will be conducted by a person with greater hearing experience. Although some arbitrators who work permanently or frequently can develop the skill to conduct good hearings, in the construction industry, hearing expertise is rarer than in other classes of arbitration. Arbitrators who decide construction disputes rarely make their living at arbitration. Not only does this result in less experienced arbitrators, but the part-time and, even more important, volunteer arbitrators must often take lengthy recesses in the hearings because of having to work at their profession or business.

One of the advantages of arbitration in

terms of hearing speed is the freedom the arbitrator has to move the hearing along unhampered by technical rules of evidence that, when accompanied by contentious attorneys, can make court hearings excessively long. The arbitrator, unlike the judge, need *not* make a transcript of the hearing, which, though helpful, is costly and time-consuming.

Looking at all aspects of this problem, arbitration is likely to be quicker than judicial dispute resolution. Speedy hearings not only should result in faster decisions but should also avoid the indirect costs of lengthy hearings, such as the unproductivity of witnesses who must attend the hearing, and direct costs, such as travel expense and attorneys' fees.

Hearing location is likely to be different. A court hearing is likely to take place at the principal city in the county where the lawsuit was commenced and whose court has jurisdiction over the dispute. This does not necessarily mean that the trial will be close to the project or convenient for witnesses. Often lawyers seek to gain tactical advantages by having the matter heard in a particular court. On the other hand, the arbitrator has more flexibility to schedule the hearing at a place that is more convenient for the witnesses. However, sometimes the arbitration clause can be deliberately designed to make it inconvenient for one of the parties to demand arbitration.

Another differentiation relates to the public nature of the hearing. Although arbitration hearings are private, the judicial process is public. Sometimes trials are reported to local newspapers. The privacy of the arbitration not only may avoid unwanted publicity but also can be a more sympathetic setting for witnesses.

One differentiation is often ignored. The courtroom and the services of judge and clerk are furnished free to the litigants. The room for the hearing and the arbitrator's fee are expenses that must be paid by the parties.

Viewing the project, often a helpful activity, is more easily accomplished in arbitration.

Quality comparisons are difficult. Arbitration has frequently been supported because of the expertise (often absent in judges) that arbitrators bring to the disputes. Doubtless, construction experience is useful. On the other hand, the dispute resolver who is *too* expert may conceive of her role not as providing a hearing and *then* making a decision but as simply deciding the dispute based upon her own knowledge. On the assumption that experience in construction is likely to produce a better decision, it is worth comparing contextual experience in the two processes.

Judges have legal training and as a rule have spent ten to twenty years practicing law before they are appointed or elected to the judiciary. This experience *may* have involved construction matters, but on the average it is likely that most judges have had little construction experience before becoming a judge. They often learn on the job. A trial judge who has been on the bench for five or ten years in a jurisdiction that has a wide range of cases may have gathered enough experience to be knowledgeable about construction matters. On the whole, trial judges do not bring great expertise to the resolution of construction disputes.

Often arbitrators possess experience in the types of matters being arbitrated before them, particularly those who arbitrate under collective bargaining agreements or who arbitrate highly specialized disputes, such as disputes between members of the stock exchange, the diamond industry, or textile trade associations. Most disputes in these industries or between members of these associations are repetitive, deal with technical matters, and are handled well through arbitration.

On the other hand, the construction industry arbitration panels, though made up of persons with experience in construction such as attorneys, architects, engineers, and contractors, do not have a sufficient volume of cases to justify full-time arbitrators or even part-time arbitrators who work with sufficient regularity to develop specialized knowledge. The construction industry, though having some elements in common, is highly specialized, and an arbitrator's experience in electrical contracting may not prepare the arbitrator to handle a dispute involving road building. On the whole, arbitrators are more likely to have had more experience in construction matters than judges have.

Another comparison of great importance is the degree of finality of the decision. A decision by a trial judge can be appealed to an appellate court. Although appeals on the whole have a low probability of success, a litigant who appeals will be successful if it can show that the trial judge has made an error of law or that factual findings were not supported by the evidence. On the other hand, for all practical purposes, an arbitrator's decision is final. To some this is a great advantage of arbitration, as it ends the dispute quickly. To others it is a disadvantage because of the immense power possessed by the arbitrator, including the power to make completely wrong decisions.

Comparing arbitration and litigation should not assume that the only alternatives are a general arbitration clause or none at all. Variations can be employed in the proper case that may be preferable to either extreme. Although the variations suggested are by no means exclusive, they do demonstrate that an arbitration clause can be "tailored" to make it preferable to either a general arbitration clause or none at all. Variations can include the following:

1. Limiting arbitration to factual disputes such as those involving technical performance standards or eliminating arbitration of other types of more "legal" disputes such as termination.

2. Specifying the place of arbitration.

3. Providing a designated person or persons as arbitrator or arbitrators.

4. Limiting arbitration to claims not exceeding a designated amount or percentage of the contract price.[96]

5. Limiting disputes to those that occur while the work is proceeding with an expedited one-person panel.

6. Permitting consolidation of separate arbitrations.

7. Providing a right to discovery.

8. Limiting the award to the most fair of the last proposals or an amount between the two final proposals of the parties.

9. Eliminating the use of attorneys.

10. Making the award "nonbinding."

Obviously, the more variables, the greater the cost of obtaining agreement and expressing it. This probably inhibits specialized arbitration clauses, at least where one party cannot dictate the clause to the other. A look at the variables and attendant complexity may persuade an exasperated drafter or negotiator that it may be simpler where possible to agree to waive a jury and have all disputes tried before the judge.

Choosing a general arbitration clause may be influenced by those who issue surety bonds and professional liability insurance. If the insurer or surety will not cover work done under a contract with a general arbitration clause, this can be a significant factor in the choice.

Whether to agree to a general arbitration clause at a particular time and place requires that the choice go beyond simply comparing abstract models of dispute resolution. For example, if the choice is between taking a dispute to an efficient and competent legal system or an unknown panel of AAA arbitrators, the former is preferable. On the other hand, an inefficient court system with questionable judgment is much

---

96. The National Society of Professional Engineers limits arbitration to claims not exceeding $200,000 in its engineer-owner agreements.

less preferable than a highly skilled arbitration system.

Many state dispute resolution systems have developed techniques that have eliminated the worst aspects of incompetence and delay that can cause parties to choose arbitration. On the other hand, sometimes arbitration can become over-judicialized and lose the advantages that arbitration was designed to achieve. For example, *Lesser Towers, Inc. v. Roscoe-Ajax Construction Co.*[97] involved arbitration of a dispute relating to the construction of a twenty-story apartment building. The arbitration took nineteen months. During that period there were 202 days of hearings and three days of oral arguments. Fifteen hundred exhibits were introduced, and the reporter's transcript was over 25,000 pages. Over $400,000 in arbitration expenses *exclusive* of attorneys' fees were incurred.

Before the arbitration hearings had even begun, each party had gone to court to obtain rulings pertaining to the arbitration. After the owner had introduced evidence to support his claim, the contractor sought to introduce evidence to support a counterclaim. At this point, the owner refused to arbitrate and went to court again to restrain the arbitrators from considering the counterclaim. Ultimately the arbitrators ruled for the owner, but the owner had to go to court again to confirm the award. While this was being done, the contractor went to federal court in an attempt to reverse the award of the arbitrator.

This case, according to the court, demonstrated that arbitration is not always a simple, expeditious, or inexpensive method of adjudicating commercial controversies. Any system will have its "horror story" cases, and perhaps the quality of the arbitrator's award could conceivably have been better than what would have resulted in court. It is unlikely, however, that arbitration in this case saved time or money.

Reference to the *Lesser Towers* case reflects another important aspect of arbitration choice. The likelihood that the parties will accept the arbitrator's decision without repeated and costly trips to the courthouse favors arbitration. This is becoming even more complicated by the expansion of jurisdiction under the U.S. Arbitration Act and the increasing likelihood that arbitration may be sought in either state or federal court and under either a state arbitration statute or under the federal act.[98] This will undoubtedly add to the complexity of legal issues and may in the long run discourage arbitration.

This section has pointed to advantages and disadvantages of arbitration. The contracting parties and their attorneys should carefully consider such factors when choosing contract language dealing with dispute resolution. The choice between a general arbitration clause, a limited arbitration clause, or no arbitration clause is an important decision that should be made with care.

SECTION 34.18
MEDIATION–ARBITRATION

Another third-party dispute resolution system that resembles arbitration but has important differences is called mediation-arbitration. This system, which has been developed in some collective bargaining contexts, places the third party in two roles. The third party seeks to mediate disputes between the parties by aiding in the negotiation process with a view toward settlement, but if parties do not settle a dispute, the third party can use normal arbitration power to decide it.

Since the third party must possess a variety of skills and would have considerable

---

97. 271 Cal.App.2d 675, 77 Cal.Rptr. 100 (1969).

98. See note 23, supra.

power, it is likely that this mediator-arbitrator would be selected in advance by the parties. For this reason, the system contemplates the use of the mediator-arbitrator throughout the performance period. Although this is technically available in ordinary arbitration, experience has shown that most disputes during performance are likely to be decided by the design professional. A mediator-arbitrator would very likely be called upon more frequently during performance and may in practice supplant the dispute resolution role of the design professional.

Mediation-arbitration is rarely used in construction disputes. One difficulty is finding a third party in whom the contracting parties have confidence and who has not only the skill but also the time to play an ongoing role during construction. Opposition can come from design professionals or construction managers who see the injection of yet another major participant as a threat to their status and as an additional complicating factor in an already complicated system.

SECTION 34.19
OTHER TECHNIQUES
TO AVOID THE COURTHOUSE

Arbitration as an alternative dispute resolution system was designed to avoid the delays, costs, and inefficiencies of the judicial system. Yet arbitration itself has developed certain defects that have caused planners to search for other techniques to handle those disputes that cannot be settled by the parties themselves. This subsection looks briefly at a few of these techniques.

(A) AAA EXPEDITED PROCEDURE

In 1984 the CI Rules created an alternative method to expedite arbitration. Section 7 requires the use of the Expedited Procedures as defined in §§ 50–54 in any claim in which the claim of any party does not exceed $15,000 exclusive of interest or arbitration costs. The parties can agree to use the Expedited Procedure in claims that exceed that amount. AAA can determine that those procedures shall not be applied to claims under $15,000.

The principal features of this process are informal notices as well as streamlined methods for selecting arbitrators and time and place of hearing. Perhaps most interesting, the hearing must be completed within one day unless the arbitrators for good cause schedule an additional hearing to be held in five days. The award must be made no later than five business days from the close of the hearing.

(B) MINI–TRIAL

The mini-trial has received attention in the popular press and is beginning to receive attention in the scholarly journals.[99] Although there can be many variations (the process being essentially a private one made by a contract), attention has been directed toward a particular mini-trial and the process it used.

After a lengthy process of negotiation, the disputants worked out an agreement for a mini-trial. The parties agreed upon a judge, or a neutral adviser. The parties were allowed an expedited discovery procedure and exchanged briefs. They were given a designated time to present their positions before executives of each disputant not *directly* connected with the dispute who have authority to settle. The hearing itself was limited to two days, moderated by the adviser. If the executives were unable to reach a settlement after the hearing, the adviser would issue a *nonbinding* opinion. The executives would meet again with the hope of settling. If they cannot, the parties can go to court, with any admissions made

---

99. For discussions of mini-trials, see Green, Marks & Olson, *Settling Large Case Litigation: An Alternative Approach*, 11 Loy.L.A.L.Rev. 493 (1978); Nilsson, *A Litigation Settling Experiment*, 65 A.B.A.J. 1818 (1979). See also *Fortune Magazine*, Feb. 26, 1979, p. 80.

or the tentative opinion of the adviser *not* admissible in any subsequent trial.

(C) REFEREES AND SPECIAL MASTERS

Attention has been directed toward the use of individuals appointed by judges (sometimes called masters or referees) who have the power to conduct hearings in disputed matters that have been brought to litigation. In California this has become known as the "rent-a-judge" system, inasmuch as the persons selected to conduct the hearings are often retired judges.

The referee's findings must stand as the findings of the court. The judge appointing the referee generally signs the judgment in accordance with the findings. This judgment, unlike an arbitration award, is subject to the same appeal as an ordinary trial court judgment. It is possible in California to designate such a referee as a temporary judge or a judge pro tem.

SECTION 34.20 PUBLIC CONTRACTS

Public construction contracts involve considerations not found in private contracts. As indicated earlier,[100] many statutes, rules, and regulations govern award of contracts, performance of contracts, and the resolution of disputes under those contracts. This treatise cannot examine the details of these legal restraints, because of both their complexity and their variations. However, the increased and often intense spotlighting of disputes in construction work necessitates some observations regarding dispute resolution in the context of public construction contracts.

(A) FEDERAL PROCUREMENT CONTRACTS

Before 1978, the disputes process in federal procurement contracts was based upon the Disputes Clause, which gave the contracting officer of the federal agency awarding the contract the power to decide disputed questions that arose during performance or thereafter. The contracting officer is usually a high administrative official of the agency awarding the contract whose decisions were conclusive unless appealed to the head of the agency within thirty days from the decision.

The agency appeals boards are appointed by the head of the agency. Their hearings and decisions are very much like those of a court. Before 1978, contractors had to go before the agency appeals boards if the dispute arose under the contract. They could appeal to the then Court of Claims if their claims were based upon a breach of contract. Alternatively, in claims under $10,000 they could appeal to a federal District Court. They could appeal from board decisions, but the board's decision was final on questions of fact if supported by substantial evidence. The Court of Claims could make its own determination on legal questions. Many federal procurement doctrines were developed that had as their objective either to keep a dispute before an agency appeals board or to allow the board to be bypassed in favor of the Court of Claims.

In 1978, Congress enacted the Federal Contract Disputes Act,[101] which gave *legislative* authorization for a disputes process that up to that time had been created solely by contract. In addition, changes were made.

Appeals can no longer be taken to the federal District Courts for claims under $10,000.

Section 605 requires contracting officers to issue decisions within sixty days on claims of $50,000 or less. Claims over that amount require a decision within sixty days or a notification of the time within which a decision will be issued. Failure to issue a

---

100. Refer to Section 9.02.

101. 41 U.S.C.A. Sections 601 et seq.

decision within the time required permits the contractor to bypass the contracting officer and go directly to higher authorities. The contractor is given sixty days from the contracting officer's decision to appeal to the agency appeals board. Section 609 allows a contractor to bring an action directly to the Court of Claims.

Congress made drastic changes in 1982 in the Federal Courts Improvement Act.[102] It created a new U.S. Court of Appeals for the Federal Circuit and a new U.S. Claims Court (to be distinguished from the earlier Court of Claims). Claims are initially presented to the Claims Court, and appellate functions go to the new Court of Appeals for the Federal Circuit.

(B) STATE AND LOCAL CONTRACTS

A detailed treatment of the many state and local laws and regulations cannot be given in this treatise. It is important, though, to recognize the importance of complying with specialized requirements for disputes under such public contracts of this type. Some states have created special courts to deal with these claims. Court decisions precluded arbitration of public contract disputes because this would place in the hands of private parties the power to decide public matters and expend public funds.[103] However, not only has hostility been fading and contracts to arbitrate future disputes upheld, but some states require the arbitration of these disputes.[104]

California's experience is interesting, and although it may not be typical, it demonstrates the continued interest and concern over the disputes resolution process. Before 1978, California state agencies followed the traditional pattern of having disputes resolved initially by a high official of the agency—the state engineer. She could issue a decision that had a substantial amount of finality, being final and conclusive unless fraudulent, capricious, arbitrary, or so grossly erroneous as to necessarily imply bad faith, a standard much like that applied to decisions by independent design professionals.

In 1977 a California appeals court decided a case that led to a drastic overhaul of the dispute resolution system. In that case the contractor appealed to the trial court after an adverse decision by the state engineer. The trial court found that errors had been committed, disregarded the decision of the state engineer, and retried the case.

The Court of Appeals concluded that the state engineer committed procedural errors,[105] but the court did not set aside the decision as had been done by the trial court but sent the claim back to the state engineer for reconsideration. In doing so, the court emphasized that the parties had "by voluntary contract" agreed that the disputes would be resolved by a designated person. This conclusion was disputed by state contractors who stated that they had no choice but to agree to such a provision.

The furor caused by this case led to an executive order and later to a statute that

---

102. 96 Stat. 25, Pub.L. 97–164. See Anthony and Smith, *The Federal Court Improvement Act of 1982*, 13 Pub. Contract L.J. 201 (1983); Miller, *The New United States Claims Court*, 32 Clev.St.L.Rev. 7 (1983–84).

103. *Pathman Construction Co. v. Knox County Hospital Association*, 164 Ind.App. 121, 326 N.E.2d 844 (1975); *E.E. Tripp Excavating Contractor, Inc. v. County of Jackson*, 60 Mich.App. 221, 230 N.W.2d 556 (1975); *City of Madison v. Frank Lloyd Wright Foundation*, 20 Wis.2d 361, 122 N.W.2d 409 (1963).

104. North Dakota requires arbitration of highway construction contract disputes. The statute was upheld in *Nelson Paving Co., Inc. v. Hjelle*, 207 N.W.2d 225 (N.D.1973). Arbitration is also required for certain public contracts in Rhode Island. See *Sterling Engineering & Construction Co. v. Town of Burrillville Housing Authority*, 108 R.I. 723, 279 A.2d 445 (1971). Likewise, Pennsylvania requires that public contracts contain arbitration clauses. See *U.S. Fidelity & Guaranty Co. v. Bangor Area Joint School Authority*, 355 F.Supp. 913 (E.D.Pa.1973).

105. *Zurn Engineers v. State Department of Water Resources*, 69 Cai.App.3d 798, 138 Cal.Rptr. 478 (1977) cert. denied 434 U.S. 985, 98 S.Ct. 612, 54 L.Ed.2d 479 (1977).

required major state procuring agencies to use a judicialized arbitration.[106]

Disputes would be decided by the persons certified as competent arbitrators by an arbitration committee composed of representatives of the state agencies and the construction industry. Those approved were placed on the State Construction Arbitration Panel. Those who wished to be placed on the panel had to submit information that indicated their education and experience and set a rate at which they were willing to serve. The disputants selected from this certified list.

Discovery is required. The arbitrator can utilize expert technical or legal advisers, depending on whether the arbitrator is an attorney or a technically trained person. Consolidation and joinder are permitted. Hearings are open to the public. The award must contain findings of fact, a summary of the evidence, and reasons underlying the award as well as conclusions of law. The award can be vacated if it is not supported by substantial evidence or based upon an error of law.

SECTION 34.21
INTERNATIONAL ARBITRATION

Earlier mention was made of the difference between domestic construction contracts and those that involve parties who are nationals of different countries, particularly contracts made by American construction companies requiring that they build projects in a foreign country.[107] Transactions of this type generate some issues of minimal or no importance in domestic contracts.

The contracts themselves may be expressed in more than one language, often generating problems that result from imprecise translation. They may also involve payment in currency that varies greatly in value. Contractors in such transactions may often have to deal with tight and often changing laws relating to export of profits and import of personnel and material. Perhaps most important, neither party may trust the legal system of the other, and the contractor may believe that it will not obtain an impartial hearing if it is forced to bring its disputes to courts in the foreign country, particularly if the owner is, as is so common in lesser-developed countries, an instrumentality of the government.

Contractors making these contracts commonly insist upon international arbitration to resolve disputes. Such arbitrations are usually held in neutral countries or in centers of respected commercial arbitration. Because of the complexities generated by international arbitration, this treatise does not discuss them.[108]

PROBLEMS

1. The arbitration clause in a contract between O and C provided that the parties agree to arbitrate all disputes under this contract.

C refused to perform under the contract, because she claimed that O defrauded her into making the contract by telling her that he would award two other construction contracts to C when he knew he had already awarded them to another contractor. C claimed that she relied on the promise by making a much lower bid than she would have ordinarily made.

---

106. See West's Ann.Cal.Pub.Contract Code, §§ 10240 et seq. See also West's Ann.Cal.Civ.Code § 1670, which requires local entities to either arbitrate or litigate. Employees of the agency cannot, as before, decide the dispute.

107. Refer to Sections 2.15 and 9.06.

108. See DeVries, *International Commercial Arbitration: A Contractual Substitute for National Courts,* 57 Tul.L. Rev. 42 (1982).

O denied making the promise and demanded arbitration. C refused to arbitrate and started a court action for fraud. O insists upon arbitration. Must this dispute be resolved by arbitration? Who decides?

2. O and C entered into a construction contract that contained a general agreement to arbitrate all future disputes. A dispute arose. O terminated the contract. C brought legal action. C sought and used the discovery process permitted by state law and examined three of O's officials under oath. Shortly after the depositions were completed, C demanded arbitration. O does not wish to arbitrate. Should O be compelled to arbitrate?

3. O and A had a dispute relating to what compensation O is required to pay for A's architectural services. A has demanded $10,000, while O has offered to pay $5,000. Under state law, all claims for under $15,000 must go to an arbitrator appointed by the court whose decision is *not* binding on the parties. A has suggested to O that their dispute be handled under the American Arbitration Association Expedited Procedures set forth in the CI Rules in Appendix F. If you were O, would you prefer to arbitrate or litigate?

35

Construction-Related Accidents: Interface Between Tort Law and Workers' Compensation

SECTION 35.01 AN OVERVIEW

This chapter deals with personal harm suffered that can be traced to the Construction Process. Mostly it deals with accidents that occur during construction, since construction is a high-risk activity. But reference is made in Section 35.09 to material dealing with persons who suffer physical harm *after* the project has been completed. Except for injuries to workers in the scope of their employment, which are accidents that fall within the jurisdiction of workers' compensation laws, tort law dominates construction accidents.

Many of the doctrines discussed in this chapter were introduced in Chapter 7, which laid the groundwork for this chapter. This chapter spotlights the construction process and applies the general principles described in Chapter 7.

Chapter 21, which described the various ways in which construction projects are organized, also bears heavily on this chapter. That chapter noted the attributes of the traditional system of construction, such as divided design and construction, prime contractor coordination and organization, and design professional review of project progress. Much of the chapter noted modern variations on the traditional methods that bear heavily upon the three principal issues described in this chapter:

1. *Statutory violations* and their effect upon liability.
2. *Vicarious liability*, the liability of one party for the failure of another to perform as required by tort law.
3. *Strict liability*, imposed on a party solely because of its status, as owner, as "person in charge," or as statutory employer.

These issues all involve control, the power to determine what is to be done, and how it is to be done. In a traditional construction organization, the owner turns over the design and site to a prime contractor and allows the prime contractor to control the site, select and coordinate subcontractors, police their work, and act as a conduit or buffer between owner and subcontractors. In this model, it is the prime contractor who has charge of the Construction Process, even though the owner, through the design professional, may reserve the right to monitor the work as it proceeds. This control is particularly emphasized in matters that relate to construction technique and the proliferating legal controls designed to protect workers and others from physical harm during construction and, on occasion, afterward.

Other methods of organization, such as separate contracts (multiple primes), coordinated and monitored by the owner through a construction manager, and design/build, affect this central assumption as to control. In the former, the owner has taken a more active role, effectively replacing the prime as the party in control. In design/build, a *sophisticated* owner may still, through its infrastructure, have either control or the right to control the process. At the other extreme—design/build—the owner may not even monitor the work through a design professional as in the traditional system, simply turning over everything to the design/build contractor.

The latter differentiation in design/build also makes it important to review Chapter 9, which classifies owners. Liability may depend on whether the owner is a large, sophisticated organization with its engineering, insurance, and legal departments that totally dominate the project or an unsophisticated owner, inexperienced in construction, that puts its trust and money in the hands of others, such as the owner's retaining an architect to design and a contractor to build or a contractor who does both. The sophisticated owner has not only a big economic stake in the project but also the capacity to protect itself and actively control the process. The inexperienced owner is often a passive participant. This is demonstrated by recent developments in New York. Section 241 of the New York Labor Law requires all owners and contractors to provide a safe workplace. This statute was the basis for holding owners strictly liable for injuries to workers.[1] In 1980, this statute was amended to exempt owners who have one- or two-family homes built for them if they do not direct or supervise construction. (Would the use of an architect deny an owner this exemption?)

The New York Labor Law was drafted originally not to place strict tort liability on *all* owners for injury to those working on their projects. However, a court reading the statute literally saw no way to avoid this result. Although it is possible to justify such a result even in the case of an unsophisticated owner by pointing to the possibility of obtaining insurance, it is clear that the legislature in 1980 chose not to place absolute liability on such owners.

This chapter must be read in light of the observations made in Chapter 7 that tort law emphasizes compensating those who suffer physical harm with less emphasis upon the need to establish moral culpability. To be sure, not all claimants recover from all those against whom they assert claims. As a rule, however, if workers' compensation recovery is taken into account, claimants will recover from someone and will rarely go totally uncompensated. The principal issue is whether they can go beyond workers' compensation to receive more generous tort awards.

SECTION 35.02
WORKER CLAIMS: WORKERS' COMPENSATION LEGISLATION

Although tort law dominates this chapter, it is important to see workers' compensation laws (described in Section 7.04(C)) as a backdrop to worker claims based upon tort law. First, workers' compensation usually provides a quick, sure recovery for a worker injured in the course of employment but one that does not provide as much compensation as a tort recovery. That factor is often influential in tort claims. Some judges view this as a reason for expanded tort recovery,[2] while others regard a workers' compensation recovery as adequate.[3] Second, workers' compensation law, with some

---

1. *Allen v. Cloutier Construction Corp.*, 44 N.Y.2d 290, 405 N.Y.S.2d 630, 376 N.E.2d 1276 (1978).

2. *Funk v. General Motors Corp.*, 392 Mich. 91, 220 N.W.2d 641 (1974) discussed in Section 35.05(B).

3. *West v. Guy F. Atkinson Construction Co.*, 251 Cal.App.2d 296, 59 Cal.Rptr. 286 (1967).

exceptions, immunizes the injured worker's employer from any tort liability. Some states extend this to co-employees of the injured worker, while a few immunize those in a common employment.[4] Because of the many entities involved in the Construction Process, immunity raises particularly difficult problems in construction accidents, both because of the uncertainty as to the extent of immunity[5] and the anomalous result of the entity most directly connected to the accident not being a defendant in the tort claim. (As Section 36.03(D) shows, the employer is often a cross-defendant through indemnification.)

The frequent creation of statutory employers in workers' compensation laws complicates the immunity issue. Many states recognize the fragmentation and financial instability of the construction industry and seek to make a party farther up the employer chain from the injured worker the statutory employer of the injured worker. The statutory employer, though not in reality employing the worker, is considered the worker's employer for workers' compensation purposes. This increases the likelihood that an injured worker will receive a compensation award. The creation of statutory employers was designed to enlarge the injured worker's right to a compensation award, with third-party claims (tort claims against all those not immune) not being common at that time. The proliferation of third-party claims has generated assertions by a party who might otherwise be liable in a tort claim that it is a statutory employer and entitled to the same immunity as the actual employer.

Predictably, in light of the variant statutory language and often different attitudes toward these third-party claims, different results occur when the issues are whether the owner is a statutory employer and whether a statutory employer receives immunity.[6]

SECTION 35.03
PRIVITY AND DUTY

Before determining whether the defendant has lived up to the standard of conduct required by tort law, two preliminary concepts, though diminishing in importance, must be examined.

The tort law requirement that persons conduct themselves in a particular way does not extend its protection in terms of liability into infinity. To avoid potentially crushing responsibility that can have an adverse effect on professional, commercial, and industrial activities, lines must be drawn to limit liability. Two concepts that can draw such a line relate to the relationship between plaintiff and defendant. One, called duty, asks whether the defendant owes a duty *to the plaintiff* to conduct itself in accordance with the tort standard of conduct. Another way of expressing this is to ask whether there is a privity between plaintiff and defendant. Usually this means the existence of a contract between the two. It also can encompass other close relationships which, though perhaps contractual in

---

4. *Bergen v. Fourth Skyline Corp.,* 501 F.2d 1174 (4th Cir.1974); *Afienko v. Harvard Club of Boston,* 365 Mass. 320, 312 N.E.2d 196 (1974). Oklahoma recently immunized design professionals from worker tort claims based upon site activities. West's Okla.Stat.Ann., tit. 85, § 12.

5. As to exceptions see Note, *Exceptions to the Exclusive Remedy Requirements of Workers' Compensation Statutes,* 96 Harv.L.Rev. 1641, 1648 (1983), which states exceptions to be dual capacity, parent-sibling corporation, intentional torts, and equitable indemnity. California dealt with this by statute, expanding some and eliminating others. See West's Ann.Cal.Lab.Code § 3602. For its handling of the now thousands of asbestosis cases, see *Johns-Manville Products Corp. v. Contra Costa Superior Court,* 27 Cal.3d 465, 165 Cal.Rptr. 858, 612 P.2d 948 (1980) noted in 33 Hastings L.J. 263 (1981).

6. Compare *Hattersley v. Bollt,* 512 F.2d 209 (3d Cir.1975) (immunity), with *Laffoon v. Bell & Zoller Coal Co.,* 65 Ill.2d 437, 3 Ill.Dec. 715, 359 N.E.2d 125 (1976) (immunity denied).

a sense, are created more by status than by contract. An illustration of such status relationships are employer and employee, parent and child, hotel keeper and guest, or common carrier and passenger. The important element is whether one person has a duty to avoid exposing the other to an unreasonable risk of harm.

Drawing lines based upon relationship and requiring either privity or a duty once had great importance in protecting defendants. Much of its effectiveness resulted from the existence of privity or a duty being a legal question to be decided by judge and not by jury.

As seen in Section 7.09, dealing with liability for defective products, and to a somewhat lesser degree in Section 7.07, dealing with misrepresentation, these limiting factors have lost much of their effectiveness, yet they still play a role in construction accidents.

The objective of compensating accident victims is often accomplished by allowing a wide range of persons who *may* be held accountable. The availability of insurance and the desire that enterprises should bear their legitimate responsibility for losses they cause have virtually eliminated privity and duty as barriers to liability where the plaintiff seeks compensation for personal harm. Elimination of these barriers when harm was caused by defective products began in 1916.[7] In construction accidents the barriers began to fall in the 1950s.[8] Court decisions have either minimized or largely eliminated the privity requirement in cases of personal harm. Similarly, any contention that the defendant owed no duty to the person suffering physical harm has had only limited success.

As a result, the worker or visitor injured during construction generally is given the chance to bring legal action against most direct participants in the Construction Process. For example, an injured worker will very likely be able to bring legal action against any contractor other than his own employer or statutory employer,[9] the owner,[10] any design professional,[11] a distributor or manufacturer of defective products,[12] possibly the lender,[13] and, where immunity does not exist, public officials whose negligent conduct may have played an important part in causing the injury.[14]

Still, the privity doctrine has had some success. For example, provisions frequently contained in policies that insure a contractor for public liability or workers' compensation give the insurer the right to inspect premises and activities of its insured as part of its loss-prevention system. Work-

---

7. *MacPherson v. Buick Motor Co.*, 217 N.Y. 382, 111 N.E. 1050 (1916).

8. *Hanna v. Fletcher*, 231 F.2d 469 (D.C.Cir.1956), certiorari denied 359 U.S. 912 (1956); *Dow v. Holly Manufacturing Co.*, 49 Cal.2d 720, 321 P.2d 736 (1958); *Inman v. Binghamton Housing Authority*, 3 N.Y.2d 137, 164 N.Y.S.2d 699, 143 N.E.2d 895 (1957); *Macomber v. Cox*, 249 Or. 61, 435 P.2d 462 (1967).

9. *Schroeder v. C.F. Braun & Co.*, 502 F.2d 235 (7th Cir.1974).

10. *Associated Engineers, Inc. v. Job*, 370 F.2d 633 (8th Cir.1966).

11. *Miller v. DeWitt*, 37 Ill.2d 273, 226 N.E.2d 630 (1967); *Evans v. Howard R. Green Co.*, 231 N.W.2d 907 (Iowa 1975). But as noted in Section 17.09(B), a differentiation is made between design and site services which, in some jurisdictions, has led to the reimposition of something like a privity defense.

12. *Morgan v. Stubblefield*, 6 Cal.3d 606, 100 Cal.Rptr. 1, 493 P.2d 465 (1972).

13. *Connor v. Great Western Savings & Loan Association*, 69 Cal.2d 850, 73 Cal.Rptr. 369, 447 P.2d 609 (1968) (property damage to tract home).

14. *Holman v. State*, 53 Cal.App.3d 317, 124 Cal.Rptr. 773 (1975). In *Morris v. County of Marin*, 18 Cal.3d 901, 136 Cal.Rptr. 251, 559 P.2d 606 (1977), an injured worker was allowed to bring legal action against the county that in violation of state law had issued a building permit to the worker's employer without requiring a workers' compensation certificate be filed.

ers have sought, with spotty success, to re-cover against these insurance companies if they can establish that the insurance company owed them a duty because the insurer *undertook* to perform a duty owed workers by their employer, the insured under the policy. If a duty *is* found because the insurer undertook the duty of its insured, the workers can show the insurer did not conduct an inspection when it should have or did not take necessary steps after having conducted one. One writer, after discussing the cases and legislation dealing with this problem, outlined several key factors as summarized in the following paragraphs.

> Focusing first on the reaction of the insured to the inspection, reliance and duty are more likely to be found when the insured discontinued his safety program after the insurer began inspecting. The same result is probable if the insured did not completely discontinue his program but simply reduced it by a sufficient degree. The reaction of the insured is the most important factor in determining the applicability of the reliance theory.

> Receiving equal attention will be the conduct of the insurer. An initial factor will be the purpose of the inspection. If the inspection is solely for internal purposes and on a limited basis, a duty is less likely to be found. Another important element will be whether open and obvious conditions were inspected. If so, a court might restrict liability to harm following from those conditions.

> A final and perhaps most important consideration deals with the character of the inspection. If the insurer makes representations as to the efficiency of its program, coupled with high safety credentials of its inspectors, liability will be more likely. The authority to enforce recommendations will also help establish the existence of a duty. By the same token, the absence of

these considerations will also lead to a finding of no duty.[15]

Those who enter another's land are not always entitled to insist that the possessor of land make the land reasonably safe for them. Here the relationship between the person who enters the land and the one who possesses (controls) it may *limit* the duty by requiring *less* of the possessor of land than is required of everyone else.

One complication of limited duty is the difficulty of determining the status of the construction worker who works on the site or at a plant while performing construction work. Some cases hold that the worker is an invitee entitled to a warning or to have the premises made reasonably safe for him.[16] Others conclude that the worker is a licensee with less protection, mainly to be warned of concealed dangers.[17] As jurisdictions begin to move toward one standard of conduct, these classification problems should be less important.[18]

The limited duty sometimes placed upon the possessor of land requires a preliminary determination of *who* is the possessor of land. The owner, before construction, is responsible for the condition of the land. As has been seen, that obligation may not be delegable.[19] In a traditional construction project—that is, where the owner turns over the site along with construction documents to the contractor and asks for a particular result—the owner's reservation of certain rights during the construction phase does not preclude a conclusion that the contractor has taken possession of the land. This transfer of responsibility is permitted

---

15. Comment, *An Insurer's Liability to Third Parties for Negligent Inspection,* 66 Ky.L.J. 910, 922–3 (1978).

16. *Hogge v. United States,* 354 F.Supp. 429 (E.D.Va.1972) (as to portions of site where he was authorized to go); *Elder v. Pacific Telephone & Telegraph Co.,* 66 Cal.App.3d 650, 136 Cal.Rptr. 203 (1977) (but no duty to sub's employee where prime contractor not "operatively present"); *Crotty v. Reading Industries,* 237 Pa.Super. 1, 345 A.2d 259 (1975) (over strong dissent); *Ferguson v. R. E. Ball & Co.,* 153 W.Va. 882, 173 S.E.2d 83 (1970) (dictum).

17. *Nagler v. United States Steel Corp.,* 486 F.2d 794 (7th Cir.1973); *Epperly v. City of Seattle,* 65 Wash.2d 777, 399 P.2d 591 (1965); *Daniel Construction Co. v. Holden,* 266 Ark. 43, 585 S.W.2d 6 (1979).

18. Refer to Section 7.08(G).

19. Refer to Section 7.08(H). See also *Weber v. Northern Illinois Gas Co.,* 10 Ill.App.3d 625, 295 N.E.2d 41 (1973).

only if the contractor is an independent contractor, (discussed in Section 35.05(B)) or if one of the many exceptions to the independent contractor rule does not apply. By and large, the question of who possesses the land for these purposes is determined by the resolution of these issues.

Another duty problem that arises in construction accidents relates to the employee-employer relationship. Frequently the law requires that an employer furnish a safe work place or safe tools for employees. Because of the multitude of entities present on a construction project, it is often difficult to determine *who* is the employer for these purposes in both traditional and nontraditional contracting methods. Here the duty concept can be invoked to determine whether a particular person responsible for furnishing a safe place of employment or tools has a duty other than to its *own* employees. For example, in the traditional process, there are prime contractors and various tiers of subcontractors. Does the prime contractor owe the employees of a subcontractor the obligation of furnishing safe tools or work place?[20] In a nontraditional system with more owner control, does primary responsibility for these matters fall upon the owner?[21] In the latter system, does the prime contractor owe a duty to employees of the owner?[22]

These questions *could* be resolved by the statutory language. However, often these statutes are drafted without construction project organization in mind. But like the question of who possesses the land, the is-sue of who is an employer for these purposes is often dominated by a determination of whether the prime contractor or subcontractor is an independent contractor. For example, suppose a prime contractor's employee is injured because of the employer's statutory violation as to work conditions. If that worker *can* bring an action against the owner, can it be based upon the statutory violation? Similar problems can arise if the injured worker is an employee of a subcontractor or even of the owner. These questions will be discussed in Section 35.05. Where the issue is faced directly, such as primary responsibility for compliance with the Occupational Safety and Health Act (OSHA), it is likely that the prime contractor will be given primary responsibility.[23] The prime contractor is the participant to whom possession is given and the participant to whom the owner looks for overall safety responsibility. This is often evidenced by prime contract provisions placing this responsibility upon the prime contractor. Where the owner in a nontraditional process has more overall control and operative presence, the principal responsibility may rest upon it.[24] It may, as seen in Chapter 36, attempt to shift any loss it suffers by obtaining indemnification from the prime contractor. This does not divest it from responsibility to those injured.

Sometimes the existence of a duty is determined by the contract that the defendant has made. Do its provisions indicate that the defendant owed a duty to the per-

20. *Caswell v. Lynch*, 23 Cal.App.3d 87, 99 Cal.Rptr. 880 (1972) held that it did. See also *Macomber v. Cox*, supra note 8, in which the dissenting judge viewed the prime contractor not as a builder but as an assembler and overseer of subcontractors and the participant most visible and responsible for the overall safety of the employees.

21. *United States v. English*, 521 F.2d 63 (9th Cir.1975) (owner an employer under California Labor Code as it controlled the premises).

22. *Valdez v. J. D. Diffenbaugh Co.*, 51 Cal.App.3d 494, 124 Cal.Rptr. 467 (1975) (contractor owed duty to owner's employee).

23. 29 U.S.C.A. § 651. For the view that the prime contractor should have primary responsibility, see *Funk v. General Motors Corp.*, 392 Mich. 91, 220 N.W.2d 641 (1974).

24. *Funk v. General Motors Corp.*, supra note 23.

son injured? For example, frequently prime contracts contain provisions placing worker safety in the hands of the prime contractor. Suppose a subcontractor's employee is injured because of a failure of the prime contractor to comply with these provisions. The contract provisions may create a duty to the injured worker if made for his benefit. Courts have reached different conclusions.[25]

Similarly, sometimes the contract is used to *deny* any duty to the injured worker. Professional societies of architects and engineers have included language in their standardized construction contracts seeking to *negate* any duty on the part of the design professional to monitor work methods. This is done to shield the design professional from any liability for accidents caused workers that are principally due to improper construction methods (covered in Section 17.09(E)).

SECTION 35.04
STANDARD OF CONDUCT

(A) NEGLIGENCE
AND STRICT LIABILITY

The negligence standard, that of avoiding unreasonable risk of harm to others, dominates construction accidents.[26] A few exceptions exist, however. Claimants against those who have made or supplied defective products need not establish negligence. On the whole, construction is considered to be furnishing services rather than supplying goods. Yet even there, there are some exceptions to the usual requirement that negligence be established. Certain construction activities can be classified as abnormally dangerous or ultrahazardous, dispensing with the negligence requirement. A few states have held developer-builders to a standard of strict liability.[27] Some states place strict liability on the party in charge of the work[28] or on owners and contractors.[29] For the most part, claims for physical harm related to construction require proof of negligence.

(B) NEGLIGENCE
FORMULAS: INADVERTENCE
AND CONSCIOUS CHOICE

Whether a person has exposed another to the unreasonable risk of harm requires an evaluation of the magnitude of the risk, the utility of the conduct, and the burden of eliminating the risk. Some negligent acts are simple inadvertence where the actors either forget to do what they know they should do or do it carelessly. For example, cases have involved acts of carelessly covering a dangerous hole[30] and placing a ladder where it could cause injury.[31]

---

25. Compare *J. D. Williams v. Fenix & Scisson, Inc.*, 608 F.2d 1205 (9th Cir.1979) (not intended beneficiary of owner-engineer contract), and *West v. Morrison-Knudsen Co.*, 451 F.2d 493 (9th Cir.1971) (not intended beneficiary of owner-contractor contract), with *Dunn v. Brown & Root, Inc.*, 455 F.2d 717 (8th Cir.1972) (intended beneficiary of prime contract).

26. *Belcher v. Nevada Rock & Sand Co.*, 516 F.2d 859 (9th Cir.1975) (rejected strict liability in action against contractor); *La Rossa v. Scientific Design Co.*, 402 F.2d 937 (3d Cir.1968) (rejected strict liability in action against design professional).

27. *Schipper v. Levitt & Sons, Inc.*, 44 N.J. 70, 207 A.2d 314 (1965). But see *Macomber v. Cox*, supra note 8. For a general discussion, see Comment, *Strict Liability to the Builder-Vendor of Homes: Schipper and Beyond?*, 10 Ohio N.U.L.Rev. 103 (1983).

28. Ill.Stat.Ann. ch. 48, ¶¶ 68, 69.

29. Refer to Section 35.01.

30. *E.L. Jones Construction Co. v. Noland*, 105 Ariz. 446, 466 P.2d 740 (1970).

31. *Hill v. Lundin & Associates, Inc.*, 260 La. 542, 256 So.2d 620 (1972). The contractor was relieved in this case because a third party moving the ladder operated as a superseding cause.

Negligence can be based upon the actor's having exercised a conscious choice, with the issue being whether that choice was reasonable under the circumstances. *Stanley v. United States* [32] involved a fatal fall by a painting subcontractor's employee that resulted in an action against the U.S. owner. The worker's widow contended that a Navy antenna tower from which the worker had fallen had been defectively designed. The court described the tower as:

> The tower in question, No. S–O, is an elongated triangular prism, its three legs arising 979 feet from the apexes of a triangular base 12 feet on a side. The sides are skeletal, with a horizontal beam about every 12 feet, with a single diagonal cross of tie-rods, or windrods, so called, in the space between. The inside is empty, except that every 70 feet there is a triangular grating, or platform, 12' × 12' × 12' with a 2' × 3' rectangular aperture, or cut-out, to afford access to a ladder which runs down the side to the platform next beneath. These apertures are staggered from platform to platform, as the ladder shifts from one side of the tower to another as one progresses from one platform to the next. Although there was no testimony on the subject, this seems an obvious safety feature, so that the platform at the base of the ascending ladder would be solid. The effect of this, of course, is to require everyone ascending or descending the tower to dismount the ladder and cross over at each platform. The principal purpose of the platform, apart from providing strength to the structure, is to serve as a place of rest. There are no guardrails around the cut-out, and none around the outside of the platform. [33]

Stanley was twenty years old and without experience in high work. Because of his inexperience he was assigned to work on the platforms, which were thought to be safer. At the time of the accident, he was using a paint roller with a six-foot rod that required both hands. He seemed to be having no difficulty but shortly thereafter fell 700 feet and died of a crushed skull.

Considerable evidence existed of negli-

gence on the part of the deceased worker's employer in permitting the worker to work without a safety belt. Negligence against the government was based upon a failure to supply guardrails to the ladder apertures or cutouts. To establish this, an expert witness testified for the plaintiff that good safety practice called for guardrails around the apertures. The expert conceded that the apertures were obvious dangers, although he stated that people engaged in painting might have their attention diverted and forget the existence of the apertures. The expert conceded that the absence of a guardrail did not constitute a danger to persons climbing the tower who could be expected to pause on the platform only for a rest and would not be as likely to be distracted from the existence of the aperture.

The government safety expert testified that guardrails would be dangerous to those servicing the antenna and cables who would be carrying equipment and who might hang the equipment up on any rail. In evaluating testimony of the design choice, the court stated:

> The platform, without rails, was a dangerous place for a painter, particularly an inexperienced painter, to work. It was perhaps more dangerous than such a painter might appreciate. On the other hand, as to all other persons, who manifestly use the tower far more frequently, and who would have to traverse apertures 26 times for a single assent, guardrails were not only not needed, but were to some degree contra-indicated. Furthermore, the special danger applicable to painters could, as the court found, be obviated by the use of safety belts with a tag line attachment. [34]

In reversing a judgment against the United States, the court stressed that the design choice had been reasonable in light of the circumstances.

The design choice arose in a different context in *George v. Morgan Construction*

---

32. 476 F.2d 606 (1st Cir.1973).

33. Id. at 607.

34. Id. at 609.

Co.[35] A worker in a steel plant was injured when a hot cobble was ejected from a rolling mill machine. Cobbles are ejected because the steel is defective or because of operational errors by the steel company. The worker recovered a workers' compensation award and then brought legal action against the designer and builder of the machine.

The worker contended that the injury could have been avoided by installation of cobblescreens as a protective guard. However, the machine designer contended that a cobblescreen would have blocked the view of employees controlling the mill from a glass enclosed pulpit twenty feet above the floor that housed operational controls. But, stated the court:

> The mill could have been so designed so that the cobblescreens could have been erected without blocking the pulpit, by putting the pulpit on the drive side of the mill instead of the working side. . . .

The judge then quoted Hume as having stated:

> . . . did I show you a house or palace, where there was not one apartment convenient or agreeable; where the windows, doors, fires, passages, stairs, and the whole economy of the building were the source of noise, confusion, fatigue, darkness, and extremes of heat and cold; you would certainly blame the contrivance, without any further examination. The architect would in vain display his subtlety, and prove to you that if this door or that window were altered, greater ills would ensue. What he says, may be strictly true: the alteration of one particular, while the other parts of the building remain, may only augment the inconveniences. But still you would assert in general, that if the

architect had had skill and good intentions, he might have formed such a plan of the whole, and might have adjusted the parts in such a manner, as would have remedied all or most of these inconveniences.

A good architect would have designed the house in such a way as to avoid these disadvantages, so that one would not have to choose between a design that was bad and one that was worse. And in the same way, a non-negligent Construction Company would have designed the steel mill in such a way as to avoid requiring the Steel Company to choose between the Scylla of no cobblescreens and the Charybdis of an obstructed view from the pulpit.[36]

(C) COMMON PRACTICE: CUSTOM

Deviation from and conformity to common practice are admissible but not conclusive on the question of negligence.[37] Such evidence can assist the finder of fact to determine what was feasible and what the defendant knew or should have known.[38]

(D) RES IPSA LOQUITUR

This concept permits negligence to be shown by circumstantial evidence, evidence of the accident coupled with inferences that the cause of the accident is more likely than not to have been the negligence of the defendant. The principal use of this doctrine is to provide sufficient evidence to submit the question of negligence to the jury.

The doctrine has uneven application in construction litigation mainly because of the many participants who were or may have been working on the project at the time the negligence occurred. As a result, it is often difficult to connect the negligence to a particular participant.

---

35. 389 F.Supp. 253 (E.D.Pa.1975).

36. Id. at 257. The Hume quotation was taken from *Dialogues Concerning Natural Religion*, Part XI, Norman Kemp Smith edition 204 (1935).

37. *Fireman's Fund Insurance Co. v. AALCO Wrecking Co.*, 466 F.2d 179 (8th Cir.1972); *George v. Morgan Construction Co.*, supra note 35; *Rivera v. Rockford Machine & Tool Co.*, 1 Ill.App.3d 641, 274 N.E.2d 828 (1971). In judging professional conduct the common standard can be the legal standard. See § 17.05.

38. *Grant v. Joseph J. Duffy Co.*, 20 Ill.App.3d 669, 314 N.E.2d 478 (1974). The industry standards considered were the American Safety Code for Building Construction, the State of Illinois Health and Safety Act, and Manual of Accident Prevention of the Associated General Contractors.

New York refused to apply the doctrine where there had been an injury from falling bricks where nineteen contractors had been working on the site.[39] Similarly, Wisconsin refused to apply the doctrine where a stone fell, injuring a subcontractor's employee.[40] The work was being performed on a state-owned building. Refusal was predicated upon the failure to show that the contractor had exclusive right to control the work since the architect and the state's representative also had control rights.

That the architect and the representative of the owner had certain powers during construction should not have defeated the plaintiff's attempt to use *res ipsa loquitur.* However, had other employees of the state been working in the area from which the stone had fallen, it would have been difficult to infer that the accident had been caused by negligence of the contractor.

The uncertainty of the application of this doctrine in construction work can be demonstrated by contrary conclusions in two cases, each of which involved injury to a worker while working at the site of a single-family home and defective stairs.[41] Yet there are cases where it has been successfully employed by the plaintiff.[42]

As tort law continues to emphasize compensating accident victims, *res ipsa loquitur* may be used more frequently in construction accidents. This is more likely to be the case in smaller construction projects or construction projects largely accomplished by one contractor. As the project becomes more complex, it will be difficult to connect the accident with the negligence of any particular participant.

(E) PRIOR ACCIDENTS

Sometimes it is necessary for the plaintiff to establish that the defendant knew of a dangerous condition and continued to allow it to exist. Evidence of this can be established by testimony of prior accidents of a reasonably similar nature.[43]

(F) DEPARTURES FROM AND CONFORMITY WITH STATUTORY STANDARDS

The Construction Process is honeycombed with laws regulating worker safety. The effect of statutes generally upon liability was discussed generally in Section 7.03(C). This section examines this important question within the context of the Construction Process.

Some statutes determine liability directly. For example, workers' compensation laws determine the employer's liability for employment accidents.[44] Similarly, as noted in Section 35.04(A), some statutes create strict liability. For example, the Illinois Structural Work Act, known as the Scaffold Act, defines the standards and places liability upon the party "in charge" who willfully

---

39. *Wolf v. American Society Tract Soc.*, 164 N.Y. 30, 58 N.E. 31 (1900). Similarly, the doctrine was not used in *Hardie v. Charles P. Boland Co.*, 205 N.Y. 336, 98 N.E. 661 (1912), where a chimney fell and the cause could have been either the contractor or the architect. But see *Schroeder v. City & County Savings Bank*, 293 N.Y. 370, 57 N.E.2d 57 (1944) (scaffold accident in which the doctrine was applied against the contractor who had erected the scaffold and another who was working nearby at the time of the accident).

40. *Goebel v. General Building Service Co.*, 26 Wis.2d 129, 131 N.W.2d 852 (1965).

41. *Moore v. Denune & Pipic, Inc.*, 26 Ohio St.2d 125, 269 N.E.2d 599 (1971), held that it could not be used, while by dictum *Ferguson v. R. E. Ball & Co.*, supra note 16, stated that it could.

42. *Smith v. General Paving Co.*, 24 Ill.App.3d 858, 321 N.E.2d 689 (1974) (roadwork that required drains be kept free followed by flooding); *Bedford v. Re*, 9 Cal.3d 593, 108 Cal.Rptr. 364, 510 P.2d 724 (1973) (decorative wall collapsed).

43. *Grant v. Joseph J. Duffy Co.*, supra note 38.

44. Refer to Section 7.04(C).

violates the statute.[45] Similarly, the New York Labor Law places strict liability upon the owner and the contractor.[46] Such direct statutory liability is relatively rare. Even the Federal Occupational Safety and Health Act of 1970 (OSHA), though relevant on the negligence issue, does not *grant* a civil remedy to employees against violators of the statute.[47] This statute, as most safety statutes, is designed to impose public law liability upon violators and a public law duty to avoid harm to workers.

Sometimes statutes state a *general* standard of conduct (such as requiring a safe place) owed by one participant, usually the contractor or owner, to other participants, usually workers or visitors to the site. More important are statutes that indirectly but significantly affect liability by providing a *detailed* standard that plays an important role in determining liability. To have any significant effect on negligence, the statute must have been designed to protect the person injured and must have dealt with the type of risk that caused the harm. Though

less important, the violation must not have been excused.

Statutory safety rules are generally for the protection of those who work on the site. For example, the Illinois Structural Work Act is designed to protect construction workers but not employees on the maintenance staff of the owner.[48] Courts on occasion have drawn an unreasonably narrow line in protection matters. For example, one court held that the statute requiring an employer to provide a safe place for its employees protected only employees of the employer and not those of an independent contractor hired to do work in the employer's plant.[49]

California, on the other hand, has given such statutes broad scope. For example, California has held that safety regulations imposed upon a prime contractor were for the benefit of employees of a subcontractor,[50] for the benefit of a physician who entered the site to perform emergency medical treatment for two injured workers,[51] and for the protection of a sixteen-year-old

---

45. Supra note 28. Illinois has struggled with the issue of who is in charge. Compare *McGovern v. Standish*, 65 Ill.2d 54, 2 Ill.Dec. 691, 357 N.E.2d 1134 (1976) (architect not in charge), with *Emberton v. State Farm Mutual Automobile Insurance Co.*, 71 Ill.2d 111, 15 Ill.Dec. 664, 373 N.E.2d 1348 (1978) (architect in charge).

46. New York has also struggled with the issue of who is strictly liable because of the statute. *Allen v. Cloutier Construction Corp.*, supra note 1, held an owner strictly liable even if the owner did not control or direct work performed by subcontractor that caused the harm. In 1980, the statute was amended to exempt owners having one- or two-family homes built if they do not direct or supervise construction. As to other participants, see *Russin v. Louis N. Picciano & Son*, 54 N.Y.2d 311, 445 N.Y.S.2d 127, 429 N.E.2d 805 (1981) (separate contractor not liable unless given authority to supervise and control portion of work where accident occurred); *Kenny v. Fuller*, 87 A.D.2d 183, 450 N.Y.S.2d 551 (1982) (construction manager *both* contractor and agent of owner); *Carollo v. Tishman Construction & Research Co.*, 109 Misc.2d 506, 440 N.Y.S.2d 437 (1981) (rejected construction manager's argument that it was merely an expediter and held it to be a contractor). However, CMs have been able to protect themselves by indemnification agreements with the subcontractors (*Carollo v. Tishman Construction & Research Co.*, supra, and *Kenny v. Fuller*, supra).

47. *Jeter v. St. Regis Paper Co.*, 507 F.2d 973 (5th Cir.1975); *Russell v. Bartley*, 494 F.2d 334 (6th Cir.1974). *Tenenbaum v. City of Chicago*, 60 Ill.2d 363, 325 N.E.2d 607 (1975) held that the violation of a city safety ordinance does not create a civil remedy. As to an OSHA violation as *evidence* of negligence, see *Industrial Tile, Inc. v. Stewart*, 388 So.2d 171 (Ala.1980), cert. denied 449 U.S. 1081, 101 S.Ct. 864, 66 L.Ed.2d 805 (1981).

48. *Bitner v. Lester B. Knight & Associates, Inc.*, 16 Ill.App.3d 857, 307 N.E.2d 136 (1974).

49. *Olds v. Pennsalt Chemicals Corp.*, 432 F.2d 1033 (6th Cir.1970), cert. denied 401 U.S. 1010, 91 S.Ct. 1257, 28 L.Ed.2d 546 (1971). Similarly, regulations of a prime contractor held not for the benefit of the subcontractor's employee in *Jones v. Indianapolis Power & Light Co.*, 158 Ind.App. 676, 304 N.E.2d 337 (1973).

50. *Alber v. Owens*, 66 Cal.2d 790, 59 Cal.Rptr. 117, 427 P.2d 781 (1967).

51. *Solgaard v. Guy F. Atkinson Co.*, 6 Cal.3d 361, 99 Cal.Rptr. 29, 491 P.2d 821 (1971).

trespasser who was taking a shortcut over the site.[52]

Expanded tort application of safety statutes gives effect to legislative concern for safety. It can aid courts in resolving the often difficult question of whether the defendant has exposed the plaintiff to an unreasonable risk of harm. Greater application of these safety statutes should induce compliance.

If the preceding requirements are met, most states hold that the violation is negligence *per se*. The trial court judge *must* take away any negligence question from the jury and rule that the plaintiffs have conclusively proved negligence. For example, *Bowman v. Redding & Co.* held that violations of local safety rules requiring that the entrance to a shaft be provided with a substantial door gate or hinged bar and that a hoist signal system be employed were negligence *per se* when a worker fell to his death.[53] Similarly, violation of a state law requiring shields over certain types of moving machinery was held to be negligence *per se* where a worker lost his balance and thrust his hand into the workings of a nearby postal conveyor belt.[54] Some jurisdictions hold that a statutory violation is evidence of negligence but not conclusive on this question.[55] One court in so concluding noted that the statute in question, one that required guardrails for steps in single-family residences, had been enacted after the home had been built and was not strictly enforced.[56]

Generally compliance with a statutory standard is evidence that legal standards have been met but not conclusive on this question.[57]

The effect of contributory negligence or assumption of risk by the injured party is discussed in Section 35.08.

SECTION 35.05
VICARIOUS LIABILITY
AND THE INDEPENDENT
CONTRACTOR RULE

Perhaps the most frequently contested issue in construction accident litigation is whether one person's negligence can be charged to another. Often the cause of the harm can be traced easily to one participant, but that participant (such as a contractor) is immune from tort action or, even if not immune, may be a small enterprise unable to pay any court judgment and may lack insurance coverage. For that reason, the injured person often seeks to charge a person's negligence to another who can be sued and can respond directly or indirectly to a court-ordered judgment.

For example, an injured employee of a subcontractor may be precluded from suing his employer and perhaps the prime contractor as well, the persons most closely connected to the accident. As a result, the injured employee often institutes legal action against the prime contractor, if the prime contractor is not a statutory employer, and the design professional. In addition, the employee may assert that the negligence of any contractor, whether immune or not, or the design professional is chargeable to the owner.

---

52. *Cappa v. Oscar C. Holmes, Inc.*, 25 Cal.App.3d 978, 102 Cal.Rptr. 207 (1972).

53. 449 F.2d 956 (D.C.Cir.1971).

54. *O'Neill v. United States*, 450 F.2d 1012 (3d Cir.1971).

55. *Fireman's Fund Insurance Co. v. AALCO Wrecking Co.*, supra note 37; *Tenenbaum v. City of Chicago*, supra note 47; *Johnson v. Salem Title Co.*, 246 Or. 409, 425 P.2d 519 (1967) (architect's violation of building code justified submitting question of negligence to jury).

56. *Cooper v. Goodwin*, 478 F.2d 653 (D.C.Cir.1973).

57. *Spangler v. Kranco, Inc.*, 481 F.2d 373 (4th Cir.1973).

(A) VICARIOUS LIABILITY

One party can be vicariously responsible for the acts of another. The most common illustration is the responsibility of the employer for the acts of its employees committed in the scope of their employment. Initially it must be determined whether the negligent person was an employee of the employer or acted as an independent contractor. The test usually applied is whether the employer controlled or had the right to control the details of how the negligent party performed its services (examined in greater detail in (B) dealing with the independent contractor rule).

(B) THE INDEPENDENT CONTRACTOR RULE

Suppose an injured worker makes a claim against an owner. A number of issues can surface. First, the claimant may assert that the owner was itself negligent. Has the owner been negligent in engaging the contractor whose negligence caused the injury?[58] Usually this relates to the competence of the person engaged. However, a recent case held that an owner who hires a financially irresponsible contractor is itself negligent.[59]

The next issue is whether the negligent party is an employee or an independent contractor. As noted in (A), the test is control or the right to control the *method* of performance, beyond the *result* to be accomplished. If the owner possesses or exercises this power, the negligent party is its employee. If the owner does not, simply asking for a result without dictating how that result is to be accomplished, the owner can assert that it engaged an independent contractor and is not responsible for its negligence.

Those supporting the rule contend that the person who has used proper care in hiring an independent contractor and has turned over the work to it should not be responsible. Those who oppose the rule assert that it allows enterprises to insulate themselves from enterprise risks. In response, defenders say that the independent contractor is usually required to carry liability insurance and the cost of insurance is transferred to the employer of the independent contractor through the contract price. Yet, assert attackers, some independent contractors do not insure, and if the employer will pay indirectly, why not compel it to pay directly by denying protection of the rule, and so the debate rages.

Undoubtedly the inconsistent cases and many exceptions to the rule reflect this controversy. Should the same rule be applied to a major addition made by General Motors to one of its plants[60] and a farmer who has furnished material and a truck to a one-person contractor who has agreed to build a chicken house?[61] In addition to the differing factual patterns, judges vary in their willingness to impose liability vicariously.

Considered swallowed up by exceptions[62] and concepts of enterprise liability, the in-

---

58. *Smith v. United States,* 497 F.2d 500 (5th Cir.1974).

59. *Becker v. Interstate Properties,* 569 F.2d 1203 (3d Cir.1977) cert. denied 436 U.S. 906, 98 S.Ct. 2237, 56 L.Ed.2d 404 (1978).

60. *Funk v. General Motors Corp.,* 392 Mich. 91, 220 N.W.2d 641 (1974). This case is discussed later in the section.

61. *Smith v. Jones,* 220 So.2d 829 (Miss.1969).

62. One court recently noted that there were so many exceptions that the rule applied only when there was no good reason for departing from it. *Walker v. Capistrano Saddle Club,* 12 Cal.App.3d 894, 896, 90 Cal.Rptr. 912, 914 (1970). An earlier court stated that the rule existed primarily as a preamble to the catalog of its exceptions. *Pacific Fire Insurance Co. v. Kenny Boiler & Manufacturing Co.,* 201 Minn. 500, 277 N.W. 226 (1937).

dependent contractor rule has displayed considerable modern vitality in many states. A large number of cases have exonerated the owner from responsibility for the negligence of the contractor despite the various powers the owner, usually through the design professional, can exercise during performance.[63] Such powers can include approval of subcontractors, inspection of the work, coordination of the work, condemnation of defective work, checking on the maintenance of schedules, and holding safety meetings. Cases that have given defenses based upon the rule have involved both traditional methods of contracting and nontraditional methods, such as awards made by sophisticated owners with technical staff competence and work often performed within the physical area of an existing facility.[64] Similarly, an owner has been exonerated for negligence of its architect[65] and prime contractors have been relieved from liability of subcontractors.[66]

Any language in the contract between employer and the party doing the work stating the latter to be an independent contractor has been considered relevant but certainly is not conclusive and often not persuasive.[67] The issue is for the jury. However, the independent contractor defense may be lost if the employer of the independent contractor assumes the latter's responsibilities.[68]

Without resort to the many exceptions to the rule, some cases have been unwilling to classify the negligent party as an independent contractor for whom the employer is not responsible. Most of these cases have been projects that have involved sophisticated, institutional owners that have never actually turned over exclusive control of the site or the project to the prime contractor.[69] Perhaps *Funk v. General Motors Corp.*[70] is typical of those cases that have not applied the rule. Funk was an employee of a subcontractor and was seriously injured in a fall on a plant construction job. He recovered workers' compensation benefits from his employer and brought an action against the prime contractor and the owner of the plant, General Motors. After stating that primary responsibility for job safety belonged to the prime contractor, the Michigan Supreme Court stated that the owner of a building under construction is not ordina-

---

63. See, e.g., *Jeter v. St. Regis Paper Co.*, 507 F.2d 973 (5th Cir.1975); *Foster v. National Starch & Chemical Co.*, 500 F.2d 81 (7th Cir.1974); *Campbell v. United States*, 493 F.2d 1000 (9th Cir.1974); *Jeffries v. United States*, 477 F.2d 52 (9th Cir.1973); *Stanley v. United States*, 476 F.2d 606 (1st Cir.1973); *Wolfe v. Bethlehem Steel Corp.*, 460 F.2d 675 (7th Cir.1972); *Fisher v. United States*, 441 F.2d 1288 (3d Cir.1971); *Olds v. Pennsalt Chemicals Corp.*, supra note 49.

64. *Jeter v. St. Regis Paper Co.*, supra note 63; *Foster v. National Starch & Chemical Co.*, supra note 63; *Wolfe v. Bethlehem Steel Corp.*, supra note 63, to name a few.

65. *Cutlip v. Lucky Stores, Inc.*, 22 Md.App. 673, 325 A.2d 432 (1974).

66. *West v. Guy F. Atkinson Construction Co.*, 251 Cal.App.2d 296, 59 Cal.Rptr. 286 (1967).

67. *Martin v. Phillips Petroleum Co.*, 42 Cal.App.3d 916, 117 Cal.Rptr. 269 (1974); *Reith v. General Telephone Co. of Illinois*, 22 Ill.App.3d 337, 317 N.E.2d 369 (1974); *Scrimager v. Cabot Corp.*, 23 Ill.App.3d 193, 318 N.E.2d 521 (1974).

68. *C.H. Leavell & Co. v. Doster*, 233 So.2d 775 (Miss.1970). Similarly, in *Hargrove v. Frommeyer & Co.*, 229 Pa. Super. 298, 323 A.2d 300 (1974) held the owner for defective design that injured an employee of the plaintiff when the plans and specifications were drawn by the owner's design department.

69. *Nagler v. United States Steel Corp.*, 486 F.2d 794 (7th Cir.1973); *Magill v. Westinghouse Electric Corp.*, 327 F.Supp. 1097 (E.D.Pa.1971); *Weber v. Northern Illinois Gas Co.*, 10 Ill.App.3d 625, 295 N.E.2d 41 (1973); *Moore v. Denune & Pipic, Inc.*, 26 Ohio St.2d 125, 269 N.E.2d 599 (1971); *Kuhns v. Standard Oil Co. of California*, 257 Or. 482, 478 P.2d 396 (1970).

70. See note 60, supra.

rily liable to construction workers for job safety. The court continued:

> In contrast with a general contractor, the owner typically is not a professional builder. Most owners visit the construction site only casually and are not knowledgeable concerning safety measures. . . . Supervising job safety, providing safeguards, is not part of the business of the typical owner.[71]

General Motors retained a significant amount of control of the project. Its internal division drafted the building plans, wrote the contractual specifications, and acted as architectural supervisor. It directly hired several of the contractors, including the employer of the plaintiff, and later assigned those contracts to the prime contractor. A General Motors' representative was the architect-engineer superintendent for one of its divisions. He could order the prime contractor to terminate any prime or subcontractor's performance within twenty-four hours. The representative was at the job site daily and interpreted the contract specifications and plans for the prime. He made on-the-spot inspections, including inspections to determine if safety requirements were being met.

The court concluded that an owner, such as GM here, will not be absolved from its failure to require observance of reasonable safety precautions when it in effect is the superintendent and has knowledge of the high degree of risk faced by construction workers. Similarly, some courts have held prime contractors for the negligence of their subcontractors.[72]

Courts refusing to allow the independent contractor defense seem motivated by hostility to an enterprise seeking to avoid responsibility for harms caused by the enterprise. Also, they wish to see victims compensated and are willing to search for a "deep pocket."[73]

(C) EXCEPTIONS TO INDEPENDENT CONTRACTOR RULE

One court stated that the rule is "primarily important as a preamble to the catalog of its exceptions."[74] Although there are many exceptions, it is important to determine the extent of their application. For example, one of the best known exceptions is for work "which the employer should recognize as likely to create during its progress a peculiar risk of physical harm to others unless special precautions are taken."[75] More commonly, courts have referred to this particular exception as work that is inherently or intrinsically dangerous or that involves a peculiarly high degree of risk.[76]

Although most jurisdictions recognize this exception, they vary in their willingness to apply it.[77] Even where the activity is

---

71. 220 N.W.2d at 646.

72. *Smith v. United States,* supra note 58; *Summers v. Crown Construction Co.,* 453 F.2d 998 (4th Cir.1972). Although some negligence seems to have been found against the prime contractor, the facts do not seem materially different from *West v. Guy F. Atkinson Construction Co.,* supra note 66, which came to a contrary conclusion.

73. Very little attention is paid to the possibility of the independent contractor carrying liability insurance, passing that cost to the employer through the contract price, and creating double insurance.

74. See note 62, supra.

75. Restatement (Second) of Torts § 416 (1965).

76. W. Prosser and P. Keeton, Torts, p. 512 (5th ed. 1984).

77. Compare *Walker v. Capistrano Saddle Club,* supra note 62, with *West v. Guy Atkinson Construction Co.,* supra note 66. The first concluded that bridge construction requiring a crane near high-voltage wires created a peculiar risk of physical harm, while the second, involving the construction of a highway overpass, concluded that the work did not. The California cases are summarized in *Castro v. State,* 114 Cal.App.3d 503, 170 Cal.Rptr. 734 (1981). For an analysis of this exception, see Comment, *A Systematic Approach to the Peculiar Risk Exception to the Independent Contractor's Rule in Iowa,* 67 Iowa L.Rev. 589 (1982).

considered inherently or intrinsically dangerous, not all accidents result in liability to the employer of the independent contractor. For example, one case held that the existence of the exception is not relevant unless a direct causal relationship exists between the peculiar risk inherent in the work and the plaintiff's injuries. The injury in this case was caused by a defective tractor scraper earth-moving machine. The court held the employer of the independent contractor not responsible.[78]

Large institutional employers who have safety expertise and a high degree of control are more likely to be held performing inherently dangerous work.[79] Courts unwilling to use this "exception" are often motivated by a desire to preserve the independent contractor rule in construction accidents.[80]

Even jurisdictions willing to recognize an exception for inherently dangerous work have difficulty coming to a common conclusion regarding the availability of this exception to employees of the independent contractor. For example, suppose that the owner hires a prime contractor to erect high-voltage electrical wires and that this is high-risk, inherently dangerous work. Suppose the prime contractor is negligent and

one of the prime's employees suffers physical harm. The employee cannot bring legal action against the prime contractor because of workers' compensation immunity granted to the employer. If the employee asserts an action against the employer of the independent contractor, that is, the owner, some jurisdictions will permit the inherently dangerous exception to apply[81] while others will not.[82]

Courts unwilling to extend the inherently dangerous exception to employees seem persuaded that the latter have an adequate remedy through workers' compensation.[83] Those that believe otherwise are likely to grant an injured worker the benefit of the inherently dangerous work exception to the independent contractor rule.[84]

The exception for inherently dangerous work has been applied vigorously in some jurisdictions but very cautiously in others. Similarly, other exceptions such as for nondelegable duties[85] and violation of building codes[86] have been applied, but jurisdictions vary in their unwillingness to actually *find* an exception to the independent contractor rule.

The *existence* of many exceptions to the rule has by no means obliterated it. It still

---

78. *Holman v. State*, 53 Cal.App.3d 317, 124 Cal.Rptr. 773 (1975).

79. *McDonough v. United States Steel Corp.*, 228 Pa.Super. 268, 324 A.2d 542 (1974).

80. *Musgrave v. Tennessee Valley Authority*, 391 F.Supp. 1330 (N.D.Ala.1975).

81. *Lindler v. District of Columbia*, 502 F.2d 495 (D.C.Cir.1974); *Walker v. Capistrano Saddle Club*, supra note 62; *McDonough v. United States Steel Corp.*, supra note 79.

82. *Campbell v. United States*, supra note 63; *Olson v. Red Wing Shoe Co.*, 456 F.2d 1299 (8th Cir.1972); *West v. Morrison-Knudsen Co.*, supra note 25; *Reber v. Chandler High School District No. 202*, 13 Ariz.App. 133, 474 P.2d 852 (1970); *Epperly v. City of Seattle*, supra note 17.

83. *West v. Guy F. Atkinson Construction Co.*, supra note 66; *Olson v. Red Wing Shoe Co.*, supra note 82.

84. Undoubtedly this played a significant part in the court's decision in *Alber v. Owens*, supra note 50.

85. *Reith v. General Telephone Co. of Illinois*, 22 Ill.App.3d 337, 317 N.E.2d 369 (1974). The employer of the independent contractor, a public utility, needed a franchise and permit from state authorities. This made its duty nondelegable and deprived it of the right to use the independent contractor rule. Similarly, in *Weber v. Northern Illinois Gas Co.*, supra note 69, the court held that the employer's duty to provide a safe work site to employees of a subcontractor was nondelegable.

86. *Johnson v. Salem Title Co.*, supra note 55. An architect was held liable for the negligence of his consulting engineer.

stands as a significant barrier to recovery in construction accidents.

SECTION 35.06
CAUSE IN FACT

To transfer the loss from the victim to the defendant, the victim must prove that the defendant's conduct has substantially caused the loss.[87] One exception relates to harm caused by two negligent persons but the plaintiff cannot prove which person caused the loss.[88]

SECTION 35.07
PROXIMATE CAUSE

Sometimes the law relieves a negligent party who has caused the loss either because the actual loss was greatly out of proportion to what could have been expected, the loss occurred to an unforeseeable plaintiff, the loss occurred in a substantially unanticipated manner, or the loss was caused by an intervening agency that should be considered a supervening cause releasing the original wrongdoer.

A few applications of proximate cause in construction accidents have been made, mainly the effect of intervening causes. For example, an architect responsible for a negligently designed wall that collapsed was not relieved by the approval of his work by a building inspector.[89] Similarly, a defendant contractor who had left a trench across a street filled with gravel was not relieved from liability when a third party placed a thin layer of blacktop across the street that ultimately wore out leaving holes.[90] The court concluded that the inter-

vention by the third party was reasonably foreseeable. However, the contractor who negligently left a ladder in a position where it might have caused an injury was relieved from liability when the ladder was unexpectedly moved by a third party.[91]

SECTION 35.08
SOME DEFENSES

(A) ASSUMPTION OF RISK AND CONTRIBUTORY NEGLIGENCE

Sometimes those against whom claims have been made point to conduct by the claimant that they assert bars or at least reduces the extent of the claim. The defenses that relate to conduct by the claimant are assumption of risk and contributory negligence. Assuming a risk is a voluntary exposure to danger with full knowledge and appreciation of its existence—or taking one's chances. Contributory negligence, on the other hand, is conduct that does not rise to the standard of conduct required by tort law and that plays a substantial part in causing the injury.

The two concepts can overlap. The claimant may have chosen voluntarily to encounter the risk, and this choice may be conduct that falls below that required by law. Yet a party can reasonably assume a risk. Before briefly looking at a case that sought to make this differentiation, it is important to recognize the difference between a claim based upon common law negligence and one based upon the violation of a statute. Claims based upon common law negligence can be defeated if it is determined that the claimant has expressly

---

87. A court seemed to relax this burden in *Fireman's Fund Insurance Co. v. AALCO Wrecking Co.*, 466 F.2d 179 (8th Cir.1972).

88. *Bowman v. Redding & Co.*, 449 F.2d 956 (D.C.Cir.1971). Refer to Section 7.03(D).

89. *Johnson v. Salem Title Co.*, supra note 55.

90. *Adams v. Combs*, 465 S.W.2d 288 (Ky.App.1971).

91. *Hill v. Lundin & Associates, Inc.*, 260 La. 542, 256 So.2d 620 (1972).

or impliedly assumed the risk.[92] Similarly, where contributory negligence by the claimant *bars* the claim (an increasingly small number of states are coming to this conclusion), a claim based upon common law negligence will be barred.[93] Complications develop when the claim is essentially based upon the violation of a statute.

In *Fonseca v. County of Orange,*[94] the plaintiff was a construction worker who fell from a bridge that was being constructed over a river by Orange County. He slipped on some wet excess cement as he was doing finish work on the concrete deck of the bridge at a height of at least twenty feet above the dry river bed. No scaffolding or railing had been installed around the perimeter of the bridge for the protection of workers as required by law. After recovering workers' compensation benefits, the plaintiff brought legal action against Orange County. The judgment in favor of the plaintiff was based principally upon the contractor's violation of state safety orders requiring installation of railings where workers must work at heights in excess of seven and one-half feet above ground level and scaffolding at heights above fifteen feet.

One assertion made by Orange County was that the plaintiff had been contributorily negligent. He had had considerable experience in cement work, and his experience should have motivated him to refuse to start work when he saw there were safety violations. The plaintiff had testified that he told the county engineer that protective railings should be installed, but in any event, the plaintiff proceeded to work on the deck from which he fell and was injured.

After defining contributory negligence, the court noted that some jurisdictions have eliminated it as a bar to recovery in actions arising from violations of safety regulations but that in California it is available as a defense.[95] On the other hand, assumption of risk does *not* bar a worker's claims against his employer, particularly where there has been a violation of safety statutes or orders. In other words, if the plaintiff's conduct were characterized as assumption of risk, it would not bar the claim, but if characterized as contributory negligence, at least at that time, it would bar the entire claim.

The court gave as a reason for this differentiation the weak bargaining position of the employee and the economic pressure upon him to work in unsafe places or with an unsafe appliance.

Noting the difference between the two doctrines and the possible overlap between them, the court concluded that the plaintiff had assumed the risk, since he knew of the danger and its magnitude and voluntarily encountered the risk. The court concluded that the plaintiff did not perform unreasonably when he continued to work despite the violation of the safety orders. This convenient conclusion, finding that the conduct did not bar the claim, reflects the tendency of the law to give wide scope to worker third-party actions.

Injured workers in other jurisdictions have not fared as well as Fonseca. Courts have held that a worker who stays on the job when asked or ordered to perform work under dangerous conditions has either assumed the risk or been contributorily negli-

---

92. *Mitchell v. Young Refining Corp.,* 517 F.2d 1036 (5th Cir.1975). But see *Gray v. Martindale Lumber Co.,* 515 F.2d 1218 (5th Cir.1975).

93. *Mundt v. Ragnar Benson, Inc.,* 61 Ill.2d 151, 335 N.E.2d 10 (1975). But see *Funk v. General Motors Corp.,* supra note 60.

94. 28 Cal.App.3d 361, 104 Cal.Rptr. 566 (1972).

95. Refer to Section 7.03(G).

gent when they bar the claim.[96] These courts have not been sympathetic to the economic compulsion arguments made by workers—that they had to work or would be sent home.

Admittedly, some arguments can be made for barring recovery in these cases. The worker *will* very likely receive workers' compensation. Barring recovery may motivate workers to refuse to work under dangerous conditions, and this may induce employers to make their work conditions reasonably safe. It can be argued that a worker who is a member of a construction trade union will often be able to find assistance there.[97]

The realities of the work place and the economic pressures upon workers support the position taken by the California court in the *Fonseca* case. Admittedly, there is an element of paternalism in concluding that workers must be protected from their own foolhardiness and from the pressures of the work place. Perhaps in the more individualistic nineteenth century, this would be appropriate, but it does not seem to be in tune with notions of modern social welfare concepts that have become predominant in tort law. In this regard, it should be noted that the modern movement toward comparative negligence would mean that conduct considered as unreasonable (and Fonseca's conduct could fall under this category) would only lower recovery rather than bar the entire claim.

Suppose the defendant seeks to establish contributory negligence by an allegation that the *worker* violated a statute. In *Morgan v. Stubblefield,*[98] Morgan and Saetelle were employees of Aaron, the electrical subcontractor retained by Associated, the prime contractor, who was building a 500-by-300-foot structure with a domed ceiling

twenty feet high. Morgan and Saetelle were affixing an electric conduit to the underside of the roof. They were working on a seventeen-foot scaffold that had been rented by Aaron from Able, an equipment supplier. Metzker, another employee of Aaron's, was assisting the two workers from the ground by moving the scaffold as they directed and obtaining material for them as needed.

Six feet from where the workers were working, Associated had created a hole in the floor about one foot square and about one foot deep to be used as the base of a stairway. The workers directed Metzker to obtain some materials. Metzker had noticed the hole in the floor but had failed to lock the wheels of the scaffold. A few minutes later, the scaffold toppled and the workers were thrown to the concrete floor below. The workers brought actions against Associated and Able, who brought cross actions against Aaron, the injured workers' employer.

The workers' claim against Able was based upon the latter's having violated safety orders issued by the Division of Industrial Safety that specified certain requirements for rolling scaffolds. The claims against Associated were based upon violations of two safety orders, one requiring adequate illumination of the building and the other requiring guarding of openings on the floor.

Each defendant contended that the workers *themselves* had violated safety regulations that prohibited riding on moving scaffolds except when the floor on which the scaffold was standing was free from holes, the scaffold itself met certain requirements not fulfilled in this case, and the scaffold was being moved by a person on the ground.

For a violation of a safety order to be

---

96. *Mitchell v. Young Refining Corp.,* supra note 92; *Demarest v. T.C. Bateson Construction Co.,* 370 F.2d 281 (10th Cir.1966); *Ralston v. Illinois Power Co.,* 13 Ill.App.3d 95, 299 N.E.2d 497 (1973).

97. *Holman v. State,* 53 Cal.App.3d 317, 124 Cal.Rptr. 773 (1975).

98. 6 Cal.3d 606, 100 Cal.Rptr. 1, 493 P.2d 465 (1972).

contributory negligence, the court held that the defendant must show that the injured employee *knew* of the safety order. The court noted that most construction safety orders were directed to the conduct of the employer. Interspersed among them were a few applicable to employees. The court noted that safety orders consist of a large volume of regulations and are amended with some frequency. The court stated:

> To presume that a workman, perhaps new to his employment, would be acquainted with these numerous and complex regulations is to permit the employer to utilize the construction safety orders for exculpatory purposes not legislatively contemplated. . . .

> Safety orders are basically intended for the protection of employees; it is the employer who is primarily responsible for complying with their provisions and it is he who has the greater opportunity and ability to obtain compliance by others. . . . These legislative aims would be subverted if an employee were deemed to have constructive knowledge of the contents of the safety orders directed to his conduct. In the absence of such a presumption, which we decline to invoke, it must follow that an employee's violation of a safety order, without more, cannot constitute negligence per se.[99]

Procedural rules make successful use of these defenses difficult for defendants even when they are permitted to assert them. First, the defendants must plead and prove that the injured party has assumed the risk or has been contributorily negligent. Second, the determination of whether the defendant has proved this by a preponderance of the evidence is typically made by juries.[100] In addition, contributory negligence is slowly disappearing in most American states and is not likely to prove to be an absolute bar. The emphasis on occupation-

al safety is likely to make successful use of assumption of risk and contributory negligence even rarer in the future.

(B) SOVEREIGN IMMUNITY

The broad outlines of sovereign immunity, past and present, were given in Section 7.06(D). Several aspects of immunity that relate more directly to the Construction Process merit brief comment. The increased participation of governmental authorities in land use and the Construction Process usually makes it necessary to obtain a building permit before construction and an occupancy permit after construction. Issuing a building permit often involves checking of drawings and specifications, while an occupancy permit usually necessitates an inspection to insure compliance with building code standards.

Public entities that issue such permits usually seek immunity from injuries or losses that may occur because of faulty design or construction. Sometimes this is accomplished by making this activity immune either by statute or case decision. Sometimes entities that issue permits seek to relieve themselves from liability by disclaimers they make at the time that such permits or approvals are given.[101]

A look at the California Design Immunity Legislation and its subsequent history is useful both as a review of design problems generally and as an illustration of the difficulties that develop because of the sharply differing policies reflected in immunity law.

Design liability has features that separate it from other forms of liability. First, it almost always involves conscious choices and tradeoffs among quality, quantity, time,

---

99. 493 P.2d at 471–2, 100 Cal.Rptr. at 7–8.

100. *Gantt v. Mobil Chemical Co.*, 463 F.2d 691 (5th Cir.1972); *Kirsch v. Dondlinger & Sons Construction Co.*, 206 Kan. 701, 482 P.2d 10 (1971); *Bennett v. Young*, 266 N.C. 164, 145 S.E.2d 853 (1966).

101. See Comment, *Municipal Tort Liability for Erroneous Issuance of Building Permits*, 58 Wash.L.Rev. 537 (1983). See also *Morris v. County of Marin*, 18 Cal.3d 901, 136 Cal.Rptr. 251, 559 P.2d 606 (1977) (injured worker allowed to sue county that in violation of state law had issued building permit without requiring a workers' compensation certificate to be filed).

and cost as well as between function and esthetics. Second, some aspects of design involve technical considerations sometimes thought to be beyond the competence of a jury. Third, events that occur *after* initial design can require reappraisal of the design. The state of the art changes and a strong danger exists that those being charged will be held to a standard that existed at the time of the accident or even at the time of the trial rather than at the time the design decision was initially made.

In 1961 the California Supreme Court abolished unlimited sovereign immunity.[102] In 1963 the legislature adopted a comprehensive statutory scheme that sought to accommodate the various interests involved. One section [103] provides:

> Neither a public entity nor a public employee is liable under this chapter for an injury caused by the plan or design of a construction of, or an improvement to, public property where such plan or design has been approved in advance of the construction or improvement by the legislative body of the public entity or by some other body or employee exercising discretionary authority to give such approval or where such plan or design is prepared in conformity with standards previously so approved, if the trial or appellate court determines that there is any substantial evidence upon the basis of which (a) a reasonable public employee could have adopted the plan or design or the standards therefor or (b) a reasonable legislative body or other body or employee could have approved the plan or design or the standards therefor.

In 1967 California granted immunity in *Cabell v. State of California,* a case where accidents had occurred before the one causing injury to the plaintiff and the claim had been made that the public entity *now* knew that the design had been improper.[104] The court emphasized the importance of judging the design at the time it was made and not

at some later date. Yet four and one-half years later the California court overruled its holding, stating that it was convinced that the legislature did not intend to permit public entities to shut their eyes to the operation of a plan or design once it had been transferred from "blueprint to blacktop." [105]

SECTION 35.09
WORK–RELATED INJURIES: SHOULD THIRD–PARTY CLAIMS BE ELIMINATED?

Originally work-place accidents were to be dealt with principally through workers' compensation claims. Although workers' compensation laws left openings for third-party actions at the time these systems originated, tort recoveries were difficult to obtain and of limited amount.

When tort law began to provide surer and larger recoveries, third-party claims proliferated, particularly in the construction work place injury with its many nonimmune participants. Once peripheral, third-party claims seem to be eclipsing the workers' compensation system. This is demonstrated not only by the increased claims but also by attempts by employers who originally sought to *avoid* being classified as statutory employers now seeking this status to avoid third-party claims.

Powerful arguments have been marshaled *against* abolition of third-party claims. One is that of economic efficiency. Those who use an economic approach to law seek to eliminate externalities, exonerating those whose activities cause harm. Externalities inhibit a *true* determination of the cost of an activity. For example, if a contractor or design professional need not pay for the physical harm its activities cause, the market cannot accurately deter-

---

102. *Muskopf v. Corning Hospital District,* 55 Cal.2d 211, 11 Cal.Rptr. 89, 359 P.2d 457 (1961).

103. West's Ann.Cal.Gov't Code § 830.6.

104. 67 Cal.2d 150, 60 Cal.Rptr. 476, 430 P.2d 34 (1967).

105. *Baldwin v. State of California,* 6 Cal.3d 424, 99 Cal.Rptr. 145, 491 P.2d 1121 (1972).

mine the cost of the activity, an important datum in determining economic efficiency.

Any attempt to bar the courtroom door to those who have suffered physical injury not only runs the risk that the efforts will generate stiff opposition in the legislature but also runs the risk that a court, as in Completion Statutes,[106] may find the legislation unconstitutional.

Yet the socialized cost in terms of courtrooms, judges' salaries, and public record-keeping, although difficult to compute, are clearly formidable. Even more difficult to quantify are the psychological cost and lost productivity generated by the lawsuit. Some costs can be determined at least in a rough way. Insurance premiums for Construction Process participants have risen sharply.

Where do these premium dollars go? Do they go into the pockets of victims as compensation for medical expenses and lost earnings? Studies of insurance premium dollar payouts reveal a sorry system under which victims receive too little and those handling claims—mainly attorneys—receive too much.

A study of the automobile insurance premium dollar showed that it was distributed in the following way:

General overhead		$ .33
Claims handling		
Defense	.13	
Claimants	.10	
		.23
Paid to claimants		.44
		$1.00 [107]

A commentator studying the product lia-

bility insurance premium dollar came up with the following breakdown:

General overhead		$ .30
Claims handling		
Defense	.185	
Claimants	.14	
		.325
Claimants		.375
		$1.00 [108]

The two studies reveal that the claims handling takes from 23 cents to 32.5 cents of each insurance dollar. If overhead costs were eliminated, the automobile accident study showed that the claims handling took 35% of the insurance payout, while the products liability study showed an even higher percentage of 46%.

The author received information supplied by a leading insurer of architects and engineers. Along with extrapolations from the other two tables it shows an even higher amount for claims handling and correspondingly lower amount to claimants:

General overhead		.28
Claims handling		
Defense	.25	
Claimants	.19	
		.44
Claimants		.28
		$1.00

Even more up-to-date information has come from a recent study of the cost of asbestos litigation.[109] That study estimated that out of a total of $400 million paid by defendants and insurers, claimants will re-

106. Refer to Section 27.03(G).

107. P. Keeton & R. Keeton, Torts: Cases and Materials 792 (2d ed. 1977).

108. O'Connell, *Bargaining for Waivers of Third-Party Claims,* [1976] Ill.L.F. 435, 437.

109. Kakalik, Ebener, Felstiner, and Shanley, Costs of Asbestos Litigation (Rand Corp. 1983).

ceive an estimated 37%. This is determined by adding the average compensation claim paid by all defendants and insurers of $60,000 and the total defense litigation costs per claim of $35,000, a total per claim of $95,000. However, from the average claim paid of $60,000, the plaintiff's legal fees and other litigation expenses averaged $25,000, leaving the plaintiffs an average of $35,000. For every $2.71 expended, defense litigation expenses are an estimated one dollar, while the plaintiff's litigation expenses are an estimated 71 cents, leaving the plaintiff one dollar, an estimated 37% of the total expended.

One approach would be to abolish third-party claims as part of an overhaul of workers' compensation laws which would improve benefits where needed. Congress went this route to a substantial degree in 1972 in legislation dealing with longshoremen.[110] Third parties cannot receive indemnification from employers. Benefits were improved. Oklahoma eliminated third party claims against design professionals for site services in 1982.[111]

It should be apparent that the partnership between the workers' compensation and tort system has developed into one with the erstwhile junior partner—the tort system—taking control through third-party actions. The result has been a return to the chaos of the tort lawsuit designed to be eliminated by the workers' compensation system. It is time to return to an efficient social insurance method of compensating work place accident victims.

SECTION 35.10
PASSAGE OF TIME:
STATUTES OF LIMITATIONS

Some claims by those who have suffered personal harm are made many years after the Construction Process has been completed. This problem in the context of owner claims was discussed in Section 27.03(G). Reference should be made to that section, since both ordinary and special completion statutes discussed there can affect claims by those who have suffered personal harm.

PROBLEM

W was an employee of S, an electrical subcontractor in a project in which P was constructing an office building for O. The site and a set of construction documents drawn by A were turned over to P. W was working on a scaffold alongside the fourth floor of a partially constructed six-story building. Safety regulations required that W wear a safety belt. On the day of the accident, W decided not to wear a belt because it was hot. A heavy plank fell from the roof, striking W, knocking him off the scaffold, and sending him to his death. The plank came from an area in which

two subcontractors, S–1 and S–2, had been working.

W's widow has brought a lawsuit against S–1, S–2, P, O, and A, as well as S's workers' compensation carrier. The claim against S–1 and S–2 was based upon the negligence of their employees which caused the plank to fall. The claim against P was based upon P's knowledge that safety orders were being violated by S. The claim against O was based upon the work being "inherently dangerous" and thereby depriving O of the independent contractor defense. The claim against A was based upon his fail-

---

110. 33 U.S.C.A. § 905(b), as amended by P.L. 98–426, Sept. 28, 1984, 98 Stat. 1639.

111. Refer to note 4, supra.

ure to insist upon safe construction methods when he knew that S had a poor reputation in worker safety. The claim against the workers' compensation carrier was based upon its unexercised power to conduct a loss-prevention program and to provide inspection. What is the likelihood of W's widow succeeding against any of these defendants?

Would your answer have been any different had W asked for safety belts on the date of the accident, was told there were none, and was told that if he did not wish to work that day he could go home?

36

Indemnification and Other Forms of Shifting and Sharing Risks: Who Ultimately Pays?

SECTION 36.01 FIRST INSTANCE AND ULTIMATE RESPONSIBILITY COMPARED

Chapter 35 demonstrated the increased likelihood that those who suffer losses incident to the Construction Process will be compensated, especially where those losses involve personal harm. Although claimants will not always receive judicial awards, the tendency has been to expand liability to insure that those who suffer losses are compensated.

Lawsuits today generally begin with claims against a number of defendants, something that is legally permissible and relatively inexpensive. Typically these defendants make claims against each other as well as claims against parties who have not been sued by the original claimant. The result is a multiparty lawsuit that generally involves as many as a half dozen interested participants as well as an almost equal number of sureties and insurers. For example, the injured employee of a prime contractor is likely to sue all contractors who are not immune, such as other separate contractors and subcontractors, the owner, the design professional, any construction manager, and, depending on the facts, those who have supplied equipment or materials. Each of the defendants will very likely bring claims against the others

based upon indemnification. If there is a building defect, such as the failure of an air conditioning system, the owner is likely to assert claims against the design professional, the contractor, and the manufacturers and sellers of the system. Again, those against whom claims have been asserted are likely to assert claims against each other.

In these lawsuits there are two levels of responsibility. There is *first instance* responsibility to the original claimant, such as the injured worker or the owner. Resolution of this issue depends upon whether any of the defendants or someone for whom they are responsible has failed to live up to the standard required by the contract or by the law. After this determination is made, the next level, that of *ultimate* responsibility must be addressed. Who among those responsible will ultimately bear the loss? This inquiry is the focal point of this chapter and one that has developed unbelievably complex and costly legal controversies.

SECTION 36.02 CONTRIBUTION AMONG WRONGDOERS

Suppose A and B, *acting together* in pursuance of a common plan or design, injure C. C can recover its loss from either A or B if each has committed a wrong. Since each has committed the wrong, and since they have acted together, neither A nor B can receive contribution from the other if either has paid more than one-half of the total judgment. In this particular instance A and B are joint wrongdoers. For example, if the design professional and owner acted together to destroy the contractor's business or reputation, the contractor could sue either or both and recover its loss from either or both. In such a case neither design professional nor owner would have a claim against the other if either paid more than one-half of the judgment.

But a number of defendants may be sued in the same legal action even though they have not acted together. They are codefendants because each may have played a substantial role in causing the injury and for procedural convenience all the claims are decided in one lawsuit. They are concurrent wrongdoers, not in the sense that their wrongdoing occurred at the same time, but in the sense that each played a substantial role in causing an indivisible loss to the claimant.

Suppose three defendants are held liable to the plaintiff and one defendant pays the award. Can the one defendant seek reimbursement or "contribution" from the other defendants? The basis for contribution is the unfairness of one defendant paying the entire judgment when all are responsible. Most American courts do not require contribution among wrongdoers. But one-half of the states have created contribution by statute. The statutes vary considerably but where they exist two factors have reduced the effectiveness of contribution statutes. First, some states require that there be a joint judgment against the defendants before contribution can be compelled. This has obvious limitations as it inhibits settle-

ment. Second, and more important for purposes of construction accidents, contribution is frequently denied against a party who was immune from the original claim. For example, it would not be available against the employer, actual or statutory as described in Section 35.02, of the injured party, if the employer were immune from liability because of Workers' Compensation laws.

Where contribution does exist, either by judicial rule or legislation, each wrongdoer generally contributes a *pro rata* share. Where there are three wrongdoers, each would contribute one-third. An increasing number of jurisdictions now compare wrongdoing quantitatively when determining the amount of contribution required.[1]

Sometimes qualitative comparisons between wrongdoers are made. But this comparison, accomplished through indemnity, can result in a total shift of the loss. Contribution, on the other hand, requires a sharing of ultimate responsibility.

SECTION 36.03 NONCONTRACTUAL INDEMNITY [2]

(A) BASIC PRINCIPLE: UNJUST ENRICHMENT

Wrongdoing can encompass a spectrum from innocent conduct to conduct intended to harm. Illustrations of the *least* culpable conduct are liability imposed upon manufacturers for defective products even though there has been no showing that they were negligent, and liability placed upon the employer for the negligence of its employees' acts committed in the scope of employment. Similarly, as seen in Section

---

1. See *Skinner v. Reed-Prentice Division Package Machinery Co.,* 70 Ill.2d 1, 15 Ill.Dec. 829, 374 N.E.2d 437 (1977); *Bielski v. Schulze,* 16 Wis.2d 1, 114 N.W.2d 105 (1962).

2. This categorization separates indemnification based upon *express* agreement from *other* forms of indemnification. The latter have various designations, such as common law indemnity, quasi-contractual indemnity, and equitable indemnity. Rather than attempt to sort out these terms and their implications, indemnification is divided into noncontractual and contractual.

35.05(C), under certain circumstances the employer of an independent contractor may be liable for the negligence of that independent contractor.

An enlarged and qualitatively wide range of liability creates the opportunity to compare wrongdoing of defendants, either original or those added by cross actions brought by original defendants. For example, suppose a worker employed by a prime contractor is injured because of deliberate safety violations by her employer. Under certain circumstances the worker can recover from the owner.[3] Liability in such a case may be based upon the nondelegable statutory duty of the owner to furnish a safe work place. Alternatively, liability may be based upon the failure to determine whether safe practices were being followed or the failure to discharge the contractor after becoming aware of the violations. Some of these reasons for owner liability would be less morally objectionable than the conduct of the prime contractor. In many states a wrongdoer who pays the claim or who would be responsible under a court judgment for the claim can receive indemnification from the other wrongdoer. Suppose the owner seeks indemnification from the contractor. Indemnification would require a comparison of the *degree* of wrongdoing between the two parties. If the owner's moral culpability is less than that of the contractor, the latter would indemnify the former. The contractor would be unjustly enriched and the owner unjustly impoverished if the owner had to pay the claim.

Similarly, suppose the prime contractor violated safety orders and a claim was made against the design professional based upon the latter's failure to detect the violation or to exercise corrective power given by the contract. The contractor's conduct can be considered active if it orders an employee to work under dangerous conditions. On the other hand, the conduct of the design professional can be considered passive, her liability based upon failure to act. Although judicial determination of active and passive conduct is sometimes at variance with ordinary meaning (a point to be explored later in this section), courts sometimes make this differentiation the basis for giving the design professional indemnity from the contractor or from someone else more directly connected with the injury or loss.[4] Indemnity in such a case is based upon the unjust enrichment of the more culpable wrongdoer that would result if it were not required to bear this loss.

(B) NONCONTRACT AND CONTRACT INDEMNITY DIFFERENTIATED

Sections 36.04 and 36.05 treat indemnification implied from or expressed in a construction contract. As indicated, noncontractual indemnity is based principally upon unjust enrichment and the concept that losses should be shifted from one wrongdoer to another based upon qualitative comparative fault. For historical reasons, however, courts have tended to treat noncontractual indemnity as more analogous to a contract claim than to a tort claim. For this reason courts tend to classify this form of indemnity as quasi-contractual (something *like* a contract) even though it is not based upon consent.[5]

(C) SOME CLASSIFICATIONS

Courts have articulated various tests to determine whether noncontractual indemnification will be awarded. Although one

---

3. Refer to Section 35.05.

4. In *Owings v. Rosé*, infra note 5, an architect was given indemnity against a consulting engineer when the architect was held liable essentially for the negligence of the engineer.

5. See *Ohio Casualty Insurance Co. v. Ford Motor Co.*, 502 F.2d 138 (6th Cir.1974); *Owings v. Rosé*, 262 Or. 247, 497 P.2d 1183 (1972). But see *J.L. Simmons Co. v. Fidelity & Casualty Co.*, 511 F.2d 87 (7th Cir.1975).

court recognized an indemnification obligation in order to impose the ultimate burden upon one who was the "active delinquent" in bringing about the injury rather than the "lesser delinquent," [6] most courts have employed the primary-secondary or passive-active differentiation, singly or together. For example, one court stated indemnity would be granted:

> to a person who, without active fault on his own part, has been compelled, by reason of some legal obligation, to pay damages occasioned by the initial negligence of another, and for which he himself is only secondarily liable.[7]

The court stated that vicarious liability, that is, liability for the wrongs of another, which arises out of a positive rule of common or statutory law,[8] as well as liability imposed for failure to discover or correct a defect or remedy a dangerous condition caused by the act of the one primarily responsible, are illustrations of *secondary* liability. The active-passive differentiation, perhaps used more frequently, looks to similar factors. Again, some positive acts create liability while sometimes negative or passive inaction, though creating liability, is less morally objectionable, though differential enough to justify indemnification.

It may be useful to look at a few cases which have dealt with noncontractual indemnity. In *Adams v. Combs,*[9] a road contractor sought indemnity from the city for whom the road work had been performed.

An accident occurred two years after completion of the contract. The court denied the claim, however, holding that any negligence on the part of the city in not maintaining the road was secondary and passive while the contractor's negligence was primary and active.

In *Barr v. Brezina,*[10] a contractor against whom legal action had been brought by a person, who was injured while crossing the site, sought indemnity from the United States, who had employed the contractor. The contractor had complained about defects in the plans but the U.S. representative ordered him to continue building according to the plans. The court held that the prime contractor would not receive indemnification. The contractor performed despite defective plans. The U.S. was only passively negligent while the contractor was actively negligent.

Architects sought indemnity in *St. Joseph Hospital v. Corbetta Const. Co.*[11] and *Owings v. Rosé.*[12] In the first case, the architect was *denied* indemnity from a supplier of a defective tile because the architect knew the tile was defective. In the second case, the architect was *granted* indemnification against a negligent consulting engineer when the architect who was *not* negligent was liable because he had a nondelegable duty. Where each party has violated the *same* duty, indemnity is usually denied. For example, in *Harris v. Algonquin Ready Mix, Inc.*[13] the electric company and the

---

6. *Miller v. De Witt*, 37 Ill.2d 273, 226 N.E.2d 630, 642 (1967).

7. *Builders Supply Co. v. McCabe*, 366 Pa. 322, 77 A.2d 368, 370 (1951).

8. For example, in *Wrobel v. Trapani*, 129 Ill.App.2d 306, 264 N.E.2d 240 (1970), the prime contractor was held liable for having violated the Illinois Structural Work Act because he was in charge of the project. He sought indemnity from a subcontractor who had violated the statute by not having the subcontractor's employee use care in putting up and using a ladder. The court held the prime contractor's liability was passive, the subcontractor's active and the former would be indemnified by the latter.

9. 465 S.W.2d 288 (Ky.1971).

10. 464 F.2d 1141 (10th Cir.1972).

11. 21 Ill.App.3d 925, 316 N.E.2d 51 (1974).

12. Supra note 5.

13. 59 Ill.2d 445, 322 N.E.2d 58 (1974).

contractor it had hired each failed to warn employees of certain dangers. The absence of any *qualitative* difference in their liability precluded indemnity.

It is generally stated that failure to discover or remedy the defect caused by another is merely passive negligence.[14] However, in *Becker v. Black & Veatch Consulting Engineers*,[15] an engineer's failure to inspect was held to be active negligence when the engineer was employed for this express purpose. This precluded the engineer from receiving indemnity from owner or contractor as all were considered actively negligent.

Contribution would have been permitted if such right existed and, generally, if *all* could have been held liable. If the contractor was immune because the injured person was its employee, however, the contractor could not be compelled to contribute to the judgment unless there was contractual indemnity. The judgment would be shared by owner and engineer unless there were contractual indemnity.

The perils of generalizations in this area are emphasized in an Illinois decision which, in discussing active and passive liability, stated:

> Determination of this question is not a matter of proceeding according to the usual dictionary definitions of the words "active" and "passive." These words are terms of art and they must be applied in accordance with concepts worked out by courts of review upon a case-by-case basis. Under appropriate circumstances, inaction or passivity in the ordinary sense may well constitute the primary cause of a mishap or active negligence. . . . It has been appropriately stated that "mere motion does not define the distinction between active and passive negligence."[16]

Noting that certain terms have become words of art is a signal that a particular doctrine has developed difficulty and, therefore, it is important to read between the lines of judicial opinions dealing with that doctrine. It is also an indication of considerable uncertainty in the law and the importance of using contract language to specifically deal with the problem rather than leave the matter to the vagaries of court decisions when either vague concepts must be applied[17] or when legal terminology varies from ordinary meaning.

One commentator, reviewing some of the criteria used for determining whether noncontractual indemnity would be granted, stated:

> Probably none of these is the complete answer, and, as is so often the case in the law of torts, no one explanation can be found which will cover all of the cases. Indemnity is a shifting of responsibility from the shoulders of one person to another; and the duty to indemnify will be recognized in cases where community opinion would consider that in justice the responsibility should rest upon one rather than the other. This may be because of the relation of the parties to one another, and the consequent duty owed; or it may be because of a significant difference in the kind or quality of their conduct.[18]

(D) EMPLOYER INDEMNIFICATION

Sometimes the propriety of shifting liability depends upon substantive law policies that tend to either protect a particular person from this responsibility or place it upon her. Many noncontractual indemnity claims are brought against employers of

---

14. See *Builders Supply Co. v. McCabe*, supra note 7.

15. 509 F.2d 42 (8th Cir.1974). See *Associated Engineers, Inc. v. Job*, 370 F.2d 633 (8th Cir.1966).

16. *Moody v. Chicago Transit Authority*, 17 Ill.App.3d 113, 117, 307 N.E.2d 789, 792–93 (1974).

17. In *Santisteven v. Dow Chemical Co.*, 506 F.2d 1216 (9th Cir.1974), the court stated that quasi-contractual indemnity is based upon whether it is "just" to shift liability.

18. W. Prosser, Torts, 313 (4th ed. 1971). A subsequent edition made marginal modifications. See W. Prosser and P. Keeton, Torts, 344 (5th ed. 1984).

injured persons. For example, suppose a subcontractor's employee is injured and the primary responsibility for the injury was the employer's failure to comply with safety rules. Because the injured employee cannot sue her own employer, suppose she institutes legal action against the prime contractor. Suppose the prime contractor can be sued as a third party; liability against the prime contractor is based upon its responsibility as the employer to provide safe working conditions and this duty cannot be delegated. This would be a classic case for granting the prime contractor noncontractual indemnity. The prime contractor's liability is based upon a nondelegable duty and clearly the subcontractor's negligence is active and primary.

Nevertheless, allowing indemnity would place *ultimate* liability upon the subcontractor. Would this deny the subcontractor immunity from tort action granted by Workers' Compensation law? One of the tradeoffs in Workers' Compensation was granting immunity to the employer from tort liability in exchange for denying the employer rights possessed prior to the enactment of Workers' Compensation statutes. Indemnification can, by indirection, frustrate this.

But there are cogent arguments for allowing indemnity in these cases. The third party, here the prime contractor, received nothing in the "trade" and arguably should not have existing rights taken away. This argument, however, ignores the likelihood that third parties are also employers who are part of some Workers' Compensation legislative trade. In addition to this argument, it would appear to be unjust to exonerate the more culpable employer when liability against the third party, such as a prime contractor or owner, may be vicarious, based upon passive negligence or a statutory violation. It might also be argued that denial of indemnity may encourage carelessness by the employer.

The difficulty of this question is reflected in the case law. It has been considered the most evenly balanced issue in Workers' Compensation law.[19] But now it appears that a slight majority precludes *noncontractual* indemnity against an employer.[20] Those that grant indemnity usually find a separate duty owed by the employer to the third party seeking indemnity.[21]

(E) COMPARATIVE NEGLIGENCE

Many states have moved to comparative negligence as a method of avoiding the all-or-nothing aspects of the contributory negligence rule.[22] Suppose an injured worker claims successfully against owner and design professional and those two defendants assert a successful cross claim against the worker's employer based upon noncontractual indemnity. Jurisdictions that grant contribution would give each of the two defendants in the original action and the additional defendant in the cross claim contribution from the others if any party paid more than one-third of the award.

Generally, no *quantitative* comparison is made when a claim for noncontractual indemnity is successful. There is a *qualita-*

19. See 2A A. Larson, Workmen's Compensation § 76.11 (1983). Compare *Dole v. Dow Chemical Co.*, 30 N.Y.2d 143, 331 N.Y.S.2d 382, 282 N.E.2d 288 (1972), which allowed indemnity, with *Galimi v. Jetco, Inc.*, 514 F.2d 949 (2d Cir.1975), dealing with the Federal Employees Compensation Act, which did not. Larson, the leading scholar in this field, advocates abolition of noncontractual indemnification against the employer. See Larson, *Third-Party Action Over Against Workers' Compensation Employer* [1982] Duke L.J. 483.

20. See *A.A. Equipment, Inc. v. Farmoil, Inc.*, 31 Conn.Sup. 322, 330 A.2d 99 (1974). California has barred noncontractual indemnity by requiring an express written indemnification. West's Ann.Cal.Labor Code § 3864. Note that *express* indemnification is permitted.

21. *Santisteven v. Dow Chemical Co.*, supra note 17; *Dole v. Dow Chemical Co.*, supra note 19.

22. See Section 31.03(G).

tive comparison, such as active-passive or primary-secondary. But this comparison is made in order to determine whether the *entire* loss will be shifted from, say, a passively negligent defendant to an actively negligent one.

To deal with this all-or-nothing approach and to eliminate the difficult active vs. passive and primary vs. secondary comparisons, New York, in *Dole v. Dow Chemical Co.,*[23] adopted quantitative comparative fault for defendants.[24] In this case an employee sued the manufacturer of a chemical claiming the package was not properly labeled and users were not warned. The manufacturer brought a cross action against the employer of the injured employee claiming that the employer did not take proper precautions, that it used untrained personnel, and that it did not follow instructions. The intermediate appellate court struck down the cross action because the manufacturer's conduct had been active. However the Court of Appeals reversed, noting the inadequacy of both active-passive and primary-secondary classification (though preferring the latter) and holding that it is more just to *share* responsibility than to either *transfer* or deny transfer of the entire loss.

New York expressed reservations a year later, however, holding the *Dole* Rule did not apply where a statute placed absolute liability on an apartment owner for another's negligence.[25] The apartment owner had made a contract with an elevator operating company that had been negligent. The court held comparative fault would not apply where one party was liable vicarious-ly or because the law had placed a nondelegable duty upon them—classic cases of passive or secondary liability. The court also noted that contracts between the defendants can distribute the ultimate loss between them.

(F) PREEMPTION

Although express indemnification will be discussed in detail in Section 36.05, one problem relating to indemnity clauses must be mentioned here. Does the presence of an express indemnification clause bar noncontractual indemnity?

If noncontractual indemnification is based on the likely intention of the parties, the matter should be resolved simply by interpreting the language of the indemnity clause. If no language deals specifically with preemption, it is likely that the parties, by focusing upon indemnification, have demonstrated an intention to exclude other forms of indemnification. If noncontractual indemnification is based upon unjust enrichment, however, the problem becomes more difficult. On the one hand it can be contended that there should be no inquiry into unjust enrichment where the parties have dealt specifically with the issue. Yet unjust enrichment can have a life of its own, based upon what the law considers fair and just.

The difficulty is reflected in case decisions that have passed upon this problem. Two recent cases have concluded that the presence of an express indemnification provision bars indemnification based upon a

---

23. Supra note 19. California followed the *Dole* case in *American Motorcycle Ass'n v. Superior Court,* 20 Cal.3d 578, 146 Cal.Rptr. 182, 578 P.2d 899 (1978), despite statutory contribution.

24. Quantitative comparisons seem more popular with contemporary courts than qualitative ones, especially all-or-nothing ones. See *Cooper Stevedoring Co. v. Fritz Kopke, Inc.,* 417 U.S. 106, 94 S.Ct. 2174 (1974); *U.S. v. Seckinger,* 397 U.S. 203, 90 S.Ct. 880, 25 L.Ed.2d 224 (1970).

25. *Rogers v. Dorchester Associates,* 32 N.Y.2d 553, 347 N.Y.S.2d 22, 300 N.E.2d 403 (1973). In 1974 New York enforced a noncontractual indemnity claim based upon primary and secondary liability brought by a prime contractor against a subcontractor when a worker was injured and brought a claim against the prime. See *Kelly v. Diesel Construction Division,* 35 N.Y.2d 1, 358 N.Y.S.2d 685, 315 N.E.2d 751 (1974).

comparison of culpability.[26] But another case, *E.L. White, Inc. v. City of Huntington Beach*,[27] not only came to a different conclusion but also demonstrated some of the complexity raised by indemnification. For that reason it may be useful to look at the facts in that case.

The city of Huntington Beach contracted with White to build a drain sewer. Under the contract, White agreed to indemnify the city. Problems developed after completion. White hired a subcontractor to make repairs. One employee of the subcontractor was killed and another injured in a cave-in caused by failure to shore or slope the trenches. The action by the workers resulted in judgments against the city and White. Before the action by the workers, the city had filed a separate action against White and its insurance company asking, among other things, that a declaration be made that the city be entitled to indemnification from White under the indemnification clause contained in the prime contract. California law does not grant indemnity when the contractual indemnity clause is written in general terms and the party seeking indemnification is *actively* negligent. The city was found to have been *actively* negligent barring its indemnification claim.

After the judgment had been rendered, White and its insurer brought a legal action against the city seeking *equitable* indemnity against the city for amounts that it paid to satisfy the judgments.

The city contended that equitable indemnity was barred by the contractual indemnification clause but the court did not agree. It held that the two forms of indemnification, contractual and equitable, are separate bases for transferring a loss. If the express contractual provision, however, does not apply to the factual setting before the court,

equitable indemnification can come into play.

The indemnity clause must be interpreted to see if it has expressed *any* intention by the parties that *all* equitable indemnity is to be eliminated. The indemnification clause protected only the *city*. The court permitted equitable indemnity when a claim was made by the *contractor,* as the clause did not preclude equitable indemnity.

SECTION 36.04
IMPLIED INDEMNITY

As mentioned earlier, one form of contractual indemnity rests not upon express language but upon the presumed intention of the parties that one party will respond for a loss it causes to the other party. For example, suppose the prime contractor negligently left a hole uncovered. A mail deliverer who entered the site during construction with permission is injured. Suppose the latter successfully asserts a claim against the owner based upon the owner's obligation as a possessor of land to keep the premises reasonably safe. Even in the absence of a specific indemnity clause in the prime contract, it can be contended that the prime contractor has impliedly promised the owner that it would indemnity the owner if its negligent conduct harmed a third party who asserted a claim against the owner and obtained a court judgment.

This form of implied indemnity has had a complicated history in admiralty law where longshoremen have been injured. The relatively uniform presence of indemnification clauses and the expansive use of noncontractual indemnity, however, have made this form of indemnification relatively unimportant in modern construction disputes.

---

26. *Wyoming Johnson, Inc. v. Stag Industries, Inc.,* 662 P.2d 96 (Wyo.1983); *Covert v. City of Binghamton,* 117 Misc.2d 1075, 459 N.Y.S.2d 721 (1983).

27. 21 Cal.3d 497, 146 Cal.Rptr. 614, 579 P.2d 505 (1978).

SECTION 36.05
CONTRACTUAL INDEMNITY

(A) INDEMNIFICATION COMPARED TO EXCULPATION, LIABILITY LIMITATION, AND AGREED DAMAGES

Frequently contracts contain language that seeks to regulate loss distribution. The simplest is language under which one party is exculpated from liability that would otherwise exist. For example, upon being admitted to a hospital, suppose a patient signs a form under which she agrees that the hospital will not be responsible if she is harmed because of the hospital's negligence. The hospital would be seeking to relieve itself from liability. While the law accords considerable freedom to contracting parties to make their own rules and distribute losses as they wish, the hospital admissions room is hardly the place for the contract process. For that reason it is likely that such an agreement would be invalid.[28] Similarly, some courts and legislatures do not allow exculpation in a residential lease[29] or one in which the tenant is a small business.[30]

On the other hand, suppose race horse owners stable their horses at a race track and sign an agreement under which they exculpate the race track from any loss or damage to the horses even if caused by negligence of the race track. Such an exculpatory clause was upheld in *Rutter v. Arlington Park Jockey Club*[31] when horses were killed in a barn fire. The court held that exculpatory clauses will be upheld if it is clear from the contract that the parties intended to accomplish this result and the evidence indicates that the parties were of relatively equal bargaining strength. After concluding that the language was sufficiently clear to include negligence by the race track the court stated:

> This was a contract between businessmen and it reflected good business judgment to place the risk of loss upon the party who could least expensively insure against it. It is unlikely that the Club would provide stable facilities free of charge if by doing so it incurred liability for damage to horses of substantial but unknown value. The cost of insurance would be greater for the Club than for the horse owners because the Club could not accurately predict the number and value of the horses that would be housed in its facilities from time to time.[32]

The court was not impressed with the argument that enforcing such an exculpatory clause would encourage the race track to be careless.

The court, while enforcing the exculpatory clause, insisted that the language of exculpation be clear, exculpatory clauses requiring a high degree of specificity.[33]

Exculpatory clauses deny the right of one party to use the legal process when it would otherwise have the opportunity of doing so. Indemnification, on the other hand, does not preclude first *instance* liability. It deals with ultimate responsibility by shifting the loss from one party to another. For example, if an injured worker recovers against the owner, permitting the owner to shift the loss to the contractor by indemnification in no way precludes the injured worker from recovering. A shift of ultimate responsibility from owner to contractor occurs. If the owner has in any way been at fault, however, risk shifting through indemnification resembles exculpation. Because of this simi-

---

28. *Tunkl v. Regents of the University of California,* 60 Cal.2d 92, 32 Cal.Rptr. 33, 383 P.2d 441 (1963).

29. *Crowell v. Housing Authority of the City of Dallas,* 495 S.W.2d 887 (Tex.1973).

30. *McLean v. L.P.W. Realty Corp.,* 507 F.2d 1032 (2d Cir.1974).

31. 510 F.2d 1065 (7th Cir.1975).

32. Id. at 1068. A case in the construction context emphasizing similar considerations is *New England Telephone & Telegraph Co. v. Central Vermont Public Service Corp.,* 391 F.Supp. 420 (D.Vt.1975).

33. *Cincinnati Gas & Electric Co. v. Westinghouse Electric Corp.,* 465 F.2d 1064 (6th Cir.1972).

larity, where one party is being indemnified despite its negligence, the judicial attitude toward indemnity clauses, both in terms of interpretation and enforcement, is likely to be similar to the judicial attitudes toward exculpatory provisions.[34]

Another contract clause which can be compared to an indemnity clause is one which seeks to limit the legal remedy. For example, sellers of machinery sometimes seek to limit their liability to repair and replace defective parts. Similarly, some design professionals seek to limit their liability to their client to a designated amount of money or a certain percentage of the fee.[35] If the actual damages in the latter illustration are *less* than the specified amount, only actual damages can be recovered. But a liability limitation sets a ceiling on the damages.

Many of the same considerations that have been discussed with regard to exculpatory clauses apply to liability limitations. Where they are determined in a proper setting by parties of relatively equal bargaining power and where the language clearly expresses an intention to limit the liability, they are given effect.[36]

Contract clauses sometimes stipulate the amount of damages in advance. As seen in Section 33.09 such clauses are frequently used for unexcused time delay. They are generally given effect as long as they are reasonable. But some of the same considerations that relate to bargaining power and appropriateness of advance agreement will be taken into account. Agreed damage clauses, unlike indemnity clauses do not shift losses.

(B) INDEMNITY CLAUSES CLASSIFIED

This section will use the prime contract as an illustration. The prime contractor is the *indemnitor*, that is, the party promising to indemnify. The owner is the *indemnitee*, that is, the party to whom indemnification has been promised. The analysis in this section also applies to a subcontract where the prime contractor is the *indemnitee* and the subcontractor the *indemnitor*.

Looking first at a claim as it relates to the *indemnitor's* conduct, the indemnity clause can be "work-related." Such a clause covers a broad variety of claims that may be asserted against the indemnitee owner by third parties such as injured workers or adjacent landowners. These claims may be predicated upon fault of the indemnitor prime. However, the only indemnification requirement is that the claim "arise out of," "be occasioned by" or "due to," to use common indemnity phrases, the work or activity of the prime.[37]

The clause can be more limited. It may cover only claims based upon wrongful conduct of the prime. This is employed by AIA Doc. A201, ¶ 4.18, which will be discussed in Section 36.05(F).

Focusing upon the *indemnitee*, the claim usually asserts one or a number of bases for owner liability. The claimant may point to acts or failures to act by the owner itself or someone for whom it is responsible, such as design professional or prime contractor. This can raise issues of active vs. passive or primary vs. secondary, a topic covered in Section 36.03.[38] The claim may be based upon *status*, such as the owner being the

---

34. *Di Lonardo v. Gilbane Building Co.*, 114 R.I. 469, 334 A.2d 422 (1975). The case is reproduced in Section 36.05(c).

35. See Section 18.03(D).

36. *Delta Air Lines, Inc. v. Douglas Aircraft Co.*, 238 Cal.App.2d 95, 47 Cal.Rptr. 518 (1965).

37. *Doloughty v. Blanchard Construction Co.*, 139 N.J.Super. 110, 352 A.2d 613 (1976).

38. California treats active and passive negligence differently when faced with the required amount of specificity to enforce such clauses. See *Rossmoor Sanitation, Inc. v. Pylon, Inc.*, 13 Cal.3d 622, 119 Cal.Rptr. 449, 532 P.2d 97 (1975), and *MacDonald & Kruse, Inc. v. San Jose Steel Co.*, 29 Cal.App.3d 413, 105 Cal.Rptr. 725 (1972).

possessor of land or the employer with common law or statutory responsibilities often strict in nature. Also, the claim against the indemnitee may be based in *whole* or *in part* on the indemnitee's own wrongful acts. Clauses that cover claims solely based upon the negligence of the indemnitee are called "broad form" while those that cover claims based in part on the negligence of the indemnitee are called "intermediate form" indemnity clauses.

The exculpatory feature of many clauses, along with work-related clauses that do not require any legal wrong by the indemnitor prime, have caused courts to scrutinize such clauses carefully. Also, the bargaining power that often enables apparently one-sided clauses to be passed "down the line" from owner to prime, from prime to subcontractor and so on, has also generated judicial concern, making litigation predictability hazardous.

(C) FUNCTION OF INDEMNITY CLAUSE: *DI LONARDO v. GILBANE BLDG. CO.*

The hostility the law has shown toward clauses under which one party has promised to indemnify another often ignores the important function served by the indemnification process. As an illustration, suppose an owner plans to make a construction contract with a prime contractor. The owner recognizes the increased likelihood that claims will be made against it that are based upon the contractor's performance, sometimes but not always negligent. With a recognition that the law increasingly makes the owner responsible to third parties or at least that the law has made it more likely that claims may be made by third parties against the owner, the owner may say to the contractor:

> I have turned over the site to you. It is your responsibility to see to it that the building is constructed properly. You must not expose others to unreasonable risk of harm. The increasing likelihood that I will be sued for what you do makes it fair that you relieve me of

ultimate responsibility for these claims by your agreeing to hold me harmless or to indemnify me.

Alternatively, a reassuring proposal may come from the contractor who may say to the owner:

> I know you may be concerned about the possibility that a claim will be made against you by a third party during the course of my performance and that you will have to defend against that claim and either negotiate a settlement or even pay a court judgment. I always conduct my work in accordance with the best construction practices and I have promised in my contract to do the work in a proper manner. I am so confident that I will do this that I am willing to relieve your anxiety by holding you harmless or by indemnifying you if any claim is made against you by third parties relating to my work. You will have nothing to worry about as I will stand behind my work. If you are concerned about my ability to pay you I will agree to back it up by public liability insurance coverage.

In this context indemnification acts to seal a deal when one party is anxious. The same scenario can be played but in a slightly different way if the architect asks the owner to obtain indemnification for her through the prime contract in the manner accomplished by AIA Doc. A201, ¶ 4.18. The architect may be saying to the contractor through the owner:

> The law may hold me accountable for injury to your workers or to employees of your subcontractor because they may connect their injury with something they claim I did or should have done. You are being paid for your expertise in construction methods and your knowledge of safety rules. These are not activities in which I have been trained or in which I claim to have great skill or experience. For that reason if a claim is made against me for conduct that is your responsibility I want you to hold me harmless and indemnify me.

In addition, either owner or architect may back up its request for indemnification from the prime contractor by noting that it is exposed to potentially open-ended tort liability if claims are made by employees of the contractor or other subcontractors while the actual employer, either the prime contractor or subcontractor, who is most directly responsible, need only pay the more limited liability imposed upon it by Workers' Compensation law.

Insurance plays a significant part in the indemnification process as the promise to indemnify may be relatively worthless unless it is backed up by a solvent insurer. Usually the indemnitor, such as the prime contractor, will have to secure a special endorsement from its insurer to cover contractually assumed liability. This will increase the cost of insurance, a cost that will ultimately be passed on to the owner. But the owner may believe that it is cheaper in the long run for it to pay this additional cost than to take the risk and insure against it. Distribution of risks through indemnification can facilitate insurance at the cheapest possible cost.[39] (An architect who is asked to indemnify, as noted in Section 18.04, may not be persuaded of this rationale if her professional liability insurance has a high deductible.)

Most of the illustrations have involved indemnification sought or demanded by the owner from the prime contractor; a similar situation arises if the prime contractor demands it from a subcontractor. Why does the prime contractor agree to indemnify? First, it may have no choice. Second, it may pass on the bulk of these risks to subcontractors. Third, and most important, it realizes that such a risk can be insured with the cost passed on to the owner.

Other problems may arise, such as the possibility that the indemnitee, such as the owner or the architect, will be indemnified if it were guilty of wrongful conduct which in part caused the harm. To that extent a broad form or intermediate indemnity clause can exculpate a party from its own wrongdoing.[40] Even more, a prime contractor who obtains a similar indemnification from a subcontractor may be encouraged to be less careful because it has received indemnification from the subcontractor despite the harm being caused at least in part by the prime contractor.

But does indemnification encourage carelessness? The factor that bears most heavily on how the contractor performs is its insurance rates, which may in part be predicated upon the number of claims that are made against it. But the insurance premium is more likely to be predicated on the type of work, the locality and the volume of the prime contractor's business. Other factors can deter careless performance, such as the possibility that public officials may take action against the prime contractor or the contractor's license may be in jeopardy. In any event, the law allows parties to provide insurance even though the existence of insurance can encourage carelessness. Indemnification is a form of insurance and there is no reason why the law should be hostile to this process.[41] Undoubtedly some of the hostility is also generated by the way in which indemnification is forced upon the weaker party who must usually accept such a clause on a "take it or leave it" basis. Again, if the risk is clearly brought to the attention of the weaker party and that party can insure against that risk and pass the cost of insurance on to the stronger party in its contract price, hostility is not justified.

A case that recognizes the functions that have been outlined is reproduced at this point.

---

39. In addition to *Di Lonardo v. Gilbane Building Co.*, reproduced in this section, see *Hicks v. Ocean Drilling & Exploration Co.*, 512 F.2d 817 (5th Cir.1975); *Buscaglia v. Owens-Corning Fiberglas*, 68 N.J.Super. 508, 172 A.2d 703 (1961), affirmed 36 N.J. 532, 178 A.2d 208 (1962).

40. *Davis v. Commonwealth Edison Co.*, 61 Ill.2d 494, 336 N.E.2d 881 (1975); but see *Kelly v. Diesel Construction Division*, 35 N.Y.2d 1, 358 N.Y.S.2d 685, 315 N.E.2d 751 (1974).

41. Hostility will be shown in Section 36.05(D) and (E) dealing with enforceability and the high degree of specificity required.

DI LONARDO v. GILBANE BUILDING CO.

Supreme Court of Rhode Island, 1975.
114 R.I. 469, 334 A.2d 422.

ROBERTS, Chief Justice. In this civil action a general contractor tort defendant, Gilbane Building Company (Gilbane), seeks indemnification from a third-party defendant subcontractor, Joseph P. Cuddigan, Inc. (Cuddigan). In October of 1970 Gilbane, a general contractor, was engaged in construction work at the Rhode Island Hospital. Cuddigan was a plumbing subcontractor on the project and was the employer of Joseph Di Lonardo, Jr., the plaintiff. It appears from the record that in the contract between Gilbane and Cuddigan, Cuddigan had agreed to indemnify Gilbane for any personal injury, death, or property damage which occurred due to the work which Cuddigan had agreed to do. The primary issue in the instant case concerns the validity of the indemnity clause.

The evidence discloses that plaintiff, Di Lonardo, was walking from the subcontractor's supply trailer at the construction site carrying certain plumbing supplies up to his specific job location. While so doing, he was injured when his left foot broke through a sheet of plywood that had been negligently placed on a stairway by Gilbane's employees. The plaintiff sued Gilbane, who in turn brought in Cuddigan as a third-party defendant. Relying on the indemnity clause, Gilbane sought to be held harmless. At the close of all the evidence, Cuddigan moved for a directed verdict. The motion was granted, thus relieving the subcontractor of any liability, the trial justice stating as his reason that the indemnity clause was unconscionable, contrary to public policy, and, therefore, void and unenforceable. The principal case was then submitted to a jury, which returned a verdict for plaintiff. At that point the trial justice granted a motion by Cuddigan for a directed verdict. Gilbane is now prosecuting an appeal from that judgment.

In *Dower v. Dower's, Inc.,* . . ., we stated that one will not be held harmless for his own negligence pursuant to an indemnity provision unless the specific and unambiguous intent of the parties is to shift liability to the non-negligent party; in such cases the rule of strict construction will be applied. In *Dower* the issue of the validity of such an indemnification provision in the face of a contention that it was violative of public policy was expressly left undecided.

In the case at bar the disputed portion of the contract provides:

"The Subcontractor agrees to indemnify and save harmless the Owner and General Contractor against loss or expense by reason of the liability imposed by law upon the Owner or General Contractor for damage because of Bodily Injuries, including death, at any time resulting therefrom accidentally sustained by any person or persons or on account of damage to property arising out of or in consequence of the performance of this work whether such injuries to persons or damage to property are due or claimed to be due to any negligence, including gross negligence, of the Subcontractor, the Owner, the General Contractor, his or their employees or agents or any other person."

We turn, first, to consider Gilbane's contention that it was error to hold the indemnity provision void as violative of public policy. We are constrained to agree with Gilbane and hold that a contract in which a subcontractor promises to indemnify and save the general contractor harmless regardless of the general contractor's negligence is a valid contract which in no way violates public policy. The law in this respect is well settled. . . . The trial justice apparently was concerned that general contractors, when armed with the subcontractor's indemnity protection, would become callously reckless and willing to sacrifice people's safety in order to avoid what might be costly precautions against injury. This reasoning fails, however, to take into consideration the fact that there is no significant distinction between the situation at bar and one in which the party has insured against his own negligence. Such practice has long been accepted. Mr. Justice Story held in 1837 that "[t]here is nothing unreasonable, unjust or inconsistent with public policy, in allowing the assured to insure himself against all losses, from any perils not occasioned by his own personal fraud." *Waters v. Merchants' Louisville Ins. Co.,* . . . We perceive the shift of liability in the instant case as simply a shift in the burden of purchasing insurance coverage. We hold that parties in adequate bargaining positions should be free to distribute this burden, by contract, as they see fit.

* * *

The appeal of the third-party plaintiff is sustained, the judgment entered below is reversed, and the cause is remanded to the Superior Court for further consideration.

(D) VALIDITY

Clauses most susceptible to attack have been broad form or work-related. These clauses can provide for indemnity not only when the indemnitor has not been negligent but, when the clause is clear enough, despite negligence by the indemnitee, that is, the party demanding indemnification.[42] As seen in *Di Lonardo v. Gilbane Building Co.,* courts tend to enforce such agreements.[43] This contractual freedom is more likely to be given where the court looks upon indemnity as an insurance facilitator. Such

---

42. *Stephens v. Chevron Oil Co.,* 517 F.2d 1123 (5th Cir.1975).

43. The following cases have upheld these clauses:

    1. *Kansas City Power & Light Co. v. United Telephone Co.,* 458 F.2d 177 (10th Cir.1972).

    2. *Kenny v. Fuller,* 87 A.D.2d 183, 450 N.Y.S.2d 551 (1982) (CM claim against prime based on work related clause);

    3. *Simon v. Corbetta Construction Co.,* 391 F.Supp. 708 (S.D.N.Y.1975) (rejecting argument that the clause was unconscionable);

courts are less concerned with inequality of bargaining power.[44] Relatively little bargaining occurs over indemnity. Owners impose indemnifications on prime contractors and the latter impose them on subcontractors. And, if they are clear enough to point to the need for insurance, they should be enforced.

Many legislatures have invalidated some indemnification clauses. These statutes were enacted mainly in the aftermath of a struggle in 1966 between the American Institute of Architects (AIA) and the Associated General Contractors (AGC) over an indemnification clause proposed by the AIA and opposed by the AGC. Before a compromise was reached, and even after, anti-indemnity legislation was enacted in many states.

Generally, legislation invalidated clauses which sought to relieve the indemnitee even if it was the sole cause of the harm, the so-called broad form indemnity clauses.[45] There are variations, however, such as the California statute, which also prohibits indemnification for design negligence [46] and the recently enacted New York legislation, which seems to require that indemnification clauses be comparative and not "all or nothing." [47] Those planning to incorporate an indemnification clause into a construction contract must consult local law.

Legislation such as that in California and New York resulted from contentions by interested groups that indemnification clauses were forced down the throats of weaker parties and could encourage careless work. If the true function of indemnification is as described in (C) such legislation is ill-advised.

Even though courts generally, in the absence of statutes, enforce indemnity clauses, the principal method of controlling them has been through a requirement of language specificity where there is negligence by the indemnitee. This will be discussed in Section 36.05(E).

(E) INTERPRETATION

Negligence of Indemnitee and Requisite Specificity

Courts that look at indemnification as an insurance facilitator are likely to interpret indemnification clauses in the same manner as any other clause.[48] The court must seek the intention of the parties. Although the principal source for determining this intention is the language of the clause, courts can look at surrounding facts and circumstances. If these processes do not prove fruitful, such courts would very likely interpret them against the party who had drafted the clause.[49]

Yet most courts still seem more impressed with indemnification as a device to exculpate a party from its own negligence and a product of bargaining oppression.

---

4. *New England Telephone & Telegraph Co. v. Central Vermont Public Service Corp.*, supra note 32 (emphasizing two substantial corporations of equal bargaining power who knew what they were doing when they made the contract);

5. *Westinghouse Electric Co. v. Murphy, Inc.*, 425 Pa. 166, 228 A.2d 656 (1967).

6. *Phillippe v. Rhoads*, 233 Pa.Super. 503, 336 A.2d 374 (1975) (requiring either legislative policy making the clause unenforceable or showing of a monopoly).

44. *Hicks v. Ocean Drilling & Exploration Co.*, supra note 39.

45. Mich.Comp.Laws Ann. § 691.991.

46. West's Ann.Cal.Civ.Code § 2782. Section 2782.5 excludes liability limitation under certain circumstances from the ambit of the statute.

47. McKinney's N.Y.Gen.Oblig.Law § 5–322.1.

48. See note 43, supra.

49. *United States v. Seckinger*, supra note 24; *Doloughty v. Blanchard Construction Co.*, supra note 37.

These courts still view such clauses with suspicion if not hostility. As a result, special rules determine whether such clauses will cover negligence by the indemnitee. These rules vary considerably as to the degree of specificity required. Most states require that the clause state clearly and unequivocally that negligence by the indemnitee is covered or the clause will not cover the loss caused by an indemnitee's negligence.[50] Other states employ similar formulas that require a high burden of clarity be placed upon the indemnitee.[51] Some even require that negligence be mentioned specifically.[52] The extent to which such a heavy burden can be placed on the drafter is demonstrated by *Chevron Oil Co. v. D. Walton Construction Co.*[53] Here the federal court applied Texas law that requires considerable clarity before such clauses will be interpreted to cover negligence of the indemnitee.

Chevron contracted with Walton for remodeling work. An employee of Walton's was injured because of the sole negligence of Chevron's employees. When Chevron demanded indemnity, it pointed to an indemnity clause that stated that Walton would indemnify Chevron and hold it harmless:

> from and against all loss, damage, liability, claims and liens of every kind arising out of or attributed, directly or indirectly, to the operations of Contractor hereunder, including without limitation and irrespective of negligence, all claims for injury to or death of persons, loss of or damage to property, and claims of workmen and materialmen.[54]

Yet despite mentioning negligence, the court held that Chevron would not be indemnified by Walton because the negligence phrase did not make clear that it was to cover Chevron's negligence. The court quoted a Texas case that worried about large and ruinous awards that might be made under such clauses. Ignoring the likelihood of insurance, the court held that the language was susceptible to two interpretations and would be interpreted against Chevron who had prepared it.

---

50. A few of the many cases are:

1. *Becker v. Black & Veatch Consulting Engineers,* 509 F.2d 42 (8th Cir.1974).
2. *Mississippi Power Co. v. Roubicek,* 462 F.2d 412 (5th Cir.1972).
3. *Warburton v. Phoenix Steel Corp.,* 321 A.2d 345 (Del.Super.1974), aff'd, 334 A.2d 225 (Del.1975).

51. *Simon v. Corbetta Construction Co.,* supra note 43, required a clear and unmistakable expression. *City of Hazard Municipal Housing Commission v. Hinch,* 411 S.W.2d 686 (Ky.1967), required that the language be clear and explicit, as did *Daniel Construction Co. v. Welch Contracting Co.,* 335 F.Supp. 303 (E.D.Va.1971), and *St. Joseph Hospital v. Corbetta Construction Co.,* 21 Ill.App.3d 925, 316 N.E.2d 51 (1974). *Laverty, Inc. v. Mel Jarvis Construction Co.,* 513 F.2d 1307 (8th Cir.1975), and *Amerco Marketing Co. v. Meyers,* 494 F.2d 904 (6th Cir.1974), required that the language be clear and unambiguous. *Wrobel v. Trapani,* 129 Ill.App.2d 306, 264 N.E.2d 240 (1970), and *Young v. Anaconda American Brass Co.,* 43 Wis.2d 36, 168 N.W.2d 112 (1969), required that the language be a clear expression of an intention to cover negligence by the indemnitee. Finally, *Gantt v. Mobil Chemical Co.,* 463 F.2d 691 (5th Cir.1972), applying Texas law, stated that the language would have to fairly "want the conclusion" that the indemnitee's negligence was to be covered. A complete collection of cases can be found in Annot., 27 A.L.R.3d 663 (1969).

52. *Cole v. Chevron Chemical Co.,* 477 F.2d 361 (5th Cir.1973); *Chesapeake & Potomac Telephone Co. v. Allegheny Construction Co.,* 340 F.Supp. 734 (D.Md.1972); *Fireman's Fund Insurance Co. v. Commercial Standard Insurance Co.,* 490 S.W.2d 818 (Tex.1972). A collection of the Texas cases which appear to require explicit reference to negligence with some exceptions is set forth in *Richmond v. Amoco Production Co.,* 390 F.Supp. 673 (E.D.Tex.1975), affirmed 532 F.2d 1373 (5th Cir.1976).

Other cases have stated that specific mention of negligence is not required. *Becker v. Black & Veatch Consulting Engineers,* supra note 50; *Hanley v. James McHugh Construction Co.,* 444 F.2d 1006 (7th Cir.1971).

53. 517 F.2d 1119 (5th Cir.1975).

54. Id. at 1121.

Complexity is confounded by some jurisdictions that attempt to distinguish between active and passive negligence, concluding that the requisite degree of specificity applies to active but not passive negligence.[55] Other courts do not make the active-passive distinction for this purpose.[56]

The plethora of cases that have interpreted indemnity clauses and have come to variant results caused the Illinois Supreme Court to state in despair:

> We have examined the authorities cited by the parties and many of those collected at 27 A.L.R.3d 663, and conclude that the contractual provisions involved are so varied that each must stand on its own language and little is to be gained by an attempt to analyze, distinguish or reconcile the decisions. The only guidance afforded is found in the accepted rule of interpretation which requires that the agreement be given a fair and reasonable interpretation based upon a consideration of all of its language and provisions.[57]

Despite this neutral approach the court held that the language did not cover negligence of the indemnitee, as such intention was not expressed clearly and explicitly.

Yet the parties who are to plan their insurance coverage upon the scope of risk and insurers who must cover particular risks must be able to know what such clauses will cover. If courts ignore existing precedents and focus too heavily upon the surrounding facts and circumstances, the uncertainty of application will make indemnification an inefficient insurance facilitator and cause over-insurance and higher rates. This is demonstrated by *American Oil Co. v. Hart*,[58] a case that involved an indemnity claim made by a large oil company against a small independent contractor who had been hired to repair a pole at a service station. The independent contractor's employee was injured while seeking to perform the repair work and successfully brought an action against the oil company based upon the latter's negligent maintenance of the pole. When the oil company sought indemnification from the independent contractor, the oil company pointed to the indemnity clause that was all-inclusive and seemed to cover this accident. In addition, the oil company pointed to an earlier case precedent that had upheld an identical clause when a similar accident had occurred. But the court, noting that the language must be "interpreted" in its setting, concluded that the indemnitor could not have intended to cover this loss when he received $40 a visit for making these repairs. Pointing to the need for a clear and unequivocal coverage of losses caused by the indemnitee's negligence, the court refused to apply the clause.

The indemnitor was a small businessman, perhaps of doubtful financial responsibility. It is likely that the indemnity claim would not have been made had he not been insured. If so, it is simply a question of whether the oil company or its insurer or the indemnitor's insurer should pay for this loss. The court's conclusion might have been proper had the language not made it clear to the indemnitor that he would be taking this risk and that he should insure against it. But if the language served this function, the court's conclusion was incorrect.

Perhaps it seems unfair for the repairman to pay a large award when he was receiving but $40 a visit. But the charge for each visit should have included that portion of the insurance that could reasonably be

---

55. *Morgan v. Stubblefield,* 6 Cal.3d 606, 100 Cal.Rptr. 1, 493 P.2d 465 (1972); *Wrobel v. Trapani,* supra note 51; *Thompson-Starrett v. Otis Elevator Co.,* 271 N.Y. 36, 2 N.E.2d 35 (1936). The standard was relaxed considerably in *Levine v. Shell Oil Co.,* 28 N.Y.2d 205, 321 N.Y.S.2d 81, 269 N.E.2d 799 (1971).

56. *Becker v. Black & Veatch Consulting Engineers,* supra note 50.

57. *Tatar v. Maxon Construction Co.,* 54 Ill.2d 64, 294 N.E.2d 272, 273–74 (1973).

58. 356 F.2d 657 (5th Cir.1966).

chargeable to the repair work. The efficiency of the indemnity clause as an insurance facilitator is destroyed by disregarding judicial precedent and interpreting identical language in similar cases differently simply because it appears to be unfair when insurance is disregarded. The court's decision reflects the older, more traditional view of indemnity, an approach different from that manifested by *Di Lonardo v. Gilbane Bldg. Co.*

The uncertainty in this area is typified by two cases involving *identical* contract language. One barred indemnity because the indemnitee had been negligent [59] while the other held negligence of the indemnitee would not preclude indemnification.[60] Variant results may be justified on the power of each jurisdiction to determine its own policy regarding indemnification and the extent to which it will be permitted to exculpate the indemnitee. But it is indefensible for courts in the same jurisdiction to reach variant results simply because one court seems more concerned about the inequity of a broad form indemnification clause while the other is willing to give the parties more room to make their own deals. Planning cannot be accomplished when courts are too willing to permit inconsistent results.

Losses and Indemnity Coverage

Clauses can cover certain losses but not others. For example, indemnity clauses can be drawn broadly enough to cover any loss, even those relating to property damage suffered by the indemnitee. Indemnity, however, is generally designed to transfer losses relating to claims third parties make against the indemnitee.[61]

Sometimes assertions of indemnity seem removed from insurance facilitation and appear to be a grasping at straws. For example, an indemnity claim was made for damage resulting to window frames faultily built by the indemnitee prime contractor and then installed by the indemnitor subcontractor.[62] The court saw no specific language covering this loss and refused to include it within the general language.

Similarly, an indemnification clause in a prime contract was held not to cover a loss incurred by the owner to a third party based upon the prime contractor's having trespassed upon the third party's land while doing the work.[63] The owner did not comply with its contract requirement to obtain an easement. The court held that the loss did not arise out of the prime contractor's performance even though the trespass was caused by its performance. A claim made based upon an injury that occurred after the work had been completed, however, falls within the ambit of the indemnity clause. Injuries, whether they occur during or after performance, are the type of loss typically covered by insurance and part of the indemnification process.[64]

Work Relatedness of Injury

The indemnitee who employs a work related indemnity clause usually seeks protection against claims that are made incident

---

59. *Daniel Construction Co. v. Welch Contracting Corp.*, 335 F.Supp. 303 (E.D.Va.1971).

60. *Davis Constructors & Engineers, Inc. v. Hartford Accident & Indemnity Co.*, 308 F.Supp. 792 (M.D.Ala.1968).

61. *Varco-Pruden, Inc. v. Hampshire Construction Co.*, 50 Cal.App.3d 654, 123 Cal.Rptr. 606 (1975); *Collins v. Montgomery Ward & Co.*, 21 Ill.App.3d 1037, 315 N.E.2d 670 (1974). In *Pacific Gas & Electric Co. v. G.W. Thomas Drayage & Rigging Co.*, 69 Cal.2d 33, 69 Cal.Rptr. 561, 442 P.2d 641 (1968), the indemnitee sought recovery for damage to his own property. The court held evidence submitted by the indemnitor that tended to show that the parties intended to cover only claims made by third parties should have been admitted into evidence.

62. *Mesker Brothers Iron Co. v. Des Lauriers Column Mould Co.*, 8 Ill.App.3d 113, 289 N.E.2d 223 (1972).

63. *Serafine v. Metropolitan Sanitary District*, 133 Ill.App.2d 93, 272 N.E.2d 716 (1971).

64. *Becker v. Black & Veatch Consulting Engineers*, supra note 50.

to or arising out of the indemnitor's performance. Interpretation problems develop when a claimant is injured while at work but the principal cause of the injury is not his doing the work but the activity of the indemnitee who later seeks indemnification from, as a rule, the indemnitor employer of the claimant. The only connection that can be made between the accident and the activity of the indemnitor is that the accident would not have happened had the indemnitor not been on the job. These problems usually involve an accident to an employee of the subcontractor, a work related clause, and a demand for indemnification from the indemnitee prime to the indemnitor subcontractor.

Work related clauses can have slightly variant language and court decisions are sometimes based upon trivial language differences. For example, in *Hanley v. James McHugh Const. Co.*,[65] the indemnity clause in the subcontract used the phrase "on account of acts or omissions of the subcontractor whether negligent or not." While unloading a truck, an employee of a subcontractor was struck and injured by an 8-foot section of a 4 × 4 that fell from the seventeenth floor because of the prime contractor's negligence. The clause was held broad enough to cover negligence caused by the prime contractor and did not require that there be negligence by the subcontractor. Yet the court fastened upon the "on account of" phrase as a basis for holding that the loss was not caused by the subcontractor's performance and did not fall within the clause. The court noted that the only connection between the subcontractor's performance and the injury was that without having made the subcontract, the worker would not have been present on the site.

This result appears to be based principally upon hostility to the clause that would require indemnification when the prime contractor was solely negligent.[66]

Other cases have interpreted work relatedness more neutrally. For example *White v. Morris Handler Co.*[67] involved the employee of a subcontractor who descended by way of an elevator from the thirtieth floor where he had been working to the third floor where he put his tools away and changed his clothes in a plumber's shanty. On his way to a stairwell he walked along a concrete walkway, slipped on gravel, and fell thirty-five feet to the ground. It was the prime contractor's responsibility to install a guardrail or handrail along the walkway.

The court enforced the indemnity clause despite it being a broad form indemnification inasmuch as it combined work relatedness and a harm caused entirely by the negligence of the indemnitee prime contractor. It also concluded that this accident arose out of the performance of the work, a conclusion quite different from that reached in the *Hanley* case. Interestingly, between the time the contract in the *White* case was made and the ensuing litigation, the Illinois legislature outlawed such broad form indemnity clauses prospectively. The court rejected an attempt by the subcontractor to argue that the enactment of the legislation showed that the broad form indemnity clause was against Illinois public policy. The court noted that such clauses were enforceable before the enactment of the statute. Opinions like that in the *Hanley* case destroy much of the value of a work related indemnity clause.

Even courts that seek neutral interpretations of work related clauses may find the task difficult. For example, in *General Ac-*

---

65. Supra, note 52.

66. Similarly, see *Martin Wright Electric Co. v. W.R. Grimshaw Co.*, 419 F.2d 1381 (5th Cir.1969); *Wilson v. Illinois Bell Telephone Co.*, 19 Ill.App.3d 47, 310 N.E.2d 729 (1974).

67. 7 Ill.App.3d 199, 287 N.E.2d 203 (1972). See also *Halverson v. Campbell Soup Co.*, 374 F.2d 810 (7th Cir. 1967).

cident Fire & Life Assurance Corp. v. Fine-gan & Burgess [68] a sign subcontract included an indemnity provision. After completion of the sign subcontractor's work the owner's project engineer went to inspect the sign, accompanied by an employee of the sign subcontractor. After inspection, the engineer indicated he wished to check a particular switch that had been installed by another subcontractor. The employee of the sign subcontractor indicated the location of the switch and then left.

On his way to inspect the switch, the engineer fell from a walkway, which had no railing because of the prime contractor's negligence. The injured engineer sued both prime and sign subcontractor. The jury found that the prime contractor had been negligent but that the sign subcontractor had not. The prime contractor's insurer paid the claim and then brought an action against the sign subcontractor based upon the indemnity provision. The purpose of inspection was to enable a tenant to move in earlier and was not directly related to the sign subcontractor's performance. The area where the injury had occurred was under the general control of the prime contractor. For these reasons the court held the clause did not cover this accident. It seems that the engineer was no longer dealing with the sign subcontractor's work.

Amount Payable

Usually the indemnitee seeks to transfer its entire loss to the indemnitor. This, as a rule, includes any money paid to the claimant, any costs of investigating the claim, any legal, investigative or expert witness costs, and interest from the time the payment was made to the third party.[69] Most clauses deal with these issues. The most troublesome questions are those that relate to amounts paid under a settlement and costs to defend when it is determined that no liability existed.

As to the first, although one court seemed to require that the indemnitee establish that it would have been liable,[70] it is better to require indemnity if a settlement were made in good faith.[71] It should not be necessary for the indemnitee to have to either litigate or settle and then establish legal responsibility.

Cost of investigating and defending the claim, the latter a particularly formidable expense, can cause difficulty. Usually indemnification clauses specifically state these costs are recoverable and this result is likely to follow even in the absence of a specific provision dealing with costs of defense.[72] If the claimant does not prevail, the indemnitor will still have to pay the defense costs, at least where the indemnity clause is work related and is not based upon liability.[73] If indemnity is based upon liability, however, courts still tend to look at such clauses as a needless exercise of bargaining strength or even incentive to carelessness. Such courts will narrowly interpret them. Those courts that see these clauses as insurance facilitators and legitimate risk shift mechanisms will give such

---

68. 351 F.2d 168 (6th Cir.1965).

69. *Larive v. United States,* 449 F.2d 150 (8th Cir.1971).

70. *Ford Motor Co. v. W.F. Holt & Sons, Inc.,* 453 F.2d 116 (6th Cir.1971). As shall be seen in Section 36.05(F), this is one of the deficiencies of the AIA clause.

71. *General Insurance Co. v. Singleton,* 40 Cal.App.3d 439, 115 Cal.Rptr. 291 (1974), based mainly upon the indemnification language.

72. *Moses-Ecco Co. v. Roscoe-Ajax Corp.,* 320 F.2d 685 (D.C.Cir.1963).

73. *Titan Steel Corp. v. Walton,* 365 F.2d 542 (10th Cir.1966); *Bethlehem Steel Corp. v. K.L.O. Welding Erectors, Inc.,* 132 N.J.Super. 496, 334 A.2d 346 (1975); *Tri-M Erectors, Inc. v. Donald M. Drake Co.,* 27 Wash.App. 529, 618 P.2d 1341 (1980).

clauses a neutral interpretation. Even the latter courts will not always find that a particular loss is covered. But they do not resolve every doubt against coverage.

(F) THE AIA INDEMNITY CLAUSE: HISTORY AND ANALYSIS

Before 1966 the AIA General Conditions of the Contract did not contain an express indemnity provision. Indemnification claims had to be based upon noncontractual indemnity.[74] But concern over an Illinois decision that signalled increased architect liability when workers were injured because of unsafe construction methods [75] caused the AIA to seek to shift the risk of certain losses to the contractor.

Section 4.18 provided for indemnity against losses, including attorney's fees, which arise from performance of the work. Indemnity included personal harm and property damage including loss of use other than to the work itself. In addition to damage to the work, economic losses such as delay damages or other losses resulting from causes specified in the indemnity clause were not covered. Indemnification would be awarded even if negligence by the owner or architect partly caused the loss.[76] But although the clause used work-related language, indemnity would not apply unless the contractor or someone for whose acts it was responsible was negligent. It was not a work related clause of the type described in Section 36.05(E). Excluded were losses caused wholly or substantially by defective drawings or specifications.

The Associated General Contractors (AGC) that had endorsed earlier documents published by the AIA refused to endorse the 1966 edition of A201 because of the indemnity clause. The AGC claimed that it would be unfair for the entire loss to be shifted to the contractor when the architect might be partially responsible for the loss. Even more important, the AGC contended that prime contractor liability insurers would not issue policies on the clause as written because it would involve them in professional liability problems. This was an insurance underwriting specialty in which they did not wish to engage.

Finally, after a year of negotiation the two groups met with a number of insurers and worked out a compromise that became the indemnity clause of 1967. There were added exclusions from indemnity which, in 1966, had been limited to defective drawings or specifications. Instead ¶ 4.18.3 was inserted which provided:

> The obligations of the Contractor . . . shall not extend to the liability of the Architect, his agents or employees arising out of (1) the preparation or approval of maps, drawings, opinions, reports, surveys, Change Orders, designs or specifications, or (2) the giving of or the failure to give directions or instructions by the Architect, his agents or employees provided such giving or failure to give is the primary cause of the injury or damage.[77]

Paragraph 4.18 required the following steps be taken to determine whether the contractor will have to indemnify owner or architect:

1. Was the type of harm (personal injury or damage to property other than the work), which was the basis of the claim, covered?

2. If so, was it caused in part by the negligence of the contractor or by someone for whom it was responsible? (sole negligence

---

74. See Section 36.03.

75. *Miller v. De Witt*, 37 Ill.2d 273, 226 N.E.2d 630 (1967). The indemnification clause appeared earlier than the decision but the ruling by the intermediate appellate court against the architect undoubtedly caused insertion of the clause in A201. The ultimate decision was not favorable to the architect on the question of liability.

76. In a dictum, the United States Supreme Court stated that negligence of the indemnitees was stated with sufficient specificity. *United States v. Seckinger*, 397 U.S. 203, 90 S.Ct. 880, 25 L.Ed.2d 224 (1970).

77. AIA Doc. A201, ¶ 4.18.3 (1967). The same language was carried forward into the 1976 edition.

of owner or architect would preclude indemnification).[78]

3. If so, was it excluded by ¶ 4.18.3?

4. If the exclusion of ¶ 4.18.3 related to the matters specified in (1), there would be no indemnity, even, evidently, if this was only a partial cause of the loss.

5. If based upon (2), there would be indemnity only if the giving or failure to give was not the primary cause of the injury or damage.

This indemnification clause would be unwieldly in the typical worker injury case where the claim by the worker is based upon a violation of safety orders and failure by the architect to detect it or take appropriate action. In such a case indemnity occurs only if it were determined that the architect's responsibility was not a primary cause of the loss. This can be a difficult factual problem.

The 1967 edition did not make clear that the architect, although not a party to the contract, could recover on the indemnification clause.[79] In 1976 A201 was modified and some changes made in the indemnification clause. Cost of investigation or expert witnesses was made somewhat easier to recover. Also, any noncontractual indemnity that would otherwise be recoverable based upon active-passive negligence would not automatically be precluded by the clause.[80] Recognition was also given to the anti-indemnity statutes that in some states might make the clause unenforceable.[81] If the statute ran afoul of such legislation, the AIA appeared to invite courts to scale it down to

make it enforceable. Finally, ¶ 1.1.2 seemed to make certain that the architect, despite not being a party to the contract, could sue for enforcement of the indemnity clause.[82]

Section 4.18 is not a work related clause. It requires negligence by the contractor. The clause would make it difficult to settle and then recover since it appears to require that liability be established. It would have been better to make clear that good faith settlements could be the basis for an indemnification recovery. Similarly, a successful defense against the claim establishing that the indemnitor was not liable, if undertaken by the indemnitee, cannot be the basis for recovery of defense costs by the indemnitee.[83]

Mention has been made of the tendency of modern law to avoid "all or nothing" solutions. Classically, noncontractual indemnity and the general practices of those drafting indemnification clauses created all or nothing solutions. Losses were transferred rather than shared. As noted earlier [84] there has been a tendency to use comparative principles in contribution and noncontractual indemnification. This tendency is beginning to make its way into express indemnification clauses. For example, New York recently revised its anti-indemnity statute. Before July 31, 1981, indemnification clauses that exculpated the indemnitee from its sole negligence were invalid. Effective July 31, 1981, "all or nothing" clauses are invalid and the only clauses that will be enforceable are those under which the indemnitor indemnifies the indemnitee only for that portion of the

78. *Cumberbatch v. Board of Trustees*, 382 A.2d 1383 (Del.Super.1978).

79. Owners have been able to recover on indemnification clauses and subcontracts to which they are not a party. *Schroeder v. C.F. Braun & Co.*, 502 F.2d 235 (7th Cir.1974); *Titan Steel Corp. v. Walton*, supra note 73. In *Pylon, Inc. v. Olympic Insurance Co.*, 271 Cal.App.2d 643, 77 Cal.Rptr. 72 (1969), the court seemed to assume the engineer could maintain an action against the prime contractor based upon the indemnity clause.

80. See Section 36.03(F).

81. See Section 36.05(D).

82. See supra note 79.

83. See Section 36.05(E).

84. See Sections 36.02 and 36.03(E).

loss caused by the indemnitor's negligence.[85] This change will undoubtedly lead to more comparative indemnification clauses.

The current indemnification clause in A201 is an all or nothing clause. In 1978, however, the AIA published A401, its standard subcontract, and changed the indemnification from all or nothing to a comparative indemnification clause.[86] This may presage a change in the next edition of A201.

(G) INSURANCE

Frequently owners require that contractors procure insurance covering the risks specified in the indemnification clause. Even without this requirement a prudent contractor will be certain that its insurance will cover this risk. Generally liability policies cover only liability imposed by law and not that imposed or assumed by contract. Courts have arrived at different results when faced with the question of whether an indemnification clause is covered under a liability policy.[87] It is usually possible for the contractor to obtain a specific endorsement covering the liability assumed by these indemnification clauses. Contractors should be certain, as should owners, that the liability policy covers this risk.

PROBLEM

O and C entered into a construction contract using AIA standard forms A101 and A201 set forth in Appendices B and C. C was to construct a twenty-unit apartment complex on land owned by O in accordance with plans and specifications drafted by A.

C decided to take steps to protect the site from theft and neighborhood children who frequently played there. It erected a six-foot barbed wire fence with a gate that was locked after work was completed during the day. It also placed a fierce dog on the site after working hours to deter vandals and others from entering the site.

The dog caused a number of problems. Neighbors complained that the dog howled during the night and made it difficult for them to sleep. Those who passed by the construction site at night were often frightened by the menacing growls of the dog. Finally, one day the dog got away from its handler when it was brought to the site after work had been completed.

It bit a subcontractor's employee who had remained on the site a little late in order to finish some work.

The neighbors, the bystanders who walked by the site, and the worker who had been bitten all made claims against the owner. The injured worker also made a claim against the architect contending that the architect knew the dog was frequently brought to the site while a few workers still remained and did not take steps to make certain that the dog could not bite or molest them. In fact the worker stated that her union steward had complained of this to the architect.

O and A have demanded that C defend these claims in accordance with ¶ 4.18 of A201. Are their demands justified? Would C have to indemnify them if they made any settlements with the neighbors, the bystanders, or the injured worker? If legal action were brought against O which terminated in O's favor, would O be able to recover its legal expenses from C?

---

85. See note 47, supra.

86. AIA Doc. A401, ¶ 11.11.1.

87. Compare *Union Paving Co. v. Thomas,* 186 F.2d 172 (3d Cir.1951), with *United States Fidelity & Guaranty Co. v. Virginia Engineering Co.,* 213 F.2d 109 (4th Cir.1954).

37

Surety Bonds: Backstopping Contractors

SECTION 37.01 MECHANICS AND TERMINOLOGY

The surety bond transaction is a peculiar arrangement, and differs procedurally from most contracts. The typical surety arrangement is essentially triangular. The "surety" obligates itself to perform or to pay a specified amount of money, if the "principal debtor," usually called the "principal," does not perform. The person to whom this performance is promised is usually called the "obligee" (sometimes called the "creditor"). In the building contract context, the surety is usually a professional bonding company. The principal is the prime contractor, or, in the case of subcontractor bonds, a subcontractor. The obligee is the owner or, in the case of a subcontractor bond, the prime contractor.

If required by the owner, the prime contractor (the principal) applies for a bond from a bonding company. Usually the cost of the bond is paid indirectly by the owner, since the bidder adds the cost of the bond to its costs when computing the bid. The bond will be issued to the owner (the obligee). The bond runs to the owner in that performance by the surety has been promised to the owner, even though the application for the bond has been made by the prime contractor.

Another large problem in dealing with surety bonds results from the antiquated way in which bonds are written. The earliest bonds were called "penal" bonds. The surety made an absolute promise to render a certain performance or to pay a specified amount of money. This would be followed by a paragraph which would state that the bond would be void if the principal promptly and properly performed all the obligations under the contract between principal and obligee. This format can cause problems. The transaction for which the surety provides financial security can be simple or complex. As illustrations of the former, the bail bondsperson will pay if the accused fails to appear at the hearing; or a banker's blanket bond requires the surety to pay if the bank official embezzles funds. However, a prime contractor backed by a surety has a wide variety of obligations set forth in the construction contract. For example, the prime contractor promises to build the project properly, not to damage the land of adjacent landowners, to perform the work in such a way as to avoid exposing workers and others to unreasonable risk of harm, and to indemnify the owner and design professional if certain claims are made. Often bond language does not state which duties are covered by the bond. For example, AIA Doc. A311 as set forth in Appendix D refers to the construction contract and simply states that ". . . if Contractor shall promptly and faithfully perform said Contract then this obligation shall be null and void; otherwise it shall remain in full force and effect." Which aspects of the contractor's performance are backed up by the surety?

Also, bond language often does not go specifically into problems of coverage, problems of notices, and other things that are normally part of any contract. Improvement in this regard has been made and some bonds today do contain specific provisions that do a better job of informing the parties of their rights and duties under the bond. See AIA Doc. A312 also contained in Appendix D.

SECTION 37.02
FUNCTION OF SURETY: INSURER COMPARED

A surety's function is to assure one party that the entity with whom it is dealing will be backed up by someone who is financially responsible. Sureties are used in transactions where persons deal with individuals or organizations of doubtful financial capacity. *Sureties* must be distinguished from *insurers*, though each serves the function of providing financial security.

An insured is concerned that unusual, unexpected events will occur that will cause it to suffer losses or expose it to liability. Although it can self-insure, that is, bear the risk itself, it usually chooses to indemnify itself against this risk by buying a promise from an insurance company in exchange for payment of a premium. The insurer distributes this risk among its policy holders.

In public liability insurance the insured itself may be at fault and cause a loss to the insurer. But the insurer does not seek to recover its loss from its own insured. Although it may seek to recover its losses

from third parties through subrogation (stepping into the position of the person it has paid, its insured, and thereby acquiring any claims of its insured against those who caused the loss), it cannot recover from those named as insureds in the policy.

Sureties, on the other hand, in addition to dealing with ordinary construction contract performance problems (not the "accidents" central to liability insurance) seek to recover any losses they have suffered from the principal party upon whom it has written a bond as well as others. In the late 1970s contractors who were liable because of defective workmanship sought to recover from their comprehensive general liability (CGL) insurers, preferring this to seeing the loss paid by their sureties. Their sureties would seek to recover from them. Their CGL insurers could not.

Although the surety furnishes, in general, financial security to the obligee, the person to whom the bond has been written, the surety's obligation is not exactly coextensive with that of the principal.

SECTION 37.03 JUDICIAL TREATMENT OF SURETIES

Judicial attitude toward sureties has changed. Before the development of professional sureties, the surety would be a private person (perhaps a relative of the principal) who sought to aid the principal to obtain a contract or to stave off a creditor by obligating himself to perform if the principal did not. Such a surety was frequently not paid and received no direct benefit for taking this risk. For these reasons the surety was considered a "favorite" of the law.

One illustration of this favored position was the Statute of Frauds, which requires that certain types of promises must be evidenced by a written memorandum.[1] The original Statute of Frauds enacted in 1677

---

1. Refer to Section 5.10.

included promises to answer for the debts, defaults, or miscarriages of another. Without a writing, the surety was not held liable. This was to protect him from the enforcement of an impulsive oral promise often made without due deliberation.[2] Also, any minor change in the contract between the principal and obligee would discharge the surety (relieve him of liability).[3] The personal, unpaid surety could not handle assurance needs in a commercial economy. The personal surety himself might not be financially responsible. For this reason, the professional, paid surety has developed as an important institution in both economic life generally and in building contracts.

The development of professional sureties casts doubt upon the protective rules that were developed largely when sureties were uncompensated. Although some protective rules are still applied, the professional surety is not regarded with the tender solicitude accorded the personal, uncompensated surety. This changeover from a legally favored position to one of neutrality, and perhaps even to one of disfavor, creates uncertainty in the law. Older cases are sometimes cited as precedents to protect sureties, but these precedents may be of limited value because being a surety is now a business.

The ambivalence toward sureties is illustrated by *Winston Corp. v. Continental Casualty Co.,*[4] which involved a claim by an owner against the surety on a performance bond. The bond incorporated all the provisions of the construction contract.

The contractor ran into financial difficulty causing delays, all known to the surety. Five months after the scheduled completion date, the owner met with the contractor and invited the surety. The surety, however, refused to attend. (The performance bond part of AIA Doc. 312, ¶¶ 2, 3.1 published in December, 1984, requires the surety to participate in conferences when the owner considers declaring the contractor "in default." See Appendix D.) At this meeting the owner and the contractor entered into an agreement designed to accelerate construction. Under the agreement the contractor assigned the construction contract to the owner, permitted the owner to take possession of the premises, and assigned the contractor's subcontracts to the owner. However, the contractor's continued participation was expected. Immediately the owner telephoned the surety notifying it of the new arrangement and mailed a copy of the letter agreement to the surety. A year after the scheduled completion date the project was completed.

The owner sued on performance and payment bonds. The surety's defense was that the agreement between owner and contractor modified the original construction contract and discharged the surety. Also the surety contended that the owner's failure to give the surety seven days written notice before terminating the contractor released the surety. The trial court, noting that sureties were favorites of the law, sustained the surety. The appellate court reversed, concluding that the owner had an absolute right to terminate the contract if the contractor was in default and that taking over the contract was merely an exercise of this right.

As for the failure to give the notice before taking over construction, the court agreed that historically any slight deviation from the contract terms would discharge the surety. But, noted the court, the application of this doctrine often caused harsh and unjust results especially when compensated sureties were relieved of their obligations because of

2. A written memorandum was not required where the main purpose or leading object of the surety was to benefit himself.

3. See Section 37.10(C) for additional applications of this doctrine.

4. 508 F.2d 1298 (6th Cir.1975). See also *Ramada Development Co. v. United States Fidelity & Guaranty Co.,* 626 F.2d 517 (6th Cir.1980). For changes discharging the surety, see Section 37.10(C).

technical breaches of the construction contract incorporated as part of the bond. This induced most courts to deviate from the doctrine under which the surety was the favorite of the law and required the compensated surety to show that the change in the original agreement was material and prejudicial to the surety. The surety must show that the change increased the surety's risk or changed it to its detriment.

In addition to no longer being considered a favorite, courts generally construe ambiguities against the surety for the same reasons that insurance policies are construed against the insurance company.[5]

The uncertainty of the surety status is also reflected by *Trinity Universal Insurance Co. v. Gould*,[6] in which the court held the surety liable despite substantial changes made in the original contract because the surety evidently knew the changes were being made and did not object. Yet in this holding against the surety the court noted that the surety was compensated and owed its patron the duty of good faith that goes beyond the morals of the market place. This language contained relics of older surety notions but in the context of a compensated surety.

Surety companies are regulated by the states in which they operate. In addition, sureties who wish to write bonds for federal projects must qualify under regulations of the United States Treasury Department. The financial capability of a surety limits the size of the projects they can bond. Bond dollar limits place a ceiling on exposure. In larger projects, there may be cosureties, or the surety may be required to reinsure a portion with another surety. Surety rates are usually regulated, and are based upon a specified percentage of the limit of the surety bond.

SECTION 37.04 SURETY BONDS IN CONSTRUCTION CONTRACTS

Surety bonds play a vital part in the Construction Process. The contracting industry is volatile. Bankruptcies are not uncommon, and a few unsuccessful projects can cause financial catastrophe. Estimation of costs is difficult and requires a great amount of skill. Fixed price contracts place many risks upon the contractor, such as price increases, labor difficulties, subsurface conditions, and changing governmental policy. Some construction companies are poorly managed and supervised. Often they are undercapitalized, and rely heavily upon the technological skill of a limited number of individuals. If these individuals become unavailable, it is likely that difficulties will arise. Credit may be difficult to obtain for many construction companies. Some contractors do not insure against the risks and calamities that can be covered by insurance. Finally, anti-inflationary government policies such as tight money policies almost always hit the building industry first.

In most construction projects, bonds are needed to protect the owner. The owner in most projects would like to be able to look to a financially solvent surety in the event that the successful bidder does not enter into the construction contract (bid bond), the prime contractor does not perform its work properly (performance bond), or the prime contractor does not pay its subcontractors or suppliers (payment bond).[7] (Bond requirements can act as preliminary screen for contractor selection.)

If these events do not occur, the amount paid for a surety bond may seem wasted. Some institutional owners believe there is

---

5. *United States v. Algernon Blair, Inc.,* 329 F.Supp. 1360 (D.S.C.1971); *School District No. 65R of Lincoln County v. Universal Surety Co., Lincoln,* 178 Neb. 746, 135 N.W.2d 232 (1965).

6. 258 F.2d 883 (10th Cir.1958).

7. As to the role of the design professional in advising on this choice, refer to Section 15.07.

no need for a surety bond system if the prime contractor is chosen carefully and if a well-administered payment system is used that eliminates the risk of unpaid subcontractors and suppliers. Such owners may choose to be self-insurers and not obtain bonds. They realize that there may be losses, but they believe that the losses over a long period will be less than the cost of bond premiums.

Even where a bond is not required at the outset it is best to include a provision in the prime contract that will require the prime contractor to obtain a bond before or during performance if the owner so requests. Usually the cost of a bond issued, after the price of the project is agreed upon, is paid by the owner.

Frequently public construction requires that performance and payment bonds be furnished. The latter are required to protect subcontractors and suppliers who have no lien rights on public work. The Miller Act requires that federal prime contractors obtain performance bonds and payment bonds based upon the contract price and similar requirements exist under state public contracting. In addition, often local housing development legislation requires that the developer furnish bonds to protect the local government if improvements promised by the developer are not made. Legal advice should be obtained to determine whether bonds are required for the project.

The invitation to bidders usually states whether the contractor is required to obtain a surety bond, the type or types of bonds, and the amount of the bonds. In some cases, the owner wishes to approve the form of the bond and the surety that is used. Also, the owner may want to have the right to refuse any substitution of surety without its express written consent given in advance of substitution.

Should the owner specify which surety bond company must be used? The choice will depend upon rates, bond provisions, and the reputation of particular bonding companies for efficient operation and for fairness in adjusting claims. If relative standardization exists on these matters, it is probably advisable to give the contractor the freedom to choose the bonding company. It may have an established relationship with a particular bonding company. In most cases it is probably sufficient to permit the bonding company to be chosen by the contractor, as long as the bonding company selected is licensed to operate in the state where the project will be built.

Occasionally a designated bonding company will not be satisfactory to the owner. This may be based upon suspicion of financial instability, or a past record of arbitrariness in claims handling. One method of exercising some control over the selection of the bonding company is to provide in the contract that the contractor submit to the owner the bond of a proposed bonding company for the owner's review. If the owner, in the exercise of its best judgment, determines that it is inadvisable to use that bonding company, the owner can veto the proposed bonding company and designate the bonding company to be used.[8]

SECTION 37.05   BID BOND

The function of a bid bond is to provide the owner with a financially responsible party who will pay all or a portion of the damages caused if the bidder to whom a contract is awarded refuses to enter into it.[9]

---

8. In *Weisz Trucking Co. v. Emil R. Wohl Construction,* 13 Cal.App.3d 256, 91 Cal.Rptr. 489 (1970) the standard for approval was held to be objective. However, a "good faith" standard can be inserted in the contract.

9. Refer to Section 22.04(E). Bid bonds are rare in Europe because of frequent prequalification of bidders. This is also the reason for European bonds being for 5% to 10% of the contract price, compared to the 50% or 100% in the United States.

SECTION 37.06
PERFORMANCE BOND

The performance bond provides a financially responsible party to stand behind some aspects of the contractor's performance. If a payment bond is furnished, the performance bond will not include payment of subcontractors, their suppliers, and suppliers of the prime contractor. Illustrations of a performance bond are reproduced in Appendix D.

Bonds usually place a designated dollar limit on the surety's liability. Typically bond limits are 50% or 100% of the contract price. Some statutory bonds are required to be 50% of the contract price.

SECTION 37.07
PAYMENT BOND

The payment bond is an undertaking by the surety to pay unpaid subcontractors and suppliers. Standard payment bonds are set forth in Appendix D. An understanding of the function of a payment bond requires a differentiation between private and public construction work. All states give unpaid subcontractors and suppliers liens if they improve private construction projects. This process has been described in Section 32.07. Although there are various ways to avoid liens, one method has been to require prime contractors to obtain payment bonds. A payment bond obligates a surety to pay subcontractors and suppliers if they are not paid by the prime contractor. The owner seeks to avoid liens filed against its property. Although the owner generally has no contractual relationship with unpaid subcontractors or suppliers, not uncommonly the latter parties assert claims against the owner. The owner would prefer directing them to the bonding company for payment.

Also, subcontractors are more likely to make bids if they can be assured of a surety if they are unpaid. The competent subcontractor who deals with a prime contractor of uncertain financial responsibility should

add a contingency to its bid for the possibility of collection costs and the risk of not collecting. The presence of a payment bond should eliminate the need for this cost factor. In addition, subcontractors and suppliers should be more willing to perform properly and deliver materials as quickly as possible when they have assurance they will be paid. Though they have a right to a mechanics' lien, the procedures for perfecting the lien and satisfying the unpaid obligation out of foreclosure proceeds are cumbersome and often ineffective. Payment bonds are preferable to mechanics' liens.

Generally subcontractors and suppliers cannot impose liens on public work. In some states they can file a "stop notice", which informs the owner that a subcontractor has not been paid and requires the owner to hold up payments to the prime contractor. Though unpaid subcontractors usually have no lien rights, they can present claims against public bodies through a request for special legislation to Congress, state legislatures, or city councils. Requiring that the prime contractor supply a payment bond provides the mechanism of relieving these legislative bodies from troublesome and time consuming claims. Some reasons for having bonds on private projects also apply to public jobs.

Competent subcontractors and willing suppliers are essential to the construction industry. These important components should have a mechanism that allows them to collect for their work. Payment bonds do this. Without a reliable payment mechanism, a substantial number of subcontractors may go out of business. Elimination of competent subcontractors can have the unfortunate effect of reducing competition and the quality of construction work.

SECTION 37.08
SUBCONTRACTOR BONDS

Sometimes the prime contractor requires that subcontractors also obtain payment bonds and performance bonds. If a subcontractor does not perform as obligated, the prime contractor (or its surety looking toward reimbursement) wants to have a financially responsible person to stand behind the subcontractor. This is the justification for a subcontractor performance bond. The justification for the subcontractor payment bond is similar to the justifications given for prime contractor payment bonds. If a subcontractor does not pay its sub-subcontractors or suppliers, the prime contractor (or its surety) is likely to be responsible, since it obligates itself to erect the project free and clear of liens. In many cases, unpaid sub-subcontractors and suppliers of subcontractors have lien rights against projects. In a large construction project, a substantial number of bonding companies who have written bonds on the various contractors in the project may exist. This makes the litigation in such cases complicated; often the principal participants in a litigation are bonding companies each seeking to shift the responsibility to the other.

SECTION 37.09 OTHER BONDS

Bonds can be used for other purposes in the Construction Process. For example, in some states owners can post a bond that can preclude a lien from being filed or dissolve a lien that has been filed. Sometimes warranty bonds are used to "back-up" the owner's claim under a warranty given by the contractor.

SECTION 37.10
SOME LEGAL PROBLEMS

(A) WHO CAN SUE ON THE BOND?

The most troublesome legal issue has been the seemingly simple question of whether unpaid subcontractors and suppliers can sue on a surety bond. Unpaid subcontractors and suppliers are not "obligees" under

the bond. That is, the bond is not written to them and the surety does not specifically oblige itself to them. Owners on prime contractor bonds or prime contractors on subcontractor bonds have no difficulty instituting legal action because they are obligees and the bonds are written to them.

Early legal problems were complicated by the use of a single bond called the Faithful Performance Bond. Since the legal standard applied to determine the right of someone other than the obligee to sue on the bond was whether the owner as obligee intended to benefit unpaid subcontractors and suppliers, some courts denied unpaid subcontractors and suppliers the right to sue on the bond. These courts reasoned that the owner must have intended to benefit itself; the bonds covered aspects of the prime contractor's performance other than nonpayment of subcontractors and suppliers. Also, the interests of owner and unpaid subcontractors and suppliers could conflict if each had claims against the prime contractor and if the amount of the bond could not satisfy all claims. As a result, the practice changed to encompass two bonds. The payment bond was to cover default consisting of nonpayment of subcontractors and suppliers while the performance bond covered all other aspects of nonperformance by the prime contractor. Yet even where two bonds are issued, some courts still deny unpaid subcontractors and suppliers the right to sue on the bond.[10]

This already complicated area was muddled further by courts that differentiated between bonds on public works and those for private projects. These courts permitted subcontractors and suppliers to sue on bonds executed for public projects. They noted that liens could not be asserted on public projects and the intention of the public agency requiring the bonds must have been to benefit the unpaid subcontractors and suppliers. But on private projects

10. See cases at infra note 12.

which were lienable an unpaid subcontractor or supplier could not sue because the intention must have been to benefit the owner.[11] Courts deciding this question often focus solely upon a private owner's desire to avoid liens being filed against its property. This ignores the other functions of surety bonds, such as those mentioned in Section 37.07.

Although the cases are not unanimous (bond language will vary) a strong modern tendency allows subcontractors and suppliers to sue directly on payment bonds if the bond seems to state that the surety will pay unpaid subcontractors or suppliers.[12] This outcome is reflected in AIA bonds in Appendix D, which clearly give unpaid subcontractors and suppliers a right to sue directly. Courts interpreting unclear language in a bond should recognize that the owner wants to be able to tell an unpaid subcontractor or supplier that it will be paid by the bonding company, and that if the bonding company wrongfully refuses to pay, the subcontractor or supplier will be able to institute legal action itself. A bond purchases this right. The party paying for the bond intends them to have this right and this should be controlling.

Where bonds are required for public work or where bonds are filed on private work under statutes that allow this as a substitute for lien rights, the language of the statute frequently determines who can sue on the bond.[13] Typically the claimant is someone who has furnished labor or material that has gone into the project. Yet recent cases have permitted a contractor's lender[14] as well as a pension fund to institute action on the payment bond.[15]

(B) VALIDITY OF BOND

Suppose the contractor misrepresents its resources when it applies for the bond. This would be misrepresentation by the applicant not by the obligee-owner. In such a case, the bond is generally valid, and the surety bond company must pursue any remedies it has against the contractor-applicant. Fraud on the part of the obligee-owner, however, gives the surety the power to avoid having to perform under the bond. If the obligee-owner participates in, or knows about, the fraudulent statements made by the applicant contractor or misleads the bonding company in some other way, the bond would not have been validly obtained, and could not be legally enforced.

The principal contract, that is, the contract between owner and contractor, may not be enforceable for various reasons. For example, suppose the building contract involved construction of a hide-out for leaders of organized crime. To permit the owner to sue upon the surety bond in such a case would further an illegal activity and would involve the court in that activity. For that reason the bond would not be enforced.

---

11. See original opinion in *Fidelity & Deposit Co. of Baltimore v. Rainer,* 220 Ala. 262, 125 So. 55 (1929) which was reversed on rehearing. The original barred a claim by the subcontractor. The final decision allowed it.

12. *Socony-Vacuum Oil Co. v. Continental Casualty Co.,* 219 F.2d 645 (2d Cir.1955) (a leading case). See also *Houdaille Industries, Inc. v. United Bonding Insurance Co.,* 453 F.2d 1048 (5th Cir.1972); *Jacobs Associates v. Argonaut Insurance Co.,* 282 Or. 551, 580 P.2d 529 (1978) (reversing earlier opinion denying right of direct action) noted in 58 Or.L.Rev. 252 (1979); *Noland Co. v. West End Realty Corp.,* 206 Va. 938, 147 S.E.2d 105 (1966). But see *State of Florida v. Wesley Construction Co.,* 316 F.Supp. 490 (S.D.Fla.1970), affirmed 453 F.2d 1366 (5th Cir.1972); *Layrite Concrete Products of Kennewick, Inc. v. H. Halvorson, Inc.,* 68 Wash.2d 70, 411 P.2d 405 (1966) (no right to sue) and *James D. Shea Co. v. Perini Corp.,* 2 Mass.App. 912, 321 N.E.2d 831 (1975); *Day & Night Manufacturing Co. v. Fidelity & Casualty Co. of New York,* 85 Nev. 227, 452 P.2d 906 (1969) (right only if language *clear*).

13. *Houdaille Industries, Inc. v. United Bonding Insurance Co.,* supra note 12.

14. *First National Bank of South Carolina v. United States Fidelity & Guaranty Co.,* 373 F.Supp. 235 (D.S.C.1974).

15. *Trustees, Fla. West Coast Trowel Trades Pension Fund v. Quality Concrete Co.,* 385 So.2d 1163 (Fla.App.1980).

Legal infirmities to contracts of a less serious nature, however, may exist. Suppose the contractor is not licensed. *Cohen v. Mayflower Corp.*[16] held that the owner can recover on a bond written on an unlicensed contractor. The court held that the licensing law was designed to protect owners and the owner could have sued the contractor even though the unlicensed contractor could not have sued the owner. The court concluded that the surety would be held on the bond because the principal debtor, the contractor, could have been held.

The stronger the public policy making the contract illegal, the less likely the bond will be enforced. If the policy is designed to protect the owner, however, such as in the *Cohen* case, it is more likely that the bond will be enforced.

(C) SURETY DEFENSES

Suppose the owner does not make progress payments despite the issuance of progress payment certificates. Suppose the design professional unjustifiably interferes with the work of the contractor or does not approve shop drawings in sufficient time to permit proper performance. The contractor's performance may be rendered impossible because of a court order, the death of the contractor, or some natural catastrophe. The performance may be rendered impracticable due to the discovery of unforeseen subsurface conditions. Most defenses the contractor has against the owner would be available to the surety.

Likewise, defenses that the prime contractor could assert against a subcontractor or supplier claimant can generally be asserted by the surety. The surety's function is to provide financial responsibility for the acts of the prime contractor. Generally the surety's obligation is coextensive with that of the principal debtor, that is, the prime contractor, but only to the extent of the bond limit.[17]

Suppose the claimant is an unpaid subcontractor or supplier under a payment bond. In *Houdaille Industries, Inc. v. United Bonding Insurance Co.*,[18] the surety was denied a defense based upon an asserted claim that the owner had breached its contract with the contractor who was the principal on the bond. The court noted that while normally the surety can assert such defenses against the owner, this defense could not be asserted against a claimant "so long as it is not participated in or authorized by the materialman." [19]

Suppose the owner and the contractor modify the construction contract, or the owner directs changes in the work. The surety's commitment can be limited by putting a fixed limit on the bond. Any change in the basic agreement, however, traditionally released the surety. The advent of the paid surety has made inroads on this rule. Frequently bonds provide that the surety "waives notice of any alteration or extension of time made by the owner." Without such a provision, the surety could be released because the principal obligation has been changed.

Suppose the end of the Construction Process concludes with claims made by the participants. Two issues that may arise are the method of resolving disputes and whether claims have been barred by the passage of time. The first issue has been

---

16. 196 Va. 1153, 86 S.E.2d 860 (1955).

17. See Section 37.10(D).

18. Supra at note 12.

19. 453 F.2d at 1053 n. 4. But see *Chicago Bridge & Iron Co. v. Reliance Insurance Co.*, 46 Ill.2d 522, 264 N.E.2d 134 (1970) (surety *was* given a defense when the claimant subcontractor submitted false lien waivers that allowed the prime to dissipate progress payments).

discussed in Section 34.15, which deals with arbitration.

The second issue, that of the passage of time barring a claim, has caused difficulty. For example, in *State v. Bi-States Construction Co.* [20] the State of Iowa brought a claim against a contractor who had abandoned a project and its surety. The contractor, a Nebraska corporation, had been dissolved under the laws of Nebraska on March 26, 1969. Iowa law required that any claim against a dissolved corporation be made within two years. The owner's claim was barred because it did not institute the claim until almost five years after the contractor had been dissolved. The court held that the claim against the bonding company was also barred. The court recognized that divided authority on the question exists, but Iowa "adheres to the rule a surety may assert as a defense the statute of limitation if available to the principal." [21]

The California Supreme Court, however, after judicial vacillation, concluded that a claim could be made against a surety even if a completion statute [22] had given a defense to the contractor.[23] The court also held that the surety could bring an action for reimbursement against the contractor because its claim did not arise until it had paid the claim of the owner. As a result the statutory protection given to the contractor was lost.[24]

Once disputes develop, both obligee and principal debtor should notify the surety and keep the surety informed as to the posture and process of the dispute. In this regard ¶ 14.2.1 of AIA Doc. A201 requires the surety to be notified in writing if the owner has terminated the contractor's performance.

(D) SURETY RESPONSIBILITY AND BOND LIMITS

If the contractor finishes the project, the surety's obligation is to pay whatever damage the owner is entitled to recover from the contractor. This is likely to be any recoverable delay damage and either the cost of correction or diminished value of the project.

More difficult questions involve contractor defaults of a sufficiently serious nature that allow the owner to terminate the contractor's performance. Usually the surety can take over the project and complete it or provide a successor contractor and sufficient funds, to the limit of the bond, to pay for a completion cost less the unpaid balance of the contract price.[25]

The surety exposure is relatively clear if a claim is made on the payment bond, that is, the amount of unpaid subcontractor and supplier claims. Exposure on a performance bond is more difficult.

The principal function of the surety is to ensure that the project is built properly within the required time. But failure by the contractor to perform as required can generate other losses to the owner. For example, one court allowed recovery for the increased cost of separate mechanical contracts and added interest costs caused by delayed performance.[26] These costs can be considered consequential damages, or damages less directly connected to the project

---

20. 269 N.W.2d 455 (Iowa 1978).

21. 269 N.W.2d at 457.

22. Refer to Section 27.03(G).

23. *Regents of the University of California v. Hartford Accident and Indemnity Co.*, 21 Cal.3d 624, 147 Cal.Rptr. 486, 581 P.2d 197 (1978).

24. The California legislature added sureties to the completion statute. West's Ann.Cal.Civ.Proc.Code § 337.15.

25. Refer to AIA Documents in Appendix D.

26. *Miracle Mile Shopping Center v. National Union Indemnity Co.*, 299 F.2d 780 (7th Cir.1962).

itself. Another court dealing with a claim of a subcontractor, however, focused principally upon the work and materials that increased the value of the project, probably a reflection of the court's view that a payment bond was a substitute for lien rights.[27] In emphasizing this limited function of the bond, the court denied recovery to the subcontractor for profits on unperformed work and delay caused by the prime contractor. Finally, another court that focused on language in the bond under which the surety would be responsible for "other costs and damages," included delay damage if the owner could prove loss of use, but not certain consequential damages because these damages were not reasonably foreseeable.[28]

Clearly the surety should not be responsible for consequential damages that would not have been chargeable to the principal debtor. But any attempt to limit liability of the surety to direct losses overemphasizes the relationship of the surety bond and mechanics' liens. The surety stands behind the prime contractor's performance. The only limit should be the liability of the principal that relates to building the project, capped by any monetary bond limit. Yet as the loss moves farther away from the work and materials that went into the project, the surety's exposure is likely to be more unclear. Generally the surety is not responsible for punitive damages that might be awarded against the obligee.[29]

Finally, the owner may be exposed to claims because the prime contractor caused damage to third parties, such as adjacent landowners. This claim should not come under the bond. It should be covered by comprehensive general liability insurance of the prime contractor. If such a claim is made against the owner, the owner will very likely have an indemnification claim against the prime contractor, which in many contracts is covered by special endorsement to the comprehensive general liability policy. Generally the surety's obligation is limited to the stipulated amount of the bond.[30] A trial court opinion, however, held that the surety's liability under certain circumstances can exceed the bond limit.[31]

(E) BANKRUPTCY OF CONTRACTOR

If, during the course of performance by the contractor, the contractor is adjudicated a bankrupt, the trustee in bankruptcy (the person who takes over the affairs of the bankrupt contractor) can determine whether or not to continue the contract.[32] Usually it does not. If the contract is not continued, the bankrupt contractor is released from any further obligation to perform under the contract. The owner has a claim against the bankrupt contractor, but is not likely to recover much. Ending the contractor's obligation, however, should not release the surety. This is the risk contemplated when the surety bond is purchased.

(F) ASSERTING CLAIMS: TIME REQUIREMENTS

The surety's obligation usually requires that certain notices be given by claimants. Likewise statutes requiring public work to be bonded, like the Federal Miller Act, often specify that notices must be given within

---

27. *Lite-Air Products, Inc. v. Fidelity & Deposit Co. of Maryland,* 437 F.Supp. 801 (E.D.Pa.1977).

28. *New Amsterdam Casualty Co. v. Bettes,* 407 S.W.2d 307 (Tex.Civ.App.1966). (AIA bond).

29. Annot. 2 A.L.R.4th 1254 (1980).

30. *Miracle Mile Shopping Center v. National Union Indemnity Co.,* supra note 26 (by implication).

31. *Continental Realty Corp. v. Andrew J. Crevolin Co.,* 380 F.Supp. 246 (S.D.W.Va.1974). As the case was settled, there was no appeal. The decision seems to be based upon improper settlement tactics of the surety. The opinion is reviewed and criticized in Wisner, *Liability in Excess of the Contract Bond Penalty,* [1976] Ins.Coun.J. 105.

32. This is discussed in Section 38.04(C).

certain periods of time to designated persons.

Frequently bonds require that legal action on the bond be brought within a designated time, usually shorter than the period specified by law. Such shortened periods to begin legal action are enforceable if reasonable.[33]

Bonds often specify the court in which legal action must be brought, usually to courts, state or federal, in the state in which the project is located.[34]

Suppose a claimant does not comply with all of the many requirements specified by bonds or statutes. A surety that denies responsibility because of some failure to comply appears to be using a technicality to avoid an obligation it was paid to perform. Courts generally interpret these requirements liberally in favor of claimants[35] and often require the surety to show prejudice caused by noncompliance before being given a defense.[36]

Also, courts are frequently faced with assertions by claimants that the surety should be estopped to assert these provisions or be found to have waived them. In *Contee Sand & Gravel Co. v. Reliance Insurance Co.,*[37] an unpaid subcontractor retained an attorney to see if it could collect from the prime contractor. The prime contractor contacted an agent for the bonding company who informed him that a performance existed but not a payment bond. This was corroborated by the surety's home office. Relying upon this corroboration the

attorney brought an action against the prime contractor but did not include the surety. The judgment obtained was not satisfied because the prime contractor was insolvent. The prime contractor then informed the attorney that there had indeed been a payment bond issued by the defendant. The earlier incorrect statement resulted from a filing mistake. The subcontractor brought an action on the bond but was met with the defense that the action had not been brought within the one-year period of limitations specified in the bond. The court held that the subcontractor had reasonably relied upon the representation that no payment bond existed by not asserting an action against the surety within the one-year period. As a result the surety was "estopped" from asserting the failure to begin the action within one year as a defense.

(G) REIMBURSEMENT OF SURETY

It is sometimes said that sureties do not expect to take a loss. Sureties do not see themselves as insurers. As a result they assert defenses that the principal debtor could have asserted. They seek bond language protection. Finally, if they *do* pay, they seek reimbursement.

Usually the surety has a valid claim against the defaulting prime contractor. This claim is often uncollectible, however, because contractors who default are rarely in a financial position to reimburse the surety. Yet in *Naylor Pipe Co. v. Murray Walter, Inc.,*[38] the prime contractor was

---

33. In *Rumsey Electric Co. v. University of Delaware,* 358 A.2d 712 (Del.1976), a one-year bond provision was held valid where the statutory period was three years. Sometimes Statutes of Limitation are excessively long. But *City of Weippe v. Yarno,* 94 Idaho 257, 486 P.2d 268 (1971) held that the period for bringing the action cannot be shortened by the parties as it would be unconscionable. The court stated that stipulating a longer period would be enforced if reasonable.

34. See Appendix D.

35. *United States v. Merle A. Patnode Co.,* 457 F.2d 116 (7th Cir.1972); *American Bridge Division United States Steel Corp. v. Brinkley,* 255 N.C. 162, 120 S.E.2d 529 (1961).

36. *Winston Corp. v. Continental Casualty Co.,* supra note 4.

37. 209 Va. 672, 166 S.E.2d 290 (1969).

38. 120 N.H. 696, 421 A.2d 1012 (1980).

forced to pay twice, a situation similar to that of the owner who sometimes pays twice under mechanic's lien laws. The prime contractor had paid the subcontractor but the latter had not paid a supplier. The court held that this was not a defense to the surety who had to pay the unpaid supplier. If the surety received reimbursement from the prime contractor the prime contractor will have paid twice. The court stated that the prime contractor could have protected itself by requiring the subcontractor to post a payment bond or to obtain waivers from suppliers before paying the subcontractors.

If the surety has obtained a guarantor for the principal debtor, such as an officer, shareholder, or friend of the prime contractor, the guarantors will be looked to for reimbursement. When the prime contractor defaults and the surety takes over, the surety usually notifies the owner that it should be paid all payments that would have gone to the prime contractor.[39] In addition, the surety usually demands any retainage at the end of the job that the owner has withheld to secure the owner against claims against the contractor.[40] In seeking the retainage the surety usually competes with other creditors of the prime contractor, the taxing authorities, and the trustee in bankruptcy, if the prime has been declared bankrupt. As a rule many more claims that can be satisfied exist and the

result is a complicated lawsuit. Typically the owner pays the retainage into court, notifies all claimants, and the court determines how the fund is to be distributed.

The surety who takes over after default usually succeeds to any claims the prime contractor may have against the owner or third parties. It is common for the prime contractor to ascribe its difficulties to the owner, the design professional, subcontractors, or other third parties. For example, in *Westerhold v. Carroll,*[41] a person who had indemnified the bonding company when the prime applied for a surety bond was forced to reimburse the bonding company when the prime defaulted. The indemnitor brought an action against an architect claiming that the architect had negligently overcertified work resulting in excessive progress payments being made to the prime contractor. The court held the indemnitor could recover against the architect despite the absence of any contractual relationship between them. Likewise, the surety on a performance bond that had to pay the obligee owner for a defective pipe installed by the prime contractor recovered from the supplier who had sold the defective pipe.[42]

Sureties make strong efforts to be reimbursed and often succeed in salvaging a substantial amount of their loss when they are called upon to respond for their principal's default.

PROBLEMS

1. C was a contractor in Idaho. He also did jobs in other nearby states. No requirement that contractors be licensed in Idaho existed, however, C undertook a job for O in Utah, which did have a licensing law. C did not apply for a

---

39. In *Gerstner Electric, Inc. v. American Insurance Co.,* 520 F.2d 790 (8th Cir.1975) the court rejected a contention of the contractor that the request for payments by the surety to the owner was a wrongful interference in the contract between contractor and owner.

40. Retainage is discussed in Section 23.03.

41. 419 S.W.2d 73 (Mo.1967).

42. *Traveler's Indemnity Co. v. Evans Pipe Co.,* 432 F.2d 211 (6th Cir.1970).

license. C applied for and received a bond from S. C defaulted, and now S refuses to perform its obligations under the bond. S has offered to refund the premium to O. But S claims that the contract between C and O was illegal, since C was not licensed. Should this defense be allowed? What would be your conclusion if Idaho had a licensing law, and C's license had been revoked for fraud and incompetence? Would your conclusion be influenced by whether O knew of the revocation? If so, in what way?

2. O engaged C to build a construction project. O required that C furnish a payment and performance bond issued by S, which it did. During the project O terminated C's contract because C had committed serious contractual breaches, including defective workmanship, unsafe construction practices, and careless work that damaged adjacent property. O has examined the bond that has been furnished by S, a bond similar to A311 set forth in Appendix D. What would O be able to recover from S? Would your answer have been different had the bond been A312 also set forth in Appendix D?

38

Termination of a Construction Contract: Sometimes Necessary But Always Costly

SECTION 38.01
TERMINATION: A
DRASTIC STEP

Termination does not occur frequently in construction contracts. One reason for this is the difficulty of determining whether a legal right to terminate exists, a point discussed principally in Sections 38.03 and 38.04. Often each party can correctly claim the other has breached. It may be difficult to determine whether a party wishing to terminate is sufficiently free from fault and can find a serious deviation on the part of the other. Another reason is the often troublesome question of whether the right to terminate has been lost, a point to be discussed in Section 38.03(E). A third reason, the serious consequences of terminating without proper cause, will be treated in this section.

Two cases will be used to illustrate the danger of an improper termination. The first, *Paul Hardeman, Inc. v. Arkansas Power & Light Co.,*[1] involved a contract under which Hardeman was to construct electric transmission lines for the Power & Light Company for a contract price of $2,700,000. Hardeman's bid had been much lower than the other bids. This was known only to the

owner because this private project did not use a public bid opening. Despite the likelihood that there had been a mistake, the Power & Light Company proceeded to award the contract without discussing the mistake possibilities.

Many difficulties developed mainly because of Hardeman's inexperience. Several months before the completion date, at a time when somewhat less than one-half of the work had been completed, Hardeman discovered that its bid had been substantially lower than the other bids because of a calculation mistake. Immediately it communicated this to the Power & Light Company and accused it of wrongfully awarding the contract while knowing of the likelihood of the mistake. It threatened to leave the project unless some equitable adjustment were made. A conference was held but no resolution was accomplished. Finally the Power & Light Company and its engineer terminated Hardeman's contract claiming that there had been unexcused delays as well as defective workmanship and material. Evidently the real reason for the termination was that the Power & Light Company sought the best tactical position for the inevitable lawsuit. Before the termination Hardeman had claimed to have

---

1. 380 F.Supp. 298 (E.D.Ark.1974).

872

spent $3,000,000 on the work. After termination a successor contractor completed performance on a cost-plus basis and received $8,000,000.

Hardeman brought legal action claiming that the Power & Light Company wrongfully accepted its bid and that the termination had been improper. It sought restitution based upon its asserted expenditures of $3,000,000 less payments made of $600,000. The Power & Light Company sought $5,800,000 on its part as damages. Although the court seemed sympathetic to Hardeman's first claim, it felt it could not grant relief because Hardeman's estimator had been grossly negligent.[2] A convenient alternative solution, however, arose. The termination had not been made in good faith. According to the court, such a drastic step had been taken prematurely at best.

Termination proved useful to Hardeman. The court concluded that improper termination entitled Hardeman to recover the reasonable value of his services, which the court found to be $2,000,000 less the amount paid of $600,000. The defendant's counterclaim for the excess cost of correction was denied.

Perhaps the case is not typical. The court seemed sympathetic to Hardeman's bidding mistake claim. Yet the case demonstrates that a precipitous termination by the owner can be very costly when the contractor has made a losing contract.

The second case generated two opinions by the Federal Court of Claims. In *J.D. Hedin Construction Co. v. United States*,[3] Hedin agreed to construct a facility for the Veterans Administration in 1955 for $2,000,000. After trouble developed, mainly delay caused by a cement shortage, the contracting officer terminated Hedin's contract 154 days after the scheduled completion date. The claim proceeded slowly through the V.A.'s administrative appeals process with the only relief given to the contractor being a reduction in the number of deficiencies from 107 to 33. Finally, *Hedin I* reached the Court of Claims. In 1969 the court concluded that the termination was improper, mainly because the cement shortage should not have been charged to the contractor and that the other 33 deficiencies were "the kind every construction contractor routinely experiences on every job."[4] This, according to the court, was not sufficient to justify the drastic sanction of termination.

Had there been a contractual power on the part of the government to terminate at its own convenience, this improper termination for default could have been converted to the less drastic convenience termination. But the absence of this termination clause entitled the contractor to recover the retained amounts for delay and expenses necessary to protect the work after termination and what the court referred to as common law damages. The court sent the case back to its trial court to determine additional amounts the contractor could recover for the improper termination.

Hedin II, decided in 1972, dealt with these additional damages. For purposes of this section, the importance of *Hedin II* is not what the contractor recovered but rather the extent of its claim. The fact that some items were granted and some were not does not change the drastic nature of the termination and the serious consequences of a wrongful one.

After Hedin's contract had been terminated, Hedin's surety obtained a successor who completed the work for $1,267,000. Hedin had to pay this amount because in its application for a surety bond it had agreed to

---

2. For a further discussion of relief from mistaken bids, refer to Section 22.04(E).

3. The first opinion was found in 187 Ct.Cl. 45, 408 F.2d 424 (1969), which will be referred to as *Hedin I*. The second, found in 197 Ct.Cl. 782, 456 F.2d 1315 (1972), will be referred to as *Hedin II*.

4. 408 F.2d at 431.

indemnify the surety. Hedin claimed it could have completed the work for $741,000 and asked for $526,000, the difference between what it paid the successor and what it claimed it could have done the work for itself. On this major item, the court awarded Hedin $420,000. As to added costs, Hedin claimed:

1. Rental value of equipment, machinery and tools taken over and used by the U.S.— $48,000 (allowed but not with interest)

2. Interest on bank loan used to pay successor—$215,000 (denied)

3. Interest on principal amounts paid to surety—$112,000 (denied)

4. Cost of employees retained on site following termination while seeking to persuade U.S. to retract termination—$4,000 (denied)

5. Legal expenses incurred by surety takeover—$1,000 (allowed)

6. Additional bond premium—$6,000 (allowed)

7. Profits (disallowed as not proven)

It may be necessary to terminate a construction contract. Performance may be going so badly and relations may be so strained that continued performance would be a disaster. But because of the reasons mentioned, the drastic step of termination should not be taken precipitously.

SECTION 38.02
TERMINATION BY
AGREEMENT OF THE PARTIES

Just as parties have the power to make a contract, they can "unmake" it. Exercise of this power in legal parlance may be described as rescission, cancellation, mutual termination or some other synonym. However described, the parties have agreed that each is to be relieved from any further performance obligations. In lay terms they have "called the deal off."

The legal requirements for such an arrangement are generally the same as those

for performing a contract, manifestations of mutual assent, consideration, a lawful purpose and compliance with any formal requirements. By and large the mutual assent requirement does not present unusual difficulty. Suppose, however, one party makes a proposal to cancel but the other party does not *expressly* accept. In most cases silence is not acceptance, but silence coupled with other acts that lead the proposer to believe there has been an agreement can be sufficient.

If each party has obligations yet to perform, the consideration consists of each party relieving the other of performance obligations. If one party has *fully* performed, however, enforcement problems may arise. For example, suppose a contractor has been paid the full contract price but has not finished performance. If the parties agree to cancel remaining obligations of the contractor, the owner is not receiving anything for its promise. Courts generally relax consideration requirements somewhat when parties adjust or cancel existing contracts. Such an agreement can be enforced by calling it a waiver or completed gift. Although problems can arise, generally agreements under which each party agrees to relieve the other are enforced.

Since most construction contracts do not have to be expressed by a written memorandum, formal requirements are rarely an impediment to enforcement of contracts of mutual termination. Desirability of proof usually, however, means that such agreements will be expressed in writing.

SECTION 38.03   CONTRACTUAL
POWER TO TERMINATE

Construction contracts frequently contain provisions giving one or both parties the power to terminate the contract. These provisions are a backdrop for the material to be discussed in Section 38.04, the common law right to terminate a contract. Although contracts are not always clear on this point, often an interrelationship between specific termination provisions and the common law exists. Specific provisions can be considered illustrations or amplifications of common law doctrines, with the common law doctrines still applicable. Alternatively, contractual termination provisions can be said to have supplanted common law doctrines.

It is likely that the common law has not been eliminated by express contract termination provisions,[5] as the intention to waive common law rights requires clear expression.[6]

(A) DEFAULT TERMINATION

Generally construction contracts drafted by an owner give explicit termination rights only to the owner. On the other hand, construction contracts published by professional associations provide that either owner or contractor can terminate for certain designated defaults by the other. Although great variations exist, it may be useful to begin discussion with AIA Documents.

AIA Doc. A201, ¶ 14.2.1, allows the owner to terminate if the contractor has designated financial problems, fails to pay his subcontractors, or

> persistently or repeatedly refuses or fails, except in cases for which extension of time is provided, to supply enough properly skilled workmen or proper materials, . . . or persistently disregards laws, ordinances, rules, regulations or orders of any public authority having jurisdiction, or otherwise is guilty of a substantial violation of a provision of the Contract Documents.

This provision does not specifically allow partial termination, which is something that may be useful if the owner wishes to retain the contractor on part of the work and find a successor for the rest.

Can the owner terminate if the contractor falls behind its schedule? Federal construction contracts permit termination if the contractor fails to prosecute the work with such diligence as will insure its completion within the time specified.[7] Suppose the owner wishes to bring in a successor contractor when the original contractor is falling far behind schedule. Paragraph 14.2.1 does permit the owner to terminate for persistent failure to supply proper workmen or materials. But delays, even unexcused ones, may be caused by other factors. Does ¶ 8.3, with its provision for time extensions, preclude termination? Does a liquidated damage clause, if present, indicate an intention that termination is not appropriate for failure to comply with schedule or completion requirements? Again, the nonexclusivity of the termination clause can support a conclusion that protracted delay, particularly if it appears that completion will not be "on time", will create a power to terminate under common law principles.[8] Any exercise of such a power should be preceded by warnings and, where appropriate, an opportunity to cure.

A termination may be wrongful if the owner has not used good faith in exercising its contractual power to terminate. In *Paul Hardeman, Inc. v. Arkansas Power & Light Co.*, the court admitted that contractual grounds to terminate existed, but held that the termination had not been made in good faith. As shall be seen in Section 38.04(A), good faith of the party in default can be a factor in determining whether the right to terminate exists at common law.

---

5. AIA Doc. A201, ¶ 7.6.1.

6. *Armour & Co. v. Nard*, 463 F.2d 8 (8th Cir.1972); *Glantz Contracting Co. v. General Electric*, 379 So.2d 912 (Miss. 1980); *Bender-Miller Co. v. Thomwood Farms, Inc.*, 211 Va. 585, 179 S.E.2d 636 (1971). See also *North Harris County Junior College District v. Fleetwood Construction Co.*, 604 S.W.2d 247 (Tex.Civ.App.1980) (dictum would permit termination for reason not specified in AIA contract).

7. 32 C.F.R. § 7–602.5. See also NSPE General Conditions, 1910–8 cited infra note 12 at § 15.2.6.

8. Whether the architect's certificate needed under ¶ 14.2.1 would be required will be discussed in Section 38.03(D).

AIA Doc. A201, ¶ 14.1.1, gives the contractor the power to terminate. Termination for nonpayment requires the contractor to stop work for 30 days and then give a seven-day notice before it can terminate. But the contractor can use common law termination, a more expansive right to terminate for nonpayment.

The sparse grounds for termination and the absence of a contractual right to terminate when the work in progress is destroyed show that AIA documents seek to continue performance and avoid the economic disruption that is caused by a contract termination.

(B) TERMINATION FOR CONVENIENCE

Pioneered by federal procurement regulations,[9] construction contracts increasingly give private owners the right to terminate for convenience, something also found in subcontracts, particularly those that are tied to prime contracts where the owner has this power.[10] Invoking such a clause requires the contractor to stop work, place no further orders, cancel those orders that had been placed, and perform other acts designed to terminate performance and protect the interests of the government. The contractor is reimbursed for work performed, unavoidable losses suffered, and expenditures incurred to preserve and protect government property. The contractor is also paid a designated profit for work performed.

In addition, federal procurement law has developed the *constructive* convenience termination.[11] Under it, a contract with a convenience termination clause converts a *wrongful* government *default* termination into a *convenience* termination.

On its face, the power given to the owner by such contracts is very broad. The owner may terminate if the project is no longer needed or if it has become outmoded or uneconomical. Yet this power is not completely unrestricted. *Torncello v. United States* [12] outlined the history of such clauses and noted the importance of the government's ability to change its procurement objectives. But the court did not permit the agency to terminate when the agency used another contractor to do the work for less. (The agency knew a cheaper source existed when it made the contract). The court stated that a termination for convenience can be used only when the circumstances of the bargain have changed.

This takes away much of the usefulness of the clause. The court should have simply concluded that the government abused its discretion because terminating for convenience in *this* case was in bad faith. Those owners who use a termination for convenience clause may have, at least judging by federal procurement, a false impression that the owner's rights are unlimited.

(C) EVENTS FOR WHICH NEITHER PARTY IS RESPONSIBLE

Some contracts specifically allow termination when events occur that have a devastating effect on performance, such as those that would justify common law relief for impracticability or impossibility. AIA documents do not specifically provide these events cause termination. The party whose

---

9. 41 C.F.R. § 1–8.703.

10. *General* Conditions 1910–8 published by the National Society of Professional Engineers, the American Consulting Engineers Council, the American Society of Civil Engineers and the Construction Specifications Institute in 1983 (known as NSPE General Conditions) allow it in § 15.4. AIA will likely follow suit in 1986. See also *Arc Electric Construction Co. v. George A. Fuller Co.*, 24 N.Y.2d 99, 299 N.Y.S.2d 129, 247 N.E.2d 111 (1969).

11. *Torncello v. United States*, note 12 infra.

12. 681 F.2d 756 (Ct.Cl.1982), overruling *Colonial Metals Co. v. United States*, 204 Ct.Cl. 320, 494 F.2d 1355 (1974), which had allowed the U.S. to terminate when it found a cheaper source it should have known of when it made the contract.

performance has been adversely affected receives time extensions and occasionally an equitable adjustment.

(D) ROLE OF DESIGN PROFESSIONAL

Despite the AIA's movement toward reducing the activities and responsibilities of the architect, the owner's power to terminate under ¶ 14.2.1 requires that the architect certify "that sufficient cause exists to justify such action." No similar requirement exists for a termination by the contractor.

Termination is a drastic step. Why require the architect to certify that there are adequate grounds? Such a decision, although involving some issues for which the architect may be trained, requires legal expertise rather than design skill. The owner may want the architect's advice. That need not require giving power to the architect to decide a sensitive and liability-exposing issue. The only possible justification is that such a preliminary step can act as a brake upon any hasty, ill-conceived decision by the owner to terminate. It would be better to have the owner's attorney perform this function, especially if the architect's conduct is itself an issue in the termination.

Would an architect's certificate be needed for common law termination? Although it is not likely that such a procedure would apply to a common law termination, the common law can add grounds but should not destroy any agreed-upon conditions precedent.

(E) WAIVER OF TERMINATION AND REINSTATEMENT OF COMPLETION DATE

Suppose one party has the power to terminate but does not exercise it. Has the power to terminate been lost? Under what conditions can it be revived?

The cases that have dealt with this problem have usually involved a performing party, usually the contractor, who has not met the completion date. But for various reasons the owner decides not to terminate. The contractor continues to perform believing that the power to terminate will not be exercised. At some point during performance the owner wishes to either terminate immediately or set a firm date for completion which, if not met, will be grounds for termination.

Clearly, a termination after the contractor has been led to believe there will be no termination would be improper.[13] The more difficult question relates to the requirements for reinstating a firm deadline that will allow termination if performance is not met by that time. The parties can agree to extend the time for completion coupled with the clear understanding that failure to comply would entitle the owner to terminate. The more difficult questions involve unilateral attempts by the owner to reinstate a firm completion date and revive the right to terminate.

In *DeVito v. United States,*[14] the contractor did not meet a revised completion date of November 29, 1960, because of performance problems. Having anticipated default, on November 25, 1960, the contracting officer requested permission from higher headquarters to terminate the contract. The request moved slowly through various offices of the agency and for unexplained reasons authority to terminate was not granted until January 16, 1961. This was communicated to the contractor on January 17, 1961. During this period, the contractor continued to perform, made commitments, and delivered some units that were accepted by the government. At the time of termination the contractor had many assemblies in various stages of completion and claimed to be on the verge of reaching full production.

---

13. *DeVito v. United States,* 413 F.2d 1147 (Ct.Cl.1969); *United States v. Zara Contracting Co.,* 146 F.2d 606 (2d Cir.1944).

14. Supra note 13.

In an action brought on behalf of the contractor the court first noted that the government is habitually lenient in granting reasonable extensions "for it is more interested in production than in litigation."[15] Moreover, the court said, "default terminations—as a species of forfeiture—are strictly construed."[16] The court noted that permitting a delinquent contractor to continue performance can preclude the government from terminating if its actions or nonactions have led the contractor to believe no termination would occur; the contractor relied upon this. The court held that the government had waived its right to terminate by allowing the contractor to continue performance under these circumstances.

As to reinstatement, the court stated:

> When a due date has passed and the contract has not been terminated for default within a reasonable time, the inference is created that time is no longer of the essence so long as the constructive election not to terminate continues and the contractor proceeds with performance. The proper way thereafter for time to again become of the essence is for the Government to issue a notice under the Default clause setting a reasonable but specific time for performance on pain of default termination. The election to waive performance remains in force until the time specified in the notice, and thereupon time is reinstated as being of the essence. The notice must set a new time for performance that is both reasonable and specific from the standpoint of the performance capabilities of the contractor at the time the notice is given.[17]

The court held that such a process would not be required if the contractor had renounced its contract or was incapable of performance.[18]

The *DeVito* case emphasized that there could be no fixed rules regarding the government's waiver of its right to terminate. *H.N. Bailey & Associates v. United States,*[19] a case decided by the Court of Claims two years after *DeVito,* illustrated this. The contractor had been awarded a contract set aside for small businesses and was relatively inexperienced in manufacturing the products sought by the procurement. The delivery date was July 27, 1966, and it seemed clear to the government that the contractor would not be able to perform. As a result, on August 10, 1966, the government notified the contractor that there had been a default and that any assistance given the contractor, or acceptance of delinquent goods, would be solely for the purpose of mitigating damages and not to be construed as an indication that the government was waiving its rights. The notice also gave the contractor ten days to advise the procuring agency of any reason why the contract should not be terminated.

On August 19, 1966, the contractor responded by stating that it was confident it could produce the goods by the end of the following month. Yet, despite this, production was not made and the government terminated for default on September 6, 1966.

The contractor claimed the conduct of the government was inconsistent with its right to terminate and constituted a waiver. The court did not agree, however, and concluded that the government did not encourage or induce the contractor to continue performance; it characterized the government's conduct as "displaying a benevolent attitude towards the defaulting contractor."[20]

---

15. 413 F.2d at 1153.

16. Id.

17. Id. at 1154.

18. A similar rule was announced in *Gamm Construction Co. v. Townsend,* 32 Ill.App.3d 848, 336 N.E.2d 592 (1975).

19. 449 F.2d 376 (Ct.Cl.1971).

20. Id. at 385.

In *John Kubinski & Sons, Inc. v. Dockside Development Corp.*,[21] promised performance by the owner essential to the contractor's performance was not accomplished by the date promised, September 1, 1968. A conference was held between the parties that appeared to extend the time for the owner's performance until approximately November 15, 1968. The contractor terminated the contract on December 18, 1968, because the owner's promised work had not been done by that time. The owner hired a replacement contractor to complete the work.

The owner contended that the contractor's failure to perform the work was a default and that the contractor had waived any owner default. The waiver claim was based upon activities engaged in by the contractor after the expiration of the original contract performance date. These activities consisted of some minor work and the delivery of a bond to receive payment for work that had been performed. The court concluded that none of the activities were inconsistent with an intention of the contractor to insist upon its contract rights.

In rejecting the contention that the contractor should have terminated at the time of the original default by the owner, the court stated:

> Whether a breach of a time provision of a contract justifies a repudiation of the agreement is a determination not to be made lightly by an injured party, since, should a court later hold his determination unwarranted, the repudiator will himself be guilty of a material breach. To hold that [the contractor] should have declared an immediate forfeiture after September 1 would defeat the public policy that encourages the extrajudicial resolution of disagreements.[22]

The court further noted that it was not clear that failure to meet the original date was a material breach. But in any event, failing to meet the time extensions along with the advent of adverse winter conditions certainly made the final breach material and terminated the contractor's obligation to perform.

(F) NOTICE OF TERMINATION: *NEW ENGLAND STRUCTURES, INC. v. LORANGER*

Termination clauses often require that a notice of termination be sent by the terminating party to the party whose performance is being terminated and, in many contracts, to the lender or surety.[23] The notice usually states that termination will become effective a designated number of days after dispatch or receipt of the notice.

In addition to problems common to all notices and communications,[24] such notice requirements create other legal problems. What are the rights and duties of the parties during the notice period? Can the defaulting party cure any defaults specified in the notice during the notice period and thereby keep the contract in effect? Can the party giving the notice to terminate demand continued performance or even accelerated performance during the notice period? Answers to such questions often depend upon the reason for requiring a notice to terminate.[25]

The notice period may be designed to allow the terminating party to "cool off." Construction performance problems often generate animosity and before the important step of termination is effective, the terminating party may wish to rethink its position. The notice period can permit a

---

21. 33 Ill.App.3d 1015, 339 N.E.2d 529 (1975).

22. 339 N.E.2d at 534.

23. Failure to notify the surety may relieve the latter from its obligation. *Winston Corp. v. Continental Casualty Co.*, 508 F.2d 1298 (6th Cir.1975).

24. Refer to Section 21.05(C).

25. Section 15.13(B) treated this problem in the context of the design professional-client relationship.

defaulting party to cure defaults in order to keep the contract in effect for the benefit of both parties. Whether the notice period is designed to "cure" may depend on the facts that give rise to termination, and the notice period.

Finally the notice period can be used to wind-down and secure the job. This allows each party to cut losses and make new arrangements. Such a position does not allow cure. But if the owner terminates it should have the option of ordering that work be done that can be completed by the effective date of termination.

Such questions should be, but rarely are, answered by the termination clause. A case will be reproduced at this point that illustrates this and other problems.

NEW ENGLAND STRUCTURES, INC. v. LORANGER

Supreme Judicial Court of Massachusetts, 1968. 354 Mass. 62, 234 N.E.2d 888.

CUTTER, Justice. In one case the plaintiffs, doing business as Theodore Loranger & Sons (Loranger), the general contractor on a school project, seek to recover from New England Structures, Inc., a subcontractor (New England), damages caused by an alleged breach of the subcontract. Loranger avers that the breach made it necessary for Loranger at greater expense to engage another subcontractor to complete work on a roof deck. In a cross action, New England seeks to recover for breach of the subcontract by Loranger alleged to have taken place when Loranger terminated New England's right to proceed. The actions were consolidated for trial. A jury returned a verdict for New England in the action brought by Loranger, and a verdict for New England in the sum of $16,860.25 in the action brought by New England against Loranger. The cases are before us on Loranger's exceptions to the judge's charge.

Loranger, under date of July 11, 1961, entered into a subcontract with New England by which New England undertook to install a gypsum roof deck in a school, then being built by Loranger. New England began work on November 24, 1961. On December 18, 1961, New England received a telegram from Loranger which read, "Because of your . . . repeated refusal . . . or inability to provide enough properly skilled workmen to maintain satisfactory progress, we . . . terminate your right to proceed with work at the . . . school as of December 26, 1961, in accordance with Article . . . 5 of our contract. We intend to complete the work . . . with other forces and charge its costs and any additional damages resulting from your repeated delays to your account." New England replied, "Failure on your [Loranger's] part to provide . . . approved drawings is the cause of the delay." The telegram also referred to various allegedly inappropriate changes in instructions.

The pertinent portions of art. 5 of the subcontract are set out in the margin.[26] Article 5 stated grounds on which Loranger might terminate New England's right to proceed with the subcontract.

---

26. [Ed. note: Court footnote renumbered.]

"The Subcontractor agrees to furnish sufficient labor, materials, tools and equipment to maintain its work in accordance with the progress of the general construction work by the General Contractor. Should the

There was conflicting evidence concerning (a) how New England had done certain work; (b) whether certain metal cross pieces (called bulb tees) had been properly "staggered" and whether joints had been welded on both sides by certified welders, as called for by the specifications; (c) whether New England had supplied an adequate number of certified welders on certain days; (d) whether and to what extent Loranger had waived certain specifications; and (e) whether New England had complied with good trade practices. The architect testified that on December 14, 1961, he had made certain complaints to New England's president. The work was completed by another company at a cost in excess of New England's bid. There was also testimony (1) that Loranger's job foreman told one of New England's welders "to do no work at the job site during the five-day period following the date of Loranger's termination telegram," and (2) that, "if New England had been permitted to continue its work, it could have completed the entire subcontract . . . within five days following the date of the termination telegram."

The trial judge ruled, as matter of law, that Loranger, by its termination telegram, confined the justification for its notice of termination to New England's "repeated refusal . . . or inability to provide enough properly skilled workmen to maintain satisfactory progress." He then gave the following instructions: "If you should find that New England . . . did not furnish a sufficient number of men to perform the required work under the contract within a reasonable time . . . then you would be warranted in finding that Loranger was justified in terminating its contract; and it may recover in its suit against New England. [T]he termination . . . cannot, as . . . matter of law, be justified for any . . . reason not stated in the telegram of December 18 . . . including failure to stagger the joints of the bulb tees or failure to weld properly . . . or any other reason, unless you find that inherent in the reasons stated in the telegram, namely, failure to provide enough skilled workmen to maintain satisfactory progress, are these aspects. Nevertheless, these allegations by Loranger of deficiency of work on the part of New England Structures may be considered by you, if you find that Loranger was justified in terminating the contract for the reason enumerated in the telegram. You may consider it or them as an

---

Subcontractor fail to keep up with . . . [such] progress . . . then he shall work overtime with no additional compensation, if directed to do so by the General Contractor. If the Subcontractor should be adjudged a bankrupt . . . or *if he should persistently . . . fail to supply enough properly skilled workmen . . . or . . .* disregard instructions of the General Contractor or fail to observe or perform the provisions of the Contract, then the General Contractor may, by *at least five . . . days prior written notice to the Subcontractor* without prejudice to any other rights or remedies, *terminate the Subcontractor's right to proceed with the work.* In such event, the General Contractor may . . . prosecute the work to completion . . . and the Subcontractor shall be liable to the General Contractor for any excess cost occasioned . . . thereby . . ." (emphasis supplied).

[Ed. note: Balance of footnotes omitted.]

element of damages sustained by Loranger." . . . Counsel for Loranger claimed exceptions to the portion of the judge's charge quoted above in the body of this opinion.

1. Some authority supports the judge's ruling, in effect, that Loranger, having specified in its telegram one ground for termination of the subcontract, cannot rely in litigation upon other grounds, except to the extent that the other grounds may directly affect the first ground asserted. See *Railway Co. v. McCarthy,* . . . ("Where a party gives a reason for his conduct and decision touching . . . a controversy, he cannot, after litigation has begun, change his ground and put his conduct upon . . . a different consideration. He is not permitted thus to mend his hold. He is estopped from doing it by a settled principle of law."

Our cases somewhat more definitely require reliance or change of position based upon the assertion of the particular reason or defence before treating a person, giving one reason for his action, as estopped later to give a different reason. See *Bates v. Cashman.* There it was said, "The defendant is not prevented from setting up this defense. Although he wrote respecting other reasons for declining to perform the contract, he expressly reserved different grounds for his refusal. While of course one cannot fail in good faith in presenting his reasons as to his conduct touching a controversy he is not prevented from relying upon one good defense among others urged simply because he has not always put it forward, when it does not appear that he has acted dishonestly or that the other party has been misled to his harm, or that he is estopped on any other ground."

We think Loranger is not barred from asserting grounds not mentioned in its telegram unless New England establishes that, in some manner, it relied to its detriment upon the circumstance that only one ground was so asserted. Even if some evidence tended to show such reliance, the jury did not have to believe this evidence. They should have received instructions that they might consider grounds for termination of the subcontract and defences to New England's claim (that Loranger by the telegram had committed a breach of the subcontract), other than the ground raised in the telegram, unless they found as a fact that New England had relied to its detriment upon the fact that only one particular ground for termination was mentioned in the telegram.

2. As there must be a new trial, we consider whether art. 5 of the subcontract . . . afforded New England any right during the five-day notice period to attempt to cure its default, and, in doing so, to rely on the particular ground stated in the telegram. Some evidence summarized above may suggest that such an attempt was made. Article 5 required Loranger to give "at least five . . . days prior written notice to the Subcontractor" of termination.

If a longer notice period had been specified, one might perhaps infer that the notice period was designed to give New England an opportunity to cure its defaults. An English text writer (Hudson's Building and Engineering Contracts, 9th ed. p. 530) says, "Where a previous warning notice of specified duration is expressly required by the contract before . . . termination [in case of dissatisfaction], the notice should be explicit as to the grounds of dissatisfaction, so that during the time mentioned in the notice the builder may have the opportunity of removing the cause of objection." . . . This view was taken of a three-day notice provision in *Valentine v. Patrick Warren Constr. Co.,* . . . without, however, very full consideration of the provision's purpose. In Corbin, Contracts, § 1266, p. 66, it is said of a reserved power to terminate a contract, "If a period of notice is required, the contract remains in force and must continue to be performed according to its terms during the specified period after receipt of the notice of termination."

Whether the short five-day notice period was intended to give New England an opportunity to cure any specified breach requires interpretation . . . of art. 5, a matter of law for the court . . . . It would have been natural for the parties to have provided expressly that a default might be cured within the five-day period if that had been the purpose.

Strong practical considerations support the view that as short a notice period as five days in connection with terminating a substantial building contract cannot be intended to afford opportunity to cure defaults major enough (even under art. 5) to justify termination of a contract. Such a short period suggests that its purpose is at most to give the defaulting party time to lay off employees, remove equipment from the premises, cancel orders, and for similar matters.

Although the intention of the notice provision of art. 5 is obscure, we interpret it as giving New England no period in which to cure continuing defaults, but merely as directing that New England be told when it must quit the premises and as giving it an opportunity to take steps during the five-day period to protect itself from injury. Nothing in art. 5 suggests that a termination pursuant to its provisions was not to be effective in any event at the conclusion of the five-day period, even if New England should change its conduct.

If Loranger in fact was not justified by New England's conduct in giving the termination notice, it may have subjected itself to liability for breach of the subcontract. The reason stated in the notice, however, for giving the notice cannot be advanced as the basis of any reliance by New England in action taken by it to cure defaults. After the receipt of the notice, as we interpret art. 5, New England had no further opportunity to cure defaults.

Exceptions sustained.

The termination clause should clearly indicate whether it permits *cure,* provides a *cooling off* period, or sets into motion a *winding down* of the project. To illustrate a cure provision, standard construction contracts used in Canada allow the owner to terminate if the contractor fails to perform after the owner notifies the contractor in writing that it is in default based upon a certificate from the architect. The notice must instruct the contractor to correct the default within five working days from receipt of the notice. If the default cannot be cured within the five working days, the owner cannot terminate if the contractor begins correction of the default within the specified time, provides the owner with an acceptable schedule for such correction, and completes the correction in accordance with that schedule. This is a sensible procedure. Termination is the last desperate step in a troubled contract.

(G) TAKING OVER MATERIALS AND EQUIPMENT

AIA Doc. A201, ¶ 14.2.1, permits the owner to:

> take possession of the site and of all materials, equipment, tools, construction equipment and machinery thereon owned by the contractor and . . . finish the work by any method he may deem expedient.

These provisions are commonly included in construction contracts. Reasons given for such provisions are:

1. to provide incentive to the contractor to take away its property from the site so that the owner can efficiently bring in a successor;

2. to provide the owner with material and equipment by which it can expeditiously continue the work with a successor (Note the conflict with (1) above); and

3. to give the owner property from which it can obtain payment for any claim it may have against the contractor.

The owner, however, must first give a seven-day termination notice. Will this encourage the contractor to take away its property before the owner does? If so, the first justification will be accomplished but the second will be frustrated. Also, suppose the successor contractor does not want to use the material and equipment. Does the contractor have the right to take away the materials and equipment during the seven-day period? (It is unlikely that the contractor will stand by passively when the owner takes possession of its property [27]). Can the owner take material or equipment owned by a subcontractor? If not, the clause may be of little value.[28] In addition, who actually owns the equipment? Much equipment is leased or purchased under conditional sales contracts. Even if the property is owned by the contractor, any attempt to use the equipment as security may force a struggle with others who contend that they have security interests that take precedence over any right created by such a clause or any other lien rights. If the contractor goes bankrupt, the trustee in bankruptcy surely will enter the fray. Also, if the contractor's property is used, a subsequent dispute may arise as to its reasonable value.

As if these problems were not enough, suppose the termination is wrongful. Taking the contractor's property constitutes the tort of conversion. The owner must pay the reasonable value of property converted, restore any gains made through the conversion of the property, and perhaps pay punitive damages.

The AIA did not expressly create a security; no express power to sell the material and equipment and retain any amounts obtained to set off against claims it has

---

27. For a case describing such an imbroglio see *Leo Spear Construction Co. v. Fidelity Casualty Co.,* 446 F.2d 439 (2d Cir.1971).

28. See AIA Doc. A201, ¶ 5.3.1.

against by the contractor exists (AIA Doc. A201, ¶ 13.2.5, *did* create such a security interest). Some justify the clause as a basis to back up emergency measures under which it is essential that the material and equipment be used to keep the project performance going. They would deal with legal subtleties later. The risks are greater if equipment, particularly construction machinery, rather than material is taken. In any event, continued inclusion of such clauses in construction contracts may indicate that they do have utility.[29]

(H) EFFECT ON EXISTING CLAIMS FOR DELAY

Termination of the contract is usually accompanied by claims by each party against the other. If the owner terminated, the contractor, as a rule, not only claims that the termination was wrongful but that the owner has committed earlier breaches causing it losses. The terminating owner also is likely to have claims for damages based upon delay and improper workmanship. Sometimes the terminated party will contend that invoking the termination remedy was a waiver of any claims that existed prior to termination. This contention was rejected in *Armour & Co. v. Nard.*[30] After pointing to a provision stating that failure to exercise a right shall not be considered a waiver, the court noted that a contractual remedy for breach generally does not exclude other remedies unless the clause shows a clear intention of the party to do so. Rarely will a terminating party implicitly give up damage claims.

(I) DISPUTED TERMINATIONS

Suppose the owner terminates the contract in accordance with the termination clause

and orders the contractor to leave the site. The contractor contends the termination is not justified and demands arbitration under a provision similar to AIA Doc. A201, ¶ 7.9.1. This provides for arbitration for all claims, disputes, and other matters arising out of or relating to the contract documents or the breach thereof. Paragraph 7.9.3 requires a contractor to continue working while the dispute is being arbitrated. The following issues can arise:

1. Does a contractor's demand for arbitration give it the right to remain on the site pending the arbitrator's decision?

2. If the *contractor* wishes to terminate, may the owner demand arbitration and insist that the contractor continue work?

3. Is the disputed termination subject to arbitration?

4. If arbitration takes place should the arbitrator order that work be resumed?

Requiring the contractor under ¶ 7.9.3 to continue working while the dispute is being arbitrated assumes performance disputes that do not involve termination. Termination is a special dispute. The *specific* provisions of a termination clause should take precedence over the *general* arbitration clause. Paragraph 7.9.3 does not require continued performance during arbitration when otherwise agreed to in writing. This can refer to the termination clause with its specific mechanism. Also, ¶ 7.9.3 can be for the owner's benefit; the owner can waive its right to continued performance during arbitration by ordering the contractor to leave the site. Finally the owner's ownership and control of the site should empower the owner to remove the contractor. Whether the termination was proper can be resolved by any applicable arbitra-

---

29. In *Northway Decking & Sheet Metal Corp. v. Inland-Ryerson Construction Products Co.,* 426 F.Supp. 417 (D.R.I.1977), the court appears to have enforced such a clause. It denied a terminated subcontractor an injunction permitting him to remove unique hanging scaffolding.

30. 463 F.2d 8 (8th Cir.1972).

tion clause.[31]   The termination has not attacked the validity of the arbitration clause.[32]

Generally arbitrators have remedial discretion. Any award to continue performance, at least in New York,[33] would be upheld. But termination is usually acrimonious and the last desperate step. Rarely will reinstatement be chosen as the remedy; this is another reason to force the contractor to leave the site even if termination is challenged. If termination were improper, a damage award would be adequate.

Suppose the *contractor* wishes to terminate and the owner seeks arbitration. The issue here is less clear. The provision for removing the contractor from the site can be considered for the benefit of the owner. This interpretation would allow the owner to order that the contractor continue working. It is more likely, however, that the conclusion will not depend upon which party terminates.

SECTION 38.04
TERMINATION BY LAW

(A) MATERIAL BREACH

If the contract does not expressly create a power to terminate, termination is allowed if the breach is material. Even with a contract clause dealing with termination, many of the factors that determine materiality can be influential when such clauses are interpreted. Also, as indicated, the termination clause may not be the exclusive source for determining when a party has the power to terminate.

Rather than establishing fixed rules, such as the importance of the clause breached,

the law examines all the facts and circumstances surrounding the breach to determine whether it would be fair to permit termination. For example, the Restatement (Second) of Contracts articulates factors in 241 that are significant in determining whether a particular breach is material. They are:

(a) the extent to which the injured party will be deprived of the benefit which he reasonably expected;

(b) the extent to which the injured party can be adequately compensated for the part of that benefit of which he will be deprived;

(c) the extent to which the party failing to perform or to offer to perform will suffer forfeiture;

(d) the likelihood that the party failing to perform or to offer to perform will cure his failure, taking account of all the circumstances including any reasonable assurances;

(e) the extent to which the behavior of the party failing to perform or to offer to perform comports with standards of good faith and fair dealing.

Looking first at breaches by the contractor, subparagraph (a) seeks to determine the importance of the deviation and the likelihood of future nonperformance. Subparagraph (b) examines whether the owner can be easily compensated for nonperformance. This would depend on whether the defect was easily correctable or whether compensation for an uncorrected defect would be easy to measure. In construction contracts the latter can be difficult. Subparagraph (c) looks for uncompensated losses which the contractor would suffer if

---

31. Many cases hold that arbitration survives termination. See *County of Middlesex v. Gevyn Construction Corp.*, 450 F.2d 53 (1st Cir.1971); *State v. Lombard Co.*, 106 Ill.App.3d 307, 60 Ill.Dec. 540, 436 N.E.2d 566 (1982); *Willis-Knighton Medical Center v. Southern Builders, Inc.*, 392 So.2d 505 (La.App.1980). But see *G&N Construction Co. v. Kirpatovsky*, 181 So.2d 664 (Fla.App.1966).

32. *Riess v. Murchison*, 384 F.2d 727 (9th Cir.1967).

33. *Grayson-Robinson Stores, Inc. v. Iris Construction Corp.*, 8 N.Y.2d 133, 202 N.Y.S.2d 303, 168 N.E.2d 377 (1960). Refer to Section 34.12.

termination occurred. If it had ordered material that had not yet been used and was not usable in other projects, this would militate against termination. Subparagraph (d) examines the likelihood that the contractor will be able to cure defects while subparagraph (e) examines the reason for the breach. An example of the latter, a breach by a subcontractor that the prime contractor could not have reasonably prevented might militate against materiality.

An owner breach would probably be nonpayment of money, failure to furnish the site, or noncooperation. Subparagraph (a) would look at the extent of the breach and the likelihood of future performance while subparagraph (b) would look at whether the breach was easily compensable, such as interest for a breach consisting of not making a progress payment.

Subparagraph (c) examines the harm termination would cause the owner, such as lost loan commitments, liability to prospective tenants, or other lost business opportunities. Subparagraph (d) seeks to determine whether the breaches are likely to be cured. Finally, subparagraph (e) examines the reasons for nonperformance. Financial reverses or a steep rise in interest rates make it less likely that the contractor can terminate the breach of nonpayment. If, on the other hand, there was "bad blood" between the design professional and the contractor, and the latter seized upon the breach by the owner to injure the design professional, termination would be less likely. Likewise, if it appears that the contractor sought an excuse to end the contract, the breach would less likely be material.

Many factors are relevant. Perhaps the most relevant factors are the particular nature of the nonperformance, the likelihood of future breaches, and the possibility of forfeiture. Some courts may look at the good faith not simply of the party failing to perform but of the party who seeks to exercise the power to terminate.[34]

Turning to construction contracts, nonpayment, perhaps the most frequently asserted justification for termination, has been discussed in Section 26.02(H). As to other breaches, one court held a prime contract terminated when the owner did not make an equitable adjustment when different underground conditions were discovered and the contractor was in financial trouble.[35] A prime contractor's failure to have the site ready for the flooring subcontractor was a material breach.[36] Hindering the subcontractor's operation was a material breach when coupled with the prime contractor having stopped payment on a check that the subcontractor was about to negotiate.[37] Another court held that the collapse of a fallout shelter was sufficient grounds for the owner to terminate the contractor's performance and recover a down payment.[38] These obviously incomplete illustrations are not designed to indicate that breaches of the type described will always be considered material. As emphasized in this subsection, the determination of whether a breach is material requires a careful evaluation of the facts and circumstances surrounding the breach as well as the effect of termination.

(B) FUTURE BREACH: PROSPECTIVE INABILITY AND BREACH BY ANTICIPATORY REPUDIATION

Subsection (A) dealt with breaches that

---

34. Refer to Sections 5.11(B) and 38.01 dealing with employment contracts.

35. *Metropolitan Sewerage Commission v. R.W. Construction, Inc.*, 72 Wis.2d 365, 241 N.W.2d 371 (1976).

36. *Great Lakes Construction Co. v. Republic Creosoting Co.*, 139 F.2d 456 (8th Cir.1943).

37. *Citizens National Bank of Orlando v. Vitt*, 367 F.2d 541 (5th Cir.1966).

38. *Economy Swimming Pool Co. v. Freeling*, 236 Ark. 888, 370 S.W.2d 438 (1963).

have occurred. This subsection deals with breaches that may occur in the future. It may appear that one party may not be able to perform when the time for performance arrives, or one party may state that it will not perform when the time for performance arrives. The contractor may discharge some of its employees, or a number may quit. The contractor may cancel orders for supplies, or its suppliers may indicate that they will not perform at the time for performance. In such cases, the owner may realize that the contractor will be unable to perform.

Such inability to perform, whether it is on the part of the owner or the contractor or subcontractors, is likely to be a breach. Each party owes the other party the duty to *appear* to be ready to perform when the time for performance arises. Even if no such promise is implied, under certain circumstances, events may permit one party to suspend its performance or terminate the contract unless the other party who appears unable to perform can give assurance or security that when the time comes for performance, it will perform.[39] Insolvency as a basis for insecurity may be affected in a construction dispute by AIA Doc. A201, ¶ 3.2.1, which gives the contractor the power to ask for evidence of financial sources from the owner before executing the contract. Failure to exercise this power may preclude the contractor's suspending its obligation to perform in the event that the owner becomes insolvent. Of course, failure to make payments as indicated in Section 26.02(H) gives the contractor a right under A201 to suspend performance.

Prospective inability deals with probabilities. In the examples given the question is whether the owner must wait to see whether actual performance or defective performance will occur, or whether it can demand assurance and, in the absence of this assurance, legally terminate any obligation to use the contractor.

Contracts often deal with termination rights. In the absence of such provisions, it will take a strong showing on the part of the owner to terminate the contractor's performance on the grounds that the contractor may not be able to perform in the future. In many cases, a combination of present and prospective nonperformance exists. If present nonperformance, such as the installation of defective materials or poor workmanship exists, the likelihood that this will continue will be a strong factor in influencing the court to allow the owner to terminate the contractor's performance. Pure prospective inability without present breach is likely to be held to be insufficient grounds unless the probabilities are very strong, or unless a contract provision giving the owner this right exists.

A breach by anticipatory repudiation occurs when one party indicates to the other that it cannot or will not perform. Each party is entitled to reasonable assurance that the other will perform in accordance with the contract. If one of the parties indicates that it will not, or cannot, perform, the other party loses this assurance. Should the latter be required to wait and see whether the threat or the indication of inability to perform will come to fruition?

The rights of a party to terminate a contract because of the other party's repudiation have expanded. A feeling of assurance is important since this is one purpose for making a contract. Sometimes a party who indicates it will not perform is jockeying for position. The owner may be trying to pay less or obtain more than the called for performance. If either party does repudiate, the other party will be given the right to terminate the obligation. The party to whom the repudiation is made need not terminate the obligation immediately. It may state that it intends to hold the other

---

39. Restatement (Second) of Contracts, §§ 251, 252 (1981).

party to the contract. If the repudiator relies upon the statement that the contract will be continued, the nonrepudiating party will lose its right to terminate based upon the earlier repudiation.

The right to continue one's performance despite repudiation by the other party is qualified by the rule against enhancing damages. One party cannot recover damages caused by the other party's breach when those damages could have been avoided by taking reasonable steps to cut down or eliminate the loss. For example, if the contractor repudiates the contract the owner cannot recover for those damages caused by breach that could have been avoided by hiring a replacement contractor.

Suppose the contractor states unequivocally without justification that it will leave the job in three days. It might be reasonable for the owner to insist that it will hold the contractor to the contract for a short period such as until the date of the walkout, or even for a few days after the walkout. But when it is clear that the contractor will not return to the project, the owner should take reasonable steps to replace the contractor if the owner wishes to continue the project. A replacement should be obtained when the repudiating contractor has committed its workers and machinery to another project and it appears that it has neither the willingness nor the capacity to return to the job. Any damages that could have been avoided by hiring a replacement will not be assessable against the repudiating contractor.

Repudiation may accompany present or prospective breach. The greater the scope of any present or prospective breach, the greater the likelihood that the court will release the innocent party by reason of both the present breach and the repudiation. Even a small breach, coupled with a repudiation, may be enough to terminate the innocent party's obligation to perform further.

(C) BANKRUPTCY

Bankruptcy can affect contracts. Two types of bankruptcy are important for the purposes of this chapter. The first is liquidation under Chapter 7 and is known as straight bankruptcy. It has two principal objectives. First, the bankrupt can wipe the slate clean, or more mostly clean, by discharging most of the bankrupt's debts. Second, bankruptcy should provide a fair and efficient liquidation of the bankrupt's estate. The trustee for the bankrupt takes over the bankrupt's estate, collects any money owed the bankrupt, compels repayment of any preferential payments made to creditors [40], turns over specific property in the hands of the bankrupt to those who have security interests in them and, after paying expenses of administering the estate, distributes any amount remaining *pro rata* to unsecured creditors. For example, if the amount of unsecured debts is one million dollars and the balance is $100,000, each unsecured creditor receives ten cents for each dollar of unsecured debt. Payments to unsecured creditors, if made at all, are usually only a small fraction of the debts.

The second type is a petition for reorganization under Chapter 11. Such a petition protects the petitioner from creditor claims while it seeks to reorganize in order to put its business on a more sound financial basis. The petitioner submits a plan that involves a reorganization of the interests of owners and creditors, a rescheduling of debts, and, one hopes, an infusion of cash and credit for judicial approval.

Two recent developments dealt with the effect of bankruptcy on existing contracts. Bankruptcy law gives the trustee an election to assume (continue performance) or reject (refuse further performance) existing con-

---

40. Broadly speaking a preference favors one creditor over the others without any legitimate business reason. For details see 11 U.S.C.A. § 547 (1979).

tracts. In 1978, effective October 1, 1979, the U.S. Congress precluded any contract clause from barring this election.[41] The trustee can assume any contract if it has the capacity to continue performance, or pay compensation if it does not.

It was common for contracts to permit automatic termination if a party filed for bankruptcy. In construction contracts the owner was usually given the power to terminate if the contractor filed for bankruptcy or had serious financial problems. Despite the change in the law, contracts often still contain such provisions, either because of ignorance as to their ineffectiveness, or simply because of inertia. In any event assumptions of construction contracts were and are rare. The trustee seldom had the resources to continue performance.

More important and more controversial were well-publicized Chapter 11 petitions by troubled airlines seeking to rid themselves of onerous labor contracts or asbestos manufacturers facing thousands of large tort claims by those who claimed that asbestos gave them cancer.

In *N.L.R.B. vs. Bildisco and Bildisco* [42] the contractor filed a Chapter 11 petition and immediately stopped paying pension, health plan, and wage increases agreed upon in a recent collective bargaining agreement with the union representing its workers. The U.S. Supreme Court held this would not be an unfair labor practice if the contractor can show the contract burdened its financial affairs and the equities balanced in favor of rejection.

Congress responded to complaints by trade unions by creating difficult procedural requirements before a collective bargaining agreement could be set aside under Chapter 11.[43]

SECTION 38.05
RESTITUTION WHERE A CONTRACT IS TERMINATED

When a contract is terminated, each party, as a rule, has conferred benefit upon the other. The owner has made progress payments and the contractor has performed the work. What effect does termination have upon the right of either to recover the value of what they have conferred upon the other?

If termination has been made for the convenience of the owner, the clause permitting termination usually provides a formula for compensating the contractor for the work performed. Terminations for default made by the owner usually involve claims by the latter that exceed the value of any unpaid work performed by the contractor. In the event they do not, the contractor, though in default, in most jurisdictions would be entitled to recover the net benefit conferred on the owner.[44] If termination was made by the contractor based upon an owner default, the contractor can use a restitutionary recovery that permits it in most cases to recover the reasonable value of the services performed.[45] If termination has been accomplished by mutual consent, the agreement will usually deal with compensation for benefits conferred. If it does not, either party is entitled to recover the net benefit it has conferred upon the other. For example, if the work performed has a value of $100,000 and the contractor has been paid $90,000, the latter should recover $10,000. Conversely, if the figures were reversed the owner would be entitled to that amount.

---

41. Id. at 365(e).

42. ___ U.S. ___, 104 S.Ct. 1188, 79 L.Ed.2d 482 (1984).

43. P.L. 98–353, Subtitle J, enacted July 10, 1984.

44. Refer to Section 26.06(D).

45. Refer to Section 31.02(E).

PROBLEMS

1. On February 1, 1985, O and C entered into a contract under which C agreed to construct a residence in accordance with designated plans and specifications in exchange for O's promise to pay $100,000. The house was to be built in Madison, Wisconsin. The site was to be furnished by the contractor no later than April 1, 1985, and the work was to be completed by November 1 of that year.

   Because of difficulty in obtaining title to the site, the contractor was not given access to the site until May 15 of that year. When O rejected a demand for additional compensation, C terminated its obligation to build the house. It contends that the delay will require that the work be done during the winter and this will substantially increase the cost of performance. Did C have legal ground to terminate? What further facts would be useful in making this decision?

   Suppose C went ahead with the work in May when it was given site access. Material and labor shortages delayed the work in the summer. By September it appeared he would not be able to finish until the following February. The delay would necessitate a substantial period of work during the winter. On November 15 it terminated the contract claiming that the delay in furnishing the site justified termination. Is this contention correct? What further facts would be important in answering this question?

2. The contract between a subcontractor and a prime contractor provided that if the owner terminates the prime contract "for any cause whatsoever at any time" the subcontract will also be terminated and the prime contractor's liability will be limited to payment for work done. The subcontractor partially performed but was not paid in accordance with the subcontract. Yet the subcontractor continued performing until it heard that the owner had terminated the prime contract. It then stopped performing, relying not upon nonpayment but rather upon the owner's termination having been caused by the prime contractor's default.

   The owner's termination was based upon its contractual power to terminate for its convenience. The owner terminated because it felt that the project would not be needed.

   1. Can the subcontractor recover from the prime contractor for work it has performed?

   2. Can the subcontractor recover its lost profits from the prime contractor?

   3. Would the subcontractor have been able to terminate for the prime contractor's failure to make payments?

3. O hired C to erect a steel building. C hired S to pour the concrete. S poured the concrete but, prior to the seven days that the concrete should have been allowed to cure in order to tolerate any heavy weight, S drove a gravel truck onto the concrete floor. The owner was present and ordered the truck off the concrete asserting that it would damage the floor.

   C met with O and promised to take reasonable steps to repair any damage caused by the truck and requested permission to continue with the work. However, O demanded that 4,000 square feet of concrete be removed and replaced before it would allow C to continue work.

   O met with S and obtained a one-year guarantee from S for all materials and workmanship on the floor. However, O continued to insist that the concrete be replaced. C contended that no evidence of damage to the floor existed, that it could, by doubling the size of its work crew, complete its work by the contract deadline without placing any weight on the floor, and that any necessary repairs could be undertaken after the building was completed.

O was not satisfied and ordered C to leave the site. A subsequent investigation revealed that the concrete was not damaged and that it exceeded requirements as to strength. O obtained another contractor who completed the project.

O has brought a claim against C for the excess costs it incurred in hiring a replacement contractor. C defended against that claim, asserting that it was justified in refusing to continue and asserted a counter claim against O for its lost profits on the contract. How would you resolve these claims?

Appendix A

Standard Form of Agreement Between Owner and Architect

INSTRUCTION SHEET *AIA DOCUMENT B141a*

FOR AIA DOCUMENT B141, STANDARD FORM OF AGREEMENT BETWEEN OWNER AND ARCHITECT — JULY 1977 EDITION

Previous Editions:

The only previous edition of AIA Document B141 was the January 1974 Edition. B141 replaced three previous Owner-Architect Agreements: B131, in which compensation was based on a Percentage of Construction Cost; B231, based on a Multiple of Direct Personnel Expense, and B331, based on a Professional Fee plus Expenses. These various compensation methods, as well as compensation based on a Fixed Fee, were all available through tear-out pages in the 1974 Edition of B141.

Other AIA Owner-Architect Agreements:

The following other AIA Owner-Architect Agreements are available for use in connection with special circumstances. These should be carefully studied and used in place of B141 where appropriate. Users should consult AIA to determine the most recent edition of each document.

B141/CM	Owner-Architect Agreement, Construction Management Edition
B151	Owner-Architect Agreement for Use on Projects of Limited Scope
B161	Owner-Architect Agreement for Designated Services
B162	Scope of Designated Services (to be used in conjunction with B161)
B171	Interior Design Services Agreement
B727	Owner-Architect Agreement for Special Services
B801	Owner-Construction Manager Agreement

Letter Forms of Agreement:

Letter forms of agreement are generally discouraged by AIA, as is the performance of a part or the whole of professional services on the bases of oral agreements or understandings. The standard AIA agreement forms have been developed through sixty years of experience, and have been tested repeatedly in the courts. In addition, the standard forms have been carefully coordinated with other AIA Documents, including the various Architect-Consultant Agreement forms, the Owner-Contractor Agreements and General Conditions of the Contract for Construction. The necessity for specific and complete correlation between these documents and any Owner-Architect Agreement used is of paramount importance. Suggested examples of letter agreements are included in Chapter 9, Owner-Architect Agreements, of the Architect's Handbook of Professional Practice.

Owner-Furnished Agreement Forms:

Many clients, especially governmental agencies, require the use of an Owner-Architect Agreement prepared by them. Such forms should be carefully compared to AIA Document B141 before an agreement is entered into. If there are any significant omissions, additions or variances from the terms of B141, both legal and liability insurance counsel should be consulted. Of particular concern is the need for consistency between the Owner-Architect Agreement and the anticipated General Conditions of the Contract for Construction in the delineation of the Architect's Construction Phase services and responsibilities. In numerous instances, AIA and its components have successfully been able to encourage clients to adopt the AIA Owner-Architect Agreements or to bring their own forms into better conformance with accepted architectural practice.

Modifications:

As with all AIA Documents, users are encouraged to consult an attorney with respect to completing the form and with respect to any modifications that are proposed. Generally, modifications can be accomplished by striking out portions of B141 which are inapplicable, and inserting any additional provisions in Article 15. Legal counsel should also be sought concerning the effect of state and local law on the terms of the Agreement, particularly with respect to registration laws, duties imposed by building codes, interest charges, and arbitration.

Changes from the 1974 Edition of B141:

Numerous changes, both in format and in content, have been made in this edition of B141, based on experience with previous editions of the Owner-Architect Agreements, the need for recognition of client concerns, and the recommendations of AIA members, committees, and legal counsel and insurance counsel. These changes are specifically identified in the side-by-side comparison of 1974 and 1977 Editions, available from AIA. A discussion of these changes is contained in the September 1977 issue of the AIA Journal. Both architect and client should have a clear understanding of the provisions of this document before entering into the Agreement.

COMPLETING THE AGREEMENT FORM:

Cover Page---

Date: The date represents the first date as of which the Agreement is entered into. It may be the date that an oral agreement was reached, the date the Agreement is submitted to the Owner, the date authorizing action was taken by the Owner, or the date of actual execution. No professional services under this Agreement should be performed prior to the date indicated.

Identification of Parties: Parties to this Agreement should be identified in the capacity in which the Agreement is to be executed, including the name of the firms and capacity of persons signing, the address of the principal office and a designation of the legal status of both parties sole proprietorship, partnership, joint venture, unincorporated association, limited partnership or corporation (general, close, or professional), etc. . Where appropriate, a copy of the resolution authorizing the individual to act on behalf of the firm or entity should be attached.

Article 14 --- BASIS OF COMPENSATION

Paragraph 14.1:

Insert the amount of the retainer, if any, and indicate whether it will be credited to the first, the last, or proportionately to all of the payments on the Owner's account.

Paragraph 14.2, Subparagraph 14.2.1:

Compensation---Multiple of Direct Personnel Expense:

In Subparagraph 14.2.1, insert: "Compensation for services rendered by Principals and employees shall be based on a Multiple of Direct Personnel Expense in the same manner as described in Subparagraph 14.4.1, and for the services of professional consultants as described in Subparagraph 14.4.2."

Compensation---Professional Fee Plus Expenses:

In Subparagraph 14.2.1, insert: "Compensation shall be based on a Professional Fee of _____ dollars ($ _____) plus compensation for services rendered by Principals and empoyees in the same manner as described in Subparagraph 14.4.1, and for the services of professional consultants as described in Subparagraph 14.4.2."

Compensation: Fixed Fee:

In Subparagraph 14.2.1, insert: "Compensation shall be a Fixed Fee of _____ _____ dollars ($ _____)."

Compensation---Percentage of Construction Cost:

In Subparagraph 14.2.1, insert: "Compensation shall be based on one of the following Percentages of Construction Cost, as defined in Article 3:

For portions of the Project to be awarded under:

A single, stipulated sum construction contract:
_____Percent (_____ %)

Separate, stipulated sum construction contracts:
_____Percent (_____ %)

A single, cost-plus construction contract:
_____Percent (_____ %)

Separate, cost-plus construction contracts:
_____Percent (_____ %)"

Paragraph 14.2, Subparagraph 14.2.2:

If applicable, insert the percentages of compensation payable for each separate Phase of Services. Percentages contained in previous Owner-Architect Agreements would be expressed as follows:

Schematic Design Phase:	fifteen percent (15%)
Design Development Phase:	twenty percent (20%)
Construction Documents Phase:	forty percent (40%)
Bidding or Negotiation Phase:	five percent (5%)
Construction Phase:	twenty percent (20%)
Total:	one hundred percent (100%)

Because phases may overlap in time, these percentages have been expressed separately for each phase, rather than cumulatively. This facilitates billing when services are being provided during more than one phase at a time.

Paragraph 14.3:

Attach AIA Document B352, Duties, Responsibilities, and Limitations of Authority of Project Representative, and the agreed compensation arrangement for such services. If this is to be determined at a later date, such as at the time of commencement of the Construction Phase, so indicate.

Paragraph 14.4, Subparagraph 14.4.1:

If billing rates are used and employees are classified in accordance with the AIA publication, Compensation Guidelines for Architectural and Engineering Services, insert:

"1. Principals' time at the fixed rate of _____ dollars ($) per hour.
     For the purposes of this Agreement, the Principals are:

                                    (list principals)

2. Supervisory time at the fixed rate of _____ dollars ($) per hour. For the purposes of this Agreement, supervisory personnel include:

          (describe supervisory personnel by job title, such as Project Architect)

3. Technical Level I time at the fixed rate of _____ dollars ($) per hour. For purposes of this Agreement, Technical Level I personnel include:

          (describe by job title, such as Senior Designer, Specifier, etc.)

4. Technical Level II time at the fixed rate of _____ dollars ($) per hour. For purposes of this Agreement, Technical Level II personnel include:

          (describe by job title, such as Junior Designer, Senior Draftsman, etc.)

5. Technical Level III and clerical time at the fixed rate of _____ dollars ($) per hour. For purposes of this Agreement, Technical Level III and clerical personnel include:

          (describe by job title, such as Junior Draftsman, Secretaries, Typists, etc.)"

If a Multiple of Direct Personnel Expense is used, insert:

"Principals' and employees' time at a multiple of _____ (   ) times their Direct Personnel Expense as defined in Article 4."

If a multiple of direct salaries is used, the term "direct salaries" should be substituted for Direct Personnel Expense above, and this should be noted in Article 15.

Paragraph 14.4, Subparagraph 14.4.2:

Insert the multiplier applied to consultant billings used to cover the costs of administration, responsibility for consultants' work, coordination and profit.

Paragraph 14.5:

Insert the multiplier, if any, applied to reimbursable expenses used to cover the costs of administration.

Paragraph 14.6:

Insert the percentage rate and basis (monthly, annual) and the time (such as a number of days) after the due date for payment on which interest charges will begin to run. This should be carefully checked against state usury laws which may set a limit on the rate of interest which may legally be charged. In addition, Federal Truth in Lending and similar state and local consumer protection laws may require setting forth the annual percentage rate and other disclosures or waivers for certain types of transactions or with certain types of clients. Advice of legal counsel should be sought on such matters.

Paragraph 14.7:

Insert the amount of time after which the compensation shall be subject to renegotiation or adjustment. If the firm requires periodic adjustments in hourly rates and multiples, this should be stated, along with any limitations on the amount of upward adjustment which may be made.

Article 15 --- OTHER CONDITIONS OR SERVICES

Here insert the following types of provisions:

Additional phases, such as Predesign, Site Analysis or Postconstruction, and the services provided in each
Identification of Additional Services, if any, provided under the Basic Compensation
Other Additional Services, and any special compensation arrangements for them
Description of consultants, if any, provided under the Basic Compensation
Preparation of multiple sets of construction documents for separate contracts
Procedure for award of Construction Contract (i.e., bidding or negotiation)
Construction delivery process (single or separate contracts, stipulated sum or cost-plus contracts)
Fixed Limit of Construction Cost
Fixed Time of Performance
Modifications to any services or conditions
Other additional conditions.

Note that any changes in the duties of the Architect during the Construction Phase must be considered with extreme care and correlated with the terms of the General Conditions of the Contract for Construction.

EXECUTION OF THE AGREEMENT:

In executing the Agreement, the parties should indicate the capacity in which they are acting.

REPRODUCTION:

AIA Document B141 is a copyrighted document, and may not be reproduced or excerpted from in substantial part without the express written permission of AIA. Purchasers of B141 are hereby entitled to reproduce a maximum of ten copies of the completed or executed document for use only in connection with the particular Project. AIA will not permit the reproduction of this document in blank, or the use of substantial portions of, or language from, this Document, except upon written request and after receipt of written permission from AIA.

THE AMERICAN INSTITUTE OF ARCHITECTS

AIA Document B141

Standard Form of Agreement Between Owner and Architect

1977 EDITION

THIS DOCUMENT HAS IMPORTANT LEGAL CONSEQUENCES; CONSULTATION WITH AN ATTORNEY IS ENCOURAGED WITH RESPECT TO ITS COMPLETION OR MODIFICATION

AGREEMENT

made as of the   day of   in the year of Nineteen
Hundred and

BETWEEN the Owner:

and the Architect:

For the following Project:
(Include detailed description of Project location and scope.)

The Owner and the Architect agree as set forth below.

> # TERMS AND CONDITIONS OF AGREEMENT BETWEEN OWNER AND ARCHITECT

ARTICLE 1

ARCHITECT'S SERVICES AND RESPONSIBILITIES

BASIC SERVICES

The Architect's Basic Services consist of the five phases described in Paragraphs 1.1 through 1.5 and include normal structural, mechanical and electrical engineering services and any other services included in Article 15 as part of Basic Services.

1.1 SCHEMATIC DESIGN PHASE

1.1.1 The Architect shall review the program furnished by the Owner to ascertain the requirements of the Project and shall review the understanding of such requirements with the Owner.

1.1.2 The Architect shall provide a preliminary evaluation of the program and the Project budget requirements, each in terms of the other, subject to the limitations set forth in Subparagraph 3.2.1.

1.1.3 The Architect shall review with the Owner alternative approaches to design and construction of the Project.

1.1.4 Based on the mutually agreed upon program and Project budget requirements, the Architect shall prepare, for approval by the Owner, Schematic Design Documents consisting of drawings and other documents illustrating the scale and relationship of Project components.

1.1.5 The Architect shall submit to the Owner a Statement of Probable Construction Cost based on current area, volume or other unit costs.

1.2 DESIGN DEVELOPMENT PHASE

1.2.1 Based on the approved Schematic Design Documents and any adjustments authorized by the Owner in the program or Project budget, the Architect shall prepare, for approval by the Owner, Design Development Documents consisting of drawings and other documents to fix and describe the size and character of the entire Project as to architectural, structural, mechanical and electrical systems, materials and such other elements as may be appropriate.

1.2.2 The Architect shall submit to the Owner a further Statement of Probable Construction Cost.

1.3 CONSTRUCTION DOCUMENTS PHASE

1.3.1 Based on the approved Design Development Documents and any further adjustments in the scope or quality of the Project or in the Project budget authorized by the Owner, the Architect shall prepare, for approval by the Owner, Construction Documents consisting of Drawings and Specifications setting forth in detail the requirements for the construction of the Project.

1.3.2 The Architect shall assist the Owner in the preparation of the necessary bidding information, bidding forms, the Conditions of the Contract, and the form of Agreement between the Owner and the Contractor.

1.3.3 The Architect shall advise the Owner of any adjust-

ments to previous Statements of Probable Construction Cost indicated by changes in requirements or general market conditions.

1.3.4 The Architect shall assist the Owner in connection with the Owner's responsibility for filing documents required for the approval of governmental authorities having jurisdiction over the Project.

1.4 BIDDING OR NEGOTIATION PHASE

1.4.1 The Architect, following the Owner's approval of the Construction Documents and of the latest Statement of Probable Construction Cost, shall assist the Owner in obtaining bids or negotiated proposals, and assist in awarding and preparing contracts for construction.

1.5 CONSTRUCTION PHASE—ADMINISTRATION OF THE CONSTRUCTION CONTRACT

1.5.1 The Construction Phase will commence with the award of the Contract for Construction and, together with the Architect's obligation to provide Basic Services under this Agreement, will terminate when final payment to the Contractor is due, or in the absence of a final Certificate for Payment or of such due date, sixty days after the Date of Substantial Completion of the Work, whichever occurs first.

1.5.2 Unless otherwise provided in this Agreement and incorporated in the Contract Documents, the Architect shall provide administration of the Contract for Construction as set forth below and in the edition of AIA Document A201, General Conditions of the Contract for Construction, current as of the date of this Agreement.

1.5.3 The Architect shall be a representative of the Owner during the Construction Phase, and shall advise and consult with the Owner. Instructions to the Contractor shall be forwarded through the Architect. The Architect shall have authority to act on behalf of the Owner only to the extent provided in the Contract Documents unless otherwise modified by written instrument in accordance with Subparagraph 1.5.16.

1.5.4 The Architect shall visit the site at intervals appropriate to the stage of construction or as otherwise agreed by the Architect in writing to become generally familiar with the progress and quality of the Work and to determine in general if the Work is proceeding in accordance with the Contract Documents. However, the Architect shall not be required to make exhaustive or continuous on-site inspections to check the quality or quantity of the Work. On the basis of such on-site observations as an architect, the Architect shall keep the Owner informed of the progress and quality of the Work, and shall endeavor to guard the Owner against defects and deficiencies in the Work of the Contractor.

1.5.5 The Architect shall not have control or charge of and shall not be responsible for construction means, methods, techniques, sequences or procedures, or for safety precautions and programs in connection with the Work, for the acts or omissions of the Contractor, Subcontractors or any other persons performing any of the Work, or for the failure of any of them to carry out the Work in accordance with the Contract Documents.

1.5.6 The Architect shall at all times have access to the Work wherever it is in preparation or progress.

1.5.7 The Architect shall determine the amounts owing to the Contractor based on observations at the site and on evaluations of the Contractor's Applications for Payment, and shall issue Certificates for Payment in such amounts, as provided in the Contract Documents.

1.5.8 The issuance of a Certificate for Payment shall constitute a representation by the Architect to the Owner, based on the Architect's observations at the site as provided in Subparagraph 1.5.4 and on the data comprising the Contractor's Application for Payment, that the Work has progressed to the point indicated; that, to the best of the Architect's knowledge, information and belief, the quality of the Work is in accordance with the Contract Documents (subject to an evaluation of the Work for conformance with the Contract Documents upon Substantial Completion, to the results of any subsequent tests required by or performed under the Contract Documents, to minor deviations from the Contract Documents correctable prior to completion, and to any specific qualifications stated in the Certificate for Payment) and that the Contractor is entitled to payment in the amount certified. However, the issuance of a Certificate for Payment shall not be a representation that the Architect has made any examination to ascertain how and for what purpose the Contractor has used the moneys paid on account of the Contract Sum.

1.5.9 The Architect shall be the interpreter of the requirements of the Contract Documents and the judge of the performance thereunder by both the Owner and Contractor. The Architect shall render interpretations necessary for the proper execution or progress of the Work with reasonable promptness on written request of either the Owner or the Contractor, and shall render written decisions, within a reasonable time, on all claims, disputes and other matters in question between the Owner and the Contractor relating to the execution or progress of the Work or the interpretation of the Contract Documents.

1.5.10 Interpretations and decisions of the Architect shall be consistent with the intent of and reasonably inferable from the Contract Documents and shall be in written or graphic form. In the capacity of interpreter and judge, the Architect shall endeavor to secure faithful performance by both the Owner and the Contractor, shall not show partiality to either, and shall not be liable for the result of any interpretation or decision rendered in good faith in such capacity.

1.5.11 The Architect's decisions in matters relating to artistic effect shall be final if consistent with the intent of the Contract Documents. The Architect's decisions on any other claims, disputes or other matters, including those in question between the Owner and the Contractor, shall be subject to arbitration as provided in this Agreement and in the Contract Documents.

1.5.12 The Architect shall have authority to reject Work which does not conform to the Contract Documents. Whenever, in the Architect's reasonable opinion, it is necessary or advisable for the implementation of the intent of the Contract Documents, the Architect will have authority to require special inspection or testing of the Work in accordance with the provisions of the Contract Documents, whether or not such Work be then fabricated, installed or completed.

1.5.13 The Architect shall review and approve or take other appropriate action upon the Contractor's submittals such as Shop Drawings, Product Data and Samples, but only for conformance with the design concept of the Work and with the information given in the Contract Documents. Such action shall be taken with reasonable promptness so as to cause no delay. The Architect's approval of a specific item shall not indicate approval of an assembly of which the item is a component.

1.5.14 The Architect shall prepare Change Orders for the Owner's approval and execution in accordance with the Contract Documents, and shall have authority to order minor changes in the Work not involving an adjustment in the Contract Sum or an extension of the Contract Time which are not inconsistent with the intent of the Contract Documents.

1.5.15 The Architect shall conduct inspections to determine the Dates of Substantial Completion and final completion, shall receive and forward to the Owner for the Owner's review written warranties and related documents required by the Contract Documents and assembled by the Contractor, and shall issue a final Certificate for Payment.

1.5.16 The extent of the duties, responsibilities and limitations of authority of the Architect as the Owner's representative during construction shall not be modified or extended without written consent of the Owner, the Contractor and the Architect.

1.6 PROJECT REPRESENTATION BEYOND BASIC SERVICES

1.6.1 If the Owner and Architect agree that more extensive representation at the site than is described in Paragraph 1.5 shall be provided, the Architect shall provide one or more Project Representatives to assist the Architect in carrying out such responsibilities at the site.

1.6.2 Such Project Representatives shall be selected, employed and directed by the Architect, and the Architect shall be compensated therefor as mutually agreed between the Owner and the Architect as set forth in an exhibit appended to this Agreement, which shall describe the duties, responsibilities and limitations of authority of such Project Representatives.

1.6.3 Through the observations by such Project Representatives, the Architect shall endeavor to provide further protection for the Owner against defects and deficiencies in the Work, but the furnishing of such project representation shall not modify the rights, responsibilities or obligations of the Architect as described in Paragraph 1.5.

1.7 ADDITIONAL SERVICES

The following Services are not included in Basic Services unless so identified in Article 15. They shall be provided if authorized or confirmed in writing by the Owner, and they shall be paid for by the Owner as provided in this Agreement, in addition to the compensation for Basic Services.

1.7.1 Providing analyses of the Owner's needs, and programming the requirements of the Project.

1.7.2 Providing financial feasibility or other special studies.

1.7.3 Providing planning surveys, site evaluations, environmental studies or comparative studies of prospective sites, and preparing special surveys, studies and submissions required for approvals of governmental authorities or others having jurisdiction over the Project.

1.7.4 Providing services relative to future facilities, systems and equipment which are not intended to be constructed during the Construction Phase.

1.7.5 Providing services to investigate existing conditions or facilities or to make measured drawings thereof, or to verify the accuracy of drawings or other information furnished by the Owner.

1.7.6 Preparing documents of alternate, separate or sequential bids or providing extra services in connection with bidding, negotiation or construction prior to the completion of the Construction Documents Phase, when requested by the Owner.

1.7.7 Providing coordination of Work performed by separate contractors or by the Owner's own forces.

1.7.8 Providing services in connection with the work of a construction manager or separate consultants retained by the Owner.

1.7.9 Providing Detailed Estimates of Construction Cost, analyses of owning and operating costs, or detailed quantity surveys or inventories of material, equipment and labor.

1.7.10 Providing interior design and other similar services required for or in connection with the selection, procurement or installation of furniture, furnishings and related equipment.

1.7.11 Providing services for planning tenant or rental spaces.

1.7.12 Making revisions in Drawings, Specifications or other documents when such revisions are inconsistent with written approvals or instructions previously given, are required by the enactment or revision of codes, laws or regulations subsequent to the preparation of such documents or are due to other causes not solely within the control of the Architect.

1.7.13 Preparing Drawings, Specifications and supporting data and providing other services in connection with Change Orders to the extent that the adjustment in the Basic Compensation resulting from the adjusted Construction Cost is not commensurate with the services required of the Architect, provided such Change Orders are required by causes not solely within the control of the Architect.

1.7.14 Making investigations, surveys, valuations, inventories or detailed appraisals of existing facilities, and services required in connection with construction performed by the Owner.

1.7.15 Providing consultation concerning replacement of any Work damaged by fire or other cause during construction, and furnishing services as may be required in connection with the replacement of such Work.

1.7.16 Providing services made necessary by the default of the Contractor, or by major defects or deficiencies in the Work of the Contractor, or by failure of performance of either the Owner or Contractor under the Contract for Construction.

1.7.17 Preparing a set of reproducible record drawings showing significant changes in the Work made during construction based on marked-up prints, drawings and other data furnished by the Contractor to the Architect.

1.7.18 Providing extensive assistance in the utilization of any equipment or system such as initial start-up or testing, adjusting and balancing, preparation of operation and maintenance manuals, training personnel for operation and maintenance, and consultation during operation.

1.7.19 Providing services after issuance to the Owner of the final Certificate for Payment, or in the absence of a final Certificate for Payment, more than sixty days after the Date of Substantial Completion of the Work.

1.7.20 Preparing to serve or serving as an expert witness in connection with any public hearing, arbitration proceeding or legal proceeding.

1.7.21 Providing services of consultants for other than the normal architectural, structural, mechanical and electrical engineering services for the Project.

1.7.22 Providing any other services not otherwise included in this Agreement or not customarily furnished in accordance with generally accepted architectural practice.

1.8 TIME

1.8.1 The Architect shall perform Basic and Additional Services as expeditiously as is consistent with professional skill and care and the orderly progress of the Work. Upon request of the Owner, the Architect shall submit for the Owner's approval a schedule for the performance of the Architect's services which shall be adjusted as required as the Project proceeds, and shall include allowances for periods of time required for the Owner's review and approval of submissions and for approvals of authorities having jurisdiction over the Project. This schedule, when approved by the Owner, shall not, except for reasonable cause, be exceeded by the Architect.

ARTICLE 2

THE OWNER'S RESPONSIBILITIES

2.1 The Owner shall provide full information regarding requirements for the Project including a program, which shall set forth the Owner's design objectives, constraints and criteria, including space requirements and relationships, flexibility and expandability, special equipment and systems and site requirements.

2.2 If the Owner provides a budget for the Project it shall include contingencies for bidding, changes in the Work during construction, and other costs which are the responsibility of the Owner, including those described in this Article 2 and in Subparagraph 3.1.2. The Owner shall, at the request of the Architect, provide a statement of funds available for the Project, and their source.

2.3 The Owner shall designate, when necessary, a representative authorized to act in the Owner's behalf with respect to the Project. The Owner or such authorized representative shall examine the documents submitted by the Architect and shall render decisions pertaining thereto promptly, to avoid unreasonable delay in the progress of the Architect's services.

2.4 The Owner shall furnish a legal description and a certified land survey of the site, giving, as applicable, grades and lines of streets, alleys, pavements and adjoining property; rights-of-way, restrictions, easements, encroachments, zoning, deed restrictions, boundaries and contours of the site; locations, dimensions and complete data pertaining to existing buildings, other improvements and trees; and full information concerning available service and utility lines both public and private, above and below grade, including inverts and depths.

2.5 The Owner shall furnish the services of soil engineers or other consultants when such services are deemed necessary by the Architect. Such services shall include test borings, test pits, soil bearing values, percolation tests, air and water pollution tests, ground corrosion and resistivity tests, including necessary operations for determining sub-soil, air and water conditions, with reports and appropriate professional recommendations.

2.6 The Owner shall furnish structural, mechanical, chemical and other laboratory tests, inspections and reports as required by law or the Contract Documents.

2.7 The Owner shall furnish all legal, accounting and insurance counseling services as may be necessary at any time for the Project, including such auditing services as the Owner may require to verify the Contractor's Applications for Payment or to ascertain how or for what purposes the Contractor uses the moneys paid by or on behalf of the Owner.

2.8 The services, information, surveys and reports required by Paragraphs 2.4 through 2.7 inclusive shall be furnished at the Owner's expense, and the Architect shall be entitled to rely upon the accuracy and completeness thereof.

2.9 If the Owner observes or otherwise becomes aware of any fault or defect in the Project or nonconformance with the Contract Documents, prompt written notice thereof shall be given by the Owner to the Architect.

2.10 The Owner shall furnish required information and services and shall render approvals and decisions as expeditiously as necessary for the orderly progress of the Architect's services and of the Work.

ARTICLE 3

CONSTRUCTION COST

3.1 DEFINITION

3.1.1 The Construction Cost shall be the total cost or estimated cost to the Owner of all elements of the Project designed or specified by the Architect.

3.1.2 The Construction Cost shall include at current market rates, including a reasonable allowance for overhead and profit, the cost of labor and materials furnished by the Owner and any equipment which has been de-signed, specified, selected or specially provided for by the Architect.

3.1.3 Construction Cost does not include the compensation of the Architect and the Architect's consultants, the cost of the land, rights-of-way, or other costs which are the responsibility of the Owner as provided in Article 2.

3.2 RESPONSIBILITY FOR CONSTRUCTION COST

3.2.1 Evaluations of the Owner's Project budget, Statements of Probable Construction Cost and Detailed Estimates of Construction Cost, if any, prepared by the Architect, represent the Architect's best judgment as a design professional familiar with the construction industry. It is recognized, however, that neither the Architect nor the Owner has control over the cost of labor, materials or equipment, over the Contractor's methods of determining bid prices, or over competitive bidding, market or negotiating conditions. Accordingly, the Architect cannot and does not warrant or represent that bids or negotiated prices will not vary from the Project budget proposed, established or approved by the Owner, if any, or from any Statement of Probable Construction Cost or other cost estimate or evaluation prepared by the Architect.

3.2.2 No fixed limit of Construction Cost shall be established as a condition of this Agreement by the furnishing, proposal or establishment of a Project budget under Subparagraph 1.1.2 or Paragraph 2.2 or otherwise, unless such fixed limit has been agreed upon in writing and signed by the parties hereto. If such a fixed limit has been established, the Architect shall be permitted to include contingencies for design, bidding and price escalation, to determine what materials, equipment, component systems and types of construction are to be included in the Contract Documents, to make reasonable adjustments in the scope of the Project and to include in the Contract Documents alternate bids to adjust the Construction Cost to the fixed limit. Any such fixed limit shall be increased in the amount of any increase in the Contract Sum occurring after execution of the Contract for Construction.

3.2.3 If the Bidding or Negotiation Phase has not commenced within three months after the Architect submits the Construction Documents to the Owner, any Project budget or fixed limit of Construction Cost shall be adjusted to reflect any change in the general level of prices in the construction industry between the date of submission of the Construction Documents to the Owner and the date on which proposals are sought.

3.2.4 If a Project budget or fixed limit of Construction Cost (adjusted as provided in Subparagraph 3.2.3) is exceeded by the lowest bona fide bid or negotiated proposal, the Owner shall (1) give written approval of an increase in such fixed limit, (2) authorize rebidding or renegotiating of the Project within a reasonable time, (3) if the Project is abandoned, terminate in accordance with Paragraph 10.2, or (4) cooperate in revising the Project scope and quality as required to reduce the Construction Cost. In the case of (4), provided a fixed limit of Construction Cost has been established as a condition of this Agreement, the Architect, without additional charge, shall modify the Drawings and Specifications as necessary to comply with the fixed limit. The providing of such service shall be the limit of the Architect's responsibility arising from the establishment of such fixed limit, and having done so, the Architect shall be entitled to compensation for all services performed, in accordance with this Agreement, whether or not the Construction Phase is commenced.

ARTICLE 4

DIRECT PERSONNEL EXPENSE

4.1   Direct Personnel Expense is defined as the direct salaries of all the Architect's personnel engaged on the Project, and the portion of the cost of their mandatory and customary contributions and benefits related thereto, such as employment taxes and other statutory employee benefits, insurance, sick leave, holidays, vacations, pensions and similar contributions and benefits.

ARTICLE 5

REIMBURSABLE EXPENSES

5.1   Reimbursable Expenses are in addition to the Compensation for Basic and Additional Services and include actual expenditures made by the Architect and the Architect's employees and consultants in the interest of the Project for the expenses listed in the following Subparagraphs:

5.1.1   Expense of transportation in connection with the Project; living expenses in connection with out-of-town travel; long distance communications; and fees paid for securing approval of authorities having jurisdiction over the Project.

5.1.2   Expense of reproductions, postage and handling of Drawings, Specifications and other documents, excluding reproductions for the office use of the Architect and the Architect's consultants.

5.1.3   Expense of data processing and photographic production techniques when used in connection with Additional Services.

5.1.4   If authorized in advance by the Owner, expense of overtime work requiring higher than regular rates.

5.1.5   Expense of renderings, models and mock-ups requested by the Owner.

5.1.6   Expense of any additional insurance coverage or limits, including professional liability insurance, requested by the Owner in excess of that normally carried by the Architect and the Architect's consultants.

ARTICLE 6

PAYMENTS TO THE ARCHITECT

6.1   PAYMENTS ON ACCOUNT OF BASIC SERVICES

6.1.1   An initial payment as set forth in Paragraph 14.1 is the minimum payment under this Agreement.

6.1.2   Subsequent payments for Basic Services shall be made monthly and shall be in proportion to services performed within each Phase of services, on the basis set forth in Article 14.

6.1.3   If and to the extent that the Contract Time initially established in the Contract for Construction is exceeded or extended through no fault of the Architect, compensation for any Basic Services required for such extended period of Administration of the Construction Contract shall be computed as set forth in Paragraph 14.4 for Additional Services.

6.1.4   When compensation is based on a percentage of Construction Cost, and any portions of the Project are deleted or otherwise not constructed, compensation for such portions of the Project shall be payable to the extent services are performed on such portions, in accordance with the schedule set forth in Subparagraph 14.2.2, based on (1) the lowest bona fide bid or negotiated proposal or, (2) if no such bid or proposal is received, the most recent Statement of Probable Construction Cost or Detailed Estimate of Construction Cost for such portions of the Project.

6.2   PAYMENTS ON ACCOUNT OF ADDITIONAL SERVICES

6.2.1   Payments on account of the Architect's Additional Services as defined in Paragraph 1.7 and for Reimbursable Expenses as defined in Article 5 shall be made monthly upon presentation of the Architect's statement of services rendered or expenses incurred.

6.3   PAYMENTS WITHHELD

6.3.1   No deductions shall be made from the Architect's compensation on account of penalty, liquidated damages or other sums withheld from payments to contractors, or on account of the cost of changes in the Work other than those for which the Architect is held legally liable.

6.4   PROJECT SUSPENSION OR TERMINATION

6.4.1   If the Project is suspended or abandoned in whole or in part for more than three months, the Architect shall be compensated for all services performed prior to receipt of written notice from the Owner of such suspension or abandonment, together with Reimbursable Expenses then due and all Termination Expenses as defined in Paragraph 10.4. If the Project is resumed after being suspended for more than three months, the Architect's compensation shall be equitably adjusted.

ARTICLE 7

ARCHITECT'S ACCOUNTING RECORDS

7.1   Records of Reimbursable Expenses and expenses pertaining to Additional Services and services performed on the basis of a Multiple of Direct Personnel Expense shall be kept on the basis of generally accepted accounting principles and shall be available to the Owner or the Owner's authorized representative at mutually convenient times.

ARTICLE 8

OWNERSHIP AND USE OF DOCUMENTS

8.1   Drawings and Specifications as instruments of service are and shall remain the property of the Architect whether the Project for which they are made is executed or not. The Owner shall be permitted to retain copies, including reproducible copies, of Drawings and Specifications for information and reference in connection with the Owner's use and occupancy of the Project. The Drawings and Specifications shall not be used by the Owner on other projects, for additions to this Project, or for completion of this Project by others provided the Architect is not in default under this Agreement, except by agreement in writing and with appropriate compensation to the Architect.

8.2 Submission or distribution to meet official regulatory requirements or for other purposes in connection with the Project is not to be construed as publication in derogation of the Architect's rights.

ARTICLE 9

ARBITRATION

9.1 All claims, disputes and other matters in question between the parties to this Agreement, arising out of or relating to this Agreement or the breach thereof, shall be decided by arbitration in accordance with the Construction Industry Arbitration Rules of the American Arbitration Association then obtaining unless the parties mutually agree otherwise. No arbitration, arising out of or relating to this Agreement, shall include, by consolidation, joinder or in any other manner, any additional person not a party to this Agreement except by written consent containing a specific reference to this Agreement and signed by the Architect, the Owner, and any other person sought to be joined. Any consent to arbitration involving an additional person or persons shall not constitute consent to arbitration of any dispute not described therein or with any person not named or described therein. This Agreement to arbitrate and any agreement to arbitrate with an additional person or persons duly consented to by the parties to this Agreement shall be specifically enforceable under the prevailing arbitration law.

9.2 Notice of the demand for arbitration shall be filed in writing with the other party to this Agreement and with the American Arbitration Association. The demand shall be made within a reasonable time after the claim, dispute or other matter in question has arisen. In no event shall the demand for arbitration be made after the date when institution of legal or equitable proceedings based on such claim, dispute or other matter in question would be barred by the applicable statute of limitations.

9.3 The award rendered by the arbitrators shall be final, and judgment may be entered upon it in accordance with applicable law in any court having jurisdiction thereof.

ARTICLE 10

TERMINATION OF AGREEMENT

10.1 This Agreement may be terminated by either party upon seven days' written notice should the other party fail substantially to perform in accordance with its terms through no fault of the party initiating the termination.

10.2 This Agreement may be terminated by the Owner upon at least seven days' written notice to the Architect in the event that the Project is permanently abandoned.

10.3 In the event of termination not the fault of the Architect, the Architect shall be compensated for all services performed to termination date, together with Reimbursable Expenses then due and all Termination Expenses as defined in Paragraph 10.4.

10.4 Termination Expenses include expenses directly attributable to termination for which the Architect is not otherwise compensated, plus an amount computed as a percentage of the total Basic and Additional Compensation earned to the time of termination, as follows:

  .1 20 percent if termination occurs during the Schematic Design Phase; or

  .2 10 percent if termination occurs during the Design Development Phase; or

  .3 5 percent if termination occurs during any subsequent phase.

ARTICLE 11

MISCELLANEOUS PROVISIONS

11.1 Unless otherwise specified, this Agreement shall be governed by the law of the principal place of business of the Architect.

11.2 Terms in this Agreement shall have the same meaning as those in AIA Document A201, General Conditions of the Contract for Construction, current as of the date of this Agreement.

11.3 As between the parties to this Agreement: as to all acts or failures to act by either party to this Agreement, any applicable statute of limitations shall commence to run and any alleged cause of action shall be deemed to have accrued in any and all events not later than the relevant Date of Substantial Completion of the Work, and as to any acts or failures to act occurring after the relevant Date of Substantial Completion, not later than the date of issuance of the final Certificate for Payment.

11.4 The Owner and the Architect waive all rights against each other and against the contractors, consultants, agents and employees of the other for damages covered by any property insurance during construction as set forth in the edition of AIA Document A201, General Conditions, current as of the date of this Agreement. The Owner and the Architect each shall require appropriate similar waivers from their contractors, consultants and agents.

ARTICLE 12

SUCCESSORS AND ASSIGNS

12.1 The Owner and the Architect, respectively, bind themselves, their partners, successors, assigns and legal representatives to the other party to this Agreement and to the partners, successors, assigns and legal representatives of such other party with respect to all covenants of this Agreement. Neither the Owner nor the Architect shall assign, sublet or transfer any interest in this Agreement without the written consent of the other.

ARTICLE 13

EXTENT OF AGREEMENT

13.1 This Agreement represents the entire and integrated agreement between the Owner and the Architect and supersedes all prior negotiations, representations or agreements, either written or oral. This Agreement may be amended only by written instrument signed by both Owner and Architect.

AIA DOCUMENT B141 • OWNER-ARCHITECT AGREEMENT • THIRTEENTH EDITION • JULY 1977 • AIA® • © 1977
THE AMERICAN INSTITUTE OF ARCHITECTS, 1735 NEW YORK AVENUE, N.W., WASHINGTON, D.C. 20006

ARTICLE 14

BASIS OF COMPENSATION

The Owner shall compensate the Architect for the Scope of Services provided, in accordance with Article 6, Payments to the Architect, and the other Terms and Conditions of this Agreement, as follows:

14.1   AN INITIAL PAYMENT of

dollars ($

shall be made upon execution of this Agreement and credited to the Owner's account as follows:

14.2   BASIC COMPENSATION

14.2.1   FOR BASIC SERVICES, as described in Paragraphs 1.1 through 1.5, and any other services included in Article 15 as part of Basic Services, Basic Compensation shall be computed as follows:

(Here insert basis of compensation, including fixed amounts, multiples or percentages, and identify Phases to which particular methods of compensation apply, if necessary.)

14.2.2   Where compensation is based on a Stipulated Sum or Percentage of Construction Cost, payments for Basic Services shall be made as provided in Subparagraph 6.1.2, so that Basic Compensation for each Phase shall equal the following percentages of the total Basic Compensation payable:

(Include any additional Phases as appropriate.)

Schematic Design Phase:	percent (	%)
Design Development Phase:	percent (	%)
Construction Documents Phase:	percent (	%)
Bidding or Negotiation Phase:	percent (	%)
Construction Phase:	percent (	%)

14.3   FOR PROJECT REPRESENTATION BEYOND BASIC SERVICES, as described in Paragraph 1.6, Compensation shall be computed separately in accordance with Subparagraph 1.6.2.

14.4   COMPENSATION FOR ADDITIONAL SERVICES

14.4.1   FOR ADDITIONAL SERVICES OF THE ARCHITECT, as described in Paragraph 1.7, and any other services included in Article 15 as part of Additional Services, but excluding Additional Services of consultants, Compensation shall be computed as follows:

*(Here insert basis of compensation, including rates and/or multiples of Direct Personnel Expense for Principals and employees, and identify Principals
and classify employees, if required. Identify specific services to which particular methods of compensation apply, if necessary.)*

14.4.2   FOR ADDITIONAL SERVICES OF CONSULTANTS, including additional structural, mechanical and electrical
engineering services and those provided under Subparagraph 1.7.21 or identified in Article 15 as part of Additional Services, a multiple of (   ) times the amounts billed
to the Architect for such services.

(Identify specific types of consultants in Article 15, if required.)

14.5   FOR REIMBURSABLE EXPENSES, as described in Article 5, and any other items included in Article 15 as Reimbursable Expenses, a multiple of (   ) times the amounts ex-
pended by the Architect, the Architect's employees and consultants in the interest of the Project.

14.6   Payments due the Architect and unpaid under this Agreement shall bear interest from the date payment is
due at the rate entered below, or in the absence thereof, at the legal rate prevailing at the principal place of
business of the Architect.

(Here insert any rate of interest agreed upon.)

*(Usury laws and requirements under the Federal Truth in Lending Act, similar state and local consumer credit laws and other regulations at the
Owner's and Architect's principal places of business, the location of the Project and elsewhere may affect the validity of this provision. Specific legal
advice should be obtained with respect to deletion, modification, or other requirements such as written disclosures or waivers.)*

14.7   The Owner and the Architect agree in accordance with the Terms and Conditions of this Agreement that:

14.7.1   IF THE SCOPE of the Project or of the Architect's Services is changed materially, the amounts of compensation
shall be equitably adjusted.

14.7.2   IF THE SERVICES covered by this Agreement have not been completed within

(   ) months of the date hereof, through no fault of the Architect, the amounts of compensation, rates and
multiples set forth herein shall be equitably adjusted.

ARTICLE 15

OTHER CONDITIONS OR SERVICES

SAMPLE

This Agreement entered into as of the day and year first written above.

OWNER   ARCHITECT

_____   _____

_____   _____

_____   _____

BY_____   BY_____

AIA DOCUMENT B141 • OWNER-ARCHITECT AGREEMENT • THIRTEENTH EDITION • JULY 1977 • AIA® • © 1977
THE AMERICAN INSTITUTE OF ARCHITECTS, 1735 NEW YORK AVENUE, N.W., WASHINGTON, D.C. 20006

Appendix B

Standard Form of Agreement Between Owner and Contractor

INSTRUCTION SHEET *AIA DOCUMENT A101a*

FOR AIA DOCUMENT A101, STANDARD FORM OF AGREEMENT BETWEEN OWNER AND CONTRACTOR — JUNE 1977 EDITION

AIA Document A101, Standard Form of Agreement Between Owner and Contractor, is for use where the basis of payment is a stipulated sum (fixed price). The 1977 Edition has been prepared for use with the 1976 Edition of AIA Document A201, General Conditions of the Contract for Construction. It is suitable for any arrangement between the Owner and the Contractor where the cost has been set in advance either by bidding or by negotiation. Although the Owner has the advantage of advance knowledge of the cost of the Work, increased efforts to assure Contract compliance may be required, in view of the fact that the price is fixed and the Contractor has a financial interest in minimizing the cost of carrying out the Work. A more complete explanation of A101 is provided in Architect's Handbook of Professional Practice, Chapter 17: Owner-Contractor and Contractor-Subcontractor Agreements.

Below is a listing of pertinent provisions revised or added to the 1977 Edition of the Stipulated Sum Owner-Contractor Agreement Form:

Article 3 — Modified to read, "Time of Commencement and *Substantial* Completion." The General Conditions, AIA Document A201, 1976 Edition, make it clear that the Contract Time runs until the Date of Substantial Completion; the Owner should be aware that an additional period of time will be required to reach final completion.

Article 4 — Revised to include reference to the Contract Documents for determination of amounts of Change Orders. Parenthetical instruction describing basis of payment now includes *base bid* and *accepted alternates*.

Article 5 — A sentence has been added at the end of the first paragraph to stipulate a specific day of the month as the end of the period for which progress payments will be made. The Agreement requires that the Owner make progress payments not later than *an agreed-upon number of days* following the end of that period covered by the Application for Payment. (Note that the General Conditions, AIA Document A201, 1976 Edition, require in Subparagraph 9.3.1 that the Contractor apply for payment at least 10 days in advance of the date payment is due.)

The provision for interest on payments due and unpaid has been revised to provide for the entry of a specific rate of interest in accordance with the changes in the interest provision of A201, Paragraph 7.8. A parenthetical statement has been added drawing attention to Truth-in-Lending and other laws which may govern the use and form of an interest provision under certain circumstances.

Article 6 — Modified to provide that final payment is due when the Work has been completed (the reference to an agreed-upon number of days after Substantial Completion of the Work has been deleted). The Certificate of Substantial Completion will provide the time period within which the Contractor will bring the Work to final completion.

Completing the form:

(NOTE: Prospective bidders should be aware of any additional provisions which may be included in A101, such as liquidated damages, retainage, or payment for stored materials, by an appropriate notice in the Bidding Documents.)

Cover Page — The names of the Owner and the Architect should be shown in the same form as in the other Project documents; include the full legal or corporate names under which the Owner and Contractor are entering the Agreement.

Article 1 — The Contract Documents

The Contract Documents must be enumerated in detail under Article 7. If unit prices are incorporated in the Contractor's bid, the bid itself may be incorporated into the Contract; similarly, other bidding documents, bonds, etc. may be incorporated, particularly in public work.

Article 2 — The Work

The general scope of the Work should be carefully defined here, since changes by Change Order, under Paragraph 12.1 of A201, must be within the general scope of the Work contemplated by the Contract. This Article should be used to describe the portions of the Project for which the Contractor is responsible, if separate contracts are used.

Article 3 — Time of Commencement and Substantial Completion

The following items should be included as appropriate:

• Date of commencement of the Work
• Provision for notice to proceed, if any
• Date of Substantial Completion of the Work
• Provision, if any, for liquidated damages if not included in the Supplementary Conditions
  (see AIA Document A511)

Date of commencement of the Work should not be earlier than the date of execution of the Contract. When time of performance is to be strictly enforced, the statement of starting time should be carefully considered.

A sample provision where a notice to proceed will be used is as follows:

The Work shall commence on the date stipulated in the notice to proceed and shall be substantially completed on _____

The Date of Substantial Completion of the Work may be expressed as a number of days (preferably calendar days) or as a specific date. The time requirements will ordinarily have been fulfilled when the Work is Substantially Complete, as defined in A201, Subparagraph 8.1.3, even if a few minor items may remain to be completed or corrected.

If liquidated damages are to be assessed because delayed construction will result in the Owner actually suffering loss, the amount per day should be entered in the Supplementary Conditions or the Agreement. Factors such as confidentiality will help determine the choice of location. Liquidated Damages are not a penalty to be inflicted on the Contractor, but must bear an actual and reasonably estimated relationship to the loss to the Owner if the building is not completed on time; for example, the cost per day of renting space to house students if a dormitory cannot be occupied when needed, additional financing costs, loss of profits, etc. This provision, which should be carefully reviewed, if not drafted, by the Owner's attorney, may be as follows:

> The Owner will suffer financial damage if the Project is not Substantially Completed on the date set forth in the Contract Documents. The Contractor (and his Surety) shall pay to the Owner the sums hereinafter stipulated as fixed, agreed and liquidated damages for each calendar day of delay until the Work is Substantially Completed: _____ dollars ($).

A provision for penalty and *bonus*, where such is appropriate, is suggested as follows:

> The Contractor agrees to pay to the Owner a sum of _____ dollars ($) for each calendar day beyond the established completion date that the Work remains uncompleted, in consideration of which the Owner agrees to pay the Contractor a sum of _____ dollars ($) for each calendar day ahead of the established completion date that the Work is determined to be Substantially Completed.

Note that a liquidated damages provision may be placed in the Supplementary Conditions in order to put Subcontractors on notice of this condition.

Article 4 — **Contract Sum**

The following items should be included as appropriate:

- The Contract Sum
- Unit prices, cash allowances, or cash contingency allowances, if any

If not covered elsewhere in the Contract Documents in more detail, the following provision for unit prices is suggested:

> The unit prices listed below shall determine the value of extra Work or changes, as applicable. They shall be considered complete including all material and equipment, labor, installation costs, overhead and profit, and shall be used uniformly for either additions or deductions.

Specific allowances for overhead and for profit on Change Orders may also be included here.

Article 5 — **Progress Payments**

The following items should be included as appropriate:

- Due dates for payments
- Retained percentage
- Payment for materials stored off the site

The due date for payment is often arbitrarily set. It should be a date mutually acceptable to both the Owner and the Contractor in consideration of the time required for the Contractor to prepare an Application for Payment, for the Architect to check and certify payment, and for the Owner to make payment, within the time limits set in Subparagraph 9.4.1, of A201, and in this Article of A101.

The last date upon which Work may be included in an Application should be normally not less than fourteen days prior to the payment date to allow seven days for the Architect to evaluate the Application and issue a Certificate for Payment and seven days for the Owner to make payment as provided in Article 9 of AIA Document A201. The Contractor may prefer an additional few days to allow time for preparation of his Application.

Retained percentage: It is a frequent practice to pay the Contractor 90 percent of the earned sum when payments fall due, retaining 10 percent to assure faithful performance of the Contract. These percentages may vary with circumstances and localities. AIA endorses the concept of reducing retainage as rapidly as possible consistent with the continued protection of all affected interests. See AIA Document A511, Guide for Supplementary Conditions, for a complete discussion.

A provision for reducing retainage should provide that the reduction will be made only if, in the judgment of the Architect, satisfactory progress is being made and maintained in the Work. If the Contractor has furnished a bond, he should be required to provide a Consent of Surety to Reduction In or Partial Release of Retainage (AIA Document G707A), before the retainage is reduced.

Payment for materials stored *off* the site should be provided for in a specific agreement and included in Article 7. Provisions regarding transportation to the site and insurance to protect the Owner's interests should be included.

Article 6 — **Final Payment**

At the time final payment is requested, the Architect should be particularly meticulous in ascertaining that all claims have been settled, in defining any claims that may still be unsettled, in obtaining from the Contractor the certification required in Article 9 of AIA Document A201 that no indebtedness against the Project remains, and in being assured that to the best of his knowledge and belief, based on the final inspection, the Contract requirements have been fulfilled.

Article 7 — **Miscellaneous Provisions**

An accurate, detailed enumeration of all Documents included in the Contract must be made in this Article.

Signatures — Subparagraph 1.2.1 of AIA Document A201, states that the Contract Documents shall be executed in not less than triplicate by the Owner and the Contractor. The Agreement should be executed by the parties in their capacities as individuals, partners, officers, etc., as appropriate.

THE AMERICAN INSTITUTE OF ARCHITECTS

AIA Document A101

Standard Form of Agreement Between Owner and Contractor

where the basis of payment is a

STIPULATED SUM

1977 EDITION

*THIS DOCUMENT HAS IMPORTANT LEGAL CONSEQUENCES; CONSULTATION WITH
AN ATTORNEY IS ENCOURAGED WITH RESPECT TO ITS COMPLETION OR MODIFICATION*

Use only with the 1976 Edition of AIA Document A201, General Conditions of the Contract for Construction.

This document has been approved and endorsed by The Associated General Contractors of America.

AGREEMENT

made as of the   day of   in the year of Nineteen
Hundred and

BETWEEN the Owner:

and the Contractor:

The Project:

The Architect:

The Owner and the Contractor agree as set forth below.

ARTICLE 1

THE CONTRACT DOCUMENTS

The Contract Documents consist of this Agreement, the Conditions of the Contract (General, Supplementary and other Conditions), the Drawings, the Specifications, all Addenda issued prior to and all Modifications issued after execution of this Agreement. These form the Contract, and all are as fully a part of the Contract as if attached to this Agreement or repeated herein. An enumeration of the Contract Documents appears in Article 7.

ARTICLE 2

THE WORK

The Contractor shall perform all the Work required by the Contract Documents for
(Here insert the caption descriptive of the Work as used on other Contract Documents.)

ARTICLE 3

TIME OF COMMENCEMENT AND SUBSTANTIAL COMPLETION

The Work to be performed under this Contract shall be commenced

and, subject to authorized adjustments, Substantial Completion shall be achieved not later than

(Here insert any special provisions for liquidated damages relating to failure to complete on time.)

ARTICLE 4

CONTRACT SUM

The Owner shall pay the Contractor in current funds for the performance of the Work, subject to additions and deductions by Change Order as provided in the Contract Documents, the Contract Sum of

The Contract Sum is determined as follows:
(State here the base bid or other lump sum amount, accepted alternates, and unit prices, as applicable.)

ARTICLE 5

PROGRESS PAYMENTS

Based upon Applications for Payment submitted to the Architect by the Contractor and Certificates for Payment issued by the Architect, the Owner shall make progress payments on account of the Contract Sum to the Contractor as provided in the Contract Documents for the period ending the   day of the month as follows:

Not later than   days following the end of the period covered by the Application for Payment percent (%) of the portion of the Contract Sum properly allocable to labor, materials and equipment incorporated in the Work and   percent (%) of the portion of the Contract Sum properly allocable to materials and equipment suitably stored at the site or at some other location agreed upon in writing, for the period covered by the Application for Payment, less the aggregate of previous payments made by the Owner; and upon Substantial Completion of the entire Work, a sum sufficient to increase the total payments to   percent (%) of the Contract Sum, less such amounts as the Architect shall determine for all incomplete Work and unsettled claims as provided in the Contract Documents.

(If not covered elsewhere in the Contract Documents, here insert any provision for limiting or reducing the amount retained after the Work reaches a certain stage of completion.)

Payments due and unpaid under the Contract Documents shall bear interest from the date payment is due at the rate entered below, or in the absence thereof, at the legal rate prevailing at the place of the Project.
(Here insert any rate of interest agreed upon.)

(Usury laws and requirements under the Federal Truth in Lending Act, similar state and local consumer credit laws and other regulations at the Owner's and Contractor's principal places of business, the location of the Project and elsewhere may affect the validity of this provision. Specific legal advice should be obtained with respect to deletion, modification, or other requirements such as written disclosures or waivers.)

ARTICLE 6

FINAL PAYMENT

Final payment, constituting the entire unpaid balance of the Contract Sum, shall be paid by the Owner to the Contractor when the Work has been completed, the Contract fully performed, and a final Certificate for Payment has been issued by the Architect.

ARTICLE 7

MISCELLANEOUS PROVISIONS

7.1 Terms used in this Agreement which are defined in the Conditions of the Contract shall have the meanings designated in those Conditions.

7.2 The Contract Documents, which constitute the entire agreement between the Owner and the Contractor, are listed in Article 1 and, except for Modifications issued after execution of this Agreement, are enumerated as follows:

(List below the Agreement, the Conditions of the Contract (General, Supplementary, and other Conditions), the Drawings, the Specifications, and any Addenda and accepted alternates, showing page or sheet numbers in all cases and dates where applicable.)

This Agreement entered into as of the day and year first written above.

OWNER   CONTRACTOR

_____   _____

_____   _____

_____   _____

Appendix C

General Conditions of the Contract for Construction

THE AMERICAN INSTITUTE OF ARCHITECTS

AIA Document A201

General Conditions of the Contract for Construction

THIS DOCUMENT HAS IMPORTANT LEGAL CONSEQUENCES; CONSULTATION WITH AN ATTORNEY IS ENCOURAGED WITH RESPECT TO ITS MODIFICATION

1976 EDITION
TABLE OF ARTICLES

This document has been approved and endorsed by The Associated General Contractors of America.

INDEX

AIA DOCUMENT A201 • GENERAL CONDITIONS OF THE CONTRACT FOR CONSTRUCTION • THIRTEENTH EDITION • AUGUST 1976
AIA® • © 1976 • THE AMERICAN INSTITUTE OF ARCHITECTS, 1735 NEW YORK AVENUE, N.W., WASHINGTON, D.C. 20006

GENERAL CONDITIONS OF THE CONTRACT FOR CONSTRUCTION

ARTICLE 1

CONTRACT DOCUMENTS

1.1 DEFINITIONS

1.1.1 THE CONTRACT DOCUMENTS

The Contract Documents consist of the Owner-Contractor Agreement, the Conditions of the Contract (General, Supplementary and other Conditions), the Drawings, the Specifications, and all Addenda issued prior to and all Modifications issued after execution of the Contract. A Modification is (1) a written amendment to the Contract signed by both parties, (2) a Change Order, (3) a written interpretation issued by the Architect pursuant to Subparagraph 2.2.8, or (4) a written order for a minor change in the Work issued by the Architect pursuant to Paragraph 12.4. The Contract Documents do not include Bidding Documents such as the Advertisement or Invitation to Bid, the Instructions to Bidders, sample forms, the Contractor's Bid or portions of Addenda relating to any of these, or any other documents, unless specifically enumerated in the Owner-Contractor Agreement.

1.1.2 THE CONTRACT

The Contract Documents form the Contract for Construction. This Contract represents the entire and integrated agreement between the parties hereto and supersedes all prior negotiations, representations, or agreements, either written or oral. The Contract may be amended or modified only by a Modification as defined in Subparagraph 1.1.1. The Contract Documents shall not be construed to create any contractual relationship of any kind between the Architect and the Contractor, but the Architect shall be entitled to performance of obligations intended for his benefit, and to enforcement thereof. Nothing contained in the Contract Documents shall create any contractual relationship between the Owner or the Architect and any Subcontractor or Sub-subcontractor.

1.1.3 THE WORK

The Work comprises the completed construction required by the Contract Documents and includes all labor necessary to produce such construction, and all materials and equipment incorporated or to be incorporated in such construction.

1.1.4 THE PROJECT

The Project is the total construction of which the Work performed under the Contract Documents may be the whole or a part.

1.2 EXECUTION, CORRELATION AND INTENT

1.2.1 The Contract Documents shall be signed in not less than triplicate by the Owner and Contractor. If either the Owner or the Contractor or both do not sign the Conditions of the Contract, Drawings, Specifications, or any of the other Contract Documents, the Architect shall identify such Documents.

1.2.2 By executing the Contract, the Contractor represents that he has visited the site, familiarized himself with the local conditions under which the Work is to be performed, and correlated his observations with the requirements of the Contract Documents.

1.2.3 The intent of the Contract Documents is to include all items necessary for the proper execution and completion of the Work. The Contract Documents are complementary, and what is required by any one shall be as binding as if required by all. Work not covered in the Contract Documents will not be required unless it is consistent therewith and is reasonably inferable therefrom as being necessary to produce the intended results. Words and abbreviations which have well-known technical or trade meanings are used in the Contract Documents in accordance with such recognized meanings.

1.2.4 The organization of the Specifications into divisions, sections and articles, and the arrangement of Drawings shall not control the Contractor in dividing the Work among Subcontractors or in establishing the extent of Work to be performed by any trade.

1.3 OWNERSHIP AND USE OF DOCUMENTS

1.3.1 All Drawings, Specifications and copies thereof furnished by the Architect are and shall remain his property. They are to be used only with respect to this Project and are not to be used on any other project. With the exception of one contract set for each party to the Contract, such documents are to be returned or suitably accounted for to the Architect on request at the completion of the Work. Submission or distribution to meet official regulatory requirements or for other purposes in connection with the Project is not to be construed as publication in derogation of the Architect's common law copyright or other reserved rights.

ARTICLE 2

ARCHITECT

2.1 DEFINITION

2.1.1 The Architect is the person lawfully licensed to practice architecture, or an entity lawfully practicing architecture identified as such in the Owner-Contractor Agreement, and is referred to throughout the Contract Documents as if singular in number and masculine in gender. The term Architect means the Architect or his authorized representative.

2.2 ADMINISTRATION OF THE CONTRACT

2.2.1 The Architect will provide administration of the Contract as hereinafter described.

2.2.2 The Architect will be the Owner's representative during construction and until final payment is due. The Architect will advise and consult with the Owner. The Owner's instructions to the Contractor shall be forwarded through the Architect. The Architect will have authority to act on behalf of the Owner only to the extent provided in the Contract Documents, unless otherwise modified by written instrument in accordance with Subparagraph 2.2.18.

2.2.3 The Architect will visit the site at intervals appropriate to the stage of construction to familiarize himself generally with the progress and quality of the Work and to determine in general if the Work is proceeding in accordance with the Contract Documents. However, the Architect will not be required to make exhaustive or continuous on-site inspections to check the quality or quantity of the Work. On the basis of his on-site observations as an architect, he will keep the Owner informed of the progress of the Work, and will endeavor to guard the Owner against defects and deficiencies in the Work of the Contractor.

2.2.4 The Architect will not be responsible for and will not have control or charge of construction means, methods, techniques, sequences or procedures, or for safety precautions and programs in connection with the Work, and he will not be responsible for the Contractor's failure to carry out the Work in accordance with the Contract Documents. The Architect will not be responsible for or have control or charge over the acts or omissions of the Contractor, Subcontractors, or any of their agents or employees, or any other persons performing any of the Work.

2.2.5 The Architect shall at all times have access to the Work wherever it is in preparation and progress. The Contractor shall provide facilities for such access so the Architect may perform his functions under the Contract Documents.

2.2.6 Based on the Architect's observations and an evaluation of the Contractor's Applications for Payment, the Architect will determine the amounts owing to the Contractor and will issue Certificates for Payment in such amounts, as provided in Paragraph 9.4.

2.2.7 The Architect will be the interpreter of the requirements of the Contract Documents and the judge of the performance thereunder by both the Owner and Contractor.

2.2.8 The Architect will render interpretations necessary for the proper execution or progress of the Work, with reasonable promptness and in accordance with any time limit agreed upon. Either party to the Contract may make written request to the Architect for such interpretations.

2.2.9 Claims, disputes and other matters in question between the Contractor and the Owner relating to the execution or progress of the Work or the interpretation of the Contract Documents shall be referred initially to the Architect for decision which he will render in writing within a reasonable time.

2.2.10 All interpretations and decisions of the Architect shall be consistent with the intent of and reasonably inferable from the Contract Documents and will be in writing or in the form of drawings. In his capacity as interpreter and judge, he will endeavor to secure faithful performance by both the Owner and the Contractor, will not

show partiality to either, and will not be liable for the result of any interpretation or decision rendered in good faith in such capacity.

2.2.11 The Architect's decisions in matters relating to artistic effect will be final if consistent with the intent of the Contract Documents.

2.2.12 Any claim, dispute or other matter in question between the Contractor and the Owner referred to the Architect, except those relating to artistic effect as provided in Subparagraph 2.2.11 and except those which have been waived by the making or acceptance of final payment as provided in Subparagraphs 9.9.4 and 9.9.5, shall be subject to arbitration upon the written demand of either party. However, no demand for arbitration of any such claim, dispute or other matter may be made until the earlier of (1) the date on which the Architect has rendered a written decision, or (2) the tenth day after the parties have presented their evidence to the Architect or have been given a reasonable opportunity to do so, if the Architect has not rendered his written decision by that date. When such a written decision of the Architect states (1) that the decision is final but subject to appeal, and (2) that any demand for arbitration of a claim, dispute or other matter covered by such decision must be made within thirty days after the date on which the party making the demand receives the written decision, failure to demand arbitration within said thirty days' period will result in the Architect's decision becoming final and binding upon the Owner and the Contractor. If the Architect renders a decision after arbitration proceedings have been initiated, such decision may be entered as evidence but will not supersede any arbitration proceedings unless the decision is acceptable to all parties concerned.

2.2.13 The Architect will have authority to reject Work which does not conform to the Contract Documents. Whenever, in his opinion, he considers it necessary or advisable for the implementation of the intent of the Contract Documents, he will have authority to require special inspection or testing of the Work in accordance with Subparagraph 7.7.2 whether or not such Work be then fabricated, installed or completed. However, neither the Architect's authority to act under this Subparagraph 2.2.13, nor any decision made by him in good faith either to exercise or not to exercise such authority, shall give rise to any duty or responsibility of the Architect to the Contractor, any Subcontractor, any of their agents or employees, or any other person performing any of the Work.

2.2.14 The Architect will review and approve or take other appropriate action upon Contractor's submittals such as Shop Drawings, Product Data and Samples, but only for conformance with the design concept of the Work and with the information given in the Contract Documents. Such action shall be taken with reasonable promptness so as to cause no delay. The Architect's approval of a specific item shall not indicate approval of an assembly of which the item is a component.

2.2.15 The Architect will prepare Change Orders in accordance with Article 12, and will have authority to order minor changes in the Work as provided in Subparagraph 12.4.1.

2.2.16 The Architect will conduct inspections to determine the dates of Substantial Completion and final completion, will receive and forward to the Owner for the Owner's review written warranties and related documents required by the Contract and assembled by the Contractor, and will issue a final Certificate for Payment upon compliance with the requirements of Paragraph 9.9.

2.2.17 If the Owner and Architect agree, the Architect will provide one or more Project Representatives to assist the Architect in carrying out his responsibilities at the site. The duties, responsibilities and limitations of authority of any such Project Representative shall be as set forth in an exhibit to be incorporated in the Contract Documents.

2.2.18 The duties, responsibilities and limitations of authority of the Architect as the Owner's representative during construction as set forth in the Contract Documents will not be modified or extended without written consent of the Owner, the Contractor and the Architect.

2.2.19 In case of the termination of the employment of the Architect, the Owner shall appoint an architect against whom the Contractor makes no reasonable objection whose status under the Contract Documents shall be that of the former architect. Any dispute in connection with such appointment shall be subject to arbitration.

ARTICLE 3

OWNER

3.1 DEFINITION

3.1.1 The Owner is the person or entity identified as such in the Owner-Contractor Agreement and is referred to throughout the Contract Documents as if singular in number and masculine in gender. The term Owner means the Owner or his authorized representative.

3.2 INFORMATION AND SERVICES REQUIRED OF THE OWNER

3.2.1 The Owner shall, at the request of the Contractor, at the time of execution of the Owner-Contractor Agreement, furnish to the Contractor reasonable evidence that he has made financial arrangements to fulfill his obligations under the Contract. Unless such reasonable evidence is furnished, the Contractor is not required to execute the Owner-Contractor Agreement or to commence the Work.

3.2.2 The Owner shall furnish all surveys describing the physical characteristics, legal limitations and utility locations for the site of the Project, and a legal description of the site.

3.2.3 Except as provided in Subparagraph 4.7.1, the Owner shall secure and pay for necessary approvals, easements, assessments and charges required for the construction, use or occupancy of permanent structures or for permanent changes in existing facilities.

3.2.4 Information or services under the Owner's control shall be furnished by the Owner with reasonable promptness to avoid delay in the orderly progress of the Work.

3.2.5 Unless otherwise provided in the Contract Documents, the Contractor will be furnished, free of charge, all copies of Drawings and Specifications reasonably necessary for the execution of the Work.

3.2.6 The Owner shall forward all instructions to the Contractor through the Architect.

3.2.7 The foregoing are in addition to other duties and responsibilities of the Owner enumerated herein and especially those in respect to Work by Owner or by Separate Contractors, Payments and Completion, and Insurance in Articles 6, 9 and 11 respectively.

3.3 OWNER'S RIGHT TO STOP THE WORK

3.3.1 If the Contractor fails to correct defective Work as required by Paragraph 13.2 or persistently fails to carry out the Work in accordance with the Contract Documents, the Owner, by a written order signed personally or by an agent specifically so empowered by the Owner in writing, may order the Contractor to stop the Work, or any portion thereof, until the cause for such order has been eliminated; however, this right of the Owner to stop the Work shall not give rise to any duty on the part of the Owner to exercise this right for the benefit of the Contractor or any other person or entity, except to the extent required by Subparagraph 6.1.3.

3.4 OWNER'S RIGHT TO CARRY OUT THE WORK

3.4.1 If the Contractor defaults or neglects to carry out the Work in accordance with the Contract Documents and fails within seven days after receipt of written notice from the Owner to commence and continue correction of such default or neglect with diligence and promptness, the Owner may, after seven days following receipt by the Contractor of an additional written notice and without prejudice to any other remedy he may make, make good such deficiencies. In such case an appropriate Change Order shall be issued deducting from the payments then or thereafter due the Contractor the cost of correcting such deficiencies, including compensation for the Architect's additional services made necessary by such default, neglect or failure. Such action by the Owner and the amount charged to the Contractor are both subject to the prior approval of the Architect. If the payments then or thereafter due the Contractor are not sufficient to cover such amount, the Contractor shall pay the difference to the Owner.

ARTICLE 4

CONTRACTOR

4.1 DEFINITION

4.1.1 The Contractor is the person or entity identified as such in the Owner-Contractor Agreement and is referred to throughout the Contract Documents as if singular in number and masculine in gender. The term Contractor means the Contractor or his authorized representative.

4.2 REVIEW OF CONTRACT DOCUMENTS

4.2.1 The Contractor shall carefully study and compare the Contract Documents and shall at once report to the Architect any error, inconsistency or omission he may discover. The Contractor shall not be liable to the Owner or the Architect for any damage resulting from any such errors, inconsistencies or omissions in the Contract Documents. The Contractor shall perform no portion of the Work at any time without Contract Documents or, where required, approved Shop Drawings, Product Data or Samples for such portion of the Work.

4.3 SUPERVISION AND CONSTRUCTION PROCEDURES

4.3.1 The Contractor shall supervise and direct the Work, using his best skill and attention. He shall be solely responsible for all construction means, methods, techniques, sequences and procedures and for coordinating all portions of the Work under the Contract.

4.3.2 The Contractor shall be responsible to the Owner for the acts and omissions of his employees, Subcontractors and their agents and employees, and other persons performing any of the Work under a contract with the Contractor.

4.3.3 The Contractor shall not be relieved from his obligations to perform the Work in accordance with the Contract Documents either by the activities or duties of the Architect in his administration of the Contract, or by inspections, tests or approvals required or performed under Paragraph 7.7 by persons other than the Contractor.

4.4 · LABOR AND MATERIALS

4.4.1 Unless otherwise provided in the Contract Documents, the Contractor shall provide and pay for all labor, materials, equipment, tools, construction equipment and machinery, water, heat, utilities, transportation, and other facilities and services necessary for the proper execution and completion of the Work, whether temporary or permanent and whether or not incorporated or to be incorporated in the Work.

4.4.2 The Contractor shall at all times enforce strict discipline and good order among his employees and shall not employ on the Work any unfit person or anyone not skilled in the task assigned to him.

4.5 WARRANTY

4.5.1 The Contractor warrants to the Owner and the Architect that all materials and equipment furnished under this Contract will be new unless otherwise specified, and that all Work will be of good quality, free from faults and defects and in conformance with the Contract Documents. All Work not conforming to these requirements, including substitutions not properly approved and authorized, may be considered defective. If required by the Architect, the Contractor shall furnish satisfactory evidence as to the kind and quality of materials and equipment. This warranty is not limited by the provisions of Paragraph 13.2.

4.6 TAXES

4.6.1 The Contractor shall pay all sales, consumer, use and other similar taxes for the Work or portions thereof provided by the Contractor which are legally enacted at the time bids are received, whether or not yet effective.

4.7 PERMITS, FEES AND NOTICES

4.7.1 Unless otherwise provided in the Contract Documents, the Contractor shall secure and pay for the building permit and for all other permits and governmental fees, licenses and inspections necessary for the proper execution and completion of the Work which are customarily secured after execution of the Contract and which are legally required at the time the bids are received.

4.7.2 The Contractor shall give all notices and comply with all laws, ordinances, rules, regulations and lawful orders of any public authority bearing on the performance of the Work.

4.7.3 It is not the responsibility of the Contractor to make certain that the Contract Documents are in accordance with applicable laws, statutes, building codes and regulations. If the Contractor observes that any of the Contract Documents are at variance therewith in any respect, he shall promptly notify the Architect in writing, and any necessary changes shall be accomplished by appropriate Modification.

4.7.4 If the Contractor performs any Work knowing it to be contrary to such laws, ordinances, rules and regulations, and without such notice to the Architect, he shall assume full responsibility therefor and shall bear all costs attributable thereto.

4.8 ALLOWANCES

4.8.1 The Contractor shall include in the Contract Sum all allowances stated in the Contract Documents. Items covered by these allowances shall be supplied for such amounts and by such persons as the Owner may direct, but the Contractor will not be required to employ persons against whom he makes a reasonable objection.

4.8.2 Unless otherwise provided in the Contract Documents:

  .1 these allowances shall cover the cost to the Contractor, less any applicable trade discount, of the materials and equipment required by the allowance delivered at the site, and all applicable taxes;

  .2 the Contractor's costs for unloading and handling on the site, labor, installation costs, overhead, profit and other expenses contemplated for the original allowance shall be included in the Contract Sum and not in the allowance;

  .3 whenever the cost is more than or less than the allowance, the Contract Sum shall be adjusted accordingly by Change Order, the amount of which will recognize changes, if any, in handling costs on the site, labor, installation costs, overhead, profit and other expenses.

4.9 SUPERINTENDENT

4.9.1 The Contractor shall employ a competent superintendent and necessary assistants who shall be in attendance at the Project site during the progress of the Work. The superintendent shall represent the Contractor and all communications given to the superintendent shall be as binding as if given to the Contractor. Important communications shall be confirmed in writing. Other communications shall be so confirmed on written request in each case.

4.10 PROGRESS SCHEDULE

4.10.1 The Contractor, immediately after being awarded the Contract, shall prepare and submit for the Owner's and Architect's information an estimated progress schedule for the Work. The progress schedule shall be related to the entire Project to the extent required by the Contract Documents, and shall provide for expeditious and practicable execution of the Work.

4.11 DOCUMENTS AND SAMPLES AT THE SITE

4.11.1 The Contractor shall maintain at the site for the Owner one record copy of all Drawings, Specifications, Addenda, Change Orders and other Modifications, in good order and marked currently to record all changes made during construction, and approved Shop Drawings, Product Data and Samples. These shall be available to the Architect and shall be delivered to him for the Owner upon completion of the Work.

4.12 SHOP DRAWINGS, PRODUCT DATA AND SAMPLES

4.12.1 Shop Drawings are drawings, diagrams, schedules and other data specially prepared for the Work by the Contractor or any Subcontractor, manufacturer, supplier or distributor to illustrate some portion of the Work.

4.12.2 Product Data are illustrations, standard schedules, performance charts, instructions, brochures, diagrams and other information furnished by the Contractor to illustrate a material, product or system for some portion of the Work.

4.12.3 Samples are physical examples which illustrate materials, equipment or workmanship and establish standards by which the Work will be judged.

4.12.4 The Contractor shall review, approve and submit, with reasonable promptness and in such sequence as to cause no delay in the Work or in the work of the Owner or any separate contractor, all Shop Drawings, Product Data and Samples required by the Contract Documents.

4.12.5 By approving and submitting Shop Drawings, Product Data and Samples, the Contractor represents that he has determined and verified all materials, field measurements, and field construction criteria related thereto, or will do so, and that he has checked and coordinated the information contained within such submittals with the requirements of the Work and of the Contract Documents.

4.12.6 The Contractor shall not be relieved of responsibility for any deviation from the requirements of the Contract Documents by the Architect's approval of Shop Drawings, Product Data or Samples under Subparagraph 2.2.14 unless the Contractor has specifically informed the Architect in writing of such deviation at the time of submission and the Architect has given written approval to the specific deviation. The Contractor shall not be relieved from responsibility for errors or omissions in the Shop Drawings, Product Data or Samples by the Architect's approval thereof.

4.12.7 The Contractor shall direct specific attention, in writing or on resubmitted Shop Drawings, Product Data or Samples, to revisions other than those requested by the Architect on previous submittals.

4.12.8 No portion of the Work requiring submission of a Shop Drawing, Product Data or Sample shall be commenced until the submittal has been approved by the Architect as provided in Subparagraph 2.2.14. All such

portions of the Work shall be in accordance with approved submittals.

4.13 USE OF SITE

4.13.1 The Contractor shall confine operations at the site to areas permitted by law, ordinances, permits and the Contract Documents and shall not unreasonably encumber the site with any materials or equipment.

4.14 CUTTING AND PATCHING OF WORK

4.14.1 The Contractor shall be responsible for all cutting, fitting or patching that may be required to complete the Work or to make its several parts fit together properly.

4.14.2 The Contractor shall not damage or endanger any portion of the Work or the work of the Owner or any separate contractors by cutting, patching or otherwise altering any work, or by excavation. The Contractor shall not cut or otherwise alter the work of the Owner or any separate contractor except with the written consent of the Owner and of such separate contractor. The Contractor shall not unreasonably withhold from the Owner or any separate contractor his consent to cutting or otherwise altering the Work.

4.15 CLEANING UP

4.15.1 The Contractor at all times shall keep the premises free from accumulation of waste materials or rubbish caused by his operations. At the completion of the Work he shall remove all his waste materials and rubbish from and about the Project as well as all his tools, construction equipment, machinery and surplus materials.

4.15.2 If the Contractor fails to clean up at the completion of the Work, the Owner may do so as provided in Paragraph 3.4 and the cost thereof shall be charged to the Contractor.

4.16 COMMUNICATIONS

4.16.1 The Contractor shall forward all communications to the Owner through the Architect.

4.17 ROYALTIES AND PATENTS

4.17.1 The Contractor shall pay all royalties and license fees. He shall defend all suits or claims for infringement of any patent rights and shall save the Owner harmless from loss on account thereof, except that the Owner shall be responsible for all such loss when a particular design, process or the product of a particular manufacturer or manufacturers is specified, but if the Contractor has reason to believe that the design, process or product specified is an infringement of a patent, he shall be responsible for such loss unless he promptly gives such information to the Architect.

4.18 INDEMNIFICATION

4.18.1 To the fullest extent permitted by law, the Contractor shall indemnify and hold harmless the Owner and the Architect and their agents and employees from and against all claims, damages, losses and expenses, including but not limited to attorneys' fees, arising out of or resulting from the performance of the Work, provided that any such claim, damage, loss or expense (1) is attributable to bodily injury, sickness, disease or death, or to injury to or destruction of tangible property (other than the Work itself) including the loss of use resulting therefrom, and (2) is caused in whole or in part by any negligent act or omission of the Contractor, any Subcontractor, anyone directly or indirectly employed by any of them or anyone for whose acts any of them may be liable, regardless of whether or not it is caused in part by a party indemnified hereunder. Such obligation shall not be construed to negate, abridge, or otherwise reduce any other right or obligation of indemnity which would otherwise exist as to any party or person described in this Paragraph 4.18.

4.18.2 In any and all claims against the Owner or the Architect or any of their agents or employees by any employee of the Contractor, any Subcontractor, anyone directly or indirectly employed by any of them or anyone for whose acts any of them may be liable, the indemnification obligation under this Paragraph 4.18 shall not be limited in any way by any limitation on the amount or type of damages, compensation or benefits payable by or for the Contractor or any Subcontractor under workers' or workmen's compensation acts, disability benefit acts or other employee benefit acts.

4.18.3 The obligations of the Contractor under this Paragraph 4.18 shall not extend to the liability of the Architect, his agents or employees, arising out of (1) the preparation or approval of maps, drawings, opinions, reports, surveys, change orders, designs or specifications, or (2) the giving of or the failure to give directions or instructions by the Architect, his agents or employees provided such giving or failure to give is the primary cause of the injury or damage.

ARTICLE 5
SUBCONTRACTORS

5.1   DEFINITION

5.1.1 A Subcontractor is a person or entity who has a direct contract with the Contractor to perform any of the Work at the site. The term Subcontractor is referred to throughout the Contract Documents as if singular in number and masculine in gender and means a Subcontractor or his authorized representative. The term Subcontractor does not include any separate contractor or his subcontractors.

5.1.2 A Sub-subcontractor is a person or entity who has a direct or indirect contract with a Subcontractor to perform any of the Work at the site. The term Sub-subcontractor is referred to throughout the Contract Documents as if singular in number and masculine in gender and means a Sub-subcontractor or an authorized representative thereof.

5.2   AWARD OF SUBCONTRACTS AND OTHER CONTRACTS FOR PORTIONS OF THE WORK

5.2.1 Unless otherwise required by the Contract Documents or the Bidding Documents, the Contractor, as soon as practicable after the award of the Contract, shall furnish to the Owner and the Architect in writing the names of the persons or entities (including those who are to furnish materials or equipment fabricated to a special design) proposed for each of the principal portions of the Work. The Architect will promptly reply to the Contractor in writing stating whether or not the Owner or the Architect, after due investigation, has reasonable objection to any

such proposed person or entity. Failure of the Owner or Architect to reply promptly shall constitute notice of no reasonable objection.

5.2.2 The Contractor shall not contract with any such proposed person or entity to whom the Owner or the Architect has made reasonable objection under the provisions of Subparagraph 5.2.1. The Contractor shall not be required to contract with anyone to whom he has a reasonable objection.

5.2.3 If the Owner or the Architect has reasonable objection to any such proposed person or entity, the Contractor shall submit a substitute to whom the Owner or the Architect has no reasonable objection, and the Contract Sum shall be increased or decreased by the difference in cost occasioned by such substitution and an appropriate Change Order shall be issued; however, no increase in the Contract Sum shall be allowed for any such substitution unless the Contractor has acted promptly and responsively in submitting names as required by Subparagraph 5.2.1.

5.2.4 The Contractor shall make no substitution for any Subcontractor, person or entity previously selected if the Owner or Architect makes reasonable objection to such substitution.

5.3   SUBCONTRACTUAL RELATIONS

5.3.1 By an appropriate agreement, written where legally required for validity, the Contractor shall require each Subcontractor, to the extent of the Work to be performed by the Subcontractor, to be bound to the Contractor by the terms of the Contract Documents, and to assume toward the Contractor all the obligations and responsibilities which the Contractor, by these Documents, assumes toward the Owner and the Architect. Said agreement shall preserve and protect the rights of the Owner and the Architect under the Contract Documents with respect to the Work to be performed by the Subcontractor so that the subcontracting thereof will not prejudice such rights, and shall allow to the Subcontractor, unless specifically provided otherwise in the Contractor-Subcontractor agreement, the benefit of all rights, remedies and redress against the Contractor that the Contractor, by these Documents, has against the Owner. Where appropriate, the Contractor shall require each Subcontractor to enter into similar agreements with his Sub-subcontractors. The Contractor shall make available to each proposed Subcontractor, prior to the execution of the Subcontract, copies of the Contract Documents to which the Subcontractor will be bound by this Paragraph 5.3, and identify to the Subcontractor any terms and conditions of the proposed Subcontract which may be at variance with the Contract Documents. Each Subcontractor shall similarly make copies of such Documents available to his Sub-subcontractors.

ARTICLE 6

WORK BY OWNER OR BY SEPARATE CONTRACTORS

6.1   OWNER'S RIGHT TO PERFORM WORK AND TO AWARD SEPARATE CONTRACTS

6.1.1 The Owner reserves the right to perform work related to the Project with his own forces, and to award separate contracts in connection with other portions of the Project or other work on the site under these or similar Conditions of the Contract. If the Contractor claims that delay or additional cost is involved because of such action by the Owner, he shall make such claim as provided elsewhere in the Contract Documents.

6.1.2 When separate contracts are awarded for different portions of the Project or other work on the site, the term Contractor in the Contract Documents in each case shall mean the Contractor who executes each separate Owner-Contractor Agreement.

6.1.3 The Owner will provide for the coordination of the work of his own forces and of each separate contractor with the Work of the Contractor, who shall cooperate therewith as provided in Paragraph 6.2.

6.2 MUTUAL RESPONSIBILITY

6.2.1 The Contractor shall afford the Owner and separate contractors reasonable opportunity for the introduction and storage of their materials and equipment and the execution of their work, and shall connect and coordinate his Work with theirs as required by the Contract Documents.

6.2.2 If any part of the Contractor's Work depends for proper execution or results upon the work of the Owner or any separate contractor, the Contractor shall, prior to proceeding with the Work, promptly report to the Architect any apparent discrepancies or defects in such other work that render it unsuitable for such proper execution and results. Failure of the Contractor so to report shall constitute an acceptance of the Owner's or separate contractors' work as fit and proper to receive his Work, except as to defects which may subsequently become apparent in such work by others.

6.2.3 Any costs caused by defective or ill-timed work shall be borne by the party responsible therefor.

6.2.4 Should the Contractor wrongfully cause damage to the work or property of the Owner, or to other work on the site, the Contractor shall promptly remedy such damage as provided in Subparagraph 10.2.5.

6.2.5 Should the Contractor wrongfully cause damage to the work or property of any separate contractor, the Contractor shall upon due notice promptly attempt to settle with such other contractor by agreement, or otherwise to resolve the dispute. If such separate contractor sues or initiates an arbitration proceeding against the Owner on account of any damage alleged to have been caused by the Contractor, the Owner shall notify the Contractor who shall defend such proceedings at the Owner's expense, and if any judgment or award against the Owner arises therefrom the Contractor shall pay or satisfy it and shall reimburse the Owner for all attorneys' fees and court or arbitration costs which the Owner has incurred.

6.3 OWNER'S RIGHT TO CLEAN UP

6.3.1 If a dispute arises between the Contractor and separate contractors as to their responsibility for cleaning up as required by Paragraph 4.15, the Owner may clean up

and charge the cost thereof to the contractors responsible therefor as the Architect shall determine to be just.

ARTICLE 7

MISCELLANEOUS PROVISIONS

7.1 GOVERNING LAW

7.1.1 The Contract shall be governed by the law of the place where the Project is located.

7.2 SUCCESSORS AND ASSIGNS

7.2.1 The Owner and the Contractor each binds himself, his partners, successors, assigns and legal representatives to the other party hereto and to the partners, successors, assigns and legal representatives of such other party with respect to all covenants, agreements and obligations contained in the Contract Documents. Neither party to the Contract shall assign the Contract or sublet it as a whole without the written consent of the other, nor shall the Contractor assign any moneys due or to become due to him hereunder, without the previous written consent of the Owner.

7.3 WRITTEN NOTICE

7.3.1 Written notice shall be deemed to have been duly served if delivered in person to the individual or member of the firm or entity or to an officer of the corporation for whom it was intended, or if delivered at or sent by registered or certified mail to the last business address known to him who gives the notice.

7.4 CLAIMS FOR DAMAGES

7.4.1 Should either party to the Contract suffer injury or damage to person or property because of any act or omission of the other party or of any of his employees, agents or others for whose acts he is legally liable, claim shall be made in writing to such other party within a reasonable time after the first observance of such injury or damage.

7.5 PERFORMANCE BOND AND LABOR AND MATERIAL PAYMENT BOND

7.5.1 The Owner shall have the right to require the Contractor to furnish bonds covering the faithful performance of the Contract and the payment of all obligations arising thereunder if and as required in the Bidding Documents or in the Contract Documents.

7.6 RIGHTS AND REMEDIES

7.6.1 The duties and obligations imposed by the Contract Documents and the rights and remedies available thereunder shall be in addition to and not a limitation of any duties, obligations, rights and remedies otherwise imposed or available by law.

7.6.2 No action or failure to act by the Owner, Architect or Contractor shall constitute a waiver of any right or duty afforded any of them under the Contract, nor shall any such action or failure to act constitute an approval of or acquiescence in any breach thereunder, except as may be specifically agreed in writing.

7.7 TESTS

7.7.1 If the Contract Documents, laws, ordinances, rules, regulations or orders of any public authority having jurisdiction require any portion of the Work to be inspected, tested or approved, the Contractor shall give the Architect timely notice of its readiness so the Architect may observe such inspection, testing or approval. The Contractor shall bear all costs of such inspections, tests or approvals conducted by public authorities. Unless otherwise provided, the Owner shall bear all costs of other inspections, tests or approvals.

7.7.2 If the Architect determines that any Work requires special inspection, testing, or approval which Subparagraph 7.7.1 does not include, he will, upon written authorization from the Owner, instruct the Contractor to order such special inspection, testing or approval, and the Contractor shall give notice as provided in Subparagraph 7.7.1. If such special inspection or testing reveals a failure of the Work to comply with the requirements of the Contract Documents, the Contractor shall bear all costs thereof, including compensation for the Architect's additional services made necessary by such failure; otherwise the Owner shall bear such costs, and an appropriate Change Order shall be issued.

7.7.3 Required certificates of inspection, testing or approval shall be secured by the Contractor and promptly delivered by him to the Architect.

7.7.4 If the Architect is to observe the inspections, tests or approvals required by the Contract Documents, he will do so promptly and, where practicable, at the source of supply.

7.8 INTEREST

7.8.1 Payments due and unpaid under the Contract Documents shall bear interest from the date payment is due at such rate as the parties may agree upon in writing or, in the absence thereof, at the legal rate prevailing at the place of the Project.

7.9 ARBITRATION

7.9.1 All claims, disputes and other matters in question between the Contractor and the Owner arising out of, or relating to, the Contract Documents or the breach thereof, except as provided in Subparagraph 2.2.11 with respect to the Architect's decisions on matters relating to artistic effect, and except for claims which have been waived by the making or acceptance of final payment as provided by Subparagraphs 9.9.4 and 9.9.5, shall be decided by arbitration in accordance with the Construction Industry Arbitration Rules of the American Arbitration Association then obtaining unless the parties mutually agree otherwise. No arbitration arising out of or relating to the Contract Documents shall include, by consolidation, joinder or in any other manner, the Architect, his employees or consultants except by written consent containing a specific reference to the Owner-Contractor Agreement and signed by the Architect, the Owner, the Contractor and any other person sought to be joined. No arbitration shall include by consolidation, joinder or in any other manner, parties other than the Owner, the Contractor and any other persons substantially involved in a common question of fact or law, whose presence is

required if complete relief is to be accorded in the arbitration. No person other than the Owner or Contractor shall be included as an original third party or additional third party to an arbitration whose interest or responsibility is insubstantial. Any consent to arbitration involving an additional person or persons shall not constitute consent to arbitration of any dispute not described therein or with any person not named or described therein. The foregoing agreement to arbitrate and any other agreement to arbitrate with an additional person or persons duly consented to by the parties to the Owner-Contractor Agreement shall be specifically enforceable under the prevailing arbitration law. The award rendered by the arbitrators shall be final, and judgment may be entered upon it in accordance with applicable law in any court having jurisdiction thereof.

7.9.2 Notice of the demand for arbitration shall be filed in writing with the other party to the Owner-Contractor Agreement and with the American Arbitration Association, and a copy shall be filed with the Architect. The demand for arbitration shall be made within the time limits specified in Subparagraph 2.2.12 where applicable, and in all other cases within a reasonable time after the claim, dispute or other matter in question has arisen, and in no event shall it be made after the date when institution of legal or equitable proceedings based on such claim, dispute or other matter in question would be barred by the applicable statute of limitations.

7.9.3 Unless otherwise agreed in writing, the Contractor shall carry on the Work and maintain its progress during any arbitration proceedings, and the Owner shall continue to make payments to the Contractor in accordance with the Contract Documents.

ARTICLE 8

TIME

8.1 DEFINITIONS

8.1.1 Unless otherwise provided, the Contract Time is the period of time allotted in the Contract Documents for Substantial Completion of the Work as defined in Subparagraph 8.1.3, including authorized adjustments thereto.

8.1.2 The date of commencement of the Work is the date established in a notice to proceed. If there is no notice to proceed, it shall be the date of the Owner-Contractor Agreement or such other date as may be established therein.

8.1.3 The Date of Substantial Completion of the Work or designated portion thereof is the Date certified by the Architect when construction is sufficiently complete, in accordance with the Contract Documents, so the Owner can occupy or utilize the Work or designated portion thereof for the use for which it is intended.

8.1.4 The term day as used in the Contract Documents shall mean calendar day unless otherwise specifically designated.

8.2 PROGRESS AND COMPLETION

8.2.1 All time limits stated in the Contract Documents are of the essence of the Contract.

8.2.2 The Contractor shall begin the Work on the date of commencement as defined in Subparagraph 8.1.2. He shall carry the Work forward expeditiously with adequate forces and shall achieve Substantial Completion within the Contract Time.

8.3 DELAYS AND EXTENSIONS OF TIME

8.3.1 If the Contractor is delayed at any time in the progress of the Work by any act or neglect of the Owner or the Architect, or by any employee of either, or by any separate contractor employed by the Owner, or by changes ordered in the Work, or by labor disputes, fire, unusual delay in transportation, adverse weather conditions not reasonably anticipatable, unavoidable casualties, or any causes beyond the Contractor's control, or by delay authorized by the Owner pending arbitration, or by any other cause which the Architect determines may justify the delay, then the Contract Time shall be extended by Change Order for such reasonable time as the Architect may determine.

8.3.2 Any claim for extension of time shall be made in writing to the Architect not more than twenty days after the commencement of the delay; otherwise it shall be waived. In the case of a continuing delay only one claim is necessary. The Contractor shall provide an estimate of the probable effect of such delay on the progress of the Work.

8.3.3 If no agreement is made stating the dates upon which interpretations as provided in Subparagraph 2.2.8 shall be furnished, then no claim for delay shall be allowed on account of failure to furnish such interpretations until fifteen days after written request is made for them, and not then unless such claim is reasonable.

8.3.4 This Paragraph 8.3 does not exclude the recovery of damages for delay by either party under other provisions of the Contract Documents.

ARTICLE 9

PAYMENTS AND COMPLETION

9.1 CONTRACT SUM

9.1.1 The Contract Sum is stated in the Owner-Contractor Agreement and, including authorized adjustments thereto, is the total amount payable by the Owner to the Contractor for the performance of the Work under the Contract Documents.

9.2 SCHEDULE OF VALUES

9.2.1 Before the first Application for Payment, the Contractor shall submit to the Architect a schedule of values allocated to the various portions of the Work, prepared in such form and supported by such data to substantiate its accuracy as the Architect may require. This schedule, unless objected to by the Architect, shall be used only as a basis for the Contractor's Applications for Payment.

9.3 APPLICATIONS FOR PAYMENT

9.3.1 At least ten days before the date for each progress payment established in the Owner-Contractor Agreement, the Contractor shall submit to the Architect an itemized Application for Payment, notarized if required, supported by such data substantiating the Contractor's right to payment as the Owner or the Architect may require, and reflecting retainage, if any, as provided elsewhere in the Contract Documents.

9.3.2 Unless otherwise provided in the Contract Documents, payments will be made on account of materials or equipment not incorporated in the Work but delivered and suitably stored at the site and, if approved in advance by the Owner, payments may similarly be made for materials or equipment suitably stored at some other location agreed upon in writing. Payments for materials or equipment stored on or off the site shall be conditioned upon submission by the Contractor of bills of sale or such other procedures satisfactory to the Owner to establish the Owner's title to such materials or equipment or otherwise protect the Owner's interest, including applicable insurance and transportation to the site for those materials and equipment stored off the site.

9.3.3 The Contractor warrants that title to all Work, materials and equipment covered by an Application for Payment will pass to the Owner either by incorporation in the construction or upon the receipt of payment by the Contractor, whichever occurs first, free and clear of all liens, claims, security interests or encumbrances, hereinafter referred to in this Article 9 as "liens"; and that no Work, materials or equipment covered by an Application for Payment will have been acquired by the Contractor, or by any other person performing Work at the site or furnishing materials and equipment for the Project, subject to an agreement under which an interest therein or an encumbrance thereon is retained by the seller or otherwise imposed by the Contractor or such other person.

9.4 CERTIFICATES FOR PAYMENT

9.4.1 The Architect will, within seven days after the receipt of the Contractor's Application for Payment, either issue a Certificate for Payment to the Owner, with a copy to the Contractor, for such amount as the Architect determines is properly due, or notify the Contractor in writing his reasons for withholding a Certificate as provided in Subparagraph 9.6.1.

9.4.2 The issuance of a Certificate for Payment will constitute a representation by the Architect to the Owner, based on his observations at the site as provided in Subparagraph 2.2.3 and the data comprising the Application for Payment, that the Work has progressed to the point indicated; that, to the best of his knowledge, information and belief, the quality of the Work is in accordance with the Contract Documents (subject to an evaluation of the Work for conformance with the Contract Documents upon Substantial Completion, to the results of any subsequent tests required by or performed under the Contract Documents, to minor deviations from the Contract Documents correctable prior to completion, and to any specific qualifications stated in his Certificate); and that the Contractor is entitled to payment in the amount certified. However, by issuing a Certificate for Payment, the Architect shall not thereby be deemed to represent that he has made exhaustive or continuous on-site inspections to check the quality or quantity of the Work or that he has reviewed the construction means, methods, techniques, sequences or procedures, or that he has made any examination to ascertain how or for what purpose the Contractor has used the moneys previously paid on account of the Contract Sum.

9.5 PROGRESS PAYMENTS

9.5.1 After the Architect has issued a Certificate for Payment, the Owner shall make payment in the manner and within the time provided in the Contract Documents.

9.5.2 The Contractor shall promptly pay each Subcontractor, upon receipt of payment from the Owner, out of the amount paid to the Contractor on account of such Subcontractor's Work, the amount to which said Subcontractor is entitled, reflecting the percentage actually retained, if any, from payments to the Contractor on account of such Subcontractor's Work. The Contractor shall, by an appropriate agreement with each Subcontractor, require each Subcontractor to make payments to his Subsubcontractors in similar manner.

9.5.3 The Architect may, on request and at his discretion, furnish to any Subcontractor, if practicable, information regarding the percentages of completion or the amounts applied for by the Contractor and the action taken thereon by the Architect on account of Work done by such Subcontractor.

9.5.4 Neither the Owner nor the Architect shall have any obligation to pay or to see to the payment of any moneys to any Subcontractor except as may otherwise be required by law.

9.5.5 No Certificate for a progress payment, nor any progress payment, nor any partial or entire use or occupancy of the Project by the Owner, shall constitute an acceptance of any Work not in accordance with the Contract Documents.

9.6 PAYMENTS WITHHELD

9.6.1 The Architect may decline to certify payment and may withhold his Certificate in whole or in part, to the extent necessary reasonably to protect the Owner, if in his opinion he is unable to make representations to the Owner as provided in Subparagraph 9.4.2. If the Architect is unable to make representations to the Owner as provided in Subparagraph 9.4.2 and to certify payment in the amount of the Application, he will notify the Contractor as provided in Subparagraph 9.4.1. If the Contractor and the Architect cannot agree on a revised amount, the Architect will promptly issue a Certificate for Payment for the amount for which he is able to make such representations to the Owner. The Architect may also decline to certify payment or, because of subsequently discovered evidence or subsequent observations, he may nullify the whole or any part of any Certificate for Payment previously issued, to such extent as may be necessary in his opinion to protect the Owner from loss because of:

.1 defective Work not remedied,

.2 third party claims filed or reasonable evidence indicating probable filing of such claims,

.3 failure of the Contractor to make payments properly to Subcontractors or for labor, materials or equipment,

.4 reasonable evidence that the Work cannot be completed for the unpaid balance of the Contract Sum,

.5 damage to the Owner or another contractor,

.6 reasonable evidence that the Work will not be completed within the Contract Time, or

.7 persistent failure to carry out the Work in accordance with the Contract Documents.

9.6.2 When the above grounds in Subparagraph 9.6.1 are removed, payment shall be made for amounts withheld because of them.

9.7 FAILURE OF PAYMENT

9.7.1 If the Architect does not issue a Certificate for Payment, through no fault of the Contractor, within seven days after receipt of the Contractor's Application for Payment, or if the Owner does not pay the Contractor within seven days after the date established in the Contract Documents any amount certified by the Architect or awarded by arbitration, then the Contractor may, upon seven additional days' written notice to the Owner and the Architect, stop the Work until payment of the amount owing has been received. The Contract Sum shall be increased by the amount of the Contractor's reasonable costs of shut-down, delay and start-up, which shall be effected by appropriate Change Order in accordance with Paragraph 12.3.

9.8 SUBSTANTIAL COMPLETION

9.8.1 When the Contractor considers that the Work, or a designated portion thereof which is acceptable to the Owner, is substantially complete as defined in Subparagraph 8.1.3, the Contractor shall prepare for submission to the Architect a list of items to be completed or corrected. The failure to include any items on such list does not alter the responsibility of the Contractor to complete all Work in accordance with the Contract Documents. When the Architect on the basis of an inspection determines that the Work or designated portion thereof is substantially complete, he will then prepare a Certificate of Substantial Completion which shall establish the Date of Substantial Completion, shall state the responsibilities of the Owner and the Contractor for security, maintenance, heat, utilities, damage to the Work, and insurance, and shall fix the time within which the Contractor shall complete the items listed therein. Warranties required by the Contract Documents shall commence on the Date of Substantial Completion of the Work or designated portion thereof unless otherwise provided in the Certificate of Substantial Completion. The Certificate of Substantial Completion shall be submitted to the Owner and the Contractor for their written acceptance of the responsibilities assigned to them in such Certificate.

9.8.2 Upon Substantial Completion of the Work or designated portion thereof and upon application by the Contractor and certification by the Architect, the Owner shall make payment, reflecting adjustment in retainage, if any, for such Work or portion thereof, as provided in the Contract Documents.

9.9 FINAL COMPLETION AND FINAL PAYMENT

9.9.1 Upon receipt of written notice that the Work is ready for final inspection and acceptance and upon receipt of a final Application for Payment, the Architect will promptly make such inspection and, when he finds the Work acceptable under the Contract Documents and the Contract fully performed, he will promptly issue a final Certificate for Payment stating that to the best of his knowledge, information and belief, and on the basis of his observations and inspections, the Work has been completed in accordance with the terms and conditions of the Contract Documents and that the entire balance found to be due the Contractor, and noted in said final Certificate, is due and payable. The Architect's final Certificate for Payment will constitute a further representation that the conditions precedent to the Contractor's being entitled to final payment as set forth in Subparagraph 9.9.2 have been fulfilled.

9.9.2 Neither the final payment nor the remaining retained percentage shall become due until the Contractor submits to the Architect (1) an affidavit that all payrolls, bills for materials and equipment, and other indebtedness connected with the Work for which the Owner or his property might in any way be responsible, have been paid or otherwise satisfied, (2) consent of surety, if any, to final payment and (3), if required by the Owner, other data establishing payment or satisfaction of all such obligations, such as receipts, releases and waivers of liens arising out of the Contract, to the extent and in such form as may be designated by the Owner. If any Subcontractor refuses to furnish a release or waiver required by the Owner, the Contractor may furnish a bond satisfactory to the Owner to indemnify him against any such lien. If any such lien remains unsatisfied after all payments are made, the Contractor shall refund to the Owner all moneys that the latter may be compelled to pay in discharging such lien, including all costs and reasonable attorneys' fees.

9.9.3 If, after Substantial Completion of the Work, final completion thereof is materially delayed through no fault of the Contractor or by the issuance of Change Orders affecting final completion, and the Architect so confirms, the Owner shall, upon application by the Contractor and certification by the Architect, and without terminating the Contract, make payment of the balance due for that portion of the Work fully completed and accepted. If the remaining balance for Work not fully completed or corrected is less than the retainage stipulated in the Contract Documents, and if bonds have been furnished as provided in Paragraph 7.5, the written consent of the surety to the payment of the balance due for that portion of the Work fully completed and accepted shall be submitted by the Contractor to the Architect prior to certification of such payment. Such payment shall be made under the terms and conditions governing final payment, except that it shall not constitute a waiver of claims.

9.9.4 The making of final payment shall constitute a waiver of all claims by the Owner except those arising from:

.1 unsettled liens,
.2 faulty or defective Work appearing after Substantial Completion,
.3 failure of the Work to comply with the requirements of the Contract Documents, or
.4 terms of any special warranties required by the Contract Documents.

9.9.5 The acceptance of final payment shall constitute a waiver of all claims by the Contractor except those previously made in writing and identified by the Contractor as unsettled at the time of the final Application for Payment.

ARTICLE 10
PROTECTION OF PERSONS AND PROPERTY

10.1   SAFETY PRECAUTIONS AND PROGRAMS

10.1.1 The Contractor shall be responsible for initiating, maintaining and supervising all safety precautions and programs in connection with the Work.

10.2   SAFETY OF PERSONS AND PROPERTY

10.2.1 The Contractor shall take all reasonable precautions for the safety of, and shall provide all reasonable protection to prevent damage, injury or loss to:

.1 all employees on the Work and all other persons who may be affected thereby;
.2 all the Work and all materials and equipment to be incorporated therein, whether in storage on or off the site, under the care, custody or control of the Contractor or any of his Subcontractors or Sub-subcontractors; and
.3 other property at the site or adjacent thereto, including trees, shrubs, lawns, walks, pavements, roadways, structures and utilities not designated for removal, relocation or replacement in the course of construction.

10.2.2 The Contractor shall give all notices and comply with all applicable laws, ordinances, rules, regulations and lawful orders of any public authority bearing on the safety of persons or property or their protection from damage, injury or loss.

10.2.3 The Contractor shall erect and maintain, as required by existing conditions and progress of the Work, all reasonable safeguards for safety and protection, including posting danger signs and other warnings against hazards, promulgating safety regulations and notifying owners and users of adjacent utilities.

10.2.4 When the use or storage of explosives or other hazardous materials or equipment is necessary for the execution of the Work, the Contractor shall exercise the utmost care and shall carry on such activities under the supervision of properly qualified personnel.

10.2.5 The Contractor shall promptly remedy all damage or loss (other than damage or loss insured under Paragraph 11.3) to any property referred to in Clauses 10.2.1.2 and 10.2.1.3 caused in whole or in part by the Contractor, any Subcontractor, any Sub-subcontractor, or anyone directly or indirectly employed by any of them, or by anyone for whose acts any of them may be liable and for which the Contractor is responsible under Clauses 10.2.1.2 and 10.2.1.3, except damage or loss attributable to the acts or omissions of the Owner or Architect or anyone directly or indirectly employed by either of them, or by anyone for whose acts either of them may be liable, and not attributable to the fault or negligence of the Contractor. The foregoing obligations of the Contractor are in addition to his obligations under Paragraph 4.18.

10.2.6 The Contractor shall designate a responsible member of his organization at the site whose duty shall be the prevention of accidents. This person shall be the Contractor's superintendent unless otherwise designated by the Contractor in writing to the Owner and the Architect.

10.2.7 The Contractor shall not load or permit any part of the Work to be loaded so as to endanger its safety.

10.3 **EMERGENCIES**

10.3.1 In any emergency affecting the safety of persons or property, the Contractor shall act, at his discretion, to prevent threatened damage, injury or loss. Any additional compensation or extension of time claimed by the Contractor on account of emergency work shall be determined as provided in Article 12 for Changes in the Work.

ARTICLE 11

INSURANCE

11.1 **CONTRACTOR'S LIABILITY INSURANCE**

11.1.1 The Contractor shall purchase and maintain such insurance as will protect him from claims set forth below which may arise out of or result from the Contractor's operations under the Contract, whether such operations be by himself or by any Subcontractor or by anyone directly or indirectly employed by any of them, or by any one for whose acts any of them may be liable:

.1 claims under workers' or workmen's compensation, disability benefit and other similar employee benefit acts;

.2 claims for damages because of bodily injury, occupational sickness or disease, or death of his employees;

.3 claims for damages because of bodily injury, sickness or disease, or death of any person other than his employees;

.4 claims for damages insured by usual personal injury liability coverage which are sustained (1) by any person as a result of an offense directly or indirectly related to the employment of such person by the Contractor, or (2) by any other person;

.5 claims for damages, other than to the Work itself, because of injury to or destruction of tangible property, including loss of use resulting therefrom; and

.6 claims for damages because of bodily injury or death of any person or property damage arising out of the ownership, maintenance or use of any motor vehicle.

11.1.2 The insurance required by Subparagraph 11.1.1 shall be written for not less than any limits of liability specified in the Contract Documents, or required by law, whichever is greater.

11.1.3 The insurance required by Subparagraph 11.1.1 shall include contractual liability insurance applicable to the Contractor's obligations under Paragraph 4.18.

11.1.4 Certificates of Insurance acceptable to the Owner shall be filed with the Owner prior to commencement of the Work. These Certificates shall contain a provision that coverages afforded under the policies will not be cancelled until at least thirty days' prior written notice has been given to the Owner.

11.2 **OWNER'S LIABILITY INSURANCE**

11.2.1 The Owner shall be responsible for purchasing and maintaining his own liability insurance and, at his option, may purchase and maintain such insurance as will protect him against claims which may arise from operations under the Contract.

11.3 **PROPERTY INSURANCE**

11.3.1 Unless otherwise provided, the Owner shall purchase and maintain property insurance upon the entire Work at the site to the full insurable value thereof. This insurance shall include the interests of the Owner, the Contractor, Subcontractors and Sub-subcontractors in the Work and shall insure against the perils of fire and extended coverage and shall include "all risk" insurance for physical loss or damage including, without duplication of coverage, theft, vandalism and malicious mischief. If the Owner does not intend to purchase such insurance for the full insurable value of the entire Work, he shall inform the Contractor in writing prior to commencement of the Work. The Contractor may then effect insurance which will protect the interests of himself, his Subcontractors and the Sub-subcontractors in the Work, and by appropriate Change Order the cost thereof shall be charged to the Owner. If the Contractor is damaged by failure of the Owner to purchase or maintain such insurance and to so notify the Contractor, then the Owner shall bear all reasonable costs properly attributable thereto. If not covered under the all risk insurance or otherwise provided in the Contract Documents, the Contractor shall effect and maintain similar property insurance on portions of the Work stored off the site or in transit when such portions of the Work are to be included in an Application for Payment under Subparagraph 9.3.2.

11.3.2 The Owner shall purchase and maintain such boiler and machinery insurance as may be required by the Contract Documents or by law. This insurance shall include the interests of the Owner, the Contractor, Subcontractors and Sub-subcontractors in the Work.

11.3.3 Any loss insured under Subparagraph 11.3.1 is to be adjusted with the Owner and made payable to the Owner as trustee for the insureds, as their interests may appear, subject to the requirements of any applicable mortgagee clause and of Subparagraph 11.3.8. The Contractor shall pay each Subcontractor a just share of any insurance moneys received by the Contractor, and by appropriate agreement, written where legally required for validity, shall require each Subcontractor to make payments to his Sub-subcontractors in similar manner.

11.3.4 The Owner shall file a copy of all policies with the Contractor before an exposure to loss may occur.

11.3.5 If the Contractor requests in writing that insurance for risks other than those described in Subparagraphs 11.3.1 and 11.3.2 or other special hazards be included in the property insurance policy, the Owner shall, if possible, include such insurance, and the cost thereof shall be charged to the Contractor by appropriate Change Order.

11.3.6 The Owner and Contractor waive all rights against (1) each other and the Subcontractors, Sub-subcontractors, agents and employees each of the other, and (2) the Architect and separate contractors, if any, and their sub-contractors, sub-subcontractors, agents and employees, for damages caused by fire or other perils to the extent covered by insurance obtained pursuant to this Paragraph 11.3 or any other property insurance applicable to the Work, except such rights as they may have to the proceeds of such insurance held by the Owner as trustee. The foregoing waiver afforded the Architect, his agents and employees shall not extend to the liability imposed by Subparagraph 4.18.3. The Owner or the Contractor, as appropriate, shall require of the Architect, separate contractors, Subcontractors and Sub-subcontractors by appropriate agreements, written where legally required for validity, similar waivers each in favor of all other parties enumerated in this Subparagraph 11.3.6.

11.3.7 If required in writing by any party in interest, the Owner as trustee shall, upon the occurrence of an insured loss, give bond for the proper performance of his duties. He shall deposit in a separate account any money so received, and he shall distribute it in accordance with such agreement as the parties in interest may reach, or in accordance with an award by arbitration in which case the procedure shall be as provided in Paragraph 7.9. If after such loss no other special agreement is made, replacement of damaged work shall be covered by an appropriate Change Order.

11.3.8 The Owner as trustee shall have power to adjust and settle any loss with the insurers unless one of the parties in interest shall object in writing within five days after the occurrence of loss to the Owner's exercise of this power, and if such objection be made, arbitrators shall be chosen as provided in Paragraph 7.9. The Owner as trustee shall, in that case, make settlement with the insurers in accordance with the directions of such arbitrators. If distribution of the insurance proceeds by arbitration is required, the arbitrators will direct such distribution.

11.3.9 If the Owner finds it necessary to occupy or use a portion or portions of the Work prior to Substantial Completion thereof, such occupancy or use shall not commence prior to a time mutually agreed to by the Owner and Contractor and to which the insurance company or companies providing the property insurance have consented by endorsement to the policy or policies. This insurance shall not be cancelled or lapsed on account of such partial occupancy or use. Consent of the Contractor and of the insurance company or companies to such occupancy or use shall not be unreasonably withheld.

11.4 LOSS OF USE INSURANCE

11.4.1 The Owner, at his option, may purchase and maintain such insurance as will insure him against loss of use of his property due to fire or other hazards, however caused. The Owner waives all rights of action against the Contractor for loss of use of his property, including consequential losses due to fire or other hazards however caused, to the extent covered by insurance under this Paragraph 11.4.

ARTICLE 12

CHANGES IN THE WORK

12.1 CHANGE ORDERS

12.1.1 A Change Order is a written order to the Contractor signed by the Owner and the Architect, issued after execution of the Contract, authorizing a change in the Work or an adjustment in the Contract Sum or the Contract Time. The Contract Sum and the Contract Time may be changed only by Change Order. A Change Order signed by the Contractor indicates his agreement therewith, including the adjustment in the Contract Sum or the Contract Time.

12.1.2 The Owner, without invalidating the Contract, may order changes in the Work within the general scope of the Contract consisting of additions, deletions or revisions, the Contract Sum and the Contract Time being adjusted accordingly. All such changes in the Work shall be authorized by Change Order, and shall be performed under the applicable conditions of the Contract Documents.

12.1.3 The cost or credit to the Owner resulting from a change in the Work shall be determined in one or more of the following ways:

.1 by mutual acceptance of a lump sum properly itemized and supported by sufficient substantiating data to permit evaluation;

.2 by unit prices stated in the Contract Documents or subsequently agreed upon;

.3 by cost to be determined in a manner agreed upon by the parties and a mutually acceptable fixed or percentage fee; or

.4 by the method provided in Subparagraph 12.1.4.

12.1.4 If none of the methods set forth in Clauses 12.1.3.1, 12.1.3.2 or 12.1.3.3 is agreed upon, the Contractor, provided he receives a written order signed by the Owner, shall promptly proceed with the Work involved. The cost of such Work shall then be determined by the Architect on the basis of the reasonable expenditures and savings of those performing the Work attributable to the change, including, in the case of an increase in the Contract Sum, a reasonable allowance for overhead and profit. In such case, and also under Clauses 12.1.3.3 and 12.1.3.4 above, the Contractor shall keep and present, in such form as the Architect may prescribe, an itemized accounting together with appropriate supporting data for inclusion in a Change Order. Unless otherwise provided in the Contract Documents, cost shall be limited to the following: cost of materials, including sales tax and cost of delivery; cost of labor, including social security, old age and unemployment insurance, and fringe benefits required by agreement or custom; workers' or workmen's compensation insurance; bond premiums; rental value of equipment and machinery; and the additional costs of supervision and field office personnel directly attributable to the change. Pending final determination of cost to the Owner, payments on account shall be made on the Architect's Certificate for Payment. The amount of credit to be allowed by the Contractor to the Owner for any deletion or change which results in a net decrease in the Contract Sum will be the amount of the actual net cost as confirmed by the Architect. When both additions and credits covering related Work or substitutions are involved in any one change, the allowance for overhead and profit shall be figured on the basis of the net increase, if any, with respect to that change.

12.1.5 If unit prices are stated in the Contract Documents or subsequently agreed upon, and if the quantities originally contemplated are so changed in a proposed Change Order that application of the agreed unit prices to the quantities of Work proposed will cause substantial inequity to the Owner or the Contractor, the applicable unit prices shall be equitably adjusted.

12.2 CONCEALED CONDITIONS

12.2.1 Should concealed conditions encountered in the performance of the Work below the surface of the ground or should concealed or unknown conditions in an existing structure be at variance with the conditions indicated by the Contract Documents, or should unknown physical conditions below the surface of the ground or should concealed or unknown conditions in an existing structure of an unusual nature, differing materially from those ordinarily encountered and generally recognized as inherent in work of the character provided for in this Contract, be encountered, the Contract Sum shall be equitably adjusted by Change Order upon claim by either party made within twenty days after the first observance of the conditions.

12.3 CLAIMS FOR ADDITIONAL COST

12.3.1 If the Contractor wishes to make a claim for an increase in the Contract Sum, he shall give the Architect written notice thereof within twenty days after the occurrence of the event giving rise to such claim. This notice shall be given by the Contractor before proceeding to execute the Work, except in an emergency endangering life or property in which case the Contractor shall proceed in accordance with Paragraph 10.3. No such claim shall be valid unless so made. If the Owner and the Contractor cannot agree on the amount of the adjustment in the Contract Sum, it shall be determined by the Architect. Any change in the Contract Sum resulting from such claim shall be authorized by Change Order.

12.3.2 If the Contractor claims that additional cost is involved because of, but not limited to, (1) any written interpretation pursuant to Subparagraph 2.2.8, (2) any order by the Owner to stop the Work pursuant to Paragraph 3.3 where the Contractor was not at fault, (3) any written order for a minor change in the Work issued pursuant to Paragraph 12.4, or (4) failure of payment by the Owner pursuant to Paragraph 9.7, the Contractor shall make such claim as provided in Subparagraph 12.3.1.

12.4 MINOR CHANGES IN THE WORK

12.4.1 The Architect will have authority to order minor changes in the Work not involving an adjustment in the Contract Sum or an extension of the Contract Time and not inconsistent with the intent of the Contract Documents. Such changes shall be effected by written order, and shall be binding on the Owner and the Contractor.

The Contractor shall carry out such written orders promptly.

ARTICLE 13

UNCOVERING AND CORRECTION OF WORK

13.1 UNCOVERING OF WORK

13.1.1 If any portion of the Work should be covered contrary to the request of the Architect or to requirements specifically expressed in the Contract Documents, it must, if required in writing by the Architect, be uncovered for his observation and shall be replaced at the Contractor's expense.

13.1.2 If any other portion of the Work has been covered which the Architect has not specifically requested to observe prior to being covered, the Architect may request to see such Work and it shall be uncovered by the Contractor. If such Work be found in accordance with the Contract Documents, the cost of uncovering and replacement shall, by appropriate Change Order, be charged to the Owner. If such Work be found not in accordance with the Contract Documents, the Contractor shall pay such costs unless it be found that this condition was caused by the Owner or a separate contractor as provided in Article 6, in which event the Owner shall be responsible for the payment of such costs.

13.2 CORRECTION OF WORK

13.2.1 The Contractor shall promptly correct all Work rejected by the Architect as defective or as failing to conform to the Contract Documents whether observed before or after Substantial Completion and whether or not fabricated, installed or completed. The Contractor shall bear all costs of correcting such rejected Work, including compensation for the Architect's additional services made necessary thereby.

13.2.2 If, within one year after the Date of Substantial Completion of the Work or designated portion thereof or within one year after acceptance by the Owner of designated equipment or within such longer period of time as may be prescribed by law or by the terms of any applicable special warranty required by the Contract Documents, any of the Work is found to be defective or not in accordance with the Contract Documents, the Contractor shall correct it promptly after receipt of a written notice from the Owner to do so unless the Owner has previously given the Contractor a written acceptance of such condition. This obligation shall survive termination of the Contract. The Owner shall give such notice promptly after discovery of the condition.

13.2.3 The Contractor shall remove from the site all portions of the Work which are defective or non-conforming and which have not been corrected under Subparagraphs 4.5.1, 13.2.1 and 13.2.2, unless removal is waived by the Owner.

13.2.4 If the Contractor fails to correct defective or nonconforming Work as provided in Subparagraphs 4.5.1, 13.2.1 and 13.2.2, the Owner may correct it in accordance with Paragraph 3.4.

13.2.5 If the Contractor does not proceed with the correction of such defective or non-conforming Work within a reasonable time fixed by written notice from the Architect, the Owner may remove it and may store the materials or equipment at the expense of the Contractor. If the Contractor does not pay the cost of such removal and storage within ten days thereafter, the Owner may upon ten additional days' written notice sell such Work at auction or at private sale and shall account for the net proceeds thereof, after deducting all the costs that should have been borne by the Contractor, including compensation for the Architect's additional services made necessary thereby. If such proceeds of sale do not cover all costs which the Contractor should have borne, the difference shall be charged to the Contractor and an appropriate Change Order shall be issued. If the payments then or thereafter due the Contractor are not sufficient to cover such amount, the Contractor shall pay the difference to the Owner.

13.2.6 The Contractor shall bear the cost of making good all work of the Owner or separate contractors destroyed or damaged by such correction or removal.

13.2.7 Nothing contained in this Paragraph 13.2 shall be construed to establish a period of limitation with respect to any other obligation which the Contractor might have under the Contract Documents, including Paragraph 4.5 hereof. The establishment of the time period of one year after the Date of Substantial Completion or such longer period of time as may be prescribed by law or by the terms of any warranty required by the Contract Documents relates only to the specific obligation of the Contractor to correct the Work, and has no relationship to the time within which his obligation to comply with the Contract Documents may be sought to be enforced, nor to the time within which proceedings may be commenced to establish the Contractor's liability with respect to his obligations other than specifically to correct the Work.

13.3 ACCEPTANCE OF DEFECTIVE OR NON-CONFORMING WORK

13.3.1 If the Owner prefers to accept defective or non-conforming Work, he may do so instead of requiring its removal and correction, in which case a Change Order will be issued to reflect a reduction in the Contract Sum where appropriate and equitable. Such adjustment shall be effected whether or not final payment has been made.

ARTICLE 14

TERMINATION OF THE CONTRACT

14.1 TERMINATION BY THE CONTRACTOR

14.1.1 If the Work is stopped for a period of thirty days under an order of any court or other public authority having jurisdiction, or as a result of an act of government, such as a declaration of a national emergency making materials unavailable, through no act or fault of the Contractor or a Subcontractor or their agents or employees or any other persons performing any of the Work under a contract with the Contractor, or if the Work should be stopped for a period of thirty days by the Contractor because the Architect has not issued a Certificate for Payment as provided in Paragraph 9.7 or because the Owner has not made payment thereon as provided in Paragraph 9.7, then the Contractor may, upon seven additional days' written notice to the Owner and the Architect, terminate the Contract and recover from the Owner payment for all Work executed and for any proven loss sustained upon any materials, equipment, tools, construction equipment and machinery, including reasonable profit and damages.

14.2 TERMINATION BY THE OWNER

14.2.1 If the Contractor is adjudged a bankrupt, or if he makes a general assignment for the benefit of his creditors, or if a receiver is appointed on account of his insolvency, or if he persistently or repeatedly refuses or fails, except in cases for which extension of time is provided, to supply enough properly skilled workmen or proper materials, or if he fails to make prompt payment to Subcontractors or for materials or labor, or persistently disregards laws, ordinances, rules, regulations or orders of any public authority having jurisdiction, or otherwise is guilty of a substantial violation of a provision of the Contract Documents, then the Owner, upon certification by the Architect that sufficient cause exists to justify such action, may, without prejudice to any right or remedy and after giving the Contractor and his surety, if any, seven days' written notice, terminate the employment of the Contractor and take possession of the site and of all materials, equipment, tools, construction equipment and machinery thereon owned by the Contractor and may finish the Work by whatever method he may deem expedient. In such case the Contractor shall not be entitled to receive any further payment until the Work is finished.

14.2.2 If the unpaid balance of the Contract Sum exceeds the costs of finishing the Work, including compensation for the Architect's additional services made necessary thereby, such excess shall be paid to the Contractor. If such costs exceed the unpaid balance, the Contractor shall pay the difference to the Owner. The amount to be paid to the Contractor or to the Owner, as the case may be, shall be certified by the Architect, upon application, in the manner provided in Paragraph 9.4, and this obligation for payment shall survive the termination of the Contract.

Appendix D

Performance and Payment Bonds, AIA Doc. A311* (1970)

---

* Reprinted with permission of American Institute of Architects, Washington, D.C. See also AIA Doc. A312.

THE AMERICAN INSTITUTE OF ARCHITECTS

AIA Document A311

Performance Bond

KNOW ALL MEN BY THESE PRESENTS: that

(Here insert full name and address or legal title of Contractor)

as Principal, hereinafter called Contractor, and,

(Here insert full name and address or legal title of Surety)

as Surety, hereinafter called Surety, are held and firmly bound unto

(Here insert full name and address or legal title of Owner)

as Obligee, hereinafter called Owner, in the amount of

Dollars ($),

for the payment whereof Contractor and Surety bind themselves, their heirs, executors, administrators, successors and assigns, jointly and severally, firmly by these presents.

WHEREAS,

Contractor has by written agreement dated   19   , entered into a contract with Owner for
(Here insert full name, address and description of project)

in accordance with Drawings and Specifications prepared by

(Here insert full name and address or legal title of Architect)

which contract is by reference made a part hereof, and is hereinafter referred to as the Contract.

AIA DOCUMENT A311 · PERFORMANCE BOND AND LABOR AND MATERIAL PAYMENT BOND · AIA ®
FEBRUARY 1970 ED. · THE AMERICAN INSTITUTE OF ARCHITECTS, 1735 N.Y. AVE., N.W., WASHINGTON, D. C. 20006

1

[B6171]

PERFORMANCE BOND

NOW, THEREFORE, THE CONDITION OF THIS OBLIGATION is such that, if Contractor shall promptly and faithfully perform said Contract, then this obligation shall be null and void; otherwise it shall remain in full force and effect.

The Surety hereby waives notice of any alteration or extension of time made by the Owner.

Whenever Contractor shall be, and declared by Owner to be in default under the Contract, the Owner having performed Owner's obligations thereunder, the Surety may promptly remedy the default, or shall promptly

1) Complete the Contract in accordance with its terms and conditions, or

2) Obtain a bid or bids for completing the Contract in accordance with its terms and conditions, and upon determination by Surety of the lowest responsible bidder, or, if the Owner elects, upon determination by the Owner and the Surety jointly of the lowest responsible bidder, arrange for a contract between such bidder and Owner, and make available as Work progresses (even though there should be a default or a succession of defaults under the contract or contracts of completion arranged under this paragraph) sufficient funds to pay the cost of completion less the balance of the contract price; but not exceeding, including other costs and damages for which the Surety may be liable hereunder, the amount set forth in the first paragraph hereof. The term "balance of the contract price," as used in this paragraph, shall mean the total amount payable by Owner to Contractor under the Contract and any amendments thereto, less the amount properly paid by Owner to Contractor.

Any suit under this bond must be instituted before the expiration of two (2) years from the date on which final payment under the Contract falls due.

No right of action shall accrue on this bond to or for the use of any person or corporation other than the Owner named herein or the heirs, executors, administrators or successors of the Owner.

Signed and sealed this   day of   19

_____
(Witness)

_____ (Principal)   (Seal)

_____ (Title)

_____
(Witness)

_____ (Surety)   (Seal)

_____ (Title)

AIA DOCUMENT A311 • PERFORMANCE BOND AND LABOR AND MATERIAL PAYMENT BOND • AIA ®
FEBRUARY 1970 ED. • THE AMERICAN INSTITUTE OF ARCHITECTS, 1735 N.Y. AVE., N.W., WASHINGTON, D. C. 20006

2

[B6172]

THE AMERICAN INSTITUTE OF ARCHITECTS

AIA Document A311

Labor and Material Payment Bond

THIS BOND IS ISSUED SIMULTANEOUSLY WITH PERFORMANCE BOND IN FAVOR OF THE
OWNER CONDITIONED ON THE FULL AND FAITHFUL PERFORMANCE OF THE CONTRACT

KNOW ALL MEN BY THESE PRESENTS: that

(Here insert full name and address or legal title of Contractor)

as Principal, hereinafter called Principal, and,

(Here insert full name and address or legal title of Surety)

as Surety, hereinafter called Surety, are held and firmly bound unto

(Here insert full name and address or legal title of Owner)

as Obligee, hereinafter called Owner, for the use and benefit of claimants as hereinbelow defined, in the

amount of
(Here insert a sum equal to at least one-half of the contract price)   Dollars ($),
for the payment whereof Principal and Surety bind themselves, their heirs, executors, administrators,
successors and assigns, jointly and severally, firmly by these presents.

WHEREAS,

Principal has by written agreement dated   19   , entered into a contract with Owner for
(Here insert full name, address and description of project)

in accordance with Drawings and Specifications prepared by

(Here insert full name and address or legal title of Architect)

which contract is by reference made a part hereof, and is hereinafter referred to as the Contract.

AIA DOCUMENT A311 • PERFORMANCE BOND AND LABOR AND MATERIAL PAYMENT BOND • AIA ®
FEBRUARY 1970 ED. • THE AMERICAN INSTITUTE OF ARCHITECTS, 1735 N.Y. AVE., N.W., WASHINGTON, D. C. 20006

3
[B6173]

LABOR AND MATERIAL PAYMENT BOND

NOW, THEREFORE, THE CONDITION OF THIS OBLIGATION is such that, if Principal shall promptly make payment to all claimants as hereinafter defined, for all labor and material used or reasonably required for use in the performance of the Contract, then this obligation shall be void; otherwise it shall remain in full force and effect, subject, however, to the following conditions:

1. A claimant is defined as one having a direct contract with the Principal or with a Subcontractor of the Principal for labor, material, or both, used or reasonably required for use in the performance of the Contract, labor and material being construed to include that part of water, gas, power, light, heat, oil, gasoline, telephone service or rental of equipment directly applicable to the Contract.

2. The above named Principal and Surety hereby jointly and severally agree with the Owner that every claimant as herein defined, who has not been paid in full before the expiration of a period of ninety (90) days after the date on which the last of such claimant's work or labor was done or performed, or materials were furnished by such claimant, may sue on this bond for the use of such claimant, prosecute the suit to final judgment for such sum or sums as may be justly due claimant, and have execution thereon. The Owner shall not be liable for the payment of any costs or expenses of any such suit.

3. No suit or action shall be commenced hereunder by any claimant:

a) Unless claimant, other than one having a direct contract with the Principal, shall have given written notice to any two of the following: the Principal, the Owner, or the Surety above named, within ninety (90) days after such claimant did or performed the last of the work or labor, or furnished the last of the materials for which said claim is made, stating with substantial

accuracy the amount claimed and the name of the party to whom the materials were furnished, or for whom the work or labor was done or performed. Such notice shall be served by mailing the same by registered mail or certified mail, postage prepaid, in an envelope addressed to the Principal, Owner or Surety, at any place where an office is regularly maintained for the transaction of business, or served in any manner in which legal process may be served in the state in which the aforesaid project is located, save that such service need not be made by a public officer.

b) After the expiration of one (1) year following the date on which Principal ceased Work on said Contract, it being understood, however, that if any limitation embodied in this bond is prohibited by any law controlling the construction hereof such limitation shall be deemed to be amended so as to be equal to the minimum period of limitation permitted by such law.

c) Other than in a state court of competent jurisdiction in and for the county or other political subdivision of the state in which the Project, or any part thereof, is situated, or in the United States District Court for the district in which the Project, or any part thereof, is situated, and not elsewhere.

4. The amount of this bond shall be reduced by and to the extent of any payment or payments made in good faith hereunder, inclusive of the payment by Surety of mechanics' liens which may be filed of record against said improvement, whether or not claim for the amount of such lien be presented under and against this bond.

Signed and sealed this   day of   19

_____   {   _____
                                                                                   (Principal)   (Seal)
              (Witness)   {
                                                                            _____
                                                                                     (Title)

_____   {   _____
                                                                                   (Surety)   (Seal)
              (Witness)   {
                                                                            _____
                                                                                     (Title)

AIA DOCUMENT A311 · PERFORMANCE BOND AND LABOR AND MATERIAL PAYMENT BOND · AIA ®
FEBRUARY 1970 ED. · THE AMERICAN INSTITUTE OF ARCHITECTS, 1735 N.Y. AVE., N.W., WASHINGTON, D. C. 20006

4
[B6174]

INSTRUCTION SHEET *AIA DOCUMENT A312a*
FOR AIA DOCUMENT A312, PERFORMANCE BOND AND PAYMENT BOND

A. GENERAL INFORMATION

1. Purpose

AIA Document A312 is a new document which combines two separate bonds into one form. This is not a single combined Performance and Payment Bond. It is customary to issue these two bonds simultaneously and to pay one premium for both. The separate procurement of one bond without the other will normally not reduce the premium.

The Performance Bond is an assurance by the Contractor and the Contractor's Surety that the work will be performed and completed in accordance with the terms of the Construction Contract. The Payment Bond is an assurance by the Contractor and the Contractor's Surety that labor and material bills incurred in connection with the Construction Contract will be paid. This assurance is limited by the amount of each bond.

Normally, these bond forms are prepared for execution by the Surety or the Surety's agent.

2. Related Documents

The 1970 edition of the Performance Bond and Labor and Material Payment Bond, AIA Document A311, will continue to be published because it complies with the federal Miller Act and various state laws, frequently called Little Miller Acts. These bonds are a required substitute under the Miller Act for mechanics lien laws which do not apply to governmental works. Caution should be exercised to assure compliance with current laws and regulations.

Other related documents are:

A201, General Conditions of the Contract for Construction
A201/CM, General Conditions of the Contract for Construction, Construction Management Edition
A201/SC, General and Federal Supplementary Conditions of the Contract for Construction
A271, General Conditions of the Contract for Furniture, Furnishings and Equipment
A501, Recommended Guide for Competitive Bidding Procedures and Contract Awards for Building Construction
A511, Guide for Supplementary Conditions
A511/CM, Guide for Supplementary Conditions, Construction Management Edition
A571, Guide for Interiors Supplementary Conditions
A701, Instructions to Bidders
A771, Instructions to Interiors Bidders

For further reference, see *Construction Bonds and Insurance Guide*, 2nd Edition, by Bernard B. Rothschild, FAIA; published by the AIA.

3. Use of Non-AIA Forms

Unlike most AIA documents, the A312 and A311 Documents are not especially interlinked by reference to the other AIA documents. They are general forms which may be used with any appropriate non-AIA document.

4. Use of Current Documents

Prior to using any AIA document, the user should consult the AIA or an AIA component chapter to determine the current edition of each document.

5. Credits

AIA Document A312 was prepared as a service to the construction and surety industries through the joint efforts of The Surety Association of America, The Engineers Joint Contract Documents Committee, The Associated General Contractors of America and The American Institute of Architects.

B. COMPLETING THE A312 FORM

1. Modifications

Users are encouraged to consult with their professional advisor (attorney or bond specialist) with respect to completing or modifying the form. Legal counsel should also be sought concerning the effect of federal, state and local laws on the terms of this Document.

Generally, modifications to the Performance Bond and Payment Bond may be made by filling in the box on the title page of each bond and stating any deletion or addition on the last page of each bond or on an additional page.

2. General

These instructions apply equally to the Performance Bond and to the Payment Bond. Both bonds require identical information on them, but each bond must be executed separately. Even though the A312 Document contains both bonds, they are still very separate bonds. The completion of one bond (e.g., the Performance Bond) is not sufficient to bind the parties to the other (e.g., the Payment Bond). Users should be careful not to mix one bond with the other. A common mistake is to fill in the cover page of the Performance Bond and to sign the signature page of the Payment Bond. In such a case, it is likely that neither bond will become binding.

3. Title Page of each Bond (Pages 1 and 4)

Identification of Parties: The Contractor and Surety should be identified along with the Owner, the Owner's Representative and the Agent or Broker. It is especially important that the Contractor and Surety be identified by using their full legal names and addresses, including the legal status of the parties: sole proprietorship, general partnership, joint venture, unincorporated association, limited partnership, corporation (general or professional), etc. The identification of the Owner's Representative and the Agent or Broker is for information only, since they are not parties to the bond agreement.

Description of the Construction Contract: The Construction Contract should be described by date and amount and by the official name and location of the Project as used in the Construction Contract. The amount of the Construction Contract should be in both written and numerical form.

Bond Amount: The dollar amount of the bond should be in both written and numerical form. Frequently, each bond (the Performance Bond and Payment Bond) will be written to equal individually 100 percent of the Construction Contract Amount.

Bond Date: This date should not be earlier than the date of the Construction Contract which is adopted by reference.

C. EXECUTION OF THE BONDS

Each bond must be separately signed by the Contractor and the Surety on the title page of each bond (pages 1 and 4). Additional space is provided on the last page of each bond (pages 3 and 6) for the signatures of additional parties. The parties executing (signing) the bond should indicate their company, print their name and title, and impress the corporate seal, if any. Where appropriate, attach a copy of the resolution or bylaw authorizing the individual to act on behalf of the firm or entity. Evidence of authority to sign on behalf of each party should be obtained. As to the Surety, this usually takes the form of a power of attorney issued by the surety company to the agent who signs on its behalf.

THE AMERICAN INSTITUTE OF ARCHITECTS

AIA Document A312

Performance Bond

Any singular reference to Contractor, Surety, Owner or other party shall be considered plural where applicable.

CONTRACTOR (Name and Address):

SURETY (Name and Principal Place of Business):

OWNER (Name and Address):

CONSTRUCTION CONTRACT
   Date:
   Amount:
   Description (Name and Location):

BOND
   Date (Not earlier than Construction Contract Date):
   Amount:
   Modifications to this Bond: ☐ None   ☐ See Page 3

CONTRACTOR AS PRINCIPAL   SURETY
Company: (Corporate Seal) Company: (Corporate Seal)

Signature: _____   Signature: _____
Name and Title: Name and Title:

(Any additional signatures appear on page 3)

(FOR INFORMATION ONLY—Name, Address and Telephone)
AGENT or BROKER: OWNER'S REPRESENTATIVE (Architect, Engineer or
                                         other party):

1 The Contractor and the Surety, jointly and severally, bind themselves, their heirs, executors, administrators, successors and assigns to the Owner for the performance of the Construction Contract, which is incorporated herein by reference.

2 If the Contractor performs the Construction Contract, the Surety and the Contractor shall have no obligation under this Bond, except to participate in conferences as provided in Subparagraph 3.1.

3 If there is no Owner Default, the Surety's obligation under this Bond shall arise after:

3.1 The Owner has notified the Contractor and the Surety at its address described in Paragraph 10 below that the Owner is considering declaring a Contractor Default and has requested and attempted to arrange a conference with the Contractor and the Surety to be held not later than fifteen days after receipt of such notice to discuss methods of performing the Construction Contract. If the Owner, the Contractor and the Surety agree, the Contractor shall be allowed a reasonable time to perform the Construction Contract, but such an agreement shall not waive the Owner's right, if any, subsequently to declare a Contractor Default; and

3.2 The Owner has declared a Contractor Default and formally terminated the Contractor's right to complete the contract. Such Contractor Default shall not be declared earlier than twenty days after the Contractor and the Surety have received notice as provided in Subparagraph 3.1; and

3.3 The Owner has agreed to pay the Balance of the Contract Price to the Surety in accordance with the terms of the Construction Contract or to a contractor selected to perform the Construction Contract in accordance with the terms of the contract with the Owner.

4 When the Owner has satisfied the conditions of Paragraph 3, the Surety shall promptly and at the Surety's expense take one of the following actions:

4.1 Arrange for the Contractor, with consent of the Owner, to perform and complete the Construction Contract; or

4.2 Undertake to perform and complete the Construction Contract itself, through its agents or through independent contractors; or

4.3 Obtain bids or negotiated proposals from qualified contractors acceptable to the Owner for a contract for performance and completion of the Construction Contract, arrange for a contract to be prepared for execution by the Owner and the contractor selected with the Owner's concurrence, to be secured with performance and payment bonds executed by a qualified surety equivalent to the bonds issued on the Construction Contract, and pay to the Owner the amount of damages as described in Paragraph 6 in excess of the Balance of the Contract Price incurred by the Owner resulting from the Contractor's default; or

4.4 Waive its right to perform and complete, arrange for completion, or obtain a new contractor and with reasonable promptness under the circumstances:

.1 After investigation, determine the amount for which it may be liable to the Owner and, as soon as practicable after the amount is determined, tender payment therefor to the Owner; or

.2 Deny liability in whole or in part and notify the Owner citing reasons therefor.

5 If the Surety does not proceed as provided in Paragraph 4 with reasonable promptness, the Surety shall be deemed to be in default on this Bond fifteen days after receipt of an additional written notice from the Owner to the Surety demanding that the Surety perform its obligations under this Bond, and the Owner shall be entitled to enforce any remedy available to the Owner. If the Surety proceeds as provided in Subparagraph 4.4, and the Owner refuses the payment tendered or the Surety has denied liability, in whole or in part, without further notice the Owner shall be entitled to enforce any remedy available to the Owner.

6 After the Owner has terminated the Contractor's right to complete the Construction Contract, and if the Surety elects to act under Subparagraph 4.1, 4.2, or 4.3 above, then the responsibilities of the Surety to the Owner shall not be greater than those of the Contractor under the Construction Contract, and the responsibilities of the Owner to the Surety shall not be greater than those of the Owner under the Construction Contract. To the limit of the amount of this Bond, but subject to commitment by the Owner of the Balance of the Contract Price to mitigation of costs and damages on the Construction Contract, the Surety is obligated without duplication for:

6.1 The responsibilities of the Contractor for correction of defective work and completion of the Construction Contract;

6.2 Additional legal, design professional and delay costs resulting from the Contractor's Default, and resulting from the actions or failure to act of the Surety under Paragraph 4; and

6.3 Liquidated damages, or if no liquidated damages are specified in the Construction Contract, actual damages caused by delayed performance or non-performance of the Contractor.

7 The Surety shall not be liable to the Owner or others for obligations of the Contractor that are unrelated to the Construction Contract, and the Balance of the Contract Price shall not be reduced or set off on account of any such unrelated obligations. No right of action shall accrue on this Bond to any person or entity other than the Owner or its heirs, executors, administrators or successors.

8 The Surety hereby waives notice of any change, including changes of time, to the Construction Contract or to related subcontracts, purchase orders and other obligations.

9 Any proceeding, legal or equitable, under this Bond may be instituted in any court of competent jurisdiction in the location in which the work or part of the work is located and shall be instituted within two years after Contractor Default or within two years after the Contractor ceased working or within two years after the Surety refuses or fails to perform its obligations under this Bond, whichever occurs first. If the provisions of this Paragraph are void or prohibited by law, the minimum period of limitation available to sureties as a defense in the jurisdiction of the suit shall be applicable.

10 Notice to the Surety, the Owner or the Contractor shall be mailed or delivered to the address shown on the signature page.

11 When this Bond has been furnished to comply with a statutory or other legal requirement in the location where the construction was to be performed, any provision in this Bond conflicting with said statutory or legal requirement shall be deemed deleted herefrom and provisions conforming to such statutory or other legal requirement shall be deemed incorporated herein. The intent is that this Bond shall be construed as a statutory bond and not as a common law bond.

12 DEFINITIONS

12.1 Balance of the Contract Price: The total amount payable by the Owner to the Contractor under the Construction Contract after all proper adjustments have been made, including allowance to the Con-

tractor of any amounts received or to be received by the Owner in settlement of insurance or other claims for damages to which the Contractor is entitled, reduced by all valid and proper payments made to or on behalf of the Contractor under the Construction Contract.

12.2 Construction Contract: The agreement between the Owner and the Contractor identified on the signature page, including all Contract Documents and changes thereto.

12.3 Contractor Default: Failure of the Contractor, which has neither been remedied nor waived, to perform or otherwise to comply with the terms of the Construction Contract.

12.4 Owner Default: Failure of the Owner, which has neither been remedied nor waived, to pay the Contractor as required by the Construction Contract or to perform and complete or comply with the other terms thereof.

MODIFICATIONS TO THIS BOND ARE AS FOLLOWS:

(Space is provided below for additional signatures of added parties, other than those appearing on the cover page.)

CONTRACTOR AS PRINCIPAL
Company: (Corporate Seal)

SURETY
Company: (Corporate Seal)

Signature: _____
Name and Title:
Address:

Signature: _____
Name and Title:
Address:

AIA DOCUMENT A312 • PERFORMANCE BOND AND PAYMENT BOND • DECEMBER 1984 ED. • AIA ®
THE AMERICAN INSTITUTE OF ARCHITECTS, 1735 NEW YORK AVE., N.W., WASHINGTON, D.C. 20006

A312-1984   3

THE AMERICAN INSTITUTE OF ARCHITECTS

AIA Document A312

Payment Bond

Any singular reference to Contractor, Surety, Owner or other party shall be considered plural where applicable.

CONTRACTOR (Name and Address):   SURETY (Name and Principal Place of Business):

OWNER (Name and Address):

CONSTRUCTION CONTRACT
  Date:
  Amount:
  Description (Name and Location):

BOND
  Date (Not earlier than Construction Contract Date):
  Amount:
  Modifications to this Bond: ☐ None   ☐ See Page 6

CONTRACTOR AS PRINCIPAL   SURETY
Company: (Corporate Seal)   Company: (Corporate Seal)

Signature: _____   Signature: _____
Name and Title: Name and Title:

(Any additional signatures appear on page 6)

(FOR INFORMATION ONLY—Name, Address and Telephone)
AGENT or BROKER: OWNER'S REPRESENTATIVE (Architect, Engineer or
                                                 other party):

1 The Contractor and the Surety, jointly and severally, bind themselves, their heirs, executors, administrators, successors and assigns to the Owner to pay for labor, materials and equipment furnished for use in the performance of the Construction Contract, which is incorporated herein by reference.

2 With respect to the Owner, this obligation shall be null and void if the Contractor:

2.1 Promptly makes payment, directly or indirectly, for all sums due Claimants, and

2.2 Defends, indemnifies and holds harmless the Owner from all claims, demands, liens or suits by any person or entity who furnished labor, materials or equipment for use in the performance of the Construction Contract, provided the Owner has promptly notified the Contractor and the Surety (at the address described in Paragraph 12) of any claims, demands, liens or suits and tendered defense of such claims, demands, liens or suits to the Contractor and the Surety, and provided there is no Owner Default.

3 With respect to Claimants, this obligation shall be null and void if the Contractor promptly makes payment, directly or indirectly, for all sums due.

4 The Surety shall have no obligation to Claimants under this Bond until:

4.1 Claimants who are employed by or have a direct contract with the Contractor have given notice to the Surety (at the address described in Paragraph 12) and sent a copy, or notice thereof, to the Owner, stating that a claim is being made under this Bond and, with substantial accuracy, the amount of the claim.

4.2 Claimants who do not have a direct contract with the Contractor:

   .1 Have furnished written notice to the Contractor and sent a copy, or notice thereof, to the Owner, within 90 days after having last performed labor or last furnished materials or equipment included in the claim stating, with substantial accuracy, the amount of the claim and the name of the party to whom the materials were furnished or supplied or for whom the labor was done or performed; and

   .2 Have either received a rejection in whole or in part from the Contractor, or not received within 30 days of furnishing the above notice any communication from the Contractor by which the Contractor has indicated the claim will be paid directly or indirectly; and

   .3 Not having been paid within the above 30 days, have sent a written notice to the Surety (at the address described in Paragraph 12) and sent a copy, or notice thereof, to the Owner, stating that a claim is being made under this Bond and enclosing a copy of the previous written notice furnished to the Contractor.

5 If a notice required by Paragraph 4 is given by the Owner to the Contractor or to the Surety, that is sufficient compliance.

6 When the Claimant has satisfied the conditions of Paragraph 4, the Surety shall promptly and at the Surety's expense take the following actions:

6.1 Send an answer to the Claimant, with a copy to the Owner, within 45 days after receipt of the claim, stating the amounts that are undisputed and the basis for challenging any amounts that are disputed.

6.2 Pay or arrange for payment of any undisputed amounts.

7 The Surety's total obligation shall not exceed the amount of this Bond, and the amount of this Bond shall be credited for any payments made in good faith by the Surety.

8 Amounts owed by the Owner to the Contractor under the Construction Contract shall be used for the performance of the Construction Contract and to satisfy claims, if any, under any Construction Performance Bond. By the Contractor furnishing and the Owner accepting this Bond, they agree that all funds earned by the Contractor in the performance of the Construction Contract are dedicated to satisfy obligations of the Contractor and the Surety under this Bond, subject to the Owner's priority to use the funds for the completion of the work.

9 The Surety shall not be liable to the Owner, Claimants or others for obligations of the Contractor that are unrelated to the Construction Contract. The Owner shall not be liable for payment of any costs or expenses of any Claimant under this Bond, and shall have under this Bond no obligations to make payments to, give notices on behalf of, or otherwise have obligations to Claimants under this Bond.

10 The Surety hereby waives notice of any change, including changes of time, to the Construction Contract or to related subcontracts, purchase orders and other obligations.

11 No suit or action shall be commenced by a Claimant under this Bond other than in a court of competent jurisdiction in the location in which the work or part of the work is located or after the expiration of one year from the date (1) on which the Claimant gave the notice required by Subparagraph 4.1 or Clause 4.2 (iii), or (2) on which the last labor or service was performed by anyone or the last materials or equipment were furnished by anyone under the Construction Contract, whichever of (1) or (2) first occurs. If the provisions of this Paragraph are void or prohibited by law, the minimum period of limitation available to sureties as a defense in the jurisdiction of the suit shall be applicable.

12 Notice to the Surety, the Owner or the Contractor shall be mailed or delivered to the address shown on the signature page. Actual receipt of notice by Surety, the Owner or the Contractor, however accomplished, shall be sufficient compliance as of the date received at the address shown on the signature page.

13 When this Bond has been furnished to comply with a statutory or other legal requirement in the location where the construction was to be performed, any provision in this Bond conflicting with said statutory or legal requirement shall be deemed deleted herefrom and provisions conforming to such statutory or other legal requirement shall be deemed incorporated herein. The intent is that this

AIA DOCUMENT A312 • PERFORMANCE BOND AND PAYMENT BOND • DECEMBER 1984 ED. • AIA ®
THE AMERICAN INSTITUTE OF ARCHITECTS, 1735 NEW YORK AVE., N.W., WASHINGTON, D.C. 20006

A312-1984 5

Bond shall be construed as a statutory bond and not as a common law bond.

14 Upon request by any person or entity appearing to be a potential beneficiary of this Bond, the Contractor shall promptly furnish a copy of this Bond or shall permit a copy to be made.

15 DEFINITIONS

15.1 Claimant: An individual or entity having a direct contract with the Contractor or with a subcontractor of the Contractor to furnish labor, materials or equipment for use in the performance of the Contract. The intent of this Bond shall be to include without limitation in the terms "labor, materials or equipment" that part of water, gas, power, light, heat, oil, gasoline, telephone service or rental equipment used in the Construction Contract, architectural and engineering services required for performance of the work of the Contractor and the Contractor's subcontractors, and all other items for which a mechanic's lien may be asserted in the jurisdiction where the labor, materials or equipment were furnished.

15.2 Construction Contract: The agreement between the Owner and the Contractor identified on the signature page, including all Contract Documents and changes thereto.

15.3 Owner Default: Failure of the Owner, which has neither been remedied nor waived, to pay the Contractor as required by the Construction Contract or to perform and complete or comply with the other terms thereof.

MODIFICATIONS TO THIS BOND ARE AS FOLLOWS:

(Space is provided below for additional signatures of added parties, other than those appearing on the cover page.)

CONTRACTOR AS PRINCIPAL
Company: (Corporate Seal)

Signature: _____
Name and Title:
Address:

SURETY
Company: (Corporate Seal)

Signature: _____
Name and Title:
Address:

Appendix E

Standard Form of Agreement Between Contractor and Subcontractor

THE AMERICAN INSTITUTE OF ARCHITECTS

AIA Document A401

SUBCONTRACT

Standard Form of Agreement Between Contractor and Subcontractor

1978 EDITION

Use with the latest edition of the appropriate AIA Documents as follows:

A101, Owner-Contractor Agreement — Stipulated Sum
A107, Abbreviated Owner-Contractor Agreement with General Conditions
A111, Owner-Contractor Agreement — Cost Plus Fee
A201, General Conditions of the Contract for Construction.

THIS DOCUMENT HAS IMPORTANT LEGAL CONSEQUENCES; CONSULTATION WITH AN ATTORNEY IS ENCOURAGED WITH RESPECT TO ITS COMPLETION OR MODIFICATION

This document has been approved and endorsed by the American Subcontractors Association and the Associated Specialty Contractors, Inc.

AGREEMENT

made as of the   day of   in the year Nineteen
Hundred and

BETWEEN the Contractor:

and the Subcontractor:

The Project:

The Owner:

The Architect:

The Contractor and Subcontractor agree as set forth below.

ARTICLE 1
THE CONTRACT DOCUMENTS

1.1 The Contract Documents for this Subcontract consist of this Agreement and any Exhibits attached hereto, the Agreement between the Owner and Contractor dated as of   , the Conditions of the Contract between the Owner and Contractor (General, Supplementary and other Conditions), the Drawings, the Specifications, all Addenda issued prior to and all Modifications issued after execution of the Agreement between the Owner and Contractor and agreed upon by the parties to this Subcontract. These form the Subcontract, and are as fully a part of the Subcontract as if attached to this Agreement or repeated herein.

1.2 Copies of the above documents which are applicable to the Work under this Subcontract shall be furnished to the Subcontractor upon his request. An enumeration of the applicable Contract Documents appears in Article 15.

ARTICLE 2
THE WORK

2.1 The Subcontractor shall perform all the Work required by the Contract Documents for

(Here insert a precise description of the Work covered by this Subcontract and refer to numbers of Drawings and pages of Specifications including Addenda, Modifications and accepted Alternates.)

ARTICLE 3
TIME OF COMMENCEMENT AND SUBSTANTIAL COMPLETION

3.1 The Work to be performed under this Subcontract shall be commenced
and, subject to authorized adjustments, shall be substantially completed not later than

(Here insert the specific provisions that are applicable to this Subcontract including any information pertaining to notice to proceed or other method of modification for commencement of Work, starting and completion dates, or duration, and any provisions for liquidated damages relating to failure to complete on time.)

3.2 Time is of the essence of this Subcontract.

3.3 No extension of time will be valid without the Contractor's written consent after claim made by the Subcontractor in accordance with Paragraph 11.10.

ARTICLE 4
THE CONTRACT SUM

4.1 The Contractor shall pay the Subcontractor in current funds for the performance of the Work, subject to additions and deductions authorized pursuant to Paragraph 11.9, the Contract Sum of
dollars ($).

The Contract Sum is determined as follows:

(State here the base bid or other lump sum amount, accepted alternates, and unit prices, as applicable.)

AIA DOCUMENT A401 • CONTRACTOR-SUBCONTRACTOR AGREEMENT • ELEVENTH EDITION • APRIL 1978 • AIA®
©1978 • THE AMERICAN INSTITUTE OF ARCHITECTS, 1735 NEW YORK AVE., N.W., WASHINGTON, D.C. 20006   **A401-1978   2**

ARTICLE 5
PROGRESS PAYMENTS

5.1   The Contractor shall pay the Subcontractor monthly progress payments in accordance with Paragraph 12.4 of this Subcontract.

5.2   Applications for monthly progress payments shall be in writing and in accordance with Paragraph 11.8, shall state the estimated percentage of the Work in this Subcontract that has been satisfactorily completed and shall be submitted to the Contractor on or before the _____ day of each month.

(Here insert details on (1) payment procedures and date of monthly applications, or other procedure if on other than a monthly basis, (2) the basis on which payment will be made on account of materials and equipment suitably stored at the site or other location agreed upon in writing, and (3) any provisions consistent with the Contract Documents for limiting or reducing the amount retained after the Work reaches a certain stage of completion.)

5.3   When the Subcontractor's Work or a designated portion thereof is substantially complete and in accordance with the Contract Documents, the Contractor shall, upon application by the Subcontractor, make prompt application for payment of such Work. Within thirty days following issuance by the Architect of the Certificate for Payment covering such substantially completed Work, the Contractor shall, to the full extent provided in the Contract Documents, make payment to the Subcontractor of the entire unpaid balance of the Contract Sum or of that portion of the Contract Sum attributable to the substantially completed Work, less any portion of the funds for the Subcontractor's Work withheld in accordance with the Certificate to cover costs of items to be completed or corrected by the Subcontractor.

(Delete the above Paragraph if the Contract Documents do not provide for, and the Subcontractor agrees to forego, release of retainage for the Subcontractor's Work prior to completion of the entire Project.)

5.4   Progress payments or final payment due and unpaid under this Subcontract shall bear interest from the date payment is due at the rate entered below or, in the absence thereof, at the legal rate prevailing at the place of the Project.

(Here insert any rate of interest agreed upon.)

(Usury laws and requirements under the Federal Truth in Lending Act, similar state and local consumer credit laws and other regulations at the Owner's, Contractor's and Subcontractor's principal places of business, the location of the Project and elsewhere may affect the validity of this provision. Specific legal advice should be obtained with respect to deletion, modification, or other requirements such as written disclosures or waivers.)

ARTICLE 6
FINAL PAYMENT

6.1   Final payment, constituting the entire unpaid balance of the Contract Sum, shall be due when the Work described in this Subcontract is fully completed and performed in accordance with the Contract Documents and is satisfactory to the Architect, and shall be payable as follows, in accordance with Article 5 and with Paragraph 12.4 of this Subcontract:

(Here insert the relevant conditions under which, or time in which, final payment will become payable.)

6.2   Before issuance of the final payment, the Subcontractor, if required, shall submit evidence satisfactory to the Contractor that all payrolls, bills for materials and equipment, and all known indebtedness connected with the Subcontractor's Work have been satisfied.

ARTICLE 7
PERFORMANCE BOND AND LABOR AND MATERIAL PAYMENT BOND

(Here insert any requirement for the furnishing of bonds by the Subcontractor.)

ARTICLE 8
TEMPORARY FACILITIES AND SERVICES

8.1 Unless otherwise provided in this Subcontract, the Contractor shall furnish and make available at no cost to the Subcontractor the following temporary facilities and services:

ARTICLE 9
INSURANCE

9.1 Prior to starting work, the Subcontractor shall obtain the required insurance from a responsible insurer, and shall furnish satisfactory evidence to the Contractor that the Subcontractor has complied with the requirements of this Article 9. Similarly, the Contractor shall furnish to the Subcontractor satisfactory evidence of insurance required of the Contractor by the Contract Documents.

9.2 The Contractor and Subcontractor waive all rights against each other and against the Owner, the Architect, separate contractors and all other subcontractors for damages caused by fire or other perils to the extent covered by property insurance provided under the General Conditions, except such rights as they may have to the proceeds of such insurance.

(Here insert any insurance requirements and Subcontractor's responsibility for obtaining, maintaining and paying for necessary insurance with limits equaling or exceeding those specified in the Contract Documents and inserted below, or required by law. If applicable, this shall include fire insurance and extended coverage, public liability, property damage, employer's liability, and workers' or workmen's compensation insurance for the Subcontractor and his employees. The insertion should cover provisions for notice of cancellation, allocation of insurance proceeds, and other aspects of insurance.)

ARTICLE 10
WORKING CONDITIONS

(Here insert any applicable arrangements concerning working conditions and labor matters for the Project.)

GENERAL CONDITIONS

ARTICLE 11
SUBCONTRACTOR

11.1   RIGHTS AND RESPONSIBILITIES

11.1.1   The Subcontractor shall be bound to the Contractor by the terms of this Agreement and, to the extent that provisions of the Contract Documents between the Owner and Contractor apply to the Work of the Subcontractor as defined in this Agreement, the Subcontractor shall assume toward the Contractor all the obligations and responsibilities which the Contractor, by those Documents, assumes toward the Owner and the Architect, and shall have the benefit of all rights, remedies and redress against the Contractor which the Contractor, by those Documents, has against the Owner, insofar as applicable to this Subcontract, provided that where any provision of the Contract Documents between the Owner and Contractor is inconsistent with any provision of this Agreement, this Agreement shall govern.

11.1.2   The Subcontractor shall not assign this subcontract without the written consent of the Contractor, nor subcontract the whole of this Subcontract without the written consent of the Contractor, nor further subcontract portions of this Subcontract without written notification to the Contractor when such notification is requested by the Contractor. The Subcontractor shall not assign any amounts due or to become due under this Subcontract without written notice to the Contractor.

11.2   EXECUTION AND PROGRESS OF THE WORK

11.2.1   The Subcontractor agrees that the Contractor's equipment will be available to the Subcontractor only at the Contractor's discretion and on mutually satisfactory terms.

11.2.2   The Subcontractor shall cooperate with the Contractor in scheduling and performing his Work to avoid conflict or interference with the work of others.

11.2.3   The Subcontractor shall promptly submit shop drawings and samples required in order to perform his Work efficiently, expeditiously and in a manner that will not cause delay in the progress of the Work of the Contractor or other subcontractors.

11.2.4   The Subcontractor shall furnish periodic progress reports on the Work as mutually agreed, including information on the status of materials and equipment under this Subcontract which may be in the course of preparation or manufacture.

11.2.5   The Subcontractor agrees that all Work shall be done subject to the final approval of the Architect. The Architect's decisions in matters relating to artistic effect shall be final if consistent with the intent of the Contract Documents.

11.2.6   The Subcontractor shall pay for all materials, equipment and labor used in, or in connection with, the performance of this Subcontract through the period covered by previous payments received from the Contractor, and shall furnish satisfactory evidence, when requested

by the Contractor, to verify compliance with the above requirements.

11.3   LAWS, PERMITS, FEES AND NOTICES

11.3.1   The Subcontractor shall give all notices and comply with all laws, ordinances, rules, regulations and orders of any public authority bearing on the performance of the Work under this Subcontract. The Subcontractor shall secure and pay for all permits and governmental fees, licenses and inspections necessary for the proper execution and completion of the Subcontractor's Work, the furnishing of which is required of the Contractor by the Contract Documents.

11.3.2   The Subcontractor shall comply with Federal, State and local tax laws, social security acts, unemployment compensation acts and workers' or workmen's compensation acts insofar as applicable to the performance of this Subcontract.

11.4   WORK OF OTHERS

11.4.1   In carrying out his Work, the Subcontractor shall take necessary precautions to protect properly the finished work of other trades from damage caused by his operations.

11.4.2   The Subcontractor shall cooperate with the Contractor and other subcontractors whose work might interfere with the Subcontractor's Work, and shall participate in the preparation of coordinated drawings in areas of congestion as required by the Contract Documents, specifically noting and advising the Contractor of any such interference.

11.5   SAFETY PRECAUTIONS AND PROCEDURES

11.5.1   The Subcontractor shall take all reasonable safety precautions with respect to his Work, shall comply with all safety measures initiated by the Contractor and with all applicable laws, ordinances, rules, regulations and orders of any public authority for the safety of persons or property in accordance with the requirements of the Contract Documents. The Subcontractor shall report within three days to the Contractor any injury to any of the Subcontractor's employees at the site.

11.6   CLEANING UP

11.6.1   The Subcontractor shall at all times keep the premises free from accumulation of waste materials or rubbish arising out of the operations of this Subcontract. Unless otherwise provided, the Subcontractor shall not be held responsible for unclean conditions caused by other contractors or subcontractors.

11.7   WARRANTY

11.7.1   The Subcontractor warrants to the Owner, the Architect and the Contractor that all materials and equipment furnished shall be new unless otherwise specified, and that all Work under this Subcontract shall be of good quality, free from faults and defects and in conformance with the Contract Documents. All Work not conforming to these requirements, including substitutions not properly approved and authorized, may be considered defective. The warranty provided in this Paragraph 11.7 shall be in addition to and not in limitation of any other warranty or remedy required by law or by the Contract Documents.

11.8 APPLICATIONS FOR PAYMENT

11.8.1 The Subcontractor shall submit to the Contractor applications for payment at such times as stipulated in Article 5 to enable the Contractor to apply for payment.

11.8.2 If payments are made on the valuation of Work done, the Subcontractor shall, before the first application, submit to the Contractor a schedule of values of the various parts of the Work aggregating the total sum of this Subcontract, made out in such detail as the Subcontractor and Contractor may agree upon or as required by the Owner, and supported by such evidence as to its correctness as the Contractor may direct. This schedule, when approved by the Contractor, shall be used only as a basis for Applications for Payment, unless it be found to be in error. In applying for payment, the Subcontractor shall submit a statement based upon this schedule.

11.8.3 If payments are made on account of materials or equipment not incorporated in the Work but delivered and suitably stored at the site or at some other location agreed upon in writing, such payments shall be in accordance with the Terms and Conditions of the Contract Documents.

11.9 CHANGES IN THE WORK

11.9.1 The Subcontractor may be ordered in writing by the Contractor, without invalidating this Subcontract, to make changes in the Work within the general scope of this Subcontract consisting of additions, deletions or other revisions, the Contract Sum and the Contract Time being adjusted accordingly. The Subcontractor, prior to the commencement of such changed or revised Work, shall submit promptly to the Contractor written copies of any claim for adjustment to the Contract Sum and Contract Time for such revised Work in a manner consistent with the Contract Documents.

11.10 CLAIMS OF THE SUBCONTRACTOR

11.10.1 The Subcontractor shall make all claims promptly to the Contractor for additional cost, extensions of time, and damages for delays or other causes in accordance with the Contract Documents. Any such claim which will affect or become part of a claim which the Contractor is required to make under the Contract Documents within a specified time period or in a specified manner shall be made in sufficient time to permit the Contractor to satisfy the requirements of the Contract Documents. Such claims shall be received by the Contractor not less than two working days preceding the time by which the Contractor's claim must be made. Failure of the Subcontractor to make such a timely claim shall bind the Subcontractor to the same consequences as those to which the Contractor is bound.

11.11 INDEMNIFICATION

11.11.1 To the fullest extent permitted by law, the Subcontractor shall indemnify and hold harmless the Owner, the Architect and the Contractor and all of their agents and employees from and against all claims, damages, losses and expenses, including but not limited to attorney's fees, arising out of or resulting from the performance of the Subcontractor's Work under this Subcontract, provided that any such claim, damage, loss, or expense is attributable to bodily injury, sickness, disease, or death, or to injury to or destruction of tangible property (other than the Work itself) including the loss of use resulting therefrom, to the extent caused in whole or in part by any negligent act or omission of the Subcontractor or anyone directly or indirectly employed by him or anyone for whose acts he may be liable, regardless of whether it is caused in part by a party indemnified hereunder. Such obligation shall not be construed to negate, or abridge, or otherwise reduce any other right or obligation of indemnity which would otherwise exist as to any party or person described in this Paragraph 11.11.

11.11.2 In any and all claims against the Owner, the Architect, or the Contractor or any of their agents or employees by any employee of the Subcontractor, anyone directly or indirectly employed by him or anyone for whose acts he may be liable, the indemnification obligation under this Paragraph 11.11 shall not be limited in any way by any limitation on the amount or type of damages, compensation or benefits payable by or for the Subcontractor under workers' or workmen's compensation acts, disability benefit acts or other employee benefit acts.

11.11.3 The obligations of the Subcontractor under this Paragraph 11.11 shall not extend to the liability of the Architect, his agents or employees arising out of (1) the preparation or approval of maps, drawings, opinions, reports, surveys, Change Orders, designs or specifications, or (2) the giving of or the failure to give directions or instructions by the Architect, his agents or employees provided such giving or failure to give is the primary cause of the injury or damage.

11.12 SUBCONTRACTOR'S REMEDIES

11.12.1 If the Contractor does not pay the Subcontractor through no fault of the Subcontractor, within seven days from the time payment should be made as provided in Paragraph 12.4, the Subcontractor may, without prejudice to any other remedy he may have, upon seven additional days' written notice to the Contractor, stop his Work until payment of the amount owing has been received. The Contract Sum shall, by appropriate adjustment, be increased by the amount of the Subcontractor's reasonable costs of shutdown, delay and start-up.

ARTICLE 12
CONTRACTOR

12.1 RIGHTS AND RESPONSIBILITIES

12.1.1 The Contractor shall be bound to the Subcontractor by the terms of this Agreement, and to the extent that provisions of the Contract Documents between the Owner and the Contractor apply to the Work of the Subcontractor as defined in this Agreement, the Contractor shall assume toward the Subcontractor all the obligations and responsibilities that the Owner, by those Documents, assumes toward the Contractor, and shall have the benefit of all rights, remedies and redress against the Subcontractor which the Owner, by those Documents, has against the Contractor. Where any provision of the

AIA DOCUMENT A401 • CONTRACTOR-SUBCONTRACTOR AGREEMENT • ELEVENTH EDITION • APRIL 1978 • AIA®
©1978 • THE AMERICAN INSTITUTE OF ARCHITECTS, 1735 NEW YORK AVE., N.W., WASHINGTON, D.C. 20006

Contract Documents between the Owner and the Contractor is inconsistent with any provisions of this Agreement, this Agreement shall govern.

12.2 SERVICES PROVIDED BY THE CONTRACTOR

12.2.1 The Contractor shall cooperate with the Subcontractor in scheduling and performing his Work to avoid conflicts or interference in the Subcontractor's Work, and shall expedite written responses to submittals made by the Subcontractor in accordance with Paragraphs 11.2, 11.9 and 11.10. As soon as practicable after execution of this Agreement, the Contractor shall provide the Subcontractor a copy of the estimated progress schedule of the Contractor's entire Work which the Contractor has prepared and submitted for the Owner's and the Architect's information, together with such additional scheduling details as will enable the Subcontractor to plan and perform his Work properly. The Subcontractor shall be notified promptly of any subsequent changes in the progress schedule and the additional scheduling details.

12.2.2 The Contractor shall provide suitable areas for storage of the Subcontractor's materials and equipment during the course of the Work. Any additional costs to the Subcontractor resulting from the relocation of such facilities at the direction of the Contractor shall be reimbursed by the Contractor.

12.3 COMMUNICATIONS

12.3.1 The Contractor shall promptly notify the Subcontractor of all modifications to the Contract between the Owner and the Contractor which affect this Subcontract and which were issued or entered into subsequent to the execution of this Subcontract.

12.3.2 The Contractor shall not give instructions or orders directly to employees or workmen of the Subcontractor except to persons designated as authorized representatives of the Subcontractor.

12.4 PAYMENTS TO THE SUBCONTRACTOR

12.4.1 Unless otherwise provided in the Contract Documents, the Contractor shall pay the Subcontractor each progress payment and the final payment under this Subcontract within three working days after he receives payment from the Owner, except as provided in Subparagraph 12.4.3. The amount of each progress payment to the Subcontractor shall be the amount to which the Subcontractor is entitled, reflecting the percentage of completion allowed to the Contractor for the Work of this Subcontractor applied to the Contract Sum of this Subcontract, and the percentage actually retained, if any, from payments to the Contractor on account of such Subcontractor's Work, plus, to the extent permitted by the Contract Documents, the amount allowed for materials and equipment suitably stored by the Subcontractor, less the aggregate of previous payments to the Subcontractor.

12.4.2 The Contractor shall permit the Subcontractor to request directly from the Architect information regarding the percentages of completion or the amount certified on account of Work done by the Subcontractor.

12.4.3 If the Architect does not issue a Certificate for Payment or the Contractor does not receive payment for any cause which is not the fault of the Subcontractor, the Contractor shall pay the Subcontractor, on demand, a

progress payment computed as provided in Subparagraph 12.4.1 or the final payment as provided in Article 6.

12.5 CLAIMS BY THE CONTRACTOR

12.5.1 The Contractor shall make no demand for liquidated damages for delay in any sum in excess of such amount as may be specifically named in this Subcontract, and liquidated damages shall be assessed against this Subcontractor only for his negligent acts and his failure to act in accordance with the terms of this Agreement, and in no case for delays or causes arising outside the scope of this Subcontract, or for which other subcontractors are responsible.

12.5.2 Except as may be indicated in this Agreement, the Contractor agrees that no claim for payment for services rendered or materials and equipment furnished by the Contractor to the Subcontractor shall be valid without prior notice to the Subcontractor and unless written notice thereof is given by the Contractor to the Subcontractor not later than the tenth day of the calendar month following that in which the claim originated.

12.6 CONTRACTOR'S REMEDIES

12.6.1 If the Subcontractor defaults or neglects to carry out the Work in accordance with this Agreement and fails within three working days after receipt of written notice from the Contractor to commence and continue correction of such default or neglect with diligence and promptness, the Contractor may, after three days following receipt by the Subcontractor of an additional written notice, and without prejudice to any other remedy he may have, make good such deficiencies and may deduct the cost thereof from the payments then or thereafter due the Subcontractor, provided, however, that if such action is based upon faulty workmanship or materials and equipment, the Architect shall first have determined that the workmanship or materials and equipment are not in accordance with the Contract Documents.

ARTICLE 13
ARBITRATION

13.1 All claims, disputes and other matters in question arising out of, or relating to, this Subcontract, or the breach thereof, shall be decided by arbitration, which shall be conducted in the same manner and under the same procedure as provided in the Contract Documents with respect to disputes between the Owner and the Contractor, except that a decision by the Architect shall not be a condition precedent to arbitration. If the Contract Documents do not provide for arbitration or fail to specify the manner and procedure for arbitration, it shall be conducted in accordance with the Construction Industry Arbitration Rules of the American Arbitration Association then obtaining unless the parties mutually agree otherwise.

13.2 Except by written consent of the person or entity sought to be joined, no arbitration arising out of or relating to the Contract Documents shall include, by consolidation, joinder or in any other manner, any person or entity not a party to the Agreement under which such arbitration arises, unless it is shown at the time the demand for arbitration is filed that (1) such person or entity is substantially involved in a common question of fact or law,

(2) the presence of such person or entity is required if complete relief is to be accorded in the arbitration, (3) the interest or responsibility of such person or entity in the matter is not insubstantial, and (4) such person or entity is not the Architect, his employee or his consultant. This agreement to arbitrate and any other written agreement to arbitrate with an additional person or persons referred to herein shall be specifically enforceable under the prevailing arbitration law.

13.3 The Contractor shall permit the Subcontractor to be present and to submit evidence in any arbitration proceeding involving his rights.

13.4 The Contractor shall permit the Subcontractor to exercise whatever rights the Contractor may have under the Contract Documents in the choice of arbitrators in any dispute, if the sole cause of the dispute is the Work, materials, equipment, rights or responsibilities of the Subcontractor; or if the dispute involves the Subcontractor and any other subcontractor or subcontractors jointly, the Contractor shall permit them to exercise such rights jointly.

13.5 The award rendered by the arbitrators shall be final, and judgment may be entered upon it in accordance with applicable law in any court having jurisdiction thereof.

13.6 This Article shall not be deemed a limitation of any rights or remedies which the Subcontractor may have under any Federal or State mechanics' lien laws or under any applicable labor and material payment bonds unless such rights or remedies are expressly waived by him.

ARTICLE 14
TERMINATION

14.1 TERMINATION BY THE SUBCONTRACTOR

14.1.1 If the Work is stopped for a period of thirty days through no fault of the Subcontractor because the Contractor has not made payments thereon as provided in this Agreement, then the Subcontractor may without prejudice to any other remedy he may have, upon seven additional days' written notice to the Contractor, terminate this Subcontract and recover from the Contractor payment for all Work executed and for any proven loss resulting from the stoppage of the Work, including reasonable overhead, profit and damages.

14.2 TERMINATION BY THE CONTRACTOR

14.2.1 If the Subcontractor persistently or repeatedly fails or neglects to carry out the Work in accordance with the Contract Documents or otherwise to perform in accordance with this Agreement and fails within seven days after receipt of written notice to commence and continue correction of such default or neglect with diligence and promptness, the Contractor may, after seven days following receipt by the Subcontractor of an additional written notice and without prejudice to any other remedy he may have, terminate the Subcontract and finish the Work by whatever method he may deem expedient. If the unpaid balance of the Contract Sum exceeds the expense of finishing the Work, such excess shall be paid to the Subcontractor, but if such expense exceeds such unpaid balance, the Subcontractor shall pay the difference to the Contractor.

ARTICLE 15
MISCELLANEOUS PROVISIONS

15.1 Terms used in this Agreement which are defined in the Conditions of the Contract shall have the meanings designated in those Conditions.

15.2 The Contract Documents, which constitute the entire Agreement between the Owner and the Contractor, are listed in Article 1, and the documents which are applicable to this Subcontract, except for Addenda and Modifications issued after execution of this Subcontract, are enumerated as follows:

(List below the Agreement, the Conditions of the Contract [General, Supplementary, and other Conditions], the Drawings, the Specifications, and any Addenda and accepted Alternates, showing page or sheet numbers in all cases and dates where applicable. Continue on succeeding pages as required.)

This Agreement entered into as of the day and year first written above.

CONTRACTOR

_____

_____

_____

SUBCONTRACTOR

_____

_____

_____

AIA DOCUMENT A401 • CONTRACTOR-SUBCONTRACTOR AGREEMENT • ELEVENTH EDITION • APRIL 1978 • AIA®
©1978 • THE AMERICAN INSTITUTE OF ARCHITECTS, 1735 NEW YORK AVE., N.W., WASHINGTON, D.C. 20006

Appendix F

Construction Industry Arbitration Rules

Construction Industry Arbitration Rules*

AMERICAN CONSULTING
ENGINEERS COUNCIL

AMERICAN INSTITUTE OF ARCHITECTS

AMERICAN SOCIETY OF
CIVIL ENGINEERS

AMERICAN SOCIETY OF
LANDSCAPE ARCHITECTS

AMERICAN SUBCONTRACTORS
ASSOCIATION

ASSOCIATED BUILDERS AND
CONTRACTORS, INC.

ASSOCIATED GENERAL CONTRACTORS

ASSOCIATED SPECIALTY
CONTRACTORS, INC.

CONSTRUCTION SPECIFICATIONS
INSTITUTE

NATIONAL ASSOCIATION OF
HOME BUILDERS

NATIONAL SOCIETY OF
PROFESSIONAL ENGINEERS

**American
Arbitration
Association**

*As amended
and in effect
February 1, 1984*

* Reprinted with permission of the American
Arbitration Association, New York, N.Y.

Construction Industry Arbitration Rules

1. Agreement of Parties

The parties shall be deemed to have made these Rules a part of their arbitration agreement whenever they have provided for arbitration under the Construction Industry Arbitration Rules. These Rules and any amendment thereof shall apply in the form obtaining at the time the arbitration is initiated.

2. Name of Tribunal

Any Tribunal constituted by the parties for the settlement of their dispute under these Rules shall be called the Construction Industry Arbitration Tribunal, hereinafter called the Tribunal.

3. Administrator

When parties agree to arbitrate under these Rules, or when they provide for arbitration by the American Arbitration Association, hereinafter called AAA, and an arbitration is initiated hereunder, they thereby constitute AAA the administrator of the arbitration. The authority and duties of the administrator are prescribed in the agreement of the parties and in these Rules.

4. Delegation of Duties

The duties of the AAA under these Rules may be carried out through Tribunal Administrators, or such other officers or committees as the AAA may direct.

5. National Panel of Arbitrators

In cooperation with the National Construction Industry Arbitration Committee, the AAA shall establish and maintain a National Panel of Construction Arbitrators, hereinafter called the Panel, and shall appoint an arbitrator or arbitrators therefrom as hereinafter provided. A neutral arbitrator selected by mutual choice of both parties or their appointees, or appointed by the AAA, is hereinafter called the arbitrator, whereas an arbitrator selected unilaterally by one party is hereinafter called the party-appointed arbitrator. The term arbitrator may hereinafter be used to refer to one arbitrator or to a Tribunal of multiple arbitrators.

6. Office of Tribunal

The general office of a Tribunal is the headquarters of the AAA, which may, however, assign the administration of an arbitration to any of its Regional Offices.

7. Initiation under an Arbitration Provision in a Contract

Arbitration under an arbitration provision in a contract shall be initiated in the following manner:

The initiating party shall, within the time specified by the contract, if any, file with the other party a notice of an intention to arbitrate (Demand), which notice shall contain a statement setting forth the nature of the dispute, the amount involved, and the remedy sought; and shall file three copies of said notice with any Regional Office of the AAA, together with three copies of the arbitration provisions of the contract and the appropriate filing fee as provided in Section 48 hereunder.

The AAA shall give notice of such filing to the other party. A party upon whom the demand for arbitration is made may file an answering statement in duplicate with the AAA within seven days after notice from the AAA, simultaneously sending a copy to the other party. If a monetary claim is made in the answer the appropriate administrative fee provided in the Fee Schedule shall be forwarded to the AAA with the answer. If no answer is filed within the stated time, it will be treated as a denial of the claim. Failure to file an answer shall not operate to delay the arbitration.

Unless the AAA in its discretion determines otherwise, the Expedited Procedures of Construction Arbitration shall be applied in any case where the total claim of any party does not exceed $15,000, exclusive of interest and arbitration costs. Parties may also agree to the Expedited Procedures in cases involving claims in excess of $15,000. The Expedited Procedures shall be applied as described in Sections 54 through 58 of these Rules.

8. Change of Claim or Counterclaim

After filing of the claim or counterclaim, if either party desires to make any new or different claim or counterclaim, same shall be made in writing and filed with the AAA, and a copy thereof shall be mailed to the other party who shall have a period of seven days from the date of such mailing within which to file an answer with the AAA. However, after the arbitrator is appointed no new or different claim or counterclaim may be submitted without the arbitrator's consent.

9. Initiation under a Submission

Parties to any existing dispute may commence an arbitration under these Rules by filing at any Regional Office two copies of a written agreement to arbitrate under these Rules (Submission), signed by the parties. It shall contain a statement of the matter in dispute, the amount of money involved, and the remedy sought, together with the appropriate filing fee as provided in the Fee Schedule.

10. Pre-Hearing Conference and Preliminary Hearing

At the request of the parties or at the discretion of the AAA, a pre-hearing conference with the administrator and the parties or their counsel will be scheduled in appropriate cases to arrange for an exchange of information and the stipulation of uncontested facts so as to expedite the arbitration proceedings.

In large and complex cases, unless the parties agree otherwise, the AAA may schedule a preliminary hearing with the parties and the arbitrator(s) to establish the extent of and schedule for the production of relevant documents and other information, the identification of any witnesses to be called, and a schedule for further hearings to resolve the dispute.

11. Fixing of Locale

The parties may mutually agree on the locale where the arbitration is to be held. If any party requests that the hearing be held in a specific locale and the other party files no objection thereto within seven days after notice of the request is mailed to such party, the locale shall be the one requested. If a party objects to the locale requested by the other party, the AAA shall have power to determine the locale and its decision shall be final and binding.

12. Qualifications of Arbitrator

Any arbitrator appointed pursuant to Section 13 or Section 15 shall be neutral, subject to disqualification for the reasons specified in Section 19. If the agreement of the parties names an arbitrator or specifies any other method of appointing an arbitrator, or if the parties specifically agree in writing, such arbitrator shall not be subject to disqualification for said reasons.

13. Appointment from Panel

If the parties have not appointed an arbitrator and have not provided any other method of appointment, the arbitrator shall be appointed in the following manner: Immediately after the filing of the Demand or Submission, the AAA shall submit simultaneously to each party to the dispute an identical list of names of persons chosen from the Panel. Each party to the dispute shall have seven days from the mailing date in which to cross off any names to which it objects, number the remaining names to indicate the order of preference, and return the list to the AAA. If a party does not return the list within the time specified, all persons named therein shall be deemed acceptable. From among the persons who have been approved on both lists, and in accordance with the designated order of mutual preference, the AAA shall invite the acceptance of an arbitrator to serve. If the parties fail to agree upon any of the persons named, or if acceptable arbitrators are unable to act, or if for any other reason the appointment cannot be made from the submitted lists, the AAA shall have the power to make the appointment from other members of the Panel without the submission of any additional lists.

14. Direct Appointment by Parties

If the agreement of the parties names an arbitrator or specifies a method of appointing an arbitrator, that designation or method shall be followed. The notice of appointment, with name and address of such arbitrator, shall be filed with the AAA by the appointing party. Upon the request of any such appointing party, the AAA shall submit a list of members of the Panel from which the party may make the appointment.

If the agreement specifies a period of time within which an arbitrator shall be appointed, and any party fails to make such appointment within that period, the AAA shall make the appointment.

If no period of time is specified in the agreement, the AAA shall notify the parties to make the appointment, and if within seven days after mailing of such notice such arbitrator has not been so appointed, the AAA shall make the appointment.

15. Appointment of Arbitrator by Party-Appointed Arbitrators

If the parties have appointed their party-appointed arbitrators or if either or both of them have been appointed as provided in Section 14, and have authorized such arbitrator to appoint an arbitrator within a specified time and no appointment is made within such time or any agreed extension thereof,

the AAA shall appoint an arbitrator who shall act as Chairperson.

If no period of time is specified for appointment of the third arbitrator and the party-appointed arbitrators do not make the appointment within seven days from the date of the appointment of the last party-appointed arbitrator, the AAA shall appoint the arbitrator who shall act as Chairperson.

If the parties have agreed that their party-appointed arbitrators shall appoint the arbitrator from the Panel, the AAA shall furnish to the party-appointed arbitrators, in the manner prescribed in Section 13, a list selected from the Panel, and the appointment of the arbitrator shall be made as prescribed in such Section.

16. Nationality of Arbitrator in International Arbitration

If one of the parties is a national or resident of a country other than the United States, the arbitrator shall, upon the request of either party, be appointed from among the nationals of a country other than that of any of the parties.

17. Number of Arbitrators

If the arbitration agreement does not specify the number of arbitrators, the dispute shall be heard and determined by one arbitrator, unless the AAA, in its discretion, directs that a greater number of arbitrators be appointed.

18. Notice to Arbitrator of Appointment

Notice of the appointment of the arbitrator, whether mutually appointed by the parties or appointed by the AAA, shall be mailed to the arbitrator by the AAA, together with a copy of these Rules, and the signed acceptance of the arbitrator shall be filed prior to the opening of the first hearing.

19. Disclosure and Challenge Procedure

A person appointed as neutral arbitrator shall disclose to the AAA any circumstances likely to affect his or her impartiality, including any bias or any financial or personal interest in the result of the arbitration or any past or present relationship with the parties or their counsel. Upon receipt of such information from such arbitrator or other source, the AAA shall communicate such information to the parties and, if it deems it appropriate to do so, to the arbitrator and others. Thereafter, the AAA shall determine whether the arbitrator should be disqualified and shall inform the parties of its decision, which shall be conclusive.

20. Vacancies

If any arbitrator should resign, die, withdraw, refuse, be disqualified or be unable to perform the duties of office, the AAA shall, on proof satisfactory to it, declare the office vacant. Vacancies shall be filled in accordance with the applicable provision of these Rules. In the event of a vacancy in a panel of arbitrators, the remaining arbitrator or arbitrators may continue with the hearing and determination of the controversy, unless the parties agree otherwise.

21. Time and Place

The arbitrator shall fix the time and place for each hearing. The AAA shall mail to each party notice thereof at least five days in advance, unless the parties by mutual agreement waive such notice or modify the terms thereof.

22. Representation by Counsel

Any party may be represented by counsel. A party intending to be so represented shall notify the other party and the AAA of the name and address of counsel at least three days prior to the date set for the hearing at which counsel is first to appear. When an arbitration is initiated by counsel, or where an attorney replies for the other party, such notice is deemed to have been given.

23. Stenographic Record

The AAA shall make the necessary arrangements for the taking of a stenographic record whenever such record is requested by a party. The requesting party or parties shall pay the cost of such record as provided in Section 50.

24. Interpreter

The AAA shall make the necessary arrangements for the services of an interpreter upon the request of one or both parties, who shall assume the cost of such services.

25. Attendance at Hearings

Persons having a direct interest in the arbitration are entitled to attend hearings. The arbitrator shall otherwise have the power to require the retirement of any witness or witnesses during the testimony of other witnesses. It shall be discretionary with the arbitrator to determine the propriety of the attendance of any other persons.

26. Adjournments

The arbitrator may adjourn the hearing, and must take such adjournment when all of the parties agree thereto.

27. Oaths

Before proceeding with the first hearing or with the examination of the file, each arbitrator may take an oath of office, and if required by law, shall do so. The arbitrator may require witnesses to testify under oath administered by any duly qualified person or, if required by law or demanded by either party, shall do so.

28. Majority Decision

Whenever there is more than one arbitrator, all decisions of the arbitrators must be by at least a majority. The award must also be made by at least a majority unless the concurrence of all is expressly required by the arbitration agreement or by law.

29. Order of Proceedings

A hearing shall be opened by the filing of the oath of the arbitrator, where required, and by the recording of the place, time, and date of the hearing, the presence of the arbitrator and parties, and counsel, if any, and by the receipt by the arbitrator of the statement of the claim and answer, if any.

The arbitrator may, at the beginning of the hearing, ask for statements clarifying the issues involved. In some cases, part or all of the above will have been accomplished at the preliminary hearing conducted by the arbitrator(s) pursuant to Section 10.

The complaining party shall then present its claims, proofs and witnesses, who shall submit to questions or other examination. The defending party shall then present its defenses, proofs and witnesses, who shall submit to questions or other examination. The arbitrator may vary this procedure but shall afford full and equal opportunity to the parties for the presentation of any material or relevant proofs.

Exhibits, when offered by either party, may be received in evidence by the arbitrator.

The names and addresses of all witnesses and exhibits in order received shall be made a part of the record.

30. Arbitration in the Absence of a Party or Counsel

Unless the law provides to the contrary, the arbitration may proceed in the absence of any party or counsel, who, after due notice, fails to be present or fails to obtain an adjournment. An award shall not be made solely on the default of a party. The arbitrator shall require the party who is present to submit such evidence as is deemed necessary for the making of an award.

31. Evidence

The parties may offer such evidence as is pertinent and material to the controversy and shall produce such additional evidence as the arbitrator may deem necessary to an understanding and determination of the controversy. An arbitrator authorized by law to subpoena witnesses or documents may do so upon the request of any party, or independently.

The arbitrator shall be the judge of the relevance and the materiality of the evidence offered, and conformity to legal rules of evidence shall not be necessary. All evidence shall be taken in the presence of all of the arbitrators and all of the parties, except where any of the parties is absent in default or has waived the right to be present.

32. Evidence by Affidavit and Filing of Documents

The arbitrator may receive and consider the evidence of witnesses by affidavit, giving it such weight as seems appropriate after consideration of any objections made to its admission.

All documents not filed with the arbitrator at the hearing, but arranged for at the hearing or subsequently by agreement of the parties, shall be filed with the AAA for transmission to the arbitrator. All parties shall be afforded opportunity to examine such documents.

33. Inspection or Investigation

An arbitrator finding it necessary to make an inspection or investigation in connection with the arbitration shall direct the AAA to so advise the parties. The arbitrator shall set the time and the AAA shall notify the parties thereof. Any party who so desires may be present at such inspection or investigation. In the event that one or both parties are not present at the inspection or investigation, the arbitrator shall make a verbal or written report to the parties and afford them an opportunity to comment.

34. Conservation of Property

The arbitrator may issue such orders as may be deemed necessary to safeguard the property which is the subject matter of the arbitration without prejudice to the rights of the parties or to the final determination of the dispute.

35. Closing of Hearings

The arbitrator shall specifically inquire of the parties whether they have any further proofs to offer or witnesses to be heard. Upon receiving negative replies, the arbitrator shall declare the hearings closed and a minute thereof shall be recorded. If briefs are to be filed, the hearings shall be declared closed as of the final date set by the arbitrator for the receipt of briefs. If documents are to be filed as provided for in Section 32 and the date set for their receipt is later than that set for the receipt of briefs, the later date shall be the date of closing the hearing. The time limit within which the arbitrator is required to make an award shall commence to run, in the absence of other agreements by the parties, upon the closing of the hearings.

36. Reopening of Hearings

The hearings may be reopened by the arbitrator at will, or upon application of a party at any time before the award is made. If the reopening of the hearing would prevent the making of the award within the specific time agreed upon by the parties in the contract out of which the controversy has arisen, the matter may not be reopened, unless the parties agree upon the extension of such time limit. When no specific date is fixed in the contract, the arbitrator may reopen the hearings, and the arbitrator shall have thirty days from the closing of the reopened hearings within which to make an award.

37. Waiver of Oral Hearings

The parties may provide, by written agreement, for the waiver of oral hearings. If the parties are unable to agree as to the procedure, the AAA shall specify a fair and equitable procedure.

38. Waiver of Rules

Any party who proceeds with the arbitration after knowledge that any provision or requirement of these Rules has not been complied with and who fails to state an objection thereto in writing, shall be deemed to have waived the right to object.

39. Extensions of Time

The parties may modify any period of time by mutual agreement. The AAA for good cause may extend any period of time established by these Rules, except the time for making the award. The AAA shall notify the parties of any such extension of time and its reason therefor.

40. Communication with Arbitrator and Serving of Notices

There shall be no communication between the parties and an arbitrator other than at oral hearings. Any other oral or written communications from the parties to the arbitrator shall be directed to the AAA for transmittal to the arbitrator.

Each party to an agreement which provides for arbitration under these Rules shall be deemed to have consented that any papers, notices or process necessary or proper for the initiation or continuation of an arbitration under these Rules and for any court action in connection therewith or for the entry of judgment on any award made thereunder may be served upon such party by mail addressed to such party or its attorney at the last known address or by personal service, within or without the state wherein the arbitration is to be held (whether such party be within or without the United States of America), provided that reasonable opportunity to be heard with regard thereto has been granted such party.

41. Time of Award

The award shall be made promptly by the arbitrator and, unless otherwise agreed by the parties, or specified by law, not later than thirty days from the date of closing the hearings, or if oral hearings have been waived, from the date of transmitting the final statements and proofs to the arbitrator.

42. Form of Award

The award shall be in writing and shall be signed either by the sole arbitrator or by at least a majority if there be more than one. It shall be executed in the manner required by law.

43. Scope of Award

The arbitrator may grant any remedy or relief which is just and equitable and within the terms of the agreement of the parties. The arbitrator, in the award, shall assess arbitration fees and expenses as provided in Sections 48 and 50 equally or in favor of any party and, in the event any administrative fees or expenses are due the AAA, in favor of the AAA.

44. Award upon Settlement

If the parties settle their dispute during the course of the arbitration, the arbitrator, upon their request, may set forth the terms of the agreed settlement in an award.

45. Delivery of Award to Parties

Parties shall accept as legal delivery of the award the placing of the award or a true copy thereof in the mail by the AAA, addressed to such party at its last known address or to its attorney, or personal service of the award, or the filing of the award in any manner which may be prescribed by law.

46. Release of Documents for Judicial Proceedings

The AAA shall, upon the written request of a party, furnish to such party, at its expense, certified facsimiles of any papers in the AAA's possession that may be required in judicial proceedings relating to the arbitration.

47. Applications to Court and Exclusion of Liability

(a) No judicial proceedings by a party relating to the subject matter of the arbitration shall be deemed a waiver of the party's right to arbitrate.

(b) Neither the AAA nor any arbitrator in a proceeding under these Rules is a necessary party in judicial proceedings relating to the arbitration.

(c) Parties to these Rules shall be deemed to have consented that judgment upon the award rendered by the arbitrator(s) may be entered in any Federal or State Court having jurisdiction thereof.

(d) Neither the AAA nor any arbitrator shall be liable to any party for any act or omission in connection with any arbitration conducted under these Rules.

48. Administrative Fees

As a not-for-profit organization, the AAA shall prescribe an administrative fee schedule and a refund schedule to compensate it for the cost of providing administrative services. The schedule in effect at the time of filing or the time of refund shall be applicable.

The administrative fees shall be advanced by the initiating party or parties in accordance with the administrative fee schedule, subject to final apportionment by the arbitrator in the award.

When a matter is withdrawn or settled, the refund shall be made in accordance with the refund schedule.

The AAA, in the event of extreme hardship on the part of any party, may defer or reduce the administrative fee.

49. Fee When Oral Hearings are Waived
Where all oral hearings are waived under Section 37, the Administrative Fee Schedule shall apply.

50. Expenses
The expenses of witnesses for either side shall be paid by the party producing such witnesses.

The cost of the stenographic record, if any is made, and all transcripts thereof, shall be prorated equally between the parties ordering copies, unless they shall otherwise agree, and shall be paid for by the responsible parties directly to the reporting agency.

All other expenses of the arbitration, including required traveling and other expenses of the arbitrator and of AAA representatives, and the expenses of any witness or the cost of any proofs produced at the direct request of the arbitrator, shall be borne equally by the parties, unless they agree otherwise, or unless the arbitrator in the award assesses such expenses or any part thereof against any specified party or parties.

51. Arbitrator's Fee
Unless the parties agree to terms of compensation, members of the National Panel of Construction Arbitrators will serve without compensation for the first two days of service.

Thereafter, compensation shall be based upon the amount of service involved and the number of hearings. An appropriate daily rate and other arrangements will be discussed by the administrator with the parties and the arbitrator(s). If the parties fail to agree to the terms of compensation, an appropriate rate shall be established by the AAA, and communicated in writing to the parties.

Any arrangement for the compensation of an arbitrator shall be made through the AAA and not directly by the arbitrator with the parties. The terms of compensation of neutral arbitrators on a Tribunal shall be identical.

52. Deposits
The AAA may require the parties to deposit in advance such sums of money as it deems necessary to defray the expense of the arbitration, including the arbitrator's fee, if any, and shall render an accounting to the parties and return any unexpended balance.

53. Interpretation and Application of Rules
The arbitrator shall interpret and apply these Rules insofar as they relate to the arbitrator's powers and duties. When there is more than one arbitrator and a difference arises among them concerning the meaning or application of any such Rules, it shall be decided by a majority vote. If that is unobtainable, either an arbitrator or a party may refer the question to the AAA for final decision. All other Rules shall be interpreted and applied by the AAA.

EXPEDITED PROCEDURES
54. Notice by Telephone
The parties shall accept all notices from the AAA by telephone. Such notices by the AAA shall subsequently be confirmed in writing to the parties. Notwithstanding the failure to confirm in writing any notice or objection hereunder, the proceeding shall nonetheless be valid if notice has, in fact, been given by telephone.

55. Appointment and Qualifications of Arbitrators
The AAA shall submit simultaneously to each party to the dispute an identical list of five members of the Construction Arbitration Panel of Arbitrators, from which one arbitrator shall be appointed. Each party shall have the right to strike two names from the list on a peremptory basis. The list is returnable to the AAA within ten days from the date of mailing. If for any reason the appointment cannot be made from the list, the AAA shall have the authority to make the appointment from among other members of the Panel without the submission of additional lists. Such appointment shall be subject to disqualification for the reasons specified in Section 19. The parties shall be given notice by telephone by the AAA of the appointment of the arbitrator. The parties shall notify the AAA, by telephone, within seven days of any objections to the arbitrator appointed. Any objection by a party to such arbitrator shall be confirmed in writing to the AAA with a copy to the other party(ies).

56. Time and Place of Hearing

The arbitrator shall fix the date, time, and place of the hearing. The AAA will notify the parties by telephone, seven days in advance of the hearing date. Formal Notice of Hearing will be sent by the AAA to the parties.

57. The Hearing

Generally, the hearing and presentations of the parties shall be completed within one day. The arbitrator, for good cause shown, may schedule an additional hearing to be held within five days.

58. Time of Award

Unless otherwise agreed to by the parties, the award shall be rendered not later than five business days from the date of the closing of the hearing.

ADMINISTRATIVE FEE SCHEDULE

A filing fee of $200 will be paid at the time the case is initiated.

The balance of the administrative fee of the AAA is based upon the amount of each claim and counterclaim as disclosed when the claim and counterclaim are filed, and is due and payable prior to the notice of appointment of the neutral arbitrator.

In those claims and counterclaims which are not for a monetary amount, an appropriate administrative fee will be determined by the AAA, payable prior to such notice of appointment.

Amount of Claim or Counterclaim	Fee for Claim or Counterclaim
$1 to $20,000	3% (minimum $200)
$20,000 to $40,000	$ 600, plus 2% of excess over $20,000
$40,000 to $80,000	$1,000, plus 1% of excess over $40,000
$80,000 to $160,000	$1,400, plus ½% of excess over $80,000
$160,000 to $5,000,000	$1,800, plus ¼% of excess over $160,000

Where the claim or counterclaim exceeds $5 million, an appropriate fee will be determined by the AAA. If there are more than two parties represented in the arbitration, an additional 10% of the administrative fee will be due for each additional represented party.

When no amount can be stated at the time of filing, the administrative fee is $500, subject to adjustment in accordance with the schedule as soon as an amount can be disclosed.

OTHER SERVICE CHARGES

$50 payable by each party for each second and subsequent hearing which is either clerked by the AAA or held in a hearing room provided by the AAA.

POSTPONEMENT FEES

Sole-Arbitrator Cases:

$50 payable by a party first causing an adjournment of any scheduled hearing.

$100 payable by a party causing a second or subsequent adjournment of any scheduled hearing.

Three-Arbitrator Cases:

$75 payable by a party first causing an adjournment of any scheduled hearing.

$150 payable by a party causing a second or subsequent adjournment of any scheduled hearing.

REFUND SCHEDULE

If the AAA is notified that a case has been settled or withdrawn before a list of Arbitrators has been sent out, all the fee in excess of $200 will be refunded.

If the AAA is notified that a case has been settled or withdrawn before the due date for the return of the first list, two-thirds of the fee in excess of $200 will be refunded.

If the AAA is notified that a case is settled or withdrawn during or following a pre-hearing conference or at least 48 hours before the date and time set for the first hearing, one-third of the fee in excess of $200 will be refunded.

Index

References are to pages